Handbook of
Plastics and Elastomers

CHARLES A. HARPER   *editor-in-chief*
Westinghouse Electric Corporation
Baltimore, Maryland

McGRAW-HILL BOOK COMPANY

New York   St. Louis   San Francisco   Auckland   Düsseldorf
Johannesburg   Kuala Lumpur   London   Mexico   Montreal
New Delhi   Panama   Paris   São Paulo   Singapore
Sydney   Tokyo   Toronto

√6152-8924

CHEMISTRY

Library of Congress Cataloging in Publication Data

Main entry under title:

Handbook of plastics and elastomers.

   Includes bibliographies and index.
   1.   Plastics—Handbooks, manuals, etc.   2.   Elas-
tomers—Handbooks, manuals, etc.   I.   Harper, Charles
A.
TP1130.H36   668.4'02'02   75-9790
ISBN 0-07-026681-6

*The editors for this book were Harold B. Crawford
and Ruth Weine, the designer was Naomi Auerbach,
and its production was supervised by Teresa F. Leaden.
It was set in Caledonia by Monotype Composition Company, Inc.*

It was printed and bound by the Kingsport Press.

I.D. 3812600Q

Handbook of
Plastics and Elastomers

OTHER McGRAW-HILL HANDBOOKS OF INTEREST

ALJIAN · Purchasing Handbook
AMERICAN SOCIETY OF MECHANICAL ENGINEERS · ASME Handbooks:
    Engineering Tables   Metals Engineering—Processes
    Metals Engineering—Design   Metals Properties
BAUMEISTER AND MARKS · Standard Handbook for Mechanical Engineers
BEEMAN · Industrial Power Systems Handbook
BRADY · Materials Handbook
BURINGTON AND MAY · Handbook of Probability and Statistics with Tables
CARROLL · Industrial Instrument Servicing Handbook
CONSIDINE · Process Instruments and Controls Handbook
CONSIDINE AND ROSS · Handbook of Applied Instrumentation
DUDLEY · Gear Handbook
EMERICK · Handbook of Mechanical Specifications for Buildings and Plants
FINK AND CARROLL · Standard Handbook for Electrical Engineers
GREENE · Production and Inventory Control Handbook
HARRIS · Handbook of Noise Control
HARRIS AND CREDE · Shock and Vibration Handbook
HICKS · Standard Handbook of Engineering Calculations
KALLEN · Handbook of Instrumentation and Controls
KING AND BRATER · Handbook of Hydraulics
KORN AND KORN · Mathematical Handbook for Scientists and Engineers
LEGRAND · The New American Machinists' Handbook
LEWIS AND MARRON · Facilities and Plant Engineering Handbook
MACHOL · System Engineering Handbook
MANTELL · Engineering Materials Handbook
MERRITT · Building Construction Handbook
MORROW · Maintenance Engineering Handbook
PERRY · Chemical Engineers' Handbook
PERRY · Engineering Manual
ROSSNAGEL · Handbook of Rigging
ROTHBART · Mechanical Design and Systems Handbook
SHAND · Glass Engineering Handbook
SOCIETY OF MANUFACTURING ENGINEERS:
    Die Design Handbook   Manufacturing Planning and
    Handbook of Fixture Design   Estimating Handbook
                            Tool Engineers Handbook
STREETER · Handbook of Fluid Dynamics
TRUXAL · Control Engineers' Handbook

Contents

Contributors

BARITO, ROBERT W. *Major Appliance Laboratories, General Electric Company:* CHAPTER 7. PLASTIC AND ELASTOMERIC FOAMS (CONTRIBUTING AUTHOR: **W. O. Eastman** *Major Appliance Laboratories, General Electric Company*)

BUCHOFF, LEONARD S. *Technical Wire Products, Inc.:* CHAPTER 8. LIQUID AND LOW-PRESSURE RESIN SYSTEMS

FENNER, OTTO *Corporate Engineering, Monsanto Company:* CHAPTER 4. CHEMICAL AND ENVIRONMENTAL PROPERTIES OF PLASTICS AND ELASTOMERS

GRAFTON, PERCY *Boonton Molding Company:* CHAPTER 12. DESIGN AND FABRICATION OF PLASTIC PARTS

HARPER, CHARLES A. *Systems Development Division, Westinghouse Electric Corporation:* CHAPTER 1. FUNDAMENTALS OF PLASTICS AND ELASTOMERS; CHAPTER 2. ELECTRICAL DESIGN PROPERTIES OF PLASTICS AND ELASTOMERS

LANDROCK, ARTHUR H. *Plastics Technical Evaluation Center, Picatinny Arsenal:* CHAPTER 11. COMMERCIAL AND GOVERNMENT SPECIFICATIONS AND STANDARDS (CONTRIBUTING AUTHOR: **Norman E. Beach,** [*deceased*] *Plastics Technical Evaluation Center, Picatinny Arsenal*)

MANSFIELD, RICHARD *United Merchants and Manufacturers, Inc.:* CHAPTER 6. PROPERTIES AND END USES FOR MAN-MADE FIBERS (CONTRIBUTING AUTHOR: **Ted V. Shumeyko** *Shumeyko Wilson Associates, Inc.*)

MOCK, JOHN A. *Materials Engineering, Reinhold Publishing Corporation:* CHAPTER 9. PLASTIC AND ELASTOMERIC COATINGS

PETRIE, EDWARD M. *Research and Development Center, Westinghouse Electric Corporation:* CHAPTER 10. PLASTICS AND ELASTOMERS AS ADHESIVES

SAMPSON, RONALD N. *Research and Development Center, Westinghouse Electric Corporation:* CHAPTER 3. MECHANICAL PROPERTIES AND TESTING OF PLASTICS AND ELASTOMERS (CONTRIBUTING AUTHOR: **George E. Rudd** *Research and Development Center, Westinghouse Electric Corporation*); CHAPTER 5. LAMINATES, REINFORCED PLASTICS, AND COMPOSITES

Preface

There are probably few who would not rate polymer technology as one of the most important growth areas in recent years. Based on a constantly increasing volume of fundamental polymer development, the impact of this technology on new and improved products has been, and continues to be, little short of phenomenal. Almost any set of statistics shows this to be true for all industries and product areas. New polymers and improvements in established polymer groups regularly extend the performance limits of plastics and elastomers. Very frequently, performance improvements are coupled with cost advantages. Polymer technology has indeed arrived at a point to justify a major single reference work to cover the field of plastics and elastomers. It is this need which prompted the publication of this *Handbook of Plastics and Elastomers*. This work is the result of years of painstaking effort by many leading experts in the polymer field. The result is the most complete single reference and text to be found—a must for the desk of anyone involved in any aspect of the development or application of plastics and elastomers, and for every reference library.

This *Handbook of Plastics and Elastomers* was prepared as a thorough sourcebook of practical data for all ranges of interest. It contains an extensive array of property and performance data, presented as a function of the most important related variables. Further, it presents every aspect of application guidelines, including cost and fabrication method tradeoffs, design and application criteria, finishing, performance limits, and all of the other important application considerations. It also fully covers chemical, structural, and other basic polymer properties. The Handbook's other major features include thorough and conveniently cross-referenced lists of standards and specifications, a completely cross-referenced and easy-to-use index, a comprehensive glossary, and end-of-chapter reference lists which have been thoroughly and exhaustively gathered and which cover years of important work not equally conveniently available in any other source.

In addition to contents, the chapter organization and coverage of the *Handbook of Plastics and Elastomers* is equally well suited to user convenience. The first four chapters cover fundamentals of plastics and elastomers, and include extremely thorough presentations of electrical properties and tests, mechanical and physical properties and tests, and chemical and environmental properties and tests. The next six chapters cover plastic and elastomeric product forms, again providing extensive data and guidelines. These product forms include laminates, reinforced plastics, composites, man-made fibers, foams, liquid and low-pressure resin systems, coatings and adhesives. The next chapter is an especially thorough treatise on specifications and standards, and the final chapter is a clearly illustrated presentation of all of the important considerations for design and fabrication of plastic products. Each chapter of the *Handbook of Plastics and Elastomers* is indeed a book in itself.

Length and coverage of a Handbook of this magnitude are necessarily measured compromises. Inevitable, varying degrees of shortages and excesses will exist, depending on the needs of the individual user. Then too, the time required to complete a major work such as this necessarily demands that some of the most recent data may not be fully covered. Further, in spite of the tremendous efforts involved, some errors or omissions may exist. While every effort has been made to minimize such shortcomings, it is my greatest desire to improve each successive edition. Toward this end, any and all comments will be welcomed and appreciated.

Charles A. Harper

Handbook of
Plastics and Elastomers

Chapter **1**

Fundamentals of Plastics and Elastomers

CHARLES A. HARPER

Westinghouse Electric Corporation,
Baltimore, Maryland

INTRODUCTION

The "Handbook of Plastics and Elastomers" will cover all the important application and use data and guidelines for the entire field of plastics and elastomers. This chapter will cover the fundamental aspects of plastics and elastomers, and subsequent chapters will present detailed coverage of the most important individual application and use subjects in the field of plastics and elastomers, as indicated by the individual chapter titles. This initial chapter will cover the nature of plastics and elastomers, including the chemical nature of polymers, testing and test significance of standard tests, primary characteristics of individual polymer classes, processing and processed forms of plastics, and related subjects which are basic to the field of plastics and elastomers.

THE NATURE OF PLASTICS AND ELASTOMERS

Practically stated, and omitting the fine points of definition, a plastic is an organic polymer, available in some resin form or some form derived from the basic polymerized resin. These forms can be liquid or pastelike resins for embedding, coating, and adhesive bonding; or they can be molded, laminated, or formed shapes, including sheet, film, or larger mass bulk shapes.

The number of basic plastic materials is large and the list is growing. In addition, the number of variations and modifications to these basic plastic materials is also quite large. Taken together, the resultant quantity of materials available is just too large to be completely understood and correctly applied by anyone other than those whose day-to-day work puts them in direct contact with a diverse selection of materials. The practice of mixing brand names, trade names, and chemical names of various plastics only makes the problem of understanding these materials the more troublesome.

Another variable that makes it difficult for those not versed in plastics to understand and properly design with plastics is the large number of processes by which plastics can be produced. Fortunately, there is an organized pattern on which an orderly presentation of these variables can be based.

While there are numerous minor classifications for polymers, depending upon how one wishes to categorize them, nearly all can be placed in one of two major classifications. These two major plastic-material classes are thermosetting materials (or thermosets) and thermoplastic materials, as shown in Fig. 1. Although Fig. 1 shows elastomers separately—and they are a separate application group—elastomers, too, are either thermoplastic or thermosetting, depending on their chemical nature. Likewise, foams, adhesives, embedding resins, and other application groups can be subdivided into thermoplastic and thermosetting plastics.

Thermosetting Plastics

As the name implies, thermosetting plastics or thermosets are cured, set, or hardened into a permanent shape. This curing is an irreversible chemical reaction known as cross linking, which usually occurs under heat. For some thermosetting materials, however, curing is initiated or completed at room temperature. Even here, however, it is often the heat of the reaction, or the exotherm, shown in Fig. 2, which actually cures the plastic material. Such is the case, for instance, with a room-temperature-curing epoxy, polyester, or urethane compound.

The cross linking that occurs in the curing reaction is brought about by the linking of atoms between or across two linear polymers, with this cross linking of atoms making a three-dimensional rigidized chemical structure. One such reaction is shown in Fig. 3. Although the cured part can be softened by heat, it cannot be remelted or restored to the flowable state that existed before the plastic resin was cured. A tabulation of most of the thermosetting plastic materials is shown in Fig. 1, and these materials are discussed in detail in a following section of this chapter.

Thermoplastics

Thermoplastics differ from thermosets in that thermoplastics do not cure or set under heat as do thermosets. Thermoplastics merely soften, when heated, to a flowable state in which under pressure they can be forced or transferred from a heated cavity into a cool mold. Upon cooling in a mold, thermoplastics harden and take the shape of the mold. Since thermoplastics do not cure or set, they can be remelted and rehardened, by cooling, many times. Thermal aging, brought about by repeated exposure to the high temperatures required for melting, causes eventual degradation of the material and so limits the number of reheat cycles. A tabulation of common thermoplastic materials is shown in Fig. 1, and these materials are discussed in detail in a following section of this chapter.

Elastomers

The term elastomers includes the complete spectrum of elastic or rubberlike polymers which are sometimes randomly referred to as rubbers, synthetic rubbers, or elastomers. More properly, however, rubber is a natural material and synthetic rubbers are polymers which have been synthesized to reproduce consistently the best properties of natural rubber. Since such a large number of rubberlike polymers exist, the broad term elastomer is most fitting and most commonly used. A tabulation of some common elastomers is shown in Fig. 1, and elastomers are discussed in detail in a following section of this chapter.

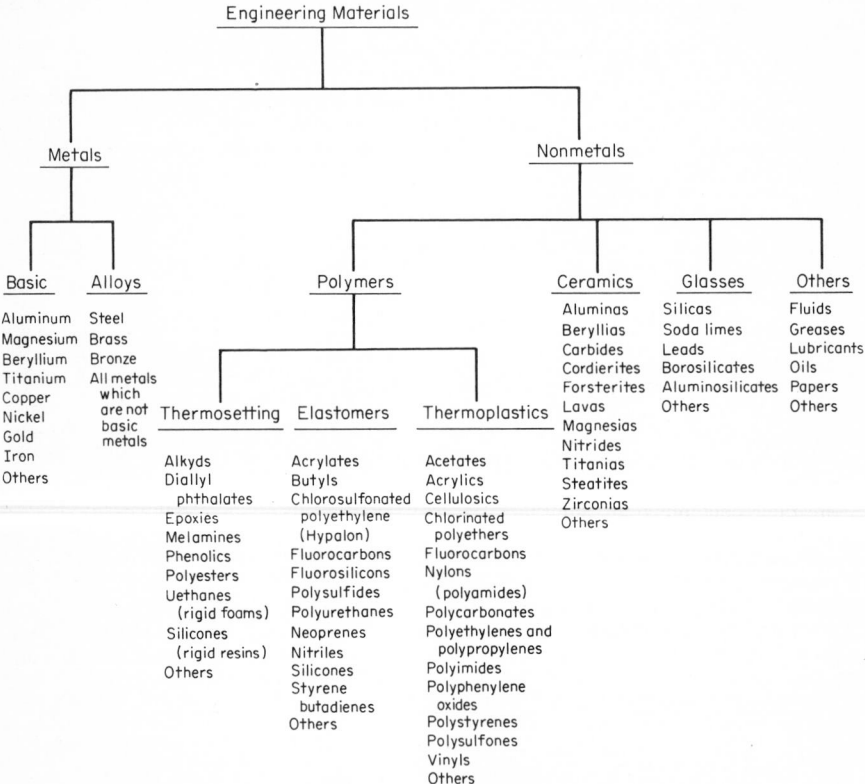

Fig. 1 Classification of engineering materials, showing polymer classes.

POLYMER STRUCTURES AND POLYMERIZATION REACTIONS

Polymer Structures

Basically, all polymers are formed by the creation of chemical linkages between relatively small molecules, or monomers, to form very large molecules, or polymers. As was mentioned above, if the chemical linkages form a rigid, cross-linked molecular structure, a thermosetting plastic results. If a somewhat flexible molecular structure is formed, either linear or branched, a thermoplastic results. An illustration of these is shown in Fig. 4.

Polymerization Reactions

One common type of polymerization reaction is the addition reaction.[1] Polymerization reactions may occur in a number of ways, with four common polymerization techniques being bulk polymerization, solution polymerization, suspension polymerization, and emulsion polymerization.[1] Basically, bulk polymerization involves the reaction of monomers or reactants among themselves without placing them in some form of extraneous media as is done in the other types of polymerization.

Solution polymerization is similar to bulk polymerization except that whereas the solvent for the forming polymer in bulk polymerization is the monomer, the solvent in solution polymerization is usually a chemically inert medium. The solvents used may be complete, partial, or nonsolvents for the growing polymer chain.

Suspension polymerization is normally used only for catalyst-initiated or free-radical addition polymerizations. The monomer is mechanically dispersed in a liquid, usually water, which is a nonsolvent for the monomer as well as for all sizes of polymer molecules which form during the reaction.

The catalyst initiator is dissolved in the monomer, and it is preferable that it does not dissolve in the water so that it remains with the monomer. The monomer and polymer being formed from it stay within the beads of organic material dispersed in the phase. Actually suspension polymerization is essentially a finely divided form of bulk polymerization. The big advantage of the suspension polymerization over bulk is that it allows the operator to cool exothermic polymerization reactions effectively and thus maintain closer control over the chain-building process. Other behavior is the same as bulk polymerization.

By controlling the degree of agitation; monomer/water ratios, and other variables, it is also possible to control the particle size of the finished polymer, thus eliminating the need to re-form the material into pellets from a melt such as is usually necessary with bulk polymerizations.

Emulsion polymerization is a technique in which addition polymerizations are carried out in a water medium containing an emulsifier (a soap) and a water-soluble initiator. It is used because emulsion polymerization is much more rapid than bulk or solution polymerization at the same temperatures and produces polymers with molecular weights much greater than those obtained at the same rate in bulk polymerizations.

In emulsion polymerization, the monomer diffuses into *micelles*, which are small, hollow spheres composed of soap film. Polymerization occurs within the micelles. Soap concentration, overall reaction-mass recipe, and reaction conditions can be varied to provide control of reaction rate and yield.

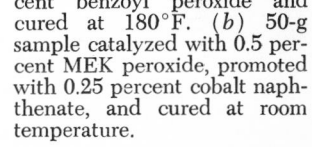

Fig. 2 Comparison of the exothermic curves for two polyester resins. (*a*) 50-g sample catalyzed with 0.5 percent benzoyl peroxide and cured at 180°F. (*b*) 50-g sample catalyzed with 0.5 percent MEK peroxide, promoted with 0.25 percent cobalt naphthenate, and cured at room temperature.

Raw-Material Sources of Major Plastic Materials

Figure 5 shows the primary raw-material sources of the major plastic materials. This knowledge will be helpful in considerations of best alternative materials to minimize the impact of the energy shortage, until such time as new raw-material sources, such as coal, are more fully developed.

PLASTIC-PROCESSING METHODS AND DESIGN GUIDELINES

Although most users of plastics buy plastic parts from plastic processors, they should still have some knowledge of plastic processing, as such information can often be helpful in optimizing product design. Also, an increasing number of user companies are doing some in-house processing. For these reasons, some guideline information in plastic processing and some guidelines on design of plastic parts are presented here. Further extensive and detailed guidelines are included in Chap. 12.

It should be mentioned that the information presented at this point applies broadly to all classes of plastics and types of processing. Most plastic suppliers will provide very specific data and guidelines for their individual products. This invaluable source of guidance is too often unused. It is strongly recommended that plastic suppliers be more fully utilized for product-design guidance. The information presented at this point will be valuable for making initial design and process decisions, however.

Table 1 explains the major ways that plastic materials can be formed into parts, and the advantages, limitations, and rough cost of each processing method. In

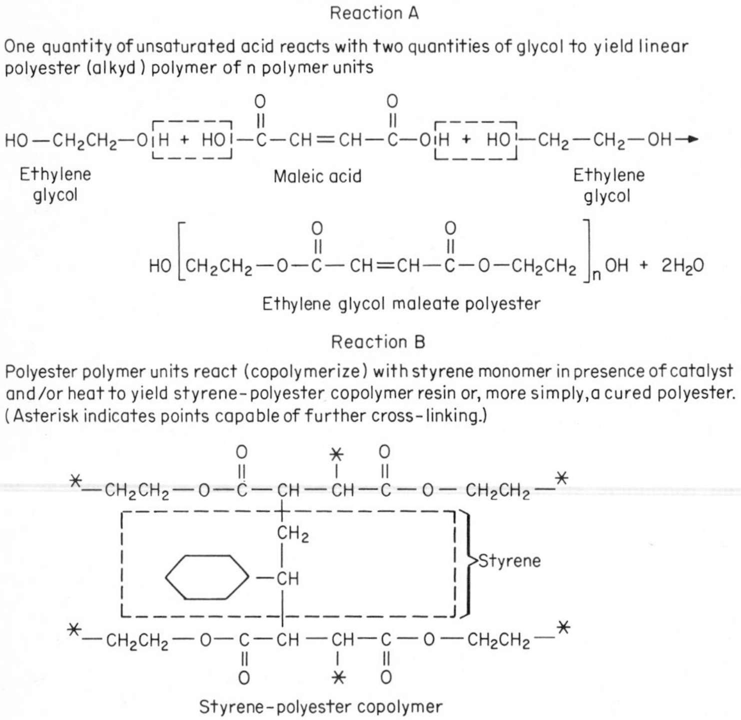

Reaction A

One quantity of unsaturated acid reacts with two quantities of glycol to yield linear polyester (alkyd) polymer of n polymer units

Ethylene glycol   Maleic acid   Ethylene glycol

Ethylene glycol maleate polyester

Reaction B

Polyester polymer units react (copolymerize) with styrene monomer in presence of catalyst and/or heat to yield styrene-polyester copolymer resin or, more simply, a cured polyester. (Asterisk indicates points capable of further cross-linking.)

Styrene-polyester copolymer

Fig. 3 Simplified diagrams show how cross-linking reactions produce polyester resin (styrene-polyester copolymer resin) from basic chemicals.[2]

general, a plastic part is produced by a combination of cooling, heating, flowing, deformation, and chemical reaction. As previously noted, the processes differ, depending on whether the material is a thermoplastic or thermoset, as discussed in more detail below.

The usual sequence of processing a thermoplastic is to heat the material so that it softens and flows, force the material in the desired shape through a die or in a mold, and chill and melt into its final shape. By comparison, a thermoset is typically processed by starting out with partially polymerized material which is softened and activated by heating (either in or out of the mold), forcing it into the desired shape by pressure, and holding it at the curing temperature until final polymerization reaches the point where the part hardens and stiffens sufficiently to keep its impressed shape.

The cost of the finished part basically depends on the material and the process used. A very rough estimate of the finished cost of a part can be obtained by multiplying the material cost by a factor ranging from 1.5 to 10. The cost factors shown in Table 1 are based on general industry experience in pricing.

Table 2 gives guidelines on part design for the various plastic-processing methods shown in Table 1. The design of a plastic part frequently depends on the processing method selected to make the part. Also, of course, selection of the best processing method frequently is a function of the part design. Listed in Table 2 are the major plastic-processing methods and their respective design capabilities, such as minimum section thicknesses and radii and overall dimensional tolerances.

The basic purpose of this guide is to show the fundamental design limits of the many plastic-processing methods. As mentioned above, it is always an advantage to review the design with the materials supplier and/or experienced plastics processor. Their suggestions on design and configuration, at an early stage, may make it possible to avoid costly production and performance problems.

Plastic-Fabrication Processes and Forms

There are many plastic-fabrication processes, and a wide variety of plastics can be processed by each of these processes or techniques. The fabrication process and the tooling determine the forms or shapes which are produced, of course. Fabrication processes can be broadly divided into pressure processes and pressureless or low-pressure processes. The former will be described at this point. Pressureless or low-pressure processes such as potting, casting, impregnating, encapsulating, and coating are reviewed in Chap. 8. Pressure processes are usually either thermoplastic-materials processes (such as injection molding, extrusion, and thermoforming) or thermosetting processes (such as compression molding, transfer molding, and laminating). There are exceptions to each, however, as mentioned below.

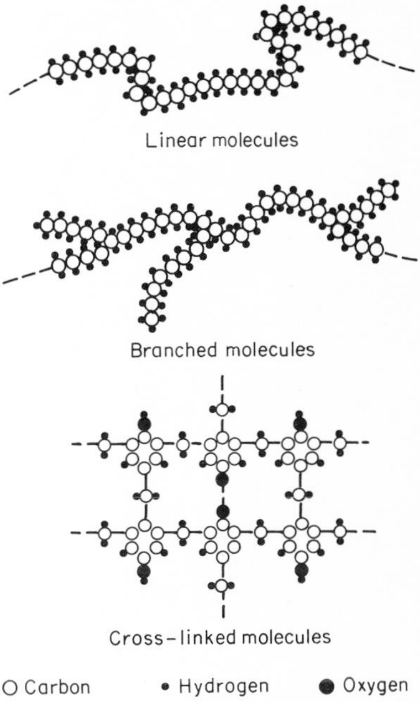

Compression molding and transfer molding Compression molding and transfer molding are the two major processes used for forming molded parts from thermosetting raw materials. The two can be

Linear molecules

Branched molecules

Cross–linked molecules

○ Carbon   • Hydrogen   ● Oxygen

Fig. 4 Some possible molecular structures in polymers.[1]

carried out in the same type of molding press, but different types of molds are used. The thermosetting materials are normally molded by the compression or transfer process, but it is possible to mold thermoplastics by these processes since the heated thermoplastics will flow to conform to the mold-cavity shape under suitable pressure. These processes are usually impractical for thermoplastic molding, however, since after the mold cavity is filled to final shape, the heated mold would have to be cooled to solidify the thermoplastic part. Since repeated heating and cooling of this large mass of metal and the resultant long cycle time per part produced are both objectionable, injection molding is commonly used to process thermoplastics.

Compression Molding. The compression-molding technique is shown in Fig. 6. In compression molding, the open mold is placed between the faces or platens of the molding press, and the mold is heated, then filled with a given quantity of molding material, and closed under pressure, causing the material to flow into the shape of the mold cavity. The actual pressure required depends on the molding material being used and the geometry of the mold. The mold is kept closed until the plastic material is suitably cured or hardened. Then the mold is opened, the molded plastic part is ejected, and the cycle is repeated. The mold is usually steel with a polished or plated cavity.

The simplest form of compression molding involves the use of a separate self-contained mold or die that is designed for manual handling by the operator. It is loaded on the bench, capped, placed in the press, closed, and cured, and then removed for opening under an arbor press. The same mold in most instances (and with some structural modifications) can be mounted permanently into the press and opened and closed as the press itself opens and closes. The press must have a positive up-and-down movement under pressure instead of the usual gravity drop found in the standard hand press.

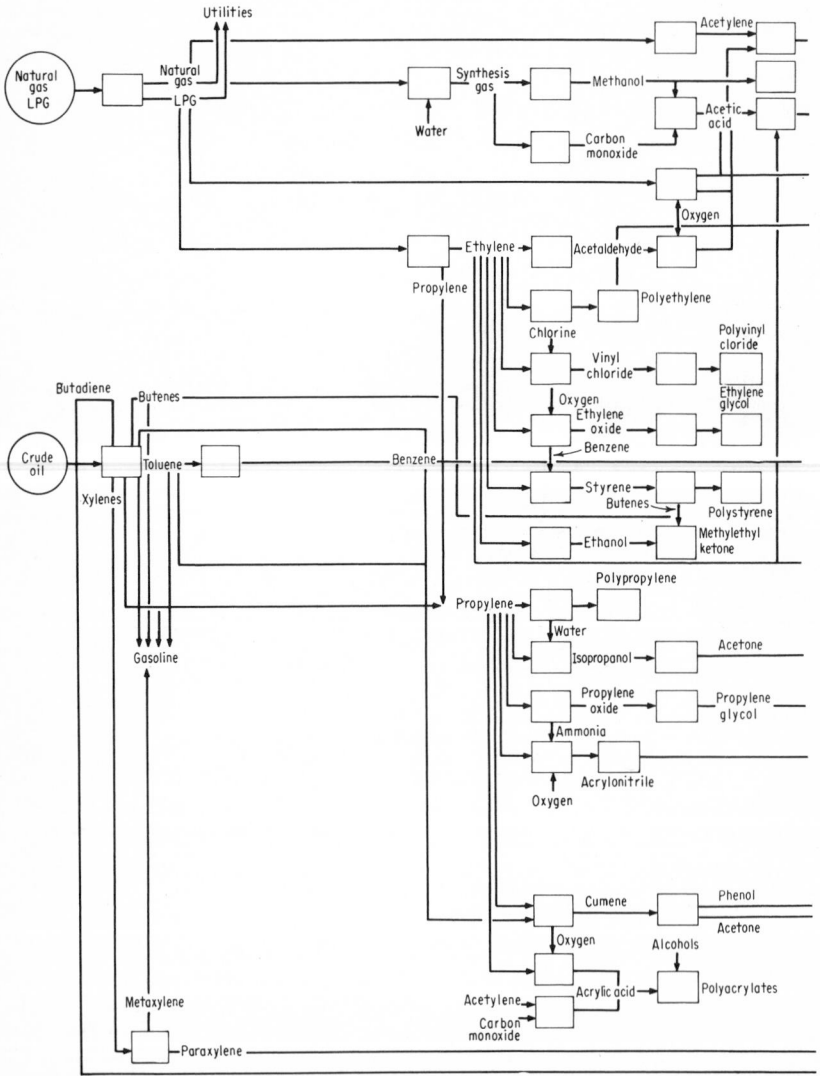

Fig. 5 Primary raw-material sources of major plastic materials.[2]

Transfer Molding.   The transfer-molding sequence is shown in Fig. 7.   The molding material is first placed in a heated pot, separate from the mold cavity.   The hot plastic material is then transferred under pressure from the pot through the runners and into the closed cavity of the mold.

The great advantage of transfer molding lies in the fact that the mold proper is closed at the time the material enters.   Parting lines that might give trouble in finishing are held to a minimum.   Inserts are positioned and delicate steel parts of the mold are not subject to movement.   Vertical dimensions are more stable than in straight compression.   Also, delicate inserts can often be molded by transfer molding, especially with the low-pressure molding compounds.

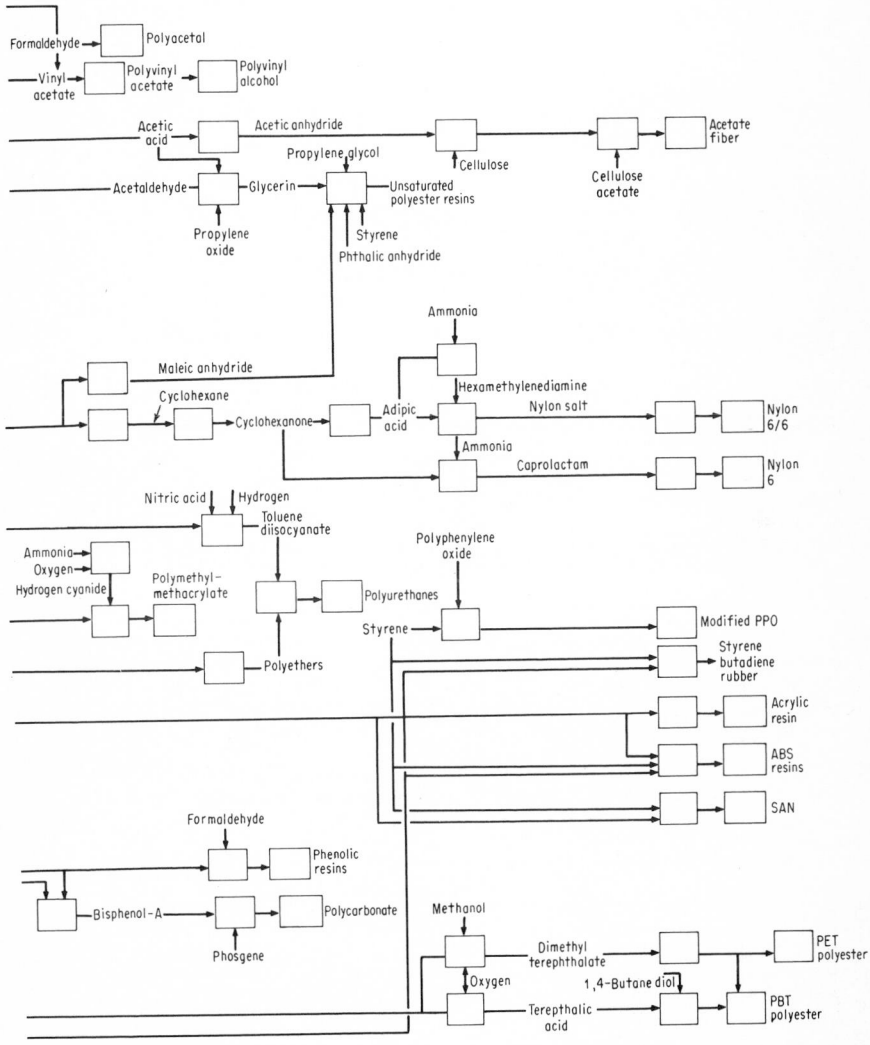

TABLE 1 Descriptions and Guidelines for Plastic-Processing Methods[3]

Process	Description	Key advantages	Notable limitations	Cost factor*
Blow molding	An extruded tube (parison) of heated thermoplastic is placed between two halves of an open split mold and expanded against the sides of the closed mold by air pressure. The mold is open, and the part is ejected	Low tool and die costs; rapid production rates; ability to mold relatively complex hollow shapes in one piece	Limited to hollow or tubular parts; wall thickness and tolerances often hard to control	1.5–5, 2–3
Calendering	Doughlike thermoplastic mass is worked into a sheet of uniform thickness by passing it through and over a series of heated or cooled rolls. Calenders also are used to apply plastic covering to the back of other materials	Low cost; sheet materials are virtually free of molded-in stresses; i.e., they are isotropic	Limited to sheet materials; very thin films not possible	1.5–3, 2–5.5
Casting	Liquid plastic (usually thermoset except for acrylics) is poured into a mold (without pressure), cured, and removed from the mold. Cast thermoplastic films are made by depositing the material, either in solution or in hot-melt form, against a highly polished supporting surface	Low mold cost; ability to produce large parts with thick cross sections; good surface finish; suitable to low-volume production	Limited to relatively simple shapes; except for cast films, becomes uneconomical at high-volume production levels; most thermoplastics not suitable	1.5–3, 2–2.5
Compression molding	A thermoplastic or partially polymerized thermosetting resin compound, usually preformed, is placed in a heated mold cavity; the mold is closed, heat and pressure are applied, and the material flows and fills the mold cavity. Heat completes polymerization, and the mold is opened to remove the part. The process is sometimes used for thermoplastics, e.g., vinyl phonograph records	Little waste of material and low finishing costs; large, bulky parts are possible	Extremely intricate parts involving undercuts, side draws, small holes, delicate inserts, etc., not practical; very close tolerances difficult to produce	2–10, 1.5–3
Cold forming	Similar to compression molding in that material is charged into split mold; it differs in that it uses no heat—only pressure. Part is cured in an oven in a separate operation. Some thermoplastic sheet material and billets are cold-formed in process similar to drop hammer–die forming of metals. Shotgun shells are made in this manner from polyethylene billets	Ability to form heavy or tough-to-mold materials; simple; inexpensive; often has rapid production rate	Limited to relatively simple shapes; few materials can be processed in this manner	

Process	Description	Advantages	Limitations	
Extrusion............	Thermoplastic or thermoset molding compound is fed from a hopper to a screw and barrel where it is heated to plasticity and then forwarded, usually by a rotating screw, through a nozzle having the desired cross-section configuration	Low tool cost; great many complex profile shapes possible; very rapid production rates; can apply coatings or jacketing to core materials, such as wire	Limited to sections of uniform cross section	2–5, 3–4
Filament winding.....	Continuous filaments, usually glass, in form of rovings are saturated with resin and machine-wound onto mandrels having shape of desired finished part. Once winding is completed, part and mandrel are placed in oven for curing. Mandrel is then removed through porthole at end of wound part	High-strength fiber reinforcements are oriented precisely in direction where strength is needed; exceptional strength/weight ratio; good uniformity of resin distribution in finished part	Limited to shapes of positive curvature; openings and holes reduce strength	5–10, 6–8
Injection molding......	Thermoplastic or thermoset molding compound is heated to plasticity in cylinder at controlled temperature; then forced under pressure through a nozzle into sprues, runners, gates, and cavities of mold. The resin solidifies rapidly, the mold is opened and the part(s) ejected. In modified version of process—runnerless molding—the runners are part of mold cavity	Extremely rapid production rates, hence low cost per part; little finishing required; good dimensional accuracy; ability to produce relatively large, complex shapes; very good surface finish	High initial tool and die costs; not practical for small runs	1.5–5, 2–3
Laminating, high pressure............	Material, usually in form of reinforcing cloth, paper, foil, etc., preimpregnated or coated with thermoset resin (sometimes a thermoplastic), is molded under pressure greater than 1,000 lb/in.^{2} into sheet, rod, tube, or other simple shape	Excellent dimensional stability of finished product; very economical in large production of parts	High tool and die costs; limited to simple shapes and cross-section profiles	2–5, 3–4
Matched-die molding............	A variation of conventional compression molding, this process uses two metal molds having a close-fitting, telescoping area to seal in the plastic compound being molded and to trim the reinforcement. The reinforcement, usually mat or preform, is positioned in the mold, and the mold is closed and heated (pressures generally vary between 150 and 400 lb/in.^{2}). Mold is then opened and part lifted out	Rapid production rates; good quality and reproducibility of parts	High mold and equipment costs; parts often require extensive surface finishing, e.g., sanding	2–5, 3–4

TABLE 1 Descriptions and Guidelines for Plastic-Processing Methods[3] (Continued)

Process	Description	Key advantages	Notable limitations	Cost factor*
Rotational molding...	A predetermined amount of powdered or liquid thermoplastic or thermoset material is poured into mold. Mold is closed, heated, and rotated in the axis of two planes until contents have fused to inner walls of mold. The mold is opened and part removed	Low mold cost; large hollow parts in one piece can be produced; molded parts are essentially isotropic in nature	Limited to hollow parts; in general, production rates are slow	1.5–5, 2–3
Slush molding.........	Powdered or liquid thermoplastic material is poured into a mold to capacity. Mold is closed and heated for a predetermined time to achieve a specified buildup of partially cured material on mold walls. Mold is opened, and unpolymerized material is poured out. Semifused part is removed from mold and fully polymerized in oven	Very low mold costs; very economical for small-production runs	Limited to hollow parts; production rates are very slow; limited choice of materials that can be processed	1.5–4, 2–3
Thermoforming........	Heat-softened thermoplastic sheet is placed over male or female mold. Air is evacuated from between sheet and mold, causing sheet to conform to contour of mold. There are many variations, including vacuum snapback, plug assist, drape forming, etc.	Tooling costs generally are low; produces large parts with thin sections; often economical for limited production of parts	In general, limited to parts of simple configuration; limited number of materials to choose from; high scrap	2–10, 3–5
Transfer molding......	Thermoset molding compound is fed from hopper into a transfer chamber where it is heated to plasticity. It is then fed by means of a plunger through sprues, runners, and gates of closed mold into mold cavity. Mold is opened and the part ejected	Good dimensional accuracy; rapid production rate; very intricate parts can be produced	Molds are expensive; high material loss in sprues and runners; size of parts is somewhat limited	1.5–5, 2–3
Wet lay-up or contact molding.....	Number of layers, consisting of a mixture of reinforcement (usually glass cloth) and resin (thermosetting), are placed in mold and contoured by roller to mold's shape. Assembly is allowed to cure (usually in an oven) without application of pressure. In modification of process, called spray molding, resin systems and chopped fibers are sprayed simultaneously from spray gun against mold surface; roller assist also is used. Wet lay-up parts sometimes are cured under pressure, using vacuum bag, pressure bag, or autoclave	Very low cost; large parts can be produced; suitable for low-volume production of parts	Not economical for large-volume production; uniformity of resin distribution very difficult to control; mainly limited to simple shapes	1.5–4, 2–3

* Material cost × factor = purchase price of a part; top figure is overall range, bottom is probable average cost.

TABLE 2 Guidelines on Part Design for Plastic-Processing Methods²

Design rules	Blow molding	Casting	Compression molding	Extrusion	Injection molding	Reinforced plastic molding			Rotational molding	Thermo-forming	Transfer molding
						Wet lay-up (contact molding)	Matched-die molding	Filament winding			
Major shape characteristics...	Hollow bodies	Simple configurations	Moldable in one plane	Constant cross-section profile	Few limitations	Moldable in one plane	Moldable in one plane	Structure with surfaces of revolution	Hollow bodies	Moldable in one plane	Simple configurations
Limiting size factor...	M	M	ME	M	ME	MS	ME	WE	M	M	ME
Min inside radius, in...	0.125	0.01–0.125	0.125	0.01–0.125	0.01–0.125	0.25	0.06	0.125	0.01–0.125	0.125	0.01–0.125
Undercuts...	Yes	Yes[a]	NR[b]	Yes	Yes[a]	Yes	NR	NR	Yes[c]	Yes[a]	NR
Min draft, degrees...	0	0–1	>1	NA[b]	<1	0	1	2–3	1		1
Min thickness, in...	0.01	0.01–0.125	0.01–0.125	0.001	0.015	0.06	0.03	0.015	0.02	0.002	0.01–0.125
Max thickness, in...	>0.25	None	0.5	6	1	0.5	1	3	0.5	3	1
Max thickness buildup, in...	NA	2–1	2–1	NA	2–1	2–1	2–1[d]	NR	NA	NA	2–1
Inserts...	Yes	Yes	Yes	Yes	Yes	Yes	Yes	Yes	Yes	NR	Yes
Built-up cores...	Yes	Yes	No	Yes	Yes	Yes	Yes	Yes	Yes	Yes	Yes
Molded-in holes...	Yes	Yes	Yes	Yes[e]	Yes	Yes	Yes	Yes	Yes	No	Yes
Bosses...	Yes	Yes	Yes	Yes	Yes	Yes	Yes	No	Yes	Yes	Yes
Fins or ribs...	Yes	Yes	Yes	Yes	Yes	Yes	No[f]	No[g]	Yes	Yes	Yes
Molded-in designs and nos...	Yes	Yes	Yes	No	Yes	Yes	Yes	No	Yes	Yes	Yes
Overall dimensional tolerance, in./in...	±0.01	±0.001	±0.001	±0.005	±0.001	±0.02	±0.005	±0.005	±0.01	±0.01	±0.001
Surface finish[h]...	1–2	2	1–2	1–2	1	4–5	4–5	5	2–3	1–3	1–2
Threads...	Yes	Yes	Yes	No	Yes	No	No	No	Yes	No	Yes

M = material. ME = molding equipment. MS = mold size. WE = winding equipment.
[a] Special molds required.
[b] NR—not recommended; NA—not applicable.
[c] Only with flexible materials.
[d] Using premix: as desired.
[e] Only in direction of extrusion.
[f] Using premix: yes.
[g] Possible using special techniques.
[h] Rated 1 to 5: 1 = very smooth, 5 = rough.

Injection molding   Injection molding is the most practical process for molding thermoplastic materials. The principle of operation is shown in Fig. 8. The operating principle is simple, but the equipment is not.

A material of thermoplastic qualities—one that is viscous at some elevated temperature and stable at room temperature without appreciable deterioration during the cycle—is maintained in a heated reservoir. This hot, soft material is forced from the reservoir into a cool mold shaped to the desired form. The cool mold is opened as soon as the material has given up enough heat to hold its shape and allow repetition.

The speed of the cycle is determined by the rapidity with which the temperature of the material used can be reduced, which in turn depends on the thermal conductivity of that material. Acrylics are slow performers, and styrenes are among the fastest.

The machine itself is usually a horizontal cylinder whose bore determines the capacity. Within the bore is a piston which, when retracted, opens a hole in the top of the cylinder through which new material can be added to replace the charge shot into the mold. The cylinder is heated by electric bands which permit temperature variation along its length. Inside the exit end of the cylinder is a torpedo over which the hot material is forced just before coming out of the nozzle into the channels leading to the cavities. This gives the material a final churning and provides thorough heating of all particles. The mold opens and closes automatically, and the whole cycle is controlled by timers. An injection machine is rated by the number of ounces it will inject per stroke of the piston, and by the square inches of working area that can be clamped against the injection pressure.

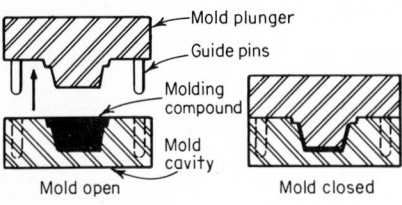

Fig. 6 Simplified illustration of compression-molding process.

Thermoset injection molding   Because of the chemical nature of the plastic materials, injection molding has traditionally been the primary molding method for thermoplastics, and compression and transfer molding have been the primary molding methods for thermosetting plastics. Because of the greater molding cycle speeds and lower molding costs in injection molding, thermoplastics have had a substantial molding-cost advantage over thermosets. Consequently, much effort has been devoted to improving the molding cycle for thermosets. As a result, advances in equipment and in thermosetting molding compounds have resulted in a rapid transition to screw-injection, in-line molding. This has been especially prominent with phenolics, but other thermosets are also included to varying degrees. The growth in screw-injection molding of phenolics has been extremely rapid. The development of this technique allows the molder to automate further, reduce labor costs, improve quality, reduce rejects, and gain substantial overall mold-cycle efficiency. The increasing use of this technique for thermosets offers the opportunity for them to make substantial gains in areas otherwise closed to them, primarily because of molding costs.

Extrusion and pultrusion   The process of extrusion consists, basically, of forcing heated, melted plastic continuously through a die which has an opening shaped to produce a desired finished cross section. The process can be envisioned as similar to that of a "heated meat grinder." Normally it is used for processing thermoplastic materials, but it can, as described below, be used for processing thermosetting materials. The main application of extrusion is the production of continuous lengths of film, sheeting, pipe, filaments, wire jacketing, or other useful forms and cross sections. After the plastic melt is extruded through the die, the extruded material is hardened by cooling, usually by air or water.

Although not widespread, extruded thermosetting materials are used increasingly. The main object here is production of shapes, parts, and tolerances not obtainable in compression or transfer molding. Pultrusion is a special, increasingly used technique for pulling resin-soaked fibers through an orifice, offering significant strength improvements. Any thermoset, B-stage, granular molding compound can be extruded.

Any type of filler may be added to the compound.   In fiber-filled compounds, the length of fiber is limited only by the cross-sectional thickness of the extruded piece.

A metered volume of molding compound is fed into the die feed zone, where it is slightly warmed.   As the ram forces the compound through the die, the compound is heated gradually until it becomes semifluid.   Before leaving the die, the

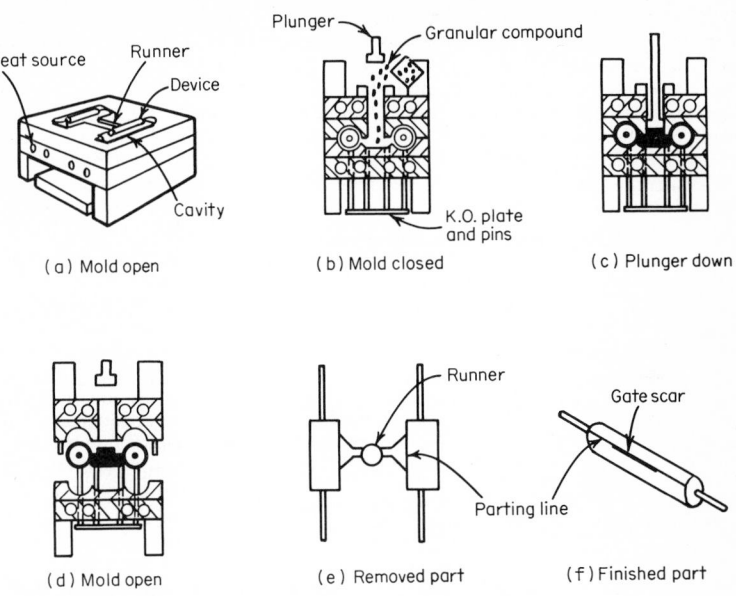

Fig. 7   Simplified illustration of transfer-molding process.

extruded part is cured by a controlled time of travel through a zone of increasing temperature.   The cured material exits from the dies at temperatures of 300 to 350°F and at rates of 3 to 6 in./min for phenolic compounds.

Thermoforming   Thermoforming is a relatively simple basic process, consisting of heating a plastic sheet, then forming it to conform to the shape of the mold either by differential air pressure or by some mechanical means.   By this processing technique, thermoplastic sheets can be rapidly and efficiently converted to a limitless number of shapes whose thickness depends on the thickness of the film being used and the processing details of an individual operation.   Although there are many variations of this process, they generally involve heating the plastic sheet and making it conform to the contour of a male or female form by air pressure or by a matching set of male and female molds.

PLASTIC PROPERTIES

Much work has been done both on standardization of measurements for properties of plastics and on analysis of plastic properties.   An understanding of plastic performance, as indicated by standard tests, is especially important to the large

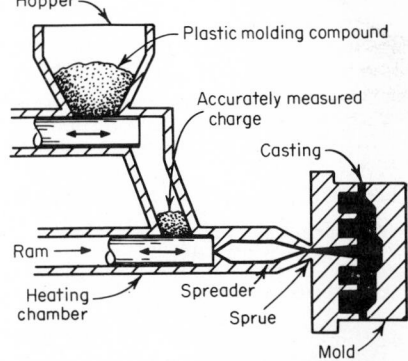

Fig. 8   Simplified illustration of injection-molding process.

percentage of non-chemically-trained users of plastic materials. While more detailed presentations are made in subsequent chapters for special areas such as mechanical design and electrical design, and Chap. 11 covers in detail the types and sources of specifications and standards, some fundamental, general-use data and guidelines will be presented at this point.

Standard Tests and Their Significance

Among the most widely used test procedures are those developed by the American Society for Testing and Materials (ASTM). These tests may be divided into categories of permanence tests, chemical tests, mechanical tests, thermal tests, analytical tests, optical tests, and electrical tests. Table 3 identifies these tests and indicates the significance of each.

A cross reference of some important ASTM tests and federal test methods is given in Table 4.

TABLE 3   ASTM Tests for Plastics and Their Significance[4]

ASTM test	Significance
Permanence tests: D 1693-70, environmental stress cracking.............	The cracking obtained in this test is indicative of what may be expected from a wide variety of other stress-cracking agents. The information cannot be translated directly into end-use service prediction but serves to rank various types and grades of PE in categories of resistance to ESC. This test can be used on high- and medium-density materials as well. In these cases it would be considered a "modified" test
G 23-69 (formerly E 42), accelerated weathering......	Twelve types of apparatus with two different exposure conditions may be employed. One condition employs continuous light with intermittent water spray and is normally used because it provides a light/moisture ratio that is generally satisfactory for simulating average natural weathering exposures likely to be encountered The alternate condition, 18 h of light at 50% relative humidity with intermittent water spray followed by 6 h of darkness at 95% relative humidity with water condensing on the test specimen, is recommended where the desire is to accentuate the effect which the moisture has had on the plastic Since weather varies from day to day, year to year, and place to place, no precise correlation exists between artificial laboratory weathering and natural outdoor weathering. However, standard laboratory test conditions produce results with acceptable reproducibility and that are in general agreement with data obtained from outdoor exposures. Fairly rapid indications of weatherability are therefore obtainable on samples of known materials which, through testing experience over a period of time, are represented by established general correlations. There is no artificial substitute for precisely predicting outdoor weatherability of those materials for which no previous weathering history has been compiled
Pipe tests (Commercial Standard CS 255-63)*.......	Sections of pipe, either thermoplastic or reinforced thermosetting in nature, are utilized in these various tests. The lengths of the sections are specified for each type, grade, and dimension of pipe sections that are to be subjected to testing
D 794-68 (1972), permanent effect of heat..............	This test is of particular value in connection with established or potential applications that involve service at

TABLE 3   ASTM Tests for Plastics and Their Significance[4] (Continued)

ASTM test	Significance
	elevated temperatures. It permits comparison of various plastics and grades of one plastic in the form of test specimens, as well as molded plastic parts that are in finished form
D 1435-69, outdoor weathering	Outdoor testing is the most accurate method for obtaining a true picture of weather resistance. The only drawback of this test is the time required for realistic exposure results. A large number of specimens is usually required to allow periodic removal and to run representative laboratory tests after measured exposure
Weight loss on heating (D 706-63, specification for cellulose acetate molding compounds).	This test for cellulosics indicates general aging stability, especially when service at elevated temperatures is anticipated. It quickly shows in weight loss or dimensional change or both whether or not a particular formulation of plastic material being tested appears to be sufficiently stable for a specific application. Plasticizer volatility is generally the cause of weight loss on heating, and this test is commonly used to help select plasticizers that give desired balances of properties and stability
D 570-63 (1972), water absorption................	The various plastics absorb varying amounts of water, and the presence of absorbed water may affect plastics in different ways. Electrical properties change most noticeably with water absorption, and this is one of the reasons that polyethylene, since it is a material that absorbs almost no water, is favored as a dielectric. Materials that absorb relatively larger amounts of water tend to change dimension in the process. When dimensional stability is required in products made of such materials, grades with less tendency to water absorption should be chosen
Chemical tests: C 619-71, chemical resistance of asbestos fiber-reinforced thermosetting resins.........	This is a rapid test to observe changes in properties of weight, appearance, and compressive strength; its results can serve as a guide in selecting asbestos-fiber reinforced-resin plastic for a self-supporting structure
C 581-68, chemical resistance of thermosetting resins......	This is a rapid test to determine changes that might have taken place in the properties of the thermosetting resins after they have been exposed to chemical involvement, and it also serves as a useful and valuable guide for selecting thermosetting resins used by glass-fiber-reinforced plastic self-supporting structures
Mechanical tests: D 790-71, flexural properties..	In bending, a beam is subject to both tensile and compressive stresses. Since most thermoplastics do not break in this test even after being greatly deflected, the flexural strength cannot be calculated. Instead, stress at 5% strain is calculated—that is, the loading in lb/in.^{2} that has been found to be necessary in order to stretch the outer surface to the extent of 5%
D 1822-68, tensile impact.....	This test is comparatively new, and relatively few industry data are available to assess its accuracy and utility adequately. However, possible advantages over the notched Izod test are immediately apparent; the notch-sensitivity factor is eliminated and energy is not used in pushing aside the broken portion of the specimen

TABLE 3   ASTM Tests for Plastics and Their Significance[4] (Continued)

ASTM test	Significance
D 747-70, stiffness in flexure..	This test does not distinguish the plastic and elastic elements involved in the measurement, and therefore a true elastic modulus is not calculable. Instead, an apparent value is obtained and called "stiffness in flexure." It is a measure of the relative stiffness of various plastics and, taken with other pertinent property data, is useful in material selection
D 256-72a, Izod impact......	The Izod impact test indicates the energy required to break notched specimens under standard conditions. It is calculated as ft-lb/in. of notch and is usually calculated on the basis of a 1-in. specimen, although the specimen used may be thinner in the lateral direction. The Izod value is useful in comparing various types or grades of a plastic. In comparing one plastic with another, however, the Izod impact test should not be considered a reliable indicator of overall toughness or impact strength. Some materials are notch-sensitive and derive greater concentrations of stress from the notching operation. The Izod impact test may indicate the need for avoiding sharp corners in parts made of such materials. For example, nylon and acetal-type plastics, which in molded parts are among the toughest materials, are notch-sensitive and are found to register relatively low resistance values when given the notched Izod impact test
D 638-72, tensile properties...	Tensile properties are the most important single indication of strength in a material. The force necessary to pull the specimen apart is determined, along with prebreak stretch. The elastic modulus ("modulus of elasticity" or "tensile modulus") is the ratio of stress to strain below the proportional limit of the material. It is the most useful item of tensile data because parts should be designed to accommodate stresses to a degree well below this For some applications where almost rubbery elasticity is desirable, a high ultimate elongation may be an asset. For rigid parts, on the other hand, there is little benefit in the fact that they can be stretched extremely long There is great benefit in moderate elongation, however, since this quality permits absorbing rapid impact and shock. Thus the total area under a stress-strain curve is indicative of overall toughness. A material of very high tensile strength and little elongation would be brittle
D 785-65 (1970), Rockwell hardness..................	Rockwell hardness can differentiate relative hardnesses of different types of a given plastic. But since elastic recovery is involved as well as hardness, it is not valid to compare hardnesses of various kinds of plastic entirely on the basis of this test. Rockwell hardness is not an index of wear qualities or abrasion resistance. For instance, polystyrenes have high Rockwell hardness values but poor scratch resistance
D 621-64, deformation under load................	This test on rigid plastics indicates their ability to withstand continuous short-term compression without yielding and loosening when fastened, as in insulators or other assemblies, by bolts, rivets, etc. This test does not indicate the creep resistance of a particular plastic for long periods of time. It measures the rigidity of plastics at service temperatures and can serve also as procurement identification

TABLE 3   ASTM Tests for Plastics and Their Significance[4] (Continued)

ASTM test	Significance
D 695-69, compressive properties of rigid plastics.......	The compressive strengths of plastics are of limited design value, since plastics products (except foams) seldom fail from compressive loading alone. The compressive-strength figures, however, may be used in specifications for distinguishing between different grades of a material and also for assessing, along with other property data, the overall strengths of different kinds of materials being tested
D 732-46 (1969), shear strength..................	Shear strength is particularly important in film and sheet products where failures from this type of load may often occur. It would seldom be a factor for the design of molded and extruded products
Thermal tests: D 648-72, deflection temperature...............	This test shows the temperature at which an arbitrary amount of deflection occurs under established loads. It is not intended to be a direct guide to high-temperature limits for specific applications. It may be useful for comparing the relative behaviors of various materials in these test conditions, but it is primarily useful for control and development purposes
D 635-72, flammability (for self-supporting materials)....	If the specimen does not ignite, it is classed "nonburning by this test." If the specimen continues to burn, it is timed until it stops or the 4-in. mark is reached. A test specimen that continues to burn to the 4-in. mark is classed as "burning by this test" and the rate is equal to (180/time) in./min. If the specimen does not continue burning to the 4-in. mark, it is classed as "self-extinguishing by this test" and the length of the burned portion is reported as the "extent of burning"
D 1238-70, flow rate (melt index) by extrusion plastometer...............	The melt-index test is primarily useful to raw-materials manufacturers as a method of controlling material uniformity. While the data from this test are not directly translatable into relative end-use processing characteristics, the melt-index value is nonetheless strongly indicative of relative "flowability" of various kinds and grades of PE. The "property" measured by this test is basically the melt viscosity or "rate of shear." In general, the materials that are more resistant to flow are those that are of higher molecular weight The flow rate obtained with the extrusion plastometer is not a fundamental polymer property and should not be so regarded. It is an empirically defined parameter critically influenced by the physical properties and molecular structure of the polymer and the conditions of measurement. The rheological characteristics of polymer melts depend upon a number of variables. Since the values of these variables occurring in this test may differ substantially from those in large-scale processes, test results may not correlate directly with processing behavior
D 569-59 (1971), flow properties................	This test is used for determining the "flow temperature" of a cellulosic, and it is not intended to correspond with temperatures that are used in commercial extrusion and molding. It serves as a control and identification test and is useful in comparing different grades or different formations of a thermoplastic material

TABLE 3 ASTM Tests for Plastics and Their Significance[4] (Continued)

ASTM test	Significance
D 1525-70, Vicat softening point.....................	The Vicat softening temperature is a good way of comparing the heat-softening characteristics of polyethylenes. It also may be used with other thermoplastics
D 746-72, brittleness temperature................	This test is of some use in judging the relative merits of various materials for low-temperature flexing or impact. However, it is specifically relevant only for materials and conditions specified in the test, and the values cannot be directly applied to other shapes and conditions. The brittleness temperature does not put any lower limit on service temperature for end products, but this property is sometimes used in specifications
Analytical tests: D 792-66 (1970), specific gravity and density.........	Specific gravity is a strong element in the price factor and thus has great importance. Beyond the price/volume relationship, however, specific gravity is used in production control in both raw-material production and molding and extrusion. Polyethylenes, for instance, may have degrees of density variation depending upon the degree of "packing" during the molding process or depending upon the rate of quench during extrusion While specific gravity and density are frequently used interchangeably, there is a very slight difference in their meaning: specific gravity is the ratio of the weight of a given volume of material at 73.4°F (23°C) to that of an equal volume of water at the same temperature. It is properly expressed as "specific gravity, 23/23°C." Density is the weight per unit volume of material at 23°C and is expressed "D23C, g/cm^3." The discrepancy enters because water at 23°C has a density slightly less than 1. In order to convert specific gravity to density, the following factor can be used: D23C, g/cm^3 = specific gravity, 23/23°C × 0.99756
D 1505-68, density by density-gradient technique..........	Density determinations by this method are very accurate and quick
Optical tests: E 308-66, spectrophotometry and description of color in CIE 1931 system...........	This recommended practice provides spectral reflectance or transmittance data, or both, for isotropic, nonluminescent and optically uniform specimens. Luminous reflectance or transmittance, tristimulus values, and chromaticity coordinates may be calculated from these data
D 1003-61 (1970), haze.......	In this test, haze of a specimen is defined as the percentage of transmitted light which, in passing through the specimen, deviates more than 2.5° from the incident beam by forward scattering. Luminous transmittance is defined as the ratio of transmitted to incident light. These qualities are considered in most applications for transparent plastics. They form a basis for directly comparing the transparencies of various grades of plastics and types of plastics. From the raw-material manufacturing standpoint, they are important control tests in the various stages of the production cycle

TABLE 3   ASTM Tests for Plastics and Their Significance[4] (Continued)

ASTM test	Significance

Electrical tests:
D 618-61 (1971), conditioning
procedures................

The temperature and moisture contents of plastics affect physical and electrical properties. The standard makes it possible to get comparable test results at different times and in different laboratories. Besides the basis test, conditions for testing at higher or lower levels of temperature and humidity are available in the "Annual Book of ASTM Standards," Part 27.

D 495-71, arc resistance......

The high-volume, low-current type of arc-resistance test is intended to simulate only approximately such service conditions as those existing in alternating-current circuits operating at high voltage and at currents limited to units and tens of milliamperes

To distinguish more easily among materials that by this test have low arc resistance, the early stages of the test are mild and the latter stages become successively more severe. The arc occurs intermittently between two electrodes resting on the surface of the specimen; the severity is increased in the early stages by successively decreasing to zero the interval between flashes of uniform duration and in later stages by increasing the current

The arc resistance of a material is described by this method by measuring the total elapsed time of operation of the test until failure occurs. Four general types of failures have been observed:
1. Many inorganic dielectrics become incandescent, whereupon they are capable of conducting the current. Upon cooling, however, they return to their earlier insulating conditions
2. Some organic compounds burst into flame without the formation of visible conducting paths in the substances
3. Others are seen to fail by "tracking"; that is, a thin wiry line is formed between the electrodes
4. The fourth type occurs by carbonization of the surface until sufficient carbon is present to carry the current
Materials often fail within the first few seconds after a change in the severity stage. When arc resistances of materials are compared, much more weight should be given to a few seconds that overlap two stages than to the same elapsed time within a stage. Thus there is a much greater difference in arc resistance between 178 and 182 s than between 174 and 178 s

D 149-64 (1970), dielectric
strength.................

This test is an indication of the electrical strength of a material as an insulator. The dielectric strength of an insulating material is the voltage gradient at which electric failure or breakdown occurs as a continuous arc (the electrical property analogous to tensile strength in mechanical properties). The dielectric strengths of materials vary greatly with several conditions such as humidity and geometry, and it is not possible to apply the standard test values directly to field use unless all conditions, including specimen dimension, are the same. Because of this, the dielectric-strength test results are of relative rather than absolute value as specification guides. The dielectric strengths of materials drop sharply if holes, bubbles, or contaminants are present in the specimens being tested. The dielectric strength varies inversely with the thickness of the specimen

TABLE 3   ASTM Tests for Plastics and Their Significance[4] (Continued)

ASTM test	Significance	
D 150-70, dielectric constant and dissipation factor.......	Dissipation factor is a ratio of the real power (in-phase power) to the reactive power (power 90° out of phase). It is defined also in other ways Dissipation factor is the ratio of conductance of a capacitor in which the material is the dielectric to its susceptance. It is the ratio of its parallel reactance to its parallel resistance. It is the tangent of the loss angle and the cotangent of the phase angle The dissipation factor is a measure of the conversion of the reactive power to real power, showing as heat. Dielectric constant is the ratio of the capacity of a condenser made with a particular dielectric to the capacity of the same condenser with air as the dielectric. For a material used to support and insulate components of an electrical network from each other and ground, it is generally desirable to have a low level of dielectric constant. For a material to function as the dielectric or a capacitor, on the other hand, it is desirable to have a high value of dielectric constant, thus a small capacitor Loss factor is the product of dielectric constant and power factor and measures total dielectric loss of the material which is being tested	
D 257—6 (1972), tests for electrical resistance.........	Insulation resistance.....	In materials used to insulate and support components of an electrical network, it is generally desirable to have insulation resistance as high as possible
	Volume resistivity.......	Knowing the volume and surface resistivity of an insulating material makes it possible to design an insulator for a specific application
	Volume resistance.......	High volume and surface resistance are desirable to limit the current leakage of the conductor that is being insulated

* This standard lists pipe-wall thicknesses required for various materials and kinds of service, general appearance and finish requirements, and tolerances, as well as the test described here.   (The pressure-test apparatus is described in ASTM D 1598-67.)   Passing all tests is a requirement for marketing pipe as meeting this commercial standard.

Rating of Plastics by Property Comparisons

Subsequent sections of this chapter will present data on plastics as a function of the most important variables.   It is frequently useful, however, to compare or rate plastics for a given property or characteristic.   Any such data must be considered approximate, of course, because of the many possible variables involved.   Some such comparative data approximations are presented in Chaps. 2 and 3 for electrical and mechanical characteristics.   Data on some other important characteristics are presented at this point.   Tables 5 through 10 show comparative data, respectively, for specific gravity, water absorption, hardness, thermal conductivity, thermal expansion, and refractive index.   The relationship of different hardness scales is shown in Fig. 9.   Frequently it is desirable to compare plastics on a cost basis for various property requirements.   Such comparison data, including property values, along with the approximate price for major plastics, are shown in Table 11.

Flammability and Flame Retardancy of Plastics

Because plastics are increasingly used in structures posing hazards to human life in the event of escalating fire, smoke, gases, etc., and because the rapid spread of

TABLE 4   Cross Reference of ASTM and Federal Tests[5]

Test method	Federal Standard Method No.	ASTM No.
Abrasion wear (loss in weight)	1091	
Accelerated service tests (temperature and humidity extremes)	6011	D 756-56
Acetone extraction test for degree of cure of phenolics	7021	D 494-46
Arc resistance	4011	D 495-61
Bearing strength	1051	D 953-54, Method A
Bonding strength	1111	D 229-63T, pars 40–43
Brittleness temperature of plastics by impact	2051	D 746-64T
Compressive properties of rigid plastics	1021	D 695-63T
Constant-strain flexural fatigue strength	1061	
Constant-stress flexural fatigue strength	1062	
Deflection temperature under load	2011	D 648-56, Procedure 6(a)
Deformation under load	1101	D 621-64, Method A
Dielectric breakdown voltage and dielectric strength	4031	D 149-64
Dissipation factor and dielectric constant	4021	D 150-64T
Drying test (for weight loss)	7041	
Effect of hot hydrocarbons on surface stability	6062	
Electrical insulation resistance of plastic films and sheets	4052	
Electrical resistance (insulation, volume, surface)	4041	D 257-61
Falling-ball impact	1074	
Flame resistance	2023	
Flammability of plastics 0.050 in. and under in thickness	2022	D 568-61
Flammability of plastics over 0.050 in. in thickness	2021	D 635-63
Flexural properties of plastics	1031	D 790-63
Indentation hardness of rigid plastics by means of a durometer	1083	D 1706-61
Interlaminar and secondary bond shear strength of structural plastic laminates	1042	
Internal stress in plastic sheets	6052	
Izod impact strength	1071	D 256-56, Method A
Linear thermal expansion (fused-quartz tube method)	2031	D 696-44
Machinability	5041	
Mar resistance	1093	D 673-44
Mildew resistance of plastics, mixed culture method, agar medium	6091	
Porosity	5021	
Punching quality of phenolic laminated sheets	5031	D 617-44
Resistance of plastics to artificial weathering using fluorescent sunlamp and fog chamber	6024	D 1501-57T
Resistance of plastics to chemical reagents	7011	D 543-60T
Rockwell indentation hardness test	1081	D 785-62, Method A
Salt-spray test	6071	
Shear strength (double shear)	1041	
Shockproofness	1072	
Specific gravity by displacement of water	5011	
Specific gravity from weight and volume measurements	5012	
Surface abrasion	1092	D 1044-56
Tear resistance of film and sheeting	1121	D 1004-61
Tensile properties of plastics	1011	D 638-64T
Tensile properties of thin plastic sheets and films	1013	D 882-64T
Tensile strength of molded electrical insulating materials	1012	D 651-48
Tensile time—fracture and creep	1063	
Thermal-expansion test (strip method)	2032	
Warpage of sheet plastics	6054	D 1181-56
Water absorption of plastics	7031	D 570-63

TABLE 5   Comparative Specific-Gravity Values for Plastics[6]

Plastic	Specific gravity
Epoxy (low density)	0.75–1.00
Methylpentene polymer (TPX)	0.83
Polypropylene copolymer	0.89–0.905
Polypropylene impact (rubber modified)	0.90–0.91
Polypropylene unmodified	0.902–0.906
Polyethylenes (low-density)	0.91–0.925
Ethylene vinyl acetate copolymer	0.925–0.95
Polyethylene (medium-density)	0.926–0.940
Ethylene ethyl acrylate copolymer	0.93
Ionomers	0.93–0.96
Styrene butadiene thermoplastic elastomers	0.93–1.10
Polyethylene, high-molecular-weight	0.94
Polyethylene (high-density)	0.941–0.965
Polyethylene cross-linkable grades	0.95–1.45
Polypropylene (inert filler)	1.00–1.30
Nylon 12 (unfilled)	1.01–1.02
ABS (high impact)	1.01–1.04
Nylon 11 (unfilled)	1.04–1.05
ABS (medium impact)	1.04–1.07
Polystyrene (general-purpose)	1.04–1.09
Polystyrene (medium and high impact)	1.04–1.10
Polystyrene heat and chemical	1.04–1.10
PVC flexible (unfilled)	1.05
ABS high (heat-resistant)	1.05–1.08
Polypropylene (glass filler)	1.05–1.24
Nylon 6/12	1.06–1.08
Polyphenylene oxides modified (Noryl)	1.06–1.10
Polyphenylene oxide (PPO)	1.06
Styrene acrylonitrile copolymer (unfilled) (SAN)	1.075–1.10
Nylon 6/10 (unfilled)	1.09
Acrylics multipolymer	1.09–1.14
Ethyl cellulose	1.09–1.17
Polycarbonate (ABS polycarbonate alloy)	1.10–1.20
Acrylic impact	1.11–1.18
Urethane elastomers	1.11–1.25
Nylon 6 (unfilled)	1.12–1.14
Nylon 6/6 (unfilled)	1.13–1.15
ABS self-extinguishing	1.15–1.21
Cellulose acetate butyrate	1.15–1.22
PVC flexible (unfilled)	1.16–1.35
Acrylics	1.17–1.20
Cellulose propionate	1.17–1.24
Phenoxy	1.18
Rubber (hard)	1.20
Polycarbonate (unfilled)	1.20
Polystyrene (20–30% glass filler)	1.20–1.33
Styrene acrylonitrile copolymer (20–33% glass filler)	1.20–1.46
Polyphenylene oxides modified (Noryl) (20–30% glass filler)	1.21–1.36
Cellulose acetate	1.22–1.34
Nylon 12 (glass filler)	1.23
ABS (20–40% glass filler)	1.23–1.36
Polysulfone	1.24
Polycarbonate (10–40% glass filler)	1.24–1.52
Alkyd polyester synthetic fiber	1.24–2.10
Polycarbonate (less than 10% glass filler)	1.25
Nylon 11 (glass filler)	1.26
Polypropylene modified vinyl chloride	1.28–1.58
Polycarbonate (unfilled)	1.28–1.43
Nylon (20–40% glass filler)	1.28–1.43

TABLE 5   Comparative Specific-Gravity Values for Plastics[6] (Continued)

Plastic	Specific gravity
Phenol-formaldehyde with butadiene—acrylonitrile copolymer—wood flour and cotton flock	1.29–1.35
Acrylic PVC alloy	1.30
Phenol formaldehyde with butadiene acrylonitrile copolymer rag	1.30–1.35
PVC flexible (filled)	1.30–1.70
Polyterephthalate 6PRO	1.31
Nylon 6/12 (glass filler)	1.32–1.46
Polyphenylene sulfides (unfilled)	1.34–1.43
Phenolic wood flour and cotton	1.34–1.45
Cellulose nitrate	0.35–1.40
PVC (rigid)	1.35–1.45
Polyester preformed chopped roving	1.35–2.30
Casein	1.35
Phenolic macerated fabric and cord	1.36–1.43
Nylon 6 (20–40% glass filler)	1.37–1.46
Polyesters (thermoplastic)	1.37–1.38
Chlorinated polyether	1.40
Acetal copolymer	1.41
Acetal homopolymer	1.42
Polyimides aromatic	1.43
Melamine cellulose	1.45–1.52
Phenolic (asbestos filler)	1.45–2.00
Melamine alpha cellulose	1.47–1.52
Urea formaldehyde alpha cellulose	1.47–1.52
Melamine formaldehyde (no filler)	1.48
Polyester (thermoplastic 18% glass filler)	1.48–1.50
Melamine (fabric filler)	1.50
Melamine fabric (phenolic modified)	1.50
Melamine (flock filler)	1.50–1.55
Melamine phenol	1.50–1.70
Polyester woven cloth	1.50–2.10
Acetal fiber reinforced (TFE)	1.54
Acetal (20% glass filler)	1.56
Phenol formaldehyde with butadiene acrylonitrile copolymer asbestos	1.57–1.65
Silicones (asbestos filler)	1.60–1.90
Epoxy (glass fiber filler)	1.60–2.00
Epoxy (mineral filler)	1.60–2.00
Cold-mold refractory (inorganic)	1.60–1.20
Alkyd granular and putty (mineral filler)	1.60–2.30
Acetal copolymer (25% glass filler)	1.61
Allyl diallyl phthalate (glass filler)	1.61–1.78
Alkyd polyester (asbestos filler)	1.65
Vinylidene chloride	1.65–1.68
Phenolic (mica filler)	1.65–1.92
Polyester premix (chopped glass filler)	1.65–2.30
Polyester low shrink	1.67
Silicones (glass filler)	1.68–2.00
Phenolic (glass filler)	1.69–1.95
Melamine (asbestos filler)	1.70–2.00
Epoxy encapsulating (glass filler)	1.70–2.00
Epoxy encapsulating (mineral filler)	1.70–2.10
Furane	1.75
Polyvinylidene fluoride (VF2)	1.77
Silicones (mineral filler)	1.81–2.82
Phenolic resin	2.00
Melamine	2.00
Chlorotrifluoro ethylene (CTFE)	2.10–2.20
Alkyd polyester (glass filler)	2.12–2.15
Fluorocarbon (FEP)	2.12–2.17
Polytetrafluoroethylene (TFE)	2.13–2.22

TABLE 6   Comparative Water-Absorption Values for Plastics[6]

Plastic	Water absorption, % by wt.
Polytetrafluoroethylene (TFE)	0
Chlorotrifluoroethylene (CTFE)	0
Fluorocarbon (FEP)	0.01
Polyallomer	>0.01
Polyethylene (medium density)	>0.01
Polyethylene (high density)	>0.01
Polypropylene (unmodified)	>0.01
Polypropylene impact (rubber modified)	>0.01
Chlorinated polyether	0.01
Methylpentene polymer (TPX)	0.01
Polybutylene	>0.01–0.026
Polypropylene (copolymer filler)	>0.01–0.030
Polypropylene (glass filler)	0.01–0.050
Phenolic (mica filler)	0.01–0.050
Polyethylene cross-linkable grade	>0.01–0.060
Furane molding compound (asbestos filler)	0.01–0.200
Polyester (preformed chopped rovings)	0.01–1.000
Polyethylene (low density)	>0.015
Polyesters (thermoplastic)	0.02
Polyesters (18% glass filler)	0.02
Polypropylene (inert filler)	0.02–0.100
Polystyrene (general-purpose)	0.03–0.100
Epoxy encapsulating (mineral filler)	0.03–0.200
Phenolic (glass filler)	0.03–1.200
Polyvinylidene (VF2)	0.04
Ethylene–ethyl acrylate copolymer (EEA)	0.04
Epoxy (mineral filler)	0.04
Epoxy encapsulating (glass filler)	0.04–0.200
Polystyrene (20–30% glass filler)	0.05–0.100
Ethylene vinyl acetate copolymer (EVA)	0.05–0.130
Epoxy (glass filler)	0.05–0.200
Polyester alkyd granular putty type (synthetic filler)	0.05–0.200
Polyester alkyd granular putty type (glass filler)	0.05–0.250
Polystyrene (chemical filler)	0.05–0.400
Polyester woven cloth	0.05–0.500
Polyester alkyd granular putty type (mineral filler)	0.05–0.500
Polystyrene (medium impact)	0.05–0.600
Polystyrene (high impact)	0.05–0.600
Polystyrene (heat-resistant)	0.05–0.600
Polyphenylene oxide modified (20–30% glass filler)	0.06
Polyphenylene oxides (modified unfilled)	0.066
Polyester (premixed chopped glass filler)	0.06–0.280
Nylon (polyamide) type 12 (glass filler)	0.07–2.0
Polycarbonate (10–40% glass filler)	0.07–0.200
PVC (rigid)	0.07–0.400
Polypropylene modified vinyl chloride (rigid)	0.07–0.400
Silicones (mineral filler)	0.08–0.130
Melamine (asbestos filler)	0.08–0.140
Styrene acrylonitrile copolymer (glass filler)	0.08–0.220
Polyester (low shrink)	0.10
Polyphenylene oxides (unfilled) (PPO)	0.10
Polyesters (sheet molding compound)	0.10–0.150
Silicones (glass filler)	0.10–0.200
Epoxy (low-density)	0.10–0.200
Acrylics	0.10–0.400
Phenol formaldehyde (asbestos filler)	0.10–0.500
Melamine alpha cellulose	0.10–0.600
Ionomers	0.10–1.400
Polycarbonate (10% glass filler)	0.12

TABLE 6   Comparative Water-Absorption Values for Plastics[6] (Continued)

Plastic	Water absorption, % by wt.
Diallyl phthalate (glass filler)	0.12–0.350
Acrylic PVC alloy (Kydene)	0.13
Polyester alkyd granular putty type (asbestos filler)	0.14
Polycarbonate (unfilled)	0.15–0.180
Phenolic acrylonitrile copolymer (asbestos filler)	0.15–0.200
Nylon polyamide type 6/12 (glass filler)	0.15–0.200
PVC flexible (unfilled)	0.15–0.750
Melamine (flock filler)	0.16–0.300
ABS (20–40% glass filler)	0.18–0.400
Styrene butadiene thermoplastic elastomers	0.199–0.390
Diallyl phthalate (synthetic-fiber filler)	0.20
ABS modified (rigid)	0.20
Styrene acrylonitrile copolymer (SAN)	0.20–0.300
ABS polycarbonate alloy	0.20–0.350
ABS (high-impact)	0.20–0.450
ABS (heat-resistant)	0.20–0.450
ABS (medium-impact)	0.20–0.450
Diallyl phthalate (mineral filler)	0.20–0.500
ABS (self-extinguishing)	0.20–0.600
Acrylics (impact)	0.20–0.700
Melamine (fabric filler phenolic modified)	0.20–1.300
Nylon (polyamide) (20–40% glass filler)	0.20–2.000
Acetal copolymer	0.22
Polysulfone	0.22
Acetal homopolymer	0.25
Nylon (polyamide) type 12	0.25
Polyaryl ether	0.25
Acetal (20% glass filler)	0.25–0.290
Acetal copolymer (25% glass filler)	0.29
Acrylics multipolymer	0.30
Melamine (fabric filler)	0.30–0.600
Melamine (phenol filler)	0.30–0.650
Phenol formaldehyde (wood-flour filler)	0.30–1.200
Phenol formaldehyde (cotton filler)	0.30–1.200
Polyimides aromatic	0.32
Melamine (cellulose filler)	0.34–0.800
Nylon (polyamide) type 6/10	0.40
Phenol formaldehyde type 6/12	0.40
Nylon (polyamide) type 11	0.40–2.900
Urea formaldehyde	0.40–0.800
Phenol (fabric and cord filler)	0.40–1.750
PVC flexible (filled)	0.50–1.000
Cold-mold refractory (inorganic) asbestos (filler)	0.50–15.000
Nylon (polyamide) type 11 (glass filler)	0.54
Cold-mold nonrefractory (organic)	0.55–2.000
Nylon (polyamide) type 6/6 (20–40% glass filler)	0.65–1.500
Cold mold (organic), asbestos, binder melamine	0.67
Urethanes elastomers	0.70–0.900
Ethyl cellulose	0.80
Phenolic resin	0.90
Cellulose acetate butyrate	0.90–2.200
Cellulose nitrate	1.00–2.000
Phenolic acrylonitrile copolymer (wood-flour and flock filler)	1.00–2.000
Phenolic acrylonitrile copolymer (rag filler)	1.00–2.000
Cellulose propionate	1.20–2.800
Nylon (polyamide) type 6	1.30–2.900
Nylons (polyamide) type 6/6	1.50
Cellulose acetate	1.70–6.500
Polyaryl sulfone	1.80

TABLE 7   Comparative Hardness Values for Plastics[6]

Plastic	Value
Rockwell M Series	
Polypropylene modified vinyl chloride (rigid)	18–55
Polystyrene (medium-impact)	20–80
Polystyrene (high-impact)	20–80
Urethanes elastomers	28
Nylon (polyamide) type 12	31
Phenolic with butadiene acrylonitrile copolymer (wood-flour and flock filler)	45–60
Phenolic with butadiene (asbestos filler)	50
Phenolic with butadiene (rag filler)	50–70
Polystyrenes (general-purpose)	65–80
Polystyrenes (heat and chemical)	65–90
ABS (20–40% glass filler)	65–100
Polyterephthalate 6 PRO	68
Polysulfone	69
Polyphenylene oxides (unfilled)	70
Polystyrenes (20–30% glass filler)	70–95
Polyester preformed chopped rovings	70–120
Silicones (glass filler)	72–90
Polycarbonate (unfilled)	73–78
Acetal (20% glass filler)	75–90
Acetal copolymer	75–80
Silicones (asbestos filler)	80
Styrene acrylonitrile copolymer (SAN)	80–90
Silicones (mineral filler)	80–95
Polyester woven cloth	80–120
Polycarbonate (10% glass filler)	85
Acrylic	85–105
Polycarbonate (up to 40% glass filler)	88–95
Phenolic resin	93
Acetal homopolymer	94
Polyesters (thermoplastic)	94
Melamine alpha cellulose (glass filler)	95–100
Polyester alkyd granular putty type (synthetic filler)	95
Phenolic (fabric and cord filler)	95–107
Polyester alkyd granular putty type (asbestos filler)	99
Styrene acrylonitrile (20–30% glass filler)	100
Phenolic (mica filler)	100–110
Epoxy molding (mineral filler)	100–110
Epoxy encapsulating (glass filler)	100–112
Epoxy encapsulating (mineral filler)	100–112
Phenolic formaldehyde (cotton filler)	100–115
Phenolic formaldehyde (wood-flour filler)	100–115
Acrylics (heat-resistant)	102–105
Epoxy molding (glass filler)	105–115
Phenolic formaldehyde (asbestos filler)	105–115
Diallyl phthalate (synthetic fiber filler)	108–115
Polyesters (18% glass filler)	109
Polyaryl sulfone	110
Melamine (asbestos filler)	110
Urea formaldehyde	110–120
Melamine (alpha cellulose filler)	115–125
Melamine (cellulose filler)	115–125
Melamine (flock filler)	115–125
Melamine (fabric filler)	120
Rockwell R Series	
Polyethylenes (low-density)	10
Cellulose propionate	10–122
Polyethylenes (medium-density)	15
Cellulose acetate butyrate	31–116

TABLE 7   Comparative Hardness Values for Plastics[6] (Continued)

Plastic	Value
Cellulose acetate	34–125
Polyallomer	50–85
Polypropylene (impact rubber modified)	50–85
Polypropylene (unmodified)	50–96
Polystyrene (medium-impact)	50–100
Polystyrene (high-impact)	50–100
Ethyl cellulose	50–115
Urethanes elastomers	60
ABS (self-extinguishing)	75–98
ABS (high-impact)	75–105
Chlorotrifluoroethylene (CTFE)	75–95
Polypropylene (unmodified)	85–110
PVC (rigid)	85–158
Polypropylene (glass filler)	90
Polypropylene (inert filler)	94–100
Cellulose nitrate	95–115
Chlorinated polyether	100
ABS modified vinyl chloride (rigid)	102
Nylons (polyamide) type 6	103–119
Acrylic PVC alloy	103
Acrylic (impact)	105–120
Nylons (polyamide) type 12	106
ABS polycarbonate alloy	106–120
ABS (heat-resistant)	107–115
ABS (medium-impact)	107–115
Polypropylene modified vinyl chloride (rigid)	107–119
Nylon (polyamide) type 11	108
Acrylics multipolymer	108–119
Nylon (polyamide) type 6/6 (unfilled)	108–120
Furane molding compound (asbestos filler)	110
Nylon (polyamide) type 6/10	111
Nylon (polyamide) type 11 (glass filler)	114
Nylon (polyamide) type 6/12	114
Polyphenylene oxides modified (unfilled)	115–119
Polycarbonate (unfilled)	115–125
Polyterphthalate 6PRO	117
Polyaryl-ether	117
Nylon type 6/12 (glass filler)	118
Acetal homopolymer	120
Polyesters (thermoplastic)	120
Polysulfone	120
Polyphenylene (25% asbestos filler)	123
Polyphenylene (40% glass filler)	123
Polyphenylene sulfides (unfilled)	124

Barcol

Polyesters (sheet molding compound)	50–70
Polyester alkyd granular putty type (glass filler)	55–80
Polyester alkyd granular putty type (mineral filler)	60–70
Polyester (premixed chopped glass filler)	60–80

Shore Durometer

Urethanes elastomers	A30–70D
Styrene butadiene thermoplastic elastomers	A40–90
PVC flexible (filled)	A50–100
PVC flexible (unfilled)	A55–90
ABS modified vinyl chloride flexible	A93
Ethylene vinyl acetate copolymer	D17–38
Ethylene–ethyl acrylate copolymer	D22–36
Polyethylenes (low-density)	D41–46D
Polyethylenes (medium-density)	D50–60D

TABLE 7   Comparative Hardness Values for Plastics[6] (Continued)

Plastic	Value
Shore Durometer (Continued)	
Fluorocarbon (FEP)..	D50–D60
Polytetrafluoroethylene (TFE)................................	D50–D65
Ionomers..	D50–D65
Polyethylenes cross-linkable grades..........................	D55–D80
Polyethylenes (high-density).................................	D60–D70
Chlorotrifluoroethylene (CTFE)...............................	D76
Polyvinylidene fluoride (VF2)................................	D80
Rockwell Series	
Nylons (polyamide) type 6/10 (20–40% glass filler)...........	40–50
Polyamides aromatic..	40–60
Nylons (polyamide) type 6 (20–40% glass filler).............	50–70
Nylons (polyamide) type 6/6 (20–40% glass filler)...........	50–80
Phenolic (glass filler)......................................	54–101
Styrene acrylonitrile copolymer (20–30% glass filler).......	60
Diallyl phthalate (mineral filler)..........................	61
Polybutylene...	D65
Polyester alkyd granular putty type (synthetic filler)......	76
Phenolic (fabric and cord filler)...........................	79–82
Diallyl phthalate (glass filler)............................	80–87
Phenolic (mica filler).......................................	88
Polyester alkyd granular putty type (glass filler)..........	95
Melamine (phenol filler).....................................	95–100
Polyester alkyd granular putty type (mineral filler)........	98

TABLE 8   Comparative Thermal-Conductivity Values for Plastics[6]

Plastic	Value, $Btu/(h)(ft^2)(°F/in.)$
Polystyrenes (medium-impact).................................	0.29–0.87
Polystyrenes (high-impact)...................................	0.29–0.87
Polystyrenes (heat-resistant)................................	0.29–0.87
Urethanes elastomers...	0.49
Polystyrenes (chemical filler)..............................	0.55–0.87
Polyallomer..	0.58–1.20
Polypropylene copolymer......................................	0.58–1.20
Polystyrenes (general-purpose)..............................	0.70–0.96
Acrylics...	0.7–1.7
Polypropylene (unmodified)...................................	0.81
Styrene acrylonitrile copolymer.............................	0.84
Polyvinylidene fluoride (VF2)................................	0.87
Polypropylene (impact rubber modified)......................	0.87–1.20
Chlorinated polyether..	0.90
PVC flexible (unfilled)......................................	0.99
PVC flexible (filled)..	0.99
Styrene butadiene thermoplastic elastomers..................	1.00
PVC (rigid)..	1.10
Nylons (polyamide) type 12 (glass filler)...................	1.10
Polyphenylene oxides (30% glass filler).....................	1.10
Ethyl cellulose..	1.10–2.00
Methylpentene polymer..	1.20
Phenolic (fabric and cord filler)...........................	1.20
Phenolic with butadiene—acrylonitrile copolymer (rag filler).......	1.20
Acrylics (impact)..	1.20–1.50
Acrylics cast..	1.20–1.70
Epoxy (low-density)..	1.20–1.70
Phenolic with butadiene—acrylonitrile copolymer (asbestos filler)...	1.20–1.70
Melamine (phenolic filler)..................................	1.20–2.00

TABLE 8   Comparative Thermal-Conductivity Values for Plastics[6] (Continued)

Plastic	Value, Btu/(h)(ft²)(°F/in.)
Cellulose acetate..	1.20–2.30
Cellulose propionate..	1.20–2.30
Cellulose acetate butyrate....................................	1.20–2.30
Phenolic formaldehyde (wood-flour filler)......................	1.20–2.40
Phenolic formaldehyde (cotton filler)..........................	1.20–2.40
Epoxy encapsulating (glass filler)............................	1.20–2.90
Epoxy (glass filler)..	1.20–2.90
Epoxy encapsulating (mineral filler)...........................	1.20–2.90
Epoxy (mineral filler)..	1.20–8.70
Polyaryl sulfone...	1.30
Polycarbonate (unfilled).....................................	1.30
ABS (high-impact)...	1.30–2.30
ABS (heat-resistant)...	1.30–2.30
ABS (medium-impact)...	1.30–2.30
ABS (self-extinguishing)......................................	1.30–2.30
Polyphenylene oxides (PPO)...................................	1.33
Polycarbonate (10% glass filler)..............................	1.40
Polycarbonate (up to 40% glass filler).........................	1.40–1.50
Chlorotrifluoroethylene (CTFE)...............................	1.40–1.50
Acrylics multipolymer..	1.50
Nylon (polyamide) type 6/10..................................	1.50
Nylon (polyamide) type 6/10 (20–40% glass filler)..............	1.50
Nylon (polyamide) type 12....................................	1.50
Polyphenylene oxides (modified) (Noryl)........................	1.50
Phenolic with butadiene—acrylonitrile copolymer (wood-flour and flock filler)................................	1.50–1.70
Diallyl phthalate (synthetic-fiber filler).......................	1.50–1.70
Diallyl phthalate (glass filler)...............................	1.50–4.40
Acetal homopolymer..	1.60
Acetal copolymer..	1.60
Cellulose nitrate..	1.60
Nylons (polyamide) type 6 (glass filler).......................	1.60
Ionomers...	1.70
Nylons (polyamide) type 6/6..................................	1.70
Nylons (polyamide) type 6....................................	1.70
Polytetrafluoroethylene (TFE)................................	1.70
Fluorocarbon (FEP)..	1.70
ABS polycarbonate alloy......................................	1.70–2.60
Phenolic formaldehyde (asbestos filler)........................	1.70–6.40
Melamine (cellulose filler)...................................	1.90–2.50
Nylons (polyamide) type 11...................................	2.00
Polyaryl ether..	2.00
Polyesters (thermoplastic)....................................	2.00
Polyesters (18% glass filler)..................................	2.00
Polyphenylene sulfides (unfilled)..............................	2.00
Urea formaldehyde..	2.00–2.90
Melamine (alpha cellulose filler)..............................	2.00–2.90
Diallyl phthalate (mineral filler)..............................	2.00–7.30
Silicones (glass filler).......................................	2.20
Polyethylenes (low-density)...................................	2.30
Polyethylenes (medium-density)...............................	2.30–2.90
Phenolic (glass filler).......................................	2.40
Nylons (polyamide) type 11 (glass filler)......................	2.60
Phenolic (mica filler) mineral................................	2.90–4.00
Melamine (fabric filler phenolic modified)......................	2.90
Polyester (premixed chopped glass)............................	2.90–4.60
Melamic (fabric filler).......................................	3.10
Polyethylenes (high-density).................................	3.20–3.60
Silicones (mineral filler)....................................	3.20–3.80
Polyester alkyd granular putty type (mineral filler).............	3.50–7.30
Polyester alkyd granular putty type (glass filler)...............	4.40–7.30

TABLE 9   Comparative Coefficient of Expansion Values for Plastics[6]

Plastic	Coefficient, 10^{-5} cm/cm/°F
Polyester (low-shrink)	0.366–0.605
Phenolic (glass filler)	0.44–1.14
Phenolic formaldehyde (asbestos filler)	0.44–2.22
Silicones (glass filler)	0.443
Melamine (phenolic filler)	0.555–2.22
Diallyl phthalate (glass filler)	0.555–2.00
Phenolic (fabric and cord filler)	0.555–2.22
Diallyl phthalate (mineral filler)	0.555–2.33
Epoxy (glass filler)	0.610–1.94
Polyester (preformed chopped rovings)	0.66–2.78
Nylons (polyamide) type 6/10 (20–40% glass filler)	0.666–1.78
Melamine (glass filler)	0.83–1.10
Nylons (polyamide) type 6/6 (30% glass filler)	0.83–1.11
Polyester alkyd granular putty type (glass filler)	0.83–1.39
Polyester woven cloth	0.83–1.66
Phenolic with butadiene acrylonitrile copolymer (wood-flour and flock filler)	0.83–1.94
Phenolic (mica filler)	1.00–1.44
Melamine (fabric filler) (phenolic modified)	1.00–1.55
Styrene acrylonitrile copolymer	1.00–2.50
Polyester (sheet molding compound)	1.11
Polyester (premixed chopped glass filler)	1.11–1.83
Silicones (mineral filler)	1.11–2.22
Melamine (asbestos filler)	1.11–2.50
Polyester alkyd granular putty type (mineral filler)	1.11–2.78
Epoxy (mineral filler)	1.11–2.78
Phenolic with butadiene acrylonitrile copolymer (rag filler)	1.16
Polyphenylene oxides (modified) (20–30% glass filler)	1.22
Urea formaldehyde	1.22–2.00
Polyesters (18% glass filler)	1.39
Melamine (fabric filler)	1.39–1.55
Styrene acrylonitrile (glass filler)	1.50–2.11
Polypropylene (inert filler)	1.61
ABS (20–40% glass filler)	1.61–2.00
Polypropylene (glass filler)	1.61–2.88
Phenolic formaldehyde (wood-flour filler)	1.66–2.50
Epoxy encapsulating (glass filler)	1.66–2.78
Epoxy encapsulating (mineral filler)	1.66–3.33
Polyester alkyd granular putty type (synthetic filler)	1.67–3.06
Nylons (polyamide) type 6 (glass filler)	1.83
Nylons (polyamide) type 11 (glass filler)	1.78
Polystyrenes (medium- and high-impact)	1.88–11.70
Polycarbonate (10% glass filler)	1.90
Styrene acrylonitrile copolymer	2.00–2.11
Acetal (20% glass filler)	2.00–4.50
Polyphenylene (40% glass filler)	2.22
Phenolic with butadiene acrylonitrile copolymer (asbestos filler)	2.22
Polyamides aromatic	2.22–2.78
Chlorotrifluoroethylene (CTFE)	2.50–3.90
Melamine alpha cellulose	2.22
Melamine (cellulose filler)	2.50
Polyaryl sulfone	2.62
PVC (rigid)	2.77–10.30
Acrylics	2.78–5.00
Polyphenylene oxides (modified) (unfilled)	2.89
Polysulfone	2.89–3.11
Polyphenylene (25% asbestos filler)	2.94
Phenoxy	3.00–3.20
Diallyl phthalate (synthetic fiber filler)	3.00–3.33
Polyphenylene sulfides (unfilled)	3.05

TABLE 9   Comparative Coefficient of Expansion Values for Plastics[6] (Continued)

Plastic	Coefficient, 10^{-5} cm/cm/°F
Polypropylene (unmodified)	3.22–5.66
Polystyrene (heat and chemical)	3.33
Polyesters (thermoplastic)	3.33
ABS (heat-resistant)	3.33–5.00
Polystyrene (general-purpose)	3.33–4.44
Acrylics (impact)	3.33–4.44
Polypropylene (impact rubber modified)	3.33–4.73
Acrylic multipolymer	3.33–5.00
ABS polycarbonate alloy	3.44–4.72
Polyaryl ether	3.61
Polycarbonate (unfilled)	3.64
PVC flexible (unfilled)	3.86–13.90
Nylons (polyamide) type 12 (glass filler)	4.16
Nylons (polyamide) type 6/6	4.44
Chlorinated polyether	4.44
Acetal homopolymer	4.50
ABS (medium-impact)	4.45–5.55
Cellulose nitrate	4.44–6.66
Acetal copolymer	4.72
Polypropylene copolymer	4.44–5.27
Polyallomer	4.60–5.55
Nylons (polyamide) type 6	4.60
Polytetrafluoroethylene (TFE)	4.60–5.80
Cellulose acetate	4.44–10.00
Nylons (polyamide) type 6/10	5.00
ABS (high-impact)	5.27–7.22
Polyethylenes cross-linkable grades	5.55–11.94
Ethyl cellulose	5.55–11.10
Polyethylene (low-density)	5.55–11.10
Urethanes elastomers	5.55–11.10
Fluorocarbon (FEP)	5.60
Nylons (polyamide) type 12	5.78
Polyethylenes (high-density)	6.10–7.22
Cellulose propionate	6.11–9.44
Cellulose acetate butyrate	6.11–9.44
Methylpentene polymer	6.50
Ionomers	6.66
Polyvinylidene fluoride (VF2)	6.70
Styrene butadiene thermoplastic elastomers	7.22–7.60
Polyethylene (medium-density)	7.77–8.89
Polybutylene	8.32
Nylons (polyamide) type 11	8.33
Ethylene vinyl acetate copolymer	8.88–11.10
Ethylene–ethyl acrylate copolymer	8.88–13.80
Polyterephthalate	4.90–13.00

some major fires has been partially attributed to plastics, the reliability of flammability testing of plastics came under serious review in the early 1970s. Perhaps the hallmark fire was the One New York Plaza fire in August 1970, when a fire in this fireproof high-rise building spread over two floors and a total of 40,000 square feet before arrival of the New York Fire Department some 20 minutes after detection of the fire.[8] Three weeks later, a 30-minute fire causing $2.5 million damage in the BOAC terminal at Kennedy International Airport was again largely attributed to foamed plastics.[8]

In review, perhaps the strongest feeling was that small-scale tests do not accurately predict results for large-scale fires. Likewise, it was felt that many descriptive terms for flammability rating such as self-extinguishing (SE) ratings tend to give the user an unjustified feeling of security. Underwriters' Laboratories is making such changes, and the plastics user should acquaint himself with the up-to-date terminology at any given time.

While large-scale tests will be developed, small-scale tests will continue to serve usefully, if judiciously used and interpreted. The test methods available for flammability are many, and they have been originated by many sources for many objectives. A summary of some of the major tests used for flammability is given in Table 12.

Flame retardants   When plastics burn, heat from an external source pyrolyzes the solid plastic to produce gases and liquids that act as fuel for the fire. Fire-retardant chemicals affect the burning rate of a solid in one or more ways:[10]

1. By interfering with the combustion reactions
2. By making the products of pyrolysis less flammable
3. By reducing the transfer of heat from the flame to the solid
4. By reducing the rate of diffusion of pyrolysis products to the flame front

TABLE 10   Comparative Refractive Index Values for Plastics[6]

Plastic	Value
Fluorocarbon (FEP)	1.34
Polytetrafluoroethylene (TFE)	1.35
Chlorotrifluoroethylene (CTFE)	1.42
Cellulose propionate	1.46–1.49
Cellulose acetate butyrate	1.46–1.49
Cellulose acetate	1.46–1.50
Methylpentene polymer	1.465
Ethyl cellulose	1.47
Acetal homopolymer	1.48
Acrylics	1.49
Cellulose nitrate	1.49–1.51
Polypropylene (unmodified)	1.49
Polyallomer	1.492
Polybutylene	1.50
Ionomers	1.51
Polyethylenes (low-density)	1.51
Nylons (polyamide) type 11	1.52
Acrylics multipolymer	1.52
Polyethylene (medium-density)	1.52
Styrene butadiene thermoplastic elastomers	1.52–1.55
PVC (rigid)	1.52–1.55
Nylons (polyamide) type 6/6	1.53
Urea formaldehyde	1.54–1.56
Polyethylene (high-density)	1.54
Styrene actylonitrile copolymer (unfilled)	1.56–1.57
Polystyrene (heat and chemical)	1.57–1.60
Polycarbonate (unfilled)	1.586
Polystyrene (general-purpose)	1.59–1.60
Polysulfone	1.633

Since various plastics burn differently and often at different temperatures and rates, there is no single universal fire retardant. Almost every application requires either a different agent or different amounts of that agent to obtain the desired flame retardancy with minimum effect on other properties.

A number of elements can act as flame retardants, but they must be incorporated in a structure that enables them to become active at the proper temperature. They include nitrogen, phosphorus, arsenic, antimony, bismuth, fluorine, bromine, chlorine, iodine, and boron. Of these, phosphorus, bromine, chlorine, and antimony are currently considered to be the most efficient.[10]

Table 13 shows the levels at which these elements are used as flame retardants in common polymers. Notice that in many cases several of these elements are used in combination to achieve a synergistic effect; that is, the combination is more effective than any one element used at the same level of loading.[10]

THERMOSETTING PLASTICS

The general nature of thermosetting materials has been described above. Plastic materials included in the thermosetting plastic category and discussed separately below are alkyds, diallyl phthalates, epoxies, melamines, phenolics, polyesters, silicones, and ureas. A list of typical trade names and suppliers is shown in Table 14. In general, unfilled thermosetting plastics tend to be harder, more brittle, and not as tough as thermoplastics. Thus, it is common practice to add fillers to thermosetting materials. A wide variety of fillers can be used for varying product properties. For molded products, usually compression or transfer molding, mineral or cellulose fillers are often used as lower-cost, general-purpose fillers, and glass-fiber fillers are often used for optimum strength or dimensional stability. There are always product and processing trade-offs, but a general guide to application of fillers is given in

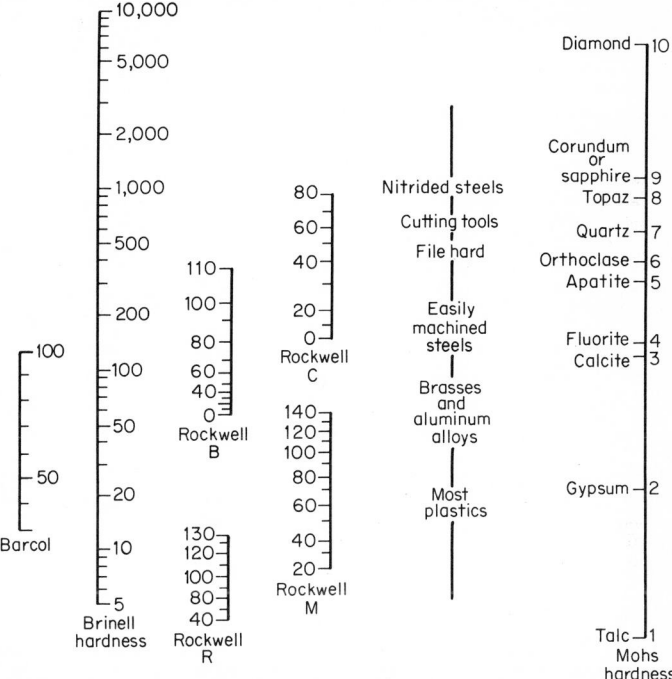

Fig. 9 Comparison of hardness scales.

Table 15. It should be added that filler form and filler surface treatment can also be major variables. Thus, it is important to consider fillers along with thermosetting material, especially for molded products. Other product forms may be filled or unfilled, depending on requirements.

A discussion of the individual groups of thermosetting plastics is given below.

Alkyds

Alkyds are available in granular, rope, or putty form, some suitable for molding at relatively low pressures, and at temperatures in the range of 300 to 400°F. They are formulated from polyester-type resins, in general reactions somewhat as shown in Fig. 3. Other possible monomers, aside from styrene, are diallyl phthalate and methyl methacrylate. Alkyd compounds are chemically similar to the polyester compounds but make use of higher-viscosity, or dry, monomers. This results in free-flowing granular and nodular types. Alkyd compounds often contain glass-fiber filler but may include clay, calcium carbonate, alumina, etc.

TABLE 11 Properties and Price per Unit Property of Important Plastics[7]

Material	Price, cents/lb	Properties							Price per unit property of plastics						
		D[a]	TM[b]	TYS[c]	FYS[d]	CS[e]	IS[f]	VR[g]	Cents/in.^{3}	Cents/TM[h]	Cents/TYS[i]	FYS[k]	Cents/CS[l]	Cents/IS[m]	Cents/VR[n]
LDPE	13.25	0.924	....	1.7	....	....	....	....	0.45	....	2.64	....	....	....	....
HDPE	14.0	0.960	1.55	4.4	....	....	1.0	....	0.49	3.16	1.11	....	....	0.49	....
PE, GR[o]	41.0	1.28	9.00	11.0	11.0	6.0	3.5	1	1.90	2.11	1.73	1.73	3.16	0.54	1.90
Polypropylene (PP)	21.0	0.902	....	5.0	....	....	0.6	1	0.69	....	1.38	....	....	1.15	0.69
PP copolymer	28.0	0.897	0.90	3.3	....	....	2.1	....	0.91	10.10	2.75	....	....	0.43	....
PP, GR	71.0	1.22	9.00	9.0	11.0	7.0	4.0	1	3.14	3.49	3.49	2.85	4.49	0.78	3.14
Ionomer	45.0	0.940	0.50	2.0	....	....	14.0	....	1.53	30.60	7.64	....	....	0.11	....
Poly(4-methylpentene-1)	105.0	0.83	2.10	4.0	....	....	0.8	1	3.15	15.00	7.89	....	....	3.94	3.15
Polystyrene (PS)	14.8	1.05	4.50	6.5	9.5	....	0.4	....	0.56	1.24	0.86	0.59	....	1.40	....
PS, GR	54.0	1.29	13.00	14.0	17.0	19.0	2.5	10	2.52	1.94	1.80	1.48	1.33	1.01	0.06
SAN	26.0	1.03	5.40	11.0	17.5	....	0.5	1	0.97	1.79	0.88	0.55	....	1.94	2.52
SAN, high-temp	23.0	1.08	5.00	9.8	15.6	....	0.5	....	0.90	1.80	0.92	0.58	....	1.80	9.70
SAN, GR	51.0	1.35	18.00	18.0	22.0	23.0	3.0	0.1	2.49	1.38	1.38	1.13	1.18	0.83	2.49
Impact styrene	17.8	1.05	....	4.2	6.2	....	1.8	1	0.68	....	1.62	1.10	....	0.38	0.68
Impact styrene, GR	56.0	1.32	13.00	14.5	17.0	16.0	2.5	1	2.68	2.06	1.85	1.58	1.68	1.07	2.68
ABS	28.0	1.07	4.10	7.9	12.0	13.0	1.1	3.7	1.08	2.64	1.37	0.90	0.83	0.98	0.29
ABS, high-impact	38.0	1.04	3.10	5.9	9.6	7.8	6.0	....	1.43	4.62	2.43	1.49	1.83	0.24	....
ABS, transparent	50.0	1.07	2.90	5.6	9.9	7.0	5.3	....	1.97	6.80	3.52	1.99	2.82	0.37	....
ABS, GR	85.0	1.36	10.00	16.0	25.0	17.0	2.4	....	4.19	4.19	2.62	1.68	2.46	1.74	....
ABS/PVC polyblend	40.0	1.21	3.30	6.0	10.2	7.4	12.5	....	1.75	5.30	2.92	1.72	2.37	0.14	....
ABS/PC polyblend	55.0	1.12	3.70	8.2	14.3	11.7	11.3	....	2.23	6.04	2.72	1.56	1.91	0.19	....
Acrylonitrile/styrene/acrylic	42.0	1.07	3.30	6.3	10.0	....	1.6	....	1.63	4.94	2.59	1.63	....	1.02	....
Rigid PVC	36.5	1.35	4.07	8.3	11.0	....	0.4	....	1.79	4.40	2.16	1.62	....	4.48	....
Rigid PVC, high-impact	40.0	1.35	3.65	5.7	....	....	15.0	....	1.95	5.33	3.42	....	....	0.13	....
PVC/acrylic polyblend	50.0	1.30	2.75	5.0	8.7	6.2	12.0	....	2.35	8.55	4.70	2.70	3.79	0.20	....
PTFE	325.0	2.20	....	3.0	....	....	3.0	100	25.90	....	86.40	....	....	8.64	0.26
PCTFE	490.0	2.13	1.80	6.0	9.0	7.4	NB[p]	10	37.80	210.00	63.00	42.00	51.00	NB	3.78
PMMA	45.5	1.18	4.00	9.5	15.0	....	0.3	0.01	1.95	4.88	2.05	1.30	....	6.50	195.00
PMMA, high-temp	65.5	1.16	4.65	10.0	12.0	18.0	0.3	....	2.75	5.92	2.75	2.29	1.53	9.20	....
Phenylene oxide-based resin	59.0	1.06	3.55	9.6	13.5	16.4	5.0	10	2.68	7.54	2.79	1.98	1.63	1.49	0.27
Phenylene oxide-based resin, GR	80.0	1.27	12.00	17.0	20.0	13.9	1.4	10	3.91	3.26	2.30	1.95	3.24	2.79	0.39
Polysulfone	100.0	1.24	3.60	10.2	15.4	13.9	1.3	5	4.50	12.50	4.41	2.92	....	3.46	0.90
Polysulfone/ABS polyblend	85.0	1.14	2.90	7.5	11.0	....	8.0	1.5	3.51	12.10	4.69	3.19	3.24	0.44	2.34
Polysulfone, GR	149.0	1.38	15.00	19.0	23.5	21.0	2.5	....	7.46	4.97	3.92	3.17	3.55	2.98	....
Polysulfone, high-temp	2,500.0	1.36	....	13.0	17.2	17.9	....	....	123.00	....	94.80	71.50	68.80	....	....

Properties and relative costs of engineering materials (per-row values). Column symbols are defined in the footnotes below; the first data column is cost in cents/lb.

Material	Cost, cents/lb	a	b	c	d	e	f	g	h	i	j	k	l	m	n
Acetal	65.0	1.42	5.20	10.0	14.1	18.0	1.4	0.1	3.34	6.44	3.34	2.35	1.86	2.39	33.40
Acetal copolymer	65.0	1.41	4.10	8.8	....	16.0	1.2	0.01	3.32	8.10	3.78	....	2.07	2.76	332.00
Acetal copolymer, high-impact	65.0	1.41	4.10	8.8	13.0	16.0	1.2	0.01	3.32	8.10	3.78	2.55	2.07	2.76	332.00
Acetal, GR	128.0	1.69	8.00	12.5	16.0	12.0	3.0	1	7.85	9.80	6.28	4.90	6.54	2.61	7.85
Cellulose acetate	52.0	1.30	....	6.4	11.2	....	1.2	0.001	2.44	....	3.81	2.18	....	2.04	2,440.00
Cellulose propionate	63.0	1.23	....	7.0	10.0	....	0.8	0.01	2.81	....	4.02	2.70	....	3.51	281.00
Cellulose acetate butyrate	62.0	1.20	....	4.5	6.9	....	4.4	0.01	2.70	....	6.00	3.95	....	0.61	270.00
Polycarbonate (PC)	75.0	1.20	3.50	9.0	12.5	10.5	13.0	1	3.25	9.30	3.61	2.60	3.09	0.25	3.25
PC, nonburning	150.0	1.35	3.43	9.9	14.2	13.3	1.0	3	7.34	21.40	7.42	5.17	5.51	7.34	2.45
PC, GR	120.0	1.52	17.00	20.0	25.0	19.0	3.5	0.15	6.61	3.89	3.30	2.64	3.48	1.89	44.00
Thermoplastic polyester	68.0	1.31	....	8.2	12.0	12.0	1.0	1	3.22	....	3.93	2.68	2.68	3.22	3.22
Phenolic, general-purpose	23.5	1.36	12.00	7.0	10.0	30.0	0.28	0.0001	1.15	0.96	1.64	1.15	0.38	4.11	11,500.00
DAP	125.0	1.71	14.00	8.0	14.0	31.0	0.50	0.01	7.72	5.51	9.65	5.51	2.49	15.44	772.00
Alkyl molding compound	39.0	2.21	20.00	7.0	17.5	25.0	3.0	0.0006	3.12	1.56	4.45	1.78	1.25	1.04	5,200.00
Polyurethane, GR	222.0	1.53	7.50	10.0	7.0	7.0	10.0	0.1	12.30	16.40	12.30	17.60	17.60	1.23	30.70
Nylon-6	75.0	1.13	4.40	11.8	18.0	....	0.9	0.01	3.07	6.98	2.60	1.70	....	3.41	310.00
Nylon-6/6	75.0	1.14	....	12.0	16.5	....	1.0	0.1	3.10	....	2.58	1.88	....	3.10	53.50
Nylon-6/6, GR	100.0	1.47	20.00	30.0	35.0	24.0	3.4	0.01	5.32	2.66	1.78	1.52	2.22	1.57	465.00
Nylon-6/10	120.0	1.07	....	8.5	....	....	0.6	0.01	4.65	....	5.48	4.69	....	7.75	28.50
Polyimide, GR	380.0	1.90	45.00	28.0	56.0	41.9	17.0	0.92	26.20	5.83	9.37	1.21	6.26	1.54	36,200.00
Urea formaldehyde	32.0	1.56	14.00	6.0	15.0	34.0	0.25	0.00005	1.81	1.29	3.02	1.90	0.53	7.24	16,300.00
Melamine formaldehyde	42.0	1.50	13.50	7.5	12.0	42.0	0.26	0.00014	2.28	1.69	3.04	....	0.54	8.78	....
Cast steel	18.0	7.77	300.00	128.0	....	....	....	....	5.07	0.17	0.40	....	....	....	....
Stainless steel	40.0	7.69	290.00	60.0	....	....	....	....	11.15	0.39	1.86	....	....	....	....
Aluminum	50.0	2.67	102.00	13.0	....	....	....	....	4.85	0.48	3.73	....	....	....	....
Magnesium	100.0	1.76	65.00	28.0	....	....	....	....	6.37	0.98	2.28	....	....	....	....
Zinc alloy	30.00	7.09	....	25.0	....	....	....	....	7.70	....	3.08	....	....	....	....
Copper	95.0	8.77	170.00	10.0	....	....	....	....	30.16	1.77	30.16	....	....	....	....

[a] Density, g/cm³.
[b] Tensile modulus, 10^5 lb/in.²
[c] Tensile yield strength, 10^3 lb/in.²
[d] Flexural yield strength, 10^3 lb/in.²
[e] Compressive strength, 10^3 lb/in.²
[f] Impact strength, ft-lb/in.
[g] Volume resistivity, 10^{16} Ω-cm.
[h] Tensile modulus, 10^{-6} cents/in.³ × in.²/lb.
[i] Tensile yield strength, 10^{-6} cents/in.³ × in.²/lb.
[j] Tensile yield strength, 10^{-4} cents/in.³ × in.²/lb.
[k] Flexural yield strength, 10^{-4} cents/in.³ × in.²/lb.
[l] Compressive strength, 10^{-4} cents/in.³ × in.²/lb.
[m] Impact strength, cents/in.³ × in./ft-lb.
[n] Volume resistivity, 10^{-16} cents/in.³ Ω-cm.
[o] GR = glass-fiber-reinforced.
[p] NB = no break.

TABLE 12 Summary of Some of the Major Flammability Tests[9]

Test	Materials	Use	Procedure
Ignition tests:			
ASTM D 2863 oxygen index test......	All	New test that gives good relative ranking of materials	Sample is burned in a measured mixture of O_2 and N_2 in which percentage of oxygen is decreased to determine minimum amount necessary to support combustion
UL hot-wire ignition.............	Plastics	Plastics parts in electrical applications certified by UL	Bar-shaped sample is wrapped with resistance wire which is heated electrically to red heat. Time for sample to ignite is measured
UL high-current arc ignition test......	Plastics	Plastics parts in electrical applications certified by UL	Electrodes resting on sample are repeatedly moved together until they arc and then apart till arc is ruptured. Number of ruptures required to ignite sample is recorded
UL high-voltage arc ignition test......	Plastics	Plastics parts in electrical applications certified by UL	Two electrodes rest on surface of sample 4.0 mm apart. Current is supplied to cause a continuous arc. Time required for ignition is measured
ASTM D 2859 methenamine pill test..	Carpets and floor coverings	For compliance with DOC standard for carpets exceeding 24 ft^2	Methenamine pill is ignited on sample and allowed to burn out. Burned area is measured (test also measures flame propagation; see below)
Flame-propagation tests:			
ASTM D 635, FTM 2021.............	Plastics sheet	Burn rate of 2.5 in./min for approval under BOCA, SP1 Model Code, and other building codes	Horizontal sample is exposed to bunsen-burner flame for 30 s and allowed to burn until fire goes out or sample is consumed. Rate is measured in in./min
ASTM 1692, UL subject 94..........	Plastics foams	Testing foams for building and furniture applications	Horizontal sample is exposed to bunsen-burner flame for 60 s. In UL 94, cotton is placed under sample. Burn rate is measured and flaming droplets must not ignite cotton
ASTM E 84, UL 723, NFPA 255, 25-ft tunnel test.................	All	Most widely used test for building materials. Model codes require a rating of 25 for interior finishes in stairwells, 75 in exit areas, and from 75 to 225 in rooms	Sample is mounted along ceiling of 25-ft-long tunnel and ignited by burner at front end of tunnel. Burn distance is noted, and compared with red oak, which is set at 100 (test also measures heat contribution and smoke generation; see below)

Test	Materials	Applications	Description
UL subject 94 test for self-extinguishing polymers............	Plastics sheet	Materials in applications certified by UL	Vertically oriented sample is exposed to a bunsen-burner flame for 10 s. If burning ceases in less than 30 s, a second 10-s application of flame is required. Flaming droplets are allowed to fall on cotton. If average burning time is less than 5 s and drips do not ignite cotton, material is self-extinguishing, group 0. If time is less than 25 s and drips do not ignite cotton, material is rated self-extinguishing group I, and if cotton is ignited, material is self-extinguishing group II
MVSS302 horizontal burn test........	All	Adopted by Department of Transportation for all materials used in automotive interiors	Horizontal specimen is ignited by a 15-s application of a bunsen-burner flame. When flame has burned 11/2 in. of the sample, time is measured until material ceases to burn or until burning has progressed 10 in., and rate must not exceed 4 in./min to pass
FAA vertical test.........	All	Required by FAA for materials including surface finishes and decorative components of aircraft crew and passenger compartments	Bunsen-burner flame is applied to a vertical specimen for 60 s. Flame time, burn length, and burn time of drips are noted. To pass, burn must be less than 6 in., flame time must not exceed 15 s, and drippings must go out before 3 s
FAA horizontal test............	All	Required by FAA for components of aircraft	Same conditions as above except samples are horizontal. Maximum acceptable burn rate is 2½ in./min for acrylic windows, instrument assemblies, seat belts, and shoulder harnesses. Small molded parts are acceptable if burn rate is less than 4 in./min
Fire endurance: ASTM E 119, UL 263, MFPA 251 fire endurance test........	All	Testing fire endurance of walls, floors, ceilings, roofs, etc., required by various building codes	Full-sized wall section is used as a partition in a room-sized furnace. In ASTM E 119 one side of the partition is exposed to gas fire with temperatures reaching 1000°F at 5 min, 1300°F after 10 min, 1700°F after 1 h, and 2000°F after 4 h. To pass, temperature on the far side of the specimen should not exceed 250°F. Similar tests specify different temperatures and time limits

TABLE 12 Summary of Some of the Major Flammability Tests[9] (Continued)

Test	Materials	Use	Procedure
Bureau of Mines flame penetration....	Plastics foams	Materials used in mines	Time required to burn through a 1-in.-thick layer of foam exposed to a continuous flame from a propane torch temperature of 2150°F
Heat contribution, factory mutual calorimeter.............	All	By insurance underwriters to test building components	Sample is burned in a gasoline-fired furnace and time-temperature curve is recorded. A non-combustible sample is substituted and time-temperature curve is reproduced using propane. Heat added to reproduce curve gives value of sample
Smoke generation:			
ASTM D 2843, Rohm & Haas XP2 smoke density chamber.............	Plastics	Testing for compliance with building codes	Sample is placed in a chamber and is ignited by a propane flame, and smoke density is measured by a photocell across a horizontal 12-in. light path. Most building codes permit materials if maximum light absorption is less than 50%. Uniform Building Code accepts up to 75%
NBS chamber.............	All	FAA for aircraft components (proposed)	Similar to above except that a radiant-heat source is used and the optical path for measurement is vertical and longer than in the ASTM chamber. Conditions in this chamber are thought to approximate more closely conditions in actual fires

These unsaturated resins are produced through the reaction of an organic alcohol with an organic acid. Selection of suitable polyfunctional alcohols and acids permits manipulation of a large number of repeating units. Formulating can provide resins that demonstrate a wide range of characteristics involving flexibility, heat resistance, chemical resistance, and electrical properties. Typical properties of alkyds are shown in Table 16.

Alkyds are easy to mold and economical to use. Molding dimensional tolerances can be held to within ±0.001 in./in. Postmolding shrinkage is small, as shown in Fig. 10. Their greatest limitation is in extremes of temperature (above 350°F) and humidity. Silicones and diallyl phthalates are superior here, silicones especially with respect to temperature and diallyl phthalates especially with respect to humidity.

Shrinkage versus time (compression-molded)

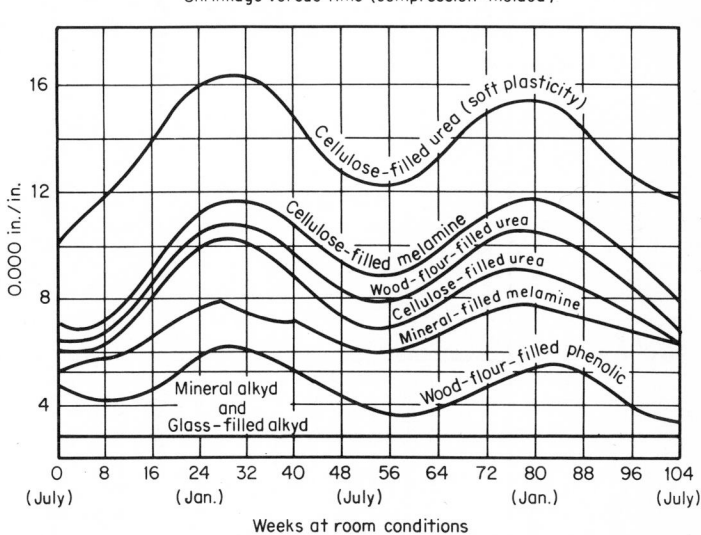

Fig. 10 Postmolding shrinkage variation of several molding compounds over a period of weeks, and showing stability of alkyds.[12]

Diallyl Phthalates (Allyls)

Diallyl phthalates, also known and often referred to as allyls, are among the best of the thermosetting plastics, with respect to high insulation resistance and low electrical losses, which are maintained up to 400°F or higher, and in the presence of high-humidity environments. Also, diallyl phthalate resins are easily molded and fabricated.

There are several chemical variations of diallyl phthalate resins, but the two most commonly used are diallyl phthalate (DAP) and diallyl isophthalate (DAIP). The primary application difference is that DAIP will withstand somewhat higher temperatures than will DAP. Typical properties of DAP and DAIP molding compounds are shown in Table 17. Outstanding electrical properties, including the retention of insulation-resistance properties after humidity conditioning, and properties of filled molding compounds are discussed further in Chap. 2.

The excellent dimensional stability of diallyl phthalates is demonstrated in Fig. 11, which compares diallyl phthalates to other plastic materials, at various temperatures.

Diallyl phthalates are extremely stable, having very low aftershrinkage, on the order of one-tenth of 1 percent, based on the MIL-M-14F dimensional-stability test. The ultimate in electrical properties is obtained by the use of the synthetic-fiber fillers; however, these materials are expensive, have high mold shrinkage, and have a

TABLE 13 Flame Retardants for Plastics and Typical Composition Ranges[10]

Plastic	Typical flame retardants	Typical percentages of flame-retardant elements for equivalent retardance							
		Br	Cl	P	P + Br	P + Cl	Sb₂O₃ + Br	Sb₂O₃ + Cl	Sb₂O₃
ABS............	Bromine and chlorine-containing organic additive compounds (such as chlorinated paraffins, brominated biphenyl, phosphate-containing aliphatic and aromatic compounds). Antimony trioxides, hydrated aluminum oxide, and zinc borate with halogenated and phosphate-containing organic compounds. Terpolymerization with halogen-containing monomers such as bis (2,3 dibromo propyl) fumerate	3	23					5 + 7	
Acrylics..........	Halogen- and phosphorus-containing organic compound. PVC to form alloy	16	20	5	1 + 3	2 + 4	7 + 5		
Cellulose acetate, cellulose butyrate, cellulose propionate	Halogenated and/or phosphate-containing organic plasticizers and halogenated compounds (such as brominated biphenyl)		24	2.5-2.5	1 + 9			12-15 + 9-12	
Epoxies..........	Halogenated and phosphate-containing compounds. Chlorinated brominated bisphenol A. Halogenated anhydrides (like chlorendic anhydride). Antimony trioxide, hydrated aluminum oxide with halogenated and phosphate-containing organic compounds	13-15	26-30	5-6	2 + 5	2 + 6		10 + 6	
Phenolic compounds	Halogen and/or phosphorus organic compounds (like chlorinated paraffins). Tris (2,3-dibromo propyl) phosphate. Hydrated aluminum oxide, zinc borate		16	6					
Polyamides........	Chlorinated and brominated biphenyls. Antimony trioxide with halogenated compounds		3.5-7	3.5				10 + 6	
Polycarbonate......	Brominated biphenyl, chlorinated paraffin. Tetrabromo bisphenol A. Antimony trioxide with halogenated compounds	4-5	10-15					7 + 7-8	

Material	Additives							
Polyesters	Halogenated phosphate-containing plasticizers. Halogenated compounds (like chlorinated paraffin and brominated biphenyl). Halogen-containing intermediates (such as tetrachlorophthalic anhydride, tetrabromophthalic anhydride, and chlorendic anhydride). Antimony trioxide, hydrated aluminum oxide, hydrated zinc borate with halogenated and phosphate-containing organic compounds	12-15	26	5	2 + 6	1 + 15-20	2 + 8-9	1 + 16-18
Polyolefins	Halogen-containing compounds (such as chlorinated paraffins and brominated biphenyl). Halogenation of polymer as chlorinated polyethylene. Antimony trioxide, hydrated aluminum oxide, zinc borate, or barium metaborate in conjunction with halogenated compounds	20	40	5	2.5 + 7	2.5 + 9	3 + 6	5 + 8
Polystyrene	Halogen- and phosphate-containing compounds like tris (2,3-dibromo propyl) phosphate and brominated biphenyls. Halogen efficiency as flame retardant is increased by incorporating small amount of free radical initiators such as peroxides. Chemical bonding of halogenated compounds into the polystyrene chain. Antimony trioxide and hydrated aluminum oxide	4-5	10-15		0.2 + 3	0.5 + 5	7 + 7-8	7 + 7-8
Polyurethanes	Halogenated phosphorus-containing plasticizers where flexibility is not a problem. Phosphorus-containing intermediates (such as phosphate, phosphite, phosphonates, and amino phosphonate polyols). Brominated isocyanates and halogen-containing prepolymers/polyols. Chlorine- and fluorine-containing compounds as blowing agents in foam. Antimony trioxide hydrated aluminum oxide, and zinc borate in conjunction with halogenated compound	12-14	18-20	1.5	0.5 + 4-7	1 + 10-15	2.5 + 2.5	4 + 4
Polyvinyl chloride	Replace other plasticizer such as DOP, with halogen- and/or phosphorus-containing plasticizer. Antimony trioxides, hydrated aluminum oxide, and zinc borate		40	2-4				5-15

Further references: C. J. Hilado, "Flammability Handbook for Plastics," Technomic Publishers; J. W. Lyons, "The Chemistry and Uses of Fire Retardants," Interscience-Wiley.

TABLE 14 Typical Trade Names and Suppliers for Thermosetting Plastics

Plastic	Typical trade names and suppliers
Alkyd.................................	Plaskon (Allied Chemical) Durez (Hooker Chemical) Glaskyd (American Cyanamid)
Diallyl phthalates.....................	Dapon (FMC) Diall (Allied Chemical) Durez (Hooker Chemical)
Epoxies..............................	Epon (Shell Chemical) Epi-Rez (Celanese) D.E.R. (Dow Chemical) Araldite (Ciba) ERL (Union Carbide)
Melamines............................	Cymel (American Cyanamid) Plaskon (Allied Chemical)
Phenolics............................	Bakelite (Union Carbide) Durez (Hooker Chemical) Genal (General Electric)
Polybutadienes........................	Dienite (Firestone) Ricon (Colorado Chemical Specialties)
Polyesters...........................	Laminac (American Cyanamid) Paraplex (Rohm and Haas) Selectron (PPG)
Silicones............................	DC (Dow Corning); see also Table 41
Ureas................................	Plaskon (Allied Chemical) Beetle (American Cyanamid)

These are but a few typical basic material suppliers.   In addition, there are many companies which formulate filled plastic systems from these basic plastics, and which are sometimes the ultimate supplier to the end-product fabricator.

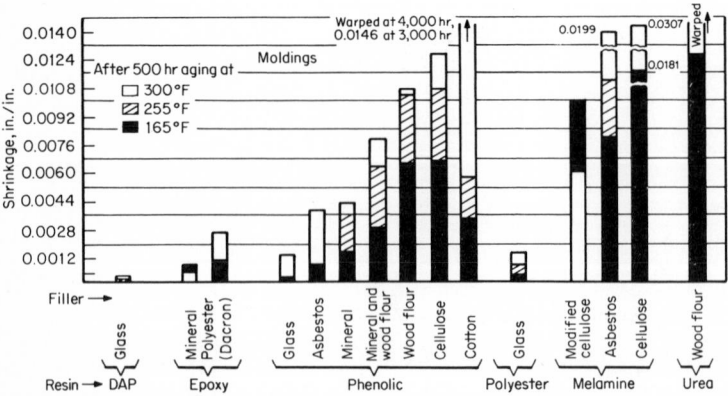

Fig. 11 Shrinkage of various thermosetting molding materials as a result of heat aging.[14]

strong flexible flash that is extremely difficult to remove from the parts. Consequently, the largest commercial volume is in the short-glass-fiber-filled materials, which combine moldability in thin sections with extremely high tensile and flexural strengths.

Epoxies

Types of epoxies   Epoxy resins are characterized by the epoxide group (oxirane rings).[15]   The most widely used resins are diglycidyl ethers of bisphenol A (Fig. 12). These are made by reacting epichlorohydrin with bisphenol A in the presence of an alkaline catalyst.   By controlling operating conditions and varying the ratio of

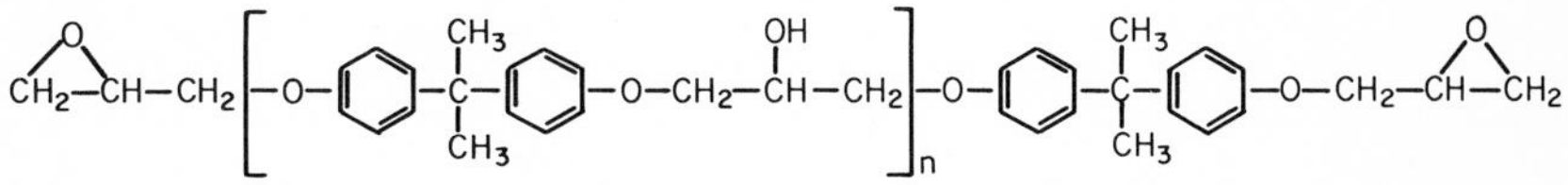

Fig. 12   General structure of epoxy (diglycidyl ether of bisphenol A).

epichlorohydrin to bisphenol A, products of different molecular weight can be made. For liquid resins the n of Fig. 12 is generally less than 1; for solid resins n is 2 or greater.   Solids with very high melting points have values for n as high as 20.

Another class of epoxy resins are the novolacs, particularly the epoxy cresols (Fig. 13) and the epoxy phenol novolacs.   These are produced by reacting a novolac resin, usually formed by the reaction of o-cresol or phenol and formaldehyde with epichlorohydrin.   These highly functional materials are particularly recommended for transfer-molding powders, electrical laminates, and parts where superior thermal properties, high resistance to solvents and chemicals, and high reactivity with hardeners are needed.

The cycloaliphatics comprise another group of epoxy resins which are particularly important when superior arc-track and weathering resistance are necessary requirements (see Fig. 13).

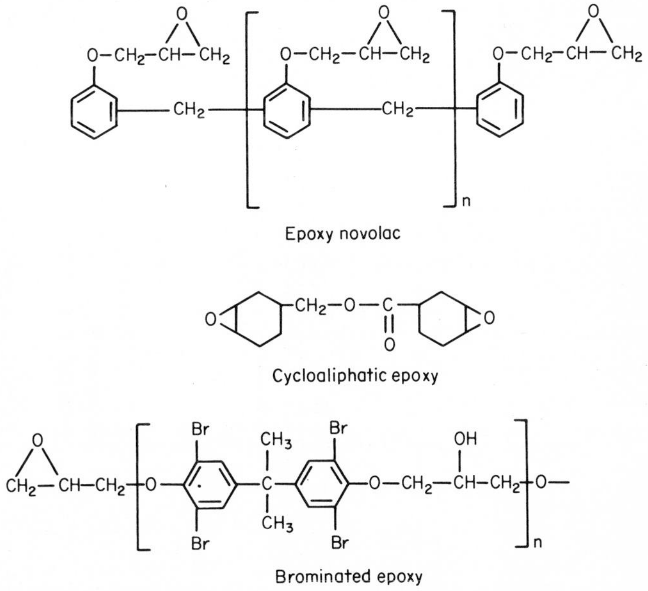

Epoxy novolac

Cycloaliphatic epoxy

Brominated epoxy

Fig. 13   General structure of novolac, cycloaliphatic and brominated epoxies.

TABLE 15 Applications of Fillers in Molding Compounds[11]

Filler or reinforcement	Chemical resistance	Heat resistance	Electrical insulation	Impact strength	Tensile strength	Dimensional stability	Stiffness	Hardness	Lubricity	Electrical conductivity	Thermal conductivity	Moisture resistance	Processibility	Recommended for use in†
Alumina, tabular	X	X				X								S/P
Alumina trihydrate, fine particle			X				X						X	P
Aluminum powder							X	X		X	X			S
Asbestos	X	X	X	X		X	X							S/P
Bronze							X	X		X	X		X	S
Calcium carbonate‡		X				X	X	X						S/P
Calcium metasilicate	X	X				X	X	X				X		S
Calcium silicate		X				X	X	X						S
Carbon black§		X				X				X	X		X	S/P
Carbon fiber							X		X	X	X			S
Cellulose			X	X	X	X	X							S/P
Alpha cellulose	X			X	X	X								S
Coal, powdered	X											X		S
Cotton (macerated/chopped fibers)			X	X	X	X	X	X						S

Filler										Symbol†
Fibrous glass	X	X		X	X	X	X	X	X	S/P
Fir bark	X				X	X	X		X	S
Graphite	X			X	X		X	X		S/P
Jute			X							S
Kaolin	X			X	X	X	X	X	X	S/P
Kaolin (calcined)	X	X		X	X	X	X	X	X	S/P
Mica	X	X		X	X		X	X	X	S/P
Molybdenum disulfide							X		X	P
Nylon (macerated / chopped fibers)	X	X		X	X	X	X	X	X	S/P
Orlon	X	X		X	X	X	X		X	S/P
Rayon	X	X		X	X	X	X			S
Silica, amorphous					X			X	X	S/P
Sisal fibers	X	X		X		X	X		X	S/P
TFE fluorocarbon						X		X	X	S/P
Talc	X	X		X	X	X	X	X	X	S/P
Wood flour						X	X	X	X	S

* The chart does not show differences in degrees of improvement; calcined kaolin, for example, generally gives much higher electrical resistance than kaolin. Similarly, differences in characteristics of products under one heading, such as talc (which varies greatly from one grade to another and from one type to another), are not distinguished.

† Symbols: P—in thermoplastics only; S—in thermosets only; S/P—in both thermoplastics and thermosets.

‡ In thermosets, calcium carbonate's prime function is to improve molded appearance.

§ Prime functions are imparting of ultraviolet resistance and coloring; also is used in cross-linked thermoplastics.

TABLE 16 Typical Properties of Alkyd Molding Compounds[12]

	ASTM test method	Putty	Granular	Glass-fiber-reinforced	Rope
Electrical					
Arc resistance, s	D 495-58T	180+	180+	180+	180+
Dielectric constant	D 150-54T				
1 MHz		5.4-5.9	5.7-6.3	5.2-6.0	7.4
1 MHz	D 150-54T	4.5-4.7	4.8-5.1	4.5-5.0	6.8
Dissipation factor	D 150-54T				
60 Hz		0.030-0.045	0.030-0.040	0.02-0.03	0.019
1 MHz		0.016-0.022	0.017-0.020	0.015-0.022	0.023
Dielectric strength, V/mil	D 149-59				
Short-time		350-400	350-400	350-400	360
Step by step		300-350	300-350	300-350	290
Physical					
Specific gravity	D 792-50	2.05-2.15	2.21-2.24	2.02-2.10	2.20
Water absorption, 24 h at 23°C, %	D 570-57T	0.10-0.15	0.08-0.12	0.07-0.10	0.05
Heat resistance max, °F:					
Long periods (continuous)		250	300	300	300
Short periods (0-24 h)		300	350	350	350
Short periods (0-1 h)		325	375	400	400
Heat-distortion temp, 264 lb/in.², °F	D 648-56	350-400	350-400	>400	>400
Coefficient of linear thermal expansion/°F	D 696-44	$10\text{-}30 \times 10^{-6}$	$10\text{-}30 \times 10^{-6}$	$10\text{-}30 \times 10^{-6}$	20×10^{-6}
Thermal conductivity, (g)(cal)/(s)(cm²)(°C/cm)		$15\text{-}25 \times 10^{-4}$	$15\text{-}25 \times 10^{-4}$	$8\text{-}12 \times 10^{-6}$	10×10^{-4}
Flammability	D 635-56T	Nonburning	Self-extinguishing	Nonburning	Self-extinguishing
Mechanical					
Impact strength, Izod, ft-lb/in. of notch	D 256-56	0.25-0.35	0.30-0.35	8-12	2.2
Comprehensive strength, lb/in.²	D 695-54	20,000-25,000	16,000-20,000	24,000-30,000	28,800
Flexural strength, lb/in.²	D 790-59T	8,000-11,000	7,000-10,000	12,000-17,000	19,500
Tensile strength, lb/in.²	D 651-48	4,000-5,000	3,000-4,000	5,000-9,000	7,100
Modulus of elasticity, lb/in.²	D 790-49T	$2.0\text{-}2.7 \times 10^{6}$	$2.4\text{-}2.9 \times 10^{6}$	$2.0\text{-}2.5 \times 10^{6}$	1.9×10^{6}
Barcol hardness		60-70	60-70	70-80	72
		MIL-M-14F type MAG	MIL-M-14F type MAG	MIL-M-14F type MAI-60	

TABLE 17 Typical Properties of Unfilled Diallyl Phthalate Molding Compounds[13]

Property by ASTM procedures	DAP	DAIP
Dielectric constant:		
At 60 Hz, 25°C	3.6	3.5
At 10^3 Hz, 25°C	3.6	3.3
At 10^6 Hz, 25°C	3.4	3.2
At 10,000 MHz, 25°C	....	3.0
At 10,000 MHz, 200°C	....	3.1
Dissipation factor:		
At 60 Hz, 25°C	0.010	0.008
At 10^3 Hz, 25°C	0.009	0.008
At 10^6 Hz, 25°C	0.011	0.009
At 10,000 MHz, 25°C	....	0.014
At 10,000 MHz, 200°C	....	0.031
Volume resistivity:		
Ω-cm at 25°C	1.8×10^{16}	3.9×10^{17}
Ω-cm at 25°C (wet)	$1.0 \times 10^{14}+*$	
Surface resistivity:		
Ω at 25°C	9.7×10^{15}	8.4×10^{12}
Ω at 25°C (wet)	$4.0 \times 10^{13}+*$	
Dielectric strength, V/mil at 25°C	450†	422†
Arc resistance	118	123–128
Moisture absorption, %—20 h at 25°C	0.09	0.1
Tensile strength, lb/in.²	3,000–4,000	4,000–4,500
Rockwell hardness (M)	114–116	119–121
Barcol hardness	43	52
Specific gravity	1.270	1.264
Izod impact, ft/in. notch	0.2–0.3	0.2–0.3
Heat-distortion temp:		
°C at 264 lb/in.²	155 (310°F)	238‡ (460°F)
°C at 546 lb/in.²	125 (257°F)	184–211 (364–412°F)
Comprehensive strength, lb/in.²	22,000–23,000	21,200–24,000
Flexural strength, lb/in.²	7,000–9,000	7,400–8,300
Refractive index at 25°C	1.571	1.569
Modulus of elasticity in flexure, lb/in.²	0.6×10^6	0.5×10^6
Chemical resistance, % gain in weight after 1 month in:		
Water at 25°C	0.9	0.8
Acetone at 25°C	1.3	−0.03
1% NaOH at 25°C	0.7	0.7
10% NaOH at 25°C	0.5	0.6
3% H_2SO_4	0.8	0.7
30% H_2SO_4	0.4	0.4
Heat-resistant properties:§		
Weight loss after aging 6 weeks at 177°C, %	7.6	1.2
Dielectric constant after aging 6 weeks at 177°C, measured at 60 Hz, 25°C	3.7	3.6
Dissipation factor after aging 6 weeks at 177°C, measured at 60 Hz, 25°C	0.01	0.006
Volume resistivity after aging 6 weeks at 177°C, Ω-cm at 25°C	2.7×10^{15}	7.1×10^{15}
Surface resistivity after aging 6 weeks at 177°C, Ω at 25°C	1.4×10^{13}	1.1×10^{14}
Flexural strength, lb/in.² at 260°C (glass-cloth laminate)	....	15,400¶

* Tested in humidity chamber after 30 days at 70°F (158°) and 100% relative humidity.
† Step by step.
‡ No deflection.
§ 50% silica filled.
¶ 12-ply, 181 glass-cloth laminate (resin = 40%).

A distinguishing feature of these resins is the location of the epoxy group(s) on a ring structure rather than on the aliphatic chain. Cycloaliphatics can be produced by the peracetic epoxidation of cyclic olefins and by the condensation of an acid such as tetrahydrophthalic with epichlorohydrin, followed by dehydrohalogenation. Cycloaliphatics are outstanding among epoxies in their resistance to electrical surface tracking.

Epoxy resins must be cured with cross-linking agents (hardeners) or catalysts to develop desirable properties. The epoxy and hydroxyl groups are the reaction sites through which cross linking occurs. Useful agents include amines, anhydrides,

TABLE 18   Curing Agents for Epoxy Resins

Curing-agent type	Characteristics	Typical materials
Aliphatic amines ..	Aliphatic amines allow curing of epoxy resins at room temperature, and thus are widely used. Resins cured with aliphatic amines, however, usually develop the highest exothermic temperatures during the curing reaction, and therefore the mass of material which can be cured is limited. Epoxy resins cured with aliphatic amines have the greatest tendency toward degradation of electrical and physical properties at elevated temperatures	Diethylene triamine (DETA) Triethylene tetramine (TETA)
Aromatic amines ..	Epoxies cured with aromatic amines usually have a longer working life than do epoxies cured with aliphatic amines. Aromatic amines usually require an elevated-temperature cure. Many of these curing agents are solid and must be melted into the epoxy, which makes them relatively difficult to use. The cured resin systems, however, can be used at temperatures considerably above those which are safe for resin systems cured with aliphatic amines	Metaphenylene diamine (MPDA) Methylene dianiline (MDA) Diamino diphenyl sulfone (DDS or DADS)
Catalytic curing agents..........	Catalytic curing agents also have a working life better than that of aliphatic amine curing agents and, like the aromatic amines, normally require curing of the resin system at a temperature of 200°F or above. In some cases, the exothermic reaction is critically affected by the mass of the resin mixture	Piperidine Boron trifluoride ethylamine complex Benzyl dimethyl-amine (BDMA)
Acid anhydrides....	The development of liquid acid anhydrides provides curing agents which are easy to work with, have minimum toxicity problems compared with amines, and offer optimum high-temperature properties of the cured resins. These curing agents are becoming more and more widely used	Nadic methyl anhydride (NMA) Dodecenyl succinic anhydride (DDSA) Hexahydrophthalic anhydride (HHPA) Alkendic anhydride

aldehyde condensation products, and Lewis acid catalysts. Careful selection of the proper curing agent is required to achieve a balance of application properties and initial handling characteristics.[15]

Curing-Agent Systems. End properties of bisphenol epoxies, as well as of other epoxies discussed here, are controlled by the type of curing agent used with the resin. Major types of curing agents are aliphatic amines, aromatic amines, catalytic curing agents, and acid anhydrides, as shown in Table 18.

Aliphatic amine curing agents produce a resin–curing agent mixture which has a relatively short working life but which cures at room temperature or at low baking temperatures, in relatively short time.   Resins cured with aliphatic amines usually develop the highest exothermic temperatures during the curing reaction; thus the amount of material which can be cured at one time is limited because of possible cracking, crazing, or even charring of the resin system if too large a mass is mixed and cured.   Also, physical and electrical properties of epoxy resins cured with aliphatic amines tend to degrade as operating temperature increases.   Epoxies cured with aliphatic amines find their greatest usefulness where small masses can be used, where room-temperature curing is desirable, and where the operating temperature required is below 100°C.

Epoxies cured with aromatic amines have a considerably longer working life than do those cured with aliphatic amines, but they require curing at 100°C or higher.   Resins cured with aromatic amines can operate at a temperature considerably above the temperature necessary for those cured with aliphatic amines.   However, aromatic amines are not so easy to work with as aliphatic amines, because of the solid nature of the curing agents and the fact that some (such as MPDA) sublime when heated, causing stains and residue deposition.

Catalytic curing agents also have longer working lives than the aliphatic amine materials, and like the aromatic amines, catalytic curing agents normally require curing of the epoxy system at 100°C or above.   Resins cured with these systems have good high-temperature properties as compared with epoxies cured with aliphatic

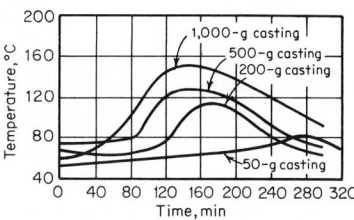

Fig. 14 Exothermic curves, as a function of resin mass, for bisphenol epoxy and 5 percent piperidene curing agent (catalytic type) cured at 60°C.

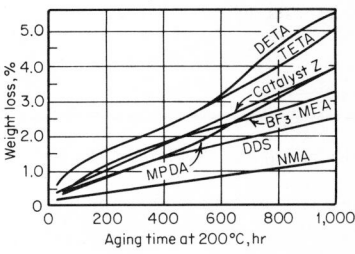

Fig. 15 Weight-loss data at 200°C for Epon 828 cured with various curing agents.   DETA and TETA are aliphatic amines, MPDA and DDS are aromatic amines, and NMA is a liquid acid anhydride.

amines.   With some of the catalytic curing agents, the exothermic reaction becomes high as the mass of the resin mixture increases, as shown in Fig. 14.

Acid anhydride curing agents are particularly important for epoxy resins, especially the liquid anhydrides.   The high-temperature properties of resin systems cured with these materials are better than those of resin systems cured with aromatic amines. Some anhydride-cured epoxy-resin systems retain most electrical properties to 150°C and higher, and are little affected physically even after prolonged heat aging at 200°C.   In addition, the liquid anhydrides are extremely easy to work with; they blend easily with the resins and reduce the viscosity of the resin system.   Also, the working life of the liquid acid anhydride systems is long compared with that of mixtures of aliphatic amine and resin, and odors are slight.   Amine promoters such as benzyl dimethylamine (BDMA) or DMP-30 are used to promote the curing of mixtures of acid anhydride and epoxy resin.   Thermal stability of epoxies is improved by anhydride curing agents, as shown in Fig. 15.

Applications of epoxies   Epoxies are among the most versatile and most widely used plastics in the electronics field.   This is primarily because of the wide variety of formulations possible, and the ease with which these formulations can be made and utilized with minimal equipment requirements.   Formulations range from flexible to rigid, in the cured state, and from thin liquids to thick pastes and molding powders in the uncured state.   Conversion from uncured to cured state is made by use of hardeners and/or heat. The largest application of epoxies is in embedding applications (potting, casting, encapsulating, and impregnating) in molded parts, and in laminated constructions such as metal-clad laminates for printed circuits and unclad laminates for various types of insulating and terminal boards. Molded parts have excellent dimensional stability, as shown in Fig. 11. The detailed properties of laminates are given in Chap. 5. Typical properties of glass-fiber-filled and mineral-filled epoxy compounds are shown in Tables 5–8 and in Chap. 2 (Table 17) and Chap. 3.

Melamines and Ureas (Aminos)

As compared with alkyds, diallyl phthalates, and epoxies, which are polymers created by addition reactions and hence have no reaction by-products, melamines and ureas (also commonly referred to as aminos) are polymers which are formed by condensation reactions and do give off by-products. Another example of this type of reaction is the polymerization reaction which produces phenolics. A typical condensation reaction is shown in the following section on phenolics. Melamines and ureas are a reaction product of formaldehyde with amino compounds containing NH_2 groups. Hence they are often also referred to as melamine formaldehydes and urea formaldehydes. Their general chemical structure is shown in Fig. 16.

Amino resins have found applications in the fields of industrial and decorative laminating, adhesives, protective coatings, textile treatment, paper manufacture, and molding compounds. Their clarity permits products to be fabricated in virtually any color. Finished products having an amino-resin surface exhibit excellent resistance to moisture, greases, oils, and solvents; are tasteless and odorless; are self-extinguishing; offer excellent electrical properties; and resist scratching and marring. The melamine resins offer better chemical, heat, and moisture resistance than do the ureas.[16]

Amino molding compounds can be fabricated by economical molding methods. They are hard, rigid, and abrasion-resistant, and they have high resistance to deformation under load. These materials can be exposed to subzero temperatures without embrittlement. Under tropical conditions, the melamines do not support fungus growth.

Amino materials are self-extinguishing and have excellent electrical-insulation characteristics. They are unaffected by common organic solvents, greases and oils, and weak acids and alkalies. Melamines are superior to ureas in resistance to acids, alkalies, heat, and boiling water, and are preferred for applications involving cycling between wet and dry conditions or rough handling. Aminos do not impart taste or odor to foods.

Addition of alpha cellulose filler, the most commonly used filler for aminos, produces an unlimited range of light-stable colors and high degrees of translucency. Colors are obtained without sacrifice of basic material properties. Shrinkage characteristics with cellulose filler are shown in Fig. 10.

Melamines and ureas provide excellent heat insulation; temperatures up to the destruction point will not cause parts to lose their shape.

Amino resins exhibit relatively high mold shrinkage, and also shrink on aging. Cracks develop in urea moldings subjected to severe cycling between dry and wet conditions.

Prolonged exposure to high temperature affects the color of both urea and melamine products.

A loss of certain strength characteristics also occurs when amino moldings are subjected to prolonged elevated temperatures. Some electrical characteristics are also adversely affected; arc resistance of some industrial types, however, remains unaffected after exposure at 500°F.

Ureas are unsuitable for outdoor exposure. Melamines experience little degradation in electrical or physical properties after outdoor exposure, but color changes may occur.

Typical physical properties of amino plastics are shown in Table 19 and typical mechanical and electrical properties in Table 20.

Phenolics

Like melamines and ureas, phenolics are formed by a condensation reaction, whose general nature is shown in Fig. 17. The general chemical structure of a phenol-formaldehyde resin is shown in Fig. 18.

Phenolics are among the oldest, best-known general-purpose molding materials.

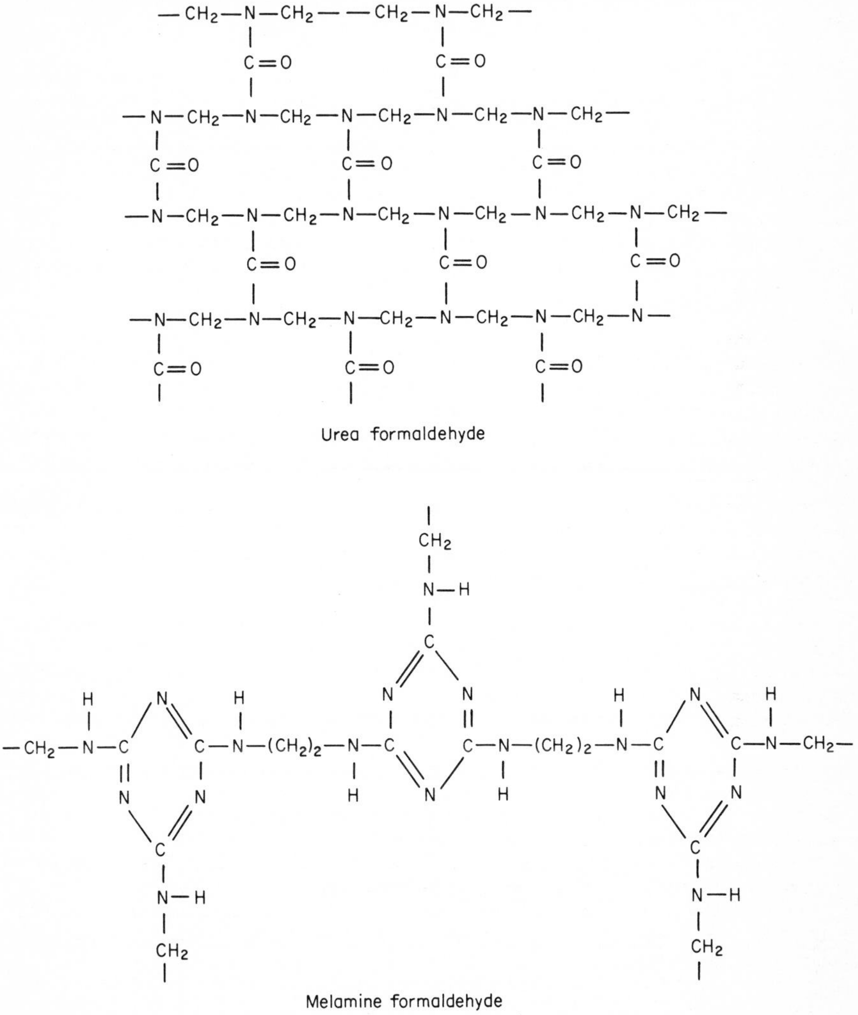

Urea formaldehyde

Melamine formaldehyde

Fig. 16 General chemical structure of urea formaldehyde and melamine formaldehyde.[16]

They are also among the lowest in cost and the easiest to mold. An extremely large number of phenolic materials are available, based on the many resin and filler combinations, and they can be classified in many ways. One common way of classifying them is by type of application or grade. Typical properties for some of these common molding-material classifications are shown in Tables 5–8. In addition to

TABLE 19 Typical Physical Properties of Amino Compounds[17]

	Urea		Melamine					
	Alpha cellulose	Wood flour	Alpha cellulose	Wood flour	Alpha cellulose, modified	Rag	Asbestos	Glass fiber
Specific gravity	1.5	1.5	1.5	1.42	1.43	1.5	1.78	1.94–2.0
Density, g/in.³	24.6	24.6	24.6	23.8	23.5	24.6	29.2	31.8–32.8
Hardness, Rockwell E	94–97	95	110	94	......	100	90	
Shrinkage,* in./in.:								
Molding	0.006–0.009	0.006–0.014	0.008–0.009	0.007–0.008	0.006–0.008	0.003–0.004	0.005–0.007	0.002–0.004
Postmold	0.006–0.012	0.006–0.012	0.009–0.011	0.004–0.007	0.001–0.002	0.004–0.008	0.002–0.003	0.002–0.005
Deflection temp, °F, at 264 lb/in.²	266	270	361	266	266	310	266	400
Heat resistance, continuous, °F	170†	170	210†	250	250	250	300	300
Coefficient of thermal expansion, per °C × 10⁻⁶	22–36	30	20–57	32–50	34–36	25–30	21–43	12–25
Thermal conductivity, (cal)(cm)/(s²)(cm)(°C) × 10⁻⁴	10.1	10.1	10.1	8.4	......	10.6	13.1	
Water absorption, 24 h, at 23°C, %	0.4–0.8	0.7	0.3–0.5	0.34–0.6	0.3–0.6	0.3–0.6	0.13–0.15	0.09–0.3
Color possibilities	Unlimited	Brown, black	Unlimited	Brown	Brown	Limited	Brown	Natural gray

* Test specimen: 4-in. diam × 1/8-in. disk.
† Based on no color change.

TABLE 20 Typical Mechanical and Electrical Properties of Amino Molding Compounds[17]

	Urea		Melamine					
	Alpha cellulose	Wood flour	Alpha cellulose	Wood flour	Alpha cellulose modified	Rag	Asbestos	Glass fiber
Mechanical:								
Tensile strength, 1,000 lb/in.²	5.5–7	5.5–10	7–8	5.7–6.5	5.5–6.5	8–10	5.5–6.5	5.9
Compressive strength, 1,000 lb/in.²	30–38	25–35	40–45	30–35	24.5–26	30–35	25–30	20–29
Flexural strength, 1,000 lb/in.²	11–18	8–16	12–15	6.5–9	11.5–12	12–15	7.4–10	13.2–24
Shear strength, 1,000 lb/in.²	11–12		11–12	10–10.5	11.4–12.2	12–14	7–8	13.0–15.6
Impact strength, Izod ft-lb/in. of notch	0.24–0.28	0.25–0.35	0.30–0.35	0.25–0.35	0.30–0.42	0.55–0.90	0.30–0.40	0.5–6.0
Tensile modulus, 10⁶ lb/in.²	1.3–1.4		1.35	1.0	1.0	1.4	1.95	
Flexural modulus, 10⁶ lb/in.²	1.4–1.5	1.3–1.6	1.1	1.0	1.1	1.4	1.8	2.4
Electrical:								
Arc resistance, s	80–100	80–100	125–136	70–106	90–120	122–128	120–180	180–186
Dielectric strength, V/mil								
Short time:								
At 23°C	330–370	300–400	270–300	350–370	350–390	250–340	410–430	170–370
At 100°C	200–270		170–210	290–330	140–190	110–130	280–310	90–350*
Step by step:								
At 23°C	220–250	250–300	240–270	200–240	200–250	220–240	280–300	170–270
At 100°C	110–150		90–130	190–210	90–190	60–90	190–210	60–250*
Slow rate of rise:								
At 23°C	250–260		210–240	240–260	280–290	210–240	270–290	170–210
At 100°C	120–170		90–120	170–200	90	70–80	170–190	70–90
Dielectric constant:								
At 60 Hz	7.7–7.9	7.0–9.5	7.9–8.2	6.4–6.6	7.0–7.7	8.1–12.6	10.0–10.2	7.0–11.1
At 10⁶ Hz	6.7–6.9	6.4–6.9	7.6–8.0	5.6–5.8	5.2–6.0	6.7–6.9	5.3–6.1	6.6–7.9
At 3 X 10, Hz							4.9	5.5
Dissipation factor:								
At 60 Hz	0.034–0.043	0.035–0.040	0.052–0.083	0.026–0.033	0.192	0.100–0.340	0.100	0.14–0.23
At 10⁶ Hz	0.029–0.031	0.028–0.032	0.026–0.030	0.034–0.035	0.044–0.12	0.036–0.041	0.039–0.048	0.013–0.016
At 3 X 10⁹ Hz							0.032	0.040
Dielectric loss factor:								
At 60 Hz	0.28–0.34	0.24–0.38	0.44–0.78	0.17–0.22	0.90–2.4	2.0–5.0	0.5–1.0	1.5–2.5
At 10⁶ Hz	0.19–0.21	0.18–0.22	0.20–0.33	0.20–0.21	0.19–0.28	0.24–0.26	0.21–0.31	0.09–0.19
Volume resistivity, Ω-cm	0.5–5.0 X 10¹¹		0.8–2.0 X 10¹²	6–10 X 10¹²	6 X 10¹⁰	1.0–3.0 X 10¹¹	1.2 X 10¹²	0.9–20 X 10¹¹
Surface resistivity, Ω	0.4–3.0 X 10¹¹		0.8–4.0 X 10¹¹	0.3–5.0 X 10¹¹	1.7 X 10¹²	0.7–7.0 X 10¹¹	1.9 X 10¹¹	3.0–4.6 X 10¹²
Insulation resistance, Ω	0.2–5.0 X 10¹¹		1.0–4.0 X 10¹⁰	1.0–3.0 X 10¹¹	2.0–5.0 X 10⁹	0.1–3.0 X 10¹⁰	1.0–4.0 X 10¹⁰	0.2–6.0 X 10¹⁰

* At 50°C.

molding materials, phenolics are used to bond friction materials for automotive brake linings, clutch parts, and transmission bands. They serve as binders for wood-particle board used in building panels and core material for furniture, as the water-resistant glue for exterior-grade plywood, and as the bonding agent for converting both organic and inorganic fibers into acoustical- and thermal-insulation pads, batts, or cushioning for home, industrial, and automotive applications. They are used to

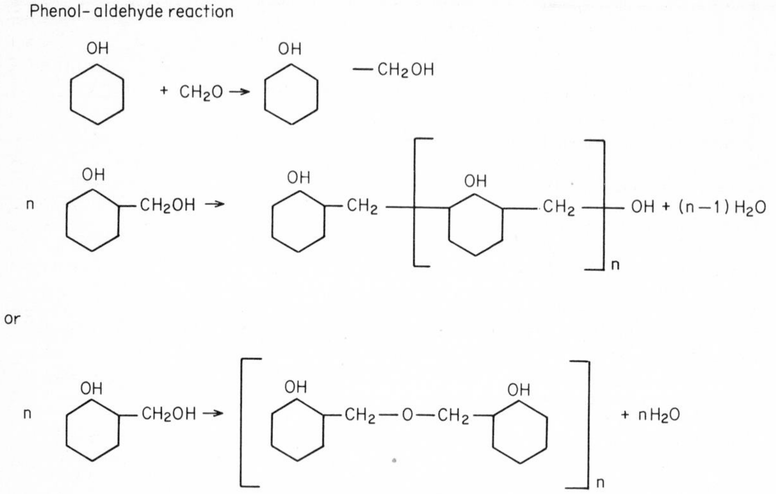

Phenol–aldehyde reaction

Fig. 17 General nature of a condensation reaction.

impregnate paper for electrical or decorative laminates and as special additives to tackify, plasticize, reinforce, or harden a variety of elastomers.

Although it is possible to get various molding grades of phenolics for various applications, as discussed above, phenolics, generally speaking, are not equivalent to diallyl phthalates and epoxies in resistance to humidity and retention of electrical properties in extreme environments. Phenolics are, however, quite adequate for a large percentage of electrical applications. Furthermore improved grades have been

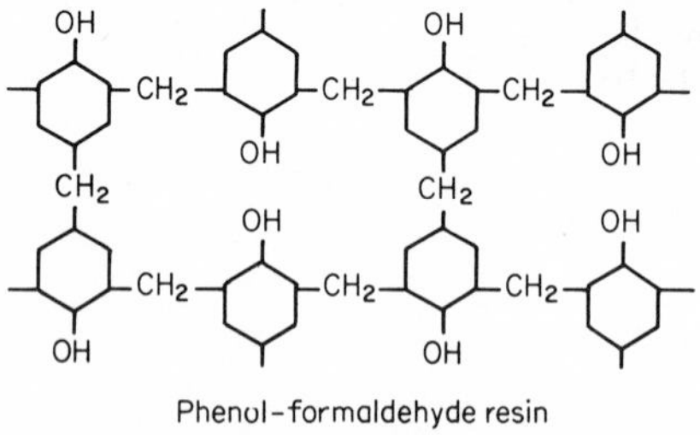

Phenol–formaldehyde resin

Fig. 18 General chemical structure of a phenol formaldehyde.

developed which yield considerable improvements in humid environments and at higher temperatures. In addition, the glass-filled, heat-resistant grades are outstanding in thermal stability up to 400°F and higher, with some being useful up to 500°F. Shrinkage in heat aging varies over a fairly wide range, depending on filler used. Glass-filled phenolics are the more stable, as shown in Fig. 11.

The growth in screw-injection molding of phenolics has been extremely rapid. The development of this technique allows the molder to automate further, reduce labor costs, improve quality, reduce rejects, and gain substantial overall mold-cycle efficiency. The increasing use of this technique for thermosets offers the opportunity for thermosets to make substantial gains in areas otherwise closed to them, primarily because of molding costs.

TABLE 21 Typical Properties of Cured Polybutadiene* Molding Compound[18]

Property	Value
Flexural strength, 10^3 lb/in.^{2}	11–14
Flexural modulus, 10^6 lb/in.^{2}	1.3–1.4
Compressive strength, 10^3 lb/in.^{2}	25
Compressive modulus, 10^5 lb/in.^{2}	5
Tensile strength, 10^3 lb/in.^{2}	7.5
Tensile modulus, 10^5 lb/in.^{2}	7
Rockwell hardness	E 70–E 95
Impact strength, ft-lb/in.	0.25–0.5
Heat-deflection temp, 264 lb/in.^{2}, °F	>500
Flammability, ASTM D 635	Nonburning
Specific gravity	2.05
Mold shrinkage, cured at 350°F, in./in.	0.015
Thermal conductivity, 10^{-4} cal/(s)(cm^2)(°C/cm)	5.2
Thermal expansion, 10^{-5} in./in./°C	4
Dielectric strength, V/mil	500–1,000+
Dielectric constant, 50 to 10^5 Hz	3.7
Dissipation factor:	
100 Hz	0.007
1,000 Hz	0.004
10,000 Hz	0.003
100,000 Hz	0.002
Volume resistivity, 10^{15} Ω-cm	1.7–7.1
Arc resistance, s	252

* Firestone FCR-1261 TC-100.

Polybutadienes

One of the recently developed groups of thermosetting polymers is the polybutadienes. The various forms offer potential as moldings, laminating resins, coatings, cast-liquid, and formed-sheet products. These materials, some being essentially pure hydrocarbon, have outstanding electrical and thermal-stability properties. Typical properties are shown in Table 21, and electrical properties are further discussed in Chap. 2. The basic chemical structure is as follows:

1,2-polybutadiene
microstructure

Polybutadienes are cured by peroxide catalysts, which produce carbon-to-carbon bonds at the double bonds in the vinyl groups.[18] The final product is 100 percent hydrocarbon except where the starting polymer is the —OH or —COOH terminated variety. The nature of the resultant product may be more readily understood if the structure is regarded as polyethylene with a cross link at every other carbon in the main chain. In a true sense, the uncured high-vinyl butadiene structure might be designated polyvinylethylene. The fully cured resin possesses dielectric properties and chemical resistance fully equal or superior to polyethylene and has a heat-deflection temperature in excess of 500°F.

Use of the high-temperature peroxides maximizes the opportunity for thermoplastic-like processing, because even the higher-molecular-weight forms become quite fluid at temperatures well below the cure temperature. Compounds thus can be injection-molded in an in-line machine with a thermoplastic screw. A low-compression-ratio thermoset screw is not required.[18]

Polyesters (Thermosetting)

Unsaturated, thermosetting polyesters are produced by addition polymerization reactions, as shown in simplified form in Fig. 3. Polyester resins can be formulated to have a wide range of physical properties. They can be brittle and hard, tough and resilient, or soft and flexible. Viscosities at room temperature may range from 50 to more than 25,000 cP. The polyesters can be employed to fabricate a myriad of products by many techniques—open-mold casting, hand lay-up, spray-up, vacuum-bag molding, matched-metal-die molding, filament winding, pultrusion, encapsulation, centrifugal casting, and injection molding.[19]

By the appropriate choice of ingredients, particularly to form the linear polyester resin, special properties can be imparted. Fire retardance can be achieved through the use of one or more of the following: chlorendic anhydride, tetrabromophthalic anhydride, tetrachlorophthalic anhydride, dibromoneopentyl glycol, and chlorostyrene. Chemical resistance is obtained by employing neopentyl glycol, isophthalic acid, hydrogenated bisphenol A, and trimethyl pentanediol. Weathering resistance can be enhanced by the use of neopentyl glycol and methyl methacrylate. Appropriate thermoplastic polymers can be added to reduce or eliminate shrinkage during curing and thereby minimize one of the disadvantages historically inherent in polyester systems.[19]

Thermosetting polyesters are widely used for moldings, laminated or reinforced structures, surface gel coatings, liquid castings, furniture products, and structures. Electrical properties are discussed in Chap. 2. Most furniture parts are made by casting. Other cast products include bowling balls, simulated marble, gaskets for vitrified-clay sewer pipe, pistol grips, pearlescent shirt buttons, and implosion barriers for television tubes. In this last use a practically water-white polyester is employed to bond a protective heavy glass shield over the television-tube face to prevent glass fragments from flying about upon accidental tube breakage.

By lay-up and spray-up techniques large- and/or short-run items are fabricated. Examples include boats of all kinds—pleasure sailboats and powered yachts, commercial fishing boats and shrimp trawlers, small military vessels—dune buggies, all-terrain vehicles, custom auto bodies, truck cabs, horse trailers, motor homes, housing modules, concrete forms, and playground equipment.[19]

Much molding is also performed with premix compounds, which are doughlike materials generally prepared by the molder shortly before they are to be molded by combining the premix constituents in a sigma-blade mixer or similar equipment. Premix, using conventional polyester resins, is employed to mold automotive-heater housings and air-conditioner components. Low-shrinkage resin systems permit the fabrication of exterior automotive components such as fender extensions, lamp housings, hood scoops, and trim rails.

Wet molding of glass mats or preforms is used to fabricate such items as snack-table tops, food trays, tote boxes, and stackable chairs.

Corrugated and flat paneling for room dividers, roofing and siding, awnings, skylights, fences, and the like is a very important outlet for polyesters. Pultrusion techniques are used to make fishing-rod stock and profiles from which slatted benches and ladders can be fabricated. Chemical storage tanks are made by filament winding.[19]

Silicones

Silicones are a family of unique synthetic polymers which are partly organic and partly inorganic.[20] They have a quartzlike polymer structure, being made up of alternating silicon and oxygen atoms rather than the carbon-to-carbon backbone which is a characteristic of the organic polymers. Silicones have outstanding thermal stability.

Typically, the silicon atoms will have one or more organic side groups attached to them, generally phenyl (C_6H_5—), methyl (CH_3—), or vinyl (CH_2=CH—) units. Other alkyl, aryl, and reactive organic groups on the silicon atom are also possible. These groups impart characteristics such as solvent resistance, lubricity and compatibility, and reactivity with organic chemicals and polymers.[20]

Silicone polymers may be filled or unfilled, depending on properties desired and application. They can be cured by several mechanisms either at room temperature (RTV) or at elevated temperatures. Their final form may be fluid, gel, elastomeric, or rigid.[20]

Some of the properties which distinguish silicone polymers from their organic counterparts are (1) relatively uniform properties over a wide temperature range, (2) low surface tension, (3) high degree of slip or lubricity, (4) excellent release properties, (5) extreme water repellency, (6) excellent electrical properties over a wide range of temperature and frequency, (7) inertness and compatibility, both physiologically and in electronic applications, (8) chemical inertness, and (9) weather resistance.

Flexible resins Flexible two-part, solvent-free silicone resins are available in filled and unfilled forms. Their viscosities range from 3,000 cP to viscous thixotropic fluids of greater than 50,000 cP.[20]

The polymer base for these resins is primarily dimethylpolysiloxane. Some vinyl and hydrogen groups attached to silicon are also present as part of the polymer.

These products are cured at room or slightly elevated temperatures. During cure there is little if any exotherm, and there are no by-products from the cure. The flexible resins have Shore A hardnesses of 0 to 60 and Bashore resiliencies of 0 to 80. Flexibility can be retained from −55°C or lower to 250°C or higher.

Flexible resins find extensive use in electrical and electronic applications where stable dielectric properties and resistance to harsh environments are important. They are also used in many industries to make rubber molds and patterns.

Rigid resins Rigid silicone resins exist as solvent solutions or as solvent-free solids. The most significant uses of these resins are as paint intermediate to upgrade thermal and weathering characteristics of organic coatings, as electrical varnishes and as glass tape and circuit-board coatings.

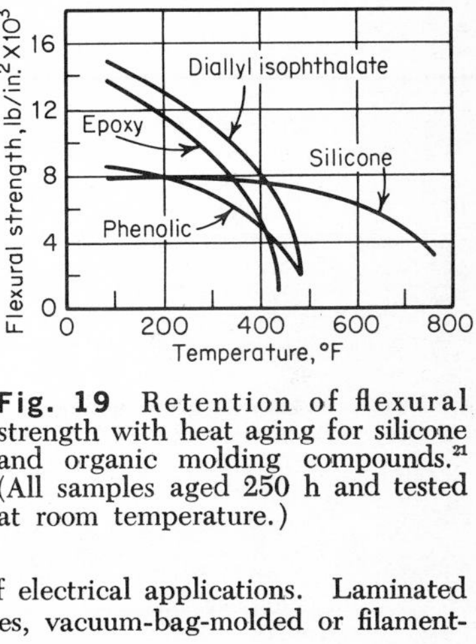

Fig. 19 Retention of flexural strength with heat aging for silicone and organic molding compounds.[21] (All samples aged 250 h and tested at room temperature.)

Glass cloth, asbestos, and mica laminates are prepared with silicone resins for a variety of electrical applications. Laminated parts can be molded under high or low pressures, vacuum-bag-molded or filament-wound.

Thermosetting molding compounds made with silicone resins as the binder are finding wide application in the electronic industry as encapsulants for semiconductor devices. Inertness toward devices, stable electrical and thermal properties, and self-extinguishing characteristics are important reasons for their use. Typical properties are given in Table 22, and thermal stability is shown in Fig. 19.

Similar molding compounds, containing refractory fillers, can be molded on conventional thermoset equipment. Molded parts are then fired to yield a ceramic article.

High-impact, long-glass-fiber-filled molding compounds are also available for use in high-temperature structural applications.

In general silicone resins and composites made with silicone resins exhibit outstanding long-term thermal stabilities at temperatures approaching 300°C and excellent moisture resistance and electrical properties, as discussed in more detail in Chap. 2.

TABLE 22   Typical Properties of Silicone Molding Compounds[21]

ASTM test method	Property	Mineral filled	Glass-fiber-filled
	Specific gravity	1.80–1.95	1.88
	Density, lb/in.^{3}	0.065–0.101	0.068
	Mold shrinkage, in./in.	0.007	0.005
	Flammability	Self-extinguishing	Self-extinguishing
	Thermal conductivity at 500°F, Btu/(h)(ft^2)(°F)(ft)	2.0–2.1	1.2
D 570	Water absorption in 24 h, ⅛-in. thickness, at 73°F, % increase in weight	0.05–0.20	0.10–0.12
D 696	Coefficient of thermal expansion, in./(in.)(°C) × 10^5, at 25 to 250°C		
	Perpendicular	2.5–6.0	Not isotropic 1.0
	Parallel	. . . .	12.1
D 648	Heat-distortion temp, °F	340–900	>900
	Max recommended intermittent-service temp, °F	500–600	700–750
	Max recommended continuous-service temp (continuous loading), °F	500–750	700–750
D 638	Tensile yield strength (avg) at 73°F, lb/in.^{2} × 10^{-3}	2.5–3.5	3.5–6.5
D 790	Flexural strength at 77°F, lb/in.^{2} × 10^{-3}	6.5–8.5	18–21
D 790	Flexural modulus at 77°F, lb/in.^{2} × 10^{-3}	1.4–1.6	2.5
D 695	Compressive strength at 77°F, lb/in.^{2} × 10^{-3}	11–18	10–12.5
D 256	Impact strength, Izod, at 73°F (notched), ft-lb/in. notch	0.30–0.40	10–24
D 785	Hardness, Rockwell M, at 73°F	70–95	83–88
D 257	Volume resistivity at 50% RH and 23°C, Ω-cm	2 × 10^{13}–5 × 10^{15}	Dry, 1.4 × 10^{14} 9 × 10^{-14}
D 150	Power factor at 1 MHz:		
	Dry	0.002	0.004
	After 24 h immersion	0.003	0.005
D 150	Dielectric constant:		
	At 60 Hz	3.4–6.3	4.35
	At 1 MHz	3.4–6.3	4.26–4.28
D 149	Dielectric strength, 500 V/s rate of rise, ⅛-in. thickness, V/mil	300–465	275–300
D 495	Arc resistance (tungsten electrodes), s	190–420	230–240
	Ultraviolet resistance	Excellent	Excellent
	Resistance to acids	To mild acids, excellent	To mild acids, excellent
	Resistance to bases	To mild bases, fair to good	To mild bases, fair to good
	Resistance to solvents	Fair to excellent	Fair to excellent
	Relative material costs, cents/in.^{3}	26–41	24

THERMOPLASTICS

The general nature of thermoplastics has been discussed above. The individual thermoplastic materials will be discussed below. In general, thermoplastic materials tend to be tougher and less brittle, so that they can be applied without use of fillers. However, while some are very tough, others do tend to craze or crack easily, so that each case must really be considered on its individual merits. Traditionally, by virtue of their basic polymer structure, thermoplastics have been much less dimensionally and thermally stable than thermosetting plastics. Hence, thermosets have offered a performance advantage, although the lower processing costs for thermoplastics have given them a cost advantage. Thus, a common decision was stable performance vs. cost. Recently, however, three major trends tend to put both thermoplastics and thermosets on a performance-consideration basis. First, much has been done in the development of reinforced, fiber-filled thermoplastics, greatly increasing stability in many areas, discussed in more detail in a later section of this chapter. Second, much has been achieved in the development of so-called engineering thermoplastics, or higher-stability, higher-performance plastics—which can also be reinforced with fiber fillers to increase their stability further. Third, and countering these gains in thermoplastics, has been the development of lower-cost processing of thermosetting plastics, especially the screw-injection-molding technology, discussed in earlier sections of this chapter. All these options should be considered in optimizing the design, fabrication, and performance of plastic parts.

A list of typical trade names and suppliers of thermoplastics is given in Table 23. Typical physical and mechanical properties of thermoplastics are presented in Table 24. Specific classes of thermoplastics are discussed below.

TABLE 23 Typical Trade Names and Suppliers for Thermoplastics

Thermoplastic	Typical trade names and suppliers
ABS	Marbon Cycolac (Borg-Warner) Abson (B. F. Goodrich) Lustran (Monsanto)
Acetals	Delrin (E. I. du Pont) Celcon (Celanese)
Acrylics	Plexiglas (Rohm and Haas) Lucite (E. I. du Pont)
Aramids	Nomex (E. I. du Pont)
Cellulosics	Tenite (Eastman Chemical) Ethocel (Dow Chemical) Forticel (Celanese)
Fluorocarbons	See Table 29
Ionomers	Surlyn A (E. I. du Pont) Bakelite (Union Carbide)
Low-permeability thermoplastics	Barex (Vistron/Sohio) NR-16 (E. I. du Pont) LPT (Imperial Chemical Industries)
Nylons (see also Aramids)	Zytel (E. I. du Pont) Plaskon (Allied Chemical) Bakelite (Union Carbide)
Parylenes	Parylene (Union Carbide)
Polyaryl ether	Arylon T (Uniroyal)
Polyaryl sulfone	Astrel (3M)
Polycarbonates	Lexan (General Electric) Merlon (Mobay Chemical)
Polyesters	Valox (General Electric) Celanex (Celanese) Celanar Film (Celanese) Mylar Film (E. I. du Pont) Tenite (Eastman Chemical)
Polyethersulfone	Polyethersulphone (Imperial Chemical Industries)

TABLE 23   Typical Trade Names and Suppliers for Thermoplastics (Continued)

Thermoplastic	Typical trade names and suppliers
Polyethylenes, polypropylenes, and polyallomers..................	Alathon Polyethylene (E. I. du Pont) Petrothene Polyethene (U.S.I.) Hi-Fax Polyethylene (Hercules) Pro-Fax Polypropylene (Hercules) Bakelite Polyethylene and Polypropylene (Union Carbide) Tenite Polyethylene and Polypropylene (Eastman) Irradiated Polyolefin (Raychem)
Polyimides and polyamide-imides......	Vespel SP Polyimides (E. I. du Pont) Kapton Film (E. I. du Pont) Pyralin Laminates (E. I. du Pont) Keramid/Kinel (Rhodia) P13N (Ciba-Geigy) Torlon Polyamide-Imide (Amoco)
Polymethyl pentene.................	TPX (Imperial Chemical Industries)
Polyphenylene oxides...............	Noryl (General Electric)
Polyphenylene sulfides..............	Ryton (Phillips Petroleum)
Polystyrenes......................	Styron (Dow Chemical) Lustrex (Monsanto) Dylene (Koppers) Rexolite (American Enka)
Polysulfones.....................	Ucardel (Union Carbide)
Vinyls...........................	Pliovic (Goodyear Chemical) Diamond PVC (Diamond Alkali) Geon (B. F. Goodrich) Bakelite (Union Carbide)

These are but a few typical basic-material suppliers. In addition, there are many companies which formulate systems from these basic materials, and which are sometimes the ultimate supplier to the end-product fabricator.

ABS Plastics

ABS plastics are derived from acrylonitrile, butadiene, and styrene. The general chemical structure is shown below. This class possesses hardness and rigidity without brittleness, at moderate costs. ABS materials have a good choice balance of tensile strength, impact resistance, surface hardness, rigidity, heat resistance, low-temperature properties, and electrical characteristics. There are many ABS modifications, and many blends of ABS with other thermoplastics. A list of trade names and suppliers is given in Table 23. The general data on the complete spectrum of ABS-based plastics are shown in Table 25.

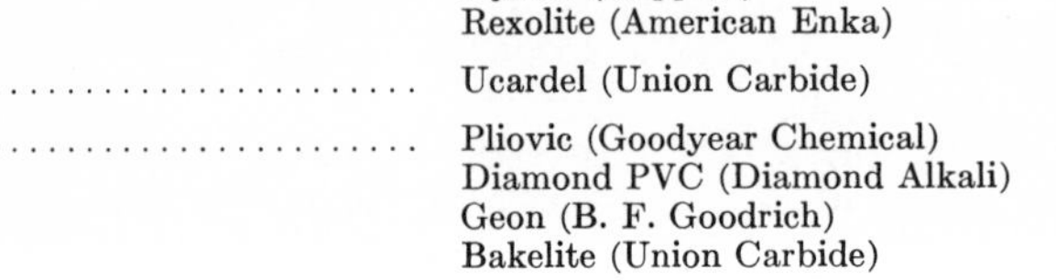

Acrylonitrile-butadiene-styrene copolymer

Polymer properties The most outstanding mechanical properties of ABS plastics are impact resistance and toughness. A wide variety of modifications are available for improved impact resistance, toughness, and heat resistance. Impact resistance does not fall off rapidly at lower temperatures. Stability under load is excellent with limited loads, shown in Fig. 20. Heat-resistant ABS is equivalent to or better than acetals, polycarbonates, and polysulfones in creep at 3,000 lb/in.² at room temperature. The Izod impact strength at 75°F is in the range of 3 to 5 ft-lb/in. of notch. This figure is gradually reduced to 1 ft-lb/in. of notch at −40°F. When impact failure does occur, the failure is ductile rather than brittle. Modulus of elasticity vs. temperature is shown in Fig. 21. Physical properties are little affected by moisture, which greatly contributes to the dimensional stability of ABS materials. ABS plastics are compared with some other materials as to flexural modulus, tensile strength, creep behavior, and other mechanical properties in Chap. 3.

ABS alloys and electroplating grades   Much work has been done to modify ABS plastics by alloying, to improve certain properties, and by modifying, to enhance adhesion of electroplated coatings. A summary of the major alloys is shown in Table 26.

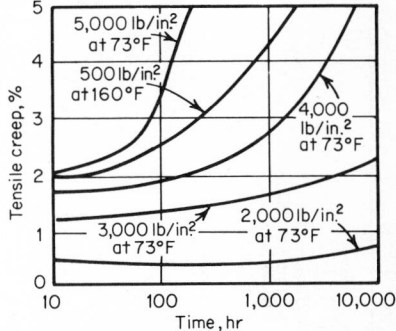

ABS alloyed or blended with polycarbonate combines some of the best qualities of both materials, resulting in a thermoplastic that is easier to process, has high heat and impact resistance, and sells for considerably less than polycarbonate.

The impact strength of the ABS-polycarbonate alloy, 10.7 ft-lb/in. of notch, is well above average for the high-impact engineering thermoplastics, but still not as high as that of polycarbonate. However, unlike polycarbonate, the alloy does not have critical thicknesses with respect to notched impact strength. Its notched

Fig. 20   Tensile creep of ABS under various loads.[24]

Izod value drops by only 2 to 4 ft-lb/in. in the ⅛- to ¼-in. range. The notched Izod value for a polycarbonate ⅛ in. thick is about 16 ft-lb/in. but only 3 to 4 ft-lb/in. at thicknesses greater than ¼ in. The flexural modulus of the ABS-polycarbonate alloy is about 15 percent greater than that of polycarbonate alone. The alloy remains more rigid than polycarbonate up to about 200°F, as shown in Fig. 22.

The 264-psi heat-deflection temperature of the alloy is 245°F, and the 66-psi value is 260°F. These values are 35 and 30°F, respectively, lower than those of polycarbonate. However, maximum recommended continuous (no-load) temperature of the alloy is only 10°F lower than that of polycarbonate.

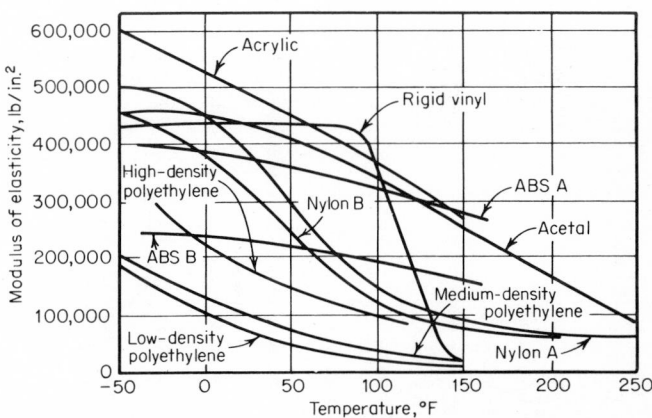

Fig. 21   Modulus of elasticity of several thermoplastic materials as a function of temperature.[25]

TABLE 24 Typical Physical and Mechanical Properties of Thermoplastics[22]

Resin material	Coefficient of thermal expansion, in./in. (°C $\times 10^{-5}$)	Thermal conductivity, (cal)/(cm²)(s)(°C)(cm) $\times 10^{-4}$	Water absorption, 24 h, %	Rockwell hardness*	Flammability, (in./min) 0.125 in.	Specific gravity	Mold shrinkage, in./in.	Clarity
Acetal	0.25	1.6	0.25	M94, R120	1.1	1.410–1.425	0.022	Translucent to opaque
ABS	3–10.5	4–9	0.2–0.5	R80–120	1.0–2	1.01–1.07	0.003–0.007	Opaque
Acrylic		1.4	0.3	M84–97	9–1.2	1.18–1.19	0.002–0.006	Transparent
Acrylic high impact	6.5–10.5	4.0	0.2–0.3	M20–67	1.1–1.2	1.11–1.18	0.004–0.008	Translucent to opaque
Cellulose acetate	8–18	4–8	1.7–4.4	R7–122	0–2	1.22–1.34	0.001–0.008	Transparent
Cellulose acetate butyrate	11–17	4–8	0.9–2.2	R17–113	0.5–1.5	1.15–1.22	0.003–0.006	Transparent
Cellulose propionate	11–16	4–8	1.2–2.8	R15–120	0.5–1.5	1.16–1.23	0.001–0.006	Transparent
Ethyl vinyl acetate	10–20	8	<0.01	R3–7	Slow-burning	0.93–0.95	0.01–0.02	Transparent
Fluorocarbons: Chlorotrifluoroethylene	5–7	4–6	Nil	R85–112	Nil	2.09–2.14	0.010–0.015	Transparent to opaque
Fluorinated ethylene propylene	8.3–10.5	5.9	<0.05	D55	Nonflammable	2.16	0.03–0.05	Transparent to opaque
Polytetrafluoroethylene	5.5 (25–60°C)	6	0.01	D60–65	Nonflammable	2.13–2.18	0.02–0.06	Transparent to opaque
Ethylene tetrafluoroethylene copolymer	5–9	......	0.029	R50, D75	None	1.7	0.030–0.040	Transparent in thin sections
Perfluoroalkoxy	12	......	0.03	D64	None	2.12–2.17	0.040	Transparent to translucent
Ethylene chlorotrifluoroethylene copolymer	8	3.8	0.01	R95	None	1.7	0.020–0.025	Transparent to translucent
Vinylidene fluoride	8.5	3.0	0.04	D80	None	1.75–1.78	0.030	Transparent to translucent
Nylon 6	4.6–5.8	5.9	1.5	R107–119	Self-extinguishing	1.13–1.14	0.007–0.011	Transparent to opaque
Nylon 6/6	8.1	5.8	1.3	R118–123	Self-extinguishing	1.13–1.15	0.007–0.015	Translucent to opaque

Material	Coef. of thermal expansion, ×10⁻⁶ in./(in.)(°F)	Thermal conductivity, Btu/(h)(ft)(°F)(in.)	Water absorption, %	Hardness	Flammability	Specific gravity		Optical properties
Nylon 6/10	10	5.5	0.4	R111	Self-extinguishing	1.07–1.09	0.015	....
Polyallomer	8–11	2–4	<0.05	R50–85	Slow-burning	0.90–0.906	0.01–0.02	Transparent to opaque
Polyaryl ether	3.6	7.13	0.25	R117	Slow	1.14	0.007	Translucent to opaque
Polyaryl sulfone	4.7	4.55	1.1	M110	Self-extinguishing	1.36	0.007–0.009	Opaque
Polycarbonate	6.7–7	4.6	0.15	M70, R112	Self-extinguishing	1.2	0.005–0.007	Transparent
Polyethylene, low-density	10–20	8	<0.05	R10	Slow-burning	0.910–0.925	0.01–0.03	Transparent to opaque
Polyethylene, medium-density	10–20	8	<0.05	R15	Slow-burning	0.926–0.940	0.01–0.035	Transparent to opaque
Polyethylene, high-density	10–20	1.9–3.3	<0.01	R30–60	Slow-burning	0.941–0.965	0.01–0.04	Translucent to opaque
Polyethylene, high-molecular-weight	13	8	<0.01	R55	Slow-burning	0.93–0.94	0.03	Translucent to opaque
Polyimide	......	......	0.32	R85–95	......	1.43	......	Opaque
Polymethyl pentene	11.7	4.0	0.01	L67–74	1.0	0.83	0.015–0.030	Transparent to opaque
Polyphenylene sulfide	5.5	6.84	0.02	R124	None	1.34	0.010	Opaque
40% glass-filled polyphenylene sulfide	4.0	......	0.01	R123	None	1.64	0.002	Opaque
Polypropylene	3.8–9	2.8–4	<0.01	R45–99	Slow-burning to nonburning	0.90–1.24	0.008–0.025	Transparent to opaque
Polystyrene	6–8	8	0.03–0.05	M65–80	0.5–2.5	1.05–1.06	0.002–0.006	Translucent to opaque
Polystyrene, high-impact	6.5–8.5	1–3	0.05–0.10	M25–69	0.5–2.5	1.04–1.06	0.003–0.005	Translucent to opaque
Polyurethane	10–20	7.4	0.60–0.80	M26, R90	Slow to self-extinguishing	1.11–1.26	0.009	Transparent to opaque
Polyvinyl chloride (flexible)	7–25	3–4	0.15–0.75		Self-extinguishing	1.15–1.80	0.002–0.004	Translucent to opaque
Polyvinyl chloride (rigid)	5–10	3–5	0.07–0.40	R100–120	Self-extinguishing	1.33–1.58	......	Transparent to opaque
Polyvinyl dichloride (rigid)	7–8	3–4	0.07–0.11	R118	Self-extinguishing	1.50–1.54	0.006–0.007	Translucent to opaque
Styrene acrylonitrile (SAN)	7	3	0.23–0.28	M30–83	0.4–0.7	1.07–1.08	0.003–0.004	Transparent
Ionomer	12–13	5.8	0.1–1.4	D60–65	0.9–1.1	0.94–0.96	0.001–0.005	Transparent
Polyphenylene oxide	5.2	......	0.06	R120	Self-extinguishing	1.06	0.006–0.008	Transparent to opaque
Polysulfone	3.1 × 10⁻⁶ in./(in.)(°F)	1.8 Btu/(h)(ft)(°F)(in.)	0.22	M69, R120	Self-extinguishing	1.24–1.25	0.0076	Transparent to opaque
Polyethersulfone	5.5	3.2–4.4	0.43	M88	......	1.37	0.007	Transparent

TABLE 24 Typical Physical and Mechanical Properties of Thermoplastics[22] (Continued)

Resin material	Impact strength notched Izod, ft-lb/in., ½-in. bar	Tensile strength, lb/in.² × 10³	Tensile modulus, lb/in.² × 10³	Elongation, %	Flexural strength, lb/in.² × 10³	Compressive strength, lb/in.² × 10³	Compressive modulus, lb/in.² × 10³	Heat-distortion temp, °F, 264 lb/in.²	Heat resistance, continuous, °F
Acetal	1.1–1.4	8.8–10	400–410	12–75	13–14	18	410	230–255	185
ABS	1.3–10.0	4.5–8.5	200–450	5–200	5–13.5	5–11	120–200	180–245	160–235
Acrylic	0.3–0.4	8.7–11.0	350–450	3–6	14–17	14–17	350–430	167–198	130–195
Acrylic high impact	0.5–2.3	5.5–8	225–330	23–38	8.5–12	7–12	250–360	169–190	140–195
Cellulose acetate	0.5–5.6	2.3–8.1	......	10–70	2.2–11.5	2.0–10.9	......	111–209	140–175
Cellulose acetate butyrate	0.4–11	2.6–6.9	......	40–88	1.8–9.3	2.1–9.4	......	113–227	140–175
Cellulose propionate	0.7–10.7	1.8–7.3	......	30–100	2.8–11	2.4–9.6	......	119–250	140–175
Ethyl vinyl acetate	No break	20–40	3.0–15	500–1,500					120–170
Fluorocarbons:									
Chlorotrifluoroethylene	3.5	6	150–190	60–190	8–10	6–12	180	160–170	390
Fluorinated ethylene propylene	No break	2–3.2	60–80	250–350	......	......	70	124	400
Polytetrafluoroethylene	No break	2–5	50–100	75–400	......	4–12	70–90	132	500
Ethylene tetrafluoroethylene copolymer	No break	6.5	120	100–400	......	7.1	......	160	300–360
Perfluoroalkoxy	No break	4.3	......	300					500
Ethylene chlorotrifluoroethylene copolymer	No break	7.0	240	200	7	8.6	120	170	330–355
Vinylidene fluoride	3.6–4.0	5.5–7.4	120	100–300	......	......	......	195	300
Nylon 6	0.9–4	9.5–12.4	200–450	25–300	9–16.6	4–11	347	150–175	250
Nylon 6/6	0.9–2	11.2–13.1	410–480	60–300	14.6	5–13	400	200	250

Nylon 6/10	0.8–3	7–8.5	160–280	50–300	10.5	4–6	....	145	220
Polyallomer	1.5–12	2.9–4.2	100–170	400–650	4–5	....	....	124–133	250
Polyaryl ether	8.0	7.5	320	25–90	11	....	....	300	250
Polyaryl sulfone	1–2	13	370	13–20	17.2	17.9	340	525	500
Polycarbonate	2–3	8–9.5	345	60–110	11–13	12.5	350	265–290	250
Polyethylene, low-density	No break	1–2.4	14–38	20–800	....	....	....	....	140–175
Polyethylene, medium density	No break	1.7–2.8	50–80	80–600	....	....	....	....	150–180
Polyethylene, high-density	0.5–23	2.8–5	75–200	10–800	1–4	0.8–3.6	50–110	110–125	180–225
Polyethylene, high molecular weight	>20	2.3–5.4	102	525–600	3.5	2.4	110	120	180–225
Polyimide	0.8–1.1	5–14.0	320–450	6–7	7–14	12–24	....	680	500–600
Polymethyl pentene	0.4–1.6	3.5–4.0	160–210	13–22	....	....	....	....	250–320
Polyphenylene sulfide	0.3	10	480	3	20	....	....	275	400–500
40% glass-filled polyphenylene sulfide	0.8	21	1,120	3	37	....	....	>425	400–500
Polypropylene	0.5–15	3.2–5	150–650	3–700	4.5–8	6–10	....	140–205	250
Polystyrene	0.25–0.40	6–8.1	400–500	1.5–2.5	9–15	11.5–16	300–560	160–215	150–190
Polystyrene, high-impact	0.7–3.5	1.9–4	200–430	10–75	5.5–12.5	8–16	....	160–205	130–180
Polyurethane	No break	4.5–8	1–3.7	400–650	0.7–1	>20	85	....	190
Polyvinyl chloride (flexible)	Varied	1–4	....	100–450	....	....	....	....	150–175
Polyvinyl chloride (rigid)	0.4–22	6–9	200–600	5–40	8–15	10–11	300–400	140–175	160–165
Polyvinyl dichloride (rigid)	1.5–7.0	7.5–9.0	360–450	10–65	14.2–17	13–22	....	212–235	195–210
Styrene acrylonitrile (SAN)	0.3–0.50	8–12	500–600	1–3.2	17	15–17.5	650	200–218	170–210
Ionomer	5.7–13	3.5–5.5	28–40	300–450	....	....	....	....	140
Polyphenylene oxide	1.5–1.9	11	380	50–80	15	15	380	375	250
Polysulfone	1.3	10.2	360	50–100	15.4	15.4	370	345	300
Polyethersulfone	1.6	12.2	350	30–80	18.6	....	....	397	300–390

* D hardness is Shore D.

TABLE 25 Typical Data for Spectrum of ABS Plastics[23]

	Marked grade	Rockwell hardness R at 73°F	Falling dart impact, ft-lb at 73°F	Flexural modulus, lb/in.² × 10⁵ at 73°F	Heat-deflection temp, °F at 264 lb/in.² (annealed)	Notched Izod, ft-lb/in. at 73°F	Notched Izod, ft-lb/in. at −40°F	Tensile strength at 73°F	Underwriters' insulating material
Standard..........	T	103	16–18	3.1	210	6.5	1.9	5,900	
High impact.......	GS	102	18–22	3.2	214	7.0	2.0	6,000	70°C
High impact.......	H	88	28–30	2.5	210	7.3	2.3	4,600	
High impact.......	L	89	28–30	2.5	210	7.5	2.6	4,900	
Controlled gloss..	CG	81	18–22	2.4	220	8.0	1.8	4,300	
Self-extinguishing.	KA*	103	5–7	3.4	...	12.5	1.6	6,000	50°C
Self-extinguishing.	KJ	96	30–32	3.1	205	2.1	0.5	5,700	70°C
High heat.........	X-17	111	.....	3.9	240	2.0	0.7	7,100	
High heat.........	X-27	111	.....	4.0	244	2.2	0.7	7,400	
Plating...........	EP	102	14–16	3.3	213	5.0	1.4	5,900	
Plating...........	EPA	110	.....	3.9	226	2.1	0.9	6,600	50°C
Medium impact.....	AH	110	8–10	3.8	219	4.3	1.1	7,100	
Medium impact.....	DH	111	6–8	3.9	221	3.4	1.0	7,500	
Cold forming......	MS	103	18–20	3.3	217	5.5	1.5	6,200	
Clear.............	CIT†	100	16–18	3.0	183	5.3	0.6	5,600	

Cycolac is a registered trademark, Borg-Warner. PPO is a registered trademark, General Electric.
* Grade KA is registered as Cycovin.
† Clear-impact thermoplastic.

TABLE 26   Summary of Major ABS Alloys

Alloys	Key characteristics	Suppliers
ABS/PVC (rigid-type).....	Virtually all the properties of high-impact grades of ABS and flame-retardant ratings up to UL's SE-O	B. F. Goodrich, Marbon, Uniroyal, A. Schulman
ABS/polycarbonate........	Very high impact strength and high heat resistance compared with other ABS	Marbon
ABS/polysulfone.........	High heat resistance and dimensional stability of polysulfone plus the low cost and processing characteristics of ABS. Chrome-plated moldings withstand thermal cycling without peeling or cracking	Uniroyal

The good creep resistance of polycarbonate, one of its biggest advantages, shown in Fig. 23, is maintained after alloying.

In addition to the polycarbonate alloy, ABS can be alloyed with other plastics to obtain special properties. Furthermore, ABS has been modified to gain improved adhesion of electroplated metals. Advances have been so great in this area that ABS is perhaps the most widely used material for producing electroplated plastic parts. Electroplated ABS is extensively employed for many electrical and mechanical products, in many forms and shapes. Adhesion of the electroplated metal is excellent. Electroplated plastics are discussed in more detail in a later section of this chapter.

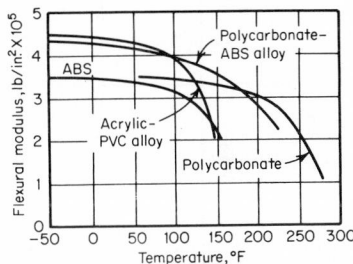

Fig. 22 Flexural modulus of ABS-polycarbonate alloy and base materials.[26]

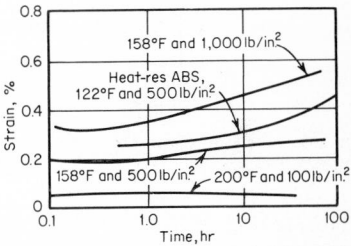

Fig. 23 Creep resistance of ABS-polycarbonate alloy and heat-resistant ABS.[26]

Acetals

Acetals are among the group of high-performance engineering thermoplastics that resemble nylon somewhat in appearance but not in properties. The general repeating chemical structural unit is shown below:

$$\left[\begin{array}{c} \text{H} \\ | \\ -\text{C}-\text{O}- \\ | \\ \text{H} \end{array}\right]_n$$

Polyacetal resin

They are strong and rigid (but not brittle) and have good moisture, heat, and chemical resistance. There are two basic types of acetals, namely, the homopolymers by du Pont, and the copolymers by Celanese. Table 23 further identifies these materials, and properties are shown in Table 24. The homopolymers are harder, have higher resistance to fatigue, are more rigid, and have higher tensile and flexural strength with lower elongation. The copolymers are more stable in long-term, high-temperature service, and more resistant to hot water. Neither type of acetal is resistant to strong mineral acids, but the copolymers are resistant to strong bases. References are frequently made to acetals, without identification of polymer type. Such references usually imply the homopolymer material.

Polymer properties The most outstanding properties of acetals are high tensile strength and stiffness, resilience, good recovery from deformation under load, and toughness under repeated impact. They exhibit excellent long-term load-carrying properties and dimensional stability and can be used for precision parts. Acetals have low static and dynamic coefficients of friction and are usable over a wide range of environmental conditions. The plastic surface is hard, smooth, and glossy. A fluorocarbon fiber-filled acetal, Delrin AF, is available and offers even better low friction and resistance properties. This is discussed in more detail below.

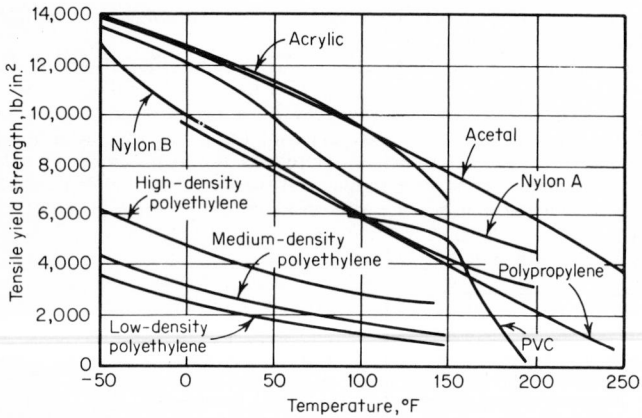

Fig. 24 Tensile yield strength of several thermoplastic materials as a function of temperature.[27]

The modulus of elasticity as a function of temperature for acetals and several other thermoplastic materials was shown earlier in Fig. 21. The tensile yield strength of acetals is compared with that of some other thermoplastics in Fig. 24. The deflection under load for Delrin acetal is compared with that of other thermoplastics in Fig. 25.

Effect of Moisture. Because acetals absorb a small amount of water, dimensions of molded parts are affected by their water content. Figure 26 presents the relationship of dimensional changes with changes in moisture content. The dimensional change resulting from an environmental change is found by subtracting the "percent in length" found at the first humidity-temperature condition from the "percent in length" at the final conditions. For example, to go from 77°F (25°C) and 0 percent water (as-molded condition) to 100°F (38°C) and 100 percent humidity, a change of 0.45 percent will occur. Figure 27 shows the rate of water absorption of acetals at various conditions.

Effect of Space and Radiation. Acetal resin will be stable in the vacuum of space under the same time-temperature conditions it can withstand in air.[28] Exposure to vacuum alone causes no loss of the engineering properties. The principal result is slight outgassing of small amounts of moisture and free formaldehyde. In a vacuum, as in air, prolonged exposure to elevated temperatures results in the liberation of increasing amounts of formaldehyde due to thermal degradation of the polymer.

Particulate radiation, such as the protons and electrons of the Van Allen radiation belts, is damaging to acetal resins and will cause loss of engineering properties.[27] For example, acetals should not be used in a radiation environment where the total electron dose is likely to exceed 1 megarad. When irradiated with 2-MeV electrons, a 1-megarad dose causes only slight discoloration while 2.3 megarads causes considerable embrittlement. At 0.6 megarad, however, acetals are still mechanically sound except for a moderate decrease in impact strength.

The regions of the electromagnetic spectrum that are most damaging to acetal resins are ultraviolet light and gamma rays.[28] In space, the deleterious effects of ultraviolet light are of prime consideration. This is due to the absence of the protective air atmosphere which normally filters out much of the sun's ultraviolet energy. Therefore, the amount of ultraviolet light in space may be 10 to 100 times as intense as on the ground.

Fluorocarbon Fiber-filled Acetal. This acetal modification, mentioned above, is a modified acetal homopolymer developed to meet the need for a thermoplastic injection-molding material to be used in moving parts in which low friction and exceptional wear resistance are the principal requirements. This resin consists of oriented TFE fluorocarbon fibers uniformly dispersed in a matrix of acetal resin. The result is an injection-molding and extrusion

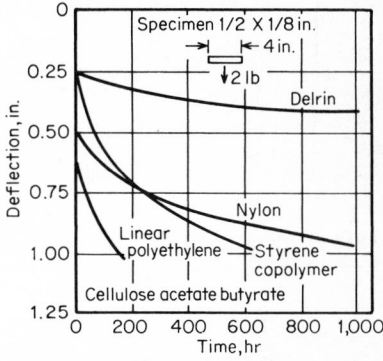

Fig. 25 Deflection of several thermoplastic materials as a function of time at 90% RH and 150°F.[27]

resin that combines the strength, toughness, dimensional stability, and fabrication economy of acetals with the unusual surface and low frictional characteristics of the fluorocarbons.

The outstanding properties of TFE fiber-filled acetal are those associated with sliding friction. Bearings made from this material sustain high loads when operating at high speeds, and show little wear. In addition, such bearings are essentially free of slipstick behavior because their static and dynamic coefficients are almost equal. Comparative properties of the filled and unfilled acetal are given in Table 27.

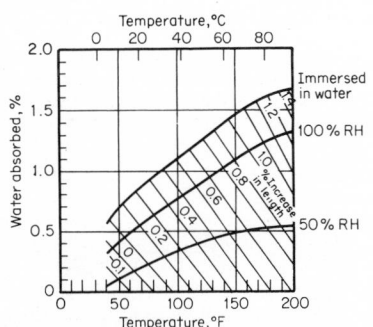

Fig. 26 Dimensional changes of Delrin acetal with variations of temperature and moisture content.[28]

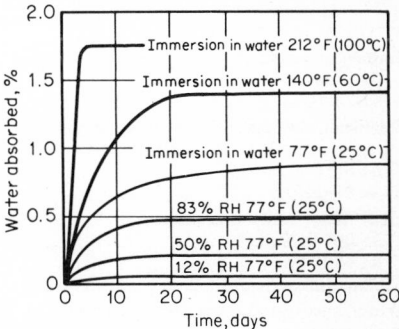

Fig. 27 Rate of water absorption of Delrin acetal at various conditions.[28]

Acrylics

The general properties of acrylics are presented in Table 24. Trade names and suppliers are listed in Table 23. Acrylics are based on polymethyl methacrylate, having the basic chemical structure shown on the next page.

$$\left[\begin{array}{c} \\ -\overset{\displaystyle H}{\underset{\displaystyle H}{\vphantom{|}C}}-\overset{\displaystyle CH_3}{\underset{\displaystyle COOCH_3}{\vphantom{|}C}}- \\ \end{array} \right]_n$$

Polymethyl methacrylate

Acrylics have exceptional optical clarity and basically good weather resistance, strength, electrical properties, and chemical resistance. They do not discolor or shrink after fabrication, have low water-absorption characteristics and a slow burning rate, and will not flash-ignite. For disadvantages, acrylics are attacked by strong solvents, gasoline, acetone, and other similar fluids.

Acrylics can be injection-molded, extruded, cast, vacuum- and pressure-formed, and machined, although molded parts for load bearing should be carefully analyzed, especially for long-term loading.

TABLE 27   Property Comparisons for Unfilled and TFE Fiber-Filled Acetal[29]

Property	ASTM No.	Average values*	
		Delrin 500 (unfilled)	Delrin AF (TFE filled)
Tensile strength and yield point,† lb/in.² $\times$ 10⁻³:			
At −68°F	D 638	14.7	9
At 73°F	D 638	10	6.9
At 158°F	D 638	7.5	4.7
Elongation,† %:			
At −68°F	D 638	13	6
At 73°F	D 638	15	12
At 158°F	D 638	330	38
Flexural modulus, lb/in.² $\times$ 10⁻³:			
At 73°F	D 790	410	400
At 170°F	D 790	190	180
At 250°F	D 790	90	85
Compressive stress, lb/in.² $\times$ 10⁻³:			
1% deformation	D 695	5.2	1.8
10% deformation	D 695	18	13
Impact strength, Izod, ft-lb/in.:			
At −40°F	D 256	1.2	0.6
At 73°F	D 256	1.4	0.7
Coefficient of linear thermal expansion, in./(in.)(°F) $\times$ 10⁵:			
At 85°F	D 696	4.5	4.5
At 85–140°F	......	5.5	5.5
Heat-distortion temperature, °F:			
At 264 lb/in.²	D 648	255	212
At 66 lb/in.²	D 648	338	329
Flammability, in./min	D 635	1.1	0.8
Water absorption after 24 h immersion, %	D 570	0.25	0.06
Rockwell hardness	D 785	M 94, R 120	M 78, R 118
Specific gravity	D 792	1.425	1.54
Taber abrasion (1,000-g load, CS-17 wheel), mg/kHz	D 1044	20	9
Coefficient of friction (no lubricant)	......	0.1–0.3	0.05–0.51

* These values represent those obtained under standard ASTM conditions and should not be used to design parts that function under different conditions. Since they are average values, they should not be used as minimums for material specifications.

† Determined at 0.2 in./min.

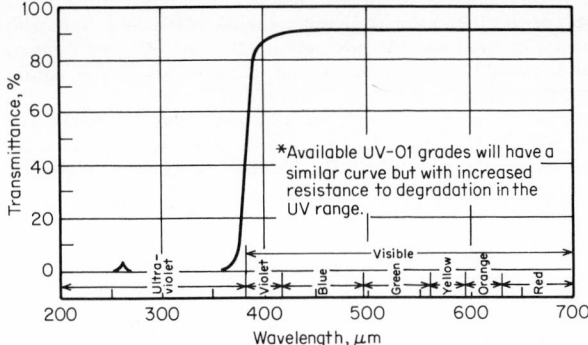

Fig. 28 Percent transmittance as a function of wavelength below 700μm for acrylic resin[30] (crystal, ⅛-in. thickness, grades 129, 130, 140, 147, 148, molded only).

Optical properties Parts molded from acrylic powders in their natural state may be crystal-clear and nearly optically perfect. The index of refraction ranges from 1.486 to 1.496. The total light transmittance is as high as 92 percent, and haze measurements average only 1 percent. Transmittance at various wavelengths is shown in Figs. 28 and 29. Light transmittance and clarity can be modified by addition of a wide range of transparent and opaque colors, most of these being formulated for long outdoor service.

Dimensional stability and aging The amount of dimensional change for a given application is determined from the coefficient of linear thermal expansion of the material. In the case of an exterior glazing unit, which is 1 by 1 ft (30.4 by 30.4 cm) and is subjected to a temperature variation of 32 to 100°F (0 to 38°C),

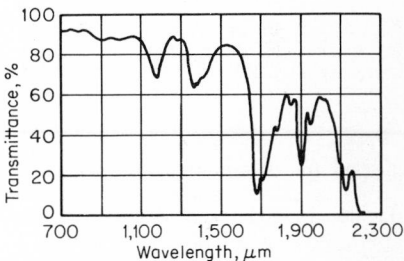

Fig. 29 Percent transmittance as a function of wavelength above 700μm for acrylic resin[30] (crystal, ⅛-in. thickness).

the space to be provided in the frame for expansion is calculated as follows (assuming no thermal expansion of frame material):

$$\Delta L = aLT$$

where ΔL = change in length, in. or cm
a = coefficient of thermal expansion = 4×10^{-5} in./in./°F
(7.2×10^{-5} cm/cm/°C)
L = initial length, in. = 12 (30.4 cm)
T = temperature variation, °F = 68 (38°C)

therefore:

$\Delta L = 4 \times 10^{-5} \times 12 \times 68 = 0.033$ in.
$\Delta L = 7.2 \times 10^{-5} \times 30.4 \times 38 = 0.083$ cm

This indicates the need for an expansion space of approximately 0.0165 in. (0.0419 cm) at each end for the direction in which this calculation was made (in this the major length).

The moisture level in acrylics will depend on the relative humidity (RH) of the environment. Figure 30 shows the equilibrium moisture content of acrylics vs. the relative humidity in the surrounding air at room temperature.

When the relative humidity of the air changes from a lower value to a higher one, acrylics will absorb moisture, which in turn causes a slight dimensional expansion. Depending on the thickness of the part, the equilibrium moisture level may be reached in a couple of weeks or in months.

Figure 31 shows the dimensional changes taking place in a ⅛-in.-thick (3.2-mm) sheet as a function of time and the environmental relative humidity at room temperature. The flat horizontal part of the curves corresponds to the equilibrium conditions at the specific environmental relative humidity.

For example, assume that a ⅛-in.-thick (3.2-mm) lighting panel as molded from dry resin is initially 24 in. long (61 cm). This panel installed in a 50 percent RH atmosphere will expand approximately 0.1 percent, or 0.024 in. (0.061 cm).

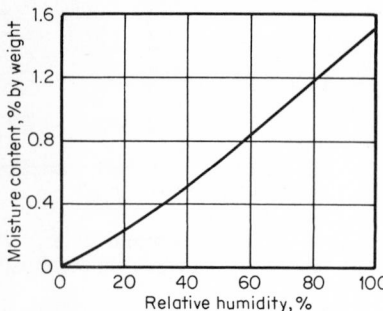

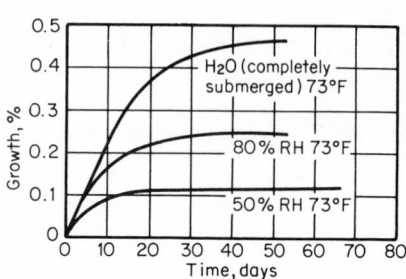

Fig. 30 Equilibrium moisture content in acrylic resins vs. percent relative humidity at 73°F.[30]

Fig. 31 Dimensional change vs. time for acrylics under various environmental moisture conditions.[30] Test samples ⅛ in. thick (3.2 mm).

Table 28 permits an easy, quick, and for most purposes, sufficiently accurate method for determining dimensional changes in acrylics affected by environmental variations in both temperature and relative humidity. The figures in the table show the dimensional changes at equilibrium in mils (0.001 in.) per foot relative to an arbitrarily chosen base condition of 70°F and 50 percent RH.

The values have been checked on actual end-use parts, specifically large lighting refractor lenses, tested and observed under environmentally controlled conditions.

The data given for calculating dimensional changes due to environmental variations in temperature and relative humidity are reproducible. Another major variable that should be considered in determining the total change in a part is the residual

TABLE 28 Dimensional Changes in Acrylics with Environmental Variations in Temperature and Relative Humidity[30]

% RH	Mils/ft*								
% moisture	10 0.18	20 0.32	30 0.46	40 0.61	50 0.76	60 0.90	70 1.05	80 1.20	90 1.35
Temp, °F									
20	−33	−31	−29	−26	−22	−18	−13	−8	−2
30	−29	−27	−25	−22	−18	−14	−9	−4	+2
40	−25	−23	−21	−18	−14	−10	−5	0	+6
50	−21	−19	−17	−14	−10	−6	−1	+4	+10
60	−16	−14	−12	−9	−5	−1	+4	+9	+15
70	−11	−9	−7	−4	0	+4	+9	+14	+20
80	−6	−4	−2	+1	+5	+9	+14	+19	+25
90	−1	+1	+3	+6	+10	+14	+19	+24	+30
100	+4	+6	+8	+11	+15	+19	+24	+29	+35
110	+9	+11	+13	+16	+20	+24	+29	+34	+40
120	+14	+16	+18	+21	+25	+29	+34	+39	+45

* 1 mil = 0.001 in.
Base is 70°F, 50% RH.

stress built into the part by processing, be it molding, extrusion, forming, or other method.

Because of their chemical structure, acrylic resins are inherently resistant to discoloration and loss of light transmission. They are unsurpassed in this respect by any other transparent plastic. The outstanding weatherability has been proved by their long-time performance in such products as automotive lenses, fluorescent street lights, outdoor signs, and boat windshields.

Cellulosics

These are a class of thermoplastics that are prepared by various treatments of purified cotton or special grades of wood cellulose. Trade names and suppliers are shown in Table 23, and data are presented in Table 24. The general cellulose structure is shown below:

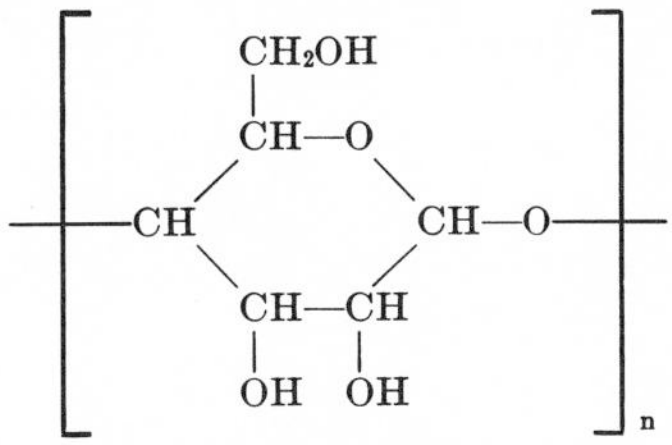

Cellulose (natural polymer)

Cellulosics are among the toughest of plastics, are generally economical, and are basically good insulating materials. However, they are temperature-limited and are not as resistant to extreme environments as many other thermoplastics. The four most prominent industrial cellulosics are cellulose acetate, cellulose acetate butyrate, cellulose propionate, and ethyl cellulose. A fifth member of this group is cellulose nitrate. Cellulose materials are available in a great number of formulas and flows and are manufactured to offer a wide range of properties. They are formulated with a wide range of plasticizers for specific plasticized properties.

Cellulose butyrate, propionate, and acetate provide a range of toughness and rigidity that is useful for many applications, especially where clarity, outdoor weatherability, and aging characteristics are needed. The materials are fast-molding plastics and can be provided with hard, glossy surfaces and over the full range of color and texture.

Butyrate, propionate, and acetate are rated in that order in dimensional stability in relation to the effects of water absorption and plasticizers. The heat deflection of cellulose acetate butyrate at 150°F is shown, compared with that of some other thermoplastics, in Fig. 25. The materials are slow-burning, although self-extinguishing forms of acetate are available. Special formulations of butyrate and propionate are serviceable outdoors for long periods. Acetate is generally considered unsuitable for outdoor uses. From an application standpoint, the acetates generally are used where tight dimensional stability, under anticipated humidity and temperature, is not required. Hardness, stiffness, and cost are lower than for butyrate or propionate. Butyrate is generally selected over propionate where weatherability, low-temperature impact strength, and dimensional stability are required. Propionate is often chosen for hardness, tensile strength, and stiffness, combined with good weather resistance.

Ethyl cellulose, best known for its toughness and resiliency at subzero temperatures, also has excellent dimensional stability over a wide range of temperature and humidity conditions. Alkalies or weak acids do not affect this material, but cleaning fluids, oils, and solvents are very harmful.

Fluorocarbons

Tetrafluoroethylene (TFE) and fluorinated ethylene propylene (FEP) It can be considered for practical purposes that there are eight types of fluorocarbons, as summarized in Table 29, and discussed further below. Suppliers and trade names are given in Table 29, and data are presented in Table 24. Like other plastics, each type is available in several grades. The original, basic fluorocarbon, and perhaps still the most widely known one, is TFE fluorocarbon. It has the optimum of electrical and thermal properties and almost complete moisture resistance and chemical inertness, but it does have the disadvantage of cold flow or creep under mechanical loading. TFE hardness as a function of temperature is shown in Fig. 32.

TABLE 29   Structure, Trade Names, and Suppliers of Fluorocarbons

Fluorocarbon	Structure	Typical trade names and suppliers
TFE (tetrafluoroethylene)....	$\left[\begin{array}{c} F \quad F \\ \mid \quad \mid \\ -C-C- \\ \mid \quad \mid \\ F \quad F \end{array}\right]_n$	Teflon TFE (E. I. du Pont) Halon TFE (Allied Chemical)
FEP (fluorinated ethylenepropylene)..........	$\left[\begin{array}{c} F \quad F \quad F \quad F \\ \mid \quad \mid \quad \mid \quad \mid \\ C-C-C-C \\ \mid \quad \mid \quad \mid \quad \mid \\ F \quad F \quad F-C-F \\ \mid \\ F \end{array}\right]_n$	Teflon FEP (E. I. du Pont)
ETFE (ethylene-tetra-fluoroethylene copolymer)...	Copolymer of ethylene (Fig. 38) and TFE, above	Tefzel (E. I. du Pont)
PFA (perfluoroalkoxy).......	$\left[\begin{array}{c} F \quad F \quad F \quad F \quad F \\ \mid \quad \mid \quad \mid \quad \mid \quad \mid \\ C-C-C-C-C \\ \mid \quad \mid \quad \mid \quad \mid \quad \mid \\ F \quad F \quad O \quad F \quad F \\ \mid \\ R_f* \end{array}\right]_n$	Teflon PFA (E. I. du Pont)
CTFE (chlorotrifluoro-ethylene).................	$\left[\begin{array}{c} Cl \quad F \\ \mid \quad \mid \\ -C-C- \\ \mid \quad \mid \\ F \quad F \end{array}\right]_n$	Kel-F (3M)
E-CTFE (ethylene-chlorotri-fluoroethylene copolymer)...	Copolymer of ethylene (Fig. 38) and CTFE, above	Halar E-CTFE (Allied Chemical)
PVF$_2$ (vinylidene fluoride)....	$\left[\begin{array}{c} H \quad F \\ \mid \quad \mid \\ -C-C- \\ \mid \quad \mid \\ H \quad F \end{array}\right]_n$	Kynar (Pennsalt Chemicals)
PVF (polyvinyl fluoride)......	$\left[\begin{array}{c} H \quad H \\ \mid \quad \mid \\ -C-C- \\ \mid \quad \mid \\ H \quad F \end{array}\right]_n$	Tedlar (E. I. du Pont)

* $R_f = C_nF_{2n-1}$.

Detailed electrical and mechanical properties are discussed in Chaps. 2 and 3. Stronger, filled modifications exist, as do newer, more cold-flow-resistant grades. FEP fluorocarbon is quite similar to TFE in most properties, except that its useful temperature is limited to about 400°F. FEP is much more easily processed, and molded parts are possible with FEP which might not be possible with TFE. Thermal-expansion curves for both are shown in Fig. 33.

Ethylene-tetrafluoroethylene (ETFE) copolymer   ETFE, a high-temperature thermoplastic fluoropolymer, is readily processed by conventional methods, including extrusion and injection molding. As a copolymer of ethylene and tetrafluoroethylene, it is closely related to TFE. It has good thermal properties and abrasion resistance, and excellent impact strength, resistance to chemicals, and electrical-insulation characteristics.

ETFE can best be described as a rugged thermoplastic with an outstanding balance of properties. Mechanically, it is tough and has medium stiffness (200,000

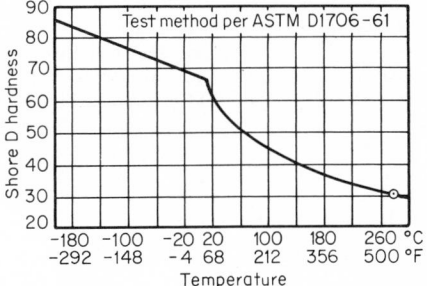

Fig. 32 Hardness of TFE fluorocarbon as a function of temperature.[31]

lb/in.²), excellent flex life, impact, cut-through, and abrasion resistance. The glass-fiber-reinforced compound has even higher tensile strength (12,000 lb/in.²), stiffness (950,000 lb/in.²), and creep resistance, but is still tough and impact-resistant.

Thermally, ETFE has a continuous temperature rating of 300°F (150°C). It can be used intermittently up to 200°C depending on exposure time, load, and environment. Glass-fiber-reinforced ETFE appears capable of useful service at 392°F (200°C).

ETFE is weather-resistant, inert to most solvents and chemicals, and hydrolytically stable. It has excellent resistance to high-energy radiation and is ultraviolet-resistant.

The fluoropolymer's dielectric constant is 2.6 and dissipation factor is 0.0006, making it an excellent low-loss dielectric. These values are not affected by changes in temperature and other environmental conditions. Dielectric strength is high; resistivity is excellent.

Molding characteristics of the resin are excellent. It exhibits good melt flow, allowing the filling of thin sections (10 mils for small parts). Cycle times are equivalent to those of other thermoplastics. It is the first fluoroplastic that can be reinforced—not merely filled—with glass fiber. Because the resin will bond to the fibers, strength, stiffness, creep resistance, heat-distortion temperature, and dimensional stability are enhanced. Electrical

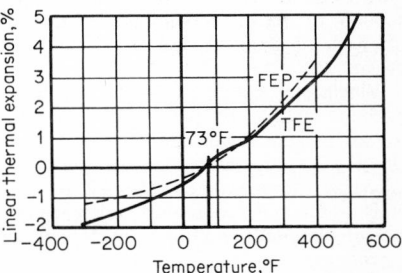

Fig. 33 Linear thermal expansion of TFE and FEP fluorocarbons as a function of temperature.[32]

and chemical properties approach those of the unreinforced resin, while the coefficient of friction is actually lower. Tests indicate the service-temperature limit is approximately 400°F. The fluoropolymer, reinforced with glass, can be molded by conventional methods with rapid molding cycles.

Perfluoroalkoxy (PFA) resins   PFA resins are a new class of melt-processible fluoroplastics. They combine the ease and economics of thermoplastic processing with high-temperature performance in the range of TFE fluorocarbon resins. PFA resins resemble FEP fluorocarbon resins in having a branched polymer chain that provides good mechanical properties at melt viscosities much lower than those of TFE. However, the unique branch in PFA is longer and more flexible, leading to improvements in high-temperature properties, higher melting point, and greater thermal stability.

In use, PFA resins have the desirable properties typical of fluorocarbons, including resistance to virtually all chemicals, antistick, low coefficient of friction, excellent electrical characteristics, low smoke, excellent flammability resistance, ability to perform in temperature extremes, and excellent weatherability. Their strength and stiffness at high operating temperatures are at least equivalent to those of TFE, while their creep resistance appears to be better over a wide temperature range. They should perform successfully in the 500°F area.

Film and sheet are expected to find use as electrical insulations in flat cables and circuitry, and in laminates used in electrical and mechanical applications.

Chlorotrifluoroethylene (CTFE)   The CTFE resins, like the FEP materials, are melt-processible and can be injection-, transfer-, and compression-molded or screw-extruded. Compared with TFE, CTFE has greater tensile and compressive strength within its service-temperature range. However, at the temperature extremes, CTFE does not perform as well as TFE for parts such as seals. At the low end of the range, TFE has somewhat better physical properties; at the higher temperatures, CTFE is more prone to stress cracking and other difficulties.

Electrical properties of CTFE are generally excellent, but dielectric losses are higher than those of TFE. Chemical resistance is poorer than that of TFE, but radiation resistance is better. CTFE does not have the low-friction and bearing properties of TFE.

Ethylene-chlorotrifluoroethylene (E-CTFE) copolymer   E-CTFE copolymer, a copolymer of ethylene and CTFE, is a strong, highly impact-resistant material that retains useful properties over a broad temperature range. E-CTFE is available in pellet and powder form and can be readily extruded, injection-, transfer-, and compression-molded, rotocast, and powder-coated. Its tensile strength, impact resistance, hardness, creep properties, and abrasion resistance at ambient temperature are comparable with those of nylon 6. E-CTFE retains its strength and impact resistance down to cryogenic temperatures.

E-CTFE also exhibits excellent ac loss properties. Its dielectric constant measures 2.5 to 2.6 over the frequency range of 10^2 to 10^6 Hz and is unaffected by temperature over the range of -80 to 300°F. Its dissipation factor is low, ranging from 0.0008 to 0.015 depending on temperature and frequency. It has a dielectric strength of 2,000 V/mil in 10-mil wall thickness and a volume resistivity of 10^{15} Ω-cm.

Vinylidene fluoride (PVF$_2$)   Vinylidene fluoride is another melt-processible fluorocarbon capable of being injection- and compression-molded and screw-extruded. Its 20 percent lower specific gravity compared with that of TFE and CTFE and its good processing characteristics permit economy and provide excellent chemical and physical characteristics. Useful temperature range is -80 to $+300$°F. Although it is stiffer and has higher resistance to cold flow than TFE, its chemical resistance, useful temperature range, antistick properties, lubricity, and electrical properties are lower.

Polyvinyl fluoride (PVF)   PVF is manufactured as a film, whose combination of excellent weathering and fabrication properties has made it widely accepted for surfacing industrial, architectural, and decorative building materials. It has outstanding weatherability, solvent resistance, chemical resistance, abrasion resistance, and color retention. It is supplied with a variety of surfaces for different bonding and antistick objectives.

Ionomers

Outstanding advantages of this polymer class are combinations of toughness, transparency, low-temperature impact, and solvent resistance.[33] Ionomers have high melt strength for thermoforming and extrusion-coating processes, and a broad processing-temperature range. There are resin grades for extrusion, films, injection molding, blow molding, and other thermoplastic processes.

Limitations of ionomers include low stiffness, susceptibility to creep, low heat-distortion temperature, and poor ultraviolet resistance unless stabilizers are added, where these properties are important.

Most ionomers are very transparent. In 60-mil sections, internal haze ranges from 5 to 25 percent. Light transmission ranges from 80 to 92 percent over the visible region, and in specific compositions high transmittance extends into the ultraviolet region.

Basically, commercial ionomers are nonrigid, unplasticized plastics. Outstanding low-temperature flexibility, resilience, high elongation, and excellent impact strength typify the ionomer resins.

Deterioration of mechanical and optical properties can occur when ionomers are exposed to ultraviolet light and weather. Some grades are available with ultraviolet stabilizers that provide up to 1 year of outdoor exposure with no loss in mechanical properties. Formulations containing carbon black provide ultraviolet resistance equal to that of black polyethylene.

Most ionomers have good dielectric characteristics over a broad frequency range. The combination of these electrical properties, high melt strength, and abrasion resistance qualifies these materials for insulation and jacketing of wire and cable. Their excellent mechanical and optical properties make them specially useful for thermoformed packaging and abrasion- and impact-resistant shoe parts. Typical materials are shown in Table 23, properties are given in Table 24, and electrical and mechanical properties are detailed further in Chaps. 2 and 3.

Low-Permeability Thermoplastics

Several grades of nitrile polymers have been found to offer outstanding low-permeability barrier properties against gases, odor, and flavor, some approaching the barrier properties of metal and glass. This strong characteristic, coupled with good transparency, good thermal and mechanical properties, and convenient processibility, makes this class of thermoplastic very useful as containers for food and beverage products. Typical materials are shown in Table 23, and properties are shown in Table 24.

Nylons

Also known as polyamides, nylons are strong, tough thermoplastics having good impact, tensile, and flexural strengths from freezing temperatures up to 300°F; excellent low-friction properties; and good electrical resistivities. The structures of four common nylons are shown in Fig. 34. Since all nylons absorb some moisture from environmental humidity, moisture-absorption characteristics must be considered in designing with these materials. They will absorb anywhere from 0.5 to nearly 2 percent moisture after 24 h water immersion. There are low-moisture-absorption grades, however; and hence moisture-absorption properties do not have to limit the use of nylons, especially for the lower-moisture-absorption grades. Typical materials are shown in Table 23, and properties are shown in Table 24. Mechanical and electrical properties are presented in more detail in Chaps. 2 and 3.

Regarding the identifications for the grades of nylon, certain nylons are identified by the number of carbon atoms in the diamine and dibasic acid used to produce that particular nylon. For instance, nylon 6/6 is the reaction product of hexamethylene and adipic acid, both of which are materials containing six carbon atoms in their chemical structure. Some common commercially available nylons are 6/6, 6, 6/10, 8, 11, and 12. Grades 6 and 6/6 are the strongest structurally; grades 6/10 and 11 have the lowest moisture absorption, best electrical properties, and best dimensional stability; and grades 6, 6/6, and 6/10 are the most flexible. Grades 6, 6/6, and 8 are heat-sealable with nylon 8 being capable of cross linking. Another grade, nylon 12, offers advantages similar to those of grades 6/10 and 11, but offers lower cost possibilities owing to being more easily and economically processed. Also a high-temperature type of nylon exists. It is discussed separately below.

In situ polymerization of nylon permits massive castings. Cast nylons are readily polymerized directly from the monomer material in the mold at atmospheric pressure. The method finds application where the size of the part required and/or the need for low tooling cost precludes injection molding. Cast nylon displays excellent bearing and fatigue properties as well as the other properties characteristic of other basic nylon formulations, with the addition of size and short-run flexibility advantages of the low-pressure casting process.

One special process exists in which nylon parts are made by compressing and sintering, thereby creating parts having exceptional wear characteristics and dimensional stability. Various fillers such as molybdenum disulfide and graphite can be incorporated into nylon to give special low-friction properties. Also, nylon can be reinforced with glass fibers, thus giving it considerable additional strength. These variations are further discussed in a later part of this chapter dealing with glass-fiber-reinforced thermoplastic materials.

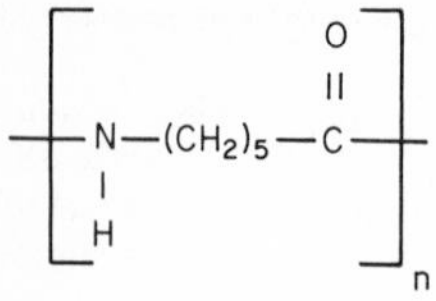

Effect of temperature of nylon One of the major property advantages of nylon resins is that they retain useful mechanical properties over a wide range of temperatures. Properly designed parts can be used successfully from -60 to $+400°F$.

Both long-term and short-term effects of temperature must, however, be considered. In the short term, there are effects on such properties as stiffness and toughness. There is also the possibility of stress relief and its effect on dimensions. Of most concern in long-term applications at high temperature is gradual oxidative embrittlement, and for such cases the use of heat-stabilized resins is recommended.

The important consideration for design work with nylon resins is that exposure to high temperatures in air for a period of time will result in a permanent change in properties due to oxidation. The permanent change in properties depends on the temperature level, the time exposed, and the composition of nylon used. The effect on the tensile strength, as measured at room temperature and 2.5 percent moisture content, is that a 25 percent strength reduction can occur in 3 months at $185°F$, and a 50 percent reduction can occur in 3 months at $250°F$. Stabilized nylon does not change appreciably at $250°F$ aging.

High-temperature oxidation reduces the impact strength even more than the static-strength properties. For instance, the impact strength and elongation of nylon are reduced considerably after several days at $250°F$.

Nylon 6

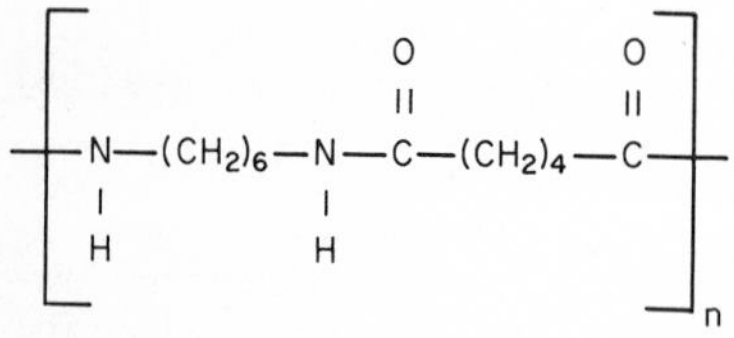

Nylon 6/6

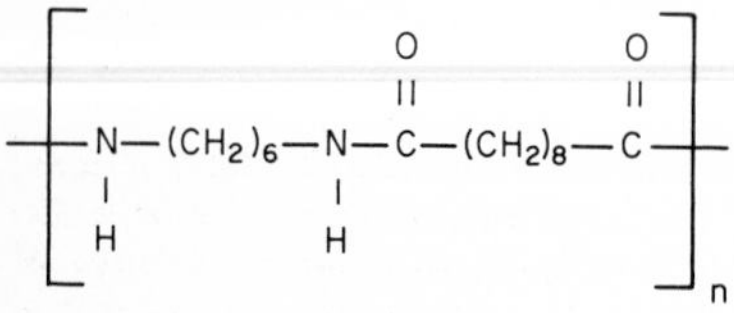

Nylon 6/10

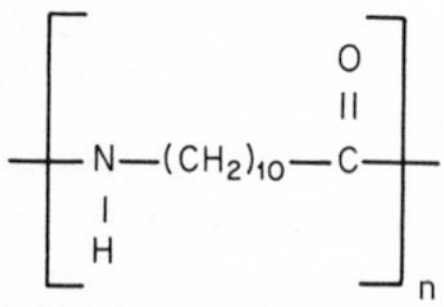

Nylon 11

Fig. 34 Chemical structures for four nylons.

A similar change in properties is encountered on exposure to high-temperature water for long periods of time. In this case a reaction with water takes place. There is no significant reaction up to $120°F$. This has been confirmed by molecular-weight measurements on 15-year-old samples.[33]

In boiling water, the tensile strength is slowly reduced until after 2,500 h it levels off at 6,000 lb/in.[2] (tested at room temperature and 2.5 percent moisture content).

The elongation drops rapidly after 1,500 h; hence, this time has been taken as the limit for the use of basic nylon. Some compositions are especially resistant to hot-water exposure, however.

Effect of moisture on dimensional control Freshly molded objects normally contain less than 0.3 percent of water, since only dry molding powder can be successfully molded. These objects will then absorb moisture when they are exposed to air or water. The amount of absorbed water will increase in any environment until an equilibrium condition based on relative humidity (RH) is reached. Equilibrium moisture contents for two humidity levels are approximately as follows:[34]

	Zytel 101	Zytel 31
50% RH air.....................	2.5%	1.4%
100% RH air (or water)..........	8.5%	3.5%

These equilibrium moisture contents are not affected by temperature to any significant extent. Thus, final water content at equilibrium will be the same whether objects of nylon are exposed to water at room temperature or at boiling temperature.

The time required to reach equilibrium, however, is dependent on the temperature, the thickness of the specimen, and the amount of moisture present in the surroundings. Nylon exposed in boiling water will reach the equilibrium level, 8.5 percent, much sooner than nylon in cool water.

When nylon that contains some moisture is exposed to a dry atmosphere, the loss of water will be the reverse of the changes described above and will take about the same length of time for a corresponding change to occur.

In the most common exposure, an environment of constantly varying humidity, no true equilibrium moisture content can be established. However, moldings of nylon will gradually gain in moisture content in such an environment until a balance is obtained with the midrange humidities. A slow cycling of moisture content near this value will then occur. In all but very thin moldings, the day-to-day or week-to-week variations in relative humidity will have little effect on total moisture content. The long period changes, such as between summer and winter, will have some effect depending on thickness and the relative-humidity range. The highest average humidity for a month will not generally be above 70 percent. In cold weather, heated air may average as low as 20 percent relative humidity. Even at these extremes, the change in moisture content of nylon is small in most cases, because of the very low rate of both absorption and desorption.

There are two significant dimensional effects that occur after molding. In some cases these oppose each other so that critical dimensions may change very little in a typical air environment. The first is a shrinkage in the direction of flow due to the relief of molded-in stresses. The second is an increase due to moisture absorption. In applications where dimensions are critical to performance and one of these effects predominates it may be necessary to anneal or moisture-condition the parts to obtain the best performance. For example, a part that will be exposed to high temperatures might require annealing, and an object in water service might be moisture-conditioned to effect dimensional changes before use.

The magnitude of the dimensional change due to molded-in stresses depends on molding conditions and part geometry. This change in size can be determined in a given case only by measuring and annealing a few pieces. In general, long dimensions (in the direction of flow) will shrink, but the short dimensions will increase in an amount that is often too small to measure.

The effect of moisture content on the dimensions of a molding can be predicted more accurately than the effect of annealing. The changes in dimension at various water contents for Zytel 101 and Zytel 31 are shown in Fig. 35. These data are for the annealed or stress-free condition. For most applications in air, the dimensions corresponding to those obtained in equilibrium with 50 percent relative humidity air are usually chosen as the average size expected when a moisture balance is established. As shown, Zytel 101 will increase 0.006 in./in. from the dry condition to equilibrium with 50 percent relative humidity. Zytel 31 under similar conditions will increase 0.0025 in./in. In many applications these changes are small enough and occur so slowly that they do not affect the operation of finished parts.

Stress-free objects of Zytel 101 and Zytel 31 will increase approximately 0.026 and 0.007 in./in., respectively, during the change from dry to completely saturated with water. In water service, if dimensions are critical, the molding can be moisture-conditioned before assembly, to allow for this change. Once a part is at or near saturation, little dimensional change can occur unless it is allowed to dry for long periods.

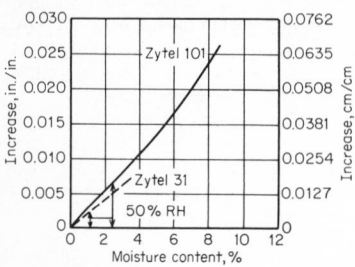

Fig. 35 Changes in dimensions with moisture content for two nylon materials in the stress-free, annealed conditions.[34]

Aramid (High-temperature, nylon) In addition to conventional nylon molding resins, there is a high-temperature nylon now called aramid. This high-temperature nylon, Nomex, retains about 60 percent of its strength at 475 to 500°F temperatures, which would melt conventional nylons. This high-temperature nylon has good dielectric strength (constant to 400°F and 95 percent relative humidity) and volume resistivity, and low dissipation factor, as reviewed in Chap. 2. Fabricated primarily in sheet, fiber, and paper form, this material is being used in wrapped electrical-insulation constructions such as transformer coils and motor stators. It retains high tensile strength, resistance to wear, and electrical properties after prolonged exposure at temperatures up to 500°F.

Parylenes

Parylene is the generic name for members of a thermoplastic-polymer series developed by Union Carbide, which are unique among plastics in that they are produced as thin films by vapor-phase polymerization.[35]

Parylene is produced by vapor-phase polymerization and deposition of para-xylylene (or its substituted derivatives). The polymers are highly crystalline, straight-chain compounds that are known as tough materials with excellent dielectric characteristics. Molecular weight is approximately 500,000.

Parylene is extremely resistant to chemical attack, exceptionally low in trace-metal contamination, and compatible with all organic solvents used in the cleaning and processing of electronic circuits and systems. Although parylene is insoluble in most materials, it will soften in solvents having boiling points in excess of 150°C.

The basic member of this thermoplastic polymer family—parylene N—is poly-para-xylylene:

$$\left(\!-CH_2-\!\!\bigcirc\!\!-CH_2-\!\right)$$

Parylene N exhibits superior dielectric strength, exceptionally high surface and volume resistivities, and electrical properties that vary remarkably little with changes in temperature. Since its melting point far exceeds that of many other thermoplastics, parylene N can even be used at temperatures exceeding 220°C in the absence of oxygen. Because parylene can be deposited in very thin coatings, heat generated by coated components is easier to control and differences in thermal expansions are less of a problem than those of conventional coatings.

The other commercially available member of the group—parylene C—is poly-mono-chloro-para-xylylene:

$$\left(\!-CH_2-\!\!\bigcirc\!\!-CH_2-\!\right)$$
Cl

Parylene C offers significantly lower permeability to moisture and gases, such as nitrogen, oxygen, carbon dioxide, hydrogen sulfide, sulfur dioxide, and chlorine, while retaining excellent electrical properties.

The parylene process   Unlike most plastics, parylene is not produced and sold as a polymer.   It is not practical to melt, extrude, mold, or calender it as with other thermoplastics.   Nor can it be applied from solvent systems, since it is insoluble in conventional solvents.

The advantages that characterize this rugged, intractable material are made possible by a fast, relatively simple vacuum-application system.   While it is a unique system, it is often less cumbersome, less complex, and easier to use and maintain than many techniques devised for more conventional plastic-coating systems.

The parylene processor starts with a dimer rather than a polymer and, in simple, compact equipment, polymerizes it on the surface of an object.   To achieve this, the dimer must first go through a two-step heating process.   The solid dimer is converted to a reactive vapor of the monomer.   When passed over room-temperature objects, the vapor will rapidly coat them with polymer.   The parylene process is shown in Fig. 36.

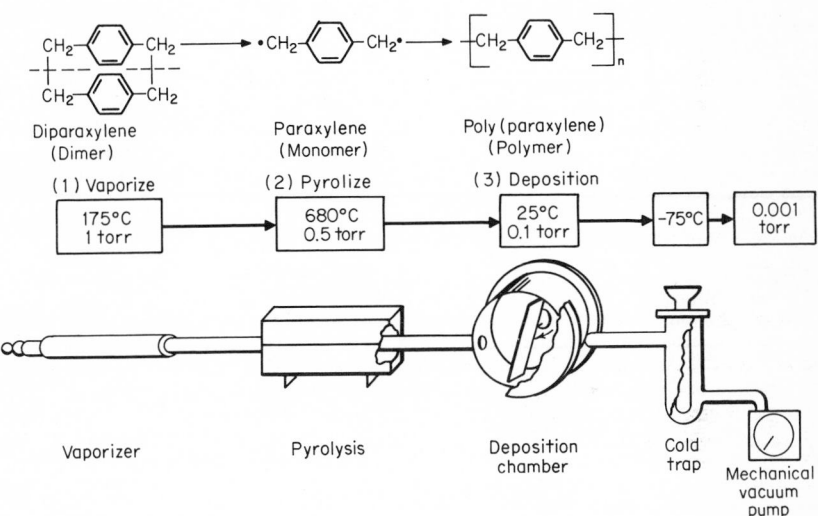

Fig. 36   Diagram of the parylene process.[35]

This process of polymerization on the object surface has much to recommend it, particularly when the goal is coating uniformity.   Unlike dip or spray coating, condensation coating does not run off or sag.   It is not "line-of-sight" as in vacuum metallizing; the vapor coats evenly—over edges, points, and internal areas.   The vapor is pervasive, but it coats without bridging so that holes can be jacketed evenly.   Plater's masking tape can be used if it is desired that some areas not be coated.   Moreover, with parylene, the object to be coated remains at or near room temperature, eliminating all risk of thermal damage.   Coating thickness is controlled easily and very accurately simply by regulating the amount of dimer to be vaporized.

Deposition chambers of virtually any size can be constructed.   Those currently in use range from 500 to 28,000 in.[3].   Large parts (up to 5 ft long and 18 in. high) can be processed in this equipment.   The versatility of the process also enables the simultaneous coating of many small parts of varying configurations.   Because of this, a time savings in labor requirements can readily be achieved.

Parylenes can also be deposited onto a cold condenser and then stripped off as a free film, or they can be deposited onto the surface of objects as a continuous, adherent coating in thicknesses ranging from 1,000 Å (about 0.004 mil) to 3 mils or more. Deposition rate is normally about 0.5 μm/min (about 0.02 mil). On cooled substrates, the deposition rate can be as high as 1.0 mil/min.

Parylene properties The material can be used at both elevated and cryogenic temperatures. The 1,000-h service life for the N and the C members, respectively, is 200 to 240°F. Corresponding 10-year service in air is limited to 140 or 175°F. Parylenes are excellent in having low gas permeability, low moisture-vapor transmission, and low-temperature ductility. Dimensional stability is reported as better than that of polycarbonate, and overall barrier properties to most gases are reported superior to those of many other barrier films. Trade names and suppliers are listed in Table 23, and important properties are listed in Tables 30 and 31.

TABLE 30   Thermal, Physical, and Mechanical Properties of Parylenes[35]

	Parylene N	Parylene C
Typical Thermal Properties		
Temp (melting), °C	405	280
Linear coefficient of expansion, $10^{-5}/°C$	6.9	3.5
Thermal conductivity, 10^{-4} cal/s/(cm²)(°C/cm)	~3	
Typical Physical and Mechanical Properties		
Tensile strength, lb/in.²	6,500	10,000
Yield strength, lb/in.²	6,100	8,000
Elongation to break, %	30	200
Yield elongation, %	2.5	2.9
Density, g/cm³	1.11	1.289
Coefficient of friction:		
Static	0.29	0.25
Dynamic	0.29	0.25
Water absorption, 24 h	0.06 (0.029 in.)	0.01 (0.019 in.)
Index of refraction n_D 23°C	1.661	1.639

Data recorded following appropriate ASTM method.

TABLE 31   Film-Barrier Properties of Parylenes[35]

Polymer	Gas permeability, cm³-mil/100 in.², 24 h-atm (23°C)						Moisture-vapor transmission, g-mil/100 in.², 24 h, 37°C, 90% RH
	N_2	O_2	CO_2	H_2S	SO_2	Cl_2	
Parylene N	7.7	39.2	214	795	1,890	74	1.6
Parylene C	1.0	7.2	7.7	13	11	0.35	0.5
Epoxies	4	5–10	8	...	...	...	1.8–2.4
Silicones	...	50,000	300,000	...	...	...	4.4–7.9
Urethanes	80	200	3,000	...	...	...	2.4–8.7

Data recorded following appropriate ASTM method.

Polyaryl Ether

Another of the relatively new thermoplastics which can be classed as engineering thermoplastics is polyaryl ether, available from Uniroyal, Inc., as Arylon T. A unique combination of three important, desirable properties makes polyaryl ether outstanding. The material has (1) a high heat-deflection temperature of 300°F measured at 264 lb/in.² on a ¼ by ¼ by 5-in. bar; (2) the highest impact strength of all engineering plastics except polycarbonate in thin sections—Izod impact strength of 8 ft-lb/in. notch at 72°F and 2.5 ft-lb/in. notch at −20°F; (3) excellent chemical resistance, resisting organic solvents except chlorinated aromatics, esters, and ketones; and it appears to resist hydrolysis, withstanding prolonged immersion in boiling water. Fabricating advantages include a low shrinkage factor, precision molding of void-free parts, and insensitivity to moisture, requiring only nominal drying before molding. Markets for Arylon include the automotive, appliance, and electrical industries. Properties are shown in Table 24.

Polyaryl Sulfone

Polyaryl sulfone offers the unique combination of thermoplasticity and retention of structurally useful properties at 500°F.[36] This material is supplied in both pellet and powder form. Its thermoplastic nature allows it to be fabricated by injection molding or extrusion techniques on standard commercially available equipment.

Polyaryl sulfone consists mainly of phenyl and biphenyl groups linked by thermally stable ether and sulfone groups. It is distinguished from polysulfone polymers by the absence of aliphatic groups, which are liable to oxidative attack. This aromatic structure gives it excellent resistance to oxidative degradation and accounts for its retention of mechanical properties at high temperatures. The presence of ether-oxygen linkages gives the polymer chain flexibility to permit fabrication by conventional melt-processing techniques. Typical materials are shown in Table 23, and typical properties are given in Table 32. Chapters 2 and 3 cover electrical and mechanical properties.

Thermal properties Polyaryl sulfone is characterized by a very high heat-deflection temperature, 525°F at 264 lb/in.^{2}, which is approximately 150°F higher than other commercially available thermoplastics, as shown in Fig. 37. This is a consequence of its high glass-transition temperature, 550°F, rather than the effect of filler reinforcement or a crystalline melting point. At 500°F it maintains a tensile strength in excess of 4,000 lb/in.^{2} and a flexural modulus of 250,000 lb/in.^{2}.

The resistance to oxidative degradation is indicated by the ability of polyaryl sulfone to retain its tensile strength after 2,000 h exposure to 500°F air-oven aging.

Chemical resistance Polyaryl sulfone has good resistance to a wide variety of chemicals, including acids, bases, and common solvents. It is unaffected by practically all fuels, lubricants, hydraulic fluids, and cleaning agents used on or around electrical components. Highly polar solvents such as N, N-dimethylformamide, N, N-dimethylacetamide, and N-methylpyrrolidone are solvents for the material.

Polycarbonates

This group of plastics is also among those which are classified as engineering thermoplastics because of their high performance characteristics in engineering designs. The general chemical structure is shown below:

Polycarbonate

Polycarbonates are especially outstanding in impact strength, having strengths several times higher than other engineering thermoplastics. Polycarbonates are tough, rigid, and dimensionally stable and are available as transparent or colored parts. They are easily fabricated with reproducible results, using molding or machining techniques. An important molding characteristic is the low and predictable mold shrinkage, which sometimes gives polycarbonates an advantage over nylons and acetals for close-tolerance parts. As with most other plastics containing aromatic groups, radiation stability is high. The most commonly useful properties of polycarbonates are creep resistance, high heat resistance, dimensional stability, good electrical properties, self-extinguishing properties, product transparency, and the exceptional impact strength which compares favorably with that of some metals and exceeds that for essentially all competitive plastics. In fact, polycarbonate is sometimes considered to be competitive to zinc and aluminum castings. Although such comparisons have limits, the fact that the comparisons are sometimes made in material selection for product design indicates the strong performance characteristics possible in polycarbonates.

In addition to their performance as engineering materials, polycarbonates are also alloyed with other plastics, to increase the strength and rigidity of these other plastics. Notable among the plastics with which polycarbonates have been alloyed are the ABS plastics. Information and data on such alloys are presented in Table 26.

TABLE 32   Typical Properties of Polyaryl Sulfone[37]

Property	Value	ASTM method
Physical:		
Specific gravity at 73°F..................	1.36	
Water absorption at 73°F, 24 h............	1.8%	D 570
Mechanical:		
Tensile strength at 73°F.................	13,000 lb/in.²	D 638
At 500°F............................	4,100 lb/in.²	D 638
Compression strength at 73°F.............	17,900 lb/in.²	D 695
At 500°F............................	7,500 lb/in.²	D 695
Flexural strength at 73°F.................	17,200 lb/in.²	D 790
At 500°F............................	8,900 lb/in.²	D 790
Tensile modulus at 73°F.................	370,000 lb/in.²	D 638
Compression modulus at 73°F.............	340,000 lb/in.²	D 695
Flexural modulus at 73°F.................	395,000 lb/in.²	D 790
At 500°F............................	252,000 lb/in.²	D 790
Tensile elongation at yield:		
At 73°F.............................	13%	D 638
At 500°F............................	7%	D 638
Notched Izod impact at 73°F.............	3 ft-lb/in.	D 256
Rockwell hardness......................	M110	D 785
Taber abrasion, CS17, 1,000-g load........	40 mg	D 1044
Thermal:		
Heat-deflection temp at 264 lb/in.²........	525°F	D 648
Glass-transition temp....................	550°F	
Coefficient of linear expansion.............	2.6×10^{-5} in./(in.)(°F)	D 696
Thermal conductivity....................	1.32 Btu/(h)(ft²)(°F/in.)	
Heat penetration, 1 mm..................	550°F	D 1525
Mold shrinkage.........................	0.008 in./in.	D 1299
Flammability..........................	Self-extinguishing	D 635
Electrical at 73°F:		
Dielectric constant, 60 Hz...............	3.94	D 150
Dielectric constant, 8.5×10^9 Hz.........	3.24	D 150
Dissipation factor, 60 Hz.................	0.0030	D 150
Dissipation factor, 8.5×10^9 Hz..........	0.010	D 150
Volume resistivity......................	3.2×10^{16} Ω/cm	D 257
Surface resistivity......................	6.2×10^{15} Ω/sq	
Dielectric strength		
(short time, 1/16 in. thick)..............	350 V/mil	D 149
Dielectric strength		
(short time, 0.011 in. thick)		
(2 in. electrode in oil)..................	1,600 V/mil	D 149
Arc resistance.........................	67 s	D 495

In addition to standard grades of polycarbonates, a special film grade exists for high-performance capacitors.[38] Typical materials are shown in Table 23, and typical properties are given in Table 24. Chapters 2 and 3 cover electrical and mechanical properties.

Moisture-resistance properties   Oxidation stability on heating in air is good, and immersion in water and exposure to high humidity at temperatures up to 212°F have little effect on dimensions. Steam sterilization is another advantage that is attributable to the resin's high heat stability. However, if the application requires continuous exposure in water, the temperature should be limited to 140°F. Polycarbonates are among the most stable plastics in a wet environment, as shown in Figs. 38 and 39.

Polyesters

Thermoplastic polyesters have been and are currently used extensively in the production of film and fibers. These materials are denoted chemically as poly-ethylene terephthalate. In the past few years a new class of high-performance molding and extrusion grades of thermoplastic polyesters has become available and is becoming increasingly competitive among plastics. These polymers are denoted chemically as poly (1,4-butylene terephthalate) and poly (tetramethylene terephthalate), which are chemically the same. These thermoplastic polyesters, identified in Table 23, are highly crystalline, with a melting point about 430°F. They are fairly translucent in thin molded sections and opaque in thick sections but can be extruded into transparent thin film. Both unreinforced and reinforced formulations are extremely easy to process and can be molded in very fast cycles. Properties are shown in Table 33.

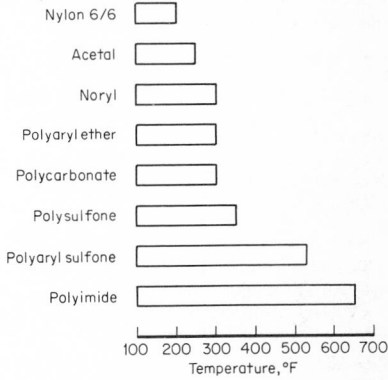

Fig. 37 Approximate heat-deflection temperature of some engineering thermoplastics at 264 lb/in.[2].

The unreinforced resin offers the characteristics: (1) hard, strong, and extremely tough; (2) high abrasion resistance, low coefficient of friction; (3) good chemical resistance, very low moisture absorption and resistance to cold flow; (4) good stress crack and fatigue resistance; (5) good electrical properties; and (6) good surface appearance. Electrical properties are stable up to the rated temperature limits. The electrical and mechanical properties are presented in more detail in Chaps. 2 and 3.

The glass-reinforced polyester resins are unusual in that they are the first thermoplastics that can compare with, or are better than, thermosets in electrical, mechanical, dimensional, and creep properties at elevated temperatures (approximately 300°F), while having superior impact properties.

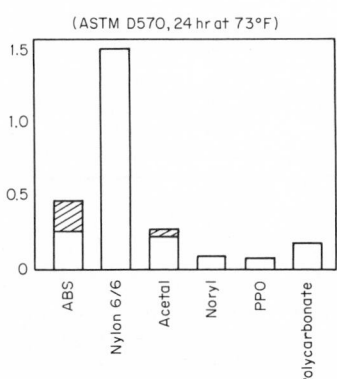

Fig. 38 Water absorption of several thermoplastics.[39,40]

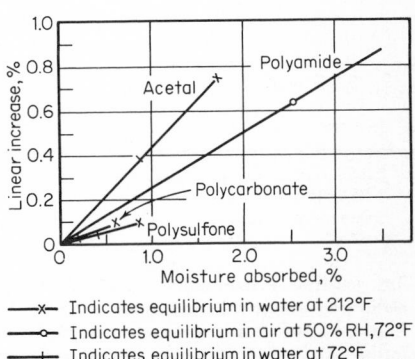

Fig. 39 Dimensional changes of several thermoplastics due to absorbed moisture.[39,40]

The glass-fiber concentration usually ranges from 10 to 30 percent in commercially available grades. In molded parts the glass fibers remain slightly below the surface so that finished items have a very smooth surface finish as well as an excellent appearance. Properties of reinforced and unreinforced polyesters are shown in Table 33.

Unreinforced resins are primarily used in housings requiring excellent impact, and in moving parts such as gears, bearings, and pulleys, packaging applications,

TABLE 33   Typical Properties of Reinforced and Unreinforced Thermoplastic Polyester[41]

Property	ASTM test	Unrein- forced	Unrein- forced SE-O*	30% glass reinforced	Glass reinforced SE-O
Physical:					
Specific gravity...............	D 570	1.31	1.42	1.52	1.58
Water absorption, 24 h, 73°F...	......	0.08	0.09	0.07	0.07
Equilibrium...................	......	0.34	0.34	0.40	0.40
Mechanical:					
Tensile strength, lb/in.^{2}........	D 638	8,000	9,000	17,000	17,000
Flexural strength, lb/in.^{2}.......	D 790	13,000	15,000	27,000	30,000
Flexural modulus, lb/in.$^2 \times 10^3$..	D 790	340	375	1,200	1,200
Izod, notched, ft-lb/in. notch...	D 256	1.2	0.9	1.4	1.4
Thermal:					
Deflection temp 264 lb/in.^{2} °F ..	D 648	130	160	416	416
Melting point.................	......	437	437	437	437
Flammability:					
Oxygen index...............	......	22.0	29.0	23.0	29.0
UL Subject 94...............	......	SB†	SE-O	SB	SE-O
Electrical:					
Volume resistivity, Ω-cm 72°F...	D 257	4×10^{16}	NA	8×10^{15}	5×10^{15}
6 weeks in 72°F water........	......	NA‡	NA	1×10^{16}	3×10^{15}
Dielectric strength, ST, V/mil 72°F.............	D 149	590	750	750	750
Dielectric constant (72°F):					
100 Hz....................	D 150	3.3	NA	3.77	3.91
10^6 Hz....................	......	3.1	NA	3.63	3.74
100 Hz (6 weeks in 72°F water)................	......	NA	NA	3.95	4.07
10^6 Hz....................	......	NA	NA	3.69	3.60
Dissipation factor:					
100 Hz, 72°F...............	D 150	0.002	NA	0.002	0.006
10 Hz.....................	......	0.02	NA	0.02	0.014
100 Hz (6 weeks in 72°F water)................	......	NA	NA	0.005	0.018
10^6 Hz....................	......	NA	NA	0.022	0.022
Arc resistance, s..............	D 495	190	NA	130	146

* SE-O, self extinguishing-O grade
† SB, slow burning
‡ NA, not available

and writing instruments. The flame-retardant grades are primarily aimed at TV/radio and electrical/electronics parts, and business machine and pump components.

Reinforced resins are being used in automotive, electrical/electronic, and general industrial areas, replacing thermosets, other thermoplastics, and metals. Electrical and mechanical properties coupled with low finished-part cost are enabling reinforced thermoplastic polyesters to replace phenolics, alkyds and DAP, and glass-reinforced thermoplastics in many applications.

Polyethersulfone

Polyethersulfone is a new high-temperature engineering thermoplastic with an outstanding long-term resistance to creep at temperatures up to 150°C.[42] It is also capable of being used continuously under load at temperatures of up to about 180°C and, in some low-stress applications, up to 200°C. Other grades are capable of operating at temperatures above 200°C and for specialized adhesive and lacquer applications.

Polyethersulfone consists essentially of a polymer of repeat unit:

$$\left[\bigcirc\!\!-\!SO_2\!-\!\bigcirc\!\!-\!O \right]$$

This structure gives an amorphous polymer which possesses only bonds of particularly high thermal and oxidative stability. While the sulfone group confers high-temperature performance, the ether linkage contributes toward practical processing by allowing mobility of the polymer chain when in the melt phase.

Polyethersulfone exhibits low creep. A constant stress of 3,000 lb/in.^{2} at 20°C for 3 years produces a strain of 1 percent, while a stress of 6,500 lb/in.^{2} results in a strain of only 2.6 percent over the same period of time.

Higher modulus values are obtained with polyethersulfone at 150°C than with polysulfone, phenylene oxide–based resins, or polycarbonate at considerably lower temperatures.

Although its load-bearing properties are reduced above 150°C, polyethersulfone can still be considered for applications at temperatures up to 180°C. It remains form-stable to above 200°C and has a heat-deflection temperature of 203°C at 264 lb/in.^{2}.

Polyethersulfone is especially resistant to acids, alkalies, oils, greases, and aliphatic hydrocarbons and alcohols. It is attacked by ketones, esters, and some halogenated and aromatic hydrocarbons.

Available materials are identified in Table 23, and properties are given in Table 24. Electrical and mechanical data are covered in Chaps. 2 and 3.

Polyethylenes, Polypropylenes, and Polyallomers

This large group of polymers is basically divided into the three separate polymer groups listed under this heading; all belong to the broad chemical classification known as polyolefins. Polyethylene and polypropylene can be considered as the first two members of a large group of polymers based on the ethylene structure, as shown in Fig. 40. Molecular changes beyond these two structures give quite different polymers and properties, of course, and are covered separately in other parts of this chapter. There are, of course, many categories or types even within each of the three polymer groups discussed in this section. Although property variations exist among these three polymer groups and among the subcategories within these polymer groups, there are also many similarities. The differences or unique features of each will be discussed below. The similarities are, broadly speaking, appearance, general chemical characteristics, and electrical properties. The differences are more notably in physical and thermal-stability properties. Basically, polyolefins are all waxlike in appearance and extremely inert chemically, and they exhibit decreases in physical strength at somewhat lower temperatures than the higher-performance engineering thermoplastics. Polyethylenes were the first of these materials developed and hence, for some of the original types, have the weakest mechanical properties. The later-developed polyethylenes, polypropylenes, and polyallomers offer improvements. The unique features of each of these three polymer groups will be mentioned at this point. Typical materials are shown in Table 23, and typical properties are given in Table 24. Chapters 2 and 3 cover electrical and mechanical properties.

Polyethylenes Polyethylenes are among the best-known plastics and come in three main classifications based on density: low, medium, and high. These density ranges are 0.910 to 0.925, 0.925 to 0.940, and 0.940 to 0.965, respectively. These three density grades are also sometimes known as types I, II, and III, respectively.

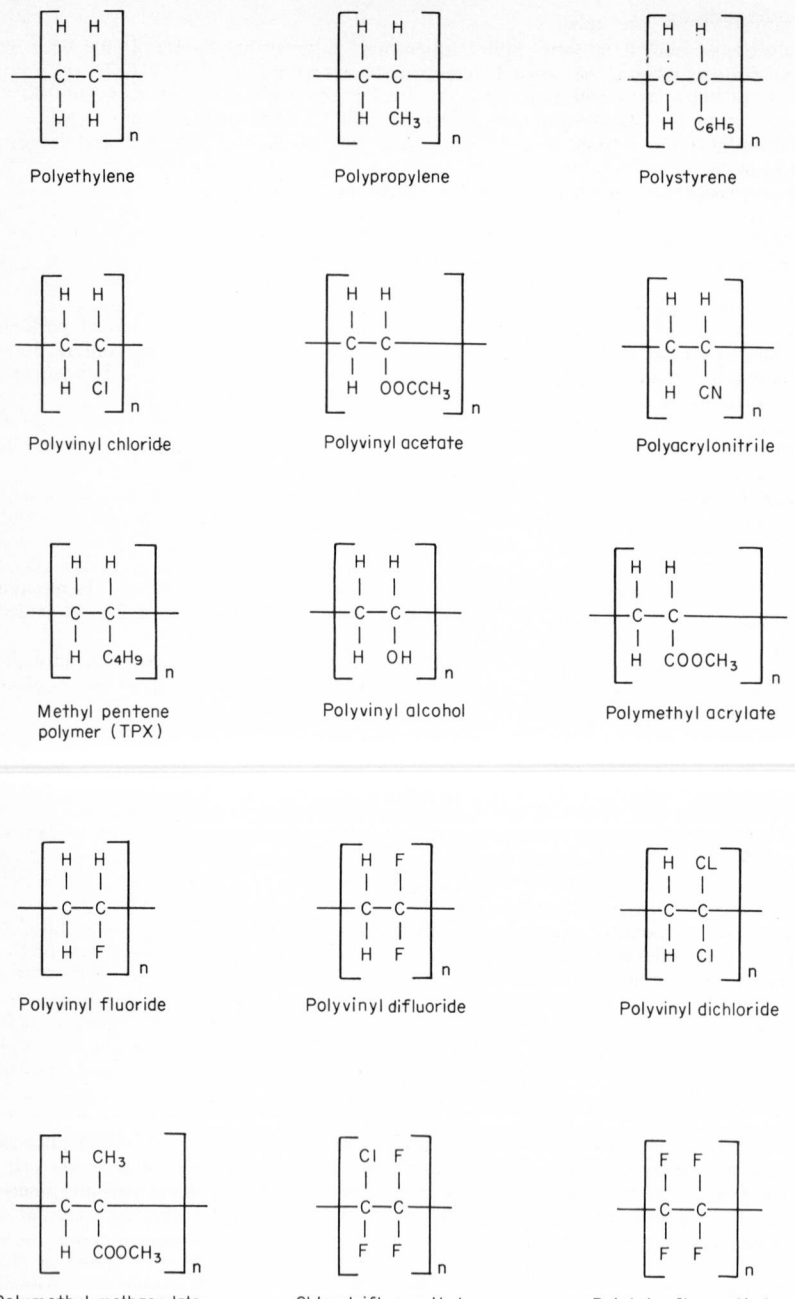

Fig. 40   Polymers based on the ethylene structure.[1]

All polyethylenes are relatively soft, and hardness increases as density increases. Generally, the higher the density, the better are the dimensional stability and physical properties, particularly as a function of temperature. Thermal stability of polyethylenes ranges from 190°F for the low-density material up to 250°F for the high-density material. Toughness is maintained to low negative temperatures.

Polypropylenes Polypropylenes are chemically similar to polyethylenes but have somewhat better physical strength at a lower density. The density of polypropylenes is among the lowest of all plastic materials, ranging from 0.900 to 0.915.

Polypropylenes offer more a balance of properties than a single unique property, with the exception of flex-fatigue resistance. These materials have an almost infinite life under flexing, and they are thus often said to be "self-hinging." Use of this characteristic is widespread as plastic hinges. Polypropylenes are perhaps the only thermoplastic surpassing all others in combined electrical properties, heat resistance, rigidity, toughness, chemical resistance, dimensional stability, surface gloss, and melt flow, at a cost less than that of most others.

Because of their exceptional quality and versatility, polypropylenes offer outstanding potential in the manufacture of products through injection molding. Mold shrinkage is significantly less than that of other polyolefins; uniformity in and across the direction of flow is appreciably greater. Shrinkage is therefore more predictable, and there is less susceptibility to warpage in flat sections.

Polyallomers Polyallomers are also polyolefin-type thermoplastic polymers produced from two or more different monomers, such as propylene and ethylene, which would produce a propylene-ethylene polyallomer. As can be seen, the monomers, or base chemical materials, are similar to those of polypropylene or polyethylene. Hence, as was mentioned above, and as would be expected, many properties of polyallomers are similar to those of polyethylenes and polypropylenes. Having a density of about 0.9, they, like polypropylenes, are among the lightest plastics.

Polyallomers have a brittleness temperature as low as −40°F and a heat-distortion temperature as high as 210°F at 66 lb/in.². Excellent impact strength plus exceptional flow properties of polyallomer provides wide latitude in product design. Notched Izod impact strengths run as high as 12 ft-lb/in. of notch.

Although the surface hardness of polyallomers is slightly less than that of polypropylenes, resistance to abrasion is greater. Polyallomers are superior to linear polyethylene in flow characteristics, moldability, softening point, hardness, stress-crack resistance, and mold shrinkage. The flexural-fatigue-resistance properties of polyallomers are as good as or better than those of polypropylenes.

Cross-linked polyolefins While polyolefins have many outstanding characteristics, they, like all thermoplastics to some degree, tend to creep or cold-flow under the influence of temperature, load, and time. In order to improve this and some other properties, considerable work has been done on developing cross-linked polyolefins, especially polyethylenes. The cross-linked polyethylenes offer thermal-performance improvements of up to 25°C or more. Cross linking has been achieved primarily by chemical means and by ionizing radiation. Products of both types are available. Radiation-cross-linked polyolefins have gained particular prominence in a heat-shrinkable form. This is achieved by cross linking the extruded or molded polyolefin using high-energy electron-beam radiation, heating the irradiated material above its crystalline melting point to a rubbery state, mechanically stretching to an expanded form (up to four or five times the original size), and cooling the stretched material. Upon further heating, the material will return to its original size, tightly shrinking onto the object around which it has been placed. Heat-shrinkable boots, jackets, tubing, etc., are widely used. Also, irradiated polyolefins, sometimes known as irradiated polyalkenes, are important materials for certain wire and cable jacketing applications.

Polyimides and Polyamide-imides

Among the commercially available plastics generally considered as having high heat resistance, polyimides can be used at the highest temperatures and they are the strongest and most rigid. Polyimides have a useful operating range to about 900°F (482°C) for short durations and 500 to 600°F (260 to 315°C) for continuous service in air.   Prolonged exposure at 500°F (260°C) results in moderate (25 to 30 percent) loss of original strengths and rigidity.

These materials, which can be used in various forms including moldings, laminates, films, coatings, and adhesives, have high mechanical properties, wear resistance, chemical and radiation inertness, and excellent dielectric properties over a broad temperature range.

Materials and sources are listed in Table 23, and properties are given in Table 24. Electrical and mechanical data are covered in Chaps. 2 and 3.   Thermal stability is compared with that of other engineering plastics in Fig. 37.

Chemical structures   Polyimides are heterocyclic polymers, having a noncarbon atom in one of the rings in the molecular chains.[3]   The atom is nitrogen and it is in the imide ring, as shown below:

The fused rings provide chain stiffness essential to high-temperature strength retention.   The low concentration of hydrogen provides oxidative resistance by preventing thermal degradative fracture of the chain.

The other resins considered as members of this family of polymers are the poly (amide-imide)s.   These compositions contain aromatic rings and the characteristic nitrogen linkages, as shown below:

There are two basic types of polyimides: (1) condensation resins and (2) addition resins.

The condensation polyimides are based on a reaction of an aromatic diamine with an aromatic dianhydride.   A tractable (fusible) polyamic acid intermediate produced by this reaction is converted by heat to an insoluble and infusible polyimide, with water being given off during the cure.   Generally, the condensation polyimides result in products having high void contents which detract from inherent mechanical properties and result in some loss of long-term heat-aging resistance.

The addition polyimides are based on short, preimidized polymer-chain segments similar to those comprising condensation polyimides.   These prepolymer chains, which have unsaturated aliphatic end groups, are capped by termini which polymerize thermally without the loss of volatiles.   The addition polyimides yield products which have slightly lower heat resistance than the condensation polyimides.

The condensation polyimides are available as either thermosets or thermoplastics, and the addition polyimides are available only as thermosets.   Although some of the condensation polyimides technically are thermoplastics, which would indicate that they can be melted, this is not the case, since they have melting temperatures which are above the temperature at which the materials begin to thermally decompose.

Properties   Polyimides and polyamide-imides exhibit some outstanding properties to their combination of high-temperature stability up to 500 to 600°F continuously and to 900°F for intermediate use, excellent electrical and mechanical properties which are also relatively stable from low negative temperatures to high positive temperatures, dimensional stability (low cold flow) in most environments, excellent resistance to ionizing radiation, and very low outgassing in high vacuum. They have very low coefficients of friction, which can be further improved by use of graphite or other fillers. Materials and properties are shown in Tables 23 and 24, and electrical and mechanical properties are further covered in Chaps. 2 and 3. Polyamide-imides and polyimides have very good electrical properties, though not as good as those of TFE fluorocarbons, but they are much better than TFE fluorocarbons in mechanical and dimensional-stability properties. This provides advantages in many high-temperature electronic applications. All these properties also make polyamide-imides and polyimides excellent material choices in extreme environments of space and temperature. These materials are available as solid (molded and machined) parts, films, laminates, and liquid varnishes and adhesives. Since data are relatively similar except for form factor, the data presented are for solid polyimides, unless indicated otherwise. Films are quite similar to Mylar, except for improved high-temperature capabilities.

Polymethylpentene

This is another thermoplastic based on the ethylene structure, as shown in Fig. 38, and having special properties due to its combination of transparency and relatively high melting point. This polymer has four combined properties of (1) a high crystalline melting point of 464°F, coupled with useful mechanical properties at 400°F and retention of form stability to near melting; (2) transparency with a light-transmission value of 90 percent in comparison with 88 to 92 percent for polystyrene and 92 percent for acrylics; (3) a density of 0.83, which is close to the theoretical minimum for thermoplastics materials; and (4) excellent electrical properties with power factor, dielectric constant (2.12) and volume resistivity of the same order as PTFE fluorocarbon. Material and source are identified in Table 23, and properties are given in Table 24. Further electrical and mechanical data are presented in Chaps. 2 and 3.

Applications for polymethylpentene have been developed in the fields of lighting and in the automotive, appliance, and electrical industries.

Polyphenylene Oxides

A patented process for oxidative coupling of phenolic monomers is used in formulating Noryl phenylene oxide–based thermoplastic resins.[43] The basic phenylene oxide structure is shown as follows:

$$\left[\begin{array}{c} CH_3 \\ | \\ \bigcirc - O \\ | \\ CH_3 \end{array} \right]_n$$

This family of engineering materials is characterized by outstanding dimensional stability at elevated temperatures, broad temperature use range, outstanding hydrolytic stability, and excellent dielectric properties over a wide range of frequencies and temperatures.

Eight standard grades are available that have been developed to provide a choice of performance characteristics to meet a wide range of engineering-application requirements.

Among their principal design advantages are (1) excellent mechanical properties over temperatures from below −40°F to above 300°F; (2) self-extinguishing, non-dripping characteristics; (3) excellent dimensional stability with low creep, high modulus, and low water absorption; (4) good electrical properties; (5) excellent resistance to aqueous chemical environments; (6) ease of processing with injection-molding and extrusion equipment; and (7) excellent impact strength.

Properties are shown in Table 24, and electrical and mechanical properties are reviewed in more detail in Chaps. 2 and 3. Thermal stability and moisture absorption are compared with other engineering thermoplastics in Figs. 37 and 38, respectively. The supplier is identified in Table 23.

Polyphenylene Sulfides

A crystalline polymer, polyphenylene sulfide (PPS), has a symmetrical, rigid backbone chain consisting of recurring para-substituted benzene rings and sulfur atoms as:

This chemical structure is responsible for the high melting point (550°F), outstanding chemical resistance, thermal stability, and nonflammability of the polymer. There are no known solvents below 375 to 400°F. The polymer is characterized by high stiffness and good retention of mechanical properties at elevated temperatures, which provide utility in coatings as well as in molding compounds. Polyphenylene sulfide is available in a variety of grades suitable for slurry coating, fluidized-bed coating, flocking, electrostatic spraying, as well as injection and compression molding.[44]

This material is identified in Table 23, and properties of unfilled and glass-filled material are shown in Table 24. Mechanical and electrical properties are discussed further in Chaps. 2 and 3.

At normal temperatures the unfilled polymer is a hard material with high tensile and flexural strength. Substantial increases in these properties are realized by the addition of fillers, especially glass. Tensile strength and flexural modulus decrease with increasing temperature, leveling off at about 250°F, with good tensile strength and rigidity retained up to 500°F. With increasing temperature there is a marked increase in elongation and a corresponding increase in toughness.

The mechanical properties of PPS are unaffected by long-term exposure in air at 450°F. Molded items have applications where chemical resistance and high-temperature properties are of prime importance. For injection-molding applications, a 40 percent glass-filled grade is recommended.

Coatings of PPS require a baking operation. Nonstick formulations can be prepared when combination of hardness, chemical inertness, and release behavior is required.

Good adhesion to aluminum requires grit blasting and degreasing treatment. Good adhesion to steel is obtained by grit blasting and degreasing, followed by treatment at 700°F in air. Polyphenylene sulfide adheres well to titanium and to bronze after the metal surface is degreased.

Polystyrenes

The basic polymer unit of polystyrene is based on the ethylene chain shown in Fig. 38, and has the following structure:

Commercial polystyrene is produced by continuous bulk, suspension, and solution polymerization techniques or by combining various aspects of these techniques.[45] The polymerization is a highly exothermic, free radical reaction. The homopolymer is characterized by its rigidity, sparkling clarity, and ease of processibility; however, the crystal polymer tends to be brittle. Impact properties are improved by copolymerization or grafting polystyrene chains to unsaturated rubbers such as polybutadiene. Rubber levels typically range from 3 to 12 percent. Commercially available impact-modified polystyrene is not as transparent as the homopolymers but it has a marked increase in toughness.

The flexibility of the styrene polymerization processes allows manufacturers to produce products with a wide variety of properties by varying the molecular-weight characteristics, additives, plasticizer content, and rubber levels. Heat resistance ranges from 170 to 200°F. Polystyrenes with tensile elongations from near zero to over 50 percent are produced. A variety of melt viscosities is also available.

Since properties can be varied so extensively, polystyrene is used in sheet and profile extrusion, thermoforming, injection and extrusion blow molding, heavy and thin-wall injection molding, direct-injection foam-sheet extrusion, biaxially oriented sheet extrusion, injection molding, and extrusion of structural foam and rotational molding. Polystyrenes can be printed; painted; vacuum-metallized and hot-stamped; sonic, solvent, adhesive, and spin welded; screwed, nailed, and stapled. Polystyrenes are most attractive when considered on a cost-performance comparison with other thermoplastics.

Limitations of polystyrene include poor weatherability, loss of clarity with impact modification, limited heat resistance, and flammability.

These polymers are identified in Table 23, and properties are shown in Table 24. Electrical and physical properties are discussed in more detail in Chaps. 2 and 3.

Polystyrenes represent an important class of thermoplastic materials in the electronics industry because of very low electrical losses. Mechanical properties are adequate within operating-temperature limits, but polystyrenes are temperature-limited with normal temperature capabilities below 200°F. Polystyrenes can, however, be cross-linked to produce a higher-temperature material, as noted in Table 23.

Cross-linked polystyrenes are actually thermosetting materials, and hence do not remelt, even though they may soften. The improved thermal properties, coupled with the outstanding electrical properties, hardness, and associated dimensional stability, make cross-linked polystyrenes the leading choice of dielectric for many high-frequency, radar-band applications.

Conventional polystyrenes are essentially polymerized styrene monomer alone. By varying manufacturing conditions or by adding small amounts of internal and external lubrication, it is possible to vary such properties as ease of flow, speed of setup, physical strength, and heat resistance. Conventional polystyrenes are frequently referred to as normal, regular, or standard polystyrenes.

Since conventional polystyrenes are somewhat hard and brittle and have low impact strength, many modified polystyrenes are available. Modified polystyrenes are materials in which the properties of elongation and resistance to shock have been increased by incorporating into their composition varying percentages of elastomers, as was described above. Hence, these types are frequently referred to as high-impact, high-elongation, or rubber-modified polystyrenes. The so-called superhigh-impact types can be quite rubbery. Electrical properties are usually degraded by these rubber modifications.

Polystyrenes are subject to stresses in fabrication and forming operations, and often require annealing to minimize such stresses for optimized final-product properties. Parts can usually be annealed by exposing them to an elevated temperature approximately 5 to 10°F lower than the temperature at which the greatest tolerable distortion occurs.

Polystyrenes have generally good dimensional stability and low mold shrinkage, and are easily processed at low costs. They have poor weatherability and are chemically attacked by oils and organic solvents. Resistance is good, however, to water, inorganic chemicals, and alcohols.

Polysulfones

In the natural and unmodified form polysulfone is a rigid, strong thermoplastic.[46] It can be molded, extruded, or thermoformed (in sheets) into a wide variety of shapes. Characteristics of special significance to the design engineer are polysulfone's heat-deflection temperature of 345°F at 264 lb/in.[2] and long-term use temperature of 300 to 340°F. This is compared with some other engineering thermoplastics in Fig. 37. These polymers are identified in Table 23, and properties are shown in

Table 24. Electrical and physical properties are discussed in more detail in Chaps. 2 and 3.

Thermal gravimetric analyses show polysulfone to be stable in air up to 500°C. This excellent thermal resistance of polysulfones, together with outstanding oxidation resistance, provides a high degree of melt stability for molding and extrusion.

Some flexibility in the polymer chain is derived from the ether linkage, thus providing inherent toughness. Polysulfone has a second, low-temperature glass transition at −150°F, similar to other tough, rigid thermoplastic polymers. This minor glass transition is attributable to the ether linkages.

The linkages connecting the benzene rings are hydrolytically stable in polysulfones. These polymers therefore resist hydrolysis and aqueous acid and alkaline environments.

Polysulfone is produced by the reaction between the sodium salt of 2,2 bis (4-hydroxyphenol) propane and 4,4'-dichlorodiphenyl sulfone.[46] The sodium phenoxide end groups react with methyl chloride to terminate the polymerization. The molecular weight of the polymer is thereby controlled and thermal stability is assisted.

The diaryl sulfone grouping is a feature of the molecular structure of polysulfones. In this highly resonating structure the sulfone group tends to attract electrons from the phenyl rings. Resonance is enhanced by having oxygen atoms para to the sulfone group. Because oxidation is a loss of electrons, tying up the electrons in resonance imparts excellent oxidation resistance to the polymer. In addition, the sulfur atom is in its highest state of oxidation.

Two additional effects are derived from the high degree of resonance. The strength of the bonds involved is increased and the grouping is fixed spatially into a planar configuration. Excellent thermal stability is provided and the polymer chain exhibits a rigidity that is retained when subjected to elevated temperatures.

Vinyls

Vinyls are structurally based on the ethylene chain, as shown in Fig. 40. Materials are identified in Table 23, and properties are shown in Table 24. Electrical and mechanical properties are reviewed in further detail in Chaps. 2 and 3.

Basically, the vinyl family is comprised of seven major types. These are polyvinyl acetals, polyvinyl acetate, polyvinyl alcohol, polyvinyl carbazole, polyvinyl chloride, polyvinyl chloride-acetate, and polyvinylidene chloride.

Polyvinyl acetals consist of three groups, namely, polyvinyl formal, polyvinyl acetal, and polyvinyl butyral. These materials are available as molding powders, sheet, rod, and tube. Fabrication methods include molding, extruding, casting, and calendering.

Polyvinyl chloride (PVC) is perhaps the most widely used and highest-volume type of the vinyl family. PVC and polyvinyl chloride-acetate are the most commonly used vinyls for electronic and electrical applications.

Vinyls are basically tough and strong, resist water and abrasion, and are excellent electrical insulators. Special tougher types provide high wear resistance. Excluding some nonrigid types, vinyls are not degraded by prolonged contact with water, oils, foods, common chemicals, or cleaning fluids such as gasoline or naphtha. Vinyls are affected by chlorinated solvents.

Generally, vinyls will withstand continuous exposure to temperatures ranging up to 130°F; flexible types, filaments, and some rigids are unaffected by even higher temperatures. Some of these materials, in some operations, may be health hazards. These materials also are slow-burning, and certain types are self-extinguishing; but direct contact with an open flame or extreme heat must be avoided.

PVC is a material with a wide range of rigidity or flexibility. One of its basic advantages is the way it accepts compounding ingredients. For instance, PVC can be plasticized with a variety of plasticizers to produce soft, yielding materials to almost any desired degree of flexibility. Without plasticizers, it is a strong, rigid material that can be machined, heat-formed, or welded by solvents or heat. It is tough, with high resistance to acids, alcohol, alkalies, oils, and many other hydrocarbons. It is available in a wide range of colors. Typical uses include profile extrusions, wire and cable insulation, and foam applications. It also is made into film and sheets.

Polyvinyl chloride raw materials are available as resins, latexes, organosols, plastisols, and compounds. Fabrication methods include injection, compression, blow or slush molding, extruding, calendering, coating, laminating, rotational and solution casting, and vacuum forming.

GLASS-FIBER-REINFORCED THERMOPLASTICS

Basically, thermoplastic molding materials are developed and can be used without fillers, as opposed to thermosetting molding materials, which are more commonly used with fillers incorporated into the compound. This is primarily because shrinkage, hardness, brittleness, and other important processing and use properties necessitate the use of fillers in thermosets. Thermoplastics, on the other hand, do not suffer from the same shortcomings as thermosets and hence can be used as molded products without fillers. However, thermoplastics do suffer from creep and dimensional-stability problems, especially under elevated-temperature and load conditions. Because of this shortcoming, most designers find difficulty in matching the techniques of classical stress-strain analysis with the nonlinear, time-dependent strength-modulus properties of thermoplastics. Glass-fiber-reinforced thermoplastics, or FRTPs, help to simplify these problems. For instance, 40 percent glass-fiber-reinforced nylon outperforms its unreinforced version by exhibiting two and a half times greater tensile and Izod impact strength, four times greater flexural modulus, and only one-fifth of the tensile creep.

Thus, glass-fiber-reinforced thermoplastics fill a major materials gap in providing plastic materials which can be reliably used for strength purposes, and which in fact can compete with metal die castings. Strength is increased with glass-fiber reinforcement, as are stiffness and dimensional stability. The thermal expansion of the FRTPs is reduced. Creep is substantially reduced, and molding precision is much greater.

Dimensional stability of glass-reinforced polymers is invariably better than that of the nonreinforced materials. Mold shrinkages of only a few mils per inch are characteristic of these products. Low moisture absorption of reinforced plastics ensures that parts will not suffer dimensional increases under high-humidity conditions. Also, the characteristic low coefficient of thermal expansion is close enough to that of such metals as zinc, aluminum, and magnesium that it is possible to design composite assemblies without fear that they will warp or buckle when cycled over temperature extremes. In applications where part geometry limits maximum wall thickness, reinforced plastics almost always afford economies for similar strength or stiffness over their unreinforced equivalents. A comparison of some important properties for unfilled and glass-filled (20 and 30 percent) thermoplastics is shown in Tables 33 and 34.

Chemical resistance is essentially unchanged, except that environmental stress-crack resistance of such polymers as polycarbonate and polyethylene is markedly increased by glass reinforcement.

PLASTIC FILMS AND TAPES

Films

Films are thin sections of the same polymers described previously in this chapter. Most films are thermoplastic in nature because of the great flexibility of this class of resins. Films can be made from most thermoplastics.

Films are made by extrusion, casting, calendering, and skiving. Certain of the materials are also available in foam form. The films are sold in thicknesses from ½ to 10 mils. Thicknesses in excess of 10 mils are more properly called sheets.

Tapes

Tapes are films slit to some acceptable size and are frequently coated with adhesives. The adhesives are either thermosetting or thermoplastic. The thermoset adhesives consist of rubber, acrylic, silicones, and epoxies, whereas the thermoplastic adhesives are generally acrylic or rubber. Tackifying resins are generally added to

TABLE 34 Comparison of Some Important Properties of Raw and Glass-Filled Thermoplastics[47]

Material	Percent loading	Tensile strength at 73°F, lb/in.² × 10⁻³ (D 638)*	Elongation at 73°F, % (D 638)	Flexural strength at 73°F, lb/in.² × 10⁻³ (D 790)	Impact strength at 73°F, notched Izod (D 256)	Heat-distortion temperature at 264 lb/in.², °F	Rockwell hardness	Specific gravity
Nylon 6/10:								
Raw........	...	8.5	85–300	......	1.2	......	R111	1.09
Short fiber.	30	17–19	3.0	22–26	1.4–2.2	400	E35-45, R118	1.30
Long fiber..	30	19.0	1.9	23	3.4	420	E70-75	1.30
Nylon 6/6:								
Raw........	...	9.0	60–300	12.5	1.0–2.0	150–186	R108-R118	1.13–1.15
Short fiber.	30	18.5–23	3.0	26.5–32	1.2–2.0	400–470	E50-55-R120	1.37
Long fiber..	30	20	1.5	28	2.5	498	E60-70	1.37
Nylon 6:								
Raw........	...	7.0	25–320	8	1.0–3.6	152–158	R103-R118	1.12–1.14
Short fiber.	30	17–24	3	22.5–32	1.3–2.0	400–420	E45-50, M90	1.37
Long fiber..	30	21.0	2.0	27	3.0	420	E55-60	1.37
Polycarbonate:								
Raw........	...	9.5	60–110	13.5	2.5	265–280	M70-R118	1.2
Short fiber.	20	12–18.5	2.5–3	17–25	1.5–2.5	285–295	M92-R118	1.35
Long fiber..	20	14–18.5	2.2–5	18.5	2.5–3.0	295	H80-90	1.35
Polypropylene:								
Raw........	...	4.3	200–700	6	1.0	135–145	R85-110	0.90–0.91
Short fiber.	20	6.0	3.0	7.5	1.0	230	M40	1.05
Long fiber..	20	8	2.2	10	3.5	283	M50	1.05

Polyacetal:								
Raw..........	..	10.0	15	14	1.4	255	M94-R120	1.425
Short fiber.......	20	10–13.5	2–3	14–15	0.8–1.4	315–325	M70-75-95	1.55
Long fiber.......	20	10.5	2.3	15	2.2	325	M75-80 (Shore)	1.55
Polyethylene:								
Raw..........	..	1.2	50–600	4.8	0.5–16	90–105	D50-60	0.92–0.94
Short fiber.......	20	6	3.0	7	1.1	225	R60	1.10
Long fiber.......	20	6.5	3.0	8	2.1	260	R60	1.10
Polysulfone:								
Raw..........	..	10.2	50–100	......	1.3	345	M69-R120	1.24
Short fiber.......	30	16	2	21	1.8	360	....	1.41
Long fiber.......	30	18.5	2.0	24	2.5	333	E45-55	1.37
PPO:								
Raw..........	..	11.6	20–40	16.5	1.2	345	M78	1.06
Short fiber.......	30	18.0	4.0–6.0	24.4	1.7	360	M94	1.27

* ASTM Test Method.

Note: Most favorable figures for short-fiber performance are based upon results with nominal ¼-in. fibers. Not included in the table are glass-reinforced styrene, SAN, ABS, and polyurethane, for which comparable data between short and long fibers are not available.

Sources: Data for raw-resin and long-glass properties for all resins—Fiberfil Inc.; short-glass polysulfone, polyethylene, polypropylene, all three nylons—Fiberfil and Liquid Nitrogen Processing Corp.; polycarbonate—Fiberfil, LNP, and General Electric Co.; polyacetal and PPO—LNP, Fiberfil, Celanese Corp., and du Pont Co., nylon 6/6—Fiberfil, LNP, and Polymer Corp.

increase the adhesion. The adhesives all deteriorate with storage. The deterioration is marked by loss of tack or bond strength and can be inhibited by storage at low temperature.

Film Properties

Films differ from similar polymers in other forms in several key properties but are identical in all others. Since an earlier section of this chapter described in detail most of the thermoplastic resins, this section will be limited to film properties. Table 35 presents the properties of common films. To aid in selection of the proper films, the most important features are summarized in Table 36.

Films differ from other polymers chiefly in improved electric strength and flexibility. Both these properties vary inversely with film thickness. Electric strength is also related to the method of manufacture. Cast and extruded films have higher electric strength than skived films. This is caused by the greater incidence of holes in the latter films. Some films can be oriented, which substantially improves their physical properties. Orientation is a process of selectively stretching the films, thereby reducing the thickness and causing changes in the crystallinity of the polymer. This process is usually accomplished under conditions of elevated temperature, and the benefits are lost if the processing temperatures are exceeded during service.

Most films can be bonded to other substrates with a variety of adhesives. Those films which do not readily accept adhesives can be surface-treated for bonding by chemical and electrical etching. Films can also be combined to obtain bondable surfaces. Examples of these combined films are polyolefins laminated to polyester films and fluorocarbons laminated to polyimide films.

PLASTIC SURFACE FINISHING

While the greatest majority of plastic parts can be and often are used either with their as-molded natural-colored surface or with colors obtained by use of precolored resins, color concentrate, or dry powder molded into the resin, increasing quantities of plastics are being surface-finished after molding to provide color or metallization. Some important points related to painting and plating will be presented below.

Painting of Plastics

Painting of plastics frequently yields lower-cost coloring than precolored resin or molded-in coloring.[49] Plastics are frequently difficult to paint, and proper consideration must be given to all the important factors involved. Table 37 is a selection guide to paints for plastics, and Table 38 shows application ratings for various paints. Some of the important considerations are given in the following sections.

Heat-distortion point and heat resistance   This determines if a bake-type paint can be used and, if so, the maximum baking temperature the plastic can tolerate.

Solvent resistance   The susceptibility of the plastic to solvent attack dictates the choice of paint system. Some softening of the substrate is desirable to improve adhesion, but a solvent that aggressively attacks the surface and results in cracking or crazing obviously must be avoided.

Residual stress   Molding operations often produce parts with localized areas of stress. Application of coatings to these areas may swell the plastic and cause crazing. Annealing of the part before coating will minimize or eliminate the problem. Often it can be avoided entirely by careful design of the molded part to prevent locked-in stress.

Mold-release residues   Excessive amounts of mold-release agents often cause adhesion problems. To assure satisfactory adhesion, the plastic surface must be rinsed or otherwise cleaned to remove the release agents.

Plasticizers and other additives   Most plastics are compounded with plasticizers and chemical additives. These materials usually migrate to the surface and may eventually soften the coating, destroying adhesion. A coating should be checked for short- and long-term softening or adhesion problems for the specific plastic formulation on which it will be used.

TABLE 35 Film Properties of Thermoplastic Films[48]

Property	ASTM test No.	Cellulose acetate	Cellulose triacetate	Cellulose acetate butyrate	FEP fluorocarbon	Polyamide — Nylon 6	Polyamide — Nylon 6/6	Polytrifluoroethylene copolymers	Polymethylmethacrylate	Polyethylene — Low density	Polyethylene — Medium density	Polyethylene — High density
Method of processing	......	Casting, extrusion	Casting	Casting, extrusion	Extrusion	Extrusion	Extrusion	Extrusion	Extrusion two-way stretch, orientation	Extrusion	Extrusion	Extrusion
Forms available	......	Sheets, rolls, tapes	Sheets, rolls	Sheets, rolls	Rolls, sheets, tubes	Rolls, tubes	Rolls	Rolls	Rolls, sheets	Rolls, sheets, tapes, tubes	Rolls, sheets, tapes, tubes	Rolls, sheets, tapes, tubes
Thickness range, in.	......	0.0005–0.250	0.0008–0.020	0.0011–0.150	0.0005–0.020	0.0005–0.030	0.005–0.020	0.0005–0.030	0.005–0.010	0.0003 and up	0.0003 and up	0.0004 and up
Max width, in.	......	40–60	40–46¼	40 cast 42 extruded	46–48	54	20	54	43	480	240	60
Area factor, in.²/(lb)(mil)	......	21,000–22,000	21,000–22,000	23,000–23,300 (1-mil)	12,900	24,500	24,200	13,000	23,400	30,000	29,500	29,000
Specific gravity	D 1505-63T	1.28–1.31	1.28–1.31	1.19–1.20	2.15	1.13	1.14	2.08–2.15	1.18–1.19	0.910–0.925	0.926–0.940	0.941–0.965
Tensile strength, lb/in.²	D 882-61T	8,000–16,400	9,000	5,000–9,000	2,500–3,000	9,000–18,000	9,000–12,000	5,000–8,000	8,200–8,800	1,500–3,000	2,000–3,500	2,400–6,100
Elongation, %	D 882-61T	15–70	10–40	50–100	300	250–550	200	50–150	4–12	100–700	50–650	10–650
Bursting strength at 1-mil thickness, Mullens points	D 774-63T	30–60	50–70	40–70	11	Elongates	......	23–31	......	10–12	......	15–300
Tearing strength, g*	D 1922-61T	4–10 (1-mil)	4–10	5–10 (1-mil)	125	50–90	......	8–29	......	100–500	50–300	......
Tearing strength, lb/mil†	D 1004-61	1–2	1–2	80–105	600	1,000–1,210	......	330–900	340–380	65–575	......	15–300
Folding endurance	D 2167-63T	500–2,000 (1-mil)	1,000–4,000 (1-mil)	800–1,200 (1-mil)	4,000	>250,000	......	Good	......	Excellent	Very high	Very high
Water absorption after 24 h, %	D 570-63	5–9 (5-mil)	2–4.5	1–2	<0.01	9.5	8.9	Nil	0.3–0.4	<0.01	<0.01	Nil
Water-vapor permeability: g/(100 in.²)(24 h)/(mil) at 25°C	E-96-635(E)	3,000	2,000	2,000	0.40	5.4–20 at 38°C	3–6	0.025–0.055	......	1.0–1.5	0.7	0.3
g/(m²)(24 h)(mm thickness) at 25°C	......	475–1,200	790	790	......	......	......	......	0.5	0.512	......	......
Permeability to gases, cm³/(100 in.²)(mil thickness)(24 h)(atm) at 25°C: CO_2	D 1434-63	860–1,000	880	6,000	1,670	9.7–45	9.1	16–40	......	2,700	2,500	580
H_2	......	835	......	......	2,200	90–250	......	220–330	......	......	1,950	......
N_2	......	30–40	30	250	320	0.9–6	0.35	2.5	......	180	315	42
O_2	......	117–150	150	950	750	2.6 (dry)–25	5.0	7–15	......	500	535	185

* Propagating.
† Initial.

TABLE 35 Film Properties of Thermoplastic Films[48] (Continued)

Property	ASTM test No.	Polystyrene	Regenerated cellulose (cellophane)	Polycarbonate	Polyimide	Polyvinyl fluoride	Polypropylene — Extrusion (cast)	Polypropylene — Biaxially oriented	Polyester (PE terephthalate)	Polytetrafluoroethylene
Method of processing	....	Extrusion oriented	Extrusion into bath	Casting, extrusion	....	Extrusion	Extrusion (cast)	Orientation	Casting, extrusion	Skiving, casting extrusion
Forms available	....	Rolls, sheets	Sheets, rolls, ribbons	Rolls (cast) Rolls (extruded)	Rolls	Rolls	Sheets, rolls, tapes	Rolls	Sheets, rolls, tapes, tubes	Sheets, tapes tubing
Thickness range, in.	....	0.001	0.0008–0.0019	0.0005–0.020	0.0005–0.005	0.0005–0.002	0.0005 and up	0.0005–0.00125	0.00015–0.014	Up to 0.010
Max width, in.	....	40	60	45 cast, 36 extruded	18	138	60	48–72	60–120	38
Area factor, in.^{2}/(lb)(mil)		26,100	11,600–25,000	23,100	19,400	17,200	30,900–31,300	30,600	19,800–22,600	12,800
Specific gravity	D 1505-63T	1.05–1.07	1.40–1.50	1.20	1.42	1.38	0.885–0.9	0.902–0.907	1.380–1.399	2.1–2.2
Tensile strength, lb/in.^{2}	D 882-64T	7,000–12,000	7,000–18,000	8,400–8,800	25,000	7,000–18,000	4,500–10,000	12,000–33,000	20,000–30,000 40,000 (type T)	1,500–4,000
Elongation, %	D 882-64T	10	10–50	85–105	70	115–250	550–1,000	50–200	70–120; 50 (type T)	100–350
Bursting strength at 1-mil thickness, Mullens points	D 774-63T	30	55–65	No break 4 mil 25–35	75	19–70	....	....	55–80	....
Tearing strength, g*	D 1922-61T	2–8	2–20	20–25	8	12–40	600-TD 25-MD	7–20	12–27	10–100
Tearing strength, lb/mil	D 1004-61	Low	110–515	1,150–1,570	510 g/mil	997–1,400	Very high	Excellent	1,000–1,300	....
Folding endurance	D 2167-63T	....	....	250–400	10,000	5,000–47,000	....	....	>100,000	....
Water absorption after 24 hr, %	D 570-63	0.04	45–115	0.35	2.9	<0.5	0.005 or less	<0.005	<0.8	0.00
Water-vapor permeability: g/(100 in.^{2})(24 h)(mil) at 25°C;	E-96-635(E)	6.2	0.4–134	11.0	5.4	3.24	0.4–1.0	0.35–0.45	1.7–1.8	....
g/(m^2)(24 h)(mm thickness) at 25°C		....	....	6.7	....	....	0.1575	0.14	....	....
Permeability to gases, cm^3/(100 in.^{2})(mil thickness)(24 h)(atm) at 25°C: CO$_2$	D 1434-63	926	0.4–6.0	1,075	45	15	800	370	16	....
H$_2$		....	1.2–2.2	1,600	250	58	1,700	....	100	....
N$_2$		42	0.5–1.6	50	6	0.25	48	....	0.7–1.0	....
O$_2$		213	0.2–5.0	300	25	3	240	120	3.5–6.0	....

TABLE 36   Film-Selection Chart

Film	Cost	Thermal stability	Dielectric constant	Dissipation factor	Strength	Electric strength	Water absorption	Folding endurance
Cellulose	Low	Low	Medium	Medium	High	Medium	High	Low
FEP fluorocarbon	High	High	Low	Low	Low	High	Very low	Medium
Polyamide	Medium	Medium	Medium	Medium	High	Low	High	Very high
PTFE polytetrafluoroethylene	High	High	Low	Low	Low	Low	Very low	Medium
Acrylic	Medium	Low	Medium	Medium	Medium	Low	Medium	Medium
Polyethylene	Low	Low	Low	Low	Low	Low	Low	High
Polypropylene	Low	Medium	Low	Low	Low	Medium	Low	High
Polyvinyl fluoride	High	High	High	High	High	Medium	Low	High
Polyester	Medium	Medium	Medium	Low	High	High	Low	Very high
Polytrifluorochloroethylene	High	High	Low	Low	Medium	Medium	Very low	Medium
Polycarbonate	Medium	Medium	Medium	Medium	Medium	Low	Medium	Low
Polyimide	Very high	High	Medium	Low	High	High	High	Medium

Other factors   Stiffness or rigidity, dimensional stability, and coefficient of expansion of the plastic are factors that affect the long-term adhesion of the coating. The physical properties of the paint film must accommodate those of the plastic substrate.

Plating on Plastics

The advantages of metallized plastics in many industries, coupled with major advances in both plateable plastic materials and plating technology, have resulted

TABLE 37   Selection Guide to Paints for Plastics[49]

Plastic	Coatability	Pretreatment*	Recommended coating types
ABS	Good	Solvent wipe	Vinyl, modified vinyl, polyurethane
Acrylic	Good	Solvent wipe	Acrylic, modified acrylic, nitrocellulose, polyurethane, epoxy
Cellulose acetate and butyrate	Good	Solvent wipe	Acrylic, vinyl, nitrocellulose
Nylon	Fair	Detergent wash	Acrylic, vinyl, polyurethane
Phenolic melamine	Difficult	Solvent wipe	Alkyds, polyurethane, epoxy, acrylic
Polycarbonate	Fair	Primer	Polyurethane, acrylic, modified acrylic
Polyester	Good	Solvent wipe	Polyurethane, epoxy
Polyolefins	Difficult	Flame solvent or primer	Polyurethane, nitrocellulose, modified acrylic
Polystyrene	Good	Primer	Polyurethane, nitrocellulose
Polyphenylene oxide-base	Fair	Primer	Polyurethane
Polyurethane	Good	Primer	Polyurethane
Polyvinyl chloride	Fair	Solvent wipe or primer	Polyurethane, nitrocellulose

* Surface pretreatment:
Solvent treatment (hot or cold).   This softens the substrate, but not to the extent of causing surface deformation, crazing, or cracking.
Etching.   Strong oxidizing agents etch the plastic surface to improve adhesion.
Flame or heat treatment.   This also provides an oxidized surface, but does so without the use of liquid agents.
Corona or arc discharge.   This is another technique for producing an oxidized surface using an ozone field.
Mechanical abrasion.   Abrasion belts or grit roughen the surface to give it "tooth" and improve adhesion.
Prime coat.   The base coatings are formulated for two properties: to adhere to the plastic substrate and provide a good bond for the topcoat.

TABLE 38   Applicating Ratings for Various Paints Used on Plastics[49]

Type	Exterior durability	Abrasion resistance	Chemical resistance	Water resistance	Cost
Acrylic.................	5	3	3	3	3
Alkyd...................	4	4	2	2	2
Cellulosic..............	3	3	2	2	3
Epoxy..................	1	5	5	4	3
Polyurethane...........	5	5	4	4	5
Vinyl..................	5	1	4	5	5

Rating scale: 0 = poor, not recommended.   5 = excellent or desirable.

in a continuing and rapid growth of metallized plastic parts. Some of the major problems have been adhesion of plating to plastic, differential expansion between plastics and metals, failure of plated part in thermal cycling, heat distortion and warpage of plastic parts during plating and in system use, and improper design for plating. The major plastics which are plated, and their characteristics for plating, are identified in Table 39. Improvements are being made continuously, especially in ABS and polypropylene, which yield generally lower product costs. Thus, the guidelines of Table 39 should be reviewed at any given time and for any given application. Aside from the commercial plastics described in Table 39, excellent plated plastics can be obtained with other plastics. Notable is the plating of TFE fluorocarbon, where otherwise unachievable electrical products of high quality are reproducibly made. Examples are corona-free capacitors and low-loss high-frequency electronic components.[51]

Design considerations   Proper design is extremely important in producing a quality plated plastic part. Since most plastic designers are not well versed in electroplating, some important design considerations are presented at this point. Figure 41 illustrates 14 design suggestions for improving designs for plating. Other key points are given below.[52]

Appearance.   Because most plated-plastic parts now being produced are decorative (i.e., washer end caps, escutcheons) rather than functional (i.e., copper-plated conductive plastic automotive distributor parts), appearance is extremely critical. For a smooth, even finish, one-piece or integral parts should be designed. Mechanical welds are difficult to plate. If they are necessary, they should be hidden on a noncritical surface.

Gates should be hidden on noncritical surfaces or should be disguised in a prominent feature. Gate design should minimize flow and stress lines, which may impair adhesion.

TABLE 39   Characteristics of Major Plated Plastics[50]

	ABS	Polypropylene	Polysulfone	Polyarylether	Modified PPO
Flow.........................	AA	AA	BA	BA	A
Heat distortion under load......	A	BA	AA	AA	A
Plateability..................	AA	A	BA	AA	BA
Thermal cycling..............	BA	A	AA	AA	AA
Warpage.....................	A	BA	A	A	A
Mold definition..............	AA	BA	A	A	A
Coefficient of expansion........	A	A	A	A	A
Water absorption.............	BA	AA	A	BA	AA
Material cost.................	A	AA	BA	BA	BA
Finishing cost...............	AA	BA	BA	AA	BA
Peel strength................	A	AA	AA	BA	BA

Polymers are rated according to relative desirability of various characteristics: AA = above average; A = average; BA = below average.

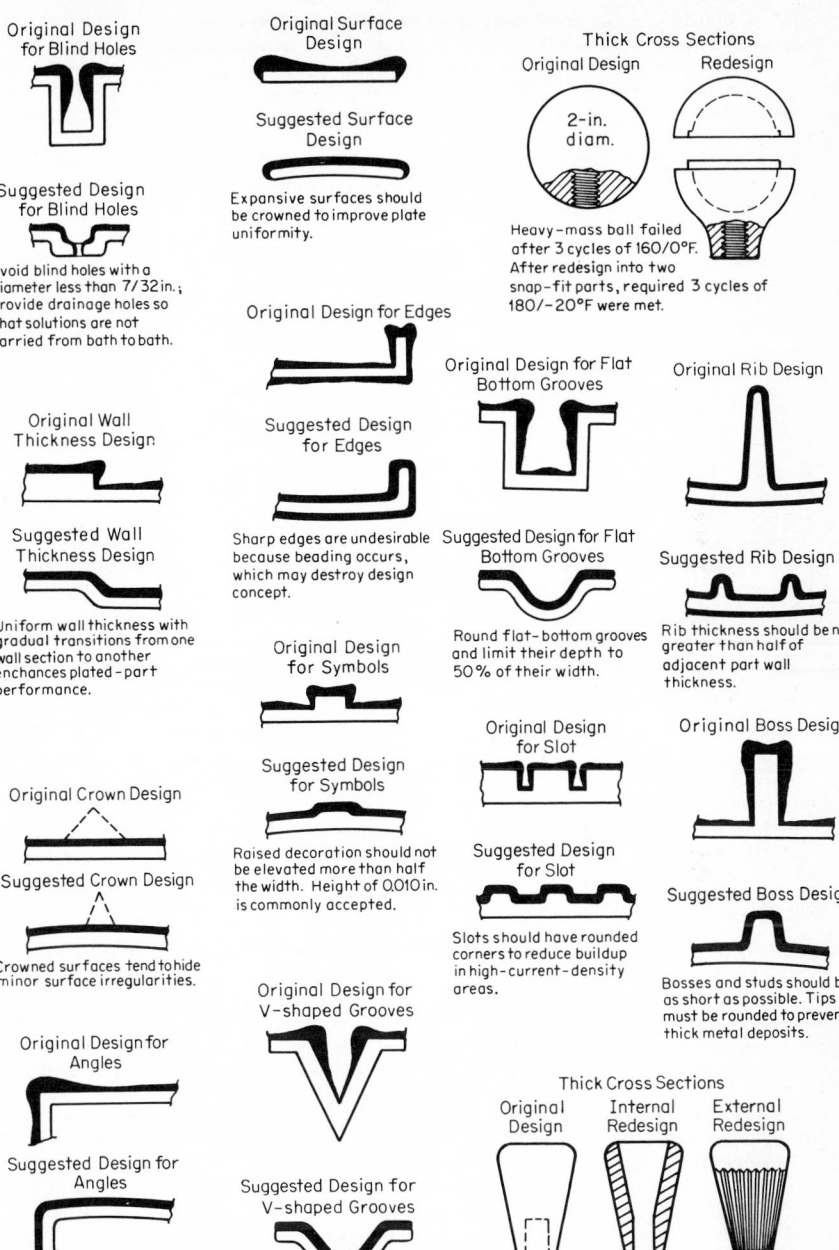

Original Design for Blind Holes

Suggested Design for Blind Holes

Avoid blind holes with a diameter less than 7/32 in.; provide drainage holes so that solutions are not carried from bath to bath.

Original Wall Thickness Design

Suggested Wall Thickness Design

Uniform wall thickness with gradual transitions from one wall section to another enchances plated-part performance.

Original Crown Design

Suggested Crown Design

Crowned surfaces tend to hide minor surface irregularities.

Original Design for Angles

Suggested Design for Angles

All angles should be as large as possible with minimum inside and outside radii of 1/32 in. and 1/16 in., respectively.

Original Surface Design

Suggested Surface Design

Expansive surfaces should be crowned to improve plate uniformity.

Original Design for Edges

Suggested Design for Edges

Sharp edges are undesirable because beading occurs, which may destroy design concept.

Original Design for Symbols

Suggested Design for Symbols

Raised decoration should not be elevated more than half the width. Height of 0.010 in. is commonly accepted.

Original Design for V-shaped Grooves

Suggested Design for V-shaped Grooves

V-shaped grooves are difficult to plate; shallow rounded grooves are better.

Thick Cross Sections
Original Design   Redesign

2-in. diam.

Heavy-mass ball failed after 3 cycles of 160/0°F. After redesign into two snap-fit parts, required 3 cycles of 180/−20°F were met.

Original Design for Flat Bottom Grooves

Suggested Design for Flat Bottom Grooves

Round flat-bottom grooves and limit their depth to 50% of their width.

Original Design for Slot

Suggested Design for Slot

Slots should have rounded corners to reduce buildup in high-current-density areas.

Original Rib Design

Suggested Rib Design

Rib thickness should be no greater than half of adjacent part wall thickness.

Original Boss Design

Suggested Boss Design

Bosses and studs should be as short as possible. Tips must be rounded to prevent thick metal deposits.

Thick Cross Sections
Original Design   Internal Redesign   External Redesign

Removing excess internal bulk provides more uniform cross section; adding external serrations acts to accommodate plastic's thermal expansion.

Fig. 41   Design suggestions for plastic parts to be plated.[52]

Avoid improperly designed ribs or bosses, which cause sink marks. Such marks will be accentuated in electroplating. Rib thickness should be no more than 50 to 70 percent of adjacent wall thickness. Check with resin supplier for individual material specifications.

Avoid large, plain surfaces where very minor defects will be especially noticeable. Use texturing wherever possible.

Avoid use of lubricants as a mold release. Design with 1° minimum draft, with 3° draft preferred. Check with resin supplier for individual material specifications.

The surface of metal inserts must be compatible with the preplate solutions used or they must be masked to avoid contact. Check with the plater before specifying.

Functional Performance. Plated parts that may be flexed either during assembly or in use should be given special consideration in the flex area.

Heavy spherical or cylindrical sections should be cored or broken up by adding flutes to reduce the total mass that is subjected to thermal stress in molding, plating, or use.

Make allowance for plating thickness when designing close-tolerance fits.

Wall thickness should be as great as is economically possible to provide maximum rigidity and adhesion. Maintain as uniform a wall thickness as possible, with 0.0075 in minimum preferable.

Design for optimum plate uniformity to provide maximum corrosion resistance.

Plating-Rack Considerations. Avoid design features that entrap solutions or make thorough rinsing difficult (such as blind holes or recesses) during processing. These will make part and bath contamination extremely critical.

Noncritical surface areas should be designated and designed for cathode contacts. These surfaces will not plate uniformly or as brightly.

Design sufficient rigidity to prevent part warpage while under tension at elevated bath temperatures.

ELASTOMERS

The term *elastomers* includes the complete spectrum of elastic or rubberlike polymers which are sometimes randomly referred to as rubbers, synthetic rubbers, or elastomers.[53] More properly, however, rubber is a natural material and synthetic rubbers are polymers which have been synthesized to reproduce consistently the best properties of natural rubber. Since such a large number of rubberlike polymeric materials now exist, the broad term *elastomer* is most fitting and most commonly used.

The Nature of Elastomers

ASTM D 1566-66T[54] defines elastomers as "macromolecular material that returns rapidly to approximately the initial dimensions and shape after substantial deformation by a weak stress and release of the stress." It also defines a rubber as: "material that is capable of recovering from large deformations quickly and forcibly, and can be, or already is, modified to a state in which it is essentially insoluble (but can swell) in boiling solvent, such as benzene, methyl ethyl ketone, and ethanol toluene azeotrope."

A rubber in its modified state, free of diluents, retracts within 1 min to less than 1.5 times its original length after being stretched at room temperature (20 to 27°C) to twice its length and held for 1 min before release.

More specifically, an elastomer is a rubberlike material that can be or already is modified to a state exhibiting little plastic flow and quick and nearly complete recovery from an extending force. Such material before modification is called, in most instances, a raw or crude rubber or a basic high polymer and by appropriate processes may be converted into a finished product.

When the basic high polymer is converted (without the addition of plasticizers or other diluents) by appropriate means to an essentially nonplastic state, it must meet the following requirements when tested at room temperature (60 to 90°F; 15 to 32°C):

1. It is capable of being stretched 100 percent.

2. After being stretched 100 percent, held for 5 min, and then released, it is capable of retracting to within 10 percent of its original length within 5 min after release (ASTM D 412-51T).[55]

The rubber definition with its swelling test certainly limits it to only the natural latex-tree source, whereas the elastomer definition is more in line with modern new synthetics.

Polymer Basis of Elastomers

The rapid rate of development of polymers can be recognized by realization that commercially available synthetics started in the late 1930s. Polymer chemistry had its start earlier, of course. The important discoveries leading to present polymer science can possibly be credited as beginning with Stardinger's work in the later 1920s on the glucose unit and the start of cellulose. The 1930s saw work by Corothers and Florey on polycondensates such as esters and polyamides (nylon). Also, the 1930s saw the important start of the addition polymers, resulting in high molecular weights of vinyl and diene polymers. The 1940s showed rapid advance of the addition polymers: copolymerization, control of molecular weights for plasticity, emulsion polymerization, development of physics in polymers, solution properties of polymers, intrinsic viscosities, effect of cis- or trans-isomerism, and mechanical behavior of polymers. In the 1950s chemists learned the detailed structure of polymers, and by means of special catalysts (Ziegler) developed stereospecific polymers and finally a true synthetic natural rubber.

Many carbon-hydrogen chain elements, it is found, tend to gather and behave in basic groups. The hydrogen in a basic group could be substituted by some other element or group to form a new group. These basic groups could react with each other in a head-to-tail fashion to make large molecules and larger macromolecules. Also, a group can react with a different group to form a new larger group which can then react with itself in a head-to-tail fashion to produce the desired new macromolecule. These basic groups are called *monomers;* and when they head-to-tail or polymerize into long chains, they then form *polymers.* The reaction between different basic groups to form a new basic recurring group is called *copolymerization.* These copolymers can also recur in a head-to-tail fashion to form long-chain molecules.

The basic monomers are usually gases or very light liquids; and, as polymerization continues, the molecular weight and viscosity both increase until solidification and the formation of the gum or solid products results. The degree of polymerization can be controlled so that the end product has the desired processing properties for the rubber industry but lacks the final properties for engineering use. The long chains in this state have a great deal of mobility between them, which also varies greatly with slight changes in temperatures. Some such materials are usable as thermoplastic materials; but generally, to obtain the properties of elasticity, this mobility must be arrested. These long chains can react further with themselves or with other chemicals. The type of reaction desired for elastomers is not a continuation of chain length at this point, but a tying or linking of the chains together at certain points to form a network and thus reduce plastic mobility. As the chains lie close to each other, especially when the reactive centers are near each other as double bonds between carbon atoms, they can be made to join to each other directly or through some element such as sulfur which will hold the two chains together. These joiners can be loosely called linking agents, and more loosely, vulcanization agents; and again the cross-linked material in the rubber industry can now be called a *vulcanized* material.

One of the basic monomers is gaseous ethylene, which is polymerized to the well-known polymer, polyethylene (Fig. 38). Ethylene is the basis for many other monomers such as styrene, acrylonitrile, isobutylene, vinyl alcohol, and vinyl chloride; which in turn can be polymerized (Fig. 38) to form polystyrene, polyacrylonitrile, polyisobutylene, polyvinyl alcohol, and polyvinyl chloride. Another basic monomer is butadiene, which is the basis for other monomers such as isoprene and chloroprene. The reactions of these two groups of monomers yield the familiar copolymers of butadiene-styrene (GR-S), butadiene-acrylonitrile (buna-N), isoprene with isobutylene (butyl), etc., which constitutes a basic elastomer series.

The mechanical properties of solid insulators are often more important than their electrical properties. Most organic materials are good insulators if they can be formed into continuous films that exclude moisture. The general principles relating mechanical properties of polymers to structures have been known for many years. Rubbers, plastics, and fibers, for example, are not intrinsically different materials. Their differences are a matter of degree rather than kind. If the forces of attraction between the molecular chains are small and the chains do not fit readily into a geometric pattern, lattice, or network, the normal thermal motion of the atoms tends to cause the chains to assume a random, more or less coiled arrangement. These conditions lead to a rubberlike character. In practical rubbers, a few cross links are added to prevent slippage of the molecular chains and permanent deformation under tension. With such polymers, when the stress is released, the normal thermal motion of the atoms causes them to return to a random coiled arrangement. If the forces between the chains are strong and the chains fit easily into a regular geometric pattern, the material is a typical fiber. In cases where the forces are moderate and the tendency to form a regular lattice is also moderate, the result is a typical plastic. Some polymers are made and used as three different materials: rubber, plastic, and fiber. Polyethylene, for example, is used as a substitute for natural rubber in wire covering, as a plastic in low-loss standoff insulators and insulating films, and as a fiber in acid-resistant filter cloths where high fiber strength is not as important as chemical resistance.

The rubber industry as such was long occupied in processing "gumlike" natural rubber. The user expected a material from this industry to be elastic in nature. Therefore, the chemical plants, in their designing of polymers for the rubber industry, kept two requirements in mind: substitutes for natural rubber with its elasticity, and also polymers that could be processed with the conventional equipment. Plastics, as such, behave quite differently from rubber in processing, and this fact led to the establishment of a new industry. Therefore, the incorporation of plastic materials into the rubber inventory was not accomplished. This separation of the two industries has resulted in separate technical groups and societies.

Elastomers, when compared with other engineering materials, are characterized by large deformability; low shape, rigidity; large energy-storage capacity; nonlinear stress-strain curves; high hysteresis; large variations with stiffness, temperature, and rate of loading; and compressibility of the same order of magnitude as most liquids. Certain of the elastomeric materials possess additional useful characteristics to a relative degree, such as corrosive chemical resistance, oil resistance, ozone resistance, temperature resistance, and other environmental conditions.

Commercially Available Elastomers

The nomenclature for elastomers, common names, chemical descriptions, and general properties are given in Table 40, and a listing of representative trade names and suppliers is given in Table 41. Further discussion of these commercially available elastomers is presented below.[53]

Natural rubber (NR) As the name implies, this is a naturally occurring product mostly derived from the *Hevea brasiliensis* tree under cultivation in large plantations around the Malay peninsula area. There are, however, other sources such as the wild rubbers of the same tree growing in Central America; guayule rubber, coming from shrubs grown mostly in Mexico; and balata. Balata is a resinous material of cis-trans isomerism and cannot be tapped like the *Hevea* tree sap. The balata tree is actually cut down and boiled to extract balata, which cures to a hard and tough product used as golf-ball covers.

Although not broadly recognized, there are many grades of natural rubber which must be selected for specific uses. When natural rubber is produced, the tree is tapped and the free-flowing latex is coagulated with an acid and coagulant, then milled or creped to reduce moisture, and further smoked or air-dried for further drying. Smoking of rubber also adds mildew protection by the introduction of creosote.

Most natural rubber is from the *Hevea* tree, but the grading which affects properties is based generally on the cleanliness of the latex before coagulation. Grade 1

TABLE 40   Nomenclature for Elastomers and General Properties[53]

ASTM D 1418	Trade names or common names	Chemical description	General properties
NR.........	Natural rubber Hevea, pure gum	Polyisoprene plus resins. Isoprene natural	Low hysteresis, high resilience, low water absorption
IR..........	Isoprene rubber, polyisoprene, Coral, rubber, Natsyn, Ameripol SN, synthetic natural	Polyisoprene, isoprene rubber synthetic	Same as NR but fewer impurities, better water resistance
SBR........	GR-S, buna-S	Styrene-butadiene rubber	General-purpose use, better heat resistance than natural, but less rubbery properties. Cheap filler insulator
CR.........	Neoprene, Denka chloroprene, Perbunan C, Petro-Tex Chloroprene	Polychloroprene	General-purpose oil-resistant and weather-resistant rubber. Widely used in cover insulation
NBR........	Buna-N nitrile rubber, Hycar, Chemigum, Paracril, FR-N, Polysar Krynac	Butadiene-acrylonitrile copolymer	General-purpose oil-resistant rubber where better oil resistance is needed than with neoprene. Needs special compounding for weather resistance. Good general-purpose chemical resistance. High power factor
IIR.........	Butyl	Isobutylene-isoprene copolymer	Low rebound, high damping, low air permeation, good weathering, good electrical properties
EPM.......	EPR, ethylene rubber Epsyn, Royalene, Vistalon, Nordel	Ethylene-propylene copolymer	Generally same as IIR but better heat resistance and higher air permeation
EPDM......	EPT, Epsyn, Royalene, Vistalon, Nordel	Terpolymer of ethylene, propylene, and a diene side chain	Generally same as EPM but faster curing. Should become the standard general-purpose rubber for all non-oil or solvent use. Good electrical properties and inexpensive. Resists hot water, steam, dry heat, ozone
CSM........	Hypalon	Chlorosulfonated polyethylene	Very similar to neoprene but good for color stability, colored covers, spark-plug boots. Good heat resistance, chemical resistance
ABR........	Acrylic, Cyanacril, Hycar, Krynac, Thiacryl, Acrylon, Paracril OHT	Acrylate butadiene	Similar to NBR but better hot-oil resistance
T...........	Polysulfide Thiokol	Organic polysulfide	Good solvent resistance, odorous, good weather resistance, tender

TABLE 40   Nomenclature for Elastomers and General Properties[53] (Continued)

ASTM D 1418	Trade names or common names	Chemical description	General properties
SI, FSI, PSI, VSI, PVSI...	Silicones, SE Rubber Si-O-Flex, Silastic	Polydimethyl siloxane variations	Outstanding at high and low temperatures, non-conductive decomposition products. Special grades for oil resistance. Good ozone and weather resistance, tender
U............	Urethane, isocyanate rubbers, Cyanaprene Vibrathene, Elasto-thane, Adiprene, Estane, Genthane, Texin, Roylar, Cas-tomer, Conothane	Reaction products of diisocyanates and polyalkylene ether glycols	High tensile strength, high abrasion resistance, high load capacity, wide hardness range, tough. More plastic than rubber in wide property range
FPM.........	Fluoroelastomer Flu-orel, Viton Kel-F	Copolymer of vinyl-idene fluoride and hexafluoropropylene Copolymer of chloro-trifluoroethylene and vinylidene fluoride	Outstanding chemical solvent resistance and good to about 450°F
BR..........	Butadiene, Cis-4, Di-ene Rubber, Ameri-pol CB, Budene, Cisdene, Taktene	Polybutadiene	Improves compression set, wear, low-temperature rebound when added to NR or SBR. Mostly used to improve wear of SBR tires
COX.........	Carboxylic, Hycar 1072	Butadiene-acryloni-trile modified with carboxylic groups	Improves wear and cold-temperature properties over NBR
CO, ECO.....	Hydrin 100, Hydrin 200	Homopolymer of epichlorohydrin, copolymer of epi-chlorohydrin with ethylene oxide	Fuel and oil resistance, chemical resistance to acids and bases, good weathering, a blend of NBR and CR properties
PVC, NBR...	Polyblend Paracril OZO, Hycar, Chemivic	A blend of polyvinyl chloride and NBR	Adds weather and ozone resistance to NBR. Allows light color covers

smoked sheets are plantation-made and generally free of any dirt. Grades 2 to 5 are from small holders, and qualities are graded on the amount of bark, sand, air inclusion, or moisture. Blankets are graded according to color only, and may contain any amount of dirt. Flat bark crepe is the scrap coagulant picked off the bark and any other pickings of small holders, and so it may contain a very high amount of bark and dirt and also may be resinous from excessive heat in smoking. Pale crepe is of the finest quality and is used for light-colored or odor-free products. River water is used for washing and at times is dirtier than the rubber. Guayule has 20 percent resin and needs to be washed to bring it up to smoked-sheet quality. Wild rubber gives long-fingered cements; otherwise, it has the same physical properties as grade 1 smoked sheets, and also needs washing to rid it of about 20 to 30 percent residual moisture.

Another rubber is plantation leaf gutta-percha. As the name implies, this material is produced from the leaves of trees grown in bush formation or on plantations. These leaves are plucked, and the rubber is boiled out as with the balata. Gutta-percha has been used successfully for submarine-cable insulation for better than 40 years, a lasting tribute to the pioneers who made this selection. The dielectric constant of gutta-percha was reported to be 2.6, while reports during the same period for *Hevea* rubber were 3.90 to 4.31.

TABLE 41   Typical Trade Names and Suppliers for Elastomers

Elastomer	Typical trade names and suppliers
Isoprene (IR)	Natsyn (Goodyear Chemical) Ameripol SN (B. F. Goodrich) Isoprene (Shell Chemical)
Neoprene (CR)	Neoprene (E. I. du Pont) Perbunan C (Farben-Fabriken Bayer)
Nitrile (NBR)	Hycar (B. F. Goodrich) Chemigum (Goodyear Chemical) FR-N (Firestone)
Butyl (IIR)	Enjay Butyl (Enjay Chemical) Petro-Tex Butyl (Petro-Tex Chemical)
EPT (EPM) copolymer	Nordel (E. I. du Pont)
EPDM terpolymer	Vistalon (Enjay Chemical) Epcar (B. F. Goodrich)
Hypalon (CSM)	Hypalon (E. I. du Pont)
Acrylic (ABR)	Cyanacril (American Cyanamid) Hycar (B. F. Goodrich) Acrylon (Borden Chemical)
Polysulfide (T)	Thiokol (Thiokol Chemical)
Silicones	Silastic (Dow Corning) SE Rubber (General Electric) Silicones (Union Carbide)
Urethane (U)	Adiprene (E. I. du Pont) Estane (B. F. Goodrich) Genthane (General Tire and Rubber)
Fluoroelastomers (FPM)	Fluorel (3M) Viton (E. I. du Pont)
Butadiene (BR)	Diene (Firestone) Ameripol CB (B. F. Goodrich)
Carboxylic (COX)	Hycar (B. F. Goodrich)
Epichlorohydrin (ECO, CO)	Hydrin (B. F. Goodrich)
Polyblend (PVC/NBR)	Chemivic (Goodyear Chemical) Paracril Ozo (Uniroyal)
Thermoplastic elastomer	Hytrel (E. I. du Pont) Kraton (Shell Chemical)

These are but a few typical basic-material suppliers. In addition, there are many companies which formulate systems from these basic materials, and which are sometimes the ultimate supplier to the end-product fabricator.

Needless to say, natural rubber, properly selected, processed, and compounded, produces very high quality rubber properties and was the original basis of electrical insulation. It is still a good basis in comparison with the new synthetic elastomers. It possesses high tensile strength, tear, resilience, wear, and cut-through. It has good electrical-insulation properties, but by special compounding can be made conductive to a specific resistance as low as 2,000 Ω-cm. It has very low compression set and good resistance to cold flow, and it can be compounded to function as low as $-70°F$. Resilience is the main superior quality over the synthetics. On the negative side, it is not as good for sunlight, oxygen, and ozone resistance as some of the synthetics, but special compounding offsets this to a degree. Its aging properties are not as good as those of the synthetics, but synthetics will harden with time while natural rubber will show a reversal and start softening. Also, natural rubber is not satisfactory in resistance to oils from petroleum, vegetable, or animal origin.

On the other hand, it has good resistance to acids and bases with the exception of oxidizing agents.

Isoprene rubber (IR)   This is the long-sought synthetic-rubber equivalent of natural rubber. As a synthetic, isoprene offers the rubber industry the purity of all synthetics and the freedom from the worry about world supply experienced during World War II. No longer is it necessary to sort, wash, screen, dry, blend, and masticate for processing qualities. Synthesizing to controlled molecular weights gives the industry uniform processing qualities and a bonus on the side—odorlessness.

Styrene-butadiene copolymer (SBR)   There are many manufacturers of SBR, but generally it is known as GR-S or buna-S. This was used for the synthetic-rubber tire that was developed because of the natural-rubber shortage during World War II; and since the government undertook the development, it became known as Government Rubber Styrene type (GR-S). There are now many variations of SBR, generally basic in the ratio of butadiene to styrene, temperature of polymerization, and type of chemicals used during polymerization. This last aspect is important in its initial choice for insulation, since the type of salts and moisture content affect the quality of insulation.

The original purpose of GR-S, of course, was to substitute for the then hard-to-obtain imported materials. In this role it became and still is a workhorse of the industry, even though it does not match the superior physical properties of natural rubber. It lacks in tensile strength, elongation, resilience, hot tear, and hysteresis; but it more than makes up for this in low cost, cleanliness, slightly better heat aging, slightly better wear than natural rubber for passenger tires, and availability at a stable price. Generally speaking, natural rubber was not employed with its ultimate properties but in a cheapened and compounded-down product to meet the desired engineering needs. In this aspect, SBR is certainly not a substitute any longer but the prime choice.

Neoprene (CR)   Neoprene was actually the first commercial synthetic rubber. This elastomer is also a major industry workhorse. Discovered in the laboratories of Notre Dame University and developed by E. I. du Pont de Nemours & Company, it was first used in the 1930s as an oil-resisting rubber. Today it is classified as a moderately oil-resisting rubber with very good weather- and ozone-resisting properties and other properties closer to those of natural rubber than SBR. Because of its crystallizing nature, it has inherent high tensile strength, elongation, and wear properties at pure gum (not extended or hardened) levels. In this respect, it compares with natural rubber, whereas SBR needs reinforcing materials for good tensile strength and thus has no pure gum properties. Electrically, neoprene ranks below natural rubber or SBR because of its polar chlorine group. It can be compounded for temperatures as low as $-67°$F but crystallizes rapidly, and it is not as good in this respect as either natural rubber or SBR. It has excellent flame resistance and is, in fact, self-extinguishing. Because of this property, it is a must in coal-mining operations and other areas where fire is extra hazardous. Also, it has good resistance to oxidative chemicals.

Nitrile rubber (NBR)   NBR is another of the four most generally used elastomers. NBR, with its nitrile group, is above neoprene in polarity and thus poorer in electrical insulation. Its big advantage is that it is considerably more resistant to oils, fuel, and solvents than neoprene. Its other properties fall in line with those of the butadiene group, and thus it closely matches the physical properties of SBR except that it has much better heat resistance. Also, considering cost (slightly above neoprene), it has the broadest resistance to chemicals with a balance of properties than any other polymer. It has a gain in oil resistance over that of neoprene but at a sacrifice to weathering and ozone resistance.

Butyl rubber (IIR)   Butyl rubber's outstanding physical properties are low air permeability (about one-fifth that of natural rubber) and high energy-absorbing qualities. It has excellent weathering and ozone resistance, excellent flexing properties, excellent heat resistance, good flexibility at low temperature, tear resistance about that of natural rubber, tensile strength about in the range of SBR, and very good insulation properties. It has very poor resistance to petroleum oils and gasoline, but excellent resistance to corrosive chemicals, dilute mineral acids, vegetable oils, phosphate ester oils, acetone, ethylene, glycol, and water. It is also very nonpolar.

EPT (EPM) copolymer and EPDM terpolymer This is a new elastomer family which may replace SBR as the general-purpose rubber. In fact, it most certainly will become the standard for wire insulation because it has so much to offer: inexpensive, excellent weathering; and excellent ozone, heat, hot water, and steam resistance. In general, it is similar to butyl rubber with the pluses just mentioned.

Because of its saturated chemical nature and its low air permeability, butyl rubber has met and still meets considerable resistance from fabricators who otherwise might use it. Slight contamination affects its cure and physical properties, and the air retention causes high defects due to air entrapment. The EPT family, with its terpolymer and air-retention properties of natural rubber, overcomes the two large objections against the use of butyl rubber. Therefore, it is foreseen that compounders may become less reluctant to use it and recommend its use. If its processing characteristics eventually match those of SBR, there will be no doubt as to the future success of EPT. Like butyl rubber, it is not oil-resistant; in fact, it is completely miscible.

Hypalon (CSM) A very close match to neoprene is Hypalon, and in many instances, substitutions can be made with hardly any difference in properties. However, Hypalon does offer some exceptional added properties over neoprene such as improved heat and ozone resistance, improved electrical properties, improved color stability, and improved chemical resistance. Light-colored neoprene will darken with age, and Hypalon will not; so if neoprene properties are desired along with color stability, Hypalon is the choice. Its very good electrical properties plus added heat resistance, color stability, and oil resistance are the reasons why it is preferred for spark-plug boots over the costlier silicone rubber and for dust covers on high-voltage ground cables for TV picture tubes, etc.

Acrylic rubber (ABR) Oils for hot applications are often fortified with sulfur-bearing chemicals; and this sulfur, in a system such as automotive transmissions, will continue to react and harden a nitrile-base rubber to a point of excessive hardening. Acrylic rubber is not affected by hot-sulfur-modified oils; and for this reason, it is used in applications of this sort. Where test temperatures are normally run at 250°F for nitrile base, 302°F is used for the acrylic rubber in conjunction with transmission oils. Acrylics, therefore, can be said to have better oil, heat, and ozone resistance than nitrile, and good sunlight resistance; but they are not as good as nitrile for low-temperature work. They are usually used in hot-oil applications.

Polysulfide rubber (T) Polysulfide is especially resistant to hydrocarbon solvents, aliphatic liquids, or blends of aliphatic with aromatic. The common alcohols, ketones, and esters used in paints, varnishes, and inks have little effect on it. It is also resistant to some chlorinated solvents, but preliminary tests should be made before it is used for this purpose. Nitrile-base rubbers will swell considerably in such solvents, and polysulfide has a special use in applications using such solvents. Compared with nitrile, it has poor tensile strength, pungent odor, poor rebound, high creep under strain, and poor abrasion resistance. Polysulfide has most of its applications in solvent-carrying hose, printer's rolls, and newspaper blankets; and because of its excellent weathering properties, it is used heavily for calking purposes.

Silicones (FSI) (PSI) (VSI) (PVSI) (SI) Silicones are an elastomer family of their own, since their polymer backbone consists of silicone and oxygen atoms rather than the carbon structure of all other elastomers. Silicon is in the same chemical group as carbon; and since it is a more stable element, chemists predicted more stable compounds from it if it could be substituted for carbon in the chain. Its outstanding property is its heat resistance (500°F+). It is also very flexible at below −100°F, and it has very good electrical properties. The outstanding property of silicone elastomers for insulation purposes is that their decomposition product on extreme heat is still an insulating silicon dioxide, whereas the decomposition product for organic compounds is conductive carbon black, which can sublime and thus leave nothing for insulation. The first user of silicone during World War II took advantage of this property in purposely undersizing motors where weight was a factor and letting them run hot. Silicone has resistance to oxidation, ozone, and weathering corona; and special modifications by inclusion of halogen groups can make oil-resistant fluorosilicones.

Silicones generally do not have high tensile strength, but their overall properties of good compression set, improved tear resistance, and stability over a wide temperature range certainly make them suitable for many engineering uses. Silicones are the most heat-resisting elastomers available today and the most flexible at low temperature. The basic structure is modified with groups such as vinyl or fluoride which enhance properties of tear and oil resistance, etc., so that there is now a family of silicone rubbers covering a wide range of physical and environmental needs.

Urethane (U)   Urethanes cross-link as well as undergo chain extension to produce a wide variety of compounds. They are available as "castables," or liquids, and as "solids," or gums, to be used on conventional rubber-processing equipment; and as thermoplastic resins to be processed similarly to polyvinyl chloride. They can be cured to tough elastic solid rubbers with outstanding load-bearing capacity, resilience, abrasion resistance, and tensile strength. The outstanding properties are exceptionally high wear and two to three times the tensile strength of natural rubber. Urethanes exhibit very good resistance to oils, solvents, oxidation, and ozone. They have poor resistance to hot water and are not recommended for temperatures above 175°F, and also are quite stiff at low temperatures. Some have even been found to depolymerize or revert at high humidities and temperatures.[56,57] They are useful where load-bearing and wear properties are desirable, and can be employed on insulation as a protective coating over an underlay.

Fluoroelastomers (FPM)   Again the halogen groups indicate stability, and these are stable. The fluorinated rubbers are exceptionally good for high-temperature service but are below silicones in this respect. They resist most of the lubricants, fuels, and hydraulic fluids encountered in aircraft and missiles; a wide variety of chemicals, especially the corrosive variety; and also, most chlorinated solvents. They have good physical properties, somewhere near those of SBR at the higher hardness levels. FPM is valuable in automotive use for its extreme heat and oil resistance and is on a much higher level in this respect than the acrylic elastomers. It has weathering properties superior to those of neoprene, but is very expensive.

Butadiene (BR)   A stereospecific controlled structure like isoprene rubber, its outstanding properties are excellent resilience and hysteresis (almost equivalent to those of natural rubber) and superior abrasion resistance compared with SBR. Butadiene is most similar to SBR and finds wide use as an admixture with SBR in tire treads to improve wear qualities. It is somewhat difficult to process, and for this reason it is hardly ever used in a larger amount than 75 percent of the total polymer in a compound.

Carboxylic (COX)   Similar in properties to nitrile rubber, the carboxylic groups result in vulcanization with outstanding abrasion resistance (approaching that of urethane), improved ozone resistance, and superior cold-temperature properties compared with nitrile.

Epichlorohydrin rubber (CO) (ECO)   Epichlorohydrin rubbers are recognized as having excellent resistance to swelling in oils and fuels, in addition to good chemical resistance to acids, bases, and water. They also have very good aging properties, including ozone resistance. The high chlorine content imparts good to fair flame retardance. Epichlorohydrin rubber appears to possess a blend of the good properties of neoprene and nitrile. However, since the rubber is somewhat new in the industry, it is too early yet to be able to determine its long-term acceptance and use.

Polyblend (PVC/NBR)   This is a nitrile rubber that has been modified by vinyl resins. The resulting elastomer gives all the rubberlike properties of nitrile rubber, in addition to excellent ozone and weather resistance. Blends are available to give nearly the same oil-resisting properties and near ozone resistance of neoprene. Color stability is excellent; and in this respect, it is superior to neoprene and nearer to Hypalon. It extrudes quite readily and matches in properties the other polymers of similar price. It is highly recommended for use in wire and cable jackets, especially where color is desired.

Polyester elastomer   Polyester elastomer is a new polymer material which can be used to fill the gaps between conventional synthetic rubbers, polyurethane rubbers, and plastics. It matches the polyurethanes in hardness and abrasion resistance and exceeds them in load-bearing capacity. It is competitive with flexible plastics in mechanical properties but exceeds them in impact and flex resistance. It matches most specialty synthetic rubbers in heat and chemical resistance.

Based on the same polyester chemistry that produced polyester fibers and polyester film, polyester elastomer possesses several distinctive characteristics. For one, it does not require vulcanization to develop its physical and chemical properties. This is an important advantage in that it simplifies and speeds manufacture. Products can therefore compete favorably in the marketplace with other specialty plastics and elastomers. For another, it has an exceptional thermal-service range for an elastomer. Still another distinction is that it offers premium resistance to chemicals.

These several design and economic advantages suggest a broad spectrum of product applications for polyester elastomer. Many items already have been field-tested, and evaluations are continuing at various stages with a number of others. To date the more promising appear to be hydraulic hose, chemical-service hose, low-pressure pneumatic tires, power and transmission belting, cable jacketing, snowmobile tracks, seals and gaskets, tubing, and molded goods.

Custom-molded or extruded articles can be conveniently produced to specification. Typical physical, thermal, and chemical properties of polyester elastomer are briefly reviewed below.

Hardness. Products can be prepared to specified hardnesses ranging from about 92 durometer A up to 63 durometer D. For familiar comparisons, this means the softer materials would be like a typewriter roll and the harder ones similar to semi-rigid plastics like flexible nylon.

Tensile Strength. The tensile strengths of specimens taken from finished goods range from 5,000 to 7,000 lb/in.^{2}.

Load-bearing Characteristics. Well-suited for load-bearing applications by virtue of its resistance to compressive stress, at hardnesses of 90 durometer A it has twice the load-bearing capacity of thermoplastic polyurethanes. At 55 durometer D its load-bearing capacity is about 40 percent higher than that of polyurethanes.

Mechanical Durability. These polymers are resilient with outstanding resistance to flex cracking at both high and low temperatures. Laboratory tests indicate excellent service under abrasive conditions.

Heat Resistance. Thermal-performance capabilities range up to 300°F and above. The harder polymers exhibit unusually high retention of physical properties at temperatures up to 350°F, both in hot air and in fluids.

Low-Temperature Behavior. Exceptionally good low-temperature properties are typical. Although they are stiffer at room temperature than polyurethanes of corresponding hardness, as the temperature reduces, the polyurethanes become much stiffer. At −40°F, the polyurethanes are two to six times stiffer and far less impact-resistant. The brittle point of polyester elastomer is below −90°F.

Resistance to Oils, Solvents, and Chemicals. In general, the fluid resistance increases with polymer hardness. Properly formulated products offer excellent resistance to lubricating oils, high-aromatic fuels and nitromethane, phosphate and chlorinated hydraulic fluids, common solvents, salt solutions, dilute alkalies and mineral acids, and dilute organic acids. They are, however, soluble in phenols, cresols, certain carboxylic acids, and some chlorinated hydrocarbons—such as chloroform, methylene chloride, and trichloroethane.

Hydrolytic Stability. Since these polymers are based on aromatic (rather than aliphatic) polyesters, they resist hydrolysis. They can be specially formulated to reinforce this resistance for prolonged periods if required for a specific product application where exposure to hot water and low-pressure saturated steam is frequently encountered.

Resistance to Flame. Being a hydrocarbon compound, the basic polymer is not inherently flame-resistant. However, where that property is required, flame-retardant compositions can be formulated.

Resistance to Weather and Ultraviolet Radiation. Products can be used successfully outdoors. However, if such exposure is required, maximum resistance to ultraviolet radiation is best achieved by special formulating.

Relative Elastomer Costs

Table 42 indicates the relative costs of the gum polymers, but it must be remembered that these represent only the raw costs to the rubber companies. Quantity purchases, freight, and credit terms will also modify these costs, as will inflation.

TABLE 42   Relative Elastomer Costs[53]

Polymer	Specific gravity	Cost/lb	Lb-vol-cost ratio (SBR = 1.0)
NR	0.92	0.24	1.0
SBR	0.94	0.23	1.0
CR	1.23	0.41	2.3
NBR	1.00	0.51	2.3
IIR	0.92	0.25	1.1
EPT	0.86	0.30	1.2
CSM	1.1	0.45	2.3
ABR	1.09	1.35	6.8
T	1.34	1.15	7.1
SI	1.10	3.60	18.2
FSI	1.38	11.98	76.3
U	1.07	2.50	11.9
FPM (hexafluor)	1.85	10.25	87.0
FPM (chlorotrifluor)	1.85	16.00	136.8
IR	0.91	0.23	1.0
BR	0.91	0.25	1.0
COX	0.98	0.66	3.0
(CO) (ECO)	1.36	1.25	7.8
Polyblend	1.07	0.44	2.2

Parameters affecting cost are many and varied:
1. Mastication to bring the rubber to proper processing plasticities
2. Compounding ingredients necessary for reinforcement, curing, processing aids, protectants, diluents, colorants, etc., many of which are costlier than the gum
3. Mixing of the batch which depends and varies as to size and type of rubber
4. Quality-control-assurance testing, depending upon extent and severity of specification
5. Processing variables of the various elastomers
6. Cure times and/or special curing techniques such as postcures
7. Defective percentages, which are known to be higher for certain elastomers in various operations
8. Type of product being made, which may entail special processing for certain degrees of physical properties
9. Possible need of clean room or refrigerated room because of contamination or pot life of mixed compounds, etc.

Engineering products are made to specific dimensions instead of weight, and for this reason a pound-volume-cost ratio of cost is given, since it is the only true basis of comparison.

Physical Properties

Proper selection and application of elastomers are difficult for design engineers in many instances, because engineering terms in conventional usage have different meanings when applied to rubber properties. Elastomers are organic materials and react in a completely different manner from metals. For this reason definitions are given below.[58]

Tensile strength in rubber refers to the force per unit of original cross section on elongating to rupture. As such, it is not really an important property in itself, since rubber is rarely required to possess tensile strength. However, it is an indication of other qualities which correlate with tensile strength. Properties that improve with tensile strength are wear and tear resistance, resilience, cut resistance, stress relaxation, creep, flex fatigue, and in some polymers such as neoprene better ozone resistance. In insulation, it should indicate better cut, abrasion, and generally a tougher and more durable shield.

Elongation is the maximum extension of a rubber at the moment of rupture. As a rule of thumb, a rubber with less than 100 percent elongation will usually break if doubled over on itself.

Hardness is an index of the resistance of rubber to deformation and is measured by pressing a ball or blunt point into the surface of the rubber. The most commonly used instrument is the durometer made by the Shore Instrument Company. There are several Shore instruments such as A, B, C, D, and O, designed to give different readings covering soft sponge up through ebonite-type materials. Shore A durometer readings are the most common, and these are the readings appearing in most specifications. On the Shore A scale, 0 would be soft and 100 hard; as a comparison, a rubber band would be about 35 and rubber tire tread 70. Hard rubber (ebonite) is much more solid and is read on the Shore D scale; as a comparison, a hard-rubber pipe stem may read about 60 Shore D, and a bowling ball about 90 Shore D. Most products will fall between 40 and 90 durometer (unless otherwise specified, Shore A is assumed).

Modulus in rubber refers to the force per unit of original cross section to a specific extension. Most modulus readings are taken at 300 percent elongation, but lower extensions can be used. It is a ratio of the stress of rubber to the tensile strain but differs from that of metal, as it is not a Young's modulus stress-strain-type curve; stress-strain values are extremely low for slight extensions but increase logarithmically with increased extension. Rubber has the useful property, for some products, of being extensible to ten times its original length. Modulus can vary from the same hardness and does affect the stiffness of insulation.

Hysteresis is energy loss per loading cycle. This mechanical loss of energy is converted into heating of the rubber product and could reach destructive temperatures.

Heat buildup is a term used to express temperature rise in a rubber product resulting from hysteresis. However, the term also can mean use of high frequencies on rubber where the power factor is too high.

Permanent set is the deformation of a rubber that remains after a given period of stress followed by release of stress.

Stress relaxation is the loss in stress that remains after the rubber has been held at constant strain over a period of time.

Creep of rubber refers to change in strain when the stress is held constant.

Compression set of rubber is the permanent creep that remains after the rubber has been held at either constant strain or stress and in compression for a given time. Constant strain is most generally employed, and is reported as a percentage of the permanent creep divided by the amount of original strain. A strain of 25 percent is most common.

Abrasion resistance is one of the remarkable properties of rubber and one most difficult to measure. The term refers to the resistance of a rubber composition to wear and is usually measured by the loss of material when a rubber part is brought into contact with a moving abrasive surface. It is specified as percent of volume loss of sample as compared with a standard rubber composition, and it is almost impossible to correlate these relative values of life expectancy. Also, since many formulations contain wax-type substances which exude to the surface, test results can be erroneous because of the lubricating effect of the waxes.

Flex fatigue is the result of rubber fracturing after being subjected to fluctuating stresses.

Impact resistance is the resistance of a rubber to abrading or cutting when hit by a sharp object.

Tear resistance is a measure of stress needed to continue rupturing a sheet of rubber, usually after an initiating cut.

Flame resistance is the relative flammability of a rubber. Some rubbers will burn profusely when ignited, whereas others are self-extinguishing when the igniting source is removed.

Low-temperature properties indicate a stiffening range and brittle point of rubber. Stiffening range is the more useful of these two but is difficult to measure. Brittle point also has little meaning unless the deforming force and rate are known. Time is also an important consideration, since some characteristics change as the temperature is lowered and held: hardness, stress-strain rate, and modulus, for example. With many materials crystallization occurs, at which time the rubber is brittle and will fracture easily. There are many tests to measure cold-temperature properties; but for a general comparison it is simplest to bend the specimens manually at the test temperature for a difference in "feel" of stiffness or actual breakage. All tests must be made in the cold box and all precautions taken so that the rubber is not warmed during the test. The specimens (0.075 standard ASTM slabs) warm very quickly and will produce false results unless extreme care is taken in the testing.

Heat resistance. No rubber is completely heat-resistant; time and temperature have their aging effects. Heat resistance is measured usually as change in tensile strength, elongation, and durometer readings from the original values, usually after a 72-h period.

Aging. Elastomeric properties can be destroyed only by further chain growth and linkage, which would result in a hard, rigid material, or a chain rupture, which would result in a plastic or resinous mass. The deteriorative agents mostly considered in this category are sunlight, heat, oxygen, stress with atmospheric ozone, atmospheric moisture, and atmospheric nitrous oxide. Chain growth or cross linkage will usually decrease elongation and increase hardness and tensile strength, whereas chain rupture will have the opposite effect. Some elastomers will continue to harden and some to soften; and some will show an initial hardening followed by softening. All are irreversible responses.

Radiation. Deteriorating effects of radiation are similar and complementary to those of aging. It has been fairly well demonstrated that damage is dependent only on dosage, or amount of radiation energy absorbed, irrespective of the form of radiation within a factor of 2. Dosage rate is an important factor only where significant degradation from other agents, such as oxygen, has access to the system. Surprisingly enough, the least heat-resistant of the elastomers displays the most radiation resistance.

TABLE 43   General Physical Properties of Rubber Compounds[53]

Compound	Tensile strength, lb/in.² (upper limit)	Durometer Shore A	% elongation	Service temp, °F	Tear resistance	Abrasion resistance	Compression set	Flame resistance
Natural	4,500	25–95	800	−80 to 175	Very good	Excellent	Good	Poor
SBR	3,000	40–95	450	−67 to 200	Good	Good to excellent	Good	Poor
Chloroprene	3,200	30–90	700	−67 to 250	Good	Excellent	Fair to good	Good
NBR	3,000	40–90	650	−67 to 275	Good	Good to excellent	Good	Poor to fair
Butyl	2,500	35–90	500	−67 to 250	Very good	Good	Poor to fair	Poor
EPT	2,800	35–90	350	−80 to 300	Good	Excellent	Fair	Poor
Chlorosulfonated polyethylene	3,300	40–90	500	−67 to 275	Good	Excellent	Fair to good	Good
Acrylic	1,800	40–95	350	−20 to 350	Fair	Good	Fair	Poor to fair
Polysulfide	1,250	20–80	400	−40 to 225	Good	Poor	Poor	Poor
Silicone	1,500	20–90	750	−140 to 550	Poor	Poor	Good to excellent	Fair to good
Urethane	8,000	60–95	700	−40 to 250	Excellent	Superior	Excellent	Poor to fair
Fluorocarbon	2,700	60–90	300	−40 to 450	Fair	Good	Good to excellent	Excellent
Isoprene	4,000	25–95	600	−80 to 175	Very good	Excellent	Good	Poor
Butadiene	3,500	35–95	550	−100 to 175	Very good	Excellent	Good	Poor
Carboxilic	2,200	50–85	400	−90 to 275	Very good	Very excellent	Fair	Poor
Epichlorohydrin	2,500	50–85	450	−40 to 325	Good	Good	Good	Fair to good
Polyblend	2,800	45–90	400	−30 to 250	Good	Excellent	Poor	Fair

Table 43 presents the physical properties of various rubber compounds. The values listed are only relatives ones, in an attempt to differentiate among the rubbers; and so extreme caution must be exercised in the use of the table.

Cold-temperature stiffening of some rubbers is shown in Fig. 42. Again, all these curves can be altered with compounding and, in fact, can easily be made to interchange their respective position.

One of the most difficult problems is the recommendation of a service temperature, and/or expected service life at specific temperatures. Without data on a specific compound for a specific application, almost all recommendations must be

TABLE 44   Comparison of Thermal and Radiation Stabilities of Various Elastomers[55]

Elastomer	Temp at which 25% of original tensile strength is lost after 8 h aging, °F	Order	Dosage at which 25% change in original properties occurs, ergs/g	Order
Silicone:				
Aliphatic side group......	480	1	4.2×10^8	7
Aromatic side group.....	. . .	1	12.0×10^8	4
Fluoroelastomer...........	450	2	4×10^8	8
Acrylic..................	425	3	3.3×10^8	9
Urethane................	350	4	4.2×10^9	1
Chlorosulfonated				
polyethylene.............	350	4	4.2×10^8	7
NBR....................	340	5	7×10^8	5
Isobutylene..............	335	6	4×10^8	8
Chloroprene.............	325	7	5.5×10^8	6
Polysulfide..............	285	8	1.5×10^8	10
SBR....................	275	9	1.5×10^9	3
Natural.................	210	10	2.5×10^9	2

based on previous experience coupled with data gathered from correlating laboratory results. Many new elastomers simply have not existed long enough for the time-span expectancy demands of some designs to be known. Most comparisons are made on a percentage of property changes from original values. The general rule of twice the speed of reaction for approximately every 15°F change in temperature follows for elastomers. Table 44 compares thermal and radiation stabilities of various elastomers.

Silicone is the best commercially available rubber for heat resistance, and Fig. 43 indicates its life expectancy at elevated temperatures.

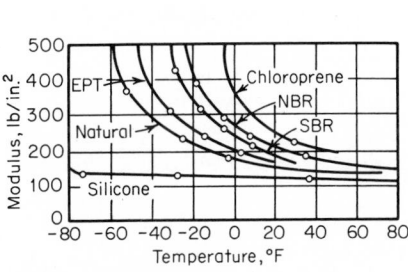

Fig. 42 Low-temperature stiffness of elastomers.[59]

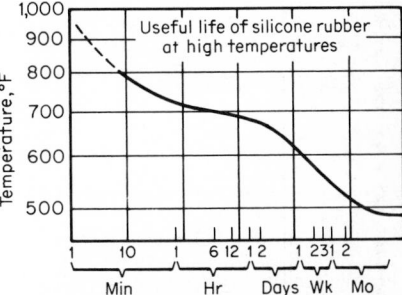

Fig. 43 Aging of silicone rubber at elevated temperatures.[60]

REFERENCES

1. Staff Report, Polymer Chemistry, *SPE J.*, June 1970.
2. The Plastics Industry and the Energy Shortage, *Celanese Plastics Co. Bull.* 1974.
3. Hauck, J. E.: Engineer's Guide to Plastics, *Mater. Eng.*, February 1967.
4. Standard Tests on Plastics, *Celanese Plastics Co. Bull.* GIC, 1970.
5. Harper, C. A.: "Handbook of Materials and Processes for Electronics," McGraw-Hill, New York, 1972.
6. Grafton, Percy: "Commonly Used Plastics," Boonton Molding Co., Inc., Boonton, N.J., 1973.
7. Deanin, R. D., and S. B. Driscoll: Cost per Unit Property of Plastics, *Modern Plastics*, April 1973.
8. Staff Report, What Are Fire Officials Squawking About, *Modern Plastics*, August 1973.
9. Staff Report, Flammability Report, *Modern Plastics*, November 1972.
10. Howarth, J. T., R. S. Lindstrom, G. Sheth, and K. R. Sidman: Flame-Retardant Additives, *Plast. World*, March 1973.
11. Seymour, R. B.: Fillers for Molding Compounds, "*Modern Plastics* Encyclopedia," 1966–1967.
12. Plaskon Plastics and Resins, Allied Chemical Corp., Plastics Division, *Tech. Bull.*
13. Dapon Diallyl Phthalate Resin, *FMC Corp. Tech. Bull.*
14. Chottiner, J.: Dimensional Stability of Thermosetting Plastics, *Mater. Eng.*, February 1962.
15. Sherman, S.: Epoxy Resins, "*Modern Plastics* Encyclopedia," 1974.
16. Petruccelli, F.: Aminos, "*Modern Plastics* Encyclopedia," 1974.
17. Sunderland, G. B., and A. Nufer: Aminos, *Machine Design* Plastics Reference Issue, June 1966.
18. Dienite Thermosetting Resins, Firestone Synthetic Rubber and Latex Co., Technical Booklet, 1972; Colorado Chemical Specialties, Inc., Golden, Colo.
19. Czarnomski, T. J.: Unsaturated Polyesters, "*Modern Plastics* Encyclopedia," 1974.
20. Kookootsedes, G. J.: Silicones, "*Modern Plastics* Encyclopedia," 1974.
21. Dow Corning Molding Compounds, *Dow Corning Corp. Tech. Bull.*
22. Staff Report, Contemporary Thermoplastics Materials Properties, *Plast. World*, Sept. 17, 1973.
23. Marbon ABS Plastics, *Borg-Warner Corp. Tech. Data Bull.* 1972.
24. Hauck, J. E.: Long-Term Performance of Plastics, *Mater. Eng.*, November 1965.
25. Patten, G. A.: Heat Resistance of Thermoplastics, *Mater. Eng.*, May 1962.
26. Hauck, J. E.: Alloy Plastic to Improve Properties, *Mater. Eng.*, July 1967.
27. Barken, H. E., and A. E. Javitz: Plastics Molding Materials for Structural and Mechanical Applications, *Elec. Mfg.*, May 1960.
28. Delrin Acetal, du Pont "Design Handbook."
29. Delrin AF, E. I. du Pont de Nemours & Co., Inc., *Tech. Data Bull.*
30. Lucite Acrylic Resins "Design Handbook." E. I. du Pont de Nemours & Co., Inc.
31. Koo, G. P., et al.: Engineering Properties of a New Polytetrafluorethylene, *SPE J.*, September 1965.
32. Mechanical Design Data for Teflon Resins, E. I. du Pont de Nemours & Co., Inc., Technical Booklet.
33. Surlyn Ionomers, E. I. du Pont de Nemours & Co., Inc., Technical Booklet.
34. Zytel Nylon Resins, E. I. du Pont de Nemours & Co., Inc., "Design Handbook."
35. Parylene Conformal Coatings, Union Carbide Corp., Technical Booklet.
36. Isaacson, W. B.: Polyaryl Sulfone, "*Modern Plastics* Encyclopedia," 1974.
37. Astrel 360 Plastic, *3M Co. Tech. Bull.*
38. Bayer Polycarbonate Films, Mobay Chemical Co., Bayer AG, *Tech. Bull.*
39. Lexan Polycarbonate Resin, General Electric Company, Technical Booklet.
40. Noryl and PPO Resins, General Electric Company Technical Bulletins.
41. Celanex, *Celanese Plastics Co. Tech. Data Bull.*
42. Leslie, V. J.: Polyethersulfone, "*Modern Plastics* Encyclopedia," 1974.
43. Eickert, J. S.: Phenylene Oxide–Based Resins, "*Modern Plastics* Encyclopedia," 1974.
44. Jones, R. V.: Polyphenylene Sulfide, "*Modern Plastics* Encyclopedia," 1974.
45. Guenin, L. R.: Polystyrene, "*Modern Plastics* Encyclopedia," 1974.
46. Walton, R. K.: Polysulfone, "*Modern Plastics* Encyclopedia," 1974.
47. Staff Report, Can RTP Meet Your Product Requirements, *Modern Plastics*, March 1966.
48. Film Properties Charts, "*Modern Plastics* Encyclopedia," 1974.

49. Kinsella, T. P., F. J. Steslow, and J. R. Weschler: Painting Plastics, *Plastics World,* April 1973.
50. Leary, F. A.: Characteristics of Plateable Plastics, *Plast. Des. and Processing,* June 1971.
51. Polyflon Plated TFE, *Polyflon Corp. Tech. Bull.*
52. Staff Report, Proper Part Design for Profitable Plating, *Plast. World,* February 1971.
53. Ceryan, J. G.: Elastomers, chap. 3 in "Handbook of Materials and Processes for Electronics," McGraw-Hill, New York, 1972.
54. ASTM standards, Part 28, April 1967.
55. Pickett, A. G., and M. M. Lemcoe: Southwest Research Institute "Handbook of Design Data on Elastomeric Materials Used in Aerospace Systems," A SDRT61-234, Aeronautical Systems Division, Air Force Systems Command, U.S. Air Force, Wright-Patterson Air Force Base, Dayton, Ohio, June 1961.
56. Gahimer, F. H., and F. W. Nieske: Hydrolytic Stability of Urethane and Polyacrylate Elastomers in Humid Environments, *J. Elastoplastics,* October 1969.
57. Gahimer, F. H., and F. W. Nieske: Testing the Hydrolytic Stability of Urethanes, *Insulation/Circuits,* August 1972.
58. Elastomers: Tailor Made Design Materials, *Design News,* Mar. 31, 1965.
59. Hauck, Jack E.: *Mater. Eng.,* April 1966.
60. Burton, Walter E.: "Engineering with Rubber," McGraw-Hill, New York, 1949.

Chapter **2**

Electrical Design Properties of Plastics and Elastomers

CHARLES A. HARPER

Westinghouse Electric Corporation,
Baltimore, Maryland

INTRODUCTION

Plastics represent one of the very important classes of materials used in the electrical and electronic industries. Plastic materials are broadly used for dielectric or insulation purposes, as well as for structural or ruggedization purposes. The useful application areas are steadily increasing owing to continuing advances in plastic technology. The upper temperature limit for plastics has risen substantially in recent years, and the number of high-performance engineering plastics has increased rapidly. In addition, chemical development has led to creation of many electronic grades of plastics, that is, plastics with improved or more stable dielectric properties. Many old industry workhorse standards, not basically suitable for high-performance electronics, can now be obtained in high-performance electronic grades. It is the purpose of this chapter to discuss the types of plastics and elastomers and properties of these materials as related to electronic and electrical industry applications. The basic types and forms of plastics were presented in Chap 1, and the reader is referred to that chapter for presentations on the chemical classes and product classes of plastics and elastomers. Their role in the field of electrical design is discussed in this chapter. Presentations will be organized into basic chemical and important product groupings, however, after an initial discussion on basic insulation properties and ratings of plastics in the functional electrical categories.

BASIC ELECTRICAL-INSULATION PROPERTIES

Plastics and elastomers usually serve some selective insulation or signal-transmission function, some structural or mechanical function in electrical systems, or frequently, a combination of these.[1-4]

Although insulating materials have many important properties and characteristics, which are discussed in detail in this chapter, there are two basic categories which should be mentioned at this time. These are (1) insulation or dielectric characteristics and (2) thermal classification.

Insulation Characteristics and Their Significance

The primary dielectric, or insulating, characteristics are dielectric strength, resistance and resistivity, dielectric constant, power factor and dissipation factor, and arc resistance. The definitions of these terms, and their significant features, are given in Table 1.

Ratings of Plastics for Insulation Characteristics

Approximate ratings of plastics in gradual or short-time breakdown per ASTM D 149 are given in Table 2. Ratings for the stepwise, or step-by-step, breakdown per ASTM D 149 are given in Table 3. Approximate ratings of volume resistivity are given in Table 4. The resistivity spectrum for materials is shown in Fig. 1.

TABLE 1   Significance of Important Electrical-Insulation Properties

Property and definition	Significance of values
Dielectric Strength	
All insulating materials fail at some level of applied voltage for a given set of operating conditions. The dielectric strength is the voltage that an insulating material can withstand before dielectric breakdown occurs. Dielectric strength is normally expressed in voltage-gradient terms, such as volts per mil. In testing for dielectric strength, two methods of applying the voltage (gradual or by steps) are used. Type of voltage, temperature, and any preconditioning of the test part must be noted. Also, thickness of the piece being tested must be recorded because the voltage per mil at which breakdown occurs varies with thickness of test piece. Normally, breakdown occurs at a much higher volt-per-mil value in very thin test pieces (a few mils thick) than in thicker sections ($\frac{1}{8}$ in. thick, for example)	The higher the value, the better the insulator. Dielectric strength of a material (per mil of thickness) usually increases considerably with decrease in insulation thickness. Material suppliers can provide curves of dielectric strength vs. thickness for their insulating materials
Resistance and Resistivity	
Resistance of insulating material, like that of a conductor, is the resistance offered by the conducting path to passage of electric current. Resistance is expressed in ohms. Insulating materials are very poor conductors, offering high resistance. For insulating materials, the term "volume resistivity" is more commonly applied. Volume resistivity is the electrical resistance between opposite faces of a unit cube for a given material and at a given temperature. The relationship between resistance and resistivity is expressed by the equation $\rho = RA/l$, where ρ = volume resistivity in ohm-centimeters, A = area of the faces, and l = distance between faces of the piece on which measurement is made. This is not resistance per unit volume, which would be ohms per cubic centimeter, although this term is sometimes erroneously used. Other terms are sometimes used to describe a specific application or condition. One such term is surface resistivity, which is the resistance between two opposite edges of a surface film 1 cm square. Since the length and width of the path are the same, the centimeter units cancel. Thus, units of surface resistivity are actually ohms. However, to avoid confusion with usual resistance values, surface resistivity is normally given in ohms per square. Another broadly used term is insulation resistance which, again, is a measurement of ohmic resistance for a given condition, rather than a standardized resistivity test. For both surface resistivity and insulation resistance, standardized comparative tests are normally used. Such tests can provide data such as effects of humidity on a given insulating-material configuration	The higher the value, the better for a good insulating material. The resistance value for a given material depends on a number of factors. It varies inversely with temperature, and is affected by humidity, moisture content of the test part, level of the applied voltage, and time during which the voltage is applied. When tests are made on a piece that has been subjected to moist or humid conditions, it is important that measurements be made at controlled time intervals during or after the test condition has been applied, since dry-out and resistance increase occur rapidly. Comparing or interpreting data is difficult unless the test period is controlled and defined

TABLE 1   Significance of Important Electrical-Insulation Properties (Continued)

Property and definition	Significance of values

Dielectric Constant

The dielectric constant of an insulating material is the ratio of the capacitance of a capacitor containing that particular material to the capacitance of the same electrode system with air replacing the insulation as the dielectric medium. The dielectric constant is also sometimes defined as the property of an insulation which determines the electrostatic energy stored within the solid material. The dielectric constant of most commercial insulating materials varies from about 2 to 10, air having the value 1

Low values are best for high-frequency or power applications, to minimize electric power losses. Higher values are best for capacitance applications. For most insulating materials, dielectric constant increases with temperature, especially above a critical temperature region which is unique for each material. Dielectric-constant values are also affected (usually to a lesser degree) by frequency. This variation is also unique for each material

Power Factor and Dissipation Factor

Power factor is the ratio of the power dissipated (watts) in an insulating material to the product of the effective voltage and current (volt-ampere input) and is a measure of the relative dielectric loss in the insulation when the system acts as a capacitor. Power factor is nondimensional and is a commonly used measure of insulation quality. It is of particular interest at high levels of frequency and power in such applications as microwave equipment, transformers, and other inductive devices.
Dissipation factor is the tangent of the dielectric loss angle. Hence, the term "tan delta" (tangent of the angle) is also sometimes used. For the low values ordinarily encountered in insulation, dissipation factor is practically the equivalent of power factor, and the terms are used interchangeably

Low values are favorable, indicating a more efficient system, with lower power losses

Arc Resistance

Arc resistance is a measure of an electrical-breakdown condition along an insulating surface, caused by the formation of a conductive path on the surface. It is a common ASTM measurement, especially used with plastic materials because of the variations among plastics in the extent to which a surface breakdown occurs. Arc resistance is measured as the time, in seconds, required for breakdown along the surface of the material being measured. Surface breakdown (arcing or electrical tracking along the surface) is also affected by surface cleanliness and dryness

The higher the value, the better. Higher values indicate greater resistance to breakdown along the surface due to arcing or tracking conditions

Similar approximate ratings for dielectric constant at 60 Hz and 1 MHz are given in Tables 5 and 6, respectively. Ratings based on 60-Hz and 1-MHz dissipation factor are listed in Tables 7 and 8. Ratings for arc resistance, per ASTM D 495, are given in Table 9. Tables 2 through 9 provide good general guides. It should be remembered, however, that individual products should be reviewed for individual design requirements, because of the many variations possible in formulating and processing the many materials involved.

TABLE 2   Short-Term Dielectric Strength of Plastics in Volts per Mil (ASTM D 149, ⅛ in. Thick)[5]

Plastic	V/mil
Cold-mold refractory (inorganic) (asbestos filler)	45
Cold-mold nonrefractory (organic)	85–115
Melamine (fabric phenolic modified)	130–370
Phenolic (glass filler)	140–400
Phenolic acrylonitrile (rag filler)	175–225
Silicones (glass filler)	200–400
Silicones (mineral filler)	200–400
Phenol (fabric and cord filler)	200–400
Melamine (phenol filler)	220–325
Melamine (fabric filler)	250–350
Epoxy encapsulating (glass filler)	250–400
Epoxy encapsulating (mineral filler)	250–400
Cellulose acetate butyrate	250–400
Polyester alkyd granular putty type (glass filler)	250–530
Cellulose acetate	250–600
PVC (flexible filled)	250–800
Polyvinylidene fluoride (VF2)	260
Phenol formaldehyde (asbestos filler)	260–360
Phenol formaldehyde (cotton filler)	260–400
Phenol formaldehyde (wood-flour filler)	260–400
Melamine alpha cellulose	270–300
Phenolic acrylonitrile copolymer (wood-flour and flock filler)	275–325
Phenolic acrylonitrile (asbestos filler)	275–350
Melamine (flock filler)	300–330
Epoxy (mineral filler)	300–400
Epoxy (glass filler)	300–400
Urea formaldehyde	300–400
Cellulose propionate	300–450
Polystyrene (medium- and high-impact)	300–600
Cellulose nitrate	300–600
PVC flexible (unfilled)	300–1,000
Urethanes elastomers	330–900
Polyester (premixed glass filler)	345–420
Polyaryl sulfone	350
Phenolic (mica filler)	350–400
Melamine cellulose	350–400
Polystyrene (glass filler)	350–425
Polyester alkyd granular putty type (mineral filler)	350–450
Polyester woven cloth	350–500
Polyester (preformed chopped rovings)	350–500
ABS polycarbonate alloy	350–500
Ethyl cellulose	350–500
ABS (high-impact)	350–500
ABS (heat-resistant)	350–500
ABS (medium-impact)	350–500
ABS (self-extinguishing)	350–500
Nylons (polyamide) type 6/6	365–480
Polyester alkyd granular putty type (synthetic filler)	365–500
Polyester alkyd granular putty type (asbestos filler)	380
Acetal homopolymer	380
Epoxy (low-density)	380–420
Diallyl phthalate (synthetic fiber)	390–400
Diallyl phthalate (mineral filler)	395–420
Diallyl phthalate (glass filler)	395–450
Acrylics PVS alloy	>400
Chlorinated polyether	400
Polycarbonate (unfilled)	400
Styrene acrylonitrile copolymer	400–500
Acrylics	400–500
Acrylics (impact)	400–500

TABLE 2   Short-Term Dielectric Strength of Plastics in Volts per Mil (ASTM D 149, ⅛ in. Thick)[5] (Continued)

Plastic	V/mil
Polyphenylene oxides (unfilled) (PPO)	400–500
Nylons (polyamide) type 6 (20–40% glass filler)	400–580
Polystyrene (heat and chemical)	400–600
Melamine (asbestos filler)	410–430
Styrene butadiene thermoplastic elastomers	420–520
Polyphenylene oxide (modified) (unfilled)	420–600
Nylon (polyamide) type 11	425
Polysulfone	425
Polyaryl ether	430
Nylon (polyamide) type 6	440–510
Polycarbonate (10% glass filler)	450
Polycarbonate (10–40% glass filler)	450
Polyethylene (high-density)	450–500
Ethylene ethyl acrylate copolymer	450–550
Polypropylene (inert filler)	450–650
Polyethylene (low-density)	450–1,000
Polyethylene (medium-density)	450–1,000
Nylon (polyamide) type 12	452
Nylon (polyamide) type 6/6 (20–40% glass filler)	470–580
Polytetrafluoroethylene (TFE)	480
Nylon (polyamide) type 11 (glass filler)	485
Polyphenylene (40% glass filler)	490
Acrylics multipolymer	493
Acetal copolymer	500
Nylons (polyamide) type 6/10 (20–40% glass filler)	500
Phenol formaldehyde type 6/12 (glass filler)	500–520
Fluorocarbon (FEP)	500–600
Chlorotrifluoroethylene (CTFE)	500–600
Polypropylene (impact rubber modified)	500–650
Polypropylene copolymer	500–660
Polypropylene (unmodified)	500–660
Polystyrene (general-purpose)	500–700
Styrene acrylonitrile (glass filler)	510
Thermoplastic polyester	560
Polyimides aromatic	560
Acetal (20% glass filler)	580
Polyphenylene sulfides (unfilled)	595
ABS modified (rigid)	600
Ethylene vinyl acetate copolymer	620–780
Polyallomer	800–950
Ionomers	900–1,000
PVC (rigid)	1,500

Thermal Classifications of Insulating Materials

Thermal classifications of electrical-insulating materials are particularly important in modern electronic systems, which often have to operate at elevated temperatures. Most electrical properties of insulators are a function of temperature. Depending on the material, these properties may vary gradually in a given direction with temperature, may alternate to some degree over a temperature range, or may change drastically beyond a critical temperature region.

For example, resistance and resistivity values generally decrease as temperature increases. Dissipation factor and dielectric constant, on the other hand, generally increase with temperature, although for many plastics they increase rapidly beyond a critical temperature region, which is different for each material.

In addition to the stability of electrical properties with temperature variation, physical stability as a function of temperature is also an important factor in analyzing the performance of insulating materials. This physical stability, together with

TABLE 3   Step-by-Step Dielectric Strength of Plastics in Volts per Mil
(ASTM D 149, ⅛ in. Thick)[5]

Plastics	V/mil
Ionomers	2.4–2.5
Cold-mold nonrefractory (organic)	50–75
Cold-mold refractory (inorganic asbestos)	50–80
Cold-mold (inorganic) asbestos, binder cement	50–80
Cold-mold (organic) asbestos, binder melamine	50–100
Phenolic (glass filler)	120–400
Silicones (glass filler)	125–300
Phenolic (fabric and cord filler)	150–300
Phenolic acrylonitrile copolymer (rag filler)	175–225
Melamine (fabric filler)	200–300
Polyester alkyd granular putty type (glass filler)	200–400
Cellulose acetate	200–520
Urea formaldehyde	220–300
Cellulose acetate butyrate	225–320
Phenolic acrylonitrile copolymer (wood-flour and flock filler)	225–275
PVC flexible (filled)	225–750
Phenol formaldehyde (asbestos filler)	230–310
Cellulose propionate	230–385
Melamine alpha cellulose	240–270
Phenolic acrylonitrile copolymer (asbestos filler)	250–300
Phenol formaldehyde (cotton filler)	250–350
Phenol formaldehyde (wood-flour filler)	250–350
Melamine cellulose	250–350
Phenolic (mica filler)	250–390
Epoxy encapsulating (glass filler)	250–400
Epoxy encapsulating (mineral filler)	260–400
Cellulose nitrate	250–550
PVC flexible (unfilled)	275–290
Polyester (premixed glass filler)	275–390
Melamine (asbestos filler)	280–320
Polyester alkyd granular putty type (asbestos filler)	290
Epoxy (glass filler)	300–400
Epoxy (mineral filler)	300–400
Polyester alkyd granular putty type (mineral filler)	300–375
Polystyrene (heat and chemical)	300–500
Ethyl cellulose	300–500
Polystyrene (medium- and high-impact)	300–600
Acetal homopolymer	320
Diallyl phthalate (mineral filler)	320–350
Polyester alkyd granular putty type (synthetic filler)	325–450
Polypropylene modified vinyl chloride (rigid)	335–380
Nylons (polyamide) type 6/6	340–410
Diallyl phthalate (glass filler)	340–400
Nylons (polyamide) type 6	340–440
Acrylics	350–400
Polyester (sheet-molding compound)	350–400
Polystyrene (glass filler)	350–430
ABS (high-impact)	350–450
ABS polycarbonate alloy	350–460
ABS (heat-resistant)	360–400
Polycarbonate (unfilled)	364
ABS (medium-impact)	370–400
Diallyl phthalate (synthetic-fiber filler)	370–410
Epoxy (low-density)	375–400
PVC (rigid)	375–750
Silicones (mineral filler)	380
Nylons (polyamide) type 6 (20–40% glass filler)	380–460
Nylons (polyamide) type 6/6 (20–40% glass filler)	390–460
Polycarbonate (10–40% glass filler)	400
Polysulfone	400

TABLE 3   Step-by-Step Dielectric Strength of Plastics in Volts per Mil (ASTM D 149, ⅛ in. Thick)[5] (Continued)

Plastics	V/mil
Nylons (polyamide) type 6/10 (30–40% glass filler)	400–450
Acrylics (impact)	400–500
Polyphenylene oxide (modified unfilled)	400–550
Polystyrenes (general-purpose)	400–600
Nylons (polyamide) type 12	409
Styrene acrylonitrile (glass filler)	410
Acrylics multipolymer	416
Styrene butadiene thermoplastic elastomers	420–540
Polyethylene (low-density)	420–700
Polytetrafluoroethylene (TFE)	430
Polyethylene (high-density)	440–600
Polycarbonate (10% glass filler)	450
Chlorotrifluoroethylene (CTFE)	450–550
Polypropylene (unmodified)	450–650
Polypropylene (copolymer filler)	450–600
Polypropylene (impact rubber modified)	450–600
Urethane elastomers	460
Phenol formaldehyde type 6/12 (glass filler)	480–490
Polyethylene (medium-density)	500–700
Ethylene vinyl acetate copolymer	620–780
Methylpentene polymer	700

the electrical stability, often is expressed by the continuous-use temperature. Although this rated value implies no specific property at the stated temperature, it is generally accepted that a material will perform without major physical or electrical instability at the specified continuous-use temperature.

One widely accepted temperature-classification system is the IEEE thermal classification of electrical-insulating materials, shown in Table 10. A material properly rated into one of these classifications can be expected to operate continuously and

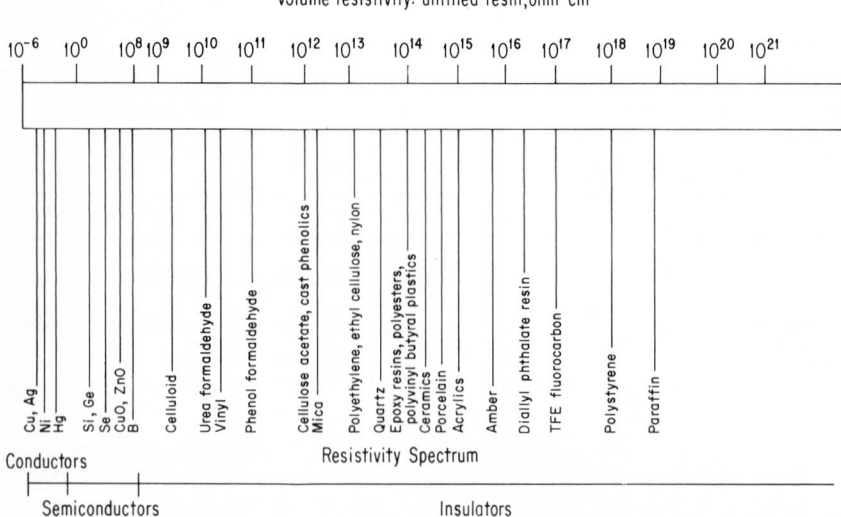

Volume resistivity: unfilled resin, ohm-cm

Fig. 1   Resistivity spectrum for a variety of engineering materials, in Ω-cm.[6]

reliably, in an electrical-insulating application, up to the maximum operating temperature of its classification for a predetermined period (usually 20,000 h but sometimes up to 100,000 h). It should be noted that this system implies performance only as electrical insulation and does not necessarily cover satisfactory mechanical performance. The IEEE classification system has especially wide use in rating wire insulation, as shown by the typical curves of Fig. 2. An approximation of the maximum continuous-use thermal ratings for plastics is shown in Table 11.

Electrical Tests for Insulation Properties

As was indicated above, ASTM tests are most commonly used for standardization of data in this subject area. Also, government-sponsored federal standards are used. Standardization is vital if data are to be compared. A listing of the federal and ASTM tests for electrical-insulation testing is given in Table 12. Some pertinent discussion on electrical-insulation tests is given below.

Dielectric strength   Dielectric strength is the voltage gradient at which electric failure results. The failure is characterized by an excessive flow of current (arc) and by partial destruction of the material. Dielectric strength is measured through the thickness of the material, as shown in Fig. 3, and is expressed in volts per unit of thickness. Two principal methods of measuring this value exist. The short-time test is performed by increasing the voltage from zero at a predetermined speed (100 to 3,000 V/s) to breakdown. The rate is specified by the investigator. The step-by-step test differs in that an initial voltage of 50 percent of the short-time voltage is applied to the material, and then the voltage is increased in equal increments and held for periods of time as determined by the investigator. Dielectric strength is influenced by temperature, humidity, voids or foreign materials in the specimen, electrode configuration, frequency, and specimen geometry. Because of this it is often difficult to compare breakdown data from different sources unless all test conditions are known.

Dielectric constant.   Dielectric constant is the ratio of the capacitance formed by two plates with a material between them and the capacitance of those same plates with a vacuum between them (Fig. 4). The value is of particular interest to designers of microwave equipment. For example, in radar the thickness of the part is dictated by the frequency, physical loads, and dielectric constant. Most plastics have dielectric constants between 2 and 10. Dielectric constant is related to the temperature and to frequency. For many plastics dielectric constant decreases with frequency and increases with temperature.

Electrical-loss properties   Dissipation factor is the ratio of the parallel reactance to parallel resistance, i.e., the tangent of the loss angle or the cotangent of the phase angle. Figure 5 shows this relationship. It should be noted that dissipation factor (tan δ) is not identical to power factor (cos θ). However, at low values of tan δ (less than 0.10) the values for power factors are nearly identical. The dissipation factor is related to the energy dissipated and hence to the efficiency of the insulation material. Some designers use another term, loss factor, which is the product of the dielectric constant and the dissipation factor. This term is related to the total loss of power (in watts) occurring in the insulation material. Dissipation factors for most plastics tend to decrease with increasing frequency and increase with increasing temperature. Since this property may vary widely through any temperature or frequency range, care should be used with reported values that do not include the test conditions.

Resistance and resistivity properties   Insulation resistance is the ratio of voltage applied to a sample to the current flowing in the sample. It is made up of two components: surface resistance and volume resistance. Measuring surface and volume resistivities (the specific resistances) involves separating these two components from the insulation resistance. This is accomplished by using three electrodes as shown in Fig. 6. The outermost (guard) electrode is at the same voltage as the measuring electrode.

Surface resistivity is the resistance to current leakage along the surface of an insulator. The units of measurements are in ohms per square. Surface resistivity is very sensitive to humidity, surface cleanliness, and surface contour.

TABLE 4   Volume Resistivity of Plastics in Ohm-Cm at 50% RH and 73°F[5]

Plastic	Ω-cm
ABS (20–40% glass filler)	7×10^4–16×10^4
Ethylene vinyl acetate copolymer	2×10^8
Polyester alkyd granular putty type (asbestos filler)	7×10^8
Polyethylenes cross-linkable grades	15×10^8–200×10^8
Ethylene ethyl acrylate copolymer	3×10^8
Melamine (fabric filler)	10^9–10^{10}
Phenolic acrylonitrile copolymer (wood-flour and flock filler)	10^{10}
Cellulose nitrate	10×10^{10}–15×10^{10}
Phenolic acrylonitrile copolymer (rag filler)	10^{10}–10^{11}
Nylons (polyamide) type 11	2×10^{11}
Phenolic acrylonitrile copolymer (asbestos filler)	3×10^{11}
Phenolic (fabric and cord filler)	10^{11}–10^{12}
Melamine alpha cellulose	10^{12}–2×10^{12}
Melamine (asbestos filler)	1.2–10^{12}
Cold-mold nonrefractory (organic)	1.3×10^{12}
Phenolic (glass filler)	10^{12}–10^{13}
Urea formaldehyde	10^{12}–10^{13}
Phenolic (asbestos filler)	10^9–10^{13}
Phenolic (cotton filler)	10^9–10^{13}
Phenolic (wood-flour filler)	10^9–10^{13}
Diallyl phthalate (mineral filler)	10^{13}
Cellulose acetate	10^{10}–10^{14}
Silicones (asbestos filler)	10^{10}–10^{14}
PVC (flexible filler)	10^{11}–10^{14}
Ethyl cellulose	10^{12}–10^{14}
Phenolic (mica filler)	10^{12}–$>10^{14}$
Urethanes elastomers	2×10^{11}–10^{14}
Epoxy (low-density)	10^{13}–10^{14}
Acetal copolymer	10^{14}
Polyester (preformed chopped rovings)	10^{14}
Polyester woven cloth	10^{14}
Silicones (mineral filler)	10^{14}
Acrylics	$>10^{14}$
Epoxy (glass filler)	$>10^{14}$
Epoxy (mineral filler)	$>10^{14}$
Epoxy encapsulating (glass filler)	$>10^{14}$
Polystyrene (general-purpose)	$>10^{16}$
Polystyrenes (medium-impact)	$>10^{16}$
Polystyrenes (high-impact)	$>10^{16}$
Styrene acrylonitrile copolymer	$>10^{16}$
Vinyl polymers	$>10^{16}$
ABS (high-impact)	10^{16}–5×10^{16}
ABS (heat-resistant)	10^{16}–5×10^{16}
ABS polycarbonate alloy	1×10^{15}–4×10^{16}
Polycarbonate (up to 40% glass filler)	4×10^{16}–5×10^{16}
Styrene butadiene thermoplastic elastomers	3×10^{16}–5×10^{16}
Acrylics impact	2×10^{16}
Acrylics multipolymer	2×10^{16}
Polyaryl ether	2×10^{16}
Polycarbonate (unfilled)	2×10^{16}
ABS (self-extinguishing)	3×10^{16}
Polyaryl sulfone	3×10^{16}

TABLE 4   Volume Resistivity of Plastics in Ohm-Cm at 50% RH and 73°F[5] (Continued)

Plastic	Ω-cm
Polycarbonate (10% glass filler)	3×10^{16}
Epoxy encapsulating (mineral filler)	$> 10^{14}$
Acetal (20% glass filler)	1.2×10^{14}
Polyvinylidene fluoride	2×10^{14}
Polyester (low-shrink)	3×10^{14}
Cellulose acetate butyrate	10^{11}–10^{15}
PVC (flexible unfilled)	10^{11}–10^{15}
Polyester (premixed glass filler)	10^{12}–10^{15}
Nylons (polyamide) type 6	10^{12}–10^{15}
Nylons (polyamide) type 6/10	10^{12}–10^{15}
Polyester alkyd granular putty type (mineral filler)	10^{13}–10^{15}
Cellulose propionate	10^{13}–10^{15}
Nylons (polyamide) type 6/6	10^{14}–10^{15}
Nylons (polyamide) type 6 (20–40% glass filler)	10^{14}–10^{15}
Polyesters sheet-molding compound	10^{14}–10^{15}
Acetal homopolymer	10^{15}
Chlorinated polyether	10^{15}
Nylons (polyamide) type 11 (glass filler)	10^{15}
Nylons (polyamide) type 6/12	10^{15}
Polyester alkyd granular putty type (glass filler)	10^{15}
Polyallomer	$> 10^{15}$
Polypropylene impact rubber modified	$> 10^{15}$
Polypropylene modified vinyl chloride (rigid)	$> 10^{15}$
PVC (rigid)	10^{15}
Nylons (polyamide) type 6/12 (glass filler)	1.4×10^{15}
Polypropylene (inert filler)	2×10^{15}–3×10^{15}
Nylons (polyamide) type 12	3×10^{15}
Polyester alkyd granular putty type (synthetic filler)	10^{8}–10^{15}
Diallyl phthalate (synthetic-fiber filler)	10^{13}–10^{16}
Diallyl phthalate (glass filler)	10^{13}–10^{16}
Nylons (polyamide) type 6/6 (20–40% glass filler)	10^{15}–10^{16}
Methylpentene polymer	10^{16}
Ionomers	$> 10^{16}$
Polyethylenes (low-density)	$> 10^{16}$
Polyethylenes (medium-density)	$> 10^{16}$
Polyethylenes (high-density)	$> 10^{16}$
Polyamide aromatic	10^{16}
Polypropylene unmodified	10^{16}
Thermoplastic polyesters	3×10^{16}
Polystyrenes (20–30% glass filler)	3×10^{16}
Polysulfone	5×10^{16}
Polyester (18% glass filler)	20×10^{16}
Polyphenylene oxides (modified) (unfilled)	$> 10^{16}$–$> 10^{17}$
Polystyrene (heat-resistant)	10^{13}–10^{17}
Polystyrene (chemical)	10^{13}–10^{17}
Polyphenylene oxides (unfilled) PPO	10^{17}
Polypropylene (copolymer filler)	10^{17}
Polyphenylene oxides modified (20–30% glass filler)	$> 10^{17}$
Polytetrafluoroethylene (TFE)	10^{18}
Chlorotrifluoroethylene (CTFE)	10^{18}
Fluorocarbon (FEP)	$> 2 \times 10^{18}$
Styrene acrylonitrile copolymer (glass filler)	1.4×10^{19}–2×10^{19}

TABLE 5   Dielectric Constant of Plastics (ASTM D 150, 60 Hz)[5]

Plastic	Constant
Fluorocarbon (FEP)	2.10
Polytetrafluoroethylene (TFE)	2.10
Methylpentene polymer	2.12
Polypropylene (unmodified)	2.20–2.60
Chlorotrifluoroethylene (CTFE)	2.24–2.80
Polypropylene copolymer	2.25–2.30
Polyethylene (low-density)	2.25–2.35
Polyethylene (medium-density)	2.25–2.35
Polypropylene (impact rubber modified)	2.30
Polyallomer	2.30
Polyethylene (high-density)	2.30–2.35
Polypropylene (glass filler)	2.37
ABS (heat-resistant)	2.40–5.00
ABS (high-impact)	2.40–5.00
ABS (medium-impact)	2.40–5.00
Ionomers	2.40–2.50
ABS self-extinguishing	2.40–5.00
ABS polycarbonate alloy	2.40–5.00
Polystyrenes (general-purpose)	2.45–2.65
Polystyrene (heat and chemical filler)	2.45–3.40
Polystyrene (medium- and high-impact)	2.45–4.75
Polypropylene (inert filler)	2.50–2.75
Styrene butadiene thermoplastic elastomers	2.50–3.40
Ethylene vinyl acetate copolymer	2.50–3.16
Polyphenylene oxides (unfilled) (PPO)	2.58
Styrene acrylonitrile copolymer	2.60–3.40
Polyphenylene oxides (unfilled) (modified)	2.64
Ethylene ethyl acrylate copolymer	2.70–2.90
Polyphenylene oxides (modified) (20–30% glass filler)	2.93
Polycarbonate (unfilled)	2.97–3.17
Ethyl cellulose	3.00–4.20
Diallyl phthalate (synthetic-fiber filler)	3.00–8.00
Chlorinated polyether	3.10
Polycarbonate (10% glass filler)	3.10
Polyaryl ether	3.14
Polysulfone	3.14
Polypropylene modified vinyl chloride (rigid)	3.10–3.70
PVC (rigid)	3.20–3.60
Polycarbonate (10–40% glass filler)	3.00–3.53
Acrylics multipolymer	3.30–3.50
Acrylics	3.30–3.90
Silicones (glass filler)	3.30–5.20
Polyimides aromatic	3.40
Silicones (mineral filler)	3.50–3.60
Acrylics (impact)	3.50–4.00
Epoxy (glass filler)	3.50–5.00
Epoxy (mineral filler)	3.50–5.00
Epoxy encapsulating (glass filler)	3.50–5.00
Epoxy encapsulating (mineral filler)	3.50–5.00

TABLE 5 Dielectric Constant of Plastics (ASTM D 150, 60 Hz)[5] (Continued)

Plastic	Constant
Cellulose acetate butyrate	3.50–6.40
Cellulose acetate	3.50–7.50
Styrene acrylonitrile (glass filler)	3.60
Thermoplastic polyesters	3.65
Polyesters (18% glass filler)	3.70
Acetal copolymer	3.70
Cellulose propionate	3.70–4.30
Polyester alkyd granular putty type (synthetic filler)	3.80–5.00
Polyester (preformed chopped rovings)	3.80–6.00
Acetal (20% glass filler)	3.90
Nylons (polyamide) type 6/10	3.90
Nylons (polyamide) type 6	3.90–5.50
Polyaryl sulfone	3.94
Acrylic-PVC alloy	4.00
Nylons (polyamide) (20–40% glass filler)	4.00–4.40
Nylons (polyamide) type 6/6	4.00–4.60
Polyester woven cloth	4.10–5.50
Nylon (polyamide) (20–40% glass filler) type 6/10	4.20
Nylons (polyamide) type 12	4.20
Diallyl phthalate (glass filler)	4.30–4.60
Nylons (polyamide) type 6 (20–40% glass filler)	4.40–4.60
Polyesters (sheet-molding compound)	4.40–6.30
Phenolic (mica filler)	4.70–6.00
PVC flexible (filled)	5.00–6.00
Phenolic (glass filler)	5.00–7.10
PVC flexible (unfilled)	5.00–9.00
Phenol formaldehyde (wood-flour filler)	5.00–13.00
Phenol formaldehyde (asbestos filler)	5.00–20.00
Celanese CDY-246	5.15
Polyester alkyd granular putty type (mineral filler)	5.10–7.50
Diallyl phthalate (mineral filler)	5.20
Phenolic (fabric and cord filler)	5.20–21.00
Polyester (low-shrink)	5.30
Polyester (premixed glass filler)	5.30–7.30
Urethanes elastomers	5.40–7.60
Polyester alkyd granular putty type (glass filler)	5.70
Melamine cellulose	6.20–7.60
Melamine (asbestos filler)	6.40–10.20
Cellulose nitrate	7.00–7.50
Melamine (phenol filler)	7.00–7.70
Urea formaldehyde	7.00–9.50
Melamine (fabric filler)	7.60–12.60
Melamine alpha cellulose	7.90–9.50
Polyvinylidene fluoride	8.40
Phenolic acrylonitrile copolymer (wood-flour and flock filler)	9.00–15.00
Melamine (fabric phenolic filler) (modified)	10.50–15.50
Phenolic acrylonitrile copolymer (rag filler)	13.00–17.00
Phenolic acrylonitrile copolymer (asbestos filler)	15.00
Cold-mold nonrefractory (organic)	17.00 (Test D 758)

TABLE 6   Dielectric Constant of Plastics (ASTM D 150, 1 MHz)[5]

Plastic	Constant
Epoxy (low-density)	2.00–3.00
Fluorocarbon (FEP)	2.10
Polytetrafluoroethylene (TFE)	2.10
Methylpentene polymer	2.12
Acrylics	2.20–3.20
Polypropylene (unmodified)	2.20–2.60
Polypropylene (copolymer filler)	2.24–2.30
Polybutylene	2.25
Polyethylene (low-density)	2.25–2.35
Polyethylene (medium-density)	2.25–2.35
Polyallomer	2.30
Polypropylene (impact rubber modified)	2.30
Polyethylene (high-density)	2.30–2.35
Chlorotrifluoroethylene (CTFE)	2.30–2.50
Polypropylene (glass filler)	2.38
Polystyrene (general-purpose)	2.40–2.65
Polystyrene (heat and chemical)	2.40–3.10
Polystyrene (medium- and high-impact)	2.40–3.80
ABS (high-impact)	2.40–3.80
ABS (heat-resistant)	2.40–3.80
ABS (medium-impact)	2.40–3.80
ABS (self-extinguishing)	2.40–3.80
ABS polycarbonate alloy	2.40–3.80
Acrylics impact	2.50–3.00
Polypropylene (inert filler)	2.50–2.60
Styrene butadiene thermoplastic elastomers	2.50–3.40
Polyphenylene oxides (unfilled) (PPO)	2.58
Styrene acrylonitrile copolymer	2.60–3.10
Ethylene vinyl acetate copolymer	2.60–3.20
Polyethylene cross-linkable grades	2.63–3.11
Polyphenylene oxide modified (unfilled)	2.64
Ethylene ethyl acrylate copolymer	2.70–2.80
Acrylics multipolymer	2.80–2.90
PVC (rigid)	2.80–3.10
Polypropylene modified vinyl chloride (rigid)	2.80–3.30
Ethyl cellulose	2.80–3.90
Chlorinated polyether	2.90
Polyphenylene oxides modified (20–30% glass filler)	2.92
Polycarbonate (unfilled)	2.92–2.93
Polycarbonate (10–40% glass filler)	3.00–3.48
Polycarbonate (10% glass filler)	3.05
Polyaryl ether	3.10
Nylon (polyamide) type 12	3.10
Polysulfone	3.10
Nylons (polyamide) type 11	3.20
Silicones (glass filler)	3.20–4.70
Cellulose acetate butyrate	3.20–6.20
Cellulose acetate	3.20–7.00
Polyphenylene sulfides (unfilled)	3.22
Diallyl phthalate (synthetic-fiber filler)	3.30–3.60
Cellulose propionate	3.30–3.80
PVC flexible (unfilled)	3.30–4.50
Polyesters (thermoplastic)	3.37

TABLE 6   Dielectric Constant of Plastics (ASTM D 150, 1 MHz)[5] (Continued)

Plastic	Constant
Polyimides aromatic	3.40
Styrene acrylonitrile (glass filler)	3.40
Acrylic-PVC Alloy	3.40
Nylons (polyamide) type 6/6	3.40–3.60
Phenol formaldehyde type 6/12 (glass filler)	3.40–4.00
Diallyl phthalate (glass filler)	3.40–4.50
Silicones (mineral filler)	3.40–6.30
Nylons (polyamide) type 6/10 (20–40% glass filler)	3.50
Nylon (polyamide) type 6/10	3.50
Polyesters (18% glass filler)	3.50
PVC flexible (filled)	3.50–4.50
Nylons (polyamide) (20–40% glass filler)	3.50–4.10
Nylon (polyamide) type 6	3.50–4.70
Epoxy (glass filler)	3.50–5.00
Epoxy (mineral filler)	3.50–5.00
Epoxy encapsulating (glass filler)	3.50–5.00
Epoxy encapsulating (mineral filler)	3.50–5.00
Polyester (preformed chopped rovings)	3.50–5.50
Polyester alkyd granular putty type (synthetic filler)	3.60–4.70
Acetal homopolymer	3.70
Acetal copolymer	3.70
Polyaryl sulfone	3.70
Polyphenylene sulfides (40% glass filler)	3.88
Acetal (20% glass filler)	3.90
Diallyl phthalate (mineral filler)	3.90–4.00
Nylon (polyamide) type 6 (20–40% glass filler)	3.90–4.00
Phenol formaldehyde type 6/12	4.00
Phenolic acrylonitrile copolymer (rag filler)	4.00–6.00
Polyester woven cloth	4.00–5.50
Phenol formaldehyde (wood-flour filler)	4.00–6.00
Phenolic (mica filler)	4.20–5.20
Urethanes elastomers	4.20–5.10
Polyesters (sheet-molding compound)	4.20–5.80
Polyester alkyd granular putty type (asbestos filler)	4.50
Phenolic (glass filler)	4.50–6.60
Phenolic (fabric and cord filler)	4.50–7.00
Polyester alkyd granular putty type (mineral filler)	4.60–6.10
Polyester (low-shrink)	4.70
Melamine cellulose	4.70–7.00
Phenolic acrylonitrile copolymer (wood-flour and flock filler)	5.00
Phenolic acrylonitrile copolymer (asbestos filler)	5.00
Phenol formaldehyde (asbestos filler)	5.00–10.00
Melamine phenol filler	5.20–6.00
Polyester (premixed glass filler)	5.20–6.40
Polyester alkyd granular putty type (glass filler)	5.20–6.80
Cold-mold nonrefractory (organic)	6.00
Melamine (asbestos filler)	6.10–6.70
Melamine (fabric filler phenolic modified)	6.10–6.70
Cellulose nitrate	6.40
Melamine (fabric filler)	6.50–6.90
Polyvinylidene fluoride (VF2)	6.60
Urea formaldehyde	6.80
Melamine alpha cellulose	7.20–8.40

TABLE 7   Dissipation Factor of Plastics (ASTM D 150, 60 Hz)[5]

Plastic	Factor
Methylpentene polymer (TPX)	0.00007
Polystyrenes (general-purpose)	0.0001–0.0003
Polypropylene (copolymer filler)	>0.0001–0.0005
Polypropylene (impact rubber modified)	>0.0003
Polyphenylene oxides (unfilled) (PPO)	0.00035
Polyphenylene oxides (modified unfilled)	0.0004
Polystyrene (medium- and high-impact)	0.0004–0.0020
Polyethylenes (medium-density)	>0.0005
Polyethylenes (high-density)	>0.0005
Polyethylenes (low-density)	>0.0005
Polypropylene (unmodified)	>0.0005
Polyallomer	$>5 \times 10^{-4}$
Polystyrene (heat-resistant)	0.0005–0.0030
Polystyrene (chemical filler)	0.0005–0.0030
Polysulfone	0.0008
Polycarbonate (10% glass filler)	0.0008
Polycarbonate (unfilled)	0.0009
Polyphenylene (oxides modified) (20–30% glass filler)	0.0009
Polycarbonate (10–40% glass filler)	0.0009–0.0013
Ionomers	0.001–0.003
Styrene butadiene thermoplastic elastomers	0.002–0.003
Polypropylene (glass filler)	0.0022
Polyaryl sulfone	0.003
ABS polycarbonate alloy	0.003–0.007
ABS (medium-impact)	0.003–0.0098
ABS (self-extinguishing)	0.003–0.008
ABS (20–40% glass filler)	0.003–0.008
ABS (heat-resistant)	0.003–0.008
ABS (high-impact)	0.003–0.008
Ethylene vinyl acetate copolymer (EVA)	0.003–0.020
Silicones (mineral filler)	0.004–0.005
Polystyrenes (20–30% glass filler)	0.004–0.014
Silicones (glass filler)	0.004–0.030
Polyesters (18% glass filler)	0.005
Ethyl cellulose	0.005–0.020
Polypropylene (inert filler)	0.0054–0.0070
Thermoplastic polyesters	0.0055
Polyaryl ether	0.005
Styrene acrylonitrile copolymer (SAN)	0.006–0.008
Styrene acrylonitrile copolymer (glass filler)	0.00645
PVC rigid	0.007–0.020
Polyesters (sheet-molding compound)	0.007–0.021
Polypropylene modified vinyl chloride (rigid)	0.008–0.010
Nylon (polyamide) type 6/6 (20–40% glass filler)	0.009–0.023
Polyester alkyd granular putty type (mineral filler)	0.009–0.06
Chlorinated polyether	0.01
Epoxy (glass filler)	0.01

TABLE 7   Dissipation Factor of Plastics (ASTM D 150, 60 Hz)[5] (Continued)

Plastic	Factor
Epoxy (mineral filler)	0.01
Epoxy encapsulating (glass filler)	0.01
Epoxy encapsulating (mineral filler)	0.01
Polyester alkyd granular putty type (glass filler)	0.010
Ethylene ethyl acrylate copolymer (EEA)	0.01–0.02
Polyester woven cloth	0.01–0.04
Polyester (preformed chopped rovings)	0.01–0.04
Cellulose acetate butyrate	0.01–0.04
Cellulose propionate	0.01–0.04
Diallyl phthalate (glass filler)	0.01–0.05
Cellulose acetate	0.01–0.06
Polyester (premixed chopped glass filler)	0.011–0.041
Polyester alkyd granular putty type (synthetic filler)	0.012–0.026
Nylons (polyamide) type 6/10 (20–40% glass filler)	0.013–0.026
Nylons (polyamide) type 6/6	0.014–0.04
Urethanes elastomers	0.015–0.048
Polyester (low-shrink)	0.019
Melamine (cellulose filler)	0.019–0.033
Nylons (polyamide) type 6 (20–40% glass filler)	0.020–0.024
Acrylics multipolymer	0.02–0.03
Melamine phenol	0.02–0.04
Polyterephthalate 6PRO	0.023
Diallyl phthalate synthetic fiber	0.026
Phenolic (mica filler)	0.03–0.05
Diallyl phthalate (mineral filler)	0.03–0.06
Melamine alpha cellulose	0.03–0.083
Urea formaldehyde	0.035–0.043
Acrylic-PVC	0.04
Nylons (polyamide) type 12	0.04
Phenolic (glass filler)	0.04–0.05
Acrylics impact	0.04–0.05
Acrylics	0.04–0.06
Nylons (polyamide) type 6	0.04–0.06
Phenolic formaldehyde (asbestos filler)	0.05–0.20
Phenolic formaldehyde (wood-flour filler)	0.05–0.30
Melamine (asbestos filler)	0.07–0.17
Melamine (fabric filler)	0.07–0.34
PVC (flexible unfilled)	0.08–0.15
Phenolic (fabric and cord filler)	0.08–0.64
Cellulose nitrate	0.09–0.12
PVC (flexible filler)	0.10–0.15
Melamine (phenolic modified fabric filler)	0.10–0.16
Phenolic acrylonitrile copolymer (wood-flour and flock filler)	0.14–0.50
Phenolic acrylonitrile copolymer (asbestos filler)	0.15
Cold-mold nonrefractory (organic)	0.20
Phenolic acrylonitrile copolymer (rag filler)	0.30–0.50

TABLE 8   Dissipation Factor of Plastics (ASTM D 150, 1 MHz)[5]

Plastic	Factor
Methylpentene polymer (TPX)	0.000025
Polystyrenes (general-purpose)	0.0001–0.0004
Polypropylene (copolymer filler)	>0.0001–0.0018
Polypropylene impact rubber modified	>0.0003
Polystyrene (medium- and high-impact)	0.0004–0.0020
Polyallomer	$>5 \times 10^{-4}$
Polyethylenes (low-density)	>0.0005
Polyethylenes (medium-density)	>0.0005
Polyethylenes (high-density)	>0.0005
Polypropylene (unmodified)	>0.0005–0.0018
Polystyrenes (heat and chemical)	0.0005–0.0050
Polyphenylene sulfides (unfilled)	0.0007
Polyphenylene oxides (unfilled) PPO	0.0009
Polyphenylene oxides (modified unfilled)	0.0009
Polyethylenes cross-linkable grades	0.001–0.002
Polystyrenes (20–30% glass filler)	0.001–0.003
Styrene butadiene thermoplastic elastomers	0.001–0.003
Polyphenylene oxides (modified 20–30% glass filler)	0.0015
Ionomers	0.0019
Silicones (mineral filler)	0.002–0.005
Silicones (glass filler)	0.002–0.020
Polypropylene (inert filler)	0.0021
Polysulfone	0.0034
Polypropylene (glass filler)	0.0035
Acrylics (impact)	0.004–0.02
Polyphenylene (40% glass filler)	0.0041
Acetal homopolymer	0.0048
Polyamides aromatic	0.005
Polybutylene	0.005
Epoxy (low-density)	0.005–0.012
Phenolic (mica filler)	0.005–0.013
Acetal copolymer	0.006
ABS polycarbonate alloy	0.006–0.013
PVC rigid	0.006–0.019
Polyester alkyd granular putty type (mineral filler)	0.006–0.04
Acetal (20% glass filler)	0.0062
Styrene acrylonitrile (glass filler)	0.00668
Polycarbonate (10–40% glass filler)	0.0067–0.0075
Polyaryl ether	0.007
Styrene acrylonitrile copolymer (SAN)	0.007–0.010
ABS (high-impact)	0.007–0.015
ABS (heat-resistant)	0.007–0.015
ABS (medium-impact)	0.007–0.015
ABS (self-extinguishing)	0.007–0.015
ABS (20–40% glass filler)	0.007–0.015
Polycarbonate (10% glass filler)	0.0075
Polyester (premixed chopped glass filler)	0.008–0.022
Polyester alkyd granular putty type (glass filler)	0.008–0.023
Diallyl phthalate (glass filler)	0.009–0.014
Chlorinated polyether	0.01
Epoxy (glass filler)	0.01

TABLE 8 Dissipation Factor of Plastics (ASTM D 150, 1 MHz)[5] (Continued)

Plastic	Factor
Epoxy (mineral filler)	0.01
Epoxy encapsulating (glass filler)	0.01
Epoxy encapsulating (mineral filler)	0.01
Phenolic acrylonitrile copolymer (asbestos filler)	0.010
Polycarbonate (unfilled)	0.010
Polyester alkyd granular putty type (synthetic filler)	0.010–0.016
Ethylene ethyl acrylate copolymer (EEA)	0.01–0.02
Polypropylene modified vinyl chloride (rigid)	0.01–0.02
Phenolic (glass filler)	0.010–0.026
Polyester woven cloth	0.01–0.03
Polyester preformed chopped rovings	0.01–0.03
Cellulose acetate butyrate	0.01–0.04
Cellulose propionate	0.01–0.05
Ethyl cellulose	0.010–0.060
Cellulose acetate	0.01–0.10
Diallyl phthalate (synthetic-fiber filler)	0.012–0.020
Polyaryl sulfone	0.013
Nylons (polyamide) type 6/10 (20–40% glass filler)	0.016–0.022
Polyester sheet-molding compound	0.016–0.024
Nylons (polyamide) type 6/6 (20–40% glass filler)	0.017–0.018
Polyester (low-shrink)	0.019
Nylons (polyamide) type 6 (20–40% glass filler)	0.019–0.021
Acrylics multipolymer	0.02
Nylons (polyamide) type 6/12	0.02
Polyesters (18% glass filler)	0.0120
Acrylics	0.02–0.03
Diallyl phthalate (mineral filler)	0.02–0.04
Thermoplastic polyesters	0.0208
Melamine alpha cellulose	0.027–0.045
Acrylic-PVC	0.03
Nylons (polyamide) type 12	0.03
Ethylene vinyl acetate copolymer EVA	0.03–0.05
Phenolic formaldehyde (cotton filler)	0.03–0.07
Phenolic formaldehyde (wood-flour filler)	0.03–0.07
Phenolic (fabric and cord filler)	0.03–0.09
Melamine (cellulose filler)	0.032–0.060
Melamine (fabric filler)	0.036–0.041
Melamine (phenolic filler)	0.04–0.06
Polyester alkyd granular putty type (asbestos filler)	0.04–0.06
Melamine (asbestos filler)	0.041–0.050
PVC (flexible unfilled)	0.04–0.14
Melamine (phenolic modified fabric filler)	0.050–0.065
Urethanes elastomers	0.050–0.075
Cellulose nitrate	0.06–0.09
Cold-mold nonrefractory (organic)	0.07
Phenolic acrylonitrile copolymer (wood-flour and flock filler)	0.08–0.10
Phenolic acrylonitrile copolymer (rag filler)	0.08–0.15
PVC (flexible filler)	0.09–0.10
Nylon (polyamide) type 6/12 (glass filler)	0.16–0.17
Urea formaldehyde	0.25–0.35
Phenolic formaldehyde (asbestos filler)	0.35–0.80

TABLE 9   Arc Resistance of Plastics in Seconds (ASTM D 495)[5]

Plastic	Value
Phenolic (glass filler)	4–190
Polycarbonate (10% glass filler)	5–120
Polycarbonate (10–40% glass filler)	5–120
Polycarbonate (unfilled)	10–120
Phenol formaldehyde (asbestos filler)	10–190
Styrene acrylonitrile copolymer (glass filler)	16
Polystyrene (medium-impact)	20–100
Polystyrene (high-impact)	20–100
Acrylic-PVC	25
Polystyrene (20–30% glass filler)	25–40
ABS (self-extinguishing)	25–50
Polyvinylidene fluoride	50–70
ABS (high-impact)	50–85
ABS (heat-resistant)	50–85
ABS (medium-impact)	50–85
Cellulose acetate	50–310
Polystyrene (general-purpose)	60–80
PVC (rigid)	60–80
Ethyl cellulose	60–80
Polyester (woven cloth)	60–120
Polystyrene (chemical filler)	60–135
Polyaryl sulfone	67
Polyphenylene oxide modified (20–30% glass filler)	70–120
ABS polycarbonate alloy	70–120
Polypropylene (glass filler)	74
Polyphenylene oxide modified (unfilled)	75
Polysulfone	75–122
Polyester alkyd granular putty type (mineral filler)	75–190
Cold-mold nonrefractory (organic)	75–200
Urea formaldehyde	80–150
Polyester alkyd granular putty type (synthetic filler)	85–185
Ionomers	>90
Type 6/10 (20–40% glass filler)	92–148
Styrene butadiene thermoplastic elastomers	95
Melamine (cellulose filler)	95–135
Styrene acrylonitrile copolymer (SAN)	100–150
Melamine (fabric filler phenolic modified)	100–200
Cold-mold refractory asbestos (inorganic)	100–500
Melamine alpha cellulose	110–140
Melamine (flock filler)	110–150
Melamine (fabric filler)	115–125
Diallyl phthalate (synthetic-fiber filler)	115–130
Epoxy (low-density)	120–150
Epoxy (glass filler)	120–180
Epoxy encapsulating (glass filler)	120–180
Epoxy encapsulating (mineral filler)	120–180
Polyester (preformed chopped rovings)	120–180
Melamine (asbestos filler)	120–180
Polyester (premixed chopped glass filler)	120–240
Polypropylene (inert filler)	121
Polyester (sheet-molding compound)	121–202
Urethanes elastomers	122
Diallyl phthalate (glass filler)	125–180
Acetal homopolymer	129
Thermoplastic polyester	130
Type 6/6 (20–40% glass filler)	130–140
Melamine (phenol filler)	130–180
Polyethylene (low-density)	135–160
Acetal (20% glass filler)	136
Polypropylene (copolymer filler)	136
Polypropylene (unmodified)	136–185
Polyester alkyd (asbestos filler)	138
Diallyl phthalate (mineral filler)	140–190

TABLE 9 Arc Resistance of Plastics in Seconds (ASTM D 495)[5] (Continued)

Plastic	Value
Epoxy (mineral filler)	150–190
Polyester alkyd (glass filler)	150–210
Silicones (glass filler)	150–250
Fluorocarbon (FEP)	>165
Cellulose propionate	175–190
Polyaryl ether	180
Polytetrafluoroethylene (TFE)	>200
Polyethylene (medium-density)	200–235
Polyimides aromatic	230
Acetal copolymer	Burn–240
Silicones (mineral filler)	250–420
Cold-mold (organic) asbestos, binder melamine	257
Chlorotrifluoroethylene (CTFE)	>360
Cold-mold (inorganic) asbestos, binder cement	400
Cold-mold (organic) asbestos, binder phenolic resin	450–500
Cold-mold (inorganic) asbestos, binder phosphoric acid	500
Acrylics	No track
Acrylics (impact)	No track
Phenol formaldehyde (wood-flour filler)	Track
Phenol formaldehyde (cotton filler)	Track
Phenol (mica filler)	Track
Phenol (fabric and cord filler)	Track
Phenolic acrylonitrile copolymer (wood-flour and flock filler)	Track
Phenolic acrylonitrile copolymer (asbestos filler)	Track
Phenolic acrylonitrile copolymer (rag filler)	Track

TABLE 10 Definitions of Insulating-Material Classes

Class	Definition
90°C (class O)	Materials or combinations of materials such as cotton, silk, and paper without impregnation. Other materials or combinations of materials may be included in this class if, by experience or accepted tests, they can be shown to be capable of operation at 90°C
105°C (class A)	Materials or combinations of materials such as cotton, silk, and paper when suitably impregnated or coated or when immersed in a dielectric liquid such as oil. Other materials or combinations of materials may be included in this class if, by experience or accepted tests, they can be shown to be capable of operation at 105°C
130°C (class B)	Materials or combinations of materials such as mica, glass fiber, and asbestos with suitable bonding substances. Other materials or combinations of materials, not necessarily inorganic, may be included in this class if, by experience or accepted tests, they can be shown to be capable of operation at 130°C
155°C (class F)	Materials or combinations of materials such as mica, glass fiber, and asbestos with suitable bonding substances. Other materials or combinations of materials, not necessarily inorganic, may be included in this class if, by experience or accepted tests, they can be shown to be capable of operation at 155°C
180°C (class H)	Materials or combinations of materials such as silicone elastomer, mica, glass fiber, and asbestos with suitable bonding substances such as appropriate silicone resins. Other materials or combinations of materials may be included in this class if, by experience or accepted tests, they can be shown to be capable of operation at 180°C
220°C	Materials or combinations of materials which, by experience or accepted tests, can be shown to be capable of operation at 220°C
Over 220°C (class C)	Insulation which consists entirely of mica, porcelain, glass, quartz, and similar inorganic materials. Other materials or combinations of materials may be included in this class if, by experience or accepted tests, they can be shown to be capable of operation at temperatures over 220°C

TABLE 11   Maximum Continuous-Use Temperature Ratings (°F) for Plastics[5]

Plastic	Value
Ethyl cellulose	115–185
Styrene butadiene thermoplastic elastomers	130–150
PVC flexible (unfilled)	130–150
PVC flexible (filled)	130–150
ABS self-extinguishing	130–180
ABS (high-impact)	140–210
Acrylics	140–200
Acrylic impact	140–195
Cellulose acetate	140–220
Cellulose acetate butyrate	140–220
Cellulose nitrate	140
Polystyrene (medium-impact)	140–175
Polystyrene (high-impact)	140–175
Styrene acrylonitrile copolymer (unfilled)	140–205
Polystyrene general-purpose	150–170
Polystyrene heat	150–170
PVC (rigid)	150–175
Polypropylene modified vinyl chloride (rigid)	150–175
Cellulose propionate	155–220
ABS (medium-impact)	160–200
Ionomers	160–200
Acrylics multipolymer	165–175
Polystyrene heat and chemical	170–200
Urea formaldehyde	170
Nylon type 6 (polyamide)	175–250
Nylon type 12 (polyamide)	175–260
Modified phenylene oxides (Noryl)	175–200
Polyesters (thermoplastic)	175–200
Nylon type 6/6 (polyamide)	180–300
Nylon type 6 (20–40%) (polyamide) (glass filler)	180–300
Nylon type 6/10 (polyamide)	180–250
Nylon type 11 (polyamide)	180–300
Polyethylene (low-density)	180–212
Polystyrene (glass filler)	180–200
Acetal 20% (glass filler)	185–220
ABS (heat-resistant)	190–230
Polypropylene copolymer	190–240
Urethanes elastomers	190
Acetal homopolymer	195
ABS (20–40% glass filler)	200–230
Acrylonitrile copolymer (wood-flour and flock filler)	200
Acrylonitrile copolymer wood-flour (rag)	200
Polypropylene impact (rubber modified)	200–250
Styrene acrylonitrile (glass filler)	200–220
Melamine alpha cellulose	210
Acetal copolymer	220
Acetal copolymer (25% glass filler)	220
Nylon type 6/12 (polyamide)	220
Phenol formaldehyde (fabric cord)	220–250
ABS polycarbonate alloy	220–250
Polyethylene (medium-density)	220–250
Acrylonitrile copolymer (asbestos filler)	225
Polypropylene (unmodified)	225–230
Modified phenylene oxides (Noryl) (20–30% glass filler)	240–265
Polycarbonate (unfilled)	240–265
Diallyl phthalate (synthetic fiber filler)	250–350
Melamine cellulose	250
Melamine (flock filler)	250–400

TABLE 11   Maximum Continuous-Use Temperature Ratings (°F) for Plastics[5] (Continued)

Plastic	Value
Melamine (asbestos filler)	250
Melamine (fabric filler)	250
Melamine (fabric phenolic modified)	250
Methylpentene polymer	250–320
Nylon type 6 copolymer (polyimide)	250
Nylon type 12 (polyimide) (glass filler)	250–340
Nylon type 11 (polyimide) (glass filler)	250
Phenol formaldehyde	250
Phenol formaldehyde (mica filler)	250–300
Polyester (thermoplastic) (18% glass filler)	250
Polyethylenes (high-density)	250
Polycarbonate (10% glass filler)	260
Furane (asbestos filler)	265–330
Melamine phenol	275–325
Polycarbonate (up to 40% glass filler)	275
Polyethylene cross-linkable grade	275
Celanex 917	275
Polycarbonate (flame-retardant)	290–300
Polypropylene (inert filler)	290–310
Diallyl phthalate (glass filler)	300–400
Polyvinylidene fluoride	300
Diallyl phthalate (mineral filler)	300–400
Epoxy (glass filler)	300–500
Epoxy (mineral filler)	300–500
Epoxy encapsulating (glass filler)	300–450
Epoxy encapsulating (mineral filler)	300–450
Nylon (polyamide) type 6/10 (20–40% glass filler)	300–400
Phenol formaldehyde (wood-flour filler)	300–350
Phenol formaldehyde (cotton filler)	300–350
Polyester premix (glass filler)	300–350
Polyester alkyd (granular)	300–450
Polyester alkyd (putty)	300–450
Polyester alkyd (mineral)	300–450
Polyester alkyd (synthetic)	300–430
Polypropylene (glass filler)	300–320
Polysulfone	300–345
Celanex 917	300
Polyvinylidene fluoride	300
Cold-mold (organic) asbestos, binder melamine	325
Cold-mold (inorganic) asbestos, binder cement	750
Phenolic resin	340
Chlorotrifluoroethylene (CTFE)	350–390
Phenolic (glass filler)	350–550
Fluorocarbon (FEP)	400
Polyphenylene sulfides (unfilled)	400–500
Polyphenylene (25% asbestos filler)	400–500
Polyphenylene (40% glass filler)	400–500
Polyester alkyd (asbestos filler)	450
Polyester alkyd (glass filler)	450
Cold-mold nonrefractory (organic)	500
Polyamide aromatic	500
Silicones (asbestos filler)	500
Silicones (glass filler)	500–750
Polytetrafluoroethylene (TFE)	550
Silicones (mineral filler)	600–750
Cold-mold (inorganic) asbestos, binder phosphoric acid	750
Cold-mold (inorganic) asbestos, binder cement	750
Cold-mold refractory (inorganic)	900–1300

TABLE 12   ASTM and Federal Tests for Electrical Insulation and Other Important Plastic Properties

Test method	Federal Standard Method No.	ASTM No.
Abrasion wear (loss in weight)	1091	
Accelerated service tests (temperature and humidity extremes)	6011	D 756-56
Acetone extraction test for degree of cure of phenolics	7021	D 494-46
Arc resistance	4011	D 495-61
Bearing strength	1051	D 953-54, Method A
Bonding strength	1111	D 229-63T, pars. 40–43
Brittleness temperature of plastics by impact	2051	D 746-64T
Compressive properties of rigid plastics	1021	D 695-63T
Constant-strain flexural fatigue strength	1061	
Constant-stress flexural fatigue strength	1062	
Deflection temperature under load	2011	D 648-56, Procedure 6(a)
Deformation under load	1101	D 621-64, Method A
Dielectric breakdown voltage and dielectric strength	4031	D 149-64
Dissipation factor and dielectric constant	4021	D 150-64T
Drying test (for weight loss)	7041	
Effect of hot hydrocarbons on surface stability	6062	
Electrical-insulation resistance of plastic films and sheets	4052	
Electrical resistance (insulation, volume, surface)	4041	D 257-61
Falling-ball impact	1074	
Flame resistance	2023	
Flammability of plastics 0.050 in. and under in thickness	2022	D 568-61
Flammability of plastics over 0.050 in. in thickness	2021	D 635-63
Flexural properties of plastics	1031	D 790-63
Indentation hardness of rigid plastics by means of a durometer	1083	D 1706-61
Interlaminar and secondary bond shear strength of structural plastic laminates	1042	
Internal stress in plastic sheets	6052	
Izod impact strength	1071	D 256-56, Method A
Linear thermal expansion (fused-quartz tube method)	2031	D 696-44
Machinability	5041	
Mar resistance	1093	D 673-44
Mildew resistance of plastics, mixed culture method, agar medium	6091	
Porosity	5021	
Punching quality of phenolic laminated sheets	5031	D 617-44
Resistance of plastics to artificial weathering using fluorescent sunlamp and fog chamber	6024	D 1501-57T
Resistance of plastics to chemical reagents	7011	D 543-60T
Rockwell indentation hardness test	1081	D 785-62, Method A
Salt-spray test	6071	
Shear strength (double shear)	1041	
Shockproofness	1072	
Specific gravity by displacement of water	5011	
Specific gravity from weight and volume measurements	5012	
Surface abrasion	1092	D 1044-56
Tear resistance of film and sheeting	1121	D 1004-61
Tensile properties of plastics	1011	D 638-64T
Tensile properties of thin plastic sheets and films	1013	D 882-64T
Tensile strength of molded electrical-insulating materials	1012	D 651-48
Tensile time—fracture and creep	1063	
Thermal-expansion test (strip method)	2032	
Warpage of sheet plastics	6054	D 1181-56
Water absorption of plastics	7031	D 570-63

Volume resistivity is the resistance to current leakage through the body of an insulator. The units of measurement are ohm-centimeters. Volume resistivity is related to the nature of the insulator, moisture in the material, and temperature.

Arc- and track-resistant properties Arc resistance is a measure of the breakdown of the surface of an insulator caused by an arc which tends to form a conducting path. Many testing methods have been developed and are useful for specific problems. However, only four have been agreed upon by ASTM. The earliest of these is ASTM D 495-61, which is a high-voltage low-current test under clean conditions.

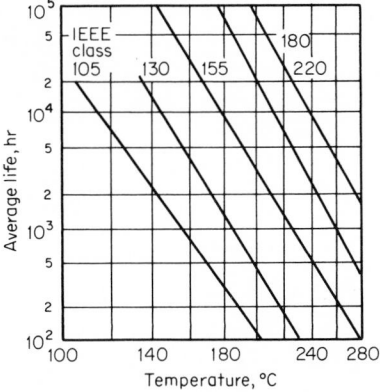

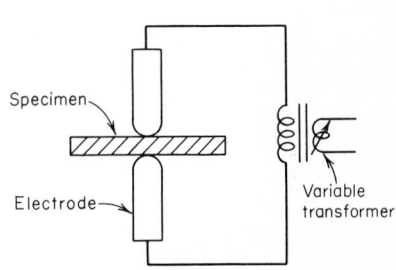

Fig. 2 Typical life curves for IEEE thermal classes of insulated magnet wire.

Fig. 3 Dielectric-breakdown test.

Figure 7 schematically shows the electrode arrangement. Other tests are specified to relate more closely with contamination and surface conditions found in practice. These tests all rely on the introduction of some contaminant in the arcing area. The dust-fog test is specified by ASTM D 2132-66T. This test is performed at 1.5 kV on a sample in a fog chamber and with a standardized dust applied to the sample surfaces. Failure is characterized by erosion of the sample or tracking. ASTM D 2302-65T describes the differential wet-track test. This test makes use of a 3-kV arc at several power levels. The sample is inclined and partially immersed in a

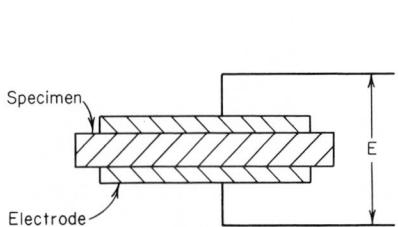

Fig. 4 Dielectric-constant test.

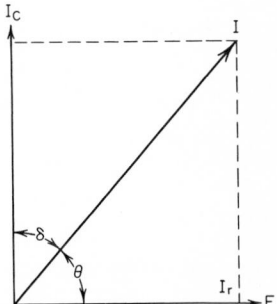

Fig. 5 Vector diagram of an equivalent parallel circuit. E = voltage; I = current; I_r = resistance component of current; I_c = capacitive component of current; θ = phase angle; and δ = loss angle.

water solution of ammonium chloride and a wetting agent. Failure is by tracking. The inclined-plane test is defined by ASTM D 2303-64T. In this test a specimen is inclined at 45°, and electrodes are placed on the underneath side. An electrolyte is fed onto the surface at a controlled rate, and voltage is increased simultaneously. Failure is by erosion and tracking.

Certain plastic materials track easily. For example, polymers based on aromatic rings such as phenolics or polystyrene track readily. Certain other plastics which are based on long straight-chain polymers are more resistant. In general the addition of mineral fillers improves arc and track resistance. Certain hydrated fillers exemplified by hydrated alumina dramatically improve the tracking properties. Fibers generally decrease wet-tracking resistance and improve dry arc resistance.

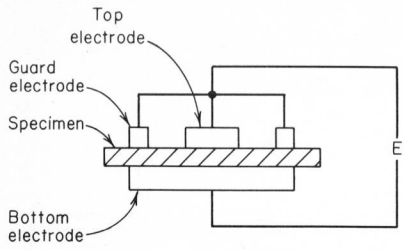

Fig. 6 Resistance test for insulating materials.

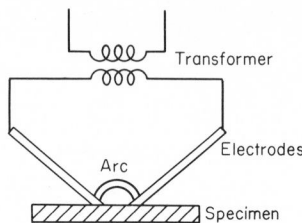

Fig. 7 Arc-resistance test.

THERMOSETTING PLASTICS

The nature of thermosetting plastics was discussed in Chap. 1. Approximate ratings of the various plastics, with respect to basic electrical-insulation properties, were presented in Tables 2 through 9 of this chapter. These ratings included both thermosetting and thermoplastic materials. Electrical design data and information on the various thermosets will be discussed in this section. The major thermosetting plastics, along with the primary design and application considerations for each, and typical commonly used names for these materials are shown in Table 13. More detailed data for each type of thermosetting material are given below.

Alkyds

Alkyds are widely used for molded electrical parts where the general application considerations given in Table 13 apply. They are chemically somewhat similar to polyester resins. Normally, alkyds are solid molding compounds, and polyesters are liquid or paste-form resins. Alkyd molding compounds are commonly available in putty, granular, glass-fiber-reinforced, and rope form (see Table 14).

Alkyds are easy to mold, and economical to use. Molding dimensional tolerances can be held to within ±0.001 in./in. Postmolding shrinkage is small, especially for glass-filled compounds. Alkyds have long been used in the electrical industry, and development has led to high-performance electronic-grade compounds. Comparisons of dielectric constant and dissipation factor for general-purpose and electronic-grade alkyd compounds are shown in Fig. 8.

The greatest limitation of alkyds is in extremes of temperature and in extremes of humidity. Silicones and diallyl phthalates are superior here, silicones especially with respect to temperature and diallyl phthalates especially with respect to humidity. The electrical-insulation resistance of alkyds decreases considerably in high, continuous-humidity conditions, as shown in Fig. 9.

Aminos (Melamines and Ureas)

Amino molding compounds can be fabricated by economical molding methods, and are very durable within their safe operating limits. They are hard, rigid, and abrasion-resistant, and have high resistance to deformation under load. These materials can be exposed to subzero temperatures without embrittlement. Under tropical conditions, the melamines do not support fungus growth.

Amino materials are self-extinguishing and have excellent electrical-insulation characteristics. They are unaffected by common organic solvents, greases and oils, and weak acids and alkalies. Melamines are superior to ureas in resistance to acids, alkalies, heat, and boiling water, and are preferred for applications involving cycling between wet and dry conditions or rough handling. Aminos do not impart taste or odor to foods.

Addition of cellulose filler produces an unlimited range of light-stable colors and high degrees of translucency. Colors are obtained without sacrifice of basic material properties. Melamines and ureas provide excellent heat insulation; temperatures up to the destruction point will not cause parts to lose their shape. Amino resins exhibit relatively high mold shrinkage, and also shrink on aging. Cracks develop

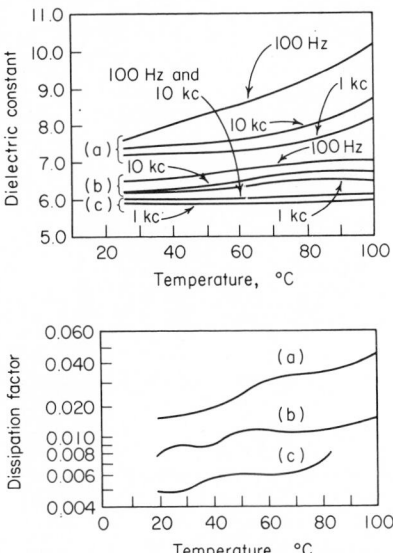

Fig. 8 Dielectric constant and dissipation factor comparisons for (a) general-purpose alkyd, (b) general-purpose electronic-grade alkyd, and (c) high-performance electronic-grade alkyd.[8]

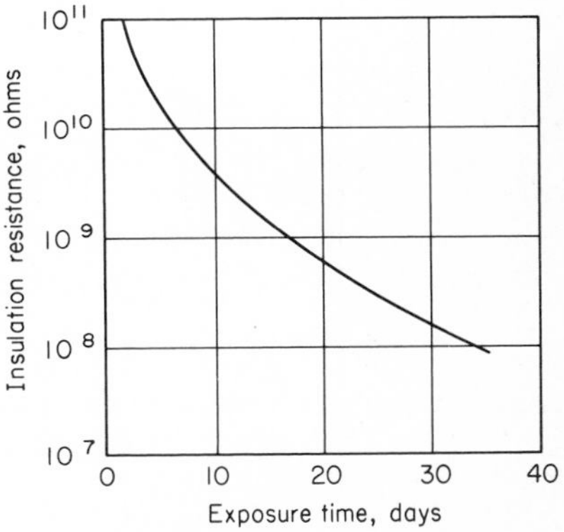

Fig. 9 Effect of 95 percent humidity at 60°C on the 500-V dc insulation resistance of an alkyd molding compound.[9]

in urea moldings subjected to severe cycling between dry and wet conditions. Prolonged exposure to high temperature can affect the color of both urea and melamine products.

A loss of certain strength characteristics can also occur when amino moldings are subjected to prolonged elevated temperatures. Some electrical characteristics are also adversely affected: arc resistance of some industrial types, however, remains unaffected after exposure at 500°F. Ureas are unsuitable for outdoor exposure. Melamines experience little degradation in electrical or physical properties after outdoor exposure, but color changes may occur. Electrical properties of various amino molding compounds are given in Table 15.

Diallyl Phthalates

Diallyl phthalates, part of the polymer class known as allylics, are among the best of the thermosetting plastics with respect to high insulation resistance and low electrical losses. Excellent properties are maintained up to 400°F or higher, and in the presence of high-humidity environments. Also, diallyl phthalate resins are easily molded and fabricated.

TABLE 13 Electrical-Application Information for Thermosetting Plastics

Material	Major application considerations	Common available forms	Typical suppliers and trade names*
Alkyds	Excellent dielectric strength, arc resistance, and dry insulation resistance. Low dielectric constant and dissipation factor. Good dimensional stability. Easily molded	Compression moldings, transfer molding	Allied Chemical Corp. (Plaskon); American Cyanamid Co. (Glaskyd)
Aminos (melamine formaldehyde and urea formaldehyde)...........	Available in an unlimited range of light-stable colors. Exhibit hard glossy molded surface, and good general electrical properties, especially arc resistance. Excellent chemical resistance to organic solvents and cleaners	Compression moldings, extrusions, transfer moldings, laminates, film	Allied Chemical Corp. (Plaskon); Monsanto Co. (Resimene); American Cyanamid Co. (Cymel for melamine; Beetle for urea)
Diallyl phthalates (DAP) (allylics)........	Rate high among thermosets in retention of properties in high-humidity environments. Also, exhibit among the highest volume and surface resistivites in thermosets. Low and relatively stable dissipation factor and dielectric constant. Heat resistance to 400°F or higher. Excellent dimensional stability. Easily molded	Compression moldings, extrusions, injection moldings, transfer moldings, laminates	FMC Corp. (Dapon); Allied Chemical Corp. (Diall)
Epoxies............	Good electrical properties, low shrinkage, excellent dimensional stability and good to excellent adhesion. Extremely easy to compound, using low-pressure or atmospheric-pressure processes, for providing a wide variety of end properties. Bisphenol epoxies are most common, but several other varieties are available for providing special properties	Castings, compression moldings, extrusions, injection moldings, transfer moldings, laminates, matched-die moldings, filament windings, foam	Shell Chemical Co. (Epon); Celanese Co. (Epi-Rez); Dow Chemical Co. (D.E.R.); Ciba Products Co. (Araldite); Union Carbide Corp. (ERL); 3M Co. (Scotchcast)

Phenolics..............	Among the lowest-cost, most widely used thermoset materials. Excellent thermal stability to over 300°F generally, and over 400°F in special formulations. Can be compounded with a broad choice of resins, fillers, and other additives. Available in general-purpose and high-electrical-performance grades	Castings, compression moldings, extrusions, injection moldings, transfer moldings, laminates, matched-die moldings, stock shapes, foam	Union Carbide Corp. (Bakelite); Hooker Chemical Corp. (Durez)
Polybutadienes.........	Excellent low-loss properties (dielectric constant and dissipation factor) at microwave frequency. Stable both electrically and physically up to 400–500°F	All forms of molded products, laminates, and liquid-resin systems	Colorado Chemical Specialties, Inc. (Ricon)
Polyesters............	Excellent electrical properties and low cost. Extremely easy to compound using low-pressure or atmospheric-pressure processes. Like epoxies, can be formulated for either room-temperature or elevated-temperature use. Not equivalent to epoxies in environmental resistance	Compression moldings, extrusions, injection moldings, transfer moldings, laminates, matched-die moldings, filament windings, stock shapes	Pittsburgh Plate Glass Co. (Selectron); American Cyanamid Co. (Laminac); Rohm & Haas Co. (Paraplex)
Silicones (rigid).......	Excellent electrical properties, especially low and stable dielectric constant and dissipation factor, which change little up to 400°F and over. Nonrigid silicones are also available as elastomers and RTV and flexible silicone potting compounds	Castings, compression moldings, transfer moldings, laminates	Dow Corning Corp. (DC Resins)

* This listing is only a very small sampling of the many possible excellent suppliers. It is intended only to orient the nonchemical reader into plastic categories. No preferences are implied or intended.

TABLE 14 Properties of Various Alkyd Molding Compounds[7]

	ASTM test method	Putty	Granular	Glass-fiber reinforced	Rope
Electrical					
Arc resistance, s	D 495-58T	180+	180+	180+	180+
Dielectric constant	D 150-54T				
1 MHz		5.4-5.9	5.7-6.3	5.2-6.0	7.4
1 MHz		4.5-4.7	4.8-5.1	4.5-5.0	6.8
Dissipation factor	D 150-54T				
60 Hz		0.030-0.045	0.030-0.040	0.02-0.03	0.019
1 MHz		0.016-0.022	0.017-0.020	0.015-0.022	0.023
Dielectric strength, V/mil	D 149-59				
Short-time		350-400	350-400	350-400	360
Step by step		300-350	300-350	300-350	290
Physical					
Specific gravity	D 792-50	2.05-2.15	2.21-2.24	2.02-2.10	2.20
Water absorption, 24 h at 23°C, %	D 570-57T	0.10-0.15	0.08-0.12	0.07-0.10	0.05
Heat resistance max, °F:					
Long periods (continuous)		250	300	300	300
Short periods (0-24 h)		300	350	350	350
Short periods (0-1 h)		325	375	400	400
Heat distortion temp, 264 lb/in.², °F	D 648-56	350-400	350-400	>400	>400
Coefficient of linear thermal expansion/°F	D 696-44	$10\text{-}30 \times 10^{-6}$	$10\text{-}30 \times 10^{-6}$	$10\text{-}30 \times 10^{-6}$	20×10^{-6}
Thermal conductivity, (g)(cal)/(s)(cm²)(°C/cm)		$15\text{-}25 \times 10^{-4}$	$15\text{-}25 \times 10^{-4}$	$8\text{-}12 \times 10^{-6}$	10×10^{-4}
Flammability	D 635-56T	Nonburning	Self-extinguishing	Nonburning	Self-extinguishing
Mechanical					
Impact strength, Izod, ft-lb/in. of notch	D 256-56	0.25-0.35	0.30-0.35	8-12	2.2
Comprehensive strength, lb/in.²	D 695-54	20,000-25,000	16,000-20,000	24,000-30,000	28,800
Flexural strength, lb/in.²	D 790-59T	8,000-11,000	7,000-10,000	12,000-17,000	19,500
Tensile strength, lb/in.²	D 651-48	4,000-5,000	3,000-4,000	5,000-9,000	7,100
Modulus of elasticity, lb/in.²	D 790-49T	$2.0\text{-}2.7 \times 10^{6}$	$2.4\text{-}2.9 \times 10^{6}$	$2.0\text{-}2.5 \times 10^{6}$	1.9×10^{6}
Barcol hardness		60-70	60-70	70-80	72
		MIL-M-14F Type MAG	MIL-M-14F Type MAG	MIL-M-14F Type MAI-60	

TABLE 15 Properties of Various Amino Molding Compounds[10]

	Urea		Melamine					
	Alpha cellulose	Wood flour	Alpha cellulose	Wood flour	Alpha cellulose modified	Rag	Asbestos	Glass fiber
Mechanical:								
Tensile strength, 1,000 lb/in.²	5.5–7	5.5–10	7–8	5.7–6.5	5.5–6.5	8–10	5.5–6.5	5.9
Compressive strength, 1,000 lb/in.²	30–38	25–35	40–45	30–35	24.5–26	30–35	25–30	20–29
Flexural strength, 1,000 lb/in.²	11–18	8–16	12–15	6.5–9	11.5–12	12–15	7.4–10	13.2–24
Shear strength, 1,000 lb/in.²	11–12	......	11–12	10–10.5	11.4–12.2	12–14	7–8	13.0–15.6
Impact strength, Izod, ft-lb/in. of notch	0.24–0.28	0.25–0.35	0.30–0.35	0.25–0.38	0.30–0.42	0.55–0.90	0.30–0.40	0.5–6.0
Tensile modulus, 10⁶ lb/in.²	1.3–1.4	1.3–1.6	1.35	1.0	1.0	1.4	1.95	
Flexural modulus, 10⁶ lb/in.²	1.4–1.5	......	1.1	1.0	1.1	1.4	1.8	2.4
Electrical:								
Arc resistance, s	80–100	80–100	125–136	70–106	90–120	122–128	120–180	180–186
Dielectric strength, V/mil								
Short time								
At 23°C	330–370	300–400	270–300	350–370	350–390	250–340	410–430	170–370
At 100°C	200–270	......	170–210	290–330	140–190	110–130	280–310	90–350*
Step by step								
At 23°C	220–250	250–300	240–270	200–240	200–250	220–240	280–300	170–270
At 100°C	110–150	......	90–130	190–210	90–100	60–90	190–210	60–250*
Slow rate of rise								
At 23°C	250–260	......	210–240	240–260	280–290	210–240	270–290	170–210
At 100°C	120–170	......	90–120	170–200	90	70–80	170–190	70–90
Dielectric constant								
At 60 Hz	7.7–7.9	7.0–9.5	7.9–8.2	6.4–6.6	7.0–7.7	8.1–12.6	10.0–10.2	7.0–11.1
At 10⁶ Hz	6.7–6.9	6.4–6.9	7.6–8.0	5.6–5.8	5.2–6.0	6.7–6.9	5.3–6.1	6.6–7.9
At 3 × 10⁹ Hz	......	......	......	......	......	......	4.9	5.5
Dissipation factor								
At 60 Hz	0.034–0.043	0.035–0.040	0.052–0.083	0.026–0.033	0.192	0.100–0.340	0.100	0.14–0.23
At 10⁶ Hz	0.029–0.031	0.028–0.032	0.026–0.030	0.034–0.035	0.044–0.12	0.036–0.041	0.039–0.048	0.013–0.016
At 3 × 10⁹ Hz	......	......	......	......	......	......	0.032	0.040
Dielectric loss factor								
At 60 Hz	0.28–0.34	0.24–0.38	0.44–0.78	0.17–0.22	0.90–2.4	2.0–5.0	0.21–1.0	1.5–2.5
At 10⁶ Hz	0.19–0.21	0.18–0.22	0.20–0.33	0.20–0.21	0.19–0.28	0.24–0.26	0.21–0.31	0.09–0.19
Volume resistivity, Ω-cm	0.5–5.0 × 10¹¹	......	0.8–2.0 × 10¹¹	6–10 × 10¹²	6 × 10¹⁰	1.0–3.0 × 10¹¹	1.2 × 10¹²	0.9–20 × 10¹¹
Surface resistivity, Ω	0.4–3.0 × 10¹¹	......	0.8–4.0 × 10¹⁰	0.3–5.0 × 10¹¹	1.7 × 10¹²	0.7–7.0 × 10¹⁰	1.9 × 10¹²	3.0–4.6 × 10¹²
Insulation resistance, Ω	0.2–5.0 × 10¹¹	......	1.0–4.0 × 10¹⁰	1.0–3.0 × 10¹¹	2.0–5.0 × 10⁹	0.1–3.0 × 10¹⁰	1.0–4.0 × 10¹⁰	0.2–6.0 × 10¹⁰

* At 50°C.

There are several chemical variations of diallyl phthalate resins, but the two most commonly used are diallyl phthalate (DAP) and diallyl isophthalate (DAIP). The primary application difference is that DAIP will withstand somewhat higher temperatures than will DAP. The retention of electrical-resistance properties after humidity conditioning is compared in Fig. 10 for several glass-filled DAP, DAIP, epoxy, and phenolic compounds, and one mineral and glass-filled epoxy. Electrical properties of various molding compounds are given in Table 16.

The excellent dimensional stability of diallyl phthalates has been mentioned above. This is demonstrated in Fig. 11, which compares diallyl phthalates with other plastic materials at various temperatures.

Likewise, the excellent electrical properties of diallyl phthalates have been stressed. The effect of frequency and temperature on dielectric constant is shown in Fig. 12, and the effect of frequency and temperature on dissipation factor in Fig. 13. The loss characteristics of DAP, epoxy, and phenolic compounds as a function of temperature are presented in Fig. 14. Diallyl isophthalate (DAIP) is also especially good in retention of dielectric strength, as indicated in Fig. 15. Diallyl phthalates rate high among plastics in resistivity levels, and the resistivity data for a variety of plastics and other materials were shown earlier in Fig. 1 and Table 4. Dielectric strengths of certain DAP and alkyds rate high, as shown in Fig. 16. Drastic reductions may occur above 50 percent humidity, however.

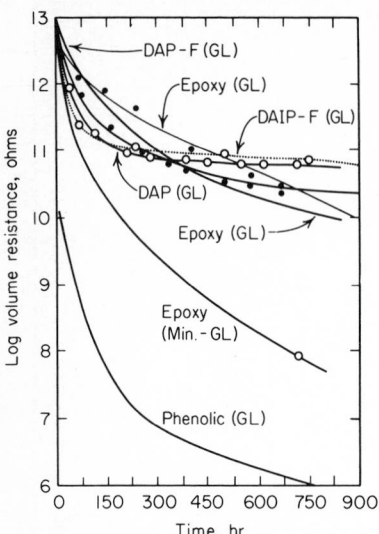

Fig. 10 Volume resistance decrease of several thermosetting materials at 70°C and 100 percent RH. (GL indicates glass filler, and Min-GL indicates mineral and glass filler.)[11]

TABLE 16 Properties of Various Diallyl Phthalate Molding Compounds

Property	Orlon	Dacron	Long glass	Asbestos	Short glass	Short glass*
Tensile strength, lb/in.²...	6,000	5,000	10,000	5,500	7,000	7,000
Compressive strength, lb/in.².................	25,000	25,000	25,000	25,000	25,000	28,000
Flexural strength, lb/in.²..	10,000	11,500	16,000	9,600	12,000	12,000
Flexural modulus, lb/in.² × 10⁻⁶..........	0.71	0.64	1.3	1.2	1.2	1.3
Impact strength, Izod, ft-lb/in. of notch........	1.2	4.5	6.0	0.4	0.6	0.5
Hardness, Rockwell M....	108	108	100	100	105	110
Specific gravity at 25°C...	1.31–1.45	1.39–1.62	1.55–1.70	1.55–1.65	1.6–1.8	1.65–1.75
Dielectric constant:						
At 1 kHz..............	3.7–4.0	3.79	4.2	....	4.4	4.1
At 1 MHz..............	3.3–3.6	3.4	4.2	4.5–6.0	4.4	3.4
Dissipation factor:						
At 1 kHz..............	0.020–0.025	0.008	0.004–0.006	0.04–0.08	0.006	0.004
At 1 MHz..............	0.015–0.020	0.012	0.008	0.04–0.08	0.008	0.008
Mold shrinkage, in./in....	0.009	0.010	0.002	0.006	0.003	0.003
Postmold shrinkage, in./in.................	0.001	0.0006	0.0007	0.001	0.0007	0.0002
Heat-deflection temperature, °F..............	265	290	392	325	400	500+
Heat resistance, continuous, °F............	300–500	350–400	350–400	350–400	350–400	450

* Based on diallyl isophthalate.

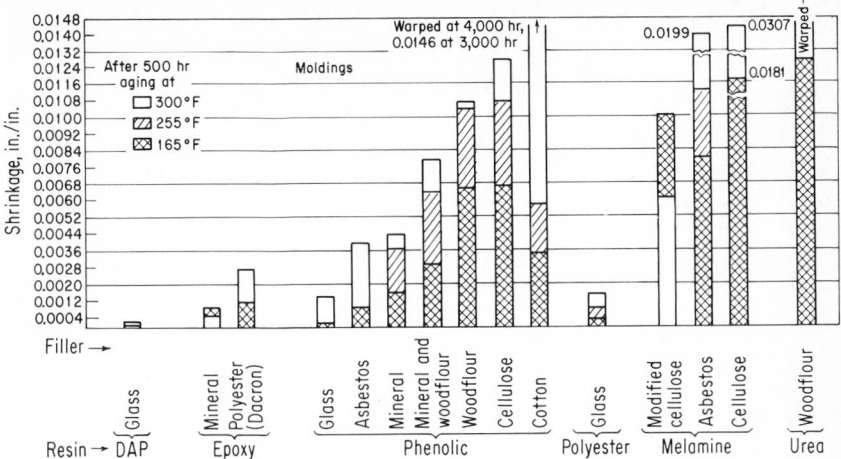

Fig. 11 Shrinkage of various thermosetting molding materials as a result of heat aging.[12]

Epoxies

Epoxies are among the most versatile and most widely used plastics in the electronic field—primarily because of the wide variety of formulations possible, and the ease with which these formulations can be made and utilized with minimal equipment requirements. Formulations range from flexible to rigid in the cured state, and from thin liquids to thick pastes in the uncured state. Conversion from uncured to cured state is accomplished by use of hardeners and/or heat. The largest applications of epoxies in electronics are in embedding applications (potting, casting, encapsulating, and impregnating), molded products, and laminated constructions such as metal-clad laminates for printed circuits and unclad laminates for various types of insulating and terminal boards.

Basically, epoxies are available as liquid or solid resins, and as powdered molding compounds. The molding compounds, while broadly used for embedment of electronic assemblies by the transfer-molding technique, are also employed for transfer and compression molding of many other types of electrical parts. Typical properties of glass-fiber-filled and mineral-filled molding compounds are given in Table 17. Physical and electrical properties are generally very stable in humid environments, although resistance properties in humidity

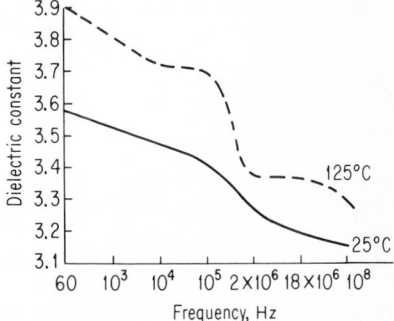

Fig. 12 Effect of frequency and temperature on the dielectric constant of unfilled diallyl phthalate.[13]

are dependent on the filler used, to some extent, as shown in Figs. 17 and 18. Also, although the general electrical properties of epoxies are good, they are not as good as those of diallyl phthalates as a function of either temperature or humidity. This is shown in Figs. 10 and 14.

In addition to their versatility and good electrical properties, epoxies are also outstanding in their low shrinkage, their dimensional stability, and their adhesive properties. Their shrinkage is often less than 1 percent, and the as-molded dimensions of an epoxy part change little with time or environmental conditions, other

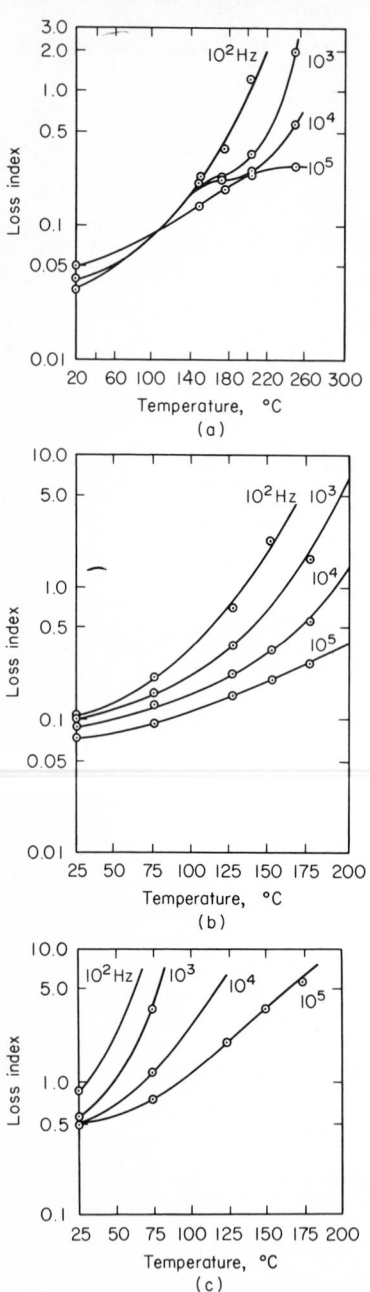

(a)

(b)

(c)

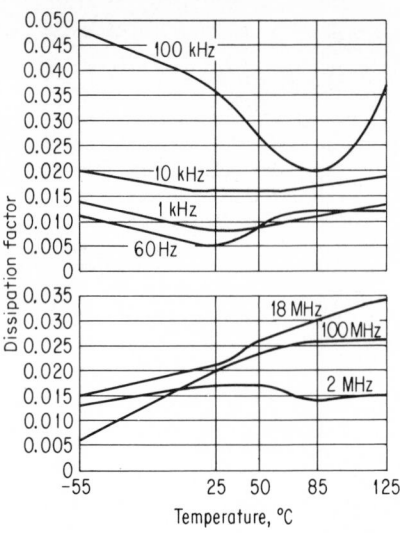

Fig. 13 Effect of frequency and temperature on the dissipation factor of unfilled diallyl phthalate.[13]

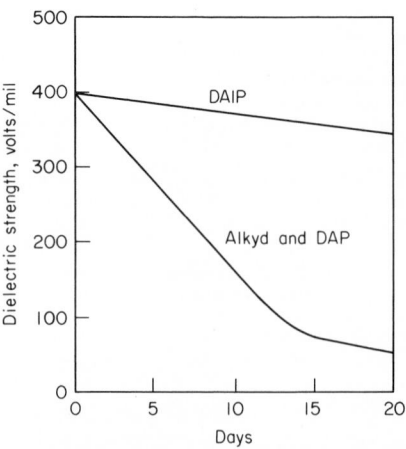

Fig. 14 Loss index vs. temperature and frequency for (a) glass-filled diallyl phthalate (DAP) compound, (b) mineral and glass-filled epoxy compounds, and (c) glass-filled phenolic compound.[11]

Fig. 15 Effect of heat aging at 400°F on the dielectric strength of diallyl phthalate (DAP), diallyl isophthalate (DAIP), and alkyd molding materials.[14]

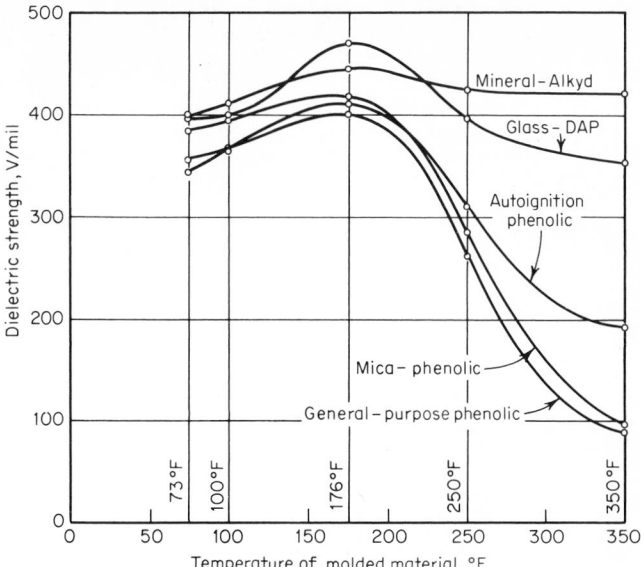

Fig. 16 Dielectric strength of various molding compounds as a function of temperature.[15]

TABLE 17 Typical Properties of Epoxy Molding Compounds with Various Fillers

	Glass-fiber filler	Mineral filler
Tensile strength, lb/in.²	14,000–30,000	5,000–7,000
Elongation, %	4	
Tensile modulus, 10^5 lb/in.²	30.4	
Compressive strength, lb/in.²	25,000–30,000	18,000–25,000
Flexural strength, lb/in.²	20,000–26,000	10,000–15,000
Impact strength, Izod, ft-lb/in. of notch	8–15	0.25–0.45
Hardness, Rockwell	M100–M108	M101
Specific gravity	1.8–2	1.6–2.06
Thermal conductivity (cal)(cm)/(s)(cm²)(°C)..	$7–10 \times 10^{-4}$	$7–18 \times 10^{-4}$
Specific heat per °C	0.19	
Coefficient of thermal expansion, per °C $\times 10^{-5}$	1.1–3	2.4–5
Heat resistance, continuous, °F	330–500	300–500
Heat-distortion temp, °F	400–500	250–450
Volume resistivity, Ω-cm	3.8×10^{15}	9×10^{15}
Dielectric strength, ⅛-in. (V/mil):		
Short-time	360	330–400
Step by step	340	350
Dielectric constant at 60, 10^3, and 10^6 Hz.....	4–5	4–5
Arc resistance, s	125–140	150–180
Burning rate	Self-extinguishing	Self-extinguishing
Water absorption, %*	0.05–0.095	0.1

* Test performed on ⅛-in.-thick piece immersed in distilled water for 24 h at room temperature.

than excessive heat. Their excellent performance in this respect is shown in Fig. 11. Because of the low shrinkage and good strength properties of epoxies, cured epoxy parts resist cracking, both upon curing and in thermal shock, better than most other rigid thermosetting materials. Based on the excellent bonds obtained with epoxy resins to most substrates, epoxy formulations are broadly used as adhesives. Even when not specifically used as adhesives, the bonding properties of epoxies often provide a better seal around inserts, terminals, and other interfaces than do most other plastic materials.

Phenolics

Phenolics are among the oldest, best-known general-purpose molding materials. They are also among the lowest in cost and the easiest to mold. An extremely large number of phenolic materials are available, based on the many resin and filler combinations. Some grades are general-purpose, medium- and high-impact, electrical-insulation, heat-resistant, moisture-resistant, arc-resistant, and injection-molding grades.

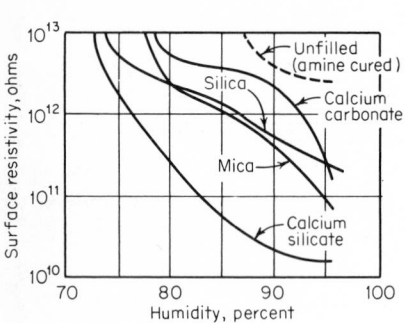

Fig. 17 Effect of humidity on surface resistivity of filled and unfilled epoxy resins at 35°C.[16]

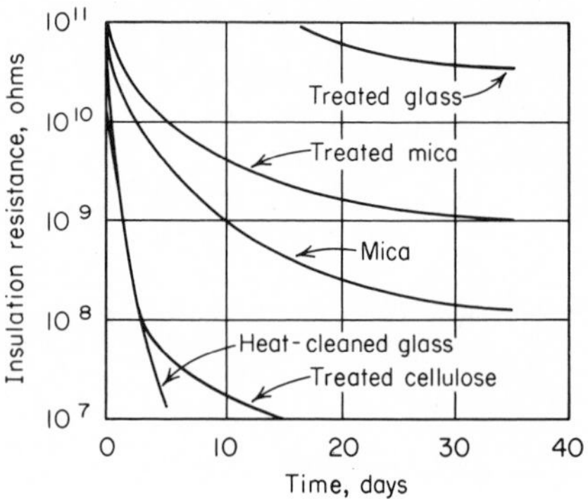

Fig. 18 Insulation resistance of epoxy compounds with various fillers, at 140°F and 95 percent RH. Treatment indicated is a chromic chloride sizing.[9]

Although it is possible to get various grades of phenolics for various applications, phenolics, generally speaking, are not equivalent to diallyl phthalates and epoxies in resistance to humidity, shrinkage, dimensional stability, and retention of electrical properties in extreme environments. Critical electrical-property comparisons are shown in Figs. 10 and 14. Phenolics are, however, quite adequate for a large percentage of electrical applications. Furthermore improved grades have been developed which yield considerable improvement in humid environments (Fig. 19) and at higher temperatures (Fig. 20). In addition, the glass-filled, heat-resistant grades are outstanding in thermal stability up to 400°F and higher, with some being useful up to 500°F. Phenolics are relatively stable, physically, as discussed in Chap. 1. Shrinkage in heat aging varies over a fairly wide range, depending on filler used. Glass-filled phenolics are the more stable, as shown in Fig. 11.

Polybutadienes

This is a versatile family of thermosetting plastic materials having excellent electrical properties which are stable at high frequencies and elevated temperatures. Further, these materials offer low moisture absorption, excellent chemical resistance, and excellent thermal stability among plastic materials. Polybutadienes are fast-curing, and can be molded, laminated, or used as casting and potting materials. They can even be injection-molded, which is the fast molding technique used for thermoplastics. This can be desirable, owing to lower processing costs for volume production. While the nature of the polymerization is such that most thermosets cannot be injection-molded, considerable effort has recently been devoted to development of injection-moldable thermosetting plastics. Table 18 shows representative properties of polybutadiene molding compounds, for a one-component system with excellent shelf life. Figures 21 and 22 show the dielectric constant and dissipation factor of glass-reinforced polybutadiene laminates up to 500°F.

Polyesters

Polyesters are versatile resins, which handle much like the epoxies. They are available in forms ranging from low-viscosity liquids to thick pastes or putties. The liquids are used for embedding applications and laminated products, much like the epoxies, and the pastes are used for molding applications. Although both epoxies and polyesters are available in formulations for room-temperature cure and formulations for heat cure, the chemical curing mechanism, or polymerization mechanism, is different for the two types of resins. Likewise, of course, the basic resins are chemically different. It is their physical forms and application forms which make them similar.

The major advantages of polyesters over epoxies are lower cost and appreciably lower electrical losses for the best electrical-grade polyesters. Some im-

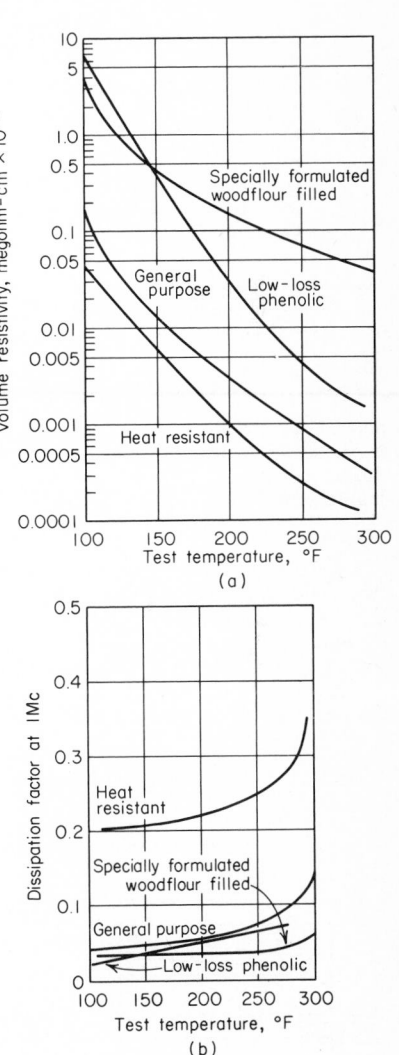

(a)

(b)

Fig. 20 Comparison of temperature effects on (a) volume resistivity and (b) dissipation factor for several grades of phenolic compounds.[17]

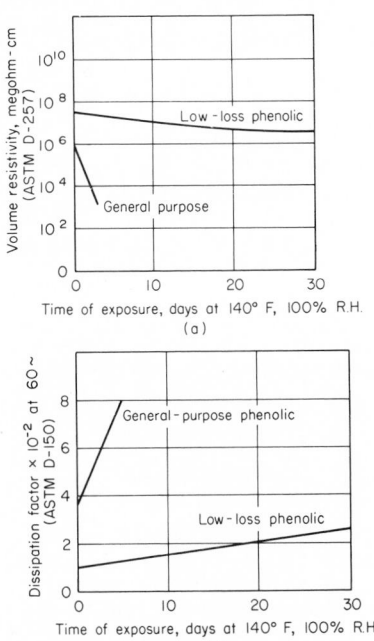

(a)

(b)

Fig. 19 Comparison of humidity effects on (a) volume resistivity and (b) dissipation factor for a general-purpose and a low-loss phenolic compound.[17]

TABLE 18   Typical Properties of Cured Polybutadiene Molding Compound (Firestone FCR-1261, TC-100)[18]

Property	Value
Flexural strength, 10^3 lb/in.^{2}	11–14
Flexural modulus, 10^6 lb/in.^{2}	1.3–1.4
Compressive strength, 10^3 lb/in.^{2}	25
Compressive modulus, 10^5 lb/in.^{2}	5
Tensile strength, 10^3 lb/in.^{2}	7.5
Tensile modulus, 10^5 lb/in.^{2}	7
Rockwell hardness	E70–E95
Impact strength, ft-lb/in.	0.25–0.5
Heat-deflection temp, 264 lb/in.^{2}, °F	>500
Flammability, ASTM D 635	Nonburning
Specific gravity	2.05
Mold shrinkage, cured at 350°F, in./in.	0.015
Thermal conductivity 10^{-4} cal/(s)(cm^2)(°C/cm)	5.2
Thermal expansion, 10^{-5} in./in./°C	4
Dielectric strength, V/mil	500–1000+
Dielectric constant, 50 to 10^5 Hz	3.7
Dissipation factor:	
100 Hz	0.007
1,000 Hz	0.004
10,000 Hz	0.003
100,000 Hz	0.002
Volume resistivity, 10^{15} Ω-cm	1.7–7.1
Arc resistance, s	252

portant disadvantages of polyesters, as compared with epoxies, are lower adhesion to most substrates, higher polymerization shrinkage, a greater tendency to crack during cure or in thermal shock, and greater change of electrical properties in a humid environment.

Silicones

Silicones are thermosetting polymers which can be classified as elastomers when the cured product is rubberlike, or as embedding materials when the basic plastic form is a castable liquid (either rubberlike or rigid), or as low-pressure transfer-molding compounds, when used in the moldable, powdered form. Furthermore, silicone polymers are widely used in electronic and electrical applications in other forms such as laminated products, coatings, and adhesives. Silicone elastomers will be discussed later in another section of this chapter which covers this material category. It is the silicone molding materials that will be discussed at this point.

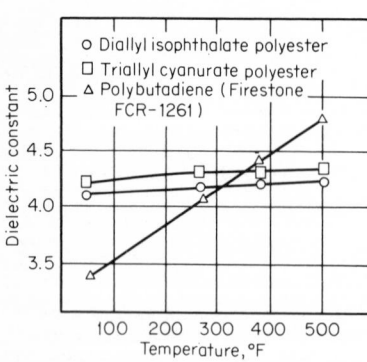

Fig. 21 Dielectric constant at 9.35 GHz of glass-fabric laminates.[18]

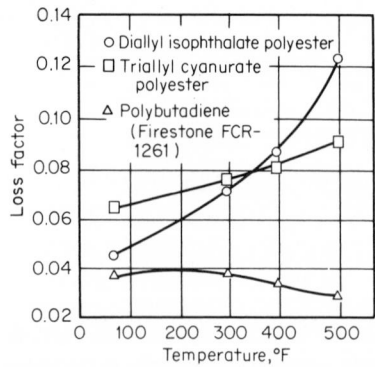

Fig. 22 Loss factor at 9.35 GHz of glass-fabric laminates.[18]

Silicone molding compounds are useful for many of the molded products required in electronic and electrical applications. These molding compounds are normally either mineral-filled or glass-fiber-filled, as described in Table 19.

The most important properties of silicones, for electronic applications, are excellent electrical properties which do not change drastically with temperature or

TABLE 19 Typical Properties of Silicone Molding Compounds[19]

ASTM test method	Property	Mineral-filled	Glass-fiber-filled
	Specific gravity	1.80–1.95	1.88
	Density, lb/in.^{3}	0.065–0.101	0.068
	Mold shrinkage, in./in.	0.007	0.005
	Flammability	Self-extinguishing	Self-extinguishing
	Thermal conductivity at 500°F, Btu/(h)(ft^2)(°F)(ft)	2.0–2.1	1.2
D 570	Water absorption in 24 h, $\frac{1}{8}$-in. thickness, at 73°F, % increase in weight	0.05–0.20	0.10–0.12
D 696	Coefficient of thermal expansion, in./(in.)(°C) $\times$ 10^5, at 25 to 250°C		
	Perpendicular	2.5–6.0	Not isotropic 1.0
	Parallel	. . . .	12.1
D 648	Heat-distortion temp, °F	340–900	>900
	Max recommended intermittent-service temp, °F	500–600	700–750
	Max recommended continuous-service temp (continuous loading), °F	500–750	700–750
D 638	Tensile yield strength (avg) at 73°F, lb/in.^{2} $\times$ 10^{-3}	2.5–3.5	3.5–6.5
D 790	Flexural strength at 77°F, lb/in.^{2} $\times$ 10^{-3}	6.5–8.5	18–21
D 790	Flexural modulus at 77°F, lb/in.^{2} $\times$ 10^{-6}	1.4–1.6	2.5
D 695	Compressive strength at 77°F, lb/in.^{2} $\times$ 10^{-3}	11–18	10–12.5
D 256	Impact strength, Izod, at 73°F (notched), ft-lb/in. notch	0.30–0.40	10–24
D 785	Hardness, Rockwell M, at 73°F	70–95	83–88
D 257	Volume resistivity at 50% RH and 23°C, Ω-cm	2 $\times$ 10^{13}–5 $\times$ 10^{15}	Dry, 1.4 $\times$ 10^{14} 9 $\times$ 10^{-14}
D 150	Power factor at 1 MHz:		
	Dry	0.002	0.004
	After 24 h immersion	0.003	0.005
D 150	Dielectric constant:		
	At 60 Hz	3.4–6.3	4.35
	At 1 MHz	3.4–6.3	4.26–4.28
D 149	Dielectric strength, 500 V/s rate of rise, $\frac{1}{8}$-in. thickness, V/mil	300–465	275–300
D 495	Arc resistance (tungsten electrodes), s	190–420	230–240
	Ultraviolet resistance	Excellent	Excellent
	Resistance to acids	To mild acids, excellent	To mild acids, excellent
	Resistance to bases	To mild bases, fair to good	To mild bases, fair to good
	Resistance to solvents	Fair to excellent	Fair to excellent
	Relative material costs, cents/in.^{3}	26–41	24

frequency over the safe operating-temperature range of silicones. This is clearly demonstrated in Fig. 23. Furthermore, silicones are among the best of all polymer materials in resistance to temperature. This is shown in Fig. 24, which compares the relative thermal stability of a group of plastic materials. Useful temperatures of 500 to 700°F are available for silicone materials. Hence, silicones are broadly used for high-temperature electronic applications, especially those applications requiring low electrical losses.

Mechanical properties of silicone molding compounds are affected by temperature aging, as shown in Fig. 25. Although silicones are not as strong as diallyl phthalates, epoxies, and phenolics upon aging at lower temperatures, Fig. 21 shows them to be stronger upon aging above 400°F. Strength at temperature will of course be lower in most cases. Generally, most silicone properties are stable in extreme environments such as humidity and vacuum. Thus, silicones are widely used for both military

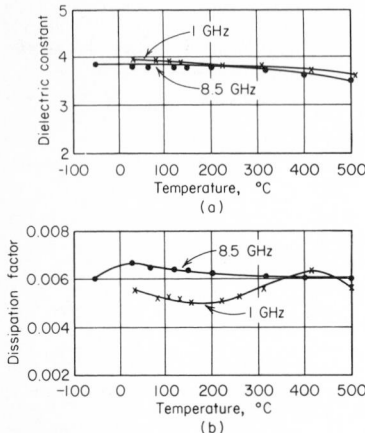

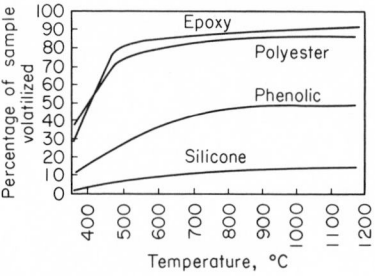

Fig. 23 Effect of heat aging at 300°C on (a) dielectric constant and (b) dissipation factor of a silicone molding compound.[20]

Fig. 24 Thermal stability of a group of plastic materials.[20]

and space applications. Weight loss at 10⁻⁶ torr, at several temperatures, is shown for silicone and epoxy materials in Fig. 26.

One other area of specific usefulness is the molding of semiconductor devices. This is particularly true when the device is expected to function or be tested under humid conditions. Ions that dissolve in water or may be transported by moisture vapors may irreversibly alter the characteristics of the device.

One method of determining the presence of ionic constituents in molding compounds is by the water-extract conductance test. This procedure involves extracting weighed samples of ground plastic with deionized water for 288 h at 71°C and measuring the electrical resistivity of the extract solution at 23°C.

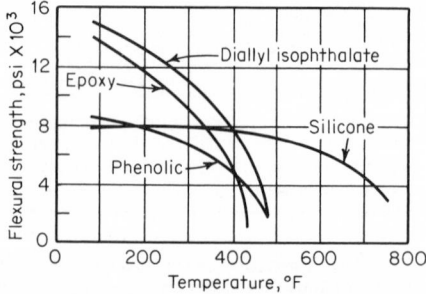

Fig. 25 Effect of temperature on flexural strength of several thermosetting compounds. (All samples were aged 250 h at temperature and tested at room temperature.)[20]

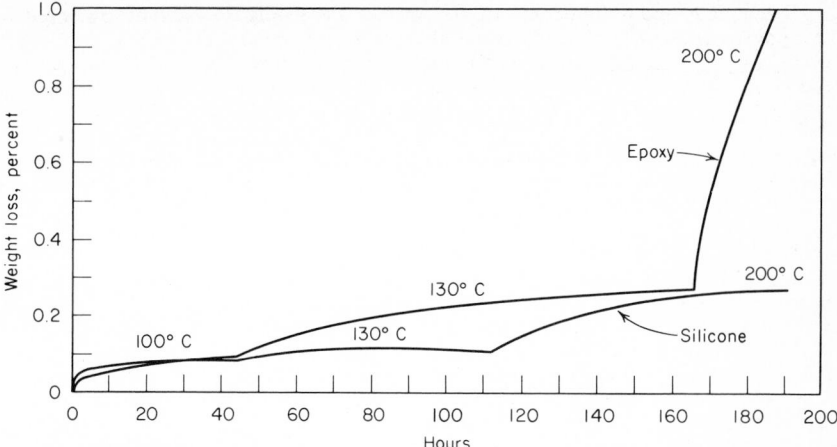

Fig. 26 Weight loss upon vacuum aging at 10^{-6} torr, at temperatures shown, for silicone and epoxy materials.[20]

Water-extract resistivity data obtained using typical electronic encapsulating grades of silicone, epoxy, and phenolic molding compounds are presented in Table 20. It can be seen that silicones excel in this test. Selection of proper curing agents and control of impurities can yield optimum-quality epoxy formulations in this respect. Also diallyl phthalates offer high water-extract resistivities, since, as with silicones, relatively nonpolar catalyst systems can be used.

TABLE 20   Water-Extract Resistivities for Silicone, Epoxy, and Phenolic Compounds[20]

Molding compound	Resistivity, Ω-cm	
	Received powder	Molded and postcured 2 h at 200°C
Silicone.................	245,000	245,000
Epoxy..................	4,180	43,000
Phenolic...............	4,580	8,370

THERMOPLASTICS

Thermoplastics differ from thermosets in that thermoplastics do not cure or set under heat as do thermosets. Thermoplastics merely soften, when heated, to a flowable state whereby under pressure they can be forced or transferred from a heated cavity into a cool mold. Upon cooling in a mold, thermoplastics harden and take the shape of the mold. Since thermoplastics do not cure or set, they can be remelted and rehardened, by cooling, many times. Thermal aging, brought about by repeated exposure to the high temperatures required for melting, causes eventual degradation of the material and so limits the number of reheat cycles. However, thermoplastics will melt to a flowable state upon heating, whereas thermosets do not melt. Essentially all thermoplastics are processed by heating to a soft state and applying pressure, such as injection molding, extruding, and thermoforming. Both thermosets and thermoplastics soften upon heating, to varying degrees. Thermoplastics are straight-chain polymers, rather than infusible, cross-linked polymers, which is the structural difference most affecting thermal properties.

Design data and information on the various thermoplastics will be discussed at this point. The major thermoplastics, along with the primary design and application considerations for each, and typical commonly used names for these materials are shown in Table 21. Basic physical and mechanical data on these thermoplastic materials are presented in Table 22 and basic electrical data in Table 23. A more detailed discussion on each is given below. In addition, further data are given in a subsequent portion of this chapter which covers glass-fiber-reinforced thermoplastics. This is of special importance because of the continuous improvements and increasing usage of reinforced thermoplastic materials.

ABS Plastics

ABS plastics are derived from acrylonitrile, butadiene, and styrene. This class possesses hardness and rigidity without brittleness, at moderate cost. ABS materials have a good balance of tensile strength, impact resistance, surface hardness, rigidity, heat resistance, low-temperature properties, and electrical characteristics. There are many ABS modifications, and many blends of ABS with other thermoplastics. This complete spectrum of ABS-based plastics offers possibilities in electrical designs, and all should be considered. The complete ABS spectrum is detailed in Chap 1.

Although ABS plastics are used largely for mechanical purposes, they also have good electrical properties that are fairly constant over a wide range of frequencies. These properties are little affected by temperature and atmospheric humidity in the acceptable operating range of temperatures.

The dielectric strength of ABS plastics is about 350 V/mil. The approximate dissipation factors of the best electrical grades are 0.004 at 60 Hz, 0.005 at 1 kHz, and 0.009 at 11 MHz. Dielectric constants range from 2.4 to 5.0, volume resistivities from 1.0 to 5.0×10^{16} Ω-cm, and arc resistance from 50 to 90 s.

Acetals

Acetals are among the group of high-performance engineering thermoplastics that resemble nylon somewhat in appearance but not in properties. They are strong and rigid (but not brittle) and have good moisture, heat, and chemical resistance. There are two basic types of acetals, namely, the *homopolymers* by Du Pont, and the *copolymers* by Celanese. Table 21 further identifies these materials. The homopolymers are harder, have higher resistance to fatigue, are more rigid, and have higher tensile and flexural strength with lower elongation. The copolymers are more stable in long-term, high-temperature service, and more resistant to hot water. Neither type of acetal is resistant to strong mineral acids, but the copolymers are resistant to strong bases. References are frequently made to acetals without identification of polymer type. Such references usually imply the homopolymer material.

Acetal resins are good insulators, having relatively low dissipation factors and dielectric constants over the operating-temperature range for these materials. The electrical properties (Table 23) of acetals are largely retained under exposure to high humidity and water immersion. Critical electrical properties do change with temperature, as shown in Table 24. These properties degrade more drastically above about 200°F. Dissipation factors vs. temperature and frequency for acetals and some other plastics are compared below under polycarbonates.

Acrylics

The general properties of acrylics are presented in Tables 21 and 22. Acrylics, or polymethyl methacrylate, provide numerous engineering advantages. They have exceptional optical clarity and basically good weather resistance, strength, electrical properties, and chemical resistance. They do not discolor or shrink after fabrication, have low water-absorption characteristics and a slow burning rate, and will not flash-ignite. For disadvantages, acrylics are attacked by strong solvents, gasoline, acetone, and other similar fluids.

Parts molded from acrylic powders in their natural state may be crystal-clear and nearly optically perfect. The total light transmittance is 92 percent, and haze measurements average only 1 percent. Light transmittance and clarity can be modified by addition of transparent and opaque colors, most of these being formulated for long outdoor surface. Acrylics can be injection-molded, extruded, cast, vacuum-and-pressure-formed, and machined, although molded parts for load bearing should be carefully analyzed, especially for long-term loading.

Acrylics are divided into the four main product groups of cast sheet, high-impact sheets, molding powder, and high-impact molding powder. Furthermore, they can be modified to improve heat resistance. Modified acrylics can provide about five times the impact strength of the standard grades. Although their outdoor weatherability is inferior to that of the straight acrylics, their indoor aging characteristics are excellent.

Acrylic-PVC alloys, in sheets of various thicknesses for thermoforming, are also commercially available. They are characterized by high impact resistance. Very high elongation at forming temperatures allows extreme draw ratios and deep parts without hot tearing, not normally possible without the PVC addition.

Acrylics have no tendency toward arc tracking. Their high arc resistance and excellent tracking characteristic make them a good choice for certain high-voltage applications such as circuit breakers. The dielectric strength is 450 to 500 V/mil. Acrylics are one of the few plastics that exhibit an essentially linear decrease in dielectric constant and dissipation factor with increase in frequency, as shown in Fig. 27.

Cellulosics

These form a class of thermoplastics that are prepared by various treatments of purified cotton or special grades of wood cellulose. Cellulosics are among the toughest of plastics, are generally economical, and are basically good insulating materials. However, they are temperature-limited and are not as resistant to extreme environments as many other thermoplastics. The four most prominent industrial cellulosics are cellulose acetate, cellulose acetate butyrate, cellulose propionate, and ethyl cellulose. A fifth member of this group is cellulose nitrate. Cellulosic materials are available in a great number of formulas and flows and are manufactured to offer a wide range of properties. They are formulated with a wide range of plasticizers for specific plasticized properties. Supply of plasticizers may be more of a problem than the base plastics in a period of petrochemical shortages.

Cellulose butyrate, propionate, and acetate provide a range of toughness and rigidity that is useful for many applications, especially where clarity, outdoor weatherability, and aging characteristics are needed. The materials are fast-molding plastics and can be provided with hard, glossy surfaces and over the full range of color and texture.

Butyrate, propionate, and acetate are rated in that order in dimensional stability in relation to the effects of water absorption and plasticizers. The materials are slow-burning, although self-extinguishing forms of acetate are available. Special formulations of butyrate and propionate are serviceable outdoors for long periods. Acetate is generally considered unsuitable for outdoor uses. From an application standpoint, the acetates generally are used where tight dimensional stability, under anticipated humidity and temperature, is not required. Hardness, stiffness, and cost are lower than for butyrate or propionate. Butyrate is generally selected over propionate where weatherability, low-temperature impact strength, and dimensional stability are required. Propionate is often chosen for hardness, tensile strength, and stiffness, combined with good weather resistance.

Fluorocarbons

Fluorocarbons are very important in electronic and electrical applications, owing to their excellent electrical properties, which are relatively unaffected by most extreme environments or operating conditions. Some fluorocarbon classes are more used than others, of course. The most widely used fluorocarbon, perhaps, is polytetrafluoroethylene, or *TFE* fluorocarbon. This was the original fluorocarbon, still known to many as Teflon. Correctly speaking, however, Teflon is the trade name for Du Pont TFE and *FEP* fluorocarbons. There are now multiple suppliers of fluorocarbon plastics. Table 21 clarifies the fluorocarbon terminology, as does further discussion below. Tables 22 and 23 give the general properties of these materials.

TABLE 21 Application Information for Thermoplastics

Material	Major application considerations	Common available forms	Typical suppliers and trade names*
ABS (acrylonitrile-butadiene-styrene)............	Extremely tough, with high impact resistance. Can be formulated over a wide range of hardness and toughness properties. Special grades available for plated surfaces with excellent pull-strength values. Good general electric properties but not outstanding for any specific electronic applications	Blow moldings, extrusions, injection moldings, thermoformed parts, laminates, stock shapes, foam	Borg-Warner Corp. (Marbon Cycolac); Monsanto Co. (Lustran); Goodrich Chemical Co. (Abson)
Acetals...........	Outstanding mechanical strength, stiffness, and toughness properties, combined with excellent dimensional stability. Good electrical properties at most frequencies, which are little changed in humid environments up to 125°C	Blow moldings, extrusions, injection moldings, stock shapes	Du Pont, Inc. (Delrin); Celanese Corp. (Celcon)
Acrylics (polymethyl methacrylate)............	Outstanding properties are crystal clarity, and resistance to outdoor weathering. Excellent resistance to arcing and electrical tracking	Blow moldings, castings, extrusions, injection moldings, thermoformed parts, stock shapes, film, fiber	Du Pont, Inc. (Lucite); Rohm and Haas Co. (Plexiglas)
Cellulosics............	There are several materials in the cellulosic family, such as cellulose acetate (CA), cellulose propionate (CAP), cellulose acetate butyrate (CAB), ethyl cellulose (EC), and cellulose nitrate (CN). Widely used plastics in general, but not outstanding for electronic application	Blow moldings, extrusions, injection moldings, thermoformed parts, film, fiber, stock shapes	Eastman Chemical Co. (Tenite); Dow Chemical Co. (Ethocel-EC); Celanese Corp. (Forticel-CAP)
Fluorocarbons: Polytetrafluoroethylene (TFE)............	Electrically one of the most outstanding thermoplastic materials. Exhibits very low electrical losses, and very high electrical resistivity. Useful to over 500°F and to below −300°F. Excellent high-frequency dielectric. Among the best combinations of mechanical and electrical properties but	Compression moldings, stock shapes, film	Du Pont, Inc. (Teflon TFE); Allied Chemical Corp. (Halon TFE)

Material	Properties	Forms/Applications	Manufacturer
	relatively weak in cold-flow properties. Nearly inert chemically, as are most fluorocarbons. Very low coefficient of friction. Nonflammable		
Fluorinated ethylene propylene (FEP)	Properties similar to those of TFE, except useful temperature limited to about 400°F. Easier to mold than TFE	Extrusions, injection moldings, laminates, film	Du Pont, Inc. (Teflon FEP)
Ethylenetetrafluoroethylene (ETFE)	A copolymer of ethylene and tetrafluoroethylene, it has excellent electricals. It is easily processable, and has a rated use temperature of about 300°F. Can be strengthened and used to 400°F by glass fiber reinforcement. Has found considerable use in wiring and cabling	Extrusion, injection molding, film	Du Pont, Inc. (Tefzel)
Perfluoroalkoxy (PFA)	Somewhat like FEP, except unique chemical branching yields improvements in high-temperature properties, higher melting point, and greater thermal stability. Useful to 500°F region	Extrusions, injection molding, film	Du Pont, Inc.
Chlorotrifluoroethylene (CTFE)	Excellent electrical properties and relatively good mechanical properties. Somewhat more stiff than TFE and FEP fluorocarbons, but does have some cold flow. Widely used in electronics, but not quite so widely as TFE and FEP. Useful to about 400°F	Extrusions, isostatic moldings, injection moldings, film, stock shapes	3M Co. (Kel-F); Allied Chemical Corp. (Plaskon CTFE)
Ethylenechlorotrifluoro-ethylene (E-CTFE)	A copolymer of ethylene and chlorotrifluoroethylene. Electrically similar to CTFE, but with easier processability. Can withstand elevated temperatures in the range of 325–350°F; strength and toughness similar to nylon	Extrusion, injection molding, film	Allied Chemical Corp. (Halar)
Polyvinyl fluoride (PVF)	Mostly used as a weatherable, architectural facing sheet. Not widely used in electronics	Extrusions, injection moldings, laminates, film	Du Pont, Inc. (Tedlar)
Polyvinylidine fluoride (PVF$_2$)	One of the easiest of the fluorocarbons to process. Stiffer and more resistant to cold flow than TFE. Good electrically. Useful to about 300°F. A major electronic application is wire jacketing	Extrusions, injection moldings, laminates	Pennsalt Chemicals Corp. (Kynar)

TABLE 21 Application Information for Thermoplastics (Continued)

Material	Major application considerations	Common available forms	Typical suppliers and trade names*
Ionomers.............	Excellent combination of toughness, solvent resistance, transparency, colorability, abrasion resistance, and adhesion. Based on ethylene-acrylic copolymers with ionic bonds. Not widely used in electronics	Film, coatings, injection moldings	Du Pont, Inc. (Surlyn A); Union Carbide Corp. (Bakelite)
Nylons..............	Good general-purpose for electrical and non-electrical applications. Easily processed. Good mechanical strength, abrasion resistance, and low coefficient of friction. There are numerous types of nylons; nylon 6, nylon 6/6, and nylon 6/10 are most common. Some nylons have limited use because of moisture-absorption properties. Nylon 6/10 is best here	Blow moldings, extrusions, injection moldings, laminates rotational moldings, stock shapes, film, fiber	Du Pont, Inc. (Zytel); Allied Chemical Corp. (Plaskon); Union Carbide Corp. (Bakelite)
Parylenes (polyparaxylene)...	Excellent dielectric properties and good dimensional stability. Low permeability to gases and moisture. Produced as a film on a substrate, from a vapor phase. Such vapor-phase polymerization is unique in polymer processing. Used primarily as thin films in capacitors and dielectric coatings. Numerous polymer modifications exist	Film coatings	Union Carbide Corp. (Parylene)
Polyallomers.........	Thermoplastic polymers produced from two monomers. Somewhat similar to polyethylene and polypropylene, but with better dimensional stability, stress-crack resistance, and surface hardness than high-density polyethylene. Electronic application areas similar to polyethylene and polypropylene. One of the lightest commercially available plastics	Blow moldings, extrusions, injection moldings, film	Eastman Chemical Products, Inc. (Tenite)
Polyaryl ether.......	High thermal stability, high impact strength, good chemical resistance, and good electrical properties	Extrusions, injection moldings, coatings	Uniroyal, Inc. (Arylon T)

Material	Properties	Forms	Manufacturers (trade names)
Polyaryl sulfone	High thermal stability, good stiffness and toughness, and good electrical properties	Extrusions, injection moldings	3M Co. (Astrel)
Polycarbonates	Excellent dimensional stability, low water absorption, low creep, and outstanding impact-resistance thermoplastics. Good electrical properties for general electronic application. Available in transparent grades	Blow moldings, extrusions, injection moldings, thermoformed parts, stock shapes, film	Mobay Chemical Co. (Merlon); General Electric Co. (Lexan)
Polyesters: Polyethylene terephthalates	Among the toughest of plastic films with outstanding dielectric-strength properties. Excellent fatigue and tear strength and resistance to acids, greases, oils, solvents. Good humidity resistance. Stable to 135–150°C	Film, sheet, fiber	Du Pont, Inc. (Mylar); Celanese (Celanar)
Thermoplastic polyesters	Excellent combination of stable electrical properties, dimensional stability, thermal stability, low surface friction, resistance to wear, flame resistance, and ease of molding	Injection moldings, extrusions, formed parts	General Electric Co. (Valox); Celanese Plastics Co. (Celanex); Eastman Chemical Co. (Tenite)
Polyethersulfone	High thermal stability, low creep, good dimensional stability, and stable electrical properties to 200°C	Extrusions, injection moldings	Imperial Chemical Industries, Ltd. (Polyethersulphone)
Polyethylenes and polypropylenes (polyolefins)	Excellent electrical properties, especially low electrical losses. Tough and chemically resistant, but weak to varying degrees in creep and thermal resistance. There are three density grades of polyethylene: low (0.910–0.925), medium (0.926–0.940), and high (0.941–0.965). Thermal stability generally increases with density class. Polypropylenes are generally similar to polyethylenes, but offer about 50°F higher heat resistance	Blow moldings, extrusions, injection molding, thermoformed parts, stock shapes, film, fiber, foam	Du Pont, Inc. (Alathon Polyethylene); U.S.I. Chemical Co. (Petrothene Polyethylene); Allied Chemical Corp. (Grex H.D. Polyethylene); Hercules Powder Co. (Hi-Fax H.D. Polyethylene); Hercules Powder Co. (Pro-Fax Polypropylene); Eastman Chemical Co. (Tenite Polyethylene and Polypropylene)

TABLE 21 Application Information for Thermoplastics (Continued)

Material	Major application considerations	Common available forms	Typical suppliers and trade names*
Polyimides and polyamide-imides..........	Among the highest-temperature thermoplastics available, having useful operating temperatures between about 400 and about 700°F or higher. Excellent electrical properties, good rigidity, and excellent thermal stability. Low coefficient of friction. Polyamide-imides and polyimides are chemically similar but not identical in all properties. Some are difficult to process, but are available in molded and block forms, and also as films and resin solutions	Films, coatings, molded and/or machined parts, resin solutions	Du Pont, Inc. (Vespel fabricated blocks, Kapton film, and Pyre-M. L. Resin); Rhodia, Inc. (Kinel); Upjohn Co. (Polyimide); Ciba-Geigy Co. (P13N); Dixon Corp. (Meldin); Amoco Chemicals Corp. (Torlon)
Polymethylpentene..........	High transparency, good form stability, low density, and excellent electrical properties	Injection moldings, extrusions, coatings	Imperial Chemical Industries, Ltd. (TPX)
Polyphenylene oxides..........	Excellent electrical properties, especially loss properties to above 350°F, and over a wide frequency range. Good mechanical strength and toughness	Extrusions, injection moldings, thermoformed stock shapes, film	General Electric Co. (Noryl)
Polyphenylene sulfide..........	Excellent thermal stability, chemical resistance, nonflammability, toughness, stiffness, and electrical properties	Injection moldings and coatings	Phillips Petroleum Co. (Ryton)

Polystyrenes..........	Excellent electrical properties, especially loss properties. Conventional polystyrene is temperature-limited, but high-temperature modifications exist, such as Rexolite or Polypenco cross-linked polystyrene, which are widely used in electronics, especially for high-frequency applications. Polystyrenes are also generally superior to fluorocarbons in resistance to most types of radiation	Blow moldings, extrusions, injection moldings, rotational moldings, thermoformed parts, foam	Dow Chemical Co. (Styron); Monsanto Co. (Lustrex); American Enka Corp. (Rexolite); Polymer Corp. (Polypenco Q-200.5)
Polysulfones..........	Excellent electrical properties and mechanical properties to over 300°F. Good dimensional stability and high creep resistance. Flame-resistant and chemical-resistant. Outstanding in retention of properties upon prolonged heat aging, as compared with tough thermoplastics	Blow moldings, extrusions, injection-mold thermoformed parts, stock shapes, film sheet	Union Carbide Corp. (Polysulfone)
Vinyls..........	Good low-cost, general-purpose thermoplastic materials, but not specifically outstanding electrical properties. Greatly influenced by plasticizers. Many variations available, including flexible and rigid types. Flexible vinyls, especially polyvinyl chloride (PVC), widely used for wire insulation and jacketing	Blow moldings, extrusions, injection moldings, rotational moldings, film sheet	Diamond Alkali Co. (Diamond PVC); Goodyear Chemical Co. (Pliovic); Dow Chemical Co. (Saran)

* This listing is only a very small sampling of the many possible excellent suppliers. The listing is intended only to orient the nonchemical reader into plastic categories. No preferences are implied or intended.

TABLE 22 Physical and Mechanical Properties of Thermoplastics

Material	Coefficient of thermal expansion, in./in. (°C × 10^{-5})	Thermal conductivity, cal/(cm²)(s)(°C)(cm) × 10^{-4}	Water absorption, 24 h, %	Flammability (in./min), 0.125 in.	Specific gravity	Price range per lb	Heat resistance, continuous, °F
Acetal	0.25	1.6	0.25	1.1	1.410–1.425	$0.65	185
ABS	3–10.5	4–9	0.2–0.5	1.0–2	1.01–1.07	$0.33–0.43	160–235
Acrylic		1.4	0.3	9–1.2	1.18–1.19	$0.455–0.75	130–195
Acrylic, high-impact	6.5–10.5	4.0	0.2–0.3	1.1–1.2	1.11–1.18	$0.525–0.70	140–195
Cellulose acetate	8–18	4–8	1.7–4.4	0–2	1.22–1.34	$0.40	140–175
Cellulose acetate butyrate	11–17	4–8	0.9–2.2	0.5–1.5	1.15–1.22	$0.62	140–175
Cellulose propionate	11–16	4–8	1.2–2.8	0.5–1.5	1.16–1.23	$0.62	140–175
Ethyl vinyl acetate	10–20	8	<0.01	Slow burning	0.93–0.95	$0.2775–0.3575	120–170
Chlorotrifluoroethylene	5–7	4–6	Nil	Nil	2.09–2.14	$4.70	390
Fluorinated ethylene propylene	8.3–10.5	5.9	<0.05	Nonflammable	2.16	$5.60–9.60	400
Polytetrafluoroethylene	5.5 (25–60°C)	6	0.01	Nonflammable	2.13–2.18	$3.25	500
Nylon 6	4.6–5.8	5.9	1.5	Self-extinguishing	1.13–1.14	$0.86–1.19	250
Nylon 6/6	8.1	5.8	1.3	Self-extinguishing	1.13–1.15	$0.84–0.875	250
Nylon 6/10	10	5.5		Self-extinguishing	1.07–1.09	$1.26	220
Polyallomer	8–11	2–4	0.4	Slow-burning	0.90–0.906	$0.28	250
Polycarbonate	6.7–7	4.6	0.15	Self-extinguishing	1.2	$0.90–3.05	250
Polyethylene, low-density	10–20	8	<0.05	Slow-burning	0.910–0.925	$0.1525–0.29	140–175
Polyethylene, medium-density	10–20	8	<0.05	Slow-burning	0.926–0.940	$0.17–0.235	150–180
Polyethylene, high-density	10–20	1.9–3.3	<0.01	Slow-burning	0.941–0.965	$0.18–0.32	180–225
Polyethylene, high-molecular-weight	13	8	<0.01	Slow-burning	0.93–0.94	$0.26–0.50	180–225
Polyimide			0.32	Slow-burning to nonburning	1.43		500–600
Polypropylene	3.8–9	2.8–4	<0.01	Slow-burning to nonburning	0.90–1.24	$0.19–0.55	250
Polystyrene	6–8	8	0.03–0.05	0.5–2.5	1.05–1.06	$0.145–0.245	150–190
Polystyrene, high-impact	6.5–8.5	1–3	0.05–0.10	0.5–2.5	1.04–1.06	$0.16–0.27	130–180
Polyurethane	10–20	7.4	0.60–0.80	Slow to self-extinguishing	1.11–1.26	$1.19–1.60	190
Polyvinyl chloride (flexible)	7–25	3–4	0.15–0.75	Self-extinguishing	1.15–1.80	$0.16–0.455	150–175
Polyvinyl chloride (rigid)	5–10	3–5	0.07–0.40	Self-extinguishing	1.33–1.58	$0.21–0.42	160–165
Polyvinyl dichloride (rigid)	7–8	3–4	0.07–0.11	Self-extinguishing	1.50–1.54	$0.50–0.53	195–210
Styrene acrylonitrile (SAN)	7	3	0.23–0.28		1.07–1.08	$0.26–0.30	170–210
Ionomer	12–13	5.8	0.1–1.4	0.4–0.7	0.94–0.96	$0.47–0.49	140
Polymethylpentene	11.7		0.01	0.9–1.1	0.83		250
Polyaryl sulfone	2.6		1.8	1.0	1.36		500
Thermoplastic polyester	5.2	3–4	0.08	Self-extinguishing	1.31	$1.15	300
Polyphenylene oxide			0.06	Slow burning	1.06		250
Polysulfone	3.1 × 10^{-6} in./in. °F	1.8 Btu/(h)(ft)(°F)(in.)	0.22	Self-extinguishing	1.24–1.25	$1.00–1.25	300
Polyethersulfone	5.5	3.2–4.4	0.43	Self-extinguishing	1.37		300–390

TABLE 23 Electrical Properties of Thermoplastics

Material	Volume resistivity, Ω-cm	Dielectric constant, 60 Hz	Dielectric strength,* $\frac{1}{8}$-in. thickness, V/mil	Dissipation or power factor, 60 Hz	Arc resistance, s
Acetal	1–10^{14}	3.7–3.8	500	0.004–0.005	129
ABS	10^{15}–10^{17}	2.6–3.5	300–450	0.003–0.007	45–90
Acrylic	$>10^{14}$	3.3–3.9	400	0.04–0.05	No tracking
Acrylic, high-impact	10^{16}–10^{17}	3.5–3.7	450–480	0.04–0.05	No tracking
Cellulose acetate	10^{10}–10^{12}	3.2–7.5	290–600	0.01–0.10	50–130
Cellulose acetate butyrate	10^{10}–10^{12}	3.2–6.4	250–400	0.01–0.04	
Cellulose propionate	10^{12}–10^{16}	3.3–4.2	300–450	0.01–0.05	170–190
Ethyl vinyl acetate	1.5×10^{8}	3.16	525	0.003	
Chlorotrifluoroethylene	10^{18}	2.65	450	0.015	>360
Fluorinated ethylene propylene	$>10^{18}$	2.1	500	0.0002	>165
Polytetrafluoroethylene	$>10^{18}$	2.1	400	<0.0001	No tracking
Nylon 6	10^{14}–10^{15}	6.1	300–400	0.4–0.6	140
Nylon 6/6	10^{14}–10^{15}	3.6–4.0	300–400	0.014	140
Nylon 6/10	10^{14}–10^{15}	4.0–7.6	300–400	0.04–0.05	140
Polyallomer	$>10^{16}$	2.3	500–1,000	0.0001–0.0005	
Polycarbonate	6.1×10^{15}	2.97	410	0.0001–0.0005	10–120
Polyethylene, low-density	10^{15}–10^{18}	2.28	450–1,000	0.006	Melts
Polyethylene, medium-density	10^{15}–10^{18}	2.3	450–1,000	0.0001–0.0005	Melts
Polyethylene, high-density	6×10^{15}–10^{18}	2.3	450–1,000	0.003–0.002	Melts
Polyethylene, high-molecular-weight	$>10^{16}$	2.3–2.6	500–710	0.0003	Melts
Polyimide	10^{16}–10^{17}	3.5	400	0.002–0.003	230
Polypropylene	10^{15}–10^{17}	2.1–2.7	450–650	0.0007–0.005	36–136
Polystyrene	10^{17}–10^{21}	2.5–2.65	500–700	0.0001–0.0005	60–100
Polystyrene, high-impact	10^{10}–10^{17}	2.5–3.5	500	0.003–0.005	60–90
Polyurethane	2×10^{11}	6–8	850–1,100	0.276	
Polyvinyl chloride (flexible)	10^{11}–10^{15}	5–9	300–1,000	0.08–0.15	
Polyvinyl chloride (rigid)	10^{12}–10^{16}	3.4	425–1,040	0.01–0.02	
Polyvinyl dichloride (rigid)	10^{15}	3.08	1,200–1,550	0.018–0.0208	
Styrene acrylonitrile (SAN)	10^{15}	2.8–3	400–500	0.006–0.008	100–150
Ionomer	$>10^{16}$	2.4–2.5	1,000	0.001	
Polymethylpentene	$>10^{16}$	2.12	700	0.001	
Polyaryl sulfone	3.2×10^{16}	3.94	350–400	0.003	67
Thermoplastic polyester	4×10^{16}	3.3	590	0.002	190
Polyphenylene sulfide	10^{16}	3.1	595	0.0004	
Polyphenylene oxide	10^{17}	2.58	400–500	0.00035	75
Polysulfone	5×10^{16}	2.82	425	0.008–0.0056	122
Polyethersulfone	10^{17}–10^{18}	3.5	400	0.001	

* Short-time.

TABLE 24 Effect of Temperature and Frequency on Dissipation Factor and Dielectric Constant of Acetals[21]

Temp, °F	Dissipation factor		Dielectric constant	
	At 1 kHz	At 100 kHz	At 1 kHz	At 100 kHz
−94	0.0160	0.0130	3.00	2.84
−76	0.0225	0.0200	3.15	2.91
−58	0.0260	0.0300	3.30	3.00
−40	0.0150	0.0360	3.36	3.14
−22	0.0030	0.0300	3.38	3.28
−4.0	0.0011	0.0170	3.40	3.36
14	0.0007	0.0080	3.41	3.39
32	0.0007	0.0036	3.41	4.41
50	0.0007	0.0018	3.41	4.43
68	0.0007	0.0012	3.43	4.43
77	0.0007	0.0010	3.44	4.43
104	0.0009	0.0018	3.52	3.50
140	0.0035	0.0029	3.58	3.55
176	0.0038	0.0060	3.68	3.62

It can be considered for practical purposes that there are eight types of fluoro-carbons, as noted in Table 21. Like other plastics, each type is available in several grades. As mentioned above, the original, basic fluorocarbon, and perhaps still the major one, is TFE fluorocarbon. It has the optimum of electrical and thermal properties but does have the disadvantage of cold flow or creep under mechanical loading. Stronger, filled modifications exist, as do newer, more cold-flow-resistant grades. FEP fluorocarbon is quite similar to TFE in most properties, except that its useful temperature is limited to about 400°F. FEP is much more easily processed, and molded parts are possible with FEP which might not be possible with TFE.

ETFE, a high-temperature thermoplastic fluoropolymer, is readily processed by conventional methods, including extrusion and injection molding. As a copolymer of ethylene and tetrafluoroethylene, it is closely related to TFE. It has good thermal properties and abrasion resistance, and excellent impact strength, resistance to chemicals, and electrical-insulation characteristics. In addition, it will tolerate rela-

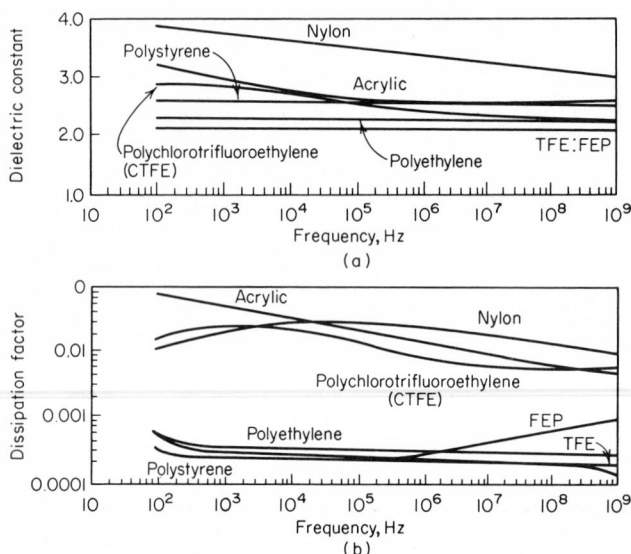

Fig. 27 Effect of frequency on dielectric constant (*a*) and dissipation factor (*b*) of several thermoplastic materials at 23°C.[22]

tively large dosages of radiation and is hot-water- and ultraviolet-resistant. It compares with TFE and acrylics in its excellent arc resistance.

The fluoropolymer's dielectric constant is 2.6 and dissipation factor is 0.0006. These values are not affected by changes in temperature and other environmental conditions, except radiation, as noted in Fig. 28. Dielectric strength is high; resistivity is excellent.

Molding characteristics of the resin are excellent. It exhibits good melt flow, allowing the filling of thin sections (10 mils for small parts). Cycle times are equivalent to those of other thermoplastics.

It is the first fluoroplastic that can be reinforced, not merely filled, with glass fiber. Because the resin will bond to the fibers, strength, stiffness, creep resistance, heat-distortion temperature, and dimensional stability are enhanced. Electrical and chemical properties approach those of the unreinforced resin, while the coefficient of friction is actually lower. Tests indicate the service-temperature limit is approximately 400°F. The fluoropolymer, reinforced with glass, can be molded by conventional methods with rapid molding cycles.

PFA resins are a new class of melt-processable fluoroplastics. They combine the ease and economies of thermoplastic processing with high-temperature performance in the range of TFE fluorocarbon resins.

PFA resins resemble FEP fluorocarbon resins in having a branched polymer chain that provides good mechanical properties at melt viscosities much lower than those of TFE. However, the unique branch in PFA is longer and more flexible, leading to improvements in high-temperature properties, higher melting point, and greater thermal stability.

In use, PFA resins have the desirable properties typical of fluorocarbons, including resistance to virtually all chemicals, antistick, low coefficient of friction, excellent electrical characteristics, low smoke, excellent flammability resistance, ability to perform in temperature extremes, and excellent weatherability. Its strength and stiffness at high operating temperatures are at least equivalent to those of TFE, while its creep resistance appears to be better over a wide temperature range.

Film and sheet are expected to find use as electrical insulations in flat cables and circuitry, and in laminates used in electrical and mechanical applications.

The *CTFE* resins, like the FEP materials, are melt-processable and can be injection-, transfer-, and compression-molded or screw-extruded. Compared with TFE, CTFE has great tensile and compressive strength within its service-temperature range. However, at the temperature extremes, CTFE does not perform as well as TFE for parts such as seals. At the low end of the range, TFE has somewhat better physical properties; at the higher temperatures, CTFE is more prone to stress cracking and other difficulties.

Electrical properties of CTFE are generally excellent, but dielectric losses are higher than those of TFE. Chemical resistance is poorer than that of TFE, but radiation resistance is better. CTFE does not have the low-friction and bearing properties of TFE.

E-CTFE copolymer, a copolymer of ethylene and CTFE, is a strong, highly impact-resistant material that retains useful properties over a broad temperature range. E-CTFE is available in pellet and powder form and can be readily extruded, injection-, transfer-, and compression-molded,

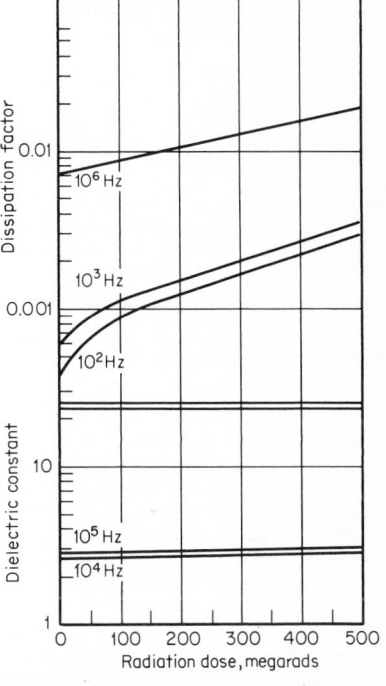

Fig. 28 Effect of radiation dose on room-temperature electrical properties (ASTM D 150) of ETFE fluoropolymer.[23]

rotocast, and powder-coated. Its tensile strength, impact resistance, hardness, creep properties, and abrasion resistance at ambient temperature are comparable with those of nylon 6. E-CTFE retains its strength and impact resistance down to cryogenic temperatures.

E-CTFE also exhibits excellent ac loss properties. Its dielectric constant measures 2.5 to 2.6 over the frequency range of 10^2 to 10^6 Hz and is unaffected by temperature over the range of −80 to 300°F. Its dissipation factor is low, ranging from 0.0008 to 0.015 depending on temperature and frequency. It has a dielectric strength of 2,000 V/mil in 10-mil wall thickness and a volume resistivity of 10^{15} Ω-cm.

Vinylidene fluoride is another melt-processable fluorocarbon capable of being injection- and compression-molded and screw-extruded. Its 20 percent lower specific gravity compared with that of TFE and CTFE and its good processing characteristics permit economy and provide excellent chemical and physical characteristics. Useful temperature range is −80 to +300°F. Although it is stiffer and has higher resistance to cold flow than TFE, its chemical resistance, useful temperature range, antistick properties, lubricity, and electrical properties are lower. Principal applications for electrical insulation have been wire insulation and heat-shrinkable tubing.

Fluorocarbons have excellent electrical properties. TFE and FEP fluorocarbons in particular have low dielectric constants and dissipation factors which change little with temperature or frequency. This is shown in Figs. 27 and 29, which also present data for some other thermoplastics.

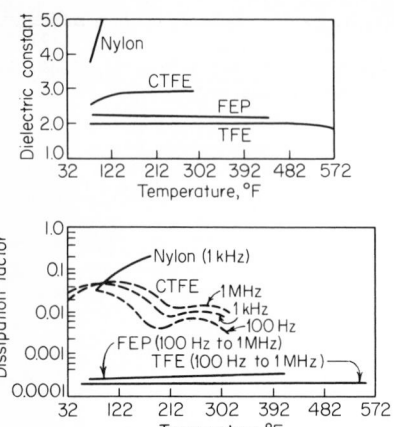

Fig. 29 Effect of temperature on dielectric constant and dissipation factor of fluorocarbons and nylon.[22]

The dielectric strength of TFE and FEP resins is high and does not vary with temperature and thermal aging. Initial dielectric strength is very high as measured by the ASTM short-time test. As with most dielectric materials, the value drops as thickness of the specimen increases, as shown in Fig. 30. Further, the dielectric strength is a function of frequency, as noted in Fig. 31. Life at high dielectric stresses is dependent on corona discharge, as shown in Fig. 32. The absence of corona, as in special wire constructions, permits very high voltage stress without damage to either TFE or FEP resins. Changes in relative humidity or physical stress imposed upon the material do not diminish life at these voltage stresses.

Surface arc resistance of TFE and FEP resins is high, and is not affected by heat aging. When these resins are subjected to a surface arc in air, they do not crack or form a carbonized conducting path. When tested by the procedure of ASTM D 495, they pass the maximum time of 300 s without failure.

The unique nonstick surface of these resins helps reduce surface arc phenomena in two ways: (1) it helps prevent formation of surface contamination, thereby reducing the possibility of arcing; (2) if an arc is produced, the discharge frequently cleans the surface of the resin, increasing the time before another arc.

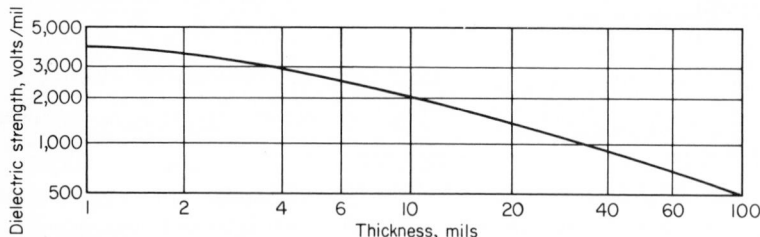

Fig. 30 Short-term dielectric strength of TFE fluorocarbon as a function of thickness.[22]

Volume resistivity (10^{18} Ω-cm) and surface resistivity (10^{16} Ω/sq) for both FEP and TFE resins are at the top of the measurable range. Neither resistivity is affected by heat aging or temperatures up to recommended service limits.

Ionomers

Outstanding advantages of this polymer class are combinations of toughness and transparency, and transparency and solvent resistance. Ionomers have high melt strength for thermoforming and extrusion-coating processes, and a broad processing-temperature range.

Limitations of ionomers include low stiffness, susceptibility to creep, low heat-distortion temperature, and poor ultraviolet resistance unless stabilizers are added, where these properties are important.

Most ionomers are very transparent. In 60-mil sections, internal haze ranges from 5 to 25 percent. Light transmission ranges from 80 to 92 percent over the visible region, and in specific compositions high transmittance extends into the ultraviolet region.

Most ionomers have good dielectric characteristics over a broad frequency range. The combination of these electrical properties, high melt strength, and abrasion resistance qualifies these materials for insulation and jacketing of wire and cable.

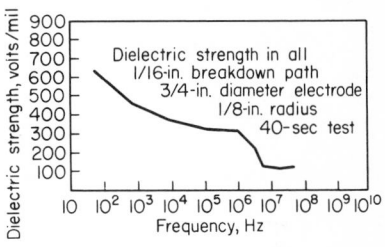

Fig. 31 Short-term dielectric strength of TFE flurocarbon vs. of frequency.[22]

Nylons

Also known as polyamides, nylons are strong, tough thermoplastics having good impact, tensile, and flexural strengths from freezing temperatures up to 300°F; excellent low-friction properties; and good electrical resistivities. They are not generally recommended for high-frequency, low-loss applications. Also, since all nylons absorb some moisture from environmental humidity, moisture-absorption characteristics must be considered in designing with these materials. They will absorb anywhere from 0.5 to nearly 2 percent moisture after 24 h water immersion. There are low-moisture-absorption grades, however; and hence moisture-absorption properties do not have to limit the use of nylons, especially for the lower-moisture-absorption grades. Application information on nylons is given in Table 21, and data for three grades are presented in Tables 22 and 23.

Regarding the identifications for the grades of nylon, certain nylons are identified by the number of carbon atoms in the diamine and dibasic acid used to produce that particular nylon. For instance, nylon 6/6 is the reaction product of hexa-

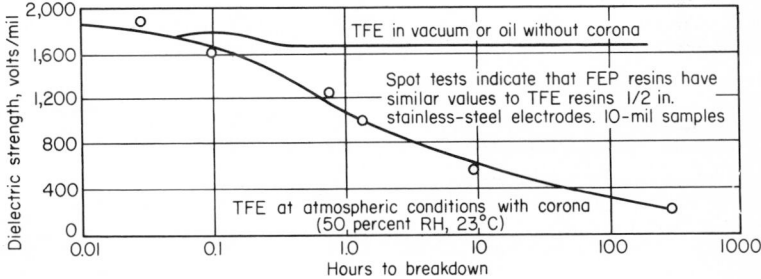

Fig. 32 Insulation life vs. continuously applied voltage stress for TFE and FEP fluorocarbons.[22]

methylene and adipic acid, both of which are materials containing six carbon atoms in their chemical structure. The commercially available nylons are 6/6, 6, 6/10, 8, 11, and 12. Grades 6 and 6/6 are the strongest structurally; grades 6/10 and 11 have the lowest moisture absorption, best electrical properties, and best dimensional stability; and grades 6, 6/6, and 6/10 are the most flexible. Grades 6, 6/6, and 8 are heat-sealable, with nylon 8 being capable of cross linking. Another grade, nylon 12, offers advantages similar to those of grades 6/10 and 11, but offers lower cost possibilities owing to being more easily and economically processed. Also a high-temperature type of nylon exists. It is discussed separately below.

Because of the combination of mechanical toughness, the ability to operate continuously at temperatures as high as 105°C, and good resistance to ambient atmospheres, nylons have become widely used in electrical applications. Examples are coil forms, plugs, switch parts, overcoatings on wire insulations, and a wide variety of molded parts.

Electrically, nylon is not outstanding, but it is adequate for most 60-Hz power applications. Although the dissipation factor may be high under some conditions of temperature, this usually means little actual power loss in 110- to 220-V applications. Mechanical requirements in insulation thickness will keep the voltage per mil value low. Power loss is proportional to (volts/mil)^2 × dielectric constant × dissipation factor × frequency.

Volume resistivity, dielectric strength, dielectric constant, and dissipation factor are all variable with moisture content and temperature of the nylon as well as with the specific kind of nylon.

The effect of frequency on the dielectric constant and dissipation factor of nylon is shown in Fig. 27. The effect of temperature on these two electrical properties is presented in Fig. 29. Resistivity decreases rapidly with moisture content, dropping as low as 10^{12} Ω-cm at 4 percent moisture, and as low as 10^9 Ω-cm at 8 percent moisture. Nylons do not compare favorably with the good electrical thermoplastics such as fluorocarbons or polystyrene in these two electrical parameters. Nylons are good general-purpose electrical materials, however, as noted in Table 12.

High-Temperature Nylon (Aramid)

In addition to conventional nylon molding resins, a high-temperature grade of nylon exists. This class of polymer has recently been reclassified as aramids. Known as Nomex, this high-temperature aramid retains about 60 percent of its strength at 475 to 500°F temperatures, which would melt conventional nylons. This high-temperature aramid has good dielectric strength (constant to 400°F and 95 percent relative humidity) and volume resistivity, and low dissipation factor, as shown in Fig. 33. Fabricated primarily in sheet, fiber, and paper form, this material is being used in wrapped electrical-insulation constructions such as transformer coils and motor stators. It retains high tensile strength, resistance to wear, and electrical properties after prolonged exposure at temperatures up to 500°F.

Parylenes

Parylene is the generic name for members of a thermoplastic polymer series developed by Union Carbide, which are unique among plastics in that they are produced as thin films by vapor-phase polymerization.

Parylenes can also be deposited onto a cold condenser and then stripped off as a free film, or they can be deposited onto the surface of objects as a continuous, adherent coating in thicknesses ranging from 1,000 Å (about 0.004 mil) to 3 mils or more. Deposition rate is normally about 0.5 micron/min (about 0.02 mil). On cooled substrates, the deposition rate can be as high as 1.0 mil/min.

Parylenes are good dielectric materials, with parylene N having a consistent dielectric constant of 2.65 and a dissipation factor that increases from 0.0002 to 0.0006, over the range of 60 Hz to 1 MHz, at room temperature. The chemical modification of parylene C, offering lower permeability to moisture and gases, raises the dielectric constant to 2.9 to 3.1, and the dissipation factor to 0.012 to 0.020 over the above frequency range. Volume-resistivity values approximate 10^{17} Ω-cm at room temperature, and short-time dielectric-strength values are 5,600 V/mil for parylene C, and 7,000 V/mil for parylene N, in 1-mil test samples. Dielectric-strength values decrease with increasing thickness, of course. The effect of temperature on key properties is shown in Fig. 34.

Polyallomers, Polyethylenes, and Polypropylenes

This large group of polymers, basically divided into the three separate polymer groups listed under this heading, all belong to the broad chemical classification known as polyolefins. There are, of course, many categories or types within each of these three polymer groups. Although property variations exist among the polymer groups and among the subcategories within the polymer groups, there are also many similarities. The differences or unique features of each will be discussed below. The similarities are, broadly speaking, appearance, general chemical characteristics, and electrical properties. The differences are more notably in physical and thermal-stability properties. Basically, polyolefins are all waxlike in appearance and extremely inert chemically, and they exhibit decreases in physical strength at somewhat lower temperatures than the higher-performance engineering thermoplastics. All these materials have nearly identical very low dissipation factors and dielectric constants,

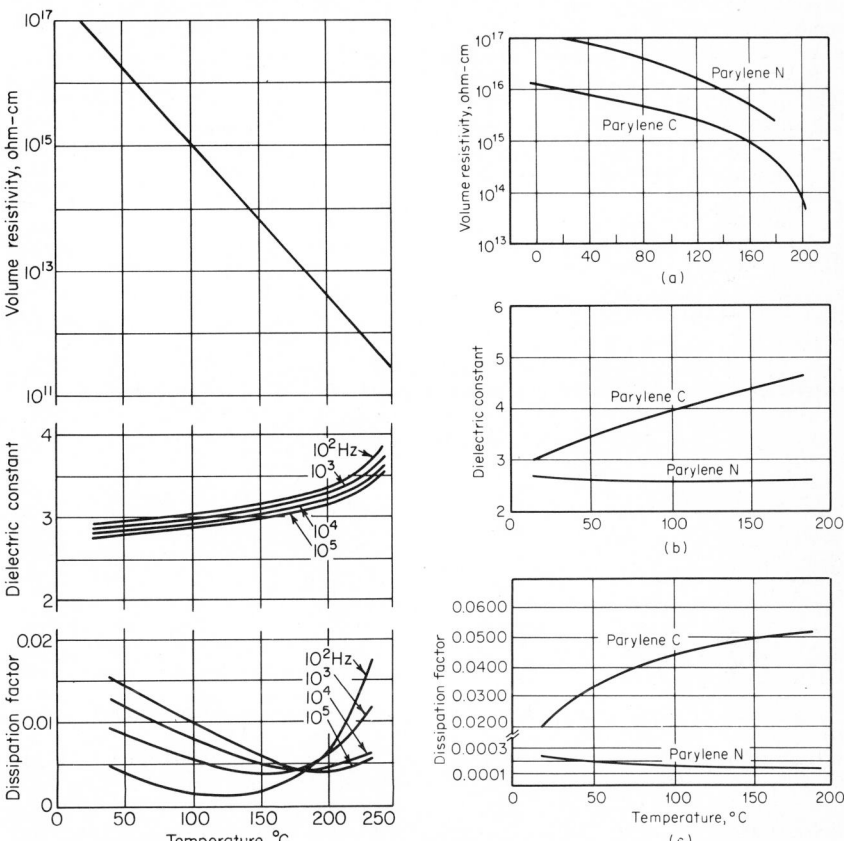

Fig. 33 Effect of temperature on electrical properties of 10-mil high-temperature aramid.[24]

Fig. 34 The effect of temperature on (a) volume resistivity, (b) dielectric constant, and (c) dissipation factor of parylene N and parylene C at 10^3 Hz.[25]

relatively independent of frequency and temperature over their operating range. These data for polyethylene are shown in Fig. 27. Resistivity values are very high among plastics. Polyethylenes were the first of these materials developed and hence, for some of the original types, have the weakest mechanical properties. The later-developed polyethylenes, polypropylenes, and polyallomers offer improvements. The unique features of each of these three polymer groups will be mentioned at this point. The application information and physical and electrical properties of each are given in Tables 10 to 12.

Polyethylenes   Polyethylenes are among the best-known plastics and come in three main classifications based on density: low, medium, and high. These density ranges are 0.910 to 0.925, 0.925 to 0.940, and 0.940 to 0.965, respectively. These three density grades are also sometimes known as types I, II, and III, respectively. All polyethylenes are relatively soft, and hardness increases as density increases. Generally, the higher the density, the better are the dimensional stability and physical properties, particularly as a function of temperature. Thermal stability of polyethylenes ranges from 190°F for the low-density material up to 250°F for the high-density material. Toughness is maintained to low negative temperatures.

Polypropylenes   Polypropylenes are chemically similar to polyethylenes but have somewhat better physical strength at a lower density. The density of polypropylenes is among the lowest of all plastic materials, ranging from 0.900 to 0.915.

Polypropylenes offer more a balance of properties than a single unique property, with the exception of flex-fatigue resistance. These materials have an almost infinite life under flexing, and they are thus often said to be "self-hinging." Use of this characteristic is widespread in plastic hinges—polypropylenes are perhaps the only thermoplastics surpassing all others in combined electrical properties, heat resistance, rigidity, toughness, chemical resistance, dimensional stability, surface gloss, and melt flow, at a cost less than that of most others.

Polyallomers   Polyallomers are also polyolefin-type thermoplastic polymers produced from two or more different monomers, such as propylene and ethylene, which would produce propylene-ethylene polyallomer. As can be seen, the monomers, or base chemical materials, are similar to those of polypropylene or polyethylene. Hence, as was mentioned above, and as would be expected, many properties of polyallomers are similar to those of polyethylenes and polypropylenes. Having a density of about 0.9, they, like polypropylenes, are among the lightest plastics.

Cross-linked polyolefins   While polyolefins have many outstanding characteristics, they, like all thermoplastics to some degree, tend to creep or cold-flow under the influence of temperature, load, and time. In order to improve this and some other properties, considerable work has been done on developing cross-linked polyolefins, especially polyethylenes. The cross-linked polyethylenes offer thermal-performance improvements of up to 25°C or more. Cross linking has been achieved primarily by chemical means and by ionizing radiation. Products of both types are available. Radiation-cross-linked polyolefins have gained particular prominence in a heat-shrinkable form. This is achieved by cross linking the extruded or molded polyolefin using high-energy electron-beam radiation, heating the irradiated material above its crystalline melting point to a rubbery state, mechanically stretching to an expanded form (up to four or five time the original size), and cooling the stretched material. Upon further heating, the material will return to its original size, tightly shrinking onto the object around which it has been placed. Heat-shrinkable boots, jackets, tubing, etc., are widely used. Also, irradiated polyolefins, sometimes known as irradiated polyalkenes, are important materials for certain wire and cable jacketing applications.

Polyaryl Ether

Another of the relatively new thermoplastics which can be classed as engineering thermoplastics is polyaryl ether, available from Uniroyal, Inc., as Arylon T. A unique combination of three important, desirable properties makes polyaryl ether outstanding. The material has (1) a high heat-deflection temperature, (2) the highest impact strength of all engineering plastics except polycarbonate in thin sections—Izod impact strength of 8 ft-lb/in. notch at 72°F and 2.5 ft-lb/in. notch at −20°F; (3) excellent chemical resistance, resisting organic solvents except chlorinated aromatics, esters, and ketones, and it appears to resist hydrolysis, withstanding prolonged immersion in boiling water. Markets for Arylon include the automotive, appliance, and electrical industries.

Polyaryl Sulfone

Yet one more of the high-thermal-stability thermoplastics is polyaryl sulfone, available from 3M Company as Astrel 360. Polyaryl sulfone offers the unique combination of thermoplasticity and retention of structurally useful properties at 500°F.

This material is supplied in both pellet and powder form. Its thermoplastic nature allows it to be fabricated by injection-molding or extrusion techniques on standard commercially available equipment.

Polyaryl sulfone is characterized by a very high heat-deflection temperature, 525°F at 264 lb/in.[2]. This is a consequence of its high glass-transition temperature, 550°F, rather than the effect of filler reinforcement or a crystalline melting point. At 500°F it maintains a tensile strength in excess of 4,000 lb/in.[2] and a flexural modulus of 250,000 lb/in.[2]. The material has been classified by Underwriters' Laboratories test procedures as self-extinguising when exposed to a flame in air.

Polyaryl sulfone has been successfully used for electrical-connector bodies that are required to meet severe structural and environmental conditions, and for parts in other critical electromechanical devices. The dielectric-constant and dissipation-factor values are shown in Fig. 35, as affected by frequency, temperature, and humidity. Room-temperature volume resistivity is 3.2×10^{16} Ω-cm, and the short-time dielectric strength is 350 V/mil for a $\frac{1}{16}$-in.-thick sample. Arc resistance is 67 s.

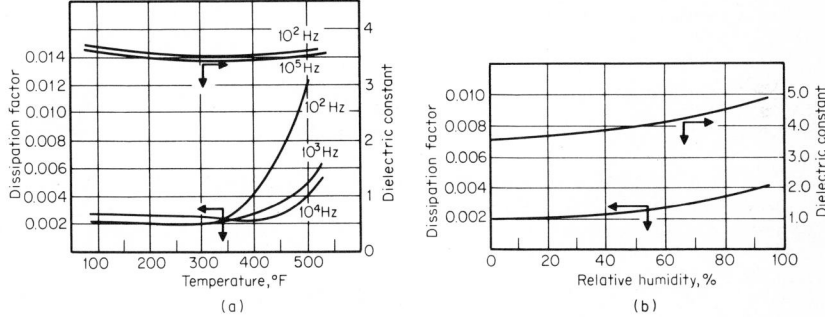

Fig. 35 Dissipation factor and dielectric constant of polyaryl sulfone as a function of (a) frequency and temperature and (b) relative humidity.[26]

Polycarbonates

This group of plastics is also among those which are classified as engineering thermoplastics because of their high performance characteristics in engineering designs. Polycarbonates are tough, rigid, and dimensionally stable and are available as transparent or colored parts. They are easily fabricated with reproducible results, using molding or machining techniques. An important molding characteristic is the low and predictable mold shrinkage, which sometimes gives polycarbonates an advantage over nylons and acetals for close-tolerance parts. As with most other plastics containing aromatic groups, radiation stability is high. The six most commonly useful properties of polycarbonates are creep resistance, high heat resistance, dimensional stability, good electrical properties, self-extinguising properties, and the exceptional impact strength, which compares favorably with that of some metals and exceeds that for essentially all competitive plastics.

In addition to the performance of polycarbonates as engineering materials, as discussed in the following paragraphs, polycarbonates are also alloyed with other plastics, to increase the strength and rigidity of these other plastics. Notable among the plastics with which polycarbonates have been alloyed are the ABS plastics. The application information on polycarbonates is given in Table 21, the physical and mechanical data in Table 22, and the electrical data in Table 23. More detailed electrical data are given below.

The basic electrical properties of polycarbonates are very good. The six most important electrical characteristics of polycarbonates are shown in Figs. 36 to 41, inclusive. The dissipation factor and dielectric constant, as a function of temperature, are shown in Figs. 37 and 39. It can be observed that there is no major change up to about 150°C. Both begin to increase beyond the 150 to 175°C temperature area, however. The dielectric-constant value remains at about 3 up to 100 MHz or higher, whereas the power factor or loss factor does increase somewhat in the frequency range. Electrical properties remain relatively stable in high-humidity environments, which also helps in many design areas.

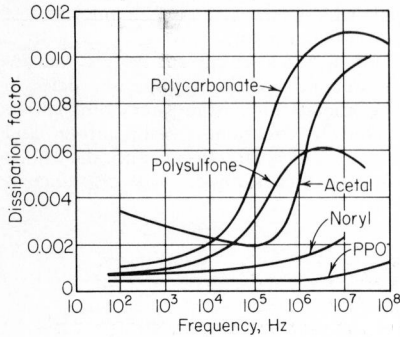

Fig. 36 Dissipation factor vs. frequency for several thermoplastics.[27]

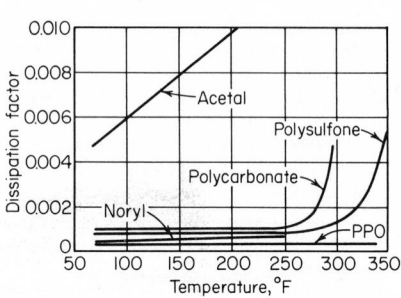

Fig. 37 Dissipation factor at 60 Hz vs. temperature for several thermoplastics.[27]

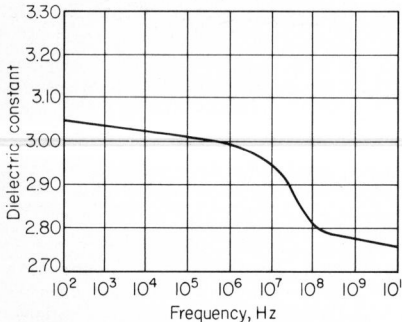

Fig. 38 Dielectric constant vs. frequency for polycarbonates at 23°C.[28]

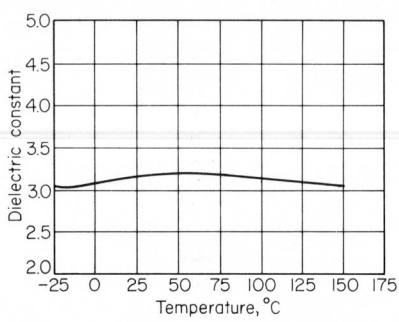

Fig. 39 Dielectric constant vs. temperature for polycarbonates at 60 Hz.[28]

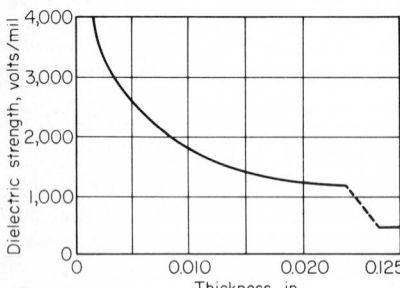

Fig. 40 Dielectric strength vs. thickness for polycarbonates.[28]

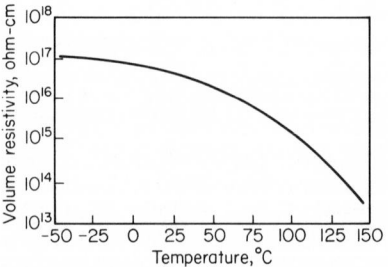

Fig. 41 Volume resistivity vs. temperature for polycarbonates.[28]

Polyesters

Thermoplastic polyesters have been and are currently used extensively in the production of film and fibers. These materials are denoted chemically as polyethylene terephthalate (see Table 21) and are discussed under Films in Chap. 1.

In the past few years a new class of molding and extrusion grades of thermoplastic polyesters has been made available. This thermoplastic polyester, identified in Table 21, is highly crystalline, with a melting point about 430°F. It is fairly translucent in thin molded sections and opaque in thick sections but can be extruded into transparent thin film. Both unreinforced and reinforced formulations are extremely easy to process and can be molded in very fast cycles.

The unreinforced resin offers the characteristics of (1) hard, strong, and extremely tough, (2) high abrasion resistance, low coefficient of friction, (3) good chemical resistance, low moisture absorption and resistance to cold flow, (4) good stress-crack and fatigue resistance, (5) good electrical properties, and (6) good surface appearance. Electrical properties are relatively stable up to the rated temperature limits of about 300°F.

The glass-reinforced polyester resins are unusual in that they are the first thermoplastics that can compare with, or are better than, thermosets in electrical, mechanical, dimensional, and creep properties at elevated temperatures (300°F), while having superior impact properties.

The glass-fiber concentration usually ranges from 10 to 30 percent in commercially available grades. In molded parts the glass fibers remain slightly below the surface so that finished items have a very smooth surface finish as well as an excellent appearance. Properties of reinforced and unreinforced polyesters are shown in Table 25.

Unreinforced resins are primarily used in housings requiring excellent impact, and in moving parts such as gears, bearings and pulleys, packaging applications, and writing instruments. The flame-retardant grades are primarily aimed at TV/radio and electrical/electronics parts, and business machine and pump components.

Reinforced resins are being used in automotive, electrical/electronic, and general industrial areas, replacing thermosets, other thermoplastics, and metals. Electrical and mechanical properties coupled with low finished-part cost are enabling reinforced thermoplastic polyesters to replace phenolics, alkyds, and DAP, and glass-reinforced thermoplastics in selected applications.

Polyethersulfone

This thermoplastic is also a new high-temperature engineering thermoplastic with an outstanding long-term resistance to creep at temperatures up to 150°C. It is capable of being used continuously under load at temperatures of up to about 180°C and, in some low-stress applications, up to 200°C.

Polyethersulfone retains good electrical properties at temperatures in excess of 200°C. For a polar material, it has a low dissipation factor which remains reasonably constant at about 0.001 over the temperature range 20 to 180°C. It is unaffected by soldering techniques. Polyethersulfone is a development of the British Imperial Chemical Industries, Ltd.

Polyimides and Polyamide-Imides

These two somewhat related groups of plastics have some outstanding properties for electronic applications. This is due to their combination of high-temperature stability up to and beyond 500°F, good electrical and mechanical properties which are also relatively stable from low negative temperatures to high positive temperatures, dimensional stability (low cold flow) in most environments, excellent resistance to ionizing radiation, and very low outgassing in high vacuum. Polyamide-imides and polyimides have very good electrical properties, though not as good as those of TFE fluorocarbons, but they are much better than TFE fluorocarbons in mechanical and dimensional-stability properties. This provides advantages in many high-temperature electronic applications. All these properties also make polyamide-imides and polyimides excellent material choices in extreme environments of space and temperature. These materials are available as solid (molded and machined) parts, films, laminates, and liquid varnishes and adhesives. Since data are relatively similar except for form factor, the data presented below are for solid polyimides, unless indicated otherwise. Films are quite similar to Mylar, except for improved high-temperature capabilities, and are discussed in Chap. 1 under Films.

The combination of very good electrical properties, high strength, and excellent thermal and radiation resistance makes polyimide parts outstanding candidates for

TABLE 25 Typical Properties of Thermoplastic Polyesters[29]

Property	ASTM test	Unreinforced	Unreinforced SE-O	30% glass reinforced	Glass reinforced SE-O
Physical:					
Specific gravity.........	D 570	1.31	1.42	1.52	1.58
Water absorption,					
24 h, 73°F...........		0.08	0.09	0.07	0.07
Equilibrium...........		0.34	0.34	0.40	0.40
Mechanical:					
Tensile strength,					
$lb/in.^2$...............	D 638	8,000	9,000	17,000	17,000
Flexural strength,					
$lb/in.^2$...............	D 790	13,000	15,000	27,000	30,000
Flexural modulus,					
$lb/in.^2 \times 10^3$.........	D 790	340	375	1,200	1,200
Izod, notched,					
ft-lb/in. notch........	D 256	1.2	0.9	1.4	1.4
Thermal:					
Deflection temp,					
$264\ lb/in.^2$ °F........	D 648	130	160	416	416
Melting point.........		437	437	437	437
Flammability:					
Oxygen index........		22.0	29.0	23.0	29.0
UL Subject 94.......		SB	SE-O	SB	SE-O
Electrical:					
Volume resistivity,					
Ω-cm 72°F...........	D 257	4×10^{16}	NA	8×10^{15}	5×10^{15}
6 weeks in 72°F water....		NA	NA	1×10^{16}	3×10^{15}
Dielectric strength,					
short-time, V/mil,					
72°F................	D 149	590	750	750	750
Dielectric constant					
(72°F)...............	D 150				
100 Hz.............		3.3	NA	3.77	3.91
10^6 Hz.............		3.1	NA	3.63	3.74
100 Hz 6 weeks in					
72°F water........		NA	NA	3.95	4.07
10^6 Hz.............		NA	NA	3.69	3.60
Dissipation factor......	D 150				
100 Hz, 72°F........		0.002	NA	0.002	0.006
10 Hz..............		0.02	NA	0.02	0.014
100 Hz 6 weeks in					
72°F water........		NA	NA	0.005	0.018
10^6 Hz.............		NA	NA	0.022	0.022
Arc resistance, s........	D 495	190	NA	130	146

SE-O is self-extinguishing.

electrical applications in most severe environments. The dielectric constant, shown in Fig. 42, decreases gradually from 3.5 at room temperature to 3.0 at 500°F. At a given temperature, dielectric constant is essentially unchanged with frequency variations in the range of 100 Hz to 100 kHz. Dissipation factor, shown in Fig. 43, is influenced by both temperature and frequency. Up to about 212°F, dissipation factor increases with increasing frequency. From 212 to 400°F, frequency has essentially no effect, while above 400°F, dissipation factor decreases with increasing frequency.

Both dielectric constant and dissipation factor increase with increasing moisture content. For example, at 1 kHz and room temperature, the dielectric constant of a

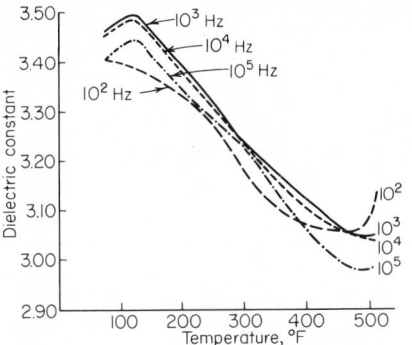

Fig. 42 Dielectric constant vs. temperature and frequency for unfilled polyimide.[30]

Fig. 43 Dissipation factor vs. temperature and frequency for unfilled polyimide.[30]

dry test bar of unfilled polyimide is 3.1, and the dissipation factor is 0.001. With a moisture content of 2.4 percent (obtained after 300 h immersion in water at room temperature), these values are 4.0 and 0.002, respectively. Drying will restore the original values.

The volume resistivity of samples molded from unfilled polyimide is 10^{17} Ω-cm at room temperature. This value decreases linearly to 10^{11} Ω-cm at 572°F (Fig. 44). Surface density is 10^{16} Ω/sq at room temperature, and decreases linearly to 5×10^{10} at 572°F.

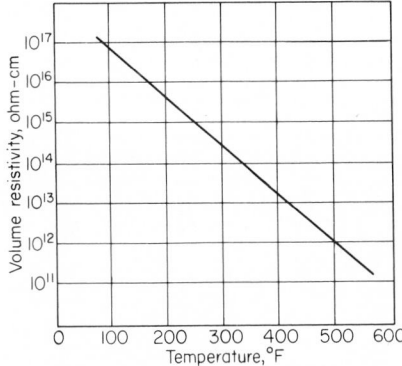
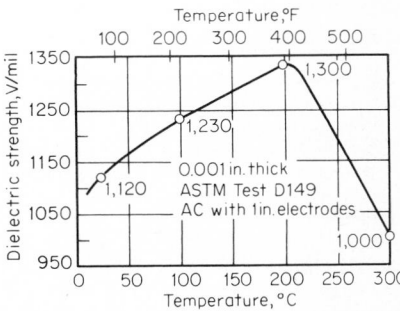

Fig. 44 Volume resistivity vs. temperature for unfilled polyimide.[03]

Fig. 45 Dielectric strength of unfilled polyimides vs. temperature in air.[30]

Polyimide parts as fabricated from unfilled resin have a very high arc resistance of 230 s. However, when exposed to arcing, a carbon track is formed. The corona resistance is superior to that attainable with fluorocarbons and polyethylenes. For example, at 200 V/mil (60 Hz and room temperature), corona life is 2,200 h. The effect of temperature on dielectric strength in air is shown in Fig. 45.

Polymethylpentene

This is another thermoplastic developed by the British Imperial Chemical Industries, Ltd. Interest in this plastic is due to item combination of transparency and relatively high melting point. Named TPX, this polymer has four combined properties of (1) a high crystalline melting point of 464°F, coupled with useful mechanical properties at 400°F and retention of form stability to near melting; (2) transparency with a light-transmission value of 90 percent in comparison with 88 to 92 percent for polystyrene and 92 percent for acrylics; (3) a density of 0.83, which is close to the theoretical minimum for thermoplastics materials; and (4) excellent electrical properties with power factor (0.001 to 0.0001), dielectric constant (2.12), dielectric strength (700 V/mil, ⅛-in.-thick sample) and volume resistivity ($> 10^{16}$ Ω-cm), of the same order as PTFE fluorocarbon.

Applications for polymethylpentene have been developed in the fields of lighting and in the automotive, appliance, and electrical industries.

Polyphenylene Oxides

This family of engineering plastics is characterized by outstanding dimensional stability at elevated temperatures, broad temperatures use range, outstanding hydrolytic stability, and excellent dielectric properties over a wide range of frequencies and temperatures. They are available in various standard grades that have been developed to provide a choice of performance characteristics to meet a wide range of engineering-application requirements.

Outstanding features are (1) excellent mechanical properties over temperatures from below −40°F to above 300°F; (2) self-extinguishing, nondripping characteristics; (3) excellent dimensional stability with low creep, high modulus, and low water absorption; (4) good electrical properties; (5) excellent resistance to aqueous chemical environments; (6) ease of processing the materials with injection-molding and extrusion equipment; and (7) excellent impact strength.

Polyphenylene oxides have excellent electrical properties which remain relatively constant with frequency and temperature over their safe operating-temperature range. The effect of frequency on dissipation factor is given in Fig. 36 and the effect of temperature on 60-Hz dissipation factor in Fig. 37. The dielectric constant is low and relatively stable up to the 220 to 225°F range. Dielectric strength is 500 to 550 V/mil in ⅛-in. sections. Electrical properties are relatively unaffected by temperature and humidity, as shown in Tables 26 to 28.

Polyphenylene Sulfides

Another high-performance thermoplastic is polyphenylene sulfide (PPS), available in both molding and coating formulations. This strong material has use temperatures to about the 350 to 450°F range, depending on use, and filled composition.

At normal temperatures the unfilled polymer is a hard material with high tensile and flexural strength. Substantial increases in these properties are realized by the addition of fillers, especially glass. Tensile strength and flexural modulus decrease with increasing temperature, leveling off somewhat between 250 and 450°F, with good tensile strength and rigidity retained up to 500°F. With increasing temperature there is a marked increase in elongation and a corresponding increase in toughness.

The mechanical properties of PPS are unaffected by long-term exposure in air at 450°F. Molded items are expected to have applications where chemical-resistance and high-temperature properties are of prime importance. For injection-molding applications, a 40 percent glass-filled grade is recommended for flexural strength.

The electrical properties of unfilled and 40 percent glass-filled polyphenylene sulfide are shown in Table 29.

Polystyrenes

Polystyrenes represent an important class of thermoplastic materials in the electronics industry because of very low electrical losses. Mechanical properties are adequate within operating-temperature limits, but polystyrenes are temperature-limited with normal temperature capabilities below 200°F. Polystyrenes can, however, be cross-linked to produce a higher-temperature material, as noted in Table 21.

TABLE 26   Resistivity of 30 Percent Glass-filled Polyphenylene Oxides at Several Temperature and Humidity Conditions[31]

Condition	Value
Volume resistivity, dry, 73°F, Ω-cm............	2.1×10^{17}
Volume resistivity, 50% RH, 73°F, Ω-cm......	1.7×10^{17}
Volume resistivity, 160°F, Ω-cm..............	1.2×10^{16}
Volume resistivity, 250°F, Ω-cm..............	2×10^{15}
Surface resistivity, 50% RH, 73°F, Ω/sq........	3×10^{17}

TABLE 27   Dielectric Constant of 30 Percent Glass-filled Polyphenylene Oxides at Several Temperature and Humidity Conditions[31]

Temp, °F	Frequency, Hz					
	60	10^2	10^3	10^4	10^5	10^6
	Dielectric constant					
73, dry..............	2.93	2.93	2.92	2.92	2.91	2.91
73, 50% RH..........	2.93	2.93	2.93	2.93	2.92	2.92
160.................	2.92	2.92	2.91	2.91	2.90	2.90
250.................	2.90	2.90	2.89	2.88	2.88	2.87

TABLE 28   Dissipation Factor of 30 Percent Glass-filled Polyphenylene Oxides at Several Temperature and Humidity Conditions[31]

Temp, °F	Frequency, Hz					
	60	10^2	10^3	10^4	10^5	10^6
	Dissipation Factor					
73, dry..............	0.0009	0.0008	0.0009	0.0009	0.0010	0.0014
73, 50% RH..........	0.0009	0.0009	0.0009	0.0009	0.0010	0.0015
160.................	0.0012	0.0011	0.0010	0.0010	0.0010	0.0012
250.................	0.0028	0.0024	0.0017	0.0013	0.0011	0.0012

TABLE 29   Electrical Properties of Unfilled and 40 Percent Glass-filled Polyphenylene Sulfide[32]

	Unfilled	40% glass-filled
Dielectric constant:		
At 1 kHz......................	3.11	3.79
At 1 MHz......................	3.22	3.88
Dissipation factor:		
At 1 kHz......................	0.0004	0.0037
At 1 MHz......................	0.0007	0.0066
Dielectric strength, V/mil..........	595	490
Volume resistivity, Ω/cm...........	10^{16}	10^{16}
Arc resistance, s...................	......	160

Cross-linked polystyrenes are actually thermosetting materials, and hence do not remelt, even though they may soften. The improved thermal properties, coupled with the outstanding electrical properties, hardness, and associated dimensional stability, make cross-linked polystyrenes the leading choice of dielectric for many high-frequency, radar-band applications. Identification of radar-band frequencies is shown in Table 30.

Conventional polystyrenes are essentially polymerized styrene monomer alone. By varying manufacturing conditions or by adding small amounts of internal and external lubrication, it is possible to vary such properties as ease of flow, speed of setup, physical strength, and heat resistance. Conventional polystyrenes are frequently referred to as normal, regular, or standard polystyrenes.

Since conventional polystyrenes are somewhat hard and brittle, having low impact strength, many modified polystyrenes are available. Modified polystyrenes are materials in which the properties of elongation and resistance to shock have been increased by incorporating into their composition varying percentages of elastomers. Hence, these types are frequently referred to as high-impact, high-elongation, or rubber-modified polystyrenes. The so-called super-high-impact types can be quite rubbery. Such compounding reduces electrical properties.

TABLE 30   Identification of Radar-Band
Frequencies

Band	Frequency, GHz*
P	0.225–0.390
L	0.390–1.550
S	1.550–5.200
C	3.900–6.200
X	5.200–10.90
K	10.90–36.00
Q	36.00–46.00
V	46.00–56.00

* G (giga) is 1,000 mega; or 1,000 million cycles per second.

The exceptionally low dissipation factor and dielectric constant for polystyrene as a function of frequency is shown in Fig. 27. The dielectric constant and dissipation factor of some polystyrenes increase rapidly above 1,000 to 10,000 MHz. Hence, the specific material and application should be checked above this frequency area. The low dissipation factor coupled with the relative rigidity of polystyrene compared with polyethylene and TFE gives polystyrene advantages in many electronic applications requiring material hardness and extremely low electrical losses, particularly at high frequencies. As was mentioned above, the cross-linked polystyrenes are especially useful here.

The dielectric strength of polystyrenes is excellent, and the resistivity properties of polystyrenes are outstanding. The excellence of the resistivity characteristic can be seen in Fig. 1. These properties coupled with the other above-mentioned excellent properties of polystyrenes make these materials most useful in high-frequency electronic applications, with the other application limitations.

Polysulfones

Polysulfones are another very useful class of engineering thermoplastics for electronic design, having excellent strength vs. temperature properties, good electrical properties (though not outstanding for high frequency), and outstanding strength retention over long periods of aging up to 300°F or over. Application guideline information is given in Table 21 and quantitative physical, mechanical, and electrical data are presented in Tables 22 and 23.

The dissipation factor vs. frequency is shown in Fig. 36, and the dissipation factor vs. temperature at 60 Hz in Fig. 37. The dielectric constant of polysulfones is approximately 3.1 up to 1 MHz, and decreases slightly at 10 MHz. The other important electrical properties of polysulfones are also good and are satisfactory for most electronic applications. The electrical properties of polysulfones are maintained to approximately 90 percent of their initial values after 1 year or more of exposure at 300°F. Also, the basic electrical properties are generally stable up to about 350°F, and under exposure to water or high humidity.

Vinyls

Vinyls are good general-purpose electrical-insulating materials but are not outstanding from the electronics-application viewpoint. The wide range of basic formulations, and the wider range of modifications, make a detailed study of specific materials desirable for a given use. General guidelines are given in Tables 21 to 23. Perhaps their widest applications in electronics are as hookup wire jacketing, and as sleeving and tubing. There are many grades and types of vinyls, among which are some special electrical grades, which should be considered for any electronic applications. Owing to the many variations available, consultation with leading suppliers is especially important. In addition to various basic vinyl classifications, vinyls may be rigid, flexible, or foamed. Further, they may be filled in many ways, alloyed with other plastics, plasticized with various plasticizers. Some vinyls are particularly outstanding in their resistance to corrosive chemicals.

Basically, the vinyl family is comprised of seven major types. These are *polyvinyl acetals, polyvinyl acetate, polyvinyl alcohol, polyvinyl carbazole, polyvinyl chloride, polyvinyl chloride-acetate*, and *polyvinylidene chloride*.

Polyvinyl acetals consist of three groups, namely, polyvinyl formal, polyvinyl acetal, and polyvinyl butyral. These materials are available as molding powders, sheet, rod, and tube. Fabrication methods include molding, extruding, casting, and calendering.

Polyvinyl chloride (PVC) is perhaps the most widely used and highest-volume type of the vinyl family. PVC and polyvinyl chloride-acetate are the most commonly used vinyls for electronic and electrical applications.

Vinyls are basically tough and strong, resist water and abrasion, and are excellent electrical insulators. Special tougher types provide high wear resistance. Excluding some nonrigid types, vinyls are not degraded by prolonged contact with water, oils, foods, common chemicals, or cleaning fluids such as gasoline or naphtha. Vinyls are affected by chlorinated solvents.

Generally, vinyls will withstand continuous exposure to temperatures ranging up to 130°F; flexible types, filaments, and some rigids are unaffected by even higher temperatures up to 200°F. These materials also are slow-burning, and certain types are self-extinguishing; but direct contact with an open flame or extreme heat must be avoided.

PVC is a material with a wide range of rigidity or flexibility. One of its basic advantages is the way it accepts compounding ingredients. For instance, PVC can be plasticized with a variety of plasticizers to produce soft, yielding materials to almost any desired degree of flexibility. Without plasticizers, it is a strong, rigid material that can be machined, heat-formed, or welded by solvents or heat. It is tough, with high resistance to acids, alcohol, alkalies, oils, and many other hydrocarbons. It is available in a wide range of colors. Typical uses include profile extrusions, wire and cable insulation, and foam applications. It also is made into film and sheets.

Polyvinyl chloride raw materials are available as resins, latexes, organosols, plastisols, and compounds. Fabrication methods include injection, compression, blow or slush molding, extruding, calendering, coating, laminating, rotational and solution casting, and vacuum forming.

Glass-Fiber-Reinforced Thermoplastics

Basically, thermoplastic molding materials are developed and can be used without fillers, as opposed to thermosetting molding materials, which are more commonly used with fillers incorporated into the compound. This is primarily because shrinkage, hardness, brittleness, and other important processing and use properties necessitate the use of fillers in thermosets. Thermoplastics, on the other hand, do not suffer from the same shortcomings as thermosets and hence can be used as molded products, without fillers. However, thermoplastics do suffer from creep and dimensional-stability problems, especially under elevated-temperature and load conditions. Because of this shortcoming, most designers find difficulty in matching the techniques of classical stress-strain analysis with the nonlinear, time-dependent strength-modulus properties of thermoplastics. Glass-fiber-reinforced thermoplastics, or FRTPs, help to simplify these problems. For instance, 40 percent glass-fiber-reinforced nylon outperforms its unreinforced version by exhibiting two and a half times greater tensile and Izod impact strength, four times greater flexural modulus, and only one-fifth of the tensile creep.

Thus, glass-fiber-reinforced thermoplastics fill a major materials gap in providing plastic materials which can be reliably used for strength purposes, and which in fact can compete with metal die castings. Strength is increased with glass-fiber reinforcement, as are stiffness and dimensional stability. The thermal expansion of the FRTPs is reduced. Creep is substantially reduced, and molding precision is much greater.

Dimensional stability of glass-reinforced polymers is invariably better than that of the nonreinforced materials. Mold shrinkages of only a few mils per inch are characteristic of these products. Low moisture absorption of reinforced plastics ensures that parts will not suffer dimensional increases under high-humidity conditions. Also, the characteristic low coefficient of thermal expansion is close enough to that of such metals as zinc, aluminum, and magnesium that it is possible to design composite assemblies without fear that they will warp or buckle when cycled over temperature extremes. In applications where part geometry limits maximum wall thickness, reinforced plastics almost always afford economies for similar strength or stiffness over unreinforced equivalents. A comparison of some important properties for unfilled and glass-filled (20 and 30 percent) thermoplastics is shown in Table 31.

One significant property to electrical-electronic engineers is temperature resistance. Glass-reinforced nylons, for example, have a deflection temperature up to 500°F, compared with 150°F for the same material in unreinforced form. This makes possible use of these materials in electrical or electronic products in which the material might be adjacent to soldering where heat might accumulate. Electrical properties of several glass-reinforced thermoplastics are shown in Table 32. Properties of reinforced and unreinforced thermoplastic polyesters are shown in Tables 33 and 25, polyphenylene oxides in Tables 26 to 28, and polyphenylene sulfides in Table 29.

In addition to the property improvements mentioned above, other properties of the base resin are also either enhanced or undergo no significant change. In most cases the electrical properties are equivalent to or better than those of the base resin. In general, they have excellent electrical-insulating properties, suiting them for connectors, insulators, and structural portions of electrical components. Electrical properties of polymers such as nylon are generally improved by glass reinforcement, particularly at high humidities. Volume resistivity of type 6/6 nylon, for example, is 10^{14} while the value for glass-filled nylon is 10^{15}. Improvement here is probably due to the reduction in moisture absorption with addition of glass fibers, as shown in Fig. 46. Chemical resistance is essentially unchanged, except that environmental stress-crack resistance of such polymers as polycarbonate and polyethylene is markedly increased by glass reinforcement.

ELASTOMERS AND RUBBERS

At one time, when natural rubber and a few synthetic rubbers constituted the primary type of rubberlike materials in use, the term rubber was predominantly used to describe this group of materials. However, with developments in the field of polymer chemistry, numerous other rubberlike materials have been developed whose chemical composition bears no resemblance to the chemical composition of the natural or the early synthetic rubbers. Also, these newer materials often exhibit vast

TABLE 31   Comparison of Some Important Properties of Raw and Glass-filled Thermoplastics[33]

Material	% glass loading	Tensile strength at 73°F, lb/in.² × 10⁻³ (D 638)*	Elonga-tion at 73°F, % (D 638)	Flexural strength at 73°F, lb/in.² × 10⁻³ (D 790)	Impact strength at 73°F, notched Izod (D 256) lb/in.², °F	Heat-distortion temp at 264 lb/in.², °F	Rockwell hardness	Specific gravity
Nylon 6/10:								
Raw...........	...	8.5	85–300	....	1.2	....	R111	1.09
Short fiber......	30	17–19	3.0	22–26	1.4–2.2	400	E35–45, R118	1.30
Long fiber......	30	19.0	1.9	23	3.4	420	E70–75	1.30
Nylon 6/6:								
Raw...........	...	9.0	60–300	12.5	1.0–2.0	150–186	R108–R118	1.13–1.15
Short fiber......	30	18.5–23	3.0	26.5–32	1.2–2.0	400–470	E50–55–R120	1.37
Long fiber......	30	20	1.5	28	2.5	498	E60–70	1.37
Nylon 6:								
Raw...........	...	7.0	25–320	8	1.0–3.6	152–158	R103–R118	1.12–1.14
Short fiber......	30	17–24	3	22.5–32	1.3–2.0	400–420	E45–50, M90	1.37
Long fiber......	30	21.0	2.0	27	3.0	420	E55–60	1.37
Polycarbonate:								
Raw...........	...	9.5	60–110	13.5	2.5	265–280	M70–R118	1.2
Short fiber......	20	12–18.5	2.5–3	17–25	1.5–2.5	285–295	M92–R118	1.35
Long fiber......	20	14–18.5	2.2–5	18.5	2.5–3.0	295	H80–90	1.35
Polypropylene:								
Raw...........	...	4.3	200–700	6	....	135–145	R85–110	0.90–0.91
Short fiber......	20	6.0	3.0	7.5	1.0	230	M40	1.05
Long fiber......	20	8	2.2	10	3.5	283	M50	1.05
Polyacetal:								
Raw...........	...	10.0	15	14	1.4	255	M94–R120	1.425
Short fiber......	20	10–13.5	2–3	14–15	0.8–1.4	315–325	M70–75–95	1.55
Long fiber......	20	10.5	2.3	15	2.2	325	M75–80 (Shore)	1.55
Polyethylene:								
Raw...........	...	1.2	50–600	4.8	0.5–16	90–105	D50–60	0.92–0.94
Short fiber......	20	6	3.0	7	1.1	225	R60	1.10
Long fiber......	20	6.5	3.0	8	2.1	260	R60	1.10
Polysulfone:								
Raw...........	...	10.2	50–100	....	1.3	345	M69–R120	1.24
Short fiber......	30	16	2	21	1.8	360	....	1.41
Long fiber......	30	18.5	2.0	24	2.5	333	E45–55	1.37
PPO:								
Raw...........	...	11.6	20–40	16.5	1.2	345	M78	1.06
Short fiber......	30	18.0	4.0–6.0	24.4	1.7	360	M94	1.27

* ASTM Test Method.

Note: Most favorable figures for short-fiber performance are based upon results with nominal ¼-in. fibers.   Not included in the table are glass-reinforced styrene, SAN, ABS, and polyurethane, for which comparable data between short and long fibers are not available.

Sources: Data for raw-resin and long-glass properties for all resins—Fiberfil Inc.; short-glass polysulfone, polyethylene, polypropylene, all three nylons—Fiberfil and Liquid Nitrogen Processing Corp.; polycarbonate—Fiberfil, LNP, and General Electric Co.; polyacetal and PPO—LNP, Fiberfil, Celanese Corp., and Du Pont Co.; nylon 6/6—Fiberfil, LNP, and Polymer Corp.

TABLE 32   Electrical Properties of Several Glass-reinforced Thermoplastics

Property	Nylon, 20–40% filled	Polycarbonate, 20–40% filled	Poly-propylene, 20–40% filled	Polyphenylene oxides, 30% filled — PPO	Polyphenylene oxides, 30% filled — Noryl	Polystyrene, 20–30% filled
Volume resistivity at 50% RH, Ω-cm...	1.53–5.5 × 10¹⁵	1.4–1.52 × 10¹⁵	1.7 × 10¹⁶	2.5 × 10¹⁷	2.1 × 10¹⁷	3.2–3.7 × 10¹⁶
Dielectric strength, ⅛ in., V/mil:						
Short time...............	408–503	475	475	500	550	350–425
Step by step.............	375–450	475	375	....	....	350–430
Dielectric constant:						
At 60 Hz.................	4.0–4.6	3.7	2.37	2.9	2.9	2.8–3.1
At 1 kHz.................	3.9–4.4	3.7	2.36	2.9	2.9	2.8–3.0
1 MHz...................	3.4–3.9	3.2–3.5	2.38	2.9	2.9	2.8–3.0
Dissipation factor:						
60 Hz...................	0.018–0.025	0.003–0.005	0.0022	0.0009	0.0009	0.004–0.014
1 kHz...................	0.020–0.025	0.002–0.004	0.0017	0.0009	0.0009	0.001–0.004
1 MHz...................	0.017–0.022	0.009	0.0035	0.0016	0.0015	0.001–0.003
Arc resistance, s............	92–148	5–120	74	120	75	25–40
Water absorption, ⅛-in. after 24 h, %..............	0.2–2.0	0.07–0.10	0.01–0.05	0.07	0.06	0.05–0.10
Max continuous-use temp, °F.................	300–400	275–300	300–320	315–365	225–250	180–200

TABLE 33   Properties of Reinforced and Unreinforced Thermoplastic Polyesters[34]

Property	Units	Test method	Unreinforced	Unreinforced and self-extinguishing	Glass-reinforced	Glass-reinforced and self-extinguishing
Specific gravity, 73°F...	....	D 792	1.31	1.41	1.52	1.58
Specific volume.........	in.³/lb	......	21.1	19.6	18.1	17.5
Water absorption, 73°F	%	D 570				
24 h................	....	......	0.08	0.08	0.06	0.07
Equilibrium..........	....	......	0.34	0.38	0.26	0.25
Mold shrinkage.........	in./in. × 10⁻³	......	17–23	17–23	2–4	2–4
Tensile strength........	lb/in.²	D 638	8,000	8,900	17,300	17,000
Elongation at break.....	%	D 638	300	100	5	5
Flexural strength.......	lb/in.²	D 790	12,800	14,700	27,500	27,000
Flexural modulus.......	lb/in.²	D 790	340,000	380,000	1,200,000	1,200,000
Compressive strength, 10% deformation......	lb/in.²	D 695	13,000	14,500	18,000	18,000
Shear strength..........	lb/in.²	D 732	7,700	7,700	8,900	9,000
Izod impact strength....	ft-lb/in.	D 256				
Notched.............	....	......	1.2	0.9	2.2	1.8
Unnotched..........	....	......	No break	No break	15	15
Gardner impact........	ft-lb	Falling dart	40–45	35–40		
Chemical resistance.....	....	......	Excellent	Excellent	Excellent	Excellent
Deformation under load, 122°F, 2,000 lb/in.²....	%	D 621	0.17	0.15	0.12	0.15
Creep, 73°F...........	%	D 674	1.1 (1,000 lb/in.²)	1.1 (1,000 lb/in.²)	0.44 (4,000 lb/in.²)	0.44 (4,000 lb/in.²)
Fatigue endurance limit, 10⁷ cycles........	lb/in.²	D 671	2,850	3,500	5,000	5,000
Coefficient of friction....	....	D 1894				
Self.................	....	......	0.17	0.16	0.16	0.16
Metal...............	....	......	0.13	0.14	0.14	0.14
Rockwell hardness, R scale...............	....	D 785	117	120	118	119
Taber abrasion, CS-17 wheel, 1,000-g load.....	mg/1,000 cycles	D 1044	6.5	10.0	9.0	11.0
Heat-deflection temp....	°F	D 648				
66 lb/in.²...........	....	......	310	325	420	420
264 lb/in.²..........	....	......	130	160	415	415
Coefficient of thermal expansion.............	in./in./°F × 10⁻⁵	D 696	5.3	4.7	3.3	3.5
Thermal conductivity...	Btu/h/ft²/°F/in.	C 177	1.1	1.2	1.3	1.3
Flammability..........	....	D 635 UL94	SB SB	SE SE-O	SB SB	SE SE-O
Volume resistivity......	Ω cm × 10¹⁶	D 257	4.0	4.0	3.2	3.4
Dielectric strength, short-time...........	V/mil	D 149	590	750	750	750
Dielectric constant:.....	....	D 150				
100 Hz..............	....	......	3.3	3.3	3.8	3.8
10⁶ Hz..............	....	......	3.1	3.1	3.7	3.7
Dissipation factor:.....	....	D 150				
100 Hz..............	....	......	0.002	0.003	0.002	0.002
10⁶ Hz..............	....	......	0.02	0.02	0.02	0.02
Arc resistance..........	s	D 495	190	63	130	80
Hot-wire ignition, ⅟₁₆ in..	s	UL55	16	20	63	34

improvements over the early rubbers in many respects while still being basically rubberlike or elastic in character. Therefore, the term elastomer came to be used to encompass the broadened range of rubberlike materials. The ASTM definition of an elastomer is: "a material which at room temperature can be stretched repeatedly to at least twice its original length and upon immediate release of the stress will return with force to its approximate original length." Currently, there are over a dozen recognized classes of elastomers, a number of which are useful in electronic assemblies.

Nature of Elastomers

Although elastomers are perhaps not so widely used in electrical structures as are thermoplastic or thermosetting plastic materials, elastomers do have wide application in many products. They are often used for wiring and cabling, for cushioning materials, for vibration damping, for gasketing and sealing, and in many other applications where rubberlike properties coupled with some selected combination of mechanical, electrical, or fluid-resistant properties are required. Elastomers are commonly used for wire and cable insulation.

Elastomers are sometimes known by their popular name, sometimes by their chemical name, and sometimes by the ASTM standards designation or some other previously used symbol. Table 34 is a cross reference of these identifications. Also commonly used are the ASTM-SAE application classifications for various elastomers, which are defined as follows:

Type R—non-oil-resistant
Type S—resistant to petroleum chemicals
  Class SA—very low volume swell
  Class SB—low volume swell
  Class SC—medium volume swell
Type T—temperature-resistant
  Class TA—resistant to high and low temperatures
  Class TB—resistant to hot air and oil

While elastomeric properties may be indicated in a manner similar to that commonly used for plastics, there are some properties of particular value that are widely used in the identification of these materials. One of these is durometer. Durometer, or hardness, is often the primary description used in identifying the characteristics of an elastomer. The hardness of an elastomer is related to its degree of vulcanization or cure and to the presence or absence of filler materials. The Shore durometer is widely used for this measurement, and the shortened term durometer is often used to specify hardness: the softer the rubber, the lower the hardness or durometer. A comparative hardness guide and detailed discussion is presented in Chap. 1.

Compression set is another property often given for elastomeric materials. Compression set, according to ASTM test D 395, is the residual decrease in the thickness of a test specimen which is observed after 30 min of test following the removal of a specified compressive loading applied under established conditions of time and temperature.

Elongation and tensile strength are also often used to describe elastomers. Elongation is the amount the material is stretched at the moment of rupture. The amount of force necessary to rupture the material is the tensile strength. Tensile strength is normally expressed in terms of the original cross section of the specimen tested.

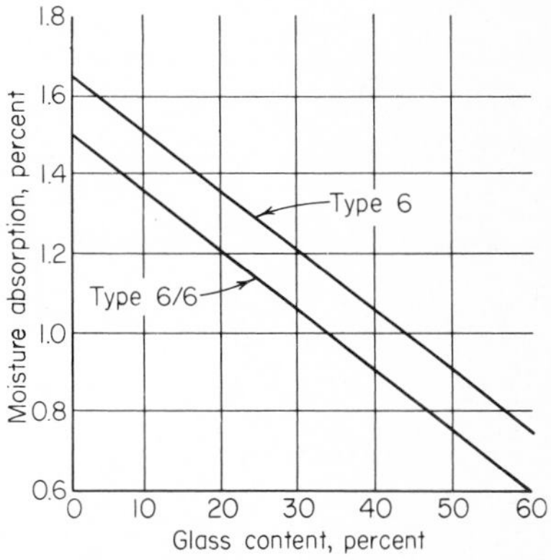

Fig. 46 Effect of glass content on moisture absorption of two nylon types.[35]

Electrical Properties of Elastomers

As with plastic materials, a broad range of properties can often be obtained with any given type of elastomer by compounding or modifying the basic material.

Electrical properties for various applications of rubber are varied and many. The products range from the large-volume usage of insulated wires to electrical-equipment parts, electrical sockets, plugs, spacers, aprons, mats, gloves, etc. Wires and cables must be designed for practically 0 to higher than 25 kV, and frequencies from direct current to 10,000 MHz. In actual service, rubber insulators are often subjected to temperature extremes, chemicals, moisture, oils, crushing abrasions, impacts, sunlight, ozone, and high operating stress. All these effects must be recognized and provided for by both proper selection of material and adequate design.

The most suitable materials for electrical purposes would be those possessing high dielectric strength and resistivity, with dielectric constant and power factor at their lowest. A combination of this sort is rarely attained, since other physical and chemical properties are also required; and so compromise values are the rule rather than exception.

In the field of wire and cable, emphasis on properties varies with the type of

TABLE 34 Designations and Application Information for Elastomers

ASTM D 1418	Trade name or common name	Chemical type	Major application considerations
Elastomer designation			
NR......	Natural rubber	Natural polyisoprene	Excellent physical properties; good resistance to cutting, gouging, and abrasion; low heat, ozone, and oil resistance. The best electrical grades are excellent in most electrical properties at room temperature
IR.......	Synthetic natural	Synthetic polyisoprene	Same general properties as natural rubber; requires less mastication in processing than natural rubber
CR......	Neoprene	Chloroprene	Excellent ozone, heat, and weathering resistance; good oil resistance; excellent flame resistance. Not so good electrically as NR or IR. However, the combination of generally good electricals for jacketing application, coupled with all the other good properties, gives this elastomer broad use for electrical wire and cable jackets
SBR......	GRS, Buna S	Styrene-butadiene	Good physical properties; excellent abrasion resistance; not oil-, ozone-, or weather-resistant. Electrical properties generally good but not specifically outstanding in any area
NBR......	Buna N, Nitrile	Acrylonitrile-butadiene	Excellent resistance to vegetable, animal, and petroleum oils; poor low-temperature resistance. Electrical properties not outstanding; probably degraded by molecular polarity of acrylonitrile constituent
IIR.......	Butyl	Isobutylene-isoprene	Excellent weathering resistance; low permeability to gases; good resistance to ozone and aging; low tensile strength and resilience. Electrical properties generally good but not outstanding in any area
IIR.......	Chlorobutyl	Chloro-isobutylene-isoprene	Same general properties as butyl
BR.......	Cis-polybutadiene	Polybutadiene	Excellent abrasion resistance and high resilience; used principally as a blend in other rubbers
PS........	Thiokol (Thiokol Chemical)	Polysulfide	Outstanding solvent resistance; widely used for potting of electrical connectors
R.........	EPR	Ethylene propylene	Good aging, abrasion, and heat resistance; not oil-resistant. Good general-purpose electrical properties

TABLE 34   Designations and Application Information for Elastomers (Continued)

Elastomer designation		Chemical type	Major application considerations
ASTM D 1418	Trade name or common name		
R.........	EPT	Ethyl propylene terpolymer	Good aging, abrasion, and heat resistance; not oil-resistant. Good general-purpose electrical properties
CSM......	Hypalon (HYP) (Du Pont)	Chlorosulfonated polyethylene	Excellent ozone, weathering, and acid resistance; fair oil resistance; poor low-temperature resistance. Not outstanding electrically, but has some special-application uses based on other properties
SIL.......	Silicone	Polysiloxane	Excellent high- and low-temperature resistance; low strength; high compression set. Among the best electrical properties in the elastomer grouping. Especially good stability of dielectric constant and dissipation factor at elevated temperatures
PU.......	Urethane	Polyurethane diisocyanate	Exceptional abrasion, cut, and tear resistance; high modulus and hardness; poor moist-heat resistance. Generally good general-purpose electrical properties. Some special high-quality electrical grades available from formulators
FLU......	Viton (Du Pont)	Fluorinated hydrocarbon	Excellent high-temperature resistance, particularly in air and oil. Not outstanding electrically
ABR......	Acrylics	Polyacrylate	Excellent heat, oil, and ozone resistance; poor water resistance. Not outstanding for or widely used in electrical applications
....	Hytrel (Du Pont)	Polyester elastomer	Relatively new elastomer combining characteristics of thermoplastics and elastomers; structurally strong, resilient, and resistant to impact and flexural fatigue; excellent resistance to oils and chemicals; good electrical insulation in the range of −55 to 150°C. Used as cable insulation and related electrical-insulation functions

circuit, as shown in Table 35.   Table 36 shows electrical properties of various elastomers, and Table 37 shows the ratings of various elastomers and plastics relative to these needs.

RECOMMENDATIONS FOR FURTHER READING

The material in this chapter has covered guidelines for use of plastics and elastomers for the broad range of parameters important to the electrical and electronics industry. Primarily, the data presented have been on bulk plastics and elastomers,

TABLE 35   Electrical-Property Requirements for Elastomers in Various Types of Service[36]

Application	Relative value of property needed			
	Dielectric strength	Dielectric constant	Power factor	Resistivity
Power:				
Low voltage............	M to L	M to H	M to H	M to L
High voltage............	H	M to L	M to L	M to L
Signal..................	M to L	M to L	M to L	M to H
Communication:				
Low frequency..........	M to L	L	M to L	M to H
High frequency..........	M to L	L	L	L

H = high; M = medium; L = low.

most commonly used for molded, extruded, machined, and other formed but bulk-type electrical and electronic products. While these data and guidelines generally apply to all types of electrical products, there are properties and considerations unique to certain plastic and elastomer forms, such as laminates, films, tapes, coatings, wiring and cabling, embedding-resin systems, and others, and detailed coverage of all these is beyond the scope of this chapter. Laminates, reinforced plastics, and composite forms are covered in Chap. 5, embedding resins in Chap. 8, and films in Chap. 1. For further extensive and detailed coverage of these and other areas, the other comprehensive handbooks in this series are recommended, namely, "Handbook of Electronic Packaging," "Handbook of Materials and Processes for Electronics," "Handbook of Wiring, Cabling and Interconnecting for Electronics," and "Handbook of Thick Film Hybrid Microelectronics." All electrical and electronic product areas, and all plastic and elastomer product forms are discussed in detail in these volumes.

TABLE 36   Electrical Properties of Various Elastomers

Elastomer	Resistivity, Ω-cm	Dielectric strength (short-time), V/mil	Dielectric constant at 1 kHz	Dissipation factor at 1 kHz
NBR....................	10^{10}	....	13.0	0.055
SBR....................	10^{15}	....	2.9	0.0032
IIR (butyl).............	10^{17}	600	2.1–2.4	0.0030
CSM (chlorosulfonated polyethylene)..........	10^{14}	500	7–10	0.03–0.07
EPR (EPM)............	10^{15}–10^{17}	900	3.17–3.34	0.0066–0.0079
EPT (EPDM)..........	10^{15}–10^{17}	900–1,050	3.0–3.5	0.004 at 60 Hz
FPM (hexafluor)........	10^{14}	613	5.9	0.053
FPM (chlorotrifluor)......	10^{13}	250–750	....	0.03–0.04
FSI (fluorosilicone).......	10^{13}–10^{14}	340–350	6.9–7.4 at 100 Hz	0.03–0.07 at 100 Hz
CR (neoprene)..........	10^{11}	150–600	9.0	0.030
NR or IR (natural or synthetic).............	10^{15}–10^{17}	....	2.3–3.0	0.0023–0.0030
Polysulfide.............	10^{12}	250–600	7.0–9.5	0.001–0.005
Urethane...............	10^{11}–10^{14}	350–525	5–8	0.015–0.09
SI (silicone).............	10^{11}–10^{17}	100–655	3.0–3.5	0.001–0.010

TABLE 37   Ratings of Various Elastomers and Plastics for Important Electrical-Service Requirements[36]

Property	Natural rubber	GR-S	Butyl	Neoprene	Nitrile rubbers	Thiokol	Silicone	Polyvinyl chloride	Polyethylene	Nylon	Teflon	Fluorothene	Mylar
Dielectric strength	H	H	MH	M	M	L	M	VH	VH	MH	VH	M	VH
Power factor	M	M	M	MH	H	VH	M	MH	VL	MH	VL	L	L
Resistivity	H	M	VH	L	L	VL	M	M	VH	L	VH	MH	MH
Dielectric constant	M	H	M	MH	M	VH	M	MH	VL	M	VL	H	L
Abrasion resistance	H	H	MH	VH	MH	M	L	H	MH	H	...	H	H
Flexibility	H	H	H	MH	MH	H	H	M	L	L	L	L	VH
Resistance to compression cutting	MH	MH	M	M	MH	M	L	H	H	VH	VH	H	VH
Cold flow at room temperature	L	L	L	L	L	M	MH	L	L	VL	VL	VH	VH
Cold cracking temperature	VL	M	L	M	MH	MH	VL	L	VL	M	VL	M	VL
Sunlight resistance	MH	MH	H	H	M	H	H	VH	H*	MH*	VH	VH	VH
Ozone resistance	L	M	H	H	MH	H	H	H	VH	H	VH	VH	H
Corona resistance	M	M	M	MH	M	...	VH	H	M	...	...	VH	H
Heat resistance	MH	H	VH	MH	MH	H	M	MH	M	H	VH	VH	M
Flame resistance	VL	VL	VL	MH	VL	L	H	H	VL	MH	VH	VH	VH
Water resistance	H	H	H	M	M	M	H	VH	VH	L	VH	VH	H
Oil resistance	VL	VL	VL	M	MH	H	M	H	M	H	VH	VH	H

* Only if suitably pigmented.
VH = very high; H = high; MH = medium high; M = medium; L = low; VL = very low.

REFERENCES

1. Harper, C. A.: "Handbook of Electronic Packaging," McGraw-Hill, New York, 1971.
2. Harper, C. A.: "Handbook of Materials and Processes for Electronics," McGraw-Hill, New York, 1972.
3. Harper, C. A.: "Handbook of Wiring, Cabling and Interconnecting for Electronics," McGraw-Hill, New York, 1973.
4. Harper, C. A.: "Handbook of Thick Film Hybrid Microelectronics," McGraw-Hill, New York, 1974.
5. Grafton, Percy: "Commonly Used Plastics," Boonton Molding Co., Inc., Boonton, N.J., 1973.
6. Dapon Diallyl Phthalates, FMC Corp. Tech. Data Bull.
7. Plaskon Plastics and Resins, Allied Chemical Corporation, Plastics Division, Tech. Bull.
8. Carpenter, R. E.: Insulating and Structural Properties of Alkyd Molding Compounds, Electro-Technol., June 1965.
9. Parry, H. L., et al.: High Humidity Insulation Resistance of Resin Systems, SPE J., October 1957.
10. Sunderland, G. B., and A. Nufer: Aminos, Machine Des., Plastics Reference Issue, December 1968.
11. Jessup, J. N., and H. H. Beacham: Effects of Temperature and Humidity on Electrical Properties of Thermosets, Modern Plastics, December 1968.
12. Chottiner, J.: Dimensional Stability of Thermosetting Plastics, Mater. Eng., February 1962.
13. Chapman, J. J., and L. J. Frisco: A Practical Interpretation of Dielectric Measurements up to 100 Mc, The Johns Hopkins University, Dielectrics Lab. Rept., Baltimore, December 1963.
14. Hauck, J. E.: Heat Resistance of Plastics, Mater. Eng., April 1963.
15. Keegan, James F.: "Durez Electrical Insulation Materials," Hooker Chemical Corp., North Tonawanda, N.Y., 1973.
16. Buchoff, L. S.: Effect of Humidity on Surface Resistance of Filled Epoxy Resins, SPE Ann. Nat. Tech. Conf., Chicago, January 1960.
17. Martino, C. F.: Phenolics Too Good to Be Outmoded, SPE Reg. Tech. Conf., Chicago, December 1966.
18. Barth, H. J., and J. J. Robertson: Fast-curing Polybutadiene Thermosetting Resins, Modern Plastics J., November 1970.
19. Silicone Molding Materials, Dow Corning Corporation Tech. Bulls.
20. Kookootsedes, G. J., and F. J. Lockhart, Silicone Molding Compounds for Semiconductor Devices, Modern Plastics J., January 1968.
21. "Delrin Acetal Resins," E. I. du Pont de Nemours & Co., Inc., Technical Booklet.
22. "Teflon for Electrical and Electronic Systems," E. I. du Pont de Nemours & Co., Inc., Technical Booklet.
23. "DuPont Tefzel Fluoropolymer," E. I. du Pont de Nemours & Co., Inc., Design Handbook, 1973.
24. Nomex High Temperature Insulation, E. I. du Pont de Nemours & Co., Inc., Tech. Bull.
25. "Parylene Conformal Coatings," Union Carbide Corporation, Technical Booklet, 1971.
26. Morneau, G. A.: Thermoplastic Polyaryl Sulfone, Modern Plastics, January 1970.
27. Gowan, A., and P. Shenian: Properties and Applications for PPO Plastic, Insulation, September 1965.
28. "Lexan Polycarbonate Resin," General Electric Company Technical Booklet.
29. Lovenguth, R. F.: Thermoplastic Polyester for Electrical/Electronic Insulation, Insulation/Circuits J., September 1972.
30. Vespel Design Handbook, E. I. du Pont de Nemours & Co., Inc., Tech. Bull.
31. "Glass-filled Polyphenylene Oxide Resins," General Electric Co. Technical Booklet.
32. "Ryton Polyphenylene Sulfide Resin," Phillips Petroleum Co. Technical Booklet.
33. Staff Report, Can RTP Meet Your Product Requirements, Modern Plastics, March 1966.
34. "Valox Thermoplastic Polyesters," General Electric Co. Technical Booklet.
35. Metz, E. A.: Fiber Glass Reinforced Nylon, Machine Des., Feb. 17, 1966.
36. McPherson and Klemin: "Engineering Uses of Rubber," Reinhold Book Corporation, New York.

Chapter **3**

Mechanical Properties and Testing of Plastics and Elastomers

GEORGE E. RUDD

RONALD N. SAMPSON
Research and Development Center,
Westinghouse Electric Corporation,
Pittsburgh, Pennsylvania

TESTING STANDARDS

Because plastic materials differ in so many ways from metals and other materials, special testing methods have been devised by governmental bodies, industry, industry associations, and professional societies. Testing standards are maintained by both the American Society for Testing and Materials and the federal government.[1,2] A selected tabulation of comparable standards is shown in Table 1. Many of these tests are similar, but this is not always the case. Standard abbreviations for terms relating to plastics are given in ASTM D 1600 and are shown in Table 2. Nomenclature relating to plastics is given in ASTM D 883. In addition, other plastic tests are described by the National Electrical Manufacturers Association.[3]

MECHANICAL-PROPERTIES TESTS

Basic Requirements

The element of time plays an important role in characterizing the mechanical properties of plastics and elastomers. For practical purposes it is safe to say that most metals at normal temperatures behave according to the classical theory of elasticity where stress is directly proportional to strain (for small strains) but independent of strain rate. This is not the case for plastics and elastomers under most conditions. They share some of the features of a viscous fluid where stress is proportional to strain rate but independent of the strain itself—that is, they are viscoelastic. This condition is often modeled mathematically as combinations of elastic springs and viscous dashpots, as shown in Fig. 1. Thus it becomes important to be concerned about how long a plastic part must sustain a load, how fast it is loaded, how far it is stretched or compressed, and so on. Obviously it would be a tremendous task to specify properties for all conceivable conditions of strain, strain rate, and creep and relaxation conditions, to say nothing of how temperature and other environmental

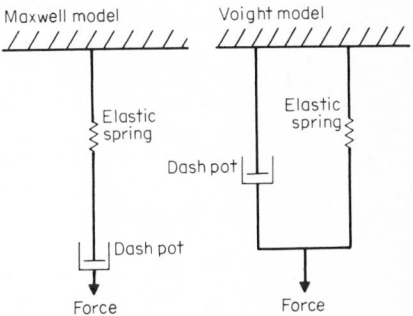

Fig. 1 Two common models used to characterize viscoelastic behavior.

TABLE 1 Standard Test Methods[1,2]

Test	Federal Standard 406 Test Method	ASTM No.
Abrasion wear (loss in weight)	1091	
Accelerated service, temperature and humidity	6011	D 756-56
Arc resistance	4011	D 495-73
Bearing strength	1051	D 953-54
Bonding strength	1111	D 229-72
Brittleness temperature of plastics by impact	2051	D 746-73
Burning rate of cellular plastics	....	D 1692-74
Compression fatigue of rubber	....	D 623-67
Compression set of rubber	....	D 395-69
Compressive properties of rigid plastics	1021	D 695-69
Compressive properties of elastomers	....	D 575-69
Crack growth in rubber	....	D 813-59

TABLE 1   Standard Test Methods[1,2] (Continued)

Test	Federal Standard 406 Test Method	ASTM No.
Cut growth in rubber by flexing	....	D 1052-55
Deflection temperature of plastics	....	D 648-72
Deformation of plastics under load	....	D 621-64
Dielectric-breakdown voltage	4031	D 149-64
Dissipation factor and dielectric constant	4021	D 150-70
Drying test, weight loss	7041	
Dynamic mechanical properties by the torsional pendulum	....	D 2236-70
Exposure of plastics, carbon-arc light and water	....	D 1499-64
Exposure of plastics, fluorescent sunlamp	....	D 1501-71
Exposure of plastics, xenon light and water	....	D 2565-70
Exposure of rubber, carbon-arc light and water	....	D 750-68
Falling-ball impact	1074	
Flame resistance	2023	
Flammability of plastics to 0.050 in. thick	2022	D 568-74
Flammability of plastics over 0.050 in. thick	2021	D 635-74
Flammability of plastics by the oxygen index	....	D 2863-74
Flammability of thin plastic sheeting	....	D 1433-74
Flexural properties of plastics	1031	D 790-71
Flexural properties of elastomers	....	D 797-64
Flexural fatigue of plastics, constant strain	1061	D 671-71
Flexural fatigue of plastics, constant stress	1062	D 671-71
Impact properties of plastics	1071	D 256-73
Incandescence resistance of rigid plastics	....	D 757-74
Indentation hardness by a durometer	1083	D 2240-68
Indentation hardness on the Rockwell scale	....	D 785-65
Indentation hardness, Barcol	....	D 2583-67
Indentation hardness, International Rubber Hardness Degrees	....	D 1415-68
Internal stress in plastic sheets	6052	
Linear thermal expansion	2031	D 696-70
Machinability	5041	
Mar resistance	1093	D 673-70
Mechanical properties of elastomers by the Yerzley oscillograph	....	D 945-72
Outdoor weathering of plastics	....	D 1435-69
Poisson's ratio	....	E 132-61
Porosity	5021	
Salt-spray test	6071	
Shear strength of plastics, double shear	1041	
Shear strength of plastics, punch	....	D 732-46
Specific gravity by water displacement	5011	D 792-66
Specific gravity by weight and volume measurement	5012	
Stress relaxation of plastics	....	D 2991-71
Surface abrasion	1092	D 1044-73
Tear strength of plastic film	....	D 1004-66
Tear strength of plastic sheeting	....	D 1938-67
Tear strength of rubber	....	D 624-73
Tensile properties of plastics	1011	D 638-72
Tensile properties of elastomers	....	D 412-68
Tensile burst strength of plastic tubes	....	D 1180-57
Tensile strength of tubing by split-disk method	....	D 2290-69
Tensile properties of thin plastic sheeting	....	D 882-73
Tensile time—fracture and creep	1063	
Tensile creep and creep rupture of plastics	....	D 2990-71
Thermal conductivity by the guarded hot plate	....	C 177-71
Thermal conductivity by the heat flow meter	....	C 518-70
Thermal-expansion test, strip method	2032	
Warpage of sheet plastics	6054	D 1181-56
Water absorption of plastics	7031	D 750-63

TABLE 2   Abbreviations for Plastics

Term	Abbreviation
Plastics and Resins:	
Acrylonitrile-butadiene-styrene plastics	ABS
Carboxymethyl cellulose	CMC
Casein	CS
Cellulose acetate	CA
Cellulose acetate butyrate	CAB
Cellulose acetate propionate	CAP
Cellulose nitrate	CN
Cellulose propionate	CP
Chlorinated polyvinyl chloride	CPVC
Cresol formaldehyde	CF
Diallyl phthalate	DAP
Epoxy, epoxide	EP
Ethyl cellulose	EC
Melamine formaldehyde	MF
Perfluoro ethylene-propylene copolymer	PFEP
Phenol formaldehyde	PF
Polyacrylic acid	PAA
Polyacrylonitrile	PAN
Polyamide (nylon)	PA
Polybutadiene acrylonitrile	PBAN
Polybutadiene styrene	PBS
Polycarbonate	PC
Polydiallyl phthalate	PDAP
Polyethylene	PE
Polyethylene terephthalate	PETP
Polymethyl-α-chloroacrylate	PMCA
Polymethyl methacrylate	PMMA
Polymonochlorotrifluoroethylene	PCTFE
Polyoxymethylene, polyacetal	POM
Polypropylene	PP
Polystyrene	PS
Polytetrafluoroethylene	PTFE
Polyurethane	PUR
Polyvinyl acetate	PVAC
Polyvinyl alcohol	PVAL
Polyvinyl butyral	PVB
Polyvinyl chloride	PVC
Polyvinyl chloride acetate	PVCA
Polyvinyl fluoride	PVF
Polyvinyl formal	PVFM
Silicone plastics	SI
Styrene acrylonitrile	SAN
Styrene butadiene	SB
Styrene-rubber plastics	SRP
Urea formaldehyde	UF
Plastic and resin additives:	
Dibutyl phthalate	DBP
Dicapryl phthalate	DCP
Diisodecyl adipate	DIDA
Diisodecyl phthalate	DIDP
Diisooctyl adipate	DIOA
Diisooctyl phthalate	DIOP
Dinonyl phthalate	DNP
Di-n-octyl n-decyl phthalate	DNODP
Dioctyl adipate	DOA
Dioctyl azelate	DOZ
Dioctyl phthalate	DOP
Dioctyl sebacate	DOS
Tricresyl phosphate	TCP
Trioctyl phosphate	TOP
Triphenyl phosphate	TPP

TABLE 2   Abbreviations for Plastics (Continued)

Term	Abbreviation
Monomers:	
Diallyl chlorendate (diallyl ester of $1,4,5,6,7,7$-hexachloro-bicyclo-$(2,2,1)$-5-heptene-2,3-dicarboxylic acid)	DAC
Diallyl fumarate	DAF
Diallyl isophthalate	DAIP
Diallyl maleate	DAM
Diallyl orthophthalate	DAP
Methyl methacrylate	MMA
Monochlorotrifluoroethylene	CTFE
Tetrafluoroethylene	TFE
Triallyl cyanurate	TAC
Miscellaneous plastic terms:	
General-purpose	GP
Single-stage	SS
Solvent-welded plastics pipe	SWP

variables affect these values.   But it is important to recognize that these factors should be of concern to the designer, and that standard material test methods often have to be modified to reflect this concern.   When these factors are considered along with the geometric complexities that often are common to many plastic designs, it is understandable that an accepted rule is to test the part itself in a manner simulating actual service conditions.

Compression Tests

Rigid plastic parts are often loaded in compression, but short-term failure is more likely to occur in some other loading mode such as tension or flexure.   However, stiffness is usually important and can be determined along with strength properties by ASTM D 695.   The standard specimen used in this method is shown in Fig. 2 with a graphical representation of the properties most often measured.   Compression rates other than the standard 0.05 in./min may be employed if appropriate.

For longer-term loadings up to 24 h, the test method described in ASTM D 621 may be useful.   Two techniques are included in this specification.   Method A can be important for determining the ability of a rigid plastic to maintain the tight clamping forces required for bolted and riveted joints, and method B measures the recovery of a nonrigid plastic after being deformed for a period of time.

Compression of elastomers is more complicated because the material "deforms" more readily and yet to a first approximation the volume is incompressible.   Therefore, when a piece is compressed between plates, the contact surfaces spread laterally unless constrained.   Figure 3 shows the effect of contact-surface constraint on a stress-strain curve.[4]   The shape of the compressed part also influences its stiffness. For situations where the loaded surfaces are parallel and the thickness does not exceed twice the smallest lateral dimension, this effect can be numerically expressed as the "shape factor."   Parts of equal hardness from the same elastomer will exhibit the same compression stiffness when the shape factor is constant even though the actual shapes are different.   Figure 4 illustrates the influence of shape factor on stiffness.[5]

The short-term compression-deflection characteristics of elastomers are often characterized by ASTM D 575, which uses a 1.129-in.-diameter (1.00 in.^{2}) specimen 0.50 in. thick.   The load required to compress the specimen a specified deflection is measured, or a specified load is applied and the resultant deflection measured. The first method is most common, and 10 percent deflection is frequently used.

Tension Tests

Tensile characteristics are the most widely reported mechanical properties of any material.   For plastics, properties are usually measured per ASTM D 638 in a testing machine capable of maintaining a constant rate of crosshead movement.

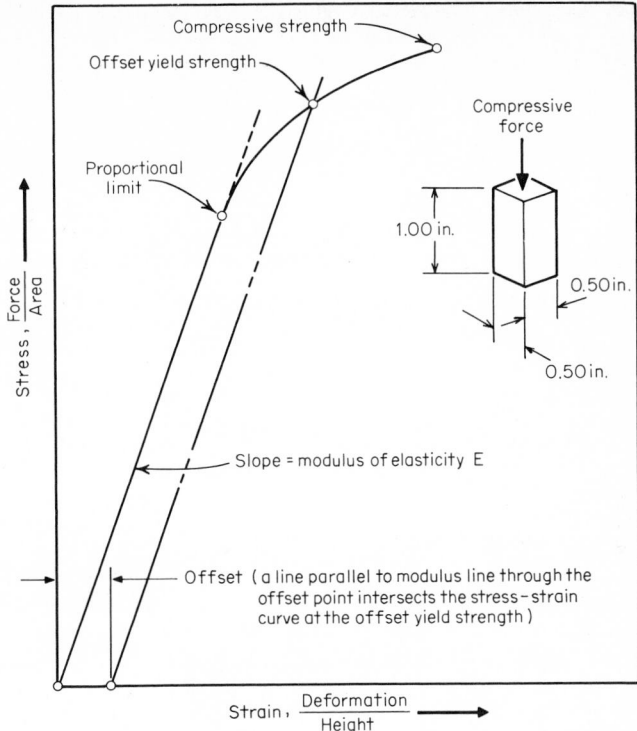

Fig. 2 Stress-strain curve compression properties often used to describe plastics.

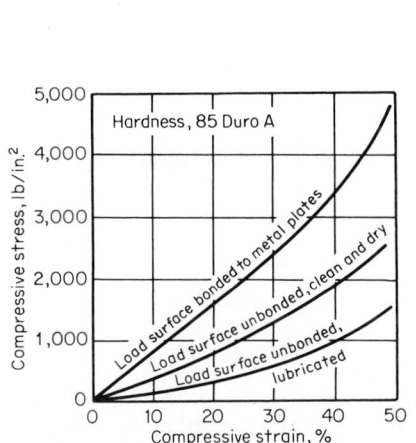

Fig. 3 The effect of surface constraint on the compressive stiffness of an elastomer.[4]

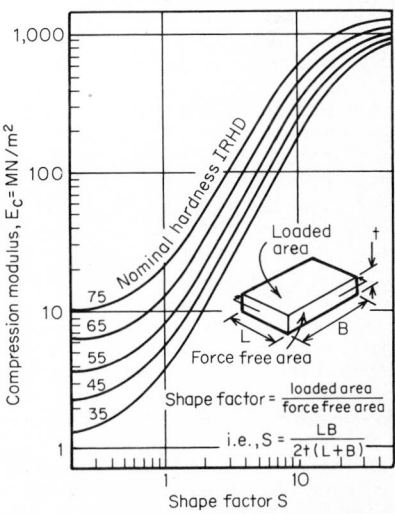

Fig. 4 The influence of shape factor on the compressive stiffness of elastomers.[5]

Specimens are usually flat but may be rod-shaped or tubular for materials produced in these shapes. Properties as defined in Fig. 2 for compression are also those most often reported in tension. The nominal tensile stress is defined as the tensile load per unit area of original unstressed cross section, but this may not be very meaningful for materials which exhibit considerable area reduction beyond yield. If this effect is of interest, the "true" stress can be calculated as the tensile load divided by the actual instantaneous area at that load. Measuring the area as the specimen "necks down" is an obvious difficulty, and this measurement is usually confined to research and development efforts.

There are two other tests for rigid tubing which produce results closely related to the tensile properties of the material. Method D 1180 bursts tube specimens by hydrostatically compressing a confined, very viscous hydraulic medium with end plugs 0.001 to 0.003 in. smaller in diameter than the tube ID. The split-disk tension test described in ASTM D 2290 gives the apparent tensile strength of rings cut from tubing with the fixture shown in Fig. 5. A slight bending moment is induced during this test by the circular shape, and tensile strength is therefore only approximated.

ASTM D 412 is used to measure the tensile properties of elastomers. The "dumbbell" specimen shown in Fig. 6 is most often used with pneumatic grips that tighten automatically to prevent slipping as the specimen tabs thin under load. Gage

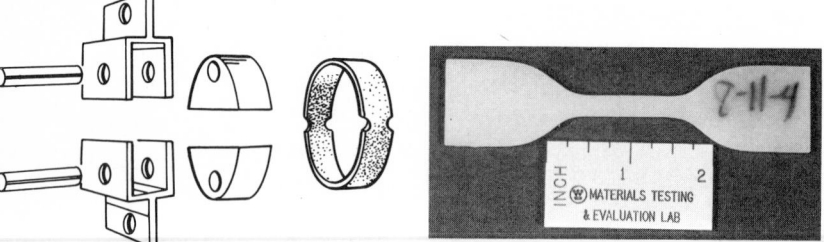

Fig. 5 Split-disk tension-test fixture and specimen as described in ASTM D 2290 for extruded or molded rings.[1]

Fig. 6 Dumbbell tensile specimen used for testing elastomer sheets per ASTM D 412.

lengths of 1 or 2 in. are defined by inked bench marks on the specimen. Elongation is determined by measuring the separation of these marks with a rule or preset calipers held close to the specimen while loading. Extensometers have been developed for use with these specimens, but care must be taken to avoid significant loading from the extensometer. This is particularly important at higher loading rates where inertia effects are significant. Tensile-modulus values for elastomers are often referred to as "100 percent modulus" or "200 percent modulus," etc. These are not true moduli but through common usage have come to mean the tensile stress at 100 or 200 percent elongation.

Flexure Tests

The simplest flexure test is conducted in three-point bending—that is, a specimen is supported as a simple beam on two supports and the load is applied midway between the supports. ASTM D 790 calls for a span/thickness ratio of at least 16:1, but this must be increased for materials of low shear strength. Many plastics do not exhibit equal tensile and compressive moduli, and therefore flexure properties show an apparent thickness effect. Many materials undergo gross flexural deformation before fracture, if in fact fracture ever occurs. Therefore, stress at 5 percent maximum fiber strain is often quoted. Flexure strength is sometimes referred to as the modulus of rupture.

Flexure properties are not often quoted for elastomers, but ASTM D 797 is sometimes used as a technique for determining glass transition temperature and measuring crystallization effects.

Shear Tests

Shear strength of plastic sheets can be determined by ASTM D 732, which uses a 1-in.-diameter punch and a fixture for clamping the sheet. Materials from 0.005 to 0.500 in. thickness can be evaluated by this procedure. Cylindrical and rectangular specimens can also be tested in double shear by a procedure outlined in Method 1041 of Federal Test Method Standard 406.

Elastomers are often used in shear where large deflections are encountered—motor and other rotating equipment mounts are an example. ASTM D 945 describes the Yerzley mechanical oscillograph, which has been in use since about 1940 and can be used to generate shear or compressive stress-strain curves on elastomers. Current methods often include special devices where loading is accomplished with servohydraulic systems capable of a wide range of loads, deflections, and loading rates. Actual parts are often tested in these systems as the only practical way to evaluate the compound effects of geometry, rate, temperature, and large deformations.

Hardness

The hardness of plastics and elastomers is often used for specification purposes and can be measured by several techniques which provide an overlap of hardness ranges, as shown in Fig. 7.[6] Methods for durometer and Rockwell hardness testing

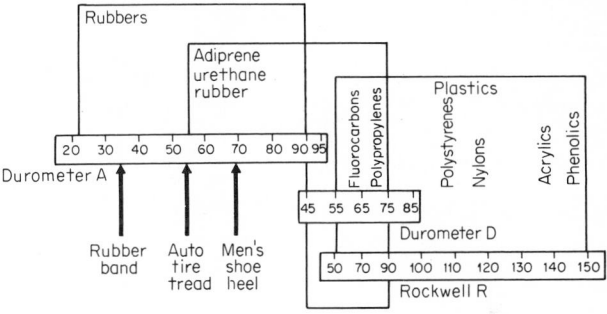

Fig. 7 Hardness of selected rubbers and plastics on a few of the common hardness scales.[6]

are given by ASTM D 2240 and D 785. Both are indentation hardness measurements, and the higher the reading the harder the material. Hardness of elastomers is also reported in International Rubber Hardness Degrees (IRHD) per ASTM D 1415. These readings are approximately the same numerically as those given by the durometer A scale. When a portable instrument is needed, the Barcol impressor can be used for rigid plastics instead of the Rockwell device. Use of this instrument is covered by ASTM D 2583. The durometer instrument is also hand-held, but reproducibility is improved if a supporting fixture is used.

Poisson's Ratio

Poisson's ratio is defined as the absolute value of the ratio of transverse strain to axial strain caused by a uniform axial stress below the material proportional limit. ASTM E 132 is a standard test method for this measurement that can be used for many materials. Poisson's ratio is seldom quoted for elastomers but is usually assumed to be between 0.4 and 0.5.

Impact Testing

As with most mechanical tests for plastics, impact-testing techniques are borrowed from methods developed for metals. The notched cantilever beam specimen for Izod testing (ASTM D 256) uses a 0.01-in. notch radius, which is quite severe for most plastic designs. Because of varying notch sensitivities among plastics, some materials are penalized more than others. Figure 8 shows that if notch sensitivity is ignored, the small radius can prejudice a material selection.[7] If these shortcomings are recognized, published Izod information can be useful, and it is often all that is available. A pendulum tester with the simple-beam Charpy impact specimen is shown in Fig. 9. The same apparatus is used for Izod specimens when the cantilever-beam mounting fixture is installed on the base and a different striker is used.

Fatigue Testing

Fatigue loading of plastics and elastomers usually leads to some heat buildup. This characteristic can be useful in certain designs where damping is a benefit, but unfortunately it causes difficulty in testing, as these materials are usually temperature-sensitive. ASTM D 671, which describes a flexural-fatigue method for rigid plastics, offers this warning: "The results are suitable for direct application in design only when all design factors such as loading, size and shape of part, frequency of stress and environmental conditions exactly parallel the test conditions." Testing under this method follows either (1) a constant-amplitude-of-deflection or (2) a constant-amplitude-of-force technique. Both approaches can be seriously affected by creep and relaxation, particularly when the initial mean stress is not zero. In fact, creep and relaxation will cause a drift toward a zero mean stress

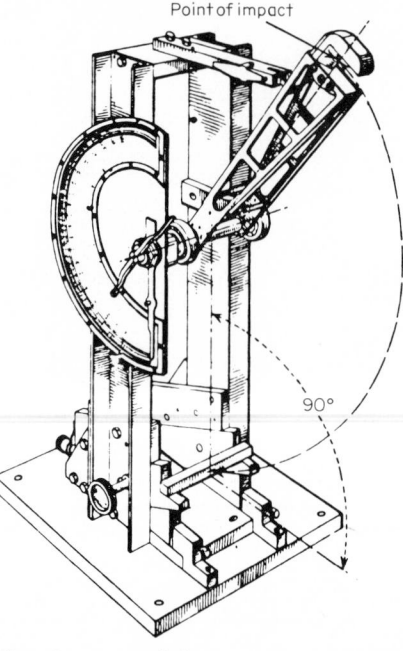

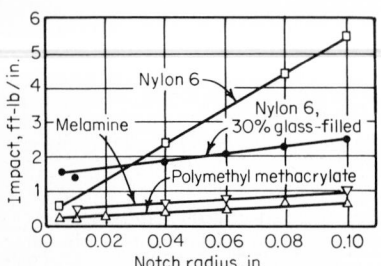

Fig. 8 The effect of notch radius on Izod impact strength for selected plastics.[7]

Fig. 9 A pendulum impact-test machine.[1]

as the test progresses. Figure 10 illustrates stress-time curves for different mean-stress conditions.

Fatigue results are usually plotted as the familiar S-N curves where stress is the ordinate and the common log of the number of cycles to failure is the abscissa, as shown in Fig. 11.[7]

There are several special-purpose tests to measure relative fatigue or fatigue-related properties of elastomer. The following ASTM methods are cited as examples of these tests:

D 623, Compression Fatigue of Vulcanized Rubber

D 813, Crack Growth in Rubber

D 1052, Measurement of Cut Growth of Rubber by the Use of the Ross Flexing Machine

The last two methods utilize a bending-type loading. Figure 12 shows a relationship to specimen thickness for the cut-growth method.[4]

Creep, Creep Rupture, Set, and Stress Relaxation

These properties are discussed together because they are all long-term tests and consequently share the same requirements regarding environmental control and measurement. These properties are highly significant in design because they involve time—an important variable for viscoelastic materials. Tensile creep tests also yield creep-rupture data which are more realistic strength parameters than short-term tensile strength. Unfortunately, long-term tests with close control are costly, especially when severe environments have to be maintained.

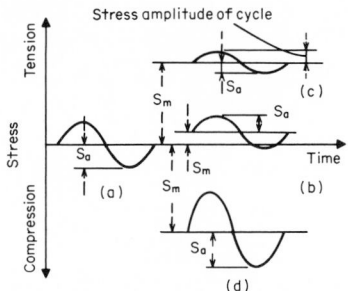

Fig. 10 Fatigue stress-time curves showing (a) a mean stress of zero, (b) and (c) a positive or tensile mean stress, and (d) a negative or compressive mean stress.[1]

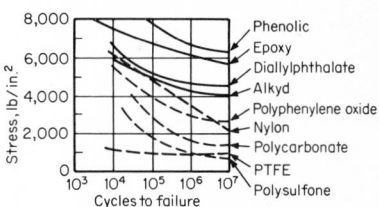

Fig. 11 Fatigue S-N curves for selected plastics.[7]

ASTM D 2990 covers the determination of tensile creep and creep rupture of plastics and defines creep strain as the total strain at any given time. It does not attempt to subtract the elastic strain induced immediately upon loading. This is an expedient move, since it is impractical to separate the two measurements. Creep data do exist, however, where creep strain is defined as the additional strain 1 min after loading. This arbitrary definition is also found in ASTM under D 674, but the procedure is scheduled for withdrawal in 1975. Other terms related to these tests are:

Creep—The time-dependent deformation resulting from an applied constant stress (constant load is another expedient compromise)

Creep modulus—Initial applied stress divided by creep strain

Creep-rupture curve—A plot of stress at rupture vs. time to rupture

Creep rate—The slope of a creep strain vs. time curve for a given creep stress

Stress relaxation—The time-dependent change in stress resulting from an applied constant strain

Creep and stress-relaxation tests run in tension can use the same specimen

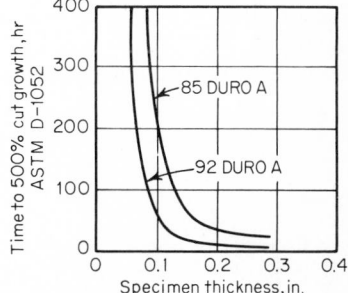

Fig. 12 The effect of specimen thickness on cut-growth rate for two hardnesses of elastomer.[4]

called for in D 638, Test for Tensile Properties of Plastics. Stress-relaxation tests under D 2991 can also be conducted in compression, and the specimen defined under D 695 can be used.

Examples of ways to present creep data are given in Figs. 13 to 15. Log coordinates are often similarly used for presenting stress-relaxation data.

The amount of set an elastomer takes after being compressed can be measured by ASTM D 395. The specimen can be the same as used for short-term measurements under D 575. Set is measured after subjecting the specimen to constant-load or constant-deflection (usually 25 percent of the original thickness) conditions, but the deflection method is most often used. The fixture shown in Fig. 16 can be used to compress the specimens for a stated time at a variety of temperatures, but the set is almost always measured after the unloaded specimen has been at room temperature for 30 min.

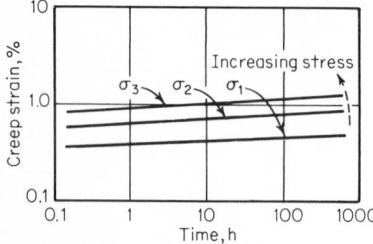

Fig. 13   Creep strain vs. time at various stress levels.[1]

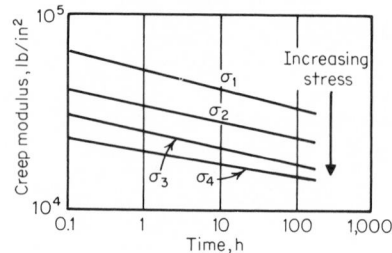

Fig. 14   Creep modulus vs. time at various stress levels.[1]

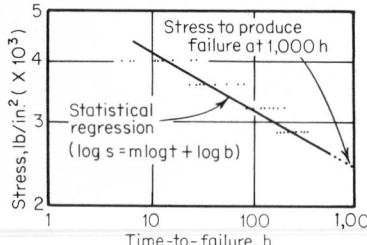

Fig. 15   Time-to-failure or stress-rupture curve.[1]

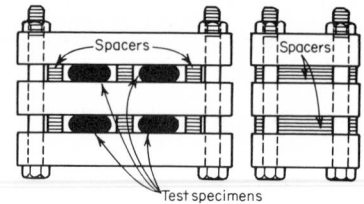

Fig. 16   Constant-deflection compression-set fixture for elastomers.[1]

Damping

Rubber and other elastomers are often used for structural damping, vibration isolation, and shock absorption. Applications cover a broad range of products from appliance-motor mounts to shock-absorbing automobile bumpers to missile-launch-tube shock-isolation pads. Some statements such as "butyl is a better damper than natural rubber" (see Fig. 17)[8] can be generalized from more or less standard test results but again the influence of temperature, rate, and environment brings one to the conclusion that almost every design is a "special case" and must be evaluated in terms of specific service conditions.

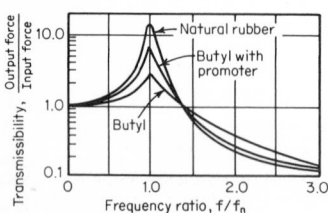

Fig. 17   Transmissibility curves for natural and butyl rubbers.[8]

Measurements describing the stiffness and damping characteristics of elastomers have been reported for some time with the Yerzley oscillograph. ASTM D 945 describes the apparatus and operation. The equipment is limited to fairly low shear and compression loads and low frequency but nonetheless has been of considerable utility. As mentioned earlier under Shear Tests, current methods of dynamically evaluating elastomers often use servohydraulic loading systems. One possible arrangement of such a system is shown in Fig. 18. Such systems can be operated over a wide range of frequencies, are readily adapted to testing at various temperatures, and can be run long enough to evaluate fatigue and self-heating effects.

Damping parameters are expressed in many ways—the damping-quality factor Q, for example, is indicated in Fig. 18. Some other relationships are:[9]

$$\frac{1}{Q} = \eta = \frac{\delta}{\pi} = \frac{2c}{c_c} = \left(\frac{\Delta f}{f_n}\right)_{3\ dB}$$

where η = loss factor
δ = logarithmic decrement
c/c_c = ratio of damping to critical damping
$(\Delta f)_{3\ dB}$ = frequency bandwidth at half-power points (3 dB down)
f_n = natural frequency

Data in the form of the last relationship can be obtained by exciting a cantilever-beam specimen with a vibration generator and varying the frequency Δf about the beam's natural frequency while measuring amplitude or strain—readily done with electric-resistance strain gages.

The logarithmic decrement is another measurement that can be made with a cantilever beam and does not involve forced vibration. A curve such as Fig. 19 is

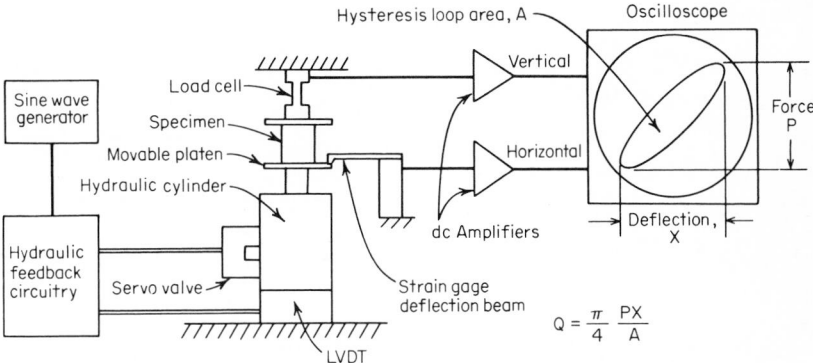

Fig. 18   Servohydraulic damping-measurement apparatus.

generated when the beam is plucked and the vibration is allowed to decay. Softer materials or heavily damped materials that do not vibrate readily are not adaptable to this technique.

The dynamic stiffness (or modulus) of a material is important in determining how much energy a damper can absorb and can be determined from the slope of the hysteresis loop shown in Fig. 18.

Damping properties of more rigid plastics (5×10^4 lb/in.$^2 < E < 2 \times 10^6$ lb/in.^{2}) can be measured by a torsional pendulum per ASTM D 2236. Bulky force-transmitting rods and grips are not required, since forces are low and generated by a freely vibrating pendulum/specimen. The fairly compact apparatus is thus adaptable to operation at different temperatures, making it useful for determining polymer transition temperatures. Figure 20 is a schematic representation of the torsional pendulum. Measuring the amplitudes of two successive oscillations permits calculation of the logarithmic decrement. With a known moment of inertia for the pendulum, the elastic shear modulus of the specimen can be calculated as a function of specimen geometry and frequency. The apparatus is limited to about 0.1- to 10-Hz operation.

Tear Strength

The tear resistance of plastics and elastomers is determined on thicknesses which are generally referred to as sheeting or film. Universal testing machines used for various other mechanical tests such as tension or compression are usually suitable for ASTM D 624, D 1004, and D 1938. The first method is appropriate for vulcanized rubber and other elastomers and is run at 20 in./min on die-cut specimens with or without razor nicks. The second and third methods cover plastic films or flexible sheeting. D 1004 is run at 2 in./min on an unnicked specimen and is designed to indicate tear-initiation characteristics, while D 1938 is run at 10 in./min on a razor-slit specimen to measure propagation tendencies. All these methods yield data of a comparative nature that are difficult to relate to design situations. Data are generally not transferrable over a very wide thickness range—in fact only one procedure, D 624, specifies that results be normalized to thickness, that is, expressed as force to tear divided by thickness. The others merely state that the thickness tested should be reported along with the tearing force. Any material grain or texture will usually affect the values considerably.

Deflection Temperature

A 0.5- by 0.5- by 5.0-in. beam specimen is used to measure the deflection temperature of plastics per ASTM D 648. The specimen is immersed in a suitable heat-transfer liquid and heated at $2°C/min$. The temperature at which the simply loaded beam deflects 0.01 in. with either a 264 or 66 $lb/in.^2$ maximum flexural fiber stress is called the deflection temperature. This is not necessarily an upper temperature limit but rather a means for comparing the relative heat resistance of plastics.

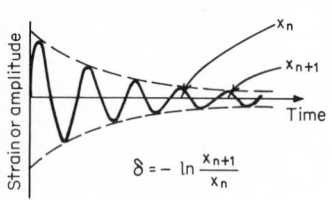

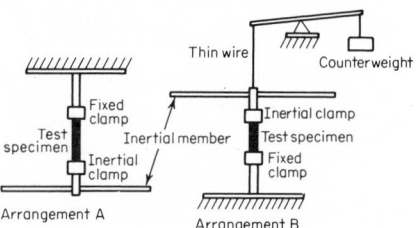

Fig. 19 Logarithmic decrement can be calculated from the decay curve of a freely vibrating cantilever beam.

Fig. 20 Two schematic arrangements of torsional-pendulum systems described in ASTM D 2236.[1]

PHYSICAL-PROPERTIES TESTS

Water Absorption

ASTM D 570 describes the procedure for determining the amount of water absorbed by molded, sheet, rod, or tubular plastics. The usual immersion time is 24 h at $23°C$ for most published data, but occasionally 2 h is used. Also tests may be run in boiling water for 0.5 or 2 h or in $50°C$ water for 48 h. The water absorbed is specified as the percent weight gained by the specimen.

Moisture and Heat Resistance

Moisture and temperature effects on plastic materials and finished parts can be evaluated by the procedures described in ASTM D 756. The exposures are not intended to be related to weathering or corrosive atmospheres but rather to extremes of sheltered storage and shipping conditions. Seven procedures are included, with exposure temperatures ranging from -40 to $+176°F$ and humidities to 100% RH for periods of 24 to 72 h. The dimensions and weight of specimens or parts are recorded before and after exposure along with a description of any qualitative changes observed such as color or odor.

Outdoor Weathering

Weathering affects all materials to some extent. With plastics volatile constituents can be lost, plasticizers can decompose, polymer links can break, hydrogen bonds can form, crystallization can take place, and so on. Damaging effects of weather can be caused by several elements and combinations thereof: sunlight, oxygen, ozone, humidity, precipitation, wind abrasion, and temperature are examples. ASTM D 1435 defines the conditions for outdoor weather exposure in the temperate zones and suggests several tests for evaluating the effects of the exposure. Samples are exposed on aluminum racks at 0, 45, or 90° to the horizontal and facing the equator. It is emphasized that exposure times of less than 1 year can be misleading and that control samples should always be used for comparison and be maintained at 23°C, 50% RH, and wrapped to exclude light. Suggested evaluation tests after exposure include those for color, gloss, and transparency in addition to tests of mechanical and electrical properties.

The Westinghouse Electric Corp. maintains a series of unique test stations at several locations across the country where insulating materials are weathered with an electric potential across the specimen. A view of one facility is shown in Fig. 21.

Fig. 21   30-kV outdoor test facility maintained by Westinghouse.

Artificial Weathering and Aging

Techniques exist for simulating and accelerating the effects of natural weathering and aging. Obviously since natural weathering cannot be rigorously defined, simulated exposures are difficult to relate to service life. However, they are necessary to evaluate new materials and methods of protection. The ASTM prescribes three light sources which may be used in combination with temperature and moisture for various exposures. For plastics the use of a carbon-arc type of light is described in D 1499, a fluorescent sunlamp in D 1501, and a xenon-arc type of light in D 2565. Use of the carbon-arc light for exposing rubber materials is described in D 750.

Two instruments frequently used for artificial weathering are the Fade-Ometer, a carbon-arc light with controlled temperature and humidity and the Weather-Ometer, an enclosed carbon arc with water sprays to wet the samples as well as control humidity. Temperature can also be controlled in the Weather-Ometer.[10]

With all these procedures the quality and type of water must be carefully noted and monitored if the exposure effects are to be reproducible. As with natural weathering, evaluation of the samples after exposure can be by a variety of standard electrical and mechanical tests. Figure 22 is an example.

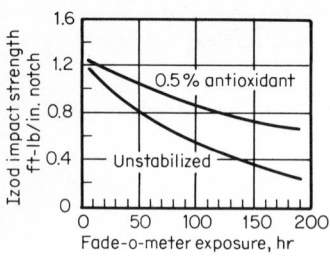

Fig. 22 An example of how impact-test results can be used to measure weathering effects on polystyrene.[10]

Thermal Expansion

The change in length of a material with temperature is indicated by the coefficient of linear thermal expansion. Plastics usually exhibit transition temperatures where the coefficient will change. In fact, determining these temperatures is one use of the measurement of thermal expansion. ASTM D 696 describes the use of a vitreous-silica or fused-quartz tube dilatometer for this measurement. A 2.0-in. (minimum)-long sample, either round or square (about 0.25 in.), is placed at the bottom of a quartz tube and a quartz rod bearing against the end of the specimen is used to transmit the change in specimen length outside the controlled-temperature zone where it can be more accurately measured. A machinist's dial gage can be used if steady-state temperature are employed, but an LVDT or potentiometer is usually used for the continuous recording of length with steadily rising or falling temperatures. A possible configuration is shown in Fig. 23.

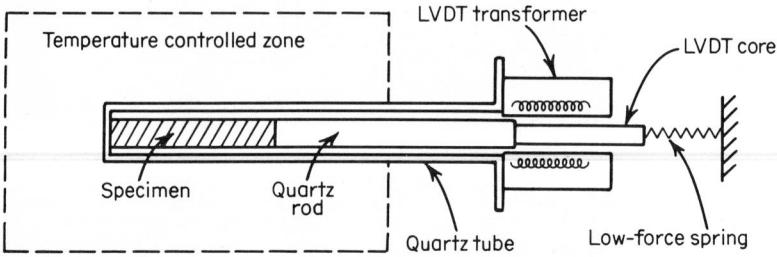

Fig. 23 Schematic configuration of a quartz-tube dilatometer.

Thermal Conductivity

Plastics and elastomers are considered to be thermal insulators—a characteristic that means they are also electrical insulators. The recognized primary technique for measuring thermal conductivity of insulating materials is the guarded-hot-plate method (ASTM C 177). A schematic of the apparatus is shown in Fig. 24. The purpose of the guard heater is to prevent heat flow in all but the axial (up and down in the schematic) direction by establishing isothermal surfaces on the specimens' hot side. With this condition established and by measuring the temperature difference across the sample, the electrical power to the main heater area, and the sample thickness, the K factor can be calculated as:

$$K = \frac{QX}{2A\,\Delta T}$$

Instruments are available for this test which use automatic means to control the guard temperature and record the sample ΔT. Unfortunately this test is fairly expensive.

Another technique uses a heat-flow sensor, which is a calibrated thermopile, in series with the heater, specimen, and cold sink. This method avoids the guard heater and requires only one specimen. This secondary technique is described in ASTM C 518.

Flammability

ASTM offers six methods for measuring the flammability of plastics. Care should be taken before declaring a material "nonflammable," because the condition of the material—dry or wet, old or new, embrittled, etc.—and its form can affect the results. These methods all use strips, sheets, or blocks and, within a given laboratory, can give fairly reliable comparative results between materials, but 40 or 50 percent variation of measured burning rates between laboratories is possible. Method D 568 for flexible plastics, D 635 for self-supporting plastics, and D 1692 for plastic sheet and cellular plastics all involve measuring the progress of burning or charring after ignition with a bunsen burner. Method D 757 can be used for materials judged self-extinguishing by D 635 and involves contacting the specimen with an incandescent silicon carbide rod at 950° C.

Method D 2863 measures flammability by determining the oxygen index (percent O_2) of an oxygen-nitrogen mixture required to support combustion of the sample. A butane flame at the end of a hypodermic needle is used to ignite flexible thin plastic sheeting, and the burning rate is measured by a procedure described in ASTM D 1433.

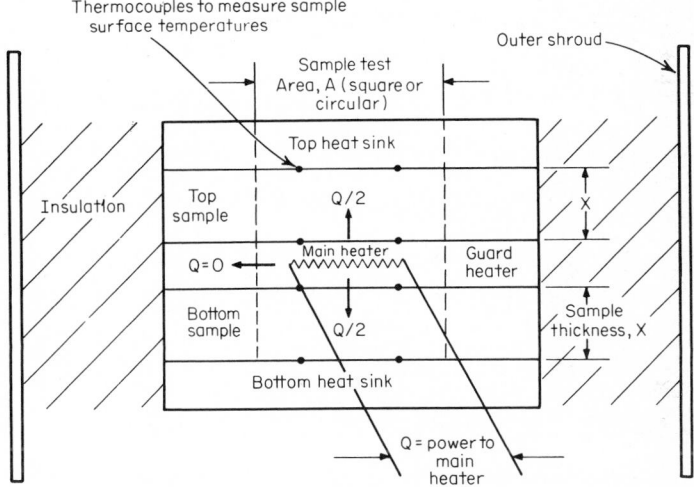

Fig. 24 Schematic assembly of the guarded hot plate.

A summary of flammability and related test results for several materials is presented in Fig. 25.[11] Note the footnote to this figure, which gives new Underwriters' Laboratories designations to describe performance.[12]

The Federal Trade Commission has recently charged that the plastics industry, among others, has been remiss in its duty by allowing the public to enjoy a false sense of security regarding the supposed inflammability of many materials. The FTC further charges that most of the small-scale material tests are not capable of predicting how a product will behave in an actual fire.[13] Where this will all lead is not clear at this time, but fire-related tests are now being carefully reevaluated, and changes most certainly will result.

THERMOPLASTIC RESINS

Thermoplastic resins are those polymers which consist of long, linear molecules accompanied by some branching but not interconnected (cross-linked) from molecule to molecule. These plastics repeatedly soften and melt with increasing temperature and harden again. Typical properties of these plastics are shown in Table 3. Each of these thermoplastic resins is described in the following section.

ABS

Originally a modification of styrene resins, these plastics now comprise a unique family on their own. They are polymers based on acrylonitrile, butadiene, and styrene. Since the weight ratio of each of the three components can be varied, the total possible number of ABS resins is unlimited. However, most common resins

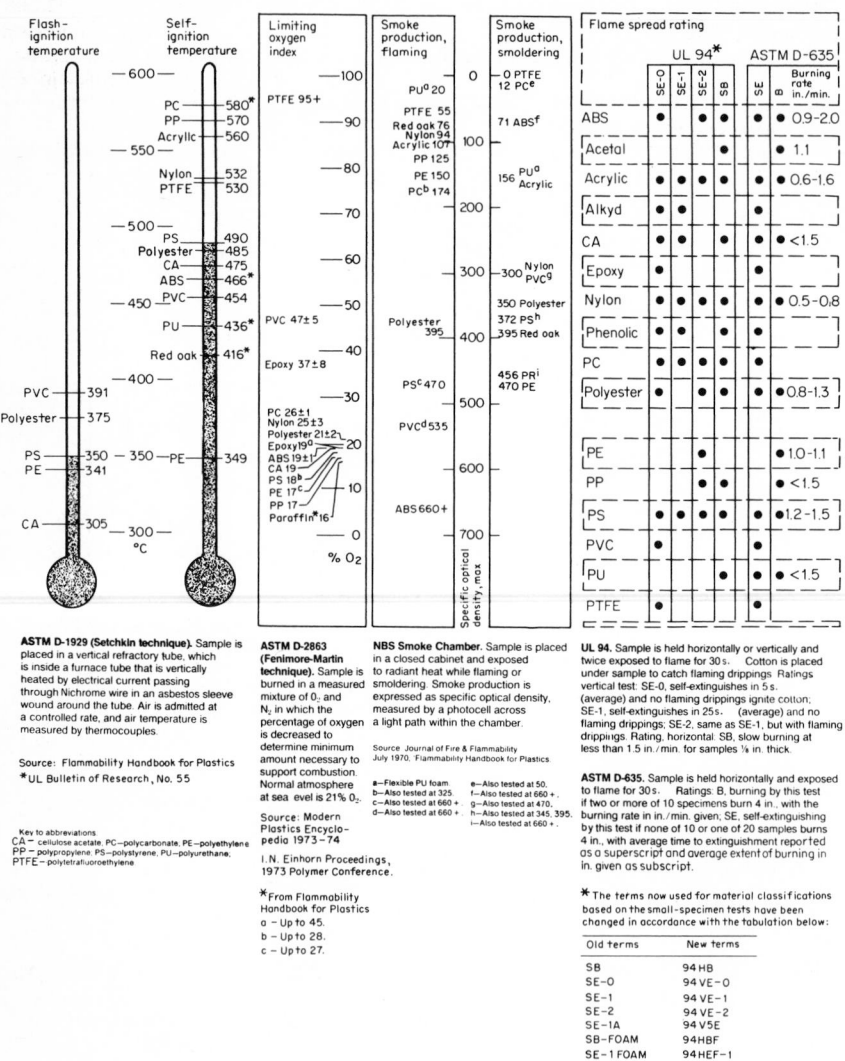

Fig. 25   Flammability test results for selected plastics.[11]

share similar properties. These are shown in Table 3. The chief advantages of ABS resins are toughness combined with chemical resistance, superior strength at elevated temperatures, and excellent impact strength. The change in tensile, compressive, and flexural properties is small up to 180°F, as shown in Fig. 26. Because of their good chemical resistance they are used in refrigerators as door and food liners and trim pieces. The chemical resistance of ABS materials is contrasted with that of impact styrene in Figs 27 and 28. The toughness of ABS leads to its use in camper bodies, trailers, appliances, and luggage. The chemical resistance permits drainpipe applications. The poor weather resistance of ABS requires its protection by other polymer films in outdoor applications. ABS can be plated and is often found in automotive parts.

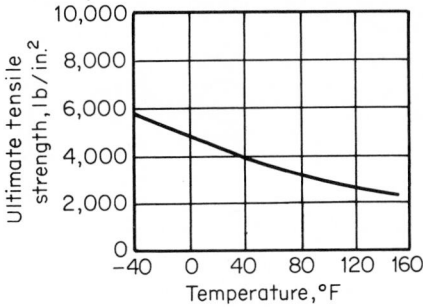

Fig. 26 Ultimate strength of ABS vs. temperature (compression moldings).[15]

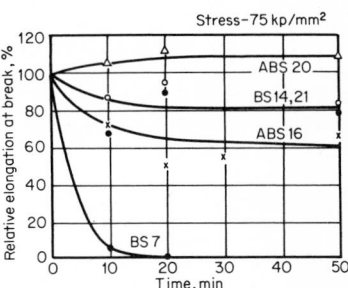

Fig. 27 Ultimate elongation of ABS and impact styrene exposed to R-11 refrigerant. ABS: acrylonitrile, butadiene, styrene polymer; BS: impact styrene polymer; BS#: butadiene content; ABS# acrylonitrile content.[16]

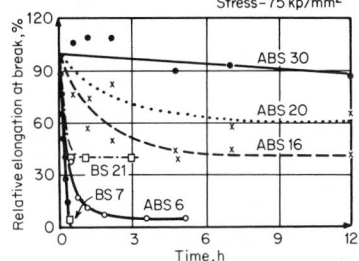

Fig. 28 Ultimate elongation of ABS and impact styrene exposed to oleic acid (salad oil). ABS: acrylonitrile, butadiene, styrene polymer; BS: impact styrene; BS#: butadiene content; ABS# acrylonitrile content.[16]

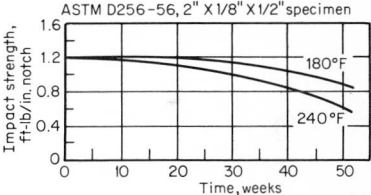

Fig. 29 Impact strength of Celcon acetal copolymer resins after aging.[17]

Acetals

Derivatives of formaldehyde, these polymers are available in two forms—homopolymers and copolymers with very similar properties, as shown in Table 3. They are characterized by high stiffness, resistance to creep, and impact strengths which are relatively unchanging with temperature. The stiffness of acetals is the highest of all thermoplastic materials, as shown in Table 4. Impact strength changes little from −40 to 230°F, as the data in Fig. 29 show. Acetals have a low surface friction in contrast with many other materials (0.1 on steel), and their abrasion is low. Hence they are used in moving parts like gears, bearings, cams, and ratchets. They also can be molded in fast cycles; so their processing cost tends to be low. Exposure to temperatures above 250°F causes the polymer to deteriorate rapidly. They have excellent fatigue properties. At room temperature the fatigue limit is 5,000 lb/in.² and they are used in zippers, springs, pump impellers, conveyors, and tape cassettes.

TABLE 3 Physical Properties—Thermoplastic Resins[14]

		ABS	ABS	ABS	ABS	ABS	ABS	ASA (acrylonitrile-styrene-acrylic elastomer)	Acetals	Acetals
Property	ASTM test method	Extrusion grades	High impact *(Molding grades)*	High heat resistant *(Molding grades)*	Medium impact *(Molding grades)*	Self-extinguishing *(Molding grades)*	Clear		Homopolymer	Copolymer
Molding qualities	....	....	Good to excellent				Good	Good to excellent	Good to excellent	Excellent	Excellent
Compression molding temp, °F	....	325–450	325–450	325–500	325–350	325–350	300–350	300–350	....	340–400
Compression molding pressure, lb/in.	....	....	1,000–8,000					1,000–5,000	4,000–8,000	....	1,000–5,000
Injection molding temp, °F	....	380–525	380–525	475–550	400–500	380–425	400–450	450–500	380–470	360–480
Injection molding pressure, lb/in.^{2}	....	....	8,000–25,000					10,000–20,000	8,000–20,000	10,000–20,000	10,000–20,000
Compression ratio	....	....	1.1–2.0					1.1–1.2	1.1–1.2	....	0.020 (avg)
Mold (linear) shrinkage, in./in.	D 792	0.004–0.009		0.004–0.009	0.004–0.009	0.004–0.008	0.006–0.008	0.004–0.008	0.020–0.025	
Specific gravity (density)	D 792	1.02–1.06	1.01–1.04	1.05–1.08	1.03–1.06	1.16–1.21	1.07	1.07	1.42	1.41
Specific volume, in.^{3}/lb		28.0	27.0–26.0	28.0	28.0–27.0		26.0	27.0	19.6	19.7
Machining qualities	....	....	Good to excellent					Good to excellent	Good to excellent	Excellent	Excellent
Tensile strength, lb/in.^{2}	D 638	4,000–7,000	4,500–7,500	6,500–8,000	6,500–7,500	5,000–6,000	5,600	6,000–8,000	10,000	8,800
Elongation, %	D 638	20.0–110.0	5.0–60.0	3.0–20.0	5.0–25.0	5.0–25.0	....	20–60	25.0–75.0	40–75
Tensile elastic modulus, 10^5 lb/in.^{2}	D 638	2.3–3.8	2.4–3.11	2.9–4.2	3.0–4.5	2.2–3.1	2.9	3.3–3.7	5.2	4.1
Compressive strength, lb/in.^{2}	D 695	5,200–10,200	4,500–8,000	7,200–10,000	1,800–12,500	6,500–7,500	7,000	9,500–21,500	18,000 (10% defl.)	16,000 (10% defl.)
Flexural yield strength, lb/in.^{2}	D 790	6,000–14,000	6,000–11,000	10,000–12,600	11,000–13,000	8,000–11,000	9,950	10,000–120,000	14,100	13,000
Impact strength, ft-lb/in. of notch (½ by ½-in. notched bar, Izod test)	D 256	2.5–10.0 at 73° 0.6–4.0 at −40° (⅛ by ½-in. bar)	5.0–8.0 at 73° 1.5–4.0 at −40° (⅛ by ½-in. bar)	2.0–6.5 at 73° 0.7–1.6 at −40° (⅛ by ½-in. bar)	2.0–6.0 at 73° 1.0–2.4 at −40° (⅛ by ½-in. bar)	1.6–5.0 at 73° (⅛ by ½-in. bar)	Notch radius 0.01 in. = 5.3 notch radius 0.11 in. = 10.9 (⅛ by ½-in. bar)	6.0–8.0 (½ by ⅛-in. bar)	1.4 (inj.), 2.3 (ext.)	1.0–1.5 (⅛ by ½-in. bar)
Hardness, Rockwell	D 785	R75–115	R75–105	R100–115	R107–115	R90–107	R-100	R102–108	M94, R120	M78–80
Flexural elastic modulus, lb/in.^{2} × 10^5	D 790	2.0–4.8	2.0–3.9	3.4–4.2	3.0–4.6	2.5–3.5	3.0	3.3–3.7	4.1	3.75
Compressive modulus, lb/in.^{2} × 10^5	D 695	1.5–3.9	1.4–2.0	1.9–2.4	2.0–2.4	1.3–2.1	1.9	1.5–2.5	6.7	4.5

Property	ASTM method									
Thermal conductivity, 10^{-4} cal/(s)(cm²) per °C/cm	C 177			4.5–8.0				4.5–8.0	5.5	5.5
Specific heat, cal/°C/g				0.3–0.4				0.3–0.4	0.35	0.35
Thermal expansion, 10^{-5} in./in./°C	D 696	6.0–13.0	9.5–13.0	6.0–9.3	8.0–10.0	6.5–9.5	8.3	6.0–10.0	8.1	8.5
Resistance to heat, °F (continuous)		140–200	140–210	190–230	160–200	130–180		160–200	195 (8,800 h)	220
Deflection temp., °F: At 264 lb/in.² fiber stress	D 648	200–220 (annealed) 210–225	200–218 (annealed) 210–225	215–245 (annealed) 225–252	199–221 (annealed) 210–227	186–215 (annealed) 199–220	170–185	215–225 (annealed)	255	230
At 66 lb/in.² fiber stress	D 648						180–195	225–230 (annealed)	338	316
Refractive index n_D	D 542						1.538		1.48	
Clarity		Translucent in thin sections to opaque					Transparent, translucent, opaque	Translucent in thin sections to opaque	Translucent to opaque	
Transmittance, %	D 1003	33.3	28.0	33.3	33.3		85.0			
Haze, %		100	100	100	100		6.0–12.0			
Water absorption, 24 h, ⅛ in. thick, %	D 570							0.5	0.25	0.22
Burning rate (flammability), in./min.	D 635	0.20–0.45 Slow				0.2–0.6, 0.050–0.086 self-extinguishing, 0.086 and above nonburning	Slow	Slow	Slow (1.1)	
Effect of sunlight		None to yellows slightly plus some embrittlement					None to slightly yellow, some embrittlement	Good resistance to prolonged exposure	Chalks slightly	
Effect of weak acids	D 543	None					None	None	Resists some	
Effect of strong acids	D 543	Attacked by concentrated oxidizing acids					Attacked by concentrated oxidizing acids	Attacked by concentrated oxidizing acids	Attacked	
Effect of weak alkalies	D 543	None					Nil	None	Resists some	None
Effect of strong alkalies	D 543	None						None	Not recommended	None
Effect of organic solvents	D 543	Soluble in ketones, esters, and some chlorinated hydrocarbons					Soluble in ketones, esters, aromatics, and chlorinated hydrocarbons	Soluble in ketones, esters, and some chlorinated hydrocarbons	Excellent resistance	

TABLE 3 Physical Properties—Thermoplastic Resins[14] (Continued)

Property	ASTM test method	Cast self-extinguishing	High-impact cast sheet	Cast	Molding	Acrylics MMA styrene copolymer	Coated sheet	Impact acrylic molding compound	MMA alpha-methylstyrene copolymer	Acrylic multipolymer	Acrylic-PVC alloy sheet
Molding qualities	....	....	....	....	Excellent	Excellent	....	Excellent	Excellent	Excellent	Good
Compression molding temp, °F	....	....	....	....	300–425	300–400	....	300–400	325–400	300–400	
Compression molding pressure, lb/in.	....	....	....	....	2,000–10,000	1,000–8,000	....	2,000–5,000	1,000–10,000	2,000–6,000	
Injection molding temp, °F	....	....	....	....	325–500	330–500	....	400–500	350–525	400–475	
Injection molding pressure, lb/in.	....	....	....	....	10,000–20,000	10,000–30,000	....	10,000–20,000	10,000–30,000	10,000–20,000	
Compression ratio	....	....	....	....	1.6–2.0						
Mold (linear) shrinkage, in./in.	....	....	....	....	0.001–0.004(C), 0.002–0.008(I)	0.002–0.006	....	0.004–0.008	0.002–0.006	0.006–0.010	
Specific gravity (density)	D 792	1.21–1.28	....	1.17–1.20	1.17–1.20	1.09	1.19	1.11–1.18	1.09	1.09–1.14	1.35
Specific volume, in.³/lb	D 792	....	....	23.7–23.1	23.7–23.1	25.2	23.4	25.6–23.3	25.2	25.6–23.3	20.5
Machining qualities	....	Good to excellent	Good to excellent	Good to excellent	Good to excellent	Good to excellent	Good to excellent	Good to excellent	Good to excellent	Good to excellent	Excellent
Tensile strength, lb/in.²	D 638	8,000–12,500	6,500	8,000–11,000	7,000–11,000	10,000	10,500	5,000–9,000	10,000	6,000–8,000	6,500
Elongation, %	D 638	3.8–5.1	143	2.0–7.0	2.0–10.0	3.0	3%	20.0–70.0	3.0	3.0–40.0	100
Tensile elastic modulus, 10^5 lb/in.²	D 638	3.5–4.8	2.65	3.5–4.5	3.8–4.5	4.3	4.50	2.0–4.0	4.3	3.1–4.3	3.35
Compressive strength, lb/in.²	D 695	11,000–12,000	....	11,000–19,000	12,000–18,000	11,000–15,000	18,000	4,000–14,000	11,000–15,000	8,000–11,500	8,400
Flexural yield strength, lb/in.²	D 790	12,000–18,000	8,900	12,000–17,000	13,000–19,000	16,000–19,000	16,000	7,000–13,000	16,000–19,000	9,500–13,000	10,700
Impact strength, ft-lb/in. of notch (½- by ½-in. notched bar, Izod test)	D 256	0.3–0.4	1.4	0.3–0.4	0.3–0.5	0.3	0.4	0.8–2.5	0.3	1.0–3.0 at 73°F, 0.9–1.6 at –40°F	15.0
Hardness, Rockwell	D 785	M61–100	M61	M80–100	M85–105	M75	M100	R105–120	M75	R108–119	R99–105
Flexural elastic modulus, lb/in.² × 10^5	D 790	3.5–4.5	3.0	3.90–4.75	4.2–4.6	2.6–3.8	4.50	2.0–3.8	2.6–3.8	3.0–4.0	4.0
Compressive modulus, lb/in.² × 10^5	D 695	4.5	....	3.90–4.75	3.7–4.6	2.4–3.7	4.50	2.4–3.7	2.4–3.7	....	4.0

Property	ASTM										
Thermal conductivity, 10^{-4} cal/(s)(cm²) per °C/cm	C 177	4.0–6.0	.....	4.0–6.0	4.0–6.0	5.0	4.0–5.0	4.0–5.0	4.0–5.0	5.3	0.293
Specific heat, cal/°C/g		0.35	.....	0.35	0.35	0.35	0.34	0.34	0.34	0.33	.....
Thermal expansion, 10^{-6} in./in./°C	D 696	.....	.....	5.0–9.0	5.0–9.0	5.2	5.0–8.0	6.0–8.0	6.0–8.0	6.0–9.0	.....
Resistance to heat, °F (continuous)		125–200	.....	140–200	140–200	180–200	140–195	180–200	180–200	165–175	.....
Deflection temp, °F: At 264 lb/in² fiber stress	D 648	155–205	172	160–215	165–210	205	165–194	205–212	205–212	185–195	160
At 66 lb/in² fiber stress	D 648	170–200	195	165–235	175–225	225	180–205	.....	.....	.....	177
Refractive index n_D	D 542	1.50	.....	1.48–1.50	1.49	1.49	.....	1.567	1.567	1.52	.....
Clarity		Transparent, translucent, and opaque		Transparent, translucent, and opaque		Transparent	Transparent, can be translucent opaque	Transparent	Transparent	Transparent	Opaque
Transmittance, %	D 1003	92	92	92	92	93–94	55–88	90	90	87	.....
Haze, %		1	2.6	1	<3	<1	5 (min)	.....	.....	6–9	.....
Water absorption, 24 h, ⅛ in. thick, %	D 570	0.3–0.4	.....	0.2–0.4	0.1–0.4	<0.4	0.2–0.8	0.15	0.15	0.06	.....
Burning rate (flammability), in./min	D 635	Self-extinguishing	1.2	Slow (0.9–1.3)	Slow (0.6–1.2)	Slow (1.1)	Slow (1.0–1.3)	Slow (1.3)	Slow	Non-burning	Nil
Effect of sunlight		.....	Nil	Nil	Nil	Nil	Nil to slight strength loss	Nil	Nil	Slight color change, slight embrittlement	Nil
Effect of weak acids	D 543	.....	Nil	Nil	None	None, except attacked by HF	Practically nil	Nil	Nil	None	None
Effect of strong acids	D 543	.....	Attacked only by high concentration oxidizing acids		None	None, except attacked by HF	Attacked only by high concentration oxidizing acids	Attacked only by high concentration oxidizing acids	Attacked by high concentration oxidizing acids	Attacked by high concentration oxidizing acids	None
Effect of weak alkalies	D 543	.....	Nil	Nil	Nil	Attacked	Practically nil	Nil	Nil	None	None
Effect of strong alkalies	D 543	.....	Attacked None except by liquid ammonia (NH_3)		Attacked	Attacked	Practically nil	None	None	Very slight	None
Effect of organic solvents	D 543	Soluble in ketones, esters, aromatic and chlorinated hydrocarbons			None	Soluble in ketones, esters, aromatic and chlorinated hydrocarbons	None	None	Soluble in ketones, esters, aromatic and chlorinated hydrocarbons	Attacked by esters, ketones, aromatic and chlorinated hydrocarbons	None

TABLE 3   Physical Properties—Thermoplastic Resins[14] (Continued)

Columns under *Cellulosic molding compounds and sheet* (and, for columns 5–6, under *Cellulose acetate*):

Property	ASTM test method	Butadiene styrene—injection, blow molding, and extrusion grades	Ethyl cellulose molding compounds and sheet	Cellulose acetate — Sheet	Cellulose acetate — Molding	Cellulose propionate molding compound	Cellulose acetate butyrate sheet	Cellulose butyrate acetate molding	Cellulose nitrate
Molding qualities	……	Excellent	Excellent	Excellent	Excellent	Excellent	Excellent	Excellent	Good
Compression molding temp, °F	……	……	250–390	……	260–420	265–400	……	265–390	185–250
Compression molding pressure, lb/in.²	……	……	500–5,000	……	100–5,000	100–5,000	……	100–5,000	2,000–5,000
Injection molding temp, °F	……	450	350–500	……	335–490	335–515	……	335–480	……
Injection molding pressure, lb/in.²	……	……	8,000–32,000	……	8,000–32,000	8,000–32,000	……	8,000–32,000	……
Compression ratio	……	……	1.8–2.4	……	1.8–2.6	1.8–2.4	……	1.8–2.4	……
Mold (linear) shrinkage, in./in.	……	……	0.005–0.009	……	0.003–0.010(C), 0.003–0.008(I)	0.003–0.009(C), 0.003–0.006(I)	……	0.003–0.009(C), 0.003–0.006(I)	……
Specific gravity (density)	D 792	1.04–1.05	1.09–1.17	1.28–1.32	1.22–1.34	1.17–1.24	1.15–1.22	1.15–1.22	1.35–1.40
Specific volume, in.³/lb.	D 792	……	25.5–23.6	21.7–21.0	22.7–20.6	23.4–22.4	24.1–22.7	24.1–22.7	20.5–19.8
Machining qualities	……	……	Good	Excellent	Excellent	Excellent	Excellent	Excellent	Excellent
Tensile strength, lb/in.²	D 638	3,400–4,000	2,000–8,000	4,500–8,000	1,900–9,000	2,000–7,800	2,600–6,900	2,600–6,900	7,000–8,000
Elongation, %	D 638	15–100	5.0–40.0	20.0–50.0	6.0–70.0	29.0–100.0	50.0–100.0	40.0–88.0	40.0–45.0
Tensile elastic modulus, 10^5 lb/in.²	D 638	180,000–200,000	1.0–3.0	3.0–6.0	0.65–4.0	0.6–2.15	……	0.5–2.0	1.9–2.2
Compressive strength, lb/in.²	D 695	……	10,000–35,000	……	2,000–36,000	2,400–22,000	……	2,100–22,000	22,000–35,000
Flexural yield strength, lb/in.²	D 790	……	4,000–12,000	6,000–10,000	2,000–16,000	2,900–11,400	4,000–9,000	1,800–9,300	9,000–11,000
Impact strength, ft-lb/in. of notch (½ by ½-in. notched bar, Izod test)	D 256	0.4	2.0–8.5, 0.3–1.7 at −40°F	……	0.4–5.2	0.5–11.5	……	0.8–6.3	5.0–7.0
Hardness, Rockwell	D 785	……	R50–115	R85–120	R34–125	R10–122	R30–115	R31–116	R95–115
Flexural elastic modulus, lb/in.² $\times 10^5$	D 790	……	……	……	……	1.5–3.4	……	……	……
Compressive modulus, lb/in.² $\times 10^5$	D 695	……	……	……	……	……	……	……	……
Thermal conductivity, 10^{-4} cal/(s)(cm²) per °C/cm	C 177	……	3.8–7.0	4.0–8.0	4.0–8.0	4.0–8.0	4.0–8.0	4.0–8.0	5.5
Specific heat, cal/°C/g	……	……	0.3–0.75	0.3–0.5	0.3–0.42	0.30–0.40	0.3–0.4	0.3–0.4	0.3–0.4
Thermal expansion, 10^{-6} in./in./°C	D 696	……	10–20	10–15	8–18	11–17	11–17	11–17	8–12
Resistance to heat, °F (continuous)	……	160	115–185	140–220	140–220	155–220	140–220	140–220	ca. 140
Deflection temp, °F: At 264 lb/in.² fiber stress	D 648	……	115–190	……	111–195	111–228	……	113–202	140–160
At 66 lb/in.² fiber stress	D 648	……	……	……	120–209	147–250	……	130–227	……
Refractive index n_D	D 542	Excellent	1.47	1.49–1.50	1.46–1.50	1.46–1.49	1.46–1.49	1.46–1.49	1.49–1.51
Clarity	……	……	……	Transparent, translucent, opaque					
Transmittance, %	D 1003	……	……	88 (⅛ in. thick)		88 (⅛ in. thick)	88 (⅛ in. thick)	88 (⅛ in. thick)	
Haze, %	……	……	……	<1 (film, thin sheet)		<1 (film, thin sheet)	<1 (film, thin sheet)	<1 (film, thin sheet)	
Water absorption, 24 h, ⅛ in. thick, %	D 570	0.08–0.09	0.8–1.8	2.0–7.0	1.7–6.5	1.2–2.8	0.9–2.2	0.9–2.2	1.0–2.0
Burning rate (flammability), in./min.	D 635	……	Slow	Slow to self-extinguishing	Slow to self-extinguishing	Slow (1.0–1.3)	Slow	Slow	Very fast
Effect of sunlight	……	……	Slight when properly stabilized	Slight	Slight	Slight	Slight	Slight	Discolors and becomes brittle
Effect of weak acids	D 543	……	Slight	Slight	Slight	Slight	Slight	Slight	Slight
Effect of strong acids	D 543	……	Decomposes	Decomposes	Decomposes	Decomposes	Decomposes	Decomposes	Decomposes
Effect of weak alkalies	D 543	……	None	Swells	Slight	Slight	Slight	Slight	Slight
Effect of strong alkalies	D 543	……	Slight	Soluble in liquid ketones and esters of lower alcohols, softened or dissolved by chlorinated hydrocarbons and aromatic hydrocarbons; little affected by aliphatic hydrocarbons		Decomposes	Decomposes	Decomposes	Decomposes
Effect of organic solvents	D 543	……	Widely soluble			……	……	……	Widely soluble

Property	ASTM test method	Fluoroplastics								Ionomers
		Polytetrafluoroethylene molding compound	PFA fluoroplastic	FEP fluoroplastic	Polyvinylidene fluoride	ETFE	ETFE glass reinforced	E-CTFE	Polychlorotrifluoroethylene	
Molding qualities			Very good	Good	Excellent	Excellent	Very good	Excellent	Excellent	Excellent
Compression molding temp, °F			625–675	600–750	400–550	575–625	575–625	500	460–550	280–350
Compression molding pressure, lb/in.²		2,000–5,000	800–3,000	1,000–2,000	500–15,000	800–2,000	800–2,000	1,000–10,000	500–15,000	100–800
Injection molding temp, °F			700–800	625–760	450–550	570–650	570–650	525–575	500–600	300–550
Injection molding pressure, lb/in.²			5,000–20,000	5,000–20,000	15,000–20,000	3,000–20,000	3,000–20,000	5,000–20,000	20,000–60,000	2,000–20,000
Compression ratio		2.5–4.5	2.0	2.0	1.8 plts. 3.6 pdr.			3.0–4.0	2.0	2.0
Mold (linear) shrinkage, in./in.	D 792		0.040	0.03–0.06	0.030	0.030–0.040	0.002–0.030	0.020–0.025	0.010–0.015	0.003–0.02
Specific gravity (density)	D 792	2.14–2.20	2.12–2.17	2.12–2.17	1.75–1.78	1.7	1.80	1.68–1.69	2.1–2.2	0.93–0.96
Specific volume, in.³/lb		12.9–12.5	13.0–12.8	13.0–12.8	15.7–15.6	16.3	14.9	16.5–16.4	13.2–12.7	30.0–29.0
Machining qualities		Excellent	Excellent	Excellent	Excellent	Excellent	Excellent	Excellent	Excellent	Fair to good
Tensile strength, lb/in.²	D 638	2,000–5,000	4,300	2,700–3,100	550–7,400	6,500	12,000	7,000	4,500–6,000	3,500–5,000
Elongation, %	D 638	200.0–400.0	300	250–330	100–300	100–400	8	200	80.0–250.0	350.0–450.0
Tensile elastic modulus, 10⁵ lb/in.²	D 638	0.58		0.5	1.2	1.2	12.0	240,000	1.5–3.0	0.2–0.6
Compressive strength, lb/in.²	D 695	1,700		2,200	8,680	7,100	10,000	7,000	4,600–7,400	
Flexural yield strength, lb/in.²	D 790								7,400–9,300	No yield
Impact strength, ft-lb/in. of notch (½- by ½-in. notched bar, Izod test)	D 256	3.0	No break	No break	3.6–4.0 (Shore)	No break	9.0	No break	2.5–2.7	6.0–15.0
Hardness, Rockwell	D 785	D50–55 (Shore)	D64	D60–65	D80 (Shore)	R50, D75	R74	R95	R75–95	D50–65 (Shore)
Flexural elastic modulus, lb/in.² × 10⁵	D 790				2.0	2.0	9.5	2.4		0.3
Compressive modulus, lb/in.² × 10⁵	D 695		1.2		1.2					
Thermal conductivity, 10⁻⁴ cal/(s)(cm²) per °C/cm	C 177	6.0		6.0	3.0			3.8	4.7–5.3	5.8
Specific heat, cal/°C/g		0.25		0.28	0.33	0.46–0.47			0.22	0.55
Thermal expansion, 10⁻⁵ in./in./°C	D 696	10.0	12	8.3–10.5	8.5	5–9	1.0–3.2	8	4.5–7.0	12.0
Resistance to heat, °F (continuous)		500	500	400	300	300–360	392	330–355	350–390	160–220
Deflection temp, °F: At 264 lb/in.² fiber stress	D 648	250		158	195	160	410	170	258	100–120
At 66 lb/in.² fiber stress	D 648				300	220	510	240		110
Refractive index n_D	D 542	1.35		1.338	1.42	1.40			1.425	1.51
Clarity		Opaque	Transparent to translucent	Transparent to translucent	Transparent to translucent	Transparent in thin sections	Opaque	Transparent to translucent	Transparent to opaque	Transparent
Transmittance, %	D 1003									75–85 (125 mil)
Haze, %										3–17 (125 mil)
Water absorption, 24 h, ⅛ in. thick, %	D 570	0.00	0.03	<0.01	0.04	0.029	0.022	0.01	0.00	0.1–1.4
Burning rate (flammability), in./min	D 635	None	Nonburning	None	Self-extinguishing	Nonburning	Nonburning	None	None	Very slow 1 in./min
Effect of sunlight		None	None	None	Slight bleaching on long exposure	None	Slight	None	None	Requires ultraviolet stabilizer
Effect of weak acids	D 543	None	None	None	None	None	None	None	None	Slow attack
Effect of strong acids	D 543	None	None	None	Attacked by fuming sulfuric	None	Slight	None	None	Attacked by oxidizing acids
Effect of weak alkalies	D 543	None	None	None	None	None	None	None	None	Very resistant
Effect of strong alkalies	D 543	None	None	None	None	None	None	None	None	Very resistant
Effect of organic solvents	D 543	None	None	None	Resists most solvents	None	None	Resists essentially all solvents	Halogenated compounds cause slight swelling	Very resistant at 75°F

TABLE 3 Physical Properties—Thermoplastic Resins[14] (Continued)

Property	ASTM test method	Nylon 6/6		Nylon 6				Nylon 6/6-6 copolymer	Nylon 6/12 unmodified
		Unmodified	Nucleated	Unmodified	Nucleated	Elastomer copolymer	Cast		
Molding qualities	.....	Excellent	Excellent	Excellent	Excellent	Excellent	.....	Fair-excellent	Excellent
Compression molding temp, °F	.....								
Compression molding pressure, lb/in.²	.....								
Injection molding temp, °F	.....	520–620	500–550	440–550	440–550	450–580		350–400	450–550
Injection molding pressure, lb/in.²	.....	Limited only by the pressure required to fill the mold and by the capacity of the machine							
Compression ratio	.....								
Mold (linear) shrinkage, in./in.	D 792	0.008–0.015	0.014–0.02	0.006–0.014		0.008–0.018		0.006–0.015	0.011
Specific gravity (density)	D 792	1.13–1.15	1.13–1.16	1.12–1.14		1.08–1.11	1.15–1.17	1.08–1.14	1.06–1.08
Specific volume, in.³/lb		24.5–24.1	24.5–23.9	24.8–24.2		24.9–25.6	24.1–23.6	25.6–24.3	26.1–25.6
Machining qualities		Excellent	Excellent	Excellent	Excellent	Excellent	Excellent	Poor–excellent	Excellent
Tensile strength, lb/in.²	D 638	12,000	13,600	11,800	13,000	7,500	11,000–14,000	7,400–12,400	8,800
Elongation, %	D 638	60	15	200	4.7	150	30–320	40–300	150
Tensile elastic modulus, 10⁵ lb/in.²	D 638		4.6				3.5–4.5	1.5–4.1	2.9–1.8
Compressive strength, lb/in.²	D 695	15,000 (yield)		13,000	16,000	5,000–12,000			
Flexural yield strength, lb/in.²	D 790	17,000	17,900		17,500		7,000–17,500		
Impact strength, ft-lb/in. of notch (½ by ½-in. notched bar, Izod test)	D 256	1.0–2.1	0.9	1.0	0.9–3.0	1.8, no breaks	0.8–3.0 (½- by ⅛-in. bar)	0.7, no break	1.0
Hardness, Rockwell	D 785	R120, M83	R123, M86	R119	M89	R81–110	R95–120	R83–119	R114
Flexural elastic modulus, lb/in.² × 10⁵	D 790	4.20	4.8	3.95	5.0	1.1	1.1	1.5	2.9
Compressive modulus, lb/in.² × 10⁵	D 695			2.5					
Thermal conductivity, 10⁻⁴ cal/(s)(cm²) per °C/cm.	C 177	5.8	5.8	5.8	5.8				5.2
Specific heat, cal/°C/g		0.4	0.4	0.4	0.4			0.4	0.3–0.4
Thermal expansion, 10⁻⁵ in./in./°C	D 696	8.0		8.3	8.0	3.11	9.0		
Resistance to heat, °F (continuous)		180–250	180–250	180–250	180–250	180–250	180–250	≤250	180–250
Deflection temp, °F: At 264 lb/in.² fiber stress	D 648	167	171	155	185	113–130	200–425	≤170	
At 66 lb/in.² fiber stress	D 648	374	425–430	365	375	260–350	400–425	≤430	410
Refractive index n_D	D 542	1.53							
Transmittance, %	D 1003	Translucent to opaque		Translucent to opaque		Translucent to opaque		Translucent to opaque	
Haze, %									
Water absorption, 24 h, ⅛ in. thick, %	D 570	1.5	1.5	1.3–1.9	1.3–1.9	1.3–1.5	0.6–1.2	1.5–2.0	0.4
Burning rate (flammability), in./min.	D 635	Self-extinguishing	Self-extinguishing	Self-extinguishing	Self-extinguishing	Slow burning	Self-extinguishing	Slow burning to self-extinguishing	Self-extinguishing
Effect of sunlight		Nylons are embrittled by prolonged exposure to sunlight but stabilized grades are available; finely dispersed carbon black is the most effective stabilizer							
Effect of weak acids	D 543	Resistant						Less resistant than 6/6 and 6	More resistant than 6/6 and 6
Effect of strong acids	D 543	Resistant		Attacked by high concentrates		Attacked		Resistant	
Effect of weak alkalies	D 543					None			
Effect of strong alkalies	D 543	Resistant	Slight			Resistant			
Effect of organic solvents	D 543	Resistant to common solvents but dissolved by phenols and formic acid							Resistant to common solvents but dissolved by phenols

TABLE 3 Physical Properties—Thermoplastic Resins[14] (Continued)

Property	ASTM test method	Nylon 6/10 unmodified	Nylon 11 unmodified	Nylon 12 unmodified	Phenylene oxide–based resins, nonreinforced	Polyallomer	Polyamide-imide Unfilled	Polyamide-imide Mineral-filled	Polyamide-imide Graphite-filled
Molding qualities	.....	Excellent	Excellent	Excellent	Excellent	Excellent	Excellent	Excellent	Excellent
Compression molding, °F	.....	.....	.....	.....	400–460	.....	630–650	630–650	630–650
Compression molding pressure, lb/in.²	.....	.....	.....	.....	500–1,000	.....	3,000–4,000	3,000–4,000	3,000–4,000
Injection molding temp, °F	.....	450–550	390–520	360–525	425–600	430–445	675–725	675–725	650–725
Injection molding pressure, lb/in.²	.....	.....	.....	.....	14,000–20,000	1,000–2,000	18,000–30,000	20,000–40,000	18,000–25,000
Compression ratio	.....	.....	.....	.....	1.3–2.2	.....	.....	.....	.....
Mold (linear) shrinkage, in./in.	.....	0.012	0.012	0.003–0.015	.....	0.01–0.02	0.006	0.005	0.005
Specific gravity (density)	D 792	1.07–1.09	1.03–1.05	1.01–1.02	1.06–1.10	0.896–0.899	1.41	1.86	1.45
Specific volume, in.³/lb	D 792	25.9–25.4	26.6–26.4	27.1–27.0	26.1–25.2	.....	19.6		
Machining qualities	.....	Good	Good		Excellent	Good	Excellent	Excellent	Excellent
Tensile strength, lb/in.²	D 638	8,500	8,000	8,000–9,250	7,800–9,600	3,050–3,850	13,300	11,300	13,000
Elongation, %	D 638	85	300	300	50.0–60.0	400.0–500.0	2.5	2.1	2.5
Tensile elastic modulus, 10^5 lb/in.²	D 638	2.8	1.85	1.8	3.55–3.80	.....			
Compressive strength, lb/in.²	D 695	.....	.....	.....	16,000–16,400	.....	35,000	46,700	33,860
Flexural yield strength, lb/in.²	D 790	.....	.....	.....	12,800–13,500	.....	23,400	21,500	22,100
Impact strength, ft-lb/in. of notch (½- by ½-in. notched bar, Izod test)	D 256	1.2	1.8	2.0	5.0 (½- by ⅛-in. bar)	.....	1.0	0.5	0.8
Hardness, Rockwell	D 785	R111	R108	R106	R115–119	R50–85	E104	E100	E98
Flexural elastic modulus, lb/in.² × 10^5	D 790	2.8	1.7	1.65	3.6–4.0	0.70–1.10	7.0	11.3	7.1
Compressive modulus, lb/in.² × 10^5	D 695	.....	1.8		3.7	.....			
Thermal conductivity, 10^{-4} cal/(s)(cm²) per °C/cm	C 177	5.2	5.2	5.2	5.16	2.0–4.0			
Specific heat, cal/°C/g	.....	0.4	0.3	0.3	0.32	0.5			
Thermal expansion, 10^{-6} in./in./°C	D 696	9.0	10.0	10.0	5.2	8.3–10.0	3.4–4.0		
Resistance to heat, °F (continuous)	.....	180–250	180–250	180–250	175–220	.....	550	550	550
Deflection temp, °F: At 264 lb/in.² fiber stress	D 648	180	130	130	212–265	124–133	540	545	565
At 66 lb/in.² fiber stress	D 648	330	300		230–280	165–192			
Refractive index n_D	D 542					1.492			
Clarity	.....	Translucent to opaque			Opaque	Transparent to translucent			
Transmittance, %	D 1003								
Haze, %	.....								
Water absorption, 24 h, ⅛-in. thick, %	D 570	0.4	0.3	0.25	0.066	<0.01	0.28	0.20	0.27
Burning rate (flammability), in./min	D 635	Self-extinguishing	Self-extinguishing	Self-extinguishing to slow-burning	Self-extinguishing, nondrip	Slow	Nonburning	Nonburning	Nonburning
Effect of sunlight	.....	Nylons are embrittled by prolonged exposure to sunlight but stabilized grades are available, finely dispersed carbon black is the most effective stabilizer			Colors may fade	Crazes; can be protected			
Effect of weak acids	D 543	More resistant than 6/6 and 6	None		None	None	Very resistant	Very resistant	Very resistant
Effect of strong acids	D 543		Resistant		None	Attacked slowly by oxidizing acids	Very resistant	Very resistant	Very resistant
Effect of weak alkalies	D 543	Resistant to common solvents but dissolved by phenols and formic acid	Resistant to common solvents but dissolved by phenols		None	None	Slight attack	Slight attack	Slight attack
Effect of strong alkalies	D 543				None	Very resistant	Attacked	Attacked	Attacked
Effect of organic solvents	D 543		Attacked		Soluble or swells in some aromatics and chlorinated aliphatics; resistant to alcohols	Resistant below 80°C	Very resistant	Very resistant	Very resistant

TABLE 3 Physical Properties—Thermoplastic Resins[14] (Continued)

Property	ASTM test method	Polyaryl ether	Polyaryl sulfone	Polybutylene	Polycarbonate — Unfilled	Polycarbonate — ABS-polycarbonate alloy	Polyester — Nonreinforced	Polyester — PCDT	Polyether sulfone
Molding qualities	....	Excellent	Excellent	Good	Good to excellent	Good	Excellent	Excellent	Excellent
Compression molding temp, °F	....	....	720–750	300–350	480–620	Not recommended	....	....	590–690
Compression molding pressure, lb/in.^{2}	....	....	1,000–2,000	500–1,000	1,000–2,000	....	....	....	1,000–1,500
Injection molding temp, °F	....	540–590	700–800	290–380	480–650	490–540	437–540	440	590–750
Injection molding pressure, lb/in.^{2}	....	10,000–20,000	20,000–50,000	10,000–30,000	10,000–20,000	8,000–25,000	8,000–23,000	....	10,000–20,000
Compression ratio	....	2.6–1.8	2.5	2.5	1.74–5.3	....	....	....	....
Mold (linear) shrinkage, in./in.	....	0.007	0.007–0.009	0.003 unaged, 0.026 aged	0.005–0.007	0.005–0.009	0.015–0.020	0.002	0.007
Specific gravity (density)	D 792	1.14	1.36	0.910–0.915	1.2	1.10–1.20	1.31–1.38	1.2	1.37
Specific volume, in.^{3}/lb	D 792	25.2	20.4	30.4–30.2	23.0	26.0–24.0	20.2–21.1	....	19.7–20.2
Machining qualities	....	Excellent	Excellent	Poor	Excellent	Good to excellent	Excellent	Excellent	Excellent
Elongation, %	D 638	25.0–90.0	13–20	300–380.0	100.0–130.0	....	50–300	210	30–80 (at break)
Tensile strength, lb/in.^{2}	D 638	7,500	13,000	3,800–4,400	8,000–9,500	7,500–8,200	8,200	7,290	12,200 (at yield)
Tensile elastic modulus, 10^5 lb/in.^{2}	D 638	3.2	3.7	0.26	3.0–3.5	3.0–3.8	2.8	....	3.5–3.54
Compressive strength, lb/in.^{2}	D 695	....	17,900	....	12,500	8,500–11,000	8,600–14,500	....	....
Flexural yield strength, lb/in.^{2}	D 790	11,000	17,200	....	13,500	10,500–14,500	12,000–16,700	10,600	18,650 (at yield)
Impact strength, ft-lb/in. of notch (½- by ½-in. notched bar, Izod test)	D 256	8.0 (½ by ¼-in. bar)	1–2 at 73°F	No break	12.0–18.0 (½ by ⅛-in. bar)	8.0–13.0 at 73°F, 1.5–3.5 at −40°F (½ by ⅛-in. bar)	0.8–1.0	1.0 at 73°F, 0.7 at −40°F	1.6
Hardness, Rockwell	D 785	R117	M110	65 (Shore D)	M70–78, R115–125	R106–120	M68, 78, 85, 98	R108	M88
Flexural elastic modulus, lb/in.^{2} × 10^5	D 790	3.0	3.95	0.49	3.2–3.5	3.0–4.2	3.3–4.01	2.97	3.7–3.75
Compressive modulus, lb/in.^{2} × 10^5	D 695	....	3.4	0.31	3.45	....	....	1.75	....
Thermal conductivity, 10^{-4} cal/(s)(cm^2) per °C/cm	C 177	7.13	4.55	....	4.6	6.0–9.0	4.2–6.9	....	3.2–4.4
Specific heat, cal/°C/g	D 696	0.35	....	0.45	0.28–0.30	....	0.28–0.55	....	0.26
Thermal expansion, 10^{-6} in./in./°C	....	3.6	4.7	15.0	6.6	6.2–8.5	6.0–9.5	4.4/°F	5.5
Resistance to heat, °F (continuous)	....	250	500	225	250	220–250	122–250	....	300–390
Deflection temp, °F: At 264 lb/in.^{2} fiber stress	D 648	300	525	130–140	265–285	220–260 (annealed)	122–185	154	397
At 66 lb/in.^{2} fiber stress	D 648	320	....	215–235	270–290	235–265	240–374	....	....
Refractive index n$_D$	D 542	....	1.67	1.50	1.586	....	....	....	1.65
Clarity	....	Translucent to opaque	Opaque	Translucent	Transparent to opaque	Opaque	Opaque	Transparent	Transparent
Transmittance, %	D 1003	....	....	....	85–91	....	....	70–5 mils	....
Haze, %	....	....	....	....	0.5–2	....	....	0.4	....
Water absorption, 24 h, ⅛ in. thick, %	D 570	0.25	1.1	<0.01–0.026	0.15–0.18	0.20–0.35	0.08–0.09	0.2	0.43
Burning rate (flammability), in./min	D 635	Slow	Self-extinguishing nondrip	1.08	Self-extinguishing, UL class I and II	Slow (0.8–1.27)	....	0.8	Self-extinguishing and nonburning
Effect of sunlight	....	None to yellows, slight embrittlement	Slight darkening	Crazes rapidly	None	None to slight yellowing, some embrittlement	....	....	Slight yellowing, strength loss
Effect of weak acids	D 543	None	None	Resistant	Attacked slowly	None	....	Resistant	None
Effect of strong acids	D 543	Resists some	None	Attacked by oxidizing	Attacked	Attacked by concentrated oxidizing acids	....	Attacked by oxidizing acids	None
Effect of weak alkalies	D 543	None	None	Very resistant	....	None	....	Resistant	None
Effect of strong alkalies	D 543	None	None	Very resistant	....	Slight	....	Slight	None
Effect of organic solvents	D 543	Soluble in chlorinated aromatics, esters, ketones	Soluble in highly polar solvents, unaffected by most common solvents, fuel, lubricants	....	Resistant to paraffinics, soluble in aromatic and chlorinated hydrocarbons	Soluble in ketones, esters, and some partly chlorinated hydrocarbons	....	Attacked by ketones, hydrocarbons	Aliphatic—none; aromatic—partly soluble

Polyethylene — "Polyethylene cross-linkable compounds" spans the *Molding grades* and *Wire and cable grades* columns.

Property	ASTM test method	Low-density	Medium-density	High-density	Rotational molding grades	Molding grades	Wire and cable grades	Ethylene-ethyl acrylate copolymer	Ethylene vinyl acetate copolymer	Ultra high molecular weight
Molding qualities	.....	Excellent	Excellent	Excellent	Good	Excellent	.....	Excellent	Excellent	Fair
Compression molding temp, °F	.....	275–350	300–375	300–450	390–450	240–450	.....	200–300	200–300	400–500
Compression molding pressure, lb/in.²	.....	100–800	100–800	500–800	500–800	100–800	.....	50–1,000	50–2,500	800–1,200
Injection molding temp, °F	.....	300–600	300–600	300–600	.....	250–300	.....	250–600	250–430	.....
Injection molding pressure, lb/in.²	.....	8,000–30,000	8,000–30,000	10,000–20,000	.....	.....	.....	8,000–20,000	8,000–20,000	.....
Compression ratio	.....	1.8–3.6	1.8–2.2	2.0	.....	.....	.....	.....	.....	.....
Mold (linear) shrinkage, in./in.	D 792	0.015–0.050	0.015–0.050	0.02–0.05	0.02–0.05	0.007–0.090	0.020–0.050	0.015–0.035	0.007–0.012	0.94
Specific gravity (density)	D 792	0.910–0.925	0.926–0.940	0.941–0.965	0.930–0.939	0.95–1.45	0.91–1.40	.....	0.920–0.950	29.4
Specific volume, in.³/lb	.....	30.4–29.9	29.9–29.4	29.4–28.7	1.0	.....	.....	.....	.....	.....
Machining qualities	.....	Good	Good	Excellent	.....	Good to excellent	Good to excellent	Fair	Fair	Good to excellent
Tensile strength, lb/in.²	D 638	600–2,300	1,200–3,500	3,100–5,500	2,500–2,700	1,600–4,600	1,500–3,100	1,600–2,100	1,440–2,800	2,500–3,500
Elongation, %	D 638	90.0–800.0	50.0–600.0	20–1,300	450–>660	10.0–4.40	180–600	700–750	550–900	300–500
Tensile elastic modulus, 10⁵ lb/in.²	D 638	0.14–0.38	0.25–0.55	0.6–1.8	.....	0.5–5.0	.....	0.041–0.075	0.02–0.12	0.20–1.10
Compression strength, lb/in.²	D 695	.....	.....	2,700–3,600	.....	2,000–5,500	.....	.....	.....	.....
Flexural yield strength, lb/in.²	D 790	.....	4,800–7,000	.....	.....	2,000–6,500	.....	3,000–3,600	.....	.....
Impact strength, ft-lb/in. of notch (½- by ½-in. notched bar, Izod test)	D 256	No break	0.5–>16.0	0.5–20.0	No break	1.0–20.0	.....	No break	No break	No break
Hardness, Rockwell	D 785	D41–50 (Shore), R10	D50–60 (Shore), R15	D60–70 (Shore)	D60–63 (Shore)	55–80 (Shore D)	33–57 (Shore D)	D27–36 (Shore)	D17–45 (Shore)	D60–70 (Shore)
Flexural elastic modulus, lb/in.² × 10⁵	D 790	0.08–0.60	0.60–1.15	1.0–2.6	1.0	0.7–3.5	.....	0.01–0.20	0.01–0.20	1.30–1.40
Compressive modulus, lb/in.² × 10⁵	D 695	.....	.....	.....	.....	0.5–1.5	.....	.....	.....	.....
Thermal conductivity, 10⁻⁴ cal/(s)(cm²) per °C/cm	C 177	8.0	8.0–10.0	11.0–12.4	.....	.....	.....	.....	.....	7.2
Specific heat, cal/°C/g	.....	0.55	0.55	0.55	.....	.....	.....	0.55	0.55	.....
Thermal expansion, 10⁻⁶ in./in./°C	D 696	10.0–22.0	14.0–16.0	11.0–13.0	.....	10.0–35.0	10.0–35.0	16.0–25.0	16.0–20.0	.....
Resistance to heat, °F (continuous)	.....	180–212	220–250	250	.....	275	275	190–200	.....	.....
Deflection temp, °F: At 264 lb/in.² fiber stress	D 648	90–105	105–120	110–130	140	105–145	100–175	.....	93	105–120
At 66 lb/in.² fiber stress	D 648	100–121	120–165	140–190	.....	130–225	.....	.....	140–147	155–180
Refractive index nD	D 542	1.51	1.52	1.54	.....	.....	.....	.....	.....	.....
Clarity	D 1003	Transparent to opaque			Transparent to opaque			Translucent	Transparent to opaque	Translucent to opaque
Transmittance, %	.....	0–75	10–80	0–40	.....	.....	.....	.....	0–80	.....
Haze, %	.....	4–50	4–50	10–50	.....	.....	.....	.....	2–40	.....
Water absorption, 24 h, ⅛ in. thick, %	D 570	<0.01 (1.04)	<0.01 (1.00–1.04)	<0.01 (1.00–1.04)	.....	<0.01–0.06	<0.01–0.06	0.04	0.05–0.13	<0.01
Burning rate (flammability), in./min.	D 635	Very slow (1.04)	Very slow (1.00–1.04)	Very slow (1.00–1.04)	Very slow	Very slow self-extinguishing	Slow	Very slow	Very slow	Very slow
Effect of sunlight	.....	Unprotected material crazes rapidly; requires black for complete protection but weather-resistant grades available in natural and colors					Same as for polyethylene	Very slight yellowing	Very slight yellowing	Very resistant
Effect of weak acids	D 543	Resistant	Very resistant			Very resistant	Very resistant	Resistant	Resistant	Attacked by oxidizing acids
Effect of strong acids	D 543	Attacked by oxidizing acids	Attacked slowly by oxidizing acids			Attacked slowly by oxidizing acids	Attacked slowly by oxidizing acids	Attacked by oxidizing acids	Attacked	Very resistant
Effect of weak alkalies	D 543	Very resistant	Very resistant		Very resistant	Very resistant	Very resistant	Resistant	Resistant	Very resistant
Effect of strong alkalies	D 543	Very resistant	Very resistant		Very resistant	Very resistant	Resistant below 80°C	Resistant	Resistant	Very resistant
Effect of organic solvents	D 543	Resistant below 60°C except to chlorinated solvents			Resistant below 80°C			Attacked by chlorinated solvents, soluble in aromatic solvents above 50°C	Soluble in chlorinated and aromatic solvents over 50°C	Resists many solvents below 80°C

TABLE 3 Physical Properties—Thermoplastic Resins[14] (Continued)

Property	ASTM test method	Polyimides Addition	Polyimides Condensation	Polymethyl-pentene	Polyphenylene-sulfide, unfilled	Polypropylene Unmodified	Polypropylene Copolymer	Polypropylene Impact (rubber-modified)
Molding qualities				Excellent	Excellent	Excellent	Excellent	Excellent
Compression molding temp, °F				540–555	600–650	340–450	340–550	340–450
Compression molding pressure, lb/in.²				500–1,000	1,000–2,000	500–1,000	500–1,000	500–1,000
Injection molding temp, °F				540–580	625–700	400–550	400–550	400–550
Injection molding pressure, lb/in.²				8,000–20,000	5,000–15,000	10,000–20,000	10,000–20,000	15,000
Compression ratio						2.0–2.4	2.0–2.4	2.0–2.4
Mold (linear) shrinkage, in./in.	D 792			0.015–0.030	0.010	0.010–0.025	0.010–0.025	0.01–0.025
Specific gravity (density)	D 792	1.43	1.43–1.51	0.83	1.34	0.902–0.910	0.890–0.905	0.890–0.91
Specific volume, in.³/lb		19.3		33.3	20.6	30.8–30.4	31.2–30.5	30.8–30.5
Machining qualities		Excellent		Good	Excellent	Good	Good	Good
Tensile strength, lb/in.²	D 638	10,500	9,000–13,000	3,500–4,000	10,000	4,300–5,500	2,900–4,500	2,800–4,400
Elongation, %	D 638	5–8	4–9	13–22	3	200.0–700.0	200–700 (at break)	350.0–>500.0
Tensile elastic modulus, 10⁵ lb/in.²	D 638	4.5	32,000–>40,000	1.6–2.1	4.8	1.60–2.25	1.0–1.7	1.0–1.7
Compressive strength, lb/in.²	D 695	>24,000				5,500–8,000	<3,700–8,000	4,000–6,500
Flexural yield strength, lb/in.²	D 790	15,000	15,000–17,000		20,000 (at break)	6,000–8,000	5,000–7,000	
Impact strength, ft-lb/in. of notch (½- by ½-in. notched bar, Izod test)	D 256	0.9	0.5–1.0 at 73°F	0.4–1.6	0.3 at 75°F 1.0 at 300°F (½- by ¼-in. bar)	0.5–2.2 at 73°F	1.1–20.0 at 73°F	1.0–15.0 at 73°F
Hardness, Rockwell	D 785	E45–58	32–58	L67–74	R124	R80–110	R50–96	R50–85
Flexural elastic modulus, lb/in.² × 10⁶	D 790	4.6		1.4–2.0	6.0	1.7–2.5	1.3–2.0	1.2–1.8
Compressive modulus, lb/in.² × 10⁵	D 695	540				1.5–3.0		
Thermal conductivity, 10⁻⁴ cal/(s)(cm²) per °C/cm.	C 177			4.0	6.84	2.8	2.0–4.0	3.0–4.0
Specific heat, cal/°C/g.				0.52		0.46	0.5	0.5
Thermal expansion, 10⁻⁵ in./in./°C.	D 696			11.7	5.5	5.8–10.2	8.0–9.5	6.0–8.5
Resistance to heat, °F (continuous)		500 in air	>470	250–320	400–500	225–260	190–240	200–250
Deflection temp, °F: At 264 lb/in.² fiber stress	D 648	680			278	125–140	115–140	120–135
At 66 lb/in.² fiber stress	D 648					200–250	185–235	160–210
Refractive index n_D	D 542			1.465		1.49		
Clarity		Opaque		Transparent, opaque	Opaque	Transparent, translucent, opaque	Transparent, translucent, opaque	Translucent
Transmittance, %	D 1003			>90		55–90 (film)		
Haze, %				<5.0		1.0–3.5 (film)		
Water absorption, 24 h, ⅛-in. thick, %	D 570	0.3	0.32	0.01	0.02	<0.01–0.03	<0.01–0.03	<0.01–0.03
Burning rate (flammability), in./min	D 635	Nonburning		1.0	Nonburning	Slow	Slow	Slow
Effect of sunlight				None	None	Unprotected material crazes rapidly, requires black for complete protection, but weather-resistant grades available in natural and colors		
Effect of weak acids	D 543	Resistant		Attacked slowly by oxidizing acids	Attacked slowly by oxidizing acids	None	None	Attacked slowly by oxidizing acids
Effect of strong acids	D 543	Resistant		Attacked slowly by oxidizing acids	Attacked slowly by oxidizing acids	Attacked slowly by oxidizing acids	Attacked slowly by oxidizing acids	Attacked slowly by oxidizing acids
Effect of weak alkalies	D 543	Slow attack		Very resistant	None	None	None	None
Effect of strong alkalies	D 543	Attack		Very resistant	None	None	None	Very resistant
Effect of organic solvents	D 543	Very resistant		Attacked by chlorinated aromatic solvents	Resistant below 375–400°F	Resistant below 80°C	Resistant below 80°C	Attacked by hydrocarbons and chlorinated hydrocarbons

Property	ASTM test method	Unfilled free-flowing general-purpose heat-resistant	Impact-resistant medium-impact high-impact heat-resistant	Special heat- and chemical-resistant type	Impact self-extinguishing	General-purpose self-extinguishing	Styrene acrylonitrile	Polysulfone	Styrene-butadiene
Molding qualities	.....	Excellent	Excellent	Good	Excellent	Excellent	Good	Excellent	Good
Compression molding temp, °F	.....	265–400	250–400	300–400	250–400	265–400	300–400	550–600	250–325
Compression molding pressure, lb/in.²	.....	1,000–10,000	1,000–10,000	1,000–5,000	1,000–10,000	1,000–10,000	1,000–10,000	1,000	100–3,000
Injection molding temp, °F	.....	325–500	350–600	350–700	350–425	350–425	375–575	650–750	300–425
Injection molding pressure, lb/in.²	.....	10,000–30,000	10,000–30,000	10,000–30,000	10,000–30,000	10,000–30,000	10,000–33,000	15,000–20,000	15,000–30,000
Compression ratio	.....	1.6–4.0	1.6–4.0	1.6–4.0	1.6–4.0	1.6–4.0	1.6–4.0	1.8–2.2	2.0
Mold (linear) shrinkage, in./in.	.....	0.001–0.006 (C), 0.002–0.006 (I)	0.002–0.006	0.001–0.008	0.002–0.006	0.002–0.006	0.002–0.007	0.007	0.001–0.005
Specific gravity (density)	D 792	1.04–1.09	1.04–1.10	1.04–1.10	1.10	1.08	1.075–1.100	1.24	0.93–1.10
Specific volume, in.³/lb	D 792	26.0–25.6	28.1–25.2	26.2–24.8	25.2	25.6	25.8–25.2	22.3	37.4–27.5
Machining qualities	.....	Fair to good	Good	Fair to good	Good	Fair to good	Good	Excellent	Poor
Tensile strength, lb/in.²	D 638	5,000–12,000	1,500–7,000	6,500–12,000	5,000	7,000	9,000–12,000	10,200 (at yield)	600–3,000
Elongation, %	D 638	1.0–2.5	2.0–90.0	1.4–2.5	13	2	1.5–3.7	50.0–100.0 (at break)	300.0–1,000.0
Tensile elastic modulus, 10^5 lb/in.²	D 638	4.0–6.0	1.4–5.0	4.0–6.0	3.0	4.5	4.0–5.6	3.6	0.008–0.500
Compressive strength, lb/in.²	D 695	11,500–16,000	4,000–9,000	11,500–16,000	....	....	14,000–17,000	13,900 (at yield)	Max stress 5,000 no break
Flexural yield strength, lb/in.²	D 790	8,000–14,000	3,000–12,000	10,000–17,000	....	....	14,000–19,000	15,400 (at yield)	Max stress 60–900 no break
Impact strength, ft-lb/in. of notch (½- by ½-in. notched bar, Izod test)	D 256	0.25–0.40 (¼-in. bar) 1.6 (⅛ by ½ in.)	0.5–8.0 (23°C) 0.4–2.5 (−25°C)	0.35–0.60 (¼-in. bar)	1.21 (⅛ by ½ in.)	0.25 (⅛ by ½ in.)	0.35–0.50	1.3 (¼-in. bar) 1.2 (⅛-in. bar) (−40°F)	No break
Hardness, Rockwell	D 785	M65–80	M10–80; R30–100	M65–90	M10	M70	M80–90	M69, R120	40–90 (Shore A)
Flexural elastic modulus, lb/in.² × 10^5	D 790	4.0–4.7	1.5–4.6	....	....	....	5.5	3.9	0.04–1.50
Compressive modulus, lb/in.² × 10^5	D 695	....	....	....	....	....	5.3	3.7	0.036–1.200
Thermal conductivity, 10^{-4} cal/(s)(cm²) per °C/cm	C 177	2.4–3.3	1.0–3.0	1.9–3.0	....	2.4–3.3	2.9	....	3.6
Specific heat, cal/°C/g	.....	0.32	0.32–0.35	0.32–0.35	....	0.32	0.32–0.34	0.31	0.45–0.50
Thermal expansion, 10^{-5} in./in./°C	D 696	6.0–8.0	3.4–21.0	6.0–8.0	....	6.0–8.0	3.6–3.8	5.2–5.6	13.0–13.7
Resistance to heat, °F (continuous)	.....	150–170	140–175	170–220	....	....	140–205	300–345	130–150
Deflection temp, °F:									
At 264 lb/in.² fiber stress	D 648	220 max	210 max	180–235	175	195	190–220	345	Subzero—150
At 66 lb/in.² fiber stress	D 648	180–230	180–220	195–240	....	....	....	358	Subzero—120
Refractive index n_D	D 542	1.59–1.60	....	1.57–1.60	....	....	1.56–1.57	1.633	1.52–1.55
Clarity	.....	Transparent	Translucent, opaque	Transparent	Translucent, opaque	Transparent	Transparent	Transparent, opaque	Transparent, opaque
Transmittance, %	D 1003	87–92	35–57	88–90	....	87	78–88		
Haze, %	.....	<0.1–3.0	>77	....	....	....	0.4–0.1	5.0	
Water absorption, 24 h, ⅛ in. thick, %	D 570	0.03–0.10	0.05–0.6	0.05–0.40	....	....	0.20–0.30	0.22	0.19–0.39
Burning rate (flammability), in./min	D 635	Slow <1.5	Slow <1.5	Slow <1.5	SE 2 (UL)	SE 2 (UL)	Slow	Self-extinguishing	Slow
Effect of sunlight	.....	Yellows slightly	Some strength loss; yellows slightly	Yellows slightly	Some strength loss; yellows slightly	Yellows slightly	Yellows slightly	Strength loss, slight yellowing	Slight color change
Effect of weak acids	D 543	None	None	None	None	None	None	None	None
Effect of strong acids	D 543	Attacked by oxidizing acids	Attacked by oxidizing acids	Attacked by oxidizing acids	Attacked by oxidizing acids	Attacked by oxidizing acids	Attacked by oxidizing acids	None	Attacked by oxidizing acids
Effect of weak alkalies	D 543	None	None	None	None	None	None	None	None
Effect of strong alkalies	D 543	None	None	None	None	None	None	None	Slight
Effect of organic solvents	D 543	Soluble in aromatic and chlorinated hydrocarbons	Soluble in aromatic and chlorinated hydrocarbons 140–200°F	Soluble in aromatic and chlorinated hydrocarbons 200–220°F	Soluble in aromatic and chlorinated hydrocarbons		Soluble in ketones, esters, some chlorinated hydrocarbons	Aliphatic—none; aromatic—partly soluble	Dissolves

TABLE 3 Physical Properties—Thermoplastic Resins[14] (Continued)

| Property | ASTM test method | Vinyl chloride and vinyl chloride-acetate molding compounds, sheets, rods, and tubes | | Vinyl polymers | | | | ABS-modified vinyl chloride | | Polypropylene-modified vinyl chloride, rigid |
		Rigid	Flexible	Vinylidene chloride molding compounds	Vinyl formal molding compound	Chlorinated polyvinyl chloride compound	Vinyl butyral molding compounds, flexible unfilled	Rigid	Flexible	
Molding qualities	....	Fair to good	Good	Excellent	Good	Good	Good	Good to excellent		Excellent
Compression molding temp, °F	....	285–400	285–350	220–350	300–350	350–400	280–320	300–350		290–360
Compression molding pressure, lb/in.²	....	750–2,000	500–2,000	250–5,000	1,000–10,000	1,500–2,000	100–3,000	1,000–2,000		1,000–2,000
Injection molding temp, °F	....	300–415	320–385	300–400	300–400	375–425	250–340	400		330–375
Injection molding pressure, lb/in.²	....	10,000–40,000	8,000–25,000	10,000–30,000	10,000–30,000	15,000–40,000	15,000–30,000	8,000–25,000		8,000–16,000
Compression ratio	....	2.0–2.3	2.0–2.3	2.0		2.0–2.5		2.0–2.5		1.7–2.0
Mold (linear) shrinkage, in./in.	....	0.001–0.005	0.010–0.050 (varies with plasticizer)	0.005–0.025	0.0015–0.0030	0.003–0.007		0.004–0.015		0.002–0.005 (inj.)
Specific gravity (density)	D 792	1.30–1.58	1.16–1.35	1.65–1.72	1.2–1.4	1.49–1.58	1.05	1.35		1.28–1.40
Specific volume, in.³/lb	D 792	20.5–19.1	23.8–20.5	16.8–16.1	23.0–19.8	18.4–17.8	26.2			21.4–18.05
Machining qualities	....	Excellent		Good	Good to poor	Excellent		Good to excellent		Excellent
Tensile strength, lb/in.²	D 638	6,000–7,500	1,500–3,500	3,000–5,000	10,000–12,000	7,500–9,000	500–3,000	5,400–7,000	2,600–3,200	5,000–8,000
Elongation, %	D 638	40–80	200.0–450.0	Up to 250.0	5.0–20.0	4.5–65.0	150.0–450.0	5.0–20.0	250.0	100.0–140.0
Tensile elastic modulus, 10⁵ lb/in.²	D 638	3.5–6.0		0.5–0.8	3.5–6.0	3.60–4.75		3.0–3.9		3.5–4.5
Compressive strength, lb/in.²	D 695	8,000–13,000	900–1,700	2,000–2,700	17,000–18,000	9,000–22,000				7,750–11,700
Flexural yield strength, lb/in.²	D 790	10,000–16,000		4,200–6,200		14,500–17,000	Varies depending on type and amount of plasticizer	7,500		10,000–15,400
Impact strength, ft-lb/in. of notch (½- by ½-in. notched bar, Izod test)	D 256	0.4–20.0	Varies depending on type and amount of plasticizer	0.3–1.0	0.8–1.4	1.0–5.6		3.0–15.0		0.36–32.00

Property	ASTM	1	2	3	4	5	6	7	8	9
Hardness, Rockwell	D 785	65–85 (Shore D)	50–100 (Shore A)	M50–65	M85	R117–122	10–100 (Shore A)	R102–113	75–95 (Shore A)	M18–55, R107–119
Flexural elastic modulus, lb/in.^{2} × 10^5	D 790	3–5	......	3.0	3.7	3.8–4.5	......	3.0–3.5	......	3.52–5.04
Compressive modulus, lb/in.^{2} × 10^5	D 695	......	......	......	......	3.35–6.00	......	......	......	......
Thermal conductivity, 10^{-4} cal/(s)(cm^2) per °C/cm	C 177	3.5–5.0	3.0–4.0	......	......	3.3	......	......	......	......
Specific heat, cal/°C/g		0.25–0.35	0.3–0.5	0.32	......	0.33	......	......	......	......
Thermal expansion, 10^{-5} in./in./°C	D 696	5.0–10.0	7.0–25.0	19.0	6.4	6.8–7.6	......	......	......	......
Resistance to heat, °F (continuous)		130–175	150–175	160–200	120–150	230	......	......	......	150–175
Deflection temp, °F:										
At 264 lb/in.^{2} fiber stress	D 648	155–170	......	130–150	150–170	202–234	......	154–170 (annealed)	......	150–170
At 66 lb/in.^{2} fiber stress	D 648	135–180	......	......	......	215–247	......	169–180	......	......
Refractive index n_D	D 542	1.52–1.55	......	1.60–1.63	1.50	......	1.47–1.49	......	......	......
Clarity		Transparent, translucent, and opaque						......	......	Transparent to opaque
Transmittance, %	D 1003	76–82	......	......	......	......	......	......	......	......
Haze, %		8–18	......	......	......	......	......	......	......	......
Water absorption, 24 h, 1/8 in. thick, %	D 570	0.04–0.4	0.15–0.75	0.1	0.5–3.0	0.02–0.15	1.0–2.0	0.20	......	0.07–0.4
Burning rate (flammability), in./min.	D 635	Nonburning to self-extinguishing	Slow to self-extinguishing	Self-extinguishing	Slow	Nonburning	Slow	Self-extinguishing	Slow	Self-extinguishing
Effect of sunlight		Varies with formulation	Varies with stabilizer	Slight	Slight	Slight	Slight	Slight	Slight	......
Effect of weak acids	D 543	None to slight	None	None	Attacked	None	Slight	None	None	None
Effect of strong acids	D 543	None	None to slight	Resistant	Attacked	None	Slight	None to slight	None to slight	None to slight
Effect of weak alkalies	D 543	None	None	Resistant	Resistant	None	Slight	None	None	None
Effect of strong alkalies	D 543	None	None	Resistant	Resistant	None	Slight	None	None	None
Effect of organic solvents	D 543	Resists alcohols, aliphatic hydrocarbons, oils, soluble or swells in ketones and esters; swells in aromatic hydrocarbons		None to slight	Attacked by some	Resists most	Resists aliphatic hydrocarbons, most oils. Swells in ketones, esters, and aromatic hydrocarbons; dissolves in alcohols	Soluble or swells in ketones, cycloalkane, esters		Same as for vinyl chloride

TABLE 4   Tensile Modulus of Thermoplastics

Plastic	Modulus, lb/in.$^2 \times 10^5$
Polyethylene, low-density	0.14
Ionomer	0.2
Polyethylene, medium-density	0.25
Fluorocarbon	0.58
Polyethylene, high-density	0.6
Cellulosic	0.65
Polymethylpentene	1.6
Polypropylene	1.6
Polyester	2.8
ABS	3.0
Polycarbonate	3.0
Polyaryl ether	3.2
Polyether sulfone	3.5
Polyvinyl chloride	3.5
Noryl	3.55
Polysulfone	3.6
Polyaryl sulfone	3.7
Acrylic	3.8
Polystyrene	4.0
Polyimide	4.5
Nylon	4.6
Polyphenylene sulfide	4.8
Acetal	5.2

Acrylics

Acrylic plastics comprise a large family of polymers dominated by methyl methacrylate (see Table 3 for properties). Methyl methacrylate is noted by its similarity to glass. In its colorless form acrylic is as transparent as the best optical glass, with a transmittance of 92 percent, an absorbance of less than 0.5 percent, and a refractive index of 1.49. The surface reflectance is greater than 4 percent for light at any angle other than perpendicular. Figure 30 shows reflectance related to the angle of incidence. The acrylic polymers also withstand weathering, retaining their optical properties after long exposure to ultraviolet light. Table 5 shows the effect of weathering on acrylics and other polymers. Less than one-half the weight of glass, acrylics have much greater resistance to breakage, as shown in Table 6. They are resistant to acids and alkalies but not to organic solvents. They are subject to creep and have low scratch and abrasion resistance. The largest market is for internally illuminated outdoor signs. Other important uses include automotive-lighting lenses, fluorescent-lamp refractors, and decorative trim

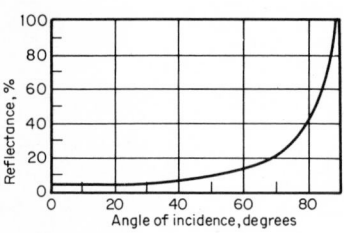

Fig. 30 Reflectance at air-acrylic interface.[18]

on appliances. The ability to act as a light pipe accounts for their use in fiber optics. Glazing applications include windscreens for boats, industrial and school buildings, and aircraft windows. The ease of machining and beauty of the polymer make it popular as a hobbyists' medium. Acrylics also have been combined with polyvinyl chloride to produce a polymer combining toughness, chemical resistance, and self-extinguishing properties. The acrylic-PVC alloy has a very high impact—15 ft-lb/in. notch—and a low water absorption.

Cellulosics

All cellulosic plastics are derived from cellulose which has been rendered moldable by reacting it with anhydrides or acids. These materials include cellulose

TABLE 5   Weathering of Acrylic Resins Compared* with Other Thermoplastics[14]

Grade	Exposure time, months	Yield, tensile (lb/in.^{2})	Ultimate elongation (%)	$\frac{1}{8}$ in. Izod (ft-lb/in.)	Gardner % light transmission	Haze	Gloss
Methyl methacrylate polymer	0	12,000	9	0.37	91.2	1.96	78
	6	......	9	0.51	91.2	3.16	78
	12	13,000	14	0.60	90.0	4.2	42
Cross-linked methacrylate polymer	0	6,200	26	3.5	87.9	11.2	65
	6	5,100	2	2.4	83.4	32.7	45
	12	6,250	...	2.3	57.7	60.0	22
Acetal	0	9,100	33	1.2	....	.....	45
	6	5,000	2	0.66	....	.....	25
	12	2,500	...	0.30	....	.....	5
Polysulfone	0	10,300	102	1.2	64.2	1.20	80
	6	9,100	...	1.3	53.4	95.3	3
	12	......	...	0.8	55.6	99.1	3
Polycarbonate ABS blend	0	8,900	13	9.6	....	.....	55
	6	8,400	4	3.7	....	.....	52
	12	6,700	...	3.6	....	.....	21
PPO resins (Noryl)	0	10,500	30	1.1	....	.....	63
	6	7,200	3	0.68	....	.....	4
	12	5,600	...	0.80	....	.....	6
Rigid PVC	0	6,700	43	22.4			
	6	7,600	21	1.7	....	.....	40
	12	8,000	14	1.2	....	.....	14
Polycarbonate	0	8,900	122	17.8	90.0	0.5	94
	6	8,900	117	17.9	88.0	3.8	83
	12	8,800	95	17.8	88.9	6.2	75

* 12 months in Pittsburgh, Pa.

acetate, cellulose acetate butyrate, cellulose propionate, cellulose triacetate, cellulose nitrate, and ethyl cellulose. Properties are given in Table 3. Most of these materials are combined with plasticizers to modify their performance.

The cellulose plastics are similar in properties in that they are lightweight materials with good impact strength. They also oxidize easily, and so have poor weather resistance. They are used for consumer goods, cosmetic containers, tools, steering wheels, and photographic film. Their good transparency is used in gages, instruments, overlay sheets, drafting equipment, and flashlights.

TABLE 6   Impact Strength of Cast-Acrylic Sheet and Glass[18]

Material	Thickness, in.	Breakage height of drop, ft	Approx speed of steel ball at moment of impact, mi/h
Double-strength window glass	0.129	2	8
Plate glass	0.250	2.5	9
Acrylic	0.060	6	13
Acrylic	0.125	14.5	21
Acrylic	0.250	36	33
Acrylic	0.375	92	52

$\frac{1}{4}$-in. steel ball dropped on 12-in. square piece.

Fluorocarbons

These plastics contain fluorine, which lends superior heat resistance, stability, and chemical resistance.   Table 7 shows a ranking of thermoplastic resins with respect to the maximum temperature at which they may be used continuously. Like many other polymers, fluorocarbons stiffen at cryogenic temperature, but these materials are still useful to −400°F, as seen in Figs. 31 to 33.   Physical properties for the resins are given in Table 3.

PTFE   The first and most widely used fluorocarbon is polytetrafluoroethylene. It has a carbon chain with two atoms of fluorine attached to each carbon.   The thermal stability is unique.   Its tensile strength at 500°F is unchanged from that at room temperature.   A phase transition occurs at 620°F; the polymer softens but does not melt.   Conventional molding methods are not possible; so the polymer must be molded with powder-metallurgy techniques.   PTFE is not affected by any chemicals except molten alkalies.   Water absorption is zero.   A slippery surface makes it useful for bearings, and it is unaffected by weather.

TABLE 7   Resistance to Heat of Thermoplastics (Continuous-Use Temperature)

Plastic	Max temp, °F
Fluorocarbon (PTFE)...........................	500
Polyaryl sulfone...............................	500
Polyimide.....................................	500
Polyphenylene sulfide..........................	400
Polyether sulfone.............................	300
Polysulfone...................................	300
Polyaryl ether.................................	250
Polycarbonate.................................	250
Polyethylene, high-density.....................	250
Polymethylpentene............................	250
Polypropylene.................................	225
Polyethylene, medium-density....................	220
Acetal..	195
Nylon..	180
Polyethylene, low-density......................	180
Noryl...	175
ABS..	160
Ionomer......................................	160
Polyester.....................................	155
Polystyrene...................................	150
Polyvinyl chloride.............................	150
Acrylic.......................................	140
Cellulosics...................................	140

The polymer is rather soft and creeps readily.   Radiation causes rapid deterioration.   Surfaces scratch and dent easily.   The material can be adhesive-bonded only with difficulty.

Uses are myriad and include cookware coatings, bridge bearings, and chemical pipes and valves.

FEP   Fluorinated ethylene-propylene is a PTFE copolymer with hexafluoropropylene.   Hence the chain is identical to PTFE except for side branching.   Properties are virtually identical except that FEP is a true thermoplastic and can be molded.   Its heat resistance is not quite as good.

CTFE   Polytrifluorochloroethylene is similar to PTFE except that it contains some chlorine.   The material is transparent up to 0.1 in. thick and has excellent chemical resistance, replacing stainless steel in many applications.   Uses include chemical ware, wiring, film, and medical devices.

E-CTFE   Polyethylene-chlorotrifluoroethylene is a copolymer of CTFE and ethylene.   Nonburning, it has an oxygen index of 60.   It compares favorably with nylon in many properties up to 350°F. Its chemical resistance makes it useful as drum liners, tanks, carboys, and laboratory ware. Fibers and fabrics are available, and film is used in electrical cables.

PVF₂ Polyvinylidene fluoride is used in chemical-process equipment as pipes and pumps. Its excellent stability in sunlight makes it popular as a decorative-finish coating on aluminum and steel building siding.

PFA Perfluoroalkoxy is a melt-processable polymer with a combination of excellent properties. After being aged at 545°F for 5,000 h it shows no decrease in tensile strength or elongation. It is more creep-resistant than PTFE. Useful as film and molded or extruded shapes, it can also be colored.

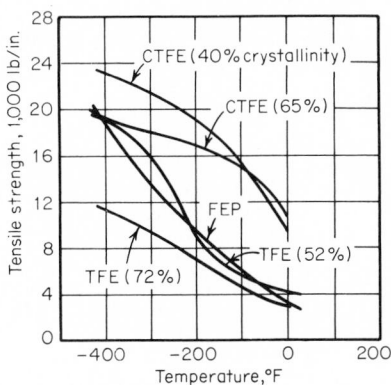

Fig. 31 Tensile strength of fluorocarbons.[20]

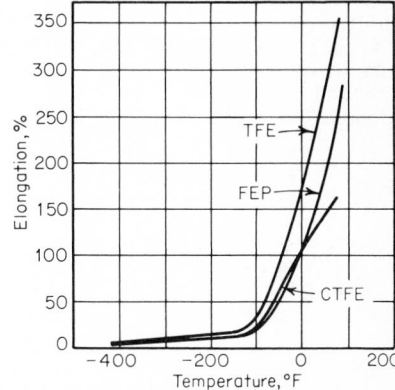

Fig. 32 Elongation of fluorocarbons.[20]

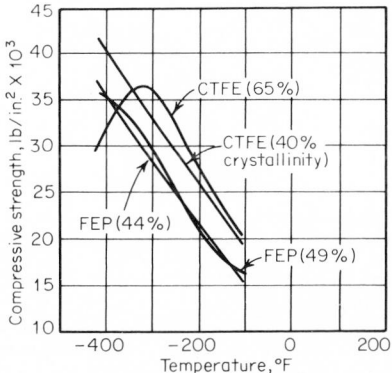

Fig. 33 Compressive strength of fluorocarbons.[20]

ETFE Ethylene tetrafluoroethylene is a copolymer of ethylene and PTFE. Good radiation resistance and improved creep make it useful to the nuclear industry. This material is the first fluorocarbon to be combined with reinforcing fibers.

Ionomers

Most plastics are linked by covalent bonds, but the ionomer resins contain both covalent and ionic bonds, resulting in polymers with improved solvent resistance and toughness. The polymer properties are given in Table 3. Used as a film or sheet, it is transparent and has high tear strength. Its chemical inertness makes it useful in food packaging and drug products. It also is used in shoes and as golf-ball covers.

Nylon (Polyamides)

This family of plastics is based on an amide group, the number in the name defining the chemical difference. The members of the family include:

Nylon 6. With a melting range of 385 to 425°F, this is the second most common nylon, with high flexibility and impact resistance.

Nylon 6/6, melting range 480 to 500°F, dominates the field, has high strength and chemical resistance.

Nylon 6/10, melting range 405 to 430°F, has a lower moisture sensitivity.

Nylon 6/12, melting range 405 to 425°F, is replacing 6/10 because it has a lower density.

Nylon 11, melting range 400 to 530°F, is usually plasticized and has low water absorption.

Nylon 12, melting range 390 to 500°F, has the lowest water absorption.

Nylons 7, 8, 9, and 1313 are used for special purposes.

TABLE 8   Tensile Strength of Thermoplastics

Plastic	Tensile strength, lb/in.^{2}
Polyaryl sulfone	13,000
Polyether sulfone	12,200
Nylon	12,000
Polyimide	10,500
Polysulfone	10,200
Polyphenylene sulfide	10,000
Acetal	10,000
Polyester	8,200
Polycarbonate	8,000
Noryl	7,800
Polyaryl ether	7,500
Acrylic	7,000
Polyvinyl chloride	6,000
Polystyrene	5,000
ABS	4,500
Polypropylene	4,300
Polymethylpentene	3,500
Ionomer	3,500
Polyethylene, high-density	3,100
Fluorocarbon	2,000
Cellulosic	1,900
Polyethylene, medium-density	1,200
Polyethylene, low-density	600

The physical properties of the various nylons are given in Table 3. Nylons have high tensile strength compared with most thermoplastics, as shown in Table 8. Stress-strain relationships are typical of thermoplastics and similar for many nylons, as seen in Fig. 34. As moisture content increases, the yield stress decreases and elongation increases. Nylons absorb moisture readily in relation to the polymer grade, humidity, and part shape. Figure 35 shows the rate of water absorption to equilibrium for nylon 6/6 in ⅛-in. sections. Nylons are notch-sensitive and so demonstrate low impact values, but dynamic tests like the falling-dart test show high values. A low coefficient of friction and good wear resistance makes nylons desirable for gears and bearings with low loads. Properties of the materials are varied widely by the addition of plasticizers, reinforcements, lubricants, stabilizers, flame retarders, and nucleating agents.

The largest market for nylon is automotive, where its resistance to gasoline and oil is useful. It is often used as battery cases. Nylon is used in contact with food and in many appliances. It also is used in kitchen utensils like ladles and spatulas. Sports applications include helmet components, putters, and shoe cleats.

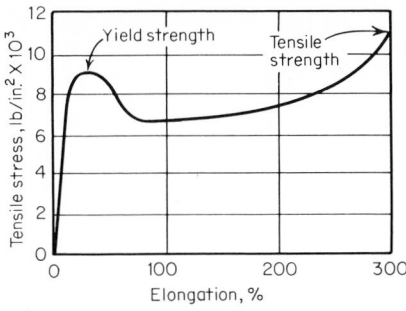

Fig. 34 Stress-strain curve for nylon 6/6 (2.5 percent moisture and 73°F: 50% RH).[21]

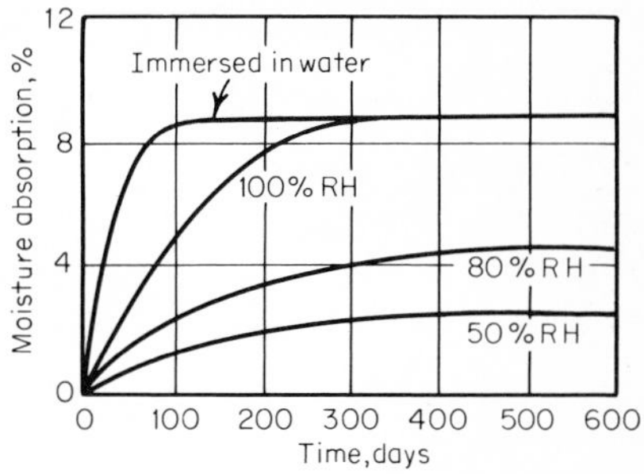

Fig. 35 Rate of water absorption to equilibrium at 73°F, nylon 6/6 (⅛-in. sample).[21]

Polyaryl Ether

This material has a combination of properties which make it unique. It has a heat-distortion temperature of 300°F at 264 lb/in.², an impact strength of 8.0 ft-lb/in., good chemical and water resistance, and is easily processable. Properties are shown in Table 3. At 200°F the tensile strength is 56 percent of the room-temperature value. Deflection temperatures at two loads are shown in Table 9, and creep strength in Fig. 36. The material can be plated and is used in automotive, appliance, and vending-machine applications, particularly in under-the-hood areas where thermal stability is important.

TABLE 9   Heat-Deflection Temperature of Polyaryl Ether (Arylon T) and Other Plastics[22]

Load, lb/in.²	Deflection temp, °F					
	Arylon T	Acetal	Nylon 6/6	Modified PPO	Poly-sulfone	Poly-carbonate
264.................	300	230	150	265	345	265
66.................	320	315	360	280	360	285

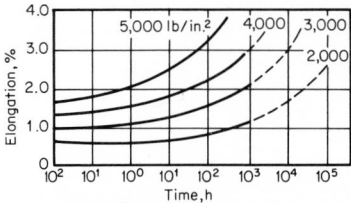

Fig. 36 Tensile creep strength of polyaryl ether (Arylon T).[22]

Polyaryl Sulfone

A strong, hard, stiff plastic which can be used from −400 to +500°F, this polymer is used mostly in aerospace applications. Table 3 shows physical properties. Figure 37 shows physical properties vs. temperature. This plastic is often used to

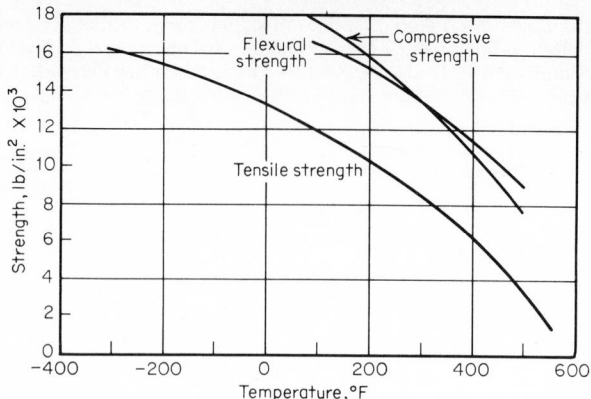

Fig. 37 Physical properties of polyaryl sulfone vs. temperature.[23]

upgrade other plastic applications with respect to temperature and to replace metals and ceramics in valve bodies, hydraulic seals, and bearings.

Polycarbonates

Polycarbonates were one of the first engineering plastics and are characterized by high impact, transparency, low flammability, and toughness, as shown in Table 3.

TABLE 10 Properties of Polycarbonate and Metals[14]

Property	Poly-carbonate	Die-cast zinc	Die-cast aluminum
Density, lb/in.³	0.043	0.24	0.98
Modulus of elasticity, lb/in.²	34×10^5	1×10^7	40×10^5
Tensile strength, lb/in.²	8,500	40,000	30,000
Ratio of tensile strength to density	198,000	160,000	204,000
Strength/weight ratio	1	0.81	1.03

The material frequently competes with zinc and aluminum, as shown in Table 10. Impact strength decreases with increasing part thickness. The water absorption of polycarbonates is low compared with that of many other plastics, as shown in Fig. 38. Polycarbonates retain their properties at lower temperatures without excessive embrittlement, as seen in Fig. 39. Polycarbonates often are used in glazing applications where high risk of breakage exists (as in Fig. 40). Table 11 lists optical properties. Compared with other optical plastics, polycarbonate does not burn readily (Table 12).

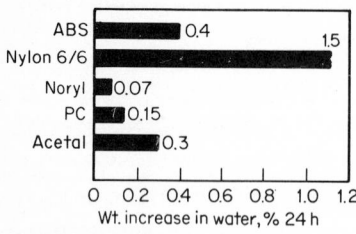

Fig. 38 Water absorption of polycarbonate and other plastics.[24]

Polycarbonates are usually selected for parts subject to sudden loads like safety helmets, tool housings, appliances, and food-dispensing equipment. Transparency applications include windows, automotive lenses, safety glasses, lighting refractors, bottles, tanks, and lids.

The resins are often influenced by chemicals, and fatigue and creep resistance at elevated temperature are marginal.

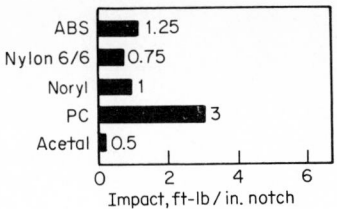

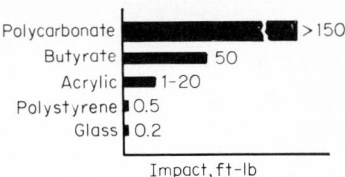

Fig. 39 Low-temperature impact of polycarbonates and other plastics.[24]

Fig. 40 Falling-ball impact strength.[24]

TABLE 11 Optical Properties of Clear Plastics and Glass[24]

	Total transmittance, %	Haze, %
Polycarbonate...................	86–89	1–3
Butyrate......................	88–90	2–4
Acrylic.......................	92	1–3
Polystyrene....................	88–92	1–3
Glass.........................	90	1–3

TABLE 12 Flammability of Polycarbonate and Other Plastics[24]

Plastic	Flammability test		
	ASTM D 635	UL 94	Oxygen index
Lexan, lighting grade.......	Self-extinguishing	Self-extinguishing, group II	25.0
Acrylic...................	Slow-burning	Burning	17.5
Polystyrene..............	Slow-burning	Burning	18.0
Butyrate................	Slow-burning	Burning	NA

Polyesters

Thermoplastic polyesters are based on polyethylene terephthalate, only recently available. The polymers are tough, chemical-resistant, and easily molded and possess a high surface gloss. Physical properties are given in Table 3. The resins are highly crystalline and are not affected by water immersion (Fig. 41). They are more dimensionally stable than acetals. Impact strengths are particularly good, as shown in Fig. 42. Because of the low moisture absorption and high stiffness,

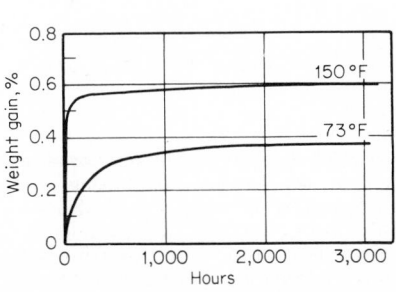

Fig. 41 Weight gain of polyester in water (24-hour soak).[25]

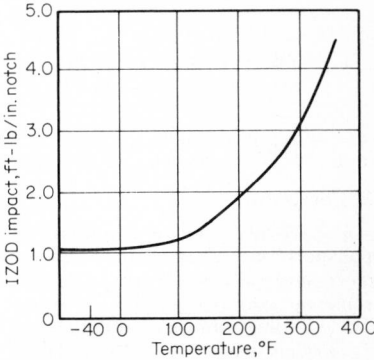

Fig. 42 Polyester impact strength vs. temperature.[25]

little creep is experienced at room temperature. Creep at 302°F is shown in Fig. 43.

Polyesters are used in the automotive industry, both under the hood and as painted-body exterior components, where they often replace thermoset materials. They also are used in machine housings, gears, and pens.

Polyether Sulfone

This resin is a new high-temperature polymer capable of continuous use at 180°C and in low-stress applications at 200°C. Polyether sulfone has a tensile strength of 9,400 lb/in.² at 180°C and has low creep, as seen in Fig. 44. Tensile yield stress with temperature is shown in Fig. 45. The material is self-extinguishing

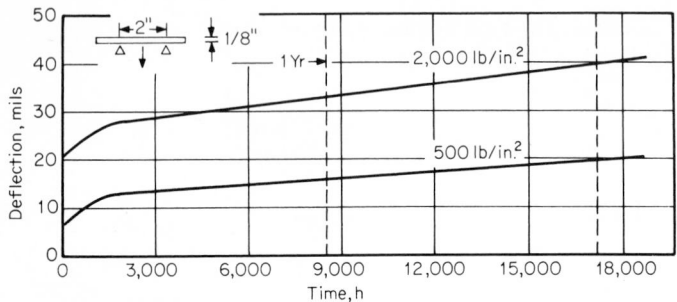

Fig. 43   PBT thermoplastic polyester flexural creep at 302°F.[26]

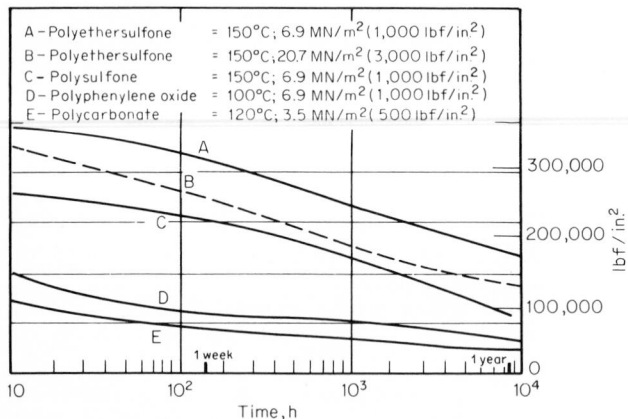

Fig. 44   Tensile creep modulus of polyether sulfone at 20°C.[27]

and emits little smoke. It is used in lamp sockets, gearboxes, dryers, irons, heaters, and aircraft components.

Polyethylene

Polyethylene polymers consist of literally hundreds of compositions based on polymers of differing molecular weight, branching, copolymers, cross-linking polymers, and polymers with additives. Basically, however, the family is divided into three categories by density, as shown in Fig. 46. Most properties vary with density; for example, Young's modulus, tensile strength, and heat distortion all decrease with decreasing density (see Table 3). Typical stress-strain curves are shown in Fig. 47 in relation to density. As density increases, elongation decreases. Low-density polyethylene has the highest elongation of all plastics (see Table 13). Adhesives do not bond to polyethylene surfaces, and this makes them useful in release coatings and separation sheets.

The low cost of the materials combined with their excellent properties makes it difficult to list all their applications. However, they are found in housewares, films, cosmetics, toys, drug and medical devices, bottles, containers, drums, laboratory equipment, kitchenware, coatings, appliances, automotive and electrical applications, and wiring.

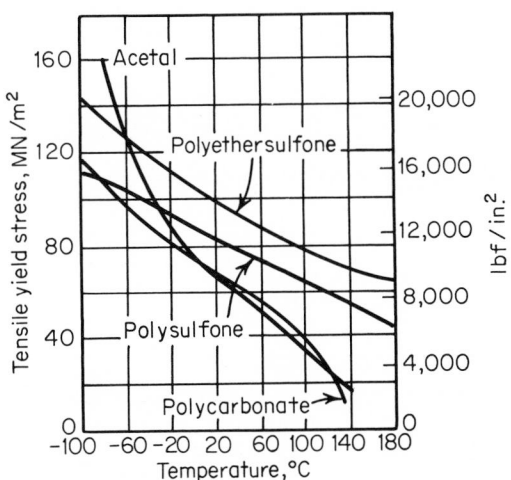

Fig. 45 Tensile yield stress of polyether sulfone vs. temperature.[27]

Polyimides

These resins exhibit outstanding thermal stability (80 percent retention of physical properties at 480°F). Two kinds are available—condensation and addition resins. The condensation molding materials are pressed and sintered like PTFE, but addition polymers are capable of being molded in either compression or transfer molds. Typical properties are shown in Table 3. A modification of this material is polyamide-imide, which performs similarly but has a somewhat lower thermal capability. As shown in Table 14, the polyimides have the lowest thermal expansion. These compounds show little creep and have excellent wear properties. They are often filled with lubricating materials such as graphite, PTFE, and molybdenum disulfide. They can be used to 500°F with little loss in properties and at 900°F for short times. Figure 48 shows tensile yield strength vs. temperature.

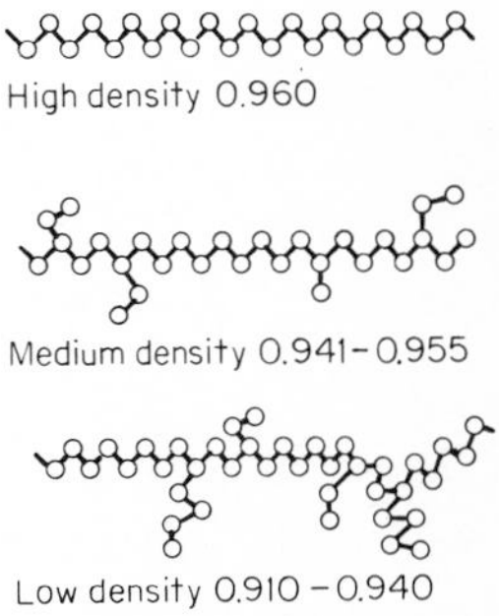

High density 0.960

Medium density 0.941–0.955

Low density 0.910–0.940

Fig. 46 Chain structure of polyethylene.[28]

Polymethylpentene

An olefin type of polymer, this plastic is characterized by a low specific gravity (0.83), high transparency (90 percent), high melting point (464°F), and chemical resistance as shown in Table 3. It creeps less than other polyolefins, as shown in Fig. 49, and has good temperature resistance, as seen in Fig. 50. At 300°F it is superior to polycarbonates. It is resistant to most chemicals, being attacked only by strong oxidizers and chlorinated solvents, thus making it useful for laboratory ware. Its largest use is in food-packaging and -handling devices such as cook-in pouches, sight glasses, trays, containers, and coffee makers.

Polyphenylene Oxide Resins (Noryl)*

PPO resins are based on polyphenylene oxide blended with styrene. The plastics have high impact strength, are dimensionally stable, and show little response to temperature up to 250°F (see Table 3). Figure 51 shows a comparison of the

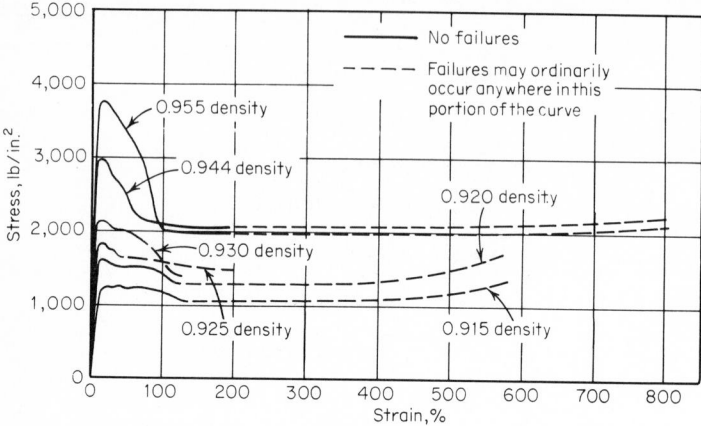

Fig. 47   Stress-strain curves for polyethylenes.[29]

TABLE 13   Tensile Elongation of Plastics

Plastic	Elongation, %
Polyethylene, low-density	600
Ionomer	350
Polypropylene	200
Fluorocarbon	200
Polycarbonate	100
Nylon	60
Polysulfone	50
Polyethylene, medium-density	50
Polyester	50
Noryl	50
Polyvinyl chloride	40
Polyether sulfone	30
Polyaryl ether	25
Acetal	25
Polyethylene, high-density	20
Polymethylpentene	13
Polyaryl sulfone	13
Cellulosic	6
Polyimide	5
ABS	5
Polyphenylene sulfide	3
Acrylic	2
Polystyrene	1

impact strength of PPO and ABS resins on both painted and unpainted samples. Painting with organic-solvent-based paints usually weakens plastics, since the solvents strain the surfaces. PPO resins are not so dramatically affected as are many materials. The fatigue strength of PPO is shown in Fig. 52 and creep strength in Fig. 53.

* Trademark, General Electric Co.

PPO is superior to nylons and to ABS in both properties. A low specific gravity (1.06) makes the volumetric cost attractive. Its chemical resistance is relatively poor.

PPO resins are used extensively in automotive applications because of their dimensional stability and impact properties. PPO can be electroplated and hence is found

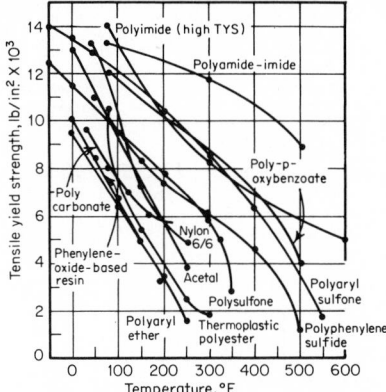

Fig. 48 Tensile yield strength vs. temperature.[29]

TABLE 14   Coefficients of Thermal Expansion

Plastic	Thermal expansion, in./in./°C, $\times 10^{-5}$
Polyimide	2.5
Polyaryl ether	3.6
Polyaryl sulfone	4.7
Acrylic	5.0
Noryl	5.2
Polysulfone	5.2
Polyether sulfone	5.5
Polyphenylene sulfide	5.5
Polyvinyl chloride	5.6
Polypropylene	5.8
Polyester	6.0
Polystyrene	6.0
Polycarbonate	6.6
ABS	8.0
Cellulosic	8.0
Nylon	8.0
Acetal	8.1
Fluorocarbon	10
Polyethylene, low-density	10
Ionomer	12
Polyethylene, high-density	11
Polymethylpentene	11.7
Polyethylene, medium-density	14

in plumbing applications such as shower heads and sprinklers. Its toughness permits use in housings, grillwork, and business machines.

Polyphenylene Sulfide

The properties of polyphenylene sulfide are shown in Table 3. This high-temperature plastic has a melting point of 550°F and has little response to temperature, as

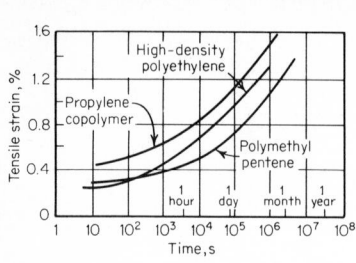

Fig. 49 Creep of polymethylpentene.[30]

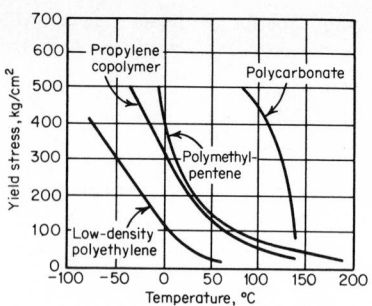

Fig. 50 Yield stress of polymethylpentene vs. temperature.[30]

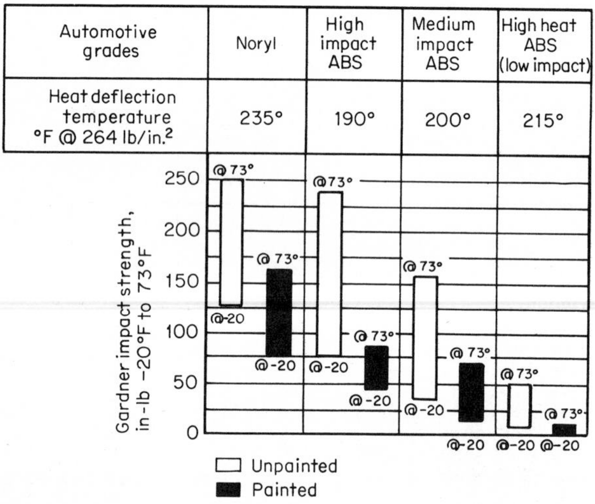

Automotive grades	Noryl	High impact ABS	Medium impact ABS	High heat ABS (low impact)
Heat deflection temperature °F @ 264 lb/in.^{2}	235°	190°	200°	215°

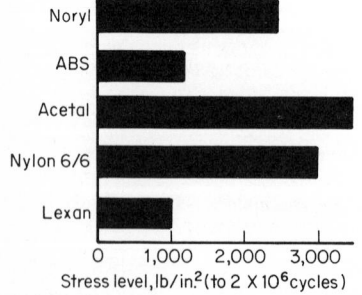

Fig. 51 Impact strength of Noryl and ABS.[31]

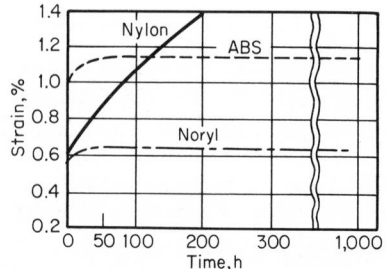

Fig. 52 Fatigue strength of Noryl.[31]

Fig. 53 Creep strength of Noryl.[31]

shown in Table 15. There are no solvents below 350°F. It has a high oxygen index and is stiff. Used in cookware, it meets FDA regulations. It also is used as piston rings, valves, and pump bodies.

Polypropylene

A polyolefin, polypropylene has an excellent balance of properties at low cost. Properties are shown in Table 3. It is superior to polyethylene in many characteristics, as shown in Table 16.[33] It has a low specific gravity (Table 17), making

TABLE 15   Effect of Temperature on Polyphenylene Sulfide Aged at 400°F in Air[32]

Property	Exposure days		
	0	3	7
Flexural modulus, lb/in.^{2}.............	2,000,000	2,020,000	2,030,000
Tensile, lb/in.^{2}......................	21,000	21,000	21,000
Elongation, %......................	3	3	3
Shore D hardness....................	92	92	92

Aging at 450°F in air		
	Control	450°F/4 months
Flexural modulus, lb/in.^{2}............................	2,000,000	2,200,000
Tensile, lb/in.^{2}.......................................	22,700	12,000
Elongation, %..	4	1
Flexural strength, lb/in.^{2}..............................	34,500	26,600
Unnotched impact strength, ft-lb.....................	2.6	1.6

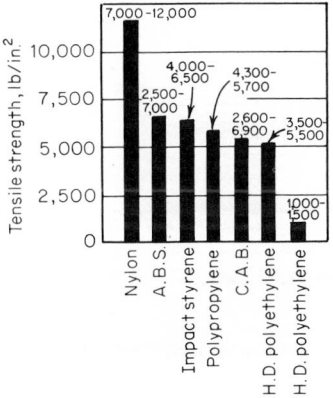

Fig. 54 Tensile strength of polypropylene compared with other plastics.[34]

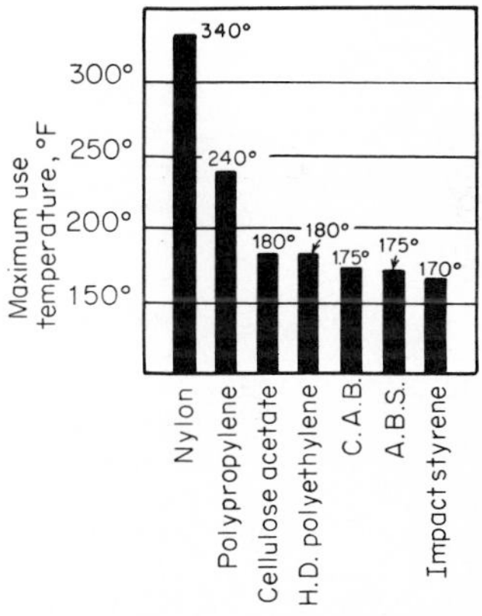

Fig. 55 Maximum use temperature of selected thermoplastics.[34]

its cost per unit volume low. It has good tensile strength, as shown in Fig. 54, and a maximum temperature of 240°F, as shown in Fig. 55. The excellent physical properties make it useful in load-bearing applications. Polypropylenes require antioxidants for optimum thermal stability, and they do not weather well. There are no solvents for polypropylene at room temperature, but aromatic and chlorinated hydrocarbons swell and soften it. This chemical stability makes it useful in dishwashers and washing machines. Capable of being plated, it is used widely in appliances. It also is used in automotive applications, furniture, housewares, carpeting, packaging, bottles, batteries, and in fiber form.

TABLE 16   Properties of Polypropylene

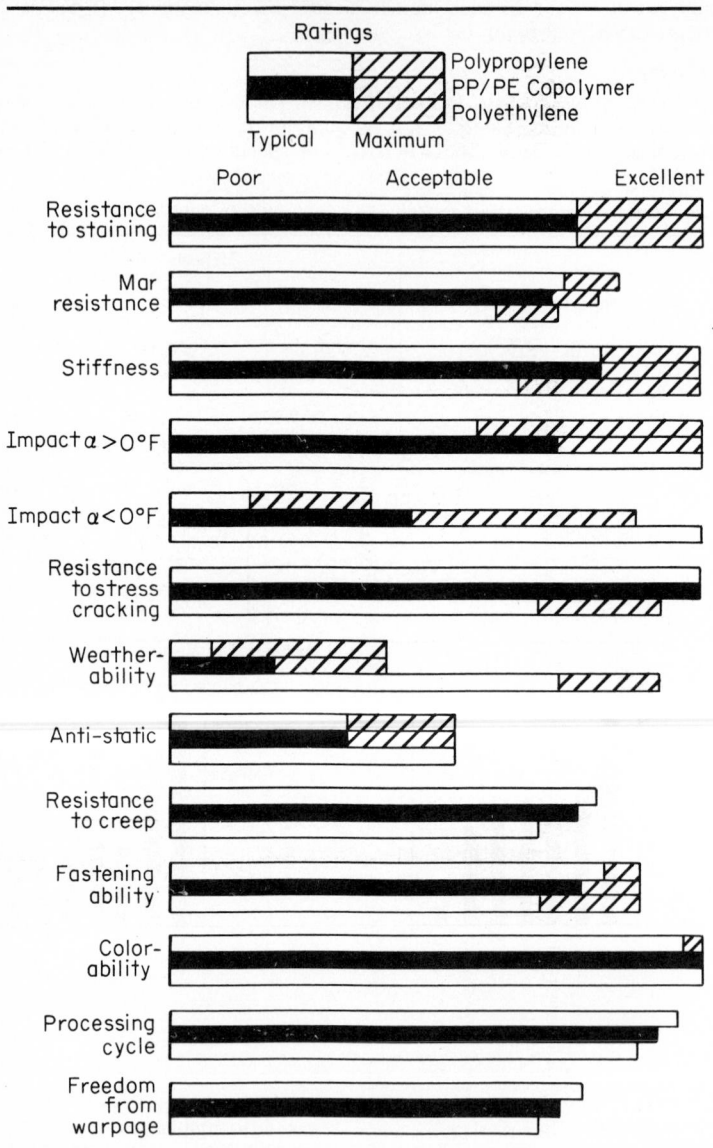

Polystyrene

The physical properties of polystyrene are given in Table 3. Many polystyrenes are available, differing in molecular structure and additives. The polymers are hard, brittle, strong, and have good gloss and high clarity. The impact resistance of polystyrene is low, as shown in Table 18. It is affected by some solvents which cause crazing (see Fig. 56), so care must be used in selecting a polystyrene when

TABLE 17   Specific Gravity of Plastics

Plastic	Specific gravity
Polymethylpentene	0.83
Polypropylene	0.902
Polyethylene, low-density	0.910
Polyethylene, medium-density	0.926
Polyethylene, high-density	0.941
Ionomer	0.93
Polystyrene	1.04
ABS	1.04
Noryl	1.06
Nylon	1.12
Polyaryl ether	1.14
Acrylic	1.17
Polycarbonate	1.2
Cellulosic	1.22
Polysulfone	1.24
Polyvinyl chloride	1.30
Polyester	1.31
Polyphenylene sulfide	1.34
Polyaryl sulfone	1.36
Polyether sulfone	1.37
Acetal	1.41
Polyimide	1.43
Fluorocarbon	2.14

TABLE 18   Izod Impact Strength of Plastics

Plastic	Impact, ft-lb/in. notch
Polyethylene, low-density	No break
Polycarbonate	12.0
Polyaryl ether	8.0
Ionomer	6.0
Noryl	5.0
Fluorocarbon	3.0
ABS	2.0
Polyether sulfone	1.6
Acetal	1.4
Polysulfone	1.3
Nylon	1.0
Polyaryl sulfone	1.0
Polyethylene, medium-density	1.0
Polyimide	0.9
Polyester	0.8
Polyethylene, high-density	0.5
Polypropylene	0.5
Cellulosics	0.4
Polymethylpentene	0.4
Polyvinyl chloride	0.4
Acrylic	0.3
Polyphenylene sulfide	0.3
Polystyrene	0.25

solvents are involved.   Polystyrenes do not weather well, mar on impact or over-stress, have severely limited heat resistance, craze in contact with many chemicals, and are flammable.

The plastics are used in toys, furniture, housewares, packaging, appliances, shoes, lighting, automotive applications, and as a modifier for many other resins.

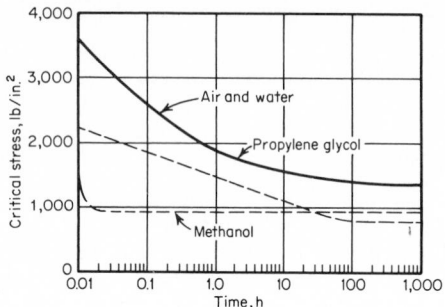

Fig. 56 Time effect of various reagents on crazing of polystyrene (compression-molded samples at 75°F).[35]

Polysulfone

Physical properties for polysulfone are given in Table 3.   The polymer is rigid and strong, as shown in Fig. 57.   Its chemical resistance is good, but it is attacked by polar organic solvents.   Resistance to water is good, as shown in Fig. 58.   Creep is better than for most other thermoplastics, particularly at temperatures over 220°F. The plastic is used in housings, appliances, and automotive, medical, and camera applications.   It can be electroplated.

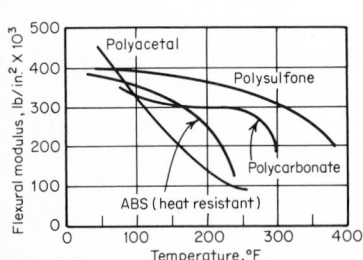

Fig. 57 Flexural modulus of polysulfone and other plastics vs. temperature.[36]

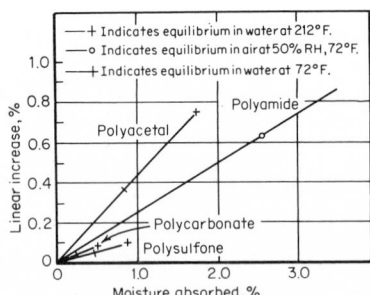

Fig. 58 Dimensional change of plastic caused by water.[36]

Polyvinyl Chloride

This hugely diversified family includes rigid polyvinyl chloride, plasticized poly-vinyl chloride, and copolymers of polyvinyl chloride, polyvinyl dichloride, and polyvinylidene chloride.   Literally thousands of formulations exist, and typical proper-ties are shown in Table 3.   Rigid materials have excellent weather stability and are self-extinguishing, as can be seen in Table 19.   Their chemical resistance is high, but some solvents dissolve them.   Extruded parts can be made to resemble wood and have similar properties.   Uses include flooring, wall covering, building com-ponents, pipes, hoods, tanks, film, sheet flooring, gasketing, packaging, waterproof clothing, upholstery, automotive applications, toys, housewares, and many others.

TABLE 19   Flammability of Thermoplastics[32]

Material	Oxygen index
Polyvinyl chloride...............................	47
Polyphenylene sulfide...........................	44
Polysulfone.....................................	30
Nylon 6/6......................................	28.7
Noryl..	28
Polycarbonate..................................	25
Polystyrene.....................................	18.3
Polyolefins.....................................	17.4
Polyacetal......................................	16.2

Reinforced Thermoplastics

The physical properties of reinforced thermoplastics are given in Table 20. Several kinds of fibers and fillers are used to reinforce thermoplastics. These include organic fibers, asbestos, talc, graphite fibers, mineral fillers, lubricants (molybdenum disulfide, PTFE, graphite), and fiberglass, the latter being most prevalent. In many thermoplastics, reinforcements act as an additive, but in some materials where the coupling agents on the fiber show full compatibility with the resin, the properties are dramatically improved—this being the case with polypropylene, polyester, polystyrene, and polyethylene. In the latter polymers the impact strengths are increased by the addition of glass fibers (see Table 21), but many other polymers show a decrease in impact. In all cases the flexural strength is increased (see Table 22). In general, 5 to 40 percent addition of glass fibers improves stiffness, tensile strength, Young's modulus, resistance to stress cracking, dimensional stability, and creep strength. Creep strength in particular is often greatly enhanced, as shown in Table 23. Disadvantages include cost and difficulty in molding, but it is often more profitable to select a reinforced polymer rather than a thermoset or a high-quality unreinforced plastic.

THERMOSET PLASTICS

Thermoset resins are those long molecules which undergo a chemical reaction, form cross links, and become infusible and resistant to further heating. Typical properties of these plastics are given in Table 24. These resins are described in the following section.

Allyl Resins

A clear resin often used for optical purposes is allyl diglycol carbonate (see Table 24), but the allyl compounds most in use are diallyl phthalate (DAP) and diallyl isophthalate (DAIP). Properties of the DAIP parallel those of DAP, but the DAIP has better thermal stability. DAP resins exhibit very little shrinkage as a result of heat aging, as shown in Table 25. They are usually filled with glass fibers, synthetic fibers, or inorganic fillers. The glass-fiber material has good flexural strength in comparison with phenolics (see Fig. 59) and often competes with phenolics in critical applications. DAP resins mostly are used in electronic applications where their dimensional stability, low water absorption, and excellent electrical properties are desired. They are beginning to find use in appliances where their good appearance and ease of coloring are needed.

Epoxy Resins

Three types of epoxies exist—bisphenol-epichlorohydrin, cycloaliphatic, and epoxy-novolac resins. The resins are reacted with amines, amides, or anhydride hardeners. The greatest variety of epoxies are used as coatings in paints, varnishes, fluid-bed powders, and chemical-resistant coatings. They are combined with mineral fillers and used as castings or with fibers and mineral fillers and used as molding

TABLE 20 Physical Properties of Reinforced Thermoplastics[14]

Property	ASTM test method	ABS, 20 to 40% glass-filled	Acetals		Nylon 6/6			Nylon 6 30-35% glass-reinforced	Nylon 6/12 30-35% glass-reinforced
			20% glass-filled homopolymer	25% glass-coupled copolymer	TFE-fiber reinforced	33% glass-reinforced	Molybdenum disulfide		
Molding qualities	...	Fair to good	Good to excellent	Good to excellent	Excellent	Excellent	Excellent	Excellent	Excellent
Compression-molding temp, °F	...	...	...	...	...	...	...	...	...
Compression-molding pressure, lb/in.²	...	...	...	...	...	...	...	...	...
Injection-molding temp, °F	...	500-550	350-480	380-480	370-410	520-560	500-600	480-550	450-550
Injection-molding pressure, lb/in.²	...	15,000-40,000	10,000-20,000	10,000-20,000	10,000-20,000	By the pressure required to fill the machine			
Compression ratio	...	...	...	...	...	...	...	...	...
Mold (linear) shrinkage, in./in.	...	0.0010-0.0020	0.009-0.012	0.004 (flow), 0.018 (trans.)	0.02-0.025	0.005	0.007-0.018	0.004	0.003
Specific gravity (density)	D 792	1.23-1.36	1.56	1.61	1.54	1.38	1.16-1.17	1.35-1.42	1.31-1.38
Specific volume, in.³/lb	D 792	22.5-20.4	18.7	17.4	18.9	20.1	...	20.5-19.5	21.1-20.0
Machining qualities	...	Good	Good to fair	Good to fair	Excellent	Excellent	Excellent	Excellent	Excellent
Tensile strength, lb/in.²	D 638	8,500-19,000	8,500-11,000	18,500	6,900 (73°F)	2,800-2,200	13,700	25,000-13,000	24,000-20,000
Elongation, %	D 638	2.5-3.0	2-7	3	12.0 (73°F)	4-5	15	3	4.5-5.0
Tensile elastic modulus, 10^5 lb/in.²	D 638	5.9-10.3	10.0	12.5	...	...	5.53	14.5-8.0	12-9
Compressive strength, lb/in.²	D 695	12,000-22,000	18,000 (10% defl.)	17,000 (10% defl.)	11,000 (10% defl.)	29,400	25,500	19,000	...
Flexural yield strength, lb/in.²	D 790	16,000-27,000	15,000 (5%)	28,000	...	41,000	29,000	24,000	...
Impact strength, ft-lb/in. of notch (½- by ½-in. notched bar, Izod test)	D 256	1.0-2.4	0.8	1.8 (⅛- by ½-in. bar)	0.7	2.2-2.6	...	3.0	2.6-2.8
Hardness, Rockwell	D 785	M65-100	M75-90	M-79	M78, R118	M100	R119	M101-78	E40-50
Flexural elastic modulus, lb/in.² $\times 10^5$	D 790	9.0-13.0	8.8	11.0	4.0	13	4.5	15.0-8.0	11.2-9
Compressive modulus, lb/in.² $\times 10^5$	D 695	...	...	...	...	...	...	...	...
Thermal conductivity, 10^{-4} cal/(s)(cm²) per °C/cm.	C 177	...	...	...	...	5.1	...	5.8	...
Specific heat, cal/°C/g	...	...	...	...	...	0.3	...	0.5	...
Thermal expansion, 10^{-6} in./in./°C	D 696	2.9-3.6	3.6-8.1	...	7.5	1.5-2.0	5.4	2.0-3.0	12.0
Resistance to heat, °F (continuous)	...	200-230	185-220	220	195 (8,800 h)	180-300	180-250	200-300	200-300
Deflection temp, °F:									
At 264 lb/in.² fiber stress	D 648	210-240	315	325	212	485	260	410	200-425
At 66 lb/in.² fiber stress	D 648	220-250	345	331	329	495	...	430	400-425

TABLE 20 Physical Properties of Reinforced Thermoplastics[14] (Continued)

Property	ASTM test method	Nylon 12, 30% glass-reinforced	Nylon 6/10, 30-35% glass-reinforced	Polycarbonate — Less than 10% glass-filled	Polycarbonate — 10 to 40% glass-filled	Polyester — 18% glass filled	Polyester — 30% glass-filled	Polyester — 36% glass-filled	PTMT, 20% asbestos-filled	Noryl 20-30% glass-reinforced
Molding qualities		Excellent	Excellent	Excellent	Very good	Good to excellent	Good to excellent	Good to excellent	Excellent	Very good
Compression molding temp, °F										
Compression molding pressure, lb/in.²										
Injection molding temp, °F		450-500	450-550	520-650	575-650	500-540	437-500	500-550	460-530	500-650
Injection molding pressure, lb/in.²				10,000-20,000	15,000-40,000	8,000-23,000	8,000-23,000	17,000	10,000-15,000	15,000-40,000
Compression ratio										
Mold (linear) shrinkage, in./in.	D 792	0.003	0.004	0.002-0.003	0.001-0.003	0.002-0.006	0.002-0.008	0.001-0.003	0.006	0.001-0.004
Specific gravity (density)	D 792	1.23	1.31-1.38	1.25	1.24-1.52	1.48-1.67	1.52	1.73	1.46	1.21-1.36
Specific volume, in.³/lb		22.5	21.1-20.0	22.2	22.4-18.2	18.2-18.7	18.2	17		22.9-20.2
Machining qualities		Excellent	Excellent	Fair to good	Fair	Excellent	Excellent	Excellent	Excellent	Excellent
Tensile strength, lb/in.²	D 638	17,400	20,000	9,600	12,000-25,000	16,000-19,000	17,000-19,000	25,000	12,000	14,500-17,000
Elongation, %	D 638	3.0	2	10.0	0.9-5.0	2.0-3.0	2-4	2	<5	4.0-6.0
Tensile elastic modulus, 10⁵ lb/in.²	D 638		10	4.5	5.0-17.0	8	13×10^5	16	10	9.25-12.0
Compressive strength, lb/in.²	D 695		18,000	14,000	13,000-21,000	16,200-18,000	18,000-23,500	30,000		17,600-17,900
Flexural yield strength, lb/in.²	D 790	22,300	23,000	15,000	17,000-32,000	26,000-26,500	26,000	38,000	19,000	18,500-20,000
Impact strength, ft-lb/in. of notch (½- by ½-in. notched bar, Izod test)	D 256	3.0	3.4	4.0-5.0 (½- by ⅛-in. bar)	1.2-4.0, 1.9-6.5 (¾ by ½ in.)	1.2-1.3	1.3-1.6	1.4	0.05	2.3 (½- by ⅛-in. bar)
Hardness, Rockwell	D 785		E40-50	M85	M88-95	M90, 109	M90	M110	M85	L106-108
Flexural elastic modulus, lb/in.² × 10⁵	D 790	10.0	10.0	5.0	5.0-14.0	8	11.0-12.0	16	9.0	7.5-11.0
Compressive modulus, lb/in.² × 10⁵	D 695			5.2	5.0-15.0	0.7×10^6	0.7×10^6			9.0-13.0
Thermal conductivity, 10⁻⁴ cal/(s)(cm²) per °C/cm.	C 177	3.9	5.1	4.8	4.9-5.2	7.0	7.0	7.1		3.8
Specific heat, cal/°C/g	D 696	0.35	0.35			0.25-0.30	0.30	0.22		
Thermal expansion, 10⁻⁵ in./in./°C		7.5	7.3	3,42	1.7-4.0	1.1-2.5	1.1-6.1	2		2.2
Resistance to heat, °F (continuous)		250-300	200-300	275	275	200-300	240-350	200-300		240-265
Deflection temp, °F: At 264 lb/in.² fiber stress	D 648	203-345	437	288	289-300	415	428	455	330	270-300
At 66 lb/in.² fiber stress	D 648		420	295	300-310	465	437	480		280-310

TABLE 20 Physical Properties of Reinforced Thermoplastics[14] (Continued)

Property	ASTM test method	Polyvinyl chloride, injection-molding grade, opaque 15% glass	Polyphenylene sulfide, 40% glass-filled	Polypropylene		Polystyrene, 20 to 30% glass-filled	Styrene-acrylonitrile 20 to 33% glass-filled
				Filled	Glass-reinforced		
Molding qualities	...	Fair	Excellent	Excellent	Excellent	Excellent	Excellent
Compression molding temp, °F	...	...	600–650	340–550	340–550	...	...
Compression molding pressure, lb/in².	...	...	1,000–2,000	500–1,000	500–1,000	...	...
Injection molding temp, °F	...	270–405	625–700	375–550	400–550	450–625	450–575
Injection molding pressure, lb/in².	...	1,300–2,000	10,000–20,000	10,000–20,000	10,000–20,000	15,000–40,000	15,000–40,000
Compression ratio.	...	1.6–2.2	...	...	...	1.6–4.0	1.6–4.0
Mold (linear) shrinkage, in./in.	D 792	0.001–0.0005	0.002	0.005–0.015	0.002–0.008	0.001–0.002	0.001–0.002
Specific gravity (density)	D 792	1.54	1.64	1.0–1.3	1.05–1.24	1.20–1.33	1.20–1.46
Specific volume, in.³/lb.	...	18.0	...	26.0	24.5	22.2–20.9	23.9–19.9
Machining qualities.	...	Good	Excellent	Fair-good	Fair	Good	Fair
Tensile strength, lb/in².	D 638	9,500	21,000	4,500–8,200	6,000–14,500	9,000–15,000	8,500–20,000
Elongation, %.	D 638	2–3	3	3.0–20.0 (at break)	2.0–3.6	0.75–1.30	1.1–3.8
Tensile elastic modulus, 10⁵ lb/in².	D 638	8.7	11.2	4.0–8.0	4.5–9.0	8.4–12.9	4.0–14.0
Compressive strength, lb/in².	D 695	9,000	...	...	5,500–7,000	13,500–18,000	22,000
Flexural yield strength, lb/in².	D 790	13,500	37,000 (at break)	8,000–9,000	7,000–11,000	10,500–20,000	22,000–26,000
Impact strength, ft-lb/in. of notch (½ by ½-in. notched bar, Izod test).	D 256	1.0	0.8 at 75°F 1.8 at 300°F (½ by ¼-in. bar)	0.4–3.0 at 73°F	1.0–5.0 at 73°F	0.4–4.5	0.4–4.0
Hardness, Rockwell	D 785	118	R123	R94–100	R110	M70–95	M100, E60
Flexural elastic modulus, lb/in.² × 10⁵.	D 790	7.5	22.0	3.0–6.5	3.8–8.5	8.0–10.0	8.0–18.0
Compressive modulus, lb/in.² × 10⁵.	D 695	...	...	2.4	...	...	...
Thermal conductivity, 10⁻⁴ cal/(s)(cm²) per °C/cm.	C 177	...	...	...	...	...	...
Specific heat, cal/°C/g.	...	...	...	...	...	0.23–0.27	...
Thermal expansion, 10⁻⁶ in./in./°C.	D 696	140	4.0	2.9	2.9–5.2	1.8–4.5	2.7–3.8
Resistance to heat, °F (continuous).	...	...	400–500	250–280	270–290	180–200	200–220
Deflection temp, °F: At 264 lb/in.² fiber stress.	D 648	155	>425	140–200	230–300	195–220	190–230
At 66 lb/in.² fiber stress.	D 648	165	...	210–290	305–310	207–231	215–240

TABLE 21   Impact Strength of Reinforced Thermoplastics

Plastic	Glass-fiber content, %	Impact strength, ft-lb/in. notch
Polypropylene	30	5.0
Polystyrene	30	4.5
Polycarbonate	40	4.0
Nylon 6/10	35	3.4
Nylon 6	35	3.0
Polypropylene	Mineral-filled	3.0
Noryl	30	2.3
Nylon 6/6	33	2.2
Polyester	36	1.4
ABS	30	1.0
Acetal	20	0.8
Polyphenylene sulfide	40	0.8

TABLE 22   Flexural Strength of Reinforced Thermoplastics

Plastic	Glass-fiber content, %	Flexural strength, lb/in.^{2}
Nylon 6/6	33	41,000
Polyester	36	38,000
Polyphenylene sulfide	40	37,000
Polycarbonate	40	32,000
Nylon 6	35	24,000
Nylon 6/10	35	23,000
Noryl	30	20,000
Polystyrene	30	20,000
ABS	30	16,000
Acetal	20	15,000
Polypropylene	30	11,000
Polypropylene	Mineral-filled	8,000

TABLE 23   Creep—Apparent Modulus of Reinforced Plastics at 72°F[37]

Resin	Applied stress, lb/in.^{2}	Creep (apparent) modulus (10^3 lb/in.^{2}) Test time, h		
		1	100	1,000
Acetal	1,500	390	280	250
Acetal, 40% glass	5,000	1,500	1,200	1,150
Polyaryl ether	2,000	309	252	185
Polycarbonate	3,000	345	320	310
Polycarbonate, 20% glass	3,000	788	755	730
Phenylene oxide base	2,000	357	339	294
Phenylene oxide base, 30% glass	2,000	1,130	1,125	1,040
Polysulfone	4,000	360	340	325
Polysulfone, 40% glass	5,000	1,600	1,450	1,450
ABS	1,500	230	211	188
ABS, 40% glass	5,000	1,800	1,650	1,600
Nylon 6/6	1,400	180	140	120
Nylon 6/6, 40% glass	5,000	1,500	1,070	1,000
Polyethylene	1,000	80	49	34
Polyethylene, 30% glass	1,500	600	470	420
Phenolic, mineral-filled	2,000	4,400	4,200	4,000
Epoxy, glass-filled	4,000	1,450	1,250	1,100
DAP, glass-filled	3,500	2,100	1,910	1,800
Polyester, 40% glass	5,000	1,870	1,760	1,680

TABLE 24 Physical Properties of Thermoset Resins[14]

Property	ASTM test method	Cast allyl diglycol carbonate	DAP molding compounds — Glass-filled	DAP molding compounds — Mineral-filled	DAP molding compounds — Synthetic-fiber-filled	Epoxy resins — Cast resins — No filler	Epoxy resins — Cast resins — Silica-filled	Epoxy resins — Cast resins — Aluminum-filled	Epoxy resins — Cast resins — Flexibilized
Molding qualities	....	....	Excellent	Excellent	Good				
Compression-molding temp, °F	....	....	290–380	270–330	270–320				
Compression molding pressure, lb/in.²	....	....	500–4,000	500–4,000	500–4,000				
Injection-molding temp, °F	....	....	300–350						
Injection-molding pressure, lb/in.²	....	....	2,000–6,000						
Compression ratio	....	....	1.9–10.0	1.2–2.3	1.2–5.2				
Mold (linear) shrinkage, in./in.	....	....	0.001–0.005	0.005–0.007	0.009–0.011	0.001–0.01	0.0005–0.008	0.001–0.005	0.001–0.01
Specific gravity (density)	D 792	1.30–1.40	1.51–1.78	1.65–1.68	1.34–1.39	1.11–1.40	1.6–2.0	1.4–1.8	1.05–1.35
Specific volume, in.³/lb	D 792	20.9–19.8	17.2–15.6	16.8–16.5	20.7–20.0	24.9–20.0	17.3–13.9		20.5
Machining qualities	....	Good	Fair	Fair	Excellent	Good		Excellent	Good
Tensile strength, lb/in.²	D 638	5,000–6,000	6,000–11,000	5,000–8,700	6,000–6,800	4,000–13,000	7,000–13,000	7,000–12,000	2,000–10,000
Elongation, %	D 638	....	....	....	....	3.0–6.0	1.0–3.0	0.5–3.0	20.0–70.0
Tensile elastic modulus, 10^5 lb/in.²	D 638	3.0	14.0–22.0	12.0–22.0	6.0	3.5			0.01–3.5
Compressive strength, lb/in.²	D 695	21,000–23,000	25,000–35,000	20,000–32,000	26,000–30,000	15,000–25,000	15,000–35,000	15,000–33,000	1,000–14,000
Flexural yield strength, lb/in.²	D 790	6,000–13,000	11,000–25,000	8,500–11,000	7,000–8,000	13,300–21,000	8,000–14,000	8,500–24,000	1,000–13,000
Impact strength, ft-lb/in. of notch (½- by ½-in. notched bar, Izod test)	D 256	0.2–0.4	0.4–15.0	0.3–0.45	0.55–8.0	0.2–1.0	0.3–0.45	0.4–1.6	3.5–5.0
Hardness, Rockwell	D 785	M95–100	80–87 (E scale)	61 (E scale)	M108–115	M80–110	M85–120	M55–85	
Flexural elastic modulus, lb/in.² $\times 10^6$	D 790	2.5–3.3	12.0–15.0						
Compressive modulus, lb/in.² $\times 10^5$	D 695	3.0							
Thermal conductivity, 10^{-4} cal/(s)(cm²) per °C/cm	C 177	4.8–5.0	5.0–15.0	7.0–25.0	5.0–6.0	4.0–5.0	10.0–20.0	15–25	
Specific heat, cal/°C/g	....	0.55				0.25	0.20–0.27		
Thermal expansion, 10^{-5} in./in./°C	D 696	8.1–14.3	1.0–3.6	1.0–4.2	5.4–6.0	4.5–6.5	2.0–4.0	5.5×10^{-6}	2.0–10.0
Resistance to heat, °F (continuous)	....	212	300–400	300–400	250–350	250–550	250–550	200–500	250–300
Deflection temp, °F: At 264 lb/in.² fiber stress	D 648	140–190	330–450	320–540+	320–400	115–550	160–550	190–600	RT–250
At 66 lb/in.² fiber stress	D 648								
Refractive index, n_D	D 542	1.50	Not applicable	Not applicable	Not applicable	1.55–1.61			
Clarity	....	Transparent	Opaque	Opaque	Opaque	Transparent	Opaque	Opaque	Transparent
Transmittance, %	....	90–92							
Haze, %	....								
Water absorption, 24 h, ⅛ in. thick, %	D 570	0.2	0.12–0.35	0.2–0.5	0.2	0.08–0.15	0.04–0.10	0.1–4.0	0.27–0.50
Burning rate (flammability), in./min	D 635	0.35 to self-extinguishing	Self-extinguishing to nonburning	Self-extinguishing to nonburning	Slow burn	Slow	Self-extinguishing		Slow
Effect of sunlight	....	Yellows very slightly	None	None	None	None	None	None	None
Effect of weak acids	D 543	None	Slight	Slight	None	None	None	None	None
Effect of strong acids	D 543	Attacked only by oxidizing acids	None to slight	None to slight	Slight	Attacked by some	Attacked by some	Attacked by some	Attacked by some
Effect of weak alkalies	D 543	None	Slight	Slight	Slight	None	None	None	None
Effect of strong alkalies	D 543	None to slight	None	None	Slight	Slight	Slight	Some	Slight
Effect of organic solvents	D 543	Resistant	None	None	None	Generally resistant	Generally resistant	Chlorinated affect	Generally resistant

TABLE 24 Physical Properties of Thermoset Resins[14] (Continued)

Epoxy resins (continued)

Property	ASTM test method	Molding compounds — Glass-fiber-filled	Molding compounds — Mineral-filled	Molding compounds — Low-density*	Encapsulating grades — Mineral-filled	Encapsulating grades — Glass-filled	Phenol novolac epoxy resins	Cyclo-aliphatic epoxy resins	Ortho cresol novolac epoxy resins	Furan, asbestos-filled
Molding qualities	....	Excellent	Excellent	Good	Good to excellent	Good	Good	....	Good	Good
Compression-molding temp, °F	....	300–330	250–330	250–320	250–330 (transf.)	280–330 (transf.)	250–330	....	250–330	275–300
Compression molding pressure, lb/in.²	....	300–5,000	100–3,000	100–2,000	50–1,000 (transf.)	50–1,000 (transf.)	100–800	....	100–800	100–500
Injection-molding temp, °F	....	....	....	250–300	....	....	....	....	....	....
Injection-molding pressure, lb/in.²	....	....	....	100–1,500	....	....	....	....	....	....
Compression ratio	....	3.0–7.0	2.0–3.0	3.0–7.0	2.0	3.0	....	....	....	....
Mold (linear) shrinkage, in./in.	....	0.001–0.005	0.002–0.008	0.006–0.010	0.004–0.010	0.004–0.008	....	....	....	....
Specific gravity (density)	D 792	1.6–2.0	1.6–2.0	0.75–1.00	1.7–2.1	1.7–2.0	1.16–1.21	1.16–1.21	1.16–1.21	1.75
Specific volume, in.³/lb	D 792	15.4–14.0	14.2–13.4	....	14.0	14.0	....	....	....	15.8
Machining qualities	....	Fair	Fair	Good	Fair	Fair	Good	Good	Good	Fair to good
Tensile strength, lb/in.²	D 638	10,000–20,000	5,000–10,000	2,500–4,000	4,000–10,000	5,000–15,000	6,000–12,000	8,000–12,000	....	3,000–4,500
Elongation, %	D 638	4	....	....	....	....	2.0–6.0	2.0–10.0	....	....
Tensile elastic modulus, 10^5 lb/in.²	D 638	30.4	....	....	....	....	....	....	....	15.8
Compressive strength, lb/in.²	D 695	25,000–40,000	18,000–40,000	10,000–15,000	18,000–30,000	18,000–30,000	....	15,000–20,000	....	10,000–13,000
Flexural yield strength, lb/in.²	D 790	10,000–60,000	8,000–15,000	5,000–7,000	6,000–15,000	8,000–20,000	....	10,000–13,000	....	6,000–9,000
Impact strength, ft-lb/in. of notch (½- by ½-in. notched bar, Izod test)	D 256	2.0–30.0	0.3–0.4	0.15–0.25	0.3–0.45	0.5–2.0	....	....	....	....
Hardness, Rockwell	D 785	M100–110	M100–110	....	M100–112	M100–112	....	....	....	R110
Flexural elastic modulus, lb/in.² $\times 10^5$	D 790	2.5–4.5 $\times 10^6$	1.4–2.2 $\times 10^6$	5.0–7.5 $\times 10^6$	....	....	....	....	....	....
Compressive modulus, lb/in.² $\times 10^6$	D 695	....	....	....	....	....	....	....	....	....
Thermal conductivity, 10^{-4} cal/(s)(cm²) per °C/cm	C 177	4.0–10.0	4.0–30.0	4.0–6.0	4.0–10.0	4.0–10.0	....	....	....	....
Specific heat, cal/°C/g	....	0.19	....	....	....	....	....	....	....	....
Thermal expansion, 10^{-5} in./in./°C	D 696	1.1–3.5	2.0–5.0	....	3.0–6.0	3.0–5.0	....	....	....	....
Resistance to heat, °F (continuous)	....	300–500	300–500	200–250	300–450	300–450	....	200–450	....	265–330
Deflection temp, °F: At 264 lb/in.² fiber stress	D 648	250–500	250–500	200–250	225–450	225–450	300–500	300–500	300–500	....
At 66 lb/in.² fiber stress	D 648	....	....	....	....	....	....	....	....	....
Refractive index, n_D	D 542	....	....	....	....	....	....	....	....	....
Clarity	....	Opaque	Opaque	Opaque	Opaque	Opaque	Transparent	Transparent	Transparent	Opaque
Transmittance, %	....	....	....	....	....	....	....	....	....	....
Haze, %	....	....	....	....	....	....	....	....	....	....
Water absorption, 24 h, ⅛ in. thick, %	D 570	0.05–0.20	0.04	0.2–1.0	0.03–0.20	0.04–0.20	Excellent	....	Good	0.01–0.20
Burning rate (flammability), in./min.	D 635	Self-extinguishing	Self-extinguishing	Self-extinguishing	Self-extinguishing to nonburning	Self-extinguishing to nonburning	....	....	....	Slow
Effect of sunlight	D 543	Slight	Slight	Slight	Slight	Slight	Darkens	None	Darkens	None
Effect of weak acids	D 543	None	None	None	None	None	None	None	None	None to slight
Effect of strong acids	D 543	Negligible	Slight	None	Slight	Slight	....	Attacked by some	Attacked by some	Attacked by oxidizing acids
Effect of weak alkalies	D 543	None	Slight to attacked	Slight to attacked	Slight to attacked	Slight to attacked	None	Attacked by some	Attacked by some	None
Effect of strong alkalies	D 543	None	Slight to attacked	Slight to attacked	Slight to attacked	Slight to attacked	None	None	None	None to slight
Effect of organic solvents	D 543	None	Slight	Slight	Slight	Slight	None	None	None	Resistant

* Filled with hollow glass spheres.

TABLE 24 Physical Properties of Thermoset Resins[14] (Continued)

Property	ASTM test method	Melamine-formaldehyde molding compounds								Melamine-phenol molding compounds
		No filler	Alpha-cellulose-filled	Cellulose-filled	Flock-filled	Asbestos-filled	Macerated-fabric-filled	Macerated-fabric-filled (phenolic-modified)	Glass-fiber-filled (including nodular)	
Molding qualities	....	Good	Excellent	Good	Good	Good	Good	Good	Good	Good
Compression-molding temp, °F	....	300–330	280–370	290–360	275–300	280–340	275–330	300–350	280–340	300–350
Compression molding pressure, lb/in.²	....	2,000–5,000	1,500–8,000	1,500–6,000	4,000–8,000	1,000–7,000	4,000–8,000	4,000–8,000	2,000–8,000	2,000–5,000
Injection-molding temp, °F	....									350–400
Injection-molding pressure, lb/in.²	....									15,000–20,000
Compression ratio	....	2.0	2.1–3.1	2.2–2.5	4.0–7.0	2.1–2.5	5.0–10.0	5.0–10.0		2.1–4.4
Mold (linear) shrinkage, in./in.	D 792	0.011–0.012	0.005–0.015	0.006–0.008	0.006–0.007	0.005–0.007	0.003–0.004	0.003–0.006	0.001–0.004	0.004–0.010
Specific gravity (density)	D 792	1.48	1.47–1.52	1.45–1.52	1.50–1.55	1.70–2.0	1.5	1.5	1.8–2.0	1.5–1.7
Specific volume, in.³/lb	....	18.7	18.8–18.2	20.0–19.1	18.5–17.8	16.3–13.8	18.5	18.5	13.8–10.0	20.9–17.8
Machining qualities	....		Fair	Good			Good	Good	Good	Good
Tensile strength, lb/in.²	D 638		7,000–13,000	5,000–9,000	7,000–9,000	5,500–7,000	8,000–10,500	6,000–10,500	5,000–10,000	6,000–8,000
Elongation, %	D 638		0.6–0.9	0.6		0.30–0.45	0.6–0.8	0.6	0.6	0.4–0.8
Tensile elastic modulus, 10⁵ lb/in.²	D 638		12.0–14.0	11.0		19.5	14.0–16.0	16.0	24.0	8.0–17.0
Compressive strength, lb/in.²	D 695	40,000–45,000	40,000–45,000	33,500	30,000–35,000	30,000	30,000–35,000	25,000–30,000	20,000–35,000	26,000–30,000
Flexural yield strength, lb/in.²	D 790	11,000–14,000	10,000–16,000	9,000–11,500	13,000	9,000–11,000	12,000–15,000	14,000–17,000	15,000–23,000	8,000–10,000
Impact strength, ft-lb/in. of notch (½- by ½-in. notched bar, Izod test)	D 256		0.24–0.35	0.27–0.36	0.40–0.45	0.28–0.40	0.6–1.0	1.1–1.4	0.6–18.0	0.27–0.38
Hardness, Rockwell	D 785		M115–125	M115–125		M110	M120	M115		E95–100
Flexural elastic modulus, lb/in.² × 10⁵	D 790									10.0–12.0
Compressive modulus, lb/in.² × 10⁵	D 695									
Thermal conductivity, 10⁻⁴ cal/(s)(cm²) per °C/cm	C 177		7.0–10.0	6.5–8.5		13.0–17.0	10.6	10.1	11.5	4.0–7.0
Specific heat, cal/°C/g	....		0.4							0.35–0.40
Thermal expansion, 10⁻⁶ in./in./°C	D 696		4.0	4.5		2.0–4.5	2.5–2.8	1.8–2.8	1.5–1.7	1.0–4.0
Resistance to heat, °F (continuous)	....	210	210	250	250	250–400	250	250	300–400	275–325
Deflection temp, °F:										
At 264 lb/in.² fiber stress	D 648	298	350–370	266		265	310	375	400	285–310
At 66 lb/in.² fiber stress	D 648									
Refractive index, n_D	D 542									
Clarity	....	Opalescent	Translucent	Opaque	Opaque	Opaque	Opaque	Opaque	Opaque	Opaque
Transmittance, %	....		29.9 (0.50 in.)							
Haze, %	....									
Water absorption, 24 h, ⅛ in. thick, %	D 570	0.3–0.5	0.1–0.6	0.34–0.80	0.16–0.30	0.08–0.14	0.3–0.6	0.2–1.3	0.09–0.21	0.30–0.65
Burning rate (flammability), in./min.	D 635	Self-extinguishing	Self-extinguishing	Self-extinguishing	None	Self-extinguishing	Self-extinguishing	Very low	Self-extinguishing	Self-extinguishing
Effect of sunlight	....	Colors may fade	Slight color change	Slight color change	Slight color change	Slight color change	Slight discoloration	Color darkens	Slight	Nil
Effect of weak acids	D 543	None	None	None to slight	None to slight	None to slight	None	None	None	Good
Effect of strong acids	D 543	....	Decomposes	Decomposes	Decomposes	Decomposes	Decomposes	Decomposes	Decomposes	Fair
Effect of weak alkalies	D 543	None	None	None	None	Very slight attack	None	None	None	Fair-good
Effect of strong alkalies	D 543	....	Attacked	Decomposes	Decomposes	Slight attack	Attacked	Attacked	None to slight	Poor-fair
Effect of organic solvents	D 543	None	None	None	None on bleedproof colors	None	None on bleedproof colors	None on bleedproof colors	None	Good

Property	ASTM test method	Phenolic cast resins			Phenol-formaldehyde and phenol-furfural molding compounds						
		No filler	Mineral-filled	Asbestos-filled	No filler	Wood-flour- and cotton-flock-filled	Asbestos-filled	Mica-filled	Glass-fiber-filled	Macerated-fabric- and cord-filled	Pulp preformed and molding board
Molding qualities					Poor	Excellent	Excellent	Good	Good to excellent	Good to fair	Good
Compression-molding temp, °F					270–320	290–380	290–380	270–350	300–380	280–380	290–350
Compression molding pressure, lb/in.					2,000–4,000	2,000–5,000	2,000–5,000	2,000–5,000	1,000–6,000	2,000–4,000	2,000–4,000
Injection-molding temp, °F					330–400	330–400	330–400	330–350	330–380	320–390	
Injection-molding pressure, lb/in.²					10,000–20,000	10,000–20,000	10,000–20,000	8,000–19,000	5,000–20,000	7,000–20,000	
Compression ratio					2.0–2.5	1.0–1.5	1.0–1.5	2.1–2.7	2.0–10.0	3.0–18.0	1.5–3.5
Mold (linear) shrinkage, in./in.					0.010–0.012	0.004–0.009	0.002–0.009	0.002–0.006	0.000–0.004	0.002–0.009	0.0018–0.008
Specific gravity (density)	D 792	1.236–1.320	1.68–1.70	1.70	1.25–1.30	1.34–1.45	1.45–2.00	1.65–1.92	1.69–1.95	1.36–1.43	1.36–1.42
Specific volume, in.³/lb	D 792	21.3–20.9	16.5–16.3	16.3	22.2–21.3	20.9–17.8	20.9–11.9	15.8–14.1	15.8–14.1	20.4–19.4	20.4–19.6
Machining qualities		Excellent	Good to fair	Fair to good	Fair to good	Fair to good	Good to poor	Poor to fair	Poor to fair	Fair to good	Fair to good
Tensile strength, lb/in.²	D 638	5,000–9,000	4,000–9,000	3,000–6,000	7,000–8,000	5,000–9,000	4,500–7,500	5,500–7,000	5,000–18,000	3,000–9,000	4,300–12,000
Elongation, %	D 638	1.5–2.0			1.0–1.5	0.4–0.8	0.18–0.50	0.13–0.50	0.2	0.37–0.57	0.3–0.7
Tensile elastic modulus, 10^6 lb/in.²	D 638	4.0–7.0			7.5–10.0	8.0–17.0	10.0–30.0	25.0–50.0	19.0–33.0	9.0–14.0	9.0–13.0
Compressive strength, lb/in.²	D 695	12,000–15,000	29,000–34,000	10,500–12,500	22,000–30,000	25,000–36,000	20,000–35,000	25,000–30,000	16,000–70,000	15,000–30,000	16,000–35,000
Flexural yield strength, lb/in.²	D 790	11,000–17,000	9,000–34,000	5,000–8,000	12,000–15,000	7,000–14,000	7,000–14,000	8,000–12,000	10,000–60,000	8,500–15,000	7,000–18,500
Impact strength, ft-lb/in. of notch (½- by ½-in. notched bar, Izod test)	D 256	0.24–0.40	0.35–0.50		0.20–0.36	0.24–0.60	0.26–3.50	0.27–0.38	0.3–18.0	0.75–8.00	0.5–4.5
Hardness, Rockwell	D 785	M93–120	M85–120	R110	M124–128, E80–95	M100–115, E77–98	M105–115, E66–97	E85–91, E88	E54–101	E79–82	E60–85
Flexural elastic modulus, lb/in.² $\times 10^5$	D 790								20.0–33.0		
Compressive modulus, lb/in.² $\times 10^6$	D 695										
Thermal conductivity, 10^{-4} cal/(s)(cm²) per °C/cm	C 177	3.5	7.5	8.4	3.0–6.0	4.0–8.2	6.0–22.0	10.0–14.0	8.2–14.5	4.0–9.0	4.0–7.0
Specific heat, cal/°C/g		0.3–0.4		0.3	0.38–0.42	0.32–0.40	0.28–0.32	0.28–0.32	0.24–0.27	0.30–0.35	0.34–0.36
Thermal expansion, 10^{-5} in./in./°C	D 696	6.8		3.3	2.5–6.0	3.0–4.5	0.8–4.0	1.9–2.6	0.8–2.05	1.0–4.0	3.0–4.5
Resistance to heat, °F (continuous)		160	160	300	250	300–350	350–500	250–300	350–550	220–250	300–400
Deflection temp, °F: At 264 lb/in.² fiber stress	D 648	165–175	150–175		240–260	300–370	300–380	300–500	300–600	250–330	260–340
At 66 lb/in.² fiber stress	D 648										
Refractive index, n_D	D 542	1.58–1.66			1.5–1.7						
Clarity		Transparent, translucent, opaque	Opaque	Opaque	Transparent to translucent		Opaque		....	Opaque	
Transmittance, %											
Haze, %											
Water absorption, 24 h, ⅛ in. thick, %	D 570	0.2–0.4	0.12–0.36	Almost none	0.1–0.2	0.3–1.2	0.10–0.5	0.01–0.05	0.03–1.20	0.40–1.75	0.6–1.8
Burning rate (flammability), in./min.	D 635	Very slow	Almost none	Almost none	Very low	Self-extinguishing	None	None	None	Almost none	Very low to self-extinguishing
Effect of sunlight		Colors may fade	Darkens	Darkens slightly	Surface darkens	General darkening					
Effect of weak acids	D 543	None	None to slight	None to slight		None to slight depending on acid; reducing and organic acids, none to slight effect					
Effect of strong acids	D 543	None	Attacked by oxidizing acids			Decomposed by oxidizing acids; reducing and organic acids, none to slight effect					
Effect of weak alkalies	D 543	Nil	Nil	None	Slight	Slight to marked, depending on alkalinity					
Effect of strong alkalies	D 543	Moderate to none	Decomposes	Decomposes	Decomposes	Attacked by strong alkalies			Fairly resistant on bleedproof materials		
Effect of organic solvents	D 543	Strong	Moderate to none	Generally resistant	Fairly resistant	Attacked					

TABLE 24 Physical Properties of Thermoset Resins[14] (Continued)

Column groups:
- *Phenol-formaldehyde and phenol-furfural molding compounds (continued)* — Compounded with butadiene-acrylonitrile copolymer: **Wood-flour and cotton-flock-filled**, **Asbestos-filled**, **Rag-filled**; **Metal-filled (iron, lead)**; **Low-shrink compound**
- *Polyesters* — Cast polyester: **Rigid**, **Flexible**; Glass-reinforced: **Preformed chopped roving**, **Premix chopped glass**, **Woven cloth**

Property	ASTM test method	Wood-flour and cotton-flock-filled	Asbestos-filled	Rag-filled	Metal-filled (iron, lead)	Low-shrink compound	Rigid	Flexible	Preformed chopped roving	Premix chopped glass	Woven cloth
Molding qualities	…	Fair to good	Fair to good	Fair to good	Good	Good	…	…	Excellent	Excellent	Hand lay-up
Compression-molding temp, °F	…	300–360	310–360	310–360	280–320	270–330	…	…	170–320	280–350	Rm. temp–250
Compression molding pressure, lb/in.²	…	2,000–4,000	2,000–4,000	2,000–4,000	2,000–5,000	>400	…	…	250–2,000	500–2,000	0–300
Injection-molding temp, °F	…	…	…	…	…	270–330	…	…	…	…	…
Injection-molding pressure, lb/in.²	…	…	…	…	…	2,000	…	…	…	…	…
Compression ratio	…	2.5–3.2	2.4	3.8–5.2	…	1.0	…	…	…	…	…
Mold (linear) shrinkage, in./in.	…	0.004–0.009	0.004–0.007	0.002–0.004	0.003–0.004	0.0001	…	…	0.000–0.002	0.001–0.012	0.000–0.002
Specific gravity (density)	D 792	1.29–1.35	1.57–1.65	1.30–1.35	2.0–4.2	1.60–2.40	1.10–1.46	1.01–1.20	1.35–2.30	1.65–2.30	1.50–2.10
Specific volume, in.³/lb	D 792	22.3–20.5	17.3–16.8	21.3–21.0	…	…	…	27.4–23.0	20.5–13.9	…	18.5–13.2
Machining qualities	…	Good	Good	Good	…	Good to excellent	Good	Fair	Good	Good	Good
Tensile strength, lb/in.²	D 638	3,500–5,000	4,000–5,000	3,000–5,000	4,000–7,000	4,500–20,000	6,000–13,000	500–3,000	15,000–30,000	3,000–10,000	30,000–50,000
Elongation, %	D 638	0.75–2.25	…	…	…	10.0–25.0	<5.0	40.0–310.0	0.5–5.0	0.5	0.5–2.0
Tensile elastic modulus, 10^5 lb/in.²	D 638	…	…	…	…	…	3.0–6.4	…	8.0–20.0	10.0–25.0	15.0–45.0
Compressive strength, lb/in.²	D 695	15,000–20,000	10,000–20,000	15,000–20,000	15,000–30,000	15,000–30,000	13,000–30,000	…	15,000–30,000	20,000–30,000	25,000–50,000
Flexural yield strength, lb/in.²	D 790	5,000–7,000	6,000–8,000	7,000–9,000	7,000–12,000	9,000–35,000	8,500–23,000	…	10,000–40,000	7,000–20,000	40,000–80,000
Impact strength, ft-lb/in. of notch (½- by ½-in. notched bar, Izod test)	D 256	0.40–0.90	0.34–0.40	2.0–2.3	0.3–0.5	2.5–15.0	0.2–0.4	>7.0	2.0–20.0	1.5–16.0	5.0–30.0
Hardness, Rockwell	D 785	M45–60	M50	M50–70	M80–115	40–70 (Barcol)	M70–115 50–75 (Barcol)	84–94 (Shore D)	50–80 (Barcol)	50–80 (Barcol)	60–80 (Barcol)

Property	ASTM									
Flexural elastic modulus, lb/in.² × 10⁵	D 790	3.0–5.0	5.0–6.0	3.5–5.5	18.0–26.0	10.0–25.0	......	10–30	10–20	......
Compressive modulus, lb/in.² × 10⁵	D 695	......	......	......	......	......	......	......	......	......
Thermal conductivity, 10^{-4} cal/(s)(cm²) per °C/cm	C 177	5.0–6.0	4.6	4.9	14.0–20.0	......	4.0	......	10.0–16.0	......
Specific heat, cal/°C/g		0.33	......	......	......	......	......	......	0.25	......
Thermal expansion, 10^{-5} in./in./°C	D 696	1.5–3.5	4.0	2.1	......	0.6–3.0	5.5–10.0	2.0–5.0	2.0–3.3	1.5–3.0
Resistance to heat, °F (continuous)		200	225	200	......	300	250	300–350	300–350	300–350
Deflection temp, °F: At 264 lb/in.² fiber stress	D 648	200–230	250	225–250	350–425	375–500	140–400	......	......	......
At 66 lb/in.² fiber stress	D 648	......	......	......	......	......	......	>400	>400	>400
Refractive index, n_D	D 542	......	......	......	......	......	1.523–1.570	1.537–1.550	1.550	......
Clarity		......	......	Opaque	......	......	Transparent to opaque	Translucent to opaque	Opaque	Translucent to opaque
Transmittance, %										
Haze, %										
Water absorption, 24 h, ⅛ in. thick, %	D 570	1.0–2.0	0.15–0.20	1.0–2.0	0.03–0.15	0.01–0.25	0.15–0.60 / 0.50–2.50	0.01–1.00	0.06–0.28	0.05–0.50
Burning rate (flammability), in./min	D 635	Slow	Very slow	Slow	None	Slow to self-extinguishing	Conventional burns fire-retardant type self-extinguishing to nonburning. Conventional 0.7–2.0 in./min; fire-retardant <0.2	Slight Depends on pigmentation	Slight	
Effect of sunlight		General darkening				Slight	Yellows slightly			
Effect of weak acids	D 543	None to slight depending on acid				None	Ortho and gen. purp. iso resis. some at RT; chem. resis. iso resis. to 170°F; bis. and HBPA to 210°F; chlor. anhyd. resis. most at RT, some to 350°F			
Effect of strong acids	D 543	Decomposed by oxidizing acids; reducing and organic acids, none to slight effect				Attacked by high concentration	Ortho and gen. purp. not resis. strong or oxidizing; chem. resis. iso resis. strong to 210°F, not resis. low-concen. oxidizing to 170°F, chlor. anhyd. resis. resis. strong to 210°F, oxidizing to 170°F			
Effect of weak alkalies	D 543	Slight to marked, depending on alkalinity				Slight	Ortho and gen. purp. resis. some at RT; chem. resis. iso to 170°F and bis. and HBPA to 210°F; chlor. anhyd. resis. some to 210°F			
Effect of strong alkalies	D 543	Attacked				Slight to considerable	Ortho and gen. purp. iso and chem. resis. iso not resis.; chem. resis.; bis. resis. to 170°F; chlor. anhyd. resis. some to 170°F			
Effect of organic solvents	D 543	Fairly resistant on bleedproof materials					Ortho and gen. purp. iso not resis.; chem. resis. iso resis. some and bis. and HBPA most at RT; chlor. anhyd. resis. most at RT, some to 200°F. Effect of salts: Ortho and gen. purp. iso resis. some at RT; chem. resis. iso resis. to 170°F; bis. and HBPA to 120°F; chlor. anhyd. resis. most to 250°F			

TABLE 24 Physical Properties of Thermoset Resins[14] (Continued)

| | | Polyesters (continued) | | | | | | | |
| | | | Alkyd | | | | | | |
Property	ASTM test method	Sheet-molding compound	Granular and putty types, mineral-filled	Asbestos-filled	Glass-filled	Synthetic-fiber-filled	Vinyl ester	Silicone, molding compounds, glass-fiber-filled	Urea formaldehyde, alpha-cellulose-filled
Molding qualities		Good	Excellent	Excellent	Excellent	Excellent	Excellent	Good	Excellent
Compression-molding temp, °F		270–350	270–350	280–320	330–350	280–350	270–300	310–360	275–350
Compression-molding pressure, lb/in.^{2}		300–1,200	100–4,000	500–2,000	2,000–6,000	500–3,000	100–200	1,000–5,000	2,000–8,000
Injection-molding temp, °F			280–390	330–380	280–380	280–380			
Injection-molding pressure, lb/in.^{2}			2,000–20,000	10,000–20,000	2,500–20,000	2,000–20,000			
Compression ratio			1.8–2.5	2.0–2.5	>1.0–11.0	>1.0–5.0		2.0–6.0	2.2–3.0
Mold (linear) shrinkage, in./in.	D 792	0.001–0.004	0.003–0.010	0.004–0.007	0.002–0.008	0.005–0.0010		0–0.005	0.006–0.014
Specific gravity (density)	D 792	1.65–2.60	1.60–2.30	1.65–2.20	2.12–2.15	1.24–2.10	1.03–1.083	1.80–1.90	1.47–1.52
Specific volume, in.^{3}/lb			17.3–12.0	16.8	13.2	22.1–13.2		15.4–14.6	18.8–18.2
Machining qualities		Good	Poor to fair	Poor to fair	Poor to fair	Fair to excellent		Fair	Fair
Tensile strength, lb/in.^{2}	D 638	8,000–20,000	3,000–9,000	4,500–9,000	4,000–9,500	4,500–8,000	10,500	4,000–6,500	5,500–13,000
Elongation, %	D 638						3.5–5.5		0.5–1.0
Tensile elastic modulus, 10^5 lb/in.^{2}	D 638		5.0–30.0	20–30	20.0–28.0	20.0			10.0–15.0
Compressive strength, lb/in.^{2}	D 695	15,000–30,000	12,000–38,000	22,500	15,000–32,000	20,000–30,000		10,000–15,000	25,000–45,000
Flexural yield strength, lb/in.^{2}	D 790	10,000–36,000	6,000–17,000	8,000–10,000	15,000	10,000–20,000	17,800–18,400	10,000–14,000	10,000–18,000
Impact strength, ft-lb/in. of notch (½ by ½-in. notched bar, Izod test)	D 256	7.0–22.0	0.30–0.50	0.45–0.50	0.60–10.0	0.50–4.50	4.7–4.9	0.3–8.0	0.25–0.40
Hardness, Rockwell	D 785	50–70 (Barcol)	60–70 (Barcol), 98 (E scale)	M99	55–80 (Barcol), 95 (E scale)	95 (M scale), 76 (E scale)		M80–90	M110–120
Flexural elastic modulus, lb/in.^{2} $\times 10^5$	D 790	10.0–22.0	20	20–30	20.0	17.0		10–25	13–16
Compressive modulus, lb/in.^{2} $\times 10^5$	D 695								
Thermal conductivity, 10^{-4} cal/(s)(cm^2) per °C/cm	C 177		12.2–25.0		15.0–25.0	7.0–16.5		7.0–9.0	7.0–10.0
Specific heat, cal/°C/g			0.25					0.19–0.22	0.4
Thermal expansion, 10^{-6} in./in./°C	D 696	2.0	–2.0–5.0		1.5–2.5	3.0–5.5		2.0–5.0	2.2–3.6
Resistance to heat, °F (continuous)		300–400	300–450	450	450	300–430	200–220	>600	170
Deflection temp, °F: At 264 lb/in.^{2} fiber stress	D 648	375–500	350–500	315	400–500	245–430	185–215	>900	260–290
At 66 lb/in.^{2} fiber stress	D 648								
Refractive index, n_D	D 542								1.54–1.56
Clarity		Opaque	Opaque	Opaque	Opaque	Opaque		Opaque	Translucent, opaque
Transmittance, %									39.4 (0.50 in.)
Haze, %									
Water absorption, 24 h, ⅛ in. thick, %	D 570	0.10–0.15	0.05–0.50	0.14	0.05–0.25	0.05–0.20	0.2	0.2	0.4–0.8
Burning rate (flammability), in./min	D 635	Slow to self-extinguishing	Slow to nonburning	Self-extinguishing	Slow to nonburning	Self-extinguishing and nonburning	Burns	Self-extinguishing	Self-extinguishing
Effect of sunlight	D 543	None	None	None	None	None		None	Pastels gray
Effect of weak acids	D 543		None	None	Poor to fair	None	Good	None to slight	None to slight
Effect of strong acids	D 543		Attacked	Slight	Poor to fair	Slight	Good	Slight	Decomposed or surface attacked
Effect of weak alkalies	D 543	Slight	Attacked	None	Poor to fair	None	Good	None to slight	Slight to marked
Effect of strong alkalies	D 543	Attacked	Decomposes	Slight	Poor to fair	Slight	Good	Slight to marked	Decomposes
Effect of organic solvents	D 543	Slight	Fair to good	None	Fair to good	Fair to good	Fair to good	Attacked by some	None to slight

TABLE 25 Shrinkage of Thermosetting Molding Materials as a Result of Heat Aging[38]

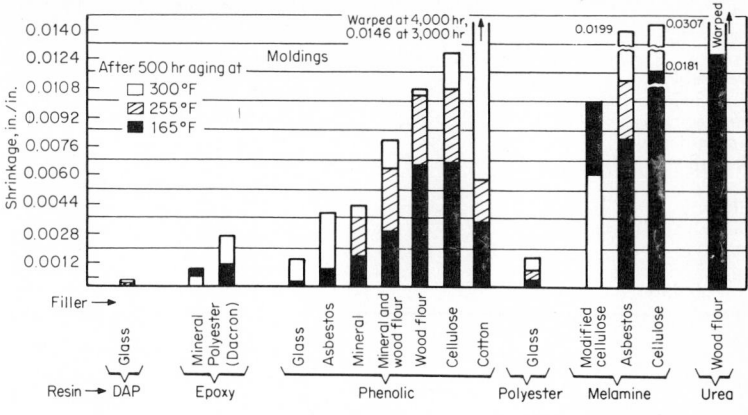

compounds. They also can be combined with hollow glass spheres as syntactic foam. Properties are shown in Table 24. Filled with glass fibers, the epoxies possess great stiffness and impact resistance, as shown in Tables 26 to 28. The upper temperature limit for epoxy molding compounds is about 300°F. Above this, they lose strength rapidly, as seen in Table 29. As moldings they have low shrinkage and are insensitive to water and many chemicals.

Melamine Resins

These resins are the reaction product of melamine and formaldehyde. Properties are given in Table 24. They usually are filled with cellulose or glass fiber. They are resistant to creep at elevated temperatures, as shown in Fig. 60. Melamines have a hard, tough surface which resists abrasion and scratching. Hence they are widely used as dinnerware and in appliance components and cutlery handles.

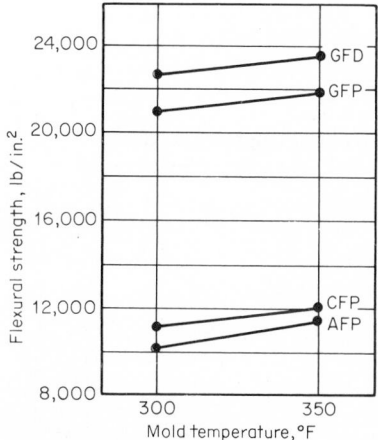

Fig. 59 Effect of mold temperature on flexural strength of DAP and phenolic moldings.[39] GFD: glass-filled diallyl phthalate; AFP: asbestos-filled phenolic; CFP: cellulose-filled phenolic; GFP: glass-filled phenolic.

TABLE 26   Flexural Strength of Molding Compounds

Plastic	Reinforcement	Flexural strength, lb/in.^{2}
Epoxy	Glass fiber	60,000
Polyester, SMC	Glass fiber	36,000
Diallyl phthalate	Glass fiber	25,000
Polyester, premix	Glass fiber	20,000
Urea formaldehyde	Wood flour	18,000
Polyester alkyd	Mineral	17,000
Epoxy	Mineral	15,000
Phenolic	Mineral	14,000
Phenolic	Wood flour	14,000
Silicone	Glass fiber	14,000
Melamine	Flock	13,000
Diallyl phthalate	Mineral	11,000

TABLE 27   Flexural Modulus of Molding Compounds

Plastic	Reinforcement	Modulus, lb/in.$^2 \times 10^5$
Epoxy	Glass fiber	28
Silicone	Glass fiber	25
Phenol formaldehyde	Mineral	22
Polyester, SMC	Glass fiber	22
Polyester alkyd	Mineral	20
Polyester premix	Glass fiber	20
Urea formaldehyde	Wood flour	16
Diallyl phthalate	Glass fiber	15
Phenol formaldehyde	Wood flour	12
Melamine	Flock	10
Diallyl phthalate	Mineral	9
Epoxy	Mineral	9

TABLE 28   Impact Strength of Molding Compounds

Plastic	Reinforcement	Impact strength, ft-lb/in. notch
Epoxy	Glass fiber	30
Polyester, SMC	Glass fiber	22
Polyester premix	Glass fiber	16
Diallyl phthalate	Glass fiber	15
Silicone	Glass fiber	8
Phenol formaldehyde	Mineral	3.5
Phenol formaldehyde	Wood flour	0.6
Polyester alkyd	Mineral	0.5
Diallyl phthalate	Mineral	0.45
Melamine	Flock	0.45
Epoxy	Mineral	0.4
Urea	Wood flour	0.4

TABLE 29 Flexural Modulus and Strength of Various Molding Materials vs. Temperature[40]

Molding material	Flexural modulus, 10^{-5} lb/in.^{2}					Flexural strength, 10^{-3} lb/in.^{2}				
	73°F	150°F	250°F	300°F	350°F	73°F	150°F	250°F	300°F	350°F
Epoxy	22.7	19.1	16.5	9.3	3.9	15.6	12.3	9.8	6.6	3.7
Phenolic	19.5	17.4	14.0	13.5	13.3	16.6	15.2	13.5	11.2	10.4
DAP	16.5	13.1	12.0	3.2	2.9	11.1	10.4	8.9	3.1	3.9
Alkyd	23.9	12.4	10.0	5.9	3.1	10.0	6.7	4.8	4.4	2.2
Polysulfone	4.2	3.8	3.9	3.7	3.3	16.2*	13.7*	10.9*	9.7*	7.0
PC, unfilled	3.9	3.3	3.5	Degraded →		14.0*	10.9*	9.1*	Degraded →	
PC, 20% glass	10.1	9.3	9.5	1.1	Degraded	22.1	18.8	16.2	1.9*	Degraded
PPO	4.0	3.8	3.3	3.0	2.7	14.5*	14.0*	10.1*	8.4*	6.2*

All values of modulus and strength are an average of five specimens for each condition as per ASTM D 790. All specimens were conditioned for 1 h at the specific temperature before testing.
* Yield strength. Other values represent strength at fracture.

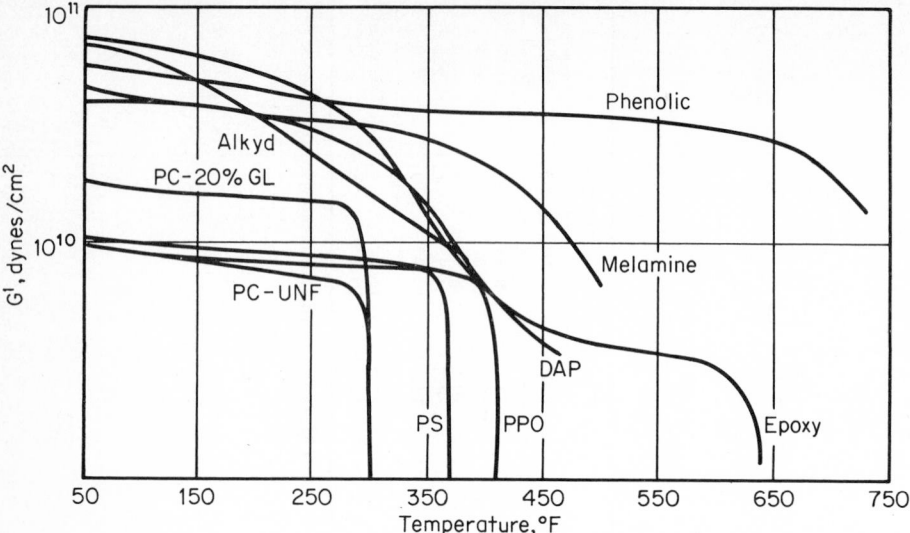

Fig. 60   Dynamic torsion modulus of melamine compared with other plastics.[41]

Phenolics

The reaction products of phenol and formaldehyde are known as phenolics; their properties are shown in Table 24. The first commercially available plastics, phenolics

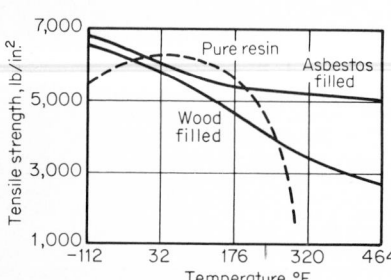

Fig. 61   The effect of filler on heat resistance of phenolics.[42]

are still used in thousands of applications. They are combined with cellulose, minerals, asbestos, chopped cloth, glass fibers, mica, or ground metals to result in molded products with widely varying properties. The effect of fillers on the tensile strength is shown in Fig. 61. Phenolics are little affected by outdoor aging (Table 30) or by exposure to water (Table 31). They also are highly temperature-resistant, as shown in Table 32. The resins burn only reluctantly, are chemical-resistant, useful to 500°F, strong, and light. Coloring and pigmenting are limited. The materials are used in automotive, housewares, appliances, toys, electrical and electronics applications, and as bearings, as knobs, and in a host of consumer applications.

TABLE 30   Tensile Strength of a Heat-resistant Phenolic in Various Climates[43]

Climate	Tensile strength, lb/in.²	
	Original test	2½-year exposure
Tropical..........................	5,780	5,130
Desert............................	4,920	3,980
Temperate........................	4,880	4,780
Arctic............................	5,380	5,220

TABLE 31   The Effect of Water on the Mechanical Properties of Phenolics[43]

Property	Flock-filled, one-step Control	Flock-filled, one-step 25 days in water at 84°C	Wood-flour-filled, two-step Control	Wood-flour-filled, two-step 10 days in water at 84°C
% change in weight (tensile specimen)	. . . . .	2.78	. . . . .	7.77
Tensile strength, lb/in.²	7,900	6,700	7,600	3,000
% change in weight (flexural specimen)	. . . . .	2.56	. . . . .	6.7
Flexural strength unnotched, lb/in.²	9,700	7,900	10,300	4,800
Flexural modulus, lb/in.² × 10⁻⁶	0.75	0.62	0.95	0.45
Flexural work to break, unnotched, ft-lb/in.³	0.69	0.50	0.50	0.23

TABLE 32   Thermal Aging of Two Phenolic Compounds[43]

Specimen thickness, in.	Flexural strength, lb/in.² * After 168 days at 150°C (305°F) 1.44 sp. gr.	1.60 sp. gr.	After 60 days at 190°C (374°F) 1.44 sp. gr.	1.60 sp. gr.
½	9,400	9,000	5,500	6,150
⅛	9,500	8,900	4,500	5,400
⅟16	8,000	. . . . .	3,700	

* Original flexural strength was 9,500 lb/in.² for both materials.

Polyesters

The polyesters and alkyds comprise a very large family of resins derived from the reaction of organic acids and anhydrides with glycols. Physical properties of molding compounds are given in Table 24. Alkyd molding materials have better flexural strength at 400°F than do the diallyl phthalate resins, as seen in Fig. 62. A recent development is sheet molding compound (SMC), which now competes with premix or bulk molding compound (BMC). The SMC materials are made by adding selected metal oxides to the resin, causing partial reaction and a dramatic increase in resin viscosity. Thus flow is controlled and molding processes are simplified. This provides parts with enhanced physical properties. Table 33 gives properties of SMC at 190°C compared with laminates and premix (BMC).

Polyester materials are used widely in automotive applications and serving trays. They often are used in sporting goods such as shuffleboard equipment, bowling balls, and billiard balls and in buttons. The resins can be made flame-retardant. Their outstanding properties include low cost, chemical resistance, low water absorption, and impact strength. They are not resistant to alkalies and are not high-temperature materials.

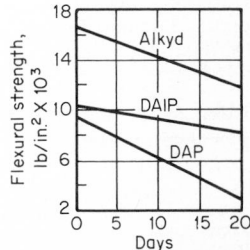

Fig. 62 Flexural strength of alkyd at 400°F vs. time.[44] DAD: diallyl phthalate; DAID: diallyl isophthalate.

TABLE 33 Properties of SMC, Laminates, and Premix after Aging at 190°C[45]

Property	Laminate	3-week premix	SMC	Laminate	6-week premix	SMC
% weight loss.........	1.0	0.8	0.6	1.3	1.3	1.0
Flexural strength, lb/in.^{2}...............	29,100	14,200	26,000	27,500	15,900	26,100
Modulus × 10^{-3}, lb/in.^{2}...............	1,534	1,560	3,214	1,543	1,727	1,284
Flexural strength, 180°C, lb/in.^{2}........	17,800	8,700	19,200	18,700	9,900	18,200
Modulus × 10^{-3}, lb/in.^{2}...............	864	701	2,023	968	926	1,547
Dielectric strength, short-term, perpendicular, V/mil........	515	369	452	503	338	498

Silicones

Silicones are true plastic materials where the silicon atom replaces some of the carbon atoms in the polymer backbone. The polymers vary from liquids to oils to rubberlike to fully thermoset, rigid plastics. Physical properties of the molding compounds are shown in Table 24. Silicones are very heat-resistant. They are capable of continuous use at 220°C for many years. Figure 63 compares the weight loss in vacuum of silicones and epoxies. Silicones are mostly used in electrical applications, but they also are used in those applications requiring thermal stability like some appliances, heaters, furnaces, aerospace devices, and automotive parts.

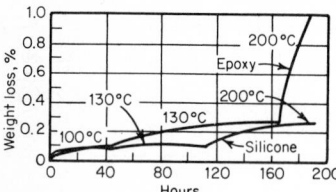

Fig. 63 Weight loss in vacuum of silicones and epoxies.[46]

Ureas

Ureas are the reaction products of urea and formaldehyde. Their physical properties are given in Table 24. Ureas usually are combined with cellulose filler, which results in materials with excellent compressive properties, as seen in Table 34. Ureas—unlike phenolics—can be highly colored, but aging in light affects colors as follows:

No effect......Red, Gray, Green, Black, Blue
Slight effect....Orange, Yellow, Coral
Fades.........Pink, Purple

Ureas are used in large moldings for business machines, toilet seats, and appliance parts. Organic solvents do not affect the material; hence it is used for chemical-resistant applications. Water absorption is high, and dimensional stability is a problem.

TABLE 34 Compressive Strength of Plastics

Plastic	Reinforcement	Compressive strength, lb/in.^{2}
Urea formaldehyde.....................	Wood flour	45,000
Epoxy.................................	Glass fiber	40,000
Epoxy.................................	Mineral	40,000
Polyester alkyd.......................	Mineral	38,000
Phenolic..............................	Wood flour	36,000
Diallyl phthalate.....................	Glass fiber	35,000
Melamine..............................	Flock	35,000
Phenol formaldehyde...................	Mineral	35,000
Diallyl phthalate.....................	Mineral	32,000
Polyester, premix.....................	Glass fiber	30,000
Polyester, SMC........................	Glass fiber	30,000
Silicone..............................	Glass fiber	15,000

ELASTOMERS

The precise definition for elastomers is provided in ASTM 1566. For the purposes of this handbook, an elastomer is defined to be a long, linear molecule with limited cross linking. Like thermoset resins, it cannot be resoftened after polymerization occurs. The polymers are soft, have high elongations, and recover from large deformations rapidly. There are many different kinds of elastomers. The properties of each are given in Table 35. Table 36 lists the definitions of these elastomers and nomenclature by ASTM and SAE. Elastomers are specified by many bodies. Such specifications are found in ASTM, MIL, AMS, and others. A reference to specifications is given in Table 37.

Natural Rubber (ASTM NR)

This elastomer is derived from the sap of the tree *Hevea brasiliensis*. Five grades depend on the purity of the product. Other natural rubbers are derived from guayule, gutta-percha, and balata. Natural rubber has excellent compression set and high physical properties. It does not age well in sunlight or ozone but, unlike synthetic polymers, softens and reverts. High temperatures result in degradation (see Fig. 64).

Butadiene Styrene (ASTM SBR)

Also known as GR-S or Buna S, this elastomer was developed as a synthetic analog to replace natural rubber during World War II. As such, its properties are not quite as good as those of natural rubber, but they are more consistent, the cost is lower, and it processes well. Thermal stability is slightly better.

Isoprene (ASTM IR)

Polyisoprene elastomers have high mechanical properties and elongation with low hysteresis. Their molecular weights can be controlled precisely, and the polymers are odorless, which makes them useful in various food-handling facilities.

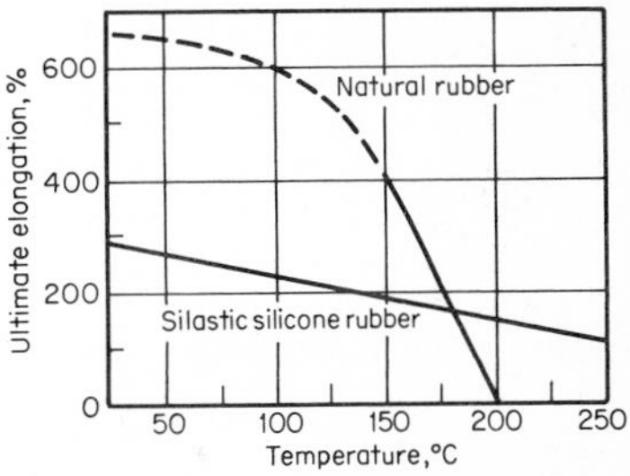

Fig. 64 Effect of temperature on elongation of natural rubber.[47]

Nitrile Rubber (ASTM NBR)

Also known as Buna N, this elastomer is made from butadiene and acrylonitrile. Chemical resistance is high, particularly with respect to oils and solvents. Heat resistance is better than that of SBR, but other properties closely parallel SBR.

Neoprene (ASTM CR)

Neoprene is polychloroprene and has moderate oil resistance but excellent weather and ozone resistance. It can be used continuously at 225°F, but crystallizes at low temperatures (below −40°F) and is not as good as SBR. It is self-extinguishing and has high tensile strength, elongation, and wear.

Butyl Rubber (ASTM IIR)

Butyl rubber is a copolymer of isobutylene and isoprene. It has extremely low permeability to air and hence is widely used for inner tubes. It weathers well and has good ozone resistance. Its heat resistance is good but is inferior to more stable materials (see Fig. 65). Its oil resistance is poor, but it withstands acids, water, and many solvents.

Polysulfide (ASTM T)

This elastomer is highly chemical-resistant and is usually found in applications where resistance to solvents is required. Chlorinated solvents may cause swelling. Its physical properties are modest. The weather resistance is excellent, and polysulfide is frequently selected as a caulking compound.

TABLE 35 Properties of Elastomers[47]

Property	Natural rubber (cis-polyisoprene)	Butadiene-styrene (GR-S)	Synthetic (polyisoprene)	Butadiene-acrylonitrile (nitrile)	Chloroprene (neoprene)	Butyl (isobutylene-isoprene)
Physical Properties:						
Specific gravity (ASTM D 792)	0.93	0.94	0.93	0.98	1.25	0.90
Thermal conductivity, Btu/(h)(ft²)(°F/ft) (ASTM C 177)	0.082	0.143	0.082	0.143	0.112	0.053
Coefficient of thermal expansion (cubical), 10^{-5} per °F (ASTM D 696)	37	37	......	39	34	32
Electrical insulation	Good	Good	Good	Fair	Fair	Good
Flame resistance	Poor	Poor	Poor	Poor	Good	Poor
Min recommended service temp, °F	−60	−60	−60	−60	−40	−50
Max recommended service temp, °F	180	180	180	300	240	300
Mechanical properties:						
Tensile strength, lb/in.²:						
Pure gum (ASTM D 412)	2,500–3,500	200–300	2,500–3,500	500–900	3,000–4,000	2,500–3,000
Black (ASTM D 412)	3,500–4,500	2,500–3,500	3,500–4,500	3,000–4,500	3,000–4,000	2,500–3,000
Elongation, %:						
Pure gum (ASTM D 412)	750–850	400–600	......	300–700	800–900	750–950
Black (ASTM D 412)	550–650	500–600	300–700	300–650	500–600	650–850
Hardness (durometer)	A30–90	A40–90	A40–80	A40–95	A20–95	A40–90
Rebound:						
Cold	Excellent	Good	Excellent	Good	Very good	Bad
Hot	Excellent	Good	Excellent	Good	Very good	Very good
Tear resistance	Excellent	Fair	Excellent	Good	Fair to good	Good
Abrasion resistance	Excellent	Good to excellent	Excellent	Good to excellent	Good	Good to excellent

Property	Material 1	Material 2	Material 3	Material 4	Material 5	Material 6
Chemical resistance:						
Sunlight aging	Very good	Very good	Poor	Fair	Poor	Poor
Oxidation	Excellent	Excellent	Good	Excellent	Good	Good
Heat aging	Excellent	Excellent	Excellent	Good	Very good	Good
Solvents:						
Aliphatic hydrocarbons	Poor	Good	Excellent	Poor	Poor	Poor
Aromatic hydrocarbons	Poor	Fair	Good	Poor	Poor	Poor
Oxygenated, alcohols	Very good	Very good	Good	Good	Good	Good
Oil, Gasoline	Poor	Good	Excellent	Poor	Poor	Poor
Animal, vegetable oils	Excellent	Excellent	Excellent	....	Poor to good	Poor to good
Acids:						
Dilute	Excellent	Excellent	Good	Fair to good	Fair to good	Fair to good
Concentrated	Excellent	Good	Good	Fair to good	Fair to good	Fair to good
Permeability to gases	Very low	Low	Very low	Low	Low	Low
Water-swell resistance	Excellent	Fair to excellent	Excellent	Excellent	Excellent	Fair
Uses	Truck and automobile tire inner tubes, curing bags for tire vulcanization and molding, steam hose and diaphragms, flexible electrical insulation, shock, vibration absorption	Wire and cable, belts, hose, extruded goods, coatings, molded and sheet goods, adhesives, automotive gaskets and seals, petroleum- and chemical-tank linings	Carburetor diaphragms, self-sealing fuel tanks, aircraft hose, gaskets, gasoline and oil hose, cables, machinery mountings, printing rolls	Same as natural rubber		Pneumatic tires and tubes; power-transmission belts and conveyor belts; gaskets; mountings; hose; chemical-tank linings; printing-press platens; sound or shock absorption; seals against air, moisture, sound, and dirt

TABLE 35 Properties of Elastomers[47] (Continued)

Property	Polybutadiene	Polysulfide	Silicone (polysiloxane)	Urethane (diisocyanate polyester)	Polyacrylate
Physical properties:					
Specific gravity (ASTM D 792)	0.91	1.35	1.1–1.6	1.25	1.09
Thermal conductivity, Btu/(h)(ft²)(°F/ft) (ASTM C 177)	......	......	0.13		
Coefficient of thermal expansion (cubical), 10^{-6} per °F (ASTM D 696)	37.5	......	45		
Electrical insulation	Good	Fair	Excellent	Fair	Fair
Flame resistance	Poor	Poor	Good	Good	Poor
Min recommended service temp, °F	−150*	−60	−178	−65	−20
Max recommended service temp, °F	200	250	600	240	350
Mechanical properties:					
Tensile strength, lb/in.²:					
Pure gum (ASTM D 412)	200–1,000	250–400	600–1,300†	>5,000	250–400
Black (ASTM D 412)	2,000–3,000	>1,000	......	......	1,000–2,500
Elongation, %:					
Pure gum (ASTM D 412)	400–1,000	450–650	100–500†	540–750	450–750
Black (ASTM D 412)	450–600	150–450	......	......	150–450
Hardness (durometer)	A40–90	A40–85	A30–90	A35–100	A40–90
Rebound:					
Cold	Excellent	Good	Very good	Bad	Fair
Hot	Excellent	Good	Very good	Good	Very good
Tear resistance	Fair	Poor	Fair	Good	Fair to good
Abrasion resistance	Excellent	Poor	Poor	Excellent	Good

Chemical resistance:					
Sunlight aging	Poor	Very good	Excellent	Excellent‡	Excellent
Oxidation	Good	Very good	Excellent	Very good	Excellent
Heat aging	Good	Fair	Excellent	Excellent‡	Excellent
Solvents:					
Aliphatic hydrocarbons	Poor	Excellent	Fair	Excellent	Excellent
Aromatic hydrocarbons	Poor	Excellent	Poor	Excellent§	Fair to good
Oxygenated, alcohols	......	Very good	Excellent	Poor	Poor
Oil, gasoline	Poor	Excellent	Poor	Excellent	Good to excellent
Animal, vegetable oils	Poor to good	Excellent	Excellent	Excellent	Very good
Acids:					
Dilute	......	Good	Very good	Fair	Fair
Concentrated	......	Good	Good	Poor	Fair
Permeability to gases	Low	Very low	High	Very low	Low
Water-swell resistance	Excellent	Excellent	Excellent	Excellent	Poor to fair
Uses	Pneumatic tires, shoe heels and soles, gaskets and seals, belting, sponge stocks; often used in blends with other rubbers to impart better resilience, abrasion resistance, and low-temperature properties	Seals, gaskets, diaphragms, valve seat disks, flexible mountings, hose in contact with solvents, balloons, boats, life vests and rafts	High- and low-temperature electrical insulation, seals, gaskets, diaphragms, ductwork, O rings	Fork-lift truck wheels, airplane tail wheels, backup wheels for turbine-blade grinders, spinning cots for glass fiber, hydraulic accumulators, shoe heels	Oil hose, searchlight gaskets, white or pastel-colored goods and automotive gaskets and O rings (especially for resistance to extreme-pressure lubricants and oils containing sulfur)

* Brittle point < −150°F; may stiffen at higher temperatures.
† Reinforced with high-temperature nonorganic fillers.
‡ Discolors, but no change in properties.
§ For up to 80% aromatics.

TABLE 35 Properties of Elastomers[47] (Continued)

Property	Epichlorohydrin homopolymer and copolymer	Fluorosilicone	Ethylene propylene (EPDM)	Chlorosulfonated polyethylene (Hypalon)	Fluorocarbon elastomers	Propylene oxides	Styrene-isoprene-styrene, and styrene-butadiene-styrene block polymers
Physical properties:							
Specific gravity	1.32–1.49	1.4	0.86	1.11–1.26	1.4–1.95	1.02	0.94–1.15
Thermal conductivity, Btu/(h)(ft²)(°F/ft)	.....	0.13	.....	0.065	0.13	.....	0.087
Coefficient of thermal expansion, 10^{-5}/°F	.....	45	.....	27	8.8	.....	7.5
Flame resistance	Fair	Poor	Poor	Good	Excellent	Poor	Poor
Colorability	Good	Good	Excellent	Excellent	Good	Good	Good
Mechanical properties:							
Hardness (Shore A)	30–95	40–70	30–90	45–95	65–90	40–80	35–90
Tensile strength, 1,000 lb/in.²							
Pure gum	.....	1	<1	4	<2	>1	0.7–4.5
Reinforced	2–3	<2	0.8–3.2	1.5–2.5	1.5–3	>2	
Elongation, %							
Reinforced	320–350	200–400	200–600	250–500	100–450	500–670	350–1,350
Resilience	Poor to excellent	Good to fair	Good	Good	Good to excellent	Very good	Good
Compression-set resistance	Very good	.....	Good	Fair to good	Fair	Fair	
Hysteresis resistance	Good	Good	Good	Good	Good	Very good	Good
Flex-cracking resistance	Very good	Good	Good	Good	Good	Very good	Good
Slow rate	Very good	Good	Good	Good	Good	Very good	Good
Fast rate	Good	Good	Good	Good	Good	Excellent	
Tear strength	Good	Fair	Poor to fair	Fair to good	Poor to fair	Good	Fair to good
Abrasion resistance	Fair to good	Poor	Good	Excellent	Good	Good	Good
Electrical properties:							
Dielectric strength	Fair	Good	Excellent	Excellent	Good	.....	Good
Electrical insulation	Fair	Good	Very good	Good	Fair to good	.....	Good
Thermal properties:							
Service temp, °F:							
Min for continuous use	−15 to −80	−90	−60	−40	−10	−80	−60 to −80
Max for continuous use	300	400	<350	<325	<500	<250	150
Low-temp stiffening, °F	−15 to −80	< −100	−20 to −60	−30 to −50	20 to −30	.....	−60 to −80

Corrosion resistance:							
Weather	Excellent	Excellent	Excellent	Excellent	Excellent	Very good	Fair
Oxidation	Very good	Excellent	Excellent	Excellent	Outstanding	Very good	Good
Ozone	Good to excellent	Excellent	Excellent	Excellent	Excellent	Very good	Fair
Radiation	....	Good	Excellent	Fair to good	Fair to good	....	Poor
Water	Good	Excellent	Good to excellent	Good	Good	Excellent	Good
Acids	Good	Very good to excellent	Good to excellent	Excellent	Good to excellent	Good	Good
Alkalies	Good	Very good	Good to excellent	Excellent	Poor to good	Very good to excellent	Good
Aliphatic hydrocarbons	Excellent	Excellent	Poor	Fair	Excellent	Poor to fair	Poor
Aromatic hydrocarbons	Very good	Excellent	Fair	Poor to fair	Excellent	Poor to fair	Poor
Halogenated hydrocarbons	Good	....	Poor	Poor to fair	Good	Poor	Poor
Alcohol	Good	....	Good	Very good	Excellent	....	Good
Synthetic lubricants (diester)	Fair to good	Excellent	Poor to fair	Poor	Fair to good	Fair to good	Poor
Hydraulic fluids:							
Silicates	Very good	Excellent	Fair to good	Good	Good		
Phosphates	Poor to fair	Excellent	Good to excellent	Poor to fair	Poor		
Uses	Diaphragms, print rolls, belts, oil seals, molded mechanical goods; gaskets, hose for petroleum handling; low-temp parts	Parts requiring resistance to high-temp solvents or oils; seals, gaskets, O rings	Electrical insulating and jacketing; footwear, sponge, proofed fabrics, automotive weather stripping, hose, belts, auto, appliance parts; parts requiring outstanding ozone and heat resistance	Flex chemical and petroleum tube and hose, rolls, tank linings; high-temp belts; wire and cable; shoe soles and heels; flooring; building products; extruded and molded parts	O rings, brake seals, shaft seals, gaskets, hose and ducting, connectors, diaphragms, carburetor needle tips, lined valves, packings, roll coverings	Electrical insulation, molded mechanical goods	Thermoplastic grades: molded mechanical goods, packaging, sports equipment, disposable pharmaceutical items. Solution grades: adhesives, coatings, calking, sealants

TABLE 36   Types of Elastomers[48]

ASTM-SAE Type	ASTM-SAE Class	Popular name	Chemical composition	Standard symbol (ASTM)	Previous symbol
R	...	Natural rubber	Isoprene	NR	
	...	SBR	Styrene/butadiene	SBR	GR-S, Buna S
	...	Butyl	Isoprene/isobutylene	IIR	GR-I
	...	Polyisoprene	Isoprene	IR	
	...	Polybutadiene	Butadiene	BR	
	...	EPR	Ethylene/aliphatic alpha olefin	EPM	
	...	EPT	Ethylene/aliphatic alpha olefin/diene monomer		
S	SA	Polysulfide	Organic polysulfide	....	GR-P
	SB	Nitrile	Acrylonitrile/butadiene	NBR	Buna N
	SB	Polyurethane	Diisocyanate/polyester or polyether		
	SC	Neoprene	Chloroprene	CR	GR-M
	SC	Hypalon*	Chlorosulfonated polyethylene	CSM	
	SC	Propylene oxide	Propylene oxide and unsaturated epoxide		
T	TA	Silicone	Polysiloxane		
	TB	Polyacrylate	Ethyl acrylate/chlorine-containing monomer	ANM	
	TB	Fluoroelastomer	1. Vinylidene fluoride/ hexafluoropropylene 2. Chlorotrifluoroethylene/ vinylidene fluoride 3. Fluorosilicone	FPM	

* E. I. du Pont de Nemours & Co., Inc.

TABLE 37   Standards and Specifications for Elastomers[48]

Title	Number
ASTM:	
Adhesives, brake lining and friction	D 1205
Battery containers	D 639
Belting, flat	D 378
Cements	D 816
Coated fabrics	D 751, D 1764, D 2136
Fire hose	D 380
Gaskets, compressibility and recovery	F 36
Gaskets, materials, gen automotive and aeronautical	D 1170
Hose, auto, air and vacuum brake	D 622
Hose, auto, air conditioning	D 1680
Hose, brake, hydraulic	D 571
Matting, electrical	D 178
Mountings, auto	D 1207
O rings	D 1414
Packing, compressed sheet	D 1330, F 39
Rings, asbestos-cement pipe	D 1869
Sleeves, insulating	D 1051
Tape, electrical	D 69
Tape, insulating	D 119, D 1373
Wire and cable, insulated	D 470, D 1350, D 1352

TABLE 37   Standards and Specifications for Elastomers[48] (Continued)

Title	Number
MIL:	
Rubber compositions, general purpose.....................	-STD-417
Boots, dust and oil seal, watertight (Navy).................	-B-19257
Cable, power, electrical, ignition..........................	-C-3702
Cable, special purpose, electrical..........................	-C-13777A
Cable, special purpose, electrical..........................	-C-23206
Connectors, general purpose electrical, miniature...........	-C-26500A
Ducting, flexible, aircraft................................	-D-7756
Duct, pneumatic start, flexible (Navy)....................	-D-22706
Duct, flexible, jet starter (USAF)........................	-D-26124
Fabricated rubber parts, general purpose..................	-R-3065
Extreme low temp resistant (class I)......................	-R-5847B
Fuel resistant (class I)...................................	-R-6855
Gaskets, cylinder liner seal, O ring.......................	-C-21569
Gaskets, heat exchanger.................................	-C-21610
High temp resistant (class II)............................	-R-5847B
High temp silicone, Dacron..............................	-C-17/92
High temp silicone, glass................................	-C-17/96
High temp and fluid resistant............................	-R-25987
Hose, air duct, flexible aircraft..........................	-H-8796
Hose assembly, low pressure, oxygen.....................	-H-6017A
Hose assembly, oxygen breathing.........................	-H-7025C
Hose assembly, oxygen breathing.........................	-H-7138C
Hose assemblies, high pressure, air and oxygen breathing.....	-H-22486
Hose assemblies, low pressure, air and oxygen breathing......	-H-22489
Mask and tubing, oxygen, pressure breathing..............	-M-6482B
Mask, oxygen...	-M-27274B
Mounts, resilient, shear.................................	-M-22322
Nonoil resistant (class III)..............................	-R-6855
Oil resistant (class II)...................................	-R-6855
O ring, hydrocarbon fuel resistant.......................	-P-5315A
Packing, rings and gaskets, hydraulic oil resistant,	
low compression set....................................	-P-5516A
Tubing, rubber, inflight feeding (USAF)..................	-T-25458
Tubes, oxygen..	-T-26385
Tear resistant (class III).................................	-R-5847B
Wire and cable, electrical (type T tape, type E extruded).....	-C-2194D
Wire, electrical, silicone................................	-W-8777B
Wire, high temp, electrical..............................	-W-16878D
AMS:	
Adhesive, Buna N type...................................	3695A
Aromatic fuel resistant (35–45)..........................	3214G
Aromatic fuel resistant (55–65)..........................	3212J
Aromatic fuel resistant (75–85)..........................	3213G
Aromatic fuel resistant (65–75)..........................	3215G
Aromatic fuel resistant, nylon fabric reinforced..............	3274C
Cork and elastomer, general purpose, soft..................	3250C
Cork and elastomer, general purpose, medium..............	3251C
Cork and elastomer, general purpose, firm.................	3252C
Dry heat resistant (35–45)..............................	3201E
Dry heat resistant (55–65)..............................	3202E
Electrical resistant (neoprene type, 65–75).................	3210A
EP lubricant resistant (65–75)...........................	3206A
Flame resistant, neoprene type (55–65)....................	3243
Flame resistant, neoprene type (65–76)....................	3244
Hose, aircraft fueling, textile reinforced, collapsing...........	3386B
Hose, aircraft fueling, textile reinforced, noncollapsing........	3387B
Hose, aircraft fueling, single wire braid reinforced,	
noncollapsing..	3388B
Hose, aircraft fueling, double wire braid reinforced,	
noncollapsing..	3389B
Hot oil and coolant resistant, low swell (45–55)..............	3226C

TABLE 37   Standards and Specifications for Elastomers[48] (Continued)

Title	Number
AMS: (continued)	
Hot oil and coolant resistant, low swell (55–65)	3227C
Hot oil and coolant resistant (65–75)	3228C
Hot oil and fuel resistant (55–65)	3220B
Hot oil and high swell (45–55)	3222C
Hot oil resistant, low swell (75–85)	3229C
Hot oil resistant, sheet, asbestos filled	3232G
Low temp resistant (25–35)	3204D
Low temp resistant (45–55)	3205D
Hydraulic fluid resistant, petroleum base (55–65)	3200C
Rings and sealing, fuel and low temp resistant (60–70)	7271C
Rings and packing, fuel and low temp resistant (70–80)	7260A
Rings and packing, phosphate ester resis butyl type (85–95)	7263
Rings and sealing, fuel resistant (65–75)	7270E
Rings and sealing, oil resistant (65–75)	7274D
Rings and sealing, synthetic lubricating oil resistant, fluorocarbon type (65–85)	7275
Rings, sealing, synthetic lubricant resistant, Buna N type (65–75)	7272B
Rings, sealing, phosphate ester fluid resistant, butyl type (70–85)	7277
Rings, sealing, high temp fluid resistant, fluorocarbon type (70–80)	7278B
Rings, sealing, high temp fluid resistant fluorocarbon type (85–95)	7279B
Rings and sealing, silicone, heat resistant (70–80)	7267A
Sheet, nylon fabric reinforced, weather resistant, neoprene type	3272A
Silicone, extreme low temp resistant (15–30)	3332
Silicone, general purpose (35–45)	3301C
Silicone, general purpose (45–55)	3302C
Silicone, general purpose (55–65)	3303E
Silicone, low compression set (65–75)	3304C
Silicone, low compression set (75–85)	3305D
Silicone sheet, glass fabric reinforced	3315B
Silicone sheet, glass fabric reinforced	3320
Silicone, extreme low temp resistant (35–45)	3334
Silicone, extreme low temp resistant (55–65)	3336
Silicone, extreme low temp resistant (75–85)	3338B
Silicone, high strength, 1800 lb/in.^{2} (45–55)	3344
Silicone, high strength, 1000 lb/in.^{2} (45–55)	3345A
Silicone, high strength, low temp (55–65)	3346A
Silicone, high strength, low temp resistant (55–65)	3356
Silicone, fuel and oil resistant (50–65)	3326A
Silicone, fuel and oil resistant (55–65)	3325
Silicone, lubricating oil and compression set resistant (65–75)	3357B
Silicone, RTV, 15,000 cps	3362
Silicone, RTV, 50,000 cps	3363A
Silicone, RTV, 1,200,000 cps	3367
Silicone, sponge, closed cell, medium	3195
Silicone, sponge, closed cell, firm	3196
Sponge, neoprene type, soft	3197F
Sponge, neoprene type, medium	3198F
Sponge, neoprene type, firm	3199F
Weather resistant, (neoprene type 25–35)	3207E
Weather resistant (neoprene type, 55–65)	3241C
Weather resistant (neoprene type, 75–85)	3242B
Weather resistant (neoprene type), cotton fabric reinforced sheet	3270E
Weather resistant (chloroprene type, 45–55)	3208E

ASTM = American Society for Testing and Materials.   MIL = military specification.
AMS = Aerospace Materials Specifications of Society of Automotive Engineers.

Silicones (ASTM FSI, PSI, VSI, PVSI, SI)

The most common silicone is polydimethylsiloxane, but many other polymers also are used. In silicone elastomers part of the carbon-backbone chain is replaced by silicon and oxygen. Its outstanding property is its heat resistance. It can be used safely up to 500°F and down to −100°F. The physical properties of silicones are not high, but they change little with increasing temperature (see Fig. 66).

Urethane (NEMA U)

This family of polymers is the reaction product of diisocyanates and polyalkylene ether glycols. Available as casting compounds, liquids, molding compounds, or gums, they have outstanding load-bearing capacity, abrasion resistance, and tensile strength (three times that of natural rubber). They resist oils, solvents, and oxidation but do not resist hot water or alkaline solutions. Low temperature causes hardening.

Polyacrylate (ASTM ABR)

Acrylics have good oil, heat, and ozone resistance—better than the nitriles—but they stiffen at low temperatures. These elastomers can be used up to 500°F.

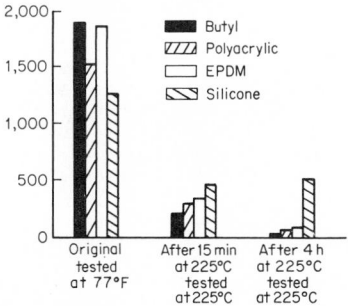

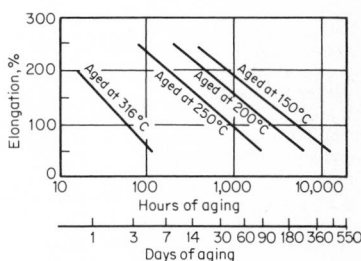

Fig. 65 Tensile strength of rubbers at various temperatures.[47]

Fig. 66 Heat aging of silicone rubber.[47]

Polybutadiene (ASTM BR)

Butadiene is similar to SBR and often is used in combination with SBR, particularly in tire treads, where it lends improved abrasion resistance.

Epichlorohydrin (ASTM CO and ECO)

These elastomers are made of epichlorohydrin as a copolymer of epichlorohydrin and ethylene oxide. They have excellent chemical resistance, particularly to oil, water, acids, bases, and ozone. The chlorine content makes them self-extinguishing.

Fluorosilicone

These polymers are silicones where some fluorine atoms have been introduced. The fluorine lends improved thermal stability and chemical resistance. They are particularly good at low temperatures, as shown in Table 38.

Ethylene Propylene (ASTM EPM, EPDM)

Based on ethylene and propylene copolymers, this family of elastomers has damping properties similar to those of natural rubber, as shown in Fig. 67. These materials—unlike butyl—do not entrap air during processing and are insensitive to contaminants. They are inexpensive, weather well, and resist ozone, steam, and hot water. Like butyl, they are not oil-resistant. They have the lowest specific gravity of all elastomers.

Chlorosulfonated Polyethylene (ASTM CSM)

This elastomer is similar to neoprene but superior in thermal stability, color stability, chemical resistance, and outdoor exposure. The polymer can be used to

TABLE 38   Low-Temperature Flexibility of Rubbers[49]

Rubber	100% modulus, km/m^{2} [aa]	Torsional modulus,[ff] km/m^2 [aa]				T_{70} [bb]	T_{100} [cc]	T_R [dd]
		0°C	−20°C	−40°C	−60°C			
Acrylate[a]	4.5	10	270	....	.....	−10	−11	−13
Acrylate modified[b]	3.4	10	17	100(−35)	.....	−30	−34	−37
Butadiene[c]	3.0	10	....	....	150	−57	−59	−59
Butyl[d]	2.3	10	....	17	560	−51	−52	−51
Chlorobutyl[e]	1.5	10	....	15	270	−53	−55	−51
Chloroprene[f]	7.3	10	....	180	.....	−37	−38	−62
Chlorosulfonated polyethylene[g]	6.8	12	300	500(−30)	.....	−14	−15	−19
EPDM[h]	2.4	10	....	15	610	−51	−52	−52
Epichlorohydrin homopolymer[i]	4.4	19	68	2000(−35)	.....	−20	−21	−22
Epichlorohydrin copolymer[j]	3.7	10	22	1900	.....	−25	−26	−27
Fluoroelastomers[k]	2.9	16	450	....	.....	−13	−14	−14
Fluorosilicone[l]	1.4	10	....	....	50	−62	−63	−66
Isoprene[m]	4.5	10	....	19	900	−51	−52	−54
Natural rubber[n]	3.5	10	....	12	600	−52	−53	−56
Nitrile[p]	4.1	10	14	700(−35)	.....	−25	−26	−27
Nitroso[q]	1.3	10	12	120	.....	−38	−40	−37
Polyester urethane[r]	0.1	10	15	700(−35)	.....	−25	−26	−27
Polyether urethane[s]	3.9	10	....	....	90	−59	−61	−63
Polyester[t]	14.8	63	77	126	211			
Polysulfide[u]	7.1	10	12	32	1500	−46	−47	−50
Propylene oxide copolymer[v]	4.1	...	....	....	.....	−18		
SBR[w]	7.0	10	....	100	.....	−38	−39	−61
Styrene butadiene thermoplastic[x]	0.4	10	....	....	12(−72)	−79	−80	−73
Silicone[y]	1.7	10	....	....	12	−63	−65	−63

[a] Elaprim AR 152X.
[b] Elaprim AR 153X.
[c] Intene 35 NS (55% trans).
[d] Butyl 301.
[e] Butyl HT 1066.
[f] Neoprene WRT.
[g] Hypalon 20.
[h] Nordel 1020.
[i] Herchlor H.
[j] Herchlor C.
[k] Viton D80.
[l] Silastic LS 2332.
[m] Cariflex IR305.
[n] Heveacrumb SMR5L.
[p] Krynac 803 (34% ACN).
[q] A copolymer of trifluoronitrosomethane and tetrafluoroethylene.
[r] Polypropylene adipate, toluene diisocyanate and hexene triol elastomer.
[s] Adiprene L140.
[t] Hytrel 55D.
[u] Thiokol ST.
[v] Parel 58.
[w] Intol 150.
[x] Cariflex TR 226.
[y] Silastomer ESP 2811U.
[aa] To convert km/m^2 to $lb/in.^2$ multiply value shown by 14.2230.
[bb] Torsional modulus is 69 km/m^2 (10,000 $lb/in.^2$).
[cc] Torsional modulus is 100 km/m^2 (14,500 $lb/in.^2$).
[dd] 100 times room temperature torsional modulus.
[ff] Torsional modulus G = Young's modulus, $E/2(1 +$ Poisson's ratio $\delta)$, where δ equals 0.5 for rubbers, and E is considered to be the 100% modulus; thus $G = 0.33E$.

$-80°F$, as shown in Fig. 68, accounting for its use in outdoor escalator handrails. It can be pigmented many colors without fading.

Fluorocarbons (ASTM FPM)

These elastomers are copolymers of vinylidene fluoride and hexafluoropropylene and of chlorotrifluoroethylene and vinylidene fluoride. They approach silicones in their thermal stability and are highly resistant to chemicals, including chlorinated solvents. Weathering properties are superior to those of all other elastomers, but they are the most expensive of all.

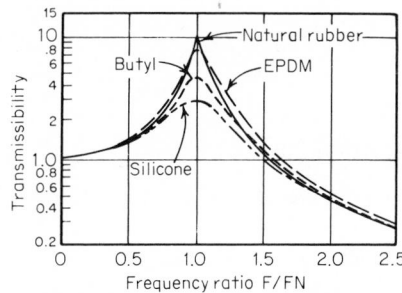

Fig. 67 Transmissibility of rubber mounts.[50]

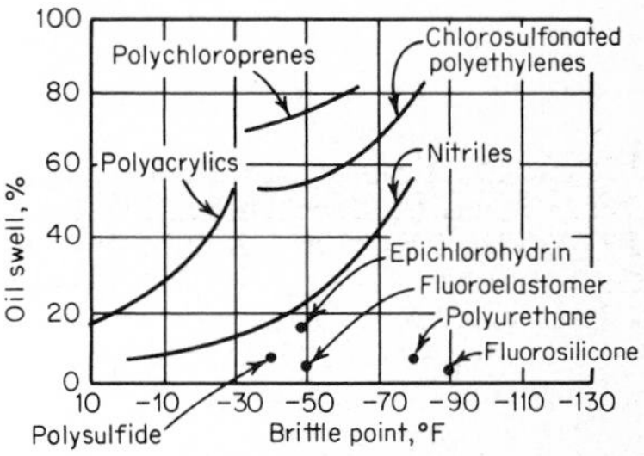

Fig. 68 Oil swell at low temperatures.[51]

Propylene Oxide

Polypropylene oxide is an elastomer which is extremely tough with low density. It weathers well but is marginal in chemical resistance.

Styrenated Block Polymers

Copolymers of styrene and isoprenes or styrene and butadiene are similar to thermoplastic resins and process similarly. They have high elongations and tensile properties and often are used as sealants.

REFERENCES

1. ASTM standards, Parts 35, 36, and 38, 1974.
2. Plastics: Methods of Testing, Federal Test Method Standard 406.
3. Industrial Laminated Thermosetting Products, NEMA Standard LI-1-1971.
4. Gallagher Corp., Glenview, Ill., advertising booklet, no date.
5. "Engineering Design with Natural Rubber," 3d ed., The Natural Rubber Producers Research Association, London, England, 1970.
6. Engineering Properties of Adiprene Urethane Rubber, E.I. du Pont de Nemours & Co., *Bull.* A88436, Wilmington, Del., 1974.
7. Riddell, M. N.: A Guide to Better Testing of Plastics, *Plast. Eng.*, April 1974.
8. Cardillo, R. M.: Dynamic Testing of Elastomer Mountings, in "High Speed Testing," vol. IV, Interscience, New York.
9. Preiss, D. M., and D. W. Skinner: Damping Characteristics of Elastomers, *Rubber Age*, August 1965.
10. Titus, J. B.: The Weatherability of Polystyrenes and Related Copolymers and Terpolymers, *Plast. Rept.* 38, Plastics Technical Evaluation Center, Picatinny Arsenal, Dover, N.J., July 1969.
11. For Flammability Tests, It's Back to Square One, *Modern Plastics*, September 1973.
12. *Mater. Eng.*, December 1973, p. 51.
13. *Mater. Eng.*, August 1973, pp. 27–36.
14. "*Modern Plastics* Encyclopedia," 1974.
15. Plastic Molding Materials Data Sheet, Dow Chemical.
16. Steinle, H.: Plastics in European Refrigerators and Dishwashers, *Appliance Eng.*, vol. 6, no. 1, 1972.

17. Celcon Acetal Copolymer Data Sheet, CIA, Celanese Plastics Co.
18. Data courtesy of Röhm & Haas Company, Manufacturers of Plexiglas.
19. Reference Manual for Polycarbonates, Mobay Chemical Co.
20. *Mater. Des. Eng.*, February 1964.
21. Nylon Bulletin, Celanese Plastics Co.
22. Arylon Data Sheet, Uniroyal, Inc.
23. Morneau, G.: A Thermoplastic That Can Be Used at 500°F, *Modern Plastics*, January 1970.
24. Lexan Polycarbonate, General Electric.
25. Valox Data Sheet, General Electric.
26. Celanex Thermoplastic Polyester *Bull*. JIA, Celanese Plastics Co.
27. Polyethersulphone Data Sheet, ICI, Ltd.
28. Fortiflex Polyethylene *Bull*. PIA, Celanese Plastics Co.
29. Raia, D. C.: Engineering Plastics—Time for a Closer Look, *Plastics World*, vol. 31, no. 9, 1973.
30. TPX Data Sheet, ICI America, Inc.
31. Noryl Thermoplastic Resins, General Electric.
32. Ryton Data Sheet, Phillips Petroleum Corp.
33. Polyolefins for Seating Applications, Plaskon HDPE, Allied Chemical Corp.
34. Polypropylene *Bull*., Avisun.
35. Ziegler, E. E., and W. E. Brown: Stress Cracking (Crazing) of Polystyrene, Dow Chemical Co., 1955.
36. Polysulfone *Bull*., Union Carbide Corp.
37. Chastain, C. F.: Plastics and Dimensional Stability, *Machine Des.*, December 9, 1971.
38. Chottiner, J.: Dimensional Stability of Thermosetting Plastics, *Machine Des.*, December 1968.
39. Sundstrom, D. W., and L. A. Walters: Effect of Molding Conditions on Strength of Thermosets, 27th ANTEC, SPE, 1969.
40. O'Toole, J. L., et al.: Thermosets vs. New Thermoplastics, *SPE Reg. Tech. Conf., The Resurgence of Thermosets*, December 1966.
41. Smack, K. G., and J. L. O'Toole: High Temperature Thermosetting Molding Compounds, *SPE Reg. Tech. Conf.*, 1967.
42. Ducca, F. W.: Phenolics—The Pioneer Insulating Plastics Make New Advances, *Insulation*, 1968.
43. Martino, C. F.: Phenolics—Too Good to Be Outmoded, *SPE Reg. Tech. Conf., The Resurgence of Thermosets*, December 1966.
44. Heat Resistance of Plastics, *Mater. Des. Eng.*, vol. 57, no. 4.
45. McNally, J., et al.: Electrical Composites; Laminates vs. Premix vs. SMC, 27th ANTEC, SPE, 1969.
46. Kookootsedes, G. J., and F. J. Lockhart: Silicone Molding Compounds for Semiconductor Devices, ACS, *Div. Org. Coatings and Plastics*, vol. 22, no. 1, 1967.
47. Rubber and Elastomers, Materials Selector, *Mater. Eng.*, 1973.
48. Specialty Elastomers, Materials and Processes Manual, no. 232, *Mater. Des. Eng.*, August 1965.
49. At Low Temperatures: Which Rubbers Are Flexible?, *Mater. Eng.*, December 1972.
50. Flanders, S.: Energy Absorbing Elastomers Soak Up Shock, Vibration, *Mater. Eng.*, June 1968.
51. King, W. H.: A Fresh Look at Elastomers, *Machine Des.*, January 1973.

Chapter **4**

Chemical and Environmental Properties of Plastics and Elastomers

OTTO H. FENNER

Monsanto Company,
Corporate Engineering,
St. Louis, Missouri

INTRODUCTION

The role of plastics and elastomers as materials of construction for all types of products has grown fantastically during the years since the end of World War II. Many of the plastics, resins, and elastomers now in common use were either relatively unknown to industry or still waiting to be discovered or developed when the

Pearl Harbor "Day of Infamy" shocked the peoples of the world. That day changed our way of living in many ways. The impact upon the availability of materials for the fabrication of untold items was frightening. Strategic materials, vitally needed for our very continued existence as a nation, were suddenly and completely cut off from import. The chemical industry rose to the situation, and its scientists brought forth from their laboratories substitute materials to solve the dilemma. The conclusion of hostilities allowed all industries to benefit from their efforts.

The importance of the chemical and environmental resistance of plastics and elastomers extends to all industries and segments of our society. Their permanence in the role of materials of construction affects all of us and our entire economy. From the synthetic turf of the football field to the moon buggy and ablative nose cone of the reentry missile, their environmental resistance is of paramount concern.

One of the fastest-growing applications of plastics and elastomers is in the fabrication of process equipment, auxiliaries, and structural members subjected to highly corrosive or aggressive chemical environments. This rapid development has resulted from the fact that design and project engineers have been faced increasingly with economic and competitive pressures. The choice of a more corrosion-resistant material of construction, which will provide good-quality products, safety, and attending low maintenance costs, frequently is the difference between success or failure in the marketplace.

Method of selection The selection of materials and the concern for their permanence or performance are generally the same throughout industry. This is true whether one is involved in the manufacture of a new steam iron, water-bed mattress, automobile bumper guard, skis, an electrical outlet plug, or an obnoxious-chemical-fumes scrubbing jet. Therefore, the known chemical- and environmental-resistance data supplied by the plastic and elastomer formulator or manufacturer can be applied readily in predicting the probable longevity of the particular product for the specific service. Where corrosion-resistance information is lacking in literature, one needs to resort to laboratory studies or specific field exposure.

Purpose of chapter It is the purpose of this chapter to perform two important functions:

1. To provide available information covering the chemical resistance and performance of the commonly employed plastics and elastomers so that reasonably accurate selections can be made to cover specific applications.

2. To furnish data on laboratory and field-testing procedures for plastics and elastomers which will enable one to obtain reliable test results that can be translated easily into predictable performance.

Plastic vs. metallic corrosion The corrosion or chemical resistance of plastics, resin systems, and elastomers is quite different from that of metals. It is pertinent to realize these main differences, as well as the things they have in common, when one is attempting to make a suitable selection. In the evaluation of the performance or acceptability of plastics or alloys subjected to specific environments there are certain general observations or measurements performed on the exposed test coupons of the material. A brief résumé of the difference between evaluations of plastics and metals or alloys is shown in Table 1. These criteria of performance and resulting acceptability are generally implemented with special tests to determine the suitability of the material for the proposed services.

The determination of the chemical resistance of plastic and elastomeric materials under specific environmental conditions is a most intriguing experience. There is both a similarity and a difference in the chemical-resistance performance of the plastics when compared with that of the metals and alloys.

Each one may experience changes in weight, color, hardness, and physical properties. However, the deterioration of metals and alloys is electrochemical in nature, and with the possible exception of high-temperature oxidation, there is always a flow of electric current. Plastics and elastomers do not experience this phenomenon during their degradation. Generally speaking, inorganic salt solutions, weak aqueous alkaline solutions, and weak aqueous acidic solutions do not have an adverse effect upon plastics, resins, or elastomers. Depending upon the metal ion in the aqueous solution, metals may be completely resistant or can experience catastrophic failure.

Most plastics and elastomers experience some degree of degradation by organic solvents. Metals and alloys are unaffected by organic solvents in most instances, one notable exception being hot chlorinated solvents under moist conditions.

Types of attack on plastics   Plastics and elastomers experience chemical attack in the following manner:

1. Swelling
2. Dissolution
3. Breaking of chemical bonds by
   a. Hydrolysis
   b. Oxidation
   c. Pyrolysis
   d. Radiation
4. Combinations of 1, 2, and 3 [1]

Variations within systems   Just as there are often vast differences in the chemical resistance of various metals and alloys to a particular environment, so too are there great differences among plastics and elastomers. There is, of course, wide variation between groups of plastics in their chemical resistivity. First there is a significant

TABLE 1   Plastics vs. Metals (Standard Evaluations)

Observation or measurement	Alloys and metals	Thermoplastic resins	Thermosetting resins
Weight change:			
As mils per year penetration (loss in thickness)	Yes	No	No
% wt. change	Rarely used	Yes	Yes
Permeation	Special cases (hydrogen blistering)	Yes	Yes
Change in hardness	Relatively few instances (graphitization, dezincification, etc.)	Yes	Yes
Dimensional changes	Yes (losses)	Yes (losses or gains)	Yes
Tensile properties	Special case. Usually not done	Yes	Generally no
Compressive properties	Special case. Usually not done	Generally no	Yes
Flexural properties	Special case. Usually not done	Generally no	Yes
Elongation	No	Yes	No
Appearance change of test sample:	Yes	Yes	Yes
Surface attack (pitting, craters, crevices, blistering, etc.)	Yes	Yes	Yes
Laminar-wall attack	Yes	Yes	Yes
Color		Yes	Yes
Deleterious effects to environment	Yes	Yes	Yes

difference in performance between thermoplastic and thermosetting materials. Second, there are extreme differences in the results of such resistances within and between the families of thermoplastics. There is just as great a chemical-resistance variation between and within the thermosetting resins. Considering this comparison still further, we find that changes in formulations, degree of or variation in the polymerization, and even the amount of specific additives can enhance or decrease the chemical resistance. [2]

Recently, plastic alloying, wherein similarly structured or vastly different polymers and resins are blended prior to use, has added to the complexity of predicting performance without some fairly extensive evaluations. It is indeed a major problem for the materials engineer charged with the responsibility of selecting a plastic material for construction from the myriad of those commercially available, identified only by trade name or by a vague generic classification.

Literature nomenclature   The current list of available plastics[3,4] reads much like an index to a textbook on organic chemistry.   It is doubly confusing to the embryonic corrosion engineer when he is faced with the herculean task of selecting or correlating compounds or materials, identified only by a trade name or product number, for a particular environmental use.   The same basic plastic may have dozens of different trade names, with each minor modification having a special trade name of its own.   Fortunately, pressure from many of the technical committees of the professional engineering societies has resulted in a vast improvement along these lines lately, and most of the companies are designating their products by specific plastic name.

CHEMICAL RESISTANCE AND PERFORMANCE

Major divisions of plastics   It is common practice to divide plastics into two main groups: thermoplastic and thermosetting.   The engineer commonly knows them as thermoplasts (also widely called thermoplastics) and thermosets.   The performance data to follow will be divided into three classifications for ease of reference.   These are thermoplastics, thermosets, and elastomers.

THERMOPLASTICS

Fundamentals

Definition of thermoplastics   The thermoplastics (thermoplasts) have been defined as resins that soften with heat and reharden to rigid materials upon cooling.   Generally there is no appreciable change in the physical properties of the thermoplastics through repeated heating and cooling cycles.   They may be remelted and reshaped as often as is practical, and as long as the procedure does not introduce charring or degradation to the reworked material.

Polymerization process   Polymerization is the secret of the production of thermoplastics.   The polymer chemist begins with single and like molecules called *monomers.*   Under heat, pressure, mechanical, or chemical persuasion, these monomers link permanently together to form long-chain molecules.   The chemist's term for these long chains is *polymer.*   A polymerized, or linked-up, plastic may contain millions of these long-chain molecules.   In many instances the research chemist will copolymerize two or more different monomers in an identical manner.   These unlike organic unsaturated molecules can be made to join in a regular, cross-linked, or even haphazard sequence to form long chains containing specific kinds of polymers with repeating monomeric links.   The copolymer produced is usually drastically different in physical properties and chemical resistance from the polymers formed from the various component monomers alone.

Physical effects caused by variation in monomer structure   The change in structure of the monomer gives rise to definite changes in the physical properties and chemical resistance of the resulting polymer.   The substitution of a portion of the available hydrogen in the ethylene molecule with other atoms or groups tends to (1) increase the density, (2) elevate the compressive and tensile strengths in most materials, and (3) increase the heat-distortion temperature.   Figure 1 diagrams the basic differences in structure between monomers of various thermoplastics and that of the ethylene monomer.   The variations in physical properties of these polymers are shown in Table 2 along with those of several other commonly employed thermoplastics.[5]

Thermoplastics considered   We will be concerning ourselves in this portion of the chapter with the chemical resistance and performance of only those plastics that are of commercial importance and currently employed as engineering plastics.   The following thermoplastics are covered:
   Acrylonitrile-butadiene-styrene (ABS)
   Acetals
   Acrylics
   Cellulosics
      Cellulose acetate butyrate
   Chlorinated polyether

Fluorocarbon plastics
  Polytetrafluoroethylene (PTFE)
  Fluorinated ethylene propylene (FEP)
  Perfluoroalkoxy (PFA)
  Polychlorotrifluoroethylene (CTFE)
  Polyvinylidene fluoride (PVF$_2$)
Nylons (polyamide resins)
Phenylene oxide resins
Polycarbonates
Polyethylene
Polyimides
Polyphenylene sulfide
Polypropylene
Polystyrene
Polysulfone
Polyvinyl chloride
  Rigid
  Flexible (plasticized)
  Chlorinated
Polyvinylidene chloride (Saran)

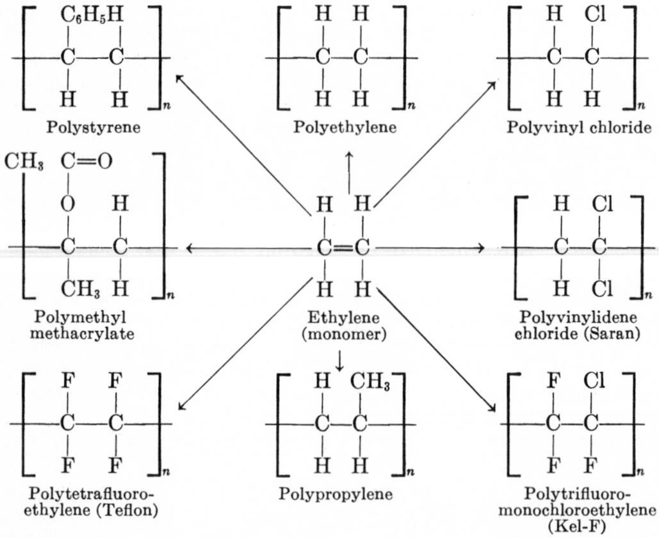

Fig. 1 Structural relationship of common thermoplastic compounds to that of the ethylene monomer.

All the thermoplastics can be produced in many forms by the conventional injection-molding, extrusion, and thermoforming techniques. Slight variations in composition for the particular operation will result in products with somewhat different properties and chemical resistances. The primary purpose of the data recorded under each specific plastic or resin system is to give the reader some idea of the main differences between these materials, in addition to a knowledge of their general chemical resistances.

Guidelines for use of corrosion-resistance tables   The accompanying chemical-resistance tables must be used with quite a degree of understanding, and with considerable reservations and precautions. These are:

1. The evaluations are *based upon* the performance of the plastic as the *pure* chemical.

2. It is assumed that the *plastic has not been degraded* through adulteration with special additives or by alloying with other plastics or resins. *It is the pure plastic* with only inert additives, if any.

3. The evaluations of the performance of the plastic as shown in the table *are not cumulative.* The fact that the plastic is identified as acceptable in several specific environments does not guarantee that it will be satisfactory in any combination of these conditions. In a similar manner, it is possible that a combination of a number of unacceptable environments may produce a compatible environment, totally acceptable for use of the plastic.

It is for the above reasons that *consideration must be given to testing procedures* if there is any question about the actual chemical composition of the proposed environment.

TABLE 2   Properties of Thermoplastic Resins[5]

Thermoplast	Specific gravity	Compressive strength, lb/in.^{2}	Tensile strength, lb/in.^{2}	Heat-distortion temp, °F
Polyethylene:				
LD	0.90–0.92	2,700–3,600	600–2,300	100–121
MD	0.926–0.940	. . . .	1,200–5,500	120–165
HD	0.941–0.965	. . . .	. . . .	144–190
Polypropylene	0.902–0.910	5,500–8,000	4,300–5,500	200–250
Polystyrene	1.04–1.10	11,500–16,000	6,500–12,000	195–240
Polyvinyl chloride (rigid)	1.30–1.45	6,000–13,000	5,000–9,000	135–180
Chlorinated polyvinyl chloride	1.20–1.40	9,000–12,000	10,000–12,000	215–247
Polyvinylidene chloride (Saran)	1.65–1.72	2,000–2,700	3,000–5,000	155
Polychlorotrifluoroethylene (Kel-F)	2.1	4,600–7,400	4,500–6,000	253
Polytetrafluoroethylene (TFE)	2.15–2.22	1,700	2,000–5,000	250
Fluorinated ethylene propylene (FEP)	2.12–2.17	2,200	2,700–3,100	158
Polymethyl methacrylate	1.17–1.28	11,000–19,000	11,000–17,000	165–235
Chlorinated polyether (Penton)	1.4	. . . .	6,000	285
Polysulfone	1.24	13,900	10,200	358
Polyphenylene sulfide	1.34	. . . .	10,800	278

ABS Resins

Preparation and composition   The ABS family of plastics is based upon the co-polymerization of acrylonitrile, butadiene, and styrene. Since each of these monomers can be readily polymerized alone to form homopolymers, as well as be copolymerized with each other, quite a variety of polymeric materials can be produced through variations in the initial ratio of the monomers. Each has its own specific chemical-resistance properties.

The ABS plastics made from the three monomers are generally manufactured by "graft" polymerization techniques. A rubber is manufactured by the polymerization of butadiene; the styrene and the acrylonitrile monomers are then added and polymerized (grafted) directly onto the rubber. The ABS polymers made by this process contain three major components: (1) a styrene acrylonitrile copolymer, (2) a styrene acrylonitrile copolymer grafted onto a polybutadiene rubber, and (3) polybutadiene rubber. There are two phases: (1) a continuous phase of the styrene acrylonitrile copolymer, and (2) a dispersed phase of polybutadiene rubber in the acrylonitrile-butadiene-styrene. There is an additional boundary layer consisting of the styrene acrylonitrile grafted onto the polybutadiene rubber, producing the required adhesion between the continuous and dispersed phases.[6]

The chemical resistance of the ABS resin depends upon the styrene-acrylonitrile continuous phase, which is the surface presented to the environment. Its performance is affected directly by the content of acrylonitrile; increased levels of acrylonitrile enhance the chemical resistance.

There are two classifications of rigid ABS. They are type I, which is the normal operating-temperature material, and type II, which has higher operating-temperature properties. The ABS resins have been divided into a number of different grades depending upon their end uses. The more important fall into the (1) basic grades consisting of general-purpose, high-impact, medium-impact, high-heat-resistant, and self-extinguishing varieties, and (2) special-purpose grades, which include clear, cold-forming, low-gloss, expandable, and plating varieties.

From the many possible variations in the composition of the ABS resins it is natural to expect chemical-resistance differences between them. In general, however, the chemical-resistance data contained in Table 3 will apply.

Chemical resistance   The ABS resins are resistant to dilute inorganic acids at ambient temperatures. They are attacked by strong acids, such as 65 percent nitric acid, perchloric acid, concentrated sulfuric acid above 50 percent, and oleum. They are destroyed by most concentrations of acetic, chloracetic, and butyric acids. They show good to excellent resistance to most alkalies up to 150°F and higher. Ammonium hydroxide does attack significantly at 150°F. Acid salts, alkaline salts, and neutral salt solutions are completely acceptable up to the maximum working temperature of the resin system. ABS is unaffected by most dry and wet gases, one lone exception being wet sulfur dioxide, which has a very significant effect. The ABS resins have very poor chemical resistance to most organic solvents. They are readily and severely attacked by aromatics (benzene, toluene, xylene), ethers, esters, chlorinated aromatics, and chlorinated aliphatics, amines, ketones, and hot alcohols.

Physical properties   The physical properties of ABS resins are listed in Table 3.

Weatherability   The ABS plastics are reported[3] to show slight yellow discoloration to none under sunlight. Continued exposure to strong ultraviolet radiation causes embrittlement.

Flammability   ABS plastics have a flammability range of 1.0 to 2.0 in./min as determined by ASTM D 635.[3]

Chief corrosive-environment services   The ABS resins find service in fume hoods, ducts, pipes, garden spray equipment, appliance housings, bases of chromeplated products, and accessory materials. ABS pipe is used in transmission and distribution of natural gas, potable water applications, irrigation and sprinkling systems, and chemical-processing equipment.

Acetal Resins

Preparation and composition   The *acetal homopolymers* are produced by the polymerization of formaldehyde. They are highly crystalline resins containing long chains of unbranched polyoxymethylene. The *acetal copolymers* are based upon trioxane. These acetal resins have excellent physical and mechanical properties. They are tough, strong materials with excellent retention of dimensional properties at elevated temperature. These characteristics permit them to bridge the gap between metals and plastics in many mechanical and structural applications.

Corrosion resistance   There is some difference in chemical resistivity of the acetal homopolymer and acetal copolymer in regard to performance toward strong alkalies. The copolymer exhibits excellent resistance to strong alkaline media, whereas the homopolymer is attacked. The acetals have excellent resistance to most organic solvents, including aliphatic and aromatic hydrocarbons, alcohols, ketones, esters, glycols, and chlorinated solvents. They are unaffected by vegetable and mineral oils. They are completely resistant to all inorganic salt solutions within the pH range of 4 to 14.

The acetal copolymer is not resistant to weak acids above 150°F and cannot be employed to handle strong acids even at ambient temperature (see Table 3). It is extremely resistant to distilled water and is completely unaffected by hot (180°F) water over long periods of exposure. Acetal pipe placed underground is unaffected by fungi, insects, or rodents. The physical properties of acetal resins are given in Table 4.

Weatherability Unpigmented, natural grades of acetal copolymer are seriously degraded by continuous exposure to ultraviolet rays. The tensile and impact properties are seriously impaired on exposure to intense ultraviolet radiation. Photostabilized compositions retain physical properties, although surface chalking occurs. Ultraviolet-stabilized and/or black-pigmented acetals exhibit very little loss in physical properties.

Flammability The acetal copolymer and homopolymer burn slowly at a rate of 1.1 in./min (ASTM D 635–56T). This rate is comparable with that of polyethylene, polystyrene, and the acrylics.[5]

Chief corrosive-environment services The exceptional resistance to long hot-water exposure makes acetal copolymers a primary plastic for plumbing fixtures. They retain their original tensile strength through 22 weeks of continuous exposure in boiling water.[3] They are used in high-purity water filters, pipes, fittings, sinks, water closets, gears, cams, bushings, housings, automotive parts, and aerosol containers.

Special considerations Ability to withstand a steam-cleaning environment and excellent food-contact chemical resistance have resulted in acceptance by FDA for such diverse uses as a coffee spigot, a milk pump, a meat hook, and a vial for antibiotics.

Acrylic Plastics

Preparation and composition The acrylic plastics and resin include not only derivatives of acrylic esters but also the polymerizable products of both acrylic and methacrylic acids, chlorides, nitriles, and amides. The acrylics are modified in some instances with elastomers and other plastics and resins to produce alloys or multiphase systems with specific properties. The chief modifications have been through the use of rubber or polyvinyl chloride. The major acrylics are esters obtained through the reaction of various alcohols with acrylic acid. The principal unmodified acrylic plastic is poly (methyl methacrylate), the hardest ester in the methacrylate family of polymers.

The methacrylate resins are of significant brilliant clarity and remarkable internal light transmission. They are of low density, are quite tough, and exhibit very good rigidity. Because of these properties they are used as structural acrylate resin sheet and as transparent panels such as airplane windows, protective-goggle lenses, and advertising signs. Table 4 shows the typical physical properties of polymethyl methacrylate, which is an excellent example of the acrylic resins.[3]

Chemical resistance The chemical resistance of poly (methyl methacrylate) is a good example of the effect of various environments upon the acrylics in general. It is essentially unaffected by weak acids, weak alkalies, inorganic salt solutions, oils, and water. It is unaffected by weak solutions of oxidizing acids, but deteriorates rapidly in highly concentrated solutions of oxidizing acids (see Table 3). Aliphatic hydrocarbons, amines, and esters containing more than 10 carbon atoms in the molecule are completely acceptable. Aromatic hydrocarbons, phenols, chlorinated aromatics, ethers, ketones, and low-molecular-weight esters and aliphatic acids have adverse solvent effects upon the acrylics.

Weatherability The acrylic resins exhibit the most outstanding weatherability of any of the commercially available thermoplastics. In outdoor exposure, clear acrylics have experienced losses of about 1 percent of their light transmission in 5 years. Accelerated exposure to radiation and heat from mercury-vapor lamps, and ultraviolet radiation from fluorescent lighting appear to have only mild effects upon the acrylics.[3]

Flammability The acrylic resins burn slowly at a rate of 0.5 to 1.8 in./min (ASTM D 635).[5]

Chief corrosive-environment services The acrylics find extensive application in outdoor signs employing internally lighted features, as well as innumerable architectural and secondary structural-support members. Other uses include dome skylights, windshields on motor vehicles and boats, windows on aircraft, and automotive taillight and stoplight lenses.

Table 3 Chemical Resistance of Thermoplastics

Substance	ABS Type I 75°F	ABS Type I 150°F	Acetal resins[d] 75°F	Acetal resins[d] 150°F	Polymethyl methacrylate 75°F	Polymethyl methacrylate 150°F	Cellulose acetate butyrate 75°F	Cellulose acetate butyrate 150°F
Acids:								
Acetic, 10–50%.	G-X	F-X	F-X	X-X	G-F	F-X	G-X	G-X
Acetic, 75%.	X	X	X	X	X	X	X	X
Acetic, glacial.	X	X	X	X	X	E	X	G
Benzoic.	E	E	F	X	E	E	G	G
Boric.	E	E	X	X	E	E	G	G
Butyric.	X	X	X	X	X	X	X	F
Chloroacetic.	X	X	X	X	X	X	X	X
Chromic.	F-X	F-X	F	X	E-F	G-X	X	X
Citric.	E	E	F	X	E	E	G	G
Fluoboric.	G	G	F	X	E	E	G	G
Hydrobromic, 25%.	G	G	X	X	G	F	X	X
Hydrochloric, 10%.	G	F	X	X	G	F	X	F
Hydrochloric, 37%.	G	F	X	X	G	F	F	F
Hydrocyanic.	G	G	X	X	E	E	X	X
Hydrofluoric.	F	X	X	X	G	F	F	X
Hypochlorous.	G	X	X	X	E	F	F	X
Lactic.	E	E	F	X	E	E	G	G
Maleic.	G	G	F	X	G	F	G	G
Nitric, 5%.	E	X	X	X	G	X	F	X
Nitric, 65%.	X	X	X	X	X	X	X	X
Oleic.	E	G	F	X	F	X	E	E
Oxalic.	G	X	X	X	X	X	X	X
Perchloric.	X	X	X	X	E	E	E	E
Phosphoric, 25%.	G	F	X	X	E	G	G	G
Phosphoric, 85%.	G	F	X	X	E	F	X	X
Phthalic.	G	G	X	X	G	F	X	X
Sulfuric, 10%.	E	E	F	X	E	G	E	E
Sulfuric, 78%.	X	X	X	X	X	X	X	X
Sulfuric, 93%.	G	G	X	X	E	E	X	X
Tannic.	G	G	F	X	E	E	E	G
Alkalies:								
Ammonium hydroxide.	F	X	E	E	G	F	X	G
Calcium hydroxide.	E	E	E	E	F	X	X	X
Potassium hydroxide.	E	G	E	E	F	X	X	G
Sodium hydroxide.	E	E	E	E	F	X	X	G

	1	2	3	4	5	6	7	8
Acid salts[a]	E	E	G	E	F-X	E	G	F
Alkaline salts[b]	E	E	E	F	G	F	G	F
Neutral salts[c]	E	E	E	E	E	E	E	G
Miscellaneous salts:								
Potassium dichromate	G	F	G	G	F-X	G	G	F
Potassium permanganate	F	F	E	G	G	G	G	F
Sodium cyanide	E	E	E	G	F	G	G	G
Sodium ferricyanide	E	E	G	E	G	G	G	G
Sodium hypochlorite	G	G	G	F	X	F	F	F
Organics and solvents:								
Acetone	X	X	E	E		X	X	X, dissolved
Alcohol, methyl	G	F	E	E		X	X	X, swollen
Alcohol, ethyl	G	F	E	E	G	X	X	X, swollen
Alcohol, butyl	F	X	G	G	G	X	X	X, swollen
Aniline	X	X	X	G	F	X	X	X
Benzene	X	X	X	E	G	X	X	X
Carbon disulfide	X	X	X	E	G	X	X	X
Carbon tetrachloride	X	X	X	E	G	X	X	X, severely swollen
Chlorobenzene	X	X	E	E	G	X	X	X
Chloroform	X	X	G	E		X	X	X
Ethyl acetate	X	X	E	E	G	X	X	X, dissolved
Ethylene chloride	X	X	G	G	F	X	X	X
Ethylene dichloride	X	X	G	G	F	X	X	X, dissolved
Ethyl ether	G	X	E	E		X	X	F
Formaldehyde, 37%	G	F	G	E	G	G	G	G
Gasoline	F	F	G	G	F	F	G	G
Heptane	X	X	G	G	G	X	X	G
Methylene chloride	X	X	G	G	F	X	X	X
Methyl ethyl ketone	X	X	E	E	G	X	X	X
Nitrobenzene	X	X	G	G	G	X	X	X
Phenol	X	X	F	F	X	X	X	X, severely swollen
Toluene	X	X	E	E	G	X	X	X
Trichloroethylene	X	X	G	G	F	X	X	X
Gases:								
Carbon dioxide	E	E	E	G	G	G	E	G
Carbon monoxide	E	E	G	G	G	G	E	G
Chlorine, dry	G	G	G	X	G	F	X	X
Chlorine, wet	G	G	X	X	X	X	X	X
Hydrogen sulfide	E	E	E	G	E	F	E	G
Sulfur dioxide, dry	G	G	G	G	G	E	E	F
Sulfur dioxide, wet	X	X	X	X	X	E	E	X

Table 3 Chemical Resistance of Thermoplastics (continued)

Substance	Chlorinated polyether			Polyvinylidene fluoride (Kynar)			Polyamide (nylon)		Polycarbonate		Polyethylene	
	75°F	150°F	250°F	75°F	150°F	Over 150°F	75°F	150°F	75°F	150°F	75°F	150°F
Acids:												
Acetic, 10–50%	E	E	E	E	E	E to 230°F	F-X	F-X	G-F	G-X	E-F	E-F
Acetic, 75%	E	E	E	E	E	E to 212°F	X	X	X	X	F	X
Acetic, glacial	E	E	E	E	G	F to 212°F	X	X	X	X	F	X
Benzoic	E	E	E	E	E	E to 250°F	X	X	X	X	E	E
Boric	E	E	E	E	E	E to 275°F	G	G	G	G	E	E
Butyric	E	E	E	E	E	E to 230°F	G	G	X	X	F	X
Chloroacetic	E	E	E	E	E	E to 212°F	X	X	G	F	X	X
Chromic	E	E	Stress cracking possible	E	G	....	X	X	G	G	F	X
Citric	E	E	E	E	E	E to 250°F	G	G	G	G	G	G
Fluoboric	E	E	....	E	E	E to 275°F	F-X	F-X			G	G
Hydrobromic, 25%	E	E	E	E	E	E to 275°F	X	X	G	G	G	F
Hydrochloric, 10%	E	E	E	E	E	E to 275°F	X	X	G	G	E	E
Hydrochloric, 37%	E	E	E	E	E	E to 275°F	X	X	X	X, crazing	G	G
Hydrocyanic	E	E	E	E	E	E to 275°F	X	X			G	G
Hydrofluoric	E	G	....	E	E	E to 250°F		X	G	X	E	G
Hypochlorous	E	E	E	E	F	E to 250°F	F		G	F	F	X
Lactic	E	E	E	E	E	E to 250°F	X	X	G	F	E	E
Maleic	E	E	E	E	G	E to 250°F	X	X	E	G	E	G
Nitric, 5%	E	E	....	E	G		X	X	E	G	E	E
Nitric, 65%	G	G	E	E	G		X	X	X	F	X	X
Oleic	F	F	....	E	E	E to 250°F	G	G	E	E	F	F
Oleum	X	X	E	X	....		X	G	X	X	X	X
Oxalic	E	E	....	E	G	F to 212°F	G-F	X	E	E	E	G
Perchloric	G	F	E	E	G		X	F			E	E
Phosphoric, 25%	E	E	E	E	E	E to 275°F	X	X	G	G	E	E
Phosphoric, 85%	E	E	E	E	E	E to 230°F	X	X	G	G	E	F
Phthalic	E	E	E	E	E	E to 212°F	G	F	G	F	E	G
Sulfuric, 10%	E	E	E	E	E	E to 230°F	X	X	E	E	E	E
Sulfuric, 78%	E	E	F	E	E	E to 212°F	X	X	G	G	F	F
Sulfuric, 93%	E	F	E	E	E	G to 212°F	X	X	X	X	F	X
Tannic	E	E	E	E	E	E to 230°F			G	G	E	E
Tartaric	E	E	E	E	E	E to 250°F	G	G	G	G	E	E
Alkalies:												
Ammonium hydroxide	E	E	E	E	E	E to 275°F	G	G	X	X	E	E
Calcium hydroxide	E	E	E	E	E	E to 275°F	G	G	X	X	E	E

4-12

Chemical resistance table (column headings appear on a preceding page). Ratings: E = excellent, G = good, F = fair, X = not recommended; G-P, F-G, F-X, G-X indicate intermediate ratings; temperature column gives service limit.

Chemical	1	2	3	4	5	6	Temp. limit	8	9	10	11	12
Potassium hydroxide	E	E	X	X	G	G	E to 212°F	E	E	E	E	E
Sodium hydroxide	E	E	X	X	G	G	E to 212°F	E	E	E	E	E
Acid salts[a]	E	E	F	E	G-P	G-P	E to 275°F	E	E	E	E	E
Alkaline salts[b]	E	E	F	F	G	G	E to 275°F	E	E	E	E	E
Neutral salts[c]	E	E	E	E	G	G	E to 275°F	E	E	E	E	E
Miscellaneous salts:												
Potassium dichromate	G	G	G	G	G	G	E to 275°F	E	E	E	E	E
Potassium permanganate	G	G	G	G	X	X	E to 275°F	E	E	E	E	E
Sodium cyanide	G	G	G	G	G	G	E to 275°F	E	E	E	E	E
Sodium ferricyanide	G	G	G	G	G	G	E to 275°F	E	E	E	E	E
Sodium hypochlorite	G	G	G	G	X	X	E to 275°F	E	E	E	E	E
Organics and solvents:												
Acetone	X	X	X	X	G	E		X	F	G	G	E
Alcohol, methyl	E	E	G	G	F-G	F-G	E to 275°F	E	E	G	E	E
Alcohol, ethyl	F	E	G	G	G	G	E to 275°F	E	E	G	E	E
Alcohol, butyl	F	F	X	X	F-X	F-X	E to 275°F	E	E	G	G	E
Aniline	X	X	X	X	G	G	F to 212°F	G	E	G	G	E
Benzene	X	F	X	X	G	E		F	E	F	F	G
Carbon disulfide	X	F	X	X	E	E		F	E	F	F	E
Carbon tetrachloride	X	X	X	X	G	F-X	E to 275°F	E	E	E	E	E
Chlorobenzene	X	X	X	X	X	E	F to 212°F	E	E	G	G	E
Chloroform	X	G	X	X	G	G	E to 212°F	E	E	G	E	E
Ethyl acetate	F	X	X	X	G	G		F	E	F	G	G
Ethylene chloride	X	E	X	X	G	G	E to 275°F	E	E	F	F	E
Ethylene dichloride	X	F	X	X	G	G		E	E	G	F	E
Ethyl ether	X	G	G	G	G	E		F	E	E	E	E
Formaldehyde, 37%	E	F	F	G	G	G	E to 275°F	G	G	G	G	E
Gasoline	E	X	F	G	E	E	E to 275°F	E	E	G	E	E
Heptane	F	F	X	X	G	G		E	E	G	G	F
Methylene chloride	X	G	X	X	G	E		F	E	X	X	F
Methyl ethyl ketone	X	X	X	X	F-G	G	E to 275°F	F	G	G	G	G
Nitrobenzene	G	X	X	X	X	F-G	E to 275°F	F	F	F	F	E
Phenol	G	X	X	X	G	X	F to 212°F	E	E	F	G	F
Toluene	F	E	X	X	G	G	E to 212°F	E	E	F	F	F
Trichloroethylene	G	F	X	X	F-G	F-G	E to 275°F	E	E	F	G	G
Gases:												
Carbon dioxide	E	E	G	G	E	E	E to 275°F	E	E	E	E	E
Carbon monoxide	E	E	G	G	E	E	E to 275°F	E	E	E	E	E
Chlorine, dry	X	X	F	F	X	X	E to 212°F	X	X	X	X	X
Chlorine, wet	X	X	F	F	X	X	E to 212°F	X	X	X	X	X
Hydrogen sulfide	G	X	G	G	G	F-X	E to 275°F	F	G	X	F	G
Sulfur dioxide, dry	G	G	G	G	X	X	E to 212°F	F	G	X	F	G
Sulfur dioxide, wet	G	G	G	G	G	G	E to 212°F	G	G	F	G	G

4-13

Table 3 Chemical Resistance of Thermoplastics (continued)

Substance	Polyimides		Polyphenylene oxide		Polyphenyl sulfide		Polypropylene		Polystyrene[e]	
	75°F	150°F	75°F	150°F	75°F	150°F	75°F	150°F	75°F	150°F
Acids:										
Acetic, 10–50%	G	G	E	E	E	E	E	E	G-F	F-X
Acetic, 75%	F	F	G	G	E	E	E	E	F	X
Acetic, glacial	F	F	G	G	G	G	F	F	X	X
Benzoic	G	G	E	E	G	G	G	G	G	G
Boric	G	G	E	E	G	G	E	E	G	G
Butyric	G	G	G	F	G	G	E	E	F	F
Chloroacetic	F	F	G	F	G	G	X	X	F	X
Chromic	X	X	G	G	X	X	G	G	G	F
Citric	G	G	G	G	G	G	E	E	G-F	G-F
Fluoboric	G	G	X	X	G	G	G	G	F	F
Hydrobromic, 25%	G	G	E	E	E	E	E	E	E	G
Hydrochloric, 10%	G	G	G	G	G	G	G	G	G	F
Hydrochloric, 37%	F	X	X	X	F	F	F	F	G	G
Hydrocyanic	G	G	X	X	X	X	G	G	E	X
Hydrofluoric	F	F	X	X	F	F	E	E	X	F
Hypochlorous	G	G	G	G	X	X	X	X	F	F
Lactic	G	G	E	E	G	G	E	E	F	F
Maleic	G	G	G	G	G	G	G	G	G	G
Nitric, 5%	X	F	G	G	F	F	G	G	F	F
Nitric, 65%	G	X	X	X	X	X	F	F	X	X
Oleic	X	X	E	E	X	X	X	X	F	F
Oleum	X	X	X	X	G	G	X	X	G	G
Oxalic	G	G	X	X	X	X	E	E	X	X
Perchloric	X	X	X	X	X	X	E	E	G	G
Phosphoric, 25%	G	G	E	E	G	G	E	E	X	X
Phosphoric, 85%	G	G	G	G	G	G	E	E	G	G
Phthalic	G	G	G	G	G	G	E	E	E	E
Sulfuric, 10%	G	G	E	E	E	E	E	E	F	F
Sulfuric, 78%	F	F	G	G	F	F	G	G	E	E
Sulfuric, 93%	X	X	G	G	X	X	X	X	X	X
Tannic	X	X	G	G	X	X	E	E	X	X
Tartaric	G	G	G	G	G	G	E	E	E	E
Alkalies:										
Ammonium hydroxide	X	X	G	G	G	G	E	E	E	E
Calcium hydroxide	X	X	E	E	G	G	E	E	G	G
Potassium hydroxide	X	X	E	E	G	G	E	E	F	F
Sodium hydroxide	X	X	E	E	G	G	E	E	F	F

Material	1	2	3	4	5	6	7	8	9	10
Acid salts[a]	G	G	E	E	G	G	E	E	G	G
Alkaline salts[b]	G	G	E	E	G	G	E	E	X	X
Neutral salts[c]	E	E	E	E	G	G	E	E	G	G
Miscellaneous salts:										
Potassium dichromate	F	F	G	E	F	F	G	G	X	X
Potassium permanganate	F	F	G	E	G	E	G	G	X	X
Sodium cyanide	F	F	E	E	G	G	G	G	G	G
Sodium ferricyanide	E	E	G	E	G	G	G	G	G	G
Sodium hypochlorite	F	F	F	G	X	F	G	G	X	X
Organics and solvents:										
Acetone	G	G	X	F	G	G	X	X	E	E
Alcohol, methyl	F	G	E	E	E	E	F	G	E	E
Alcohol, ethyl	F	G	E	E	E	E	G	G	E	E
Alcohol, butyl	G	G	E	E	E	E	G	G	E	X
Aniline	X	X	E	E	F	F	X	X	X	E
Benzene	X	X	X	F	G	G	X	X	E	E
Carbon disulfide	X	X	X	X	F	G	X	X	E	E
Carbon tetrachloride	X	X	X	F	G	G	X	X	E	E
Chlorobenzene	X	X	X	X	F	G	X	X	E	E
Chloroform	X	X	X	G	E	E	X	X	E	E
Ethyl acetate	X	X	X	F	G	G	G	G	E	E
Ethylene chloride	X	X	X	F	G	E	X	X	E	E
Ethylene dichloride	X	X	X	X	E	G	X	X	E	E
Ethyl ether	X	X	X	E	G	E	X	X	G	G
Formaldehyde, 37%	X	X	X	X	E	E	X	X	E	E
Gasoline	X	X	G	F	E	G	X	X	E	E
Heptane	F	F	X	X	G	G	X	X	E	E
Methylene chloride	F	F	X	F	G	F	X	X	E	E
Methyl ethyl ketone	X	X	X	F	F	G	G	G	G	G
Nitrobenzene	X	X	X	G	X	X	X	X	E	E
Phenol	X	X	X	F	G	F	X	X	E	E
Toluene	X	X	X	X	G	G	X	X	E	E
Trichloroethylene	X	X	X	X	X	G	X	X	E	E
Gases:										
Carbon dioxide	E	E	E	E	E	E	E	E	G	G
Carbon monoxide	G	G	E	E	G	G	E	E	G	G
Chlorine, dry	X	X	X	F	G	G	F	F	G	G
Chlorine, wet	X	X	X	F	X	F	G	G	G	G
Hydrogen sulfide	X	X	E	E	G	G	G	G	G	G
Sulfur dioxide, dry	F	F	F	G	G	G	G	G	G	G
Sulfur dioxide, wet	F	F	F	G	F	G	G	G	G	G

Table 3 Chemical Resistance of Thermoplastics (continued)

Substance	Polysulfone 75°F	Polysulfone 150°F	Polyvinyl chloride (type I) 75°F	Polyvinyl chloride (type I) 150°F	Polyvinyl chloride (type II) 75°F	Polyvinyl chloride (type II) 150°F	Chlorinated polyvinyl chloride 75°F	Chlorinated polyvinyl chloride 150°F	Vinyl chloride-vinylidene chloride copolymer (Saran)[f] 75°F	Vinyl chloride-vinylidene chloride copolymer (Saran)[f] 150°F
Acids:										
Acetic, 10–50%	G	G	E	G	F	F	G-F	G-X	G	F
Acetic, 75%	G	G	F	F	F	X	F	X	F	X
Acetic, glacial	F	F	F	X	X	X	F	X	X	X
Benzoic	G	G	G	F	E	E	E	E	G	G
Boric	G	G	G	E	E	E	E	E	E	E
Butyric	G	G	G	F	F	X	G-F	F-X	X	X
Chloroacetic	G	X	G	F	E	X	E	G	F	F
Chromic	G	G	E	E	F	E	G	E	G	G
Citric	G	X	E	E	E	E	E	E	E	E
Fluoboric	X	X	E	E	E	G	E	X	G	G
Hydrobromic, 25%	X	X	E	E	E	G	G	X	G	G
Hydrochloric, 10%	G	G	E	E	E	G	E	G	G	G
Hydrochloric, 37%	G	G	E	E	E	E	G	G	G	G
Hydrocyanic	X	X	E	E	G	X	G	F	G	G
Hydrofluoric	X	X	E	E	E	E	F	X	G	G
Hypochlorous	F	F	E	E	E	E	E	G	G	G
Lactic	E	E	E	E	E	E	E	G	E	E
Maleic	G	G	E	E	G	X	G	G	G	G
Nitric, 5%	G	G	E	E	F	E	E	E	E	E
Nitric, 65%	F	X	E	G	E	X	G	X	F	F
Oleic	E	X	E	E	X	G	E	E	X	X
Oleum	X	X	X	X	X	F	X	X	E	E
Oxalic	E	E	E	E	G	G	E	E	E	E
Perchloric	F	F	E	G	G	E	G	X	G	G
Phosphoric, 25%	G	G	E	E	E	G	G	G	G	G
Phosphoric, 85%	G	G	E	E	E	E	G	G	G	G
Phthalic	G	G	E	E	E	G	E	E	G	G
Sulfuric, 10%	G	F	E	E	F	F	G	G	G	G
Sulfuric, 78%	G	X	E	E	F	F	F	X	F	X
Sulfuric, 93%	F	G	E	G	X	X	X	E	E	E
Tannic	G	G	E	E	E	E	E	E	E	E
Tartaric	G	G	E	E	E	E	E	E	E	E

Chemical resistance ratings (E = excellent, G = good, F = fair, X = not recommended)

Material	1	Remarks	2	3	4	5	6	7	8	9
Alkalies:										
Ammonium hydroxide	G		E	E	E	E	E	G	F	X
Calcium hydroxide	G		E	E	E	E	E	E	G	G
Potassium hydroxide	G		E	E	E	E	E	E	F	X
Sodium hydroxide	G		E	E	E	E	E	G	F	X
Acid salts[a]	G		E	E	E	E	E	E	G	G
Alkaline salts[b]	G		E	E	E	E	E	E	G	F-G
Neutral salts[c]	G		E	E	E	E	E	E	G	G
Miscellaneous salts:										
Potassium dichromate	G		E	E	E	E	E	E	G	G
Potassium permanganate	G		E	E	G	X	F	F	G	G
Sodium cyanide	G		E	E	E	G	G	E	G	G
Sodium ferricyanide	G		E	E	E	E	E	E	G	G
Sodium hypochlorite	G		E	E	G	G	G	G	G	F
Organics and solvents:										
Acetone	F-X	Cracking	X	X	X	X	X	X	F	X
Alcohol, methyl	X	Crazing	E	E	G	G	E	E	E	E
Alcohol, ethyl	F	Crazing	E	G	X	X	G	G	E	E
Alcohol, butyl	G	Crazing	G	F	X	X	X	X	G	G
Aniline	X	X	X	X	X	X	X	X	X	X
Benzene	X	Fracture	X	X	X	X	X	X	G	F
Carbon disulfide	F	X	F	X	X	X	X	X	X	X
Carbon tetrachloride	F	Cracking	X	F	X	X	X	F	F	X
Chlorobenzene	X	Fracture	X	X	X	X	X	X	X	X
Chloroform	X	Cracking	X	X	X	X	X	X	F	X
Ethyl acetate	X	X	X	X	X	X	X	X	F	X
Ethylene chloride	X	X	X	X	X	X	X	X	X	X
Ethylene dichloride	F	X	X	X	X	X	X	X	X	X
Ethyl ether	G	X	G	F	X	X	X	G	X	X
Formaldehyde, 37%	G	G	F	F	F	F	F	G	G	F-G
Gasoline	F	G	F	F	X	X	X	F	F	G
Heptane	X	Crazing	E	G	X	X	X	X	X	X
Methylene chloride	F	X	X	X	X	X	X	X	X	X
Methyl ethyl ketone	F	Fracture	X	X	X	X	X	X	X	X
Nitrobenzene	X	Fracture	F	X	X	X	X	F	X	X
Phenol	X	X	X	X	X	X	X	F	G	X
Toluene	X	Cracking	X	X	X	X	X	X	X	F
Trichloroethylene	X	X	X	X	X	X	X	X	X	X

Table 3 Chemical Resistance of Thermoplastics (continued)

Substance	Polysulfone 75°F	Polysulfone 150°F	Polyvinyl chloride (type I) 75°F	Polyvinyl chloride (type I) 150°F	Polyvinyl chloride (type II) 75°F	Polyvinyl chloride (type II) 150°F	Chlorinated polyvinyl chloride 75°F	Chlorinated polyvinyl chloride 150°F	Vinyl chloride-vinylidene chloride copolymer (Saran)ᶠ 75°F	Vinyl chloride-vinylidene chloride copolymer (Saran)ᶠ 150°F
Gases:										
Carbon dioxide	G	G	E	E	E	E	G	G	G	G
Carbon monoxide	G	G	E	E	E	E	G	G	G	G
Chlorine, dry	F	F	G	G	G	G	X	X	G	G
Chlorine, wet	F	F	G	F	F	F	X	X	G	G
Hydrogen sulfide	F	F	E	E	E	E	G	G	G	G
Sulfur dioxide, dry	F	F	E	E	E	E	X	X	G	G
Sulfur dioxide, wet	F	F	E	E	X	X	X	X	G	G

E—excellent, no discernible attack.   F—fair, mild attack, limited use.   P—poor.
G—good, no significant attack.   X—unacceptable, prohibited attack.

ᵃ Acid salts: aluminum chloride, aluminum sulfate, copper chloride, copper sulfate, ferric chloride, ferric sulfate, stannic chloride, zinc chloride, zinc sulfate, etc.
ᵇ Alkaline salts: potassium bicarbonate, potassium carbonate, sodium bicarbonate, sodium carbonate, trisodium phosphate, etc.
ᶜ Neutral salts: calcium chloride, calcium nitrate, calcium sulfate, magnesium chloride, potassium nitrate, potassium sulfate, sodium chloride, sodium nitrate, sodium sulfate, etc.
ᵈ The alkali ratings are for acetal copolymer resin.   The homopolymer is destroyed by concentrated caustics and suffers significant attack by hot, weak caustics.
ᵉ These values are based upon the unadulterated polystyrene, not upon any copolymers.
ᶠ Stannic chloride has a deteriorating effect on Saran.

TABLE 4 Physical Properties of Thermoplastics

Thermoplastic	Specific gravity	Tensile strength, lb/in.²	Flexural strength, lb/in.²	Flexural modulus, lb/in.² × 10⁵	Compressive strength, lb/in.²	Heat-distortion temp, °F	Continuous heat resistance, °F	Coefficient of thermal expansion, in./in./°F × 10⁻¹	Flammability
ABS resins	1.05	7,800	10,000	3.4	11,000	215	180	3.8	Very slow-burning
Acetal resins:									
Copolymer	1.41	8,800	13,000	3.8	4,500	316†	220	4.7	Slow-burning
Homopolymer	1.42	10,000	14,000	4.1	5,200	338†	195	4.5	Slow-burning
Polymethyl methacrylate	1.17–1.20	8,000–11,000	12,000–17,000	4.0–4.8	11,000–19,000	150–220	140–200	4.5	Slow-burning
Cellulose acetate butyrate*	1.18–1.20	5,000–6,000	5,600–6,700	1.2–1.4	5,300–7,100	171–184†	250–275	6–9	Slow-burning
Chlorinated polyether (Penton)	1.4	6,000	5,000	1.3	9,000	285†	250–275	6.6	Self-extinguishing
TFE fluorocarbon	2.14–2.20	2,000–5,000	....	....	1,700	250†	550	6.0	Nonburning
FEP fluorocarbon	2.12–2.17	2,700–3,100	....	....	2,200	215†	400	5–6	Nonburning
PFA fluorocarbon	2.12–2.17	4,300‡	....	1.0‡	....	....	....	6.7§	Nonburning
CTFE fluorocarbon (Kel-F)	2.1–2.2	4,500–6,000	7,400–9,300	....	4,600–7,400	265†	360–390	2.5–4.0	Nonburning
PVF₂ fluorocarbon (Kynar)	1.75–1.78	5,500–7,400	5,000	2.0	8,600	195†	300	4.5–5.0	Self-extinguishing
Nylon 6,6	1.14	12,000	....	4.1	4,900	470†	275–300	5.5	Self-extinguishing
Polycarbonate	1.20	9,500	13,500	3.4	12,500	280–290†	250	3.8	Self-extinguishing
Polyethylene:									
LDPE	0.910–0.925	600–2,300	....	0.08–0.60	....	100–121	186	8.9–11.0	Very slow-burning
MDPE	0.926–0.940	1,200–3,500	4,800–7,000	0.60–1.15	....	120–155	220	8.5–16.0	
HDPE	0.941–0.965	3,100–5,500	....	1.0–2.6	3,600	144–190	250	10.0–13.0	
Polyimides	1.43	10,000	15,000	4.5	25,000	300¶	500	....	Nonburning
Polyphenylene oxide	1.06	9,600	13,500	4.0	16,400		190–220	3.3	Self-extinguishing
Polyphenylene sulfide	1.34	10,800	20,000	6.0	16,000	280¶	500	3.0	Nonburning
Polypropylene	0.900	4,800–5,500	6,000–7,000	1.7–2.5	5,500–6,500	200	230	3.8–5.8	Slow-burning
Polystyrene (GP grade)	1.04–1.09	5,000–12,000	8,000–14,000	4.0–4.7	11,000–16,000	220 max¶	150–170	10.8–14.0	Fast-burning
Polysulfone	1.24	10,200	15,400	3.9	13,900	358	345	3.1	Self-extinguishing
Polyvinyl chloride:									
Type I	1.38	8,500	13,500	5.0	8,000	165	150	3.0	Self-extinguishing
Type II	1.35	6,000	12,000	3.5	....	155	140	5.0	Very slow-burning
Chlorinated polyvinyl chloride (CPVC)	1.50	8,000	16,000	4.0	17,000	234¶	230	12.0	Nonburning
Vinyl chloride-vinylidene chloride copolymer (Saran)	1.68	4,200	5,000	0.75	2,500	140	180	8.8	Self-extinguishing

* ASTM grade MH.
† 66 lb/in.².
‡ At 73°F.
§ At 70 to 212°F.
¶ 264 lb/in.².

Cellulosic Plastics

Preparation and composition   The cellulosic plastics, unlike those plastics manufactured by the polymerization of monomers or condensation of resinous products, are produced from the naturally occurring polymer cellulose through chemical modification. The most important derivatives include cellulose nitrate, the very first such product of commercial significance, and the acetates, butyrates, and propionates. These materials all have excellent dimensional stability and low water absorption, and they are tough. They all have much the same chemical resistance.

Cellulose acetate butyrate   Cellulose acetate butyrate is typical of the cellulose ester plastics in its chemical resistance and applications. It is the esterification product of alpha cellulose with acetic and butyric acids and anhydrides in the presence of sulfuric acid as a catalyst in most processes. The esterification is carried almost to the triester stage, and then the product is hydrolyzed to an acetyl content of about 12 percent (acetic acid content of 16.7 percent) and a butyryl content of approximately 37 percent (butyric acid content of 45.9 percent). The molecular weight of the cellulose acetate butyrate is controlled within the range yielding viscosities 8,000 to 12,000 centipoises of 20 percent solution by weight of the ester and acetone. Formulations may contain plasticizers, pigments, and dyes as desired.

The chief characteristics of cellulose acetate butyrate compositions are their toughness, high impact strength, dimensional stability, and ease of molding (see Table 4).

Since the properties of cellulose acetate butyrate can be varied significantly through plasticizer proportions, a limited number of types have arbitrarily been chosen in accordance with the primary requirement. ASTM D 707 should be consulted in the selection of the formulation most suited for the particular application.

Cellulose acetate butyrate structure:

Cellulose Acetate butyrate

Chemical resistance   Cellulose acetate butyrate is very resistant to water. It is highly resistant to most aqueous solutions. Both strong acids and strong caustics attack it severely. Its solvent resistance is very good toward the aliphatic hydrocarbons and ethers. However, it is readily swelled or in some instances completely dissolved by low-molecular-weight alcohols, esters, ketones, aromatic hydrocarbons, and chlorinated hydrocarbons (see Table 3).

The cellulose propionate follows the same line of chemical resistance as the cellulose acetate butyrate. The cellulose acetate is slightly better in its chemical resistance to organic solvents. It is not as resistant to water and aqueous solutions.

Weatherability   Cellulose acetate butyrate is unaffected by ultraviolet radiation and resists the effects of outdoor exposures over long periods.

Flammability   Cellulose acetate butyrate is classified as slow-burning by Underwriters' Laboratories. It has a burning rate of 0.5 to 1.5 in./min (ASTM D 635). However, it may be modified to provide varieties exhibiting self-extinguishing properties.

Chief corrosive-environment services   Cellulose acetate butyrate has been employed successfully as pipe for the transportation of natural gas and the handling of sour crudes, oils, and brine. The low frictional-resistance properties permit a flow of fluids 30 to 40 percent greater than that through steel pipe.

A second family of compounds of cellulose is typified by ethyl cellulose. This compound is produced by the reaction of ethyl chloride with cellulose in on alkaline medium. It is readily plasticized by many plasticizers, and is dissolved by most of the common solvents and hydrocarbons.   It has very good resistance to moisture and is unaffected by the majority of aqueous salt solutions.   Since these types of compounds have ether linkages, they are not as chemically resistant to acids as the cellulose esters such as cellulose acetate butyrate.   Ethyl cellulose exhibits much better resistance to alkalies than the cellulose esters.

Chlorinated Polyether

Preparation and composition   Chlorinated polyether, an oxatane polymer of 3,3 bis (chloromethyl)-1-oxacyclobutane, may be prepared from pentaerythritol via the trichloride or trichloride monoacetate.[7]   Chlorine constitutes approximately 46 percent by weight of the polymer.   The position of each chlorine atom in a chloromethyl group attached to the central carbon atom makes for excellent chemical resistance.   It is a thermoplastic resin of high molecular weight (250,000 to 350,000) and, because of its regular structure, is crystalline in nature and extremely resistant to thermal degradation at molding and extrusion temperatures.   Two crystalline forms have been obtained.   One is formed by cooling the melt very slowly, and the other from the slow heating of the amorphous polymer.   Because there are decided differences in the physical properties of the two crystalline forms, the properties of commercial products will depend somewhat upon their applied heat experience (see Table 4).   The glass transition temperatures are in the range of 7 to 32°C.

Chlorinated polyether (Penton) structure:

$$\left[-O-CH_2-\overset{\displaystyle CH_2Cl}{\underset{\displaystyle CH_2Cl}{C}}-CH_2-O- \right]_n$$

Chemical resistance   Chlorinated polyether thermoplastic is highly resistant to chemical attack from mineral acids, strong alkalies, and most common solvents.   The resin is resistant to all ranges of most inorganic salt solutions up to 120°C (250°F). It is unacceptable for use in highly oxidizing acids such as fuming nitric acid, oleum, or chlorosulfonic acid.   It is not recommended for services involving fluorine or many chlorinated solvents.   The maximum continuous operating temperature for use with aggressive environments is about 110 to 120°C, although intermittent operations to as high as 140 to 150°C frequently have been acceptable[8] (see Table 3).

Weatherability   Articles fabricated of chlorinated polyether exhibit excellent weathering characteristics.

Flammability   Chlorinated polyether is classified as self-extinguishing by ASTM D 635.

Chief corrosive-environment services   The chief chemical applications involve the use of pipes, fittings, valves, pumps, and tank linings.   Of frequent use also are coatings applied to metal surfaces by means of fluidized-bed, water-suspension, and organic-dispersion, and flocking techniques.   Both rigid pipe and lined steel pipe find much service for high-temperature corrosive environments.

Special considerations   The heat-distortion temperatures are well above the range normal for most thermoplastic materials, and chlorinated polyether effectively retains its tensile-strength properties, as well as allied mechanical properties at elevated temperatures[7] (see Table 4).

Water absorption is negligible, resulting in no change in molded shapes between dry and wet environments.

The low thermal conductivity—0.91 Btu/(lb)(ft²)(°F/in.)—makes it an effective heat insulator.

The extremely good chemical-resistance properties of chlorinated polyether make it an ideal material for use in many aggressive environments.   The fact that it can be readily molded or machined to produce parts that have tolerances approaching those provided by high-cost metals and alloys makes it highly desirable for tough precision parts.[9]

It is applied as a coating, as mentioned above, to transform low-cost substrates into equipment capable of handling highly corrosive chemicals at elevated temperatures. These coatings can be particularly useful and economical where difficult configurations and custom-fabricated parts require the polymer's excellent chemical resistance along with its high-temperature resistance. Extruded chlorinated polyether sheet can be applied directly to the surface of metal substrates, i.e., tank walls, using a standard heat-activated adhesive backing, rolling in place, and butt strapping and sealing the seams by conventional hot-air welding.

Fluorocarbons

Preparation and composition These plastics may be considered as polymers or copolymers of saturated hydrocarbons in which hydrogen atoms have been replaced with fluorine. There are many molecular possibilities in the fluorocarbon plastics family, but the most useful and fruitful polymeric compounds appear to be derived from ethylene and propylene molecules. Those types most generally of greatest application include these five major resins:

1. Polytetrafluoroethylene (PTFE or TFE).
2. Polyfluoroethylenepropylene (PFEP or FEP). This is a copolymer of tetrafluoroethylene and hexafluoropropylene which has better thermoplastic properties than TFE resin, and improved processability.
3. Polychlorotrifluoroethylene (PCTFE or CTFE).
4. Polyvinylidene fluoride (PVF$_2$ or VF$_2$).
5. A new class of perfluoropolymers that combine the carbon-fluorine backbone with a perfluoroalkoxy (PFA) side chain, hence their terminology as PFA fluorocarbons.[10]

Physical and mechanical properties The outstanding characteristic of the fluorocarbon resins is their almost complete inertness to most chemicals and their ability to withstand higher operating temperatures than most plastic or resinous compounds. The fluorocarbons have extremely low coefficients of friction and as a consequence are employed in nonlubricated bearings and gliding surfaces. Their nonsticking surface properties minimize bridging of wet cake or fine powders in hoppers or chutes with high angles of repose. Their excellent dielectric properties are relatively unaffected by extremes in operating temperature and power frequency (see Table 4). In the unfilled condition, the fluorocarbons evidence only fair resistance to deformation under load. The lack of resistance of the fluorocarbons to cold flow is substantially solved through the incorporation of fillers and reinforcements. Among those materials most employed are graphite, molybdenum disulfide, asbestos fibers, and glass fibers and flakes.

Chemical resistance In considering the chemical compatibility (i.e., inertness or chemical resistance), it has been reported[11] that both Teflon TFE and FEP resins are chemically inert to essentially all industrial chemicals and solvents, even at elevated temperatures and pressures. This nearly universal chemical compatibility stems from three causes:

1. The very strong interatomic bonds between the carbon-carbon and carbon-fluorine atoms
2. The almost perfect shielding of the polymer's carbon background by fluorine atoms
3. The very high molecular weight (or long polymer chain length)

TFE has outstanding chemical resistance. It is not compatible with certain environments, however. These fluorocarbon resins will react with molten alkali metals (such as metallic sodium), fluorine, strong fluorinating agents (such as chlorine trifluoride, and oxygen difluoride), and molten sodium hydroxide at temperatures above 300°C. TFE has exceptional thermal stability and is rated for continuous service to 500°F. It is stable also at cryogenic temperatures.

FEP is considered to be generally as chemically resistant as TFE. Its maximum recommended service temperature for continuous application is 400°F.

PFA is reported[10] to be as chemically resistant as either TFE or FEP fluorocarbon resin. As in the case of FEP, PFA is a truly thermoplastic resin and can be fabricated by common thermoplastic techniques not applicable to TFE. PFA has been found to be serviceable from cryogenic temperatures to as high as 545°F.

CTFE has less chemical resistance than TFE or FEP. It is swelled (not attacked) by some chlorinated solvents at high temperatures and is deteriorated by the same group of fluorinated compounds, fluorine, and molten alkali metals as TFE. It is unaffected in general by most other chemicals. The maximum recommended continuous service temperature for CTFE is 390°F. CTFE has better mechanical, rheological, and optical properties than TFE. CTFE film has the lowest water-vapor-transmission rate among thermoplastics.

The chemical resistance of TFE, FEP, PFA, and CTFE resins is so excellent to most environments, with the few exceptions previously noted, that environmental-performance tables like those shown for the other thermoplastics are superfluous and are not presented.

Polyvinylidene fluoride is not as chemically resistant as TFE or FEP fluorocarbon resins. PVF_2 is resistant to most inorganic acids and bases. Fuming nitric, oleum, and other sulfonating agents, and concentrated caustics show some effect at high temperatures. Among the organics, nearly complete resistance is shown to aliphatic and aromatic hydrocarbons, alcohols, acids, and chlorinated solvents. Strong reducing agents such as the primary amines have an effect depending upon the temperature. Strongly polar solvents such as ketones and esters cause partial solvation, particularly at elevated temperatures. PVF_2 exhibits good resistance to oxidizing agents and the halogens.

Weatherability   The fluorocarbons are extremely resistant to ultraviolet radiation and are unaffected by prolonged exposures to direct sunlight and severe outdoor environments. One of the outstanding properties of polyvinylidene fluoride is its stability to weather. Its combination of low moisture absorption, high chemical resistance, and stability to ultraviolet radiation makes it attractive for applications requiring long service life under severe atmospheric conditions.

Flammability   The fluorocarbon resins (TFE, FEP, and CTFE) are classified nonburning as determined by ASTM D 635. Polyvinylidene fluoride is classified "Self Extinguishing Group 1" by Underwriters' Laboratories, Inc., and by ASTM D 635.

Chief corrosive-environment services   For the fluorocarbons these include:[12,13]
1. Envelope-type gaskets for flanged pipe.
2. Thin-film tape substitute for pipe-thread sealant in joining threaded pipe.
3. Expansion bellows and flexible couplings in glass-lined steel piping and between fragile equipment.
4. Lined hoppers for nonsticking, nonbridging properties in handling dry powders or wet crystals.
5. Flexible hose in plasticizer- and solvent-transfer lines, nonbinding feed lines off weigh tanks, and for relief of vibrational stresses on centrifuge effluent lines.
6. Fluorocarbon-lined steel and rigid fluorocarbon pipe for highly corrosive environments.
7. Solid fluorocarbon pipe employed as a standpipe or distribution feed pipe within a glass-lined steel vessel.
8. Nonsticking, nonlubricated plug-seat and ball valves, and fluorocarbon diaphragm valves for multiple services.
9. TFE board, in conjunction with resinous or silica-cement overlay and tantalum-metal backup plate. This finds extensive use in repair of spalled glass-lined steel equipment.
10. Suspended liners of FEP sheet within columns, horizontal and vertical storage tanks, and various chemical reactors.
11. Miscellaneous other fluorocarbon corrosive-environment applications of splash shields, spray nozzles, steam spargers, heat-exchange tube inserts, coated fabrics, filter clothes, filter cartridges, flowmeter accessories, roll covers, and transparent sight-glass inspection plates.
12. FEP tubing heat exchangers.

Polyvinylidene fluoride finds applications in the aerospace industry for expulsion bladders, missile wire and cable, valve seats, lip seals, and frangible diaphragms. In the chemical industry it is used as solid and lined pipe and fittings; solid and lined valves and pumps; seals and gaskets; coated and lined process vessels, fans, scrubbing columns, storage tanks, and filters; tower packing, and filter cloth; containers for corrosive chemicals, bags for radioactive waste, and specialty packages for pharmaceuticals; primary insulation for computer wire, jacketing for aircraft wire and cable, and injection tube for waste disposal.

Permeation of fluorocarbons Gases and vapors will permeate TFE and FEP resins at a considerably lower rate than most other plastics. There is still some permeation, however, and concern must always be observed in the use of fluorocarbon-lined steel equipment in any highly corrosive environment. Since permeation is a "transport" phenomenon (molecular migration),[11] its rate is highly dependent upon temperature. Permeation of TFE polymers occurs by passing molecules of gas through microvoids in the semiporous molding; alternatively, these molecules may in some instances truly migrate between the polymer molecules; both phenomena are strictly physical. In the case of corrosive environments such permeation can result in serious chemical attack to the substrate steel in TFE- or FEP-lined equipment even though no deterioration is experienced by the fluorocarbon resins.[14]

Polyamide Resins (Nylon)

Preparation and composition The polyamides (nylon) are basically linear polymers with repeating amide linkages as an integral part of the molecular chain.[15] They may be produced by two methods:

1. Through amidation of diamines with dibasic acids. Typical of this process is the condensation between hexamethylenediamine plus adipic acid to produce nylon 6/6:

$$H_2N(CH_2)_6NH_2 + NOOC(CH_2)_4COOH \xrightarrow[-H_2O]{\Delta} [NH(CH_2)_6NHCO(CH_2)_4CO]_n$$
$$(nylon\ 6/6)$$

and the reaction between hexamethylenediamine and sebacic acid giving nylon 6/10:

$$H_2N(CH_2)_6NH_2 + HOOC(CH_2)_8COOH \xrightarrow[-H_2O]{\Delta} [NH(CH_2)_6NHCO(CH_2)_8CO]_n$$
$$(nylon\ 6/10)$$

2. Through polymerization of amino acids or their derivatives. In such reactions polycaprolactam results in the product

$$[NH(CH_2)_5CO]_n$$
$$(nylon\ 6)$$

while polymerization of 11-aminoundecanoic acid yields

$$[NH(CH_2)_{10}CO]_n$$
$$(nylon\ 11)$$

Among the myriad of possibilities of other commercially available nylons is nylon 12, characterized by its low-moisture-absorption properties, obtained through the polymerization of the cyclic laurolactam. Many others have been reported or are in various stages of development.

The nylons have many similar properties; they are strong, tough, have extremely good impact resistance, are abrasion-resistant, and have excellent fatigue resistance (see Table 4). All exhibit about the same degree of chemical resistance. Their chief differences lie in melting points, stiffness, resistance to moisture absorption, and processability.

Chemical resistance The nylons have very good resistance to most chemical compounds (see Table 3). They are biologically inactive and resistant to fungal and bacterial attack. They resist hydrolysis. They exhibit extremely good solvent resistance to most of the common organic solvents. They are particularly applicable in exposures to aromatic and aliphatic hydrocarbons, ketones, and esters. They are essentially unaffected by lubricants, oils, fuel, many phosphate esters, and refrigerants. The nylons show very poor chemical resistance toward strong mineral acids, oxidizing agents, and certain salts such as potassium thiocyanate, calcium chloride, and zinc chloride. Nylon 6/6 is generally considered to have the best chemical-resistance properties of the nylons. They are unaffected by alkalies and most salt solutions.

Weatherability Unstabilized nylons do not weather well in outdoor applications under long-term exposures. Where they are exposed to direct sunlight and ultra-violet radiation, it is necessary to employ ultraviolet absorbers or fill with carbon black.

Flammability The nylons are classified as self-extinguishing or slow-burning by ASTM D 635.

Chief corrosive-environment services These include gasoline environmental ex-posures, automotive-industry grease, oil, cams, gears, bushings, rod, tubing, bearings, wear plates, high-strength high-impact components, rollers, wire jacketing, tough molded products subjected to abusive services, hot-water valves, fasteners, electrical parts, and accessories, packaging film, fibers, kitchen utensils, and glass-fiber-rein-forced molded parts.

Polycarbonates

Preparation and composition Polycarbonates are linear aromatic polyesters of carbonic acid. The reaction of polyhydroxy compounds with a carbonic acid deriva-tive produces polymers with carbonate ($-O-\overset{\overset{O}{\|}}{C}-O-$) linkages. Polycarbonates can be described as polymeric combinations of bifunctional phenols or bisphenols, linked together through carbonate linkages.[16]

There are four possible production procedures for linear polycarbonates. These are:

1. Ester exchange of dihydroxy compounds with diesters of carbonic acid and monofunctional aromatic or aliphatic hydroxy compounds

2. Ester exchange of bis-alkyl or bis-aryl carbonates of dihydroxy compounds with themselves or with other dihydroxy compounds

3. Reaction products of dihydroxy compounds with phosphene in the presence of acid acceptors

4. Reaction of the bis-chlorcarbonic acid ester of dihydroxy compounds with dihydroxy compounds in the presence of acid acceptors

While many variations in the final structure are possible, the principal polycar-bonate of commercial interest is that based on bisphenol A (4,4'-dihydroxy-diphenyl propane). This polycarbonate has the following molecular structure:

$$\left[-O-\!\!\bigcirc\!\!-\overset{\overset{CH_3}{|}}{\underset{\underset{CH_3}{|}}{C}}-\!\!\bigcirc\!\!-O-\overset{}{\underset{\underset{O}{\|}}{C}}- \right]_n$$

Physical and mechanical properties Polycarbonate polymers are one of the better thermoplastics where strength and elastic moduli are concerned, offering both excel-lent impact strength and rigidity. They are relatively insensitive to changes in their properties with increases in temperature. The deflection temperature under load (heat-distortion point) of the polycarbonates of approximately $270°F$ is one of the highest among thermoplastics (see Table 4).

Dimensional stability is one of the outstanding properties of polycarbonates. Pre-cision moldings are readily prepared within tolerances of plus or minus 0.001 in. as a result of its low and predictable mold shrinkage.

Chemical resistance[17] The unstressed carbonate resin is unaffected by dilute mineral and organic acids. It is insoluble in aliphatic hydrocarbons, petroleum ether, and most alcohols. It is, however, partially soluble in aromatic hydrocarbons such as benzene and toluene, and ketones and esters. It is extremely soluble in chlori-nated hydrocarbons, such as methylene chloride, and materials such as dioxan 1,4. Ammonia and amines readily attack it, and it is slowly decomposed by strong alkaline solutions (see Table 3).

Molded polycarbonate items often experience stress cracking or crazing when exposed to certain chemical compounds or vapors. This appears to be due to residual stresses locked in during the molding operations being relieved when contacted by compounds exhibiting partial solvency. This susceptibility to stress cracking and crazing can be eliminated or minimized by proper mold design or special annealing treatments in the temperature range of 250 to 300°F.

Weatherability Polycarbonates exhibit outstanding resistance to sunlight, atmospheric chemicals, severe temperature changes, and rain-erosion effects. Polycarbonates are good ultraviolet absorbers and consequently experience surface effects but little subsurface deterioration. These surface effects can be minimized through use of ultraviolet stabilizers.

Flammability Polycarbonates are classified as self-extinguishing by ASTM D 635. They are listed by Underwriters' Laboratories, Inc., in their Class II category of self-extinguishing materials.

Chief corrosive-environment services These include structural parts, electrical parts, safety helmets, portable tool housings, aircraft components, food-service equipment, containers, indoor and outdoor lighting fixtures, medical equipment, signs and building panels, glazing, automotive parts, aircraft hot-air ducts, doors, windows, partitions, and skylights, as well as shields and storm-door panels.

Special consideration Without reinforcement, polycarbonate resins rank extremely high as engineering materials. However, addition of glass-fiber reinforcement raises the level of many of their properties. For example, a 40 percent reinforced glass-filled polycarbonate has a tensile modulus five times as great and flexural compressive and tensile strength twice as great as those of the unreinforced polycarbonate.

Polyethylene

Preparation and composition Polyethylene is not the simple-structured polyolefin arising from the polymerization of multiple ethylene molecules as one might think. Polyethylene varies from type to type according to each specific molecular structure. This depends upon the crystallinity, molecular weight, and molecular-weight distributions. The final molecular structure is dependent upon the process variables. These include operating temperature, employed pressures, types of catalyst, and additives. Depending upon these variables the polyethylene may vary in its physical properties from soft to hard, flexible to rigid, and weak to tough.

Polyethylene is obtained by the linking together of numerous molecules of ethylene $CH_2{=}CH_2$ into very long molecules of varying lengths. However, the polyethylene does not consist of bundles of long molecules bunched together. There is generally some cross linkage of these long-chain molecules across branches along the chain. The manufacturer controls the synthesis and production techniques to produce polyethylene which may vary according to (1) average size of molecules or molecular weight, (2) distribution of size of molecules within the product, and (3) extent of branching or nonlinearity of these molecules.

Polyethylene is covered by ASTM D 1248. This specification provides for three types of polyethylene:

Type 1 is polyethylene in a density range of 0.910 to 0.925 g/cm³. This is termed "low-density" or "branched" polyethylene, or LDPE.

Type 2 is polyethylene in a density range of 0.926 to 0.949 g/cm³. This is known as "medium-density" polyethylene, or MDPE.

Type 3 is polyethylene in a density range of 0.941 to 0.965 g/cm³. This is termed "high-density" or "linear" polyethylene, or HDPE.

Physical and mechanical properties There is a difference in the physical properties of the polyethylene types (see Table 4). The chief differences among these three types are in rigidity, heat resistance, and resistance to loading. Increased density results in greater stiffness, strength, hardness, and heat resistance, while resulting in a lowering of the impact strength and resilience. The rate of attack of environments aggressive to the polyethylenes is somewhat more rapid with the type 1 polyethylene plastics in comparison with that with types 2 and 3. In general, however, chemical resistances for long-term exposures are essentially the same.

Chemical resistance The polyethylenes have excellent chemical resistance to many chemicals and solvents (see Table 3). At ambient temperature, they are resistant to acids and alkalies except for oxidizing acids such as nitric, chlorosulfonic, and fuming sulfuric. They are unaffected by hydrofluoric acid. The polyethylenes are generally insoluble in organic solvents at temperatures below 122°F. However, at higher temperatures they are soluble to varying degrees in hydrocarbons and halogenated hydrocarbons. They are appreciably affected by chlorinated solvents, aliphatic and aromatic hydrocarbons, certain esters, and oils. Polyethylene can be dissolved in hot (160°F) toluene, xylene, amyl acetate, trichloroethylene, petroleum ether, paraffin, turpentine, and lubricating oils.

Weatherability Rapid ultraviolet degradation resulting in severe crazing is experienced by the unprotected material. A black-pigmented variety is recommended for exposure to direct sunlight. It is reported that as little as 3 percent carbon black will absorb all the ultraviolet radiation in sunlight and will prevent its deleterious effects. Other ultraviolet absorbers may be employed if carbon-black additions cannot be tolerated.

Flammability The polyethylene resins burn slowly at a rate of about 1.0 in./min (ASTM D 635). Flame retardants, usually antimony compounds, may be employed at the expense of certain physical properties to meet specific nonburning or self-extinguishing demands.

Chief corrosive-environment services Uses of polyethylene plastic include pipe for handling chemicals, natural-gas transmission, irrigation systems, and fertilizer and agricultural application; rigid and lined steel tanks for processing chemicals; carboys and packaging containers; and battery parts.

Special considerations Polyethylene is susceptible to stress cracking in certain environments. Rapidity of failure is dependent upon the amount of stress and temperature. Environments most conducive to stress cracking include surface-active agents such as metallic soaps and sulfated and sulfonated alcohols. Environments such as aliphatic and aromatic hydrocarbons and specific organic acids introduce stress cracking to a somewhat less active degree. In most instances there is an incubation period, and stress cracking frequently does not occur for several months. Stress cracking can be minimized in many instances by the correct composition of polyethylene, or redesign of the part to eliminate residual stresses.

Polyimide

Preparation and composition Polyimides are produced through the reaction of a dianhydride and a diamine to give a soluble polyamic acid polymer, which is converted thermally to the polyimide by loss of water. Typical are those polymers derived from the condensation of pyromellitic dianhydride and aromatic diamines, such as m-phenylene diamine, benzidine, and 4.4'-diaminodiphenylether.[16]

The basic chemical structures of the polyimides may be represented by

Physical and mechanical properties The polyimide resins are very strong and tough, highly resistant to radiation and abrasion, have excellent electrical properties, and are unaffected by many chemicals (see Table 4).

Probably the greatest virtue of the polyimides is that they have the highest heat resistance of all commercially available organic polymers. They have been used continuously in air at 500°F and under intermittent operations at temperatures as high as 900°F. This high heat resistance is attributed to the highly aromatic structure, in addition to the nitrogen-containing ring. An example of its exceptional heat resistance is the fact that under test the polymer retained 90 percent of its original tensile strength after 1,000 h at 575°F. Polyimides are reported to be serviceable in an operating range of 850 to 900°F. They have very high *PV* values for bearing applications, high dielectric strength, excellent mechanical strength and solvent resistance, low outgassing, and high resistance to ionizing radiation. It is indicated that they will perform satisfactorily in cryogenic systems at −320 to −420°F, and as a matter of fact tensile strengths at −320°F show a 30 percent increase above the strength at room temperature.

Chemical resistance   The polyimides exhibit good resistance to mild concentrations of inorganic or organic acids, and are essentially unaffected by grease and oils and aliphatic and aromatic solvents. Concentrated mineral acids cause severe embrittlement of polyimide parts in a relatively short time (see Table 3). They are, however, severely attacked by strong alkalies and oxidizing agents. They are also severely attacked by strong bases such as aqueous ammonia, hydrazine, nitrogen dioxide, and primary or secondary amines. Exposure to steam or hot water at 212°F produces a decrease in tensile and flexural strength. However, most of the reduced tensile values can be restored by drying, which indicates the reduced properties are not due to chemical changes. At high temperatures some solvents containing functional groups such as *m*-cresol and nitrobenzene can cause swelling of polyimides without substantially reducing their mechanical strength. Chemicals which act as powerful oxidizing agents can cause oxidation of polyimide parts even under mild conditions.

Weatherability   Polyimide parts undergo losses in tensile strength after prolonged outdoor exposure. They are completely resistant to fungus attack.

Flammability   Polyimides are classified as nonburning by ASTM D 635.

Chief corrosive-environment services   The polyimide family is used currently as films, adhesives, castings, moldings, enamels and varnishes, fibers, coated glass fabrics, and binders for reinforced plastics. The conventional techniques employed in the reinforced-plastics industry are directly applicable to the polyimides.

Structural products have been produced from glass-fiber rovings highly impregnated with polyimide using vacuum-bag or autoclave techniques. The glass-fiber reinforced polyimide can be filament-wound to produce high-strength items with excellent heat and chemical resistant properties.

Applications in corrosive environments include nonlubricated bearings, ball-bearing retainers, thrust washers, compressor vanes, seals, valve seats, and piston and wear rings through special compounding of the polyimide resin. They are extremely good for electrical insulation at extremes of temperature.

Special considerations   Polyimides have unusual resistance to ionizing radiation. Mechanical and electrical properties do not show an appreciable change until the absorbed energy exceeds 10 rd, a higher threshold than most plastics.

Polyphenylene Oxide

Preparation and composition   Polyphenylene oxide is currently prepared by the copper-catalyzed oxidation of 2,6-xylenol.[16] It has the following structural formula:

$$\left[\begin{array}{c} CH_3 \\ \\ \\ CH_3 \end{array} \right]_n$$

Physical and mechanical properties   Polyphenylene oxide resin is characterized by excellent impact strength, high tensile strength and modulus, excellent dimensional stability for long periods of time under heavy loads, and a practical application-temperature range from a brittle point of approximately −275°F to a heat-distortion point of 375°F at 264 lb/in.^{2}. Its linear coefficient of thermal expansion, 2.9×10^{-5} in./in./°F, is extremely low for a thermoplastic. It is identified as a self-extinguishing resin (see Table 4). The polyphenylene oxide resins have excellent processability on conventional extrusion and injection-molding equipment.

Chemical resistance   The polyphenylene oxide resins show excellent chemical resistance to dilute acids, alkalies, and aqueous salt solutions (see Table 3). They are severely attacked by strong oxidizing acids.   It has been reported that polyphenylene oxide resins under stressed conditions will crack when exposed to some organic liquid environments such as aliphatic hydrocarbons, ketones, esters, and petroleum derivatives; stress-relieved polyphenylene oxide is unaffected in such environments.   It is readily dissolved by chlorinated aliphatics (chloroform) and aromatics such as benzene and toluene.

The resistance of the polyphenylene oxide resin to hydrolytic breakdown is emphasized by its excellent performance when exposed to stema at 270°F for long periods of time.   This characteristic suggests its possible application for the fabrication of steam traps and exhaust jets handling high-temperature steam.

Flammability   The polyphenylene oxide resins are classified as self-extinguishing, nondripping by ASTM D 635.

Weatherability   The resin experiences mild discoloration by direct sunlight, and is unaffected in general by moisture and outdoor environmental exposures.

Polyphenylene Sulfide

Preparation and composition   Polyphenylene sulfide (PPS) is produced from p-dichlorobenzene and sodium sulfide in a polar solvent.[18]   PPS is highly crystalline, with a melting point near 550°F.   Thermogravimetric analysis of PPS in nitrogen or in air indicates no appreciable weight loss below 900°F.   Since thermogravimetric analysis as such is inherently a short-term indication of thermal stability, Short and Hill[19] conducted tests that demonstrated high-temperature (260°C) resistance of thin film coatings (1 mil thick) on steel for a period of 1,686 h in air.   Degradation in air is essentially complete at 1300°F.   PPS has outstanding resistance to heat and can be used at temperatures as high as 450°F for prolonged periods of time.

Physical and mechanical properties   The unfilled resin is characterized by high tensile strength, high flexural modulus, high-temperature heat deflection, and modest impact strength (see Table 4).

The electrical properties of PPS are excellent.   Its low dielectric constant, along with a low dissipation factor and a high dielectric strength, combined with its ease of molding, nonflammability, high-temperature stability, and corrosion resistance make it a most attractive polymer in the electrical industry.

Polyphenylene sulfide has the following structural formula:

$$\left[\bigcirc\!\!\!\!-\!S \right]_n$$

The recurring para-substituted benzene rings and sulfur atoms of polyphenylene sulfide are considered responsible for its outstanding chemical inertness and thermal stability.

Chemical resistance   Polyphenylene sulfide possesses unusual chemical resistance.[20]   The cured polymer is essentially unaffected by aliphatic hydrocarbons, alcohols, ketones, chlorinated aliphatic compounds, esters, aliphatic ethers, toluene, liquid ammonia, ammonium hydroxide, aqueous sodium hydroxide, organic acids, and most inorganic salt solutions (see Table 3).   Dilute hydrochloric and nitric acids, concentrated sulfuric acid, and strong oxidizing agents such as chromic acid, bromine water, and sodium hypochlorite cause deterioration, particularly at elevated temperatures.   No change in weight was noted[18] after 4 months of exposure in xylene, dimethyl formamide, chlorobenzene, 10 percent nitric acid, concentrated hydrochloric acid, 48 percent sulfuric acid, and concentrated sulfuric acid at ambient temperature.

There are no known solvents for polyphenylene sulfide below 375°F.

Weatherability   Polyphenylene sulfide has good resistance to prolonged exposure to direct sunlight and is unaffected by high-humidity conditions.

Flammability   Polyphenylene sulfide is classified as nonburning according to ASTM D 635.

Chief corrosive-environment services   Polyphenylene sulfide has been finding extensive service as corrosion-resistant pump components, valves, pipes, metering devices, elevated-temperature solventless protective coatings, automotive parts subjected to heat exposure, electrical insulators, and electronic components.

Polypropylene

Preparation and composition   Polypropylene is prepared as a linear polymer with regular steric or spatial features as a result of the Ziegler/Natta coordination catalyst system employed.

$$
\left[
\begin{array}{c}
\overset{\displaystyle H}{\underset{\displaystyle H}{\overset{|}{\underset{|}{-C}}}}=\overset{\displaystyle H}{\underset{|}{C}}-CH_3
\end{array}
\right]_n
$$

This stereoregular, isotactic structure of polypropylene accounts for the relatively good combination of its mechanical properties.

Physical and mechanical properties   The fairly high crystalline melting point of polypropylene gives rise to the fact that it retains its mechanical strength up to as high a temperature as $140°C$.   Commercially available polypropylene has the lowest density of any of the commonly employed thermoplastics.   Polypropylene is somewhat similar to higher-density polyethylene in many ways, but it has a greater rigidity and heat resistance, and items made from it act much lighter in weight. Unlike polyethylene, it appears to be quite resistant to environmental stress cracking. Polypropylene exhibits very low moisture-absorption characteristics and excellent retention of its electrical properties over a wide temperature range.   It is an excellent electrical insulator.   Polypropylene has high impact strength (see Table 4).

Chemical resistance   The chemical resistance of polypropylene toward many environments is quite similar to that exhibited by high-density polyethylene (see Table 3).   The two polymers tend to be swollen by the same solvents.   Polypropylene is highly resistant to most acids, alkaline media, and inorganic salt solutions at elevated temperatures.   Polypropylene differs from polyethylene in its chemical resistivity because of the presence of tertiary carbon atoms occurring alternately on the chain skeleton.   The most significant effect is the susceptibility of polypropylene to oxidation at elevated temperatures.   This may be minimized through the addition of selective antioxidants, which are generally contained in all commercially available polypropylenes.

Weatherability   Because polypropylene is susceptible to ultraviolet degradation in strong sunlight, ultraviolet absorbers are generally added before processing.

Flammability   Polypropylene burns slowly at a rate of 0.7 to 1.0 in./min under test conditions of ASTM D 635.

Chief corrosive-environment services   Polypropylene is used in automotive parts; electrical equipment and appliances; chemical piping, pumps, valves, fittings, and tanks; filter cloths; packing for scrubbing towers; blowers; and molded parts.

Polystyrene

Preparation and composition   The production of styrene monomer is essentially a two-step process.   The first reaction is the production of ethylbenzene from ethylene and benzene in the presence of a Friedel-Crafts catalyst such as aluminum chloride. The ethylbenzene is separated by fractional distillation from the residual benzene and polyethylbenzenes:

$$
\bigcirc + C_2H_4 \rightarrow \bigcirc\!\!-C_2H_5
$$

Styrene is produced by dehydrogenation of the ethylbenzene:

$$
\bigcirc\!\!-C_2H_5 \xrightarrow[\text{+ steam}]{630°\ C} \bigcirc\!\!-CH{=}CH_2 + H_2\uparrow
$$

Polystyrene is available in a number of grades as follows:
1. General-purpose types
2. Impact grades, consisting of polystyrene and rubber
3. Chemical-resistant grades—copolymers of styrene and acrylonitrile
4. Specialty grades—light-stabilized styrene–methyl methacrylate copolymers for weathering and glass-reinforced polystyrenes

Physical and mechanical properties The unfilled polystyrene is a hard, rigid, transparent plastic. It is odorless and tasteless. It has good physical/mechanical properties, varying somewhat according to molecular weight (see Table 4). One of its most important properties is its high transmission of all wavelengths of visible light and its high refractive index (1.592), which gives it a particularly high brilliance. Its main mechanical limitation is brittleness, unless it is strengthened by a reinforcing agent.

Chemical resistance The chemical resistance of polystyrene is somewhat less than that of polyethylene (see Table 3). It is dissolved by a number of hydrocarbons such as benzene, toluene, and ethylbenzene and by chlorinated hydrocarbons such as carbon tetrachloride, chlorobenzene, chloroform, and methylene chloride. It is attacked by ketones (with the exception of acetone) and esters. Certain other materials such as acids, alcohols, oils, cosmetic creams, and foodstuffs often foster crazing and cracking. Polystyrene has good resistance to many ordinary chemicals such as weak acids, all concentrations of alkalies, and aqueous solutions of many salts. It is readily attacked by oxidizing agents. Polystyrene has a low moisture absorptivity. It also has good resistance to water-vapor transmission.

Weatherability Nonstabilized polystyrene turns yellow and becomes brittle when exposed to ultraviolet radiation. The incorporation of about 1 percent saturated aliphatic amines, cyclic amines, or amino alcohols greatly improves resistance to weathering.

Flammability Polystyrene ignites easily and burns horizontally at 0.5 to 2.5 in./min by ASTM D 635. It burns with a heavily sooty flame.

Chief corrosive-environment services Polystyrene applications include houseware, appliances, containers, film, toys, accessories, cabinets, auto parts, tape reels, foams, and liners for refrigerator doors. The polyester resins are diluted to 50 percent with styrene monomer for use as the conventional glass-fiber-reinforced polyester laminar structures used in fabrication of chemical-processing equipment.

Polysulfone

Preparation and composition Polysulfone is the reaction product of the sodium salt of 2,2-bis (4-hydroxy phenol) propane and 4,4'-dichlorodiphenyl sulfone. The sodium phenoxide end groups react with methyl chloride to terminate the polymerization, controlling the molecular weight of the polysulfone.[21]

The basic chemical structure of the polysulfone is represented by the repeating diphenylene sulfone group, an ether linkage, and an isopropylidene group:

Physical and mechanical properties The polysulfone has an exceptionally high heat-deflection temperature of $345°F$ at 264 lb/in.^{2}. It has excellent thermal stability and can be used under continuous service as high as 330 to 450°F. It is reported that thermal gravimetric analyses show polysulfone to be stable in air up to 500°C, a valuable characteristic for melt stability in molding and extrusion (see Table 4).

Its toughness is attributed to the ether linkage in the polymer, which makes for a flexible chain. Its physical properties are good down to −150°F, and the polymer is recommended for applications where low-temperature toughness is required. The electrical properties of polysulfone are excellent and are unchanged over a wide range of temperatures and frequencies.

Chemical resistance   Polysulfone is highly resistant to mineral acids, alkali, and salt solutions. Resistance to detergents and hydrocarbon oils is good, even at elevated temperatures under moderate levels of stress (see Table 3). For example, specimens have withstood over 500 h in 2 percent Cascade solution at 140°F and 1,500 lb/in.[2] stress.[21] It is unaffected by aliphatic hydrocarbons. Polysulfones will be attacked by polar organic solvents such as esters and ketones, chlorinated hydrocarbons, and aromatic hydrocarbons. Polysulfone is not attacked by water through chemical reaction (hydrolysis). Products made from polysulfone can be steam-sterilized repeatedly without any significant degradation of physical properties or chemical resistance.

Weatherability   The polysulfones experience mild degradation of their mechanical properties and slight yellowing on prolonged exposure to sunlight.

Flammability   Polysulfone is classified as self-extinguishing, either SE1 or SE0 by Underwriters' Laboratories test, depending upon the scientific formulation. Polysulfone is classified nonburning per ASTM D 635.

Chief corrosive-environment services   The polysulfones are used in electrical and electronic parts, housings for meters, switches, insulating bushings, alkaline battery cases, coffee makers, humidifiers, sight glasses, clothes steamers, automatic washer and dryer parts, ironers, hair dryers, automotive and aircraft parts, dairy and food-processing equipment, water-heater dip tubes, medical instruments, microwave-oven racks and trays, and chemical-processing equipment (pumps, valves, piping, tanks).

Special considerations   Polysulfone products containing residual locked-in stresses are susceptible to stress cracking when exposed to aggressive environments such as carbon tetrachloride, butyl cellosolve, 2-ethyl ethanol, trichloroethane, and ethyl acetate. This stress cracking can be relieved by special postfabrication annealing techniques at 330°F or by incorporation of about 10 percent short glass fibers as reinforcement in the product.

Polyvinyl Chloride (PVC)

Preparation and composition   Polyvinyl chloride is one of the more widely used thermoplastic resins as a result of its ease of fabrication and extremely good acid and alkali resistance. Polyvinyl chloride is available commercially as the rigid unplasticized chemical-resistant normal-impact type I, as the slightly less corrosion-resistant optimum-impact plasticized type II, and as high-impact (greater than 5.0 Izod impact) type III.

Type I rigid PVC is generally the formulation used for process equipment and piping in the chemical industry. It is used chiefly for the fabrication of extruded pipe, molded fittings, valves, and heat-welded thermoplastic equipment and structures. Copolymerized with vinyl acetate, it serves as a solvent-applied protective coating for many aggressive environments.

It is also available as low-viscosity, high-solids dispersions in organic liquids, either as the lone polymer or copolymerized. These are termed "organosols." If the organic liquid is a relatively nonvolatile material, such as a plasticizer, the dispersion is termed a "plastisol." Dioctylphthalate is a typical plasticizer which is employed as the continuous phase in which the particles of the polyvinyl chloride resin are suspended. The object to be plastisol-coated is first primed and preheated to 350 to 375°F, and then dipped into the vat containing the heated plastisol. The deposited coating must be heated further at about the same temperature in order to fuse the resin into a tough, flexible, void-free layer. Plasticized polyvinyl chloride sheet is used extensively for the lining of plating tanks, for pickling tanks used in processing stainless steel, and for the handling of corrosives such as nitric acid, chromic acid, and other strong oxidizing agents incompatible with rubber.

To take advantage of the greater chemical resistance of the unplasticized polyvinyl chloride, a laminar, composite sheet of both plasticized and unplasticized polyvinyl chloride is available. The softer, plasticized layer is cemented to the steel tank, minimizing thermal-expansion problems inherent with the more rigid, unplasticized polyvinyl chloride, which is faced toward the corrosive environment within the tank. Polyvinyl chloride has this structural formula:

$$\left[\begin{array}{cc} H & Cl \\ -C-C- \\ H & H \end{array} \right]_n$$

Special considerations Major difficulties experienced with the use of polyvinyl chloride have been the inadvertent introduction of excessive heat (about 165°F), which may cause plastic flow and deformation. The presence of residual specific solvents may deteriorate the surface or cause prohibitive swelling or loss of strength.

Chemical resistance Rigid type I polyvinyl chloride is completely resistant to inorganic alkalies in all concentrations (see Table 3). It is resistant to all inorganic and organic salts. PVC is extremely resistant to inorganic acids such as hydrochloric, nitric, phosphoric, and concentrated sulfuric acid. The fatty acids, stearic, oleic, and linoleic, have little effect upon PVC at ambient temperatures. Prolonged exposure to hot fatty acids effects a softening at the liquid/vapor phase juncture. It is believed this is the result of oxidation of the fatty acids to ketonic and similar oxygenated molecules. Acetic and formic acids mildly attack PVC at ambient temperatures and are quite aggressive at 140 to 150°F. PVC is severely attacked by glacial acetic acid. Rigid PVC is extremely resistant to oxidizing agents such as strong sulfuric acid, nitric acid, chromic acid, hydrogen peroxide, and sodium hypochlorite. The common solvents such as ketones, esters, ethers, chlorinated hydrocarbons, and aromatic hydrocarbons soften and in some instances dissolve rigid PVC. Alcohols and aliphatic hydrocarbons have little effect upon PVC. It is completely resistant to all common animal, mineral, and vegetable oils.

Weatherability Polyvinyl chloride darkens on exposure to ultraviolet radiation because of the formation of polythene structures; it also becomes brittle. Surface cracking and chalking are experienced in polymers not protected with adequate ultraviolet absorbers and antioxidants. Polyvinyl chloride exhibits long service life on outside installations below ground, since it is completely unaffected by the action of acid soils, fertilizers, bacteria, or fungi.

Chief corrosive-environment services PVC is used for chemical-processing equipment including piping, valves, ducts, hoods, stacks, pumps, rigid tanks, lined steel tanks and equipment, hoppers, metering devices, scrubbers, absorbers, columns, jets and eductors, dip and feed pipes, screw feed troughs, filters, transfer carts, blowers, plastisol-coated steel units, floor gratings, tape pipe wrappings, mist eliminators, and protective guards for motors, belts, and rotating equipment.

Flammability Polyvinyl chloride type I is classified as self-extinguishing by ASTM D 635.

Chlorinated PVC (CPVC)

Preparation and composition Polyvinyl chloride has been modified through chlorination to obtain a vinyl plastic with improved corrosion resistance and the ability to withstand operating temperatures that are 50 to 60°F higher. Chlorinated polyvinyl chloride has this structural formula:

$$\left[\begin{array}{cc} Cl & Cl \\ -C-C- \\ H & H \end{array} \right]_n$$

Physical properties of chlorinated PVC are given in Table 4.

Chemical resistance CPVC has about the same range of chemical resistance as rigid PVC (see Table 3). It has been reported[3] that marked improvement in residual tensile strength is noted with CPVC in comparison with PVC on exposure for 30 days in such environments as 20 percent acetic acid, 40 to 50 percent chromic acid, 60 to

70 percent nitric acid, 80 percent sulfuric acid, hexane, and 50 percent sodium hydroxide at 73, 140, and 185°F.

Weatherability   CPVC darkens on exposure to direct sunlight.

Flammability   CPVC is classified as self-extinguishing by ASTM D 635.

Chief corrosive-environment services   CPVC has been chiefly used as pipe, fittings, tanks, ducts, and bands for handling highly corrosive environments in the chemical and allied industries.

Polyvinylidene Chloride Copolymers (Saran)

Preparation and composition   Vinylidene chloride monomer is obtained from trichloroethane by dehydrochlorination:

$$CH_3CHCl_2 \rightarrow CH_2{=}CCl_2$$

Vinylidene chloride monomer is a highly inflammable liquid which polymerizes very readily into its polymer.   It must be protected from exposure to light, air, and water to prevent acceleration of this polymerization.   The polyvinylidene chloride, however, is thermally unstable and is not able to withstand melt processing (see Table 4).   Vinylidene chloride is copolymerized with vinyl chloride:

$$\left[\begin{array}{cccc} H & Cl & H & Cl \\ -C{-}C{-} & & C{-}C{-} \\ H & Cl & H & H \end{array}\right]_n$$

or with acrylonitrile:

$$\left[\begin{array}{cccc} H & Cl & H & CN \\ -C{-}C{-} & & C{-}C{-} \\ H & Cl & H & H \end{array}\right]_n$$

to obtain polymers that are of commercial value and sold under the name of Saran.   The plasticized copolymers of vinylidene chloride–vinyl chloride are used in the manufacture of extruded pipe, valves, molded fittings, and pumps.   The copolymers of vinylidene chloride and acrylonitrile are used in tank linings, Saran fibers, and monofilaments.   The vinylidene chloride–vinyl chloride copolymers have better vapor-barrier characteristics than PVC or vinyl chloride–vinyl acetate copolymers.

Chemical resistance   The vinylidene chloride–vinyl chloride copolymers exhibit greater chemical resistance than polyvinyl chloride (see Table 3).   They are un-affected by weak inorganic and organic acids.   They show excellent resistance to concentrated acids.   They are unattacked by weak and strong alkaline solutions with the exception of ammonium hydroxide, which deteriorates them rather rapidly.   They are resistant to most aliphatic and aromatic solvents, oils, greases, gasoline, kerosene, esters, and alcohols.   They are, however, readily attacked by solvents such as dioxane, cyclohexanone, dimethyl formamide, and certain chlorinated aromatic hydrocarbons.   They are extremely impervious to water, and their films have exceptional resistance to water-vapor transmission.

Weatherability   The vinylidene chloride copolymers with vinyl chloride or acry-lonitrile are slightly degraded in direct sunlight.   Ultraviolet absorbers such as 5-chlor-2-hydroxybenzophenone minimize such degradation.

Flammability   Polyvinylidene chloride and copolymers with vinyl chloride or acrylonitrile are classified as self-extinguishing or slow-burning by ASTM D 635.

Chief corrosive-environment services   Saran is used in chemical-processing and waste-disposal equipment and piping, heat-sealable sheet film, wrapping, brushes, screening, valves, pump liners, marine applications, automotive parts and accessories, protective coatings, and filaments.

THERMOSETS

Fundamentals

Definition of thermosets   The thermosetting resins generally are liquids or low-melting solids which under the influence of heat, pressure, ultraviolet light, radio-frequency excitation, catalysts, and accelerators are transformed into rigid products that are relatively infusible and insoluble. They cannot be reshaped or remelted after they have been cured to their final state.

Energy of activation   The heat required to activate the molecule to an energy level high enough to initiate the cross-linkage and polymerization of the thermosets can be furnished by a number of sources. There are two main methods used in industry to initiate the cure of thermosetting resins. The first employs external heating by way of electrical, steam, or hot-gas elements. The second method relies upon the use of catalysts and accelerators that react exothermically and provide thereby sufficient heat to promote the setting of the resin.

Thermosets considered   As with the thermoplastics, we will be concerning ourselves in this portion of the chapter with the chemical resistance and performance of only the more important, commercially employed thermosets. The chemical resistance of the thermosets more generally used for chemical-processing equipment is noted in Table 5. The following thermosets are covered:

Allylics
  Diallyl maleate
  Diallyl phthalate
  Diallyl isophthalate
  Diethylene glycol bis(allyl carbonate)
  Triallyl cyanurate
Amino resins
  Melamine formaldehyde
  Urea formaldehyde
Epoxies
Furans
Phenolics
Silicones
Unsaturated polyesters
Urethanes
Vinyl esters

TABLE 5   Resistivities of Thermosetting Plastics at Ambient Temperature

	Epoxy (amine-cured)	Furan	Phenolic (unmodified)	Polyester (chemical-resistant type)
Weak acids, bases and salts.....	A	A	B	B
Strong bases.................	A	B	D	C
Strong acids.................	B–C	A	A	A
Strong oxidants..............	X	X	D	C–D
Aliphatic solvents.............	A	A	A	A
Aromatic solvents.............	A	A	A	A
Esters and ketones............	A–B	A	C	B–C
Chlorinated solvents...........	A	A	A	A

A = excellent, relatively inert.
B = good, very slight attack, acceptable.
C = fair, mild attack, acceptable in many cases.
D = poor, attacked, softened, swollen, very limited service.
X = unacceptable, severe attack, delamination.

Allylic Resins

Preparation and composition   Monomers of the allylic resins are prepared by the esterification of allyl alcohols $CH_2 = CHCH_2OH$ with organic polybasic acids. The monomers of greatest commercial interest include diallyl phthalate (DAP); diallyl isophthalate (DAIP); diallyl maleate (DAM), a highly reactive trifunctional monomer containing two types of polymerizable double bonds; diallyl chlorendate

(DAC), a highly fire-retardant variety; and triallyl cyanurate (TAC), a high-temperature-resistance-imparting member of the family. A polymer of diethylene glycol bis (allyl carbonate) has exceptionally good optical properties; it can be obtained in colorless, optically clear coatings. It is prepared in the following manner:

$$
\begin{array}{ccc}
CH_2-CH_2OH & CH_2-CH=CH_2 & CH_2-CH_2-O-COOCH_2-CH=CH_2 \\
| & | & | \\
O & +2\ O & \rightarrow O \\
| & | & | \\
CH_2CH_2OH & COOH & CH_2-CH_2-O-COO-CH_2-CH=CH_2 \\
\text{Diethylene} & \text{Allyl acid carbonate} & \\
\text{glycol} & &
\end{array}
$$

Physical and mechanical properties The allylics exhibit exceedingly good electrical properties under elevated temperatures and high-humidity conditions (see Table 6). They have high temperature-resistance levels. For example, diallyl phthalate resin is resistant to 300 to 350°F, and diallyl isophthalate resin resists continuous exposure to temperatures in the range 400 to 450°F. The allylics have low-moisture-absorption characteristics and good abrasion resistance.

Chemical resistance The allylic resins exhibit good resistance to common solvents, acids, aqueous inorganic salt solutions, and alkaline media (see Table 7).

Flammability The allylic resins are classified as self-extinguishing to nonburning by ASTM D 635.

Weatherability These resins are essentially unaffected by exposure to direct sunlight and high-humidity environments.

Applications Major use of allylic resins is in molded and laminated products, electrical and electronic parts, appliances, tubing, ducting, radomes and aircraft windows, and missile parts. The allylic monomers are often substituted for a portion of the commonly employed styrene in the cross-linking of polyester resins to obtain resin systems with good performance at higher temperatures.

Amino Resins

Preparation and composition The main amino resins are the reaction products of formaldehyde $\left(\begin{array}{c}H \\ \diagdown \\ C=O \\ \diagup \\ H\end{array}\right)$ with urea $\left(\begin{array}{c}O \\ \| \\ H_2N-C-NH_2\end{array}\right)$ or melamine:

$$
\begin{array}{c}
N \\
\diagup\ \diagdown \\
H_2N-C \quad\quad C-NH_2 \\
\| \quad\quad\quad \| \\
N \quad\quad\quad N \\
\diagdown\ \diagup \\
C \\
| \\
NH_2
\end{array}
$$

Physical and mechanical properties Melamine-formaldehyde resins have good electrical-resistance properties. They are rated as the hardest of the thermosets, with a Rockwell hardness range of M 118 to 124. They exhibit good heat resistance, low moisture absorption, and good dimensional stability. The melamines can be readily molded by all the common processing procedures (see Table 6).

Urea-formaldehyde resins have high strength, and exhibit good surface hardness. They have good electrical properties. The urea-formaldehyde resins have only fair temperature resistance, with 170°F being the top continuous operating temperature. These plastics should be considered rather brittle.

Chemical resistance The melamines and ureas are severely attacked by strong oxidizing acids and strong alkalies. Weak acids and weak caustics cause mild attack. Commercial-grade sodium hypochlorite solutions deteriorate the amino resins. Both the melamine- and urea-formaldehyde resins are unaffected by ketones, alcohols,

TABLE 6 Physical Properties of Thermosetting Resins

Resin	Specific gravity	Tensile strength, lb/in.²	Flexural strength, lb/in.²	Flexural modulus, lb/in.² × 10⁵	Compressive strength, lb/in.²	Heat-distortion temp, °F	Continuous heat resistance, °F	Coefficient of thermal expansion, in./in./°F × 10⁻¹	Flammability
Allylic resins:									
DAP	1.27	4,000	9,000	12.0	24,000	311	300	5.4	Self-extinguishing
DAIP	1.26	4,000	8,000	15.0	24,000	464	400	3.6	Self-extinguishing
Amino resins:									
Urea	1.5	6,000	15,000	14.0	38,000	266	170	4.0	Self-extinguishing
Melamine	1.45	6,500	11,000	12.0	45,000	298	210	4.5	Self-extinguishing
Epoxy resin (cast)	1.10–1.40	4,000–13,000	13,300–21,000	....	15,000–25,000	115–550	250–550	8.1–11.4	0.3 in./min
Furfuryl alcohol polymer (GFR)	1.45	11,500	18,000	6.5	12,500	420	330	....	Slow-burning
Phenolic resin (asbestos-filled)	1.70	3,500–5,500	5,500–7,500	....	12,000	....	300	6.0	Self-extinguishing
Silicone resin (glass-fiber-filled)	1.9	4,300	12,500	9.5	14,000	900	600	3.2	
Polyester (GFR)	1.8–2.3	4,000–10,000	12,000–20,000	7–10	20,000–30,000	400	300	4.5–6.0	Self-extinguishing
Urethane (cast)	1.25	7,500	3,500	0.8	20,000	....	225	9.1	
Vinyl ester resin (clear casting)	1.06	11,900	19,000	4.5	16,700	215	200	....	Slow-burning

Table 7 Chemical Resistance of Thermosetting Resins

Substance	Epoxy 75°F	Epoxy 150°F	Furan resin 75°F	Furan resin 150°F	Phenolic resin§ 75°F	Phenolic resin§ 150°F	Polyesters¶ 75°F	Polyesters¶ 150°F	Polyesters¶ 220°F	Vinyl ester 75°F	Vinyl ester 150°F
Acids:											
Acetic, 10–50%	E-F	E-F	E	E	E	E	E	G	F	E	G
Acetic, 75%	F	X	G	G	E	G	F	F	X	E	G
Acetic, glacial	F	X	G	G	G	G	X	X	X	G	G
Benzoic	E	E	E	E	E	E	E	E	G	E	E
Boric	G	E	E	E	E	E	E	E	E	E	E
Butyric	G	F	E	E	E	E	E	G	F	E	E
Chloroacetic	G	G	E	X	G	E	E	F	X	G	F
Chromic	F	X	X	E	G	G	G	G		G	F
Citric	G	G	E	E	E	E	E	E	E	E	E
Fluoboric	G	G	E	E	E	E	E	E		E	E
Hydrobromic, 25%	E	E	E	E	E	E	E	E	E (Dynel)	E	G
Hydrochloric, 10%	E	E	E	E	E	E	E	E	G	E	G
Hydrochloric, 37%	E	E	E	E	E	E	E	G	E	E	G
Hydrocyanic	E	E	F	X	X	X	E	E	F	G	F
Hydrofluoric	E	E	X	X	E	E	E	G		F	X
Hypochlorous	F	X	E	E	E	E	E	G	G	G	G
Lactic	E	E	E	E	F	X	E	E	F (Dynel)	E	E
Maleic	E	E	E	E	X	X	E	E	F	E	E
Nitric, 5%	F	X	F	X	E	E	G	G	G	G	E
Nitric, 65%	X	X	X	X	X	X	E	E	F	X	F
Oleic	X	E	E	E	E	E	E	X	X	E	X
Oleum	E	X	X	X	X	X	X	E		X	E
Oxalic	E	E	E	E	E	E	E	E	X	E	X
Perchloric	E	E	G	G	G	F	X	X	E	G	E
Phosphoric, 25%	E	E	E	E	E	E	E	E	G	E	X
Phosphoric, 85%	E	E	E	E	E	G	E	E	F	E	E
Phthalic	E	E	E	E	E	E	E	E	G	E	E
Sulfuric, 10%	X	X	G	X	F	E	E	X	G	E	E
Sulfuric, 78%	X	X	X	X	F	X	F	X	X	G	F
Sulfuric, 93%	E	X	X	X	X	X	X	X	X	F	X
Tannic	E	E	E	E	E	E	E	E	E	E	E
Tartaric	E	E	E	E	E	E	E	E	E	E	E

	C1	C2	C3	C4	C5	C6	C7	C8	C9	C10	C11
Alkalies:											
Ammonium hydroxide	X	G	F	G	G	E	E	E	E	E	E
Calcium hydroxide	E	E	F	G	G	E	E	E	E	E	E
Potassium hydroxide	F	X	X	F	F	E	E	E	E	E	E
Sodium hydroxide	E	E	X	X	X	E	E	E	E	E	E
Acid salts*	G-F	F	E	G	G	E	E	F	E	E	E
Alkaline salts†	E	E	E	E	E	E	E	E	E	E	E
Neutral salts‡											
Miscellaneous salts:											
Potassium dichromate	E	E	G	E	E	E	E	E	F	G	G
Potassium permanganate	E	E	G	E	E	E	E	E	X	X	X
Sodium cyanide	E	E	G	E	E	E	E	E	G	G	G
Sodium ferricyanide	E	E	X	E	E	E	E	E	E	E	E
Sodium hypochlorite	F	G	X	X	G	X	G	G	X	X	X
Organics and solvents:											
Acetone	X	X	X	X	X	E	G	E	E	G	G
Alcohol, methyl	F	G	G	G	E	E	G	E	E	E	E
Alcohol, ethyl	F	G	G	G	E	E	G	E	E	E	E
Alcohol, butyl	F	X	E	E	E	E	G	E	E	F	E
Aniline	X	X	X	X	X	E	E	E	X	G	X
Benzene	X	F	X	X	X	E	E	E	F	F	G
Carbon disulfide	X	F	X	X	X	E	E	T	X	F	F
Carbon tetrachloride	X	F	X	X	X	E	E	E	F	F	G
Chlorobenzene	X	F	X	X	X	E	E	E	X	F	F
Chloroform	X	X	X	X	X	E	E	C	X	G	G
Ethyl acetate	X	X	X	X	X	E	E	E	G	E	E
Ethylene chloride	X	X	X	E	E	E	E	E	E	E	E
Ethylene dichloride	X	X	X	E	E	E	E	E	E	E	G
Ethyl ether	X	X	X	X	X	E	E	G	X	G	F-X
Formaldehyde, 37%	G	E	E	E	E	E	E	E	E	E	E
Gasoline	X	G	G	E	E	E	E	E	G	G	G
Heptane	X	G	X	G	E	E	E	G	G	G	G
Methylene chloride	G	X	X	X	X	E	E	E	X	X	X
Methyl ethyl ketone	X	X	X	X	X	E	E	G	F	G	G
Nitrobenzene	X	X	X	X	X	E	E	E	F	F	F
Phenol	X	F	X	X	X	E	E	E	F	F	G
Toluene	X	F	X	X	X	E	E	E	X	F	G
Trichloroethylene	X	X	X	X	X	E	E	E	X	F	G

Table 7 Chemical Resistance of Thermosetting Resins (continued)

Substance	Epoxy 75°F	Epoxy 150°F	Furan resin 75°F	Furan resin 150°F	Phenolic resin§ 75°F	Phenolic resin§ 150°F	Polyesters¶ 75°F	Polyesters¶ 150°F	Polyesters¶ 220°F	Vinyl ester 75°F	Vinyl ester 150°F
Gases:											
Carbon dioxide	G	G	E	E	E	E	E	E	E	E	E
Carbon monoxide	G	G	E	E	E	E	E	E	G	E	E
Chlorine, dry	F	X	G	F	F	X	G	(Formation of chlorinated protective surface barrier)		E	G
Chlorine, wet	X	X	E	G	X	X	G			E	G
Hydrogen sulfide	E	E	E	E	E	E	E	E	G	E	G
Sulfur dioxide, dry	E	E	E	E	E	E	E	E	E	E	G
Sulfur dioxide, wet	E	E	E	E	E	E	E	E	E	G	G

E—excellent, no discernible attack. F—fair, mild attack, limited use.
G—good, no significant attack. X—unacceptable, prohibited attack.

* Acid salts: aluminum chloride, aluminum sulfate, copper chloride, copper sulfate, ferric chloride, ferric sulfate, stannic chloride, zinc chloride, zinc sulfate, etc.

† Alkaline salts: potassium bicarbonate, potassium carbonate, sodium bicarbonate, sodium carbonate, trisodium phosphate, etc.

‡ Neutral salts: calcium chloride, calcium nitrate, calcium sulfate, magnesium chloride, potassium nitrate, potassium sulfate, sodium chloride, sodium nitrate, sodium sulfate, etc.

§ These values are based upon a straight heat-cured phenol-formaldehyde resin. Modified phenolics may be expected to have less chemical resistance and should be tested.

¶ Note that this is a composite of various polyesters and should be used only as a guide. The resistance of each polyester should be checked before use.

chlorinated aliphatics, aliphatic hydrocarbons (heptane, hexane), aromatic hydrocarbons (benzene, toluene, xylene), kerosene, gasoline, oils, and greases. They are
completely resistant to most aqueous inorganic salt solutions (see Table 7).

Flammability   The melamines and ureas are classified as self-extinguishing by
ASTM D 635.

Weatherability   Melamine-formaldehyde and urea-formaldehyde resins suffer significant discoloration when subjected to direct sunlight for prolonged periods.

Epoxy Resins

Preparation and composition   Epoxy resins are basically thermosetting resins
which can be reacted with curing agents or catalytically homopolymerized to form
a cross-linked polymeric structure.   Their most outstanding property is their excellent adhesion to both metallic and nonmetallic surfaces.   In addition, cured epoxy
resins are characterized by their good mechanical and electrical properties, superior
dimensional stability, and good resistance to heat and chemical attack.

The epoxy resins may be divided into two broad categories: (1) glycidyl ether
resins and (2) epoxidized olefins.

Delmonte[22] has prepared an excellent concise "Summary of Miscellaneous Epoxy
Systems."

The glycidyl ether resins are manufactured by reacting epichlorohydrin and bisphenol A in the presence of aqueous caustic soda.   The reaction is always carried
out with an excess of epichlorohydrin so that the resin which is produced has terminal epoxy groups.   By varying the process conditions, as well as the amount of
excess epichlorohydrin, it is possible to obtain epoxy resins of low, medium, or high
molecular weight.   Chlorinated or brominated bisphenol A may be used to react
with epichlorohydrin to produce flame-resistant resins.

Substitution of the bisphenol A with polyglycols or dimerized unsaturated fatty
acids produces epoxy resins with greater flexibility and toughness.   Since these are
of lower viscosity than the conventional glycidyl ether resins, they are often used
to modify the final resin.

The epoxidized olefins may be produced through several routes using (1) oxygen
and selective metal catalysts, (2) perbenzoic acid, (3) acetaldehyde monoperacetate,
or (4) peracetic acid.[23]

The epoxy resin is characterized by the presence in the molecule of highly reactive
ethoxyline groups, which serve as terminal linear polymerization points.   Cross-
linkage through these groups and hydroxyl or other reactive groups in the molecule
results in tough, infusible thermosetting-resin systems.

The simplest form of the epoxy resin is represented by the diglycidyl ether of
bisphenol A:

Catalyst effect   The properties of cured epoxy resins may be varied over a wide
range through the selection of specific curing agents and variation in the curing
cycle.   Incorporation of reactive diluents, organic and inorganic fillers, and modification with other synthetic resins imparts even greater differences.[26]   Flexibilizing
epoxies are obtained by introducing long-chain glycols or fatty acids between the
epoxy groups in the molecule.   Similar flexibility can be accomplished through use of
special curing agents or compound modifiers.   However, such flexibilizing generally
results in lower solvent and chemical resistance.[26]   Brenner and Singer[27] reported on
the superior chemical resistance of epoxy-furane resin blends in acidic environments.

Physical and mechanical properties   The heat-distortion temperature of the
epoxies varies considerably among the various resin formulations.   Delmonte[22] summarizes thus:

Aliphatic polyamide–cured standard epoxies............................... 200–220°F
Aromatic amine–cured standard epoxies................................... 300–320°F
Aromatic resinous amine–cured standard epoxies.......................... 420–450°F
NADIC methyl anhydride–cured epoxies.................................. 400–500°F
Cycloaliphatic epoxies–optimum heat cure................................ 450–525°F

Other physical properties of epoxy resins are given in Table 6.

Chemical resistance   The epoxy resins are unaffected by most weak acids (see Table 7). They are attacked by acetic acid of concentrations in excess of 50 percent, chromic acid, hypochlorous acid, nitric acid, sulfuric acid in excess of 35 percent, and oleum. They are very resistant to the organic acids such as benzoic, citric, lactic, oxalic, maleic, oleic, and phthalic. They exhibit good resistance to weak and concentrated (38 percent) hydrochloric acid, dilute 15 percent hydrofluoric acid, 85 percent phosphoric acid, and 20 percent sulfuric. They are completely resistant to most salt solutions and are unaffected in aqueous solutions containing potassium permanganate or sodium hypochlorite. The chemical resistance of epoxy resins is influenced by the hardener used to cure them. Studies reported by Pschorr[25] demonstrated the effects of three chemically different types of hardeners (anhydride, amine, boron fluoride complex) upon chemical resistance. He reported, "Strong organic solvents such as chloroform, acetone and tetrahydrofuran had an adverse effect upon all systems. Weak solvents, such as hydrocarbons, alcohols or glycols affected them only slightly. Anhydride-cured systems exhibited better resistance to diluted acids while amine-cured systems showed better resistance to diluted bases. The best overall chemical resistance was obtained with deaminodephenyl sulfone."

Weatherability   Clear and light-colored epoxies will tend to yellow or discolor in sunlight. Observations on coatings after 15 years of exposure are not characterized by excessive cracking-shrinkage lifting, or loss of adhesion. Of all the epoxies, cycloaliphatic epoxies exhibit the best weathering.[22]

Flammability   The cast-epoxy resins range from self-extinguishing to slow-burning according to ASTM D 635, depending upon the resin system.

Chief corrosive-environment services   Epoxies are used for piping and ducts for handling highly corrosive liquids and vapors; rigid glass-fiber-reinforced storage tanks, receivers and feed tanks, hoppers, wet-crystal and dry-powder bins; coated-steel tanks, epoxy-resin-impregnated-graphite processing equipment, heat exchangers, and spargers; acidproof floor toppings and mortars for acidproof brick linings; pumps, valves, and blowers; and adhesives and structured laminates.

Furans

Preparation and composition   The furan resins are derivatives of:

$$
\begin{array}{cc}
\text{HC}\!\!-\!\!-\!\!-\!\!\text{CH} & \text{HC}\!\!-\!\!-\!\!-\!\!\text{CH} \\
\text{HC} \quad\; \text{C}\!\!-\!\!\text{C}\!\!=\!\!\text{O} \quad \text{or} \quad \text{HC} \quad\; \text{C}\!\!-\!\!\text{CH}_2\text{OH} \\
\text{O} \qquad \text{H} & \text{O}
\end{array}
$$

Furfuryl aldehyde   Furfuryl alcohol

which resinify rapidly under strongly acidic conditions.

Of the two, furfuryl alcohol has become the more important. The chief furan resins involve the alcohol as the sole polymeric compound, although there are resins of lesser commercial importance derived from combinations with formaldehyde or urea formaldehydes. Other combinations in the past have employed mixtures of phenol and furfural.

Furfuryl alcohol is readily resinified in the presence of acid catalysts. In practice, the resinification is checked at an intermediate liquid-resin stage by cooling and proper neutralization to a pH of 5 to 8. The dehydrated liquid resin is recatalyzed at the time of use by acidification, and passes through a rubbery stage to the final stage of infusibility.

The resinification is one of intermolecular dehydration with linear structure buildup:

$$
\begin{array}{l}
\text{HC}\text{——}\text{CH} \qquad\qquad \text{HC}\text{——}\text{CH} \\
\text{HC} \qquad \text{C—CH}_2\text{:OH} \quad \text{H:C} \qquad \text{C—CH}_2\text{OH} \xrightarrow{\text{—H}_2\text{O}} \\
\qquad \diagdown_{\text{O}}\diagup \qquad\qquad\qquad \diagdown_{\text{O}}\diagup
\end{array}
$$

$$
\begin{array}{l}
\text{HC}\text{——}\text{CH} \qquad \text{HC}\text{——}\text{CH} \\
\text{HC} \qquad \text{C—CH}_2\text{—C} \qquad \text{C—CH}_2\text{OH} \\
\qquad \diagdown_{\text{O}}\diagup \qquad\qquad \diagdown_{\text{O}}\diagup
\end{array}
$$

5-furfurylfurfuryl alcohol.

The reaction is allowed to continue to form the desired length of linear polymer:

$$
\begin{array}{l}
\text{HC}\text{——}\text{CH} \quad \Bigg[\; \text{HC}\text{——}\text{CH} \quad \Bigg]\; \text{HC}\text{——}\text{CH} \\
\text{HC} \qquad \text{C—CH}_2\text{—} \quad \text{—C} \qquad \text{C—CH}_2 \;\Bigg]_n\; \text{—C} \qquad \text{C—CH}_2\text{OH} \\
\qquad \diagdown_{\text{O}}\diagup \qquad\qquad \diagdown_{\text{O}}\diagup \qquad\qquad \diagdown_{\text{O}}\diagup
\end{array}
$$

This liquid is usually neutralized, and the water is removed prior to final cure of the resin. These fairly stable compounds are then recatalyzed with acid media and cross-linked to form infusible resinous masses that are extremely resistant to chemical attack.

It has been suggested[16] that loss of unsaturation during cross-linking indicates that the reaction is essentially a form of double-bond polymerization:

The fully cured furfuryl alcohol resins are dark-colored, thermally stable solids, extremely inert to most acids, alkalies, moisture, and solvents. The furan resins have extremely good thermal shock resistance, and will withstand temperatures as high as 265°F (see Table 6).

The speed with which cure occurs is dependent upon the concentration of the acidic catalyst as well as on the temperature. The catalysts most frequently employed include isopropyl sulfuric acid, p-toluene sulfonyl chloride, p-toluene sulfonic acid, and phosphoric acid.

The reaction which occurs during the linkage of the resin is quite exothermic and generates enough steam, from the water of formation, to result in severe blowholes through the resin. It is generally advisable to catalyze only that amount of furan which can be used immediately, since the catalyzed resin has a very short shelf life. Shallow mixing pans are recommended because deep pans permit a buildup of heat which results in eruptive blowing. Frequently, a prohibitively rapid cure is prevented by immersing the mixing container in ice to precool the catalyzed furan resin.

Chemical resistance The cured furan resins have excellent chemical resistance (see Table 7). The furan resins have better chemical resistance and heat resistance than the phenolics and polyesters. They are extremely resistant to alkalies and exhibit excellent solvent resistance. The furan resins are unaffected by boiling chloroform, chlorobenzene, toluene, acetone, ether, and esters. They show good resistance to water penetration. They are severely deteriorated by strong oxidizing agents, and are rapidly attacked by bromine, iodine, hydrogen peroxide, nitric acid, sulfuric acid above 60 percent concentration, sodium hypochlorite solutions, and concentrated chromic acid.

Weatherability   The furans exhibit excellent resistance to ultraviolet radiation and outdoor environments.

Flammability   The furans are classified as slow-burning by ASTM D 635.

Chief corrosive-environment services   Typical applications include:

1. Asbestos-filled or glass-fiber-reinforced furan-resin tanks, process or sewer piping, ducts, vent stacks, and pumps.

2. Furan-resin-impregnated graphite valves, pump-tube-and-shell heat exchangers, plate- and bayonet-type heat exchangers, cubical or block heat exchangers, spargers, falling-film absorber towers, bubble-cap or Raschig-ring-packed columns and piping, scrubbing towers, and vacuum jets.

3. Encapsulation of process equipment with hand lay-up of furan resin and glass tape to produce corrosion-resistant properties. Agitators, pH blocks, baffles.

4. Glass-fiber-reinforced furan-resin hand-lay-up lining in acidic service.

5. Repair patching of spalled-glass areas in glass-lined steel equipment.

Special considerations   Since the furans are acid-catalyzed to cure, it is impossible for their protective coatings to be applied directly to steel surfaces. It is necessary to employ a prime coat over the steel. It has been the practice in many cases to employ a modified phenolic prime coat for this service. Since the residual alkalinity of this prime coat would react with the acid catalyst in the furan coating and interfere with the generally good adhesive properties of the furan, it is imperative that complete removal of this residual alkalinity be accomplished before any attempt is made to apply the furan top coat.

Phenolics

Preparation and composition   The phenolics are usually considered to be the condensation products of phenol with aldehydes or ketones. In the simplest conception, although many types of end products may be produced through variations in the catalyzing medium and reaction conditions, phenol alcohols are produced and subsequently polymerized to the desired resinous material. The short-chain two-ring methylene derivatives, known as dihydroxydiphenylmethanes, and the phenol alcohols are the starting points from which the long-chain resins are derived. The curing is usually accomplished with heat or by acid catalyst with the splitting out of water.[1]

The reactions proceed along these general lines:

Phenol   Formaldehyde   Phenol alcohol

Phenol alcohol   Phenol   Dihydroxydiphenylmethane

While formaldehyde is the most widely used aldehyde, others, such as furfural, may be employed to produce resins with excellent molding properties. Similarly, other phenols, such as butyl phenol, the cresols, and resorcinol, may be substituted for the phenol to produce resins with special desirable properties.

Physical and mechanical properties   The unmodified phenol-formaldehyde resins exhibit high dimensional stability over a wide temperature range; they may be used at temperatures as high as 350 to 400°F for long periods of time and are non-flammable. Phenolic-board stock, prepared from impregnated macerated fabric and cotton and nylon flock, or cloth, finds many elevated-temperature applications in industry. Phenolic-impregnated graphite has tensile strength in the range of 800 to 1,500 lb/in.². The tensile strengths of the asbestos-filled-phenolic cast materials are quite low (3,000 to 6,000 lb/in.²) (see Table 6).

Chemical resistance   The unmodified, heat-cured phenol-formaldehyde resins are unaffected by most concentrations of inorganic acids (see Table 7). They are, however, severely attacked by concentrated sulfuric acid, oleum, nitric acid, and hot chromic acid. They are insoluble in most aliphatic, aromatic, and halogenated hydrocarbons. No deleterious effects are experienced on exposure to esters, ethers, glycols, ketones, and alcohols. They are resistant to gasoline, mineral and vegetable oils, and lubricants. They are unaffected by refrigerants such as Freon, methyl chloride, and ammonia. They are extremely resistant to acidic and neutral inorganic salt solutions. They are resistant to most organic carboxylic acids, organic fatty acids, and their salt solutions. Unless modified, they are poorly resistant to alkaline media and strong oxidizing agents. They have high resistance to moisture absorption.

Weatherability   The phenol-formaldehyde resins exhibit darkening and minor degradation under direct sunlight. They experience very low moisture absorption.

Flammability   The unmodified phenol-formaldehyde resins are classified as self-extinguishing by ASTM D 635.

Chief corrosive-environment services   The phenolic resins are employed extensively as baked coatings for linings in tank cars, storage tanks, process equipment, valves, and piping. Glass fibers and other natural and synthetic fibers are employed as laminates in many applications. The phenolic resins are used as chemical-resistant impregnants for carbon and graphite fabrications, as well as with asbestos in specifically molded process piping, valves, pumps, blowers, and tanks. These fabrications have the undesirable quality of being fragile and easily fractured under tensional stresses or by impact. Some specific applications of phenolics are:

1. Asbestos-filled phenol-formaldehyde resin crystallizers, quenchers, boiling tubs, hold tanks, bins, stacks, and ducting
2. Fume and acid-vapor scrubbing towers and jets
3. Phenolic-resin-impregnated graphite valves, pumps, heat exchangers, distillation (bubble-cap or packed) columns, and split-flow controllers
4. Coated tray and shelf vacuum dryers, rotary dryers, tanks, pressure filters, tank cars, and containers
5. Impregnated-wood surfaces: filter plates, plywood structures, filter bins
6. Doctor and scraping blades on flaker drums
7. Ring-seal gaskets in gas and oil industry for Christmas-tree manifolds and piping
8. Photographic-development tanks
9. Rayon-spinning brackets and accessories
10. Electrical equipment, housings, ignition parts, blocks, etc.
11. Pulleys, wheels, gears, etc.

Special considerations   Particular care must be taken to maintain all nozzles and other appendages under compression to ensure against cleavage or cracking. Heavy piping hung on the end of valves leading off tanks or flanges on equipment must be supported independently to prevent fractures of the associated nozzles. Reinforcement by armoring the nozzle with epoxy-impregnated glass cloth may serve as a substitute for supports in many cases.

Silicone Resins

Preparation and composition   The silicone resins are prepared by the hydrolysis of the organochlorosilane ($RSiCl_3$, R_2SiCl_2, R_3SiCl) monomers into the polymers by cross-linking.

$$\left[\begin{array}{cc} CH_3 & CH_3 \\ | & | \\ -O-Si-O-Si-O- \\ | & | \\ CH_3 & CH_3 \end{array} \right]_n$$

Polydimethylsiloxane

The silicone resins owe their high heat stability to the strong silicon-oxygen-silicon bonds. The resin systems vary significantly in their physical properties as a result of the degree of their cross-linkage and the type of radical (R) within the monomer molecule. In this regard the chief radicals are methyl, phenyl, or vinyl groups.

Physical and mechanical properties The silicone resins have extremely good thermal stability, excellent electrical-insulating properties, and high water repellency; they also exhibit good chemical resistance. Employed in the form of glass-fiber-reinforced laminates, they find many uses in temperatures as high as 500°F (see Table 6).

Chemical resistance The silicone resins are unaffected by weak acids, weak caustics, and aqueous inorganic salt solutions. They offer very good resistance to most solvents, aliphatic hydrocarbons, gasoline, greases, and lubricating oils. They are severely attacked by hot concentrated alkalies and strong acid solutions. Since the chemical resistance varies with the molecular structure of the resin, it is advisable to test the specific silicone resin in the intended environment and not rely upon the general chemical-resistance tables.

Flammability The silicone resins are classified as slow-burning to self-extinguishing as determined by ASTM D 635.

Weatherability The high moisture resistance and inertness to prolonged periods of direct sunlight emphasize the excellent weatherability of these silicone polymers.

Application The silicone resins find use as high-temperature-resistant protective coatings for exhaust stacks; tubing; auxiliary parts for the electromotive, jet aircraft, and electrical industries; laminated structural members; radomes; thermal barriers, and sizing on glass cloth for polyesters.

Unsaturated Polyesters

Preparation and composition Polyester resins are the polycondensation products of dicarboxylic acids with dihydric alcohols, either or both of which contain a double-bonded pair of carbon atoms. The initial unsaturated polyester formed is cross-linked with a polymerizable monomer (e.g., styrene, vinyl toluene, methyl methacrylate) under the influence of a peroxide catalyst to form a thermoset, which when fully cured is transformed into a rigid product that is relatively infusible and insoluble. Polyester resins cannot be reshaped or remelted after they have been cured to their final stage. The heat required to activate the molecule to an energy level high enough to initiate the cross-linkage and polymerization of the polyester can also originate from externally applied heat, pressure, or ultraviolet radiation. Thus these materials should be kept in relatively cool storage areas and not stored in translucent containers.

Linear repeating molecules of polyester are partially interlinked during the initial production. The monomer used in cross-linking, usually styrene, is added to this mixture along with a small amount of inhibitor; the peroxide catalyst and other curing accelerators are added just prior to the application.

The structural formula of the polyester resin prepared from ethylene glycol maleate and styrene may be represented by the following:

$$-CH_2-CH_2O-C-\underset{|}{CH}-CHC-O-CH_2-CH_2O-C-\underset{|}{CH}-CH-C-O-$$

Although the general-purpose polyester above has some fairly good chemical-resistance properties, it is not considered acceptable for severely aggressive environments.

Polyester-resin differences In the industry one is aware that there are pronounced differences in the chemical or solvent resistance of the various polyester resins on the market. The general-purpose resins, usually employed for such structural items as windows, doors, corrugated awnings, carts, boats, and decorative materials, are generally obtained as condensation products from phthalic anhydride, maleic anhydride, ethylene or propylene glycol, and styrene. They are not usually considered acceptable as corrosion-resistant materials of construction. They are somewhat less expensive than the chemical-resistant polyesters and are employed in those areas in which they would be exposed only to mildly corrosive environments.

However, for highly corrosive services, these general-purpose resins yield to the chemical-resistant varieties. The chemical-resistant polyesters are either general-purpose formulas which have been modified to enhance their chemical and solvent resistance or specially formulated products. The latter group, the specially formulated polyesters, includes propoxylated bisphenol A–fumarate, special isophthalic acid resins, those based upon hydrogenated bisphenol A, and those formulated with chlorendic anhydride.

Flammability Because the polyester resins, being organic compounds, will burn, some fire-retardant formulations have been introduced. The fire-retardancy arises either through incorporation of additives into the standard polyester resin or through internal modifications of the molecule. Additives generally take such a form as 5 percent by weight of antimony trioxide or similar materials. Chlorine-bearing plasticizers as well as others, such as tricresyl phosphate, have been employed as diluents in some instances, but they lessen the chemical resistance significantly. Internal modification is exemplified by polyesters made with hexachloroendomethylene or chlorendic anhydride; these resins have excellent chemical resistance, too.

Further modification of the polyester resin can be accomplished through selection of specific monomers to instill higher-temperature resistance in the final product. An example of this technique is the substitution of a portion of the styrene monomer with materials such as triallyl cyanurate or diallyl phthalate.

Reinforced polyester physical properties With only rare exceptions the polyesters are not employed without some reinforcing media or fillers. This is due basically to an inherent shrinkage during final curing, which can be in the range of 1 to 6 percent. For this reason it has been found advisable to use glass or synthetic fibers as reinforcement. The many forms of glass fiber available include continuous strands, chopped strands, reinforcing mats, surfacing mats, woven rovings, and fabric. Recently a glass flake has been added to this group.

The surface-treatment specifications for the reinforcement will depend upon the ultimate performance requirement. Special sizes for protection of the strands are available for use depending upon the specific resin system.

Chemical resistance General-purpose polyester resins are resistant to weak inorganic acids, weak alkalies, and many salt solutions. They are not resistant to strong acids, strong alkalies, or organic solvents (see Table 7).

Polyesters made from isophthalic acid, bisphenol A, hydrogenated bisphenol A, and the adduct of hexachlorocyclopentadiene and maleic anhydride also have good to excellent chemical resistance.

These chemical-resistant polyester resins are unaffected by most weak and strong acids, strong alkalies, aliphatic hydrocarbons, most alcohols, and most salt solutions. They are not resistant to attack by concentrated nitric acid, oleum, 93 percent sulfuric acid and other highly oxidizing media, aromatic hydrocarbons, esters, ketones, or halogenated solvents. They are highly resistant to moisture absorption.

Weatherability The polyesters experience yellowing and mild degradation under the influence of direct sunlight. Generally ultraviolet absorbers such as 2-hydroxy-benzophenones are employed in the formulation of the resin if the structure is to experience outdoor exposures.

Flammability Conventional rigid thermosetting polyester coatings burn at a rate of 0.7 to 2.0 cm/min by ASTM D 635. Fire-retardant varieties are classified as self-extinguishing to nonburning, with rates of 0.2 to 0.8 cm/min.

Chief corrosive-environment services

1. Reactors, storage tanks, catch tanks, boiling tubs, receivers, ground tanks, separators
2. Reclaimed water funnels, sumps, laboratory sinks
3. Ducts, vent stacks, hoods
4. Process piping, sewer piping, standpipes, dip pipes

   5. Dry-powder bins, feeders, chutes, carts
   6. Wet-cake hoppers, screw-feed troughs, conveyor covers
   7. Chlorine absorbers, off-gas scrubbers, mist eliminators
   8. Doors, windows, canopies, paneling, structural members
   9. Equipment safety shields, motor covers
  10. Fans, blowers, eductors, jets, exhausters
  11. Filter-press frames, slide-blast gates, check valves
  12. Waste-disposal troughs, settling basins
  13. Guttering, downspouts
  14. Alcohol columns
  15. Encapsulated steel-saddle supports for horizontal tanks
  16. Chopped fiber-spray and/or glass-flake linings for in-place rehabilitation of process equipment
  17. Polyester-resin-coated laminated plywood evaporators and boiling tubs
  18. Boats and marine equipment
  19. Electrical equipment and parts
  20. Structural members
  21. Automotive industry
  22. Laundry equipment
  23. Sporting goods

Polyurethanes

Preparation and composition[1]   The reaction of unsaturated polyesters or polyethers with isocyanates is the general source for polyurethanes.   The isocyanates react with those polyesters containing hydroxyl groups to form urethane linkages and with those polyesters containing carboxyl groups to form amide linkages.   In either case, CO_2 is evolved to produce the polyurethane, as shown below.

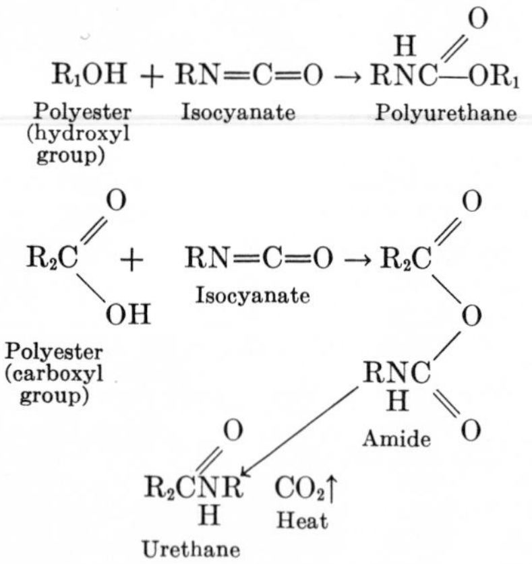

$$R_1OH + RN{=}C{=}O \rightarrow RNC\overset{O}{\underset{}{\Big/\!\!/}}OR_1$$

Polyester (hydroxyl group)   Isocyanate   Polyurethane

$$R_2C\overset{O}{\underset{OH}{\diagup\!\!/}} + \;RN{=}C{=}O \rightarrow R_2C\overset{O}{\underset{O}{\diagup\!\!/}}$$

Polyester (carboxyl group)   Isocyanate

$$RNC\overset{H}{\diagup}$$

Amide

$$R_2C\overset{O}{\underset{H}{N}R} \quad CO_2\uparrow$$

Urethane   Heat

   Curing is obtained through further reaction with bifunctional compounds containing active hydrogen, such as water, glycols, diamines, or amino alcohols.

   Depending upon the reaction procedure, the urethane can be obtained as rigid or flexible forms, castings, elastomers, protective coatings, or adhesives.

Physical and mechanical properties   The flexible forms have densities in the range of 1.0 to 6.0 lb/ft³. The rigid urethane forms may be prepared with densities from as low as 1.0 lb/ft³ to as high as 50 lb/ft³. They have good load-bearing properties, extremely low thermal conductivity, high-temperature resistance, good high tensile properties, excellent acoustical properties, and good erosion resistance (see Table 6).

Chemical resistance   The urethanes are swelled by many common solvents. However, the urethane returns to its original dimensions when removed from the solvent and air-dried.   Urethanes are fairly resistant to aliphatic solvents, alcohols, ethers, and fuel oils. They are attacked by hot water and polar solvents. They exhibit good oxygen resistance. Strong oxidizing acids, strong inorganic acids, and concentrated caustics cause severe attack.

Flammability   The urethanes are classified as slow-burning to self-extinguishing by ASTM D 635.

Weatherability   Urethanes experience mild deterioration under the influence of prolonged direct sunlight. They have low moisture absorption and are unaffected by high-humidity environments.

Application   The major application of the polyurethanes has been in protective coatings, synthetic elastomers, and rigid and flexible foams. A typical foam formulation may incorporate the reaction of a polyester—prepared from adipic acid, diethylene glycol, and trimethylol propane—with tolylene diisocyanate in the presence of a suitable activator mixture containing the adipic acid ester of N-ethyl aminoethanal. Such foams have extremely good K factors, which together with their comparatively low cost, make them excellent choices for insulation coverings. They serve also as cushioning to absorb vibration. An outstanding characteristic of these polyurethane foams is the fact that they may be foamed in place, without the application of external heating or pressure. Polyurethane protective coatings show excellent resistance to oxidation and solvents and thermal stability to 240 to 260°F. They are quite abrasion-resistant.

Vinyl Esters

Preparation and composition   A new family of resins, designated as vinyl ester resins, has found application recently as a substitute for polyesters in the fabrication of glass-fiber-reinforced structures. These are polymerizable resins in which the terminal portions of the resin are vinyl ester groups

$$\left[-CH_2=CH-\overset{\overset{\displaystyle O}{\|}}{\underset{\underset{\displaystyle R}{|}}{C}}-O- \right]_n$$

and in which the main polymeric chain between the terminal portions comprises the residue of a polyepoxide resin.

An example of the molecular structure of one typical ester might be that resulting from the reaction product of two moles of methacrylic acid with a conventional polyepoxide:

$$\left[CH_2=\overset{\overset{\displaystyle}{}}{\underset{\underset{\displaystyle CH_3}{|}}{C}}-\overset{\overset{\displaystyle O}{\|}}{C}-OCH_2CHCH_2-O-\bigcirc \right]_2 \rightarrow \overset{\overset{\displaystyle CH_3}{|}}{\underset{\underset{\displaystyle CH_3}{|}}{C}}$$

Polyesters vs. vinyl esters   The main differences between vinyl esters and polyesters have been stated to be the following:

1. Vinyl esters do not necessarily have regularly recurring ester groups as part of the main polymeric chain.

2. Vinyl esters terminate in vinyl groups, while polyesters terminate in carboxylic or hydroxyl groups.

3. Vinyl esters are produced from a different set of raw materials than polyesters (polyester variations involve only the use of various dibasic acids and polyhydric alcohols).

4. Polyester-resin manufacture is a condensation polymerization, while vinyl-ester manufacture is an addition polymerization.

5. Vinyl esters are low-molecular-weight, precision polymers, while polyesters are high-molecular-weight, random polymers.

The vinyl ester is diluted, like polyesters, with styrene to about 45 to 50 percent and the resin mixture is cured by the use of peroxides, cobalt naphthenate, and dimethyl aniline.

Chemical resistance   Vinyl esters are unaffected by aliphatic hydrocarbons, strong bases, and dilute acids, bases, and salt solutions. They exhibit fair resistance to strong oxidants and strong acids. They are attacked by aromatic solvents, chlorinated solvents, esters, and ketones (see Table 7).

Chief corrosive-environment services   The vinyl esters are employed to build all types of chemical-processing equipment and are used in about all the corrosive environments in which the epoxy and polyester resins see service. As glass-fiber-reinforced resin structures they are used in piping, ducts, tanks, hoods, vent stacks, bins, feeders, chutes, scrubbing towers, fans, blowers, jets, pumps, valves, etc. Reinforced with glass flake, the vinyl esters are used as protective linings in steel tanks.

Flammability   The vinyl-ester resins are classified as burning resins by ASTM D 635. Some modified vinyl esters are fire-retardant, however.

ELASTOMERS

Fundamentals

Relationship to plastics   Closely related to the thermoplastic and thermosetting polymeric compounds are the synthetic elastomers. In many instances the same building-block monomers, such as styrene and acrylonitrile, may be found as part of their structure.

ASTM definition of elastomer   The term *elastomer* has been defined by the American Society for Testing and Materials as meaning "a polymeric material which at room temperature can be stretched to at least twice its original length and upon immediate release of the stress will return quickly to approximately its original length."

In this section we will be reviewing the general physical properties and chemical resistivity of natural rubber, for comparison, and the chief synthetic elastomers of commercial value.

Natural Rubber

Preparation and composition   Natural rubber is obtained principally from the latex of *Hevea brasiliensis*, a tree. This elastomer is based upon natural *cis*-polyisoprene. The principal repeating unit is $(-CH_2-C=CH-CH_2-)_n$.

$$\overset{\displaystyle |}{CH_3}$$

Natural rubber is generally classified into soft, semihard, and hard rubber depending upon the degree of its curing. Vulcanization is accomplished by reaction with sulfur and/or sulfur-bearing compounds and additional organic accelerators in the presence of certain inorganic oxides at temperatures in the range of 212 to 400°F. The degree of vulcanization (hardness) is dependent upon the amount of sulfur contained in the final product. Soft rubbers have a maximum of about 0.5 to 4.0 percent sulfur, with a durometer Shore A reading of 20 to 30. Hard rubber (Ebonite) contains about 32 percent sulfur with a Shore D durometer between 50 and 100. Semihard rubber falls somewhere between these ranges. Neither semihard nor hard rubber meets the ASTM definition of elastomers; only soft rubber may be so considered. Thus chemical-resistance information is included only for soft rubber (see Table 8).

Natural rubber has outstandingly high tensile strength and excellent resiliency, and it retains its mechanical properties to a great extent at both elevated and low temperatures. However, it is susceptible to high-temperature degradation under prolonged exposure (see Table 9).

Chemical resistance   Natural rubber is affected by mineral and vegetable oils, gasoline, benzene, toluene, and chlorinated hydrocarbons. It is severely attacked by strongly oxidizing materials such as nitric acid, concentrated sulfuric acid, dichromates, permanganates, sodium hypochlorites, and chlorine dioxide. It is not affected by most inorganic salt solutions, alkalies, and nonoxidizing acids. Hydrochloric acid reacts with soft rubber to form rubber hydrochloride. The reaction of hydrochloric acid with semihard and hard rubber is quite slow, and they are acceptable for such service (see Table 8).

Synthetic Rubber (IR)

This is synthetic *cis*-polyisoprene, which is chemically and physically similar to natural rubber. Its chemical resistance is identical to that of natural rubber.

Acrylonitrile-Butadiene (NBR)

Preparation and composition These elastomers are familiar as nitrile rubber or Buna N. Their properties and chemical resistance can be varied by changing the ratio of acrylonitrile to butadiene. These polymers usually contain between 20 and 40 percent acrylonitrile. A decrease in acrylonitrile results in a decrease in fuel and oil resistance. However, lowering the acrylonitrile content results in an elastomer with improved low-temperature flexibility and resiliency. Proper compounding can result in products with high tensile strength, excellent abrasion resistance, and good aging characteristics (see Table 10).

Chemical resistance The nitrile rubbers have good resistance to oils and solvents. They exhibit good resistance to alkalies and aqueous salt solutions. They swell very slightly in aliphatic hydrocarbons, fatty acids, alcohols, and glycols. The reduction in physical properties as a result of swelling is small, making the NBR elastomers especially suitable for gasoline and oil-resistant applications. NBR is attacked by strong oxidizing agents, ketones, ethers, and esters (see Table 8).

Polyacrylic Rubber

Preparation and composition The class of elastomers known as polyacrylic rubbers are polymers of acrylic acid esters prepared from alcohols of intermediate weight. Typical of this group are polyethyl acrylate and polybutyl acrylate, as well as the copolymer of ethyl acrylate and 2-chloroethyl vinyl ether. Vulcanization of these polymers is quite difficult inasmuch as they are saturated materials. Sulfur and sulfur-bearing compounds in fact act as curing retardants, and perform more as aging deterrents. These polyacrylic rubbers are cured with amines, such as triethylenetetramine. Alkaline reagents such as sodium metasilicate, potassium hydroxide, or lead oxide have been used used satisfactorily too. Cross linking is said to occur by a Claisen-type condensation reaction with the splitting-out of alcohol:

$$
\begin{array}{ccc}
\text{—CH}_2\text{—CH—} & & \text{—CH}_2\text{—CH} \\
| & & | \\
\text{C=O} & & \text{C=O} \\
| & & | \\
\text{OC}_2\text{H}_5 \rightarrow \text{—CH}_2\text{—C} & & + \text{ C}_2\text{H}_5\text{OH}\uparrow \\
\text{—CH}_2\text{—CH—} & & | \\
| & & \text{C=O} \\
\text{C=O} & & \text{OC}_2\text{H}_5 \\
| & & \\
\text{OC}_2\text{H}_5 & &
\end{array}
$$

Chemical resistance The fully cured polyacrylic rubbers exhibit good resistance to hot oils and air up to 350°F. They have good resistance to petroleum products, aliphatic hydrocarbons, and animal and vegetable fats and oils. They will swell in aromatic hydrocarbons, alcohols, and ketones. Solvents frequently employed with polyacrylic rubbers are methylethyl ketone, toluene, xylene, or benzene. The polyacrylic rubbers deteriorate when subjected to aqueous media, steam, glycols, and caustic environments. Copolymers of acrylates and acrylonitrile produce acrylic rubbers with improved water and steam resistance. The acrylic rubbers are unaffected by oxygen and ozone (see Table 11).

Physical properties The polyacrylic rubbers have good flex resistance and low permeability to hydrogen, helium, and carbon dioxide. They have fairly poor weathering characteristics, since they are affected by water. Discoloration in direct sunlight is negligible.

Butyl Rubber (IIR)

Preparation and composition A general-purpose synthetic rubber is obtained through the copolymerization of isobutylene with 1 to 35 percent of isoprene in the

Table 8 Chemical Resistance of Elastomers

Substance	Natural soft rubber		Butadiene-acrylonitrile (Buna N)		Butyl rubber		Chloroprene rubber (Neoprene)	
	75°F	150°F	75°F	150°F	75°F	150°F	75°F	150°F
Acids:								
Acetic, 10–50%	G-F	F-X	X	X	G	G	E	G
Acetic, 75%	X	X	X	X	G	F	F	X
Acetic, glacial	X	X	X	X	G	F	F	X
Benzoic	G	G	G	G	G	G	E	E
Boric	G	G	E	E	X	X	E	E
Butyric	G	G	X	F	G	G	X	X
Chloroacetic	X	X	X	X	F	X	G	F
Chromic	X	X	X	X	E	X	X	X
Citric	G	G	G	F	E	E	E	E
Fluoboric	G	G	X	G	X	G	E	E
Hydrobromic, 25%	G	F	X	X	X	G	X	F, severe permeation
Hydrochloric, 10%	G	G	G	F	X	X	G	X, severe permeation
Hydrochloric, 37%	G	F	X	X	X	X	G	X, severe permeation
Hydrocyanic	G	G	X	X	G	X	E	G
Hydrofluoric	F	X	X	X	X	X	G	X
Hypochlorous	G	F	X	X	E	G	X	X
Lactic	G	F	G	F	E	G	E	E
Maleic	F	X	X	X	X	X	G	X
Nitric, 5%	X	X	X	X	F	X	F	F
Nitric, 65%	X	X	X	X	X	X	X	X
Oleum	X	X	X	G	X	X	X	X
Oxalic	G	G	G	G	G	G	F	X
Perchloric	G	G	F	X	G	G	X	E
Phosphoric, 25%	G	G	X	X	E	E	E	E
Phosphoric, 85%	G	G	X	X	G	F	E	G
Phthalic	F	X	G	G	X	X	E	G
Sulfuric, 10%	E	G	X	X	E	E	E	E
Sulfuric, 78%	X	X	X	X	G	F	X	X
Sulfuric, 93%	X	X	X	X	X	X	X	X
Tannic	G	F	G	G	E	E	E	E
Tartaric	E	G	G	G	E	E	E	E

Chemical	1	2	3	4	5	6	7	8
Alkalies:								
Ammonium hydroxide	E	E	E	E	E	G	G	G
Calcium hydroxide	E	E	E	E	E	G	G	G
Potassium hydroxide	E	E	E	E	E	G	G	G
Sodium hydroxide	E	E	E	E	G	G	G	G
Acid salts*	G	E	E	E	G	G	G	G
Alkaline salts†	E	E	E	E	E	G	G	G
Neutral salts‡	E	E	E	E	G	G	G	G
Miscellaneous salts:								
Potassium dichromate	E	X	E	G	E	X	X	X
Potassium permanganate	E	E	G	G	E	X	G	X
Sodium cyanide	E	E	G	E	G	G	G	G
Sodium ferricyanide	E	G	G	G	X	X	G	G
Sodium hypochlorite	E	X	F	G	X	X	X	X
Organics and solvents:								
Acetone	X	X	F	X	X	X	X	X
Alcohol, methyl	F	E	G	G	G	G	G	G
Alcohol, ethyl	F	E	G	G	G	E	E	E
Alcohol, butyl	F	G	G	G	X	G	E	E
Aniline	X	X	X	F	X	X	X	X
Benzene	X	X	X	X	X	X	X	X
Carbon disulfide	X	X	X	X	X	X	X	X
Carbon tetrachloride	X	X	X	X	X	X	X	X
Chlorobenzene	X	X	X	X	X	X	X	X
Chloroform	X	X	X	F	X	X	X	X
Ethyl acetate	X	X	X	F	X	X	X	X
Ethylene chloride	X	X	X	F	X	X	X	X
Ethylene dichloride	X	X	X	X	X	X	X	X
Ethyl ether	E	E	F	G	F	F	F	F
Formaldehyde, 37%	F	E	X	X	F	X	X	F
Gasoline	G	G	X	X	G	G	X	X
Heptane	G	G	X	X	F	X	X	X
Methylene chloride	X	X	X	X	X	X	X	X
Methyl ethyl ketone	F	F	F	F	X	X	X	X
Nitrobenzene	F	G	F	F	X	X	X	X
Phenol	X	X	X	F	X	X	X	X
Toluene	X	X	X	F	X	X	X	X
Trichloroethylene	X	X	X	X	X	X	X	X
Gases:								
Carbon dioxide	G	G	E	E	E	E	G	G
Carbon monoxide	G	G	F	G	X	X	X	X
Chlorine, dry	X, severe permeation	X, severe permeation	X	X	X	X	X	X
Chlorine, wet	X	X	X	X	X	X	X	X
Hydrogen sulfide	E	E	G	E	X	X	X	X
Sulfur dioxide, dry	E	E	G	E	X	X	X	X
Sulfur dioxide, wet	E	E	G	E	X	X	X	X

Table 8 Chemical Resistance of Elastomers (continued)

Substance	Chlorosulfonated polyethylene (Hypalon)		Ethylenepropylene dimer		Polysulfide rubber (Thiokol)		Polyurethane elastomer	
	75°F	150°F	75°F	150°F	75°F	150°F	75°F	150°
Acids:								
Acetic, 10–50%	E	G	G	G	G	F	F	F
Acetic, 75%	G	G	G	G	G	F	X	X
Acetic, glacial	G	F	G	G	G	F	X	X
Benzoic	G	G			G	G	G	G
Boric	E	E	G	G	G	G	E	E
Butyric	F	X	G	G	G	F	X	X
Chloroacetic	E	G			X	F	X	X
Chromic	E	G	F	F	G	X	X	X
Citric	E	E	G	G	G	G	G	G
Fluoboric	E	E	G	G	E	G	E	E
Hydrobromic, 25%	G	G	G	G	E	F	X	X
Hydrochloric, 10%	E	E	G	G	G	E	F	F
Hydrochloric, 37%	G	F	G	G	G	E	F	F
Hydrocyanic	E	E			X	F	X	X
Hydrofluoric	E	G	F	F	G	G	G	G
Hypochlorous	G	F			G	G	X	X
Lactic	E	E	F	F	X	X	G	G
Maleic	G	G			G	G	G	F
Nitric, 5%	E	E	G	G	X	G	X	X
Nitric, 65%	F	X	X	X	G	X	X	X
Oleic	G	G	G	G	X	G	F	F
Oleum	X	X	X	X	G	X	X	X
Oxalic	G	G	G	G	X	G	G	F
Perchloric	G	F			G	X	F	F
Phosphoric, 25%	E	E	G	G	E	G	F	X
Phosphoric, 85%	E	G	G	G	E	F	X	F
Phthalic	G	G			G	G	G	G
Sulfuric, 10%	E	E	G	G	G	G	F	F
Sulfuric, 78%	E	G	F	F	X	X	X	X
Sulfuric, 93%	G	F	F	X	G	G	X	X
Tannic	E	E	G	F	G	G	G	G
Tartaric	E	E	F	F	G	G	G	G
Alkalies:								
Ammonium hydroxide	E	E	G	G	E	G	E	E
Calcium hydroxide	E	E	G	G	E	E	E	E
Potassium hydroxide	E	E	G	G	E	E	E	E
Sodium hydroxide	E	E	G	G	E	E	E	E
Acid salts*	E	E-G	G-F	G-F	E	G-F	G	G

Material						
Alkaline salts†	E	E	E	G	E	E	E
Neutral salts‡	E	E	E	G	E	E	E
Miscellaneous salts:							
Potassium dichromate	E	E	E	G	G	G	F
Potassium permanganate	G	G	G	G	G	G	F
Sodium cyanide	G	G	G	G	G	E	F
Sodium ferricyanide	E	E	E	E	E	E	G
Sodium hypochlorite	G	G	G	G	G	G	F
Organics and solvents:							
Acetone	F	F	F	G	G	G	X
Alcohol, methyl	G	G	E	G	E	E	X
Alcohol, ethyl	E	E	E	E	E	E	X
Alcohol, butyl	G	G	E	G	E	E	X
Aniline	F	X	X	X	G	G	X
Benzene	X	X	X	X	G	G	X
Carbon disulfide	X	X	X	X	G	G	X
Carbon tetrachloride	X	X	X	F	X	X	X
Chlorobenzene	X	X	X	G	G	G	X
Chloroform	X	X	X	F	G	G	X
Ethyl acetate	X	X	X	X	G	G	X
Ethylene chloride	X	X	X	F	F	F	X
Ethylene dichloride	X	X	X	X	X	X	X
Ethyl ether	X	X	X	G	G	G	X
Formaldehyde, 37%	E	E	E	E	E	E	F
Gasoline	F	F	G	F	G	E	F
Heptane	G	G	X	G	X	G	X
Methylene chloride	X	X	X	X	G	X	X
Methyl ethyl ketone	X	X	X	X	G	X	X
Nitrobenzene	X	X	X	X	F	F	X
Phenol	X	X	X	X	G	X	X
Toluene	X	X	X	X	F	F	X
Trichloroethylene	X	X	X	X	F	F	X
Gases:							
Carbon dioxide	E	E	E	G	G	E	E
Carbon monoxide	E	E	E	G	G	E	E
Chlorine, dry	F	F	F	F	F	X	X
Chlorine, wet	F	F	F	F	G	X	X
Hydrogen sulfide	E	E	E	G	F	G	G
Sulfur dioxide, dry	E	E	F	F	F	G	G
Sulfur dioxide, wet	E	E	F	F	F	G	G

E—excellent, no discernible attack.   F—fair, mild attack, limited use.
G—good, no significant attack.   X—unacceptable, prohibited attack.

* Acid salts: aluminum chloride, aluminum sulfate, copper chloride, copper sulfate, ferric chloride, ferric sulfate, stannic chloride, zinc chloride, zinc sulfate, etc.

† Alkaline salts: potassium bicarbonate, potassium carbonate, sodium bicarbonate, sodium carbonate, trisodium phosphate, etc.

‡ Neutral salts: calcium chloride, calcium nitrate, calcium sulfate, magnesium chloride, potassium nitrate, potassium sulfate, sodium chloride, sodium nitrate, sodium sulfate, etc.

TABLE 9 General Properties of Natural Soft Rubber

Property	Value
Specific gravity	0.93
Tensile strength, lb/in.^{2}	2,500–3,500
Elongation, %	750–850
Hardness (durometer)	A30–90
Tear resistance	Excellent
Abrasion resistance	Excellent
Recommended operating temp:	
Min °F	−60
Max °F	180
Chemical resistance:	
Acids, dilute	Good
Acids, concentrated	Fair
Aliphatic solvents	Poor
Aromatic solvents	Poor
Chlorinated solvents	Poor
Alcohols	Good
Alkalies	Good
Esters, ethers, ketones	Poor
Resistance to (heat) aging	Good
Resistance to oxidation	Good

presence of aluminum chloride as a catalyst, maintaining the reaction temperature at −125 to −140°F. It has the following molecular structure:

$$\left(-CH_2-\underset{\underset{CH_3}{|}}{\overset{\overset{CH_3}{|}}{C}}-CH_2-\overset{\overset{CH_3}{|}}{C}=CH-CH_2- \right)_n$$

Physical properties Butyl rubber exhibits exceedingly low permeability to gases such as oxygen and nitrogen. It was this characteristic that enabled it to replace natural rubber for pneumatic-tire inner liners. Butyl rubber is about equal in tear

TABLE 10 General Properties of Butadiene-Acrylonitrile (Buna N)

Property	Value
Specific gravity	1.00
Tensile strength, lb/in.^{2}	500–900
Elongation, %	450–700
Hardness (durometer)	A40–95
Tear resistance	Good
Abrasion resistance	Excellent
Recommended operating temp:	
Min °F	−60
Max °F	250
Chemical resistance:	
Acids, dilute	Good
Acids, concentrated	Good
Aliphatic solvents	Excellent
Aromatic solvents	Good
Chlorinated solvents	Fair
Alcohols	Good
Alkalies	Good
Esters, ethers, ketones	Poor
Resistance to (heat) aging	Excellent
Resistance to oxidation	Good

TABLE 11   General Properties of Polyacrylate

Property	Value
Specific gravity	1.09
Tensile strength, lb/in.^{2}	250–400
Elongation, %	450–750
Hardness (durometer)	A40–90
Tear resistance	Fair
Abrasion resistance	Good
Recommended operating temp:	
Min °F	−20
Max °F	350
Chemical resistance:	
Acids, dilute	Fair
Acids, concentrated	Fair
Aliphatic solvents	Good
Aromatic solvents	Fair
Chlorinated solvents	Fair
Alcohols	Poor
Alkalies	Fair
Esters, ethers, ketones	Poor
Resistance to (heat) aging	Excellent
Resistance to oxidation	Good

resistance and abrasion resistance to natural rubber. Its electrical properties are about the same as those of natural rubber also. While the resilience of butyl rubber is relatively low at room temperature, significant increase in resilience is noted at elevated temperature (212°F) (see Table 12).

Chemical resistance   Butyl rubber is quite resistant to many acidic and alkaline media. Unlike natural rubber it is very resistant to swelling by animal and vegetable oils (see Table 8). Vulcanized butyl rubber swells and deteriorates rapidly when exposed to aliphatic and aromatic solvents. Butyl rubber may be used at service temperatures in the range of 200 to 240°F. Heat has a tendency to soften the material over a period of time rather than to harden it as with natural rubber.

TABLE 12   General Properties of Butyl Rubber

Property	Value
Specific gravity	0.90
Tensile strength, lb/in.^{2}	2,500–3,000
Elongation, %	750–950
Hardness (durometer)	A40–90
Tear resistance	Good
Abrasion resistance	Good
Recommended operating temp:	
Min °F	−50
Max °F	300
Chemical resistance:	
Acids, dilute	Excellent
Acids, concentrated	Excellent
Aliphatic solvents	Poor
Aromatic solvents	Poor
Chlorinated solvents	Poor
Alcohols	Good
Alkalies	Excellent
Esters, ethers, ketones	Poor
Resistance to (heat) aging	Excellent
Resistance to oxidation	Excellent

Chloroprene Rubber (CR)

Preparation and composition More familiarly known as neoprene, this compound is prepared by polymerization of chloroprene, 2-chlorobutadiene-1, 3. It has the following repeating structure:

$$\left[\begin{array}{cccccccc} H & Cl & H & H & H & Cl & H & H \\ -C & -C= & C & -C & -C & -C= & C & -C- \\ H & & H & H & & & & H \end{array} \right]_n$$

Neoprene has excellent resistance to permeation by gases. Depending upon the gas, permeability is one-fourth to one-tenth that of natural rubber (see Table 13).

Chemical resistance Neoprene is unaffected by aliphatic hydrocarbons, alcohols, glycols, and fluorinated hydrocarbons. It is not resistant to chlorinated hydrocarbons, organic esters, aromatic hydrocarbons, phenols, and ketones (see Table 8). It is unaffected by dilute mineral acids, concentrated caustics, and aqueous inorganic salt solutions. It is severely attacked by concentrated oxidizing acids like nitric acid and sulfuric acid, as well as strong oxidizing agents such as potassium dichromate

TABLE 13 General Properties of Chloroprene (Neoprene)

Property	Value
Specific gravity	1.25
Tensile strength, lb/in.^{2}	3,000–4,000
Elongation, %	800–900
Hardness (durometer)	A40–95
Tear resistance	Fair
Abrasion resistance	Good
Recommended operating temp:	
Min °F	−40
Max °F	240
Chemical resistance:	
Acids, dilute	Excellent
Acids, concentrated	Good
Aliphatic solvents	Good
Aromatic solvents	Fair
Chlorinated solvents	Fair
Alcohols	Good
Alkalies	Good
Esters, ethers, ketones	Good
Resistance to (heat) aging	Excellent
Resistance to oxidation	Excellent

and peroxides. Neoprene is slightly inferior to nitrile rubber in oil resistance but is markedly better than natural, Buna S, or butyl rubber. This elastomer has good resistance to sunlight and general weathering.

Abrasion resistance Neoprene synthetic rubber is a tough, strong, resilient elastomer with excellent resistance to abrasive wear under severe service conditions. It has better abrasive-resistant properties than natural rubber at elevated temperatures.

Hardness range Hardness of products made of neoprene generally ranges from 45 to 95 durometer A. Neoprene products are also available in open- or closed-cell sponge form.

Temperature resistance Neoprene is able to withstand continuous-exposure conditions at 200 to 225°F for long periods. There are specially compounded formulas of neoprene which can operate in intermittent service at temperatures up to 250°F. Conventional neoprene compounds can operate at temperatures down to −40°F, while specially formulated compounds have performed satisfactorily at temperatures as low as −67°F.

Flammability Neoprene will not propagate a flame. It burns in the presence of a flame but is self-extinguishing when the flame is removed.

Chlorosulfonated Polyethylene (CSM)

Preparation and composition   More familiarly known as Hypalon, this elastomer is prepared by reacting polyethylene with chlorine and sulfur dioxide. The process converts the thermoplastic polyethylene into an elastomer which can be compounded and vulcanized. The final product can be slightly varied in its properties; these depend upon the degree of chlorination, which is generally in the range of 25 to 30 percent chlorine. A typical molecular structure may be considered as follows:

$$\left[-\left(CH_2-CH_2-CH_2-\underset{\underset{Cl}{|}}{\overset{\overset{H}{|}}{C}}-CH_2-CH_2-CH_2 \right)_{12} -\underset{\underset{\underset{Cl}{|}}{SO_2}}{C}- \right]_n$$

Chemical resistance   Extremely resistant to most chemicals, oils, and greases (see Table 8), Hypalon is particularly unaffected by aqueous salt solutions, alcohols, weak and concentrated alkalies, and concentrated sulfuric acid. It experiences good performance with hypochlorites. The resistance is poor to aliphatic and aromatic

TABLE 14   General Properties of Chlorosulfonated Polyethylene (Hypalon)

Property	Value
Specific gravity	1.12–1.25
Tensile strength, lb/in.^{2}	4,000
Elongation, %	250–500
Hardness (durometer)	A45–95
Tear resistance	Fair
Abrasion resistance	Excellent
Recommended operating temp:	
Min °F	−40
Max °F	325
Chemical resistance:	
Acids, dilute	Excellent
Acids, concentrated	Excellent
Aliphatic solvents	Fair
Aromatic solvents	Poor
Chlorinated solvents	Poor
Alcohols	Good
Alkalies	Excellent
Esters, ethers, ketones	Poor
Resistance to (heat) aging	Excellent
Resistance to oxidation	Excellent

hydrocarbons, gasoline, jet fuels, chlorinated solvents, aldehydes, ketones, and fuel oils. The elastomer is resistant to microorganisms and underground environments. It has excellent resistance to moisture, high-humidity conditions, and ozone.

Flammability   Hypalon is classified as self-extinguishing by ASTM D 635.

Weatherability   Hypalon exhibits excellent resistance to ultraviolet radiation and atmospheric oxidation. It has excellent resistance to outdoor weathering by the elements and is unaffected by prolonged immersion in water.

Physical properties   This elastomer has excellent abrasion resistance and good retention of physical properties to temperatures as low as 0°F and as high as 300°F. Special formulations can resist temperatures to 350°F. Hardness may be varied from 50 to 95 durometer A. The elastomer's excellent electrical properties make it well suited for insulation in low-voltage applications. Hypalon has excellent resistance to fatigue, cracking, and impact (see Table 14).

Ethylene Propylene (EPM)

Preparation and composition   There are two ethylene-propylene compounds of commercial value. The first, known as EPR rubber, is a completely saturated ethylene-propylene copolymer made by solution polymerization. This product employs a peroxide or peroxide-sulfur modified curing system. EPR has excellent resistance to ozone and weathering. It can be used at temperatures up to 325°F without appreciable loss in properties (see Table 15).

A second type of ethylene-propylene elastomer, designated EPDM, is a sulfur-curing terpolymer.

Chemical resistance   The ethylene-propylene elastomers exhibit good to excellent chemical resistance to water, acids, and caustics. They are essentially unaffected by alcohols. They are quite resistant to the phosphate-based hydraulic fluids. They show only mild resistance to the aromatic hydrocarbons, and are attacked by aliphatic and halogenated hydrocarbons. They are poorly resistant to diester-type synthetic lubricants. They exhibit excellent resistance to oxidation and ozone (see Table 8).

TABLE 15   General Properties of Ethylene Propylene

Property	Value
Specific gravity	0.86
Hardness (durometer)	A30–90
Tear resistance	Fair
Abrasion resistance	Good
Recommended operating temp:	
Min °F	−60
Max °F	300
Chemical resistance:	
Acids, dilute	Excellent
Acids, concentrated	Good
Aliphatic solvents	Poor
Aromatic solvents	Fair
Chlorinated solvents	Poor
Alcohols	Good
Alkalies	Good
Esters, ethers, ketones	Poor
Resistance to (heat) aging	Good
Resistance to oxidation	Excellent

Weatherability   EPDM exhibits excellent weathering properties, suffering no significant deterioration under prolonged exposure to direct sunlight.

Flammability   The ethylene propylene elastomers are classified as burning by ASTM D 635.

Fluoroelastomers

Preparation and composition   Typical members of this elastomer family are produced as:

1. A copolymer of chlorotrifluoroethylene and vinylidene fluoride
2. A trifluoropropyl siloxane polymer
3. A perfluorobutyl acrylate elastomer
4. A copolymer of vinylidene fluoride and hexafluoropropylene
5. Hexafluoropentamethylene adipate

There are a number of other compositions of merit.

Chemical resistance   The fluoroelastomers all have excellent chemical and solvent resistance. They are extremely resistant to aliphatic hydrocarbons, chlorinated solvents, animal, mineral, and vegetable oils, gasoline, jet fuels, dilute acids, alkaline media, and aqueous inorganic salt solutions. They exhibit fair to poor resistance

TABLE 16 General Properties of Fluorosilicone

Property	Value
Specific gravity	1.4
Tensile strength, lb/in.^{2}	1,000
Elongation, %	200–500
Hardness (durometer)	A40–75
Tear resistance	Fair
Abrasion resistance	Poor
Recommended operating temp:	
Min °F	−90
Max °F	400
Chemical resistance:	
Acids, dilute	Excellent
Acids, concentrated	Very good
Aliphatic solvents	Excellent
Aromatic solvents	Excellent
Chlorinated solvents	Excellent
Alcohols	Good
Alkalies	Good
Esters, ethers, ketones	Good
Resistance to (heat) aging	Good
Resistance to oxidation	Excellent

to oxygenated solvents, alcohols, aldehydes, ketones, esters, and ethers (see Tables 16 and 17).

Physical properties These elastomers have a practical application and retain their physical properties over a wide temperature range of from −90 to 400°F. They have low permeability rates with air and extremely low water absorption. This resistance to moisture makes them highly desirable for use in electrical and electronic equipment requiring excellent insulation. They have excellent aging characteristics. They exhibit good tensile strength and tear resistance but generally are compounded with such fillers as finely divided silica or carbon black when these are of added importance.

TABLE 17 General Properties of Vinylidene Fluoride Hexafluoropropylene

Property	Value
Specific gravity	1.6
Tensile strength, lb/in.^{2}	2,000
Elongation, %	100–400
Tear resistance	Poor
Abrasion resistance	Good
Recommended operating temp:	
Min °F	−10
Max °F	<500
Chemical resistance:	
Acids, dilute	Good
Acids, concentrated	Good
Aliphatic solvents	Excellent
Aromatic solvents	Excellent
Chlorinated solvents	Good
Alcohols	Excellent
Alkalies	Fair
Esters, ethers, ketones	Poor
Resistance to (heat) aging	Good
Resistance to oxidation	Excellent

Silicone Rubber (VMQ)

Preparation and composition   Silicone rubber is an organo-silicone oxide polymer characterized by exceptional heat resistance. Chemically it is a linear-chain dimethyl siloxane polymer containing several thousand units of this structure:

$$\left[\begin{array}{cccc} CH_3 & CH_3 & CH_3 & CH_3 \\ | & | & | & | \\ -Si-O-Si-O-Si-O-Si-O- \\ | & | & | & | \\ CH_3 & CH_3 & CH_3 & CH_3 \end{array}\right]_n$$

Although the polymer is saturated, it can be vulcanized by heating with a source of free radicals such as benzoyl peroxide. It is postulated that the free radicals derived from the benzoyl peroxide abstract hydrogen atoms from the polymer molecules. The unsatisfied valences on the polymer molecules then give rise to cross links which result in a vulcanized product (see Table 18).

Temperature range   The silicone rubbers have an extremely broad useful temperature range from −100 to 500°F. Their flexibility, resilience, and tensile strength are retained in this temperature range to an amazing degree.

TABLE 18   General Properties of Silicone Rubber (Polysiloxane)

Property	Value
Specific gravity	1.1–1.6
Tensile strength, lb/in.^{2}	600–1,300
Elongation, %	100–500
Hardness (durometer)	A30–90
Tear resistance	Fair
Abrasion resistance	Poor
Recommended operating temp:	
Min °F	−178
Max °F	600
Chemical resistance:	
Acids, dilute	Good
Acids, concentrated	Good
Aliphatic solvents	Excellent
Aromatic solvents	Excellent
Chlorinated solvents	Good
Alcohols	Good
Alkalies	Good
Esters, ethers, ketones	Good
Resistance to (heat) aging	Fair
Resistance to oxidation	Good

Chemical resistance   Silicone rubbers are quite resistant to lubricating, animal, and vegetable oils, alcohols, dilute acids, and alkalies. They are excessively swollen by aromatic solvents, such as benzene and toluene, gasoline, and chlorinated solvents. They are not resistant to steam at elevated temperatures. The silicone polymers exhibit excellent resistance to ozone and weathering. They are particularly desirable in many aircraft applications such as jet-engine components, aircraft ducting, gaskets, seals, and diaphragms.

Styrene-Butadiene (SBR)

Preparation and composition   This product, known also as GR-S rubber and Buna S rubber, is made by copolymerizing 75 parts of butadiene to 25 parts of styrene in soap water at 50°C. It is compounded in much the same manner as natural rubber and can be vulcanized to soft, semihard, or hard rubber. It has the following molecular structure:

$$\left[\begin{array}{c} \overset{H}{\underset{H_2}{-C}}-C\!\!=\!\!C-\overset{H_2}{C}-\overset{H}{\underset{\displaystyle \bigcirc}{C}}-\overset{H_2}{C}- \\ \end{array} \right]_n$$

Chemical resistance Its compounds are similar to natural rubber in physical properties and in chemical resistance. Like natural rubber, SBR formulations deteriorate quickly in contact with oils and solvents. They are unaffected by water, alcohol, dilute acids, or alkalies. However, they are soluble in ketones, esters, and many hydrocarbons. They resist atmospheric deterioration slightly better than natural rubber (see Table 19).

Physical and mechanical properties SBR has lower tensile strength, is less resilient, and has lower heat resistance than natural rubber. The SBR products are more flexible than natural rubbers, especially at temperatures down to $-120°F$.

TABLE 19 General Properties of Styrene-Butadiene

Property	Value
Specific gravity	0.94
Tensile strength, lb/in.^{2}	200–300
Elongation, %	400–600
Hardness (durometer)	A40–90
Tear resistance	Fair
Abrasion resistance	Good
Recommended operating temp:	
Min °F	−60
Max °F	180
Chemical resistance:	
Acids, dilute	Good
Acids, concentrated	Fair
Aliphatic solvents	Poor
Aromatic solvents	Poor
Chlorinated solvents	Poor
Alcohols	Good
Alkalies	Good
Esters, ethers, ketones	Poor
Resistance to (heat) aging	Good
Resistance to oxidation	Good

Polysulfide Rubbers (T)

Preparation and composition These elastomers are prepared by reacting different organic dihalides with polysulfides. Typical is Thiokol B, which is prepared from di-2-chloroethyl ether, $Cl-(CH_2)_2-O-(CH_2)_2Cl$, and sodium polysulfide, Na_2S_x.

Chemical resistance Thiokol polymers are outstanding in solvent, gasoline, and oil resistance. The resistance to solvent swelling depends significantly upon the weight percent of sulfur in the polymer. The polysulfides have good resistance to aging and exceptional resistance to ozone, oxidation, and weathering. The polysulfides can be used within the temperature range of -60 to $250°F$; it also has short-term uses up to $300°F$ without loss of properties (see Tables 8 and 20).

Polyurethane Diisocyanate Elastomers (AU)

Preparation and composition These are commonly referred to by the names urethane and/or polyester rubbers. The urethane solid elastomers are made with various isocyanates. The chief ones employed are tolylene diisocyanate (TDI) 4,4′-diphenylmethane diisocyanate (MDI), and 1,5-naphthalene diisocyanate. Typical starting point is the polyester obtained from the reaction of ethylene or propylene

TABLE 20   General Properties of Polysulfide (Thiokol)

Property	Value
Specific gravity	1.35
Tensile strength, lb/in.^{2}	250–400
Elongation, %	450–650
Hardness (durometer)	A40–85
Tear resistance	Poor
Abrasion resistance	Poor
Recommended operating temp:	
Min °F	−60
Max °F	250
Chemical resistance:	
Acids, dilute	Good
Acids, concentrated	Good
Aliphatic solvents	Excellent
Aromatic solvents	Excellent
Chlorinated solvents	Very good
Alcohols	Good
Alkalies	Good
Esters, ethers, ketones	Fair
Resistance to (heat) aging	Fair
Resistance to oxidation	Good

glycol with adipic acid. This product is reacted with 1,5-naphthalene diisocyanate to produce a prepolymer with a structure along these lines:

P = polyester group

This prepolymer can be chain-extended with water, glycols, or amines through linkage across terminal isocyanate groups:

Chemical resistance   The solid polyurethane elastomers are clear, flexible plastics with excellent resistance to most mineral and vegetable oils, greases, and fuels. They exhibit good to excellent resistance to aliphatic, aromatic, and chlorinated hydrocarbons. Resistance to esters, ethers, and ketones is generally only fair. They are softened and swollen by alcohols. They have limited service in weak acid solutions and are completely unacceptable for use in concentrated acids. Caustics severely degrade the urethane rubbers. They are not resistant to steam. They show good resistance to oxygen and ozone (see Tables 8 and 21).

Weathering   Urethanes have good weathering properties, being unaffected by extremes of weather. Prolonged exposure to ultraviolet light will darken the elastomers and somewhat diminish their physical properties. Pigmentation or use of ultraviolet absorbers can minimize such effects. Standard urethane elastomers do not support fungus growth and are generally quite resistant to microbiological attack. Some of the urethane rubbers offer outstanding resistance to the damaging effects of gamma-ray radiation.

TABLE 21   General Properties of Urethane Elastomer

Property	Value
Specific gravity	1.1–1.3
Tensile strength, lb/in.^{2}	4,000–8,000
Elongation, %	400–750
Hardness (durometer)	Shore A50–D70
Tear resistance	Good
Abrasion resistance	Excellent
Recommended operating temp:	
Min °F	−65
Max °F	240
Chemical resistance:	
Acids, dilute	Fair
Acids, concentrated	Poor
Aliphatic solvents	Excellent
Aromatic solvents	Good
Chlorinated solvents	Good to fair
Alcohols	Good to fair
Alkalies	Fair to poor
Esters, ethers, ketones	Poor
Resistance to (heat) aging	Excellent
Resistance to oxidation	Good

Physical properties   These elastomers have excellent abrasion resistance, impact resistance, and load-bearing properties. They can be processed by conventional thermoplastic methods, i.e., injection molding, extrusion, and solution application. They can be used safely for prolonged periods at temperatures above the boiling point of water, and retain much of their physical properties as low as −45 to 60°F (see Table 21).

Flammability   The urethane rubbers are classified as slow-burning to self-extinguishing by ASTM D 635. Compounding with flame retardants can result in nonburning products.

PLASTIC TESTING AND EVALUATION

General Considerations

It is particularly interesting to note that plastic evaluations in many laboratories, as recorded throughout the literature, generally do not follow typical cookbook standards or set patterns of procedures which we associate with the testing of alloys. The standard expression "mils per year attack" referring to metals and alloys exposed to corrosive environments is obviously lacking from reports on plastics studies.

The choice of plastic testing is frequently dictated by the proposed application or end use and the design of the experiment is usually directed by the ingenuity of the materials engineer, whereas corrosion of alloys and metals is an electrochemical phenomenon, with the material deteriorating only, with the possible exception of high-temperature oxidation, when there is actual electric-current flow. Plastics, of course, are not susceptible to such deterioration.

It is unfortunate that a double standard of acceptance has been applied to metals vs. plastics in the fabrication of chemical-processing equipment. Steel, or even a premium-priced alloy, may fail when subjected to highly corrosive environments, but often the material is not criticized and may even be replaced with like construction. Blame is placed upon the environment, not the metal, and studies are initiated to determine whether changes in the operating process may not be required.

Generally, the case is different when similar difficulties are experienced with plastics. Often the plastic is just condemned automatically with little or no effort made to ferret out the true cause of the failure. The inadvertent exposure or misapplication of a particular plastic in a highly aggressive environment may lead to a blanket condemnation of the entire family of plastics and resin materials for such service. Further testing and selection of the proper plastic would have presented no more difficulty than a similar upgrading with a metallic material.

All of which leads us to several questions:

1. What constitutes a good testing or evaluation procedure for plastics subjected to highly corrosive environments?

2. What kind of studies, laboratory or field, can we conduct which will give us the best chance of evaluating plastics and resins as materials of construction for chemical-processing equipment?

3. How compatible are the proposed plastic and the actual environment? In the selection of a particular plastic for fabrication of a piece of equipment several major factors need to be considered initially. These include a thorough knowledge of the environmental conditions to which the plastic will be subjected. Trace or residual percentages of detrimental chemicals are too frequently disregarded, with resultant ill effects.

4. Will the physical and mechanical properties of the plastic meet the engineering-design requirements of pressure, temperature, vibration, flexing, tensional, and torsional or compressive loads, etc.?

5. Does the use of the particular plastic system introduce a potential safety hazard through possible ultraviolet degradation, static-charge buildup, poor fire-retardant properties, etc.?

6. Does the long-range cost-analysis picture reflect the plastic as a competitive cost item in comparison with other choices of materials?

All these questions can be satisfactorily answered through the judicious testing of the proposed plastics in laboratory or actual field exposures, followed by thorough visual examinations and evaluations of their residual physical and mechanical properties.

One of the major considerations is to remember that there is a wide variation among generic groups of plastics in their chemical resistance as well as in their physical and mechanical properties. Even within their own generic classification, the performance of differently formulated plastic systems may evince wide variations in a particular environment. For this reason it is imperative that an acceptable performance in the specific-process environment be established for any of the plastics being considered before it is recommended as a material of construction.

An additional point of caution should be considered in selecting a material of construction based upon performance-rating data obtained from reference manuals tabulated by plastics manufacturers or equipment fabricators. Many times these ratings have been derived from testing of the plastic in a "pure" product. The presence of contaminants can make such ratings questionable. When the environment consists of several components, or is not known precisely, reliable test data can be obtained only through actual exposure. As a matter of fact, tests in the several components are not individually additive as to acceptance or rejection of the plastic or resin system. Many a failure of a plastic or reinforced-resin unit, which was deemed acceptable for a specific service after exposure to each of the individual chemicals involved alone, has been traced to the detrimental effects resulting from their combination.

Initial Evaluations

The testing of plastics either in the laboratory or under actual plant operations by the normal static-submersion method is frequently sufficient for detecting the inapplicability of the plastic or reinforced resin when it is readily attacked. However, when there is little visual change in the test sample, determination of its compatibility with the particular environment becomes somewhat more difficult. It is possible, for example, that the presence of undesirable trace contaminants in the process stream may be selectively absorbed over a long period of time by the plastic with no visible evidence of harmful effects. A gradual degradation of the plastic's physical and mechanical properties may, however, indicate that the plastic is developing a weakened structure. Under those conditions of no apparent visual change, a prime concern may still be whether or not some insidious or adverse conditioning of the plastic is occurring which could promote future difficulties. That is the reason why performance data based upon changes in physical and mechanical properties are so meaningful.

Visual Changes

Incompatibility of a plastic or resin with the environment is usually visually apparent by such occurrences as the following:

1. *Abnormal surface changes,* which may be identified by charring, chalking, blistering, cracking, and crazing, all of which can result in a weakened structure, prohibiting the material's extended use in the particular service.

2. *Color changes.* Although any significant changes in color of the plastic usually indicate some interaction of the plastic and the environment, it does not necessarily follow that such changes prohibit the use of the plastic. Many times a color change is limited to either the immediate surface or at best penetration of a few mils. Permeation sometimes causes the development of a color change throughout the entire body of the plastic. However, often the discolored plastic will not have experienced any significant degradation in its physical or mechanical properties and may be entirely acceptable for the proposed service.

3. *Dimensional changes.* Swelling or shrinkage to a minor degree is no reason for rejection provided such dimensional changes do not prohibit the use of the plastic in its intended service. Minor swelling in the plastic lining of a tank is not sufficient cause for alarm or replacement as long as its adhesion to the tank wall is unaffected. Swelling to the same extent within the volute of a pump, body of a valve, or throat of a flowmetering device might be quite unsatisfactory. Prohibited swelling is often only "in the eyes of the beholder" and quite difficult to define in terms of percentage-growth numbers as to rejection or acceptance. Excessive shrinkage is more easily defined by evidence of shrinkage, cracks, distortion, or loss of adhesive qualities.

4. *Hardness changes.* Softening or hardening of a plastic or a resin system may or may not be evidence of prohibitive attack. Softening, in many instances, is due to the permeation of a solvent into a thermoplast. Upon removal from exposure to the environment, the solvent may evaporate, and the plastic may resume its original hardness. Tackiness accompanying a softening of the plastic generally denotes its unacceptability. However, again this may be only a surface phenomenon and a very reasonable, practical life could be experienced by material in the environment. Brittleness and losses in tensile or flexural strength are usually associated with surface hardening of the plastic. Certain media, such as strong acids, may increase the surface hardness of catalytically cured resins without detrimental effects.

5. *Laminar-wall attack.* This type of attack is suffered by fiber-reinforced-resin structures in environments to which the resin is unacceptable. Should the resin system not have been properly cured, failures of this nature can also occur in an otherwise compatible environment. In other instances, delamination occurs because

the woven-glass-cloth reinforcement has not been sufficiently wetted during fabrication of the structure, resulting in poor adhesion between the glass mat and glass cloth. Failure to remove paraffin wax from the surface of a glass-fiber-reinforced polyester laminate prior to the addition of any overlay may cause delamination also. Exposure of glass fibers, not protected by the encapsulating corrosion-resistant resin, to environments such as strong caustic or hydrofluoric acid which are incompatible with glass can cause rapid deterioration of the glass reinforcement.

6. *Weight changes.* Many laboratories report performance of plastics with regard to changes in weight through exposure to the environment. These weight changes result from the (a) absorption of liquids, (b) production of corrosion products, (c) leaching of resin, (d) dissolution of plastic, (e) deterioration of fiber through chemical attack, or (f) combination(s) of these. Generally, a single weight change at the completion of an exposure period is satisfactory for evaluating the plastic for the specified environment. However, false or questionable conclusions or inferences may result from relying solely on such measurements. To avoid this pitfall, a series of physical- and mechanical-properties tests and weight-change evaluations should be made on each plastic at definite time intervals during the exposure period. These measurements are then plotted, so that a performance over time may be prognosticated from the curves. A fairly satisfactory analytical procedure has been developed which is based upon the percentage weight change (PWC) and change in flexural strength of a plastic material with time.[28] This system is not only valuable for predicting the probable life of a particular plastic or resin system, but is exceptionally good for comparing the rates of attack of several plastic or resin materials in the same environment, or the performance of a specific plastic in different environments.

Two sets of curves are employed in this evaluation. The first, the cumulative PWC curve, as the name indicates, shows the actual percentage weight changes of the sample at a particular time as compared with its initial weight, and can be extrapolated to give an estimated annual percentage weight change. An arbitrary rating system, based upon the observation of a great number of exposures in various environments, serves to determine the suitability of the plastic or reinforced resin for the particular application.

The second, the differential PWC curve, is a plot of the individual weight changes between the successive weighings, and serves to tell whether the rate of change is speeding up, slowing down, or remaining fairly constant for any particular increment of time. The slope of this differential curve determines the predominating trend. Unfortunately ambiguous results frequently occur with reference to changes in weight or dimension, and most testing laboratories have therefore resorted to evaluations which include changes in specific physical properties of the plastic test sample with time of exposure.

Standard Test Procedures

Many of the technical societies, including the American Society for Testing and Materials, National Association of Corrosion Engineers, The Society of Plastics Engineers, and the Society of Plastic Industry, Inc., have special committees and task groups concerned with standard chemical resistance and testing procedures for plastics. Final acceptance has generally been through the preparation of an ASTM Standard or Tentative Method of Test covering the specific evaluation. Those in effect are contained in the current ASTM "Book of Standards" dealing with Plastics —General Methods of Testing, Nomenclature.[29] Among these ASTM standards dealing specifically with the general testing of plastics in chemical environments are the following.

ASTM D 543–65, Standard Method of Test for Resistance of Plastics to Chemical Reagents This method is intended for the immersion testing of all plastic materials, including cast, hot-molded, cold-molded, and laminated resinous products, and sheet materials, for resistance in any of 51 standard or other chemical reagents or compounds. This method includes provisions for reporting, after 7 days' exposure, changes in weight, dimensions, appearance of evidence of loss of gloss, developed texture, decomposition, discoloration, swelling, clouding, tackiness, rubberiness, crazing, bubbling, cracking, solubility, etc., and mechanical-properties changes in tensile strength, flexural strength, compressive strength, etc., as prescribed in the ASTM methods of test for the specific properties being determined.

ASTM C 581–65T, Tentative Method of Test for Chemical Resistance of Thermosetting Resins Used in Glass-Fiber-Reinforced Structures   This method is intended for use as a relatively rapid test to evaluate the chemical resistance of thermosetting resins used in the fabrication of glass-fiber-reinforced self-supporting structures under anticipated service conditions. Its primary purpose is to promote uniformity of testing and comparison of different resin/fiber reinforcement systems throughout the industry. This method provides for the determination of changes in Barcol hardness, flexural strength, color and surface of specimen, and appearance of immersion medium. The results obtained by this method serve as a guide in, but not as the sole basis for, the selection of a thermosetting resin used in a glass-fiber-reinforced plastic self-supporting structure. This method provides for the procedure to employ in the fabrication of standard laminates for comparison of various thermosetting-resin systems. It includes a basis for the selection of the standard test specimen, as well as the recommended standard reagents and testing temperatures for the basic evaluation of resin-glass systems.

Specially Designed Test Procedures

The ASTM tests are adequate as far as their basic scope and intent are concerned. However, such evaluations do not take into consideration the possible performance of plastics in a particular environment as a consequence of fabrication or design characters. In many instances no standard or appropriate test is readily available. Under these circumstances, ingenuity is demanded and more sophisticated or extensive testing procedures are devised by the materials engineer which more nearly approximate the actual conditions under which the plastic equipment will be exposed. Literature is replete with special articles dealing with such significant studies as the effect upon a particular plastic or reinforced resin of compressive, flexural, or tensional loads under actual environmental conditions. Care must be taken in the evaluation of the plastic that such loads applied during the test are in the same direction as those which the piece of equipment will experience in actual service.

Some plastics, in addition to those possessing crystalline orientation, are found to react differently to external conditions, depending on the direction in which the test is applied. This is termed *anisotropy*. Materials in which the properties do not vary with the direction of test are termed *isotropic*. A material that is anisotropic must be tested in two directions, e.g., calendered sheet in the direction of calendering (longitudinal) and at right angles to the direction of calendering (transverse). Similar testing is required with fiber-reinforced resin structures depending on the direction in which the fiber or cloth is laid down.

When a nonuniform cure is encountered, e.g., in thick sections of thermosetting plastics, it is usual for the surfaces to be more highly polymerized than the interior. The skin is stressed in tension and the interior is compressed. Loss of plasticizer during curing may give a similar effect.

Frequently the best testing consists in the substitution of a short spool of pipe or section of duct directly into the process stream. After what may be considered a suitable period of exposure (usually 90 to 120 days minimum), the section is removed for visual examination, and physical and mechanical tests are conducted to determine residual properties.

Stress Cracking

Many thermoplastics, such as polyethylene, polymethyl methacrylate, polypropylene, polystyrene, and polysulfone, experience stress-corrosion cracking which is quite similar to that which affects metals and alloys. However, annealing or stress-relieving techniques as well as certain design precautions may be employed to remedy such conditions and prevent the possibility of catastrophic failures. It is advisable whenever possible to conduct tests upon stressed and stress-relieved coupons in the contemplated environment on those plastics susceptible to stress cracking.

Polysulfone materials, for example, may be annealed at 330°F for a period depending upon the thickness of the polysulfone part and varying roughly from 1 h for a thickness of 65 mils to as much as a 4½-h soak period for a 400-mil-thick section. A properly annealed polysulfone part will not crack or craze when tested in aggressive ethyl acetate solvent.

A procedure has been developed[30] employing certain solvents for use in determining the amount of residual stress present in a polysulfone part. The solvents and approximate stress levels are listed in Table 22.

Polysulfone parts to be tested are conditioned at room temperature before being subjected to a 1-min dip, allowed to dry, and then examined for crazing or cracking. If no crazing occurs after application of a given solvent, the stress level is less than the amount (in pounds per square inch) indicated for that particular test solvent. The test is conducted in steps whereby higher stresses are checked first.

Frequently fiber reinforcement or use of specific fillers in the thermoplast will eliminate such stress-cracking tendencies. It is reported[30] that polysulfone will stress-crack in MEK without glass reinforcement, yet the product formed by addition of 10 percent short glass fibers compounded into the material will not crack.

A recent publication by Broutman[31] does a fine job of reviewing the problem of stress-cracking in thermoplastics as well as outlining excellent testing procedures, remedial measures, and corrective engineering design. Kambour[32] discusses the role of organic agents in inducing stress-cracking of polymers.

TABLE 22

Test solvent	Approximate stress level, lb/in.^{2}
Carbon tetrachloride	2,400–2,600
Butyl cellosolve	1,400
Cellosolve solvent (2-ethoxy ethanol)	1,200
1,1,1,-Trichloroethane	500–600
Ethyl acetate	200

Flexural Modulus–Flexural-Strength Criteria

Other investigations, such as those of VanDelinder[33] employing Barcol techniques, have been based on changes in surface hardness in different environments with time. Still another technique by Atkinson[34] advances a chemical-resistance test involving a "quality index schedule" for changes in appearance, hardness decrease, thickness increase, and flexural-strength decrease. This test can be very effective for determining the performance of reinforced plastics subjected to highly corrosive environments.

Examination of test data obtained after many exposures of plastics and resins shows that either initial loss or gain in weight, or even a decrease in flexural strength or modulus, is only part of the performance picture. Rate of change or even stability of these with additional exposure is a much more important criterion of acceptance for use. Many investigations such as those of Fisher and Gackenbach[35] and Cass and Fenner[36] disclose rapid decrease in physical properties during the first few hours or days of exposure, with a subsequent flattening out of the curve indicating no appreciable further lowering of the strength of the material.

Such studies and results lead to the conclusion that changes in flexural strength and flexural modulus are the best criteria for evaluating the chemical resistance of thermosets in both field and laboratory exposures. The performance of thermoplastics, on the other hand, is more frequently determined by means of changes in their tensile properties. In either instance, changes in tensile strength and flexural modulus of the exposed coupon are evaluated with time. These evaluations are rapidly becoming the standard reference, since they appear to be the most indicative fundamental check on the physical behavior of plastic and resin materials.

It has been recorded[37] that special test coupons or standard ASTM tensile dog-bone bars are often employed in immersion testing. These are edge-sealed with the base resin to minimize any chances of wicking and laminar-wall attack during exposure. These coupons are removed after definite intervals for physical-property determinations. An alternate procedure employs larger laminate panels which are suspended in the process stream of the process equipment. At inspection time, the samples are removed, the bottom ¼ in. is trimmed off with an abrasive wheel, and the next ¾ in. is cut off for testing. The cut edge of the panel is resealed with the base resin and returned to the environment for additional exposure.

Such samples may be put through tests to determine tensile strength, flexural

TABLE 23   37% Formaldehyde, 25°C

Resin	Exposure time				
	0	1 month	9 months	15 months	36 months
Flexural modulus $\times 10^6$ lb/in.^{2}					
Epoxy laminate.......	1.96	2.10	2.14	2.20	2.02
Polyester A..........	1.94	1.80	1.86	1.87	1.70
Polyester B..........	1.68	1.87	1.90	1.85	1.83
Polyester C..........	1.84	1.77	1.97	1.93	1.81
Polyester D..........	1.92	1.71	1.77	1.90	1.93
Flexural strength $\times 10^3$ lb/in.^{2}					
Epoxy laminate.......	38.3	42.0	41.5	43.9	39.4
Polyester A..........	35.7	33.5	35.1	31.2	30.8
Polyester B..........	32.4	30.8	36.0	32.9	33.3
Polyester C..........	33.7	27.8	33.4	30.5	31.8
Polyester D..........	36.6	33.1	34.1	35.1	35.4

TABLE 24   Steam Condensate, pH = 4, 90°C

Resin	Exposure time				
	0	1 month	10 months	17 months	36 months
Flexural modulus $\times 10^6$ lb/in.^{2}					
Epoxy laminates.......	1.96	1.98	1.84	1.48	1.58
Polyester A..........	1.94	1.84	1.61	2.10	0.86
Polyester B..........	1.68	1.74	1.29	1.08	0.95
Polyester C..........	1.84	1.42	1.35	1.01	1.30
Polyester D..........	1.92	1.80	1.86	1.57	1.49
Flexural strength $\times 10^3$ lb/in.^{2}					
Epoxy laminates.......	38.3	34.6	27.3	18.6	17.0
Polyester A..........	35.7	21.9	20.2	11.1	8.3
Polyester B..........	32.4	23.5	20.4	12.7	10.9
Polyester C..........	33.7	19.5	16.6	13.7	13.0
Polyester D..........	36.6	31.9	30.2	21.8	19.3

strength, or flexural modulus. In some instances, these evaluations should be made at actual operation temperatures, and heating chambers are employed. Many laboratories use either the Tinius Olsen tensile tester or the Instron unit tester for these determinations. The advantage in the use of either of these units is their high degree of accuracy, reproducibility of test results, and the fact that each automatically plots the data as a permanent record.

Exposure data on a group of glass-fiber-reinforced resins exposed for 36 months in 37 percent formaldehyde at ambient temperature are shown in Table 23. A similar exposure for 36 months is shown in Table 24 for samples tested in a hot

(90°C) steam-condensate storage tank, pH = 4. It is obvious in examining these tables that there is a leveling off of the loss in residual flexural strength of the various reinforced resins with time. Fisher and Gackenbach[35] have concluded from their experiments "that chemical resistant polyester resins, if they are resistant to specific chemicals, will lose strength properties up to a maximum and then level off at a constant strength." Cass and Fenner[1,2,36] and others have reported similar results in which glass-fiber-reinforced polyester-resin laminates experience losses in flexural strength to a constant value. The exposure period to reach this particular leveling-off point is in the range of 90 to 120 days. Figures 2 and 3 are typical curves of such performance. Note the flattening out of the curve, and retention of residual flexural strength with time.

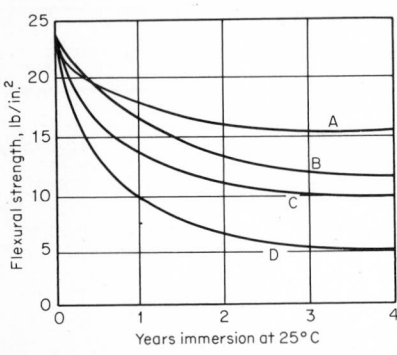

Fig. 2 Changes in flexural strength of FRP polyesters with time of exposure in acids at 25°C. $A = 50$ percent H_2SO_4, $B = 85$ percent H_3PO_4, $C = 10$ percent H_3PO_4, $D = 20$ percent HCl.

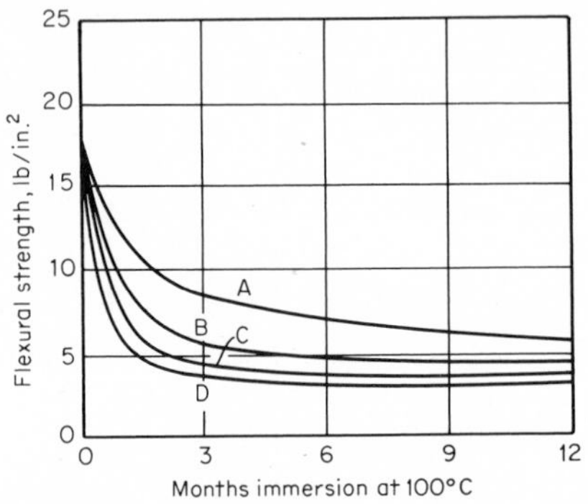

Fig. 3 Changes in flexural strength of FRP polyester with time of exposure in acids at 100°C. $A = 50$ percent H_2SO_4, $B = 85$ percent H_3PO_4, $C = 10$ percent H_3PO_4, $D = 20$ percent HCl.

Nondestructive Evaluations

Any determination of tensile, flexural, or compressive strength results in the destruction of the test sample. However, a definite repetitive pattern is observed in the performance of fiber-reinforced plastics when residual percent of flexural strength is plotted against the flexural modulus (Fig. 4). This is extremely valuable, since test measurement of the modulus of elasticity of a plastic may be nondestructive. The relationship between flexural strength and modulus thus may be employed to predict the probable strength of the plastic. The modulus of elasticity can be measured on a tank under actual operating conditions and yields significant engineering-design data; it also furnishes an excellent means of predicting the longevity of existing chemical processing equipment.

The Deflectron has been employed as a nondestructive test unit for this type of testing (Fig. 5). This portable unit is placed within slots in lugs fastened to the outer wall of the tank, duct, or other plastic structure. By applying a load through special weights at a specific area, a measurement of the deflection of the wall at that point is obtained. Performance of the plastic tank is determined through subsequent readings which indicate whether greater deflection occurs under the same applied load. Significant increases or decreases in the deflection under identical operating conditions of temperature and fill volume are indications of attack upon the plastic.

One-Side Exposure Panel Testing

The standard static-submergence test makes interpreting of exposure data difficult, and it has become apparent to many researchers and corrosion engineers that such testing is inadequate for good design. It is interesting to note that many companies and individuals have simultaneously arrived at the decision that "one-side" testing of the plastic panel under conditions simulating the inner wall of a tank or duct is more desirable.   Although such test equipment is slightly different in design, all the investigators employ similar means of one-side panel testing.   The Deflectron Corrocell is a typical laboratory unit for such testing.   It consists of a 4-in.-diameter by 8½-in.-long glass cylinder, with several ground-glass joint openings placed in a

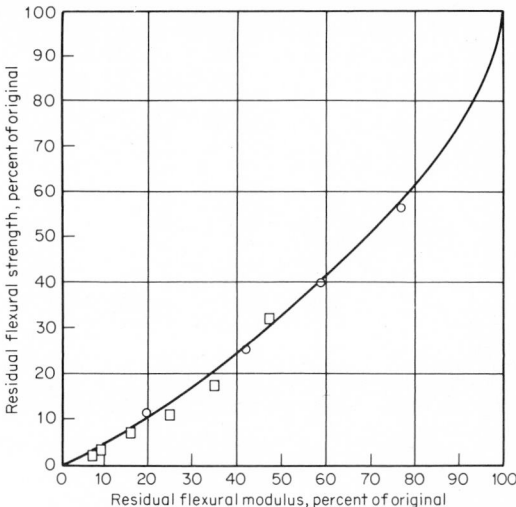

Fig. 4   Plot of percent residual flexural strength vs. percent residual flexural modulus for FRP resins.

Fig. 5   Field application of Deflectron nondestructive instrument measuring flexural modulus of wall of FRP polyester tank.

line perpendicular to the axis of the cylinder (Fig. 6).   The ends of the glass cylinder are closed off with the plastic panel to be tested, and are sealed with suitable gaskets, allowing one-side exposure to the environment.   The plastic panel is backed up by a suitable flange that serves as a stand and a mount for the testing instrument.

Fig. 6 Deflectron Corrocell open-end test unit with Deflectron instrument in place, weights applied, measuring the deflection of the panel in place during exposure.

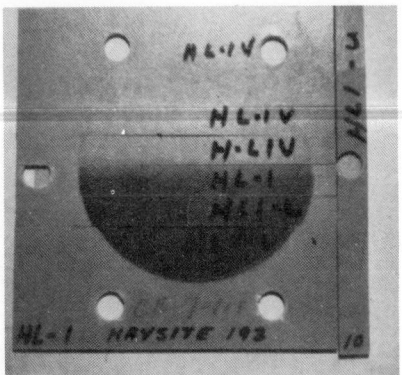

Fig. 7 Deflectron Corrocell one-side test panel. This panel shows strips cut through those areas exposed to the vapor phase and liquid phase of the test environ. A section is cut also from the unexposed portion of the test panel and used as a comparative control. Residual flexural-modulus and flexural-strength values are obtained on these strips using the Instron instrument to evaluate the performance of the FRP laminate in the specific environment.

This vessel is filled with the test solution and heated with a heating mantle to maintain the required temperature. Liquid/vapor conditions permit the test panels to be subjected to one-side exposures in much the same manner as that experienced by process equipment in the plant (Fig. 7).

As previously detailed in literature,[38] the Deflectron may be placed within slots provided in the flange mount, and the changes in the modulus of the plastic panel can then be followed under the actual operating temperature of the corrosive medium.

It has been possible by such means to obtain experimental data relating the temperature to the modulus of elasticity. This has been particularly useful in extrapolating modulus data between various temperature points as desired for design calculations, or in predicting the probable hazards involved in elevating the operating temperatures in existing plastic equipment. The decrease in modulus is generally 15 to 20 percent between 100 and 25°C. Studies have shown that there is also a general lowering of the modulus of a glass-fiber-reinforced resin laminate if it is held at an elevated temperature for an extended period of time, even in such relatively noncorrosive environments as those normally associated with distilled water or steam condensate.

A vast number of precise evaluations arising from programmed variations in both operating conditions and sample compositions is made economically possible through this technique. It is particularly valuable for comparing one-side exposures of two different laminate resin systems simultaneously under identical environmental conditions, as well as comparison of plastic piping (Fig. 8).

Use of the open-end one-side testing apparatus, through group testing, allows comparative evaluations of more than two plastic or fiber-reinforced resin systems with a minimum of effort on the part of the investigator. On completion of the test procedure the test panels are removed and visually examined, and their compressive strength, flexural modulus, flexural strength, and tensile strength are determined under precise conditions on an Instron, Tinius Olsen, or similar apparatus for testing physical properties.

Fig. 8 Evaluation of glass-fiber-resin pipe. Deflectron Corrocell open-end test units group evaluating sections of glass-fiber-reinforced-resin pipe along with one-side panels. Electric-heating mantles are supplying heat through wall of FRP pipe.

Permeation

One of the most destructive conditions experienced by plastics is that occurring through permeation of the environment into the material. The seriousness of this condition is attested to by the many instances of its contribution to catastrophic failure of equipment.

Many researchers have directed their studies in this direction. Lebovits[14] has reviewed the mechanism of the permeability of polymers to gases, vapors, and liquids in detail. Delmonte and Sama[30] have discussed at length a technique to determine the rates of penetration of moisture or chemical solutions into plastics reinforced with glass fibers and the effects of these substances. They have observed significant changes taking place in the dielectric properties of the plastic material against length of exposure in the aggressive environment. This method of observation is very convenient inasmuch as it permits rapid examination of the relative porosity of the laminated structure or rate of permeation of the particular plastic by the specific liquid or vapor.

This type of plastic evaluation is useful in predicting the probable service longevity of chemical-processing equipment. Knowledge of the penetration rate of chemicals into plastic structures, particularly glass-fiber-reinforced-resin structures, is extremely important to the engineer engaged in the design of storage tanks to be employed under vacuum or pressure conditions. The permanence of the physical properties of the reinforced-plastic structure is dependent upon the extent of penetration by the environment. Dielectric-measurement techniques are generally nondestructive procedures which are sensitive to early changes taking place within the reinforced plastic. Consequently their employment can serve as an accelerated method of predicting performance of the structure in the proposed service.

To determine the rate of penetration of an aggressive electrolyte through the wall of a glass-fiber-reinforced polyester tank, Cass and Fenner[2] embedded probes within the laminate wall at depths of 10, 60, and 125 mils. Electrical-conductivity or -resistivity measurements between the electrolyte in the tank and the embedded probes were monitored to determine the length of exposure or service period before penetration of the electrolyte to the probes was accomplished. Similar laboratory evaluations have been made with one-side exposure panels containing embedded probes, employing open-end Deflectron test cells to determine rates of penetration or attack.

Cass and Fenner[36] developed a technique whereby thin layers of the exposed surface of a glass-fiber-reinforced polyester-resin laminate were removed through successive "skinnings" with a diamond cutter and the flexural modulus of the remaining laminate was measured after each "skinning" until the original flexural modulus was obtained. This technique permits a determination of the effective depth of penetration of the corrosion into the structure and is somewhat comparable with that reported as mils/year penetration for metals and alloys.

A special word or two appears apropos in considering the testing of reinforced plastics. Here we are dealing with complex laminates or composites which pose problems that are somewhat different from those of the unreinforced or unfilled thermoplastics or thermosets.

The major difficulty with the chemical testing of reinforced plastics is that their performance and final acceptance or rejection are influenced by the fact that one is dealing with entire systems and mixtures of fiber reinforcements, fillers, catalysts, hardeners, resins, or reactive diluents. The effects of percentage variations in the specific ingredients, as well as the proportions (ratios) of these components; whether a postcure has been employed; significance of temperature and length of cure; type of fiber, weave, and cloth form; and techniques employed in fabrication of reinforced plastics all contribute to laminate performance.

In the evaluation of reinforced resins under corrosive environments, two additional, very important, precautions must be observed to minimize the possibility of reaching wrong conclusions as to their resistance or applicability.

The first deals with the need to make certain that any solutions which are rapidly weakened in their corrosiveness through attack upon the glass-fiber-reinforced-resin laminate must be replaced periodically to ensure that proper environmental conditions are maintained. Such materials as sodium or calcium hypochlorite, hypochlorous acid, bromine water, hydrogen peroxide, and deionized water are specific examples that come to mind.

The second is concerned with the fact that occasionally materials which show no significant attack when exposed to strong alkalies and/or acids will rapidly bloom and deteriorate some short period after removal from the aggressive environment. The phenomenon appears to be associated with a permeation of the concentrated caustic or acid into the subsurface of the laminate while under test, followed by a hygroscopic phase wherein the caustic or acid picks up water from the atmosphere. During this diluting period it appears that either excessive hydraulic pressures are introduced within the structure, excessively high, very localized heat of dilution temperatures may be reached, or severely corrosive concentrations are bridged. This condition has been experienced with certain reinforced plastics subjected to both 50 percent sodium hydroxide and 60 percent sulfuric acid. In any event it is recommended practice to reexamine test coupons after a 24-h period on removal from highly corrosive liquids to ascertain any possibility of adverse aftereffects which could occur in FRP equipment during idle periods or under shutdown conditions.

Plastographic-Analysis Techniques

Procedures for obtaining additional performance data after plastic materials have been subjected to aggressive environments, termed plastographic-analysis techniques, have been developed and reported recently.[40]

These techniques, employing the stereomicroscope, metallographic optical microscope, and scanning electron microscope, open a new, intriguing, and effective avenue for:

1. The plastics or fiber research chemist involved with determining the compatibility of new materials of construction
2. The corrosion engineer engaged in the selection of materials for highly corrosive environments
3. The design engineer concerned with the structural limitations and laminar integrity of fiber-reinforced-resin equipment
4. The plastic process-equipment manufacturer and user in securing meaningful specifications and standards

The technique has been extremely helpful in performance and failure analyses which in themselves have led to the development of better engineering materials of construction and design considerations for plastic materials of construction. In this procedure plastic samples and glass-fiber-resin laminates were prepared and subjected to "known-conditions" exposures. This technique has had its greatest employment in permanence studies with fiber-reinforced-resin materials. Typical "known-condition" samples for comparison against SPI standard control laminates were prepared by the following methods:

1. Physical fractures of the SPI standard control sample through fold-over, torsional breakage, flexural-fatigue cracking, and hydrostatic rupture
2. Undercured laminates arising from insufficient additions of catalysts or accelerators, excessively low cure temperatures, air inhibition, or high-humidity conditions
3. Overcured laminates resulting from excessive additions of catalyst or accelerators, poor heat dissipation, high exothermic conditions
4. Improperly resin-wetted laminates produced as a consequence of insufficient mechanical impregnation, incompatible sizings on fibers, or presence of resin-phobic contaminants
5. Chemically degraded SPI standard control laminates in which the resin matrix, glass fibers, or both have been attacked by selective solvents, oxidizing agents, acids, and strongly caustic compounds

Examination of these and similarly prepared special samples under magnifications ranging from 100 to 13,000X has proved to be extremely valuable in determining the resistivity of fiber reinforcements and the particular resin system toward the specific environments under consideration. Figures 9 and 10 are examples of such examination under the Stereoscan electron microscope made at 2,500X and 2,375X magnifications. These figures were photographs taken of a portion of a standard SPI hand-lay-up panel of a glass-fiber-reinforced chemical-resistant polyester resin after its suspension in the vapor phase about 48 percent hydrofluoric acid for 15 days in the laboratory under room-temperature conditions. This environment was chosen as one which would show attack on the glass fibers and yet leave the resin matrix essentially unaffected. The chemical resistance of the polyester and severe deterioration of the glass fibers indicate the effectiveness of plastographic-analysis techniques as a tool in evaluating plastics for corrosive environments.

Infrared Spectroscopy

Plastic construction materials, and wherever possible commercially available resins or proprietary plastics, should be identified for plastographic analysis. Such identification assists in determining the cause of poor performance in failure-analysis studies, establishing good standards, and promoting the design of better corrosion-resistant chemical-processing equipment and structures.

By means of infrared spectroscopy a very small segment of the plastic or resin material can be rapidly and accurately identified. Additives such as plasticizers, waxes, flameproofing materials, reactive diluents, and copolymers can also be determined by this means. It has been somewhat more difficult to identify antioxidants, accelerators, catalysts, and stabilizers because of their relatively small amounts.

Many excellent reference texts are available on the subject,[41] and more specific reports on the application of infrared analysis are too numerous to list here. However, Kraft[42] has prepared a very compact, interesting article in which he discusses the use of "attenuated total reflectance" (ATR) techniques of infrared spectroscopy

Fig. 9 Stereoscan at 2,500 × magnification showing specimen exposed to 48 percent hydrofluoric acid for 15 days at room temperature. The glass fibers are severely attacked. The bisphenol A fumarate polyester resin experienced no significant attack.

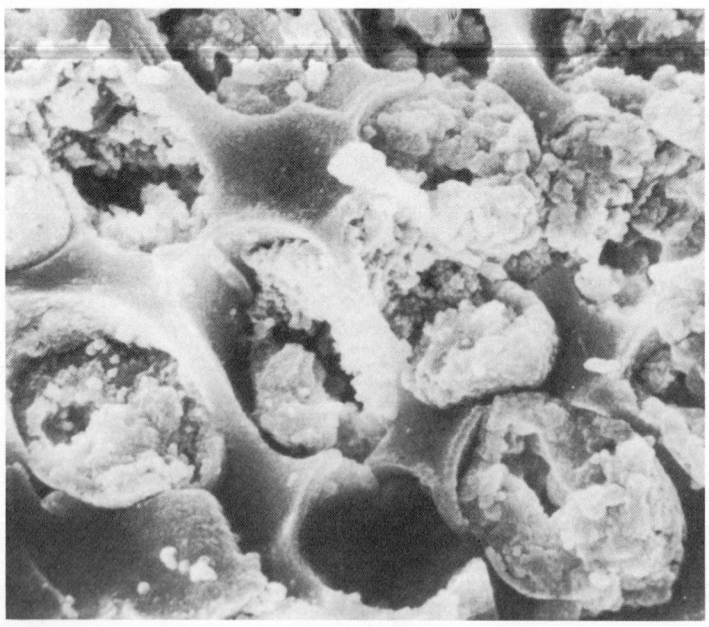

Fig. 10 Stereoscan at 2,375 × magnification showing severe attack to glass-fiber reinforcement by 48 percent hydrofluoric acid. The bisphenol A fumarate polyester-resin matrix discloses no appreciable attack.

in the analysis of plastics. Whetsel[43] has covered the developments in instrumentation, techniques, data handling, and applications of infrared spectroscopy in an excellent, recently featured article. Hannah et al.[44] have also recently reviewed continuing and future growth applications of infrared spectroscopy in polymer and molecule identification which are of current interest. Most sophisticated plastics research laboratories have prepared or acquired a complete set of each characteristic infrared-spectrum curve for each of the major polymers for systematic polymer identification, as well as the necessary equipment to perform infrared studies.

Nondestructive Testing

Various investigators have employed nondestructive tests including ultrasonic techniques for detecting voids with air interfaces, such as delaminations, porosity, cracks, and poor wetting of fibers in fiber-reinforced-resin laminates. Adams[45] reports a procedure in which immersed through-transmission methods of inspection with automatic scanning and facsimile C-scan recording were used for detection of discontinuities. Epstein and Rowland[46] describe nondestructive tests, including visual, ultrasonic, sonic, radiographic, electrical, microwave, and thermal, in a recent excellent résumé. Botsco[47] has prepared an excellent two-part discussion on the use of microwaves in the nondestructive testing of plastics; the discussion covers not only theory and instrumentation but thickness gaging, flow detection, fiber orientation, measurement of degree of cure, density, moisture content, and chemical composition.

Again, these references serve only to refresh one's memory that these and many more applications of nondestructive techniques are applicable to the testing of plastics and resins. Literature is replete with excellent specific references which are uncovered readily through a library survey.

REFERENCES

1. Cass, R. A., and O. H. Fenner: Evaluating Performance of Fiberglass Laminates, *Ind. Eng. Chem.*, vol. 8, August 1964.
2. Cass, R. A., and O. H. Fenner: Effect of Chemical Formulation of GFR Resin Laminates upon Their Performance in Corrosive Environments, *18th Ann. Conf., National Association of Corrosion Engineers*, Kansas City, Mo., Mar. 19–23, 1962.
3. "Modern Plastics Encyclopedia," vol. 49, no. 10A, 1972–1973.
4. Brydson, J. A.: "Plastic Materials," Iliffe Books, Ltd., London, 1966.
5. 1971 Plastics/Elastomers Reference Issue, *Machine Des.*, Feb. 11, 1971.
6. Zimmerman, R. L.: Engineers Guide to ABS Plastics, *Mater. Eng.*, October 1971, pp. 34–39.
7. "Penton—An Engineering Polymer Information on Hercules Chlorinated Polyether," Her. 500–325 15M, Hercules Powder Co., Wilmington, Del.
8. "The New ABC's of Penton for Corrosion Resistance," Her. 500–498, Hercules Powder Co., Wilmington, Del.
9. "Designing with Penton," Her. 500–420A, Hercules Powder Co., Wilmington, Del.
10. "DuPont Teflon PFA," Fluorocarbon Div., Plastics Dept., Brochure PIB no. 2, E. I. du Pont de Nemours & Co., Wilmington, Del., Aug. 31, 1972.
11. *Journal of Teflon*, vol. 5, no. 9, December 1964.
12. Fenner, O. H.: Designing with Fluorocarbons for Chemical Applications, *Modern Plastics*, October 1964.
13. Fenner, O. H.: Plastics for Process Equipment, 21st Biennial Materials of Construction Report, *Chem. Eng.*, Nov. 4, 1968.
14. Lebovits, A.: Permeability of Polymers to Gases, Vapors and Liquids, *Modern Plastics*, March 1966.
15. Floyd, Donald E.: "Polyamide Resins," Reinhold Plastics Application Series, Reinhold Publishing Corporation, New York, 1958.
16. Lee, Henry, Donald Stoffey, and Kris Neville: "New Linear Polymers," McGraw-Hill Book Company, New York, 1967.
17. Lexan Polycarbonate Chemical Resistance Data, *Tech. Rept.* CDC-392, General Electric, Pittsfield, Mass.
18. Jones, R. Vernon: Newest Thermoplastic—PPS, *Hydrocarbon Processing*, November 1972.
19. Short, J. W., and H. Wayne Hill, Jr.: Polyphenylene Sulfide Coating and Molding Resins, *Chemtech*, August 1972, pp. 481–485.

20. Ryton Polyphenylene Sulfide for Injection Molding, *Tech. Serv. Mem.* TSM-258, rev. June 1972, Phillips Petroleum, Bartlesville, Okla.
21. Walton, R. K.: Polysulfone Resin, Resin and Molding Compounds, "*Modern Plastics* Encyclopedia," pp. 259–260, 1968.
22. Delmonte, John: Guidelines in the Selection and Use of Epoxy Resin Systems, *Plast. Des. and Processing*, March 1967, p. 29.
23. Burge, Robert E.: Epoxy Resins, Resins and Molding Compounds, "*Modern Plastics* Encyclopedia," pp. 166–171.
24. Epoxy Plastics–Materials and Processes Manual 227, *Mater. Des. Eng.*, January 1965, p. 113.
25. Pschorr, Frank E.: The Effect of Hardener Composition upon the Chemical Resistance of Epoxy Resins, Ciba Products Corporation, Fair Lawn, N.J.
26. Lee, Henry, and Kris Neville: "Epoxy Resins—Their Application and Technology," McGraw-Hill Book Company, New York, 1957.
27. Brenner, W., and E. J. Singer: *Mater. and Methods*, vol. 41, no. 6, p. 102, 1955.
28. Cass, R. A., and O. H. Fenner: Accelerated Test Procedure for Evaluation of Fiber Reinforced Resin Equipment in the Chemical Industry, *Corrosion*, vol. 17, no. 1, January 1961.
29. ASTM "1967 Book of Standards," American Society for Testing and Materials, Philadelphia, Pa.
30. Union Carbide Chemicals Corporation, Polysulfone Plastics Div., private communications.
31. Broutman, L. J.: How to Minimize Stress Cracking in Plastics, *Mater. Des. Eng.*, September 1965.
32. Kambour, R. P.: "The Role of Organic Agents in the Stress Crazing and Cracking of Glassy Polymers," General Electric, Research and Development Center Publication.
33. VanDelinder, L. S.: Evaluation of Rigid Plastics for Chemical Service, *17th Ann. Conf., National Association of Corrosion Engineers*, Buffalo, N.Y.
34. Atkinson, H. E.: Reinforced Plastics for Chemical Process Equipment, ASME paper 61-WA-267, 1961.
35. Fisher, J. J., and R. E. Gackenbach: Reinforced Plastics in the Chemical Industry, 15th Annual Meeting of Reinforced Plastics Div., Society of Plastics Industries, February 1960.
36. Cass, R. A., and O. H. Fenner: Engineering Data on Glass Fiber Reinforced Resin Materials, *Mater. Protection*, vol. 4, no. 10, October 1965.
37. Fenner, O. H.: Evaluation of Reinforced Plastics for Corrosion Series, Reinforced Plastics Technology, May 31–June 4, 1965.
38. Fenner, O. H.: Evaluating Plastics and Resins, Materials Engineering Forum, *Chem. Eng.*, Nov. 18, 1968.
39. Delmonte, John, and E. Sama: Use of Dielectric Measurements to Determine the Penetration of Chemicals into Glass Laminates, *Proc. 18th Ann. Tech. Management Conf.*, Reinforced Plastics Div, SPE, February 1963.
40. Fenner, O. H.: Plastographic Analysis Techniques in the Evaluation of FRP Structures and Equipment, *SAMPE Quart.*, April 1970.
41. Hausdorff, H. H.: Analysis of Polymers by Infrared Spectroscopy, The Perkin-Elmer Corporation, Norwalk, Conn., Pittsburgh Conference on Analytical Chemistry.
42. Kraft, E. A.: Analysis of Plastics by ATR Spectroscopy, *Modern Plastics*, March 1968.
43. Whetsel, Kermet: Infrared Spectroscopy, *Chem. Eng. News Feature*, Feb. 5, 1968.
44. Hannah, R. W., J. G. Frasselli, and C. L. Angell: IR Spectroscopy—Even More Useful Now, *Ind. Res.*, November 1968.
45. Adams, C. J.: Ultrasonic Testing of Resin-Fiber Composites, Testing Techniques for Filament Reinforced Plastics, *Tech. Rept.* AFML-TR-66-274, Air Force Materials Laboratory and American Society for Testing and Materials Test Methods Symposium, September 1966, Wright-Patterson Air Force Base, Ohio.
46. Epstein, George, and Richard Rowland: NDT Methods for Composites, Reinforced Plastics, *Mater. Eng.*, October 1969.
47. Botsco, Ron J.: Non-destructive Testing of Plastics with Microwaves, *Plast. Des. and Processing*, part 1, November 1968; part 2, December 1968.

General References

Simulated Service Testing in the Plastics Industry, *ASTM Spec. Tech. Pub.* 375.
Baldanze, Nicholas T.: A Review of Non-destructive Testing for Plastic Methods and Applications, *Plastic Rept.* 22, Picatinny Arsenal, Dover, N.J., August 1965.

Bergen, R. L., Jr.: Tests for Selecting Plastics, *Metal Progr.*, November 1966.
Castle, J. E., and H. G. Masterson: Applications of the Scanning Electron Microscope to Oxide Surfaces, *Anti-Corrosion*, December 1964.
Cohen, Lester A.: Are You Evaluating Thermoplastics Properly, *Mater. Des. Eng.*, June 1964.
Delmonte, John: Photomicrographs Show the Way to Better Parts, *Modern Plastics*, August 1968.
Fenner, O. H.: Where to Use Plastic Materials, *Loss Prevention, a CEP Technical Manual*, 1968.
Fenner, O. H.: Use of Reinforced Plastics for Corrosion Resistance, paper presented at Plastic Institute of America/Washington University Symposium on Reinforcement of Plastics, May 27, 1969.
Fenner, O. H.: Application of Plastic Materials in Handling Hazardous Chemicals, paper delivered before 61st National Meeting of the American Institute of Chemical Engineers, Rice Hotel, Houston, Tex., Feb. 22, 1967.
Fenner, O. H.: Chemical Resistance of Polymeric Materials, paper presented Nov. 13, 1969, Plastic Design Institute, Symposium on Permanence of Polymeric Materials, Cherry Hill, N.J.
Gackenbach, R. E.: Reinforced Polyesters Enjoy Continued Success, 18th Annual Conference, National Association of Corrosion Engineers, Kansas City, Mo.
Hitt, W. C., and J. E. Ramsey: Ultrasonic Inspection and Evaluation of Plastic Materials, *Ultrasonics*, January–March 1963.
Katlafsky, Bernard, and R. E. Keller: ATR and Transmittance Spectra of Solvents, *Instrument News*, vol. 15, no. 2, spring 1964, The Perkin-Elmer Corporation, Norwalk, Conn.
Kellam, Barney: Evaluating and Specifying Polymers for an Electric Utility, *SPE J.*, December 1966.
Kerle, E. J., and J. J. Fisher: Hydrogenated Bisphenol in Chemical Resistant Polyester Resins, *Proc. 19th Ann. SPI Tech. Management Conf.*, Reinforced Plastics Division, Feb. 4–6, 1964, Chicago, Ill.
Mc Neil, D. W., and B. Bennett: Light Microscopy and Reinforced Plastics, *Plastics Technical Section, Modern Plastics*, December 1964.
Miller, J. C., et al.: Factors Affecting Performance of Reinforced Plastics in Sodium Hypochlorite Environments, Sec. 4D, *Proc. SPI 25th Anniv. Conf.*, Reinforced Plastics/Composites Division, February 1970.
Nerren, B. H.: Fractography of PTFE, *Modern Plastics*, December 1969.
Raphall, Thomas: Predicting the Service Life of Plastics, *ASM Tech. Rept.* D5–1.3, 1965 Metals/Materials Congress, Detroit, Mich.
Regester, R. F.: Accelerated Corrosion Testing of Glass Reinforced Polyesters, *Proc. 22nd Ann. Tech. Conf. Reinforced Plastics Composites Division.*
Seymour, R. B.: Better Tests for Chemical Resistance, *Plastics World*, May 1966.
Sherr, A. E., Preliminary Evaluation of Polymer Properties, *SPE J.*, January 1965.
Smith, W. E., and W. A. Severance: Visual Standards for Reinforced Plastics, *Mater. Des. Eng.*, March 1959.
Vetters, C. M.: Derakane Vinyl Ester Resins Show Improved Chemical Resistance for Corrosion Control, Spec. 4B, *Proc. SPI 25th Anniv. Conf.*, Reinforced Plastics/Composites Division, February 1970.
Webster, R. M.: Reinforced Plastics for Corrosive Service, *Chem. Eng.*, July 10, 1961.
Zolin, Byron: Evaluating Polyester Laminates for Chemical Resistance, *Mater. Des. Eng.*, April 1964.
Peering into the Wonderland of Minutiae with the Scanning Electron Microscope, a staff article, *Machine Des.*, July 24, 1969.
Methods for Evaluating Protective Coatings Used as Lining Materials in Immersion Service, Proposed NACE Standard Test Method, T-6A *Unit Committee Rept.*, NACE, Houston, Tex.
Standard Tests on Plastics, Celanese Plastics Co., *Bull.* G1C, Newark, N.J.
Simplified Explanations of Common ASTM Tests Mentioned in the Plastics Properties Chart, 1969–1970 "*Modern Plastics* Encyclopedia," vol. 46, no. 10A, October 1969.
Bellamy, L. J.: "The Infra-red Spectra of Complex Molecules," 2d ed., Methuen, London, 1958.
Colthup, N. B., L. H. Daly, and S. E. Weberly: "Introduction to Infrared and Raman Spectroscopy," Academic, New York, 1964.

Chapter **5**

Laminates, Reinforced Plastics and Composites

RONALD N. SAMPSON

Chemical Sciences Department, Research and
Development Center, Westinghouse Electric Corporation,
Pittsburgh, Pennsylvania

LAMINATES

Definition

Laminates are plastic materials made by bonding together two or more sheets of reinforcing fibers, usually with heat and pressure. The fibrous sheets or webs may be fabric, paper, or mat made of cellulose, asbestos, glass fibers, or synthetic plastics. The binders—called *resins*—can be almost any polymeric material but more often are based on thermosetting resins such as a phenolic, melamine, polyester, epoxy, or silicone. These resins are dissolved in suitable solvents and used to impregnate the reinforcing sheets in a treating tower, as shown in Figs. 1 and 2. In the treater

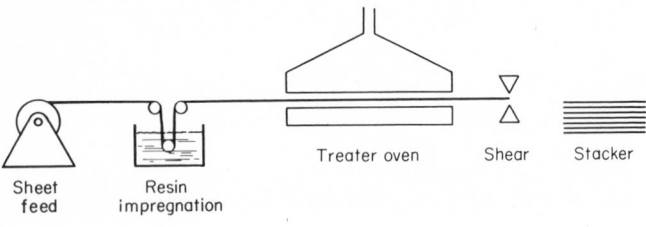

Sheet   Resin   Treater oven   Shear   Stacker
feed   impregnation

Fig. 1 Horizontal treater tower.

Fig. 2 Web treater. (*Westinghouse Electric Corp.*)

oven the solvents are removed, and in some cases the resins are advanced to a semi-cured condition called the *B stage* in which the resins are dry and hard but yet capable of further flow. The sheets are then placed into a molding press and pressed at high temperature (250 to 400°F) and high pressure (200 to 3,000 lb/in.²) until they are cured. In the curing (themosetting) process the resins become permanently infusible and insoluble. The finished laminates are usually cooled under pressure, removed from the press, and trimmed to size.

Design Criteria

The presence of the reinforcing media (paper, cloth, etc.) differentiates laminates from other plastics and results in a material with substantially improved physical properties. These high mechanical strengths make it possible to use laminates in structural applications where regular plastics could not be used. Because the high strength/weight ratio of certain laminates rivals even alloy steels, one can find these reinforced-plastic materials being used in such applications as rocket-motor casings. However, when comparing laminates with metals, one should remember that laminates differ in behavior quite dramatically, and the designer must allow for these differences. The biggest difference between metals and plastics is that plastics are not ductile. Whereas metals can be designed to be used up to their yield point, plastics (lacking a well-defined yield) can be used almost to their ultimate strength. The lack of ductility also means that reinforced plastics will not change greatly in shape when loaded, and stress risers are not easily formed, as is the case with metals.

The first step in design selection of a laminate is to define carefully the service that the laminate is expected to meet. Are the requirements physical or electrical? What specific properties are important, and what values are to be assigned to these properties?

The second step is selection of the laminate. The choice of the laminate should be made—if possible—from standard available materials to avoid costly and time-consuming development programs. Many property standards for laminates have been established by the National Electrical Manufacturers Association, ASTM, and the federal government. NEMA standards are described later. Military and other specifications are listed in Table 1. When consulting standard-property tables, one should remember that the values listed are usually conservative, and perhaps a lower-cost grade should be considered for a particular application. At this point one may also find that several laminates are candidates and that—particularly when cost is a vital factor—compromises in requirements may have to be made to obtain a material with the best balance of properties.

Many properties of laminates are conflicting. For example, materials with high resin content usually have improved electrical properties and reduced physical properties. Cloth-based materials are more easily punched than paper-based materials, but they have poorer electrical properties. Cloth-based materials are of higher cost than paper materials, but finished parts may be less expensive because punching costs less than machining. Flame-retardant materials are often lower in physical properties than similar non-flame-retardant grades.

The next step in the design process is to ensure that the material selected is available in the form and size required. Not all materials are available in all forms (for example, many materials are not available in tubes), and each laminate has certain thickness limits. If you have selected incorrectly and the form or size is critical, you may have to go back and select a laminate which compromises the properties required.

The next step is to determine whether the tolerances you require are available. If you have selected nonstandard tolerances, you can expect the cost of your laminate to be substantially increased. The use of nonstandard tolerances should be carefully considered. Can some element of the design be modified so that sheet with standard tolerances can be used?

The last design step is to check your results. Have you under- or overspecified? For example, have you specified a glass-based laminate to obtain water resistance when sufficient resistance is available in a paper-based material? Have you carefully considered machining or processing cost in relation to the cost of the laminate?

Have you specified accurately in relation to the following features—all of which affect costs?

Fabric base
Flame resistance
Nonstandard sizes or tolerances
Sanded surfaces
Thermal stability
Punchability
High mechanical strength
Machinability
Improved electrical properties
Arc and track resistance

TABLE 1 Specifications Relating to Laminates

Specification No.	Type
Military specifications:	
MIL-Y-1140	Yarn, cord, sleeving, cloth and tape, glass
MIL-R-7575	Resin, polyester, low-pressure laminating
MIL-C-9084	Cloth, glass, finished, for polyester-resin laminates
MIL-R-9299	Resin, phenolic, low-pressure laminating
MIL-R-9300	Resin, epoxy, low-pressure laminating
MIL-R-24607	Resin, polyester, low-pressure laminating, fire-resistant
MIL-R-25042	Resin, polyester, high-temperature-resistant, low-pressure laminating
MIL-P-25395	Plastic materials, heat-resistant, low-pressure laminated, glass-fiber base, polyester resin
MIL-P-25421	Plastic materials, glass-fiber base, epoxy resin, low-pressure laminated
MIL-R-25506	Resin, silicone, low-pressure laminating
MIL-P-25515	Plastic materials, phenolic-resin, glass-fiber base, low-pressure laminated
MIL-P-25518	Plastic materials, silicone resin, glass-fiber base, low-pressure laminated
MIL-P-25770	Plastic materials, asbestos base, phenolic resin, low- or high-pressure laminates
MIL-R-60346	Roving, glass, fibrous (for filament-winding applications)
Federal specification:	
L-P-383	Plastic material, polyester resin, glass-fiber base, low-pressure laminated
Aerospace material specifications (SAE):	
AMS 3822	Cloth—type E glass, B-stage epoxy resin impregnated, style 181-75DE
AMS 3823	Fabric, glass—style 181-75DE, chrome complex finish No. 550
AMS 3828	Glass roving, epoxy resin preimpregnated, type E glass
AMS 3830	Cloth—silica, B-stage phenolic resin impregnated, high-pressure molding

Reinforcing Fibers

Many reinforcing fibers are used in laminates. The selection of the reinforcing medium is related to the properties desired, cost, and to a certain extent, the resin system desired. Reinforcements include fiber glass, asbestos, cellulose, and synthetic fibers. Table 2 lists common reinforcements and the special properties they contribute to a laminate. Special fibers such as graphite and boron are described in a later section on composites.

TABLE 2   Laminate Properties in Relation to Reinforcing Media

Reinforcing medium	Mechanical strength	Electrical properties	Impact resistance	Chemical resistance	Machinability	Heat resistance	Moisture resistance	Abrasion resistance	Low cost	Punchability	Appearance
Glass fabric	X	X	X	X	...	X	X				
Glass mat	...	...	X	X	...	X	X	...	X		
Woven roving	...	...	X	X	...	X	...	...	X		
Asbestos paper	...	X	...	...	...	X	...	...	X		
Asbestos fabric	X	X	X	...	...	X					
Kraft paper	X	...	...	...	X	...	...	...	X	X	
α-Cellulose, paper	...	...	...	...	X	...	...	...	...	X	X
Rag paper	...	X	...	...	X	...	...	...	...	X	X
Cotton fabric	X	X	X	...	X						
Nylon	...	X	X	X	...	...	...	X			
Polyester	...	X	...	X	...	...	X	X			
Acrylic	...	...	...	X	...	...	X				

Glass fibers   Glass fibers are drawn continuously from a melt in special fiber-drawing furnaces. Three types of glass are used. The most common is E glass, which is moisture- and alkali-resistant with good electrical and physical properties. This fiber is most widely used in plastics. C glass is especially compounded for acid resistance; it is used mostly as surfacing mat or veil. A recent glass is S glass, which has higher strength and modulus than the others and is used in aerospace applications, usually in nonwoven form. Table 3 lists the composition of these glasses, and Table 4 contains physical and electrical properties. The glasses are produced continuously as filaments with diameters from 20×10^{-5} to 55×10^{-5} in., but staple fibers can also be made. A strand of fiber glass is composed of 204 filaments or some multiple of 204. Strands are then processed into yarns or rovings. Strands intended for weaving are coated with a protective binder (size), which is subsequently removed from the finished fabric. These strands are used as is or twisted with others to make yarns. Yarns have been standardized and are shown in Table 5. In yarns the first letter describes the glass composition. The second letter describes the fiber type, i.e., C for continuous and S for staple. The third letter defines the filament diameter. The first number series designates the weight of the yarn in

TABLE 3   Glass Compositions[1]

	SiO_2	$Al_2O_3 + Fe_2O_3$	CaO	MgO	$Na_2O + K_2O$	B_2O_3	BaO	F_2
E-type glass	53.2	14.8	21.1	0.3	1.3	9.0	0.3	
C-type glass	64.6	4.1	13.4	3.3	9.6	4.7	0.9	Trace

	SiO_2	Al_2O_3	Fe_2O_3	CaO	MgO	Na_2O	B_2O_3
S-type glass	64.3	24.8	0.2	<0.01	10.3	0.27	<0.01

TABLE 4 Properties of E Glass and S Glass[2]

Property	E glass	S glass
Specific gravity	2.54	2.48
Density, lb/in.^{3}	0.092	0.090
Tensile, 1,000 lb/in.^{2}:		
Monofilament	500	665
S. D.	58	79
Roving	372*	550*
S. D.‡	12	12
Modulus, 10^6 lb/in.^{2}	10.5	12.4
Elongation, %	4.8	5.4
Thermal:†		
Coefficient of expansion (80–212°F), 10^{-6} in./(in.)(°F)	2.8	1.6
Specific heat	0.192	0.176
Conductivity, Btu/(ft)(h)(°F)	0.56	
Electrical:†		
D. C.‡ 10 GHz, 75°F	6.13	5.21
D. C. 10 GHz, 460°F	6.25	5.30
L. T.‡ 10 GHz, 75°F	0.0039	0.0068
L. T. 10 GHz, 460°F	0.0055	0.0084
Velocity of sound, ft/s:		
Calculated	17,500	19,000
Measured	18,000	
Index of refraction	1.547	1.523

* 12-end roving.
† Measured from bulk glass.
‡ S. D. stands for standard deviation; D. C. is dielectric constant; L. T. is loss tangent.

TABLE 5 Continuous Fiber-Glass Strands[2]

Strand No.	No. of filaments	Nominal yd/lb	Filament diam, 10^{-5} in.
E Glass			
ECD 1800*	51	180,000	20–25
ECD 900*	102	90,000	20–25
ECD 450*	204	45,000	20–25
ECE 225*	204	22,500	25–30
ECDE 75*	816	7,500	23–28
ECG 150†	204	15,000	35–40
ECG 75†	408	7,500	35–40
ECH 68‡	408	6,800	40–45
ECH 30‡	612	3,000	40–45
ECK 75†	204	7,500	50–55
ECK 37‡	408	3,700	50–55
ECK 25‡	612	2,500	50–55
ECK 18‡	816	1,800	50–55
ECK 6.75‡	2,040	675	50–55
S Glass			
SCG 135†	204	13,500	35–40
SCG 140†	204	14,000	35–40
SCG 150†	204	15,000	35–40

* Used for fabric yarns.
† Used for roving and fabric yarns.
‡ Used for roving and woven rovings.

hundreds of yards per pound of basic strand.   The last number within the parentheses defines the number of strands in the yarn.

Rovings are continuous reinforcements made by combining strands.   Strands intended for roving are protected with a resin-compatible binder which often remains on the glass.   The rovings are designated by the number of strands, which in this form are called *ends*.   Hence, typical rovings have 12, 20, 30, or 60 ends.   Rovings are used for filament winding and for certain nonwoven products.

Glass-fabric reinforcements are cloths made in a variety of woven patterns.   The weaves commonly used are plain, leno, satin, crowfoot satin, and twill (shown in

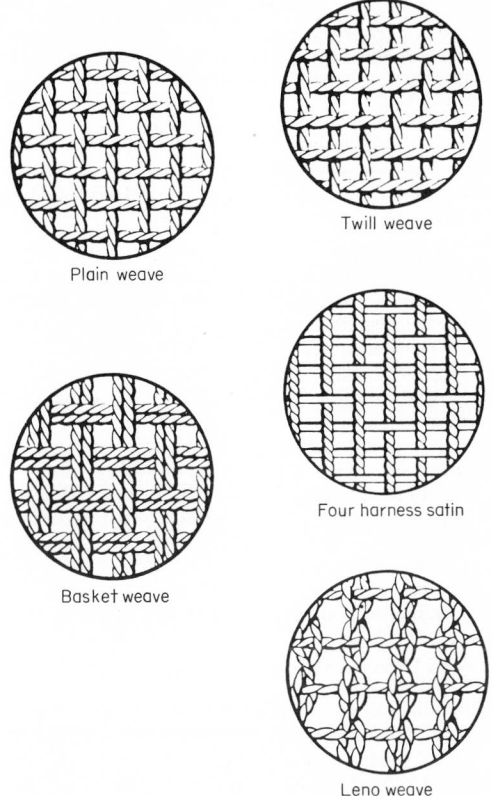

Plain weave

Twill weave

Basket weave

Four harness satin

Leno weave

Fig. 3   Weave patterns for glass cloth.[3]

Fig. 3).*   With over one hundred weaves to choose from, the fabrics are available in thicknesses from 0.001 to 0.050 in.   Other fabric variables include weight, number of warp and fill picks per inch, yarn size, and yarn twist.   Fabrics and their use are defined in MIL-C-9084B and MIL-Y-1140C.   All fabrics (and most glass-fiber products) to be used with polymers are coated with a coupling agent called a *finish*. Before the finish is applied, the yarn size is removed by heat cleaning.   The purpose of these finishes is to increase the adhesion of the plastics to the glass surface.   Hence, the type of finish to be specified is dependent on the type of polymer to be used. Table 6 describes common glass finishes and their uses.   Glass-fiber finishes are defined in MIL-F-9118A.

* See Glossary, pages 2, 5, 6, 8, 9, and 13.

TABLE 6   Glass-Fiber Finishes

Finish	Use
111............	For high-pressure melamine laminates.
516............	Similar to 111, but with improved electrical properties.
210............	With silicone rubber.
112............	Designates fabric that has been heated to high temperature to remove all organic constituents. Also known as "heat-cleaned." Used with silicone laminates.
Neutral pH.....	Fabric for silicones which is washed in deionized water to remove alkali.
Volan A*.......	Based on methacrylate chromic chloride. Widely used with polyester, epoxy, and phenolic resins.
Garan†........	Used with polyesters where white color and high wet strength is required.
A-172..........	A vinylsilane material used with polyesters.
A-1100.........	Used with phenolic, epoxy, melamine, phenolic, and polyimide resins, this material is an aminosilane.
A-1100 Soft.....	Similar to A-1100, but treated fabrics are much more drapable.

* Trademark of E. I. du Pont de Nemours & Company, Inc., Wilmington, Del.
‡ Trademark of Johns-Manville Corp., New York, N.Y.

Glass-mat reinforcement is composed of randomly deposited strands of glass from ¼ to 1 in. long. The strands are held together with a light coating of some polymer called a *binder*. Mats are available in a range of thicknesses and with many of the same glass treatments described in the previous section. Mat laminates do not possess physical properties comparable with fabric laminates, but they are less costly and can be molded to more complex shapes. For molded shapes a mat preform is sometimes made. This is a formed-in-place mat made in a special machine and also consisting of randomly deposited fibers with a binder. Another special case of a mat laminate is the process used primarily by boat manufacturers called *spray-up*. The spray-up process consists of chopping fibers of glass into small lengths, mixing them with a resin, and then spraying them onto a mold surface. In this fashion a mat with pre-added resin is assembled at the mold. All these methods produce products with similar properties.

Woven roving laminates are a compromise in physical properties between mat and cloth. This material is made from roving—continuous strands of fiber-glass roving which is very simply woven in an over-under pattern. The mat is easily handled and can be molded in rather complex shapes. The cost is intermediate between those of mat and cloth. Such constructions are available in 15 styles comprising a variety of strengths in warp and fill directions. A special example of this material is unidirectional roving. In this material the roving strands are laid parallel and held in position with a light fabric or a crosstie thread. The plies are highly directional in properties, and laminates can be made with a wide range of properties by varying the direction of each ply. Table 7 summarizes the forms of fiber glass and their uses.

Asbestos fibers   Asbestos paper is used in higher-temperature laminates. Asbestos is a naturally occurring mineral consisting of $3MgO \cdot 2SiO_2 \cdot 2H_2O$ with varying quantities of silicates of iron and calcium. Another asbestos form, crocidolite, is also used. The asbestos fibers are separated from the rock strata, purified, and made into a paper in a conventional papermaking machine. Various resins are used as binders. The papers are fairly weak compared with cellulose.

Asbestos fabric is cloth made from asbestos thread. The fibers in the thread are staple fibers, that is, discontinuous. The fabrics are available in several woven patterns and thicknesses.

Cellulose   Cellulose paper is widely used for a reinforcing sheet. The three most common papers are kraft, alpha cellulose, and rag paper. Kraft paper is the cheapest and is made by the sulfate process involving hydrolysis of the lignin in cellulose. Kraft-paper laminates have good physical properties. Alpha-cellulose papers are made from purified pulp blended with waste printed stock such as magazine or letterhead.

TABLE 7 Characteristics and Uses of Different Forms of Fiber Glass[1]

Form	General description	Process	Glass content	Typical applications
Rovings................	Ropelike bundle of strands of continuous glass fibers	Filament winding, continuous panels, matched die molding, spray-up, pultrusion, centrifugal casting	25–85	Pipe, automobile bodies, rod stock, rocket-motor cases, ordnance
Chopped strands.........	Strands cut to lengths of ⅛ to 2 in., sold in bulk	Premix molding, wet slurry preforming, thermoplastic screw injection molding	15–45	Electrical and appliance parts, ordnance, components, industrial parts
Reinforcing mats.........	Chopped or continuous strands in nonwoven random matting	Matched die molding, hand lay-up, centrifugal casting	20–45	Marine applications, translucent sheets, truck and automobile body panels
Surfacing and overlay mats..	Low-reinforcing monofilaments in random mat	Matched die molding, hand lay-up, filament winding	5–15	Automobile bodies, housings, etc., where smooth surfaces are required
Yarns.................	Twisted yarns with afterfinish	Fabric weaving, filament winding, unidirectional reinforcement	50–70	Fishing rods
Woven fabrics.........	Woven cloths from fibrous glass yarns with afterfinish	Hand lay-up, vacuum bags, autoclave high-pressure laminating	45–65	Aircraft structures, marine, ordnance, electrical flat sheet and tubing, radomes, aircraft skin
Woven roving........	Woven fibrous glass roving—coarser and heavier than fabrics	Hand lay-up	40–70	Large containers, marine
Nonwoven fabrics..........	Unidirectional or parallel rovings in sheet form	Hand lay-up, filament winding	60–80	Aircraft structures
Sheet-molding compound.. (SMC)	Mat or chopped fibers, semisolid filler resin mix, catalyst in relatively dry flowable sheet	Matched die molding	20–35	Automobile bodies and components, housings

Laminates from this paper have improved appearance, machinability, and electrical properties. The paper is more expensive than kraft. Rag paper is made from pure cotton fibers obtained from scrap textiles. The fibers are purified by being cooked in caustic solutions which remove all waxes, dirt, and polymers. Rag-paper laminates have the best electrical properties and machinability and are not badly affected by water. This paper is the most costly of the three grades.

Cellulose fabrics are often used in laminates and are available in weights from 2 to 10 oz/yd² and in many weave patterns. The fabrics are used in laminates for mechanical rather than electrical purposes.

Synthetic fibers   Synthetic fibers are used in certain laminates for special purposes. These fibers most often are used in textile form, but mats and nonwoven batts are sometimes useful. Nylon (polyamide) fabric-based laminates are high in electrical properties, impact strength, and chemical resistance. Not all resin binders may be used with nylon because of the difficulty in obtaining a good resin-to-fiber bond. Polyester-based fabrics such as Dacron° result in laminates with excellent water and chemical resistance. Abrasion resistance is also improved. Dynel† and Orlon° are acrylic fibers. Laminates containing these fabrics are characterized by mildew resistance, acid resistance, and dimensional stability. Some other synthetic fibers such as Nomex° (aromatic polyamide), polyvinyl alcohol, and polypropylene are used for special applications, but the cost of such laminates is high and their use is limited.

Laminated Materials

The large number of reinforcing materials combined with an equally large selection of resin binders could theoretically result in an infinite number of laminated materials. The next sections will present the properties and uses of the most commonly used materials. Table 8 lists the most important laminate properties to aid in proper selection.

Phenolic laminates   Phenolic resins are among the oldest plastics. They are available in a large number of variations depending on the nature of the reactants, the ratios used, and the catalysts, plasticizers, lubricants, fillers, and pigments employed. The properties of phenolic laminates based on papers, fabrics, and impregnated wood are shown in Table 9. Military Specifications MIL-R-9299 and MIL-P-25515B describe phenolics. These laminates are excellent general-purpose materials. They have a wide range of mechanical and electrical properties, and by selecting the correct reinforcement, products with many properties can be obtained. Phenolics are used widely for both high- and low-temperature service. Figure 4 presents tensile strength, elongation, modulus, and Izod impact at −60, −40, +10, and +77°F for phenolic laminates based on glass cloth, cotton cloth, and paper. Figure 5 shows the room-temperature flexural, tensile, and compressive strength of a high-temperature glass-cloth-reinforced phenolic laminate after 200 h at various temperatures.

Phenolics are often used because their cost is lower than that of laminates with other resins. They have excellent water resistance and are frequently chosen for marine applications. Their wear and abrasion resistance make them suited for gears, wheels, and pulleys. Their stability under a variety of environmental conditions makes them useful in printed circuits and terminal blocks. Two disadvantages limit their use. Because of the nature of the resin the laminates are available only in dark colors, usually brown or black. The second disadvantage is the rather poor resistance to electric arcs. High filler loading can improve the low-power arc resistance as measured by ASTM D 495, but the presence of moisture, dirt, or higher voltages usually results in complete arcing breakdown.

Melamine laminates   Melamine-formaldehyde polymers are widely used in the home in the form of high-quality molded dinnerware. This application takes advantage of their heat resistance and water-white color. Also used in the home are melamine-surfaced decorative laminates described in a later section. Typical properties of melamine laminates are shown in Table 9. Melamine resin can be combined with a variety of reinforcing fibers, but properties are best when glass cloth is the

---

° Trademark of E. I. du Pont de Nemours & Company, Inc., Wilmington, Del.
† Trademark of Union Carbide Corporation, New York, N.Y.

TABLE 8 Laminate-Selector Chart

Laminate	Property								
	Cost	Electrical properties	Chemical resistance	Thermal stability	Arc resistance	Flame resistance	Humidity resistance	Dimensional stability	Mechanical stability
Phenolic	X			X				X	X
Melamine		X			X	X	X	X	X
Epoxy			X		X	X	X	X	
Polyester	X		X						
Diallyl phthalate		X	X				X	X	
Silicone			X	X	X	X	X		
Teflon*		X		X	X	X			
Polyimide		X		X			X		
Diphenyl oxide		X		X			X		X
Thermoplastic	X	X		X				X	X

* Trademark of E. I. du Pont de Nemours & Company, Inc., Wilmington, Del.

TABLE 9   Laminates Property Chart[1]

Properties	Acrylic laminates Glass fabric base	Acrylic laminates Glass mat base	Diallyl phthalate laminates Glass fabric base	Diallyl phthalate laminates Cotton fabric base	Polyester laminates Glass fabric base	Polyester laminates Glass mat base[4]	Polyester laminates Paper base	Polyester laminates Cotton base
Physical								
1. Laminating temp., °F.	200–250	200–250	200–350	200–300	R.T.–300	R.T.–300	R.T.–300	R.T.–300
2. Laminating pressure, p.s.i.	0–200	50–200	10–1500	10–1500	0–120	0–500	0–10	0–10
3. Specific gravity, D792	1.34–1.45	1.25–1.38	1.65–1.8	1.38	1.5–2.1	1.4–2.1	1.2–1.5	1.2–1.4
4. Specific volume, cu. in./lb., D792	21–19	22–20	16.8	20.1	18.5–13.2	18.5–14.4	23.0–18.5	23.0–19.7
5. Tensile strength, D638, p.s.i.	20000–60000	10000–25000	21000–62000	8500–15000	18000–65000	10000–25000	6100–14300	7000–9000
6. Modulus of elasticity in tension, D638, 10^5 p.s.i.	12–35	10–19	—	—	10–28	10–19	8–12	—
7. Compressive strength, D695, p.s.i.	25000–60000	20000–50000	25000–55000	40000	20000–60000	20000–45000	19700–25000	23000–24000
8. Modulus of elasticity in compression, D695, 10^5 p.s.i.	—	—	—	—	30–40	15–25	—	—
9. Flexural strength, D790, p.s.i.	25000–80000	15000–40000	30000–75000	12000–21000	12500–90000	15000–40000	13000–28000	13000–18000
10. Modulus of elasticity in flexure, D790, 10^5 p.s.i.	14–38	12–20	6–30	11–13	6–35	10–18	8–13	5–12
11. Shear strength, D732, p.s.i.	12000–20000	10000–20000	17800–20000	14000–14700	12000–23000	10000–20000	—	—
12. Modulus of elasticity in shear, 10^5 p.s.i.	—	—	—	—	5–7	—	—	—
13. Bearing strength, D953, p.s.i.	—	—	—	—	40000–60000	15000–35000	24000–30000	—
14. Impact strength, Izod, D256, ft. lb./in. of notch	15.0–35.0	6.0–20.0	4.3–20.0	0.8–1.5	19.0–35.0	4.5–18.0	—	1.5–2.3
15. Bond strength, D952, lb.	2000–3000	800–1800	1300	1100–1500	2000–3000	850–1700	1000–2000	1000–1500
16. Hardness, Rockwell, D785	M100–M110	M80–M110	M120	M104–M110	M100–M110	M70–M110	—	M50–M90
17. Water absorption, 24 hr., $\frac{1}{8}$-in. thickness, D570, %	0.3–1.0	0.3–1.0	0.12	0.60	0.15–2.50	0.1–0.8	0.10–5.00	1.3–3.5
18. Effect of sunlight	Nil	Nil	Nil	Nil	Approx. nil	Approx. nil	Slight darkening	
19. Machining qualities	Good	Good	Fair to good	Good	Fair to good	Good	Excellent	Excellent
Thermal								
20. Thermal conductivity, C177, 10^{-4} cal./sec./sq. cm./1° C./cm.	1.7–2.0	1.7–2.0	—	6.3–6.6	—	1.7–8.7	—	—
21. Specific heat, cal./°C./gm.	0.3	0.3	—	—	0.25–0.26	—	—	—
22. Thermal expansion, D696, 10^{-5}/in./in./°C.	1.8–2.2	2.0–2.5	—	2.0–2.7	—	1.0–4.2	3.1	—
23. Resistance to heat (continuous), °F.[5]	180–240	180–240	300	250	200–350	125–400	220–250	250
24. Burning rate, D635	Slow	Slow	Slow to nil	—	Slow to nil Depending on resin		Moderate to self-extinguishing	
Electrical								
25. Insulation resistance (96 hr. at 90% R.H. and 35° C.), D257, megohms	—	100–10^7	150	80000	—	100–10^6	—	—
26. Volume resistivity (50% R.H. and 25° C.), D257, ohm-cm.	—	10^6–10^{12}	—	—	—	10^6–10^{12}	10^{13}	—
27. Dielectric strength, short-time, $\frac{1}{8}$-in. thickness, D149, volts/mil	250–700	300–500	530	390	250–700	300–700	600–800	300–500
28. Dielectric strength, step-by-step, $\frac{1}{8}$-in. thickness, D149, volts/mil	220–600	200–400	—	—	220–600	200–400	400–600	270–400
29. Dielectric strength, step-by-step, parallel to lamination, KV	45–50	40–85	75	40–70	45–50	40–85	40–60	—
30. Dielectric constant, 60 cycles, D150	4.0–6.0	4.4–6.0	4.3	5.1	4.0–6.0	4.4–6.0	5.1	—
31. Dielectric constant, 10^3 cycles, D150	—	4.2–5.0	—	—	—	4.2–5.0	—	—
32. Dielectric constant, 10^6 cycles, D150	3.0–4.0	3.0–6.0	4.1	4.8	3.0–4.0	3.0–6.0	3.0–4.2	2.9–3.6
33. Dissipation factor, 60 cycles, D150	0.02–0.04	0.005–0.050	0.013	0.015	0.02–0.04	0.005–0.050	0.1	—
34. Dissipation factor, 10^3 cycles, D150	—	0.004–0.040	—	—	—	0.004–0.040	—	—
35. Dissipation factor, 10^6 cycles, D150	0.007–0.030	0.007–0.040	0.015	0.018–0.040	0.007–0.030	0.007–0.040	0.02–0.03	0.02–0.04
36. Arc resistance, D495, sec.	80–140	70–185	150	130	80–140	65–185	28–75	70–85
Chemical								
37. Effect of weak acids, D543	None	None	None	None	None	None	None	None
38. Effect of strong acids, D543	Slight	Slight	Slight	Decomposes	Some attack	Slight to some	Some attack	Some attack
39. Effect of weak alkalies, D543	None	None	None	None	Slight to none	Slight to none	Slight	Slight
40. Effect to strong alkalies, D543	Slight	Slight	Slight	Attacked	Some to severe	Some to severe	Attacked	Attacked
41. Effect of organic solvents, D543	Attacked by some	Attacked by some	None	None	Some to severe	Some to severe	Generally nil	Generally nil

TABLE 9 Laminates Property Chart[1] (Continued)

	Epoxy				Melamine-formaldehyde laminates				
	Nonwoven continuous glass filament base	Nonwoven noncontinuous fibrous glass base	Glass fabric base	Cellulose paper base	Cellulose paper base	Cotton fabric base	Asbestos paper or fabric base	Glass fabric base	Glass mat base
1.	250–330	250–310	R.T.–370	250–300	270–320	260–320	270–320	270–300	270–320
2.	10–100	10–1000	10–1800	600–1000	500–1800	1000–1500	1000–1800	1000–1800	1000–1800
3.	1.75–2.00	1.50–1.70	1.7–2.0	1.4–1.52	1.40–1.55	1.30–1.52	1.75–1.95	1.82–1.98	1.75–1.85
4.	15.8–14.9	18.5–16.5	16.3–14.5	19.5–18.0	19.1–17.8	20.5–18.5	15.8–14.9	15.2–13.9	16.9–15.0
5.	110000–215000	—	35000–85000	10000–19000	10000–25000	7000–17000	6000–12000	25000–63000	16000–25000
6.	55–89	—	20–35	7–12	—	10–19	16–22	20–25	—
7.	90000–120000	—	35000–80000	23000–30000	30000–48000	33000–50000	27000–50000	25000–85000	60000
8.	53–85	—	—	—	—	—	—	47	—
9.	120000–210000	20000–25000	40000–105000	16600–24000	14000–20000	14000–28000	12000–24000	35000–85000	21000–30000
10.	53–80	10–14	20–45	10–13	—	16–20	—	15–41	13
11.	—	—	14000–25000	9000–12200	—	16500–18000	13000–17000	19000–35000	—
12.	—	—	—	—	—	—	—	—	—
13.	—	—	—	—	—	—	—	—	—
14.	36.0	—	5.5–25.0	0.5–0.8	0.3–1.5	0.5–2.4	0.75–4.0	5.0–15.0	5.0–6.5
15.	—	—	1600–3200	1200	1000	1500–2500	800–1700	1500–2300	1400–2000
16.	M100–M108	—	M105–M120	M97–M110	M110–M125	M110–M120	M110–M118	M115–M125	M115–M125
17.	0.04–0.30	—	0.04–0.30	0.15–0.50	1.0–2.0	0.9–3.5	1.0–5.0	0.20–2.5	1.25–2.8
18.	Slight	Slight	Slight color change		Slight color change				—
19.	Excellent	Excellent	Excellent	Excellent	Fair	Fair	Fair	Fair	Fair
20.	—	—	7.1–7.4	7.0	—	—	—	7.12	—
21.	0.217	0.2	0.35–0.40	0.35–0.40	—	—	—	0.23–0.40	—
22.	0.9–2.5	—	1.0–1.5	2.0–3.8	—	0.7–2.5	—	0.7–12.0	—
23.	325	—	225–300	300	210–260	210–275	225–275	300	290
24.	0.1 in./min.	Nil	Slow to nil	Slow	Approximately nil			Nil	Nil
25.	52000	100000	50000–1000000	100000–500000	—	100–400000	5	100000	10–500
26.	3.3×10^{13}	$>10^{12}$	1×10^{15}	4.5×10^{10}	—	—	—	—	—
27.	450	450	400–750	475–800	400–700	200–450	50–150	200–600	300–450
28.	—	—	300–600	325–540	200–450	150–350	60–120	150–450	250–350
29.	36	—	35–100	60–80	—	12–75	7	23–75	35–55
30.	5.9	—	4.2–5.3	4.0–6.0	—	7.5–8.6	—	6.5–10.0	4.5–6.5
31.	5.6	—	—	—	7.9	7.3–8.3	—	6.1–9.5	4.2–6.0
32.	5.4	3.9	4.5–5.3	4.0–5.0	6.4–8.5	6.2–8.0	8.0–9.6	6.0–9.0	4.2–7.5
33.	0.005	—	0.003–0.015	0.010–0.025	—	0.06–0.15	—	0.04–0.10	0.004–0.05
34.	0.007	—	—	—	0.057	0.03–0.09	—	0.012–0.03	0.004–0.05
35.	0.018	0.020	0.010–0.030	0.030–0.040	0.035–0.05	0.03–0.07	0.12–0.22	0.011–0.025	0.006–0.08
36.	60	120	15–180	30–120	100	120–180	—	175–200	170–200
37.	None	None	None		Slight to marked	Slight to marked	None	None	None
38.	Slight	Slight	Slight		—	—	Decomposes	—	—
39.	Slight	Slight	Slight		None	None	None	None	None
40.	Slight	Slight	Attacked		Attacked	Attacked	None	Attacked	Attacked
41.	Slight	None	Slight to none	Slight	None on bleed-proof materials				

TABLE 9   Laminates Property Chart[1] (Continued)

Properties	Phenol-formaldehyde laminates (see also facing page)						
	Cotton web base	Nylon fabric base	Glass fabric base	Asbestos fabric base	Asbestos felt base	Asbestos paper base	Wood base
Physical							
1. Laminating temp., °F.	275–350	275–325	275–350	300–350	260–325	300–350	390–320
2. Laminating pressure, p.s.i.	1000–1800	15–1800	15–2000	300–1800	2–1000	1000–1800	1000–2000
3. Specific gravity, D792	1.31–1.37	1.15–1.19	1.40–1.95	1.55–1.92	1.30–1.90	1.55–1.83	1.30
4. Specific volume, cu. in./lb., D792	21.1–20.2	24.0–23.2	19.7–15.4	17.8–15.4	21.3–14.6	16.8–15.1	21.3
5. Tensile strength, D638, p.s.i.	14000–16000	5000–9500	9000–50000	10000–13000	11000–50000	5000–15000	16000–32000
6. Modulus of elasticity in tension, D638, 10^5 p.s.i.	6–11	3.5–5.0	12–25	10–17	17–52	16–25	37
7. Compressive strength, D695, p.s.i.	32000–47000	28000–36000	34000–75000	30000–55000	6000–85000	40000	12000–21000
8. Modulus of elasticity in compression, D695, 10^5 p.s.i.	4–8	—	—	7	4–53	5	—
9. Flexural strength, D790, p.s.i.	18000–25000	9000–22000	16000–80000	10000–35000	16000–60000	11000–30000	25000–40000
10. Modulus of elasticity in flexure, D790, 10^5 p.s.i.	6–10	5–10	10–40	10–21	15–54	14–23	—
11. Shear strength, D732, p.s.i.	12600–14000	10000–13700	17200–24000	12000–24000	10000–29000	7500–13500	—
12. Modulus of elasticity in shear, D953, 10^5 p.s.i.	—	—	—	—	—	—	—
13. Bearing strength, D953, p.s.i.	—	—	—	—	16000–47000	—	—
14. Impact strength, Izod, D256, ft. lb./in. of notch	1.4–2.5	2.0–5.8	4.0–18.0	1.5–5.0	4–12	0.6–1.5	4.0–8.0
15. Bond strength, D952, lb.	1000–2200	1000–1500	800–2000	1500–2200	800–2000	600–2000	—
16. Hardness, Rockwell, D785	M100–M110	M100–M110	M100–M110	M70–M115	M70–M120	M90–M111	M90–M105
17. Water absorption, 24 hr., $\frac{1}{8}$-in. thickness, D570, %	1.2–1.6	0.15–0.60	0.12–2.70	0.3–2.5	—	0.4–2.0	2.5–11.0
18. Effect of sunlight	Lowers surface resistance and general darkening				No effect	Lowers surface resist., gen'l darkening	
19. Machining qualities	Fair to excellent		Fair to good				Good
Thermal							
20. Thermal conductivity, C177 10^{-4} cal./sec./sq. cm./1° C./cm.	—	—	1.4–1.8	4.6	3.4–6.9	—	—
21. Specific heat, cal./°C./gm.	—	0.35–0.40	0.23–0.30	0.30	0.31	0.30	0.2–0.4
22. Thermal expansion, D 696, 10^{-5}/in./in. °C	1.3–1.6	4.0	1.5–2.5	1.3–2.5	0.30–0.60	1.0–2.2	0.6–6.5
23. Resistance to heat (continuous), °F.[4]	225–255	165	250–500	275–325	275–350	275	150–200
24. Burning rate, D635	Very low	Slow	Nil	Approx. nil	Approx. nil	Approx. nil	Very low
Electrical							
25. Insulation resistance (96 hr. at 90% R.H. and 35° C.), D257, megohms	5–200	30000–1000000	25–5000	25	—	—	0.25–335
26. Volume resistivity (50% R.H. and 25° C.), D257, ohm-cm.	—	7×10^{13}	2.5×10^{10}	—	1.4×10^{11}	—	—
27. Dielectric strength, short-time, $\frac{1}{8}$-in. thickness, D149, volts/mil	200	360–600	300–700	50–100	55–85	160–250	75–500
28. Dielectric strength, step-by-step, $\frac{1}{8}$-in. thickness, D149, volts/mil	150	300–450	250–650	50–75	—	95–200	—
29. Dielectric strength, step-by-step, parallel to lamination, KV	10–40	40–100	30–70	1–15	—	3–25	—
30. Dielectric constant, 60 cycles, D150	—	3.7–6.0	4.0–10.0	—	39	—	—
31. Dielectric constant, 10^3 cycles, D150	—	3.6–4.1	4.8–6.3	7.5	20	7.0	—
32. Dielectric constant, 10^6 cycles, D150	5.5–7.0	3.3–4.5	3.7–6.6	5.5–10.0	4.5	5.2	5.0
33. Dissipation factor, 60 cycles, D150	0.10–0.50	0.02–0.06	0.01–0.10	—	0.3	—	0.20
34. Dissipation factor, 10^3 cycles, D150	—	0.01–0.02	0.015–0.042	0.06–0.10	0.3	0.15–0.20	—
35. Dissipation factor, 10^6 cycles, D150	0.06–0.50	0.015–0.040	0.005–0.050	0.10–0.15	0.2	0.08–0.14	0.05
36. Arc resistance, D495, sec.	Tracks	10	Tracks	4	—	4	Tracks
Chemical							
37. Effect of weak acids, D543	None to slight depending on acid						
38. Effect of strong acids, D543	Decomposed by oxidizing acids; reducing and organic acids none to slight effect						
39. Effect of weak alkalies, D543	Slight to marked depending on alkalinity and grade						
40. Effect of strong alkalies, D543	Attacked by strong alkalies unless a special alkali-resistant resin is used						
41. Effect of organic solvents, D543	None on bleed-proof materials						

TABLE 9   Laminates Property Chart[1] (Continued)

	Phenol-formaldehyde laminates (see also facing page)		Lignin plastic	Silicone laminates		Teflon laminates			Vulcanized fiber
	Cellulose paper base	Cotton fabric base	Laminated	Glass fabric base	Asbestos fabric base	Glass nonwoven fiber base	Ceramic nonwoven fiber base	Glass fabric base	(In general, low thicknesses give high values)
1.	275–350	275–350	275–365	325–480	350–400	—	—	—	—
2.	1000–1800	1000–1800	2000–2500	30–2000	30–2000	—	—	—	—
3.	1.28–1.40	1.30–1.40	1.35–1.45	1.6–1.9	1.75	2.15–2.25	1.9–2.2	2.2	0.95–1.45
4.	21.8–19.7	21.3–20.4	21.6–19.0	17.3–14.5	15.8	12.5–13.1	14.6–12.8	12.8–12.6	28.0–19.0
5.	8000–20000	7000–16000	5000–8000	10000–37600	—	3000–7500	1500–4000	12000–20000	6000–14000
6.	8–20	5–15	—	15–20	—	—	—	—	8–12
7.	20000–40000	30000–44000	25000–30000	25000–46000	40000–50000	10000	—	20000	25000–35000
8.	8	6	—	—	—	—	—	—	—
9.	10500–30000	14000–30000	9000–15000	10000–38000	12000–16000	7500–15000	—	11000–17000	13000–20000
10.	8–15	8–10	13	10–35	—	2.4–3.7	—	—	7–10
11.	6000–14000	7000–18000	—	16500–20000	—	—	—	11000	—
12.	—	—	—	—	—	—	—	—	—
13.	—	—	—	—	—	—	—	—	—
14.	0.3–1.10	0.8–3.0	0.5–1.5	5.0–13.0	6.0–9.0	1.3–2.5	1.1–2.5	1.5–15.0	1.0–3.0
15.	700–1400	1000–2500	450–600	600–1300	—	—	—	600–900	1000
16.	M70–M120	M90–M115	M90	M100	—	L50	—	M30	R50–R80
17.	0.2–6.0	0.7–4.4	2.5 ($\frac{1}{4}$ in.)	0.070–0.65	1.0–1.5	0.005–0.25	0.5–2.0	0.02	18.0–61.0
18.	Lowers surface resis., gen. darkening		Slight	Nil	Nil	Nil	Nil	Nil	Slight
19.	Fair to excellent		Good	Fair to good	Fair to good	Excellent	Fair to good	Good	Good to excellent
20.	4.8–7.0	2.1–7.0	—	7.0	—	5.0–7.0	—	1.0–2.0	3.0–5.0
21.	0.38–0.41	0.35–0.40	—	0.25–0.27	—	0.23	0.25	0.2	0.4
22.	1.4–3.4	1.8–3.2	0.42	0.5–1.0	—	—	3.7–3.8	—	2.3
23.	225–250	225–250	150	400–700	>480	500	500	500	225
24.	Very low	Very low	Slow	Nil	Nil	Nil	Nil	Nil	Slow
25.	35–1000000	10–5000	—	2500–3000000	—	400000	30	20000 to 10^5	—
26.	10^{10}–10^{13}	10^{10}–10^{12}	1.6×10^{12}	—	—	$>10^7$	$>10^7$	$>10^6$	10^6–10^{13}
27.	300–1000	150–600	135 ($\frac{1}{4}$ in.)	180–480	50–150	300	150–250	300–800	150–300
28.	260–800	120–440	105 ($\frac{1}{4}$ in.)	150–350	50–100	250	—	250–700	125–225
29.	15–90	10–60	—	25–60	—	>55	—	40–80	—
30.	4.5–7.5	5.0–10.0	6.0 ($\frac{1}{4}$ in.)	4.3–5.0	—	2.2	2.5–2.7	2.6–2.8	—
31.	4.2–6.0	4.2–6.5	—	3.7–4.3	—	2.2	—	2.6–2.8	4.7
32.	3.6–6.0	5.0–7.5	—	3.7–4.3	—	2.2–2.4	2.4–2.5	2.4–2.7	
33.	0.02–0.10	0.04–0.50	0.11	0.005–0.01	—	0.004	0.002–0.003	0.0005	—
34.	0.03–0.07	0.04–0.09	—	0.005–0.01	—	0.001	—	0.0004	0.03–0.08
35.	0.02–0.08	0.05–0.10	—	0.005–0.01	—	0.00035	0.001–0.003	0.0008	
36.	4–75	4	135	150–250	—	180	190	180	110–150
37.	None	None	None to slight	None	Very slight	Nil	None	None	Embrittled by some
38.	Slight	Slight	Poor	Some attack	Some attack	Attacked by HF only		None	Embrittled
39.	Slight	Slight	Fair	Very slight	Very slight	None	None	None	Swells
40.	Attacked	Attacked	Poor	Attacked	Attacked	Slight	Slight	Slight	Swells
41.	None	None	Slight	Attacked by some		None	None	None	None

reinforcement. Table 10 shows that electrical properties are maximized when the filler is glass. Melamines retain their properties at low temperatures. Table 11 shows the physical properties of melamine with several reinforcements down to −65°F. For many years melamines were used exclusively where electric-arc resistance was desired. Recently, however, filled polyester and epoxy materials have begun to compete with melamine on both a cost and an arc-resistance basis. Melamines are often used where resistance to caustic is required. Other uses include switchgear, terminal blocks, circuit-breaker parts, slot wedges, and bus-bar supports. Melamine laminates have poor dimensional stability, particularly when the part is exposed to alternating cycles of high and low humidity.

Epoxy laminates Epoxies were first synthesized in 1931, but they did not become popular until 1955. The family consists of many resins based on a large variety of chemical modifications. Typical properties are given in Table 9. Note the extremely high physical-strength values of laminates made from unidirectional fiber mat as shown in the first column. This laminate approaches filament-wound materials in properties. Epoxies are frequently used in cryogenic applications where resistance to resin crazing is desired. Figure 6 shows the tensile properties of laminates to −423°F. Military Specifications MIL-R-9300A and MIL-P-25421 describe epoxy laminates. Epoxies are frequently used in electronic applications where their excellent dielectric properties are of value. In general, however, these properties increase with temperature, limiting the electrical application of epoxies to about 250°F. They are used also in marine applications, where their resistance to water is utilized. With the correct choice of hardener,

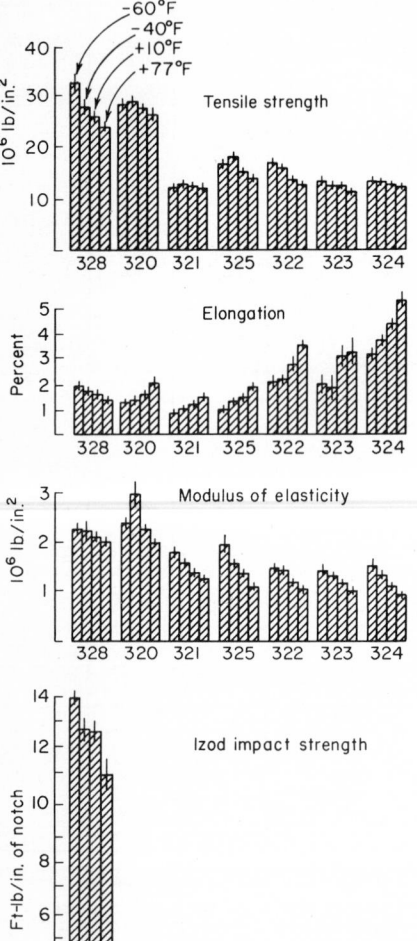

Fig. 4 Low-temperature properties of phenolic laminates.[4]

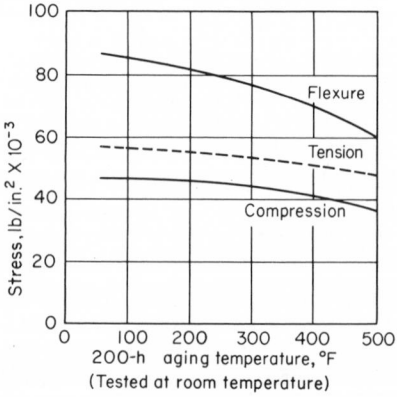

Fig. 5 Physical properties of high-temperature phenolic laminates after aging.[5]

TABLE 10   Electrical Properties of Melamine Laminates Related to Reinforcements[6]

Property	Reinforcement		
	Asbestos	Glass	Cellulose
Insulation resistance, ohms..............	10^8	10^9	10^7
Arc track resistance, volts (D 2303).......	1,500	1,500	1,000
Dielectric strength, volts/mil.............	50	300	250

laminates can be made with good retention of physical properties up to 400°F. One such glass cloth–epoxy laminate is shown in Figs. 7 to 10 along with the changes in flexural strength, water absorption, and electric strength from aging at 329, 392, 436, and 482°F.   The changes in volume resistivity, dielectric constant, and power factor vs. temperature and at several frequencies for the same material are given in Figs. 11 to 13.   Epoxy laminates are used in printed circuits and chemical-resistant applications.   The cost of these materials is higher than that of phenolic, melamine, or polyester composites.

Table 11   Mechanical Properties of Melamine at Varying Temperatures[4]

Property*	Temperature, °F	Melamine† glass	Melamine‡ asbestos	Electrical melamine-formaldehyde §	α-Melamine-formaldehyde ¶
Tensile strength, psi $\times 10^{-3}$	77	32.5	13.5	5.4	7.8
	10	32.9	14.5	6.7	6.9
	−40	38.1	15.4	5.7	6.9
	−65	37.2	14.1	5.6	6.7
Modulus of elasticity, psi $\times 10^{-6}$	77	2.130	2.146	1.060	1.270
	10	2.290	3.040	1.610	1.640
	−40	1.430	3.390	1.540	1.730
	−65	1.580	3.060	1.390	1.880
Elongation at break, %	77	2.15	0.82	0.54	0.62
	10	2.17	0.64	0.50	0.44
	−40	2.36	0.60	0.40	0.39
	−65	2.75	0.55	0.38	0.37
Work to produce failure, ft-lb/in.³	77	31.0	......	1.32	2.08
	10	36.7	......	1.40	1.33
	−40	44.3	......	1.02	0.86
	−65	52.2	......	0.97	1.13
Proportional limit, psi $\times 10^{-3}$	77	15.8	14.5	33.2	25.9
	10	7.1	9.5	10.5	9.5
	−40	5.4	6.7	5.7	5.6
	−65	7.8	6.9	6.9	6.7
Izod impact strength, ft-lb/ in. notch (D 256)	77	11.12	0.98	0.33	0.31
	10	12.64	0.90	0.37	0.28
	−40	13.43	0.96	0.32	0.28
	−65	14.68	0.93	0.28	0.29

* Average values.
† Laminated melamine, glass fabric base.
‡ Laminated melamine, asbestos paper base.
§ Melamine-formaldehyde, cellulose filler, electrical grade.
¶ Melamine-formaldehyde, α-cellulose filler.

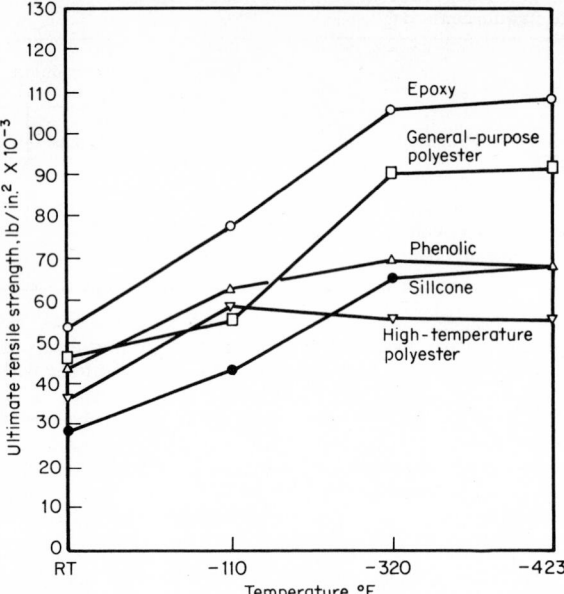

Fig. 6 Cryogenic tensile strength of various laminates.[7]

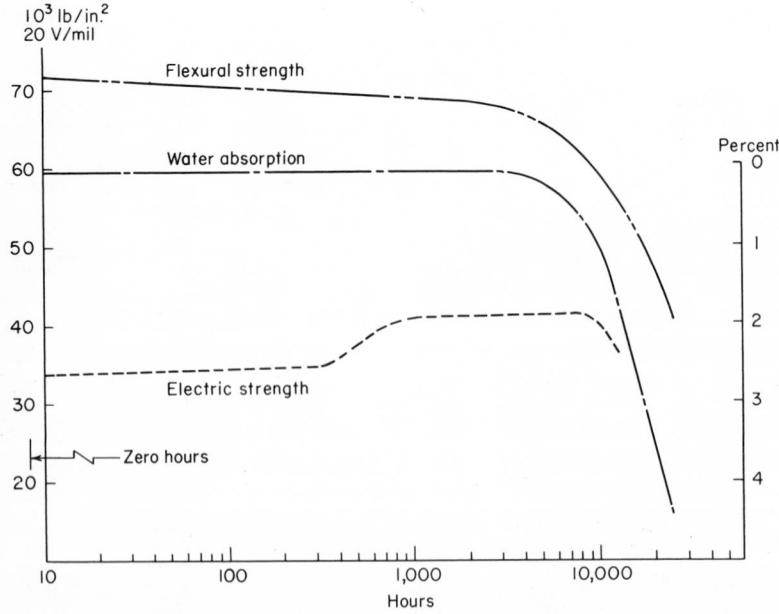

Fig. 7 Properties vs. time at 329°F for high-temperature epoxy laminate.

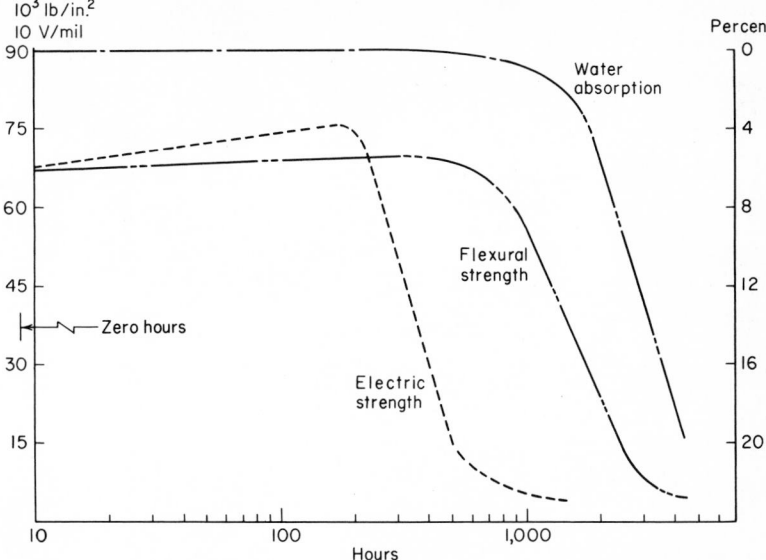

Fig. 8 Properties vs. time at 392°F for high-temperature epoxy laminate.

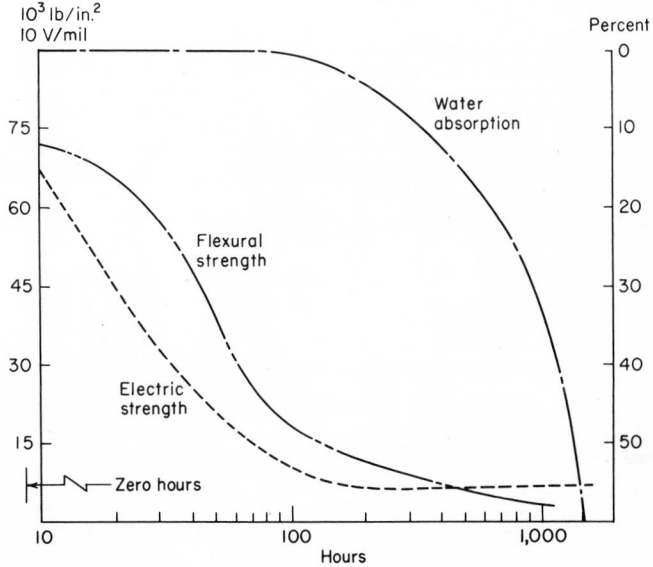

Fig. 9 Properties vs. time at 436°F for high-temperature epoxy laminate.

Polyester laminates   Polyester laminates are described in MIL-R-7575B and MIL-P-8013C. The resins were first used during World War II for a variety of structural and insulating applications. Commercial polyesters usually consist of the polyester polymer dissolved in a monomer such as styrene. Typical properties are shown in

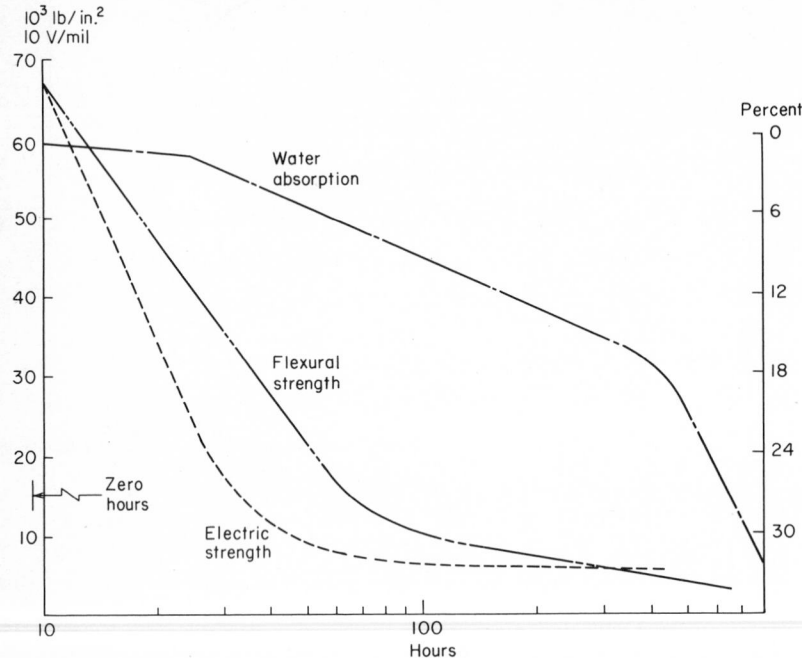

Fig. 10   Properties vs. time at 482°F for high-temperature epoxy laminate.

Table 9. The resins are usually combined with glass fibers, since many cellulosic reinforcements inhibit their cure. Table 12 shows the effect of reinforcement media on several properties. One of the chief virtues of these resins is the excellent chemical resistance of the laminates, and hence they are used for corrosion-resistant tanks, exhaust ducts, scrubbers, and plating equipment. Note that special grades of polyester are used for chemical service. These materials resist acids and bases for long times

TABLE 12   **Relationship of Properties to Reinforcement for Polyester Laminates**

Reinforcement material	Moisture absorption, %	Ultimate tensile strength, psi × 10⁻³				Ultimate flexural strength, psi × 10⁻³			
		At room temperature		At −65°F (frozen)		At room temperature		At −65°F (frozen)	
		Dry	Wet	Dry	Wet	Dry	Wet	Dry	Wet
Glass cloth......	0.25	52.9	48.7	64.7	62.8	64.8	47.7	72.9	69.4
Glass mat.......	0.28	13.0	11.8	16.2	13.3	31.1	23.1	31.9	29.4
Cotton cloth....	2.6	5.5	6.0	5.9	7.2	13.2	....	12.8	

even at 200°F. They are attacked by most chlorinated solvents. Table 13 shows the effect of a variety of chemicals on a bisphenol-fumarate polyester. The chemical resistance of these laminates also can be influenced by the type of glass reinforcement, the fabrication method, and the amount of stress in the product. Polyesters are useful over a fairly wide temperature range below 250°F. Table 14 shows physical properties for typical polyester laminates to −60°F. Polyester laminates are widely used in the switchgear industry. They can be made highly arc-, track-, and flame-resistant. The polyester resins themselves possess good arc and track resistance, which is further enhanced by the addition of mineral fillers, particularly aluminum oxide trihydrate. Flame resistance is obtained by modifying the basic resin, usually by the incorporation of chlorine, although bromine and phosphorus also are used. Polyesters are moderate in cost. They are used in applications requiring dimensional stability, such as furniture, and water resistance, such as boats and bathtubs. They are used in automotive applications such as grills, hoods, fenders, and interior parts. A large use for resins stabilized to sunlight is in corrugated sheet used as screens, patio roofs, and dividers. Low-profile polyesters are resins to which have been added any of several thermoplastic polymers. This addition decreases the resin shrinkage and, thereby, produces exceptionally smooth surfaces. These resins are used in automotive and recreational-vehicle body parts. Surfacing resins are polyesters to which have been added waxes to eliminate air inhibition and thixotropic agents for viscosity control. They mostly are used in hand lay-up applications like boat hulls.

A closely related polymer family, the vinyl esters, has been introduced recently. This family has even superior chemical resistance and certain processing advantages.

Diallyl phthalate laminates   Properties of diallyl phthalate laminates are given in Table 9. This resin has two principal

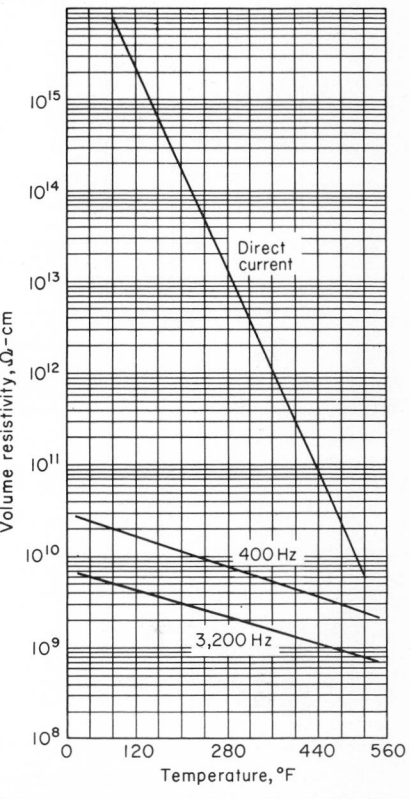

Fig. 11   Volume resistivity vs. temperature at 400 and 3,200 Hz and direct current for high-temperature epoxy laminate.[8]

forms: diallyl orthophthalate (DAP) and diallyl isophthalate (DAIP). The isophthalate is more thermally stable, being useful up to 350°F, whereas the orthophthalate is useful only to about 300°F. In most other respects the resins are similar. Diallyl phthalate laminates are used almost exclusively as electronic insulations. Their low-loss characteristics are attractive over a range of frequencies and up to their maximum temperature limits. Radar and microwave applications make use of DAP materials because of these low losses. Figures 14 and 15 show the influence of frequency on dielectric constant and loss factors of DAIP and DAP laminates in comparison with various laminates.

In these curves the epoxy MPDA and CL are epoxy-cured with *m*-phenylenediamine, epoxy DICY is epoxy-cured with dicyandiamide, and TAC polyester refers to a high-temperature polyester resin incorporating triallyl cyanurate monomer. All materials were laminated with style 181 glass cloth.

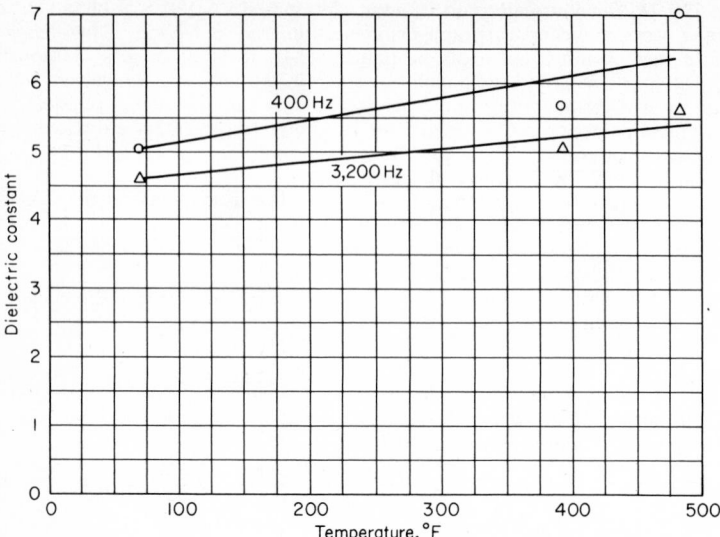

Fig. 12 Dielectric constant vs. temperature at 400 and 3,200 Hz for high-tempera-ture epoxy laminate.[8]

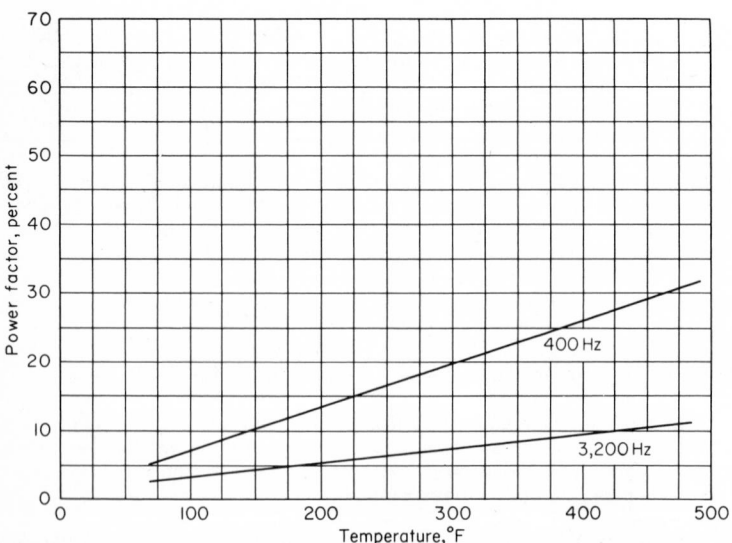

Fig. 13 Power factor vs. temperature at 400 and 3,200 Hz for high-temperature epoxy laminate.[8]

TABLE 13   Resistance of a Bisphenol A–Fumarate Polyester to Various Chemicals[10]

Acids	Concentration, %	Max temp, °F		Alkalies	Concentration, %	Max temp, °F
Acetic	10	220		Ammonium hydroxide	5	180
Acetic	25	200		(Aqueous ammonia)	10	140
Acetic	50	160		Ammonium hydroxide	20	140
Acetic	75	160		(Aqueous ammonia)	29	100
Acrylic	100	Amb		Calcium hydroxide	25	160
Benzene sulfonic	All	220		Potassium hydroxide	25	160
Benzoic	All	220		Sodium hydroxide	5	200
Boric	All	220		Sodium hydroxide	10	160
Butyric	50	220		Sodium hydroxide	25	160
Chloroacetic	25	200		Sodium hydroxide	50	Amb
Chloroacetic	50	140				
Chromic	5	200				
Chromic	10	140				
Chromic	30	NR		**Bleaches**		
Citric	All	220				
Fatty acids	All	220		Calcium hypochlorite	20	200
Fluoboric	All	220		Chlorine dioxide	15	220
Fluosilicic*	25	Amb		Chlorine water	Sat'd	160
Formic	All	Amb		Chlorite	...	210
Gluconic	50	Amb		Hydrogen peroxide	30	140
Hydrobromic	48	160		Hydrosulfite	...	220
Hydrochloric	10	220		Peroxide	...	220
Hydrochloric	20	180		Sodium hypochlorite	5¼	†
Hydrochloric	37	160		Sodium hypochlorite	15	†
Hydrochlorous	20	160		Textone	...	220
Hydrocyanic	10	160				
Hydrofluoric*	10	180				
Hydrofluoric*	20	NR		**Gases**		
Hypochlorous	20	160				
Lactic	All	220		Bromine	...	NR
Maleic	All	220		Carbon dioxide	...	220
Nitric	5	160		Carbon monoxide	...	220
Nitric	60	Amb		Chlorine, dry	...	220
Nitrous oxide fumes	...	140		Chlorine, wet	...	220
Oleic	All	220		Sulfur dioxide, dry	...	220
Oleum	All	NR		Sulfur dioxide, wet	...	220
Oxalic	Sat'd	220		Sulfur trioxide	...	220
Perchloric	30	Amb				
Phosphoric	80	220				
Stearic	All	220		**Organic Materials**		
Sulfuric	25	220				
Sulfuric	50	220		Acrylonitrile	100	NR
Sulfuric	70	160		Diethyl ether	100	NR
Superphosphoric 115%				Diethyl ketone	100	NR
P_2O_5	...	200		Dimethyl formamide	100	NR
Tannic	All	220		Dimethyl phthalate	100	Amb
Tartaric	All	220		Dimethyl sulfoxide	100	NR
Trichloroacetic	50	220		Dioxyl phthalate	100	Amb
				Diphenyl ether	100	120
				Ethyl acetate	100	NR
Alcohols				Ethylene chlorohydrin	100	220
				Ethylene glycol	All	220
Amyl	All	210		Formaldehyde solution	All	†
Benzyl	All	Amb		Furfural	5	160
Butyl	All	Amb		Furfural	20	Amb
Ethyl	All	Amb		Gasoline	100	Amb
Methyl	All	Amb				

TABLE 13   Resistance of a Bisphenol A–Fumarate Polyester to Various Chemicals[10] **(Continued)**

Organic Materials (Continued)

	Concentration, %	Max temp, °F
Isopropyl palmitate......	100	220
n-Heptane..............	100	Amb
Kerosene..............	100	Amb
Orange oil.............	100	100
Phenol................	10	NR
Phthalic anhydride......	100	220
Pyridine...............	100	NR
Sour crude oil..........	100	220
Toluene diisocyanate.....	100	Amb
Triethanolamine........	100	Amb

Solvents

	Concentration, %	Max temp, °F
Acetone...............	10	180
Acetone...............	100	NR
Benzaldehyde..........	100	NR
Benzene..............	100	NR
Carbon disulfide........	100	NR
Carbon tetrachloride.....	100	Amb
Chlorobenzene.........	100	NR
Chloroform............	100	NR
Dichlorobenzene........	100	NR
Ethylene chloride.......	100	NR
Ethylene dichloride......	100	NR
Ethyl ether............	100	NR
Methylene chloride......	100	NR
Methyl ethyl ketone.....	100	NR
Monochlorobenzene......	100	NR
Naphtha..............	100	Amb
Nitrobenzene..........	100	NR
Tetrachloroethylene......	100	Amb
Toluene...............	100	NR
Trichloroethylene........	100	NR
Trichloromonofluoromethane..............	100	Amb
Water, distilled..........	...	210
Xylene................	100	NR

Salts

	Concentration, %	Max temp, °F
Aluminum chloride......	All	220
Aluminum potassium sulfate..............	All	220
Aluminum sulfate.......	All	220
Ammonium bicarbonate..	10	160
Ammonium bicarbonate..	50	160
Ammonium carbonate....	50	Amb
Ammonium chloride.....	All	220
Ammonium nitrate......	All	220
Ammonium persulfate....	All	180
Ammonium sulfate......	20	220
Aniline sulfate..........	All	220
Antimony trichloride.....	All	220
Barium carbonate.......	All	220
Barium chloride........	All	220

Salts (Continued)

	Concentration, %	Max temp, °F
Barium sulfide..........	All	140
Calcium chlorate........	All	220
Calcium chloride........	All	220
Calcium sulfate.........	All	220
Copper chloride.........	All	220
Copper cyanide.........	All	220
Copper sulfate..........	All	220
Ferric chloride..........	All	220
Ferric nitrate...........	All	220
Ferric sulfate...........	All	220
Ferrous chloride.........	All	220
Ferrous nitrate..........	All	220
Ferrous sulfate..........	All	220
Lead acetate............	All	220
Magnesium carbonate....	All	140
Magnesium chloride.....	All	220
Magnesium sulfate.......	All	220
Mercuric chloride........	All	220
Mercurous chloride......	All	220
Nickel chloride..........	All	220
Nickel nitrate...........	All	220
Nickel sulfate...........	All	220
Potassium aluminum sulfate...............	All	220
Potassium bicarbonate...	All	160
Potassium carbonate.....	10	140
Potassium carbonate.....	25	100
Potassium carbonate.....	50	Amb
Potassium chloride.......	All	220
Potassium dichromate....	All	220
Potassium ferricyanide...	All	220
Potassium ferrocyanide...	All	220
Potassium nitrate........	All	220
Potassium permanganate.	All	220
Potassium persulfate.....	All	220
Potassium sulfate........	All	220
Silver nitrate............	All	220
Sodium acetate..........	All	220
Sodium bicarbonate......	10	160
Sodium bisulfate........	All	220
Sodium carbonate.......	10	180
Sodium carbonate.......	32	160
Sodium cyanide.........	All	220
Sodium chloride.........	All	220
Sodium chlorite.........	10	Amb
Sodium ferricyanide......	All	220
Sodium nitrate..........	All	220
Sodium nitrite..........	All	220
Sodium silicate.........	All	220
Sodium sulfate..........	All	220
Sodium sulfide..........	All	220
Sodium sulfite..........	All	220
Stannic chloride.........	All	220
Stannous chloride........	All	220
Trisodium phosphate.....	25	Amb
Zinc chloride............	All	220
Zinc sulfate.............	All	220

TABLE 13 Resistance of a Bisphenol A-Fumarate Polyester to Various Chemicals[10] (Continued)

Plating Solutions	Concentration, %	Max temp, °F	Other	Concentration, %	Max temp, °F
Cadmium cyanide.......	...	200	Aluminum chloro-		
Chrome*...............	...	160	hydroxide............	50	Amb
Gold..................	...	200	Glycerine..............	100	220
Lead..................	...	200	Linseed oil..............	100	220
Nickel................	...	200	Sorbitol solutions........	All	120
Silver................	...	200	Succinonitrile, aqueous...	...	Amb
Tin fluoborate*.........	...	200	Sulfonated detergents....	100	160
Zinc fluoborate*........	...	200	Urea-ammonium nitrate fertilizer mixture......	...	100
			8-8-8 fertilizer...........	...	100

NR = not recommended. Amb = ambient temperature. Sat'd = saturated.

* 10 to 20 mils of synthetic surfacing mat such as Dynel or Orlon should be used to reinforce surface in contact with chemical.

† Satisfactory up to maximum stable temperature for product.

TABLE 14 Low-Temperature Properties of Polyester Laminates[11]

Property*	Degrees centigrade						
	−60	−40	−20	0	23	50	80
Modulus of elasticity, primary:							
High rate, psi × 10³...	4,770	4,480	4,540	4,460	5,545	3,500†	3,360†
Static, psi × 10³......	3,760	2,910	2,690	2,820	2,825	2,500†	2,240†
Increase, %..........	27	54	69	58	96	40	50
Modulus of elasticity, secondary:							
High rate, psi × 10³...	1,930	2,090	1,870	1,990	2,352		
Static, psi × 10³......	1,840	1,890	1,610	1,890	2,086		
Increase %..........	5	11	16	5	13		
Tensile strength:							
High rate, psi × 10⁻³.	73.6	72.5	68.8	66.47	62.34	58.00	54.2
Static, psi × 10⁻³.....	51.58	44.84	41.4	39.0	36.74	34.13	29.4
Increase, %.........	43	62	66	70	70	70	84
Elongation, %:							
High rate............	3.20	3.24	2.91	2.78	2.41	2.42	2.10
Static...............	2.46	2.21	2.01	1.83	1.66	1.65	1.48
Increase.............	30	47	45	52	45	47	42
Work to produce failure:							
High rate, ft-lb/in.³...	111.5	111.2	97.2	90.1	74.6	70.4	52.4
Static, ft-lb/in.³.......	60.7	45.3	38.6	32.9	27.4	25.4	19.6
Increase, %..........	84	145	152	174	172	177	167
Time to failure:							
High rate, msec.......	10.7	10.7	10.1	10.1	9.9	9.8	9.0
Static, msec × 10⁴....	7.8	6.8	6.5	6.2	5.9	5.8	5.8

* ASTM D 638, type I specimen: ⅛-in.-thick hand lay-up panels formulated with Stypol 25 resin, vacuum pressure >12 psi; gelled under heat lamps; postcured 3 hr at 250°F.

† Value at 0.5% strain.

DAIP has the lower dielectric constant and retains it virtually unchanged up to 10^{10} Hz. DAIP does not show an inflection point in the loss-factor curve as is the case with many other laminates. Figures 16 and 17 show the effect of temperature. DAIP again demonstrates flat curves, with only the more expensive silicones rivaling its performance. However, as shown in Fig. 18, when moisture is present, DAIP is better than the silicone. These data are for laminates exposed to 100 percent relative humidity for 24 h.

Diallyl phthalate laminates are used for microwave and radar applications, where they appear as radomes, terminal boards, feedomes, and connectors. The cost of these materials is fairly high. They are respected for their dimensional stability, being

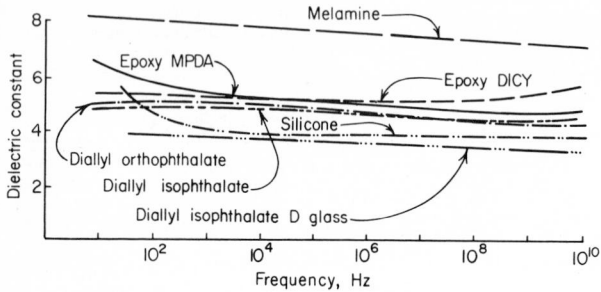

Fig. 14   Dielectric constant of various laminates vs. frequency.[9]

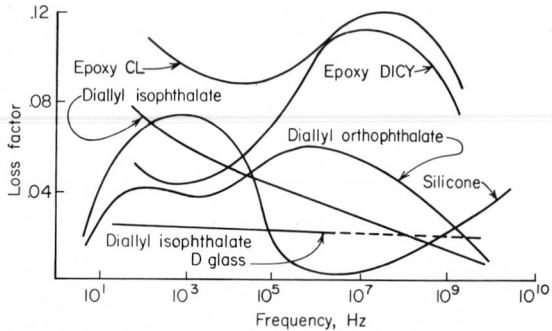

Fig. 15   Loss factor of various laminates vs. frequency.[9]

little affected by humidity changes. In hermetic and high-vacuum applications, care must be taken to ensure that these materials are well cured and postbaked because undercured laminates outgas DAP monomer, which can be detected as crystals.

Silicone laminates   Table 9 gives typical values of silicone laminates, and these are further defined by Military Specifications MIL-R-25506A and MIL-P-25518A. The silicone resins were the first of the true high-temperature plastics. Silicone polymers are similar to organic polymers except that certain carbon atoms are replaced by silicon and oxygen atoms. Silicone polymers analogous to the corresponding organic polymers are available in the form of liquids, gels, elastomers, and brittle solids. The latter polymers when combined with glass fabrics or asbestos constitute the silicone laminates known to the electronics and electrical industries. The materials are universally acceptable for class 180°C (class H) operation. As might be expected, these materials are useful from cryogenic temperatures to about 500°F. Cryogenic properties of the silicone laminates are given in Table 15, where their physical properties are compared with those of other laminating resins. Dielectric properties are compared with those of other laminating resins. Dielectric properties of silicones are particularly useful. The dissipation factor is low at room temperature and stays relatively constant to 300°F. Figure 19 compares glass-cloth-reinforced silicones with epoxy laminates. In this curve and the others following, two silicones are shown, one designated for low-pressure laminating and the other for high-pressure, and they are compared with NEMA G-10 and G-11 epoxy-glass laminates. Notice the flat slope of the silicones. The epoxies characteristically show an inflection point near the glass transition temperature, and their losses get progressively worse. The response of dissipation factor to frequency is shown in Fig. 20. Thermal aging does little to influence these properties. Table 16 shows the effect of aging at 572°F on

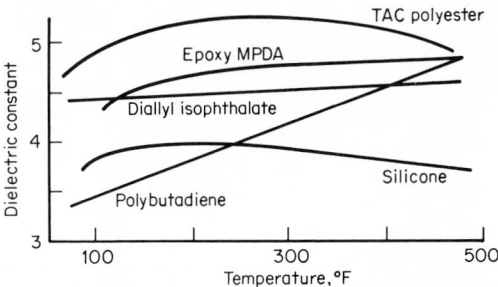

Fig. 16 Dielectric constant at 9,375 MHz of various laminates vs. temperature.[9]

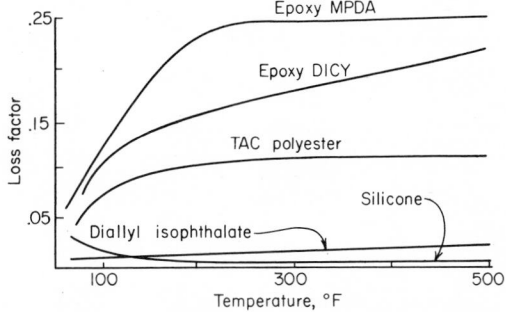

Fig. 17 Loss factor at 9,375 MHz of various laminates vs. temperature.[9]

the dielectric constant, dissipation factor, surface resistivity, and electric strength of a silicone laminate ⅛ in. thick formed from 116 glass cloth and Dow Corning 2105A resin. Figure 21 shows the effect of aging at 300°C on the insulation resistance. Because of the presence of the silicon atom, silicone laminates have good arc and track resistance. Table 17 shows typical tracking properties. Physical properties of silicones are not greatly influenced by aging, but compared with laminates based on other resins, the flexural and tensile strengths are not unusually high. Figures 22 and 23 show the influence of heat aging at 250°C on flexural and compressive strength. Silicone laminates are used as radomes, structures in electronic heaters, rocket components, slot wedges, ablation shields, coil forms, terminal boards, and printed circuits. As printed-circuit boards they have bond strengths to copper much less than those of epoxies (epoxy bond strength about 10 lb/in.). Physical properties at elevated temperature do not compare favorably with those of other high-tempera-ture laminates. The cost of silicones is high but not as high as some of the other specialty materials described in later sections. The use of silicones should be care-fully considered in applications where they may be adjacent to sliding electrical surfaces (such as carbon brushes) or moving contacts (such as switches). When exposed to elevated temperatures, certain silicones give off volatile fragments which condense on the surfaces in question, insulating these surfaces with a thin layer of silicon dioxide.

Phenylsilane laminates   The phenylsilane resins are a hybrid version of phenolics and silicones.   Phenylsilane laminates are used primarily for structural purposes at

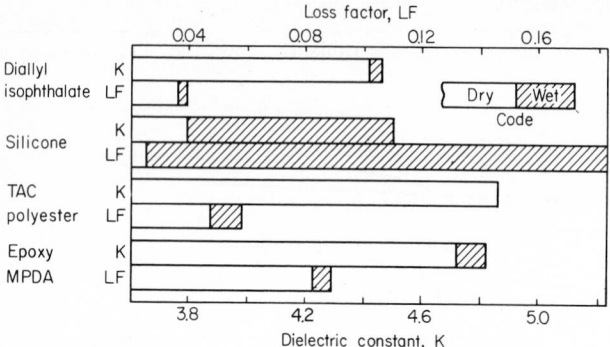

Fig. 18 Dielectric properties of various laminates as a function of humidity at 9,375 MHz.[9]

elevated temperatures.   Electrical properties are relatively poor compared with those of the silicones or DAP, for example, but they change very little with changing temperature.   Table 18 presents physical properties of a ⅛-in. style 181 glass cloth-

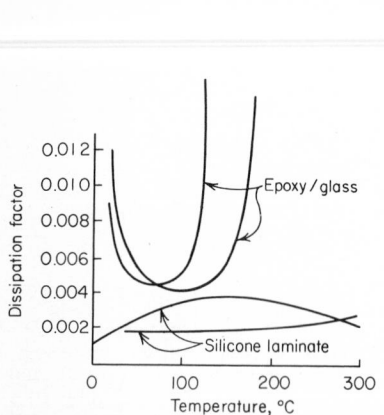

Fig. 19 Dissipation factor vs. temperature for silicone and epoxy laminates at 60 Hz.[12]

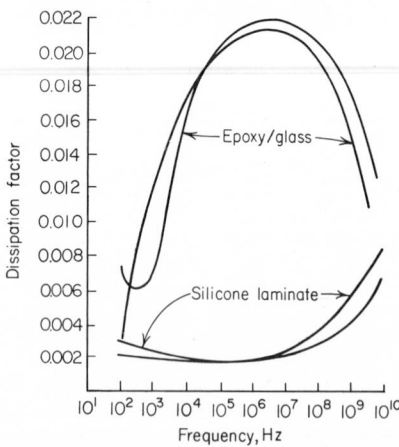

Fig. 20 Dissipation factor vs. frequency for silicone and epoxy laminates at room temperature.[12]

phenylsilane laminate based on Monsanto's SC 1013 resin and compares these properties with the requirements of MIL-R-9299A.

Teflon laminates   The excellent properties of Teflon resins are described elsewhere in this handbook.   Teflon laminates were developed to improve the creep, cold flow, and other physical properties of unmodified Teflon without disturbing the electrical, chemical, and thermal properties.   Physical properties for Teflon laminates are shown in Table 9.   The physical properties are much improved over the virgin polymer at

TABLE 15 Strength of Glass-reinforced Plastic Laminates Laminated with 181 Glass Cloth[7] (In psi $\times 10^{-3}$)

Resin system	Ultimate tensile strength				Ultimate flexural strength				Ultimate compressive strength			
	RT*	−110°F	−320°F	−424°F	RT*	−110°F	−320°F	−424°F	RT*	−110°F	−320°F	−424°F
Epoxy												
Epon[a] 828	40.9	62.3	91.6	97.9	77.9	99.3	150.4	145.0	43.5	85.4	109.9	108.8
Epon 1001	49.0	71.4	94.9	100.7	83.2	105.2	154.0	122.0	58.6	86.2	89.4	78.9
Epon 826 NMA	58.5	82.0	127.2	121.1	86.0	116.5	166.7	171.2	50.5	84.4	113.9	129.1
Epon 826 DDS	61.0	98.1	115.2	118.0	94.9	129.6	167.5	172.4	71.7	107.4	138.1	135.2
Narmco 522 (prepreg)	54.0	75.1	104.4	108.6	75.6	99.1	152.6	163.7	55.8	93.2	122.9	135.5
Phenolic												
Narmco 506	38.1	58.3	79.4	72.2	63.4	76.2	84.5	83.0	46.5	55.1	81.8	76.7
CTL 91 LD	51.7	65.7	65.1	64.9	77.4	100.1	109.0	110.0	70.4	88.1	99.1	91.6
Polyester (general purpose)												
Hetron 92[b]	39.2	63.2	71.7	84.9	71.4	82.1	87.2	81.9	22.9	35.8	43.6	41.4
Paraplex P43[c]	47.9	58.5	81.3	78.2	68.8	79.4	84.2	84.9	31.5	43.2	51.3	47.9
Narmco 527 (prepreg)	49.5	66.9	102.1	104.1	75.2	80.9	93.8	106.3	36.6	55.1	66.8	65.8
Hetron 31	50.0	68.1	105.3	100.1	71.2	104.1	126.7	129.0	19.9	41.3	68.5	66.4
Polyester (high-temperature)												
Laminac[d] 4232	41.0	60.7	56.9	55.5	63.3	74.7	70.0	65.4	28.4	42.0	47.6	44.1
Vibrin[e]	32.7	57.6	52.0	54.6	52.0	76.2	78.1	75.1	26.3	33.5	38.1	36.1
Silicone												
Narmco 513	29.6	47.8	76.3	80.5	34.1	40.7	52.1	52.8	25.3	28.7	39.1	43.3
Trevarno F 130[f]	25.7	46.7	67.9	74.6	36.1	43.5	63.7	72.8	17.7	23.6	34.9	30.8
Trevarno F 131[f]	29.4	47.4	70.4	76.2	37.9	45.8	66.9	65.0	21.0	39.0	42.7	46.1
Dow-Corning 7146	31.1	32.3	50.7	42.4	23.8	29.4	18.9	13.4	22.9	35.9	19.0	No test
Phenylsilane												
Narmco 534 (prepreg)	34.4	55.3	63.4	61.2	69.0	102.1	105.4	110.4	45.7	59.6	59.5	55.8
Nylon-epoxy												
Metlbond 406[g]	29.5	57.2	78.8	76.4	33.8	76.0	96.0	81.7	21.6	76.1	82.4	82.2
Polybenzimidazole	37.4	50.9	77.1	77.6	78.9	103.7	129.8	144.0	65.4	45.2	71.1	69.1
Flexible polyurethane												
Adiprene L-100[h]	11.6	76.1	73.8	70.2	Too flexible to test	51.8	58.7	48.5	4.9	48.6	90.1	91.1
Teflon												
FEP	24.5	43.0	65.7	81.3	19.4	37.4	67.3	66.3	7.3	11.4	22.6	33.3

* RT Room temperature.
[a] Trademark of Shell Chemical Co., New York.
[b] Trademark of Hooker Chemical Corporation, Niagara Falls, N.Y.
[c] Trademark of Rohm & Haas Co., Philadelphia, Pa.
[d] Trademark of American Cyanamid Co., Wayne, N.J.

[e] Trademark of United States Rubber Co., Naugatuck, Conn.
[f] Trademark of Coast Manufacturing and Supply Co., Livermore, Calif.
[g] Trademark of Whittaker Corp., Costa Mesa, Calif.
[h] Trademark of E. I. du Pont de Nemours & Co., Wilmington, Del.

TABLE 16   Effect of Accelerated Heat Aging on Silicone-Glass Laminates[12]

Hours aging at 572°F	Condition A	Condition D 48/50
Dielectric constant at 1 MHz		
0	3.78	3.82
24	3.72	3.76
48	3.77	3.80
96	3.75	3.79
192	3.71	3.90
Dissipation factor at 1 MHz		
0	0.0009	0.0027
24	0.0013	0.0024
48	0.0012	0.0030
96	0.0012	0.0033
192	0.0012	0.0230
Surface resistivity, ohms		
0	3.4×10^{13}	8.5×10^{13}
24	4.2×10^{14}	4.6×10^{13}
48	7.9×10^{14}	2.6×10^{14}
96	3.3×10^{14}	1.5×10^{14}
192	3.6×10^{14}	5.5×10^{13}
Perpendicular electric strength in air, volts/mil		
0	>328	>337
24	>379	>361
48	>375	349
96	351	315
192	106	100

only slight sacrifice in dielectric properties. These laminates are used almost exclusively in electrical insulation, where they meet the requirements of class 180°C. They are used as slot wedges, coil separators, terminal blocks, and printed circuits. One other use is in bearing pads on such structures as bridges. They are described in NEMA Standards Publication LI-1-1971.

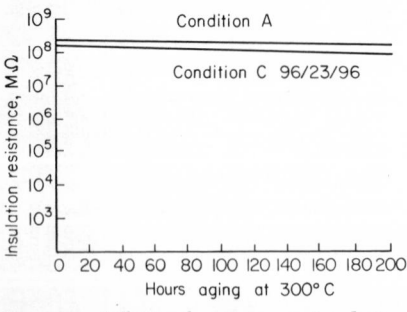

Fig. 21 Effect of aging on insulation resistance of silicone laminates.[12]

Polyimide laminates Polyimide resins first appeared in the early 1960s, and they represent a new concept in thermally stable polymers. Also available as copolymer amide-imides, the resins do not oxidize quickly and the rate of pyrolysis is slow to about 650°F. The polyimides have certain similarities to nylon. They are available in the form of films, molding materials, wire enamels, adhesives, and laminates. The polyimides are really thermoplastic resins, their molecules being linear and not cross-linked. However, like Teflon, their melting point is virtually identical with the decomposition point,

TABLE 17   Track Resistance of
Silicone-Glass Laminate[13]

IEC CTI,* volts	320
Dust fog (ASTM D 2132-62T) hr	1.0
Inclined plane, liquid contaminated, kv	1.5
Differential wet track:	
Discharge power, watts	1.3
Time to track, sec	8.0

*International Electrotechnical Commission—
Critical Tracking Index.

TABLE 18   Physical Properties of Phenylsilane Laminates[14]

Property	Phenylsilane, press grade				MIL-R-9299A, Class 2		
	At room temperature		At 500°F after:		At room temperature		At 500°F after ½ hr
	Dry	Wet	½ hr	200 hr	Dry	Wet	
Flexural strength, psi × 10⁻³	64	62.6	53.8	49	50	45	40
Flexural modulus, psi × 10⁻⁶	3	2.5	3.01	2.6	3	2.5	3
Compressive strength, psi × 10⁻³	59	59	49.9	48.9	35	30	30
Tensile strength, psi × 10⁻³	42	39.5	33.1	33.6	40	38	30
Laminate resin content, %	29.8						
Barcol hardness	76				55 min.		
Specific gravity	1.8						
Laminate thickness, in	0.12						
Flammability, in./min	Self-extinguishing				1.0 (max.)		
Water absorption after 24 hr immersion, % change in weight	0.15				+1¼ (max.)		

and the resin is thermally destroyed before it melts. This means that in the molding process laminates do not flow and that the various plies become bonded together by a cold-flow or creep process. Therefore, the molding process calls for pressures of at least 200 lb/in.² and temperatures reaching 700°F.

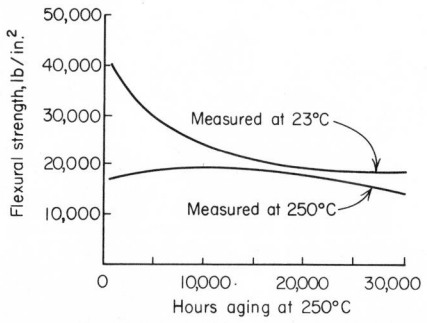

Fig. 22 Effect of long-term aging on flexural properties of silicone laminates.[12]

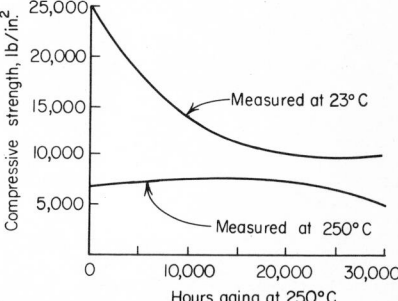

Fig. 23 Effect of long-term aging on compressive properties of silicone-glass laminate.[12]

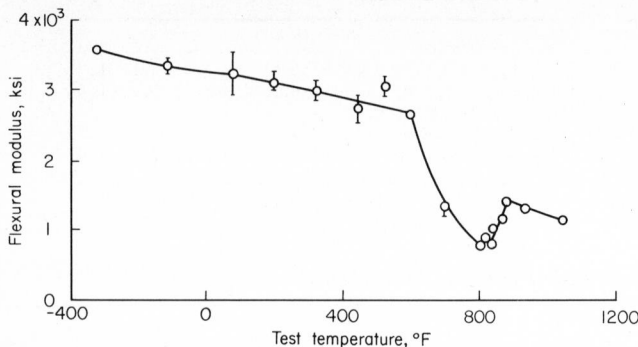

Fig. 24   Variation of flexural modulus with temperature for polyimide-glass laminate.[15]

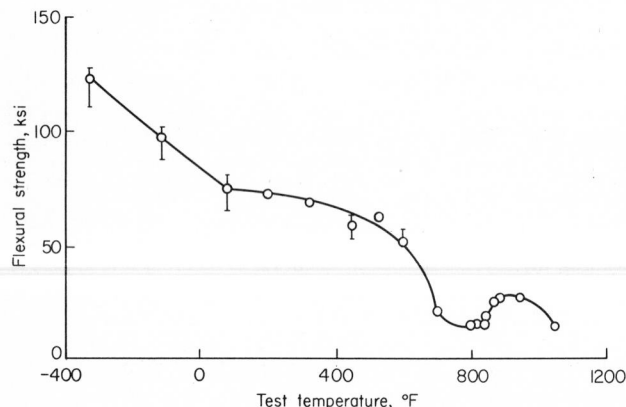

Fig. 25   Variation of flexural strength with temperature for polyimide-glass laminate.[15]

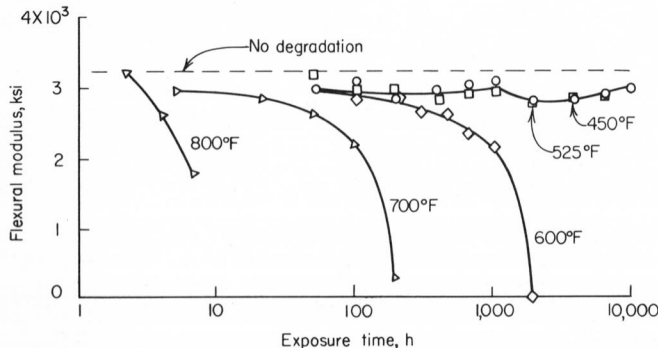

Fig. 26   Variation of room-temperature flexural modulus with exposure time for polyimide laminate.[15]

The physical properties of polyimides are good to about 600°F. Flexural strength and modulus for a du Pont polyimide–glass cloth laminate with A-1100 finish are shown in Figs. 24 to 27. Polyamide-imide laminates exhibit a similar response of flexural properties with aging, as shown in Fig. 28. This curve compares the aging at 300°C of several different polyamide-imides and silicones and phenolic laminates. Complete data for 12-ply laminates of style 181 glass fabric with A-1100 finish are shown in Table 19. Retention of the flexural strength is shown in Table 20. Polyimides are used in special aerospace applications where their thermal and radiation resistance are required. Their cost is higher than that of silicones.

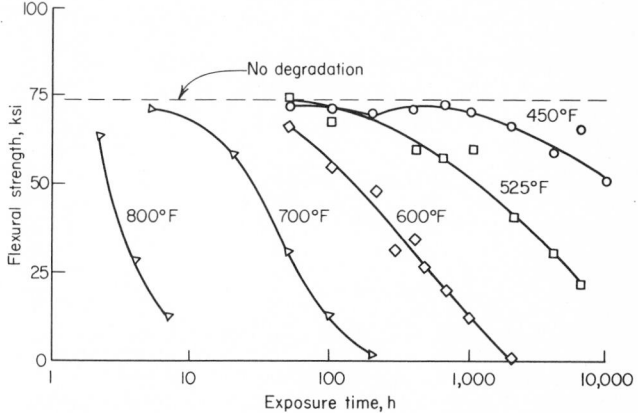

Fig. 27 Variation of room-temperature flexural strength with exposure time for polyimide laminate.[15]

Diphenyl oxide laminates Another new high-temperature laminate is based on diphenyl oxide (DO) resins. These materials have good chemical resistance and retain their properties well up to 500°F.

The changes in flexural and dielectric strength of diphenyl oxide–resin laminate are shown in Fig. 29, while dielectric constant, power factor, and volume resistivity as a function of temperature and frequency are given in Figs. 30 to 32. DO laminates have been used as terminal boards, slot wedges, and coil forms.

Thermoplastic laminates Laminates with a reinforcing base of glass or asbestos and containing a thermoplastic-resin binder are just becoming available. Mostly aimed at the automotive markets, these laminates can be easily shaped on simple tools by heating and pressing. Some of the materials can be stamped on metal-forming equipment. Tables 21 and 22 present physical properties for several kinds of

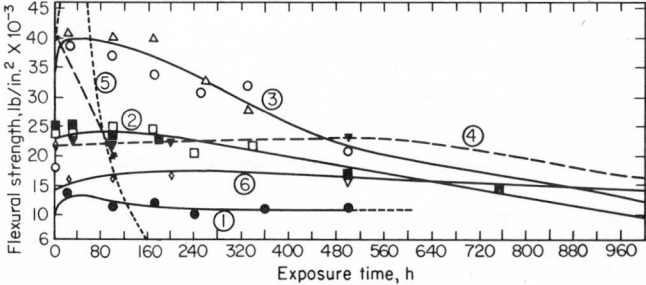

Fig. 28 Flexural strength of laminates aged and tested at 300°C in air.[16] (1) and (6) are silicone, (2), (3), and (4) are aromatic amide-imide, (5) is phenolic MIL-R-9299.

thermoplastic laminates. Table 23 presents relative cost and the maximum long-time service temperature for all the laminating materials described in the preceding sections.

NEMA Industrial Laminates

Industrial laminates are those materials in sheet, tube, and rod form used chiefly by the electrical and electronic industries for primary insulation and for dielectric

TABLE 19   Properties of Polyimide Laminates[17]

Laminate properties:	
Resin content, %	23
Specific gravity	1.60
Flexural strength, 1,000 lb/in.^{2}:	
RT	64
550°F	43
Flexural modulus, 10^6 lb/in.^{2}:	
RT	2.9
550°F	2.2
Tensile strength, 1,000 lb/in.^{2}:	
RT	52
550°F	40
Tensile modulus, 10^6 lb/in.^{2}:	
RT	2.8
Compressive strength, 1,000 lb/in.^{2}:	
RT	56
Compressive modulus, 10^6 lb/in.^{2}:	
RT	3.6
Interlaminar shear strength, 1,000 lb/in.^{2}:	
RT	2.8
550°F	2.3
Short-beam shear strength, 1,000 lb/in.^{2}:	
RT	5.6
Elongation, %, RT	1.74
Poisson's ratio	0.20
Notched Izod impact, ft-lb/in.	15.7
Heat-distortion temperature, °F	480
Thermal conductivity, cal/(cm)(s)(°C)	3.46×10^{-4}
Coefficient of linear thermal expansion, in./(in.)(°C):	
−70–90°C	0.74×10^{-5}
90–170°C	0.34×10^{-5}
170–400°C	0.69×10^{-5}
Dielectric constant	3.9
Loss tangent	0.015

TABLE 20   Strength Retention of Polyimide Laminate[17]

Temp, °F	Time to 40% strength retention, h
500	30,000–40,000
550	5,000–7,000
600	1,500–2,000
700	100–150

purposes where structural requirements must be met. Standards for the laminates belonging to this category have been developed by the National Electrical Manufacturers Association (NEMA).[20]

The NEMA laminates are described below. All resins are phenolic except where specified differently. Table 24 lists the NEMA laminates and shows the outstanding properties of each grade. In addition to NEMA, many specifications relating to

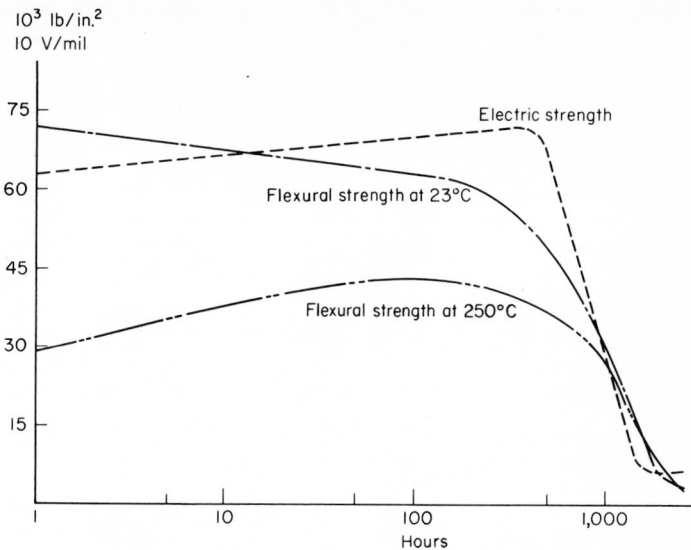

Fig. 29 Effect of temperature on flexural strength and electrical strength for diphenyl oxide–glass cloth laminate.

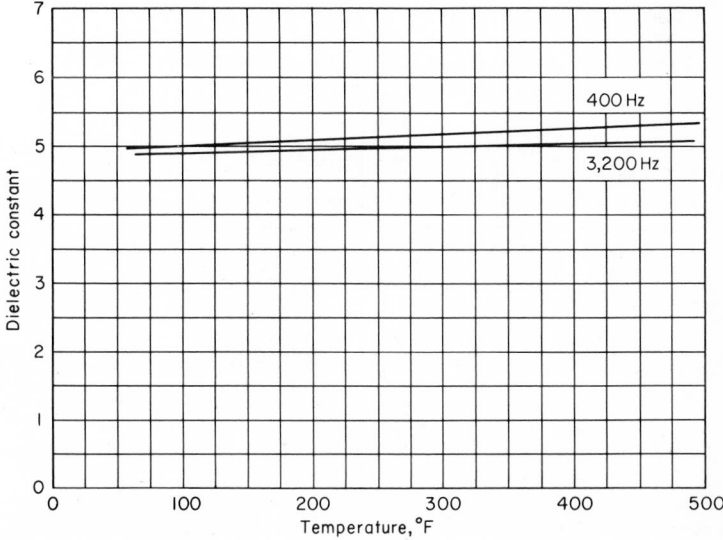

Fig. 30 Dielectric constant at 400 and 3,200 Hz vs. temperature for diphenyl oxide–glass cloth laminate.[8]

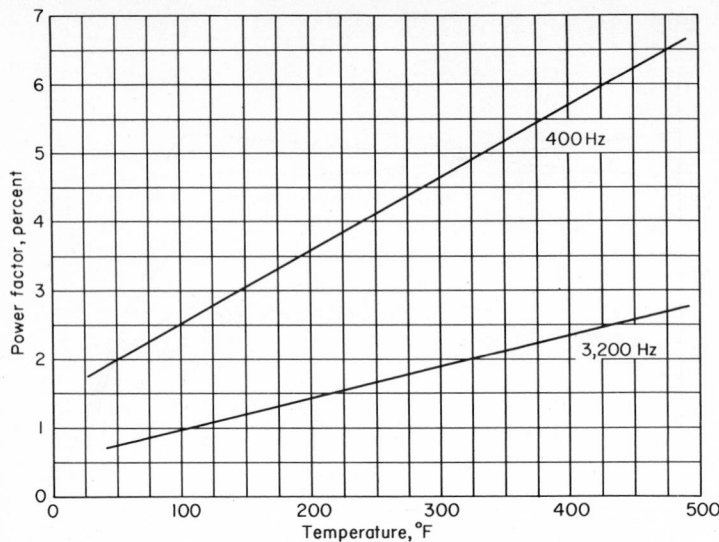

Fig. 31 Power factor at 400 and 3,200 Hz vs. temperature for diphenyl oxide–glass cloth laminate.[8]

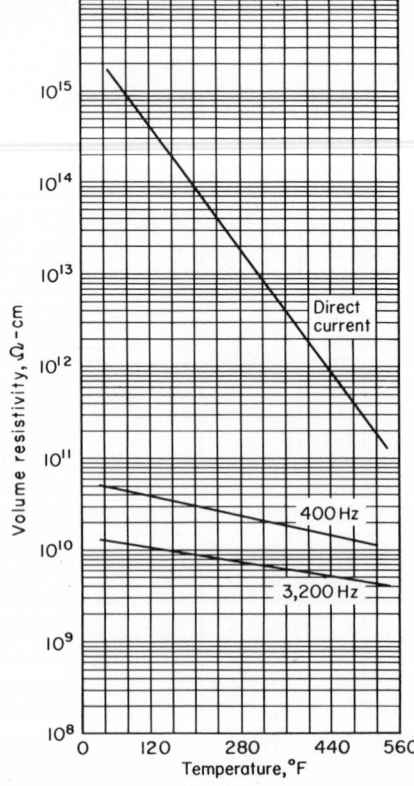

Fig. 32 Volume resistivity vs. temperature for diphenyl oxide–glass laminate.[8]

laminates have been developed by the military. A comparison of NEMA and military grades is given in Table 25.

Grade X Paper-based sheet and tubes intended for mechanical applications where electrical properties are of little importance. These materials are humidity-sensitive

TABLE 21   Properties of Thermoplastic Laminates[18]

Resin used	Reinforcement	Tensile strength, psi	Flexural strength, psi	Flexural modulus, psi $\times 10^{-6}$	Flatwise impact strength, ft-lb/in.
Polymethylmeth-	Nil	10,000	15,000	0.45	0.5
acrylate	Glass mat	15,000	26,000	1.1	11
Polystyrene........	Nil	7,000	13,000	0.5	0.5
	Glass mat	15,000	25,000	1.2	10
	Glass woven roving	34,000	18,000	....	....
	Asbestos felt	20,000	28,000	3.0	7.0
Polyvinylchloride...	Nil	7,000	14,000	0.5	2.0
	Glass mat	17,000	24,000	1.1	14
	Glass woven roving	30,000	19,000	....	30
	Asbestos felt	18,000	21,000	2.5	8.0

TABLE 22   Properties of Glass-mat-reinforced Thermoplastics[19]

Property	Poly-propylene	Styrene-acrylo-nitrile copolymer	Polyvinyl chloride
Flexural strength at RT, psi......................	26,900	32,400	31,300
Flexural modulus at RT, psi.....................	1,030,000	1,300,000	1,260,000
Tensile strength at RT, psi......................	16,500	18,400	19,100
Tensile modulus at RT, psi......................	860,000	1,100,000	950,000
Flexural strength at 180°F, psi..................	14,000	22,000	
Flexural modulus at 180°F, psi..................	720,000	1,200,000	
Heat distortion temperature at 264 psi, °F.........	327	255	221
Maximum short-term-use temperature, °F.........	300	200	200
Notched Izod impact, ft-lb/in....................	11.2	15.7	27
Falling-dart impact, in...........................	16	36	
Wet flexural strength, psi........................	24,600	26,900	
Wet flexural modulus, psi........................	1,000,000	1,300,000	
Coefficient of thermal expansion, in./(in.)(°F) $\times 10^{-5}$	1.5	1.3	1.61
Glass fiber content, %...........................	44	42	36
Lb/ft² at 100 in.................................	0.65	0.70	0.85
Weight for equivalent stiffness....................	0.7	0.7	
Specific gravity.................................	1.26	1.36	

and are not equal to fabric-based materials in impact strength. The rolled tubes have good power factor and electric strength when dry, and they can be punched well.

Grade XP Paper-based sheet used for hot punching. More flexible and not as strong as grade X. The moisture resistance and electrical properties are better than those of grade X and less than those of grade XX.

Grade XPC Paper-based sheet similar to grade XP but can be punched cold. Slightly more flexible but with lower flexural strength and higher cold flow than grade XP.

Grade XX Paper-based sheet, tube, and rods with good machinability and good electrical properties. Not as strong as grade X but better in moisture resistance. Electric strength of thin-walled tubes may be low at seam areas.

Grade XXP Paper-based sheet with better electrical and moisture resistance than grade XX. Punching and cold-flow qualities are intermediate between those of grades XP and XX.

Grade XXX Paper-based sheet, tube, and rods for radio-frequency applications. Has good resistance to humidity and possesses low cold flow. Rolled tubes are superior to all other paper-based grades when exposed to high humidity.

TABLE 23 Cost and Thermal Rating of Laminates

Resin	Reinforcement	Relative cost (paper phenolic = 1)	Max use temp, °F
Phenolic	Paper	1	220
Phenolic	Glass cloth	2.8	450
Phenolic	Asbestos cloth	3.9	350
Phenolic	Asbestos paper	1.1	350
Phenolic	Cotton cloth	1.7	250
Phenolic	Nylon cloth	4.0	250
Melamine	Glass cloth	2.7	300
Epoxy	Glass cloth	3.0	250
Epoxy	Paper	2.3	220
Polyester	Glass cloth	2.7	250
Polyester	Glass mat	1.1	250
Diallyl phthalate	Glass cloth	5.0	300
Silicone	Glass cloth	6.0	500
Phenylsilane	Glass cloth	6.0	500
Teflon	Glass cloth	6.5	400
Polyimide	Glass cloth	25.0	600
Diphenyl oxide	Glass cloth	5.0	450
Thermoplastic	Glass mat	1.0	180

Grade XXXP Paper-based sheet intermediate between grades XXP and XX in punching. Has high insulation resistance and low losses at high humidity.

Grade XXXPC Paper-based sheet similar to XXXP but with lower punching temperatures.

Grade FR-2 Paper-based sheet modified chemically so as to be self-extinguishing after ignition source is removed. Similar in all other properties to grade XXXPC.

Grade FR-3 Paper-based epoxy-resin sheet. The epoxy resin contributes excellent stability of electrical properties at high humidity. The laminate is self-extinguishing, has high flexural strength, and is punchable.

Grades ES-1, ES-2, and ES-3 Paper-based sheets used for nameplate applications.

Grade C Cotton-fabric-based sheet, tube, and rod made from a cloth weighing more than 4 oz/yd². Mechanical grade; not recommended for electrical applications. Used for gears and other applications requiring high impact strength.

Grade CE Cotton-fabric-based sheet, tube, and rod. Similar to grade C. Used for electrical applications requiring greater toughness than that possessed by grade XX or for mechanical applications requiring better moisture resistance than that possessed by grade C. Not recommended for primary insulation at voltages exceeding 600 V.

Grade L Cotton-fabric-based sheet, tube, and rod made from a cloth weighing less than 4 oz/yd². This grade machines better than grades C or CE but is not as tough. A mechanical material and not recommended for electrical applications. Has good moisture resistance.

Grade LE Cotton-fabric-based sheet, tube, and rod similar to grade L. Not recommended for primary insulation at voltages over 600 V. Used for applications requiring greater toughness than grade XX or better machining than grade CE. Available in thin sheets.

Grade CF Cotton-fabric-based sheet made from a cloth weighing more than

TABLE 24 NEMA Laminate-Selector Chart

Grade	X	XP	XPC	XX	XXP	XXX	XXXP	XXXPC	FR-2	FR-3	C	CE	L	LE	CF	A	AA	G-2	G-3	G-5	G-7	G-9	G-10	G-11	FR-4	FR-5	GPO-1	GPO-2	GPO-3	N-1	GT	GX
Mechanical application	X	X	X								X	X	X	X	X	X	X		X	X		X	X	X	X	X	X	X	X	X	X	X
Electrical application		X	X												X			X			X	X	X	X	X	X	X	X	X			
Punching qualities																				X		X										
Tubes		X	X	X	X	X	X	X	X	X	X	X		X	X	X	X		X		X	X	X	X	X	X						
Rods	X				X	X	X	X	X		X	X		X					X			X	X	X	X	X						
Flame resistance				X		X							X			X	X	X		X	X	X						X	X	X	X	X
Impact resistance				X		X			X	X	X	X			X		X	X		X	X	X	X	X	X	X	X	X	X			
Flexural strength											X	X			X		X		X	X	X	X	X	X	X	X	X	X	X			
Thermal stability																X	X		X	X	X	X	X	X		X				X	X	X
Cold flow											X	X	X					X	X	X	X	X	X		X		X	X	X	X	X	X
Humidity resistance						X			X	X											X			X		X				X	X	X
Electronic application						X	X	X					X			X						X										
Machinability				X		X	X	X			X	X		X																		
Arc resistance																				X	X				X	X	X	X	X		X	X

5-39

TABLE 25   Cross Reference of NEMA and Military Designations for Plastic Laminates[21]

Resin: reinforcement	NEMA grade	MIL type	Military (MIL) Specification
Sheets			
Epoxy:			
Glass cloth..............	G-10	GEE	
Glass cloth..............	G-11	GEB	
Paper...................	FR-3	PEE	P-3115C
Glass (copper-clad)........	G-10, G-11	GE, GB	P-13949B
Glass (copper-clad)........	....	GF	P-13949B
Melamine:			
Paper..................	ES-1, ES-3	NDP	P-78A
Glass cloth.............	G-5	GMG	P-15037C
Glass (copper-clad)........	G-5	GM	P-13949B
Phenolic:			
Paper..................	X, P		
Paper..................	XP	HSP	P-78A
Paper..................	XPC		
Paper..................	XX	PBG	P-3115C
Paper..................	XXX, XXP	PBE, PBE-P	P-3115C
Paper..................	XXXP, XXXPC	PBE-P	P-3115C
Paper..................	ES-2	NDP	P-78A
Paper..................	FR-1	PBG	P-3115C
Paper..................	FR-2	PBE-P	P-3115C
Cotton fabric............	C	FBM, FBG	P15035A
Cotton fabric............	CF	....	P-8655A
Cotton fabric............	L	FBI	P-15035C
Cotton fabric............	LE	FBE	P-15035C
Asbestos paper...........	A, AA	PBA, FBA	P-8059A
Glass fiber..............	G-2		
Glass fiber..............	....	II-2	P-25515A
Glass cloth..............	G-3		
Nylon cloth.............	N-1	NPG	P-15047B
Glass (copper-clad)........	XXXP, XXXPC	PP	P-13949B
Polyester:			
Glass fiber..............	GPO-1		
Glass fiber..............	GPO-1	1, 2, 3	P-8013C
PTFE:			
Glass cloth..............	GTE	GTE	P-19161A
Silicone:			
Glass cloth..............	G-6, G-7	GSG	P-997C
Glass (copper-clad)........	G-6	GS	P-13949B
Rods and tubes			
Phenolic:			
Paper..................	X	PBM	P-79C
Paper..................	XX	PBG	P-79C
Paper..................	XXX	PBE	P-79C
Cotton fabric............	C	FBM	P-79C
Cotton fabric............	CE	FBG	P-79C
Cotton fabric............	L	FBI	
Cotton fabric............	LE	FBE	P-79C
Glass cloth..............	G-3		
Melamine:			
Glass cloth..............	G-5	GMG	P-79C

4 oz/yd^2. The laminates are postforming, or capable of being shaped by the application of heat and pressure.

Grade A   Asbestos-paper-based sheet and tube. More resistance to flame and heat than cellulose-based materials. Limited to service at voltages less than 250 V. Moisture causes small dimensional changes.

Grade AA   Asbestos-fabric-based sheet and tube.   Similar to grade A but with better physical properties and thermal resistance.   Not recommended for primary insulation.

Grade G-2   Glass-cloth-based sheet.   Cloth is made of a staple (noncontinuous) fiber.   Good electrical properties under wet conditions.   Weakest of all glass-based grades.   Lower in dielectric losses than all other glass-based grades except the silicone.

Grade G-3   Glass-cloth-based sheet, tube, and rod.   Cloth is made of a continuous fiber.   High mechanical properties except for bond strength, which is low.   Good electrical properties under dry conditions.

Grade G-5   Glass-cloth-based sheet, rod, and tube with a melamine-resin binder. Cloth is made of continuous fiber.   Maximum physical properties.   Hardest of all grades.   Good flame resistance, arc resistance, dimensional stability.   Tubing has excellent burst strength.

Grade G-7   Glass-cloth-based sheet and tube with a silicone-resin binder.   Cloth is made of continuous fiber.   Extremely good dielectric loss and insulation resistance when dry.   Good electrical properties when wet.   Excellent heat and arc resistance. Second only to grade G-5 in flame resistance.   Meets IEEE requirements for class 180°C insulation.

Grade G-9   Glass-cloth-based sheet, tube, and rod with a heat-resistant melamine-resin binder.   Cloth is made of continuous filaments.   Similar to grade G-5 except more thermally stable.   Second only to grade G-7 in arc and heat resistance.

Grade G-10   Glass-cloth-based sheet, tube, and rod with an epoxy-resin binder. Cloth is made of continuous fiber.   Extremely high mechanical strength at room temperature.   Good dielectric loss and electric-strength properties both wet and dry. Insulation resistance at high humidity is better than that of grade G-7.

Grade G-11   Glass-cloth-based sheet, tube, and rod with an epoxy-resin binder. Cloth is made of continuous fibers.   Similar to grade G-10 except that it retains at least 50 percent of its room-temperature flexural strength at 150°C after 1 h at 150°C.

Grade FR-4   Glass-cloth-based sheet, rod, and tube with an epoxy-resin binder. Similar to grade G-10 but self-extinguishing when ignition source is removed.

Grade FR-5   Glass-cloth-based sheet, rod, and tube with an epoxy resin binder. Similar to grade G-11 but is self-extinguishing when ignition source is removed.

Grade FR-6   Glass-mat-based sheet with polyester resin and fillers.   Used as a flame-resistant printed-circuit board.

Grade GPO-1   Glass-mat-based sheet with a polyester-resin binder.   General-purpose mechanical and electrical grade.

Grade GPO-2   Glass-mat-based sheet with a polyester-resin binder. Similar to grade GPO-1 except has low flammability and is self-extinguishing.

Grade GPO-3   Glass-mat-based sheet with a polyester-resin binder.   Similar to grade GPO-1 but has resistance to carbon tracking and low flammability and is self-extinguishing.

Grade N-1   Nylon-cloth-based sheet.   Excellent electrical properties at high humidity.   Good impact strength but subject to creep and cold flow.

Grade GT   Glass-cloth-based sheet with polytetrafluoroethylene binder.   Has controlled dielectric constant for general use.

Grade GX   Glass-cloth-based sheet with polytetrafluoroethylene binder.   Similar to grade GT, but the dielectric constant is controlled within closer limits.   Used for X-band electrical applications.

Properties of NEMA industrial laminates   The properties of each grade standardized by NEMA are fully described in Ref. 20.   Table 26 gives typical values for these sheets.   These properties should not be used as standards but can be used as guides for the selection of materials. NEMA standards also describe tubing and rods. Tubing is made by rolling impregnated sheet on a mandrel between heated pressure rollers and then either curing the tubes in an oven (rolled tubing) or pressing them in a mold (molded tubing).   The properties of the finished tube depend to some extent on the curing method used.   Table 27 gives properties of molded tubing, and Table 28 gives properties of rolled tubing.   Table 29 shows the properties of laminated rods.

TABLE 26 Typical Values for NEMA Laminates (Condition A)[20]

ASTM and NEMA grades	X	XP	XPC	XX	XXP	XXX	XXXP	XXXPC	ES-1	ES-2	ES-3
Equivalent MIL-P or Federal (LP) specification No.	LP 509	LP 509	LP 509	LP 513	LP 509	LP 513	LP 513	LP 513	LP 509	LP 509	LP 509
MIL-P or federal type	...	...	...	PBG	...	PBE	PBE-P	PBE-P	NDP	NDP	NDP
NEMA temperature indexes (see NEMA Publication LI-5):											
$\frac{1}{32}$ to $\frac{1}{16}$ in. Electrical test index	130	130	130	130	130	130	125	125			
Mechanical test index	130	130	130	130	130	130	125	125			
$\frac{1}{16}$ in. and over Electrical test index	130	130	130	140	140	140	125	125			
Mechanical test index	130	130	130	140	140	140	125	125			
Electric strength perpendicular to laminations, V/mil:											
Short time											
$\frac{1}{32}$-in. thickness	950	900	850	950	950	900	900	900			
$\frac{1}{16}$-in. thickness	700	650	600	700	700	650	650	650	750	550	
$\frac{1}{8}$-in. thickness	500	470	425	500	500	470	470	470	...		
Step by step											
$\frac{1}{32}$-in. thickness	700	650	625	700	700	650	650	650			
$\frac{1}{16}$-in. thickness	500	450	425	500	500	450	450	450	550	400	
$\frac{1}{8}$-in. thickness	360	320	290	360	360	320	320	320	...		
Insulation resistance, condition C-96/35/90, MΩ	...	...	...	60	500	1,000	20,000	50,000			

Property											
Tensile strength, lb/in.²:											
Lengthwise	20,000	12,000	10,500	16,000	11,000	15,000	12,400	12,400	12,000	13,000	15,000
Crosswise	16,000	9,000	8,500	13,000	8,500	12,000	9,500	9,500	8,500	9,000	12,000
Modulus of elasticity, lb/in.²:											
In tension											
Lengthwise	1,900,000	1,200,000	1,000,000	1,500,000	900,000	1,300,000	1,000,000	1,000,000			
Crosswise	1,400,000	900,000	800,000	1,200,000	700,000	1,000,000	800,000	800,000			
In flexure											
Lengthwise	1,800,000	1,200,000	1,000,000	1,400,000	900,000	1,300,000	1,000,000	1,000,000			
Crosswise	1,300,000	900,000	800,000	1,100,000	700,000	1,000,000	700,000	700,000			
Compressive strength, lb/in.²											
Flatwise	36,000	25,000	22,000	34,000	25,000	32,000	25,000	25,000			
Edgewise	19,000			23,000		25,500					
Rockwell hardness (M)	M-110	M-95	M-75	M-105	M-100	M-110	M-105	M-95	M-118	M-118	M-120
Deformation and shrinkage, cold flow at 4,000 lb/in.² for ⅛-in. thickness, % change				0.90		0.80	0.80				
Density, g/cm³	1.36	1.33	1.34	1.34	1.32	1.32	1.30	1.30	1.45	1.40	1.38
Specific volume, in.³/lb	20.4	20.8	20.6	20.6	21.0	21.0	21.3	21.3	19.1	19.8	20.1
Thermal expansion, cm/(cm)(°C)	2.0×10^{-5} →										
Thermal conductivity, cal/(s)(cm²)(°C)(cm)	7.0×10^{-4} →										
Specific heat	0.35–0.40										
Effect of acids	All grades except G-5 and G-9 are resistant to dilute solutions of most acids.										
Effect of alkalies	Not recommended for use in alkaline solutions except for melamine grades G-5 and G-9, which are resistant to dilute alkaline solutions.										
Effects of solvents	Unaffected by most organic solvents except acetone, which may soften the punching-grade stocks. Aromatic hydrocarbons and chlorinated aliphatics may affect silicone grade G-7.										

TABLE 26 Typical Values for NEMA Laminates (Condition A)[20] (Continued)

ASTM and NEMA grades	C	CE	L	LE	G-3	G-5	G-7	G-9	G-10	G-11
Equivalent MIL-P or Federal (LP) specification No...... MIL-P or federal type...	15035 FBM	15035 FBG	15035 FBI	15035 FBE	LP 509 ...	LP 509 GMG	997 GSG	15037 GME	18177 GEE	18177 GEB
NEMA temperature indexes (see NEMA Publication LI-5):										
1/32 to 1/16 in. Electrical test index	85	85	85	85	140	...	170	...	130	140
Mechanical test index......	85	85	85	85	170	140	220	140	140	160
1/16 in. and over Electrical test index	115	115	115	115	140	...	170	...	130	170
Mechanical test index......	125	125	125	125	170	140	220	140	140	180
Electric strength perpendicular to laminations, V/mil:										
Short time										
1/32-in. thickness....	...	...	...	700	750	...	450	450	750	750
1/16-in. thickness....	...	500	...	500	700	350	400	400	700	700
1/8-in. thickness....	...	360	...	360	600	260	350	350	550	550
Step by step										
1/32-in. thickness....	...	...	...	450	550	...	400	400	500	500
1/16-in. thickness....	...	300	...	300	500	220	350	350	450	450
1/8-in. thickness....	...	220	...	220	450	160	250	275	350	350
Insulation resistance, condition C-96/35/90, MΩ......	...	...	...	30	...	100	2,500	10,000	200,000	200,000

Property										
Tensile strength, lb/in.^{2}:										
Lengthwise	10,000	9,000	13,000	12,000	23,000	37,000	23,000	37,000	40,000	40,000
Crosswise	8,000	7,000	9,000	8,500	20,000	30,000	18,500	30,000	35,000	35,000
Modulus of elasticity, lb/in.^{2}:										
In tension										
Lengthwise	1,000,000	900,000	1,200,000	1,000,000	2,000,000	2,300,000	1,800,000	2,300,000	2,500,000	2,500,000
Crosswise	900,000	800,000	900,000	850,000	1,700,000	2,000,000	1,800,000	2,000,000	2,000,000	2,000,000
In flexure										
Lengthwise	1,000,000	900,000	1,100,000	1,000,000	1,500,000	1,700,000	1,400,000	2,500,000	2,700,000	2,800,000
Crosswise	900,000	800,000	850,000	850,000	1,200,000	1,500,000	1,200,000	2,000,000	2,200,000	2,300,000
Compressive strength, lb/in.^{2}:										
Flatwise	37,000	39,000	35,000	37,000	50,000	70,000	45,000	70,000	60,000	60,000
Edgewise	23,500	24,500	23,500	25,000	17,500	25,000	14,000	25,000	35,000	35,000
Rockwell hardness (M)	M-103	M-105	M-105	M-105	M-100	M-120	M-100	M-120	M-111	M-112
Deformation and shrinkage, cold flow at 4,000 lb/in.^{2} for $\frac{1}{8}$-in. thickness, % change	...	...	...	...	0.30	0.30	0.30	0.25	0.25	0.20
Density, g/cm^3	1.36	1.33	1.35	1.33	1.65	1.90	1.68	1.90	1.80	1.80
Specific volume, in.^{3}/lb	20.4	20.8	20.5	20.8	16.8	14.6	16.5	14.6	15.3	15.3
Thermal expansion, cm/(cm)(°C)	←—— 2.0 × 10^{-5} ——→				←—— 1.8 × 10^{-5} ——→		←—— 1.0 × 10^{-5} ——→		←—— 0.9 × 10^{-5} ——→	
Thermal conductivity, cal/(s)(cm^2)(°C)(cm)	←—— 7.0 × 10^{-4} ——→				←— 12.0 × 10^{-4} —→		←——— 7.0 × 10^{-4} ———→			
Specific heat	←—— 0.35–0.40 ——→				0.30	0.26	0.25	←—— 0.35–0.40 ——→		
Effect of acids	All grades except G-5 and G-9 are resistant to dilute solutions of most acids.									
Effect of alkalies	Not recommended for use in alkaline solutions except for melamine grades G-5 and G-9, which are resistant to dilute alkaline solutions.									
Effects of solvents	Unaffected by most organic solvents except acetone, which may soften the punching-grade stocks. Aromatic hydrocarbons and chlorinated aliphatics may affect silicone grade G-7.									

TABLE 26 Typical Values for NEMA Laminates (Condition A)[20] (Continued)

ASTM and NEMA grades	A	AA	FR-2	FR-3	FR-4	FR-5	FR-6	N-1
Equivalent MIL-P or Federal (LP) specification No.	LP 509	LP 509	LP 509	22324	18177	18177	...	15047
MIL-P or federal type	...	...	PBE-P	PEE	GEE	GEB	...	NPG
NEMA temperature indexes (see NEMA Publication LI-5):								
1/32 to 1/16 in.								
Electrical test index	140	140	75	90	130	140		
Mechanical test index	140	140	75	90	140	160		
1/16 in. and over								
Electrical test index	155	155	105	110	130	170		
Mechanical test index	155	155	105	110	140	180		
Electric strength perpendicular to laminations, V/mil:								
Short time								
1/32-in. thickness	...	...	800	650	750	750	...	850
1/16-in. thickness	225	...	650	600	700	700	400	600
1/8-in. thickness	160	...	470	475	550	550	...	450
Step by step								
1/32-in. thickness	...	...	550	500	500	500	...	650
1/16-in. thickness	135	...	450	450	450	450	...	450
1/8-in. thickness	95	...	320	325	350	350	...	300
Insulation resistant, condition C-96/35/90, MΩ	...	...	20,000	100,000	200,000	200,000	20,000	50,000

Property								
Tensile strength, lb/in.²:								
Lengthwise	10,000	12,000	12,500	14,000	40,000	40,000	7,000	8,500
Crosswise	8,000	10,000	9,500	12,000	35,000	35,000	7,000	8,000
Modulus of elasticity, lb/in.²:								
In tension								
Lengthwise	2,500,000	1,700,000	1,000,000	1,200,000	2,500,000	2,500,000	…	400,000
Crosswise	1,600,000	1,500,000	800,000	1,000,000	2,000,000	2,000,000	…	400,000
In flexure								
Lengthwise	2,300,000	1,600,000	1,100,000	1,300,000	2,700,000	2,800,000	…	600,000
Crosswise	1,400,000	1,400,000	900,000	1,000,000	2,200,000	2,300,000	…	500,000
Compressive strength, lb/in.²								
Flatwise	40,000	38,000	25,000	29,000	60,000	60,000	30,000	
Edgewise	17,000	21,000	…	…	35,000	35,000	…	…
Rockwell hardness (M)	M-111	M-103	M-97	M-100	M-111	M-114	M-90	M-105
Deformation and shrinkage, cold flow at 4,000 lb/in.² for ⅛-in. thickness, % change	…	…	0.85	0.50	0.25	0.20	0.5	…
Density, g/cm³	1.72	1.70	1.33	1.42	1.85	1.85	1.8	1.15
Specific volume, in.³/lb	16.1	16.3	21.0	19.5	14.9	14.9	16.0	24.1
Thermal expansion, cm/(cm)(°C)	← 1.5×10^{-5} →		← 2.0×10^{-5} →		1.0×10^{-5} →	0.9×10^{-5} →	← 2.0×10^{-5} →	
Thermal conductivity, cal/(s)(cm²)(°C)(cm)	← 0.30 →		← 7.0×10^{-4} →				← 15.0×10^{-4} →	
Specific heat				← 0.35 to 0.40 →				
Effect of acids	All grades except G-5 and G-9 are resistant to dilute solutions of most acids.							
Effect of alkalies	Not recommended for use in alkaline solutions except for melamine grades G-5 and G-9 and polyester grade FR-6, which are resistant to dilute alkaline solutions.							
Effects of solvents	Unaffected by most organic solvents except acetone, which may soften the punching-grade stocks. Aromatic hydrocarbons and chlorinated aliphatics may affect silicone grade G-7.							

TABLE 27 Properties of Molded Tubing[22]

Grade	Filler	Resin	Colors available	Specific gravity	Compressive strength, axial, psi	Tensile strength, psi	Power factor at 1 MHz	Dielectric constant at 1 MHz	Water absorption, in 24 hr, % — Wall thickness, in.				Dielectric strength, volts/mil (short time) perpendicular to laminations — Wall thickness, in.		
									$1/32$	$1/16$	$3/32$	$1/8$	Over $1/16$ to $1/8$, incl.	Over $1/8$ to $1/4$, incl.	Over $1/4$ to $1/2$, incl.
X	Paper	Phenolic	Natural and black	1.25	18,000	12,000	......	...	6.0	4.5	4.0	3.6	270	180	100
XX	Paper	Phenolic	Natural and black	1.25	18,000	11,000	0.040	5.5	...	2.0	1.8	1.6	220	150	110
XXX	Paper	Phenolic	Natural and black	1.22	20,000	9,000	0.040	5.5	...	1.4	1.2	1.1	220	150	110
C	Coarse-weave cotton fabric	Phenolic	Natural and black	1.2	19,000	9,000	......	...	...	3.5	2.6	2.2	160	110	75
CE	Coarse-weave cotton fabric	Phenolic	Natural and black	1.25	19,000	8,500	......	...	...	3.0	2.2	2.0	175	125	90
C-M	Coarse-weave cotton fabric	Melamine	Grayish brown	1.30	19,000	8,500	......	...	...	6.2	4.7	3.9	150	100	80
L	Fine-weave cotton fabric	Phenolic	Natural and black	1.25	18,000	9,000	......	...	6.5	3.5	2.2	1.8	160	110	75
LE	Fine-weave cotton fabric	Phenolic	Natural and black	1.25	19,000	8,500	......	...	4.5	2.2	1.8	1.5	175	125	90
A	Asbestos paper	Phenolic	Natural	1.55	18,000	10,000	......	...	...	2.0	1.8	1.6	85	70	55
AA	Asbestos fabric	Phenolic	Natural	1.55	18,000	9,000	......	...	...	2.0	1.8	1.6	90	75	60
G-3	Continuous-filament glass fabric	Phenolic	Grayish brown	1.55	20,000	30,000	......	...	6.5	3.7	3.2	3.0	170	120	85
G-5	Glass fabric	Melamine	Grayish brown	1.80	20,000	30,000	......	...	5.0	3.9	3.7	3.0	155	105	75
N-1	Nylon fabric	Phenolic	Natural	1.18	6,500	6,000	0.040	5.0	...	1.0	0.8	0.6	200	160	80

TABLE 28 Properties of Rolled Tubing[23]

Grade	Filler	Resin	Colors available	Specific gravity	Compressive strength, axial, psi	Tensile strength, psi	Power factor at 1 MHz	Dielectric constant at 1 MHz	Water absorption, in 24 hr, % — Wall thickness, in.				Dielectric strength, volts/mil (short time) perpendicular to laminations — Wall thickness, in.		
									1/32	1/16	3/32	1/8	Over 1/16 to 1/8, incl.	Over 1/8 to 1/4, incl.	Over 1/4 to 1/2, incl.
X	Paper	Phenolic	Natural and black	1.10	10,000	8,500	.....	...	8.0	5.0	4.3	4.0	325	200	145
XX	Paper	Phenolic	Natural and black	1.10	13,000	8,000	0.040	5.0	6.0	3.0	2.5	2.0	290	200	145
XXX	Paper	Phenolic	Natural and black	1.12	13,000	7,000	0.040	5.3	...	1.5	1.3	1.0	...	...	...
C	Coarse weave cotton fabric	Phenolic	Natural and black	1.12	12,000	6,000	.....	...	...	5.0	3.6	3.0	120	110	75
CE	Coarse weave cotton fabric	Phenolic	Natural and black	1.12	13,000	6,000	.....	...	...	4.7	3.3	2.7	140	120	85
C-M	Grayish brown	Melamine	Grayish brown	1.12	13,000	6,000	.....	...	...	7.0	5.0	4.2	150	100	75
L	Fine weave cotton fabric	Phenolic	Natural and black	1.12	12,000	6,000	.....	...	8.0	5.0	3.3	2.8	120	110	75
LE	Fine weave cotton fabric	Phenolic	Natural and black	1.12	13,000	7,000	.....	...	7.5	5.0	3.0	2.5	140	120	85
A	Asbestos paper	Phenolic	Natural	1.25	10,000	7,000	.....	...	...	2.2	2.0	1.8	90	75	60
G-3	Continuous filament glass fabric	Phenolic	Grayish brown	1.50	13,000	25,000	.....	...	7.0	4.0	3.5	3.2	175	125	90
G-5	Continuous filament glass fabric	Melamine	Grayish brown	1.70	13,000	25,000	0.012	7.0	5.0	3.9	3.7	3.0	160	110	85
G-7	Continuous filament glass fabric	Silicone	Grayish brown	1.55	6,000	.....	.....	...	1.0	1.0	0.8	0.8	125	140	140
G-10	Continuous filament glass fabric	Epoxy	Grayish brown	1.55	25,000	15,000	0.005	7.0	1.0	0.6	0.4	0.3	300	250	200
N-1	Nylon fabric	Phenolic	Natural	1.15	5,500	5,000	0.040	5.0	...	1.5	1.2	1.0	220	180	120

TABLE 29   Properties of Laminated Rods[24]

Grade	Filler	Resin	Range of diameters available, in., incl.	Colors available	Specific gravity (min.)	Flexural strength, psi (min.)	Compressive strength, psi (axial)	Tensile strength, psi (average)	Water absorption, in 24 hr, % Diameter, in.			
									1/8	1/4	1/2	1
X	Paper	Phenolic	1/8–1¾	Natural and black	1.30	16,000	20,000	10,000	3.0	2.0	1.5	1.5
XX	Paper	Phenolic	1/8–1¾	Natural and black	1.30	15,000	20,000	10,000	2.5	1.5	1.0	1.0
XXX	Paper	Phenolic	1/8–1¾	Natural and black	1.25	13,000	20,000	9,000	1.5	1.0	0.75	0.75
C	Coarse weave cotton fabric	Phenolic	1/4–4	Natural and black	1.28	16,000	19,000	9,000	...	2.5	2.0	2.0
CE	}	Phenolic	1/4–4	Natural and black	1.26	13,000	20,000	8,000	...	1.7	1.3	1.0
L	Fine weave cotton fabric	Phenolic	3/16–4	Natural and black	1.28	16,000	19,000	11,000	2.5	2.0	1.5	1.2
LE	}	Phenolic	3/16–4	Natural and black	1.26	12,000	20,000	10,000	2.2	1.4	1.1	1.0
A	Asbestos paper	Phenolic	3/8–2	Natural	1.70	15,000	18,000	6,000	2.5	1.5	1.0	1.0
AA	Asbestos fabric	Phenolic	3/8–4	Natural	1.60	17,000	20,000	8,000	...	1.5	1.0	1.0
G-3	Continuous filament glass fabric	Phenolic	1/4–4	Grayish brown	1.65	30,000	22,000	30,000	...	4.2	2.5	2.5
G-5	}	Melamine	1/4–4	Grayish brown	1.80	30,000	22,000	30,000	...	4.5	3.0	3.0
N-1	Nylon fabric	Phenolic	1/4–4	Natural	1.15	10,000	20,000	6,000	0.8	0.6	0.5	0.5

Polyester-glass-mat laminates grades GPO-1, GPO-2, and GPO-3 have been standardized by NEMA and the Society of the Plastics Industry, Inc. These materials are made of random fiber-glass mat and polyester resins and are used both mechanically and electrically, particularly in electrical-switchgear applications. Table 30 illustrates the properties of these laminates.

Polytetrafluoroethylene-glass-cloth laminates grades GT and GX have also been standardized by NEMA, and their properties are shown in Table 31. These laminates are used in the radar and microwave industries where the exceptionally low electrical-loss characteristics and their stability and resistance to heat are needed.

TABLE 30   Properties of NEMA Grades GPO-1, GPO-2, and GPO-3

Property	GPO-1	GPO-2	GPO-3
Flexural strength, lb/in.$^2 \times 10^{-3}$..........	18*	18*	18*
Flexural modulus, lb/in.$^2 \times 10^{-6}$:			
Lengthwise.........................	1.2	1.2	1.2
Crosswise...........................	1.0	1.0	1.0
Tensile strength, lb/in.$^2 \times 10^{-3}$:			
Lengthwise.........................	12	10	10
Crosswise...........................	10	9	9
Tensile modulus, lb/in.$^2 \times 10^{-6}$:			
Lengthwise.........................	1.0	1.0	1.0
Crosswise...........................	0.9	0.9	0.9
Compressive strength, lb/in.$^2 \times 10^{-3}$:			
Flatwise...........................	30	30	30
Edgewise...........................	20	20	20
Shear strength, lb/in.$^2 \times 10^{-3}$.........	14	14	14
Impact strength, ft-lb/in. notch...........	8*	8*	8*
Bond strength (Condition A), lb...........	850*	850*	850*
Water absorption, %, $\frac{1}{8}$-in. thickness......	0.7*	0.6*	0.6*
Thermal expansion per °C................	2.0×10^{-5}		
Flame resistance, s:			
Ignition...........................	......	75*	75*
Burning............................	......	85*	85*
Electric strength, V/mil.................	400	400	400
Dielectric constant:			
At 1 MHz...........................	4.3		
At 60 Hz...........................	4.5	4.5	4.5
Dissipation factor:			
At 1 MHz...........................	0.03		
At 60 Hz...........................	0.05	0.05	0.05
Arc resistance, s......................	100*	100*	150*

* Values shown with an asterisk are guaranteed maximum or minimum values. Other properties are not to be considered as standards but rather as design guides. All values quoted are Condition A—as received.

Thermal aging of NEMA grade laminates Materials used in electrical devices often are exposed to elevated temperatures. Both electrical and physical properties of plastic materials degrade as a function of time and temperature. The IEEE has classified insulation materials in various classes related to temperature, as shown in Table 32. In designing electrical devices, the effect of temperature on electrical strength and flexural strength is often important. The decrease in these properties after 1,168 and 1,000 h at a variety of temperatures is shown in Figs. 33 to 40 for several NEMA laminates.

Impact strength, insulation resistance, and power factor for NEMA grade laminates from −100 to 200°F are shown in Figs. 41 to 43. Other conditions influence the thermal properties of laminates. For example, storage of laminates at high relative humidity often decreases the physical properties. This is caused by the plasticizing effect of water on the polymers. Figures 44 to 46 illustrate the effect of high (90%

TABLE 31   Properties of NEMA Grades GT and GX[20]

Property to be tested	Conditioning (see LI-1-7.03)	Grade GT	Grade GX	Test method
Flexural strength, min average, lb/in.^{2}:				
$\frac{1}{32}$ to $\frac{1}{8}$ in., inclusive				
Lengthwise............	A	15,000	15,000	ASTM D 790
Crosswise..............	A	10,000	10,000	ASTM D 790
$\frac{1}{4}$ in.				
Lengthwise............	A	11,000	11,000	ASTM D 790
Crosswise..............	A	8,000	8,000	ASTM D 790
Impact strength (Izod), ft-lb/in.:				
$\frac{1}{8}$ to $\frac{1}{4}$ in., inclusive				
Lengthwise............	E-48/50	10	10	ASTM D 256
Crosswise..............	E-48/50	6	6	ASTM D 256
Water absorption, max average, %:				
$\frac{1}{32}$ in...................	D-24/23	0.20	0.20	ASTM D 229
$\frac{1}{16}$ in...................	D-24/23	0.10	0.10	ASTM D 229
$\frac{3}{32}$ in...................	D-24/23	0.08	0.08	ASTM D 229
$\frac{1}{8}$ in...................	D-24/23	0.08	0.08	ASTM D 229
$\frac{1}{4}$ in...................	D-24/23	0.08	0.08	ASTM D 229
Electric strength, min average, V/mil, short time:				
0.005 in..................	C-40/23/50	700	700	ASTM D 229
0.010 in..................	C-40/23/50	700	700	ASTM D 229
0.015 in..................	C-40/23/50	600	600	ASTM D 229
0.025 in..................	C-40/23/50	600	600	ASTM D 229
$\frac{1}{32}$ in..................	C-40/23/50	600	600	ASTM D 229
$\frac{1}{16}$ in..................	C-40/23/50	500	500	ASTM D 229
Dielectric breakdown parallel to laminations, kV, step by step:				
$\frac{1}{32}$–$\frac{1}{2}$ in................	C-40/23/50	60	60	ASTM D 229
$\frac{1}{32}$–$\frac{1}{2}$ in................	D-48/50 + D-$\frac{1}{2}$/23	20	20	ASTM D 229
Arc resistance, min average, s, tungsten electrodes.........	D-48/50 + D-$\frac{1}{2}$/23	180	180	ASTM D 495
Dielectric constant at 1 MHz, max:				
Less than $\frac{1}{32}$ in...........	C-40/23/50	2.8	...	Preferred method— see LI-1-7.10
$\frac{1}{32}$ in. and over...........	C-40/23/50	2.8	...	
Less than $\frac{1}{32}$ in...........	D-24/23	2.8	...	Alternate method— see ASTM D 150, Fig. 9
$\frac{1}{32}$ in. and over...........	D-24/23	2.8	...	
Dielectric constant at 8–12.4 GHz, $\frac{1}{32}$ or $\frac{1}{16}$ in. only....	C-40/23/50	...	2.4–2.6*	MIL-P-13949E
Dissipation factor, max:				
At 1 MHz				
Under $\frac{1}{32}$ in............	C-40/23/50	0.0018	...	Preferred method— see LI-1-7.10
$\frac{1}{32}$ in. and over.........	C-40/23/50	0.0018	...	
Under $\frac{1}{32}$ in............	D-24/23	0.005	...	Alternate method— see ASTM D 150, Fig. 9
$\frac{1}{32}$ in. and over.........	D-24/23	0.005	...	
At 8–12.4 GHz,				
$\frac{1}{32}$ or $\frac{1}{16}$ in.............	C-40/23/50	...	0.0022	MIL-P-13949E

* Tolerance is plus or minus 0.05 on any nominal value within this range.

RH) humidity exposure on the hot flexural properties and hardness of a typical G-10 laminate.

Tolerances of NEMA grade laminates   NEMA grade laminates are available in a wide variety of sheet sizes from different manufacturers, depending on the molder's available presses. In most cases sheets 48 by 96 in. can be obtained. Thickness tolerances for sheets are given in Table 33.

Vulcanized Fiber

Vulcanized fiber is made by soaking rag paper in a zinc chloride bath. Zinc chloride causes the cellulose to gelatinize, thereby making it formable.

The term *vulcanize* is meant to describe this softening process and not the thermal vulcanization normally associated with rubber. The soft cellulose is then soaked in

TABLE 32   Definitions of Insulating Materials[25]

Class	Definition
Class 90.....	Materials or combinations of materials such as cotton, silk, and paper without impregnation. Other materials or combinations of materials may be included in this class if by experience or accepted tests they can be shown to have comparable thermal life at 90°C
Class 105....	Materials or combinations of materials such as cotton, silk, and paper when suitably impregnated or coated or when immersed in a dielectric liquid. Other materials or combinations may be included in this class if by experience or accepted tests they can be shown to have comparable thermal life at 105°C
Class 130....	Materials or combinations of materials such as mica, glass fiber, and asbestos with suitable bonding substances. Other materials or combinations of materials may be included in this class if by experience or accepted tests they can be shown to have comparable thermal life at 130°C
Class 155....	Materials or combinations of materials such as mica, glass fiber, and asbestos with suitable bonding substances. Other materials or combinations of materials may be included in this class if by experience or accepted tests they can be shown to have comparable thermal life at 155°C
Class 180....	Materials or combinations of materials such as silicone elastomer, mica, glass fiber, and asbestos with suitable bonding substances such as appropriate silicone resins. Other materials or combinations of materials may be included in this class if by experience or accepted tests they can be shown to have comparable thermal life at 180°C
Class 220....	Materials or combinations of materials which by experience or accepted tests can be shown to have the required thermal life at 220°C

water to leach out the zinc chloride, and the sheet is dried and calendered, resulting in a hard, dense, regenerated cellulose.

Vulcanized fiber is made in many grades, the grade and properties thereof being determined by the quality of the rag paper and by the processing. Thirteen grades have been standardized by the National Electrical Manufacturers Association in Publication VU-1-1971. Certain grades are available in sheets, rolls, tubes, and rods; they consist of the following:

1. *Commercial grade* (sheets, rolls, and rods): This grade is considered as the general-purpose grade and is sometimes referred to as "mechanical and electrical grade." It possesses good physical and electrical properties and can be fabricated satisfactorily by punching, turning, and forming operations. It is made through the entire range of thicknesses from 0.012 to 2 in. Rods are turned from commercial-grade sheets.

2. *Electrical insulation grade* (sheets and rolls): This grade is primarily intended for electrical applications and others involving difficult bending or forming operations. It is made in all standard thicknesses from 0.004 to ⅛ in. Thin material of this grade is sometimes referred to as "fish paper."

Bone grade (sheets, tubes, and rods): This grade is characterized by the greater hardness and stiffness associated with higher density. It machines smoother and with less tendency to separate the plies in difficult machining operations.   It is made on the limited thickness range of $\frac{1}{32}$ to $\frac{1}{2}$ in.   Rods are turned from bone-grade sheets.

4. *Abrasive grade* (sheets and rolls): This grade is designed as the supporting base for abrasive grit for both disk and drum sanders.   It has exceptional tear re-

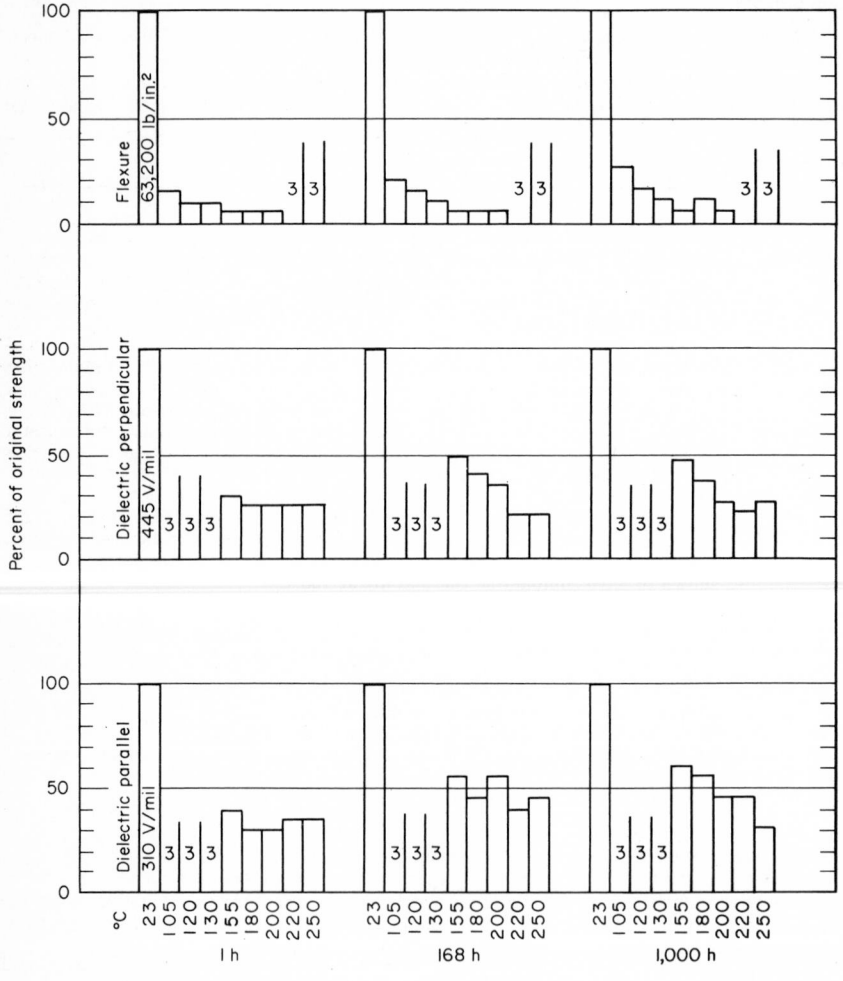

3 insufficient material

Fig. 33   Elevated-temperature properties of NEMA grade G-10.[26]

sistance, ply adhesion, resilience, and toughness.   The surface allows uniform adhesive distribution.

5. *Bobbin grade* (sheets): This grade is used for the manufacture of textile bobbin heads.   It is soft enough to punch under proper conditions, yet firm enough to resist edge denting in use.   It machines to a very smooth finish.

6. *Flexible grade* (sheets): This grade is made sufficiently soft and resilient by

incorporating a plasticizer to make it suitable for gaskets, packing, and similar applications.   It is not recommended for electrical use.

7. *Hermetic grade* (sheets and rolls): This grade is used as electric-motor insulation in hermetically sealed refrigeration units.   High purity and low methanol extractables are essential because it is immersed in the refrigerant.

8. *Pattern grade* (sheets): This grade is made to have best dimensional stability

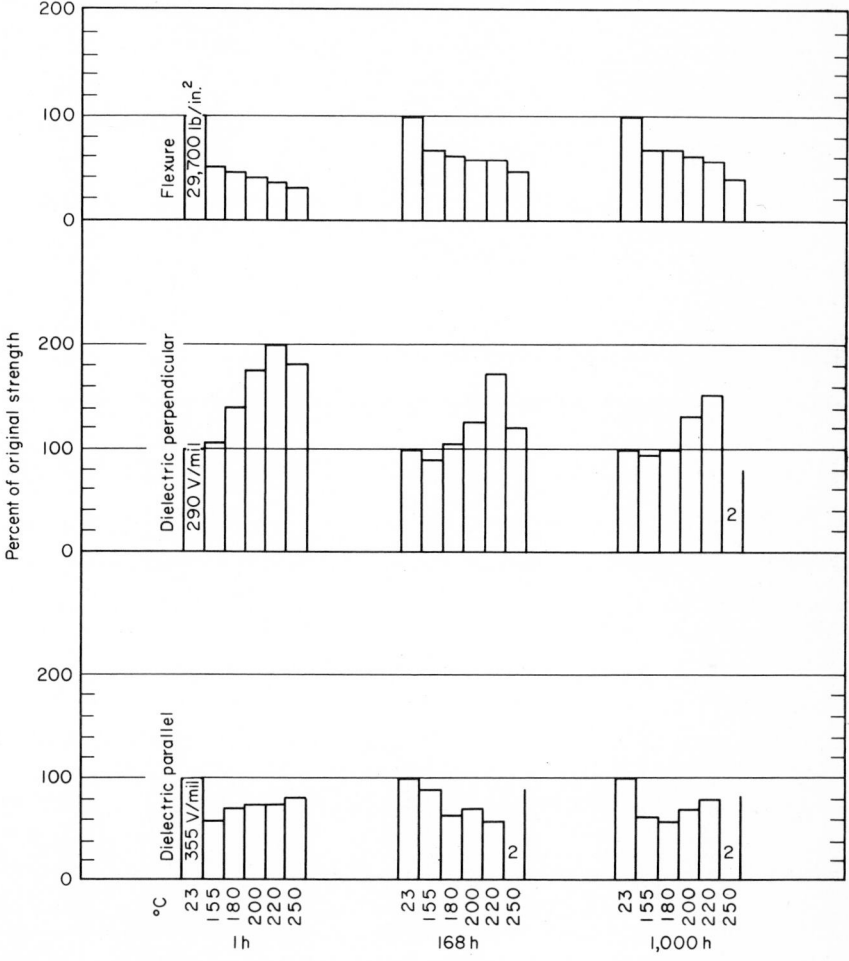

Fig. 34   Elevated-temperature properties of NEMA grade G-7.[26]

and least warpage for use as patterns in cutting cloth, leather, and other materials.

9. *Railroad grade* (sheets): This grade is for use in railroad track joints, switch rods, and other insulation applications for track circuits.   It conforms to Parts 58 and 178 of the AAR (Association of American Railroads) Signal Section "Specifications for Hard Fiber."   Its toughness and high compressive strength enable it to withstand the repeated shock loads occurring in track joints.

10. *Shuttle grade* (sheets): This grade is designed for gluing to wood shuttles to

withstand the repeated pounding received in power looms. Its dense structure permits machining to the very smooth finish required to prevent snagging fine yarn.

11. *Trunk and case grade* (sheets): This grade is especially adapted for trunks, cases, receptacles, and athletic guards. It conforms to the mechanical requirements of commercial grade but possesses better bending qualities and smoother and cleaner surfaces.

12. *White grade* (sheets and reels): This grade is recommended for applications

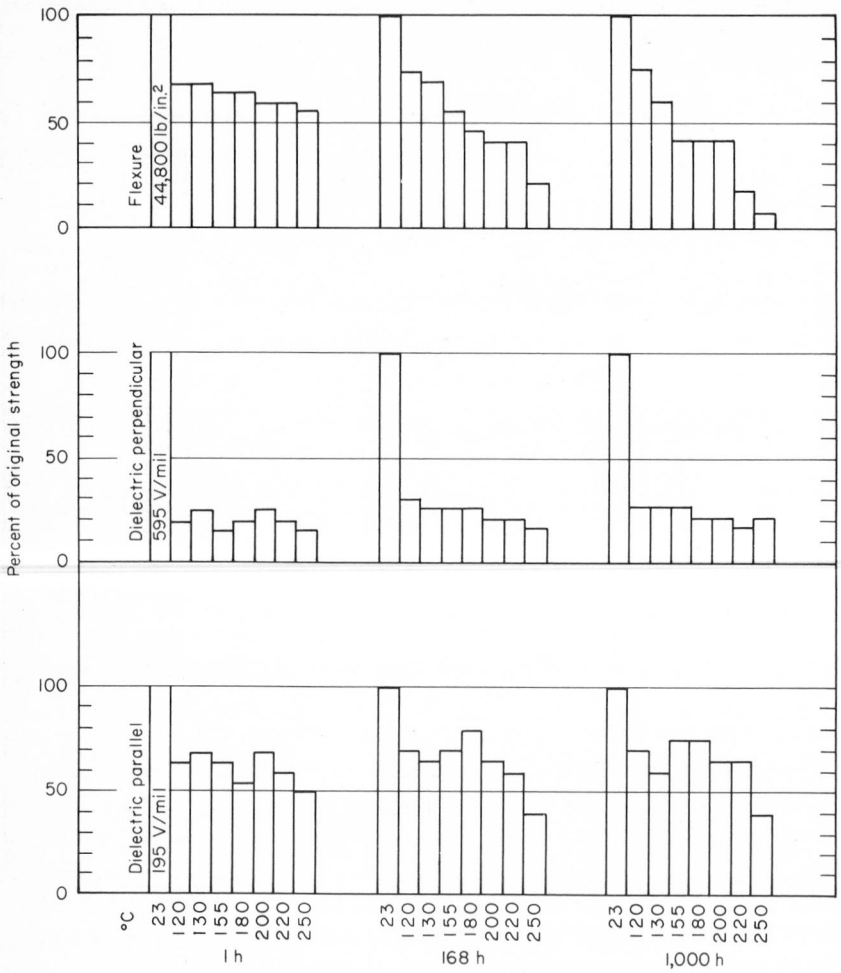

Fig. 35   Elevated-temperature properties of NEMA grade G-5.[26]

where whiteness and cleanliness are essential, such as spoons, suture reels, file index cards, and advertising novelties. It is suitable also for most mechanical and electrical applications.

13. *White-tag grade* (sheets and rolls): This grade has smooth, clean surfaces and is capable of being printed and written upon without feathering. The resistance of vulcanized fiber to organic chemicals is important where tags are used in dry-cleaning solutions. For certain tag applications, red or black fiber is available.

Moisture plasticizes fiber and permits deep drawing. For example, if a sheet of fiber is soaked in hot water, it can be drawn in conventional shaping equipment in thickness up to ³⁄₃₂ in. Wet fiber can be stretched up to 25 percent of the original dimension in tension and 50 percent in compression. Dry fiber can be formed also after being heated to 350°F.

Fiber is noted for being hard and tough with good electrical strength. It is low in cost and possesses unusual arc resistance. It is resistant to abrasion, gasoline, oil, and

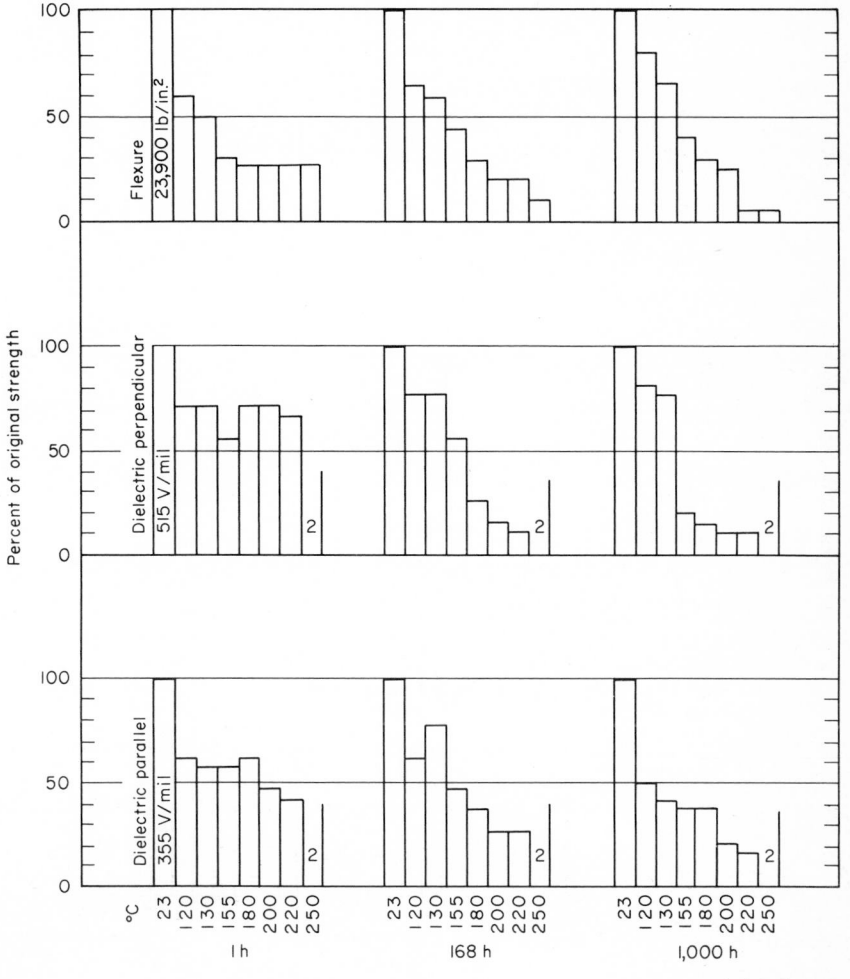

Fig. 36 Elevated-temperature properties of NEMA grade GPO-1.[26]

many solvents and it machines like wood. Table 34 presents some of the properties of fiber. Testing standards are given in ASTM D 619-63, and fiber is further described in ASTM D 710 and MIL-F-1148.

Typical uses for fiber include slot liners, washers, relay parts, lightning arresters, suitcases, golf clubs, loom parts, grinders, and railroad-track insulation. One disadvantage of fiber exists in applications where precise dimensional stability is required. In such cases the swelling of fiber accompanying water absorption would exclude its use. Figure 47 shows the absorption of water by vulcanized fiber after being exposed to various humidities.

Decorative Laminates

Decorative laminates are high-pressure laminates with a paper base, similar in many respects to industrial laminates except for the special attractive surfaces. These laminates have a core of sheets of phenolic-resin-impregnated kraft paper. On top of

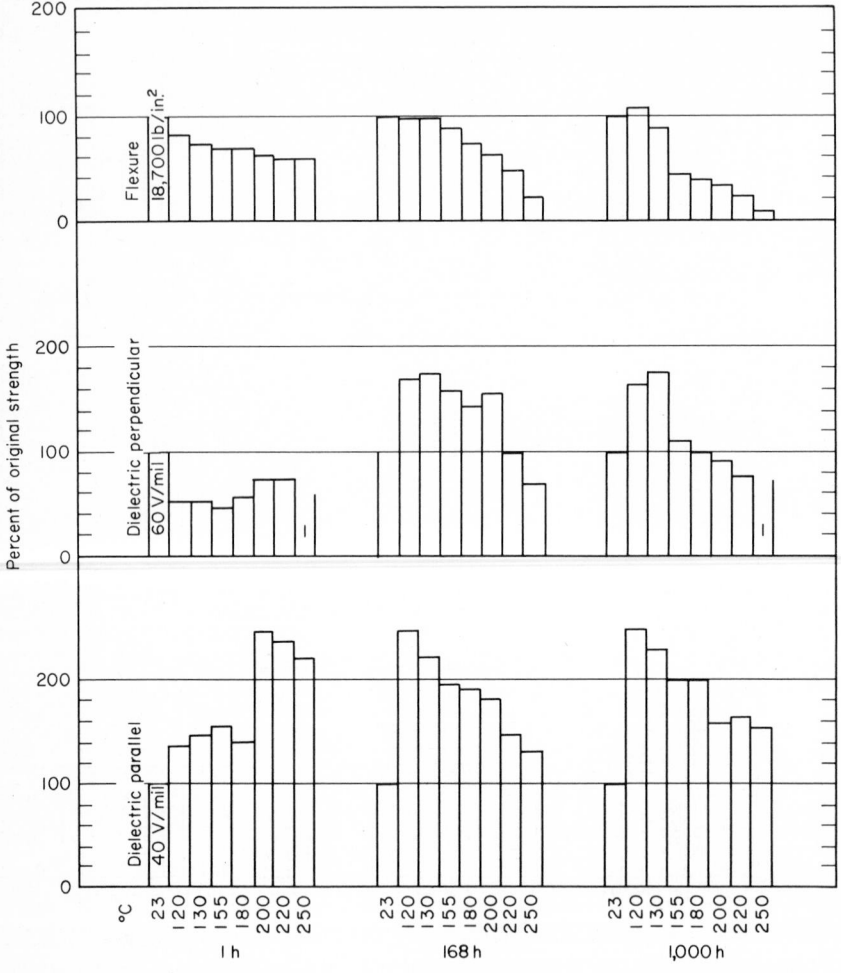

Fig. 37   Elevated-temperature properties of NEMA grade AA.[26]

this is placed a special grade of paper with a decorative pattern printed on the surface and impregnated with a clear melamine resin. The decorative pattern may be wood grain, solid color, or any other design. On top of this is placed another sheet of paper called an *overlay* which is impregnated with a melamine so chosen that the index of refraction is similar to that of the cellulose in the paper. This overlay protects the decorative sheet and provides the unique abrasion and stain resistance of these laminates. The stack then goes into large, multiple-opening presses similar to that shown in Fig. 48. Here the sheets are bonded together at temperatures of 300°F and pressures up to 1,500 lb/in.² The sheets are trimmed and the back is sanded to achieve tolerances and provide for adhesive bonding.

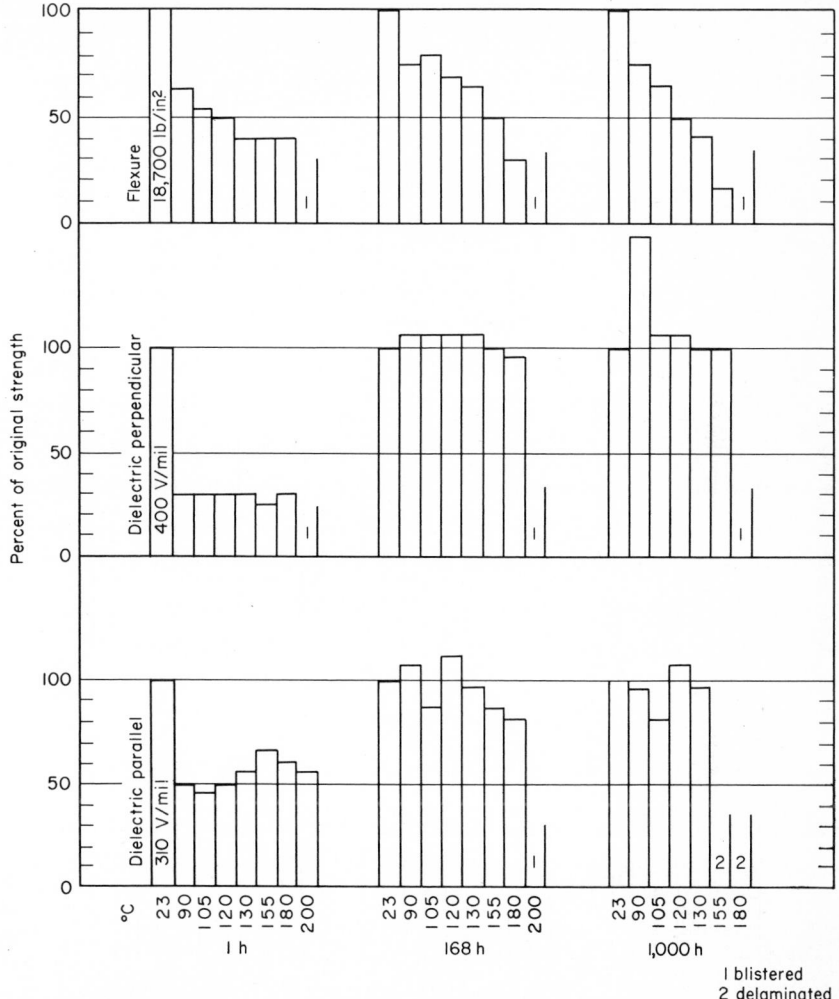

Fig. 38 Elevated-temperature properties of NEMA grade LE.[26]

The top surface of the decorative laminate may be any of several finishes with varying gloss. The gloss finish has a mirrorlike surface with a gloss-meter reading of 80 to 100. Furniture finish with a reading of 36 to 55 is often used in dinette sets. Satin finish (15 to 34) is the most popular of all laminates. Velvet finish with a gloss of 7 to 14 is a new finish popular on walls and partitions. Low-glare finish with a gloss of 4 to 10 is especially attractive on wood-paneled walls, and the very recent oil-rub finish (2 to 5) can be treated with furniture polish or oil to result in a fine craftsmanlike wood-grain surface.

Decorative laminates are available in several grades intended for differing service. Table 35 shows the grades available and describes typical applications. Postforming grades are capable of being bent to produce curved surfaces. They must be heated to 315 to 325°F rapidly and pressed to shape in a form or mold with moderate pressure. Typical properties of decorative laminates are shown in Table 36. The resistance of these laminates to chemicals and household preparations is excellent. Chemical resistance is shown in Table 37.

These laminates are usually bonded to some substrate like plywood, chipboard, or

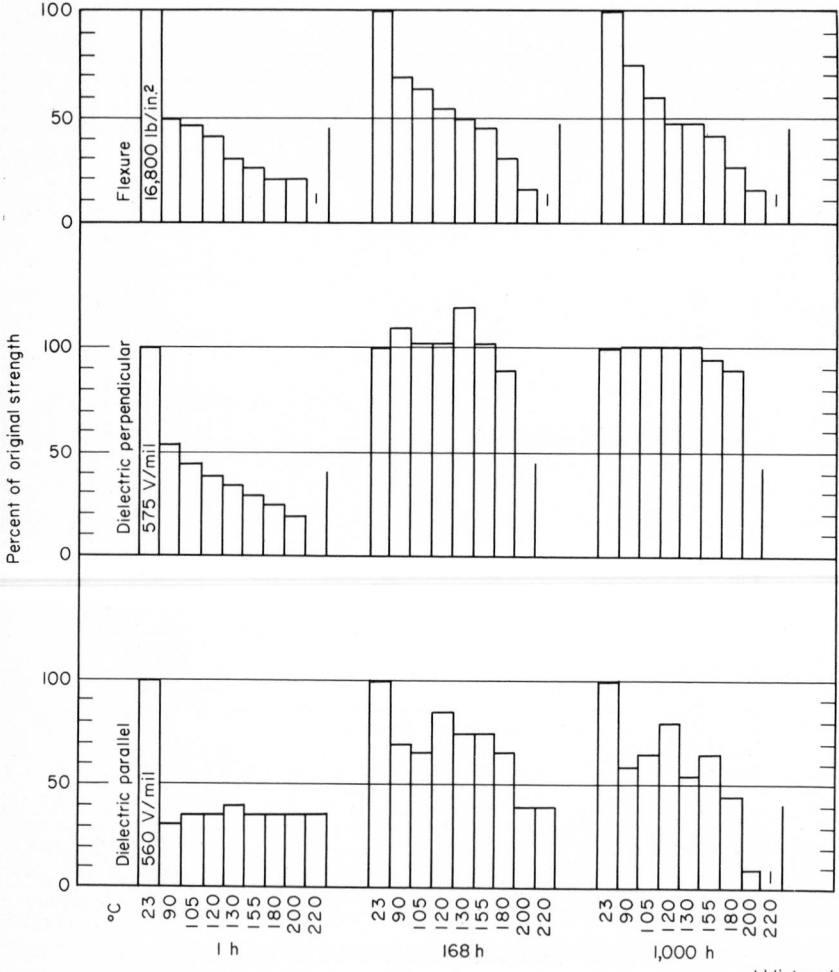

Fig. 39 Elevated-temperature properties of NEMA grade XXXP.[26]

composition board. When bonding, a balancing sheet is recommended. This sheet consists of several layers of phenolic-impregnated paper similar to the core in the decorative laminate. When bonded between the laminate and the back side of the laminate, the balance sheet prevents moisture absorption and minimizes warpage of the structure. Adhesives used include urea formaldehydes, phenolic, casein, epoxy, polyvinyl acetate, and contact cements.

Decorative laminates are not structural materials and should be treated as veneers except where they are bonded to structural cores. Although the face wears well in abrasive environments, the materials are handled as brittle, rigid, thin structures. A laminate has grain direction much like wood, and its dimensional stability is similar to that of wood. With varying humidity the width of the laminate changes two times as much as the length; so in critical applications the machine or grain direction should be established.

Decorative laminates have been standardized by the National Electrical Manufac-

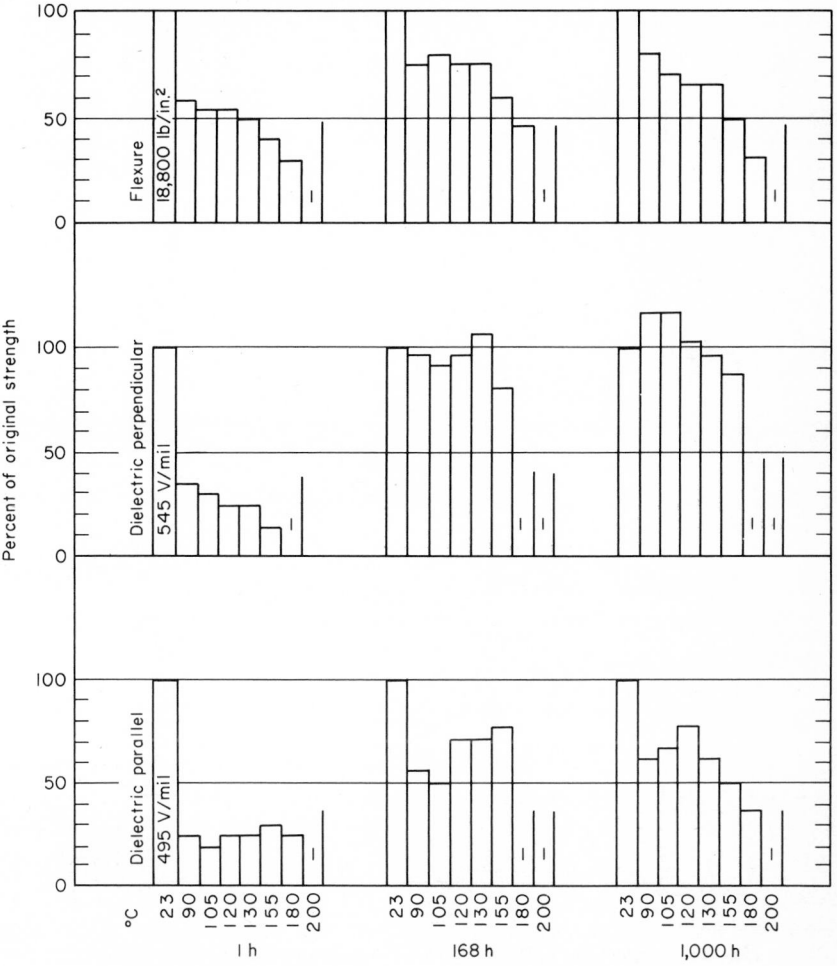

Fig. 40 Elevated-temperature properties of NEMA grade XX.[26]

turers Association in Publication LD-1-1971; NFPA Building Exits Code; U.S. Coast Guard; Public Health Service Regulations, Part 53; MIL-T-1717-B (ships); FHA Minimum Property Standards; and Federal Specifications L-T-0041C, AA-H-625, FF-H-111A (kick plate).

Printed-Circuit Laminates

The many advantages of photofabrication have resulted in its ever-increasing use in electronic applications. A printed circuit made by photofabrication is defined as an electrical conducting path reproduced onto the surface of an insulation medium. In the most common case the insulating medium consists of a plastic laminate with copper foil bonded to the surface. The desired electric circuit is printed onto the copper foil by any one of many reproduction techniques such as photography, offset printing, or silk screening, the object being to deposit, in a precise pattern, some polymer (resist) which is insoluble in a copper-etching media. The printed board is then passed through the etchant (often hot ferric chloride) where all unprinted copper foil is dissolved, leaving behind the desired circuit. The board is then passed through another

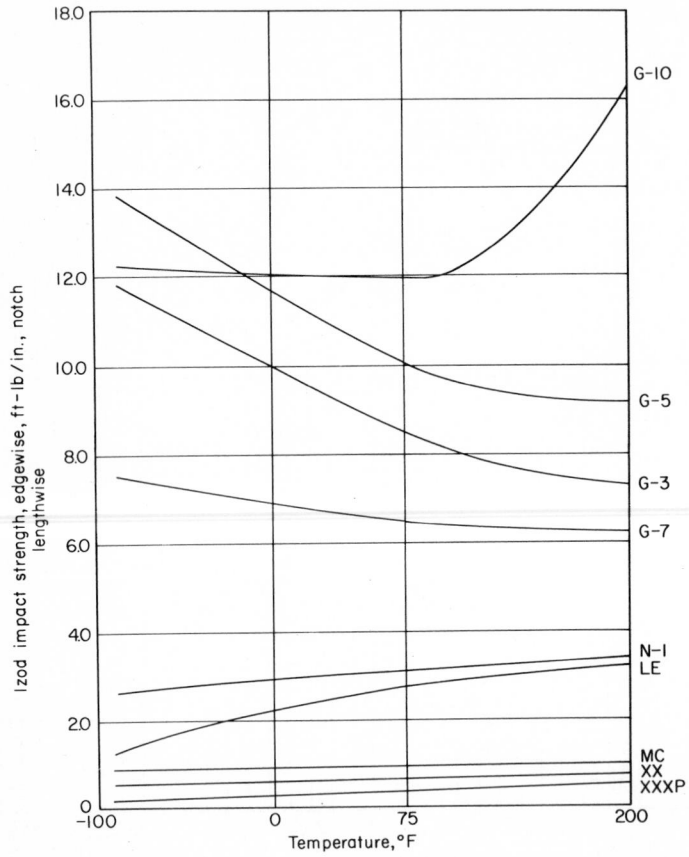

Fig. 41 Impact strength of NEMA grade laminates vs. temperature.[27]

bath, where the resist is removed. Washing, drying, and in some cases, plating conclude the process. Printed circuits offer the designer weight reductions, decreased cost, and increased reliability, and are admirably suited for mass-production methods.

A vital part of a printed circuit is the plastic (copper-clad) laminate, for it is this material that dictates the board's physical and insulation properties. Theoretically any plastic or ceramic material can serve as the base for a printed circuit, and special applications have required the use of many unusual combinations. However, certain materials are more useful and have been standardized by NEMA (ASTM standards are similar) and the federal government. A list of the NEMA grades of copper-clad is given in Table 38. Many copper-clad materials also are specified in military specifications. Frequently the military specification is not identical to commercial standards but requires a higher level of performance.   However, most commercial copper-clad laminates can qualify under both commercial and military standards.   Table 39 presents military standards for copper-clad laminates in comparison with similar NEMA specifications.   Military Specification MIL-P-13949D referenced in the table defines the copper-clad laminates and their properties, and is similar in many respects to NEMA LI-1-1965, Part 10, and ASTM D 709-67.   Design standards are defined in MIL-275B, and quality control and reliability standards are given in MIL-P-55110A.

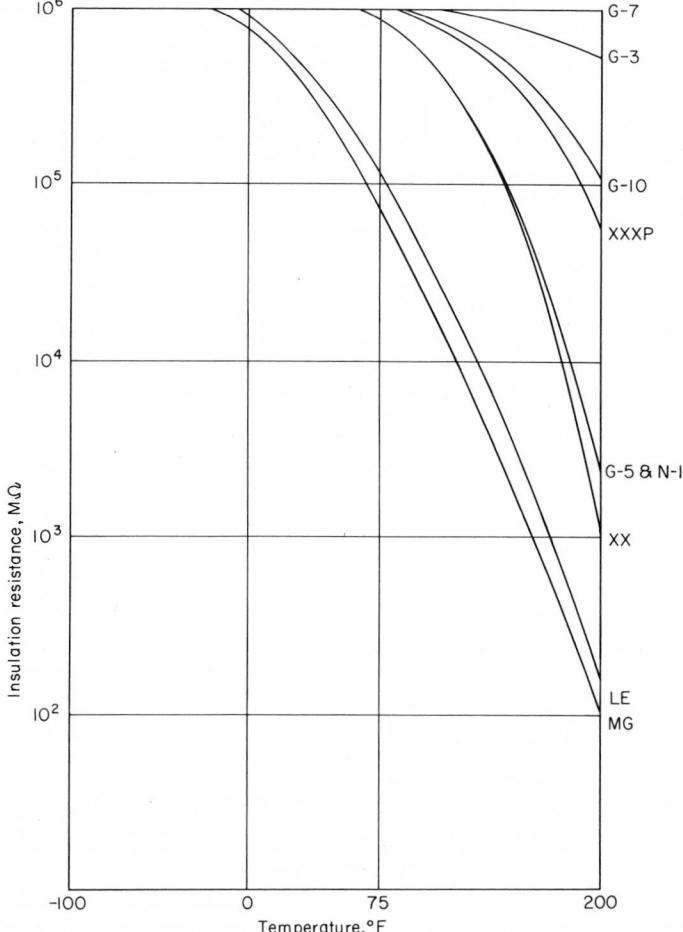

Fig. 42   Insulation resistance of NEMA grade laminates vs. temperature.[27]

Phenolic paper copper-clad is often used for commercial applications where great strength is not required and extreme environments are not encountered.   NEMA XXXPC is similar to NEMA XXXP but can be punched at about 140°F.   FR-2 laminates have the properties of the XXXPC but are flame-retardant as well.   All materials are medium brown in color and are opaque, with an identifying code molded into the surface by the manufacturer.

Epoxy paper copper-clad is NEMA grade FR-3.   This material has physical properties superior to those of NEMA XXXPC and is flame-retardant.   Type PH in MIL-P-13949D is similar in properties to NEMA XXXP but is flame-retardant and, like

NEMA XXXP, must be punched hot. These materials are light yellow to white in color and partly translucent.

Epoxy glass NEMA G-10 and G-11 are made with glass cloth and epoxy resin and are usually light green and semitransparent. They are used when optimum physical properties are required. Moisture resistance and dielectric properties are also excellent. Both grades behave similarly except that NEMA G-11 retains more than 50 percent of its room-temperature flexural strength when measured at 150°C after 1 h aging at 150°C. Thus this grade has better thermal stability and should be employed when

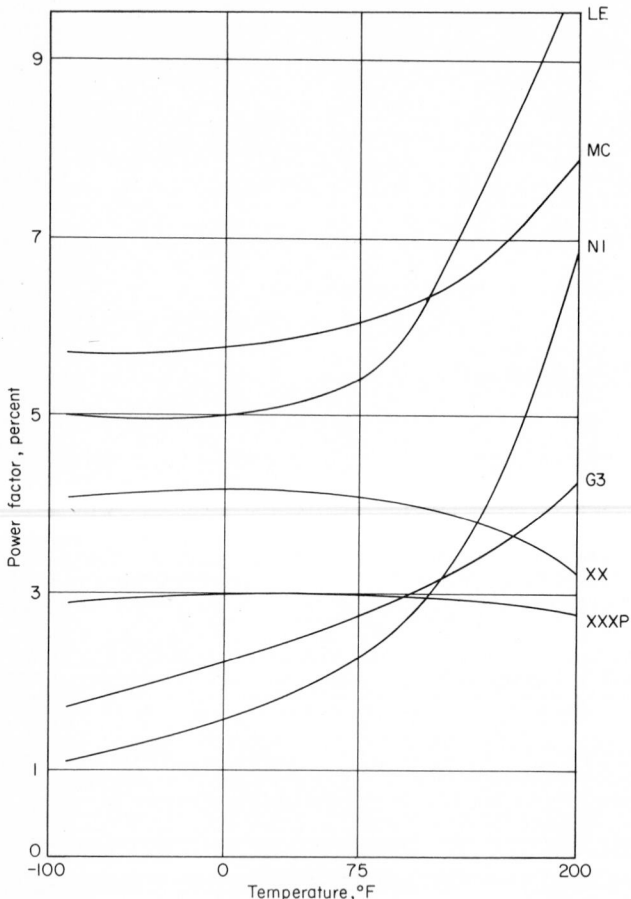

Fig. 43 Power factor of NEMA grade laminates vs. temperature.[27]

physical properties at elevated temperatures are critical. NEMA grades FR-4 and FR-5 are flame-retardant versions of NEMA G-10 and G-11, respectively. The NEMA FR-4 is yellow-green, while NEMA FR-5 varies in color from green to brown depending on the manufacturer. Table 40 lists the properties of the NEMA copper-clad laminates.

Testing copper-clad Certain properties relating to interactions of laminate and copper foil are unique to copper-clad, and specific test methods have been devised to measure them.

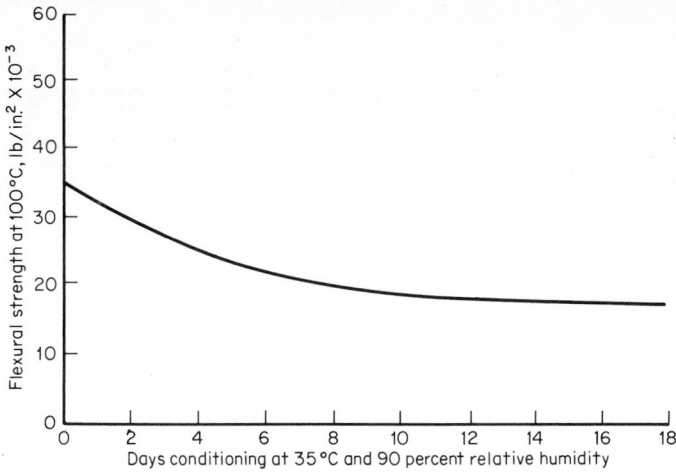

Fig. 44 Effect of high-humidity storage on elevated-temperature flexural strength for G-10 laminate.

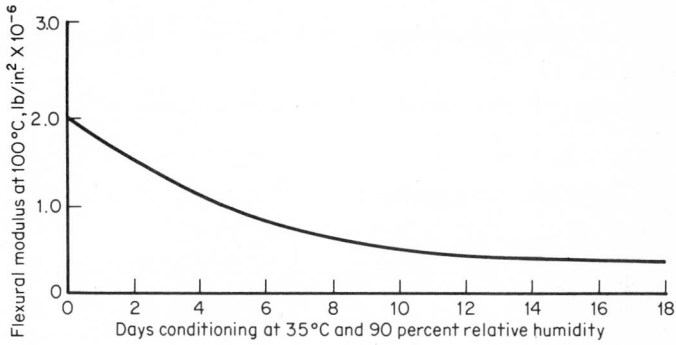

Fig. 45 Effect of high-humidity storage on elevated-temperature flexural modulus for G-10 laminate.

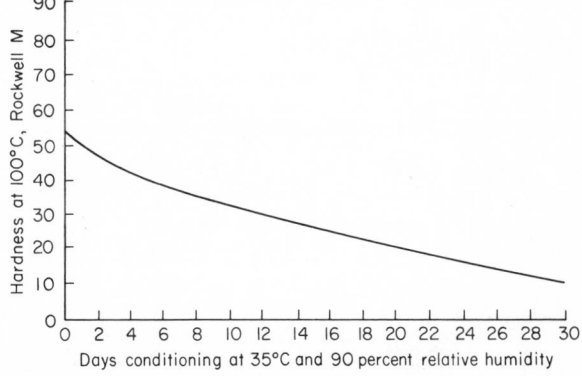

Fig. 46 Effect of high-humidity storage on elevated-temperature hardness for G-10 laminate.

TABLE 33 Thicknesses and Thickness Tolerances of NEMA Grade Laminates (in inches)

Nominal thickness	X, XP, XPC, XX, XXP, XXX, XXXP, XXXPC, FR-2, FR-3	ES-1, ES-2, ES-3, CE, A	C	L	LE	AA	G-2	G-3, G-5, G-7, G-9, G-10, G-11, FR-4, FR-5	N-1
	Plus or minus	Plus or minus	Plus or minus	Plus or minus	Plus or minus	Plus or minus	Plus or minus	Plus or minus	Plus or minus
0.010	0.002	.....	.....	0.003	.....	.....	.....	0.002	0.003
0.015	0.0025	.....	.....	0.0035	0.0035	.....	.....	0.003	0.0035
0.020	0.003	.....	.....	0.004	0.004	.....	.....	0.004	0.004
0.025	0.0035	0.005	.....	0.0045	0.0045	.....	.....	0.005	0.0045
1/32	0.0035	0.0065	0.0065	0.005	0.005	.....	0.008	0.0065	0.0065
3/64	0.0045	0.0075	0.0075	0.0055	0.0055	.....	0.010	0.0075	0.0075
1/16	0.005	0.0075	0.0075	0.006	0.006	0.018	0.010	0.0075	0.0075
3/32	0.007	0.009	0.009	0.007	0.007	0.018	0.012	0.009	0.009
1/8	0.008	0.010	0.010	0.008	0.008	0.020	0.012	0.012	0.010
5/32	0.009	0.011	0.011	0.009	0.009	.....	0.015	0.015	0.011
3/16	0.010	0.0125	0.0125	0.010	0.010	0.024	0.019	0.019	0.0125
7/32	0.011	0.014	0.014	0.011	0.011	.....	0.021	0.021	0.014
			(Plus only)	*(Plus only)*					
1/4	0.012	0.015	0.030	0.024	0.012	0.028	0.022	0.022	0.015
5/16	0.0145	0.0175	0.035	0.029	0.0145	0.034	0.026	0.026	0.024
3/8	0.017	0.020	0.040	0.034	0.017	0.038	0.030	0.030	0.032
7/16	0.019	0.022	0.044	0.038	0.019	0.044	0.033	0.033	0.040
1/2	0.021	0.024	0.048	0.042	0.021	0.048	0.036	0.036	0.048
5/8	0.024	0.027	0.053	0.048	0.024	0.058	0.040	0.040	0.054
3/4	0.027	0.029	0.058	0.054	0.027	0.068	0.043	0.043	0.058
7/8	0.030	0.031	0.062	0.060	0.030	0.076	0.046	0.046	0.062
1	0.033	0.033	0.065	0.065	0.033	0.086	0.049	0.049	0.066

Nominal thickness, in.							G-5 and G-9 only
1⅛	0.035	0.069	0.069	0.035	......	0.053	0.053
1¼	0.037	0.073	0.073	0.037	......	0.055	0.055
1⅜	0.039	0.077	0.077	0.039	0.106	0.058	0.058
1½	0.041	0.081	0.081	0.041	0.124	0.061	0.061
1⅝	0.043	0.085	0.085	0.043	......	0.064	0.064
1¾	0.045	0.089	0.089	0.045	......	0.067	0.067
1⅞	0.047	0.093	0.093	0.047	0.144	0.070	0.070
2	0.049	0.097	0.097	0.049	0.160	0.073	0.073
2¼	......	0.105	......	......	......	......	0.079
2½	......	0.113	......	......	......	......	0.085
2¾	......	0.121	......	......	......	......	0.090
3	......	0.130	......	......	......	......	0.097
3½	......	0.146	......	......	......	......	0.110
4	......	0.163					
4½	......	0.179					
5	......	0.190					
5½	......	0.210					
6	......	0.230					
6½	......	0.240					
7	......	0.260					
7½	......	0.280					
8	......	0.290					
8½	......	0.310					
9	......	0.320					
9½	......	0.340					
10	......	0.360					

NOTES:
1. For a sheet having a nominal thickness not listed in the table, the tolerances shall be the same as for the next larger nominal thickness.
2. The minimum and maximum thickness for each grade shall be in accordance with LI 1-4.05.

TABLE 34   Typical Properties of Vulcanized Fiber

Property	Value
Dielectric strength, volts/mil:	
Short time	150–300
Step by step	125–225
Dielectric constant at 60 Hz	4.7
Dissipation factor at 60 Hz	0.03–0.08
Arc resistance, sec	100
Tensile strength, psi $\times$ 10^{-3}:	
Crosswise	6.5–8
Lengthwise	13–18
Tensile modulus, psi $\times$ 10^{-6}:	
Crosswise	0.8
Lengthwise	1.2
Flexural strength, psi $\times$ 10^{-3}:	
Crosswise	12–14
Lengthwise	14–18
Flexural modulus, psi $\times$ 10^{-6}:	
Crosswise	0.7
Lengthwise	1.0
Izod impact strength, ft-lb/in. notch:	
Crosswise	1.1–1.2
Lengthwise	1.4–1.6
Water absorption, % in 2 hr	30–52
Specific gravity	0.9–1.55
Hardness, Rockwell R	50–80
Bond strength, lb	1,000
Thermal conductivity, Btu/(hr)(ft²)(°F/in.)	0.168
Thermal expansion per °F $\times$ 10^{-5}:	
Crosswise	1.1
Lengthwise	1.7
Specific heat, Btu/(lb)(°F)	0.4
Burning rate	Slow

TABLE 35   Types and Applications of Decorative Laminates

Grade	Application	Data
$\frac{1}{16}$-in. general-purpose	Most widely used. For vertical and horizontal use requiring tough, long-wearing properties such as counters, tables, sinks, furniture, bars, and doors	For bonding to other structures. Supplied oversize. Should be bonded to plywood or Novoply
0.050-in. post-forming	Used where inside or outside radii are required, such as sink and counter tops, and restaurant tables	Should be bonded to a solid base. Formable with inside radii of $\frac{3}{16}$ in. and outside radii of $\frac{3}{4}$ in.
$\frac{1}{32}$-in. vertical	For furniture, cabinets, and walls or other vertical applications. Also where small radii postformability is needed	Similar to postforming grade. Minimum outside radius is $\frac{3}{8}$ in.
0.050-in. fire-resistant	Underwriters' label. Used on buses, ships, and public buildings	On incombustible core has flame spread of 15, fuel contributed −10, smoke developed −5
$\frac{1}{16}$-in. cigarette-proof	Will withstand lighted cigars and cigarettes. Used on tables and bars	Similar to general-purpose but contains an aluminum foil for heat dissipation. Meets MIL-T-17171B type II. Withstands 550°F for 10 min without blister

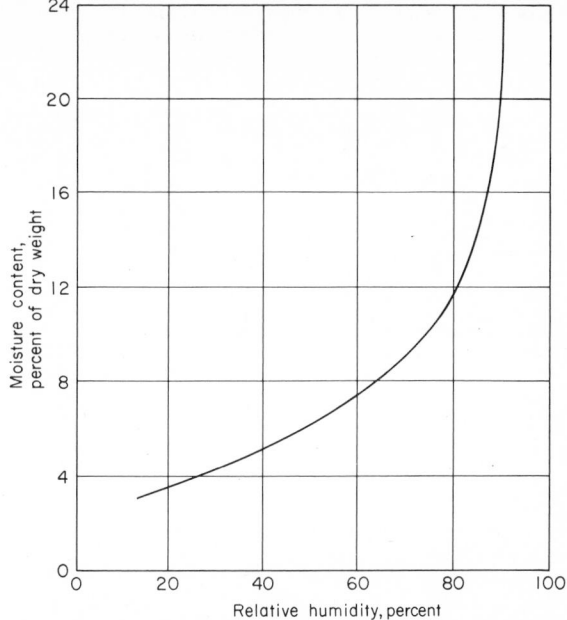

Fig. 47 Equilibrium moisture content of vulcanized fiber.[28]

Fig. 48 Multiple-opening press. (*Westinghouse Electric Corp.*)

Oven Blister Test. This NEMA test is to evaluate the delamination of copper foil from the surface of the laminate. It consists of exposing the laminate to elevated temperature for various times and inspecting for blisters in the foil. No blistering should occur in the foil under the following conditions:

Grade	Temp, °C	Time, min
XXXP, XXXPC, FR-2	120 ± 2	30
FR-3	120 ± 2	60
FR-4, FR-5, G-10, G-11	140 ± 2	60

Solder-Float Test. This test is made on the copper-clad both before and after a pattern is etched. The test identifies the resistance of the copper bond to soldering

TABLE 36   Typical Properties of Decorative Laminates

Property; NEMA test	$\frac{1}{16}$-in. General-purpose	0.050-in. Postforming	$\frac{1}{32}$-in.
Thickness, in.	0.062 ± 0.004	0.050 ± 0.003	0.031 ± 0.002
Wear resistance 2–2.01:			
Wear value, cycles min	475	430	416
Rate of wear, g/100 cycles	0.058	0.061	0.062
Boiling-water resistance, 2–2.03	No effect	No effect	No effect
High-temperature resistance, 2–2.03	No effect	Slight dulling	Slight dulling
Cigarette-burn resistance, 2–2.04, s	142	90	79
Stain resistance 2–2.05:			
Reagents 1–20	No effect	No effect	No effect
Reagents 21–29	Slight–none	Slight–none	Superficial stains
Color fastness, 2–2.06	No change	No change	Slight–none
Moisture resistance 2–2.07:			
Weight increase, %	4.16	8.89	9.3
Thickness increase, %	4.55	10.72	11.0
Dimensional stability 2–2.08:			
With grain, %	0.25	0.69	0.29
Cross grain, %	0.63	0.51	0.61
Flexural strength, 2–2.09:			
Lb/in.², face in tension	22,775	18,330	23,800
Modulus, face in tension	1,979,500	1,317,000	2,110,000
Lb/in.², face in compression	31,110	19,400	30,000
Modulus, face in compression	1,885,000	1,284,000	2,500,000
Formability, 2–2.11		$\frac{1}{2}$-in. radius	$\frac{3}{8}$-in. radius

conditions. In the case of unetched samples, the laminate is floated copper side down in a solder bath for the times and temperatures given in Table 41. At the end of the test no blistering or delamination should be seen in either foil or laminate. For etched samples, a pattern like that shown in Fig. 49 is made. Etching processes are detailed in ASTM D 1825-65T. The process just described is performed, and a similar inspection is made. If no blistering has occurred, the etched specimens are tested for peel strength as described next.

TABLE 37   Chemical Resistance of Decorative Laminates

Chemicals not affecting laminates:

Gasoline	Olive oil
Naphtha	Ammonia solution (10%)
Water	Citric acid
Alcohols	Coffee
Amyl acetate	Mustard
Acetone	Sodium bisulfite
Carbon tetrachloride	Wax and crayons
Moth spray	Urine
Fly spray	Shoe polish
Soaps	Trisodium phosphate
Detergents	

Chemicals which may stain laminates—stain can be removed by buffing with mild abrasive cleanser:

Tea	Ink
Beet juice	Iodine solution
Vinegar	Mercurochrome solution
Bluing	Phenols (Lysol)
Dyes	

Chemicals which may damage laminates:

Hypochlorite bleaches	Potassium permanganate
Hydrogen peroxide	Berry juices
Mineral acids	Silver nitrate
Lye solutions	Gentian violet
Sodium bisulfate	Silver protein (Argyrol)

After exposure to solder the laminates should be inspected for color change and for "measling." Measling, a phenomenon sometimes observed in glass-fabric-based laminates, is caused by separation of individual glass fiber bundles at their crossover points and is oberved as white spots. The Institute of Printed Circuits in the publication "Acceptability of Printed Circuit Boards" defines measling and provides rejection criteria. It should be noted that measling is purely cosmetic, and no impairment of any property has been attributed to this condition.

Peel-Strength Test. The sample from the solder-float test is clamped horizontally, and about 1 in. is hand-peeled. A suitable testing machine then peels the foil from the surface at about 2 in./min. Peel strength is also often measured after oven aging.

Solvent Resistance. To remove the resist from the completed wiring, the circuit boards are generally exposed to trichloroethylene. Thus, NEMA and the military specifications suggest evaluating the boards for delaminations and blisters after exposure for 2 min over boiling trichloroethylene. Table 42 is a chart designed to aid

TABLE 38   NEMA Copper-clad Laminates

NEMA grade	Resin	Reinforcement	Description
XXXP	Phenolic	Paper	Hot-punching grade for general use
XXXPC	Phenolic	Paper	Room-temperature punching for boards with small, close holes
G-10	Epoxy	Glass cloth	General-purpose glass base. Excellent electrical and physical properties. Good moisture resistance
G-11	Epoxy	Glass cloth	Similar to G-10 but more thermally stable
FR-2	Phenolic	Paper	Similar to XXXPC but flame-retardant
FR-3	Epoxy	Paper	Room-temperature punching, flame-retardant, better physical properties than XXXPC
FR-4	Epoxy	Glass cloth	Similar to G-10 but flame-retardant
FR-5	Epoxy	Glass cloth	Similar to G-11 but flame-retardant

TABLE 39 Specifications for Military Copper-clad

Military Specification	Similar NEMA standard	Description
MIL-P-13949D, Type PH	. . . . .	Paper base, epoxy resin, hot punch, flame-retardant
MIL-P-13949D, Type PX	FR-3	Paper base, epoxy resin, flame-retardant
MIL-P-13949D, Type GB	G-11	Glass fabric base, epoxy resin, temperature-resistant
MIL-P-13949D, Type GC	. . . . .	Glass fabric and nonwoven fabric base, polyester resin, flame-retardant
MIL-P-13949D, Type GE	G-10	Glass fabric base, epoxy resin, general purpose
MIL-P-13949D, Type GF	FR-4	Glass fabric base, epoxy resin, flame-retardant
MIL-P-13949D, Type GH	FR-5	Glass fabric base, epoxy resin, temperature resistant and flame-retardant
MIL-P-13949D, Type GP	. . . . .	Glass fabric base, polytetrafluoroethylene resin
MIL-P-13949D, Type GT	. . . . .	Glass fabric base, polytetrafluoroethylene resin
MIL-P-22324A, Type PEE	FR-3	Paper base, epoxy resin, flame-retardant
MIL-P-19161A, Type GTE	. . . . .	Glass fabric base, polytetrafluoroethylene resin
MIL-P-997, Type GS	. . . . .	Glass fabric base, silicone resin
MIL-P-15037, Type GM	. . . . .	Glass fabric base, melamine resin
MIL-P-8013C, Types 1, 2, 3	. . . . .	Glass fabric base, polyester resin

in the selection of the proper copper-clad material. The value judgments offered are comparative only and not absolute.

Copper foil Copper foil used in printed circuits may be either rolled or electroformed, but most of it is electroformed. Most manufacturers treat the copper to oxidize the surface, which results in higher bond strength. Adhesives are used on the foil on phenolic copper-clad but not on epoxy copper-clad. The copper foil is supplied in thicknesses of 0.0014 in. (1 oz/ft^2) and 0.0028 in. (2 oz/ft^2). The standard tolerances for foil are as follows:

	Tolerance, in.	
Thickness, in.	Plus	Minus
0.0014	0.0004	0.0002
0.0028	0.0007	0.0003

Other foil thicknesses are available in special orders. The copper surface in relation to pits and dents is described in MIL-P-13949D, which assigns point values to the dimensions of the pit. A score exceeding 30 points per square foot is sufficient cause for rejection. Another method is by comparison with photographic overlays which provide a standard with which to measure each defect in terms of area. The areas are summed, and the board is rejected at some critical value. The user should remember that in the typical printed circuit 95 percent of the copper is etched away; therefore, the probability of a defect's occurring in the circuit is small. Even when such a defect does occur within the pattern, it may not be of significance unless it is in a critical area. The purchaser must carefully examine his needs to avoid paying a premium price for specially prepared boards free from defects.

Other metal foils are available such as aluminum, nickel, and steel, but these materials are special-order items with attendant premiums.

Dimensions and tolerances Since a copper-clad is a composite made of layers of insulation and of foil, the total thickness tolerance is determined by the tolerances of the individual layers and the care that the laminator exercises during the molding process. Table 43 presents standard tolerances prepared by NEMA.

Copper-clad laminates are cut to shape either by sawing on a band or circular saw or by shearing. The area tolerances depend on which method is used and are presented in Tables 44 and 45.

TABLE 40 Properties of NEMA Copper-clad[29]

Property	Conditioning procedure by ASTM Methods D 618*	XXXP	XXXPC	FR-2	FR-3	FR-4	FR-5	G-10	G-11
Peel strength, min., lb/in. width									
1-oz copper:									
After solder dip..........	A	6	6	6	7	7	7	7	7
After elevated temperature	E-1/140†	6	6	6	7	7	7	7	7
2-oz copper:									
After solder dip..........	A	7	7	7	9	9	9	9	9
After elevated temperature	E-1/140†	7	7	7	9	9	9	9	9
Volume resistivity (min.),									
megohm-cm..............	C-96/35/90	10,000	10,000	10,000	100,000	100,000	100,000	100,000	100,000
Surface resistance (min.),									
megohms	C-96/35/90	1,000	1,000	1,000	1,000	1,000	1,000	5,000	5,000
Dielectric breakdown parallel to laminations, min., kv,									
step by step.............	D-48/50	15	15	15	30	30	30	30	30
Dielectric constant (avg. max.) at 1 MHz.‡........	D-48/50	5.3	5.3	5.3	5.0	5.8	5.8	5.8	5.8
Dissipation factor (avg. max.) at 1 MHz.‡..............	D-48/50	0.05	0.05	0.05	0.045	0.045	0.045	0.045	0.045
Flexural strength (avg. min.), psi									
Lengthwise§................	A	12,000	12,000	12,000	20,000	55,000	55,000	55,000	55,000
Crosswise§................	A	10,500	10,500	10,500	16,000	45,000	45,000	45,000	45,000
Lengthwise¶	....	......	......	......	....	50,000	50,000	50,000	50,000
Crosswise¶................	....	......	......	......	....	40,000	40,000	40,000	40,000
At elevated temperature, % of Condition A value retained, lengthwise:	E-1/150 T-150								
½₂-in. thickness.........	....	......	......	......	....	......	50	....	50
¹⁄₁₆- and ³⁄₃₂-in. thicknesses	....	......	......	......	....	......	40	....	40
⅛- and ¼-in. thicknesses..	....	......	......	......	....	......	50	....	50
Flammability (avg. max.), seconds to extinguish......	A	......	......	15	15	15	15		
Water absorption (avg. max.), %									
½₂-in. thickness.........	....	1.30	1.30	1.30	....	0.80	0.80	0.80	0.80
¹⁄₁₆-in. thickness.........	....	1.00	0.75	0.75	0.65	0.35	0.35	0.35	0.35
³⁄₃₂-in. thickness.........	....	0.85	0.65	0.65	....	0.25	0.25	0.25	0.25
⅛-in. thickness..........	....	0.75	0.55	0.55	0.50	0.20	0.20	0.20	0.20

* Methods of Conditioning Plastics and Electrical Insulating Materials for Testing (ASTM Designation D 618).
† For grades XXXP, XXXPC, FR-2, and FR-3, use condition E-1/120.
‡ Applies only to ³⁄₃₂- and ⅛-in. thicknesses.
§ Applies only to ½₂- to ³⁄₃₂-in. thicknesses.
¶ Applies only to ⅛- and ¼-in. thicknesses.

TABLE 41 Solder-Float Test[29]

Grades	Composite thickness, in.				
	½₂ up to ¹⁄₁₆			¹⁄₁₆ and over	
	Time, sec	Temperature, °F		Time, sec	Temperature, °F
XXXP, XXXPC and FR-2	10	475 ± 3.6 (246.1 ± 2°C)		5	500 ± 3.6 (260 ± 2°C)
FR-3	10	500 ± 3.6 (260 ± 2°C)		10	500 ± 3.6 (260 ± 2°C)
FR-4, FR-5, G-10, and G-11	20	500 ± 3.6 (260 ± 2°C)		20	500 ± 3.6 (260 ± 2°C)

Copper-clad laminates received from the manufacturer invariably demonstrate a certain warp or twist. This is caused by strains of molding and differences in thermal expansion between copper and laminate. NEMA identifies the maximum twist of copper-clad as shown in Table 46. The change in dimension of a copper-clad is always a problem to the designer. When the copper is etched from the surface of a laminate, locked-in strains are removed, and certain dimensional changes occur. These changes may vary from 0.001 to 0.010 in./in. and are rarely predictable. It is always wise to investigate new patterns for distortion. It is equally wise to identify carefully every batch of copper-clad throughout the process so that manufacturing problems can be referred to the copper-clad manufacturer when appropriate.

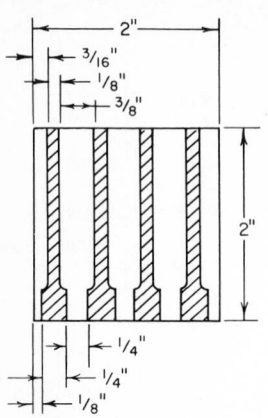

Fig. 49   Solder-float-test sample.[29]

Multilayer Printed Circuits

Development of multilayer circuits   Multilayer printed circuits are individual circuit layers which have been bonded together and interconnected. First used experimentally in the early 1960s, they frequently are employed with integrated circuits. Multilayer circuits are extremely precise because of the complicated miniaturized patterns required. Circuits 1 ft square with 10 layers and a total thickness of 0.100 in. are common. Such a circuit may contain many plated-through holes 0.020 in. in diameter on 0.050-in. centers. Circuits have been made with as many as 40 layers.

Manufacturing process   As many manufacturing techniques for multilayer circuits exist as there are manufacturers. In general, however, the following basic process is used. The thin copper-clad sheets have the desired circuit impressed onto the copper surface, often by photographic methods. The copper is etched using conventional etching practices. The individual layers are laminated together using B-stage prepreg. Holes are then drilled at the appropriate places and to the correct depth. The internal connections from layer to layer are then made by plating copper through the holes, resulting in the completed multilayer circuit.

Multilayer copper-clad   The copper-clad laminates used in multilayer circuits are

TABLE 42   Copper-clad Selector Chart

Property	XXXP	XXXPC	G-10	G-11	FR-2	FR-3	FR-4	FR-5
Cost index	1.2	1.0	2.6	2.6	1.4	2.1	2.6	2.6
Copper bond strength......	Good	Good	Excellent	Excellent	Good	Excellent	Excellent	Excellent
Flexural strength	Low	Low	High	High	Low	Medium	High	High
Water absorption	High	High	Low	Low	High	High	Low	Low
Flammability...	Burns	Burns	Burns	Burns	Self-extin-guishing	Self-extin-guishing	Self-extin-guishing	Self-extin-guishing
Thermal stability	Low	Low	Medium	High	Low	Low	Medium	High
Punchability (room temperature)........	Poor	Excellent	Good	Good	Excellent	Good	Good	Good
Dip solder......	Good	Good	Excellent	Excellent	Good	Good	Excellent	Excellent
Electric strength	Good	Good	Excellent	Excellent	Good	Excellent	Excellent	Excellent
Insulation, resistance........	Low	Low	High	High	Low	Low	Low	Low
Impact, strength	Low	Low	High	High	Low	High	High	High

the same materials employed in conventional printed circuits except they are much thinner. These laminates are available in 0.002-in. increments from 0.002 to $\frac{1}{32}$ in. in thickness. Thickness tolerances for the base copper-clad as specified by NEMA are shown in Table 47, and tolerances for cut sheets of copper-clad less than 18 in. on a side are given in Table 48. Copper tolerances are the same as those for conventional printed circuits presented in the previous section.

The condition of the copper surface is of more concern in multilayer laminates than in regular printed circuits because of the need for exact alignment of circuit lines. NEMA has provided a variety of copper quality guides including pinholes, roughness, scratches, pits, and indentations.

TABLE 43   Thicknesses and Tolerances of Copper-clad Sheet[30]

Nominal overall thickness, including copper, in.	Plus or minus thickness tolerances of copper-clad sheet, in.				
	Class I tolerances			Class II tolerances	
	Grades XXXP, XXXPC, FR-2, and FR-3		Grades FR-4, FR-5, G-10, and G-11, All weights of copper, one and two sides	All weights of copper, one and two sides	
	1-oz copper, one side	1-oz copper, two sides; 2-, 3-, 4- and 5-oz copper, one and two sides		Grades XXXP, XXXPC, FR-2, and FR-3	Grades FR-4, FR-5, G-10, and G-11
$\frac{1}{32}$ (0.031)............	0.004	0.0045	0.0065	0.003	0.004
$\frac{3}{64}$ (0.046)............	0.005	0.0055	0.0075	0.0035	0.005
$\frac{1}{16}$ (0.062)............	0.0055	0.006	0.0075	0.004	0.005
$\frac{3}{32}$ (0.093)............	0.007	0.0075	0.009	0.005	0.007
$\frac{1}{8}$ (0.125)............	0.0085	0.009	0.012	0.006	0.009
$\frac{5}{32}$ (0.156)............	0.0095	0.010	0.015	0.007	0.010
$\frac{3}{16}$ (0.187)............	0.010	0.011	0.019	0.008	0.011
$\frac{7}{32}$ (0.218)*..........	0.011	0.012	0.021	0.009	0.012
$\frac{1}{4}$ (0.250)*..........	0.012	0.012	0.022	0.009	0.012

* These values do not apply to grades XXXPC and FR-2.

At least 90 percent of the area of a sheet shall be within the tolerances given and at no point shall the thickness vary from the nominal by a value greater than 125 percent of the specified tolerance.

Cut panels less than 18 by 18 in. shall meet the applicable thickness tolerances in 100 percent of the area of the panel.

For a sheet having a nominal thickness not listed in the table, the tolerances shall be the same as for the next higher nominal thickness.

Although any polymer could be used as the base in multilayer circuits, the excellent electrical properties and bond strength of epoxy resins make the choice of epoxies particularly attractive. Two grades have been standardized by NEMA: G-10-UT and FR-4-UT. These materials—except for their thicknesses—are similar to the previously discussed NEMA grades G-10 and FR-4. Table 49 shows the properties of these two materials.

Multilayer prepreg materials Prepregs, as used in multilayer circuits, consist of a glass-fiber fabric saturated with a resin—usually epoxy—and partly reacted to a dry tack-free condition. With heat and pressure the resin softens, flows, and acts as an adhesive to bond adjacent circuit layers together. Many epoxy prepreg materials are available from the same suppliers who make prepreg materials for laminates and also from suppliers of the copper-clad laminates. In use, the prepreg is cut to the correct shape and inserted between the various circuit layers. Locating pins in a molding frame hold the various layers in register. The frame and lay-up are placed into a molding press with platens aligned accurately to within 0.001 in. in parallel. The lay-up is then heated at a rate and to a temperature as specified by the prepreg sup-

TABLE 44   Tolerances in Length and Width of Fabricated Panels Cut by Sawing[30]

Nominal thickness, in.	Plus or minus tolerance in length or width, in.		
	6 in. and under	Over 6 up to 18 in.	18 in. and over
$\frac{1}{32}$–$\frac{1}{4}$, incl..................	0.010	0.015	$\frac{1}{32}$

TABLE 45   Tolerances in Length and Width of Fabricated Panels Cut by Shearing in Lengths Less than 48 In.[30]

Squaring Sheared

Nominal thickness, in.	Plus or minus tolerances in length or width, in.			
	2 in. and under	Over 2 to 6 in., incl.	Over 6 to 24 in., incl.	Over 24 in.
$\frac{1}{32}$–$\frac{1}{16}$, incl.............	0.010*	0.015	$\frac{1}{32}$	$\frac{3}{64}$

Rotary Sheared (Not Applicable to Grades GT and GX)

Nominal thickness, in.	Plus or minus tolerance in length or width, in.		
	$\frac{5}{64}$–$\frac{3}{4}$ inch, incl.	Over $\frac{3}{4}$–3 in., incl.	Over 3–6 in., incl.
$\frac{1}{32}$–$\frac{1}{16}$, incl.............	0.005	0.005	0.010

* 0.015 in. for grades GT and GX.

TABLE 46   Maximum Warp or Twist of Copper-clad Sheet[30]

Range of thickness, in.	Paper grades—XXXP, XXXPC, FR-2, and FR-3 Glass grades—G-10, G-11, FR-4, FR-5, GT, and GX; Max warp or twist (based on 36-in. dimension), %					
	Class A warp or twist			Class B warp or twist		
	1- and 2-oz copper, one side	1- and 2-oz copper, two sides		1- and 2-oz copper, one side	1- and 2-oz copper, two sides	
		Glass	Paper		Glass	Paper
$\frac{1}{32}$–$\frac{3}{64}$, incl...........	12	5	6	10	2	5
Over $\frac{3}{64}$–$\frac{1}{16}$, incl......	10	5	6	5	1	$2\frac{1}{2}$
Over $\frac{1}{16}$–$\frac{1}{8}$, incl.......	8	3	3	5	1	$2\frac{1}{2}$
Over $\frac{1}{8}$–$\frac{1}{4}$, incl........	5	1.5	1.5	5	1	$1\frac{1}{2}$

plier. Careful control of pressure is needed. Pressing temperatures vary from 300 to 450°F, and molding pressures average 500 lb/in.² Molding times may reach 1 h. The molded circuit must be cooled under pressure to 120°F, and then it can be removed from the press.

Most of the rejections of multilayer circuits result from an improper molding process.

TABLE 47 Thicknesses and Tolerances of Multilayer Base Laminates After Etching[31]

Nominal thickness, excluding copper, in.	Plus or minus thickness tolerance of base laminate after etching, in.	
	Class I	Class II
0.002–0.0045, incl.................	0.0010	0.0007
Over 0.0045–0.006, incl.............	0.0015	0.0010
Over 0.006–0.012, incl.............	0.0020	0.0015
Over 0.012–0.020, incl.............	0.0025	0.0020
Over 0.020–0.030, incl.............	0.0030	0.0025

Causes of rejects include blisters, voids, delamination, measling, displaced layers, and board distortions. The last-named condition is very common and results from stresses induced in the molding process. Warped circuits can often be straightened by clamping them between flat plates and heating them at 250°F for 1 h.

TABLE 48 Thicknesses and Tolerances in Length and Width of Cut Sheets of Multilayer Copper-clad[31]

Thickness, in.	2 in. and under	Over 2 to 6 in.	Over 6 to 18 in.
Under 0.010	0.015	0.031	0.046
0.011 to 0.015	0.015	0.020	0.031
0.016 to under 0.031	0.010	0.020	0.031

TABLE 49 Property Values for NEMA Grades G-10-UT and FR-4-UT Laminates[31]

Property to be tested	Conditioning procedure	Grade		Test methods
		G-10-UT	FR-4-UT	
Peel strength, min., lb/in. width:				
1-oz copper...............	A	6	6	LI 1-11.08
2-oz copper...............	A	8	8	
Volume resistivity (min.), meg-ohm-cm..................	C-96/35/90	10^6	10^6	LI 1-11.09
Surface resistance (min.), meg-ohms.....................	C-96/35/90	10^4	10^4	LI 1-11.09
Dielectric constant (max. avg.) at 1 MHz (1/32-in. laminate)..	D-24/23	5.4	5.4	LI 1-11.10
Dissipation factor (max. avg.) at 1 MHz (1/32-in. laminate)....	D-24/23	0.035	0.035	LI 1-11.10
Flammability................	A	......	Self-extinguishing	LI 1-11.11

REINFORCED PLASTICS

Definition

Reinforced plastics are materials made by bonding together two or more sheets of reinforcing fibers. Heat and pressure may or may not be used. They are distinguished from laminates in that they may have a varying cross section and the reinforcing sheets need not be parallel.

Prepreg Materials

Prepregs can be defined to be any resin treated on a reinforcing web and sold to the purchaser in an uncured form. The user of any prepreg can then mold his own laminated structure. The resin on a prepreg may be "B-staged," which means partly cured to a non- or semitacky state, or in certain resin systems, the resin may normally be a solid at room temperature.

Prepregs have several advantages over wet lay-up molding:

1. For making small runs and prototypes this form of the molding is most convenient—particularly for smaller molders.

2. By pretreating the reinforcement the prepreg supplier removes most of the cost of resin treatment from the user. Better housekeeping results for the molder by the elimination of the sticky, wet lay-up resins. The prepreg supplier can often be more efficient, and treating costs are lower as a result.

3. The molder is relieved of the necessity of maintaining extensive inventories of many items.

4. A prepreg is often of higher quality than wet molding material. Resin contents accurate to 2 percent are standard.

5. The prepreg supplier will supply die-cut material, thereby saving the molder labor expense.

6. Prepreg sometimes permits more efficient mass-production runs because of its uniformity and the ability to handle it with automated equipment.

7. Molding lay-up can often be made external from the press, thereby saving valuable press time. Various layers of prepreg can be tacked together with a hot iron to form complicated lay-ups.

8. In the special case of filament-winding prepregs, better tension, pattern accuracy, and quality are possible.

These advantages must be weighed against the following disadvantages:

1. Room-temperature cures are impossible, which may limit the material and applications possible.

2. Prepregs are much more expensive than the cost of the raw materials. Depending on the molding volume and other factors, the molder may find that economics dictate making his own resin treatment.

3. Prepregs are often most useful to vacuum-bag molders to whom the efficiencies of prepregs are not important because of the slow nature of this molding process.

4. Prepregs are best kept under refrigeration, which necessitates expensive storage equipment. Sometimes under adverse conditions or if the molder grows careless, the prepreg may spoil, resulting in poor-quality parts and extra costs.

Prepreg materials are available with most of the reinforcing webs described in a previous section. Most common resins are phenolic, epoxy, diallyl phthalate, polyester, silicone, and polyimide. Other resins can be obtained in special order. Many suppliers of prepreg exist, but fewer than ten supply a full range of prepregs—the remainder supply special markets. The prepregs are sold in 38-, 44-, 50-, 54-, and 60-in. widths on a wide range of reinforcing bases. Style 181 glass cloth is popular because of its formability. The prepregs consist of 30 to 55 weight percent resin with carefully controlled flow and gel time. All the prepregs have some volatiles, with the epoxies and silicones averaging less than 2 percent and the other materials less than 10 percent.

Many of the prepregs are available in two forms—a vacuum-bagging grade and a press-curing grade. Typical vacuum-bagging cure cycles consist of molding for 240 min at 300 to 375°F. Press cycles average 90 min at 350°F and 200 to 1,000 lb/in.^{2} Moldings destined for high temperature must be postcured, for which a typical schedule is: preheat to 200°F for a fairly long time and then gradually increase the temperatures to 400°F. The exact time and temperature cycle has been established by the manufacturer for each prepreg requiring postcure and is related to the nature of the resin and the thickness of the molding.

An unusual prepreg material is being introduced by one supplier. Called prepromoted mat, it consists of chopped strand fiber-glass mat which has been impregnated with the accelerator portion of a room-temperature-curing polyester system. In causing polyesters to cure at room temperature, accelerators such as cobalt naphthanate or dimethyl aniline are added to the resin followed by an organic peroxide catalyst. Hence in this prepromoted mat the accelerator is added to the fiber. When a catalyzed resin is then added to the mat, the polymerization occurs, providing a laminate or lay-up which cures at ambient condition. This results in several advantages to the molding shop:

- No weighing, dispensing, or mixing is required.
- Gel times are carefully controlled.
- Room-temperature cures are available without stability problems.
- Materials cure at low temperatures and in the presence of water.
- The mat can easily be used in the field for rapid, high-quality repair of reinforced-plastic parts.

Prototypes

Often during the design process it is desirable to produce parts to check dimensions or obtain more design information. Several methods of making such prototypes are available, and these methods—with others—are also used in production. Table 50 compares the features for each molding method to aid in process selection. Table 51 lists design rules for the molding methods used on reinforced plastics.

Hand lay-up The simplest method of making prototypes is the hand lay-up or contact lay-up method. It is best used to make large parts and is frequently employed by boat manufacturers as shown in Fig. 50. Hand lay-up can be used on either male or female molds. The molds can be wood, metal, plastic, or plaster. When wood or plaster is used, the pores must be sealed with varnishes or lacquers. After the sealing compound is dry, the mold must be coated or covered with a release agent such as floor wax or a film-forming polymer like polyvinyl alcohol (PVA). At this point the reinforcing web (glass cloth, mat, or woven roving) is placed on the mold, and a low-viscosity resin is painted onto the reinforcement, the resin generally being a room-temperature-curing polyester or epoxy. The air from the reinforcement is removed carefully with a paint roller or a rubber squeegee. Additional layers of reinforcement and resin are added until the correct thickness is obtained. The exposed topmost layer is sometimes covered with cellophane or Mylar* to make a smooth surface. Some polyesters are slightly inhibited by air, and thus film overlays help to complete the cure. When the resin has hardened, the part is removed and given a postbake. Although this process is cheap and easy to perform, the moldings are not of the highest quality and may not be completely representative of the properties expected. The laminates contain more resin than is obtained from other methods, and more voids may be present. Little control can be exercised over the uniformity of wall thickness.

Spray-up A special case of hand lay-up is the spray-up process shown in Fig. 51. This is accomplished with a special type of spray gun. Molds similar to those used in hand lay-up are used. The gun chops glass fiber into predetermined lengths, mixes them with resin, and deposits them onto the mold surface. The mold surface is often pretreated with a mineral-filled resin which has been permitted to cure partially. This surface is called a *gel coat*. When enough glass has been deposited, an impregnated glass mat has been formed. From this point on, the process is identical to the hand lay-up process. The process is attractive because it makes use of glass fiber in its least expensive form. Parts with complicated shapes can be made, and size is not critical. Skilled operators are required if control over wall thickness is required.

Vacuum-bag molding Vacuum-bag molding is one of the most popular methods of making prototype or short-run parts. Molds similar to those just discussed are used. Either wet resins or prepregs can be employed. Figure 52 shows the molding process. Over the mold surface is placed a release sheet such as PVA or release-treated Mylar. The reinforcement is placed onto the release film and impregnated by brushing or spraying if prepregs are not being used. Next bleeders are installed along the top of the mold. These are made of burlap, muslin, osnaburg cloth, or felt, and are intended to provide a space from which the air in the mold can be evacuated. The impregnated reinforcement is sometimes covered with a perforated release film: PVA, Teflon, or treated Mylar. Special release fabrics are used also.

---

* Trademark of E. I. du Pont de Nemours & Company, Inc., Wilmington, Del.

TABLE 50 Molding Methods for Prototypes

Feature	Hand lay-up	Spray-up	Vacuum-bag molding	Pressure-bag molding	Autoclave molding	Vacuum-injection molding	Matched metal die	Cold-press molding
Mold cost.............................	Low	Low	Medium	Medium	Medium	High	Very high	Medium
Facility cost.........................	Low	Medium	Low	Low	High	Low	Very high	Low
Production rate......................	Low	High	Low	Low	Low	Medium	Very high	Medium
Physical properties.................	Very low	Medium	Low	Low	Medium	Low	Very high	Medium
Quality of molding..................	Very low	Medium	Low	Medium	Medium	Medium	High	High
Surface appearance.................	Poor	Good	Poor (1 side)	Poor (1 side)	Poor (1 side)	Good	Excellent	Good
Resin content........................	High	Medium	High	High	Medium	Very high	Low	Medium
Finishing required..................	Much	Little	Much	Much	Much	Little	Very little	Little
Operator skill required............	Low	High	Medium	Medium	Medium	Medium	Low	Medium

TABLE 51 Design Rules for Reinforced Plastics[32]

Design rules	Spray-up, hand lay-up (contact)	Matched-metal-die molding preform mat	Sheet molding compound	Cold molding
Min inside radius, in.	$\frac{1}{4}$	$\frac{1}{8}$	$\frac{1}{16}$	$\frac{1}{4}$
Molded-in holes..	Large	Yes, parallel or perpendicular to ram action	Yes	No
Trimmed-in mold.	No	Yes	Yes	No
Undercuts.......	Yes	No	Yes	No
Min draft recommended	0°	1–3°	1–3°	2°
¼- to 6-in. depth		3° + or as required	3° + or as required	3°
6-in. + depth				
Min practical thickness, in.	0.060	0.030	0.050	0.080
Max practical thickness, in.	0.500+	0.250	1	0.500
Normal thickness variation, in.	±0.020	±0.008	±0.005	±0.010
Max thickness buildup	As desired	2:1 max	As desired	2:1
Corrugated sections	Yes	Yes	Yes	Yes
Metal inserts....	Yes	Yes	Yes	NR
Surfacing mat....	Yes	Yes	No	Yes
Max size part to date, ft²	3,000	200	50	50
Limiting size factor	Mold size	Press dimension	Press capacity	
Metal edge stiffness	Yes	Yes	No	No
Bosses..........	Yes	NR	Underside only	NR
Ribs............	Yes	NR	As required	No
Molded-in labels	Yes	Yes	Yes	Yes
Raised numbers..	Yes	Yes	Yes	Yes
Gel-coat surface..	Yes	Yes	No	Yes
Shape limitations.	None	Moldable	Moldable	Moldable
Translucency....	Yes	Yes	No	Yes
Finished surfaces.	One	Two	Two	Two
Strength orientation	Random	Random	Random	Random
Typical glass loading, % by wt	20–30	25–40	15–35	20–35

NR = not recommended.

Over this is often placed dry cloth to assist in air evacuation. In other cases the bag can contact the lay-up directly, depending on the shape and size of the part. Finally the vacuum bag is clamped into place.

Vacuum bags are made from many materials. Cellophane is cheap and easy to install, its chief limitations being tear resistance and temperature. Polyvinyl alcohol

Fig. 50   Hand lay-up of reinforced plastics.   (*Molded Fiberglass Boat Co.*)

(PVA) can be used up to about 150°F, and it is tougher and more tear-resistant than cellophane.   PVA is highly sensitive to humidity, being water-soluble.   Polyethylene has been used where temperatures permit (room temperature only).   Polyvinyl chloride (PVC) bags are useful up to 150°F and are less expensive than PVA.   Certain PVC polymers inhibit polyester resins and prevent a tack-free cure.   When high curing temperatures or steam-autoclave curing are required, the bag should be made of neoprene rubber or some similar elastomer.   Neoprene and some other rubbers also

Fig. 51   Spray-up of reinforced-plastic part.   (*Owens Corning Fiberglas.*)

inhibit polyester resins, and if polyesters are to be cured, the bag should have a film such as Mylar or Kapton* between the lay-up and the bag.

The edge of the bag must be well sealed to prevent loss of vacuum and introduction of air into the lay-up. A common method is to use zinc chromate paste, which is supplied in ribbons or beads. The paste is pressed into intimate contact with the mold, and the bag is pressed into the paste. This material holds a vacuum well and is easily removed from the assembly after molding. Other rubber gaskets can be used along with hold-down or clamp rings, either to replace or to supplement the paste. Getting a good seal is critical to the success of the vacuum-bag process, and care exercised at this step will avoid much difficulty. Vacuum is then applied, bringing air pressure to bear on all surfaces. The use of a resin trap in the vacuum line as shown in Fig. 53 is essential. The trap plays the same role that an oil separator plays in an air line. As the resin softens and flows during the molding, it will enter the vacuum

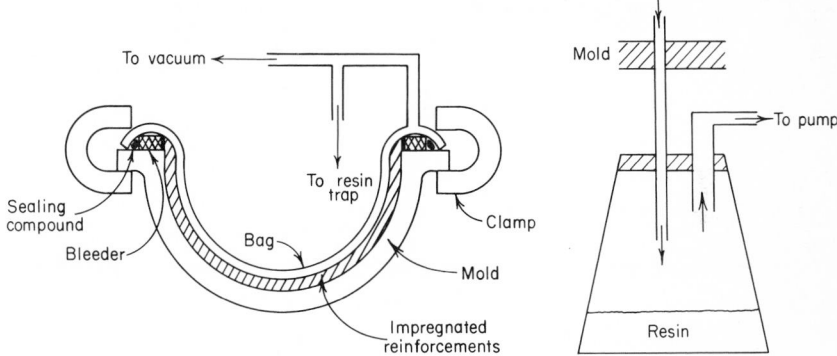

Fig. 52 Vacuum-bag molding.   **Fig. 53** Resin trap.

line. If not removed, it will decrease the vacuum, clog and ruin the lines, and if permitted to enter the pump, will cause pump failure. Mechanical working of the bag with a roller or squeegee aids in air removal when wet resins are used.

If prepregs are used, the whole assembly is placed in an oven for cure. If prepregs requiring temperatures greater than 300°F are employed, all release sheets should be made of Mylar or Teflon. A typical cure schedule consists of increasing the temperature to 250°F in 30 min, to 300°F in 15 min, and to 350°F in 20 min. When the part is cured, temperatures should be reduced to about 150°F before the vacuum is released.

Advantages of the vacuum-bag process are:

1. The molds are relatively inexpensive.
2. The process is faster than hand lay-up.
3. Prepregs can be used.
4. The product is not as rich in resin as in hand lay-up, and properties are improved.
5. The process does not require highly skilled personnel.

Disadvantages are:

1. Only the surface adjacent to the mold surface is smooth. The other surface may be duller and show folds and marks from the bag.
2. The part still contains more resin than those made by high-pressure techniques, and hence properties are not optimized.
3. Certain shapes with complicated surfaces or sharp corners are difficult to form.
4. Dry spots and voids are common.

Pressure-bag molding A vacuum-bag molding process augmented with an externally supplied pressure is known as pressure-bag molding. A pressure plate is

* Trademark of E. I. du Pont de Nemours & Company, Inc., Wilmington, Del.

added to the top of the mold, as shown in Fig. 54. The bleeder strips may need to be repositioned to accommodate the top plate. The molding is the same as in the previous case except that pressures up to 50 lb/in.^{2} can be introduced into the mold. Care must be taken to ensure that the mold is capable of withstanding this extra load. The advantage obtained by pressure-bag molding is the extra consolidation of the lay-up and the decrease in resin content with attendant increase in properties.

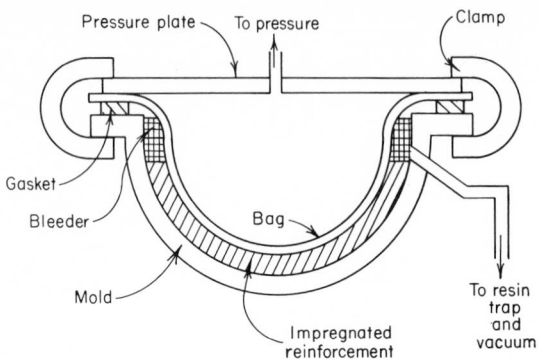

Fig. 54 Pressure-bag molding.

Autoclave molding Autoclave molding is similar to the preceding methods except that the pressure plate is removed and the entire mold and the lay-up are placed into an autoclave. Steam or hot pressurized air up to 100 lb/in.^{2} is admitted, supplying both pressure and heat. If steam is used, the covering bag must have no holes or leaks because steam will enter the lay-up and ruin the molding. The chief advantages of this method are faster cycles and increased properties.

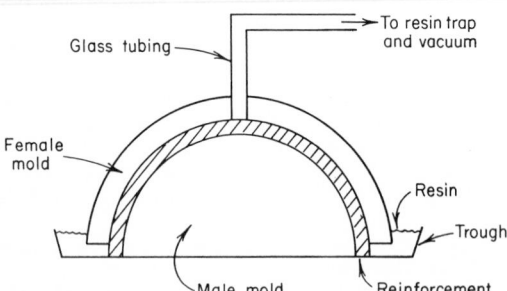

Fig. 55 Vacuum-injection-molding process.

Vacuum-injection molding This method is very useful in making prototype moldings because of its simplicity and low cost. Figure 55 shows the process. Both male and female molds are required, although the female mold may be a bag. The mold surfaces must be very smooth and very well coated with release agents. Dry reinforcements (mat or cloth) are arranged over the male mold to the thickness desired. The female section is carefully—to avoid moving the reinforcements—lowered into place over the male. Catalyzed room-temperature-curing resin (epoxy or polyester) is poured into the trough surrounding the bottom of the male mold, and vacuum is applied. As the air is pumped from the molds, the resin will rise through and saturate the reinforcements. When the resin is seen to reach the glass tubing (which must be at the highest place in the mold) and become bubble-free, the vacuum line is clamped

and the pump turned off. After cure the part is usually stuck to the male mold and may be removed by blowing compressed air between the mold and the part. The advantages for the vacuum-injection process are:
1. Both surfaces are glossy and good in appearance.
2. Little finishing is required.
3. The process is neat and clean.
Disadvantages include:
1. The parts are high in resin content.
2. It is more costly to manufacture two molds.
3. Dry spots can occur and will not be detected until the part is finished.
4. Poor selection and application of mold release may cause sticking.

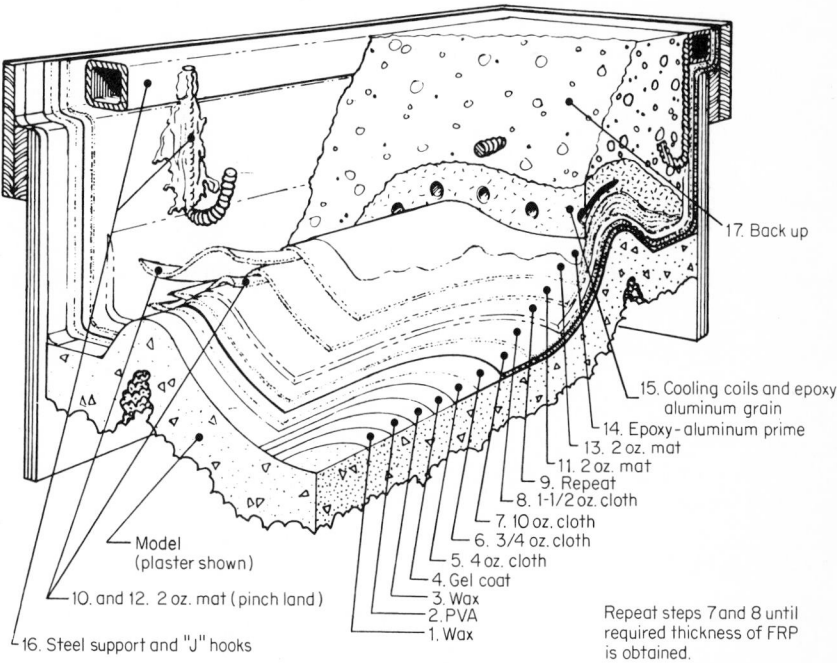

17. Back up
15. Cooling coils and epoxy aluminum grain
14. Epoxy-aluminum prime
13. 2 oz. mat
11. 2 oz. mat
9. Repeat
8. 1-1/2 oz. cloth
7. 10 oz. cloth
6. 3/4 oz. cloth
5. 4 oz. cloth
4. Gel coat
3. Wax
2. PVA
1. Wax

Model (plaster shown)

10. and 12. 2 oz. mat (pinch land)

16. Steel support and "J" hooks

Repeat steps 7 and 8 until required thickness of FRP is obtained.

Fig. 56 Mold for cold-press molding.[32]

Cold-press molding This method is similar in many respects to vacuum-injection molding and is becoming increasingly popular. Cold molding is a method for manufacturing intermediate-volume products (from 200 to 8,000) with low pressures, room-temperature-curing resins, and inexpensive molds. The mold is made from a wood pattern duplicating the desired part. A box is built around the pattern, and guide pins are installed. A gel coat is sprayed over the pattern and backed up with glass-reinforced polyester-resin laminate. Figure 56 shows a typical mold section. Both a top and bottom mold are made and installed in a simple press. Glass-fiber reinforcement (usually mat) is cut to shape and placed in the mold. Resin (either polyester or epoxy) with a room-temperature-curing system is poured into the mold and the mold is closed, distributing the resin throughout the reinforcement. The advantages of this process are:
1. Plastic-mold costs are low compared with machined metal molds and can be fabricated in-house with only a few weeks lead time instead of the months required for metal tools.

2. Uniform parts can be produced more rapidly with minimum supervision using cold molding than with spray-up or hand lay-up, thus accelerating market testing for structural performance and consumer acceptance.

3. Cold-molded parts are more dimensionally controlled than spray-up or hand lay-up parts. This permits the use of more mechanized assembly methods and reduces variation in product cost.

4. The quality of cold-molded parts is comparable with those made with matched metal die molding. Appearance is similar and physical properties approach those of matched die-molded parts at less cost in intermediate volume.

5. Parts made by cold molding have a higher flexural modulus than those made with spray-up or hand lay-up.

6. Molding personnel can be more easily trained for this process than for spray-up or hand lay-up and generally require less supervision to assure quality.

7. Low press costs result from simpler, lighter-duty requirements compared with presses used in matched metal die molding.

Disadvantages include:

1. Mold-preparation cost is moderate.
2. Molds are easily damaged.
3. Molds are difficult to break in.
4. Some scrap loss due to imperfect parts must be expected.

Matched metal die molding   This method is not usually a candidate for prototype manufacture. Instead it represents the ultimate in molding speed and part quality. The cost of the molded part is directly related to the number of parts desired, since expensive molds must be amortized and high-cost molding facilities are used. Molds are generally made of high-quality alloy steels and are machined to exacting tolerances. Aluminum-, cast-iron-, and beryllium-based alloys are sometimes used for special purposes. The molds usually contain hardened cutoff edges which automatically trim the parts, eliminating costly finishing operations. The mold surface is polished to a mirror finish and is usually chrome-plated to aid in part ejection. Figure 57 shows a matched metal die in use.

Filament Winding

Filament winding refers to the process of wrapping resin-impregnated continuous fibers around a mandrel, forming a surface of revolution. When enough layers have been wound, the part is cured and the mandrel is removed. The continuous filaments are generally tapes or rovings of fiber glass, but other reinforcements are used also. Epoxies are popular impregnating resins, although others are used. Typical properties of glass-epoxy filament-wound parts are shown in Table 52. Of special notice are the high tensile and modulus values. Filament-wound parts contain a high percentage of reinforcing fiber and approach the properties of the fiber. The role of the resin is to bond adjacent fibers together and to transmit stress from fiber to fiber. The resin content is determined by the amount of resin on the glass and the winding tension. Table 53 shows the effect of tension on the properties of wound parts, and Table 54 compares filament-wound pressure vessels with metal ones.

In designing filament-wound parts, one attempts to orient the fibers in the direction of stress. If, because of the geometry of the pair, it is impossible to do so to align the fibers, a balanced structure is used wherein all fibers regardless of direction have equal stress. Careful design consideration must be given to holes and flanges to avoid interrupting the continuous nature of the fibers.

Many other variables must be controlled to optimize the structure. These include the winding angles, the relative location of the layers of differing winding angle, the winding tension, the width of the roving bond, the type of fiber, and the number of ends of fiber in the roving. The type of mandrel used contributes to the quality of the product. Most mandrels are machined and polished steel with a slight taper (about $\frac{1}{2}°$). Other mandrels can be made of plaster, salt, low-melting alloys, and plastic. When closed-end structures are wound, water-softening plasters are often used.

Winding machines are of two types—helical and polar. Helical winders consist of a rotating mandrel and a reciprocating roving delivery system. Both motions are controlled to fix the winding pattern and fiber angle. It takes two layers to cover the mandrel with fiber-containing crossovers, except for those structures with a wind angle of 90° (called circumferential or hoop wraps). Helical winding produces open-ended structures such as a cylinder.

To produce closed-end structures like bottles, a polar winder is used. In this machine the mandrel is generally stationary and the delivery arm revolves around it inclined to the proper angle. After each rotation the mandrel is indexed one fiber

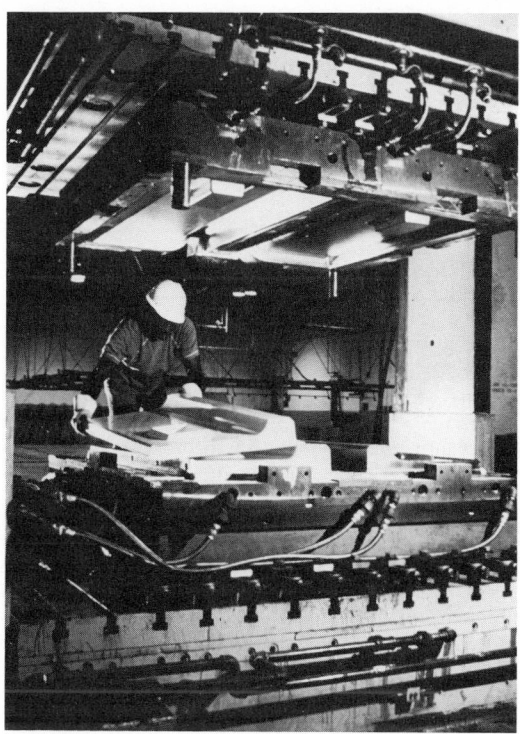

Fig. 57 Matched metal die molding. (*Owens Corning Fiberglas.*)

band width, resulting in a two-layer application without fiber crossovers. Hoop wraps can also be added to the center portion of the structure.

The resin used can be applied at the time of the winding process (wet winding), or prepreg rovings can be used. Such prepreg rovings are available from C glass, E glass, S glass, boron, and graphite in 12, 20, 30, and 60 ends. The resin constitutes 17 to 23 percent by weight of the roving. In most cases the materials are shipped frozen and must be kept refrigerated. Resins available include epoxy, phenolic, diallyl phthalate, and polyester.

Filament-wound tubes used in the electrical industry have been standardized by NEMA in Publication LI-10-1971. Two grades are listed, FW-G-10, which is a general-purpose grade, and FW-G-11, which is more thermally stable, retaining at least 50 percent of its room-temperature axial compression strength measured at 150°C after 1 h at 150°C.

Filament-wound structures are used in rocket-motor cases, pipelines, cherry-picker booms, lightning arresters, chemical-storage tanks, sporting equipment, shotgun barrels, and circuit-breaker parts.

TABLE 52   Properties of Filament-wound Products

Property	Value
Specific gravity...........................	2.0
Density, lb/in.³............................	0.072
Thermal conductivity, Btu/(hr)(ft²)(°F/in.)......	2.2
Thermal expansion, per °F × 10⁻⁶..............	7.0
Specific heat, Btu/(lb)(°F)...................	0.227
Maximum use temperature, °F.................	<400
Hoop tension, psi × 10⁻³:	
Unidirectional windings.....................	230
Helical windings...........................	135
Compressive strength, psi × 10⁻³..............	70
Flexural strength, psi × 10⁻³.................	100
Bearing strength, psi × 10⁻³.................	35
Shear strength, psi × 10⁻³:	
Interlaminar..............................	6
Cross....................................	18
Modulus of elasticity, tension, psi × 10⁻⁶........	6.0
Modulus of rigidity, torsion, psi × 10⁻⁶.........	2
Dielectric strength, step-by-step, volts/mil.......	400

Pultrusion

Pultrusion is a continuous method of making reinforced plastics wherein reinforcing rovings are pulled through a resin bath and into a die where the composite is cured. Since the process is continuous, there is no theoretical limit on the length of the product. Figure 58 shows the process schematically, and Fig. 59 shows the delivery and impregnation sections of a typical pultrusion facility.

The reinforcing fibers used determine the physical properties of a pultruded product. Because of its cost and physical properties, fiber glass is usually used—particularly E glass. With fiber glass the product has highly directional properties. The tensile and flexural strengths in the fiber direction are extremely high, exceeding 100,000 lb/in.², while the properties transverse are much lower. Transverse properties can be enhanced somewhat by incorporating glass mat, glass cloth, or chopped fibers, but with attendant reduction in longitudinal property. For this reason, the direction of prime stress must be carefully considered when one is designing a pultruded part.

TABLE 53   Hoop-Strength Data—Biaxial Test Specimens[33]

Spec. No.	Tension at mandrel, lb/end	Band-width per 30 ends, in.	Density, lb/in.³	Glass content by wt. %	Voids volume, %	Burst pressure, psi	Ultimate hoop stress, psi	Ultimate stress per single glass end, psi × 10⁻⁵	Strength-weight ratio, in. × 10⁻⁶
1A	0.12	0.180	0.0694	73.87	3.25	4,100	126,000	3.09	1.82
1B	0.12	0.180	0.0694	73.87	3.25	4,600	143,000	3.48	2.06
1C	0.12	0.180	0.0694	73.87	3.25	3,800			
Avg.	....	.....	......	......	....	4,350	134,000	3.28	1.92
2A	0.20	0.180	0.0740	79.70	1.92	3,900			
2B	0.20	0.180	0.0740	79.70	1.92	5,200	163,000	3.80	2.21
2C	0.20	0.180	0.0740	79.70	1.92	5,150	161,600	3.76	2.18
Avg.	....	.....	......	......	....	5,175	162,250	3.78	2.19
6A	0.5	0.180	0.0756	80.752	3.31	4,600	155,000	3.24	2.05
6B	0.5	0.180	0.0756	80.752	3.31	4,300	145,000	3.05	1.92
6C	0.5	0.180	0.0756	80.752	3.31	4,200	142,000	2.96	1.88
Avg.	....	.....	......	......	....	4,360	147,000	3.08	1.95

TABLE 54   Properties of Internal-Pressure-Vessel Materials[34]

Material	Ultimate hoop (tensile) strength, psi × 10⁻³	Density, lb/in.³	Specific strength,* in. × 10⁻⁶	Thermal conductivity†	Tensile modulus of elasticity, in. × 10⁻⁶	Compressive strength, psi × 10⁻³
Glass-resin‡ (unidirectional)..	130–170	0.077	1.6–2.1	2.0–5.0	6.0–9.0	70–175
Glass-resin (bidirectional)....	60	0.072	0.7	2.0–5.0	2.0–3.5	40–60
Steel wire–resin (bidirectional)	150	0.166	1.0	. . . .	12	
Titanium (homogeneous)....	50	0.163	0.9	. . . .	16.5	135
Steel (homogeneous)........	280	0.280	0.9	314	30.0	
Aluminum (homogeneous)...	80	0.097	0.8	1,416	10.0	40
Magnesium (homogeneous)..	32	0.064	0.5			

* Specific strength equals ultimate tensile strength/density (approx.).
† Btu/(hr)(ft²)(°F)/in.
‡ Interlaminar shear strength = 6,000 to 8,000 psi (parallel to laminations); 20,000 psi (perpendicular to laminations).
    Axial bearing strength   = 20,000 to 40,000 psi
    Compression strength   = 50,000 to 70,000 psi

Physical properties of a 60 percent by weight fiber glass–polyester pultruded part are shown in Table 55.

Polyesters, acrylics, vinyl esters, and silicones have all been used in pultrusion, but polyesters comprise 90 percent of the usage because of their favorable cost, desirable properties, and easy handling characteristics. Liquid systems with low viscosity (about 2,000 cP) are used to ensure rapid saturation of the fiber bundles. The resin determines the weather resistance, the thermal resistance, the burning characteristics, the moisture sensitivity, the chemical resistance, and some of the electrical properties. The curing rate of the resin in the heated die must be carefully controlled. Most of these resins are exothermic (give off heat while curing), and too rapid curing results in voids and cracks while too slow curing produces a poorly cured product with poor shear and compression strength.

Dies used in pultrusion are usually quite simple and inexpensive. Steel dies up to 6 ft long often are used with production rates of 5 ft/min. Dies are heated electrically or with rf generators up to 10 kW at 100 MHz. In the cases where rf heating is used, the dies are Teflon.

Advantages and disadvantages of pultruded shapes are shown in Table 56. Typical uses of pultruded shapes are fishing rods, ladders, I beams, electrical pole-line hardware, tool handles, and electric-motor wedges. In applications where pultrusions are to be used like cables, special termination devices must be used to avoid interrupting the fiber continuity and to achieve maximum properties.

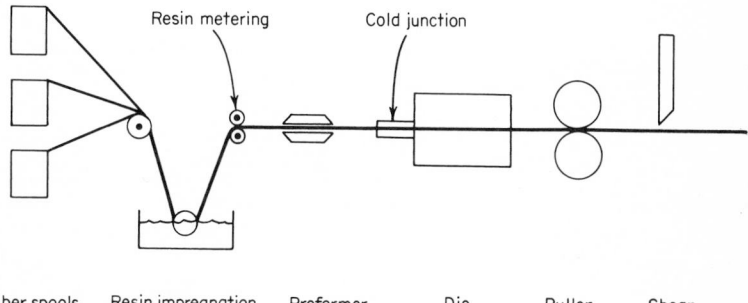

Resin metering   Cold junction

Fiber spools   Resin impregnation   Preformer   Die   Puller   Shear

Fig. 58 Pultrusion process.

Fig. 59   Pultrusion equipment.   (*Westinghouse Electric Corp.*)

SMC and BMC

Sheet molding compound (SMC) and bulk molding compound (BMC) are specially thickened polyester materials with exceptional molding characteristics.   The discovery that magnesium oxide[35] thickens polyester resins, increasing their room-temperature viscosities and improving their molding parameters, permitted a change in the way polyester materials could be prepared and molded.   Since the materials could be made nontacky and easily handled, they could be used with automated equipment and without the housekeeping problems normally associated with polyester materials.   SMC is composed of glass mat impregnated with a thickened polyester resin.   It differs from BMC in that its fiber length exceeds ½ in. and it is made in flat sheets.   For molding, the material is cut into the proper size and placed in a compression mold.   Many complex shapes are possible.   A typical press cycle is 300°F at 800 lb/in.² for 2 min.   Design parameters of SMC are shown in Table 51. The properties of SMC are compared with those of BMC in Table 57.   SMC is used for automotive parts, electrical components, and appliances.

TABLE 55   Typical Properties of Pultruded Products

Property	Test method	Value
Flexural strength	ASTM D 790-63	100,000 lb/in.²
Tensile strength	ASTM D 638-64T	120,000 lb/in.²
Compressive strength	ASTM D 695-63T	40,000 lb/in.²
Impact strength	ASTM D 256-56	40 ft-lb/in. notch
Electric strength:		
Perpendicular	ASTM D 149-64	450 V/mil
Parallel	ASTM D 14-64	60 kV
Arc resistance	ASTM D 495-61	130 s
Water absorption	ASTM D 570-63	0.1%

TABLE 56   Advantages and Disadvantages of Pultrusions

Advantages	Disadvantages
High strength	Unidirectional strengths
Chemical resistance	Difficult to terminate
Unlimited lengths	Low shear strength
Low cost	Limited binder selection
Inexpensive tools and facilities	
Excellent dimensional tolerances	
Smooth—glossy surfaces	
Dimensional stability	

BMC is similar to SMC except that $\frac{1}{2}$-in. chopped fibers are combined with the thickened resins. It is then usually extruded into "ropes" or "logs" for ease of handling. It is molded by compression molding, transfer molding, or injection molding. In the latter two processes, automatic feed equipment is used to charge the press. Press-cycle conditions are similar to those of SMC. BMC is used for low-cost laundry equipment, housings, fan blades, and electrical devices.

Wall and Panel Systems

Plastic materials and laminates lend themselves well to wall or panel structures. The decorative ability, low thermal conductivity, low cost, and strength make them exceedingly popular for both internal and external use. In general, a wall or panel system consists of a plastic skin or sheet supported on a core or frame. Skins may be high-pressure decorative laminates, reinforced plastics, or thermoplastic sheets or films. Cores include honeycomb, gypsum board, asbestos, balsa wood, foams, metal angles, plywood, or particle board. Often the panels are movable and come equipped with suitable hardware to permit rapid, simple installations. Specific examples follow.

Interior panels   These wall and partition systems are usually surfaces with high-pressure decorative laminates, but polyvinyl chloride film and polyester-impregnated paper are used also. Several cores can be specified. For fire-retardant applications, the core is usually cement-asbestos or fire-resistant chipboard. For areas of high humidity, such as bathrooms or kitchens, the core should be polystyrene foam. For general use in hotels, stores, and offices, the most popular core is particle board. The panels are supplied with aluminum extrusion which holds the panels in place at top and bottom. Special moldings fit between panels and in corners. Integrated wiring and telephone systems are available. The cost of a panel system varies from $1.50 to $8.00/ft^2, depending on patterns and sophistication.

TABLE 57   Properties of SMC and BMC[36]

Property	SMC	BMC
% glass fiber by weight	15–30	15–35
Flexural strength, lb/in.^{2} $\times 10^3$	18–30	10–20
Flexural modulus, lb/in.$^2 \times 10^5$	14–20	14–20
Tensile strength at yield, lb/in.$^2 \times 10^3$	8–20	4–10
Tensile modulus, lb/in.$^2 \times 10^5$	16–25	16–25
Compressive strength, lb/in.$^2 \times 10^3$	15–30	20–30
Ultimate tensile elongation, %	0.3–1.5	0.3–0.5
Impact strength, Izod, ft-lb/in. of notch	8–22	2–10
Thermal conductivity Btu/(h)(ft^2)(°F/in.)	1.3–1.7	1.3–1.7
Flammability	Slow to nonburning	Slow to nonburning
Rockwell hardness	H 50–H 112	H 80–H 112
Specific gravity	1.7–2.1	1.8–2.1
Heat distortion, °F at 264 lb/in.^{2}	400–500	400–450

Thermal panels Panels with a thermally insulating core of foam—either polystyrene or polyurethane—are often used for refrigerated buildings. In those cases the exterior side is often plywood, aluminum, or steel, while the inside is often glass-reinforced polyester. Most of the foams used have densities of about 2 lb/ft^3 and a thermal conductivity of about 0.26 Btu/(h) (ft^2) (°F/in.). Such panels or specially molded sections also are used as spandrel and column covers, where they provide decorative appearances and thermal insulation. One supplier furnishes an ABS foam core surfaced with an ABS sheet on each side; the composite can be vacuum-formed into small cabins, trailers, or campers. Certain trailer trucks are molded glass polyester with foam cores.

Aircraft panels Lightweight panels are required for aircraft usage, where they comprise floors, partitions, bulkheads, and doors. The core material most often selected is honeycomb where the light weight and stiffness are required. Face materials can be decorative high-pressure laminates or thermoplastics, but strict fire-retardant requirements exist.

Special masonry effects Glass-reinforced polyester laminates can be molded against appropriate mold surfaces and made to appear like stone or brick walls. Made in thicknesses of about $\frac{1}{32}$ in., the specially pigmented materials are almost identical to real masonry. They are bonded to any suitable core or even bonded directly to studs on inside partitions. Inexpensive, they provide a lightweight structure which saves on supporting members and avoids the high cost of in-place masonry work.

Reinforced-plastic panels These panels also are known as corrugated fiber-glass sheet. They are made in matched metal molds or by hand lay-up on large open molds. Several types are available. The most common are polyester-resin-based materials with about 30 percent by weight of fiber-glass mat. They are available in a wide range of colors and several surface finishes consisting of smooth or satin. Panels come in many sizes up to 4 by 12 ft and in thicknesses from 30 to 125 mils (3 to 12 oz/ft^2). Many styles of corrugations exist. Widely differing shapes, pitches, and depths can be found from one manufacturer to the next. Typical physical properties of corrugated sheets are shown in Table 58.

TABLE 58 Properties of Corrugated Fiber-Glass Sheet

Property	Value
Flexural strength, lb/in.^{2}	30,500
Flexural modulus, $\times 10^{-6}$, lb/in.^{2}	1.6
Tensile strength, lb/in.^{2}	14,000
Tensile modulus, $\times 10^{-6}$, lb/in.^{2}	1.4
Hardness (Barcol)	50
Coefficient of expansion, in./(in.)(°F)	1.8×10^{-5}
Thermal conductivity	1.1 Btu/(h)(ft^2)(°F/in.)
Flame spread*	35
Specific gravity	1.4
Water absorption (24 h), %	0.2
Light transmission,† %	95
Solar transmission,† %	85

* Fire-resistant Grades only—others are slow-burning.

† Clear grade—Light transmission varies with color and may be as low as 4%. Solar transmission may be as low as 15%.

Several special grades of corrugated sheet can be selected. For outdoor use with maximum weather protection one can select panels covered with a thin film—2 to 5 mils of polyvinyl fluoride (Tedlar*) or acrylic. A different type of sheet with excellent weather protection contains an all-acrylic binder resin; others contain acrylic-modified polyester. Some suppliers offer guarantees up to 10 years on these products. One grade of sheeting known as "security panel" is a flat sheet of glass polyester with

* Trademark of E. I. du Pont de Nemours & Company, Inc., Wilmington, Del.

expanded steel in the center. Giving the appearance of old-fashioned beaded glass, it provides maximum protection against unauthorized entry or vandalism. In yet a different approach to this problem, steel hardware cloth is sandwiched inside transparent polyvinyl chloride sheets, producing a laminate with excellent chemical and weather resistance.

To specify a corrugated product, one must consider the weather resistance needed, the stiffness required, the nature of the loads to be applied, and the decorative effect desired. One should ensure that the corrugations are uniform in pitch and depth to avoid leaks between panels. Thickness must also be constant, because thin sections may break and flashing may not fit. Soft materials indicate inadequate cure with premature discoloration sure to occur.

Corrugated sheets are used as patio roofs, garage doors, greenhouses, factory glazing, wind screens, skylights, and area dividers. Standards and approvals include U.S. Department of Commerce Standard 214-57, ASTM D 1919-65T, Military and Federal Specification L-P-505b GSA-FSS, and Underwriter's Guide 4OU8.16.18.

COMPOSITES

Definition

A composite is a homogeneous material created by the combination of two or more materials to obtain specific characteristics or properties. More specifically, it is the union of a reinforcing element and a compatible resin binder. Both reinforced plastics and laminates by this definition are composites, but the general usage in the aerospace industry has focused the term composite on those high-performance constructions obtained from high strength–high modulus fibers and specialty resin systems.

High-Performance Fibers

The need for lighter weight, higher temperature resistance, and greater stiffness has compelled the aerospace industry to develop new fibrous reinforcements to replace fiber glass. These fibers have been made from both conventional metal alloys and specially prepared materials. Table 59 lists new reinforcements evaluated in composites. These reinforcements can be used in filament winding, but they are generally combined into unidirectional tapes or sheets in prepreg form and then laminated into sheets or formed structures. The fibers are oriented to optimize the strength in the direction of greatest stress. The cost of these structures remains much higher than fiber-glass composites, but price reductions of the fibers have continued.

Boron composites Boron fibers are made in continuous form by decomposing boron trichloride and depositing the elemental boron onto thin (0.0005-in.) tungsten wire. The final diameter of the wire is 0.004 in. Boron filaments have a modulus approaching 60×10^6 lb/in.^{2} as shown in Table 60. Boron filaments are supplied as a preimpregnated tape with either an epoxy or polyimide binder at 50 percent fiber volume. The tapes are $\frac{1}{8}$-in.-wide single ply or 3-in.-wide single ply combined with fiber-glass scrim cloth. Boron fibers are easily oxidized above 600°F and must be well coated with the binder or with other compounds like silicon carbide.

Obtaining good adhesion of the resin to boron fibers is difficult and has been solved by acid etching the fiber surface. Physical properties of an epoxy-boron laminate related to etching with nitric acid are shown in Table 61. Boron fibers combined with polyimides also are available where the superior thermal capability is needed. Table 62 lists properties of this type of laminate. Boron composites have been used primarily by aircraft designers where the stiffness and low specific gravity are needed. Such structures as vertical stabilizers, flaps, helicopter blades, and drive shafts are in use.

Graphite composites Carbon and graphite fabrics have been available for many years based on graphitized rayon cloth. Only recently, however, has it become possible to control the graphitization of organic fibers and induce sufficient orientation to result in greatly improved strengths and stiffnesses. Pioneered by the Royal Aircraft Establishment in Great Britain, filaments of graphite now are available with strengths approaching 400,000 lb/in.^{2} and moduli up to 60×10^6 lb/in.^{2}. The graphite fibers result from the pyrolytic degradation of synthetic organic fibers (precursor)—either

TABLE 59 Properties of Fibrous and Wire Reinforcements[37]

Fiber/wire	Density, lb/in.³	Specific gravity	Melting point, °F	Tensile strength		Modulus of elasticity	
				Ultimate, lb/in.² ×10⁻³	Ratio to density, ×10⁻⁶	Lb/in.² ×10⁻⁶	Ratio to density, ×10⁻⁶
Aluminum............	0.097	2.70	1220	90	0.9	10.6	110
Aluminum oxide........	0.144	3.97	3780	100	0.7	76	530
Aluminum silica........	0.140	3.90	3300	600	4.3	15	110
Asbestos............	0.090	2.50	2770	200	2.2	25	280
Beryllium............	0.067	1.84	2343	250	3.7	44	660
Beryllium carbide.......	0.088	2.44	3800	150	1.7	45	510
Beryllium oxide........	0.109	3.03	4650	75	0.7	51	470
Boron..............	0.093	2.59	3812	500	5.4	60	650
Carbon.............	0.051	1.40	6700	250	4.9	27	530
Glass:							
E glass...........	0.092	2.55	2400	500	5.4	10.5	110
S glass...........	0.090	2.49	3000	700	7.8	12.5	140
Graphite............	0.051	1.40	6600	250	4.9	37	720
Molybdenum..........	0.367	10.20	4730	200	0.5	52	140
Polyamide...........	0.041	1.14	480	120	2.9	0.4	10
Polyester............	0.050	1.40	480	100	2.0	0.6	10
PRD-49-III...........	0.053	1.45	—	400	7.5	19	360
Quartz (fused silica)......	0.079	2.20	3500	1,000	12.7	10	130
Steel..............	0.282	7.87	2920	600	2.1	30	110
Tantalum............	0.598	16.60	5425	90	0.2	28	50
Titanium............	0.170	4.72	3035	280	1.6	16.7	100
Tungsten............	0.695	19.30	6170	620	0.9	58	80
Tungsten monocarbide.....	0.565	15.70	2500	106	0.2	104	200

TABLE 60   Properties of Boron Filaments[2]

Property	Value
Diameter, mils.	4.0
Density, lb/in.^{3}.	0.095
Specific gravity.	2.63
Tensile strength, 1,000 lb/in.^{2}.	450
Tensile modulus, 10^6 lb/in.^{2}.	58
Shear modulus, 10^6 lb/in.^{2}.	24
Coefficient of thermal expansion, 10^{-6} in./(in.)(°F).	2.7

rayon or polyacrylonitrile (PAN). Rayon is in the form of continuous two-ply yarns containing 720 filaments per ply, while the PAN is supplied as untwisted tow of 10,000 filaments. The fibers pass through carefully controlled ovens where they are pyrolized to graphite. The higher the temperature reached, the higher the degree of graphitization and the resulting modulus. However, the strength of the fiber decreases as modulus increases. Hence in graphite fibers (unlike boron fibers) many combinations of strength and modulus can be obtained. The fibers, after leaving the oven, often are coated with either polyvinyl alcohol (PVA) or a resin-compatible size to protect the fiber from damage. The PVA must be removed before the fiber is laminated. Table 63 lists graphite fibers now available. The fibers can be either continuously made or produced in batch fashion in small lengths. Ultimately, as the technology matures, the small fibers will no doubt be discontinued. Epoxy resins often are selected as the binder in graphite composites because of their handling ease and compatibility with the fiber. Table 64 shows strengths of a variety of epoxy-graphite composites. Table 65 lists the properties of several polyimide-graphite laminates. In this table, the polyimide was Du Pont's PI 3301, and PVA on or off refers to the presence or absence of the PVA size on the fibers.

In designing a high-performance composite, one must select between boron or graphite in relation to the requirements of the structure. Table 66 lists those design factors which relate to each type of fiber. Graphite composites have been used in many of the same aircraft and aerospace applications as boron composites. In addition, graphite is beginning to find uses in industrial or commercial products. Archery equipment and tennis rackets have been made of graphite-epoxy and golf-club shafts

TABLE 61   Properties of Epoxy-Boron Laminates as a Function of
Etching Process[38] (In lb/in.^{2} × 10^{-3})

Property	Not etched	Cold-etched	Hot-etched
Fiber content, volume %	65.2	60.1	67.3
Tensile strength.	50.5	37.5	71.7
Tensile standard deviation.	4.4	5.3	3.8
Flexural strength.	73.8	65.7	107.0
Flexural modulus.	24.6	22.8	23.9
Flexural standard deviation.	8.4	9.0	16.8
Compressive strength.	140.0	124.5	148.2
Compressive modulus.	.....	21.9	22.2
Compressive standard deviation.	25.3	11.6	7.0

NOTES:
1. ERLA 0400/CL resin system used throughout.
2. Data reported are averages of eight replicates.
3. Compressive results based on edgewise compression of 0.5 × 2.0-in. specimens; flexural tests based on four-point loading; 32:1 span-depth ratio on 0.5 × 4.0-in. specimens.

are now in production.   Certain parts of looms are now in graphite, and the stiffness has permitted increased loom speeds.

Beryllium-fiber composites   Beryllium fibers also are being used as reinforcements. Beryllium has a high modulus/density ratio and satisfactory tensile properties.   These laminates demonstrate a yielding-type stress-strain diagram with the proportional limit 40 percent of ultimate and a yield at 96 percent of ultimate.   As with other exotic fibers, coupling the resin to the fiber surface is difficult and acid etches seem to be best.   Properties of a beryllium laminate are compared with other fibers in Table 67, with an epoxy resin used as a binder.

TABLE 62   Mechanical Properties of Boron-Polyimide Laminates[39]

Property	Value
Flexural strength, lb/in.²	192,330
Flexural modulus, lb/in.²	23.4×10^6
Compressive strength, lb/in.²	62,464
Compressive modulus, lb/in.²	28.4×10^6
Tensile strength, lb/in.²	123,750
Tensile modulus, lb/in.²	37.4×10^6
Horizontal shear strength, lb/in.²	5,350
Transverse flexural strength, lb/in.²	6,210
Tensile strength, lb/in.², at RT after 100 h at 600°F	114,000
Tensile modulus, lb/in.², at RT after 100 h at 600°F	34.4×10^6
Average boron content, % by vol.	58
Average resin content, % by wt.	13
Average void content, % by vol.	15
Molding pressure, lb/in.²	100

Whisker-reinforced composites   Whiskers are short, single-crystal fibers with extremely high physical properties.   Since they are in reality grown crystals, the cost is very high.   Most common are aluminum oxide (sapphire), boron carbide, silicon carbide, and silicon nitride.   The fibers are available as loose fibers or as papers. Modulus values as high as 50×10^6 lb/in.² have been obtained for epoxy-sapphire whisker laminates.   One use of whiskers is to add small amounts to a resin matrix being used to impregnate some other fiber.   The interlaminar shear strength is greatly improved thereby, making possible more efficient structures.   Table 68 shows the influence of adding silicon carbide to epoxy-boron and other laminates.   It is also possible to grow these crystals onto the surface of other reinforcing fibers, the only limit being that the parent fiber be able to withstand the processing temperatures.

PRD-49 composites   A new fiber thought to be an aromatic polyamide has been introduced by E. I. du Pont de Nemours & Co. and designated PRD-49.*   Three versions exist: PRD-49-I, the first laboratory fiber; PRD-49-III, a yarn for reinforcing plastics; and PRD-49-IV, which is intended for cables and fabrics.   Only the PRD-49-III is discussed here.   This organic fiber has a surprisingly high modulus (19 × 10^6 lb/in.²) and high tensile strength (see Table 59).   The fiber is much more handleable than graphite or boron and can withstand abrasion without fraying.   With a round cross section, the fiber withstands acids, bases, organic solvents, fuels, and lubricants.   The thermal stability is high.   Epoxy resin-coated strands lose only 18 percent in modulus and 25 percent in tensile strength after 1 min at 400°F in air. Maximum temperature is limited to 465°F.   The fiber is self-extinguishing, with an oxygen index of 29.

In unidirectional laminates with epoxy binder, flexural properties differ from those based on boron or graphite in that there is a yield point at 50,000 lb/in.² and a yielding area to 125,000 lb/in.² similar to aluminum.

In shear stress, the fiber shows excellent resistance to water, as shown in Table 69. Impact strengths based on unnotched Charpy are 250 ft-lb/in.², which is 11 times that of graphite composites, 6 times that of boron composites, and about the same as

* Kevlar, Trademark of E. I. du Pont de Nemours & Company, Inc., Wilmington, Del.

TABLE 63 Graphite Fibers

Supplier	Type	Price, $/lb	Precursor	Tensile, lb/in.² × 10³	Modulus, lb/in.² × 10⁶	Length	Process
Hercules	HMS	250	PAN	250–350	50–60	3,000 ft	Continuous
Hercules	HTS	230	PAN	300–350	32–40	3,000 ft	Continuous
Hercules	A	150	PAN	250–300	25–32	3,000 ft	Continuous
Hercules	HMS	100	PAN	250–325	50–60	48 in.	Batch
Hercules	HTS	92	PAN	350–450	35–42	48 in.	Batch
Hercules	A	53	PAN	275–325	28–35	48 in.	Batch
Celanese	Celion GY-70	315	PAN	310	80	1,000 ft	Continuous
Union Carbide	Thornel 50s	300	Rayon	285	57	21,000 ft	Continuous
Union Carbide	Thornel 50	250	Rayon	315	57	21,000 ft	Continuous
Union Carbide	Thornel 75s	350	Rayon	345	79	21,000 ft	Continuous
Union Carbide	Thornel 75	300	Rayon	380	79	21,000 ft	Continuous
Great Lakes Carbon	Fortafil 4T	120	PAN	350	37	800 ft	Continuous
Great Lakes Carbon	Fortafil 5T	120	PAN	400	48	800 ft	Continuous
Great Lakes Carbon	Fortafil 6T	120	PAN	420	59	800 ft	Continuous
Great Lakes Carbon	Fortafil 5Y	275	PAN	250	50	21,000 ft	Continuous
Great Lakes Carbon	Fortafil 3-T (L)	80	PAN	200	28	90 ft	Continuous
Narmco	Modmor I	315	PAN	250	50	1,000 ft	Batch
Narmco	Modmor I	120	PAN	250	60	39 in.	Batch
Narmco	Modmor II	295	PAN	350	35	1000 ft	Batch
Narmco	Modmor II	120	PAN	400	40	39 in.	Batch
Hitco	HMG-50	315	Rayon	250	50	21,000 ft	Continuous

TABLE 64   Strengths of Graphite-Epoxy Laminates

Fiber manufacturer....	Hercules	Hercules	Hercules	Hercules	Hercules	Hercules
Fiber type...........	HMS	HTS	A	HMS	HTS	A
Fiber length.........	3,000 ft	3,000 ft	3,000 ft	48 in.	48 in.	48 in.
Fiber volume, %......	56.5	56.6	55.0	50.5	55.0	55.0
Longitudinal (0°) properties:						
Tensile strength, lb/in.^{2}............	126,300	143,400	208,000	95,800	183,000	*
Tensile modulus, lb/in.^{2} $\times$ 10^6.......	30.0	20.9	27.8	26.4	20.9	*
Flexural strength, lb/in.^{2}............	150,700	193,900	189,000	131,200	230,000	220,000
Flexural modulus, lb/in.^{2} $\times$ 10^6.......	29.5	17.2	15.0	27.1	19.1	19.5
Compressive strength, lb/in.^{2}............	110,000	126,600	185,000	*	125,400	152,000
Compressive modulus, lb/in.^{2} $\times$ 10^6.......	27.7	21.1	*	*	19.5	20.0
Transverse (90°) properties:						
Tensile strength, lb/in.^{2}............	6,000	4,850	*	*	*	*
Tensile modulus, lb/in.^{2} $\times$ 10^6.......	0.84	0.90	*	*	*	*
Interlaminar shear, lb/in.^{2}..............	10,700	12,600	11,300	6,000	13,000	8,500

* Not available.

TABLE 64   Strengths of Graphite-Epoxy Laminates (Continued)

Fiber manufacturer....	Celanese	Union Carbide	Union Carbide	Union Carbide	Union Carbide
Fiber type...........	Celion GY-70	Thornel 50s	Thornel 50	Thornel 75s	Thornel 75
Fiber length.........	1,000 ft	21,000 ft	21,000 ft	21,000 ft	21,000 ft
Fiber volume, %......	60	60	59	57	....
Longitudinal (0°) properties:					
Tensile strength, lb/in.^{2}............	90,000	148,000	142,000	214,000	200,000
Tensile modulus, lb/in.^{2} $\times$ 10^6.......	45	33	25	44	40
Flexural strength, lb/in.^{2}............	....	137,000	*	115,000	*
Flexural modulus, lb/in.^{2} $\times$ 10^6.......	....	32	*	41	*
Compressive strength, lb/in.^{2}............	110,000	100,000	*	97,000	*
Compressive modulus, lb/in.^{2} $\times$ 10^6.......	40	33	*	44	*
Transverse (90°) properties:					
Tensile strength, lb/in.^{2}............	....	4,500	3,000	4,600	*
Tensile modulus, lb/in.^{2} $\times$ 10^6.......	....	1.0	1.13	0.9	*
Interlaminar shear, lb/in.^{2}..............	9,000	8,000	3,500	6,000	9,500

* Not available.

TABLE 64   Strengths of Graphite-Epoxy Laminates (Continued)

Fiber manufacturer....	GLC†	GLC	GLC	GLC	GLC
Fiber type............	Fortafil 4T	Fortafil 5T	Fortafil 6T	Fortafil 5Y	Fortafil 3-T (L)
Fiber length.........	800 ft	800 ft	800 ft	21,000 ft	90 ft
Fiber volume, %......	55	55	55	55	55
Longitudinal (0°) properties:					
Tensile strength, lb/in.²..........	170,000	170,000	150,000	61,000	*
Tensile modulus, lb/in.² × 10⁶......	*	*	32	24.5	*
Flexural strength, lb/in.²..........	200,000	220,000	170,000	125,000	180,000
Flexural modulus, lb/in.² × 10⁶......	20	25	30	24.5	17
Compressive strength, lb/in.²..........	160,000	150,000	120,000	55,000	*
Compressive modulus, lb/in.² × 10⁶......	*	*	*	*	*
Transverse (90°) properties:					
Tensile strength, lb/in.²..........	*	*	*	*	*
Tensile modulus, lb/in.² × 10⁶......	*	*	*	*	*
Interlaminar shear, lb/in.²..........	14,000	11,000	9,000	8,300	11,000

† Great Lakes Carbon Corporation.

TABLE 64   Strengths of Graphite-Epoxy Laminates (Continued)

Fiber manufacturer....	Narmco	Narmco	Narmco	Narmco	Hitco
Fiber type............	Modmor I	Modmor I	Modmor II	Modmor II	HMG-50
Fiber length.........	1,000 ft	39 in.	1,000 ft	39 in.	21,000 ft
Fiber volume, %......	60	53	60	60	55
Longitudinal (0°) properties:					
Tensile strength, lb/in.²..........	120,000	110,000	180,000	165,000	120,000
Tensile modulus, lb/in.² × 10⁶......	28	26.5	22	23	25.5
Flexural strength, lb/in.²..........	120,000	120,000	250,000	255,000	116,000
Flexural modulus, lb/in.² × 10⁶......	25	25	20	20	25
Compressive strength, lb/in.²..........	*	86,000	160,000	145,000	70,000
Compressive modulus, lb/in.² × 10⁶......	*	*	*	*	*
Transverse (90°) properties:					
Tensile strength, lb/in.²..........	7,000	*	8,000	*	*
Tensile modulus, lb/in.² × 10⁶......	*	*	*	*	*
Interlaminar shear, lb/in.²..........	10,000	6,000	16,000	13,000	7,500

* Not available.

S glass composites and aluminum. Fatigue life is better than that of other composites, and PRD-49-III composites can support higher continuous loads than glass composites. Table 70 compares fabric laminates of PRD-49-III and E glass. Dielectric properties are excellent, with a dielectric constant of 3.4 and loss tangent of 0.005 at 9×10^9 H.

The cost of the fiber is high, but PRD-49 should find uses in pressure vessels, missile cases, aircraft panels, rotor blades, and radomes.

Machining and Joining Laminates and Reinforced Plastics

Machining Plastic materials present unique machining requirements because of their basic property differences from wood and metals. The resin systems are all similar in their hardness and machining properties, the chief difference in machinability

TABLE 65 Properties of Polyimide-Graphite Laminates[39]

Material or process variable	Thornel 25 (250 psi, PVA on)	Thornel 40 (250 psi, PVA on)	Thornel 40 (250 psi, PVA off)	Thornel 40 (PVA off, vacuum bag)	Thornel 40 (PVA off, 500 psi)	Thornel 40 (PVA off, 1,000 psi)
Horizontal shear strength, psi:						
At RT.................	1,775	1,350	1,700	1,850	1,875	1,600
At 600°F.............	2,775	1,300	1,950			
At 600°F after 100 hr at 600°F...............	1,225	1,050	1,425			
At 600°F after 250 hr at 600°F...............	1,175	575	475			
At 700°F...............	....	1,175	1,650			
At RT after 2-hr H₂O boil.	....	1,250	1,700			
Flexural strength (psi $\times 10^{-3}$)/ modulus (psi $\times 10^{-6}$):						
At RT.................	35.3/8.5	56.3/18.1	59.0/16.4	63.5/18.7	62.5/16.0	59.8/16.5
At 600°F after 100 hr at 600°F...............	19.0/6.5	31.0/16.8	30.2/12.7	26.6/13.3	30.5/13.9	27.0/13.6
At 600°F after 250 hr at 600°F...............	10.9/5.1	....	17.6/10.1			
Tensile modulus at RT, psi $\times 10^{-6}$..................	....	....	20.5	22.0	19.4	17.9
Physical properties:						
Resin content, % vol......	38	36	32	27	33	33
Fiber content, % vol......	52	59	57	63	57	58
Void content, % vol......	10	6	11	10	10	9
Specific gravity..........	1.24	1.42	1.34	1.35	1.33	1.36

being caused by the type of reinforcement. Reinforcements can be listed in order of increasing difficulty of machining as:

Paper
Fabrics
Asbestos
Glass

The cellulose-paper- and fabric-based materials machine much like wood. Glass- and asbestos-reinforced laminates should be treated like hard brass. The glass-based materials, and laminates with certain mineral fillers, tend to wear tools rapidly because of the abrasiveness of the fibers. Thus, if glass-based laminates are to be machined, hardened tools should be used.

The differences in machinability of laminates are caused by several property peculiarities:

1. The materials are in laminar form. Laminates can be considered to be sheets of reinforcement glued together. The bond strength of the glue (resin) is much lower than the strength of the reinforcements. Hence, machining forces tend to separate (delaminate) the plies of the reinforcements. For example, holes should be drilled perpendicular to the plane of the reinforcements and not parallel. If a hole is to be drilled parallel, clamps and slow speeds are required.

TABLE 66 Fiber Selection for High-Performance Composites[40]

Boron composites		Carbon composites	
Design requirement	Property	Design requirement	Property
High-compression loading	High compressive strength per unit density (high specific compressive strength)	Thin-wall construction under low compressive loading	High specific (per unit density) modulus provides high buckling stability
Large sheet goods; little or no machining	Large-diameter fiber; collimates well; forms straight, high-quality sheet	Structures with cutouts, holes, or tapers	Easily machined by conventional metal-removal techniques
Combination of high stiffness and high strength	High specific (per unit) density, strength and modulus in same fiber	Sandwich structures; thin skins	Individual ply thickness can be varied from 0.001 to 0.010 in.; ability to produce specific gages in laminate
Minimum thermal distortion when combined with metals	Thermal coefficient of expansion for boron closer to metals (such as titanium) than carbon	Material must be formed around tight radius	Small fiber diameter permits small radius of curvature without incurring high stresses in the fiber
		Exact modulus, or more than one modulus required in the same structure	Available in a range of moduli from 2 to 80 million $lb/in.^2$

5-101

TABLE 67   Properties of Beryllium-Wire Laminate* Compared With Other Fibers[37] (Epoxy Resin Used as Binder†)

Property	Composite fiber/wire					
	Beryllium	Boron	Graphite	Carbon	S-glass‡	None
Density, lb/in.^{3}........	0.058	0.074	0.067	0.054	0.074§	0.043
Specific gravity........	1.61	2.05	1.87	1.50	2.05	1.20
% volume matrix......	35	35	35	35	35	100
Ultimate tensile strength, psi $\times$ 10^{-6}.............	97	320	160	80	260	12
Ratio to density, psi $\times$ 10^{-6}........	1.67	4.32	2.39	1.48	3.53	20.8
Tensile modulus of elasticity, psi $\times$ 10^{-6}..	28.0	36.0	20.0	4.3	7.6	0.5
Ratio to density, psi $\times$ 10^{-6}........	482	486	299	80	103	11

* Unidirectional orientation.

† Epoxy resin: tensile strength 12,000; tensile modulus of elasticity 0.5 $\times$ 10^6 psi; elongation 4.47 %; compressive strength 18,000 psi; compressive modulus of elasticity 0.5 $\times$ 10^6 psi; deformation 6 %.

‡ S-glass epoxy composite: tensile elongation 3.2 %, compressive strength 175,000 psi, compressive modulus of elasticity 7.5 $\times$ 10^6 psi; interlaminar shear 11,000 psi; flexural strength 190,000 psi; Poisson's ratio 0.2.

§ 20 wt. %.

2. The brittleness of the resins is severe. All resins have similar moduli. They are considered to be brittle materials with low elongations. Hence machining with dull tools causes chipping and gouging. This condition is at its worst in glass-reinforced materials, since the glass fibers also are brittle. Glass laminates require hard, sharp tools and carefully controlled cutting speeds.

3. Resilience of laminates is much greater than that of metals. As plastics are machined, they yield away from the tool only to spring back when the deforming forces are removed. Dull tools emphasize the condition. Tools require greater clearance and less rake to minimize the condition. All holes and machined areas will end out smaller that the tool which formed them. Hence tools must be oversized to obtain correct dimensions.

4. The thermal conductivity of plastics is much lower than that of metals. If forced hard during cutting, both laminate and tools may become hot, with decomposition of

TABLE 68   Addition of SiC to Resin Matrices[41]

Reinforcing fibers	Matrix	Whisker content of matrix, vol. %	Interlaminar shear strength, psi $\times$ 10^{-3}	As % of control	Flexural modulus of matrix, psi $\times$ 10^{-6}
4-mil boron....	Epoxy resin	0	9.27		0.41
4-mil boron....	Epoxy resin	4½	11.97	129	0.72
6-mil steel.....	Epoxy resin	0	6.5		0.41
		4½	10.4	160	0.72
6-mil steel.....	Phenolic resin	0	4.10		0.75
		2½	4.56	111	1.1
		4½	5.23	128	1.3

TABLE 69   PRD-49/Epoxy Interlaminar Shear Strength[42]

Epoxy resin (3-day water boil) ERLA 4617/MPDA

	Fiber vol. %	Interlaminar shear strength		
		Before, lb/in.^{2}	After, lb/in.^{2}	% change
PRD-49-III.....................	60	9,000	8,000	11
Graphite No. 2..................	48	9,600	5,300	44
Graphite No. 1..................	60	3,000	1,300	57
E glass........................	60	9,200	5,500	36
S glass........................	60	11,800	8,100	31
Boron.........................	60	13,300	9,400	29

the material and dulling of the tool as a result. Carbide tips are superior because of their heat resistance. Coolants—either air or liquids—should be used wherever possible. If coolants are not available, the cutting should be interrupted periodically to permit natural cooling.

5. The densities and hence the weights of plastics are lower than those of metals. This means that larger parts can be machined on the equipment, and higher speeds are often possible.

6. Tolerances are different. In general, the tolerance obtainable in metals cannot be achieved in plastics. Mold stresses, water absorption, and thermal effects prevent the precision obtained with metal parts.

The following sections describe particular machining processes and the best methods of machining.

Shearing. All laminates may be sheared, the limit being—as in metals—the thickness of the part. Paper-based laminates up to $\frac{1}{16}$ in. and fabric-based materials up to $\frac{1}{8}$ in. may be sheared on conventional guillotine shears. For thicknesses up to $\frac{1}{4}$ in., rotary shears should be used. Some grades must be heated to 250°F to be sheared, but dimensional stability then may introduce tolerance problems. For glass-based materials, shearing is not possible over $\frac{1}{16}$ in., and thicker laminates should be sawed.

Sawing. Paper- and fabric-based grades can be cut with ease on conventional band saws having teeth with 4 to 10 points per inch. The speeds can reach 9,000 ft/min. Circular saws should have hollow-ground no-set teeth and can be operated up to 13,000 surface ft/min.

Glass-based laminates require hardened blades. A regular high-speed-steel saw blade will be useless after 3 min use. Carboloy*-tipped teeth should be used on saw

* Trademark of General Electric Co., Schenectady, N.Y.

TABLE 70   PRD-49 Fabric Laminate Performance[42]

$\frac{1}{8}$-in. panels, 181 style cloth, autoclave molded epoxy-resin matrix

	E glass	PRD-49-III
Laminate density, lb/in.^{3}...........................	0.065	0.048
Modulus (tension-compression) (10^6 lb/in.^{2}):		
RT dry...	3.6	4.5
RT wet...	3.2	4.0
300°F ($\frac{1}{2}$ h)..	3.4	4.6
Strength (tension) (10^3 lb/in.^{2}):		
RT dry...	63	75
RT wet...	55	73
300°F ($\frac{1}{2}$ h)..	50	51
Strength (compression) (10^3 lb/in.^{2}):		
RT dry...	62	37
300°F ($\frac{1}{2}$ h)..	61	14

blades. These should have 10 to 12 teeth per inch, and speeds should be limited to about 2,000 ft/min. Cutoff wheels are useful but should be equipped with diamond blades. Air or water coolants should be used. If air is used, provisions should be made to collect the dust, which may cause dermatitis.

Drilling. Holes should be drilled perpendicular to laminations. If parallel drilling is required, the work should be clamped securely between plates, and feeds should be slow. Frequent back-out of the drill will keep the hole clean and avoid chipping. Air cooling should be used. Special drills for plastic laminates are available from all suppliers. These have 70 to 90° points, "slow spirals," and a high helix angle. In all cases, holes will decrease in size after the tool is removed. Drills should be 0.002 to 0.004 in. oversized, and tolerances of ±0.001 in. can be obtained.

Paper and fabric laminates can be drilled with high-speed drills; surface speeds of 350 ft/min at 10,000 r/min can be maintained.

Glass laminates should be drilled with polished and chrome-plated tungsten-carbide drills at speeds up to 250 ft/min. Feed rates should approach 0.003 in./revolution. Diamond drills can be run at high speeds up to 35,000 r/min.

All laminates can be countersunk or counterbored. Speeds one-half those of drilling should be used. Tolerances can be held to ±0.005 in.

Punching. Laminates can be punched with standard presses and dies. Because of the resilience of the materials, all holes should be punched 4 percent oversize. The dies must be equipped with a spring-loaded stripper plate to avoid lifting the punched part on the pins and to minimize the delamination forces. The stripper plate should have 0.001-in. clearance from the die. All hole spacing should be at least 1.5 times the laminate thickness. Die clearances should be one-half those for steel. Best results have been found with Carpenter No. 610 steel hardened to Rockwell C 60. Minimum hole size should be 75 percent of laminate thickness. Laminates up to $\frac{3}{32}$ in. can be punched with sharp, clean holes. Holes in thicker laminates get rough and chipped. The number of holes capable of being punched in a laminate with a given press is a function of the shear area and the stripper-plate spring force. Holes should be no smaller in diameter than the laminate thickness and no larger than three times the laminate thickness. All grades can be cold-punched, but maximum punchable thickness is related to both resin type and reinforcements. The following thicknesses are the maximum for cold punching:

Paper-based laminates...................	$\frac{1}{16}$ in.
Fabric-based laminates..................	$\frac{1}{8}$ in.
Glass-polyester laminates................	$\frac{5}{32}$ in.
Glass-silicone laminates.................	$\frac{1}{4}$ in.

Heating the laminates makes punching easier. Most laminates can be heated to 150°F, but some grades require 250°F. Best punching of glass-based grades results when carbide tools are used. Quality criteria for punched phenolic laminates are presented in ASTM D 617-44.

One is often faced with the decision as to whether to punch or drill a laminate. The correct economic answer is a function of the number of units to be punched, the dimensions of laminate and holes, the shape of the part, and the facilities available. Table 71 lists the factors to be considered in making a correct decision.

Tapping and Threading. Maximum accuracy for threads is obtained with a 75 percent thread and a class 2 fit. The tools should be made oversized to allow for the resilience of the laminates. Tapholes need to be deep enough for at least three threads, and the edges should be chamfered to avoid chipping of the walls. Tools should have a 5 to 10° negative rake on front of lands.

The taps for paper- and fabric-based grades can be made of high-speed plated steel with four flutes and should be about 0.002 in. oversize. Taps for glass-based laminates should be nitrided and plated with two or three flutes. The same rake should be used, and they should be 0.002 to 0.008 in. oversize. Speeds of no more than 80 ft/min are used.

Dies for paper- and fabric-based materials should have a chamfer of about 33° on the lead and a negative rake of 10°. Cutting speeds can approach 400 ft/min. External threads on glass-based laminates are not recommended.

Lathe Work. Laminates cut much like brass. Critical to success are highly sharpened tools and air cooling. When cutting laminates, never stop the tool, or a mark will remain on the surface of the work. Excessive pressure or dull tools may cause delaminations to occur.

Paper- and fabric-based laminates can be machined with high-speed-steel tools with a 10° negative rake and about a 35° clearance. Rough cuts ⅛ in. deep can be made at surface speeds up to 500 ft/min.

Glass-based laminates should be cut with tools made of carbide or diamond. The rake should be 10° negative, and a 30° clearance is needed. Surface speeds should be less than 400 ft/min, and feeds of 0.010 in./revolution can be maintained. Dust must be collected to avoid dermatitis problems.

TABLE 71　Punching and Drilling Laminates

Factor	Punching	Drilling
Machine tool cost...................	High	Low
Die or jig cost.....................	High	Low
Labor cost.........................	Low	High
Production rate....................	High	Low
Maintenance needed................	Little	Much
Tool life..........................	Long	Short
Size of hole.......................	Limited	Not limited
Ability to work curved surfaces......	No	Yes
Irregular-shaped holes..............	Yes	No

Milling. Laminates must be handled carefully on a milling machine, since milling-tool forces act to delaminate the work. Tools should be of carbide, should be operated at surface speeds of less than 1,000 ft/min, and should have a negative rake. Feed speeds of about 30 in./min are common. The materials cut like brass.

Joining

Laminates can be joined with any of the conventional joining methods such as rivets, bolts, and screws. One must always remember their tendency to delaminate and their anisotropic nature.

Rivets Riveting practices should be those normally used. Riveted joints should be designed to act in shear rather than tension if possible. If tension loads must be used, they should amount to no more than one-half the shear loads defined for the rivets. Rivet holes should be large enough to permit the rivets to be slid in easily. Remember in drilling the holes to allow for the resiliency of the plastic materials. Holes that are too large will permit the rivets to buckle or bend, thereby reducing the rivet strength. Holes that are too small will cause damage to the hole walls as the rivet is driven. In general, the rivet hole should be about 1/64 in. larger than the rivet diameter. Rivet holes should be at least 4½ rivet diameters from the edge of the laminate. The rivet diameter should be larger than the laminate thickness. Best results are found with type 56S or 2S aluminum rivets with truss heads. These larger heads distribute the loads on the plastic surface better than the normal buttonheaded rivets. The length of the rivets should be equal to the thickness of the parts being joined plus one rivet diameter. It is better to have a rivet too long than too short.

In addition to the standard truss-headed rivets described, other special types are useful in joining plastic laminates. These include blind rivets, "Pop" rivets, drive rivets, and plastic rivets.

Bolts Care must be exercised in bolting laminates so that the laminate is not crushed or placed under such high loads that creep will cause failure. As in riveting, the bolt hole should be 1/64 in. larger than the bolt diameter, and the bolt diameter should be larger than the laminate thickness. Always use washers or load strips for the head and nut to bear against. All plastic materials are notch-sensitive. Glass-mat laminates, in particular, exhibit stress sensitivity at holes, and hence bolts should be located at least 5 times the bolt diameter in from the side of the laminates. Fabric and paper laminates can have bolts located only 4½ times the bolt diameter in from the side.

Screws Screws work about as well in plastic laminates as in wood. Tapping of holes for screws was described in a previous section. Screw joints possess excellent resistance to axial loads (loads in the direction of the screw axis) and lateral loads (loads perpendicular to the direcion of the screw axis). These forces are related to both the type of resin and the type of reinforcement. Tables 72 to 74 give the holding forces for polyester laminates based on several types of glass reinforcements.

TABLE 72   Holding Forces of Screws in Glass-reinforced Plastics[33]

Fastener size threads	Plastic material polyester—181 cloth, axial holding force				Type of fastener machine screw, lateral holding force			
	Minimum		Maximum		Minimum		Maximum	
	DoP, 16th in.	Force, lb	DoP, 16th in.	Force, lb	DoP, 16th in.	Force, lb	DoP, 16th in.	Force, lb
4–40	2	130	5	500	1	210	2	350
6–32	2	170	6	820	1	250	3	700
8–32	3	320	7	1,140	2	440	3	810
10–32	3	420	8	1,500	2	850	4	1,200
¼–20	4	700	10	2,800	2	1,200	5	1,900
$\frac{5}{16}$–18	4	820	12	4,200	3	2,100	6	2,800
⅜–16	5	1,160	14	6,000	4	3,300	7	3,900
$\frac{7}{16}$–14	5	1,300	16	8,000	4	3,800	8	5,200
½–13	6	1,660	17	9,800	5	5,300	10	7,500
$\frac{9}{16}$–12	6	1,820	18	12,000	5	5,800	11	9,000
⅝–11	7	2,200	20	15,300	6	7,300	13	11,900
¾–10	7	2,550	22	20,600	7	8,900	15	15,000

Plastic mechanical properties; psi:
Tensile strength: 30,000–45,000.
Edge compressive strength: 18,000–27,000.
Shear strength (Johnson): 14,000–17,000.

Self-tapping screws are becoming increasingly popular. Their strengths approach those of screws in tapped holes, and they eliminate one machining step. Frequently used with adhesives, self-tapping screws are used to join either two laminates together or one laminate to some other material. Literally dozens of sheet-metal designs can be chosen for use with laminates, and some screws have been developed specifically for laminates. These screws are of two types: thread forming (in which the material is forced aside) and thread cutting (in which the material is cut or machined). Thread-forming screws are available in hardened alloy steel and stainless. They can be inserted into punched or drilled holes and resist tension and shear forces well. Tables 75 and 76 present pullout strengths for type Z screws in cloth and mat laminates.

Thread-cutting screws cut a thread that is similar to a class 2 tolerance. They are available in coarse and fine threads, both of which have similar holding power, but the fine thread causes less delamination. Tables 77 and 78 present pullout strengths for type F screws in cloth and mat laminates.

Since the strength of a screw is proportional to the shear area and hence the screw diameter, fasteners of increased diameters represented by inserts are useful. They are invaluable when a screw is to be removed frequently, because threads in a laminate do not wear nearly so well as threads in a metal. These inserts can be placed in a molded laminate or, where design permits, molded in place. The inserts are self-tapping or can be pushed into a properly sized hole. Bonding of inserts with appropriate adhesives is possible also.

Adhesive Bonding

One widely accepted method of fastening laminates together is adhesive bonding. Adhesive bonding presents several distinct advantages to the designer. One of the most important is distribution of stresses. Plastic laminates are notch-sensitive, and

TABLE 73   Holding Forces of Screws in Glass-reinforced Plastics[33]

Fastener size threads	Plastic material polyester—1000 cloth, axial holding force				Type of fastener machine screw, lateral holding force			
	Minimum		Maximum		Minimum		Maximum	
	DoP, 16th in.	Force, lb	DoP, 16th in.	Force, lb	DoP, 16th in.	Force, lb	DoP, 16th in.	Force, lb
4–40	1	90	3	500	1	180	2	350
6–32	1	110	4	700	1	220	2	450
8–32	2	230	5	1,000	1	280	3	880
10–32	2	260	6	1,500	2	750	3	1,100
1/4–20	2	320	8	2,700	2	1,000	4	1,800
5/16–18	3	520	10	4,000	2	1,350	6	2,000
3/8–16	3	620	11	4,850	3	2,600	7	3,300
7/16–14	4	950	13	6,800	3	3,000	9	4,500
1/2–13	4	1,100	15	7,720	4	4,500	10	6,000
9/16–12	5	1,600	16	9,000	5	6,000	12	9,000
5/8–11	5	1,750	18	11,600	5	6,500	13	10,000
3/4–10	6	2,500	22	24,000	6	8,000	16	15,000

Plastic mechanical properties, psi:
  Tensile strength: 30,000–45,000.
  Edge compressive strength: 18,000–27,000.
  Shear strength (Johnson): 14,000–17,000.

therefore conventional mechanical fasteners cause stress risers. Adhesive bonding provides a large area over which the stresses are distributed. Other advantages and disadvantages are given in Table 79.

The veracity of an adhesive joint is a function of the adhesive, the quality of the joint, joint design, and the environment. Before an adhesive is selected, the joint should be carefully designed. Any joint design should take into consideration the following factors:

1. Joints can be loaded in shear, tension, or compression, but most loads are combinations of these.

2. The joint should keep the stresses on the bond line to a minimum.

3. Rigid adhesives should be used for shear loads and flexible adhesives for peel loads.

4. Tensile loads should be minimized.

5. As large a bond area as possible should be used.

TABLE 74   Holding Forces in Glass-reinforced Plastics[31]

	Plastic material polyester— 1½-oz mat, axial holding force				Type of fastener machine screw, lateral holding force			
	Minimum		Maximum		Minimum		Maximum	
Fastener size threads	DoP, 16th in.	Force, lb	DoP, 16th in.	Force, lb	DoP, 16th in.	Force, lb	DoP, 16th in.	Force, lb
4–40	2	40	5	450	1	150	2	290
6–32	2	60	6	600	1	180	2	380
8–32	2	100	7	1,150	1	220	3	750
10–32	2	150	8	1,500	2	560	4	1,350
¼–20	3	300	10	2,300	3	1,300	5	1,900
⁵⁄₁₆–18	3	400	12	3,600	3	1,600	7	2,900
⅜–16	4	530	14	5,000	4	2,600	10	4,000
⁷⁄₁₆–14	4	580	16	6,500	5	3,800	12	5,000
½–13	4	620	18	8,300	6	5,500	14	6,000
⁹⁄₁₆–12	4	650	20	10,000	7	6,500	15	8,000
⅝–11	4	680	22	12,000	7	6,800	16	11,000
¾–10	4	700	24	13,500	7	7,000	17	17,000

Plastic mechanical properties, psi:
Tensile strength: 6,000–25,000.
Edge compressive strength: 10,000–22,000.
Shear strength (Johnson): 10,000–13,000.

6. The adherends should be stressed in the direction of greatest strength.   In a laminate this is parallel to the laminations.

The simplest joints to make are butt joints.   Butt joints cannot withstand bending loads, and thick bond lines induce internal stresses which result in poor performance. The most popular modification of joint design is the lap joint.   Tensile loading causes the lap joint to be stressed in shear.   This induces an unbalance of strain in the adhesive and adherends, and hence a bending of the joint.   These factors result in stresses at the ends of the lap joint substantially greater than the stresses elsewhere.

TABLE 75   Pullout Strength of Type Z Thread-forming Screws[33]
Style 1000 cloth—66.5 wt. % glass

	Fastener size threads				
Property	4–24	6–20	8–18	14–14	⅜–12
Initial penetration, 16th in..................	2	4	4	5	4
Pullout strength at initial penetration, lb.....	70	200	120	300	100
Pullout strength per ¹⁄₁₆ in., lb..............	150	165	170	300	410
Range, 16th in............................	2–6	4–9	4–15	5–18	2–16
Approximate break load of fastener, lb.......	650	1,100	2,000	4,300	

TABLE 76   Pullout Strength of Type Z Thread-forming Screws[33]

$1\frac{1}{2}$-oz mat—37 wt. % glass

Property	Fastener size threads				
	4–24	6–20	8–18	14–14	$\frac{3}{8}$–12
Initial penetration, 16th in.................	2	4	4	5	4
Pullout strength at initial penetration, lb.....	110	110	180	100	200
Pullout strength per $\frac{1}{16}$ in., lb..............	115	85	145	265	250
Range, 16th in...........................	4–6	4–10	4–14	5–20	4–8

TABLE 77   Pullout Strength of Type F Thread-cutting Screws[33]

$1\frac{1}{2}$-oz mat—37 wt. % glass

Property	Fastener size threads			
	4–40	6–32	8–32	$\frac{1}{4}$–20
Initial penetration, 16th in..............	2	4	4	5
Pullout strength at initial penetration, lb..	10	130	360	400
Pullout strength per $\frac{1}{16}$ in., lb..........	150	90	230	300
Range, 16th in.......................	2–6	4–12	4–12	5–17

TABLE 78   Pullout Strength of Type F Thread-cutting Screws[33]

Style 1000 cloth—66.5 wt. % glass

Property	Fastener size threads			
	4–40	6–32	8–32	$\frac{1}{4}$–20
Initial penetration, 16th in.............	2	4	4	5
Pullout strength at initial penetration, lb..	70	250	400	350
Pullout strength per $\frac{1}{16}$ in., lb..........	130	185	195	365
Range, 16th in.......................	2–6	4–9	4–12	5–12
Approximate break load of fastener, lb...	650	1,300	2,200	4,000

TABLE 79   Advantages and Disadvantages of Adhesive Bonding of Laminates

Advantages	Disadvantages
1. Large stress-bearing area.	1. Surfaces must be cleaned.
2. Excellent fatigue strength.	2. Inspection of joint is difficult.
3. Adhesive tends to damp vibration.	3. Jigs and fixtures may be needed.
4. Adhesive may absorb shock.	4. Long cure times may be needed.
5. Eliminates galvanic corrosion when metals used.	5. Special facilities required.
6. Joins all thicknesses.	
7. Appearance improved—smooth contours.	
8. Seals joint.	
9. May be less expensive.	

Breaking strength of lap joints increases linearly with overlap length up to some limiting value. Strengths of lap joints are proportional to overlap area only at constant lap length. Modifications of joint design which result in stronger joints include:
1. Reducing joint bending by stiffening the adherends.
2. Making the adherends thinner.
3. Using a tougher adhesive.
4. Using a thinner bond line.
5. Decreasing the overlap length.
Other joint designs useful in bonding reinforced plastics are defined in Chap. 10.

REFERENCES

1. "*Modern Plastics* Encyclopedia," vol. 48, no. 10A, McGraw-Hill, New York, 1972.
2. "Plastics for Aerospace Vehicles," Part 1, Reinforced Plastics, MIL-HDBK-17A, Department of Defense, January 1971.
3. "Industrial Woven Glass Fabrics and Tapes," Hess Goldsmith & Co.
4. Tish, H. A.: Mechanical Properties of Rigid Plastics at Low Temperature, *Modern Plastics*, vol. 33, no. 11, 1955.
5. Brenner, W., et al.: "High Temperature Plastics," p. 60, Reinhold, New York, 1962. Reprinted by permission of Van Nostrand Reinhold Co.
6. Laurie, W. A.: New Concepts for Thermosets, *Reg. Tech. Conf. Plastics in Electrical Insulation*, Society of Plastics Engineers, 1964.
7. Landrock, A. H.: Properties of Plastics and Related Materials at Cryogenic Temperatures, *Plastic Rept.*, no. 20, Picatinny Arsenal, 1965.
8. Kueser, P. E., et al.: Electric Conductor and Electrical Insulation Materials, *Topical Rept., NASA Rept.* CR-54092, October 1964.
9. Raech, H., and J. M. Kreinik: Prepreg Materials for High Performance Dielectric Applications, *20th Ann. Tech. Conf. Reinforced Plastic Division*, Society of Plastics Industry, 1965.
10. "A Guide to Corrosion Control with Atlac 382," Atlas Chemical Industries, 1967.
11. McAbee, E., and M. Chmura: Effect of Rate and Temperature on the Tensile Properties of 181 Glass Cloth Reinforced Polyester, *21st Ann. Tech. Conf.*, Society of Plastics Engineers, 1965.
12. Herberg, W. F., and E. C. Elliott: Reinforced Silicone Plastics Retain Physical and Electrical Properties with Little Regard to Temperature or Frequency, *Plast. Des. and Processing*, vol. 6, no. 1, 1966.
13. Korb, L. L.: Arc and Track Resistance Tests, *Reg. Tech. Conf. Plastics in the Electrical Industry*, Society of Plastics Engineers, 1964.
14. Hexcel Trevarno, Inc., *Data Bull.* 13b, 1966.
15. Pride, R. A., et al.: Mechanical Properties of Polyimide Resin/Glass Fiber Laminates for Various Time, Temperature and Pressure Exposures, *23d Ann. Tech. Conf., Reinforced Plastics/Composites Division*, Society of Plastics Industry, 1968.
16. Freeman, J. H., et al.: New Organic Resins for High Temperature Reinforced Laminates, *Soc. Plast. Eng. Trans.*, vol. 2, no. 3, 1962.
17. Pyralin Data Sheet, E. I. du Pont de Nemours & Co., Inc., 1970.
18. Wicker, G. L.: A New Concept in Reinforced Thermoplastic Sheet, *20th Ann. Tech. Conf. Reinforced Plastics Division*, Society of Plastics Industry, 1965.

19. "Azdel Product Sheet," GRTL Company.
20. "Industrial Laminated Thermosetting Products," Part 20, NEMA Standards Publication, LI-1-1971.
21. Harper, C. A.: "Plastics for Electronics," Kiver Publications, Chicago, 1964.
22. "*Modern Plastics* Encyclopedia," p. 546, McGraw-Hill, New York, 1967.
23. "*Modern Plastics* Encyclopedia," p. 547, McGraw-Hill, New York, 1967.
24. "*Modern Plastics* Encyclopedia," p. 548, McGraw-Hill, New York, 1967.
25. "General Principles upon Which Temperature Limits Are Based in the Rating of Electrical Equipment," IEEE Publication 1, Institute of Electrical and Electronics Engineers, 1962.
26. "High Temperature Properties of Industrial Thermosetting Laminates," NEMA Standards Publication LI-3-1961, 1961.
27. Skow, N. A.: Performance Properties of Industrial Laminates at High Temperature, *1st Nat. Conf. Appl. Elec. Insul.*, Institute of Electrical and Electronics Engineers, 1958.
28. "Vulcanized Fibre," NEMA Standards Publication VU-1-1971.
29. Copper-clad Thermosetting Laminates for Printed Wiring, ASTM D 1867-68.
30. "Industrial Laminated Thermosetting Products," Part 10, Copper-clad Laminates, NEMA Standards Publication LI-1-1971.
31. "Industrial Laminated Thermosetting Products," Part 11, Thin Copper-clad Laminates, NEMA Standards Publication LI-1-1971.
32. "Fiberglas-reinforced Plastic Cold Press Molding," Owens/Corning Fiberglas Corporation Manual 5-PL-5361.
33. Oleesky, S. S., and J. G. Mohr: "Handbook of Reinforced Plastics of the SPI," Litton Educational Publishing, New York (reprinted by permission of Van Nostrand Reinhold, New York, 1964).
34. Rosato, D. V., and C. S. Grove, Jr.: "Filament Winding," Interscience, New York, 1968.
35. U.S. Patent 2,628,209.
36. Keown, J. A.: Fiberglas Plastic Design Guide, *26th Ann. Tech. Conf. Reinforced Plastic/Composites Div.*, Society of Plastics Industry, 1971.
37. "*Modern Plastics* Encyclopedia," McGraw-Hill, New York, 1968.
38. Jaffe, E. H.: Boron Fibers in Composites, *21st Ann. Tech. Conf. Reinforced Plastics Div.*, Society of Plastics Industry, 1966.
39. Copeland, R. I., et al.: Polyimides—Reinforced and Unreinforced, *23d Ann. Tech. Conf., Reinforced Plastics/Composites Div.*, Society of Plastics Industry, 1968.
40. Berg, K. R., and F. J. Filippi: Advanced Fiber-Resin Composites, *Machine Design*, Apr. 1, 1971.
41. Economy, J., and L. C. Woher: The Potential of Whisker Reinforced Plastics, *Plast. Des. and Processing*, vol. 7, no. 7, 1967.
42. Moore, J. W.: PRD-49, A New Organic High Modulus Reinforcing Fiber, *27th Ann. Tech. Conf., Reinforced Plastics/Composites Div.*, Society of Plastics Industry, 1972.

Chapter **6**

Properties and End Uses for Man-Made Fibers

TED V. SHUMEYKO
Shumeyko Wilson Associates, Inc.
New York, New York

RICHARD G. MANSFIELD
United Merchants and Manufacturers, Inc.,
New York, New York

INTRODUCTION

The Forms of Fibers

Textile fibers are the building blocks of yarns and fabrics. Traditionally, the term fiber referred to a discontinuous structure whose length was several hundred times its diameter and which possessed a balance of functional and aesthetic properties sufficient to suit it to yarn manufacture, fabric formation, dyeing, mechanical and chemical finishing, and end-use service.

Today, these parameters of processing and use are still largely valid but the term fiber is no longer limited by the concept of discontinuity; it is used to describe any material of natural or man-made origin which can be used to manufacture textile-mill products. And the advances made in fabric-forming technology and end-use development have broadened and made obsolete many of the property criteria formerly required to qualify a material as a textile fiber.

The physical forms which the textile fibers of today take fall into three broad categories: staple, tow, and continuous-filament. The significant usage of fibers in all three of these forms is attributable, in large measure, to the efforts of man-made-fiber producers.

Staple Among the natural fibers, most have been in discontinuous form, which, as noted, was part of the definition of the textile fiber. The term staple is synonymous with this type fiber, and in fibers of natural origin is commonly used to preface the mean fiber length as well.

In man-made-fiber terminology, the term staple refers to that discontinuous form which has been purposefully manufactured during continuous extrusion of fibers via cutting to controlled lengths. This is one of a multiplicity of controls which separate the processing and use-function aesthetics of man-made fibers from those of the natural fibers. Together they provide the textile technologist with raw materials which more closely resemble engineering materials than any of the natural fibers and have changed the economic profile of the textile industry.

For the most part, natural and man-made-fiber staple is the raw material of the spinning process whose output is a yarn. After one or more of several possible aftertreatments such as lubrication, twisting, and dyeing, such spun yarns become raw materials for fabric-forming processes or for handcrafts.

Tow Again the history of natural-fiber processing has furnished a term commonly used in man-made-fiber technology, but in the case of tow, the meaning has been altered considerably.

Originally, bast fibers such as flax were buffeted to separate relatively short fibers from long ones. The dividing criterion was 10 in., and these "short" fibers were known as tow.

In man-made-fiber manufacturing, tow is a ropelike strand of continuously extruded fibers in which the fibers are generally parallel but the individual fibers or fiber aggregate exhibit no twist on themselves or among each other. Tow is the raw material for production of staple, flock, and most importantly, yarns which resemble those conventionally spun. The tow to yarn process is peculiar to man-made-fiber technology and will be described in a following section.

Filament In common textile usage, the term filament is an abbreviation for continuous-filament yarn. In nature, the two spinnerets in the head of the silkworm, *Bombyx mori,* are the principal exemplification of a mechanism to generate filaments continuously. And when the cocoon produced by the silkworm from silk, so extruded, is manually joined with others and unwound, a continuous-filament textile yarn is produced.

Man has simulated this natural extrusion-through-spinneret process to form the interface between chemistry and textile-fiber manufacture. When the man-made spinneret has one orifice, a monofilament yarn is necessarily extruded through it. Multifilament yarns are formed when more than one filament, controlled by the number of orifices in the spinneret, is extruded and gathered into a smooth parallel aggregate of sufficient diameter to suit it to fabric-forming processes. While tow is technically a continuous-filament structure, it is generally much greater in diameter than those termed yarns. As noted, tow is a raw material for yarn production.

The design of the spinneret used to form continuous-filament yarn is a sophisticated, often proprietary science. Aside from controlling the number of filaments in the cross section of this fiber form, it determines the cross section of the filament itself. While most commonly round, man-made-fiber producers have developed many filamental-cross-section variations such as trilobal or pentalobal, which, when combined with the amount of "producer twist" imparted to the yarn and the way in which this is accomplished, have profound influence on the physical performance of filament yarns in fabric formation and performance aesthetics in end use.

Aside from manufacturing the filament form via extrusion, another technique employed to a lesser extent involves the slitting of films extruded or cast in wide-roll form. The slitting, performed on the entire roll width simultaneously, can produce a ribbonlike continuous product narrow enough to be usable for selected textile-fabric-forming modes. This technique generates a fiber form which is classifiable as monofilament.

Processing of Fibers

The use of the staple, tow, or filament fiber forms as raw materials to produce a yarn or a yarn modification is a vital link between fiber and fabric. Generally, staple and tow are processed by the textile-mill-products industry, filament by the man-made-fiber producer.

Historically, the fiber producer completed the processing of filament yarns, except for the use of silk-system "throwsters." Their main function was to add additional twist so that filaments broken during fabric formation were minimized and the textile mill could, in part, vary the surface texture, feel (hand), and drape of the fabrics it made. But with the advent of the texturing or bulking process after World War II, the throwster changed his role to that of texturizer. With the rapid growth of importance of textured filament yarns commencing in the mid 1960s, the fiber producer turned his eye to texturing as an extension of the filament-yarn-manufacturing process. The significance of the texturing process for modifying filament yarns lies in achieving more spun-like qualities (bulk, warmth, coverage, and a pleasing sensation next to the skin) in filament yarns at more filament-yarn-like economics and availability (rate of production). And using certain thermoplastic man-made generic fibers, the texturing mode of processing can impart the property of "comfort" or "action" stretch, sometimes true elasticity.

Common to all fiber processing is the potential of coloration. While most fibers are formed into fabric in their "natural," "gray," "grey," or "greige," state and colored "in the piece" by dyeing or printing in roll form, "stock dyeing" and yarn dyeing are commonly performed by commission dyers or integrated textile mills in staple (stock) or filament-yarn form. Certain producers of man-made fibers precolor staple, tow, and filament yarns by adding durable pigments to fiber-forming substances before they are extruded through the spinneret. Filament yarns so colored are referred to as "solution-dyed" or "dope-dyed."

Staple to yarn   This mode of spinning has the purpose of cleaning, paralleling, and reducing the size of product through a variety of mechanical processes from baled staple to finished yarn. Fiber-to-fiber cohesion and "workability" during the processing are accomplished by the inherent cross-sectional configuration of the fiber, control of atmospheric conditions in spinning departments, and addition of lubrication during or before the start of the process and twist during its latter phases.

Again, most spinning systems have evolved from a technological base designed for natural fibers. Thus, man-made fibers and blends made with other man-made fibers or natural fibers are commonly spun on the cotton system, the wool system, or the worsted system or modifications of them. When spinning predominantly man-made-fiber yarns, modifications usually are designed to preclude the need to remove trash and other foreign matter from the raw staple (stock) and to remove short fiber lengths. In natural-fiber spinning, combed cotton is produced on a system designed to remove the short fibers. Run-of-the-mill (carded) cotton yarns are produced from shorter fibers; worsted yarn is produced on a system designed to remove short fibers from woolen yarns. The shorter wool fibers can be spun on the woolen system. Since man-made fibers are delivered devoid of foreign matter at a controlled distribution level of staple length, those processes designed to bring about such control in natural-fiber spinning will not be considered here.

The cotton system of spinning is by far the most common. Mixing, blending, and opening (loosening and separation of fibers) are the first functions in the process flow and are performed continuously; the three functions are combined into a continuous operation, generally referred to simply as *opening*. The input to opening consists of layers of fiber from the bale; the output is a loose, separated mixture or blend of fibers. Opening inputs to *picking*, in which the fibers are further opened and formed into a flat, thick batting of randomly oriented fibers known as a *lap*. The picker lap inputs to the *card*, which further opens and separates fibers, forming them into a thin web of partially oriented fibers which is condensed into a ropelike strand known as *sliver*. The sliver inputs to *drawing* (the draw frame), which serves to "average out" sliver weight uniformity via multiple sliver end feeding and drawing down or drafting, and to continue the function of paralleling fiber which persists through all remaining stages of spinning. Sliver output from drawing is input to the *roving* or *slubbing* operation(s). The purpose of this operation is to draw the sliver down (or draft it) to a soft, relatively small diameter product form known as roving which contains a low number of turns per inch and which is wound up on a bobbin. Roving output is input to the spinning frame, which dramatically reduces the diameter of the roving, adds considerable twist (turns per inch), and winds it on a bobbin. This output is called *yarn;* if desired, it may be combined with one or more other similar or dissimilar yarns to form a plied yarn in an operation known as *twisting*. The purpose of twisting is to increase diameter or strength or to develop other mechanical properties. While machine development has increased the efficiency of drafting in roving and spinning, past picking, cotton-system spinning is still a discontinuous process. There is one exception to this statement; i.e., one continuous, automated bale-to-yarn spinning system is operational in this country, but its details have not been disclosed.

The woolen system of spinning is functionally similar to the cotton system. The first step in the process, which mixes, blends, and opens fiber, is known as *picking*. Picked stock is metered to a system of carding which outputs a sliver, if the process is to proceed to the worsted system, or to a system of cards which produces a roving-like product if the process is to remain on the woolen system. The woolen-system card output is spun to yarn much like cotton roving.

The worsted-system card output is doubled, drafted, and paralleled in an indigenous process known as *gilling*. This process is analogous to cotton-system drawing. If wool fiber has been used, short fibers are removed from sliver output of gill boxes via combing. In man-made-fiber processing, this is not necessary—indeed the gill box has been replaced by the modern pin drafter, which inputs card sliver and outputs roving ready for the spinning frame. Functionally, this modernized, man-made, fiber-oriented system may be compared with the efficient cotton-system roving and spinning operations which have been developed. The terms "long draft" and "American system" are frequently used in connection with these developments.

While unification of spun-yarn numbering systems, indeed all yarn-formation systems, has met with success overseas, domestic practice is to designate the three systems of spun yarns according to weight/length ratios established for the natural fibers. These so-called "direct" yarn-designation systems are based on the relationship: count (yarn number or size) $\times$ standard = yards per pound, where the standard for the cotton system is 840, for wool 1,600 (run system) or 300 (cut system), and for worsted 560. As adoption of the metric system evolves in this country, the relationship should change to: count $\times$ 1,000 = meters per kilogram.

Tow to yarn   This system of fiber processing is more compatible with man-made fibers and the industries which produce and process them. Tow is processed to achieve blending and the aesthetics of spun yarns and at the same time bypass nonconventional or modified conventional natural-fiber-spinning systems. This is accomplished by random breakage of crimped or uncrimped continuous filaments in the tow to produce a sliver. The sliver is then drafted into roving and spun via conventional spinning techniques. The so-called "Perlock system" is usually used to break tow as described. The Pacific converter accomplishes the same purpose

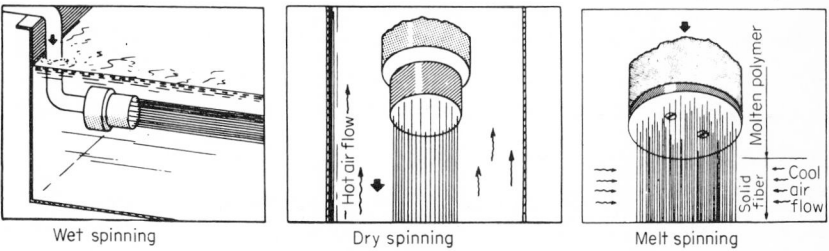

Fig. 1   The three methods of extrusion.[1]

by cutting rather than breaking. The Saco-Lowell system, known as *a direct spinning system,* inputs lightweight tow and outputs yarn with no need for conventional roving and spinning; yarns so produced are characteristically stronger, lower in elongation, and higher in shrinkage in temperature than other tow-to-yarn systems and are used in special applications.

Filament   Continuous-filament yarns are applicable to fabric formation by the textile mill with the least amount of processing. As filaments are produced via extrusion of a fiber-forming substance through spinnerets, they are oriented and wound up with zero twist in one operation. When additional twist is required of the fiber producer, a separate operation is used to insert about three turns per inch. One process, Du Pont's Rotoset, precludes addition of twist by entangling filaments during filament-yarn spinning.

Because much of this chapter will deal with the development, properties, and usage of the major generic man-made fibers, it is fitting at this point to differentiate among the three methods of extruding a fiber-forming substance: wet, dry, and melt spinning. This may be simply done on a graphic basis by reference to Fig. 1.

Similarly, it is useful to differentiate among three so-called "two-constituent" fiber structures in which two chemically and physically different fiber-forming substances can cohere during continuous-filament-yarn spinning. Known as the bicomponent, skin-core, and matrix-fibril systems, this may also be simply done on a graphic basis by reference to Fig. 2.

Finally, the numbering system used for any man-made fiber in its extruded form should be described. It is an indirect system and is based on the original silk system of yarn numbering, i.e., the so-called "denier" system. It may best be understood by expression of the (arbitrary) relationship:

$$\frac{4,464,528}{\text{denier}} = \text{yards per pound}$$

While tow and staple forms of man-made fibers often are assigned direct numbering systems, i.e., when at some stage after delivery they are spun by the fiber producer on the cotton, wool, or worsted systems, expression of denier is used at the stage of extrusion to define the extruded aggregate and denier per filament. Denier per filament is a specification of man-made fibers which is preserved in all manufacturing stages subsequent to extrusion. It is common to express filament yarns' fiber-breaking strength in terms of tenacity or grams per denier.

Texturing The economics, aesthetics, function, and processability balance potential of bulking continuous-filament thermoplastic man-made fiber have been realized; and textured polyester, nylon, and acetate are in volume usage in fabric-formation systems, notably knitting.

The first textured-yarn system was devised by Heberlein & Co. A.G. Known as Helanca, it yields a true processed stretch yarn and is based on multistage twisting, heat setting, and untwisting—usually of a two-ply continuous-filament yarn.

The development of the so-called "false-twist" method of texturing followed the same principle but was performed in one operation via a unique series of patent developments. As commercialization of this kind of texturing developed, more and

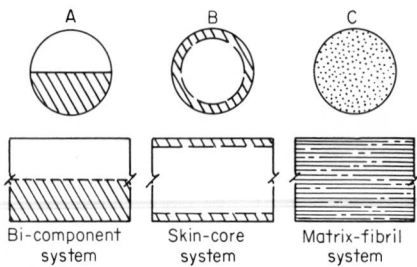

Fig. 2 Two-constituent fiber systems.

more emphasis was placed on development of bulk, less on stretch. These modified stretch yarns were produced by fiber selection, reducing the plies from two to one, and "setting" the yarn after texturing in an autoclave or by a heating stage incorporated in the final stage of the false-twist machine.

These twist-bulking techniques are generally broken down into an output nomenclature according to the amount of stretch which can and is allowed to remain in the yarn: stretch yarns are termed "torque bulk yarns" or "torque yarns."

Other basic texturing methods which produce a nontorque bulked or bulked and stretch yarn include Deering Milliken's Agilon (edge-crimped), Du Pont's Taslan (air-jet-bulked), and Joseph Bancroft's Ban-Lon (applied only to fabrics made from this Textralizing method based on stuffer tube bulking).

While generally not considered a method applicable to filament yarn texturing, texture can be developed in man-made fibers in staple, tow, or filament bicomponent extruded form. When the two polymeric components in the fiber shrink to different degrees, bulk and stretch properties may be developed when such fibers are subjected to dry or wet heat. This technique is used largely in developing bulk in yarns made from acrylic staple and tow.

Fabric Formation

Because of the development of methods which produce textile cloth by unconventional means, there is no up-to-date accepted definition for the term fabric. Description is usually based on the fabric-forming mode. But it may be advanced that a fabric is a form of nondurable soft goods made from textile fibers or yarns which possess suitable aesthetics, function, and economics for its use to manufacture articles of apparel, home furnishings, and recreational goods; and for its use as interior trim in transport vehicles and as parts and components in surgical and health-care procedures and parts, and in manufacturing of other nondurable and durable industrial and consumer products.

Weaving This oldest of fabric-forming techniques may be defined as interlacing two systems of threads running at right angles to each other. The machine used to produce woven fabrics is known classically as the *loom,* and while modifications have been made which suggest that they now be called weaving machines, the basic principles are essentially the same as those used in biblical times. The parallel system of threads or ends wound on a spool-like beam and running lengthwise in the loom is called the *warp.* The *warp* is prepared from yarns in a separate operation (certain looms are equipped to input warps directly from a creel at the loom). The system of threads or filling which interlace at right angles to the warp is known as *filling*—individual threads are known as picks. In the classic fly-shuttle loom, filling is wound onto a bobbin or quill, placed in a shuttle, and propelled across the loom negatively (the shuttle is struck but not controlled in flight), leaving a "trail" of filling or one pick. The order in which warp ends are raised for each pick so as to define a "repeat" of a given weave form and the order in which picks are inserted are known as *weave formation.* There are three orders of weaves: plain, twill, and satin—variations on them, their combination, their use in conjunction with variations and combination of warp and filling yarns (fiber content, yarn count, textural configuration, and the method by which they were processed mechanically and chemically, including coloration and coloration potential) define the economic/aesthetic/functional suitability of the fabric so woven for a given end use.

The most significant technological advancement in loom design is the shuttleless loom, which has, after some 25 years of applications development among textile mills which weave broad goods (more than 18 in. wide), found widespread acceptance. The concept replaces the quill and the shuttle with a weft-insertion device such as a dart or rapier which carries the filling across the loom or on a jet of air or water. These developments reduce noise, vibration, and power requirements of weaving machines.

Most fabrics formed by the weaving mode or any other to be summarized here are not suited for use by the customer of the textile mill directly after formation. Common to most modes is, at least, a separate operation of inspection, sometimes mending of defects, and usually rerolling into a presentable, shippable, packaged-to-protect form.

But an entire science of wet and mechanical processing to develop coloration (dyeing), design (printing), and such function and aesthetics as "hand" or feel, raised surface effects, water repellency, soil release, and flame retardance (mechanical and chemical finishing) exists. It is beyond the scope of this summary to detail the science, but certain brief observations should be advanced concerning coloration, since the history and current development of man-made fibers are intimately bound up with the class of dyestuffs usable on man-made generics and variants thereof.

Fabrics may be yarn-dyed (known as ingrain in knitting) or piece-dyed. In the former case, fibers or yarns are precolored by the producer or are dyed in yarn form prior to fabric formation. Color needs of the market can be more flexibly met by storing fabrics in the "greige" and dyeing them only when those needs are commercially defined.

Dyestuffs used for natural and man-made fibers may be classified according to their dyeing mechanism or their chemical structure. Kaswell[1] suggests that the former classification is the more important. Following is an abstract of this classification as it applies to man-made fibers and as it interfaces with the classification according to chemical structure.

1. *Physical Dyes:* Direct or substantive water-soluble dyes are selectively absorbed by fibers via a reversible physical chemical reaction (chemical classes—direct dyes for cellulosics; basic dyes for acrylics).

2. *Chemical Dyes:* Preferential absorption, probable reaction with fiber (chemical classes—acid dyes for polyamides; reactive dyes for cellulosics).

3. *Insoluble Dyes:* Synthesization of color within the fiber (chemical classes—vat dyes for cellulosics; sulfur dyes for cellulosics; acetate or dispersed dyes for acetate, polyamide, and polyester; pigment color with binders or resins for all fibers).

4. *Mordant Dyes:* Have no affinity for the fiber but can be anchored to it by a third substance which has affinity for both fiber and dyestuff (chemical classes—metallized for polyamides; basic for cellulosics and polyamides).

It should be stressed that the dyeing of textile fibers in fiber, yarn, or piece form is often looked upon as an art as well as a science, and development of the correct shade to various criteria of color fastness depends on control of many things other than mere dyestuff-class selection. Among these factors are time, temperature, concentration, pH, and addition of auxiliary dyeing chemicals.

Knitting   Knitting is a mode of fabric formation which depends on the formation of a series of loops from textile yarns which hang on and support one another. There are two useful means (among many) of classifying knitting machines and the fabrics which they produce: by the type of needle employed, i.e., latch, spring, and —more recently—compound; and by the configuration of the "hardware" which houses the needles. The following outline summarizes a combination of the two methods.

1. *Large-diameter Circular Machines:* Needles are housed around the circumference of a cylinder with a diameter in excess of 9 in. Camming mechanisms (or other modes) have the capability of raising the needle or fabric looped on it. When an end of yarn is served to the rotating or stationary cylinder at a preselected point on the cylinder (a feed or feeder) from a stationary or rotating yarn rack above the cylinder or from a side creel, the first step in loop formation results: the yarn is held in the hook of the needle. When the needle is lowered, the loop forms, and loops on adjacent needles define a characteristic serpentine path. When this happens, the latch (of a latch needle) closes or the beard (of a spring needle) is pressed closed and the "old" loop retained on the needle rides over it, casting it off. Series of loops formed in this manner in the vertical direction are called *wales*, in the horizontal direction *courses*. The number of feeders and the speed of the machine define production capability, the specifications of the yarns used and the patterning mechanism (selective controller of needle motion) define pattern, and the type of needle used defines fabric fineness capability. Usually, large-diameter circular knitting machines which employ spring needles make fine fabrics in terms of needle per inch capability, and those which employ latch needles make relatively coarse fabrics. They are both loosely defined as single knits.

When two systems of needles are contained in the same machine at right angles to each other, they may readily knit the so-called "rib structure," and when they are capable of forming interactive stitches, a double cloth may be knitted. When the second set of needles, contained in a dial, is in alignment with those in the cylinder, the so-called "interlock construction" is formed. When cylinder and dial needles are offset, the important double-jersey (double-knit) construction is formed. Again, the parameters of feeds, speeds, yarn input, and patterning mechanism define output appearance, function, and economics. All these double cloths are knitted with latch needles; interlock machines are most commonly used to produce stretch fabrics from torque-textured nylon, and double-jersey machines are most commonly used to produce "action stretch" fabrics from set false-twist textured polyester. The latter concept has accounted for the rapid rise of the use of textured polyester knits in apparel. Common to all double-cloth knitting is firmness, bulk with fine-stitch definition, and dimensional stability (except those fabrics made with stretch yarns). With proper yarn and stitch selection, these double cloths are inherently easy-care, with the most marketable feature being "packability."

2. *Small-diameter Circular Machines:* With few exceptions, machines of this class are of the latch-needle type and are used to make so-called "seamless" hosiery for men, women, and children. Patterning capability and production efficiency have been developed to a fine degree in these machines, but they still form a stitch in

the same basic manner as in large-diameter knitting. Both large- and small-diameter circular knitting are frequently referred to as "weft knitting."

Both large- and small-diameter circular or weft-knitting machines may be used to knit "yard goods," i.e., roll goods delivered to customers who cut and sew the cloth into a consumer product. Both may be used to partially "fashion" or form a product for apparel or industrial end use. In large-diameter work, an example is the sweater strip; in small-diameter work, an example is hosiery per se but particularly pantyhose. In all such cases product-configuration knitting with minimum afterprocessing has led to verticalization among knitting mills possessing this hardware: they receive yarn and deliver packaged consumer products. In the case of pantyhose, the concept, executed usually with torque-textured nylon so as to produce supportive, snug, "one-size-fits-all" garments, has led to worldwide acceptance and manufacturing.

3. *Warp Knitting:* Often misnamed flat knitting, warp knitting inputs, as the name implies, a warp prepared as in weaving but to more critical specifications. Each end in the warp is served to one or more needle bars (a horizontal housing for latch, spring, or compound needles arranged side by side in a straight line) via one or more guide bars (a horizontal housing for eyelet-shaped guides arranged side by side in a straight line). The number of bars of each type used, the manner in which patterning mechanisms permit the guide bar to interact with reciprocating needles, the pattern selected, and the yarns used define the economics, function, and aesthetics of machine output. But basically, stitches are formed by drawing loop series through newly formed loop series as in all knitting. In circular knitting, it is generally required to slit the knitted tube after knitting to form yard goods. In warp knitting this is not necessary.

The type of needle used in warp-knitting machines plays a large part in defining the machine subclass and the fabrics which it can make: latch-needle machines are known as "Raschel" and are used for a variety of fabrics including lacelike lingerie, outerwear trimming fabrics, men's and women's outerwear, drapery, and upholstery; spring-needle machines are known as "tricot" and are generally used for lingerie fabrics and substrates for bonded or laminated fabrics; and compound-needle machines are usually used for apparel outerwear fabrics.

Because of some rapid machine-development effort, these classifications are no longer strictly valid: Rockwell International Corp. has developed a machine which is essentially Raschel in terms of output capability but which uses spring needles, and several weft-insertion machines of both the latch- and spring-needle type have been commercialized. Most of the development on existing and new machines has been directed at broadening the end-use bases of machine output.

4. *Flat-Bed Knitting:* As the name implies, fabrics are produced on this machine class in a single (horizontal) plane. Yarn(s) fed to the machines from creels reciprocate when they reach each of the two extremities of the needle bed.

When spring needles are employed and when the machine is equipped with mechanisms to add to or decrease the number of needles knitting in the bed according to a predetermined pattern, the machine is known as a "Cotton's patent" or a "full-fashioned" machine. Before the advent of torque-textured stretch yarns, these machines were used almost exclusively for the manufacture of ladies' full-length hosiery. Today, they are used essentially for production of full-fashioned men's, women's, and children's sweaters.

When latch needles are employed, they are usually arranged in two beds at right angles to each other and are used to produce coarse accessories such as scarves. Because these V-bed machines are capable of readily producing plain and patterned rib stitches and because these structures are desirable for sweaters, V-bed machines are finding increasing acceptance for production of cloth for this end use.

Tufting This form of multiple-needle sewing has had profound influence on the way in which many fabrics for the home, notably floor coverings (carpets and rugs), are made. In the process, numerous rapidly moving needles positioned side by side in the tufting machine stitch the face or pile yarns through a base-fabric input to the machine. Usually woven, these "primary-backing" fabrics serve as the first anchoring mechanism for the pile. In a separate operation, the U's of the tufted face yarns are secured to the primary backing by latex coating its reverse side. A more open, less costly fabric is then often adhered to this anchored reverse side. This additional reinforcement is known as secondary backing.

Needle punching   While animal-hair fibers have for many years been "felted" in web form to produce textile-mill products, the peculiar chemical nature and scalelike surface structure of fibers such as wool have not been duplicated in man-made fibers. Thus the simplified fabric-forming mode of developing felt by subjecting these fibers to heat, moisture, and mechanical action has fallen into relative disuse. In its place a process known as needle punching has been developed for man-made fibers. In the process, a web made from shrinkable man-made fibers is subjected to insertion of numerous reciprocating barbed needles. Unlike tufting, the needles carry no stitching thread or yarns; rather the barbs themselves engage and entangle the fibers in the web within the web. After being subjected to wet or dry heat, the entangled web shrinks to form a dense feltlike structure. This technique is used to make filter fabrics, papermaker felts, and inexpensive carpeting.

Nonwovens   Technically, stitch bonding produces a form of nonwoven or "non-knit" textile. But the technology for forming fabrics without using yarns or conventional machines to manipulate them has been developed far beyond this technique.

Basically, a nonwoven fabric is produced by bonding webs of man-made fibers with chemical adhesives or by taking advantage of the thermoplastic nature of selected man-made fibers to effect self-bonding. The principal fiber used, however, is a regenerated cellulosic—rayon—which is not thermoplastic and which requires application of a bonding resin and usually aftercuring of the resin.

There are two general basic approaches to the manufacture of nonwovens: textile industry and paper industry. The textile approach depends on forming webs in one of two ways:

1. Use of conventional cotton- or wool-system cards to produce the web, and then impregnation with resin and curing in a separate operation.

2. Use of a cardlike machine to produce the web, collecting it (doffing it) via an air system. The fibers in webs so produced are characterized by their random orientation. Random nonwoven substrates were first produced by a machine known as a Curlator, and this hardware is still in active use.

In either case, tensile and other mechanical properties of nonwoven fabrics may be manipulated by assembling multiple web layers at right angles or greater (lesser) angles to each other prior to bonding. This is a common practice, and nonwovens so produced are known as "cross-laid." With the growing importance of polyester fibers, the thermoplastic method of bonding is more important than formerly, but the growing importance of nonwoven fabrics developed for such uses as medical-surgical and sanitary consumer products has made disposability an important facet of the industry. Ergo, biodegradability is also important, and rayon remains the most important fiber in use for nonwovens.

Webs formed by textile processes are often referred to as "dry-laid" because that is the way they are formed. When webs are formed from textile fibers on the screens of conventional papermaking machines, they are in a necessarily wet state, and they form substrates for fabrics known as "wet-laid nonwovens." The sophisticated technology required for bonding, and enhancing functions such as strength by cross laying and aesthetics of porosity by various means are essentially the same for dry-laid (textile) and wet-laid (paper) nonwoven-fabric manufacture. Additionally, both basic methods may be made to produce reinforced nonwovens by "gridding" filament yarns in one or two directions. These reinforcing yarns need not be prefabricated: they are applied without interlacing during nonwoven-fabric manufacture.

End Uses

One of the most useful ways in which to classify textile-mill products is by end use. For these industries and their suppliers, it has provided a means by which to structure manufacturing technology and marketing. End-use development will be considered for each generic fiber discussed in the following section of this chapter.

Table 1 introduces the reader to the scope of end-use programming and shows how man-made-fiber consumption in the United States has penetrated markets formerly held by natural fibers. Generally, the textile industries classify textiles according to apparel, home furnishings, and industrial end usage, i.e., the consumer-product class into which textile yarns and fabrics are manufactured.

Physical Distribution

Until the mid-1950s it was relatively easy to classify the United States textile industries into four distinct categories:
1. Cotton broad-woven-fabric producers
2. Man-made-fiber fabric (ex-silk) producers
3. Woolen- and worsted-fabric producers
4. Knit-fabric producers

Since then, the adoption of man-made fibers by all these machine-class sectors has blurred the lines among these categories, as has the broadening of machine classes in common use, i.e., the expansion of the knitting industries and the addition of such processes as tufting and nonwovens.

Simultaneous with this change has come a change in the stature of traditional middlemen, notably the commission spinner, weaver, knitter, dyer, finisher and printer, and the converter. For the most part this change has been one of attrition due to vertical integration by large textile companies or mill groups. Indeed some of them have integrated back as far as fiber production per se and as far forward as ownership of retail outlets. But this is a generalization which bears a paradoxical qualification: in certain facets of physical distribution, the middleman has gained in importance if he has been able to grasp the new fiber and fabric technologies as they have rapidly emerged. A classic example of this is the apparel knit-goods converter and the supplier of spun and/or textured sales yarns to the independent or commission knitting mill, notably the double-knitting mill.

The flow chart in Fig. 3 illustrates the complex evolution of physical distribution among the textile industries.

THE MAN-MADE FIBERS

Each class of man-made fibers to follow will be headed by a definition of that class as given by the Federal Trade Commission under the Textile Fiber Products Identification Act along with the generic name given pursuant to the provisions of Section 7(c) of the Act.

Acrylic

A manufactured fiber in which the fiber-forming substance is any long-chain synthetic polymer composed of at least 85 percent by weight of acrylonitrile units $(-CH_2-CH-)$.
$$| \atop CN$$

History As early as the 1920s, polymers from acrylonitrile appeared in the German patent literature. In 1938, H. Rein of I.G. Farbenindustrie described fibers based on acrylonitrile polymer. The development of acrylic fibers was made possible by the discovery of suitable solvents for the polymer. Du Pont announced the development of an acrylic fiber, Fiber A, in 1944. A pilot plant for this fiber was started in 1945. Du Pont produced the first commercial acrylic fiber under the trademark Orlon in 1949. The first Orlon produced was in filament form. The staple fiber was produced from a commercial plant in 1952.

In 1952, Chemstrand Co., a joint venture of American Viscose Co. and Monsanto Chemical Co., commercially produced their acrylic fiber called Acrilan.

Dow Chemical Co. introduced their acrylic fiber Zefran in 1958, and American Cyanamid introduced their acrylic fiber Creslan in 1959.

All the major manufacturers of acrylic fibers in the United States encountered major problems in the commercialization of their products. In some cases dyeability was the major problem, in others "fibrillation" (the tendency of the fiber to splinter either in processing or in actual use in a fabric).

TABLE 1 Fiber-Consumption Comparison[1]
(Pounds)

Article	1964		1969	
	Man-made fibers	Cotton plus wool	Man-made fibers	Cotton plus wool
End-Use Consumption*				
Men's and boys' wear	269,300,000	1,444,900,000	622,500,000	1,270,000,000
Women's, misses', children's, and infants' wear	734,600,000	884,000,000	1,203,200,000	686,900,000
Home furnishings	943,500,000	1,340,100,000	1,700,000,000	1,336,200,000
Other consumer-type products	382,900,000	495,300,000	661,100,000	543,100,000
Industrial uses	752,900,000	598,300,000	1,286,400,000	551,400,000
Fiber Consumption in Selected Articles				
Men's and boys' wear:				
Suits	16,400,000	33,500,000	25,100,000	38,400,000
Separate slacks	62,300,000	137,800,000	174,500,000	114,600,000
Overcoats, topcoats, and rainwear	8,100,000	17,500,000	11,100,000	17,100,000
Outdoor jackets and athletic uniforms	23,200,000	48,500,000	26,500,000	64,800,000
Sweaters	19,800,000	39,200,000	20,700,000	40,700,000
Utility clothing, except shirts	13,700,000	361,100,000†	54,800,000	311,300,000†
Work and uniform shirts	3,300,000	66,500,000	26,000,000	57,900,000
Business and dress shirts	13,900,000	101,400,000†	60,800,000	56,200,000†
Sport shirts	46,500,000	232,100,000	101,700,000	162,800,000
Underwear and nightwear	9,500,000	230,900,000	22,100,000	241,200,000
Hose, all types	29,000,000	66,200,000	49,500,000	41,700,000
Other uses	23,600,000	110,200,000	49,700,000	123,300,000
Sum	269,300,000	1,444,900,000	622,500,000	1,270,000,000
Women's, misses', children's, and infants' wear:				
Suits	16,500,000	15,500,000	21,000,000	10,600,000
Skirts	30,400,000	65,300,000	57,500,000	36,300,000
Slacks	67,100,000	45,000,000	93,400,000	54,000,000
Dresses	252,100,000	188,000,000	434,000,000	117,000,000
Coats and jackets	15,700,000	94,000,000	28,900,000	91,400,000
Rainwear	9,900,000	10,300,000†	12,200,000	6,100,000†

Playsuits, sunsuits, and shorts	65,700,000†	27,600,000	88,500,000	13,100,000
Sweaters	32,300,000	62,900,000	35,900,000	52,200,000
Swimwear	5,900,000†	8,600,000	6,800,000†	7,900,000
Loungewear	28,000,000	29,200,000	34,000,000	15,500,000
Washable service apparel	2,700,000	13,600,000	13,800,000†	8,600,000
Blouses and shirts	57,900,000	100,700,000	79,200,000	49,400,000
Brassieres and foundation garments	7,200,000†	33,500,000	16,000,000†	24,900,000
Underwear	34,900,000†	119,700,000	42,600,000†	90,300,000
Nightwear	62,300,000	45,700,000	83,200,000	24,200,000
Full-length hosiery		78,000,000		33,900,000
Anklets and socks	19,800,000	11,900,000	24,400,000	8,200,000
Other uses	54,800,000†	24,800,000	41,500,000†	14,800,000
Sum	686,900,000	1,203,200,000	884,000,000	734,600,000
Home furnishings:				
Bedspreads and quilts	126,500,000†	38,600,000	108,000,000†	42,500,000
Blankets	30,600,000	101,600,000	32,000,000	84,800,000
Sheets and other bedding	441,700,000	88,100,000	464,300,000†	25,500,000
Towels and toweling	282,600,000†	4,500,000	271,200,000†	3,600,000
Carpet and rug face yarns	105,600,000	1,007,100,000	161,900,000	448,900,000
Carpet and rug backing	51,100,000†	79,300,000	61,900,000†	9,100,000
Curtains—lace, cottage, etc.*	27,800,000†	41,000,000	22,900,000†	32,200,000
Drapery, upholstery, and slip covers*	232,200,000	327,100,000	174,700,000	285,700,000
Other uses*	38,100,000†	12,700,000	43,200,000†	11,200,000
Sum	1,336,200,000	1,700,000,000	1,340,100,000	943,500,000
Other consumer-type products:				
Apparel linings	115,900,000	187,300,000	114,400,000	138,000,000
Retail piece goods	134,000,000	125,000,000	113,000,000	59,000,000
Narrow fabrics—elastic and nonelastic	74,000,000†	66,700,000	71,000,000†	44,600,000
Luggage, handbags, etc.	13,200,000†	19,500,000	13,100,000†	18,000,000
Toys	7,000,000†	22,000,000	4,500,000†	10,500,000
Medical, surgical, and sanitary	106,700,000†	68,000,000	82,400,000†	25,300,000
Tricot for laminating; stuffing fibers; sports equipment; hand umbrellas and other miscellaneous	3,200,000	154,500,000	2,300,000	75,000,000
Shoes and slippers	62,100,000†	6,500,000	72,100,000†	5,000,000
Handwork yarns	27,000,000	11,600,000	22,500,000	7,500,000
Sum	543,100,000	661,100,000	495,300,000	382,900,000

TABLE 1 Fiber-Consumption Comparison¹
(Pounds)
(Continued)

Article	1964		1969	
	Man-made fibers	Cotton plus wool	Man-made fibers	Cotton plus wool
Industrial uses:				
Transportation upholstery	16,300,000	61,300,000	25,200,000	66,000,000†
Tires, all types*	423,900,000	10,700,000†	584,800,000	3,900,000†
Hose—fire, fuel, etc.*	16,200,000	12,500,000†	26,200,000	7,500,000†
Belting—V and flat belts*	15,600,000	43,000,000†	23,000,000	48,000,000†
Laundry supplies	6,700,000	9,800,000†	6,000,000	7,000,000†
Electrical applications*	21,200,000	21,400,000†	27,800,000	12,800,000†
Filtration*	12,100,000	9,300,000†	25,700,000	6,300,000†
Sewing thread	13,800,000	94,100,000†	27,500,000	81,200,000†
Rope, cordage, fishline, etc.*	39,600,000	64,100,000†	84,600,000	49,600,000†
Tarps, tents, parachutes, etc.*	13,000,000	63,700,000†	26,100,000	85,000,000†
Paper and tape reinforcing*	27,300,000	3,800,000†	44,100,000	9,500,000†
Reinforced plastics*	114,900,000	28,000,000†	282,900,000	24,000,000†
Bags and bagging	400,000	77,900,000†	41,900,000	38,300,000†
Felts	6,700,000	18,100,000	12,600,000	17,000,000
Other uses*	25,200,000	80,600,000	48,000,000	95,300,000
Sum	752,900,000	598,300,000	1,286,400,000	551,400,000

* Including textile glass fiber.
† Mainly cotton. Little or no wool used.

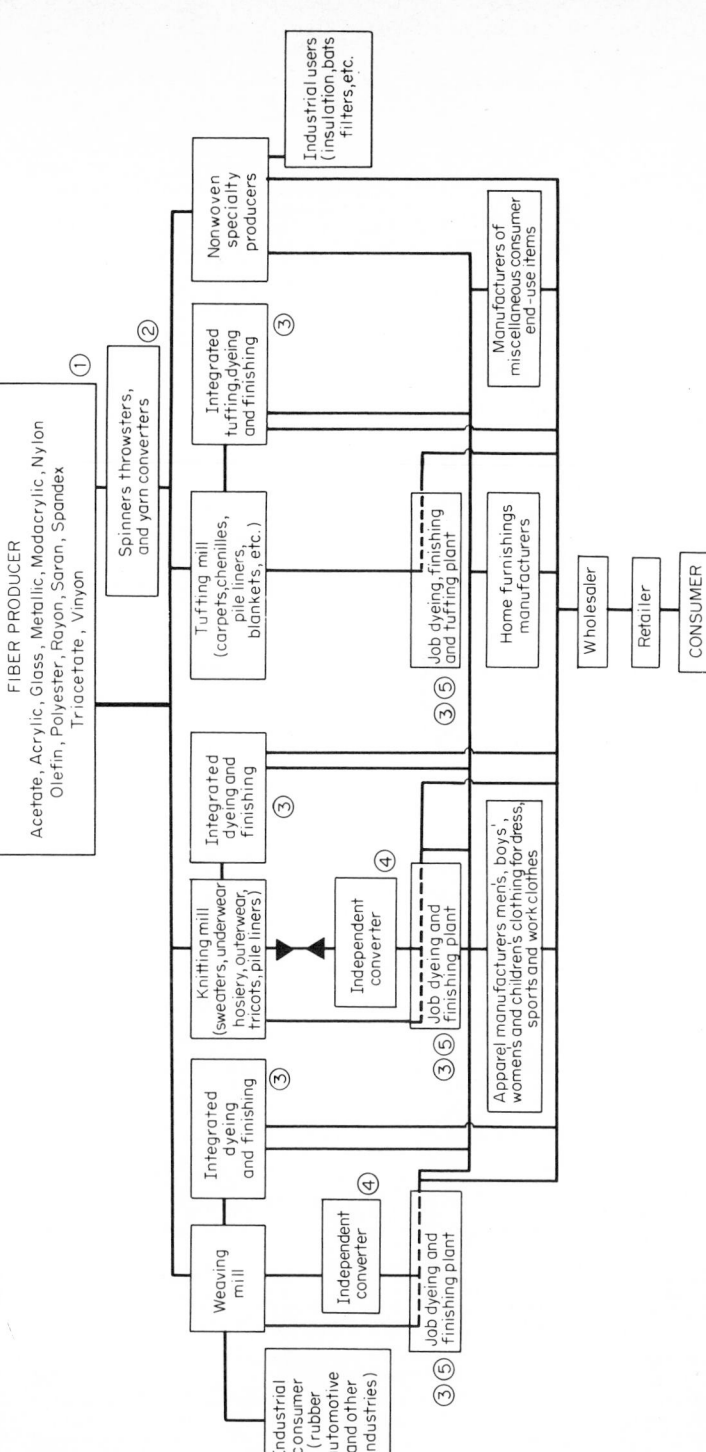

Fig. 3 Flow of man-made fibers from producer to consumer.[1] (1) Man-made fibers are used in the textile process either alone or in blends with each other or with natural fibers. (2) Spinners produce yarns from short fibers. Throwsters, by the application of twisting machines, can alter filament yarns to make them suitable for such fabric constructions as crepe, or can add to filament yarns such special characteristics as bulk or stretch. Yarn converters, who may also be throwsters, dye or otherwise prepare yarns for fabrication, as do integrated weaving and knitting mills. (3) Dyeing and finishing plants dye or print fabrics and apply finishes such as durable press, crush resistance, water and soil repellence, and flameproofing. Independent converters own stock in process, but the processing is done by others. (5) Job dyeing, finishing, and tufting plants contract work from weaving, knitting, and tufting mills and from independent converters. Note: Companies which combine two or more of the above functions are integrated operations.

6-15

Raw materials and manufacturing process   Acrylic fibers can be made by either wet or dry spinning. The following are the major steps in producing an acrylic fiber:

1. In addition to the base monomer acrylonitrile, other monomers are incorporated in the mixture along with the catalyst and delusterants. The mixture is stirred in a reactor until the proper degree of polymerization is obtained.

2. The polymer is filtered, dried, and blended, then dissolved in a suitable solvent depending on whether the fiber is to be dry-spun or wet-spun. In the case of Orlon acrylic fiber, which is dry-spun, a volatile solvent such as dimethyl formamide is used. In the case of fibers that are wet-spun such as Acrilan, Creslan, or Zefran acrylic fiber, solvents such as dimethyl acetamide or zinc or thiocyanate salts can be used.

3. The fiber is extruded through a spinneret either into warm air if it is dry-spun or into a coagulating bath if it is wet-spun. The fiber is also stretched during the spinning process. In the wet-spun fibers the impurities are also washed out.

4. Finish is applied to the fiber, and it is then crimped and cut into fiber form or packaged as tow.

Figure 4 illustrates a flow chart for production of a typical wet-spun acrylic fiber.

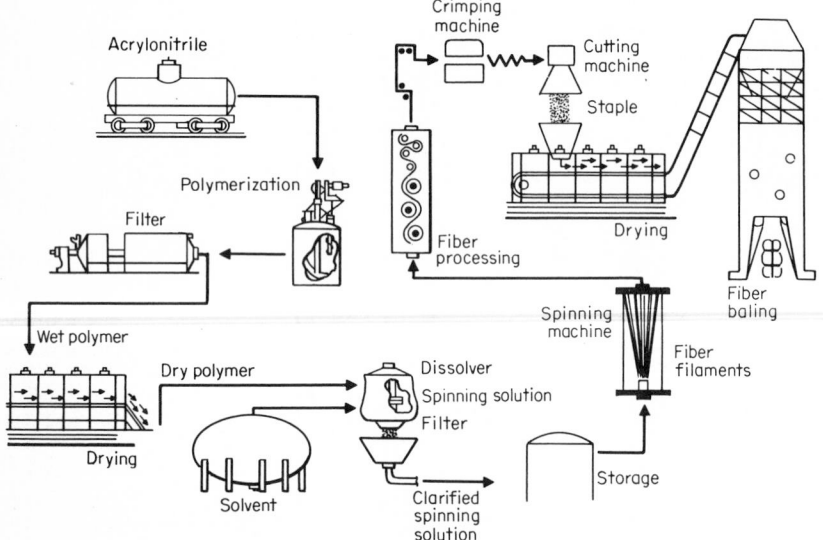

Fig. 4   Flow chart for production of a typical wet-spun acrylic fiber.

Properties of acrylic fibers[2]   See Table 2.

The product types of acrylics and other genera developed in recent years and currently under development are so numerous in themselves and as generated by the man-made-fiber manufacturer under branding programs that summarizing them is impractical. The reader is advised to consult with individual fiber producers for details of their type-number assignation system for a generic. Also useful is reference to "Textile Fibers and Their Properties" (Burlington Industries, Inc., Greensboro, N.C.) and *Textile Industries* for August 1972 and August 1973 for detailed analyses of these man-made-fiber product types or, as they are more frequently known, variants.

It should be noted, however, that the principal reasons for variant development include cross-sectional modification to improve aesthetics of hand and luster as well as soil-hiding properties; basic chemical modification to introduce producer coloration, dyestuff receptivity, versatility, low rate of burning, and antistatic properties; and extrusion and orientation modifications to customize denier-per-filament tensile and percent-of-shrinkage properties. Few variants are developed for the convenience of the fiber producer. For the most part, they contribute to the aesthetics, versatility, function, and economics of fiber users—from fabric-forming mill to the ultimate consumer.

End uses   The staple and tow shipments of acrylic and modacrylic man-made fibers for 1970 listed in Table 3 illustrate the scope of end-use development commercialized for these fiber generics. It should be noted that they are not manufactured in this country in continuous-filament form.

Modacrylic

A manufactured fiber in which the fiber-forming substance is any long-chain synthetic polymer composed of less than 85 percent but at least 35 percent by weight of acrylonitrile units ($—CH_2—CH—$).
$$\underset{\displaystyle CN}{|}$$

History   Dynel was the first modacrylic fiber commercialized in the United States. It was introduced by Union Carbide in 1948.

Verel was the second modacrylic fiber commercialized in the United States. It was introduced by Tennessee Eastman in 1956. The polymer was originally developed by Dow Chemical Co. and subsequently licensed to Eastman for commercialization.

SEF was introduced as a modacrylic fiber by Monsanto in 1971.

Raw materials and manufacturing process   In general the polymerization methods are similar to those used in the acrylic fibers. The spinning methods are also similar to those used for acrylic fibers, although in some cases annealing steps are needed to obtain some of the required physical properties.

Figure 5 illustrates a flow chart for production of Dynel modacrylic.

Properties of modacrylic fibers   See Table 2.

End uses   The uses for modacrylics are grouped with data given for end uses in the acrylic section, but it should be noted that modacrylics found first wide acceptance in high-pile knitted fabrics for outerwear shell ("fake fur") and liner fabrics. This was due largely to the ability to develop an animal-hair-like luster and character during the so-called "electrifying" or fiber-straightening and polishing process originally used for furs and adapted to finishing of knitted imitation furs. More recently, the self-extinguishing properties of the fiber have made it important in development of flame-safe fabrics which comply with current FTC Flammable Fabrics Act regulations.

Polyester

A manufactured fiber in which the fiber-forming substance is any long-chain synthetic polymer composed of at least 85 percent by weight of an ester of a dihydric alcohol and terephthalic acid ($p—HOOC—C_6H_4—COOH$).

History   Carothers and Hill worked with aliphatic polyesters but did not consider them as fiber-forming candidates because of their low melting points and poor hydrolytic stability. Whinfield and Dickson, in the laboratories of the Calico Printers Association in England, discovered that polyethylene terephthalate had fiber-forming properties and a high melting point together with good hydrolytic stability.

The development of polyester fiber in England was slowed down by World War II. Calico Printers entered into an agreement with I.C.I. for the latter to develop the polyester fiber which became known as Terylene. Du Pont bought the rights from I.C.I. to manufacture polyester fibers in the United States under the trademark Dacron. In 1953, a full-scale polyester plant was built by Du Pont in Kinston, N.C. Previously Du Pont was producing developmental quantities of Dacron fibers and yarns in a pilot plant at Seaford, Del.

Raw materials and manufacturing process   Ethylene glycol was already an item of commerce, particularly for antifreeze, in the United States when polyester fibers were being developed. It was only necessary to provide this material in a more

TABLE 2 Properties of Fibers

Fiber	Acrylic	Modacrylic	Polyester
Cross-sectional appearance of fibers			
	Round, bean-shaped, dog bone or lobal	Dog bone, irregular dog bone, and ribbon	Round
Longitudinal description	Smooth, twisted, or wide striations	Twisted ribbon, broad striations	Smooth
Burning test	Ignites and burns readily. Forms hard black bead. Odors range from sweet to acrid or burnt meat	Difficult to ignite. Shrinks away from flame and forms irregular bead. Self-extinguishing. Sharp, sweet odor	Shrinks from flame. Forms hard round bead. Pungent odor
Treatment with 60% H_2SO_4	Insoluble	Insoluble	Insoluble
Treatment with 5% caustic	Insoluble	Insoluble	Partially soluble
Solubility reagent	Dimethyl formamide	Warm acetone	M-Cresol

Effects of heat................................	420–490°F sticking temperature	275–300°F sticking temperature	Sticks 445–455°F, melts 482–554°F
Specific gravity................................	1.16–1.18	1.31–1.37	1.38
Moisture regain (commercial), %..............	1.5	0.4–3.5	0.4
Effects of acids, alkalies, and solvents.........	Generally good resistance to mineral acids. Fair to good resistance to weak cold alkalies. Good resistance to common solvents	Excellent resistance to highly concentrated acids. Unaffected by alkalies. Soluble in warm acetone	Good to excellent resistance to mineral acids. Some affected by concentrated sulfuric acid. Decomposes in alkalies at the boil. Generally insoluble in common solvents but soluble in some phenolic compounds
Dyestuffs used................................	Acid, basic, cationic, chrome, direct, disperse, naphthol, neutrals, premetallized, and sulfur	Basic, cationic, disperse, neutral, premetallized, and vat	Cationic, developed, and disperse
Stress-strain curve............................			

TABLE 2 Properties of Fibers (Continued)

Fiber	Rayon	Acetate	Triacetate
Cross-sectional appearance of fibers			
Longitudinal description	Serrated to round	Cloverleaf to slightly irregular	Irregular
	Striated to smooth	Striated	Striated
Burning test	Ignites readily. Burns rapidly with paper odor. Does not bead	Ignites readily. Forms black brittle bead. Acetic odor	Ignites readily. Forms black brittle bead. Acetic odor less pronounced than acetate
Treatment with 60% H_2SO_4	Soluble	Soluble	Soluble
Treatment with 5% caustic	Swells, does not dissolve	Partially soluble	Partially soluble
Solubility reagent	60% H_2SO_4 concentrated hydrochloric acid	80% acetone	Methylene chloride

Effects of heat.................................	Decomposes 300–464°F	Sticks 350–375°F, melts 500°F	Sticks 480°F, melts 572°F
Specific gravity................................	1.50–1.54	1.31	1.30
Moisture regain (commercial), %...............	11.0	6.5	3.5
Effects of acids, alkalies, and solvents.........	Disintegrates in cold or hot concentrated acids. Resistant to cold weak alkalies, and organic solvents. Attacked by strong oxidizing agents. High-modulus rayons can be mercerized	Unaffected by weak acids but disintegrates in concentrated acids. Cold weak alkalies have little effect. Soluble in acetone, phenol, and glacial acetic acid	Unaffected by weak acids but disintegrates in concentrated acids. Cold weak alkalies have little effect. Swells and partially dissolved by acetone. Soluble in methylene chloride
Dyestuffs used	Azoics, direct, reactive, sulfur and vats	Azoics and some disperse	Azoics and some disperse
Stress-strain curve.............................			

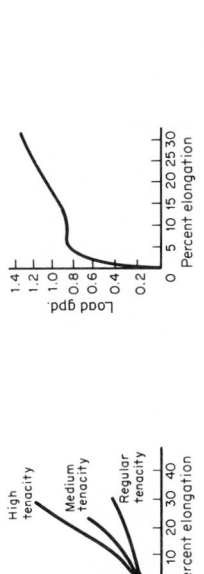

TABLE 2 Properties of Fibers (Continued)

Fiber	Nylon 6	Nylon 6/6	Spandex
Cross-sectional appearance of fibers...........			
	Round or trilobal	Round or trilobal	Dog bone or teardrop
Longitudinal description..................	Smooth or few striations	Smooth or few striations	Dark, no markings
Burning test.............	Difficult to ignite. Shrinks from flame. Forms hard pale brown bead. Celery-like odor fainter than nylon 6/6	Difficult to ignite. Shrinks from flame. Forms hard pale brown bead. Celery-like odor	Does not shrink from flame. Burns with melt. Leaves soft fluffy residue
Treatment with 60% H_2SO_4..........	Soluble	Soluble	Soluble
Treatment with 5% caustic...........	Insoluble	Insoluble	Insoluble
Solubility reagent............	Formic acid	Formic acid	Dimethyl formamide

Effects of heat	Sticks 320–375°F, melts 425°F	Sticks 445°F, melts 490°F	Sticks 345–445°F, melts 445–554°F
Specific gravity	1.14	1.14	1.20–1.25
Moisture regain (commercial), %	4.5	4.5	1.30
Effects of acids, alkalies, and solvents	Resistant to weak acids but decomposes in strong mineral acids. Alkalies have little or no effect. Soluble in phenol and formic acid	Same as nylon 6	Generally good resistance to most acids and alkalies. Good resistance to most common solvents
Dyestuffs used	Acid, direct, disperse, and premetallized	Same as nylon 6, plus cationic	Acid, cationic, chrome, disperse, direct, and premetallized
Stress-strain curve	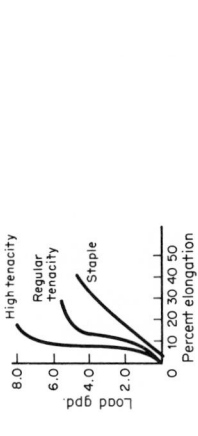		

TABLE 2 Properties of Fibers (Continued)

Fiber	Fiber glass	Polyolefin
Cross-sectional appearance of fibers		
Longitudinal description	Round	Round and other
	Smooth	Smooth
Burning test	Glows—Does not burn	Shrinks from flame. Forms hard bead. Odors: polyethylene—paraffin; polypropylene—burning asphalt
Treatment with 60% H$_2$SO$_4$	Insoluble	Insoluble
Treatment with 5% caustic	Insoluble	Insoluble
Solubility reagent	Insoluble in all commercial solvents	Xylene

Effects of heat	Softens at 1350–1560°F	Melts 250–333°F
Specific gravity	2.49–2.55	0.90–0.96
Moisture regain (commercial), %	0.0	0.0
Effects of acids, alkalies, and solvents	Resists most acids and alkalies. Unaffected by common solvents	Excellent resistance to most acids and alkalies. Generally soluble above 160°F in chlorinated hydrocarbons
Dyestuffs used	Resin-bonded pigments	Usually pigmented before extrusion. When modified, can be dyed with selected dyestuffs
Stress-strain curve		

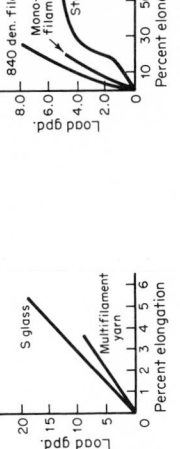

TABLE 3 Shipments of Acrylic and Modacrylic for Major End Uses, 1972*

End use	Staple tow, million lb
All apparel..	349
Five largest apparel uses:	
Hosiery, sweaters, sweater shirts, and hard knitting yarns..............	115
Knit dresses..	67
Pile and fleece linings...	51
Suits, slacks, and coats (women's)..................................	30
Hosiery (men's and women's)......................................	24
All home furnishings..	173
Three largest home furnishings uses:	
Carpets and rugs...	131
Blankets..	24
Drapery and upholstery...	9
All industrial and other...	14
Three largest industrial uses:	
Doll hair, toys, etc., sand bags....................................	3
Filtration, woven and nonwoven...................................	1
Coated and protective fabrics......................................	2

* No acrylic filament made in the United States.

highly purified form for fiber use. This is the glycol used by all the fiber manufacturers in the United States with one exception. Tennessee Eastman uses 1,4-cyclohexanedimethanol in Kodel II. Kodel II has a somewhat higher melting point and is a bulkier fiber and deeper-dyeing than the conventional polyester fibers.

Dimethyl terephthalate is still used extensively as an intermediate for polyester fibers, although terephthalic acid is being used more frequently. The major route

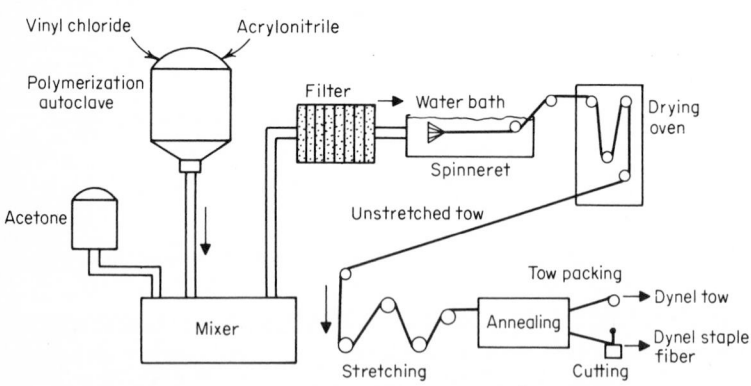

Fig. 5 Flow chart for production of Dynel modacrylic.

to dimethyl terephthalate is with *p*-xylene as a starting material, which is oxidized with nitric acid. The crude terephthalic acid is esterified in methanol to give dimethyl terephthalate, which is then purified.

End uses The 1972 shipments of staple, tow, and filament for polyester fiber shown in Table 4 illustrate the scope of end-use development commercialized for this fiber generic.

Terephthalic acid is also manufactured by the oxidation of *p*-xylene. The key to the use of terephthalic acid as a more direct route to the intermediate has been the ability to obtain the material with a high degree of purity. Initial attempts to make polyester fibers with terephthalic acid failed until purification technology was developed.

The batch process for the manufacture of polyester fibers may be described in a six-step flow:

1. Dimethyl terephthalate is reacted with a slight excess of ethylene glycol over a temperature range of 150 to 210°C in the presence of a catalyst. The process of ester interchanging takes place and yields the monomer. Since it is a reversible reaction, methanol is fractionated off. Titanium dioxide is added as a dispersion at this stage, acting as a delustrant.

TABLE 4   Shipments of Polyester for Major End Uses, 1972

End use	Staple and tow, million lb	End use	Filament, million lb
All apparel.................	741.0	All apparel.................	685.0
Five largest apparel uses:		Largest apparel uses:	
Men's suits, slacks, and		Textured yarn for circular	
coats...................	160.0	knit or woven use........	558.0
Men's woven shirts........	123.0		
Woven dresses...........	84.0		
Women's suits, slacks, and			
coats...................	74.0		
Work clothing, men's and			
women's...............	68.0		
All home furnishings........	497.0	All home furnishings........	
Four largest home furnishings uses:		Figures not available. Primarily used in curtains/	
Carpets.................	174.9	draperies and bedspreads	
Sheets and pillowcases......	153.0		
Fiberfill.................	121.4		
Draperies...............	21.0		
All industrial and other uses..	48.5	All industrial and other uses..	250.0
Two largest industrial uses:		Three largest industrial uses:	
Nonwovens..............	7.5	Tires...................	229.0
Sewing thread............	5.0	Other rubber industries....	15.0
		Rope and cordage........	9.0

2. The monomer is transferred to a polymerization autoclave, and the temperature is reduced to 280°C. Generally an antimony derivative is used as a polymerization catalyst. The glycol which is formed is removed.

3. When the desired viscosity of polymer has been reached, the polymer is extruded, cooled, formed into chips, and blended.

4. The blended chip is remelted, filtered, and extruded through a spinneret. The spinning and melting equipment is similar to that used for nylon manufacture.

5. The spun filaments are amorphous and weak and must be drawn to obtain the desired strength and elongation properties in the yarn or tow. Nylon is cold-drawn, but it is necessary to heat polyester above 80°C (its glass-rubber transition temperature) in order to draw it successfully. The flow charts for making polyester staple and filament are shown in Fig. 6.

6. The yarn or tow is treated with a combination antistatic and lubricating finish before it is packaged.

Properties of polyester fibers   See Table 2.

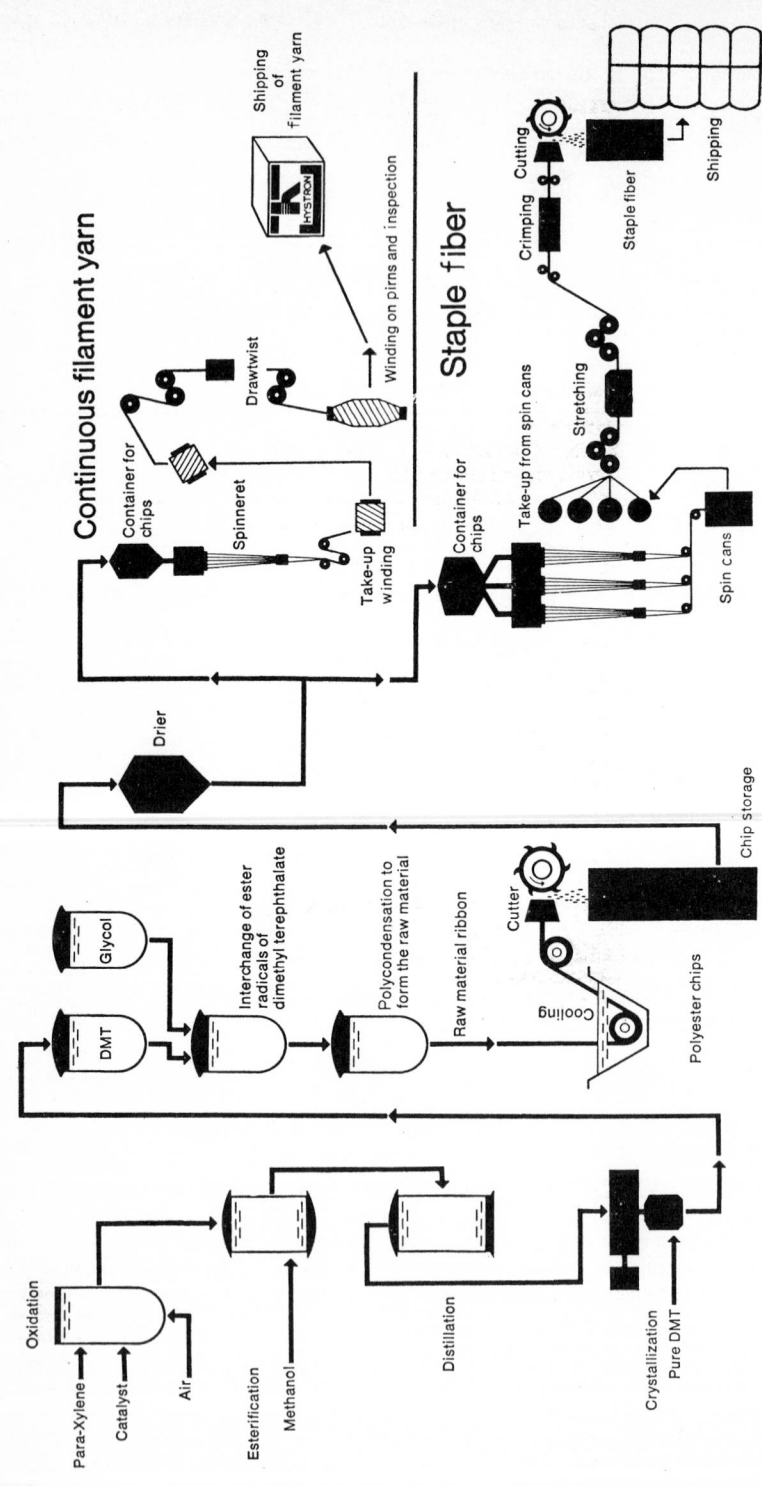

Fig. 6 The manufacture of Trevira polyester fiber by the melt-spun process.

Rayon

A manufactured fiber composed of regenerated cellulose, as well as manufactured fibers composed of regenerated cellulose in which substituents have replaced not more than 15 percent of the hydrogens of the hydroxyl groups.

History The history of the man-made-fiber business really starts with the development of fibers from cellulose. The viscose process, the process most commonly used to make rayon, is based on the work of Cross and Bevan in 1891 to 1892. The pioneering work on the viscose-rayon process was carried out by Courtauld, Ltd., of England. Samuel Courtauld and Company in England was successful, and soon yarn was being exported to the United States. Courtauld's was represented in the United States by Samuel A. Salvage, who soon realized the great future for viscose yarns and persuaded Samuel Courtauld to buy the American rights to produce rayon in the United States. The Viscose Company of America was formed and started its first plant at Marcus Hook, Pa. in 1910.

Raw materials and manufacturing processes

The Viscose Process. The basis for the process is the discovery by Cross and Bevan that alkali cellulose, obtained by treatment with caustic soda, can be made into cellulose xanthate by reaction with carbon disulfide. The cellulose xanthate is soluble in dilute caustic soda and can be regenerated to cellulose by the action of an acid bath.

The following are the major steps in the production of viscose rayon:

1. The wood pulp, which contains between 87 and 98 percent alpha cellulose, is steeped in 18 percent caustic soda to form the alkali cellulose.

2. The alkali cellulose is shredded into crumbs and aged under very carefully controlled conditions.

3. The alkali cellulose is reacted with carbon disulfide in a xanthation churn.

4. The orange cellulose xanthate crumb is dissolved in caustic soda to produce a solution containing 6 to 8 percent cellulose and 6 to 8 percent sodium hydroxide.

5. The viscose spinning solution is carefully matured, deaerated, and filtered.

6. The viscose solution, along with suitable additives such as delusterants, is spun into filaments through a metallic spinneret, into an acid bath to regenerate the cellulose.

7. The wet acid-containing filament rayon is washed, desulfurized, rinsed, has a finish applied, and is dried.

The Cuprammonium Process. This process was introduced at the end of the nineteenth century by the German firm of Glanzstoff. It was abandoned in a few years in favor of the cheaper viscose process.

The cuprammonium process was revived commercially in 1919, when J. P. Bemberg A.G. started to produce rayon yarn by this process, hence the name Bemberg rayon. Bemberg's success with this yarn was due to the invention of stretch spinning, which entailed the replacement of caustic soda with a milder coagulant and stretching the newly formed filaments in the spinning bath. In 1926, the American Bemberg Corp. began to produce Bemberg rayon at Elizabethton, Tenn. The following are the major steps in the production of cuprammonium rayon:

1. The purified wood pulp, with an alpha cellulose content of at least 96 percent, is prepared with a slurry of basic copper sulfate and ammonia at a low temperature. Caustic soda and stabilizers are subsequently added.

2. The impurities are filtered out by passing the solution through fine-mesh nickel screens, and the solution is deaerated.

3. The spinning process produces a mild coagulation as the solution is moved through the spinnerets.

4. The threads are subjected to the action of dilute sulfuric acid, which completes the hardening of the thread and converts the rest of the ammonia and copper salts to the sulfate form.

5. The yarn is washed to remove all acid and copper salts, and suitable lubricants are added prior to drying.

The Polynosics and High-Wet-Modulus Rayons. In general, rayon fibers and yarns have a good balance of properties for use in textiles. Almost all the shortcomings of rayon's properties relate to the hydrophilic nature of the cellulose molecule.   Early attempts to reduce water uptake and water retention involved the use of cross-linking agents for the cellulose.   The cross-linking caused severe loss in tenacity and abrasion resistance.

In the early 1950s, Cox and coworkers at Du Pont made a breakthrough by the use of organic spin-bath "modifiers" in combination with zinc ions to retard dexanthation and to enable greater stretching.   This breakthrough allowed the production of tire yarns with tenacities up to 8 g/denier, yet retaining 10 percent elongation at the break.   These developments have enabled rayon tire yarns to maintain a competitive position much longer than was expected.

As a by-product of the rayon-tire-yarn developments, American Viscose and Chemiefaser Lenzing in Europe developed "high-wet-modulus" fibers.   These fibers have good strength combined with a medium-range elongation to break and a high modulus.   This well-balanced combination of properties makes the "high-wet-modulus" rayons ideal blending fibers for use with polyesters.

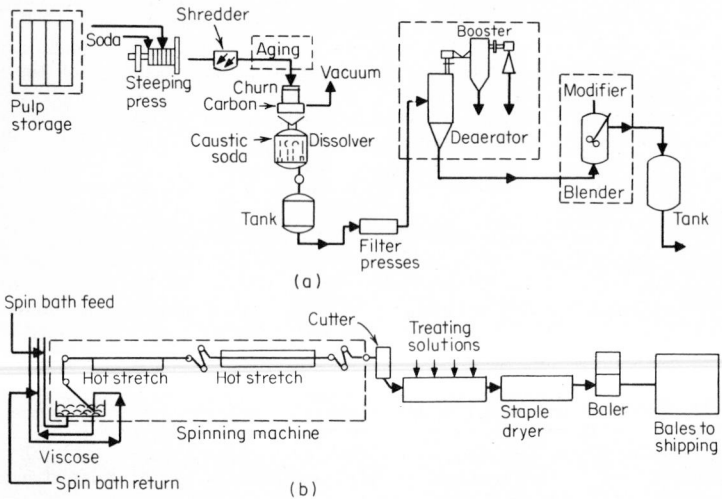

Fig. 7   Flow chart for production of high-wet-modulus rayon.   (*a*) High-wet-modulus viscose manufacture.   (*b*) High-wet-modulus staple-fiber spinning.

The "polynosics" are based on technology developed by Tachikawa in Japan. These high-modulus rayons depend on the spinning of viscoses with a high degree of xanthation and high viscosity into spin baths having a high percentage of acid. Polynosics have high strength, elongation to the break below 11 percent, low water swelling, high modulus both wet and dry, and better alkali resistance than "high-wet-modulus" rayons.   The polynosics, however, have a greater tendency toward fibrillation than the "high-wet-modulus" rayon fibers.   Figure 7 illustrates a typical flow chart for high-wet-modulus rayon.

Fire-retardant Rayon.   American Viscose Corp. introduced a permanent fire-retardant rayon fiber (PFR) in December 1967.   PFR is manufactured by metering a water-insoluble organophosphorus compound into the viscose dope before it reaches the spinning jets.   The fire-retardant agent is encapsulated within a cellulose matrix as a dispersion of fine droplets throughout the fiber.

American Viscose states that this fire-retardant system can be incorporated in all regular rayon staple and filament products with only minor effects on fiber properties.

One of the major difficulties in commercializing the PFR products has been the high cost of the organophosphorus additive.   It is hoped that the cost will decrease as the volume of production increases.

Defosse and Welch give the following summation of the characteristics of the American Viscose fire-retardant rayon:

1. Aesthetics are unchanged.
2. Normal dyeing properties result with dye fasteners unimpaired.
3. Fiber is chemically inert to bleaching, alkali, etc., likely to be encountered in fabric preparation.
4. Fiber accepts chemical finishing agents.
5. Conversion of fiber into yarn and fabric requires no major machine changes.
6. Fiber is nontoxic and nonallergenic to skin.

Properties of rayon fibers See Table 2.

TABLE 5   Shipments of Rayon for Major End Uses, 1972

End use	Staple and tow, million lb	End use	Filament, million lb
All apparel..................	191.1	All apparel................	41.0
Five largest apparel uses:		Four largest apparel uses:	
Dresses...................	91.1	Apparel linings, men and	
Women's suitings and		women..................	20.0
slacks..................	23.1	Dresses and rainwear......	9.4
Men's suitings and slacks...	20.8	Suits, slacks, coats, men's	
Shirts, woven and knit.....	22.0	and women's............	2.4
Apparel linings...........	22.2	Other apparel, men and	
		women..................	1.0
All home furnishings.........	192.5	All home furnishings.........	32.8
Four largest home furnishings uses:		Four largest home furnishings uses:	
Drapery, upholstery, and		Drapery and upholstery....	28.3
slipcovers...............	109.7	Bedspreads, quilts, and	
Bed ticking...............	16.3	other bedding...........	2.3
Bedspreads and quilts......	12.5	Carpets and rug backing...	1.2
Curtains.................	7.9	Curtains.................	0.7
All industrial and other uses..	307.2	All industrial and other uses..	301.5
Five largest industrial uses:		Five largest industrial uses:	
Towels, medical, surgical,		Total tire use............	102.1
and sanitary............	148.6	Narrow fabrics and braids..	17.5
Coated and protective		Hose....................	21.9
fabrics..................	28.2	Rope twine and cordage....	6.5
Flock....................	17.9	Flat belting..............	4.6
Filtration and sewing			
thread..................	22.2		
Mop yarns...............	14.9		

End uses Table 5, which lists staple, tow, and filament shipments for rayon fiber for 1972, illustrates the scope of end-use development commercialized for this fiber generic. It may be safely assumed that fibers for industrial uses are made from high-tenacity rayon.

Acetate and Triacetate

A manufactured fiber in which the fiber-forming substance is cellulose acetate. Where not less than 92 percent of the hydroxyl groups is acetylated, the term *triacetate* may be used as a generic description of the fiber.

History The first successful commercialization of acetate yarns occurred in 1921, and was based mainly on the work of Camille and Henri Dreyfuss of Basel, Switzerland. In 1901, a process for production of acetate was patented in the United States, but it was not until 1924 that Celanese Corporation bought the 1901 patent and began production. In 1930, American Viscose Corporation entered the field. Today, acetate is also produced in textile-fiber form by Eastman Chemical Products, Inc., and Du Pont.

In 1954, Celanese Corporation introduced Arnel triacetate, which exhibits greater thermal stability than conventional acetate. It is also more hydrophobic.

Raw materials and manufacturing Acetate is derived from cellulose in combination with acetate from acetic acid and acetic anhydride. The cellulose acetate is dissolved in acetone and extruded. As the filaments emerge from spinnerets, the solvent is evaporated in warm air—dry-spun—leaving a fiber of almost pure cellulose acetate. Figure 8 is a flow chart for the production of acetate.

Triacetate is similarly derived, but the cellulose acetate is dissolved in a mixture of methylene chloride and methanol for extrusion. As filaments emerge from spinnerets, the solvent is evaporated in warm air—again dry-spun—leaving a fiber of almost pure cellulose acetate. Triacetate contains a higher ratio of acetate to cellulose than do acetate fibers.

Properties of acetate and triacetate fibers See Table 2.

End uses Table 6, which lists 1972 filament shipments of acetate fiber, illustrates the scope of end-use development commercialized for this fiber generic. Currently, the fiber is being used only in filament form for textile purpose.

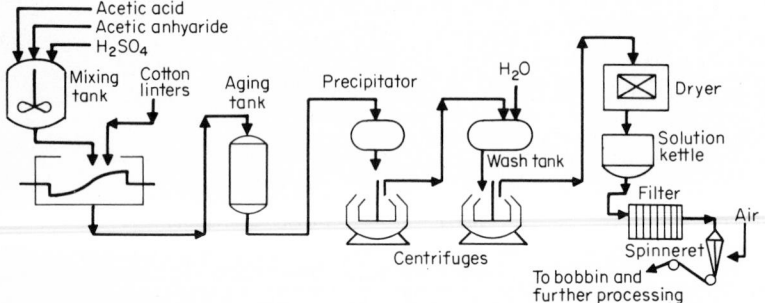

Fig. 8   Flow chart for production of acetate fiber.

Nylon

A manufactured fiber in which the fiber-forming substance is any long-chain synthetic polyamide having recurring amide groups (—C—NH—) as an integral part of the polymer chain.

$$\underset{O}{\overset{\|}{}}$$

History The basic research which Dr. Wallace H. Carothers and Dr. Julian H. Hill undertook at Du Pont to make "superpolymers" was reported in 1931 at an American Chemical Society meeting in Buffalo, N.Y. They described an interesting polymer which could be drawn or extruded into a fiber. "A clump of filaments rolled into a ball and compressed," they reported, "showed remarkable springiness resembling wool." In their elastic properties, these fibers are much superior to any known synthetic silk.

The basic patent for Carothers' work was filed in 1931, and the patent covering "linear condensation polymers" was issued in 1937. The material to be known later as nylon 6/6 was developed in a Wilmington laboratory in 1935.

One of the first commercial uses developed for nylon was for toothbrush bristles in 1936.

The pilot plant for nylon yarn was set up in Wilmington in 1938. In 1938, Du Pont appropriated $8,000,000 (one-sixth of Du Pont's 1938 earnings) for a commercial plant. Dr. Charles Stine of Du Pont made the first public announcement of nylon in New York in 1938.

On Dec. 12, 1939, the first commercial pound of nylon was produced at Seaford, Del. The total investment in nylon at this point was $27,000,000. The first commercial use for nylon was in women's hosiery.

Raw materials and manufacturing processes

Type 6/6 Nylon. Originally nylon was explained to the public as a material made from coal, air, and water. The oversimplification was based on the manufacture of the raw materials from coal-tar-derived chemicals and ammonia derived from nitrogen from air and hydrogen from water.

As our prime source of organic chemicals in this country changed from coal tar to petrochemicals, other routes to the intermediates for nylon 6/6 were developed.

In making nylon 6/6, the base components hexamethylenediamine and adipic acid are ammoniated and dehydrated to adiponitrile, which is then hydrogenated

TABLE 6   Shipments of Acetate for Major End Uses, 1972

End use	Staple and tow, million lb	End use	Filament, million lb
All apparel and home furnishings uses............	28.0	All apparel.................	325.7
Cigarette filter tow..........	144.0	Five largest apparel uses:	
		Underwear and nightwear..	57.0
		Suits and dresses..........	140.8
		Apparel linings............	49.7
		Robes, ties, and lounge- wear....................	30.9
		Blouses and shirts.........	27.7
		All home furnishings.........	480.0
		Three largest home furnishings uses:	
		Drapery, upholstery, and slipcovers...............	36.0
		Curtains.................	4.7
		Bedspreads and quilts......	3.9
		Industrial and other uses.....	12.2
		Electrical applications......	0.4
		Other industrial and consumer uses...........	11.8

to hexamethylenediamine, the other base component of the nylon salt. Monsanto has developed an electrochemical process of producing the components for nylon 6/6 through the use of acrylonitrile as raw material.

The starting material for nylon 6/6 is the nylon salt (formed by reacting hexamethylenediamine and adipic acid).

1. The nylon salt in 60 percent aqueous solution and acetic acid to help regulate chain lengths are pumped into an oxygen-free stainless-steel autoclave. The temperature is raised in several steps over several hours, and the steam from the condensation reaction is bled off until the mixture reaches 525°F, where it is held until equilibrium is reached. At an early stage of the polymerization, delusterants and heat and light stabilizers are added.

2. The finished polymer is extruded in ribbon form onto a casting wheel by applying a pressure of oxygen-free nitrogen. The chilled ribbon passes to a chipping unit and is batch-blended.

3. Unlike nylon 6, it is not necessary to wash nylon 6/6 polymer to remove any low-molecular-weight impurities, since the percentage of these is quite low.

4. The basic steps involved in spinning and processing nylon 6/6 are similar to those outlined under nylon 6, except that higher melting temperatures are required for nylon 6/6. See Fig. 9 for production flow chart.

Type 6 Nylon. Caprolactam, the starting material for nylon 6, can start with phenol, which is oxidized to cyclohexanol, then to cyclohexane, then to cyclohexanone oxime, which is then rearranged to caprolactam.

This polymerization is a chain-forming reaction with no splitting-off of water. The steps in making nylon 6 are as follows:

1. A 95 percent aqueous solution of caprolactam and a 0.2 percent solution of acetic acid (for molecular-weight stabilization) are heated to 500°F for 12 h in the absence of oxygen.

2. The high-viscosity polymer is forced out of the polymerization unit as a ribbon and is cooled and chopped into small chips.

3. The chips are washed to remove unreacted monomer, dried in the absence of oxygen, and then sent to the spinning hoppers.

4. In grid spinning, polymer chips are melted on a grid where the temperature is very carefully controlled. The polymer chips are melted in an inert atmosphere,

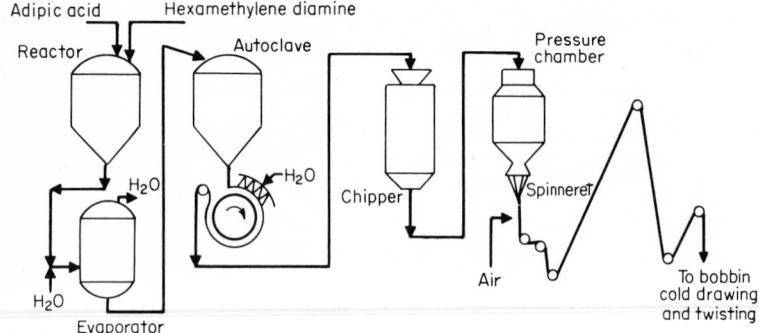

Fig. 9   Flow chart for production of nylon 6/6.

usually nitrogen. The molten polymer falls through to a pool, where it is precisely metered by a gear pump through a sand-pack and wire-screen assembly to remove any impurities and polymer gels. The spinneret is part of the sand pack; it is made of a high-temperature alloy and is 0.25 to 0.50 in. thick. Precisely drilled holes in the spinneret determine the size of the yarn and the shape of the fiber cross section.

5. As it leaves the spinneret holes the molten polymer is cooled by a counter-current of cool air and solidifies.

6. The individual filaments are brought together into a yarn and are given a water-emulsion spun finish to control static and frictional properties. The yarn is then wound up on a bobbin.

7. The next operation is a very critical one—that of drawing. In drawing, the yarn is stretched four to five times its length or more, depending on the end-use properties desired. This drawing process orients the polymer molecules along the fiber length. Drawing increases fiber strength and reduces elongation. It also controls the dyeing properties of the yarn.

Properties of nylon fibers   See Table 2.

In addition to nylon 6, and nylon 6/6, various other nylon types have been and are being developed. The properties of these types, unlike variants on a single type, can be dramatically different from one another. Table 7 lists these types along with a brief description of each.

End uses   Table 8 lists staple, tow, and filament shipments of all types of nylon fiber for 1972. It illustrates the scope of end-use development for this fiber.

Spandex

A manufactured fiber in which the fiber-forming substance is a long-chain synthetic polymer comprised of at least 85 percent of a segmented polyurethane.

History Du Pont first introduced an elastomeric polyurethane into commercial production in 1958 and called it Fiber K. After commercialization, it was marketed under the trade name Lycra.

Raw materials and manufacturing process Spandex fibers owe their high extensibility, just as rubber does, to the fact that their major ingredient is a soft elastic material. In place of isoprene or rubber, spandex uses polyethers or polyesters as segments of a block copolymer chain which must be so designed as to be easily deformable at room temperature. For rapid and good recovery, the soft segment

TABLE 7 Types and Properties of Nylon

Nylon	Made from	MP, °C	Moisture regain, %	Specific gravity	Status
6	Caprolactam	218	4.5	1.14	Commercial: fibers and molding
6/6	Hexamethylene-diamine and adipic acid	256	4.5	1.14	Commercial: fibers and molding
6/10	Hexamethylene-diamine and sebacic acid	215	2.6	1.09	Commercial: brush bristles
11	Aminodecanoic acid	189	1.2	1.04	Commercial (France), fibers and molding
4	Pyrollidone	265	. . .	. . . .	Developmental (U.S.)
7	Aminoheptanoic acid or the ester	225	2.4	1.10	Developmental (U.S. and Russia): fiber
8	Capryllactam	195	1.4	1.08	Developmental (Germany)
9	9-Aminononanoic acid	200	1.2	1.06	Developmental (Russia): fiber
MXD-6	m-Xylylenediamine and adipic acid	234	4.0	1.22	Developmental (U.S.): fiber and molding
Polyoxamides .	Oxalic esters and diamines	235–250	1.0	. . . .	Developmental (U.S. and England)
HT-1, Nomex .	Aromatic diamine and di-acid	375	4.9	1.38	Developmental (U.S.): high-temp fibers
Polyazine	Copolymer of adipic, and isocinchomeronic esters with diamine	260	7.7	1.23	Developmental (Japan)

must have little interchain bonding. It should also be highly flexible and have low internal viscosity. The spandex soft segment not only governs the recovery efficiency but also largely determines the sorption properties and the oxidative and hydrolytic stability of the fiber. Recent trends in soft-segment improvement have been centered on achieving the desired characteristics through copolyesters, specially selected to resist hydrolysis during wet finishing.

Spandex diverges basically from rubber in the mechanism by which the flexible chains are anchored to limit the plastic flow which would result in permanent set.

Rubber relies on vulcanization for cross-linking individual molecular chains. Most spandex fibers, on the other hand, are not cross-linked but contain discrete inflexible domains of material which reinforce the system somewhat as carbon black does in rubber. These domains are agglomerates of the so-called "hard" segments which have been built into the spandex chains.

The hard-segment domains are composed of polyureas typically melting above 200°C. Their surprising efficiency in effecting recovery may be due to their very small size, about 20 to 100 Å, and their thorough anchorage in the soft matrix. The hard and soft phases are chemically bound, the soft-polymer segments having been attached to the hard via urethane groups when the polymer was synthesized. In the relaxed state the structure is not highly oriented. If the structure is stretched, at first the modulus is low since not much force is required to distort the soft material; as the hard segment domains begin to be oriented, however, the soft chains begin to become parallelized, forming low-melting crystalline areas. These crystalline

TABLE 8   Shipments of Nylon for Major End Uses, 1972

End use	Staple and tow, million lb	End use	Filament, million lb
All apparel..................	18.0	All apparel.................	411.6
Three largest apparel uses:		Three largest apparel uses:	
Men's suits, slacks, and coats...................	5.0	Flat knitting..............	171.2
		Circular knitting..........	104.4
Work clothing, men's and women's................	5.0	Hosiery..................	88.8
Socks and knitting.........	1.0		
All home furnishings.........	442.8	All home furnishings.........	569.3
Two largest home furnishings uses:		Two largest home furnishings uses:	
Carpets..................	432.6	Broadloom nylon and transportation rugs.......	549.5
Upholstery and other......	10.1	Upholstery..............	10.0
All industrial and other uses..	16.5	All industrial and other uses..	449.8
Three largest industrial uses:		Four largest industrial uses:	
Miscellaneous industrial uses...................	9.0	Tire cord and fabric.......	284.9
		Rope and cordage.........	25.0
Felts, woven and nonwoven...............	4.0	Auto seat belts............	23.0
Flock...................	2.0	Sewing thread............	13.0

areas are typical of both rubber and spandex. Crystallization continues to occur with increasing stretch until the whole system acts essentially as a hard fiber. This stress-induced crystallization accounts for a final upsweep of the spandex stress-strain curve and the high tenacity (3 to 6 g based on denier at instant of break). When the fiber is allowed to retract, the soft-segment crystallites melt out and revert to the random coil configuration, but the hard-segment crystallites persist.

The combination of a structure of hard and soft segments has several advantages. Not only does it have the regularity of structure which leads to high power, but the hard segments, free of interchain covalent bonds, are sufficiently labile that with heat they can be broken and reformed into new physical configurations. This is necessary for heat-settability, which is very important in developing optimum fabric structure and behavior.

Because elastomerics such as spandex may be used by the textile-mill products industries in three distinct and unique forms, all stemming from fiber-producer output, some mention must be made of them. The technology at the mill level evolved during some 40 years of formation of elastic fabrics using rubber yarns. These forms include bare filament, covered yarns, and core-spun yarns.

The use of bare filament is simple to define: it is fabric formation directly from the producer package. But the control of tensions during weaving and knitting is quite another matter and is considered somewhat of a textile-mill art. Covered yarns are produced by textile mills on so-called "hollow-spindle" machines. The elastomer is threaded up through a hollow spindle, and supporting and stretch-controlling yarns of either spun or filament yarns are wrapped around the elastomer in either one or two (opposite) directions. Core-spun yarns are made on conventional spinning frames: the core of elastomer is predrafted and joined with conventional natural or man-made fibers in the normal draft zone of the spinning frame.

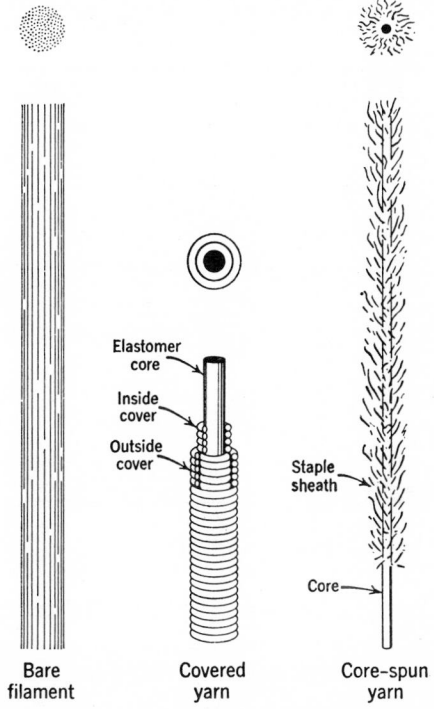

Elastomer
core

Inside
cover

Outside
cover

Staple
sheath

Core

Bare
filament

Covered
yarn

Core-spun
yarn

Fig. 10 The three forms of elastomeric yarns.[3]

Figure 10 illustrates the three forms of elastomer yarns. Figure 11 illustrates a typical core-spinning setup.

Properties[3]   See Table 2.

End uses   The following list of end uses for spandex illustrates the scope of development and the form in which this fiber is used. In 1970, the total man-made-fiber industry capacity for spandex was 28 million pounds.

End Use	*Form of Spandex Yarn*
Foundation garments	Bare, covered
Form-persuasive garments	Bare
Swimwear	Covered, bare, core-ply
Hosiery	Covered, core-ply, bare
Outerwear	Core-spun, core-ply

Glass

A manufactured fiber in which the fiber-forming substance is glass.

History   In 1893, Edward D. Libbey made fiber-glass yarn which was used as a warp with a silk filling for a dress fabric at the Columbian Exposition in Chicago. Mr. Libbey was the founder of Libbey Glass Co. in Toledo, Ohio. His fiber-glass efforts, however, were premature from a commercial standpoint.

During World War I, the Germans made thermal insulation from fiber glass by the Gossler process. This process drew the fibers from the molten glass and collected them on a revolving drum.

Only very slow progress was made in fiber-glass technology until 1931. At about this time, Owens Illinois Co. in Columbus, Ohio, developed steam-blowing techniques for making glass fiber. The resulting material was used for air-filter media. During

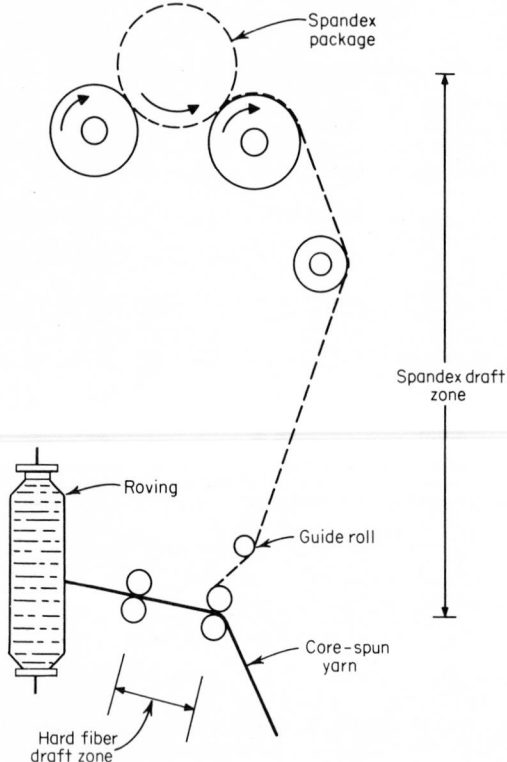

Fig. 11   Typical spandex core-spinning setup.[3]

the 1930s, air-drawing techniques were used for staple fiber-glass production. In 1936, Owens Illinois developed a continuous-filament drawing process for fiber glass. This process drew the streams of molten glass from small heated furnaces called "bushings," at speeds exceeding 2 mi/min.

The Corning Glass Co. was also producing glass fibers, and in 1938, Owens Illinois Co. and Corning Glass Co. pooled their efforts and formed the Owens Corning Fiberglas Corp. The products they produced were sold under the Fiberglas trademark. The first products sold were for air filtration, thermal insulation, and textile yarns for industrial use.

The first two types of yarns developed for textile use were the E type (glass suitable for electrical purposes) and the C type (glass resistant to chemical attack).

In the mid-1960s, Owens Corning Fiberglas achieved a breakthrough in the development of very fine filaments for yarns in the range of 3.8 μm or ¼ denier. These yarns, designated "beta," enabled fiber glass to be used in textile structures which require softer fabrics, higher strengths, and resistance to flexing. They also enable items such as bedspreads and special products for aerospace applications such as astronauts' uniforms to be made from fire-resistant fiber glass.

Raw materials and manufacturing process The major ingredients for the E and C types of glass are listed in the following comparison. E glass is characterized by superior electrical properties and higher heat resistances; C glass can vary in composition according to the need for a variant to resist chemical action.

Composition	E Glass, %	C Glass, %
Silicon dioxide	52–56	60–65
Calcium oxide	16–25	
Aluminum oxide	12–16	2–6
Boron oxide	8–13	2–7
Sodium and potassium oxide	0–1	8–12
Magnesium oxide	0–6	
Magnesium oxide and calcium oxide	.....	15–20

The following five steps outline the basic methods of manufacturing glass yarns:

1. The basic ingredients are mixed and blended according to the type of yarn or fiber to be made.

2. The blended raw materials are fed into the melting and glass-refining unit.

3. The molten glass can either be made into marbles or spun directly from the melt. Originally almost all fiber glass was made into marbles. Now, however, much more processing is being done by direct melt.

4. The difference between the staple-fiber process and the continuous-filament process is in the drawing and collecting techniques. The staple-fiber process employs jets of compressed air to attenuate the molten glass into fine fiber after it passes through orifices in the base of furnaces. The compressed air drives the fibers onto a revolving drum, where they resemble a cobweb. A lubricant is applied by spraying, and the material can be collected as sliver.

5. In the continuous-filament process the fibers are drawn mechanically after being extruded through small orifices. A sizing is applied after the filaments are gathered together and are wound on a high-speed winder at a rate of over 1 mi/min.

Properties of fiber-glass fibers See Table 2.

End uses In 1970, domestic producers of glass fiber shipped 434.1 million pounds. The principal textile use for glass fiber is in multifilament form for such articles as draperies, electrical insulation, and industrial filtration cloths.

Olefin

A manufactured fiber in which the fiber-forming substance is any long-chain synthetic polymer composed of at least 85 percent by weight of ethylene, propylene, or other olefin units, except amorphous (noncrystalline) polyolefins qualifying under category (1) of Paragraph (j) of Rule 7.

History Polyethylene, the first of the olefin fibers, was introduced in 1941. Polyethylene resin was developed commercially by I.C.I. just before World War II. It was based on polymerizing ethylene at what were then very high pressures of 1,000 to 2,000 atmospheres. The low melting point of 128°C, however, has severely limited its use as a textile fiber. It is mainly made in heavy-denier monofilaments used in webbing and cordage applications.

However, the major breakthrough in the preparation of polyolefins came in 1954 with the development of organometallic catalysts by Professor Ziegler. Ziegler's work was based in part on work carried out by G. Natta and coworkers at Milan Polytechnic. The organometallic catalysts enabled polyethylene and polypropylene to be made under normal pressures. The polymerization of the polypropylene into the isostatic form provided a polyolefin with suitable fiber-forming properties. Development of polypropylene fibers started in the United States in 1957.

Raw materials and manufacturing processes Olefin fibers are products of the petroleum industry and are derived from propylene and ethylene gases. Because one olefin—polypropylene—is used mostly for woven primary carpet backing in the specific form of ribbon-shaped monofilament, description of methods for manufacturing it will be limited here to that form.

1. The major methods of making undrawn film in order to produce a 2- by 100-mil ribbon are as follows:

   a. A flat monofilament die where rectangular orifices produce individual narrow ribbons which are quenched in water.

   b. A flat film die where wide sheets several mils thick are extruded, cooled or chill-rolled, and the edges are trimmed and slit into undrawn ribbons.

   c. The film can also be extruded through a circular film die as a tube. The tube is collapsed and forms a double-lay flat sheet which is edge-trimmed and slit into two individual layers.

2. The slit film produced by any of the above methods must be drawn and oriented by heat. The heat source can be heated rolls, radiant heat, or a hot-air chamber. The ribbons are stretched six to eight times.

3. An annealing step is needed to give better dimensional stability and release strains in the yarn. This is usually done by passing ribbon yarn through an annealing oven where the ribbon yarn is overfed by several percent to allow for relaxation.

4. The ribbon yarns are then put up in individual packages and placed on loom beams for use by the weaver.

When polypropylene tapes or ribbons are highly stretched in a lengthwise direction, they have little transverse strength. Controlled mechanical action can shatter or "fibrillate" these yarns into a web of filaments much lower in denier than the original tape.

The steps in making yarns by the fibrillation techniques are as follows:

1. A slit film or tape is produced by any of the three methods previously cited.

2. After the tape is oriented in an oven, it can be split by several techniques such as a rotary knife, razor blades, or a porcupine-like roller.

Fibrillated film yarns can be used as carpet backing, face yarn for carpets, or upholstery fabric.

Conventional multifilament yarns can be spun from polypropylene by techniques similar to those employed in making nylon.

Properties of polyolefin fibers   See Table 2.

End uses   Table 9, which lists staple, tow, and filament shipments of all olefin fibers for 1972, illustrates the scope of end-use development for these fibers.

Other Generics

Because of their developmental status in this country or limited use, several generics will be described in more general and less comprehensive terms than those covered to this point.

Saran   A manufactured fiber in which the fiber-forming substance is any long-chain synthetic polymer composed of at least 80 percent by weight of vinylidene chloride units ($-CH_2-CCl_2-$).

First domestic commercial production was in 1941 by Firestone Plastics Company, the predecessor company of Firestone Synthetic Fibers and Textiles Company.

General fiber properties include good resistance to wear, common chemicals, sunlight, staining, mildew, and weather. It is self-extinguishing, comparatively stiff and dense, and has low softening temperatures.

End Uses.   Transportation fabrics (seating upholstery), outdoor furniture.

Rubber   A manufactured fiber in which the fiber-forming substance is comprised of natural or synthetic rubber, including the following categories:

1. A manufactured fiber in which the fiber-forming substance is a hydrocarbon such as natural rubber, polyisoprene, polybutadiene, copolymers of dienes and hydrocarbons, or amorphous (noncrystalline) polyolefins.

2. A manufactured fiber in which the fiber-forming substance is a copolymer of acrylonitrile and a diene (such as butadiene) composed of not more than 50 percent but at least 10 percent by weight of acrylonitrile units ($-CH_2-CH-$).

CN

The term "lastrile" may be used as a generic description for fibers falling within this category.

3. A manufactured fiber in which the fiber-forming substance is a polychloroprene or a copolymer of chloroprene in which at least 35 percent by weight of the fiber-forming substance is composed of chloroprene units (—CH$_2$—C=CH—CH$_2$—).

$$Cl$$

First domestic commercial production was in 1930 by U.S. Rubber Company, now Uniroyal, Inc.

The general fiber properties are characteristically elastomeric. See the section on spandex.

End Uses. Foundation garments, surgical supports, and elastic webbing (narrow woven or braided fabrics).

TABLE 9 Shipments of Olefin for Major End Uses, 1972

End use	Staple and tow,* million lb	End use	Filament,† million lb
All apparel.................	1.0	All apparel................	Very little used, no data available
Pile fabric linings...........	1.0		
Total home furnishings.......	58.0	Total home furnishings.......	209.4
Three largest home furnishings uses:		Four largest home furnishings uses:	
Needle-punched carpets, rugs, and backing........	48.0	Carpet backing...........	126.1
Tufted broadloom.........	6.0	Broadloom carpets.........	45.3
Upholstery...............	4.0	Upholstery..............	18.8
		Outdoor furniture (webs, saran and olefin)........	9.2
All industrial and other uses..	4.0	All industrial and other uses..	103.7
Largest industrial uses:		Largest industrial uses:	
Filtration woven and nonwoven..............	3.0	Rope, twine, and cordage................	53.4
Other nonwoven..........	1.0		

* Includes a small amount of vinyon in industrial uses.
† Includes monfil, split and slit film.

Vinyon   A manufactured fiber in which the fiber-forming substance is any long-chain synthetic polymer composed of at least 85 percent by weight of vinyl chloride units (—CH$_2$—CHCl—).

First domestic commercial production was in 1939 by American Viscose Corporation.

General fiber properties include low softening temperatures and high general chemical resistance.

End Uses.   Industrials such as thermoplastic bonding agents for nonwoven fabrics. In some parts of the world vinyon is known as polyvinyl chloride fiber.

Metallic   A manufactured fiber composed of metal, plastic-coated metal, metal-coated plastic, or a core completely covered by metal.

First domestic commercial production was in 1946 by The Dobeckmun Company, now absorbed into Dow-Badische Company.

General fiber properties include nontarnishability of the plastic-coated type (metallized polyester or polyester–aluminum foil laminates).   When suitable adhesives and films are used in these plastic metallizations or laminations with metal, they have good general chemical resistance but poor resistance to alkalies, good general color fastness, and processable tensile properties.

End Uses. Most apparel and home furnishings uses, as a decorative yarn.

SIGNIFICANCE OF PROPERTY DATA

In his classic and prophetic Edgar Marburg Lecture,[4] Harold DeWitt Smith proposed various criteria for evaluation and use of textile fibers like those in the durables industries for steel and other rigid materials. Until that time, much of the textile technology for the "limp structures"—fibers—was subjective. Fabric-forming and end-use parameters were for the most part empirical. While the "limp" characteristics, among others, of textile fibers often preclude the optimum in true engineering, the ability to construct textile fabrics predictably from known fiber properties (see Table 10) which Dr. Smith presented in his lecture gave us the basis for modern textile engineering.

TABLE 10   Fundamental Fiber Characteristics[4]

Form	Physical	Chemical

Length:
  Mean length..............
  Length distribution (stable
  diagram)................
Cross section:
  Mean area................
  Area distribution..........
  Shape....................
Crimp:
  Frequency (crimps per
  inch)....................
  Amplitude................
Surface character............
	Density (specific gravity)
	Moisture absorption:
	  Relation between relative
	  humidity and moisture
	  regain
	  Swelling phenomena
	Stress-strain behavior
	  Strength
	  Elongation
	  Stiffness
	  Toughness
	  Elasticity
	  Resilience
	Interfiber friction
	Color
	Luster
	Refractive index
	Scorching or melting point
	Specific heat
	Specific conductivity:
	  Heat
	  Electricity
		Composition
		Structure
		Behavior toward:
		  Acids
		  Alkalies
		  Oxidation
		  Dyes
		  Solvents

Many of the measurable criteria have already been discussed and quantified for various fibers. Some bear further discussion because of their importance.

Density (Specific Gravity)

The specific gravity of fibers has both aesthetic and economic implications. Bulk with light weight is a desirable fiber characteristic for fabrics intended for apparel; and specific gravity has a direct bearing on fiber money value, since all fibers are sold on the basis of price per pound for all textile end uses. Table 11 lists the density of both man-made and natural fibers.

Stress-Strain Behavior

From the time a fiber enters a textile mill until the fiber product is worn out or discarded, the fiber is being pulled and stretched at frequent intervals. To realize this, one has only to think of the nature of the mechanical actions that occur in spinning, throwing, fabric finishing, bending of a trouser knee, or towing with a rope. These pulls on the fiber and the stretching that results are often referred to as stresses and strains. Any pull or stress that tends to stretch the fiber in only one direction is called a tensile stress. Thus, how a fiber behaves when it is pulled or stretched is the same as its tensile stress-strain properties.

The conditions under which a fiber is subjected to tensile stress greatly affect the behavior of the fiber. Generally, a higher stress is necessary to increase the length of a fiber when it is dry or at low temperatures than when it is moist or at high temperatures.

The stress-strain chart   If a fiber is clamped vertically by one end and sufficient weight is attached to the other, the fiber will increase in length. As the weight is increased, the fiber will continue to increase in length until it breaks. A plot of weight added against increase in length gives a graph, commonly called a stress-strain chart. Examination of the shape, length, and height of the stress-strain curve can answer many questions about a fiber such as:

1. How well does it resist stretching?
2. How much will it stretch before breaking?
3. How strong is it?
4. How tough is it?

TABLE 11   Density (Specific Gravity) of Textile Fibers

Generic and type of natural and man-made fibers	Lb/in.³	Specific gravity
Acetate	0.0476	1.3
Acrylic	0.0419–0.0430	1.15
Azlon	0.0471	1.3
Bast	0.0535	1.5
Cotton	0.0559–0.0560	1.5
Polyester	0.0441–0.0500	1.2–1.4
Fluorocarbon	0.0760–0.0795	2.2
Glass	0.0900–0.0920	2.5
Modacrylic	0.0469–0.0470	1.15
Nylon	0.0410–0.0520	1.14
Olefin	0.0392–0.0343	0.9
Rayon	0.0540–0.0550	1.5
Saran	0.0400–0.0621	1.7
Silk	0.0469	1.34
Spandex	0.0361	1.0
Triacetate	0.0469	1.3
Vinyl	0.0466	1.3
Vinyon	0.0495	1.4
Wool	0.0471–0.0473	1.3

The above description is a simple way of obtaining a stress-strain curve. In practice, any one of a number of different types of laboratory instruments might be used. Such apparatus is precision-built, and most types have automatic devices for plotting the stress-strain curve. A typical curve from the Instron tester is shown in Fig. 12.

In this stress-strain curve, distance between the fiber clamps is measured along the horizontal axis and the load on the fiber along the vertical axis. Line *AB* represents the length of a "crimped" fiber without any load on it. As the clamps holding the fiber separate, crimp is pulled out of the fiber, and as the load on the fiber gradually increases, the fiber increases in length, plotting the curve *BGF*.

For a short while after the crimp is removed, the plot of load against increase in length gives a reasonably straight line. If a line *CH* is drawn along this initial "straight-line" portion of the curve through the line of zero load, the line *AC* will represent the approximate length of the straightened, uncrimped fiber.

Resistance to stretch   In the sample curve, line *CD* represents any increase in length of the fiber in centimeters and line *DG* represents the load in grams at that

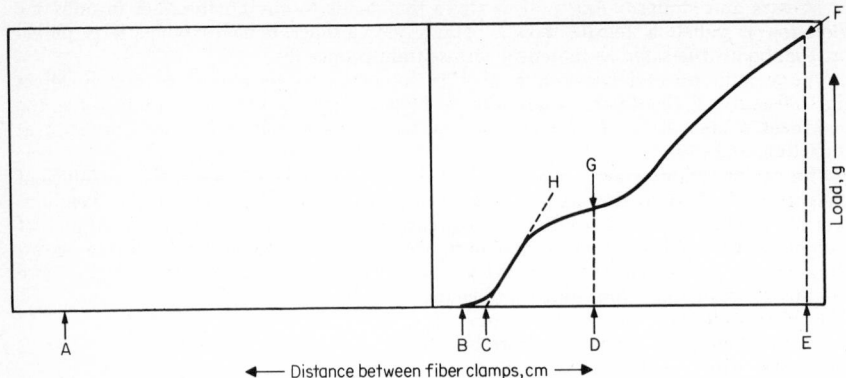

Fig. 12 Stress-strain curve from an Instron tensile-tester chart. (*Instron Engineering Corporation.*)

Point A = point of zero distance between Instron fiber clamps
Line AB = length of "crimped" fiber
Line AC = test length of straightened fiber at estimated zero elongation
Line CD = any specified elongation of fiber
Line CE = elongation at breaking point
Line CH = line along the initial "straight-line" portion of the stress-strain curve BGF
Line DG = load required to elongate by the amount CD
Point F = breaking point of fiber
Line EF = breaking load

$$\text{Breaking tenacity, g/denier} = \frac{EF}{\text{denier at length } AC}$$

$$\text{Breaking elongation, \%} = \frac{CE \times 100}{AC}$$

$$\text{Work to break, g-cm/denier-cm} = \frac{\text{area under curve } BGF}{\text{denier at length } AC \times AC}$$

$$\text{Tenacity at any specific elongation, g/denier} = \frac{DG}{\text{denier at length } AC}$$

length. Any increase in fiber length caused by a load or stress on the fiber is called elongation. Elongation is expressed as the percentage increase in length $\dfrac{CD \times 100}{AC}$.

If the load DG in grams is divided by the fiber denier before elongation, the result is the tenacity required to elongate the fiber by the amount CD. Tenacity is the pull or stress on the fiber measured in grams per denier.

Tenacity to elongate* by a given percentage is a measure of the fiber's stretch resistance or its ability to resist elongation. Since this is a ratio of grams per denier to percent, the slope or direction of the stress-strain curve gives an indication of stretch resistance. The more nearly vertical the curve, the greater is the fiber's resistance to stretching.

Resistance to stretching is an important factor in processing and is related to such fabric properties as drape and hand.

For the most part fibers in apparel, household items, and many industrial uses are elongated during use by only a small amount. In such cases only the initial portion of the stress-strain curve needs to be considered.

* "Tenacity at various elongations" is used synonymously. Where the initial "straight-line" portion of the stress-strain curve extends out to an elongation of 1 percent or more, the tenacity at 1 percent elongation multiplied by 100 equals the "initial modulus" in grams per denier.

Stretch at break point In the stress-strain chart, line CE represents the elongation at the point of fiber rupture or the breaking elongation of the fiber. Breaking elongation is expressed as the total percent increase in length. This is an important property, since a fiber with a high breaking elongation is needed in fabrics and yarns that must withstand large deformations without breaking.

Breaking tenacity Breaking tenacity is the tenacity or stress at the point of fiber rupture. It is determined by dividing the breaking load in grams EF by the fiber denier before any elongation. Obviously a fiber with a high breaking tenacity is needed where high strength is a requirement in fabrics or yarns.

The term breaking tenacity should not be confused with the popular terms breaking strength and tensile strength. Breaking strength is the load in pounds required to break a specimen of textile material. Tensile strength is expressed as stress per unit cross-sectional area, such as pounds per square inch or grams per square centimeter. To convert breaking tenacity in grams per denier to tensile strength in pounds per square inch, multiply tenacity by the appropriate factor as given below:

Acetate	17021	Rayon	19453
Dacron	17661	Cotton	19197
Nylon	14590	Silk	15998
Orlon	14590	Wool	16893

For a better understanding of the nomenclature that is used to describe fiber properties, consult "ASTM Standards on Textile Materials" and "Textile Terms and Definitions."

Figure 13 lists breaking tenacities of various man-made and natural fibers.

Toughness A tough fiber is one that can take a severe amount of punishment up to the point of breaking. Toughness is measured by the work required to break the fiber. Since work is the product of force times distance, work to break can be determined from the product of load times elongation or the total area $BFEB$ under the stress-strain curve BGF. The larger this area the greater is the work to break or toughness. Since work to break increases as the dimensions of the fiber become greater, it must be expressed in terms of unit denier and unit length. Otherwise results will not be comparable.

Work to break is indicative of the flex life, abrasion resistance, and shock absorbency of the fiber. It is also an important property where factors such as brittleness are concerned.

Application of Stress-Strain Properties

1. *Yarn Manufacturing:* In opening, picking, and other early stages of yarn spinning, fibers are subjected to rugged treatment. Tough fibers can withstand this vigorous working without damage.

With fibers of low stretch resistance, high roving twists and close ratch settings can cause tight picks and ends in finished fabric.

In blend yarns, wide differences in the stress-strain properties of the component fibers will reduce the breaking strength of the yarn.

Cockling is more likely to be a problem with high-tenacity staples having a high breaking elongation and low stretch resistance.

2. *Throwing:* Throwing consists primarily of yarn transfer from one package to another while twisting. The yarn is under tension; and, as is well known, variations in tension in filament yarns can cause such quality problems as warp streaks, filling bands, barré, or quill junctions in finished fabric. Tension levels can be chosen based on a knowledge of the stretch-resistance properties of the fiber being handled. This is especially useful if fibers with relatively low stretch resistance, such as nylon, are being used. For example, consider two fibers a and b, fibers having about the same breaking tenacity and breaking elongation. To stretch fiber a by the amount E would require a tenacity about three times as great as that required to stretch fiber b by the same amount. With the introduction of heat or moisture, as in sizing, it is usually necessary to handle the yarns more carefully, since these two conditions both reduce the stretch resistance of fibers (see Fig. 14).

3. *Dyeing and Finishing:* With so much of this work taking place under variable conditions of heat and moisture, fabrics and yarns are susceptible to distortion. Therefore, careful consideration must be given to the fibers in each item coming into the area. Here it should be emphasized that the stress-strain properties at the dyeing conditions must be known to visualize how well the fiber will withstand the necessary mechanical handling. For example, Orlon* acrylic fiber stretches quite readily in a hot bath, and skein-dyed yarns must be cooled before being lifted from the bath; otherwise, they will lose their bulkiness.

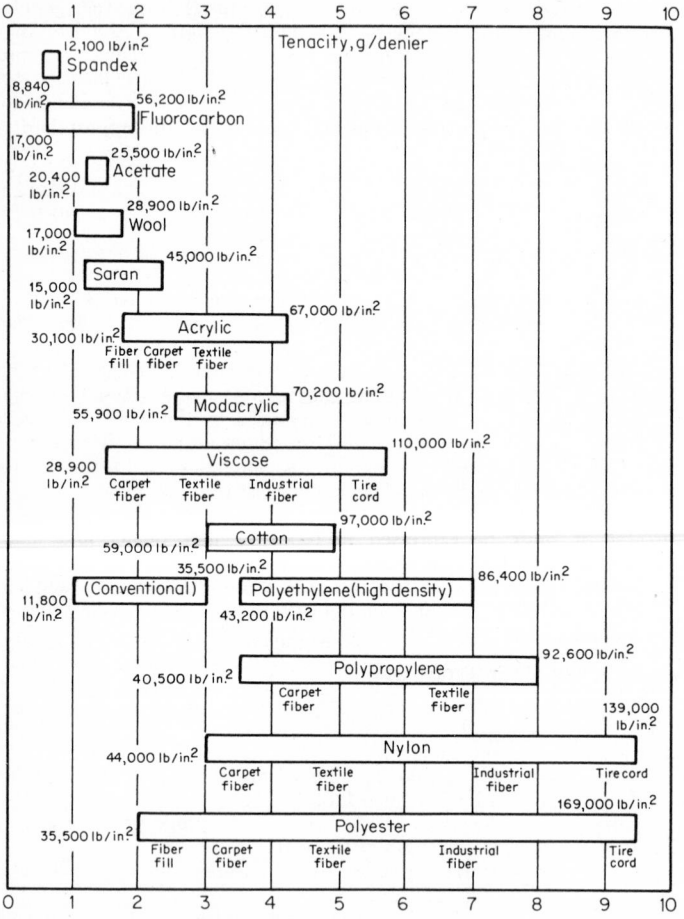

Fig. 13   Range of breaking tenacities in grams per denier of textile yarns.

4. *Weaving:* Close attention to the adjustment of tensions is necessary in controlling the tightness of the ends and picks in finished fabric. The fact that the stretch resistance of fibers varies should be constantly considered in quill winding, warp preparation, and loom adjustment. It is also important to emphasize that the tenacity required to elongate a given fiber varies depending upon the amount of elongation. Therefore, better tension adjustments can be made by a study of how stretch resistance varies with tension level.

* Trademark of E. I. du Pont de Nemours & Co., Inc.

5. *Knitting:* Proper tension control is the key to quality knitting. Stitch formation, quality of dyeing, and fabric stability are all dependent to a great extent upon the fiber's tensile properties, namely, resistance to stretching. Because of the low tension involved and the wide variation in the stretch resistance of fibers at low elongations, critical attention must be given to the knitting machine when changing from one fiber to another.

Recovery of Fibers from Deformation

During the production and processing of fibers and the use of fiber products, the fibers are constantly being subjected to the action of external forces which deform or distort them. The types of deformation or distortion encountered are elongation, compression, bending, shear, and various combinations of these. When the deformations occur, internal stresses are introduced which tend to return the fibers to their original shape and dimensions. If the deformations are small and the external forces are applied for only a short time, the fibers may return almost immediately to their original form, and the work of deformation may be almost completely recoverable. If, however, the fibers are held in the deformed state, internal stresses caused by the deformation diminish with time, and the tendency of the fibers to return to their original form is reduced. A very rapid decay of internal stresses is obtained by subjecting the fibers, while they are held in the deformed state, to heat, moisture, or other swelling agents.

This is done in the twist setting, crimp setting, and hot setting of fibers or yarns, as well as in the heat setting, embossing, and pressing of fabrics.

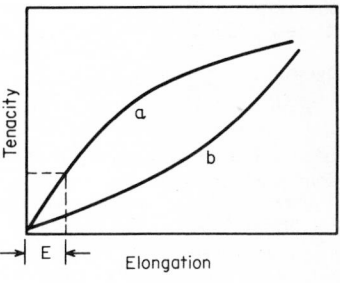

Fig. 14

The extent and manner of recovery of fibers from deformations differ considerably from one type of fiber to another and play a major part in determining the processing characteristics of fibers and the end-use performance of fiber products. For example, high recovery from low stretch is an important factor affecting the use of Dacron polyester fiber in the suiting market because suiting fibers are normally subjected to very small deformations. It is of particular significance that the recovery properties of Dacron* are only slightly affected by moisture conditions. The low moisture sensitivity of Dacron contributes to the good wrinkle resistance, crease retention, and "wash-and-wear" characteristics of suiting fabrics made from this fiber.

Some important factors to consider when relating fiber-recovery behavior to processing problems or end-use performance are the following:

1. Whether small or large deformations are encountered in the processing operations and in the normal use of the fiber product
2. How long the fiber is held in the deformed state by the external forces before these forces are removed to permit recovery
3. How fast the fiber recovers after removal of the deforming force
4. What the environmental conditions (dry, hot, wet) are during deformation, holding, and recovery
5. How much the recovery of the fiber is prevented or delayed by retarding factors such as interfiber friction

The following examples will indicate how recovery can affect the behavior of the fiber during mill processing and during use of the fiber product.

Yarn and Fabric Manufacturing

In various processing stages—such as (1) opening, picking, and other staple-fiber-processing steps through spinning; (2) transferring of spun or filament yarns from one package to another; and (3) warping of spun or filament yarns for weaving or knitting—the fibers or yarns may be subjected to high and nonuniform tensions

* Trademark of E. I. du Pont de Nemours & Co., Inc.

that cause large and nonuniform deformations of the fibers. Residual internal stresses resulting from these deformations, if not relieved before the fibers are spun into yarn or before the yarns are woven into fabric, can manifest themselves in many undesirable forms in subsequent processing steps. Cockled spun yarns, crushed yarn packages, distorted beams, warp streaks, tight picks, barré, and other defects have frequently been traced to erratic tensions encountered by fibers or yarns in some previous stage of the textile process.

In the manufacture of some high-bulk yarns, staple fibers with a high residual shrinkage (introduced by large deformations in previous processing) are blended with fibers having a low residual shrinkage and then spun into yarn. When the yarn is subjected to heat and moisture while under little or no tension, the high-shrinkage fibers shrink and cause buckling of the other fibers to produce the desired high-bulk yarn.

In the preparation of some stretch yarns, continuous-filament yarns are crimped, coiled, looped, or otherwise deformed and then heat-treated to relieve internal stresses caused by the deformations. As a result, any subsequent attempt to restore the yarn to its original length creates a multitude of new fiber stresses which resist change and tend to return the stretched yarn to its heat-set length.

Twisting of yarns often results in excessive twist liveliness, caused by the internal stresses introduced during the twisting operation. Although these stresses can be relieved by twist setting, nonuniform penetration of the heat and/or moisture into the yarn on the twist-setting package will cause differences in residual internal stresses which could result in streaky fabric or nonuniform fabric dimensions.

Dyeing and Finishing

With so many of the dyeing and finishing operations taking place under hot and/or wet conditions, and because of the lower modulus of fibers under these conditions, fabrics and yarns are susceptible to large deformations. Careful consideration must therefore be given to each of the fibers in each item being processed. It should be emphasized that the recovery properties at the dyeing and finishing conditions (i.e., environmental temperature and medium) and the setting effects of these conditions must be known in order to visualize how well the fiber will withstand the necessary mechanical handling. Boiling water, for example, is a fairly severe setting medium for nylon, and for this reason most woven fabrics of filament nylon yarn should be held smooth during dyeing at temperatures near the boil. This prevents the setting of wrinkles which would later be very difficult to remove.

Wrinkle Resistance of Fabrics

Normally, only small fiber deformations are involved in the casual wrinkling of suiting fabrics during wear; hence the ability of the fiber to recover from small deformations is important. Interfiber friction and the setting effects of moisture at high relative humidities are other factors affecting wrinkle resistance.

In some sheer fabrics, especially those made either from monofilament yarns or from multifilament yarns with the filaments bonded together with a resin, the fibers may be subjected to large bending deformations during normal wear. Here a different fiber may be required for maximum wrinkle resistance than was required for the suiting fabric.

Fiber recovery from deformation is also an important factor in the wrinkling of fabrics during laundering; however, of even greater importance for the hydrophilic fibers may be the tendency of water to "set" wrinkles in the fabric being laundered, and to overcome the effects of any previous pressing treatments.

Crease or Pleat Retention of Fabrics

When a crease or pleat is formed in a fabric, internal stresses are introduced into the fibers. These stresses are relieved by heat and/or moisture during pressing, and the crease is "set" in the fabric. When an external force is applied to remove the crease and straighten the fibers, new internal stresses are introduced which tend to return the fibers to their "set" shape. However, these new stresses may also be relieved by heat and/or moisture, and when this occurs the crease will be either removed or reduced in sharpness.

Casual wear wrinkles and wash wrinkles can be removed from an "automatic wash-and-wear" fabric, without destroying either crease or pleats, by tumbling the fabric in a clothes dryer. The tumbling action tends to straighten the fibers in the wrinkles while the fibers are exposed to the mild setting conditions in the dryer. This procedure will usually remove the casual wear wrinkles, which were not "set" in the fabric, while having only a minor effect on the creases which were "set" in the fabric by pressing under more drastic conditions than those normally found in a dryer. New wrinkles may be introduced during drying, however, if the dryer is overloaded or if a hot fabric is allowed to remain in a wrinkled condition in the bottom of the dryer after the tumbling action has stopped.

Fabric Liveliness or Resiliency

A rapid rate of recovery from deformations is important for fibers used in the manufacture of resilient fabrics. Such fibers should also be able to overcome inter-fiber friction and other retarding factors. This factor of resilience is related to the amount of work the fibers can perform during rapid recovery from bending deformations.

Other Fabric Characteristics

Some other fabric characteristics which are greatly influenced by the recovery properties of the fiber are:

Shape and form retention in woven and knit fabrics and garments
Bulk and thickness retention of fabrics
"Automatic wash-and-wear" behavior
Abrasion resistance and flex life
Fabric hand

Recovery Testing

A test study of the load-elongation behavior of a typical fiber during recovery from deformation is illustrated and outlined in Fig. 15. Definitions of various quantified recovery properties are suggested in this figure.

Chemical Resistance

Textile fibers are frequently described in property-data listings in terms of degradation or discoloration resistance to various chemicals. The reasons for this center on the numerous exposures to chemically concentrated wet media to which fibers are subject during yarn and fabric manufacture and end use, principally home care (dry cleaning, laundering).

Most often listed are alkalies, acids, and oxidizing and reducing agents. Some of the wet media in which these chemicals are used in fiber, yarn, fabric, and made-up end-use form include scouring (washing), bleaching, rinsing, dyeing, printing, and chemical finishing (such as water-repellent, flame-resistant, and soil-release finishing).

Table 12 lists the resistance of certain man-made fibers to these frequently used textile chemicals in descending order of resistance.

STATISTICAL PROFILE OF THE TEXTILE INDUSTRY

Numerous government agencies, trade associations, trade journals, and private consultants provide statistical data relative to the textile industry. Some typical examples of important data are given in the following tables and figures:

Fiber consumption:
World fiber consumption (Fig. 16)[5]
U.S. fiber consumption (Fig. 17)[6]
U.S. per capita fiber consumption (Table 13)[7]
Fiber production:
U.S. man-made fiber production (Table 14)[8]

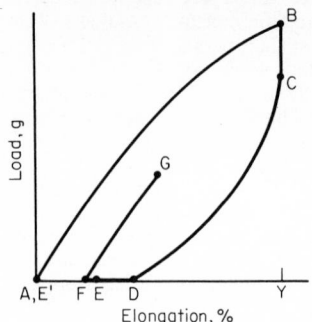

Fig. 15 Test study and definitions of load-elongation behavior of a typical fiber during recovery from deformation.

Outline of recovery test:

Point A	Start of first cycle of test. Elongation phase begins.
Curve AB	Specimen of fiber is stretched at constant rate of speed.
Point B	Specified elongation AY is reached. Holding phase begins.
Line BC	Specimen is held at constant length for a specified time. Stress decay becomes apparent.
Point C	Recovery phase begins.
Curve CD	Fiber specimen is allowed to recover at same rate of speed used for elongation phase.
Point D	Fiber under no load. Rate of fiber recovery changes.
Line DE	Recovery continues but at a progressively slower rate. Slack in specimen becomes increasingly evident as fiber clamps of tester return to their initial position.
Point E	Second cycle of test begins when elapsed time of recovery phase equals elapsed time of initial elongation phase. (For a perfectly elastic fiber, point E would be located at E', recovery would be complete, and reelongation would begin.)
Line EF	Some recovery still taking place. Slack in specimen becomes less evident as fiber clamps of tester retrace their movements of the initial elongation phase.
Point F	Slack in specimen disappears. Recovery stops. Reelongation of fiber begins.
Curve FG	Specimen is restretched at same constant rate of speed as before.

Definition of recovery properties:

$$\text{Initial stress, g/denier} = \frac{load\ BY}{\text{fiber denier at zero elongation}}$$

$$\text{Stress decay, \%} = \frac{\text{load } BC}{\text{load } BY} \times 100$$

$$\text{Tensile form recovery, \%} = \frac{\text{recovered elongation } FY}{\text{original elongation } AY} \times 100$$

$$\text{Work to elongate, g-cm/denier-cm} = \frac{\text{area under elongation curve } AB}{\text{fiber denier} \times \text{length of specimen at } A}$$

$$\text{Work recovery, \%} = \frac{\text{area under recovery curve } CD}{\text{area under elongation curve } AB} \times 100$$

Production and price index of Noncellulosic man-made fibers (Fig. 18)[8]
Estimated man-made fiber sales by U.S. producers (Table 15)[8]
Estimated time required for profitable production of some man-made fibers (Table 16)
Fabric manufacturing:
The 38 largest U.S. textile corporations, and their rank in U.S. industry (Table 17)[9]

TRADEMARKS

See Table 18.

TABLE 12   Chemical Resistance of Certain Fibers in Descending Order of Resistance

Resistance to alkalies	Resistance to acids	Resistance to oxidizing and reducing agents
Teflon	Teflon	Teflon
Polypropylene	Polypropylene	Polypropylene
Modacrylics	Modacrylics	Nomex nylon
Nylon 6/6 and 6	Acrylics	Modacrylics
Nomex nylon	Polyesters	Polyesters
Polyesters	Nomex nylon	Acrylics
Rayon	Nylon 6/6 and 6	Nylon 6/6 and 6
Acrylics	Rayon	Rayon

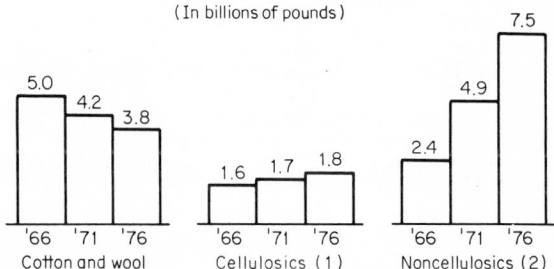

Fig. 16   World fiber consumption.[5]   (1) Rayon and acetate.   (2) Polyester, nylon, and acrylics.

TABLE 13   U.S. per Capita Consumption by Fiber Type[7]

Year	Lb per capita			
	Man-made	Cotton	Wool	Total
1960	10.1	23.3	3.1	36.5
1970	27.7	19.8	1.7	49.2
1976	36.0	18.0	1.0	55.0

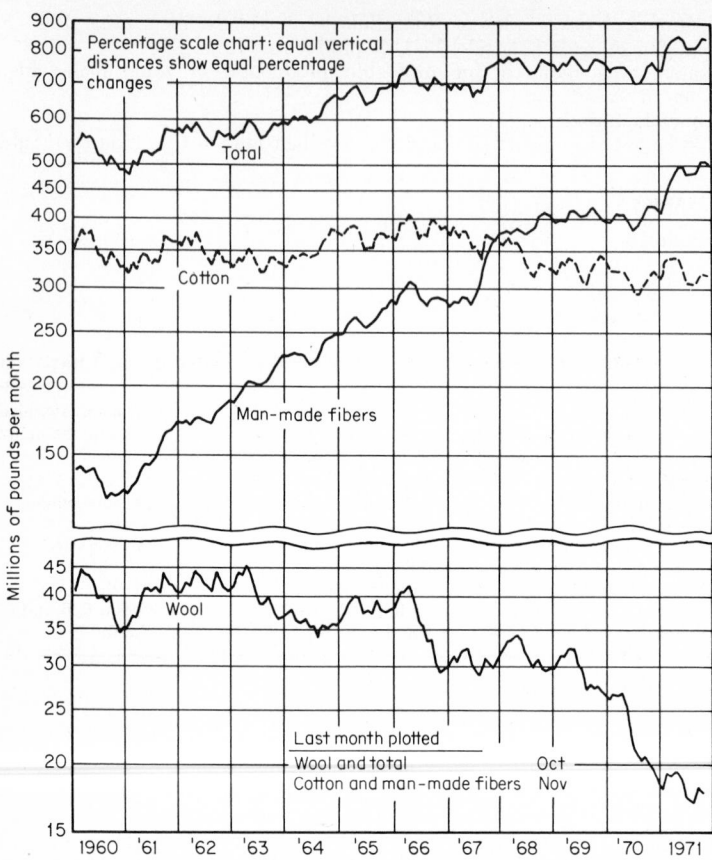

Fig. 17   U.S. textile fiber consumption.[6]   Data are 3-month moving averages, plotted at the middle month.   For cotton and wool, consumption means fiber put into process at spinning mills.   The man-made fiber data are U.S. producers' domestic shipments plus imports for consumption of (a) rayon staple and rayon-acetate yarn and (b) the noncellulosic fibers.   Acetate staple and textile glass fiber are not included.

TABLE 14   U.S. Man-Made-Fiber Production, 1961 and 1970[8]

	Million lb		% increase, 1961–1970
	1961	1970	
Noncellulosic:			
Nylon:			
Filament...................................	442	1,137	157
Staple.....................................	34	221	550
	476	1,358	185
Polyester:			
Filament...................................	27	440	1,530
Staple.....................................	77	1,022	1,227
	104	1,462	1,306
Acrylic:			
Staple.....................................	140	492	251
Olefins:			
Filament...................................	19	196	932
Staple.....................................	1	59	5,800
	20	255	1,175
Other:			
Filament...................................	10	15	50
Staple.....................................	1	3	200
	11	18	64
Total, noncellulosic.......................	751	3,585	377
Cellulosic:			
Rayon:			
High-tenacity...............................	251	147	−41
Other filament.............................	142	121	−15
Staple.....................................	400	607	52
	793	875	10
Acetate:			
Filament...................................	249	463	86
Staple.....................................	53	35	−34
	302	498	65
Total, cellulosic..........................	1,095	1,373	25
Total.................................	1,846	4,958	169

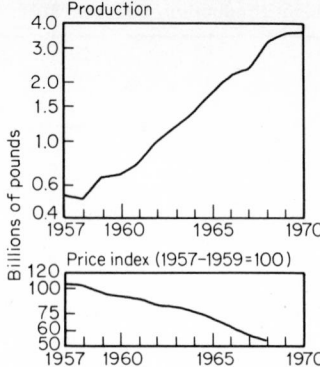

Fig. 18 Production and price index of noncellulosic fibers, 1957–1970.[8]

TABLE 15   Estimated Man-Made-Fiber Sales by U.S. Firms in 1970[8]

Rank	Producer	Sales ($ million)	Product code‡
1	Du Pont	1,100	A-N-P-Ac-O-S-X
2	Celanese	550	A-N-P-O
3	Monsanto	420	N-P-Ac
4	American Enka	230	R-N-P
5	Eastman	220	A-P-Ac-X
6	FMC	210	R-A-P-X
7	Allied Chem	160	N-P
8	American Cyanamid	100	R-P-Ac
9	Beaunit	100	R-N-P
10	Dow Badische	60	N-P-Ac
11	Hercules	40	O
12	Phillips	35	N-P-O
13	Courtaulds*	30	R-N-Ac
14	Firestone	30	N-P
15	Hoechst†	30	R-P-Ac
16	Röhm & Haas	25	N-X
	All others	30	
		3,370	

* Sells acrylic fiber manufactured by parent company in United Kingdom.
† Sells acrylic and rayon fibers manufactured by parent firm in Germany.
‡ A = acetate and triacetate   P = polyester
  Ac = acrylic and modacrylic   R = rayon
  N = nylon   S = spandex
  O = olefin   X = others

TABLE 16   Estimated Time Required for Profitable Production of Some Man-Made Fibers

Company	Product	Year of first full commercial production	Years before profitable operations
Allied Chemical Co	Nylon 6	1955	8
Chemstrand (now Monsanto Textile Division)	Nylon 6/6	1954	2
Chemstrand	Acrylic	1952	12
American Cyanamid	Acrylic	1959	7
Celanese	Polyester	1960	4
Hercules	Polypropylene	1961	6

Textile rank 1971	Textile rank 1970	Overall rank 1971	Overall rank 1970		Sales ($000)	Net profit ($000)	Profit as % of invested — Sales	Profit as % of invested — Capital	Employees	Sales per employee
1	1	65	56	Burlington	1,727,045	40,141	2.3	5.3	82,000	$21,061
2	2	133	145	J. P. Stevens	861,066	(642)*	..	6.2	42,100	20,453
3	3	170	166	United Merchants	738,388	16,400	2.2	6.2	36,000	20,510
4	4	248	226	Kayser-Roth	476,884	9,395	2.0	5.9	29,500	16,166
5	5	260	270	M. Lowenstein	444,078	8,932	2.0	6.3	17,524	25,341
6	6	305	277	West Point-Pepperell	355,243	6,232	1.8	3.3	19,500	18,218
7	7	340	329	Springs Mills	325,639	9,321	2.9	3.7	19,600	16,614
8	8	342	331	Cannon Mills	323,478	16,833	5.2	7.2	24,000	13,478
9	9	346	338	Cone Mills	319,124	6,229	2.0	4.6	14,300	22,316
10	10	351	344	Dan River	312,405	(2,560)*	..	..	19,553	15,977
11	11	362	360	Jonathan Logan	301,548	17,537	5.8	15.2	13,000	23,196
12	12	366	402	Kendall	294,388	11,547	3.9	9.8	13,000	22,645
13	13	385	416	VF	276,076	14,721	5.3	14.6	16,000	17,255
14	14	427	420	Collins & Aikman	239,182	12,410	5.2	15.0	8,300	28,817
15	15	437	433	Fieldcrest Mills	227,275	10,640	4.7	12.7	11,801	19,259
16†	18	258	257	Indian Head	192,300	9,800	5.1	..		
17	17	500	480	Hanes	176,031	3,385	1.9	4.0	13,018	13,526
18	19	503	504	Reeves Brothers	175,202	5,244	3.0	8.8	7,000	25,029
19	20	506	527	Ludlow	173,346	5,101	2.9	8.8	11,325	15,306
20	22	514	538	Riegel Textile	171,112	2,827	1.7	4.5	8,900	19,226
21	28	531	713	Texfi Industries	164,183	11,294	6.9	24.5	4,660	35,232
22	16	541	475	E. T. Barwick	158,123	(1,621)*	..	..	4,675	33,823
23	23	561	588	Chadbourn	150,997	(17,585)*	..	..	12,300	12,276
24	25	584	633	Avondale	144,082	5,720	4.0	9.1	5,218	27,612
25	21	596	531	Duplan	141,431	5,606	4.0	13.1	5,300	26,685
26	26	611	660	Graniteville	135,204	3,717	2.7	6.8	5,853	23,100
27	27	644	675	Chelsea Industries	126,528	2,763	2.2	10.0	4,000	31,632
28	24	686	632	Bibb	113,224	(7,516)*	..	..	6,800	16,651
29	29	734	763	Dixie Yarns	99,363	2,325	2.3	7.5	4,000	24,841
30	30	743	771	Belding Heminway	95,258	2,829	3.0	10.4	2,800	34,021
31	35	807	892	Textiles, Inc.	85,279	3,537	4.1	11.8	4,300	19,832
32	31	863	837	National Spinning	76,347	1,927	2.5*	7.8	2,400	31,811
33	34	877	859	Bates	73,660	2,275	3.1	5.8	3,900	18,887
34	32	885	835	American Thread	72,570	2,607	3.6	7.4	4,176	17,378
35	33	891	855	Ozite	71,672	401	0.6	2.8	1,633	43,890
36	36	938	969	Chatham	65,939	3,688	5.6	12.5	3,100	21,271
37	37	966	981	C. H. Masland	63,580	1,020	1.6	4.1	1,450	43,848
38	..	990	..	Crompton	60,411	4,551	7.5	10.2	3,000	20,137

From The Fortune Directory of the 500 Largest Industrial Corporations and The Fortune Directory of the Second 500 Largest Industrial Corporations. Does not include privately owned companies, such as Deering Milliken, which do not publish financial results, or companies, e.g., Bangor Punta, to whose total sales textiles contribute a relatively minor part. Figures include textile sales only.

* Loss.

† Sales by Indian Head's specialty textiles group during 1971 made up only 42 percent of the firm's total sales.

TABLE 18 Trademarks[10]

Trademark	Generic name	Member company
Acele..............	Acetate	E. I. du Pont de Nemours & Company, Inc.
A-Acrilan...........	Acrylic, modacrylic	Monsanto Textiles Company
Acrilan..............	Acrylic, modacrylic	Monsanto Textiles Company
Actionwear..........	Nylon	Monsanto Textiles Company
Anavor..............	Polyester	Dow Badische Company
Angelrest............	Polyester	Fiber Industries Inc., marketed by Celanese Fibers Marketing Co., a Division of Celanese Corp.
Anim/8.............	Anidex	Röhm and Haas Company, Fibers Division
Anso................	Nylon	Allied Chemical Corporation, Fibers Division
Antron..............	Nylon	E. I. du Pont de Nemours & Company, Inc.
Ariloft..............	Acetate	Eastman Kodak Company, Tennessee Eastman Company Division
Arnel...............	Triacetate	Celanese Fibers Marketing Co., Celanese Corp.
Arnel Plus...........	Triacetate, nylon	Celanese Fibers Marketing Co., Celanese Corp.
Astroturf............	Nylon, polyethylene	Monsanto Company, New Enterprise Division
Avceram............	Inorganic	FMC Corporation, American Viscose Division
Avicolor.............	Acetate, rayon	FMC Corporation, American Viscose Division
Avicrimp............	Rayon	FMC Corporation, American Viscose Division
Aviloc..............	Rayon	FMC Corporation, American Viscose Division
Avisco..............	Acetate, rayon, vinyon	FMC Corporation, American Viscose Division
Avistrap.............	Rayon	FMC Corporation, American Viscose Division
Avlin...............	Polyester	FMC Corporation, American Viscose Division
Avril...............	Rayon (high-wet-modulus)	FMC Corporation, American Viscose Division
Ayrlyn..............	Nylon	Röhm and Haas Company, Fibers Division
Beau-Grip...........	Rayon	Beaunit Corporation
Bi-Loft..............	Acrylic	Monsanto Textiles Company
Blue C..............	Nylon, polyester	Monsanto Textiles Company
Bodyfree............	Nylon	Allied Chemical Corporation, Fibers Division
Briglo...............	Rayon	American Enka Company
Cadon..............	Nylon	Monsanto Textiles Company
Cantrece............	Nylon	E. I. du Pont de Nemours & Company, Inc.
Caprolan............	Nylon	Allied Chemical Corporation, Fibers Division
Captiva.............	Nylon	Allied Chemical Corporation, Fibers Division
Cedilla.............	Nylon	Fiber Industries Inc., marketed by Celanese Fibers Marketing Co., a Division of Celanese Corp.
Celacloud............	Acetate	Celanese Fibers Marketing Co., Celanese Corp.
Celaloft.............	Acetate	Celanese Fibers Marketing Co., Celanese Corp.
Celanese............	Acetate	Celanese Fibers Marketing Co., Celanese Corp.
Celanese............	Nylon, polyester, triacetate	Fiber Industries Inc., marketed by Celanese Fibers Marketing Co., a Division of Celanese Corp.
Celaperm............	Acetate	Celanese Fibers Marketing Co., Celanese Corp.
Celara..............	Acetate	Celanese Fibers Marketing Co., Celanese Corp.
Chromspun..........	Acetate	Eastman Kodak Company, Tennessee Eastman Company Division
Coloray.............	Rayon	Courtaulds North America Inc.
Cordura.............	Nylon	E. I. du Pont de Nemours & Company, Inc.
Courtaulds Nylon.....	Nylon	Courtaulds North America Inc.
Crepeset............	Nylon	American Enka Company
Creslan.............	Acrylic	American Cyanamid Company

**TABLE 18   Trademarks[10]
(Continued)**

Trademark	Generic name	Member company
Cumuloft............	Nylon	Monsanto Textiles Company
Dacron..............	Polyester	E. I. du Pont de Nemours & Company, Inc.
Dye I...............	Nylon	Monsanto Textiles Company
Elura...............	Modacrylic	Monsanto Textiles Company
Encron..............	Polyester	American Enka Company
Englo...............	Rayon	American Enka Company
Enka................	Rayon, nylon, polyester	American Enka Company
Enkaloft.............	Nylon	American Enka Company
Enkalure............	Nylon	American Enka Company
Enkalure II..........	Nylon	American Enka Company
Enkasheer...........	Nylon	American Enka Company
Enkrome............	Rayon	American Enka Company
Esterweld...........	Polyester	American Cyanamid Company, IRC Fibers Company
Estron..............	Acetate	Eastman Kodak Company, Tennessee Eastman Company Division
Estron SLR..........	Acetate	Eastman Kodak Company, Tennessee Eastman Company Division
Fiber 40.............	Rayon (high-wet-modulus)	FMC Corporation, American Viscose Division
Fiber 410............	Rayon	FMC Corporation, American Viscose Division
Fiber 700............	Rayon (high-wet-modulus)	American Enka Company
Fibro...............	Rayon	Courtaulds North America Inc.
Fibro DD............	Rayon	Courtaulds North America Inc.
Fibro FR............	Rayon	Courtaulds North America Inc.
Fortrel..............	Polyester	Fiber Industries Inc., marketed by Celanese Fibers Marketing Co., a Division of Celanese Corp.
Fortrel 7............	Polyester	Fiber Industries Inc., marketed by Celanese Fibers Marketing Co., a Division of Celanese Corp.
Fybrite..............	Polyester	Celanese Fibers Marketing Co., Celanese Corp.
Golden Touch........	Polyester	American Enka Company
Herculon............	Olefin	Hercules Incorporated, Fibers Division
I.T.................	Rayon	American Enka Company
Jetspun.............	Rayon	American Enka Company
Kodel...............	Polyester	Eastman Kodak Company, Tennessee Eastman Company Division
Kolorbon...........	Rayon	American Enka Company
Krispglo............	Rayon	American Enka Company
Loftura.............	Acetate	Eastman Kodak Company, Tennessee Eastman Company Division
Loktuft.............	Olefin	Phillips Fibers Corporation, Subsidiary of Phillips Petroleum Company
Lurex...............	Metallic	Dow Badische Company
Lycra...............	Spandex	E. I. du Pont de Nemours & Company, Inc.
Marvess.............	Olefin	Phillips Fibers Corporation, Subsidiary of Phillips Petroleum Company
Marvess III BCF......	Olefin	Phillips Fibers Corporation, Subsidiary of Phillips Petroleum Company
Monvelle............	Biconstituent nylon/spandex	Monsanto Textiles Company
Multisheer...........	Nylon	American Enka Company
Nandel..............	Acrylic	E. I. du Pont de Nemours & Company, Inc.
Nomex..............	High-temperature-resistant nylon (DPO1*)	E. I. du Pont de Nemours & Company, Inc.

**TABLE 18   Trademarks[10]
(Continued)**

Trademark	Generic name	Member company
Orlon	Acrylic	E. I. du Pont de Nemours & Company, Inc.
Phillips 66 Nylon	Nylon	Phillips Fibers Corporation, Subsidiary of Phillips Petroleum Company
Phillips 66 Nylon, BCF	Nylon	Phillips Fibers Corporation, Subsidiary of Phillips Petroleum Company
Purilon	Rayon	FMC Corporation, American Viscose Division
Qiana	Nylon	E. I. du Pont de Nemours & Company, Inc.
Quintess	Polyester	Phillips Fibers Corporation, Subsidiary of Phillips Petroleum Company
Rayflex	Rayon	FMC Corporation, American Viscose Division
SEF	Modacrylic	Monsanto Textiles Company
Serene	Polyester	Fiber Industries Inc., marketed by Celanese Fibers Marketing Co., a Division of Celanese Corp.
Shareen	Nylon	Courtaulds North America Inc.
Skybloom	Rayon	American Enka Company
Skyloft	Rayon	American Enka Company
Softglo	Rayon	American Enka Company
Source	Biconstituent fiber†	Allied Chemical Corporation
Spectran	Polyester	Monsanto Textiles Company
Spectrodye	Nylon	American Enka Company
Strialine	Polyester	American Enka Company
Stryton	Nylon	Phillips Fibers Corporation, Subsidiary of Phillips Petroleum Company
Super Bulk	Nylon	American Enka Company
Super L	Rayon	FMC Corporation, American Viscose Division
Super Rayflex	Rayon	FMC Corporation, American Viscose Division
Suprenka	Rayon	American Enka Company
Teflon	Fluorocarbon	E. I. du Pont de Nemours & Company, Inc.
Textura	Polyester	Röhm and Haas Company, Fibers Division
Trevira	Polyester	Hoechst Fibers Incorporated
Tricocel	Acetate	Celanese Fibers Marketing Co., Celanese Corp.
Tyweld	Rayon	American Cyanamid Company, IRC Fibers Company
Ultron	Nylon	Monsanto Textiles Company
Variline	Nylon	American Enka Company
Verel	Modacrylic	Eastman Kodak Company, Tennessee Eastman Company Division
Villwyte	Rayon	American Cyanamid Company, IRC Fibers Company
Vivana	Nylon	Dow Badische Company
Vycron	Polyester	Beaunit Corporation
Wear-Dated	Acrylic, modacrylic, nylon, polyester	Monsanto Textiles Company
Xena	Rayon (high-wet-modulus)	Beaunit Corporation
Zantrel	Rayon (high-wet-modulus)	American Enka Company
Zefkrome	Acrylic	Dow Badische Company
Zefran	Acrylic, nylon, polyester	Dow Badische Company
Zefstat	Metallic	Dow Badische Company

* Temporary designation assigned by FTC pending action on application for a new generic name and definition.
† A biconstituent fiber of nylon and polyester polymers.

REFERENCES

1. "Man-made Fiber Fact Book," Man-made Fiber Producers Association, Inc., Washington, D.C., 1971.
2. *Modern Textiles,* March 1972.
3. Ibrahim: Spandex Fibers, Du Pont.
4. Smith, Harold DeWitt: Textile Fibers, An Engineering Approach to Their Properties and Utilization, ASTM 47th Annual Meeting, 1944.
5. *Textile Organon,* Textile Economics Bureau.
6. *Textile Organon,* Textile Economics Bureau, January–February 1972.
7. *Textile Organon,* Textile Economics Bureau.
8. Liang: *Textile Industries,* August 1971.
9. *Textile Industries,* August 1972.
10. "Man-made Fibers," Man-made Fiber Producers Association, Inc., 1973.

Chapter **7**

Plastic and Elastomeric Foams

R. W. BARITO
and
W. O. EASTMAN

Major Appliance Laboratories,
General Electric Company,
Louisville, Kentucky

INTRODUCTION

Over the past two decades polymer foams have emerged in an ever-increasing array of forms from technology centers of the plastics and rubber industry. They have been spawned either in answer to a materials need in industry or as a developmental expansion of commercialized precursors.

To generalize typical characteristics for such a large family of materials would seem overambitious. However, a few general attributes summarize the key reasons why a designer or engineer might specify a cellular polymeric material.

Foams fall into that rather narrow portion of physical chemistry called *colloid chemistry*. The world colloid (Greek: gluelike) was used by T. Graham (1861) to describe solutions that passed through a membrane at a very slow rate, e.g., gelatin, glue, and albumin. Colloidal systems are between true solutions, such as sugar dissolved in water, and suspensions or emulsions where the particles are large enough to be visible to the naked eye. Thus the term *colloid state* can be applied broadly to any sparingly soluble substance.

A colloidal system is composed of two phases: the disperse phase (phase forming particles) and the dispersion medium (phase in which the particles are distributed). The total system is called a *disperse system*. The dispersion medium can be liquid, solid, or gas; similarly the disperse phase can be liquid, solid, or gas. Obviously, this leads to a number of possibilities for disperse systems; for example, smokes and dusts are solids dispersed in a gas; fog and mist are liquids dispersed in a gas. There are many examples in nature of gases, liquids, and solids that are dispersed in a solid medium. The two phases in a foam system are the gaseous phase and the liquid or solid phase, even though all foams at some time in their formation consist of a gas dispersed in a liquid. The foams we discuss in this chapter will be confined to gases dispersed in solids. In the foams we discuss the dispersion medium will consist of a plastic material and the gas phase or disperse phase will be one or more gases.

Virtually every polymer that has ever been made has at some time or other been foamed. Some of these foams have found widespread use in a variety of applications. Many have limited usage in specialized applications. A large number of plastic foams have never been taken past the laboratory bench. Those which have been most successful have been so because of a balance of favorable economics, good properties, and versatility.

The dispersion medium of plastic foams is composed of either thermoset or thermoplastic materials. A thermoset material is defined as one that cannot be melted or dissolved once the chemistry of the system has been carried to completion. Some examples of thermoset materials that have been foamed are polyurethanes, epoxies, phenolics, and natural and synthetic rubber. Thermoplastics are materials that can always be melted and dissolved. Polymers in this category that have been foamed and have achieved commercialization are polystyrene, polyethylene, polypropylene, and polyvinyl chloride.

The disperse phase in a foam is gaseous. The gas can be either totally encapsulated in discrete cells within the plastic or can be unconfined so it can, through interconnected passages in the plastic, move unrestricted through the plastic medium. The general terms for these two types of foam are closed-cell and open-cell. This generalization suffers, as all generalizations do, from being not totally accurate. Closed-cell foams are not completely closed and open-cell foams are not usually completely open. The type of foam used in a given application is generally dictated by the application.

A special type of foam which has found applications in the high-performance and electrical fields is called a *syntactic foam*. Syntactic foams are produced by dispersing hollow spheres of glass or plastic in a polymer matrix. These foams, which will be described in more detail in a later section, allow the engineer another degree of freedom in his design.

Foams are used in a wide variety of applications.[1] To name a few: rigid foams are used for thermal and electrical insulation, structural and decorative parts, packaging and flotation devices; flexible foams are used for cushioning, filtration, noise alleviation, packaging, and clothing.

The basic physical/chemical description of foams has already been given. What sort of properties emerge as we add gaseous materials such as air, nitrogen, or organic materials to rigid or flexible thermoplastic or thermoset polymers? (The long list of polymers which can be foamed and the methods by which they are foamed are covered later.)

Since most gaseous substances are considerably less dense than polymers, the density or unit weight of a composite of gas and polymer will be considerably less than that of the solid polymer. The extent of weight reduction is dependent on the amount of gas-filled cells. Thus, the weight of material per unit volume can be drastically reduced along with associated materials cost (as long as the gas used is not expensive). This property is quite important when one is primarily interested in space filling, such as packaging applications.

This reduced density also offers a range of buoyancy properties which can be useful in flotation applications.

Another general property of many gases is low thermal conductivity. Should the thermal conductivity of the gas in a polymer foam be considerably less than that of the polymer matrix, the composite thermal conductivity will be in direct proportion to the volume of incorporated gas. In addition then to a reduction in weight, considerable improvement in thermal conductivity is available from foamed materials. Aside from packaging, insulation constitutes the greatest market use of rigid foamed materials, the leading materials being polystyrene and polyurethane. A comparison of the thermal conductivity of polymers with other insulating materials is found in Fig. 1.

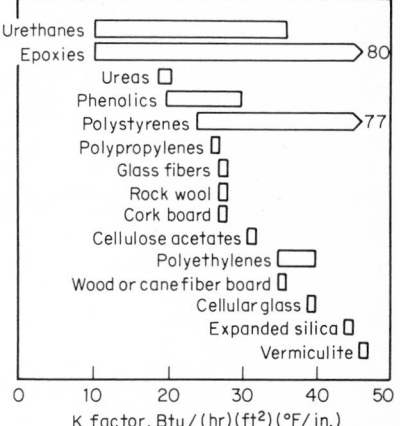

Expanding our analysis to mechanical behavior, we can see that adding cells of a "strengthless" gas to a polymer greatly reduces the strength properties of the composite. On the other hand, classical beam formulas show that the increased thickness at lower weights, possible with foams, results in remarkable improvements in beam stiffness and highly attractive weight/stiffness ratios. W. O. Weber

Fig. 1 Comparison of thermal insulating properties of plastic foams with other insulating materials.[3]

of Hercules, Inc., provides an excellent mathematical summary of this property relationship (see Fig. 2).

Coupled with the aforementioned beam stiffness/weight advantages of foam, one can expand the structural capabilities of cellular materials through the lamination of high-strength, high-stiffness skins to cellular plastic cores.

The use of this "I-beam" effect in structures has been covered in countless articles on structures. An excellent engineering review is presented in a *Modern Plastics* article by Robert Darvas of the University of Michigan.[4]

Gas possesses another attribute which makes it a unique partner with polymers— compressibility. A compressible gas dispersed in small cells in a noncompressible but ductile polymer enables the resulting composite to absorb high energy levels without transmitting the load to a substrate material or component. Plastic foams are rapidly growing in the packaging industry for this very reason. If this compressible gas is added to a flexible or semirigid matrix, interesting aesthetic and "feel" properties develop. Thus, soft and resilient foams are used in highly decorative and functional crash pads, sun visors, etc., in the automotive industry.

A benefit derived from the cellular structure of a foam is acoustical damping. Although too complex a subject to explore, the labyrinth and thickness effect of foams can be of considerable advantage in designing sound shields for equipment

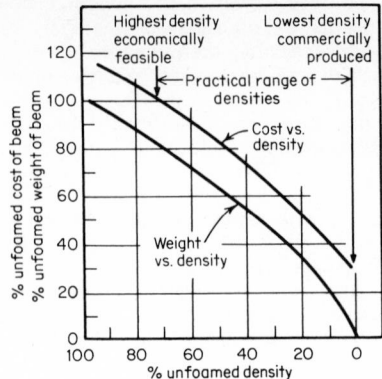

Fig. 2   Relationship of density to foamed plastic properties.[6]

The derivation of relationships between weight of solid beam and form beam of equal stiffness follows: The simple beam-deflection formula is

$$e = \frac{WL^3}{48EI}$$

where e = deflection, in.
 W = load, lb, concentrated at center of span
 L = length of span, in.
 E = modulus of elasticity
 I = moment of inertia of beam

If we assume the beam to be rectangular in cross section, its moment of inertia about the central axis is shown by

$$I = \frac{bh^3}{12}$$

where b = width of beam, in.
 h = depth of beam, in.

Substituting for I in the beam-deflection formula, we get

$$e = \frac{WL^3}{48E(bh^3/12)} = \frac{WL^3}{4Ebh^3}$$

and

$$e_f = \frac{WL^3}{4E_f bh_f^3}$$

$$e_s = \frac{WL^3}{4E_s bh_s^3}$$

where e_f = deflection, in. of foam-material beam
 e_s = deflection, in. of solid-material beam
 E_f = modulus of elasticity of foam material
 E_s = modulus of elasticity of solid material
 h_f = beam depth (or thickness) of foam beam
 h_s = beam depth (or thickness) of solid beam

To obtain beams of equal stiffness, the total deflection of each beam must be the same; so

$$e_f = e_s$$

or

$$\frac{WL^3}{4E_f bh_f^3} = \frac{WL^3}{4E_s bh_s^3},$$

where W, L, and b are constants for any given beam with a fixed load.

or shutting out surrounding noises from a living or working space. Typical acoustical properties of several foams are found in Table 1.

The electrical properties of the dispersed gas can also be used to great advantage, particularly if the desired properties of the composite and the gas are the same. An example is the use of expanded polyolefins in the communications industry to reduce attenuation caused by high capacitance.

In many cases one has the choice between open and closed cells, thus allowing for additional avenues of application. Open or interconnecting cell structures find use in filtration and in absorption applications, while closed-cell or unicellular structures find their use in structural, insulating, or flotation applications.

The degrees of freedom available in gas/polymer composites just reviewed are, of course, only part of the story. One must also consider the many aspects of polymer behavior in assessing requirements for a cellular material. The selection of a polymer system from a list of almost two dozen available polymers will depend upon the balance of performance criteria, some of which are chemical resistance, temperature rating, permeation, flammability, cost, weatherability, rigidity/flexibility, and strength. Recently, polymer technology has focused on modifications of the classical polymers. This list has been expanded to include fillers and fiber blends and alloys and a host of additive modifications.

Some idea as to the breadth of properties available with typical foams is found in Table 2. Shown in Table 3 are examples of the general usage pattern of foams.

Clearing the equation of constants, we get

$$E_s h_s{}^3 = E_f h_f{}^3$$

or

$$h_f{}^3 = \frac{E_s}{E_f} h_s{}^3$$

Assuming that the modulus of the material E is proportional to its density, and that the proportionality constant is 1, which can be shown to be conservative, the equation below follows:

$$\frac{E_s}{E_f} = \frac{d_s}{d_f}$$

where d_f = density of foam material
d_s = density of solid material

or substituting d_s/d_f for E_s/E_f in the preceding equation:

$$h_f{}^3 = \frac{d_s}{d_f} h_s{}^3$$

and

$$h_f = \sqrt[3]{\frac{d_s}{d_f}} h_s$$

Now, for example, consider a foam material containing 90 percent air by volume. The thickness of this foam beam for stiffness equivalent to that of the solid material is determined as follows:

$$h_f = \sqrt[3]{\frac{d_s}{d_f}} h_s$$

$$= \sqrt[3]{\frac{1}{0.1}} h_s$$

$$= \sqrt[3]{10}\, h_s = 2.15 h_s$$

or 2.15 times the depth or thickness of a solid beam. However, since the weight of material per unit thickness in the foam beam would be one-tenth that of solid material, the total weight of the foam beam is $1/10 \times 2.15$ or 0.215 times the weight of the solid beam. This represents a 78.5 percent savings in weight, and to a lesser degree in cost.

POLYURETHANE

The variety of chemicals available to produce urethane foams make these foams, as a class, perhaps the most versatile foams available today. By selection of the chemicals used, it is possible to produce urethane foams that vary from rigid, board-like foams to supersoft, resilient materials. It is possible to vary the density from extremely high density to extremely low density, to produce foams with open or closed cells.

Chemistry The chemistry of urethanes has been treated extensively by other authors, and the reader is referred to these publications[9,10] for an in-depth treatment. Here we will only touch on some of the more important reactions that occur when the urethane foam is produced.

TABLE 1 Typical Acoustical Properties of Plastic Foams[3]

Material	Frequency, Hz	Sound-absorption coefficient
Expanded polystyrene (2.5 lb/ft³)...............	250	0.03
	500	0.05
	1,000	0.14
	2,000	0.49
	4,000	0.19
	NRC†	0.18
Rigid urethane (2 lb/ft³)......................	250	0.18
	500	0.27
	1,000	0.19
	2,000	0.62
	4,000	0.22
	NRC	0.32
Flexible urethane (1.9 lb/ft³)..................	250	0.02
	500	0.80
	1,000	0.99
	2,000	0.79
	4,000	0.88
	NRC	0.60–0.70
Phenolic (2–4 lb/ft³).........................	NRC	0.50–0.75

† Noise-reduction coefficient.

The predominant reaction, and the one from which the foams derive their name, is the urethane reaction. When an alcohol is reacted with an isocyanate, a urethane is produced.

Expressing this in chemical language:

$$R{-}OH + R{-}N{=}C{=}O \rightarrow R{-}O{-}\overset{\displaystyle O}{\overset{\|}{C}}{-}NH{-}R$$

Alcohol   Isocyanate   Urethane

When the alcohol is replaced by a *polyol* and the isocyanate is replaced by a *di*isocyanate or a polyisocyanate, the result is a polyurethane. The polyols used are either polyethers or polyesters which contain two or more terminal hydroxyl (OH) groups. Similarly, the isocyanate used contains two or more isocyanate groups. When a polyol containing two hydroxyl groups (diol) and a diisocyanate containing two isocyanate groups are used, the result is a thermoplastic polymer. Chemically:

$$HO{-}R{-}OH + O{=}C{=}N{-}R{-}N{=}C{=}O$$

Diol   Diisocyanate

$$HO{-}R{-}O{-}\overset{\displaystyle O}{\overset{\|}{C}}{-}NH{-}R{-}NH{-}\overset{\displaystyle O}{\overset{\|}{C}}{-}O{-}R{-}O{-}\overset{\displaystyle O}{\overset{\|}{C}}{-}NH{-}R{-}N{=}C{=}O$$

Polyurethane

TABLE 2 Typical Properties of Various Foamed Plastics[2]

	Density, lb/ft^3	K factor, equilibrium 70°F	Compressive strength, lb/in.^{2}	Impact strength, lb/in.^{2}	Tensile strength, lb/in.^{2}	Burning rate	Max service temp, °F	Water absorption, lb/ft^2	Dielectric constant
Urethane (flexible)	1.5–20	0.20–0.30	0.13–4.5 (at 25%)	....	12–150	Self-extinguishing	200–300	High	1.1
Urethane (rigid)	1.5–70	0.16–0.50	10–10,000+ (at 10%)	....	15–7,000+	Self-extinguishing	250–300	Low	1.5
Polystyrene	1.25–6.0	0.25	14–130	0.14–0.21	33–300	Self-extinguishing	175	Extremely low	1.05
Polyvinyl chloride	3–25	Low	....	....	10–200	Self-extinguishing	150–170	Low to high	1.1
Epoxy	5–20	0.12	13–17	....	26–31	Self-extinguishing	250–300	Low	1.1
Polyethylene	1.8–2.2	0.35	10 (at 25%)	....	20–30	2.5	160	Low	1.05
Silicone	14–16	0.3	200–325 (at 10%)	....	....	Non-flammable	650	2.2	1.25
Cellulose acetate	6–8	0.30	125	0.12	170	4.5	350	....	1.12
Phenolic	1–22	0.20–0.28	2–1,200	....	3–130	Self-extinguishing	300	....	1.19
Urea formaldehyde	0.8–1.2	0.18–0.21	5	....	....	....	120	....	
Syntactic	20–32	....	750–5,000	....	....	....	800	....	1.4–1.7

TABLE 3   Typical Applications of Foamed Plastics[2]

Structural insulation.........	Urethane, polystyrene, rigid PVC, phenolic, urea formaldehyde, epoxy, silicone, cellulose acetate
Appliance insulation.........	Polystyrene, urethane, urea formaldehyde
Structural reinforcement......	Urethane, polystyrene, epoxy, cellulose acetate, syntactic
Flotation medium...........	Polystyrene, urethane, cellulose acetate, phenolic, syntactic
Structural void filler.........	Urethane, polystyrene, cellulose acetate, epoxy, syntactic
Decorative material..........	Polystyrene, acrylic
Packaging..................	Polystyrene, polyethylene, urethane, epoxy
Cushioning.................	Flexible PVC, flexible urethane, polyethylene
Garment insulation..........	Flexible PVC, flexible urethane
Shock-absorbent padding.....	Flexible urethane, flexible PVC, polyethylene
Electronic encapsulation......	Urethane, epoxy, silicone, syntactic
Electrical insulation.........	Urethane, polyethylene, epoxy, silicone, syntactic
Gaskets and filters...........	Polyethylene, urethane, urea formaldehyde
Carpet backing..............	Flexible urethane, flexible PVC, polyethylene
Mattresses.................	Flexible (polyether-based) urethanes
Upholstery.................	Flexible urethanes, flexible PVC, polyethylene
Acoustical insulation.........	Phenolic, urea formaldehyde
Floral displays..............	Urea formaldehyde, phenolic polystyrene
Toys, novelties..............	Polystyrene, PVC, polyethylene
Food containers.............	Polystyrene
Leather substitute...........	Flexible PVC
Chemical-resistant filler......	Urethane, polyethylene, epoxy, silicone, urea formaldehyde, phenolic

If a polyol or isocyanate is used which contains more than two hydroxyls or isocyanate groups, a thermoset polymer results.

Again chemically:

If the foam being made utilizes carbon dioxide as either part or all of the blowing agent, the following carbon dioxide–producing reaction takes place:

$$R{-}N{=}C{=}O + H_2O \rightarrow R{-}NH_2 + CO_2$$

Isocyanate   Water   Amine   Carbon dioxide

Further, the amine produced above can undergo reaction with an isocyanate:

$$RNH_2 + R{-}N{=}C{=}O \rightarrow R{-}NH{-}\overset{\overset{\displaystyle O}{\|}}{C}{-}NH{-}R$$

Amine   Isocyanate   Urea

Other reactions that take place are:

$$\text{R—N=C=O} + \underset{\text{Urethane}}{\text{R—NH—}\overset{\displaystyle O}{\overset{\|}{\text{C}}}\text{—OR}} \rightarrow \underset{\text{Allophanate}}{\overset{\displaystyle R}{\overset{|}{\text{NH}}}\text{—}\overset{\displaystyle O}{\overset{\|}{\text{C}}}\text{—}\overset{\displaystyle R}{\overset{|}{\text{N}}}\text{—}\overset{\displaystyle O}{\overset{\|}{\text{C}}}\text{—OR}}$$

Isocyanate

and

$$\underset{\text{Isocyanate}}{\text{R—N=C=O}} + \underset{\text{Urea}}{\text{R—NH—}\overset{\displaystyle O}{\overset{\|}{\text{C}}}\text{—NH—R}} \rightarrow \underset{\text{Biuret}}{\text{R—NH—}\overset{\displaystyle O}{\overset{\|}{\text{C}}}\text{—}\overset{\displaystyle R}{\overset{|}{\text{N}}}\text{—}\overset{\displaystyle O}{\overset{\|}{\text{C}}}\text{—NH—R}}$$

and

$$3\text{—N=C=O}$$

3 isocyanate groups   Isocyanurate

As the above polymerization reactions proceed, the materials are converted from low- to medium-viscosity liquids to solids.

In addition to the above chemistry there is the chemistry associated with the catalysis of the above reactions. The most commonly used types of catalysts are tertiary amines and organometallic materials. For the chemical mechanisms involved in catalysis and the reaction rates of the various catalysts the reader is referred to other publications.

The physical and colloid chemistry of nucleation, bubble stability, and polymer rheology in polyurethane foams will not be discussed in this chapter. For an in-depth treatment of these subjects the reader is referred to other publications.

Composition   The following materials are common to all urethane foams (rigid, semiflexible, and flexible):

Polyol(s)
Isocyanate
Surfactant
Catalyst(s)
Blowing agent(s)

By the selection and formulation of these materials a wide range of properties can be produced. For example, depending upon the polyol selected, foams can be made that vary from supersoft to rigid. Further, by varying the type and amount of these materials, other properties are affected such as density, open- and closed-cell content, and rate of cure.

The surfactants, which are generally silicone block copolymers, play an important role in compatibilizing the liquid foam ingredients, stabilizing the foam as it rises, controlling cell size, and producing the desired type of cell structure (open or closed).

The type and amount of catalyst(s) used in a formulation will affect the rise time and cure time of the foam.

The most widely used blowing agent in flexible foams is carbon dioxide. This is produced when the isocyanate reacts with water in the formulation. Fluorocarbons are sometimes used as supplemental blowing agents in flexible foams. These blowing agents are volatilized by the exothermic heat of the polyol-isocyanate reaction. In rigid foams, fluorocarbons are the primary blowing agents used, with carbon dioxide occasionally being used as a supplemental blowing agent. The amount of blowing agent used in a given formulation will affect the density of the finished foam. Figure 3 illustrates the effect of the amount of blowing agent on the density of the resultant foam.

In addition to the basic ingredients listed above, urethane foams sometimes include fillers, dyes, pigments, and flame retardants to modify or improve the finished foam.

Flexible urethane foams Flexible polyether urethane foams are made as slab stock or as molded-foam parts. Slab stock is made by a continuous process where the resultant foam is in the form of a bun 2 to 8 ft wide, 1 to 5 ft high, and 10 to 60 ft long. The slab stock differs from the molded foam in that before it can be used it has to be cut to the desired shape or size depending on the end use, whereas the molded foam has the dimensions and contour of the end-use product. Molded foams are made in molds (usually cast aluminum) that are preheated to 80 to 200°F, whereas slab-stock foam is made at ambient conditions.

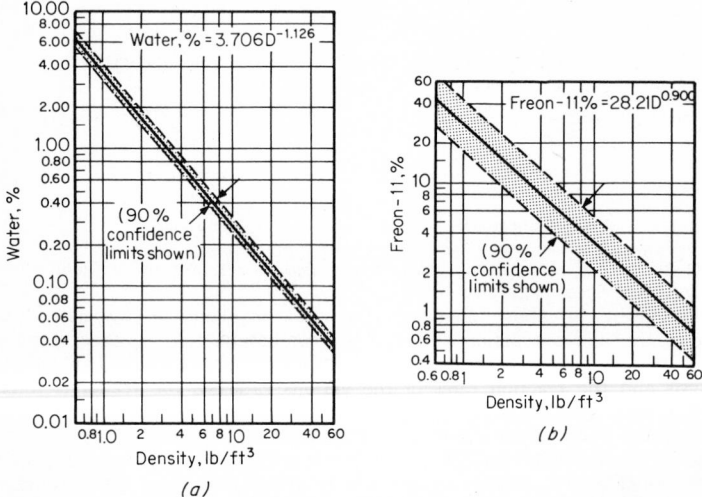

Fig. 3 Effect of blowing-agent concentration on density.[20] (a) Effect of water concentration on density. (b) Effect of Freon-11 concentration on density.

Table 4 shows the formulation, reactivity, processing conditions, and properties of a typical high-resiliency flexible-foam bun. As can be seen from the table, this formulation requires a multicomponent (four) machine in order to process the foam. This gives the foam processor an added degree of freedom in that the catalyst level can be adjusted to accommodate different processing conditions. Further, the isocyanate index can be easily adjusted to produce foams with varying degrees of firmness. The foam is suitable for furniture, mattresses, and cushioning applications because of its high modulus and hysteresis.

Using the same isocyanate (PAPI 901) and a different polyol, Table 5 demonstrates the properties obtained on molded flexible foam. This foam is termed a resilient "cold-molded" flexible foam—resilient because of its high modulus and hysteresis, "cold-molded" because of the low mold temperature required when it is foamed. "Cold-molded" foams are produced in molds that have been preheated to 80 to 100°F. They are made best in nonmetal molds such as plastic or wood. Foams of this type can be demolded in 5 to 10 min.

The selection of a release agent for any urethane-molding application (flexible or rigid) is extremely important. Most release agents are either waxes, soaps, or silicones dissolved in low-boiling solvents. They are usually applied by spraying them onto the surfaces of the mold. The foam will stick to any part of the mold that is

TABLE 4 Formulation and Properties of High-Resiliency Flexible-Foam Bun[14]

Formulation (parts by wt)		Conveyor Processing Data	
A: PAPI-901	Vary (depending on firmness)	Cream time, s	10
		Rise time, s	41
B:		Tack-free time, s	110
6000 mw triol	97	Bun exotherm, °F	275
X-90	6	Bun height, in	20
Water	3	Conveyor speed, ft/min	7.4
C:		Conveyor angle, deg	5.0
XV-10	2.4	Mixer type	Medium shear C-type
6000 mw triol	1		
D:		Mixer speed, r/min	6,600
Dibutyltindilaurate	0.1	Nucleation, ft³/h	0.02
6000 mw triol	2.0		

Component	A	B	C	D
Temp, °F	70	70	75	75
Pump capacity	3.0 gal/min	6.0 gal/min	11 cm³	11 cm³
Orifice size, in	0.25	0.25	0.039	0.0625
Line pressure, lb	61	118	50	52
Isocyanate index	1.00	1.10	1.20	
Component mix ratio:				
A	59.4	65	71.7	
B	100	100	100	
C	3.2	3.2	3.2	
D	1.98	1.98	1.98	
Throughput, lb/min:*				
A	34.0 (257)	37.1 (281)	40.9 (310)	
B	57.1 (432)	57.1 (432)	57.1 (432)	
C	1.83 (13.9)	1.83 (13.9)	1.83 (13.9)	
D	1.13 (8.5)	1.13 (8.5)	1.13 (8.5)	

Foam Properties			
Parts isocyanate	63	69	76
Isocyanate index	1.00	1.10	1.20
Density, lb/ft³	2.92	2.81	2.64
ILD,‡ lb/50 in.²:			
At 25% deflection	20	28	38
At 65% deflection	52	74	76
Modulus (65/25)	2.60	2.64	2.26
Hysteresis, 25 retained/25 original, %	80	79	74
Split tear, pli†	1.04	1.10	1.33
Die C tear, pli	2.70	3.20	4.00
Tensile, lb/in.²	15.5	24.2	28.6
Elongation, %	115	105	110
Compression set, % deflection:			
At 50% deflection	14	14	13
At 90% deflection	61	13	19
Autoclave set, % deflection:			
At 50% deflection	17	20	17
At 90% deflection	86	62	46
Auto flame test 302, in./min	2.7	2.6	2.5

* Figures in parentheses are grams per second.
† Pounds per lineal inch.
‡ ILD stands for indentation load deflection.

inadequately sprayed and cause the foam to tear when the part is removed from the mold. One of the major problems with all release agents is the buildup of the mold release on the surfaces of the mold after repeated use. This requires that the molds be cleaned frequently; otherwise part definition and surface appearance will be impaired. On appearance parts this becomes a very critical factor.

Processing   All urethane foams (flexible, microcellular, semiflexible, rigid) are processed in much the same manner. Two or more components are accurately metered into a mixing chamber, mixed, and dispensed. Three basic types of metering equipment are available: positive-displacement gear pumps, piston- or rod-displacement units, and high-pressure units. With the gear pumps and piston- or rod-displacement types, mixing is accomplished by a mechanical mixer (impeller) or by frothing. Frothing is a technique whereby a gas is introduced, at high pressure, into the mixer housing at the same time the liquid foam ingredients are metered into the housing. This gas, coming in at high pressure, causes turbulence which mixes the liquid foam ingredients. Mixing is accomplished in the high-pressure metering units by the liquid ingredients impinging on each other at very high velocities in a small mixing chamber.

TABLE 5   Formulation and Properties of Molded Flexible Foam[14]

Formulation (parts by weight):	
A: PAPI 901	61
B:	
CP-4701	100
Water	3.6
DABCO WT	1.4
Properties:	
Density, lb/ft^3	2.37
Tensile, lb/in.^{2}	20
Elongation, %	98
Split tear, pli	0.9
Compression set, %:	
50% deflection	15.8
90% deflection	14.2
ILD, lb/50 in.^{2}:	
25% deflection	51
50% deflection	86
65% deflection	136
Hysteresis, 25 retained/25 original, %	71
Modulus, 65/25	2.67
Mold temp, °F	85
Mold type	Aluminum

Integral-skin flexible foam   Foam parts of this type derive their name from the continuous, dense skin on the surfaces of the finished molded part. The thickness of the skin, depending on the formulation and processing conditions, varies between 0.040 and 0.100 in. If the liquid ingredients are pigmented, the finished part will have the desired shape, surface texture, and color. Color can also be achieved by painting the molded part with paints that have been specially developed for urethane foams.

Mold design and the proper selection of release agent are the two most important elements in making good integral-skin foam parts. The mold must be vented adequately, so as not to entrap air and thus restrict the filling of the mold. Second, the mold design must be such that once the mold is filled, sufficient back pressure can be developed to give a part without large voids. The release agent should give adequate release and not build up on the mold with repeated use. A build up mold release agent causes a loss of surface texture.

Table 6 demonstrates the properties that can be obtained from a typical integral-skin foam. As the ratio of the two polyols changes, the properties of the final foam are altered. This change in the ratio of polyols results in a higher cross-link density in the cured foam. Formulation A has the lowest cross-link density, and formulation C has the highest. The higher the cross-link density the stiffer the foam. This results in higher tensile strength, modulus, tear strength, and compression set while lowering the elongation.

Some applications of integral-skin foam are for interior automotive parts: armrests, headrests, sun visors. Other applications include toys, seats, and padding for recreational vehicles, furniture seats and arms, and appliance trim.

Microcellular urethane foam Microcellular urethane foams are high-density urethane foams with an integral skin of a desired thickness. These materials are characterized by their outstanding wear resistance and toughness at the surface and

TABLE 6   Formulation, Processing Characteristics, and Typical Properties of Integral-Skin Flexible Foams[15]

	A	B	C
Foam formulations:			
NIAX polyol 34-28	90	85	80
NIAX polyol 50-810	10	15	20
Silicone L-5302	0.75	0.75	0.75
Dabco* 33-LV	0.5	0.5	0.5
NIAX catalyst D-22	0.1	0.1	0.1
NIAX isocyanate SF-50 (105 index)	26.2	35.9	45.7
NIAX blowing agent-113	15	15	15
NIAX blowing agent-11	10	10	10
Foam-processing characteristics:			
Stream ratio, part A/part B	1.98	1.66	1.43
Cream time, s	12	12	12
Rise time, s	70	58	55
Tack-free time, s	150	90	75
Core-foam properties:			
Density,† lb/ft³	5.1	4.9	4.4
25% CLD,§ lb/in.²	1.0	1.6	2.4
Tensile strength, lb/in.²	44.1	60.7	63.8
Elongation, %	165	153	134
Tear resistance, lb/in.	2.7	3.7	5.5
50% compression set, %	3.5	10.5	33.0
Humid aging load loss (3 h at 220°F), %	20.7	16.7	17.7
Skin properties:			
100% modulus, lb/in.²	279	795	1,011
Tensile strength, lb/in.²	442	1,004	1,096
Elongation, %	153	148	130
Die C tear, pli	58	122	158
Puncture resistance,‡ lb	17.3	20.5	18.9

* Houdry Process and Chemical Co. trademark for triethylenediamine.
† The overall density of molded test samples 7 by 7 by 2 in. was ~12.5 lb/ft³.
‡ Bic (RM-25) ball-point pen driven at 20 in./min; skin attached to foam.
§ CLD stands for compression load deflection.

their flexible, lower-density inner core. Microcellular urethanes can be cast into complex shapes to give mirrorlike surfaces. Colored parts can be made by either color coating the mold prior to filling or color coating the finished part.

In Table 7 the properties of three microcellular urethane foams are shown. In each formulation the isocyanate was kept constant and the polyols were varied to demonstrate the range of properties that can be obtained.

One of the most important properties to control is density. This can be controlled by the amount of material poured into the mold. Overpacking the mold results in higher density. If a lower-density material is desired, additional blowing agent can be added (see Fig. 3). The effect of density on properties can be seen in Fig. 4 *a, b,* and *c.*

Another important process variable is mold temperature. A cold mold will give

TABLE 7 Properties of Microcellular Urethane Elastomers at 45 lb/ft³ Core Density[16]

	A	B	C
NIAX activator.........................	W-902	W-902	W-902
NIAX resin............................	W-202	W-207	W-208
Hardness, Shore A.....................	71	78	85
100% modulus, lb/in.².................	510	890	1,180
Tensile strength, lb/in.²..............	920	1,140	1,400
Ultimate elongation, %.................	187	149	142
C tear, lb/in.........................	130	165	212
B compression set, %..................	16	15	30
Zwick rebound, %......................	28	26	29
50% compression modulus, lb/in.²......	780	1,620	2,210

a foam with a thicker skin, whereas a hotter mold will give a foam with a thinner skin.

Applications for this type of foam have been in automotive exterior components such as bumpers. Another application for microcellular foams is shoe soles.

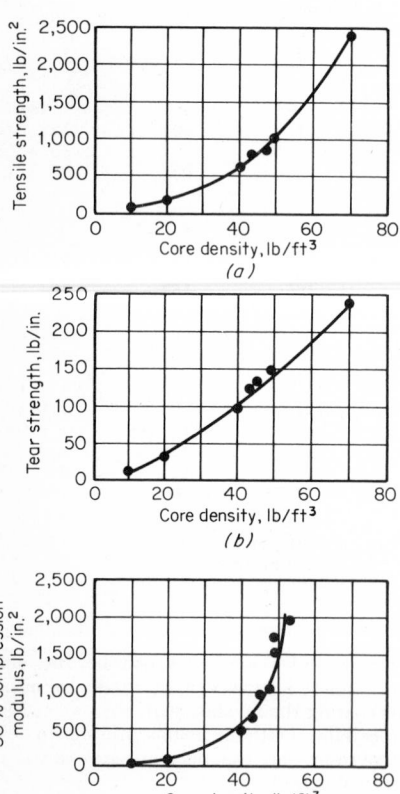

Fig. 4 Effect of core density on tensile strength (*a*), tear strength (*b*), and compression modulus (*c*), of microcellular urethane elastomers (formula A).[16]

Semiflexible urethane foam This type of foam is sometimes referred to as semi-rigid. These foams are characterized by their low rebound, or bounceback. Their densities can range from 4 to 10 lb/ft³.

Semiflexible foams are used principally by the automotive industry for such applications as dashboard covers, armrests, head restraints, and crash pads. In these applications it is important that the foam maintain its firmness and impact properties over a wide temperature range.

Table 8 shows a typical semiflexible-foam formulation, its processing characteristics, and properties. Note that as the isocyanate index is varied, the foam hardness is altered. The hardness can also be affected by the amount of material packed into the mold. This, of course, will also affect the density of the finished part.

Integral-skin rigid urethane furniture foam These foams are generally produced in two density ranges, depending upon the application. For nonstructural applications such as ceiling beams, pilasters, and bas-relief items, lower-density foams are used: 4 to 12 lb/ft³. In structural furniture applications higher-density (20 to 30 lb/ft³) foams are required. These include such applications as chairs, tables, and television cabinets.

Urethane furniture foams will reproduce complex design characteristics with authentic woodlike texture, weight, and handleability. The foam surface can be stained, painted, or varnished. Special surface effects can be produced by the proper selection of stains and varnishes.

TABLE 8 Formulation, Processing Characteristics, and Properties of Semiflexible Foam[14]

Formulation				

Materials	Parts by Weight			
B:				
6000 M wt triol.................	100.0			
Isonol 550....................	10.0			
Triethanolamine...............	4.0			
Dibutyltindilaurate.............	0.5			
XV-26.......................	0.7			
Water.......................	2.0			
Fluorocarbon 11-B.............	10.0			
DC 200 (5 cstk)...............	0.03			
A: Papi 901				
Parts by weight..............	59.5	62.5	65.5	68.5	71.5
Isocyanate index.............	1.00	1.05	1.10	1.15	1.20

Processing Characteristics				
Mix ratio A...................	46.8	49.2	51.5	54.0	56.2
Parts by weight B.............	100	100	100	100	100
Index.......................	1.00	1.05	1.10	1.15	1.20
Component viscosities A at 25°C, cP........	60–110			
Component viscosities B at 25°C, cP........	800–1,000			
A. Bench mix data:				
Component temp, °F...........	$77 \pm 2°$			
Cream, s....................	8–12			
Rise, s.....................	45–55			
Tack-free, s.................	55–65			
Free rise cup density, lb/ft³..............	4.0–4.5			
Free rise core density, lb/ft³.............	3.5–4.0			
B. Machine mix data:				
Component temp, °F...........	80–85			
Cream, s....................	8–10			
Rise, s.....................	30–40			
Tack-free, s.................	55–65			
Free rise cup density, lb/ft³..............	3.5–4.0			
Mixer type..................	3 stage			
Mixer speed, r/min............	5,600			
Mold temp, °F...............	105–120			
Mold composition.............	Aluminum			
Demold time, min.............	5			

Isocyanate index.............	1.00	1.05	1.10	1.15	1.20
Density (core), lb/ft³............	4.01	4.12	4.12	4.01	4.11
Split tear, pli................	1.18	1.26	1.41	1.40	1.41
Tensile, lb/in.²..............	25.5	27.7	28.8	29.5	31.3
Aged tensile 22 h at 285°F........	13.0	13.2	14.0	15.0	18.3
Elongation, %................	98	96	90	86	83
Aged elongation 22 h at 285°F........	76	73	71	68	65
50% compression set, % original deflection....	26.9	28	29.1	30.5	31.5
CLD, lb/in.²:				
At 25% deflection.............	0.90	1.05	1.10	1.15	1.22
At 50% deflection.............	1.90	2.70	2.70	2.70	3.00
At 75% deflection.............	8.10	11.50	11.70	12.0	12.4

Tables 9 and 10 give typical data for rigid furniture foams.

Foam density affects the physical properties of rigid foam more than any other single factor. Figures 5 and 6 demonstrate the effect of density on compressive strength and tensile strength. Affected in a similar manner by density are flexural strength, shear strength, and modulus.

TABLE 9 Formulation, Properties, and Processing Data for Integral-Skin Rigid Furniture Foam[14]

Formulation	
Materials	Parts by Weight
A: PAPI 901	86.5
B:	
ISONOL 550	60
6000 mol. wt. triol	40
Surfactant	2
Dibutyltindilaurate	0.1
Fluorocarbon 11-B	10

Polymer Properties (Machine Foams)		
Density, lb/ft³:		
Molded	21	31
Skin	62	71
Core	13	24
Skin thickness, in. ($\times 10^{-2}$)	6	8
Hardness:		
Shore D	70	74
Rockwell S	45	65
Falling-ball impact ($\perp$ to skin surfaces), ft-lb	7.4	10.8
Pendulum impact ($\perp$ to skin surfaces), ft-lb	7.7	9.9
Screw holding, lb	115	130
Heat-distortion temp (264 lb/in.² load), °C	71	80
Compressive strength, lb/in.²	310	510
Compression modulus, lb/in.²	10,530	15,280
Tensile strength, lb/in.²:		
Core	240	850
Skin/core*	1,070	1,520
Elongation, %		
Core	5	6
Skin/core*	6	6
Tensile modulus, lb/in.²:		
Core	2,900	10,000
Skin/core*	14,850	20,000
Flexural strength ($\perp$ to skin surfaces), lb/in.²	580	1,740
Flexural modulus, lb/in.²	30,400	45,100

Processing Data	
Mix ratio A, parts by weight	77
Mix ratio B, parts by weight	100
Component temp A, °F	80
Component temp B, °F	85
Component viscosities A, cP	90
Component viscosities B, cP	2,500
Free-rise density, lb/ft³	8–9
Foam reactivity:	
Cream, s	35
Gel, s	60
Rise, s	85
Firm, min	4
Mixer type	3 stage
Mixer speed, r/min	5,500
Mold temp, °F:	
Rubber molds	70–80
Large metal molds	100–110
Small metal molds	135–140
Demold time, min	4–5

* Property determinations on ¼-in.-thick specimen.

TABLE 10   Formulation and Properties of Water-blown Rigid Furniture Foam

Formulation (parts by weight)				
A: PAPI 901	93	107	104	97
B:				
G-2450	100			
G-2451	...	100		
RH-360	...	...	100	
Poly G-340 DM	...	...	...	100
Water	0.6	0.6	0.5	0.5
Surfactant	2.0	2.0	2.0	2.0
DABCO R-8020	1.2	1.1	1.1	1.0
Reactivity				
Cream time, s	48	35	40	
Rise time, s	150	140	130	
Firm time, s	420	300	300	
Polymer Properties				
Density, lb/ft³:				
Molded	14.8	14.3	16.9	16.1
Core	14.1	13.6	16.1	15.3
Compression strength at 10% deflection ($\perp$ to skin surface), lb/in.²	500	513	667	625
Compression modulus ($\perp$ to skin surface), lb/in.²	11,620	11,820	14,560	13,260
Flexural strength ($\perp$ to skin surface), lb/in.²	810	780	890	860
Flexural modulus ($\perp$ to skin surface)	29,150	23,380	37,980	32,060
Screw holding, lb	57.1	68.9	80.8	83.5
Falling-ball impact ($\perp$ to skin surface), ft-lb	1.6	1.1	1.5	1.8

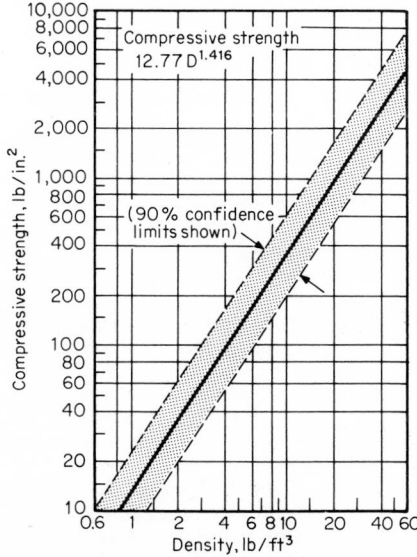

Fig. 5 Effect of density on compressive strength of rigid urethane foam.[20]

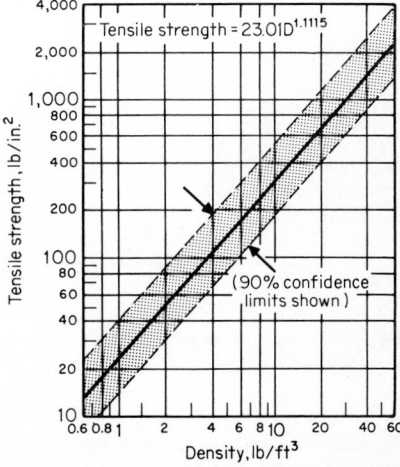

Fig. 6 Effect of density on tensile strength of rigid urethane foam.[20]

Urethane foams in flotation and structural applications Urethane foams are uniquely suited for marine applications where low density, buoyancy, and structural strength are required. Typical applications for urethane foam where buoyancy is a prime requisite are boats, floating docks, and buoys. In these applications the foam is generally poured between metal or plastic skins that are impervious to water. The skins with the foam between give structural integrity to the total part.

Table 11 shows the properties of a typical urethane foam used in flotation devices.

TABLE 11   Properties of Rigid Urethane Foam for Flotation and Structural Applications[17]

Property	Machine	
	Pour	Froth
Density:		
Free rise, core, lb/ft³..	3.2	2.4
Molded, core, lb/ft³..	3.4	2.8
Molded, overall, lb/ft³..	3.9	3.1
Closed-cell content, %...	94	94
K factor, Btu/(h)(ft²)(°F/in.)...................................	0.16	0.13
Dimensional stability, average volume change (ASTM D 2126-62T):		
158°F, 100% RH, 1 week, %....................................	+8	+22
158°F, 100% RH, 2 weeks, %...................................	+9	+28
158°F, 100% RH, 4 weeks, %...................................	+13	+14
−16°F, ambient humidity, 2 weeks, %..........................	−1	−1
Compressive strength, parallel to rise, lb/in.² (ASTM D 790-61)........	53	35
Compressive strength, perpendicular to rise, lb/in.² (ASTM D 790-61)...	51	34
Water absorption, 4 weeks, g/1,000 cm³ (ASTM D 2127-62T)..........	45	50

Rigid, low-density urethane foam The principal application of rigid low-density (1.5 to 2.5 lb/ft³) urethane foam is for thermal insulation in household refrigerators, freezers, picnic coolers, and other related products where low thermal conductivity is important. The primary reason for its use in these applications is the very low thermal conductivity of fluorocarbon-blown urethane foam. Further, these foams maintain this low conductivity over a period of many years. These foams have low thermal conductivity because of the high-molecular-weight fluorocarbon gas which is entrapped within the closed-cell structure of the foam. The urethane matrix of the foam is unique in its ability to contain the fluorocarbon gas.

In addition to the fluorocarbon blowing agent, the thermal conductivity of a foam is greatly influenced by the cell size, the closed-cell content, and the density. These three properties can be controlled by the formulation and processing technique. The effect of density on K factor (thermal conductivity) is shown in Fig. 7.

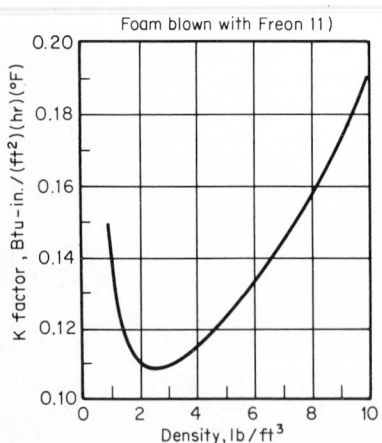

Foam blown with Freon 11)

Fig. 7 Effect of density on thermal conductivity of rigid urethane foam.[20]

The replacement of other insulation materials with fluorocarbon-blown urethane foam permits the engineer the freedom to design a product that has less insulation space, thus increasing the usable volume within the device. Or, if the space occupied by the other insulation is directly replaced with foam, the total heat leakage through the device will be considerably reduced.

A secondary benefit that is derived from the use of foam in these applications is the structural strength it renders to the device. When the foam is encased between metal or plastic skins, it greatly increases the rigidity of the whole product.

The properties of a typical rigid low-density urethane foam are shown in Table 12.

Spray foaming of urethanes Urethanes easily lend themselves to spray foaming. They fill such application needs as thermal insulation for storage tanks, pipelines, ducts, and buildings. Urethane foams are ideally suited for these applications because of their low thermal conductivity, excellent dimensional stability, durability, and the ease with which they can be applied. In addition, because of their fast set times, they can be applied to vertical surfaces without sagging, as well as to horizontal surfaces. A further advantage is the ability to apply them in thin layers.

Spray foams are generally applied under uncontrollable ambient conditions. This can cause problems if the ambient conditions under which the foam is applied are not specified. Since the urethane-foam reaction is exothermic, the reaction rate,

TABLE 12   Properties of a Frothed Rigid Urethane Foam[18]

Property	Value
Core density, lb/ft³	1.91
Overall density, lb/ft³	2.10
Closed-cell content, % (ASTM D 2856)	95
Dimensional stability, average volume change, %:	
176°F, dry heat, 100 h	+6
158°F, 100% RH, 7 days	+10
110°F, 100% RH, 7 days	+5
−16°F, 100 h	No change
Water-vapor transmission, perm inch (ASTM C 355)	3.1
Water absorption, % by volume, 2 weeks	3.1
Compressive strength, 10% deflection, lb/in.² (ASTM D 1621)	20
Tensile strength, lb/in.² (ASTM D 1623)	39
Thermal conductivity at 77°F, Btu/(h)(ft²)(°F/in.):	
Initial	0.11
Aged 1 month at 110°F, 75% RH	0.15

adhesion to the substrate, and the physical properties of the foam will vary with the ambient temperature. Generally, urethanes are spray-foamed at ambient temperatures between 65 and 100°F.

Since the isocyanate in the formulation can react with water, it is important that the surface to which the foam is being applied be completely dry. Another environmental condition that affects the properties of the foam is wind velocity. It is therefore recommended that foam not be sprayed when wind velocities are greater than 10 to 5 mi/h.

If the foam is to be exposed to outdoor weathering, barrier or protective coatings are necessary.

The properties of a typical fire-retardant spray foam are shown in Table 13.

POLYSTYRENE

Since its introduction in 1938 polystyrene has grown constantly in volume of pounds produced. It enjoys the position of being one of the "big three" polymers in terms of worldwide production and usage (the others being vinyls and olefins). The reason for the polymer's widespread usage is the balance of characteristics in the key areas of:

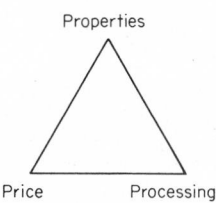

TABLE 13   Properties of a Fire-Retardant Urethane for Spray Application[19]

Property	Value
Density, ASTM D 1622:	
Core, lb/ft³	1.8–2.2
Overall, applied, lb/ft³	2.6–3.0
Closed-cell content, % (ASTM D 2856)	91
Compressive strength, ASTM D 1621:	
10% deflection, lb/in.²	34
No definite yield	
Tensile strength, lb/in.² (ASTM D 1623)	43
Shear strength, lb/in.² (ASTM D 732)	92
Friability, % weight loss (ASTM C 421):	
After 2 min	0.4
After 10 min	0.4
Water absorption (ASTM D 2127), lb/ft² (surface area)	0.07
Water-vapor transmission, perm inch (ASTM C 355)	2
Thermal conductivity (ASTM C 518):	
Initial	0.12
Design (aged)	0.16
Flammability (ASTM D 1692)	SE 0.14 in., 32 s
UL Guide BRYX, foamed plastic (UL-723 tunnel test):	
Flame spread	25
Fuel contribution	5
Smoke developed	115–150

Since polystyrene and its properties are well covered in another part of this book, a brief summary will suffice at this point.

The basic structure of polystyrene

$$\left[\begin{array}{c} -C-C- \\ \bigcirc \end{array} \right]_n$$

of course dictates its properties. It is a fairly simple hydrocarbon and is nonpolar.

The benzene rings $\bigcirc$ cause the carbon backbone (C—C—C) to be rigidized, rendering the polymer rather stiff. Some of the less obvious basic properties are excellent optical clarity, good tensile strength, a transition of 180°F, and rather low ductility.

The basic polymer can be modified in several ways to overcome deficiencies or to achieve specific attributes. Some of the chemical modifications of styrene are:

1. *Adding divinyl benzene:* The addition of this monomer increases the heat resistance by tying together or cross-linking the polymer chains.

2. *Adding α-methyl styrene:* This monomer improves the temperature characteristics of the polymer (transition temperatures of the copolymer).

3. *Adding acrylonitrile:* The addition of acrylonitrile, a highly polar monomer, increases the solvent resistance of the copolymer. These copolymers are called SAN.

4. *Adding polybutadiene, or styrene butadiene copolymer:* These rubbers are added to the polymer by a grafting mechanism to improve the ductility of the polymer. The result is high-impact polystyrene (HIPS) or medium-impact styrene (MIPS).

5. *A combination of 3 and 4* results in a very ductile solvent- and chemical-resistant plastic known as acrylonitrile-butadiene-styrene (ABS).

The styrene family of plastics is produced in the United States at a rate greater than 3.4 billion pounds per year and is used in a myriad of applications. This chapter will concern itself with the portion which is foamed—estimated at 10 percent or 340 million pounds. It is second only to polyurethane in usage as a foamed plastic.

Styrenic plastic foams are available in several basic forms—extruded foams, expanded-bead-molded foams, and injection-molded foams.

Extruded foams   This form of styrene foam is quite familiar to most consumers as the board-, plank-, or block-shaped items readily available at variety stores. The foam is fabricated into its final form by cutting, sawing, laminating, and bonding these rigid shapes together.

The mechanical and physical properties of foamed plastics, as discussed in the introduction, are dependent upon the general properties of the matrix polymer and the degree to which it is expanded with the gas or blowing agent. Thus the properties of foams in a given class should vary directly with the density. Such is the case with extruded polystyrene foam, as is shown in Table 14.

TABLE 14   Variation of Expanded Polystyrene (Extruded) Properties with Density (ASTM D 2125-62T)

Property	Class	Grade 1.5	Grade 2	Grade 3	Grade 4.5
Nominal density	1, 2				
kg/m³		24	32	48	72
lb/ft³		1.5	2.0	3.0	4.5
Compressive strength, min	1, 2				
kg/cm²		0.70	1.27	3.52	8.44
lb/in.²		10	18	50	120
Thermal conductivity, avg, max:					
kg-cal m/(h)(cm²)(°C)	1	0.041	0.041	0.031	0.031
Btu in./(h)(ft²)(°F)		0.33	0.33	0.25	0.25
kg-cal m/(h)(cm²)(°C)	2	0.037	0.035		
Btu in./(h)(ft²)(°F)		0.30	0.28		
Water-vapor permeability, max	1,2				
Metric perm-cm		5.0	3.3	3.3	3.3
Perm-in.		3.0	2.0	2.0	2.0
Flammability	1		Burns		
	2		SE or nonburning		

Since Dow is the dominant factor in this country in extruded polystyrene foams, they are generally known as Styrofoam.° Although detailed process techniques are not released to the public, the extruded foams are manufactured by injecting a blowing agent such as methylene chloride into molten polystyrene which contains additives such as nucleating agents, stabilizers, and process aids. This molten mixture is conveyed in an extruder to an orifice die which is in the form of a long, narrow slit. Upon exposure to atmospheric pressure the polymer expands rapidly because of the volatilization and gas expansion of the methylene chloride, whose boiling point is considerably below room temperature. Since each small volume of gas expands around a point at which the gas cell was nucleated, the resultant foam is closed-cell. Depending upon the amount of methylene chloride introduced, the expansion can be as high as forty times the original volume. The foam is now cooled in a controlled pattern to prevent cell rupture and cut to the final dimensions. Although the process is basically simple, the entire system must be precisely controlled to produce a consistent product.

Styrofoam is used primarily in areas where lightweight, easy-to-install insulations are needed and where environments do not exceed the performance of the basic polymer. Styrene board stock is also laminated between sheets of other materials of construction from kraft paper to metal skins to provide relatively low cost building panels. Fire-resistant grades are available where required by specification.

Another major application area makes use of the lightweight, closed-cell structure, and low water absorption. Thus flotation applications such as rafts and small boats consume large quantities of Styrofoam.

Registered trademark of Dow Chemical Company.

Finally a highly visible but relatively low tonnage application for Styrofoam is the hobby, florist, and do-it-yourself category where most consumer exposure is found. Approximately 50 million pounds of extruded polystyrene foam, or Styrofoam, is produced each year.

Expanded-bead foams The general public is perhaps most familiar with this form of foam through many cups of coffee or other beverages consumed in cups molded from expanded bead. Expanded-bead foams start with the same basic polymer as extruded foams, and thus ultimate properties (at equivalent densities) are comparable.

Since considerably more expanded-bead styrene is consumed (over 200 million pounds) than extruded foams, the applications are much more diverse and the properties much more dependent upon design and processing techniques. Data are available on these materials from raw-material suppliers such as Arco and Dow.

One of the more important aspects of foam usage is its insulating value. The thermal conductivity of expanded-bead foam (Dylite°) at various foam densities is shown in Fig. 8. The effect of temperature on this property is shown in Fig. 9.

° Trademark of Arco Polymers, Inc.

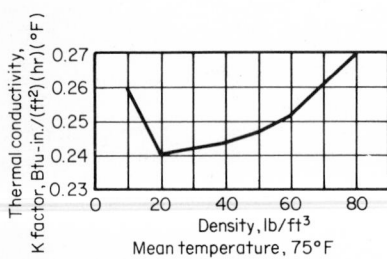

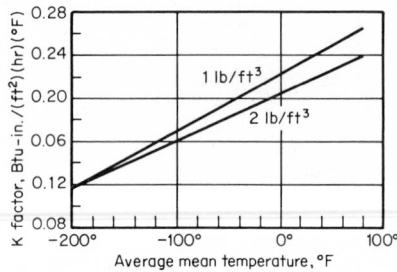

Fig. 8 Thermal conductivity of foamed dylite expandable polystyrene at various densities.[41]

Fig. 9 Thermal conductivity of foamed dylite expandable polystyrene.[41]

Many of the insulating applications of styrene bead foam involve water and water-vapor environments (i.e., cold and dry on one side, humid and hot on the other).

Thus the pickup and diffusion of water and water vapor through these foams becomes rather important. Table 15 and Fig. 10 describe these properties in a Pelaspan° sample.

TABLE 15   Water Absorption vs. Density[42]

Density, lb/ft³	1.0	1.5	2.0	3.0
Water absorption lb/ft²	0.04	0.03	0.03	0.03

In addition to density, a major factor in determining the performance properties of bead foam is the consistency of fusion of the expanded beads to one another. Poor fusion can drastically affect strength properties and can even change water-absorption and transmission properties to a large degree. All properties shown in the tables and graphs in this section are based on maximum fusion of the beads. The subject will be covered in more detail in the processing section.

° Trademark of Dow Chemical Company.

Another aspect of physical strength not obvious to the casual observer is the fact that foams fabricated from polystyrene by an internal blowing agent, whether it be by extrusion or by pre-expanding and molding beads, exhibit unusually high strength. This phenomenon (i.e., $1 + 1 = > 2$) is primarily due to the fact that polystyrene is easily oriented. That is, the polymer molecules can be lined up.

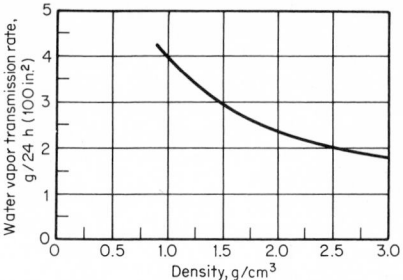

Fig. 10 Water-vapor transmission of a sample molded of Pelaspan (ASTM E 96-53T, Procedure E).[42]

This orientation increases the strength of the polymer in the direction of the orientation. When a nucleated cell begins to form and expand, it grows in a spherical manner (until it meets other cells and becomes a dodecahedron). The polystyrene is thus oriented in several directions simultaneously—this is generally known as biaxial orientation. The resultant polymer comprising the cell structure is thus considerably stronger (tensile strength) and more ductile (elongation to rupture) than

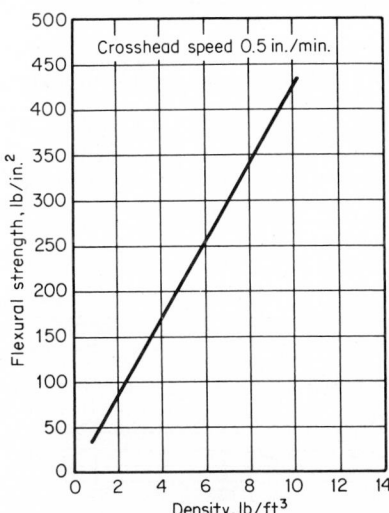

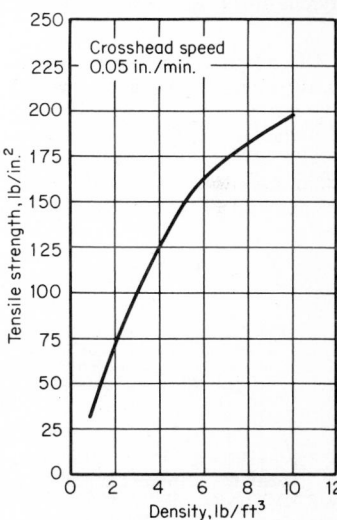

Fig. 11 Flexural strength of polystyrene foam vs. density at 73°F.[43]

Fig. 12 Tensile strength of polystyrene foam vs. density at 73°F.[43]

the base polymer. In addition the gas-transmission and chemical-resistance properties are improved.

Figures 11 through 14 describe the mechanical-strength properties of expanded styrene foam as a function of density.

The need for specific physical properties depends upon the design of the article to be used. This factor is particularly important in package design where the package must support the article to be shipped. In addition to basic strength properties just covered, the change in properties with time—particularly dimensions—is vitally important to a packaging or structural engineer.

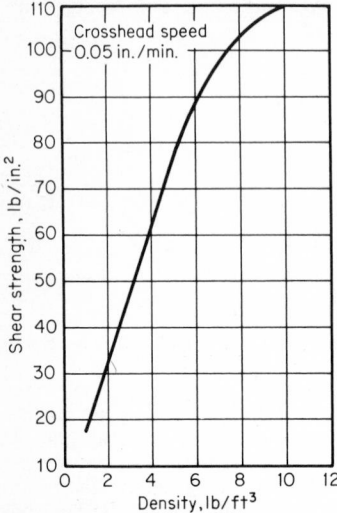

Fig. 13 Shear strength of polystyrene foam vs. density at $73°F$.[43]

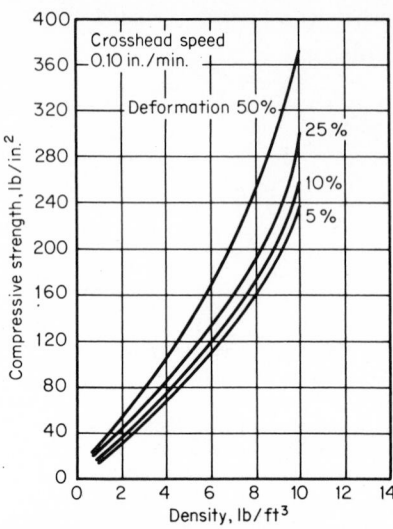

Fig. 14 Compressive strength of polystyrene foam vs. density at $73°F$.[43]

One term for this property which is particularly appropriate to packaging is *compressive creep*. Data on expanded-bead foam at two densities and varying loads are shown in Figs. 15 and 16.

It should be pointed out, however, that these data were collected at room temperature. As the ambient temperature rises, the creep rate increases, becoming quite rapid and possibly catastrophic at temperatures approaching the heat-deflection temperature of these foams (170 to 180°F).

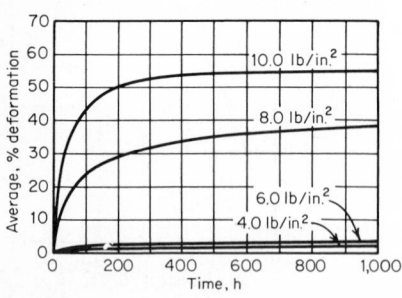

Fig. 15 Foamed Dylite polystyrene static compressive creep density, 1.0 lb/ft³.[41]

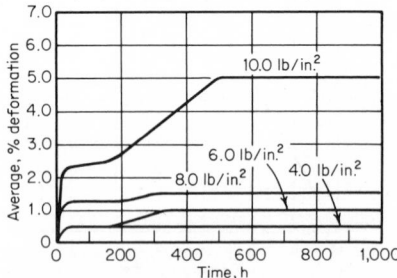

Fig. 16 Foamed Dylite polystyrene static compressive creep. Density, 1.5 lb/ft³.[41]

In addition to static physical properties, the dynamic cushioning qualities of packaging foams must be considered when items are to be protected from shock. Energy absorption of foams is an area where considerable technical information is available, and the reader is directed to some of the technical bulletins issued by Koppers and Dow on this subject.

Basically data are developed on foams by dropping a weighted load (static stress) onto the proposed cushioning material. The most effective cushioning material will absorb energy and thus require more time for complete deceleration.

Thus if

$$G = \frac{\text{measured acceleration}}{\text{acceleration of gravity}}$$

G is the portion of energy not absorbed by the cushion, and the lower the G value the better the energy absorption and protection. The curves in Figs. 17 through 20, obtained by dynamically testing the material at different drop heights, show a convenient method of expressing the cushioning characteristics of materials. (Similar information is presented in Sinclair-Koppers *Bulletin* 9–272, Principles of Packaging

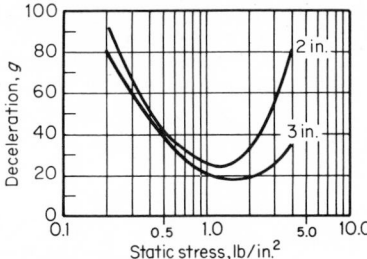

Fig. 17 Cushioning properties of polystyrene at varying densities and energy levels.[43] (0.9 lb/cu ft density)

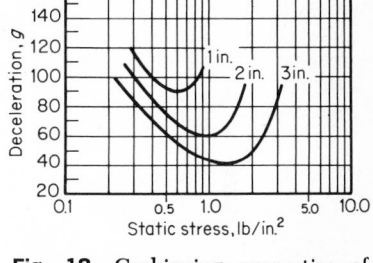

Fig. 18 Cushioning properties of polystyrene foam at varying densities and energy levels.[43] (1.5 lb/cu ft density)

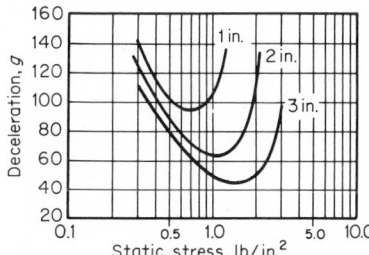

Fig. 19 Cushioning properties of polystyrene foam at varying densities and energy levels.[43]. (2.0 lb/cu ft density)

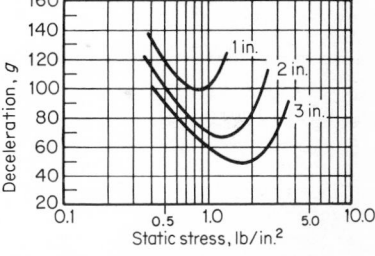

Fig. 20 Cushioning properties of polystyrene foam at varying densities and energy levels.[43] (2.5 lb/cu ft density)

Design.) A different curve is obtained for each specimen thickness for the various densities tested at 24-, 30-, and 36-in. drop heights. The curves are a good indication of the type of protection to be expected from a given material. The adjacent curves indicate what variations in dynamic cushioning might be expected with changes in thickness and density from a drop height of 24 in. A more complete set of curves may be found in Ref. 42, pages 11 and 12, along with information on their use in designing packages.

Molded expanded polystyrene parts or slabs are generally produced by the following sequence of events:

1. A volatile organic liquid, usually neopentane, is forced into beads of crystal (unmodified) polystyrene with some heat and pressure. These beads are then shipped to the foam processor.

2. The fabrication of styrene-foam beads uses two expansion steps in moving from the dense polymer (>60 lb/ft^3) to the final density of 1.0 to 10 lb/ft^3. The beads are first pre-expanded using steam heat (steam is preferred because of its heat capacity) to the approximate density of the final product. The pre-expanded beads, sometimes called *pre-puff*, are then aged up to 24 to allow the foam system to come to equilibrium. The use of pre-expanded beads allows for more uniform density and filling of the thick-walled parts usually produced by this method.

3. The pre-expanded beads are then placed in a mold called a *steam-chest mold*, which is then heated through direct injection of steam into the mold cavity containing the pre-expanded beads. The injection of steam through the cavity walls causes the beads to enlarge in size and to begin to flow in a viscous manner. Thus the beads fuse against one another, creating a uniform cell structure throughout the whole part. The temperature in the part reaches a maximum of 230 to 240°F. Because of this temperature and the fact that considerable gas (organic vapor, air, and water vapor) is contained within the system, the mold must be securely clamped to prevent part distortion. While under clamp pressure, the part must be cooled to below its distortion temperature so that deformation will not occur upon removal of the part from the mold.

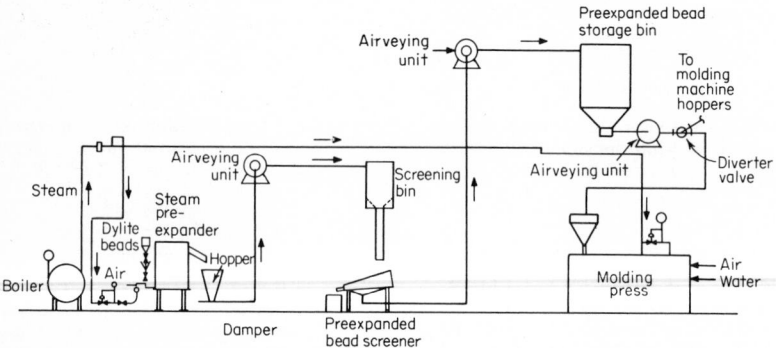

Fig. 21   Schematic of typical expandable polystyrene manufacturing facility.[41]

Since there is a great range of part shape and thickness variations to contend with, expandable beads are supplied in a range of sizes from 0.012 to 0.079 in. particle size (unexpanded).

A product category which lies somewhere between extruded foam and molded expandable bead is extruded film and sheet from expandable beads. In this process the expandable beads containing a blowing agent such as neopentane and usually a nucleating agent are fed directly to an extruder and extruded through film sheet, tube, or rod dies into the desired end product. The resultant film or sheet is uniform, of relatively low density, and exhibits a satin sheen which makes it attractive as a consumer-item packaging material. In many cases the film or sheet is laminated with nonexpanded film to develop better structural or aesthetic properties.

These foam films and sheets with or without laminated skins are used extensively in packaging and are generally converted to their final shapes by a vacuum-forming method. Typical properties of a foam sheet and laminated foam sheet are shown in Table 16.

A typical schematic of the process involved in converting expandable beads to finished parts is shown in Fig. 21. A more detailed view of the steam-chest molding process is shown in Fig. 22.

A variation of the expanded-bead process which has been adapted to ABS polymers by Marbon is worth noting. In this process ABS pellets containing a heat-decomposable chemical blowing agent are supplied by the plastics manufacturer.

TABLE 16 Typical Properties of Foam Sheet and Polystyrene-Foam Laminates[41]

Property	0.035-in. foam sheet		0.050-in.[f] foam sheet and 0.010-in. crystal polystyrene laminate		0.050-in.[f] foam sheet and 0.010-in. modified polystyrene laminate	
Density,[a] lb/ft^3	6.5					
Tensile strength[b] at break, lb/in.^{2} (73°F)	570 (MD)	300 (TD)	1,200 (MD)	1,000 (TD)	720 (MD)	430 (TD)
Ultimate elongation,[b] % (73°F)	3.6 (MD)	4.7 (TD)	2.5 (MD)	2.1 (TD)	6.2 (MD)	6.4 (TD)
Tensile strength at break,[b] lb/in.^{2} (0°F)	570 (MD)	300 (TD)	1,200 (MD)	970 (TD)	690 (MD)	430 (TD)
Ultimate elongation,[b] % (0°F)	3.6 (MD)	4.8 (TD)	3.3 (MD)	1.3 (TD)	8.0 (MD)	6.4 (TD)
Falling-weight impact at 50% failure,[c] in.-oz.[d]	3.5		7		81	
Water-vapor transmission,[d] g/(24 h)(100 in.^{2})(100°F, 90% RH)	4.75		0.42		0.43	
Rate of temp increase through sheet,[e] °F/min	2		2		2	

MD = machine direction.   TD = transverse to machine direction.
[a] MIL-P-19644.
[b] ASTM D 882-56T (crosshead speed 2 in./min).
[c] Gardner tester.
[d] ASTM E 96-53T.
[e] The sheet materials are placed between two aluminum plates.   One plate is held at a constant temperature of 180°F; the second plate is set at 73°F. The temperature of the second plate is allowed to increase by conduction through the sheet material from the initial temperature of 73°F.   The rate of temperature increase of the second plate is a relative measure of the insulating property of the sheet material.   For comparison: 12 mils double-wall kraft-paper insulated bag = 11°F/min; 9 mils 120-lb cup board = 29°F/min.
[f] Foam sheet was 0.035 in. thick before lamination.   Additional thickness represents expansion during lamination.

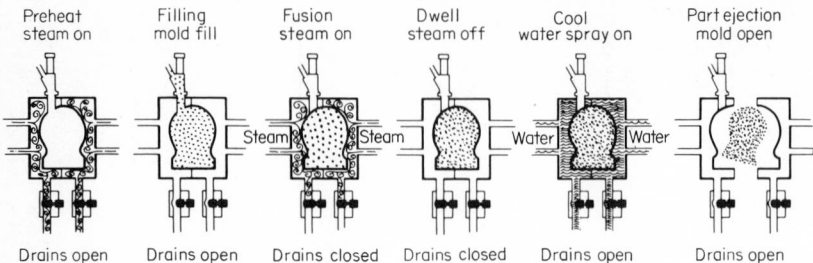

Preheat steam on	Filling mold fill	Fusion steam on	Dwell steam off	Cool water spray on	Part ejection mold open
Drains open	Drains open	Drains closed	Drains closed	Drains open	Drains open

Fig. 22 Representation of typical molding operation for expandable polystyrene.[43]

These unfoamed pellets are poured into a mold with agitation or vibration. The mold is then closed and heated in an oven, causing the blowing agent to decompose and the ABS pellets to soften and expand. The process is similar to steam-chest molding of expandable beads except that steam is not used and the degree of expansion and thus the lowering of density are significantly reduced.

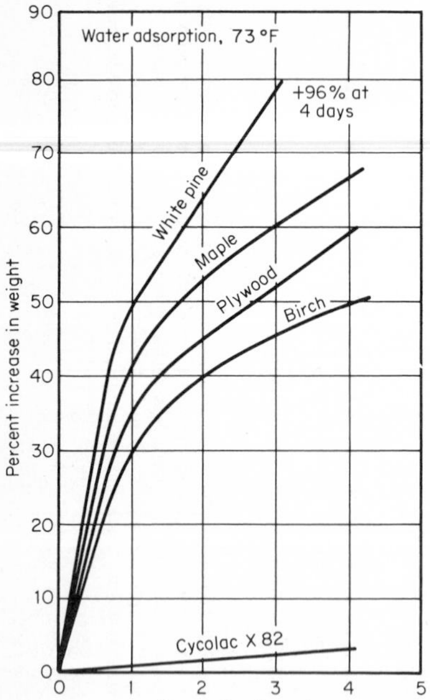

The process was developed by Marbon to fill a need for a low-cost tooling and investment method for large structures. The total system has been called "expansion casting" in a "foam foundry" by Marbon, who also developed the equipment to put the product into operation. Expansion-cast ABS foam is currently being used in specialty pallets for materials handling, furniture structures, utility boxes for underground applications, and as precast concrete forms in the construction industry.

Since the reduction in density is only about 50 percent or 32 lb/ft³, the cast ABS foams perform more like structural materials. Typical properties of expanded ABS are shown in Table 17. Since wood is one of the replacement targets for ABS foam, data are also shown for typical strength/weight ratios (Table 18), screw and staple pull-out strengths (Table 19), and weight change with water immersion (Fig. 23). Again the properties are highly dependent upon the density achieved (although a rather narrow range is available by this process) and the consistency and degree of fusion of the pellets during expansion.

A similar approach taken by Hercules with polypropylene will be discussed under Polyolefins.

A rapidly growing technology in foamed plastics is in the realm of injection-molded structural foam. This subject will be covered in another section.

Fig. 23 Swelling of wood vs. foamed ABS immersed in room-temperature (73°F) water. (*Marbon Chemical Division, Borg Warner Corp. product literature.*)

[Graph: Percent increase in weight vs. Immersion time, days. Water adsorption, 73°F. Curves labeled White pine (+96% at 4 days), Maple, Plywood, Birch, and Cycolac X 82.]

TABLE 17 Properties of Expanded ABS Foams*

Mechanical	
Tensile properties:	
Tensile strength, lb/in.^{2}	
At 160°F	800
At 73°F	1,400
At −40°F	2,200
Flexural properties:	
Flexural strength, lb/in.^{2}	
At 160°F	900
At 73°F	2,400
At −40°F	3,600
Flexural modulus, lb/in.^{2}	
At 160°F	60,000
At 73°F	90,000
At −40°F	140,000
Impact properties:	
Notched Izod impact, ft-lb/in.	
At 73°F	1.2
At −40°F	.6
Hardness, Shore D	60

Thermal	
Coefficient of linear thermal expansion, in./(in.)(°C)	9.7 × 10^{-5}
Deflection temp (unannealed, 264 lb/in.^{2}), °F	159
Thermal conductivity, Btu/(h)(ft^2)(°F/in.)	0.58
Flammability, in./min	1.04

Analytical	
Specific gravity, 31 lb/ft^3	0.50
Water absorption, %	0.60
Mold shrinkage, in./in.	0.008

Electrical	
Dielectric constant:	
50 Hz	1.63
10^3 Hz	1.56
10^6 Hz	1.59
Power factor:	
50 Hz	0.002
10^3 Hz	0.0035
10^6 Hz	0.007
Arc resistance, s	66
Dielectric strength, V/mil	

* Marbon Chemical Division, Borg Warner Chemicals product literature.

TABLE 18 Stiffness Weight Ratio for Various Materials*

Material	Weight for equal stiffness
Steel	7.95
Linear polyethylene	6.40
Polypropylene	4.38
ABS	4.29
Polystyrene	3.90
Aluminum	3.66
Expanded ABS	3.06
Wood:	
Maple	1.62
White pine	1.00

* Marbon Chemical Division, Borg Warner Chemicals product literature.

TABLE 19   Screw and Staple Pull-out Strengths of Expanded ABS and Various Woods (In pounds)

Material	Screw		Staple	
	1 in. long	¾ in. long	$\frac{3}{16}$ in. long No. 903	½ in. long No. 908
JS.........................	270	193	11.3	38.0
Birch.....................	642	442	13.5	46.3
Maple.....................	668	463	22.8	*
White pine.................	239	168	3.1	17.0

SOURCE: Marbon Chemical Division, Borg Warner Chemicals product literature.

Removal speed 0.1 in./min. Number 10 gage steel flathead wood screws embedded to depth of threads.

* Unable to staple for test purposes.

POLYOLEFINS

Introduction   Although low-density polyethylene, high-density polyethylene, polypropylene, and their copolymers differ significantly in their individual characteristics, they will be considered together as a single class of materials in this chapter.

Polyolefins are a rather young family in the commodity-resins field, low-density polyethylene being born during World War II, high-density polyethylene emerging in the middle fifties, and polypropylene being commercially introduced in the late fifties. These three olefins (and their copolymers) comprised 38 percent of the total plastics consumption in 1972.

Chemistry   As with all other polymers the structure of the major building blocks of the polymer dictates the performance, processing, and to a large degree the cost of the resultant polymer.

$$
\text{Ethylene} \quad
\begin{array}{c} \text{H} \quad \text{H} \\ | \quad\quad | \\ \text{C} = \text{C} \\ | \quad\quad | \\ \text{H} \quad \text{H} \end{array}
\quad \text{and propylene} \quad
\begin{array}{c} \text{H} \quad\quad \text{H} \\ | \quad\quad\quad | \\ \text{C} = \text{C} \\ | \quad\quad\quad | \\ \text{H H—C—H} \\ \quad\quad | \\ \quad\quad \text{H} \end{array}
$$

are relatively low-cost monomers derived from petrochemical feedstocks. When these monomers are polymerized, the following generalized properties result (the reader is referred to the chapters on these polymers for a detailed examination of performance):

1. The polymers are extremely nonpolar and thus have excellent electrical properties.

2. The nonpolar nature gives the polymers outstanding resistance to polar solvents and most inorganic chemicals and solutions. Water transmission and absorption are extremely low.

3. The degree that the polymers can pack together into crystalline regions provides for a wide range of stiffness and melting points.

Comonomers can be polymerized along with these basic monomers to generate even more variation in performance. Examples of comonomers and their effect are listed below.

1. *Vinyl Acetate or Acrylates:* reduces crystallinity and thus stiffness in low-density polyethylene.

2. *Butene-1:* reduces crystallinity somewhat in high-density polyethylene—more importantly greatly improves stress-crack resistance.

3. *Acrylic Acid:* reduces crystallinity and stiffness; creates sites for "quasi" cross-link with metal ions to form ionomers.

4. *Ethylene:* improves impact as a comonomer for polypropylene (at low levels), eliminates crystallinity, and becomes a cross-linkable elastomer (at high levels).

5. *Chlorine:* not a comonomer; however, polyethylene in several densities can be chlorinated to varying degrees to accomplish flexibility, flame resistance, or compatibility with other polymers.

Processing and properties   Polyolefin foams can be produced from the numerous types of polymers outlined above by a process similar to those described for polystyrene.   In summary these are:

One process is addition or injection of a gas or volatile liquid into the polymer melt during (usually) an extrusion process.   Upon exposure to atmospheric pressure and having been shaped by the extrusion die, the polyolefin foam expands with a closed-cell structure to the desired shape and density.   Typical shapes can be rods, sheets, boards or planks, or ovals.   Depending on the amount of gas or blowing agent used, foams can be produced in densities ranging from 2 to 9 lb/ft³.

Several varieties of polyethylene foam are available from Dow Chemical under the trade name Ethafoam.   Typical properties of Ethafoam are shown in Table 20.

Two additional products which are presumably produced by the extrusion process are Microfoam sheeting available from Du Pont (polyethylene) and Minicel-PPF

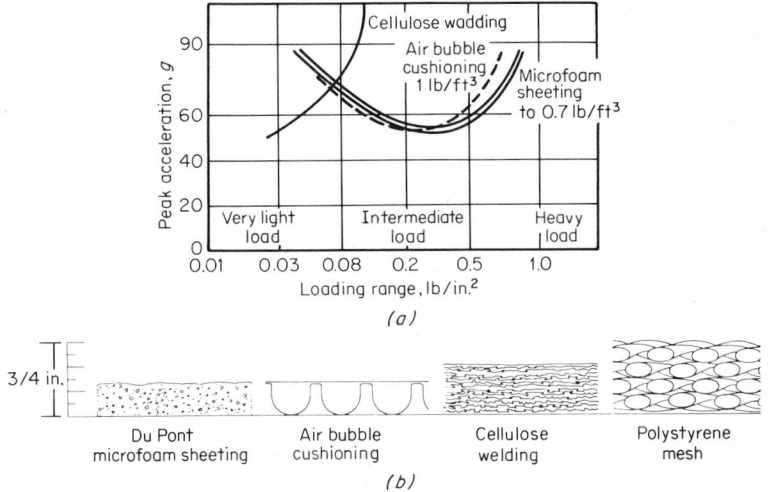

Fig. 24   Shock protection of various cushioning materials.   (*Du Pont product literature.*)

sheet from Haveg Industries (polypropylene).   Both products are aimed at the cushioning market, with Microfoam focusing on protective packaging and Minicel-PPF aiming at the broader cushioning market.   Typical properties of Minicel-PPF appear in Table 21.   A comparison of the cushioning properties of Microfoam is shown in Fig. 24.

A particular advantage of foamed olefins over polystyrene-bead packaging lies in the fact that the lower modulus and higher toughness of the polyolefin foams give rise to higher resilience and resistance to repeated impact.   The low density coupled with the negligible water pickup and closed-cell structure make polyolefin foams extremely popular in personal flotation devices.

Polyolefin foams are also produced by adding chemical blowing agents to pellets followed by injection molding of structural parts.   The blowing agent is added in several ways:

1. By direct addition or dusting of the chemical on the pellets just prior to molding
2. By direct incorporation of the blowing agent by the material supplied into the pellet
3. By adding the blowing agent via a concentrate

TABLE 20 Typical Properties of Ethafoam Polyethylene Foam

Typical properties	Test procedure	Ethafoam 220 sheet	Ethafoam 220 plank	Ethafoam 400	PZ 4139.22	Ethafoam 900 plank
Density, lb/ft³	Dow CSP-3A	2.4	2.2	4	6	9
Cell size (mm avg)	Dow FP-156	⅛ in. = 1.0 ¼ in. = 1.5 ½ in. = 1.7	1.0	1.0	1.0	1.0
Compressive deflection, lb/in².:	ASTM D 1056					
At 5%		2.5	3.5	6.5	15	44
At 10%		3	5	8.5	17	48
At 25%		6	8	12	20	54
At 50%		15	15	21	33	75
Compression set, % original deflection	50% compression for 22 h; 24 h recovery period	10–15	10–15	10–15	10–15	10–15
Compressive creep, % deflection	Loaded at specified lb/in².² static load for 1,000 h:					
75°F		......	<5 at 1.5 lb/in².²	<5 at 3 lb/in².²	<5 at 4 lb/in².²	<4 at 20 lb/in².²
160°F		......	<5 at 0.25 lb/in².²	<6 at 1 lb/in².²	<5 at 2 lb/in².²	<7 at 7 lb/in².²
Tensile strength, min lb/in².²	ASTM D 1564	40	20	35	60	100
Tensile elongation, %	ASTM D 1564	50	60	60	65	65
Tear strength, lb/in.	ASTM D 624	15	6	10	15	29
Water absorption, % by volume	ASTM C 272	<.5	<.5	<.5	<.5	<.5

Property	Test method					
WVTR, perm. in.	ASTM C 355	0.2	0.4	<0.4	<0.4	<0.4
Buoyancy, lb/ft³		55	55	53	51	48
Thermal conductivity, Btu-in./(h)(ft²)(°F)	Dow FP-4	1/8 in. = 0.28 1/4 in. = 0.30 1/2 in. = 0.34	0.37	0.40	0.40	0.40
Thermal stability, % shrinkage	Conditioned at specified temp. with no load:					
24h		−0.3 at 180°F	−0.3 at 180°F	−1.0 at 190°F	−1.0 at 200°F	−1.0 at 200°F
48 h		−1.5 at 180°F	−2.4 at 180°F	−1.5 at 190°F	−1.0 at 200°F	−1.0 at 200°F
Burning characteristics, in./min	ASTM D 1692	2.1	2	2	2	1.2
Dielectric strength, V/mil; 1/4-in. thickness		52	52	59	86	133
Dielectric constant at 10^6 Hz		1.05	1.05	1.06	1.07	1.15
Volume resistivity, Ω/cm		$>10^{16}$	$>10^{16}$	$>10^{16}$	$>10^{16}$	$>10^{16}$
Dissipation factor at 10^9 Hz		$1 - 2 \times 10^{-4}$	$1 - 2 \times 10^{-4}$	$1 \times 2 \times 10^{-4}$	$1 - 2 \times 10^{-4}$	$1 - 2 \times 10^{-4}$
Chemical resistance		Good	Good	Good	Good	Good
Odor		None	None	None	None	None

SOURCE: Dow Chemical product literature.

Injection-molded parts produced from these systems are usually thick-walled semi-structural parts where low part cost and tool cost are desirable. The inevitable surface splay or "frostlike" appearance has restricted the use of this approach in areas where surface appearance is of critical importance.

Low densities have not been achieved with foams produced by this method. Typical densities for most polyolefins range from 25 to 50 lb/ft³.

A variation of this process has been offered by Hercules, Inc. In this Hercocel process pellets of polypropylene which have been precompounded with a chemical

TABLE 21   Typical Properties of Haveg Minicel-PPF

Property	Test method	Minicel-PPF type	
		5UM*	5B-1†
Density, lb/ft³.		5	5
Compression deflection, lb/in.²			
25%.	ASTM D 1056	120	55
10%.		80	40
Compression set, % of original deflection:			
70°C.	ASTM D 1056	85	85
100°C.	D 395 Method B	89	90
Compression modulus:			
25°C, lb/in.² to 10%.		1,200	500
100°C, lb/in.² to 10%.		300	175
Tensile strength, lb/in.².	ASTM D 638	160	145
Tensile modulus, lb/in.².	ASTM D 638	2,500	
Elongation on tension, %.	ASTM D 638	25	65
Flexural strength, lb/in.².	ASTM D 790	230	
Flexural modulus, lb/in.².	ASTM D 790	9,550	
Shrinkage, %.	7 days at 158°F	0.6	
Volume loss, %.	8 days at 220°F	0.7	
Odor after aging.		None	None
K factor, Btu/(h)(ft²)(in./°F).		0.27	0.27
Cell structure.		Extremely fine	Extremely fine

SOURCE: Haveg Industries, Inc. FPO/65-10 (Rev. 1).
* 5UM—unmodified polypropylene.
† 5B-1—modified polypropylene.

blowing agent are fed into a closed mold. The mold is then heated to expand the pellets to fill the mold. The mold is then cooled and the parts removed.

Data on several polyolefins foamed by the injection-molding process appear in Table 22. Data from Hercules on the Hercocel process are found in Table 23.

The electrical industry and in particular the telecommunications segment of that industry have long capitalized on the excellent electrical properties of polyolefins for power, control, and communications cable. It has been found that the capacitance of a telephone line could be reduced (thus reducing attenuation) to a very low level of foaming the polyethylene or polypropylene insulation. In most plants this is accomplished by the use of chemical blowing agents, although direct use of gas has been proposed. Considerable work has been done in the nucleation of foam cells in polyolefins at the Bell Telephone Laboratories.

Some attention has also been given to the foaming of cross-linked polyethylene foam by the addition of a chemical blowing agent and a peroxide cross-linking agent. Although the thermal properties of the foam are enhanced, the added processing complication and added cost have restricted its growth in the United States. (This process has apparently enjoyed significant commercialization in Japan, however.)

TABLE 22 Typical Properties of Foamed Polyolefins[47]

Property	High-density polyethylene, Marlex 6009*		Low-density polyethylene, Marlex 1478		Ethylene butene-1 copolymer Marlex TR-880		Polypropylene	
	Foamed	Unfoamed	Foamed	Unfoamed	Foamed	Unfoamed	Foamed	Unfoamed
Density, g/cm³	0.83	0.960	0.77	0.914	0.76	0.957	0.80	0.905
Part weight, g	615	740	609	......	564	760	650	740
Wall thickness, mils	183	159	160	......	165	160	170	159
Rim stiffness†, lb	12	14.1	......	......	11	9.4		
Flex modulus, lb/in²	127,000	163,000	......	......	119,000	150,000		
Impact strength,‡ ft-lb	10–12	>44	2	>44	2.1	>44	2.3	8
Tensile strength, lb/in²:								
Machine direction	1,100	3,750	920	1,320	......	......	2,600	4,400
Transverse direction	1,100	3,900	640	1,530	......	......	2,000	4,600
Elongation, %:								
Machine direction	95	50	175	150	......	......	10	11
Transverse direction	95	640	70	460	......	......	15	190
Tensile strength across gate, lb/in²	820	1,800	650	1,240	......	......	1,610	2,920
Elongation across gate, lb/in²	10	20	20	60	......	......	8	2
Thermal conductivity, Btu/(h)(ft²)(°F/in.)	0.9 (at 0.5 density) 0.8 (at 0.4 density)	3.7						

* Trademark of Phillips Petroleum Co.
† Force in pounds to deflect rim 1 in.
‡ Falling-ball test.

TABLE 23   Typical Properties of Hercocel Structural Polypropylene Foam

Property	ASTM method	Value
Density at 23°C, lb/ft^3	D 792	33
Tensile yield strength at 2 in./min (⅛-in. specimen), lb/in.^{2}	D 638	1,480
Tensile yield elongation at 2 in./min (⅛-in. specimen), %	D 638	15.5
Elongation at break at 2 in./min (⅛-in. specimen), %	D 638	22.1
Flexural modulus (1% secant) (⅛-in. specimen), lb/in.^{2}	D 790	80,000
Stiffness in flexure (⅛-in. specimen), lb/in.^{2}	D 747	59,000
Izod impact (⅛-in. specimen), ft-lb/in. notch	D 256	
23°C		0.7
−18°C		0.5
Deflection temp (⅛-in. specimen), °C	D 648	
66 lb/in.^{2}		71
264 lb/in.^{2}		45
Deformation under load (composite of ⅛-in. specimens) 1,000 lb/in.^{2} and 50°C, %	D 621	>25
Shrinkage from mold dimensions, in./in.	......	0.014
Water absorption, 24 h, %	D 570	1.21

SOURCE: Hercules Technical Information Report.

POLYVINYL CHLORIDE

Introduction Polyvinyl chloride, or PVC, is another of the plastics world's big three. The reasons for its position (production in 1972 was 4.3 billion pounds) are its tremendous versatility, balance of properties, and economics.

The myriad uses of solid or unfoamed PVC will be covered in another chapter of this book. Almost every form taken by PVC in its solid state has been reproduced in a foam, the exception being thin packaging film and blow-molded bottles and containers. The result is that in our normal lives we are totally surrounded by PVC, or "vinyl," application from wall coverings to upholstery covering to clothing to shoes to the floor we walk on. Some of us may even have plasticized PVC tubing in our bodies as prosthetic devices.

Chemistry The structure of PVC

$$-(\overset{\overset{\displaystyle H}{|}}{\underset{\underset{\displaystyle H}{|}}{C}}-\overset{\overset{\displaystyle H}{|}}{\underset{\underset{\displaystyle Cl}{|}}{C}})-_n$$

gives an indication of its inherent properties, namely:

1. It is derived from low-cost feedstocks, ethylene and chlorine, and is therefore relatively low-cost.

2. The presence of the chlorine atom gives rise to resistance to solvents and chemicals.

3. The chlorine atom also results in a polymer that is normally of low flammability when ignited.

4. The bulk of the chlorine atom renders the carbon chain relatively stiff. This property gives rise to relatively high modulus, which is important in structural applications.

5. The polarity of the molecule allows semisolvents or plasticizers to be added to the polymer, which gives rise to new families of semirigid and flexible products.

PVC exhibits some less than desirable qualities not immediately apparent from the basic structure. A few of the more critical of these are:

1. Processing difficulty of rigid systems due to polymer stability

2. Relatively low heat deflection (and therefore use) temperatures

3. A tendency of plasticized materials to creep or deform excessively under high load, particularly at elevated temperatures

Rigid foams Like most thermoplastic polymers, PVC can be foamed by mixing it with or injecting volatile liquids or gases which expand upon processing or by intimately mixing the molten polymer with a chemical blowing agent which decomposes into a gas (usually nitrogen) at desired temperatures. Both open-cell and closed-cell structures are thus possible.

One of the most rapidly growing segments of the rigid-vinyl-foam business in recent years has been the profile extrusion of extremely fine cell foam. The technology, originally developed in Europe, is known as the *Cekula process* (see Fig. 25). The integral-skin foamed profiles produced by this method are called *microcellular vinyl*. Considerable inroads have been made into the prefinished-molding wood market in relatively small cross sections. As soon as compound and processing technology develop, penetration into additional construction applications will occur.

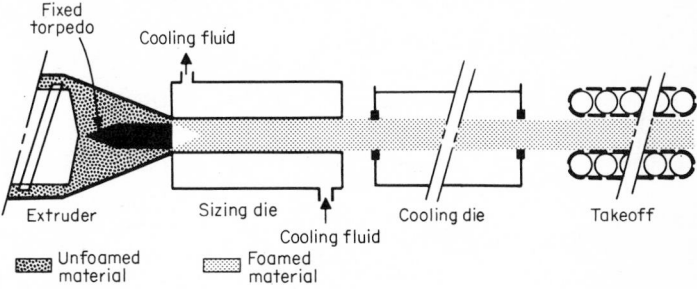

Fig. 25   The Cekula process for foam-profile extrusion.[54]

Advantages of microcellular-vinyl extrusions over classical wood profiles such as pine include:
1. Less splitting
2. Dent resistance
3. Impact resistance
4. Moisture/chemical resistance
5. Less flammability
6. Integral color

Properties of cellular products, as previously discussed in the introduction, are highly dependent upon density, matrix composition, and processing. Thus it is dangerous to generalize the types of properties one would obtain from these products. In order to put them in a general framework of performance, however, the data in Tables 24 and 25 are presented.

A unique rigid PVC foam system which resembles the chemical processes of polyurethane more than the hot-melt thermoplastic was developed by a French company, Kleber Colombes S.A., about a decade ago. This cross-linked, fluorocarbon-gas-filled foam exhibited excellent thermal-insulation properties, higher strength/weight values than most rigid polyurethane foams, very low water pickup and permeability, and inherent nonburning characteristics. Its growth rate was anticipated to be quite rapid; however, its somewhat higher cost and processing limitations have prevented this system from making large inroads into rigid polyurethane foam.

Johns Manville has been promoting such a foam for insulation purposes, and Plyfoam, Inc. (Plainview, Long Island, New York) has proposed the use of rigid PVC foams as cores with GRP skins for marine construction. Comparative properties of rigid foams most generally used for insulation, structural applications, and flotation are shown in Table 26. These data clearly indicate the areas of superiority of a cross-linked PVC foam.

It is expected that cross-linked, rigid, cast PVC foam will grow at a moderate rate, capturing those markets where its high strength/weight ratio and low flammability are needed. Extruded microcellular thermoplastic PVC foam profiles, on the other hand, are expected to achieve an extremely high growth rate as wood and the labor to shape wood become scarcer. In this latter market PVC must share its growth with other low-cost polymers such as high-impact polystyrene and ABS.

Flexible foams   The largest use of PVC foams lies in flexible-foam applications. Because of PVC's previously mentioned great versatility coupled with several diverse processing methods which have been developed over the years, plasticized or flexible vinyl has moved into a great number of products over the last 10 to 15 years. The addition of a suitable plasticizer to polyvinyl chloride renders the polymer soft and pliable. The expansion of thin softened material by gases or chemical blowing agents creates a resilient material which in film or sheet foam has a luxurious hand, or feel. This quality, coupled with the other attributes previously discussed, has created

TABLE 24   Properties of a Typical Foamed PVC Compound[53]

Property	ASTM test method	Value
Flexural strength, lb/in.^{2}	D 790	4,400
Modulus of elasticity in flexure, lb/in.^{2}	D 790	135,000
Deflection temp, °C	D 648	60.0
Flammability	D 635	Nonburning
Coefficient of thermal expansion and contraction, in./(in.)(°C $\times$ 10^{-5})	D 696	3.7
Specific gravity	D 792	0.7
Coefficient of thermal conductivity	Equivalent to C 177	
Btu/(h)(ft^2)(°F/in.)	....	0.44
cal/s/cm^2/(°C/cm)	....	1.5 $\times$ 10^{-4}
Gardner drop impact, in.-lb:		
At 85 ± 2°F	SPI method	80*, 272†
At 32 ± 2°F	SPI method	60; 100†

All tests performed on extrusion expanded strips.
* Surface scissions visible only on underside of specimen.
† First surface scissions visible on impact side of specimen.

TABLE 25   Properties of a Foamed PVC Compound

Property	Test method	Value
Specific gravity (expanded)	ASTM D 792	0.6
Tensile strength, lb/in.^{2}	ASTM D 638	2,400–2,700
Modulus of elasticity, lb/in.^{2}	ASTM D 638	150,000
Heat-deflection temp, °F at 264 lb/in.^{2}	ASTM D 648	149
Izod impact, ft-lb/in. notch	ASTM D 256	0.6–0.7
Coefficient of linear expansion, in./(in.)(°C $\times$ 10^{-5})	ASTM D 696	7.5
Thermal conductivity, cal/(s)(cm^2)(°C/cm) $\times$ 10^{-4}	ASTM C 177	2.01
Flammability	ASTM D 635	Self-extinguishing

SOURCE: B. F. Goodrich Chemical Product Data Sheet, Geon 85721.

such diverse markets that all cannot be covered here. A survey of processing techniques and general properties should direct the reader to proper materials selection for use in specific designs. Table 27 illustrates the large number of processes and applications that are possible using chemically blown vinyl foam.

Most flexible PVC foams are produced from a plastisol. A plastisol is an emulsion-polymerized PVC resin dispersed in a liquid plasticizer. The surface characteristics of the polymer particle prevent solution or gelling of the polymer by the plasticizer until appropriate elevated temperatures are reached—usually 275 to 350°F. As ungelled viscous liquids, plastisols can be pumped, roller-coated, knife-coated, cast, and laminated quite easily at high speeds.

Foaming is accomplished by two basic methods:
1. By incorporating a chemical blowing agent into the liquid plastisol.
2. By forcing or beating a gas such as air or nitrogen into the plastisol. This method, known as *mechanical frothing*, was pioneered by the R. T. Vanderbilt Co. and uses an Oakes foamer.

The chemically blown systems offer more degrees of freedom in that their cell size and amount and ratio of open cell/closed cell can be controlled. Thus properties can be designed into the foam to a large degree. A disadvantage is the fact that much more control must be exercised in the formulation and the choice of an oven (for both plastisol fusion and chemical decomposition of the blowing agent and resultant foaming must take place at the proper time).

The mechanical-froth system, although somewhat simpler, can produce only open-cell foam structures. Table 28 summarizes these processes.

TABLE 26 Comparative Properties of Plastic Foams[51]

Property	Cross-linked PVC		Polystyrene		Poly-urethane
Density, lb/ft^3................	2.0	4.0	1.0	1.8	2.0
Compressive strength, ultimate or yield........................	50	160	16–18	30	30
Shear strength, ultimate.........	40	100	33	30–40	22
Water absorption, % by vol......	1.9	1.0	2.0	0.25	3.0
Coefficient of thermal expansion, in./(in.)(°F $\times$ 10^{-6})...........	9.0	9.0	25	40	30–45
Flame spread (ASTM E 84), 1-in.-thick material.................	3	25	10–20	10–25	45
Max continuous service temp (hot face exposure).................	225–250	225–250	175	165	200–275
Thermal conductivity					
K at 75°F mean..............	0.18	0.20	0.26	0.28	0.17
K at 0°F mean..............	0.16	0.18	0.22	0.23	0.17
Water-vapor permeability, perm-in.....................	0.10	0.04	2.0	1.2–1.5	2.2
Dimensional stability, dry heat, after 100 h at 212°F, % vol. change....................	−5	−2	−100	−100	6
Dimensional stability, moist heat, after 28 days at 100% RH, 158°F, % vol. change.................	−4	−1	−5	−5	30
Solvent resistance..............	Good	Good	Poor	Poor	Good

The properties of flexible vinyl foams vary greatly with the unlimited compounding combinations that are possible. Tables 29 and 30 show typical formulations for both mechanical froth and chemically blown foams. In addition to the range of amounts shown in this figure one must keep in mind the large number of types of materials in each ingredient category. Figure 26 shows the range of a typical property—compression—over a range of plasticizer content.

Typical properties of flexible vinyl foams are shown in Table 51 at the end of the chapter. In general vinyl foams are higher in density and lower in resilience than polyurethane foams. On the other hand their availability in a liquid or plastisol form allows usage in thin films that are impractical for polyurethane systems.

Flexible PVC foams can also be produced by incorporating a chemical blowing agent or a volatile liquid into a compounded pellet or powder, with subsequent conventional extrusion into profiles, sheet, film, or wire insulation.

In summary PVC foams have a highly desirable balance of properties such as:
1. Softness, hand, or feel
2. Thermoplastic processibility, leading to easy processing, heat sealing, and embossing

TABLE 27 Summary of Processes and Applications for Foamed Polyvinyl Chloride[55]

Construction	Application	Method of manufacture							
		Calender	Plasticol coating	Slush molding	Injection molding	Rotational molding	Plastisol spray	Extrusion	Compression molding
		Existing Uses							
Skin/foam/fabric	Automotive	X	X						
	Clothing	X	X						
	Footwear	X	X						
	Furniture	X	X						
	Luggage	X	X						
	Flooring		X						
Skin/foam/felt	Flooring	X							
Skin/fabric/foam	Footwear			X					
Skin/foam	Place mats		X						
Foam/fabric	Handbag		X						
	Sock liner		X						
Foam (sheet)	Automotive		X						
	Padding		X						
	Tape		X						
	Underlays		X						
	Closures		X						
Foam (bulk)	Soles and heels				X				

7-40

Developing Uses

Foam (bulk)							
Dolls							
Toys			X	X			
Packaging					X	X	
Profiles						X	X

Potential Uses

Skin/foam/fabric			X				
Skin/foam			X				
Foam (bulk)	X		X			X	X
Wall covering							
Shower curtains							
Draperies							
Wire coating							
Floor tiles							
Records							
Gaskets							

TABLE 28   Plastisol-Foam Processes

Type	Technique	Cell structure	Process or blowing agent
I	Mechanical atmospheric	Open	Whipped (Oakes-Vanderbilt) Solvent evaporation (e.g., methylene chloride) Pressurized-gas incorporation (Dennis, elastomer)
II	Chemically blown (liquid-paste) atmospheric	Open	Sodium borohydride Nitrosan
III	Chemically blown (hot-melt) atmospheric	Partially closed	Celogen OT Azodicarbonamide Celogen RA Expandex 177
IV	Chemically blown high pressure	Closed	Chemical blowing agents, post-expanded

SOURCE: B. F. Goodrich Chemical Co.

Plastisol foam can be classified by method of gas incorporation (mechanical or chemical), pressure on the paste during blow (atmospheric or elevated), and/or the proportion of closed cells formed (mostly open, to all closed).

TABLE 29   Typical Formulation for a Mechanically Frothed PVC Foam

Material	phr*
PVC dispersion resin..............	50–100
PVC blending resin................	50–0
Plasticizer.......................	75–120
Silicone surfactant................	2–4
Stabilizer........................	1–3
Filler............................	0–50
Pigment..........................	As desired

SOURCE: "*Modern Plastics* Encyclopedia," 1972–1973.
* Parts per 100 resins.

TABLE 30   Formulation for a Chemically Blown PVC Foam

Material	phr*
PVC dispersion resin..............	60–100
PVC blending resin................	40–0
Plasticizer.......................	40–120
Blowing agent....................	0.25–5
Stabilizer........................	2.5
Filler............................	0.50
Pigment..........................	As desired

SOURCE: "*Modern Plastics* Encyclopedia," 1972–1973.
* Parts per 100 resins.

3. Resistance to most chemical environments and weather
4. Desirable electrical properties
5. Insulating properties for heat and sound
6. Ability to absorb energy

These have led to hundreds of uses in industry, some of which are listed in Table 31.

A special version of PVC foam which could be characterized as an elastomer foam is the foamed PVC/nitrile rubber blend. The nitrile elastomer is added to the rigid PVC polymer along with selected plasticizers to lower the viscosity. The nitrile (acrylonitrile/butadiene copolymer) rubber is then cured within the plasticized PVC matrix using normal rubber curatives. Considerable care must be

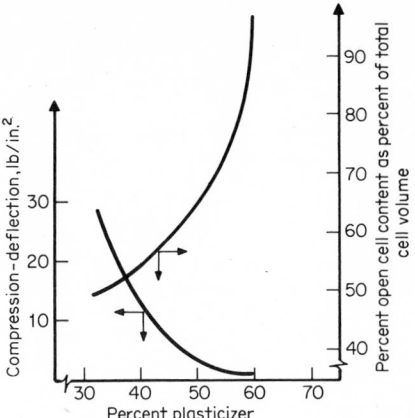

Fig. 26 Foam properties vs. plasticizer content of a typical PVC foam. (*VIPAG Newsletter*, vol. II, no. 2, p. 42.)

exercised in both the compounding and curing of these systems to balance the cure rate with expansion rate.

The resultant polymer foam achieves an attractive balance of properties in the areas of flame resistance, chemical, oil, and solvent resistance, low compression set, weather assistance (when pigmented black), and fairly good insulating characteristics. As such it has found widespread use in insulating of refrigeration systems, gasketing, and seals.

The properties of a typical nitrile/PVC foam are shown in Table 32.

TABLE 31 Uses of PVC Foam

Type*	Applications
I. Mechanical atmospheric.....................	Rug backing, Oakes
	Sound-deadening wall covering, Oakes
	Cushioning, elastomer
	Skived sheet, elastomer
II. Chemically blown atmospheric (liquid-paste)...	Molded toys
	Automotive padding
	Carpet backing
	Gasketing
III. Chemically blown atmospheric (hot-melt)......	Upholstery
	Flooring
	Apparel
	Handbags, luggage
	Shoe fabric
IV. Chemically blown high pressure..............	Fish floats
	Marine fenders
	Surfboards
	Insulation
	Athletic mats
	Life rings and jackets

SOURCE: B. F. Goodrich.
* From Table 28.

TABLE 32 Typical Properties of PVC/Nitrile Rubber Foams[61]

Property	Min	Max
Density, lb/ft³...	3.5	7.0
Compression to 25%, lb/in.²......................................	1.5	3.5
50% compression set at room temp, 24-h recovery, % of original height..	15	20
Dimensional change, linear, 24 h at 158°F, %.........................	1.5	2.5
Flammability, strip ¼ by ½ in., time until flame is out, s..............	0	5
Water absorption, lb/ft² of cut surface............................	0	0.1
Thermal conductivity, Btu/(hr)(ft²)(°F/in.).........................	0.24	0.28
Cold-crack temperature, pass, 180° bend over ½-in. mandrel, °F.........	−55	

EPOXY

Introduction Epoxy resins are formed from the condensation of coreactants such as bisphenol-A and epichlorohydrin and cured through cross-linking reactions of anhydrides and diamines. Many variations are possible in the reactants which determine strength and electrical properties, while the choice of curing agent determines cure rate and ultimate temperature resistance. The resultant polymer exhibits excellent mechanical and electrical properties coupled with outstanding chemical resistance. The reader is referred to the chapter covering these resins for full information on performance.

Several attempts have been made to produce low-density thermal insulations using epoxy resins and fluorocarbon gases. Although the properties and densities have been comparable with polyurethane foams, the ease of processing polyurethanes and the economic advantage of these foams have prevented epoxy foams from attaining any significant level of penetration into the thermal-insulation field.

One important application where epoxies show a distinct advantage is the insulation of large outdoor gas-fuel-holding tanks where the superior weathering of epoxy foams is needed.

The major use of epoxy foams has been in the syntactic-foam field covered in the next section of this chapter.

Properties of various types of epoxy foams are covered in the general properties table at the end of the chapter (Table 51).

SYNTACTIC

Introduction Syntactic foams, like syntactic crystalline polymers, are characterized by their high degree of order or structure. Unlike the foams discussed in other sections of this chapter, the gas cells are not produced within the foam matrix but are synthetically produced from glass, ceramic, or polymers in a separate operation and are then introduced into the liquid-polymer matrix.

Some polymer cells or microspheres differ significantly from the glass in that they are formed from a thermoplastic polymer. Thus during the heat-curing cycle of epoxies or polyesters these thermoplastic spheres, which contain a trapped organic gas, expand several-fold, reducing the composite density to relatively low levels. Some polymer spheres have also been produced from a thermosetting phenolic resin.

Glass or ceramic hollow spheres are available from several manufacturers; however, their use in foams is somewhat low, primarily because of the cost.

The major use for inorganic microballoons has been for high-performance syntactic foams for tooling or flotation buoyancy in deep-submergence vehicles where the high compressive strength of these foams can be fully utilized. Typical uses of syntactic foams are illustrated in Fig. 27. Some glass microballoon/epoxy foams have also been employed in the packaging of delicate electronic instruments. Of the organic or polymer spheres the predominant material is Dow Chemical's Saran microspheres. This product has been used mostly in the reinforced and unreinforced polyester field for electrical applications and to reduce the weight and improve the ductility of decorative and structural castings. The properties of Saran microspheres are shown in Table 33.

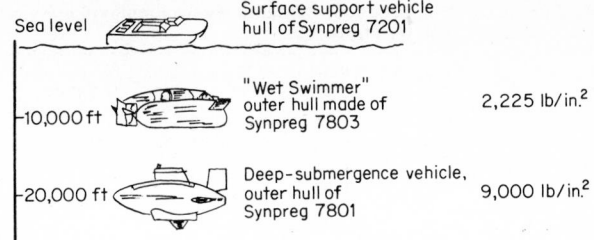

Fig. 27 Typical usage of an epoxy/glass sphere prepreg (Synpreg-Whittaker Corp) in buoyancy application.[63]

TABLE 33 Typical Properties of Thermoplastic Microspheres[67]

	Precursor particle	Foamed sphere
Composition....................	68 vinylidene chloride/22 acrilonitrile/10 isobutane	75 VeCl$_2$/25 VCN
Bulk density, lb/ft^3.................	45	<1.0
Particle size, μm:		
No. average......................	8	30
Range...........................	4–20	10–100
Foaming temp, °C.................	90–120	
K, Btu(in.)(h)(ft)(°F)...............	......	0.25
Hydraulic pressure, lb/in.^{2}:		
Buckling.........................	......	10–50
Rupture..........................	......	50–250
Solvent resistance:		
Nonpolar organics................	......	Excellent
Polar organics....................	......	Fair
Mineral acids.....................	......	Excellent
Alkali............................	......	Excellent

TABLE 34 Typical Properties of Syntactic Foams Using Plastic and Silica Spheres[67]

Formulation 50% TBS/50% DPGM*	Sphere loading, %	Density, lb/ft^3	Shrink-age, %	Compression properties, modulus, lb/in.^{2} 0.0625 in. compression × 10^4	0.125 in. compression × 10^4	0.250 in. compression × 10^4	Uniaxial compressive strength, lb/in.^{2}
Control...........	....	63.8	1.75	....	....	....	19,098
Thermoplastic.....	1	52.5	....	....	16.5	10.2	9,549
	2	42.7	1.60	....	10.5	7.3	6,366
	3	36.7	1.50	....	8.5	5.6	4,902
	4	33.0	1.25	....	7.5	4.5	3,788
	5	28.8	1.25	....	....	3.7	2,897
Silica............	15	52.0	1.00	24.2	20.8	Broke	6,800
	30	43.0	0.50	25.2	18.4	....	6,100
	40	39.5	0.45	25.7	20.8 at 0.100 in.	....	5,540
Phenolic..........	10	53.8	1.25	20.8	17.4	11.7	8,800
	20	45.2	1.15	16.7	14.0	9.2	6,460
	30	39.2	0.75	14.0	11.7	7.5	5,160

Test cylinders: 2 in. diam × 4 in. length.
* 50% tertiary butyl styrene/50% dipropylene glycol maleate.

TABLE 35   Typical Electrical Properties for Syntactic Foam Employing Saran Microspheres and a Polyester Matrix[67]

	Frequency, Hz	Control, 63.8 lb/ft³	1% spheres, 52.5 lb/ft³	3% spheres, 36.7 lb/ft³	5% spheres, 28.8 lb/ft³
Dielectric constant........	10^2	3.17	2.59	2.04	1.75
	10^3	3.12	2.55	2.02	1.73
	10^4	3.08	2.51	2.00	1.71
	10^5	3.02	2.48	1.97	1.69
	10^6	2.93	2.40	1.95	1.67
Dissipation factor.........	10^2	0.0078	0.0094	0.0058	0.0056
	10^3	0.0120	0.0103	0.0067	0.0068
	10^4	0.0133	0.0111	0.0073	0.0077
	10^5	0.0158	0.0139	0.0095	0.0092
	10^6	0.0205	0.0180	0.0130	0.0103

Composition: 50% tertiary-butylstyrene/50% DPGM plus various concentrations of thermoplastic microspheres.

TABLE 36   Typical Physical Properties for Glass-Reinforced Polyester Syntactic Foams[67]

Formulation				Flexural properties, 4-in. span	
50% TBS/50% DPGM*	Sphere loading, %	Density, lb/ft³	Tensile, lb/in.²	Modulus, lb/in.² ×10⁵	Flexural strength, lb/in.²
Control....................	....	75.6	6,880	6.32	14,420
Thermoplastics..............	1	63.1	5,960	5,23	11,980
	2	52.4	.....	3.86	9,500
	3	44.9	.....	3.40†	8,115†
	4	39.4	.....	2,76†	7,061†
	5	37.4	.....	2.71†	8,289†
Silica......................	15	62.4	6,000	5.96	12,100
	30	54.3	5,790	5.81	11,820
	40	51.2	.....	5.21	9,720
Phenolic....................	5	68.2	.....	5.42	12,600
	10	60.3	.....	4.91	10,500
	15	53.2	.....	3.82	8,500
	20	52.4	4,035	3.72	7,520
	30	48.7	4,000	3.34	8,460‡

Laminates: Contained two layers of 1¼-oz continuous filament glass-fiber mat.
* 50% tertiary butyl styrene/50% dipropylene glycol maleate.
† 5-in. span.
‡ Thickness 0.165 to 0.170 in.

From the basic description of the hollow spheres it is obvious that the processes used to produce foams must be of extremely low shear intensity in order to prevent complete destruction of the fragile spheres. Thus only low-viscosity reactive-liquid systems are used and hot-melt processes are not used.

Tables 34 through 36 describe the general properties of Saran microsphere polyester-resin composite foams. The data are compared with phenolic and silica hollow spheres.

Figure 28 illustrates the differences of the strength/weight ratio of the hollow spheres discussed in this section.

Tables 37 through 39 describe several of the mechanical and physical properties of syntactic foams constructed from glass microballoons and epoxy-resin matrices.

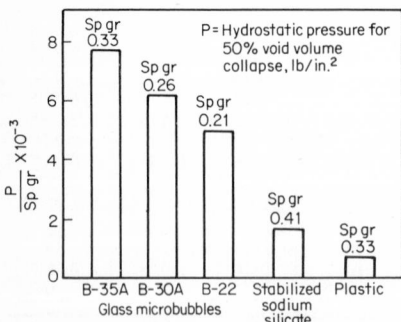

Fig. 28   Strength/weight comparison of strength of glass spheres with plastic and silica spheres.[65]

TABLE 37 Typical Properties of a Syntactic Foam Employing Glass Microballoons and Epoxy-Resin Matrix[65]

Nom. density, lb/in.²	Nom. buoyancy in sea water	Water absorption, % water	Uniaxial strength, lb/in.²	Compressive yield, lb/in.²	Properties modulus, lb/in.²	Hydrostatic crush point, lb/in.²	Tensile strength, lb/in.²	Shear strength, lb/in.²	Bulk modulus, lb/in.²
34	30	1.5*	7,500	6,500	300,000	10,000	3,000	3,500	275,000
36	28	2.1†	9,500	8,000	370,000	11,000	3,000	3,500	325,000
42	22	1.5‡	14,500	12,000	450,000	19,000	5,000	5,000	450,000

Tests conducted on 1-in. diam by 2-in. cylinders (ASTM D 2735-68T).
* After 1 week at 4,500 lb/in.².
† After 6 weeks at 5,000 lb/in.².
‡ After 1 week at 13,500 lb/in.².

TABLE 38   Properties of a High-Performance Syntactic Foam[65]

	Density, lb/ft^3	Uniaxial strength, lb/in.^{2}	Compressive properties		Hydro-static crush point, lb/in.^{2}	Water absorp-tion, % water*
			Yield, lb/in.^{2}	Modulus, lb/in.^{2}		
Standard epoxy resin....	33.2	11,700	10,300	450,000	15,200	2.1
Bis (2, 3 epoxy-cyclo-pentyl) ether resin.....	34.4	13,800	12,500	536,000	17,500	2.4
Alternate high-strength resin.................	33.1	13,400	12,300	540,000	17,000	0.67

* After 12 weeks at 9,000 lb/in.^{2}.

TABLE 39   Properties of Prepreg Used in Buoyancy Application[63]

Properties	Synpreg 7801 (for use at ocean depths to 20,000 ft)	Synpreg 7802 (for use at ocean depths to 10,000 ft)	Synpreg 7803 (for use at ocean depths to 1,000 ft)	Synpreg 7804 (com-patible 181-glass fabric prepreg for surface finishing)	Synpreg‡ 7201 (fire-resistant syntactic prepreg primarily for surface use)
Specific gravity...............	0.7–0.9	0.7–0.9	0.7–0.9	1.7–1.9	0.7–0.9
Thickness, in..................	0.365	0.400	0.420	0.100	0.400
Tensile properties:					
Strength, lb/in.^{2}.............	13,000	13,000	13,000	60,000	16,000
Modulus, 10^6 lb/in.^{2}........	1.0	1.0	1.0	3.6	1.3
Flexural properties:					
Strength, lb/in.^{2}.............	20,000	20,000	19,000	92,000	
Modulus, 10^6 lb/in.^{2}........	1.0	1.0	1.0	3.7	
Compression, flatwise:					
Yield strength, lb/in.^{2}........	19,000	17,000	15,000	....	6,000
Ultimate strength, lb/in.^{2}.....	22,000	20,000	16,000		
Compression, edgewise:					
Strength, lb/in.^{2}.............	23,000	23,000	23,000	55,000	12,000
Modulus, 10^6 lb/in.^{2}........	1.0	1.0	1.0	3.8	1.1
Water absorption, %..........	<3*	....	<1†	0.5	
Izod impact, ft-lb/in...........	7	7	6	16	6
Dielectric constant at 9.375 GHz.	2.3	2.3	2.3	4.4	2.5
Loss tangent at 9.375 GHz......	0.014	0.014	0.010	0.020	0.017
Thermal conductivity, 10^{-4} cal cm/(cm^2)(s)(°C).......	2.6	2.6	2.6	....	2.6
Thermal expansion, 10^{-6} in./(in.)(°F).............	9.1	9.1	9.1	5	9.1

Test panels 10 by 10 in. were fabricated from 10 layers of syntactic foam prepreg (Synpreg 7804 laminated 12-ply).

The laminates were bagged with top and edge bleeder on caul sheets and cured 9 min/ply at 300°F and 50 lb/in.^{2}.

All specimens were tested per Federal Test Method Standard 406.

* After cycling 1,500 times to a hydrostatic pressure of 13,500 lb/in^2.

† After cycling 1,500 times to a hydrostatic pressure of 300 lb/in.^{2}.

‡ Trademark of American Cyanamid Company.

ELASTOMERS

Introduction   Although there are almost as many families of polymers in the category of elastomers as in plastics, they will be considered as a group in this section. These materials are basically similar in several aspects, which distinguishes them from plastics foams. Some of these characteristics are:

1. They are highly flexible, compressible, and soft.
2. They have a cross-linked network which resists permanent deformation under loads.
3. They have varying degrees of heat resistance, which, coupled with deformation resistance, allows usage with applied loads at elevated temperatures.

With the correct elastomer system or rubber compound based on specific performance criteria, the above characteristics have led to the usage of elastomer foams in the following general areas:

1. Sealing
2. Gasketing
3. Cushioning
4. Vibration isolation
5. Packaging
6. Insulation

Use of plastic foams in many of the above applications has increased in the past few years, for example, in flexible urethane foam in cushioning and bedding and PVC flexible foam in carpet underlay.

Although considerably more detail on elastomers is available in another section to help the engineer select the proper foam system, the general characteristics of some of the more common elastomers which can be foamed are listed in Tables 40 and 41.

The performance of foams produced from these elastomers can be qualitatively related to the general performance in the unfoamed or solid state.

Processing   The processing of elastomeric foams involves a complication not present with thermoplastic foams, for as the foam is being generated the elastomer is curing or vulcanizing. Should expansion occur too rapidly, the foam will collapse upon complete vulcanization; however, should the cure rate advance too rapidly, the cellular-structure growth will be restricted. Generally plasticizers or oils are used to soften the vulcanizates to render them more foamable.

Several methods of processing have evolved over the years for rubber foam, or sponge rubber, as it is sometimes called.

The classical method for foam or sponge preparation is to incorporate low-cost inorganic salts which decompose with heat and organic-acid acceleration into carbon dioxide. Because of the rapid rate at which carbon dioxide diffuses through the walls, the cell walls rupture and an open-cell structure results. In addition to the use of rubber in the solid or bulk form the elastomer system can be compounded in a latex or water-emulsion system. This liquid system can then be handled like the previously discussed PVC plastisol system. Thus fabric coatings and rug backing are large uses for elastomeric foam.

Nitrogen diffuses much less readily through these polymers and enables the production of closed-cell foams. The chemical blowing agents employed are similar to those used in plastics and are discussed in the blowing-agent section of this chapter.

Direct nitrogen injection is also used in elastomers, although the process is somewhat more complex than with thermoplastics. The rubber process is primarily used to fabricate closed-cell flat foam-rubber sheets which are die-cut for use as seals and gaskets. The process involves the following steps:

1. The elastomeric compound is molded into large flat sheets and the state of cure is advanced to a "B" stage, or semicured state.
2. The sheet is rolled into a cylindrical shape and placed in a high-pressure chamber, where the nitrogen pressure is raised to 5,000 lb/in.^{2} This pressure can be maintained for as long as 40 h to complete the pressure diffusion of the nitrogen into the partially cured rubber.
3. The rubber sheet is then placed in the final mold, and the temperature is

TABLE 40 Typical Property Comparisons of Major Elastomers[70]

Type	Isoprene rubbers NR and IR	SBR	Neoprene	Butyl	Nitrile copolymers	Ethylene propylene	
						EPDM	EPM
Specific gravity	0.93	0.94	1.23	0.92	1.00	0.86	0.86
Tensile strength, lb/in.²:							
Pure gum	>1,000	<500	>3,000	>1,500	>500	<400	<300
Reinforced	>4,000	>3,000	>3,000	>2,000	>3,000	>3,000	>3,000
Practical Shore A:							
hardness range	30–100	40–100	40–95	40–75	30–100	45–95	45–95
Resilience:							
Room temp	E	G	G	P	F	G	G
Hot	E	G	G	G	F	G	G
Tear resistance	G–F	F	G	G	F	G	G
Abrasion resistance	E	G–E	E	F–G	F	G–E	G–E
Atmospheric aging resistance	P–E*	P–E*	E	G–E	F	E	E
Oxidation resistance	G	G	G	G–E	F	E	E
Heat resistance	G	G	G	G–E	G	G–E	G–E
Low-temperature flexibility	E	G	F	F	F	G–E	G–E
Compression set	F–G	F	P(GN)‡ G(W)‡	F	G	E	E
Permeability	F	F	G	E	E	F	F
Flame resistance	P–G*	P–G*	E	P	P	P–E*	P–E*
Acid resistance:							
Dilute	G	F–G	E	E	G	E	E
Concentrated	F	F–G	G	E	F	G	G
Moisture resistance	F–G	G–E	F	F	E	E	E
Solvent resistance:							
Hydrocarbons	P	P	G†	P	F	P	P
Oxygenated solvents	G	G	P	G	P	G	G
Oil and gasoline	P	P	G	P	E	P	P
Animal and vegetable oils	P–G	P–G	G	E	E	G	G
Dielectric properties	G–E	G	P	G–E	P	G–E	E

E = excellent. G = good. F = fair. P = poor.
* Specially compounded.
† Except aromatics.
‡ A type of neoprene.

raised to cure the elastomer. The nitrogen gas expands and creates a closed-cell structure.

Typical properties for various elastomeric foams are shown in Tables 42 and 43.

BLOWING AGENTS

Introduction Throughout this chapter references are made to blowing agents. These materials can be classified into several categories:

1. Gaseous materials injected directly into the molten polymer under pressure
2. Volatile liquids incorporated into the polymer
3. Chemical agents which decompose into a gas and other by-products

Although this chapter is limited to the comparative performance of the finished products rather than the chemistry of formulation and reactions, the tables that follow will help to acquaint the reader with the many possible variations when foaming a polymer. The reader must also realize that foaming a polymer adds yet another important variable to the complex interrelationship of composition, process, performance, and economics.

Gases. Gases are a relatively simple method of creating foams as long as one takes care to inject the gas at a sufficiently high pressure to keep its volume low prior to expansion and to inject the gas at a point in the melt where it cannot "backflow" out the polymer entrance point.

TABLE 41 Typical Property Comparisons of Specialty Elastomers[70]

Type	Thiokol		Sili-cone	Fluoro-carbon	Hyp-alon	Acrylic
	FA	ST				
Specific gravity	1.34	1.25	.98	1.85	1.1	1.09
Tensile strength, lb/in.²:						
Pure gum	170	200	0	1,800–3,000	2,100	400
Reinforced	1,300	1,300	1,000	2,500–3,500	2,800	1,800
Practical Shore A hardness range	20–80	20–80	20–90	55–80	45–95	40–90
Resilience:						
Room temp	F	F	G	P	G	P
Hot	F–G	F–G	F	F	E	E
Tear resistance	G	G	F	F–G (gum) G–E (reinf.)	F	F
Abrasion resistance	F	F	F	G–E	E	G
Atmospheric aging resistance	E	E	E	E	E	E
Oxidation resistance	E	E	E	E	E	E
Heat resistance	F	F–G	E	E	E	E
Low-temperature flexibility	G	E	E	F	F	P
Compression set	P	F	E	F–G	F	G
Permeability	E	E	F	G–E	G	G
Flame resistance	P	P	F	E	G	P
Acid resistance:						
Dilute	G	G	G	E	E	F
Concentrated	P	P	P	E	E	P
Moisture resistance	F	F	E	E (perox.) F (amine)	E	F
Solvent resistance:						
Hydrocarbons	E	E	F	E	F	G
Oxygenated solvents	G	G	G	P* G†	F	P
Oil and gasoline	E	E	F	G	F	E
Animal and vegetable oils	E	E	G	G	F	E
Dielectric properties	G	G	E	G	G	P

E = excellent. G = good. F = fair. P = poor.
* Ketones, esters.
† Alcohols.

Liquids. Liquids are handled much in the same way as gases. Examples of some of the more common liquids are shown in Table 44. Care must be exercised with flammable liquids.

Chemical Blowing Agents. The simpler species of chemical blowing agents are simple inorganic salts which decompose into carbon dioxide upon the addition of heat or an acid. The reaction is not dissimilar to the chemistry involved in baking a cake. Thus sodium bicarbonate has found considerable use over the years as a blowing agent for plastics and rubber.

TABLE 42 Specification Properties for ASTM Type O Open-Cell Foam Rubber[71]

Grade No.	Basic requirements				Requirements added by suffix letters		
	Compression deflection (limits), 25% deflection (limits), kg/m² (lb/in.²)	Oil-aged 22 h at 70°C (158°F), change in volume in ASTM oil No. 3 (limits), %	Oven-aged 7 days at 70°C (158°F), change from original deflection values (limits), %	Compression set, 22 h at 70°C (158°F), 50% deflection, max, %	Suffix B — Compression set, 22 h at 70°C (158°F), 50% deflection, max, %	Suffix F1 — Low-temp test at −40°C (−40°F), change from original deflection values, max, %	Suffix F2 — Low-temp test at −55°C (−67°F), change from original deflection values, max, %
Type R, Non-Oil-Resistant							
RO 10	3.5-14 (0.5-2)	...	±20*	15	...	25	25
RO 11	14-35 (2-5)	...	±20	15	...	25	25
RO 12	35-63 (5-9)	...	±20	15	...	25	25
RO 13	63-91 (9-13)	...	±20	15	...	25	25
RO 14	91-119 (13-17)	...	±20	15	...	25	25
RO 15	119-168 (17-24)	...	±20	15	...	25	25

Type S, Class SB, Oil-Resistant, Low Swell

SBO 10.....	3.5–14 (0.5–2)	.....	±20*	.....	.....	50
SBO 11.....	14–35 (2–5)	−25 to +10	±20	40	.....	50
SBO 12.....	35–63 (5–9)	−25 to +10	±20	40	.....	50
SBO 13.....	63–91 (9–13)	−25 to +10	±20	40	.....	50
SBO 14.....	91–119 (13–17)	−25 to +10	±20	40	.....	50
SBO 15.....	119–168 (17–24)	−25 to +10	±20	40	.....	50

Type S, Class SC, Oil-Resistant, Medium Swell

SCO 10.....	3.5–14 (0.5–2)	+10 to +60	±20*	50	25	50
SCO 11.....	14–35 (2–5)	+10 to +60	±20	50	25	50
SCO 12.....	35–63 (5–9)	+10 to +60	±20	50	25	50
SCO 13.....	63–91 (9–13)	+10 to +60	±20	50	25	50
SCO 14.....	91–119 (13–17)	+10 to +60	±20	50	25	50
SCO 15.....	119–168 (17–24)	+10 to +60	±20	50	25	50

* If this grade after aging still falls within the compression-deflection requirement of $0.070\ {}^{+0.070}_{-0.035}$ kg/cm² $(1\ {}^{+1}_{-\frac{1}{2}}$ lb/in.²$)$, it will be considered acceptable even though the change from the original load deflection is greater than ±20%.

Recently the technology has become highly sophisticated, and researchers and formulators continually strive for improvements and refinements in creating cellular polymer systems.

The most active field has been the nitrogenous materials which, depending upon

TABLE 43   Specification Properties for ASTM Type E Closed-Cell Foam Rubber[71]

Grade No.	Basic requirements				Requirements added by suffix letters
	Compression deflection, 25% deflection (limits), kN/m² (lb/in.²)	Fluid immersion 7 days at 23°C (73.4°F) change in weight in ASTM reference fuel B, % max*	Oven-aged 7 days at 70°C (158°F) change from original deflection values (limits), %	Water absorption, max, weight %†	suffix B: compression set, 22 h at room temp; 50% deflection, after 24 h recovery at room temp max %
Type R, Non-Oil-Resistant					
RE 41	14–35 (2–5)	....	±30	5	25
RE 42	35–63 (5–9)	....	±30	5	25
RE 43	63–91 (9–13)	....	±30	5	25
RE 44	91–119 (13–17)	....	±30	5	25
RE 45	119–168 (17–24)	....	±30	5	25
Type S, Class SB, Oil-Resistant, Low Swell					
SBE 41	14–35 (2–5)	50	±30	5	25
SBE 42	35–63 (5–9)	50	±30	5	25
SBE 43	63–91 (9–13)	50	±30	5	25
SBE 44	91–119 (13–17)	50	±30	5	25
SBE 45	119–168 (17–24)	50	±30	5	25
Type S, Class SC, Oil-Resistant, Medium Swell					
SCE 41	14–35 (2–5)	50	±30	5	25
SCE 42	35–63 (5–9)	50	±30	5	25
SCE 43	63–91 (9–13)	50	±30	5	25
SCE 44	91–119 (13–17)	50	±30	5	25
SCE 45	119–168 (17–24)	50	±30	5	25

† This test of weight change in reference fuel B is used in place of the usual oil resistance test of volume change in No. 3 oil for the following reason. Oil or solvent immersion of flexible closed-cellular materials usually causes loss of gas, by diffusion through the softened cell walls, which results in some shrinkage of the test sample. This shrinkage counteracts the swell which would normally occur, thus invalidating test data based on volume change. Reference fuel B is used because it produces a wider and more consistent differentiation among the R, SB, and SC grades than does the No. 3 oil.

* For cellular materials with densities 160 kg/m³ (10 lb/ft³) or less, the value of water absorption allowed is 10% max by weight. For densities of more than 160 kg/m³ (10 lb/ft³), the value of water absorption is 5% max by weight.

structure, will decompose at the desired temperature and rate for the polymer selected. A series of chemical blowing agents is shown in Table 45.

Care must be exercised in selecting a chemical blowing agent when one is formulating a foamed-polymer "recipe." Some of the items which must be considered are:

1. The blowing agent should not decompose into products which are disagreeable in odor or color.

TABLE 44 Designation and Critical Properties of Several Blowing Agents[74]

Usual designation	Compound	Gas available/g	Working range, °C	Typical applications	Typical concentrations, %*
ABFA (AC) (AZ)......	Azodicarbonamide (azobisformamide)	220 ml at 200°C	165–215	PP, PS, ABS, PVC	0.1–2
AZDN............	Azodiisobutyronitrile	135 ml at 100°C	110–125	PVC	0.1–2
BSH............	Benzenesulfonhydrazide	120 ml at 100°C	90–100	PVC	0.1–2
OBSH (OT)......	PP'-oxybis (benzonesulfonhydrazide)	120 ml at 140°C	165–210	PE, PS	0.1–2
TSSC (RA)......	p-Toluene sulfonyl semicarbazide	146 ml at 263°C	213–225	ABS, PP, nylon	0.1–2
DMDNTA........	N,N'-dimethyl N,N'-dinitrosoterephthalamide	170 ml at 140°C	80–160	PVC	0.1–2
THT†..........	Trihydrazino triazine	225 ml at 320°C	250–320	Nylon, PP, ABS, PE, PC	0.1–5
N₂...........	Nitrogen	....	Any	PE, PS	1–2
CO₂..........	Carbon dioxide	....	.....	PVC	1–5
	Pentane	....	.....	PS	5–10
	Fluorocarbon 11	....	.....	PS	5–10
	Fluorocarbon 12	....	.....	PS	5–10
	Fluorocarbon 114	....	.....	PS, PE	5–10
	Methylene chloride	....	.....	.....	5–10

* Higher concentrations for lower densities.
† The high-temperature agents are 1 OBSC (4,4-oxybis benzene sulfonyl-semicarbazide), TSSC (p-toluene-sulfonyl-semicarbazide), and barium azodicarboxylate.

TABLE 45 Typical Blowing Agents for Various Thermoplastics[74]

Foam	Typical blowing agent	%	Method of production	Density	Typical product
PS...........	Pentane, isopentane (and water)	7	Expandable bead	1.5–2.5	Closed-cell rigid board and shapes
	Pentane, fluorocarbons	7	Extrusion	2–6	Closed-cell rigid film and sheet
	Methyl chloride, fluorocarbons	10–15	Extrusion	2–5	Closed-cell rigid board stock
PVC...........	Air, N_2, CO_2	.....	Mix and cast or calender	3–6	Open or closed cells, flexible shapes and sheets
	AZ, DMDNTA	10–18	Calender	4–25	Open- or closed-cell flexible shapes and sheets
	Acetone plus chemical agent	.....	Extrusion	7–25	Closed-cell flexible profiles
	Isocyanate, (CO_2) solvent, H_2O	50	Extrusion, molding	7–25	Closed-cell rigid shapes
PE...............	Fluorocarbon (114) (and hydrocarbon)*	10–20	Molding	3–6	Closed-cell rigid board and shapes
			Extrusion	2–4	Closed-cell semiflexible insulation
	OBSH	.....	Extrusion	2–4	Open- and closed-cell flexible films
	AZ	2	Extrusion	19–20	Closed-cell semiflexible wire insulation
	AZ, nitrogen	1–2	Injection, extrusion	25–55	Open- and closed-cell rigid shapes
	AZ, dicumyl peroxide	2–7	Cross-linked, compression and extrusion	1.5–12	Closed-cell semirigid blocks, sheet, and film
PP.........	AZ, TSSC	4–2	Injection, extrusion	25–55	Closed-cell rigid shapes
	AZ, peroxides	2–7	Cross-linked compression and extrusion	2–6	Closed-cell semirigid sheet, film

2. The blowing agent should not be toxic.
3. The blowing agent should be stable under normal storage conditions.
4. The blowing agent should be relatively compatible with the polymer matrix.

Several methods are used to incorporate chemical blowing agents into polymer matrices depending down upon dispersion requirements, manufacturing flexibility, and economics.

1. The blowing agent can be added as a powder to pellets of the polymer or to a polymer-powder mix just prior to processing. There is some danger of desegregation using this method. At times additives such as mineral oil are used to make the blowing-agent powder stick to the pellets or powder.

2. The blowing agent can be precompounded into the base polymer (10 to 15 percent) and this mixture used as a concentrate.

3. The blowing agent can be compounded directly into the final composition by hot-melt mixing methods.

TABLE 46   Effect of Acids and Bases on the Thermal-Decomposition Temperature of a Typical Blowing Agent—Celogen AZ (Partial Table)[72]

Material	Decomposition temp, °C
Celogen AZ	180–210
Acids:	
Oxalic	100–150
p-Toluene sulfonic	100–180
Glycol	110–150
Lactic	115–180
Acetic	140–195
Citric	145–165
Succinic	155–190
Salicylic	165–200
Acetyl salicylic	165–200
Sulfamic	160–205
Loralkyl phosphoric	155–195
Maleic anhydride + water	100–180
Phosphoric	90–160
Butyl phosphoric	130–170
Malic	155–185
Bases:	
Guanidine carbonate	135–180
Potassium carbonate plus water	100–145
Potassium carbonate anhydrous	155–210
Borax	100–185
Ethanolamine	85–100

In the above methods, extreme care must be taken not to allow the mix to reach the decomposition point of the blowing agent. Moreover early (lower-temperature) decomposition of many blowing agents is induced by the presence of many organic and inorganic species. A partial list of these activators is found in Table 46.

Finally the cost of the blowing agent, the density of the final product, and the fabrication method must be considered when selection is made. Figure 29 compares the costs of gaseous, liquid, and chemical blowing agents used in most public foams.

STRUCTURAL FOAM

Introduction During the past few years there has been considerable growth, particularly in the furniture field, of a technology which has acquired the term *structural foam*. This term was initially used by Union Carbide to promote the

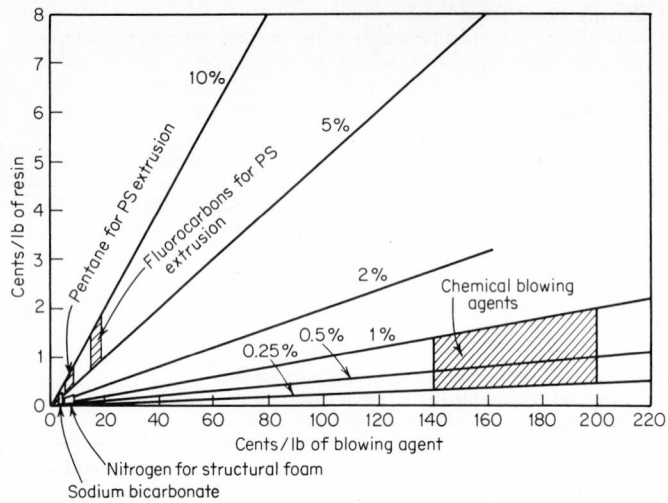

Fig. 29   Cost curves for several blowing agents at different concentrations.[74]

licensing of a patented technology involving a process for injection molding a thermoplastic structure with a low-density (relatively) core and a high-density skin. There was one notable precursor—Dow's Frostwood—and there have been several variations of this basic process since that time. "Structural foam" has now been applied (or misapplied) to essentially any thermoplastic foamed structure produced in a variety of ways. These processes are briefly summarized below:

Process	Characteristics
Frostwood	Dow patent 3,058,161. Discusses in general terms use of blowing agents in styrenic materials for molding foam structures. Discusses cam opening of the mold to produce a high-density skin
Union Carbide	UCC patent 3,268,636. Injects nitrogen gas into melt and thence into an accumulator. Mold (cast) is filled by expanding melt at relatively low pressure
USM	USM patents. Discloses injection molding at high pressures using gas or chemical blowing agent. Mold expands after high-pressure injection to allow molten plastic to expand. Shorter cycles than low pressure is claimed. Tool costs are considerably higher
Hoover	A joint development of Hoover and General Motors, details of which are vague. A two-stage high-pressure injection system incorporates the blowing agent (part of the development) just before injection
Phillips	Conventional low-pressure molding using a blowing-agent concentrate. Promoted primarily for polyethylene
Phillips/Engelit	Pellets containing blowing agent are fluxed on a rotating table and scraped into a conveying screw by a knife or doctor blade
Allied	Patents 3,211,506 and 3,384,691. A high-pressure system wherein the hot melt containing blowing agent is allowed to egress from the tool via the manifold, thus reducing the pressure and allowing expansion
ICI	A novel process called sandwich molding wherein two polymer melts, one without blowing agent and one with blowing agent, are injected sequentially into a mold; the solid material encapsulating the foamable, creates a dense skin. Opening the mold an appropriate amount allows foaming of the core. U.S. Patent 3,599,290

Processing   General considerations for injection molding structural foam from thermoplastics are quite adequately covered in an article from *Fisons Industrial Chemicals* reprinted from the British publication *Plastics*. A portion of that article follows.

Injection Moulding

All types of machines can be used for the production of cellular mouldings. The degree of control over melt temperature and pressure found necessary in extrusion applies to the injection moulding of foam. To obtain this control, the following additional equipment is necessary—

(a) A nozzle shut-off to maintain sufficient pressure on the melt to prevent foaming in the cylinder/barrel.

(b) A non-return valve is required for the same reason. Whilst it is possible to use all types of machines, an in-line screw machine is preferred to a plunger type.

Injection moulding of foam differs from that of homogeneous material in that injection speed must be high to eliminate foaming in both the gates and the nozzle, and injection and locking pressure are reduced. The pressure required to completely fill the mould is derived almost entirely from the gas.

Three moulding techniques have proven successful for producing cellular articles.

(a) Standard molding techniques.

(b) Flow moulding (i.e., mould closes as shot enters).

(c) 'Dropping-plate' technique.

Both (a) and (b) are established techniques; flow moulding may offer advantages in improved mould filling with cellular compounds. In neither case is a thick homogeneous skin of controllable thickness obtained. The surface effect is that of woodgrain due to turbulence and flow lines of entrained gas.

The 'dropping-plate' technique utilizes normal injection pressures, preventing cell formation within the mould. One wall of the mold retracts, increasing mould volume and reducing melt pressure, resulting in cell formation in the centre of the moulding. Skin thickness can be controlled by mould temperature and dwell time before increasing the mould volume.

It is in the field of injection moulding that chemical blowing agents appear to have the greatest potential application, and their inclusion in moulding compounds would assist in over-coming a number of problems inherent in injection moulding.

A greater platen area can be utilized in view of the reduced locking pressures required.

Mould volume can be increased proportional to density without increasing shot weight.

The reduction in injection pressures permits the use of cheap moulds.

Internal mould pressure reduced sinkage problems.

Chemical blowing agents can be used in three applications: anti-sink, density reduction and structural foam.

Anti-sink or anti-void applications for chemical blowing agents occur in thick-section mouldings having poor flow characteristics, where void formation may impair physical properties. Incorporation of chemical blowing agent at levels of 0.02%–0.075% will disperse voids, give sufficient internal pressure to eliminate 'sink marks' without loss of physical properties in the moulding.

Density reductions of 15% in mouldings can be obtained by the inclusion of 0.15%-0.4% chemical blowing agent, with little loss of physical properties. The only difference in injection technique for mouldings having 15% density reduction is the reduction in back-up pressure to permit cell formation.

Structural foams with density reductions of less than 30% are achievable on conventional machinery, but using equipment specifically designed for the production of cellular mouldings, the reduction is greater than 40%.

Operating hints when injection moulding cellular polymers which may prove useful:

(i) Injection rates must be very high; a ten times increase in the injection rate will not induce melt fracture.

(ii) Control of cell structure is a function of mould pressure.

(iii) Mould temperatures will control surface finish; i.e., hot moulds will give a glossy surface, cold moulds a pearlescent finish.

(iv) To prevent swelling, cooling times may need to be increased by a factor of two.

(v) Foam density and cell structure will be controlled by the diameter of the gates and runners, i.e., small openings with high injection rates give small cells, the converse will give large cells. In the case of ABS the large cells will coalesce to give a 'wood grain' finish.

Properties of the foamed polymers are of course dependent upon the degree of foaming or density reduction. This dependence on density is shown for several typical polymers in Figs. 30 through 32. More recently the higher-performance engineering thermoplastics have received considerable attention, thus accelerating the anticipated growth rate of structural thermoplastic foam.

The structural-foam boom has been well documented in the literature, and few polymers have not been foamed. The greatest growth appears to be in furniture, where wood simulation presents an extremely attractive economic opportunity. Tables 47 and 48 illustrate this growth.

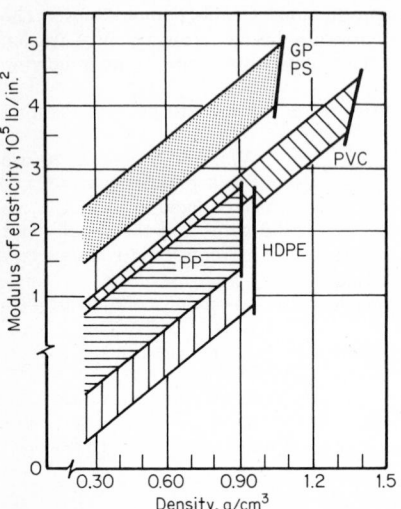

Fig. 30 Modulus of elasticity of several foamed thermoplastics at varying densities.[77]

MISCELLANEOUS FOAMS

Phenolics Phenolic foams, like epoxy foams, approach polyurethane foams in thermal-conductivity performance and excel in thermal endurance and dimensional stability. Their highly corrosive catalyst systems and high friability (brittleness), however, have prevented significant growth. In addition water transmission and absorption are relatively high for phenolic foams because of the cell structure, which averages around 90 percent open cells.

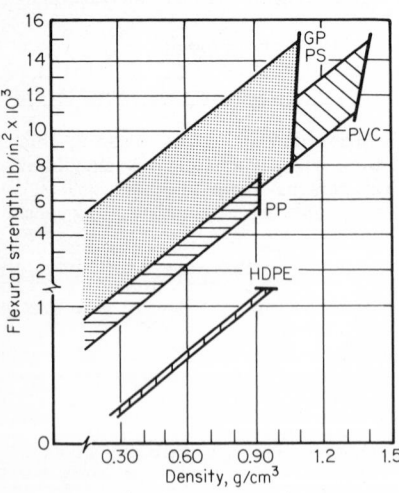

Fig. 31 Flexural strength of several foamed thermoplastics at varying densities.[77]

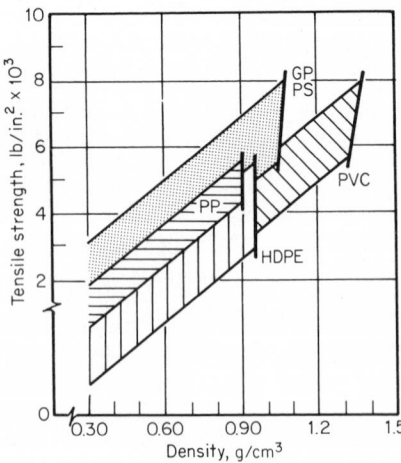

Fig. 32 Tensile strength of several foamed thermoplastics at varying densities.[77]

TABLE 47   Growth Pattern for Structural Foam by Polymer Type, 1968–1980[79]

Material	Consumption, million lb			
	1968	1971	1975	1980
PS............................	0.35	10.5	75	300
HDPE......................	0.10	2.5	40	200
PP...........................	0.25	1.7	20	100
Other*......................	N†	0.3	15	400
Total.....................	0.70	15.0	150	1,000

* Includes ABS, PVC, nylon, polycarbonate, new resins developments, etc.
† Negligible.

TABLE 48   Growth Pattern for Structural Foam by Application, 1968–1980[79]

Market	Consumption, million lb			
	1968	1971	1975	1980
Automotive...................	0.1	1.5	20	100
Building.....................	N*	1.0	20	300
Furniture†...................	0.2	8.0	50	200
Materials handling‡...........	0.1	1.5	10	100
Miscellaneous§...............	0.3	3.0	50	300
Total.....................	0.7	15.0	150	1,000

* Negligible.
† Includes TV cabinets, stereo components, etc.
‡ Includes tote boxes, filed lugs, pallets, etc.
§ Includes business machine housings, appliance parts, water-meter housings, recreational products, toys, novelties, instrument bases, military applications, etc.

Phenolic foams are produced by reaction casting similar to polyurethanes. The reactants are phenol formaldehyde, an acid catalyst, surfactants, and a volatile organic-liquid blowing agent (usually methylene chloride, pentane, or fluorocarbon).

Some phenolic foams have been used in sandwich panel structures where their high-temperature resistance and flame resistance are desirable; however, their major use has been as moisture-retaining flower holders.

Figure 33 shows the variation in density possible by varying the reactants in a phenolic foam. Table 49 illustrates typical mechanical properties of a phenolic foam at two densities. Figures 34 and 35 illustrate the variation of thermal conductivity with density and with temperature.

Urea formaldehyde   Urea-formaldehyde foams resemble phenolic foams in many aspects. They are formed by foaming a urea-formaldehyde polymer with compressed air and curing through the addition of phosphoric acid. Like phenolics, urea-formaldehyde foams are low in mechanical strength and are friable or brittle. Unlike phenolics, urea-formaldehyde foams have a maximum service life of 120°F, whereas phenolics can operate successfully above 300°F continually.

Since urea-formaldehyde foams can be

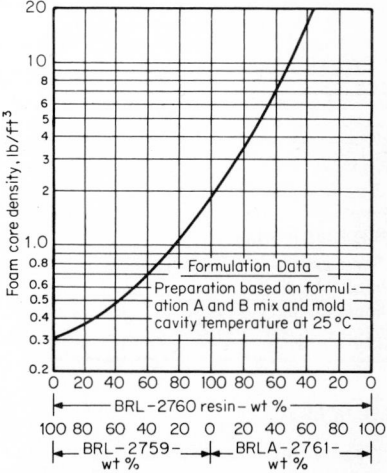

Fig. 33   Variation in phenolic foam density as a function of reactant ratios.[96]

TABLE 49   Typical Strength Properties of Rigid Phenolic Foam
at Two Densities[96]

Strength, lb/in.^{2}	2 lb/ft^3	4 lb/ft^3
Compressive...................	25	60
Flexural.......................	25	65
Shear.........................	14	30
Tensile........................	20	42
Shear modulus.................	400	750

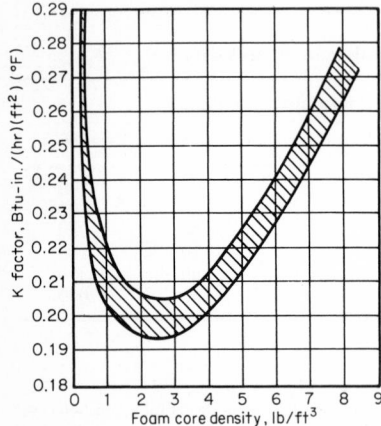

Fig. 34 Insulating qualities (K factor)
of phenolic foam at varying densities.[96]

Fig. 35 Variation in K factor of phenolic foam with mean temperature.[96]

produced continuously with relatively simple equipment, they have been used extensively in Europe as dwelling thermal and acoustical insulation.   Because of their high
water absorption they are also used for floral-display bases.   Typical properties of a
urea-formaldehyde foam used for thermal insulation are shown in Table 50.

TABLE 50   Typical Properties of a Urea-Formaldehyde Foam[94]

Property	Value
K factor (at 80°F), Btu/(ft^2)(h)(°F) for 1 in...............	0.2
Sound-transmission class (STC), dB.......	5–7 (improvement)
Fuel contribution.....................	Noncombustible (ASTM E 136)
Flame spread........................	25 (ASTM E 84)
Smoke density.......................	0–5 (ASTM E 84)
Smoke toxicity.......................	Nontoxic (Federal Hazardous Substances Act)
Weight, lb/ft^3.........................	0.7

SUMMARY

The foregoing discussion shows that the multitude of factors which describe an
application-requirement profile, and the many choices of polymer types and process
methods currently available, make the choice of polymer foam for a specific use

bewildering. This chapter will not make instant materials engineers of mechanical or electrical engineers but is intended to establish a working vocabulary and knowledge of the polymers and techniques currently used so that the design engineer can work with materials people on site or representatives from materials or parts suppliers. If terms, nomenclature, and background are on a more equal basis, better materials decisions can be made.

Table 51 presents a summary of properties for most of the plastic foams discussed in this chapter. Unfortunately elastomeric foams are not compared on the same basis, primarily because of their historically different usage patterns (and thus specifications). The development of thermoplastics with elastomeric properties and the interdiffusion of new process techniques between these two technologies should result in a gradual unification of evaluation techniques and specification. This event should prove of great value to the design or specification engineer.

Also included at the end of the chapter is a listing of foam and foam-raw-material suppliers (Table 52). A technically capable materials supplier is of immeasurable help in selecting the correct foam system. A more extensive listing which includes raw-material suppliers and equipment manufacturers appears in *Foamed Plastics 73,* a publication of Technomics, Inc.

Economics   The final cost of a foam structure or part is dependent upon a great many factors, making a simple list of typical cost/board foot potentially misleading. A few of the factors which enter the cost analysis are:

1. Polymer raw-material cost
2. Processing method
3. Density of the final part
4. Part fabrication (molding, casting, cutting)
5. Volume purchased
6. Source (captive, prime manufacturer, distributor)
7. Specification requirements

Approximate costs of the unfoamed polymer systems discussed in this chapter are listed below as a starting point. The reader may then use the "rule of thumb" to determine initial processed cost by multiplying by a factor of 2 to 2½. Additional factors should be added for fabrication and distribution costs.

Polymer	Approximate Cost, Cents/lb (Unfoamed)
Polyurethane	37–65
Polystyrene	33–36
Polyolefins	25–40
PVC	30–40
Epoxy	68–90
Elastomers	25–350
Phenolics	41–50
Urea formaldehyde	34–45

Flammability   Although terms such as "self-extinguishing" and "nonburning" have been used* to describe the performance of foam specimens when tested in accordance with certain industry and regulatory-agency specifications, it must be remembered that all the materials discussed in this chapter are predominantly organic and will burn (albeit to highly varying degrees) when exposed to an ignition source. Foams are further complicated by their varying density, which causes large surface areas to be exposed to the flame.

The reader is urged to use specimen specification testing for screening purposes, but to employ sound application engineering tests to determine final-product safety.

Testing   Since polymers are employed in widely diverse applications, test methods and specifications tend to be centered around end uses. A summary of test methods used for most plastic foams appears in Table 53.

---

* No longer allowed per FTC consent decree.

TABLE 51 Typical Properties of Several Classes of Polymer Foams

Type of material	ABS (acrylonitrile-butadiene-styrene) Injection molding type	ABS Self-ext.	ABS Expansion cast type	Cellulose acetate	Rigid closed cell			Epoxies Syntactic (partly open cell)	Epoxies Syntactic (partly open cell)	Epoxies Syntactic	Epoxies Spray applied	Epoxies Foam-in-place
Forms available	Pellets	Pellets	Pellets	Boards and rods, (rigid, closed cell foam)	Precast blocks, slabs, sheet			One-part, free flowing powder	Two part, moldable like damp sand	Rigid sheet or pack in place	*2-package systems (liquid)	*2-package systems (liquid)
Density, lb./cu. ft.	31.0–56.0	31.0–46.0	31.0–35.0	6.0–8.0	5.0	10.0	20.0	14.0–20.0	20.0	23.0	1.8–2.0	2–2.3
Tensile strength, p.s.i., ASTM D1623	1800–4100	1100–3700	1400	170	51	180	650	—	—	—	26–31	—
Compression strength at 10% deflection, p.s.i., ASTM D1621	2300–3700 (1000 h)	—	1000	125	90	260	1080	125	1000	2100	13–17	20–26
Impact strength, ft. lb./in.	0.8–1.5	0.8	1.2	0.12	—	—	—	—	—	—	—	—
Maximum service temp., °F. Dry:	180	180	180	350	350	350	350	350–400	—	500	160	200
Maximum service temp., °F. Wet:	—	—	—	—	—	—	—	—	—	—	—	—
Flammability, ASTM D1692	Slow burning; can be self exting.			Slow burning	Can be formulated to be self-extinguishing			S.E. or N.B.	—	—	S.E.	—
Thermal conductivity, Btu/sq. ft./hr./°F./in., ASTM D2326	0.58		0.58	0.31	0.26	0.28	0.32	<0.5	0.38	0.36	0.11–0.12	0.11–0.13
Coefficient of linear expansion, 10^{-5} in./in./°F., ASTM D696	4.9–9.5	—	9.7	2.5	—	—	—	—	—	—	—	—
Dielectric constant, ASTM D1673	—	—	1.63	1.12	1.19–1.08 1.36–1.24 @ 10^6 and 10^{10}	—	1.55–1.41	1.38	1.45	—	—	—
Dissipation factor @ 28° C. and 1 meg., ASTM D1673	—	—	—	—	—	—	—	0.006	0.01	—	—	—
Water-vapor trans., perm-in. ASTM C355	—	—	—	—	—	—	—	—	—	—	—	—
Water absorption, % by volume (96 hr.), ASTM D2842	0.4–0.6	—	Nil	13–17 @ 100% R.H. 1.9–2.5 @ 50% R.H.	—	—	—	—	—	1.8	—	—

Type of material	Ethylene copolymer	Ionomers	Phenolics (foam in place) Liquid resin				Syntactic castable	Syntactic polybenzimidazole
Forms available	Extruded sheet	Sheet, rod					Two-component: liquid and paste	Molding powder
Density, lb./cu. ft.	35.0	2.0–20.0	⅓–1½	2–5	7–10	10–22	50–60	25–38
Tensile strength, p.s.i., ASTM D1623	600–800	57–105	3–17	20–54	80–130	—	>1000	1800–2000
Compression strength at 10% deflection, p.s.i., ASTM D1621	—	24–15.2	2–15	22–85	158–300	300–1200	8000–13,000 p.s.i.	3780–4310
Impact strength, ft. lb./in.	—	Resilient	—	—	—	—	.19–21	—
Maximum service temperature, °F. Dry:	130	150–155	—	Continuous service at 300			275	600 continuous in air
Wet:	—	—	—				—	
Flammability, ASTM D1692	Slow burn	Slow burn or S.E.	Self-extinguishing				—	Non-flammable
Thermal conductivity, Btu/sq. ft./hr./°F./in., ASTM D2326	—	0.27–0.34	0.21–0.28	0.20–0.22	0.24–0.28	—	1.0	0.42–0.78
Coefficient of linear expansion, 10^{-5} in./in./°F., ASTM D696	—	—	—	0.5	—	—	10	—
Dielectric constant, ASTM D1673	—	1.5 @ 10^6	—	—	1.19–1.2	1.19–1.2	2.1	1.82–2.40
Dissipation factor @ 28° C. and 1 meg.	—	0.003	—	—	0.028–0.031	0.028–0.031	0.03	—
Water-vapor trans, perm-in. ASTM C355	—	0.34–2.00	—	2 lb./ft.³ 5 lb./ft.³ 2074 g 1844 g per day/m²	—	—	—	—
Water absorption, % by volume (96 hr.), ASTM D2842	<0.5	0.40–1.0	—	13–51 @ 100% R.H. 1–4 @ 50% R.H.	10–15 @ 100% R.H. 1–5 @ 50% R.H.	—	>0.5	—

TABLE 51 Typical Properties of Several Classes of Polymer Foams (Continued)

	Polyethylene								Polypropylene	
	Low density foam			Intermediate density foam			High density foam	Cross-linked foam	High density foam	Cross-linked foam
Type of material	Planks, rods, rounds, ovals	Sheet	Tubing	Plank						
Forms available	Planks, rods, rounds, ovals	Sheet	Tubing	Plank			Molded parts and shapes with solid integral skin; density typically 25–50 lb./cu. ft.	Sheets	Molded parts and shapes with solid integral skin	Sheets
Density, lb./cu. ft.	2.2	2.4	2.1–3.3	4.0	6.0	9.0	35.0	0.9–6.5	35.0	3
Tensile strength, p.s.i., ASTM D1623	20–30	40	—	35	60	100	1200	46–210	1600	118–147
Compression strength at 10% deflection, p.s.i., ASTM D1621	3.5	—	—	10.0	15.0	28.0	1300	—	2100	175–1200
Impact strength, ft. lb./in.	—	—	—	—	—	—	—	—	260	—
Maximum service temperature, °F. Dry:	180	180	180	180	200	210	230	—	—	275° F.
Wet:	—	—	—	—	—	—	230	—	—	275° F.
Flammability, ASTM D1692	Slow burn²	Slow burn²	S.E.	Slow burn	Slow burn	Slow burn	Slow burn	—	Slow burn	Slow burn
Thermal conductivity, Btu/sq. ft./hr./°F./in., ASTM D2326	0.35–0.40	0.29	0.26 @ 40° F. mean	0.40	0.40	0.40	0.92	0.30–0.37	0.87	0.27
Coefficient of linear expansion, 10^{-5} in./in./°F., ASTM D696	9.5	—	—	—	—	—	4.18	—	4.2	—
Dielectric constant, ASTM D1673	0.05 @ 10^9	1.05 @ 10^9	—	—	—	—	—	1.1	—	—
Dissipation factor @ 28° C. and 1 meg.	0.0002 @ 10^9	0.0002 @ 10^9	—	—	—	—	—	—	—	—
Water-vapor trans., perm-in. ASTM C355	<0.40	<0.40	0.13	<0.40	<0.40	<0.40	<0.40	—	—	—
Water absorption, % by volume (96 hr.), ASTM D2842	<0.50	<0.50	<0.50	<0.50	<0.50	<0.50	0.22	0.1–0.5	—	0.5

Polystyrene

Type of material	Products or shapes molded from expandable beads; finished boards				Extruded boards and billets			Extruded film and sheet		
Forms available										
Density, lb./cu. ft.	1.25	1.0	2.0	5.0	1.5–2.0	2.0–2.6	2.0–5.0	6.0	8.0	10.0
Tensile strength, p.s.i., ASTM D1623	26–34	21–28	42–68	148–172	55–70	60–105	180–200	300–500	400–700	600–1000
Compression strength at 10% deflection, p.s.i., ASTM D1621	17–24	13–18	25–40	85–130	25–45 @ 5%	25–50 @ 5%	100–140 @ 5%	42.5	52.5	68.0
Impact strength, ft. lb./in.	—	0.14	0.21	0.20	—	—	—	—	—	—
Maximum service temperature, °F. — Dry:	167–185 (aged)				165–175 (aged)			170–175 (aged)	175	175
Maximum service temperature, °F. — Wet:	—				—			—	—	—
Flammability, ASTM D1692	All self-extinguishing when compounded									
Thermal conductivity, Btu/sq. ft./hr./°F./in., ASTM C177	0.26	0.25–0.26 @ 70° F. mean temp.	0.23–.24 @ 70° F. mean temp.	.246	0.18–0.24 @ 40° F. mean temp.			0.24 @ 70° F. mean temp.		
Coefficient of linear expansion, 10^{-5} in./in./°F., ASTM D696	3.0–4.0				3.0–4.0			—		
Dielectric constant, ASTM D1673	1.06–1.02 @ 10^2 to 10^6 cycles				<1.05 @ 10^2 to 10^8 cycles			1.27	—	1.28
Dissipation factor @ 28° C. and 1 meg.	0.0001–0.0007 @ 10^2 to 10^6 cycles				<0.0004 @ 10^3 to 10^8 cycles			0.00011	—	0.00015
Water-vapor trans., perm/in. ASTM C355	—	1.2–3.0	0.6–1.2	0.4–0.6	0.3–1.5			1.50	1.25	1.00
Water absorption, % by volume (96 hr.), ASTM D2842	Nil to <3				Nil to <0.5			Nil		

TABLE 51 Typical Properties of Several Classes of Polymer Foams (Continued)

Type of material	Flexible — Free rise (slab stock)	Flexible — Molded (foam-in-piece)	Polyurethane Rigid (closed cell)							Isocyanurate foams	Semi-rigid (foamed-in place)	Integral skin molded (flexible)	Skinned molded (rigid)	Micro-cellular elastomer
Forms available	Slabs, sheets, blocks, custom shapes	2 and 3 package or one-shot systems for mixing on job	Molded parts; boards, blocks, slabs, pipe covering; one-shot, two- and three-package systems for foam-in-place; for spray, pour, or froth-pour techniques							Boards, blocks, slabs, two and three component systems	—	2 and 3 package systems	2 and 3 package systems	2 and 3 package systems
Density, lb./cu. ft.	0.8-26.0	1.0-40.0	1.5-3.0	4-8	9-12	13-18	19-25	28-40	41-70	1.8-6.0	1.2-26.0	25-65 (skin) 5-20 (core)	25-50 (skin) 3-30 (core)	21-65
Tensile strength, p.s.i., ASTM D1623	8-50	10-1350	15-95	90-290	230-450	475-700	775-1300	1350-2500	3000-8000	28-160	20-1350	20-100	100-1200	100-2500 @ 50% 10-100 @ 10%
Compressive strength at 10% deflect., p.s.i., ASTM D1621	0.2-2.0	0.25-2100	15-60	70-275	290-550	650-1100	1200-2000	2100-4000	5000-15000	16-100	20-2100	1-5 @ 25%	40-1700	100-2500 @ 50%
Impact strength, ft. lb./in.	—	—	<0.05	0.32	0.45	0.45-1+				—	1.0-5.0	—	20-100	Flex
Max. service temp., °F. (Dry / Wet)	260-275	250	250	300	300	300	300			300	175-350	150-175	250	250 / 212
Flammability, ASTM D1692	Burnable or self-exting.	Burnable or self-exting.	Burnable, but can be formulated to be self-extinguishing. Also, 25 or less rating in tunnel test (ASTM E-84)							S.E.	Burnable or self-exting.	Burnable or self-exting.	Can be S.E.	Slow to S.E.
Thermal conductivity, Btu/sq. ft. hr./°F./in., ASTM D2326	0.2-0.25 @ 2 lb./cu. ft. density	0.3 @ 2 lb./cu. ft. density	Blown with fluorocarbon: 0.11-0.17	0.15-0.21	0.19-0.25	0.26-0.34	0.34-0.42			0.14-0.20	0.11-0.30	—	0.12-0.45	—
			Blown with CO$_2$: 0.23	0.21-0.29	0.31-0.35	0.36-0.40	0.42-0.52		0.57					
Coefficient of linear expansion, 10^{-5} in./in./°F., ASTM D696	—	—	3-8	4	4	4	4			5 × 10^5	—	—	4	—
Dielectric constant, ASTM D1673	1.0-1.5	—	1.05	1.10	1.2	1.3	1.4	1.5		1.05-1.10	2.2	—	—	—
Dissipation factor @ 28° C. and 1 meg.	—	—	0.0018	0.0032	0.0055					—	—	—	—	—
Water-vapor trans., perm-in. ASTM C355	0.9-4.0	—	0.9-2.0	0.08						4.0-1.5	—	—	—	—
Water absorption, % by volume (96 hr.), ASTM D2842	—	—	<0.1-5.0	0.5	0.4	0.2				0.5-1.0	—	—	—	—

Type of material	Plastisol — Mechanically frothed	Plastisol — With blowing agent	Polyvinyl chloride Flexible — Open cell	Polyvinyl chloride Flexible — Closed cell	Rigid closed cell	Silicones — Liquid (10% closed cell)	Silicones — Liquid (closed cell) (open cell)	Silicones — Sheet (open cell)	Styrene acrylonitrile		Urea formaldehyde
Forms available	Liquid or paste	Liquid or paste	Sheet and rolls; cored cushions & other molded shapes	Sheets and molded shapes	Boards and billets	Liquid (10% closed cell)	Liquid (closed cell) (open cell)	Sheet (open cell)	Products or shapes molded from expandable beads; finished boards		Block and shred
Density, lb./cu. ft.	13-60	3.0-4.0	10-up	4.0	2-6	9.6 (unrestricted)	21.0-31.0	10.0	0.8	1.0	0.8-1.2
Tensile strength, p.s.i., ASTM D1623	—	50-3000	10-200	—	70 and up	100-150	100-150	45	20	30	Poor
Compression strength at 10% deflect., p.s.i., ASTM D1621	—	—	—	0.5-40.0	45-250	10 @ 75%	10 @ 75%	5 @ 75%	1.5 @ 5% 6.0 @ 5%	6.0 @ 5%	5
Impact strength, ft. lb./in.	—	—	—	—	—	—	—	—	—	—	Poor
Maximum service temp., °F. (Dry / Wet)	150-175	150-175	125-225	S.E.	200 / 200	400	500-650	500	170-190		120
Flammability, ASTM D1692	S.E.	S.E.	S.E.	S.E. @ 70° F.	Non-burning	—	—	—	S.E. when compounded		—
Thermal conductivity, Btu/sq. ft. hr./°F./in., ASTM D2326	Depends on density	Depends on density	Depends on density	0.24-0.28 @ 70° F.	0.18-0.23 @ 70° F.	—	0.36	0.6	0.29	0.29	0.18-0.21
Coefficient of linear expansion, 10^{-5} in./in./°F., ASTM D696	—	—	—	—	4.0-6.0	—	—	—	0.32		0.29
Dielectric constant, ASTM D1673	—	—	—	—	—	1.42 @ 10^5	1.3-1.4	1.2	—	—	—
Dissipation factor @ 28° C. and 1 meg.	—	—	—	—	0.003-0.15	0.001 @ 10^5	<0.01	0.007	—	—	—
Water-vapor trans., perm-in. ASTM C365	—	—	—	—	—	—	—	—	2.0-4.0		—
Water absorption, % by volume (96 hr.), ASTM D2842	—	—	Nil	Nil	Nil	0.1	—	—	Nil		Nil

SOURCE: "*Modern Plastics* Encyclopedia," 1972–1973, p. 295.

TABLE 52   Commercial Sources of Foamed Plastics

ABS Foam

Farbenfabriken Bayer AG, Leverkusen, Germany
Marbon Div., Borg-Warner Corp., Washington, W. Va.
Mitsubishi Rayon Co., Tokyo, Japan
Ube Cycon, Ltd., Tokyo, Japan

Cellular Cellulose Acetate Foam

Strux Corp., Lindenhurst, N. Y.

Epoxy Foam

1. Ingredients for (base resins)
2. Slab stock

Ablestik Adhesive Co., Div. Ablestik Laboratories, Gardena, Calif. (1)
Atlas Minerals & Chemicals Div., ESB Inc., Mertztown, Pa. (1)
CIBA Products Co., Ciba-Geigy Corp., Summit, N. J. (1)
DeBell & Richardson, Inc., Enfield, Conn. (2)
Emerson & Cuming, Inc., Canton, Mass. (1)
Isochem Resins Co., Lincoln, R. I. (1)
Leepoxy Plastics, Inc., Fort Wayne, Ind. (1)
Ren Plastics, Inc., Lansing, Mich. (1)

Ionomer Foam

Arlon Products, Inc., Harbor City, Calif.
Gilman Brothers Co., Gilman, Conn.

Phenolic Foam

Dynamit Nobel, AG, Troisdorf-Koln, Germany
Dynamit Nobel of America Inc., Norwood, N. J.
Marblette Co., Long Island City, N. Y.
Smithers Co., Kent, Ohio

Polyethylene Foam

1. Low density
2. High density
3. Cross-linked

Airex AG, Sins, Switzerland (1)
Air-O-Plastik Corp., Carlstadt, N. J. (1, 2, 3)
Arlon Products, Inc., Harbor City, Calif. (1)
Badische Anilin- & Soda-Fabrik AG, Ludwigshafen, Germany (1, 3)
Bagdad Plastics Co., Sub. Bagdad Copper Corp., Phoenix, Ariz. (1)
Bakelite Xylonite Ltd., London, England (1, 3)
Bernel Foam Products Co., Buffalo, N. Y. (1, 3)
Bontec Corp., W. Haven, Conn. (1, 2, 3)
Cellu Products Co., Patterson, N. C. (1)
Dew-Foam Co., Van Nuys, Calif. (1, 2)
Dow Chemical Co., Midland, Mich. (1)
Dynamit Nobel, AG, Troisdorf-Koln, Germany (3)
Emerson & Cuming, Inc., Canton, Mass. (1, 2)
Foam Products Corp., Maryland Heights, Mo. (1)
Foam Specialties, Inc., Rye, N. Y. (1, 2, 3)
Foamade Industries, Royal Oak, Mich. (1, 2, 3)
Hercules Incorporated, Wilmington, Del. (1, 3)
Kristal Kraft, Inc., Palmetto, Fla. (1)
Marine Plastics, Div. Northern Petrochemical Co., Clinton, Mass. (1)
National Distributing Co., Northvale, N. J. (1, 2)
Paramount Industries, Inc., Piscataway, N. J. (1, 2, 3)
Polymer Materials, Inc., Farmingdale, N. Y. (1, 2)
Rogers Foam Corp., Somerville, Mass. (1, 2, 3)
Rubbermaid Industrial Products Corp., Statesville, N. C. (1)

TABLE 52   Commercial Sources of Foamed Plastics (Continued)

Polyethylene Foam (Continued)

Stallman, M. H., Co., Providence, R. I. (1, 2)
Toyad Corp., Latrobe, Pa. (1, 2)
UCB-Sidac, Ghent, Belgium (1)
Voltek, Inc., Lawrence, Mass. (1, 2, 3)

Polypropylene Foam

Hercules Incorporated, Wilmington, Del.
Paramount Industries, Inc., Piscataway, N. J.
Rogers Foam Corp., Somerville, Mass.
Toray Industries, Inc., Tokyo, Japan

Polystyrene Foam

1. Expandable beads
2. Molded board, slab, etc.
3. Expanded board, slab, etc.
4. Extruded foam film and sheet

Acer Industries, Inc., Baltimore, Md. (2, 3)
BASF Canada Ltd., Station St. Laurent, Canada (1)
BASF Wyandotte Corp., Wyandotte, Mich. (1)
Badische Anilin- & Soda-Fabrik AG, Ludwigshafen, Germany (1)
Bagdad Plastics Co., Sub. Bagdad Copper Corp., Phoenix, Ariz. (2)
Bakelite Xylonite Ltd., London, England (3, 4)
Baxenden Chemical Co., Accrington, Lancs., England (3)
Cadillac Plastic & Chemical Co., Detroit, Mich. (2)
Cellu Products Co., Patterson, N. C. (4)
Cincinnati Foam Products, Inc., Cincinnati, Ohio (2, 3)
Construction Products Div., W. R. Grace & Co., Cambridge, Mass. (2)
Cosden Oil & Chemical Co., New York, N. Y. (1)
Crown Products Corp., St. Louis, Mo. (3)
Dew-Foam Co., Van Nuys, Calif. (2, 3)
Dow Chemical Co., Midland, Mich. (2, 3, 4)
Emerson & Cuming, Inc., Canton, Mass. (2)
Expanda-Foam, Inc., Antioch, Ill. (2, 3, 4)
Falcon Mfg. of Michigan, Inc., Byron Center, Mich. (2)
Farbwerke Hoechst AG, Frankfurt/Main-Hoechst, Germany (1)
Foam Products Corp., Maryland Heights, Mo. (3)
Foam Products, Inc., York Haven, Pa. (2)
Foam Specialties, Inc., Rye, N. Y. (2, 3)
Foam-Lite Plastics, Inc., Knoxville, Tenn. (2)
Formex Corp., Elkhart, Ind. (2, 3)
Foster Grant Co., Leominster, Mass. (1, 4)
General Box Co., Plastics Div., St. Charles, Ill. (2)
General Foam Plastics Corp., Norfolk, Va. (2)
Gilman Brothers Co., Gilman, Conn. (3, 4)
Grace, W. R., & Co., Cryovac Div., Duncan, S. C. (4)
Hitachi Chemical Co., Tokyo, Japan (1)
Hydra-Matic Packing Co., Bethayres, Pa. (3)
Montedison S.p.A., Milano, Italy (1)
National Distributing Co., Northvale, N. J. (2, 3)
Paramount Industries, Inc., Piscataway, N. J. (2, 3, 4)
Plastifoam Corp., Rockville, Conn. (2)
Plasti-Kraft Corp., Ozona, Fla. (3)
Plastronic Packaging Corp., Stevensville, Mich, (2, 3)
Poly Chemical Co., Tainan, Shien, Taiwan (1, 2)
Poly Foam, Inc., Lester Prairie, Minn. (2, 3)
Poly Molding Corp., Haskell, N. J. (2, 3)
Polymer Materials, Inc., Farmingdale, N. Y. (1)
Radva Plastics Corp., Radford, Va. (2, 3)
Robinson Industries, Inc., Coleman, Mich. (2, 3)

TABLE 52   Commercial Sources of Foamed Plastics (Continued)

Polystyrene Foam (Continued)

Sekisui Chemical Co., Tokyo, Japan (1, 2, 4)
Showa Denko K.K., Tokyo, Japan (1, 3)
Sinclair-Koppers Co., Pittsburgh, Pa. (1, 4)
Stallman, M. H., Co., Providence, R. I. (3)
Sterling Container Corp., Manchaug, Mass. (4)
Styro-Molders Corp., Colorado Springs, Colo. (3)
Swedish Crucible Steel Co., Packaging & Insulation Material Div., Detroit, Mich. (2)
Topper Plastics, Inc., Covina, Calif. (2)
Toyad Corp., Latrobe, Pa. (2)
Tufflite Plastics, Inc., Ballston Spa, N. Y. (3)
Tuscarora Plastics, Inc., New Brighton, Pa. (2)
United States Mineral Products Co., Stanhope, N. J. (2, 3)

Polyvinyl Chloride and Copolymers Foam

---

 1. Ingredients for (base resins)
 2. Slab stock
 3. Cross-linked
 4. Modifiers

A. & S. Corp., Verona, N. J. (1, 4)
Abbey Plastics Corp., Hudson, Mass. (1)
Air Products & Chemicals, Inc., Wayne, Pa. (4)
Airex AG, Sins, Switzerland (2)
Arlon Products, Inc., Harbor City, Calif. (2)
Bagdad Plastics Co., Sub. Bagdad Copper Corp., Phoenix, Ariz. (2)
Balsa Ecuador Lumber Corp., Northvale, N. J. (2, 3)
Cadillac Plastic & Chemical Co., Detroit, Mich. (2)
Crest-Foam Corp., Moonachie, N. J. (2)
Crown Products Corp., St. Louis, Mo. (2)
Dayton Polymer Industries, Inc., Dayton, Ohio (2)
Dew-Foam Co., Van Nuys, Calif. (2)
Diamond Shamrock Chemical Co., Plastic Div., Cleveland, Ohio (1)
Dynamit Nobel, AG, Troisdorf-Köln, Germany (2)
Dynamit Nobel of America Inc., Norwood, N. J. (2)
Ferro Corp., Composites Div., Norwalk, Conn. (2)
Firestone Plastics Co., Div. Firestone Tire & Rubber Co., Pottstown, Pa. (1)
Flexible Products Co., Marietta, Ga. (1)
Franklin Fibre-Lamitex Corp., Wilmington, Del. (2)
Goodrich, B. F., Chemical Co., Cleveland, Ohio (1)
Goodrich, B. F., General Products Co., Akron, Ohio (2)
Goodyear Tire & Rubber Co., Chemical Div., Akron, Ohio (1)
Howell Industries, Inc., Paterson, N. J. (1)
ICI America Inc., Stamford, Conn. (4)
Interstab Ltd., Kirby, Liverpool, England (4)
M R Plastics & Coatings, Inc., Maryland Heights, Mo. (1)
No-Ark Inc., Oakville, Conn. (2)
Plyfoam Inc., Plainview, L. I., N. Y. (2, 3)
Polytex Industrial Finishes Corp., Long Island City, N. Y. (3)
Porter, W. H., Inc., Holland, Mich. (2)
Rhone-Progil, Direction Commerciale Matières Plastiques, Neuilly, France (1)
Rogers Foam Corp., Somerville, Mass. (2)
Rubatex Corp., Bedford, Va. (2)
Ruco Div., Hooker Chemical Corp., Hicksville, N. Y. (1)
Scott Bader Co., Wellingborough, Northants, England (1)
Shinetsu Chemical Co., Tokyo, Japan (1)
Stallman, M. H., Co., Providence, R. I. (2)
Stauffer Chemical Co., Plastics Div., Westport, Conn. (1)
Tenneco Chemicals, Inc., Foam & Plastics Div., New York, N. Y. (2)
Union Carbide Corp., Chemicals & Plastics, New York, N. Y. (1)
Uniroyal, Inc., New York, N. Y. (1)

TABLE 52   Commercial Sources of Foamed Plastics (Continued)

Silicone Foam

Arlon Products, Inc., Harbor City, Calif.
Dow Corning Corp., Midland, Mich.
Emerson & Cuming, Inc., Canton, Mass.
General Electric Co., Silicone Products Dept., Waterford, N. Y.
Rhone-Poulenc, Société des Usines Chimiques, Paris, France
Shinetsu Chemical Co., Tokyo, Japan
3M Co., St. Paul, Minn.
Wacker-Chemie GmbH, Muenchen, Germany

Urea Foam

1. Ingredients for
2. Slab stock

Badische Anilin- & Soda-Fabrik AG, Ludwigshafen, Germany (1)
Isochem Resins Co., Lincoln, R. I. (1)
Mitsubishi Gas Chemical Co., Tokyo, Japan (1)

Urethane-Foam Film, Sheet, or Slab Stock

1. Flexible
2. Rigid

Acushnet Co., New Bedford, Mass. (1, 2)
Air-O-Plastik Corp., Carlstadt, N. J. (1)
Albis Corp., Houston, Tex. (1)
Almac Plastics, Inc., Long Island City, N. Y. (1, 2)
American Rubber & Plastics Corp., La Porte, Ind. (1)
Arlon Products, Inc., Harbor City, Calif. (1)
Bagdad Plastics Co., Sub. Bagdad Copper Corp., Phoenix, Ariz. (1, 2)
Baxenden Chemical Co., Accrington, Lancs., England (2)
Bernel Foam Products Co., Buffalo, N. Y. (1)
Burnett, William T., & Co., Baltimore, Md. (1)
Califoam Div., Mobay Chemical Co., Santa Ana, Calif. (1)
Canadian Industries Ltd., Industrial Chemicals Div., ICI Products Group, Montreal, Que., Canada (1, 2)
Carpenter, E. R., Co., Richmond, Va. (1)
Crest-Foam Corp., Moonachie, N. J. (1)
Crown Products Corp., St. Louis, Mo. (1, 2)
Dacar Chemical Products Co., Pittsburgh, Pa. (2)
Dew-Foam Co., Van Nuys, Calif. (1, 2)
Diamond Shamrock Chemical Co., Nopco Chemical Div., N. Arlington, N. J. (1)
Elastolabs Corp., Brooklyn, N. Y. (1)
Elliott Co. of Indianapolis, Inc., Indianapolis, Ind. (2)
Expanda-Foam, Inc., Antioch, Ill. (1, 2)
Firestone Foam Products Co., Fall River, Mass.
Foam Products Corp., Maryland Heights, Mo. (2)
Foam Products, Inc., York Haven, Pa. (1, 2)
Foam Specialists, Inc., Rye, N. Y. (1)
Foamabrix Corp., New York, N. Y. (1)
Foamade Industries, Royal Oak, Mich. (1)
Foamcraft, Inc., Chicago, Ill. (1, 2)
Franklin Fibre-Lamitex Corp., Wilmington, Del. (1, 2)
Future Foam, Inc., Omaha, Neb. (1)
General Plastics Mfg. Co., Tacoma, Wash. (1, 2)
General Tire & Rubber Co., Chemical/Plastics Div., Akron, Ohio (1)
Hoodfoam Industries, Inc., Marblehead, Mass. (1)
Hydra-Matic Packing Co., Bethayres, Pa. (1, 2)
International Foam Div., Holiday Inns of America, Inc., Chicago, Ill. (1)
Luzerne Rubber & Plastics Co., a unit of Beisinger Industries Corp., Taunton, Mass. (1, 2)
Matsushita Electric Works, Ltd., Kadoma-shi, Osaka-fu, Japan (2)
Millmaster Onyx Corp., New York, N. Y. (2)
Mobay Chemical Co., Div. Baychem Corp., Pittsburgh, Pa. (1, 2)
Morristown Foam Corp., Morristown, Tenn. (1)
National Distributing Co., Northvale, N. J. (1, 2)

TABLE 52   Commercial Sources of Foamed Plastics (Continued)

Urethane-Foam Film, Sheet, or Slab Stock (Continued)

North American Urethanes, Inc., Troy, Mich. (1, 2)
North Carolina Foam Industries, Inc., Mount Airy, N. C. (1, 2)
Owens-Corning Fiberglas Corp., Toledo, Ohio (2)
Paramount Industries, Inc., Piscataway, N. J. (1)
Phillips-Foscue Corp., High Point, N. C. (1)
Plastics Consulting & Mfg. Co., Camden, N. J. (1, 2)
Polymer Building Systems, Inc., Riverside, Calif. (2)
Porter, W. H., Inc., Holland, Mich. (2)
Precision Insulation Co., Houston, Tex. (2)
Reeves Brothers, Inc., New York, N. Y. (1)
Reeves Hardifoam Ltd., Toronto, Ont., Canada (1)
Rubba, Inc., Bronx, N. Y. (1, 2)
Scott Paper Co., Foam Div., Chester, Pa. (1)
Sekisui Chemical Co., Tokyo, Japan (1)
Stallman, M. H., Co., Providence, R. I. (1)
Strux Corp., Lindenhurst, N. Y. (2)
Techcraft, Inc., Wolfeboro, N. H. (1, 2)
Tenneco Chemicals, Inc., Foam & Plastics Div., New York, N. Y. (1)
Tiffin Enterprise, Inc., Tiffin, Ohio (2)
Unipoint Industries, Inc., High Point, N. C. (1)
United Foam Corp., Compton, Calif. (1)
United States Mineral Products Co., Stanhope, N. J. (2)
Upjohn Co., CPR Div., Torrance, Calif. (1, 2)

Urethane-Foam Ingredients

1. Complete foaming systems (e.g., prepolymers)
2. Diisocyanate
3. Modifiers
4. Polyesters for
5. Polyether polyols
6. Polymeric isocyanates

ARCO Chemical Co., Philadelphia, Pa. (3)
Air Products & Chemicals, Inc., Wayne, Pa. (3).   Catalysts, stabilizers
Allied Chemical Corp., Specialty Chemicals Div., Morristown, N. J. (2)
Atlas Chemicals Div., ICI America, Inc., Wilmington, Del. (5)
BASF Wyandotte Corp., Wyandotte, Mich. (1, 2, 3, 5)
Baker Castor Oil Co., Bayonne, N. J. (4)
Baxenden Chemical Co., Accrington, Lancs., England (1)
Berger Chemicals, Resinous Chemicals Div., Newcastle upon Tyne, England (4)
Callery Chemical Co., Div. Mine Safety Appliances Co., Callery, Pa. (1)
Canadian Industries Ltd., Industrial Chemicals Div., ICI Products Group, Montreal, Que., Canada (1, 2, 3, 4, 5)
Canadian Industries Ltd., Plastics Div., Industrial Chemicals Div., Montreal, Que., Canada (1, 2)
Chase Chemical Corp., Pittsburgh, Pa. (1)
Chemical Coatings & Engineering Co., Media, Pa. (1)
Cincinnati Milacron Chemicals, Inc., New Brunswick, N. J. (3)
Compo Industries, Inc., Waltham, Mass. (1)
Cook Paint & Varnish Co., Kansas City, Mo. (1)
Dacar Chemical Products Co., Pittsburgh, Pa. (1)
Diamond Shamrock Chemical Co., Nopco Chemical Div., N. Arlington, N. J. (1, 4)
Dow Chemical Co., Midland, Mich. (1)
Dow Corning Corp., Midland, Mich. (3). Surfactants for cell control
Du Pont de Nemours, E. I., & Co., Wilmington, Del. (2)
Durez Div., Hooker Chemical Corp., N. Tonawanda, N. Y. (1)
Eastman Chemical Products, Inc., Sub. Eastman Kodak Co., Kingsport, Tenn. (3)
Emerson & Cuming, Inc., Canton, Mass. (1, 2)
Farbenfabriken Bayer AG, Leverkusen, Germany (1, 2, 3, 4, 5)
Flexible Products Co., Marietta, Ga. (1, 2, 3, 5)
Freeman Chemical Corp., Div. H. H. Robertson Co., Port Washington, Wis. (1)
Fuller-O'Brien Corp., Technical Coatings Div., S. San Francisco, Calif. (1)

TABLE 52   Commercial Sources of Foamed Plastics (Continued)

Urethane-Foam Ingredients (Continued)

Furane Plastics, Inc., Los Angeles, Calif. (1)
General Electric Co., Silicone Products Dept., Waterford, N. Y. (3). Silicone surfactants
General Latex & Chemical Corp., Cambridge, Mass. (1)
Hightemp Resins, Inc., Stamford, Conn. (1)
ICI America Inc., Stamford, Conn. (4)
Insta-Foam Product Co., Addison, Ill. (1)
Isochem Resins Co., Lincoln, R. I. (1, 3, 5)
Isocyanate Products Div., Witco Chemical Corp., New Castle, Del. (1, 2, 3)
Jefferson Chemical Co., Houston, Tex. (3, 5, 6)
Kaiser Chemicals, Oakland, Calif. (2)
Kenrich Petrochemicals, Inc., Bayonne, N. J. (1, 4, 5)
Lankro Chemicals, Ltd., Eccles, Manchester, England (1, 2, 4, 5)
M R Plastics & Coatings, Inc., Maryland Heights, Mo. (1)
M & T Chemicals Inc., Sub. American Can Co., Rahway, N. J. (3)
Magnolia Plastics, Inc., Chamblee, Ga. (1)
Midwest Mfg. Corp., Burlington, Iowa (1)
Millmaster Onyx Corp., New York, N. Y. (4)
Mobay Chemical Co., Div. Baychem Corp., Pittsburgh, Pa. (1, 2, 3, 4, 5, 6)
Mobil Chemical Co., Industrial Chemical Div., Richmond, Va. (3).   Flame retardants
Mol-Rez Div., Whittaker Corp., Minneapolis, Minn. (1, 4)
Montedison S.p.A., Milano, Italy (2, 5)
Nippon Soda Co., Tokyo, Japan (2, 5)
North American Urethanes, Inc., Troy, Mich. (1, 4, 5)
North Carolina Foam Industries, Inc., Mount Airy, N. C. (1)
Olin Corp., Plastics Operations, Stamford, Conn. (1, 2, 5)
Owens-Corning Fiberglas Corp., Toledo, Ohio (1)
PPG Industries, Inc., Resin Products Sales, C&R Div., Pittsburgh, Pa. (1, 5)
Pelron Corp., Lyons, Ill. (1, 4, 5)
Pfizer Inc., Special Chemicals Dept., Greensboro, N. C. (4)
Podell Industries, Inc., Saddle Brook, N. J. (3).   Cell stabilizers
Polymer Building Systems, Inc., Riverside, Calif. (1)
Precision Insulation Co., Houston, Tex. (1)
Purethane Div., Easton RS Corp., Brooklyn, N. Y. (1, 3, 4, 5)
Reichhold Chemicals, Inc., White Plains, N. Y. (1, 4)
Ren Plastics, Inc., Lansing, Mich. (1)
Reynolds Chemical Products Div., Hoover Ball & Bearing Co., Ann Arbor, Mich. (1)
Rhone-Poulenc, Société des Usines Chimiques, Paris, France (1, 2, 3, 4, 5, 6)
Rubba, Inc., Bronx, N. Y. (1)
Ruco Div., Hooker Chemical Corp., Hicksville, N. Y. (4)
Schramm Fiberglass Products Div., High Strength Plastics Corp., Chicago, Ill. (1, 2, 4, 5)
Scott Bader Co., Wellingborough, Northants, England (4)
Unifoam Inc., Toledo, Ohio (1)
Uniglass Industries, Glascoat Div., Miami, Fla. (1, 2)
Union Carbide Corp., Chemicals & Plastics, New York, N. Y. (1, 2, 3, 5)
Uniroyal, Inc., New York, N. Y. (2, 5, 6)
United States Gypsum Co., Chicago, Ill. (1)
Upjohn Co., CPR Div., Torrance, Calif. (1, 2, 4, 5)
Upjohn Co., Polymer Chemicals Div., Kalamazoo, Mich. (1, 2, 5)
Urecon Corp., Pontiac, Mich. (1)
Verona Div., Baychem Corp., Union, N. J. (1, 2)
Whittaker Coatings & Chemicals, Lenoir Coatings & Resins, Div. Whittaker Corp., Lenoir,
    N. C. (1, 2, 3, 4, 5)
Witco Chemical Corp., Organics Div., New York, N. Y. (3, 4, 5).   Couping agents, tin
    catalysts

Foaming Agents

1.   Decomposable solids
2.   Gaseous blowing agents
3.   Volatile liquids

AMSCO Div., Union Oil Co. of Calif., Palatine, Ill. (3)
A. & S. Corp., Verona, N. J. (1)

TABLE 52   Commercial Sources of Foamed Plastics (Continued)

Foaming Agents (Continued)

Allied Chemical Corp., Specialty Chemicals Div., Morristown, N. J. (2)
Canadian Industries Ltd., Industrial Chemicals Div., ICI Products Group, Montreal, Que.,
  Canada (1, 3)
Custom Chemicals Co., E. Paterson, N. J. (1)
Dow Chemical Co., Midland, Mich. (2)
Du Pont de Nemours, E. I. & Co., Wilmington, Del. (3)
Fairmount Chemical Co., Newark, N. J. (1)
Farbenfabriken Bayer AG, Leverkusen, Germany (1)
Farbwerke Hoechst AG, Frankfurt/Main-Hoechst, Germany (2)
Kaiser Chemicals, Oakland, Calif. (2)
National Polychemicals, Inc., Wilmington, Mass. (1)
Noury & van der Lande N.V., Deventer, Holland (1, 2, 3)
Phillips Petroleum Co., Chemical Dept., Plastics Div., Bartlesville, Okla. (2)
Polymer Materials, Inc., Farmingdale, N. Y. (2)
Sobin Chemicals Inc., Boston, Mass. (1)
Union Carbide Corp., Chemicals & Plastics, New York, N. Y. (2)
Uniroyal, Inc., New York, N. Y. (1, 2)
Vulcan Materials Co., Chemicals Div., Wichita, Kans. (3)

SOURCE: "*Modern Plastics* Encyclopedia," 1972–1973, p. 298.

TABLE 53   Specifications and Test Methods for Polymer Foams[3]

ASTM test	Title
Tests for Rigid Cellular Plastics	
C 177-45	Method of Test for Thermal Conductivity of Materials by Means of the Guarded Hot Plate
C 273-61	Method of Shear Test in Flatwise Plane of Flat Sandwich Constructions or Sandwich Cores
C 355-63T	Tentative Methods of Test for Water Vapor Transmission of Materials Used in Building Construction
C 393-62	Tentative Method of Flexure Test of Flat Sandwich Constructions
C 518-63T	Tentative Method of Test for Thermal Conductivity of Materials by Means of a Heat Flow Meter
D 790-61 . . . . (Proced. A)	Method of Test for Flexural Properties of Plastics
D 1621-64	Tentative Method of Test for Compressive Strength of Rigid Cellular Plastics
D 1622-63	Tentative Method of Test for Apparent Density of Rigid Cellular Plastics
D 1623-64	Tentative Method of Test for Tensile Properties of Rigid Cellular Plastics
D 1673-61	Method of Test for Dielectric Constant and Dissipation Factor of Expanded Cellular Plastics Used for Electrical Insulation
D 1692-59T	Tentative Method of Test for Flammability of Plastics Foams and Sheeting
D 1940-62T	Tentative Method of Test for Porosity of Rigid Cellular Plastics
D 2126-62T	Tentative Method of Test for Resistance of Rigid Cellular Plastics to Simulated Service Conditions
D 2127-62T	Tentative Method of Test for Water Absorption of Rigid Cellular Plastics
	Proposed Tentative Method of Test for Thermal Conductivity of Cellular Materials by Means of a Line Heat Source
	Proposed Tentative Method of Test for Porosity of Rigid Cellular Plastics by Means of the Air Pyrometer

TABLE 53   Specifications and Test Methods for Polymer Foams[3] (Continued)

ASTM test	Title
Specifications for Rigid Cellular Plastics	
D 2125-62T......	Tentative Specification for Cellular Polystyrene
	Proposed Tentative Specification for Rigid Urethane Foam
	Proposed Tentative Specification for Rigid Cellular Urethane Thermal Insulation
	Proposed Tentative Specifications for Preformed Block Type Thermal Insulation, Polystyrene, Cellular, Molded Type
Methods and Specifications for Flexible Cellular Plastics	
D 1564-64T......	Tentative Specifications and Methods of Test for Flexible Urethane Foam
D 1565-60T......	Tentative Specifications and Methods of Test for Flexible Foam from Polymers or Copolymers of Vinyl Chloride
D 1596-59T......	Tentative Method of Test for Dynamic Properties of Package Cushioning Materials
D 1667-64........	Tentative Specifications and Methods of Test for Sponge Made from Closed Cell Polyvinyl Chloride or Copolymers Thereof
D 2221-63T......	Tentative Method of Test for Creep Properties of Package Cushioning Materials
Other Applicable Tests but Not Specifically Approved for Cellular Plastics	
D 149-61.........	Methods of Test for Dielectric Breakdown Voltage and Dielectric Strength of Electrical Insulating Materials at Commercial Power Frequencies
D 257-61.........	Method of Test for Electrical Resistance of Insulating Materials
D 150-59T........	Methods of Test for AC Capacitance, Dielectric Constant and Loss Characteristics of Electrical Insulating Materials
E 162-62T........	Method of Test for Surface Flammability of Materials Using a Radiant Heat Energy Source
D 1929-62T......	Method of Test for Ignition Properties of Plastics
C 351-61.........	Method of Test for Mean Specific Heat of Thermal Insulation
C 272-53.........	Method of Test for Water Absorption of Core Materials for Structural Sandwich Constructions
C 384-58.........	Methods of Test for Impedance and Absorption of Acoustical Materials by the Tube Method
C 423-60T........	Method of Test for Sound Absorption of Acoustical Materials in Reverberation Rooms
D 543-60T........	Method of Test for Resistance of Plastics to Chemical Reagents
D 696-44.........	Method of Test for Coefficient of Linear Thermal Expansion of Plastics
D 1924-63........	Recommended Practice for Determining Resistance of Plastics to Fungi
C 367-57.........	Methods of Test for Strength Properties of Prefabricated Architectural Acoustical Material
C 421-61.........	Method of Test for Weight Loss by Tumbling of Preformed Thermal Insulation

Federal and military Spec. Nos.	Title
HH-I-524.........	Insulation Board, Thermal (Polystyrene)
HH-I-00530....... (Interim)	Insulation Board, Thermal (Urethane)
MIL-P-15280D....	Plastic Foam, Unicellular, Sheet and Tubular Form
MIL-P-19644A....	Plastic Foam, Molded Polystyrene (Expanded Bead Type)
MIL-P-21929..... (Ships)	Plastic Material, Cellular Polyurethane, Rigid, Foam-in-Place
MIL-S-27332......	General Specification for Seat Cushion Insert, Polyurethane Plastic Foam
MIL-P-40619.....	Plastic Material, Cellular Polystyrene
MIL-P-43110..... (MR)	Plastic Foam Insulation, Thermal

REFERENCES

1. The Latest in Foam Technology, *Modern Plastics*, January 1969.
2. Foamed Plastics Now, *Plast. World*, March 1964.
3. Plastic Foams, *Mater. Des. Eng.*, May 1966, no. 236.
4. Darvas, R.: How to Design Foam Core Sandwich Laminates, *Modern Plastics*, June 1964.
5. Matonis, V. A.: "Elastic Behavior of Low Density Rigid Foams in Structural Applications," *SPE J.*, September 1964.
6. Weber: "Why Buy Air," *Hercules Chem.*
7. Glasstone, S.: "Textbook of Physical Chemistry," Van Nostrand, Princeton, N.J., 1940, 1946.
8. Benning, C. J.: "Plastics Foams," vols. I, II, Interscience-Wiley, New York, 1969.
9. Saunders, J. H., and K. C. Frisch: "Polyurethanes: Chemistry and Technology," vols. I, II, Interscience-Wiley, New York, 1962, 1964.
10. Dombrow, Bernard A.: "Polyurethanes," Reinhold, New York, 1957.
11. Britain and Gemeinhardt: *J. Appl. Polymer Sci.*, vol. 4, p. 207, 1960.
12. Smith: *J. Appl. Polymer Sci.*, vol. 7, p. 85, 1963.
13. Baker, et al.: *J. Chem. Soc.*, vol. 9, pp. 19, 24, 27, 1949; p. 4649, 1957.
14. *Upjohn Tech. Bull.* 202.
15. High Performance Integral Skin Urethane Foams, *Union Carbide Tech. Service Rept.*
16. High Performance Microcellular Urethane Elastomers, *Union Carbide Tech. Rept.*
17. Selectrofoam G403–6568, *Pittsburgh Plate Glass Industries Tech. Service Bull.*
18. Selectrofoam 6596, *Pittsburgh Plate Glass Industries Tech. Service Bull.*
19. Selectrofoam 6445–65058, *Pittsburgh Plate Glass Industries Tech. Service Bull.*
20. Properties of Rigid Urethane Foam, *Du Pont Tech. Rept.*
21. Humbert, W. E., and R. N. Kennedy: Expanded Polystyrene, vol. 18, no. 2, Paper 25, Division of Paint, Plastics, and Printing Ink Chemistry, American Chemical Society, September 1958.
22. Collins, F. H.: Controlled Density Polystyrene Foam Extrusion, *SPE J.*, July 1960, p. 705.
23. New License-free Moldable ABS Foams, *Mater. Eng.*, May 1967.
24. Injection-molded Expandable Polystyrene Simulates Wood, *Plast. Des. and Processing*, December 1963.
25. Davis, D. R.: Trouble-shooting Expandable Polystyrene Molding, *Plast. Technol.*, July 1962.
26. Smart Technology Comes to EPS molding, *Modern Plastics*, March 1972.
27. Allflatt, P. A.: How to Set Up a Plant for EPS Custom Molding, *Plast. Technol.*, November 1971.
28. Whedon, P. G.: Expandable ABS—A New Product and a New Conversion Method, *SPE Ann. Tech. Conf.*, vol. XIV, May 6–10, 1968.
29. Ingram, A. R.: Recent Progress in Expandable Polystyrene Foam Technology, *21st Ann. Tech. Conf. SPE*, Mar. 4, 1965.
30. Economics of Polystyrene Foams, *SPE J.*, February 1967.
31. Skinner, S. J., S. Baxter, S. D. Eagleton, and P. J. Grey: How Polystyrene Foam Expands, *Modern Plastics*, January 1965.
32. Stefanides, E. J.: Automated System Speeds Production of ABS Foam Parts, *Des. News*, Apr. 27, 1970.
33. Now: Expansion-cast ABS Using Structural Foam "Foundry," *Plast. Technol.*, March 1970.
34. Polystyrene Foam Contributes to Compactness of Room Air Conditioners, *Plast. Des. and Processing*, February 1964.
35. Molding Expandable Polystyrene, *Plast. Des. and Processing*, November 1970.
36. Waechter, C. J.: Processing Foamed Styrene for Thermoformed Packages, *Plast. Des. and Processing*, December 1970.
37. Zaslawsky, M.: Creep of Polystyrene Bead Foam under Uniaxial Compression, *SPE J.*, vol. 24, June 1968.
38. DiVerdi, J. M.: How to Buy Polystyrene Bead Molding Equipment, *Plast. Technol.*, October 1968.
39. Kreinik, J. M.: ABS Foam, "*Modern Plastics* Encyclopedia," 1971–1972.
40. Franson, G. R.: Expandable Polystyrene Fabrication Methods and Applications, *SPE Ann. Tech. Conf.*, Jan. 19, 1956.
41. Dylite Expandable Polystyrene, *Sinclair Koppers Corp. Tech. Manual* 9–273.
42. Packaging with Pelaspan, *Dow Chemical Co. Product Data Book* 171-192-10M-463.

43. Pelaspan Expandable Polystyrene, *Dow Chemical Co. Product Data Book Form* 171–414.
44. "*Modern Plastics* Encyclopedia," 1971–1972.
45. New Polypropylene Foam, *Mater. Des. Eng.*
46. Davidson, N. V.: How to License Polyolefin Foam Systems, *Plast. Technol.,* October 1968, p. 189.
47. Now Foam with Concentrates, *Modern Plastics,* May 1968, p. 88.
48. Polypropylene Pellets Developed for Expansion Molding, *Plast. Technol.,* January 1969, p. 15.
49. Kato, Y., and H. Ohshima: Foaming Process for LDPE Wire Insulation, *Modern Plastics,* May 1971.
50. Lanthier, P. F., and H. R. Lasman: Preparation of Low Density Cellular Vinyl and Other Thermoplastics with Azobisformamide, *SPE VIPAG Newsletter,* vol. II, no. 2, p. 34.
51. A New PVC Foam through Cross-linking, *Modern Plastics,* October 1967, p. 94.
52. Roggi, P. E.: What Are the Prospects for Expanded Vinyl? *Modern Plastics,* March 1971, p. 50.
53. Extruded PVC Foam Profiles Take on Wood Moldings, *Modern Plastics,* November 1971, p. 50.
54. Structural Foam Goes Extruded, *Modern Plastics,* January 1969, p. 100.
55. What's Ahead for PVC Markets, *Modern Plastics,* April 1971, p. 86.
56. Lasman, H. R.: Foam Vinyl and Polyolefins, *SPE 21st Ann. Tech. Conf.,* Mar. 1–4, 1965.
57. Cellular Vinyl, *B. F. Goodrich Chemical Co. Service Bull.* G-26.
58. Hopton, T. E.: Rigid Vinyl Profiles That Compete with Decorative Woods, *SPE J.,* vol. 27, p. 42, February 1972.
59. Renshaw, J. T., J. A. Cannon, and J. D. Hinchen: How Formulation and Process Variables Affect Vinyl Foams, *SPE J.,* vol. 20, p. 47, August 1970.
60. Stickman, S. W.: Vinyl Foam, "*Modern Plastics* Encyclopedia," 1971–1972.
61. Benning, C. J.: "Structure Properties and Application, in "Plastic Foams," vol. II, p. 113, Wiley, New York, 1969.
62. Thermoplastic Microspheres Are on Scene for Syntactic Foam, *Mater. Des. Eng.,* April 1969, p. 67.
63. Kausen, R. C., and F. F. Corse: Syntactic Foam Prepreg, *Modern Plastics,* December 1969, p. 146.
64. Doris, J. W., and R. W. Johnson: Syntactic Foam, *Modern Plastics,* September 1967, p. 25.
65. Johnson, R. W.: Here's What's Happening in High Performance Syntactic Foams, *Plast. Des. and Processing,* April 1973, p. 14.
66. Luoma, E. J.: Syntactic Foam, "*Modern Plastics* Encyclopedia," 1971–1972.
67. Morehouse, D. S., and H. A. Walters: Foamed Thermoplastic Microspheres in Reinforced Polyesters, *SPE J.,* vol. 25, p. 45, May 1969.
68. Expandable Filler Catches On, *Modern Plastics,* August 1969.
69. Design for Foam Growth, *Chem. Week,* Oct. 26, 1963.
70. Winspear, G. G. (ed.): "The Vanderbilt Rubber Handbook," R. T. Vanderbilt, Inc., New York, 1968.
71. "1969 Book of ASTM Standards," Part 28, American Society for Testing and Materials, Philadelphia, Pa., 1969.
72. Celogen AZ-A Nitrogen Blowing Agent for Sponge Rubber and Expanded Plastics, *Uniroyal Product Literature Form* 200R76.
73. Callington, K. T.: Chemical Blowing Agents in Plastics Conversion, *Plastics,* June 1969.
74. Naturman, L. I.: How to Select Blowing Agents for Thermoplastics, *Plast. Technol.,* October 1969, p. 41.
75. LaClair, R. C.: Foaming Agents, "*Modern Plastics* Encyclopedia," 1972–1973, p. 292.
76. Structural Foam in Your Future? It's Closer Than You Think, *Plast. World,* April 1973, p. 40.
77. Weir, C. L.: These Data Show Why the Action Is Swinging to Structural Foams, *Plast. Technol.,* April 1972, p 37.
78. Craayleek, G., Jr.: Now You Can Make Structural Styrene Foam Components That Look Like Genuine Wood, *Plast. Des. and Processing,* August 1972, p. 10.
79. If You Think Structural Foam Is Set to Boom, You're Right, *Modern Plastics,* December 1971, p. 34.
80. Weir, C. L.: New Foam Molding Process Offers New Product Versatility, *Modern Plastics,* March 1969, p. 68.

81. New Foam for New Markets, *Modern Plastics,* December 1968, p. 66.
82. Smooth-Surface Foam Molding, *Modern Plastics,* June 1974, p. 78.
83. Wilson, M. G.: Glass Reinforced Foam for Injection Molding, *Plast. Des. and Processing,* January 1972, p. 12.
84. Chabot, D. G.: Injection Molded Foam, "*Modern Plastics* Encyclopedia," 1971–1972.
85. Rutherford, H.: Union Carbide Method, "*Modern Plastics* Encyclopedia," 1971–1972.
86. Angell, R. G.: The Structural Foam Molding Process, *Union Carbide Tech. Paper.*
87. New Way to Make a Sandwich: Injection Mold, *Modern Plastics,* October 1970, p. 66.
88. Chabot, D. G.: Furniture Parts: What Does Structural Foam Offer, *Modern Plastics,* April 1971, p. 72.
89. Enter the All Foam Bed, *Modern Plastics,* August 1971, p. 45.
90. Scheiner, L. L.: Structural Thermoplastic Foams: An Emerging Technology, *Plast. Technol.,* December 1968, p. 35.
91. Injection Molding of Plastics Foams Shapes Up for 1971, *Plast. Des. and Processing,* February 1971, p. 18.
92. Molded Structural Foams Are Starting to Move, *Modern Plastics,* April 1967, p. 76.
93. Rigid Foam Gains Momentum as 100-Lb Parts Become Feasible, *Plast. World,* October 1970, p. 49.
94. Riley, C. P., Jr.: "Foam Phenolics and Ureas," unpublished talk, National Poly Chemicals, Inc., March 1965.
95. Cellular Plastics (Theory, Properties, Applications, and Economics), *Univ. Michigan Eng. Summer Conf.,* no. 6424-103, 1964.
96. Einhorn, E. N.: A Review of Cellular Plastics, *Univ. Michigan Eng. Summer Conf.,* no. 6424-103, 1964.
97. Collington, K. T.: Chemical Blowing Agents in Plastics Conversion, *Plastics,* vol. 34, no. 380, p. 724, June 1969.

Chapter **8**

Liquid and Low-Pressure Resin Systems

LEONARD S. BUCHOFF

Technical Wire Products, Inc.,
Cranford, New Jersey

INTRODUCTION

The materials covered in this chapter are resins that are liquids at room temperature or solids that melt up to 200°C. The chapter will not cover materials that are liquids by virtue of being dissolved in solvents, i.e., solvent-based adhesives, or suspended in water, e.g., rubber. Their common attribute is conversion to a solid by the action of heat and/or a curing agent. Most of the materials are cross-linked during the curing, resulting in an infusible, insoluble thermoset. Some, like cast acrylic, are converted to a thermoplastic. In a third group, typified by vinyl plastisols, the conversion from liquid to solid is accomplished by a physical process, with no chemical change. In general, if solvents are used with these resins, they play a minor role in the conversion of the formulation, or function other than as a solvent. They may enter into the reaction, as styrene does in polyester.

Liquid resins are useful in the fields of encapsulation, art, adhesives, laminates, foams, coatings, construction, resurfacing, centrifugal casting, embedments, tooling, and model making. Other chapters in this handbook deal with adhesives, foams, laminates, coatings, and electrical insulation. In this chapter, properties and guides to the selection of materials and compounds will be presented.

Advantages of Liquid Resins

Liquid resins are important because of their versatility and ease of application. Compounding is often accomplished by dispersing curing agents, fillers, and other ingredients at room temperature with a simple propeller mixer. Fillers are wetted better by the liquid resins than by molten polymer, and the fillers can be chemically bound. A variety of properties are readily available by changing formulations; stiffness, color, strength, and electrical properties can be tailored to specific needs by changing proportions and types of ingredients.

A range of cure temperature and times is available. Most classes of liquid resins can be formulated to cure at room temperature, eliminating the need for ovens. Equipment in general can be very simple and inexpensive; although automated, highly sophisticated machinery is available for metering, mixing and dispensing, and curing. Often, disposable containers can be used for hand mixing to prepare material for repairs or use in remote areas, such as in making field repairs on electrical equipment, in resurfacing highways and concrete floors, or in installing sealants or tiles.

Liquid resins also allow a wide range of mold construction. Inexpensive, one-shot paper molds and wooden molds for constructing entire walls and boats can be used. Plaster of paris, plastic, sheet metal, foam, and many other materials have been used in mold construction for castings.

There is no theoretical limitation to the size of castings that can be produced. Insulators weighing several tons have been cast from epoxies, and nylon parts much too large to be molded have been cast from caprolactams. Liquid resins can also be poured around small electrical components, flowers, or other delicate objects without altering their shape or changing their positions. Proper compounding can produce castings with negligible shrinkage and optical clarity. Castings can be produced with a minimum of internal stress. They will therefore be stronger, more dimensionally stable, and less subject to stress cracking than parts made by other processes.

Selection of the Resin

Many of the materials covered in this chapter are used in electrical and electronic parts. A comparison of properties of potting resins is given in Tables 1 and 2. Table 3 identifies resin characteristics that influence the choice of materials in electrical application. Table 4 is included to aid in selecting materials for non-electrical applications, as the properties requirements are not necessarily the same as in electrical applications.

Elastomers In recent years, liquid elastomers have gained wide acceptance. Table 5 lists the reactive sites, trade names, and sources for these materials. The price, specific gravity, and viscosity are given in Table 6.

Processing Liquid Resins

Electrical components and circuits are embedded in liquid resins by casting, potting, liquid injection molding, encapsulation, or impregnation.

Fabricated parts are products from liquid resins by casting, rotational molding, centrifugal casting, liquid injection molding, and pultrusion. Features of these processes are presented in the following paragraphs.

Casting Prepared liquid-resin compound is poured into a stationary mold and cured. If objects are to be embedded, they are prepositioned in the mold or introduced after the resin is poured.

Advantages
Minimum equipment required
Delicate parts can be embedded
Wide choice of optical effects possible
No limit to part size
Wide choice of mold construction

Disadvantages. Bubbles and voids are problems, high-viscosity formulations difficult to handle.

Material Used. Epoxies, polyesters, acrylics, allylics, polyurethanes, silicones.

Equipment. Simple molds to elaborate metering and mixing machines and automated lines.

Applications. Electrical embedding, casting art statuary, embedding biological specimens, building facades.

Rotational molding The catalyzed resin is poured into a mold, and the mold is rotated biaxially in a heated chamber. The resin is distributed solely by gravitational force; centrifugal force is not a factor.

Advantages. Large hollow parts can be produced, molds are inexpensive, stress-free parts.

Disadvantages. Heating and cooling cycles of heavy molds are time-consuming. Room-temperature-setting materials would reduce this problem.

Material Used. Plastisols, presently; epoxies, polyesters, silicones, and polyurethanes in the future.

Equipment. Heavy machinery is necessary for handling large molds; carousel machines having multiple molds are used to speed production; heating is done with hot air, hot salt spray, or heating oil.

Applications. Toys, novelty items, tank bodies.

TABLE 1 Comparison of Properties of Liquid Resins

Material	Cure shrinkage	Adhesion	Thermal shock	Electrical properties	Mechanical properties	Handling properties	Cost
Epoxy:							
Room temp. cure	Low	Good	Fair*	Fair	Fair	Good	Moderate
High temp.	Low	Good	Fair*	Good	Good	Fair to good	Moderate
Flexible	Low	Excellent	Good	Fair	Fair	Good	Moderate
Polyesters:							
Rigid	High	Fair	Fair*	Fair to good	Good	Fair	Low to moderate
Flexible	Moderate	Fair	Fair	Fair	Fair	Fair	Low to moderate
Silicones:							
Rubbers	Very low	Poor	Excellent	Good	Poor	Good	High
Rigid	High	Poor	Poor	Excellent	Poor	Fair	High
Polyurethanes:							
Solid	Low	Excellent	Good	Fair	Good	Fair	Moderate
Foams	Variable	Good	Good	Fair	Good for density	Fair	Moderate
Butyl LM	Low	Good	Excellent	Good	Poor	Fair	Low to moderate
Butadienes	Moderate	Fair	Poor*	Excellent	Fair	Fair	Moderate

* Depends on filler.

Centrifugal Casting   The catalyzed resin is introduced into a rapidly rotating mold where it forms a layer on the mold surfaces and hardens.   Vacuum is sometimes employed to aid in bubble removal.

Advantages. Reinforcing elements can be accurately placed; high-viscosity, rapidly curing material can be used; dense bubble-free parts; precise wall thickness possible; finished inside diameter.

Disadvantages. Fillers separate rapidly, limited to toroidal shapes.

Material Used. Polyurethane, plastisols, other liquid elastomers, epoxy.

Equipment. High-speed rotating machinery, vacuum system sometimes used.

Applications. Elastomer sheets, reinforced pipe, rubber track.

Liquid injection molding   Catalyzed resin is metered into closed molds, curing rapidly and automatically ejecting the finished parts.

Advantages. Liquids can be accurately metered; low injection pressure; rapidly curing materials can be automatically mixed immediately before entering the molds; flash is easily avoided.

Disadvantages. Low-viscosity resin necessary; expensive equipment, precision molds needed.

Material Used. Polyurethane, epoxies, silicones, polyesters, vinyl plastisols.

Equipment. Injection presses, automatic mixing and dispensing machine, multiple molds.

Applications. Polyurethane shoe soles and heels, transparent caps for light-emitting diodes, large structural polyester parts.

Pultrusion   Reinforced-plastics profiles are produced in continuous lengths. Fibrous glass in the form of roving, mat, or cloth is impregnated with catalyzed resin, drawn through a heated die, and the cured shapes are cut to appropriate lengths.

Advantages. High-strength products; continuous lengths available; potentially inexpensive.

Disadvantages. Expensive equipment, complex specialized dies, speed limited.

Material Used. Polyester and epoxy—low viscosity, internally lubricated.

Equipment. Impregnation tank, heated die, rf generator and/or oven, puller, diamond-faced cutoff saw.

Application. Fishing rods, ladders, electrical-pole-line hardware, standard structural shapes.

EPOXIES

Epoxy resins have gained wide acceptance in the industrial field in the past 20 years in adhesives, coatings, castings, potting, building construction, chemical-resistant equipment, boats, etc.   The properties that have made the epoxies popular in so many fields are their versatility, excellent adhesion, low cure shrinkage, good electrical properties, compatibility with a great number of materials, resistance to chemicals and weathering, dependability, and ability to cure under adverse conditions.

Epoxies can be compounded to produce a wide range of handling, curing, and final part properties by choice of the basic resin(s), curing agent(s), filler(s), and modifier(s).   Each of these topics is discussed separately in this section.   As the curing agent becomes an integral part of the cured compound, its choice is a controlling influence on the curing and final properties of the mixture.   Fillers and modifiers are used to tailor the liquid viscosity and cured properties to the application.

Basic Types of Epoxy Resin

Epi-bis epoxies   Most of the epoxies used today are the liquid reaction product of *epi*chlorohydrin and *bis*phenol A, often called epi-bis epoxies.   Table 7 shows trade names of some of the major epi-bis resins, and the companies that produce them.   Others in this group range from resins of increasing viscosity to solids melting up to 175°C.   Generally, the higher the melting point, the less curing agent is needed.   The cured properties of all these resins are similar, but the toughness increases as the melting point increases.   Although most of the epi-bis resins are light amber, transparent, colorless epoxies are available for optical embedments.

TABLE 2 Properties of Various Liquid Resins[1]

Property	ASTM	Rigids (silica-filled)			Flexibilized epoxies			Liquid elastomers				Rigid polyurethane
		Styrene-polyester	Epoxy	Silicone	Epoxy-polysulfide (50-50)	Epoxy-polyamide (50-50)	Epoxy-polyurethane (50-50, diamine cure)	Polysulfide	Silicone	Polyurethane (diamine cure)	Plastisol	Prepolymer
Electrical Properties												
Dielectric strength, V/mil*	D 149	425	425	350	350	430	640	340	400	350	200	
Dielectric constant:												
60 Hz	D 150	3.7	3.8	3.7	5.6	3.2	......	7.2	3	8.2	5.6	1.05
10^3 Hz	D 150	3.7	3.6	3.6	5.4	3.2	4.9	7.2	3	7.3	4.9	1.06
10^6 Hz	D 150	3.6	3.4	3.6	4.8	3.1	......	7.2	3	......	3.6	1.04
Dissipation factor:												
60 Hz	D 150	0.01	0.02	0.008	0.02	0.01	......	0.01	0.005	0.08	0.12	0.004
10^3 Hz	D 150	0.02	0.02	0.004	0.02	0.01	0.04	0.01	0.004	0.09	0.1	0.003
10^6 Hz	D 150	0.02	0.03	0.01	0.06	0.02	......	0.02	0.003	......	0.12	0.003
Surface resistivity, Ω/square:												
Dry	D 257	10^{15}	10^{15}	10^{15}	10^{12}	10^{14}	......	10^{11}	10^{13}	......	10^{10}	$>10^{12}$
After 96 h at 95°F, 90% RH	D 257	10^{11}	10^{13}	10^{15}	10^{10}	......	......	10^{10}	10^{12}	......	10^{10}	$>10^{12}$
Volume resistivity (dry), Ω-cm	D 257	$>10^{15}$	$>10^{15}$	10^{15}	10^{12}	10^{14}	10^{14}	10^{11}	10^{14}	10^{12}	10^{10}	$>10^{14}$
Mechanical Properties												
Tensile strength, lb/in.²	D 638, D 412	10,000	9,000	4,000	1,800	4,600	6,000	800	275	4,000	2,400	110
Elongation, %	D 412	......	......	......	30	10	10	450	200	450	300	
Compression strength, lb/in.²	D 695, D 575	25,000	16,000	13,000	......	7,000	......	......	......	......	......	85

Physical Properties

Property	Test method											
Flexural modulus of rupture, lb/in.^{2}	D 790	10,000	12,000	8,000	……	……	8,300	……	……	……	……	……
Izod impact strength (notched), ft-lb	D 256	0.3	0.4	0.3	……	……	……	……	……	……	……	……
Shore hardness	D 676, D 1484	……	……	……	40D	80D	……	45A	35A	90A	80A	……
Penetration at 77°F, mils	D 5	……	……	……	……	……	……	1	2	4.5	……	……
Shrinkage on cooling or curing, % by vol		6	3.5	8	3	3	6	……	……	……	……	……
Specific gravity	D 71, D 792	1.6	1.6	1.8	1.2	1.0	1.2	1.2	1.1	1.1	1.2	0.1†
Min cold flow, °F												
Softening or drip point, °F	D 36											
Heat-distortion temp, °F	D 648	230	165	300	……	……	100	……	……	……	……	……
Max continuous service temp, °F		250	250	480	225	175	250	200	350	210	150	165
Coefficient of thermal expansion per °F $\times 10^{-6}$	D 696	26	22	44	44	44	110	……	128	110	……	19
Thermal conductivity, Btu/(ft^2)(h)(°F/ft)		0.193	0.29	0.23	……	……	0.77	……	0.13	……	0.09	0.02
Cost, $/lb		0.25	0.50	2.75	0.90	0.80	1.10	0.85	4.25	1.30	0.30	1.75

* Short-time test on ⅛-in. specimen.
† 6 lb/ft^3.

Table 8 lists properties of a typical cured epi-bis resin with several types of curing agents.

Cycloaliphatic epoxies These resins are characterized by the saturated ring present in the structure. They are free of the hydrolyzable chlorine, sometimes present in the epi-bis epoxies, which adversely affects some electronic devices. The cycloaliphatics have superior arc-track resistance, good electrical properties under adverse conditions, good weathering properties, high heat-deflection temperatures,

TABLE 3   Resin Characteristics That Influence the Choice of Resins in Electrical Applications

Resin	Cure and handling characteristics	Final part properties
Epoxies........	Low shrinkage, compatible with a wide variety of modifiers, very long storage stability, moderate viscosity, cure under adverse conditions	Excellent adhesion, high strength, available clear, resistant to solvents and strong bases, sacrifice of properties for high flexibility
Polyesters......	Moderate to high shrinkage, cure cycle variable over wide range, very low viscosity possible, limited compatibility, low cost, long pot life, easily modified, limited shelf life, strong odor with styrene	Fair adhesion, good electrical properties, water-white, range of flexibilities
Polyurethanes...	Free isocyanate is toxic, must be kept water-free, low cure shrinkage, solid curing agent for best properties	Wide range of hardness, excellent wear, tear, and chemical resistance, fair electricals, excellent adhesion, reverts in humidity
Silicones .... (flexible)	Some are badly cure-inhibited, some have uncertain cure times, low cure shrinkage, room temp cure, long shelf life, adjustable cure times, expensive	Properties constant with temperature, excellent electrical properties, available from soft gels to strong elastomers, good release properties
Silicones........ (rigid)	High cure shrinkage, expensive	Brittle, high-temperature stability, electrical properties excellent, low tensile and impact strength
Polybutadienes..	High viscosity, reacted with isocyanates, epoxies, or vinyl monomers, moderate cure shrinkage	Excellent electrical properties and low water absorption, lower strength than other materials
Polysulfides.....	Disagreeable odor, no cure exotherm, high viscosity	Good flexibility and adhesion, excellent resistance to solvents and oxidation, poor physical properties
Butyl LM rubber	High viscosity, low cure shrinkage, low cost, variable cure times and temperature, one-part material available	Low strength, flexible, low vapor transmission, good electrical properties
Allylic resins....	High viscosity for low-cure-shrinkage materials, high cost	Excellent electrical properties, resistant to water and chemicals

and good color retention. Some members of this group are low in viscosity and serve as reactive diluents in laminate structures.

While most cycloaliphatics are cured only with anhydrides, others do react with amines. Table 9 shows how the properties of these materials can be modified by blending a flexible and a rigid resin. Massive high-voltage insulators weighing 5,000 lb have been cast with cycloaliphatics.[5,6]

Extensive testing has shown that bushings cast from cycloaliphatic resins resist high-voltage breakdown and tracking better than bushings cast from other epoxies. Figure 1 shows typical high-voltage cycloaliphatic epoxy bushings.

TABLE 4 Resin Characteristics of Liquid Resins for Nonelectrical Applications

Resin	Handling characteristics	Final part properties	Typical applications
Cast acrylics	Low to high viscosity, long cure times, bubbles a problem, special equipment necessary for large parts	Optical clarity, excellent weathering, resistance to chemicals and solvents	In glazing, furniture, embedments, impregnation
Cast nylon	Complex casting procedure, very large parts possible	Strong, abrasion-resistant, wear-resistant, resistance to chemicals and solvents, good lubricity	Gears, bushings, wear plates, stock shapes, bearings
Phenolics	Acid catalyst used, water given off in cure	High density unfilled, brittle, high temperature, brilliance	Billiard balls, beads
Vinyl plastisol	Inexpensive, range of viscosities, fast set, cure in place, one-part system, good shelf life	Same as molded vinyls, flame-resistant, range of hardness	Sealing gaskets, hollow toys, foamed carpet backing
Epoxies	Cures under adverse conditions, over a wide temp range, compatible with many modifiers, higher filler loadings	High strength, good wear, chemical and abrasion resistance, excellent adhesion	Tooling, fixtures, road and bridge repairs, chemical-resistant coating, laminates, and adhesives
Polyesters	Inexpensive, low viscosity, good pot life, fast cures, high exotherm, high cure shrinkage, some cure inhibition possible, wets fibers easily	Moderate strength, range of flexibilities, water-white available, good chemical resistance, easily made fire retardant	Art objects, laminates for boats, chemical piping, tanks, aircraft, and building panels
Polyurethanes	Free isocyanate is toxic, must be kept water-free, low cure shrinkage, cast hot	Wide range of hardnesses, excellent wear, tear, and chemical resistance, very strong	Press pads, truck wheels, impellers, shoe heels and soles
Flexible soft silicone	High-strength materials, easily inhibited, low cure shrinkage, adjustable cure time, no exotherm	Good release properties, flexible and useful over wide temperature range, resistant to many chemicals, good tear resistance	Casting molds for plastics and metals, high-temperature seals
Allylic resins	Long pot life, low vapor pressure monomers, high cure temperature, low viscosity for monomers	Excellent clarity, abrasion-resistant, color stability, resistant to solvents and acids	Safety lenses, face shields, casting impregnation, as monomer in polyester
Butyl LM	High viscosity, low cost, adjustable cure time	Low strength, low vapor transmission, resistant to reversion in high humidity	Roofing coating, sealant, in reservoir liners
Polysulfide	High viscosity, characteristic odor, low cure exotherm	Good flexibility and adhesion, excellent resistance to solvents and oxidation, poor physical properties	Sealants, leather impregnation

Novolac epoxies   Phenolic or cresol novolacs are reacted with epichlorohydrin to produce these resins.   The novolacs are high-viscosity liquids or semisolids, and are often mixed with other epoxies to improve handling properties.   Figure 2 shows the viscosities of blends of a novolac resin with an epi-bis resin vs. temperature.   They cure more rapidly than epi-bis epoxies and have higher exotherms.   The cured novolacs have higher heat-deflection temperatures than epi-bis resins (Table 10).

TABLE 5   Liquid Elastomers—Types and Suppliers[2]

Chemical description	Reactive site	Trade name	Supplier
Isoprene, natural........	Unsaturation	DPR	Hardman Inc.
Isoprene, synthetic.......	Unsaturation	Isolene D	Hardman Inc.
Isobutylene.............	Unsaturation	Enjay LM 430	Enjay Chemical Co.
Butadiene/styrene.......	Unsaturation	Ricon 100	The Richardson Co.
Butadiene/styrene.......	—OH	Poly bd CS-15	Arco Chemical Co.
Butadiene..............	Unsaturation	Lithene A,P,Q	Lithium Corp.
Butadiene..............	Unsaturation	Ricon 150	The Richardson Co.
Butadiene..............	—OH	Butarez HTS	Phillips Petroleum Corp.
Butadiene..............	—OH	Poly bd R-45 (M&HT)	Arco Chemical Co.
Butadiene..............	—OH	Hystl G	Hystl Dev. Co.
Butadiene..............	—COOH	Hystl C	Hystl Dev. Co.
Butadiene..............	—COOH	Butarez CTL II	Phillips Petroleum Corp.
Butadiene..............	—COOH	Hycar CTB	B. F. Goodrich Chemical Co.
Butadiene..............	—COOH	HC-434	Thiokol Chemical Corp.
Butadiene..............	—Br	RTV liquid rubber	Polymer Corp.
Polysulfide.............	—SH	LP	Thiokol Chemical Corp.
Butadiene/acrylonitrile...	—COOH	Poly bd CN-15	Arco Chemical Co.
Butadiene/acrylonitrile...	—COOH	Hycar CTBN	B. F. Goodrich Chemical Co.
Butadiene/acrylonitrile...	—SH	Hycar MTBN	B. F. Goodrich Chemical Co.
Butadiene/acrylonitrile...	Unsaturation	Hycar 1312	B. F. Goodrich Chemical Co.
Chloroprene.............	—Cl	Neoprene FB, FC	E. I. du Pont de Nemours & Co.
Polyester..............	—NCO (urethane)	Many	American Cyanamid Co., Thiokol Chemical Corp., Uniroyal, Inc., Witco Chemical Corp., North American Urethanes, Inc., Upjohn Co., Hooker Chemical Corp., Mobay Chemical Co., Conap, Inc., Hysol Div., Dexter Corp., etc.
Polyether..............	—NCO (urethane)	Many	E. I. du Pont de Nemours & Co., Uniroyal, Inc., North American Urethanes, Inc., etc.
Siloxane...............	Unsaturation, silane	Many	Dow Corning Corp., Union Carbide Corp., General Electric Corp., Stauffer-Wacker Silicone Corp., Farbenfabriken Bayer AG.

The novolacs have excellent resistance to solvents and chemicals.   This property is compared with that of an epi-bis resin in Table 11.

Other epoxy resins   Resorcinol diglycidyl ether (Kopper's Kopoxite 159 and Ciba-Geigy's ERE 1359) has a viscosity of 400 cP and one of the highest epoxy concentrations available in a casting resin.   This resin exhibits low cure shrinkage, rapid curing, good resistance to moisture and chemicals, and good adhesive qualities. It is often used to reduce the viscosity of epi-bis compounds.

Chlorinated or brominated bisphenol A is used to produce flame-resistant epoxies.

As the halogen content increases, the viscosity generally increases. Table 12 compares the properties of systems prepared from a chlorinated epi-bis resin and a standard epi-bis resin.

Resins that flexibilize epoxy formulations are straight-chain compounds terminated in epoxy groups. Other polyhydric compounds such as glycerin are used as starting points for epoxy resins. These generally have low viscosities, high reactivity, and poorer physical, electrical, and chemical resistance properties than epi-bis resins.

Curing Agents

The curing agent used in a formulation plays a major role in determining the handling properties, curing schedule, and final part properties. The curing agents

TABLE 6 Liquid Elastomers—Price, Specific Gravity, and Viscosity[2]

Trade name*	Midrange price, $/lb	Specific gravity	$/lb vol	Viscosity, poises 25°C	Other
Lithene P & Q.......	0.35	0.89	0.31	1–2,000	
Lithene A...........	0.35	0.93	0.32	1–2,000	
Poly bd 45HT.......	0.48	0.90	0.43	50	
Poly bd 45M........	0.50	0.90	0.45	50	
Poly bd CS-15.......	0.50	0.90	0.45	225	110 at 50°C
Poly bd CN-15......	0.56	0.90	0.505	525	100 at 70°C
Hycar 1312.........	0.51	1.00	0.51	High	225 at 50°C
Isolene D..........	0.59	9.02	0.54	High	140–1,700 at 50°C
DPR..............	0.61	0.92	0.56	High	160–1,300 at 50°C
Ricon.............	0.70	0.89	0.62	400	
Ricon 100..........	0.70	0.92	0.64	3,800	100 at 60°C
Enjay LM 430......	0.70	0.92	0.64	High	800 at 120°C
Hystl G & C........	0.76	0.88	0.67	650–2,600	80–335 at 45°C
Neoprene FB & FC..	0.65	1.23	0.80	High	750–1,500 at 90°C
RTV liquid rubber...	1.00	0.95	0.95	750–2,000	200 at 65°C
Ester urethane......	0.87	1.14–1.25	0.99–1.09	High	120 at 60°C
Ether urethane......	0.97	1.06–1.11	1.03–1.08	60–200	
Thiokol LP.........	0.83	1.27	1.04	7–1,400	
Hycar MTBN.......	1.15	0.98	1.13	400	
Hycar CTBN.......	1.25	0.95	1.19	1,150	
Hycar CTB (162)....	1.50	0.91	1.36	250	
Baysilone E & K.....	1.65	0.97	1.60	20–500	
HC-434............	3.18	0.91	2.90	250	
Butarez CTL II.....	4.25	0.90	3.83	300	
Butarez HTS........	4.25	0.905	3.85	130	

* Suppliers are listed in Table 5.

will be discussed as to their effect on epi-bis resins, but the data will be generally applicable to most other types of epoxies.

The major classes of epoxy curing agents are (1) aliphatic amines, (2) aromatic amines, (3) acids and anhydrides, (4) catalysts, (5) latent curing agents, and (6) flexibilizing curing agents. Mixtures of curing agents are used to modify the curing schedule of the compound and cured properties.

Aliphatic amines Aliphatic amines such as diethylenetriamine (DETA) are widely used for curing epoxies at room temperature in small masses or in films. These materials are highly exothermic so that a large mixture may overheat during cure. Disadvantages of these materials include a short pot life and tendency to pick up moisture from the air, which inhibits the cure. DETA and their adducts are used in potting, adhesives, patching, coatings, and wet-lay-up laminates.

Slower curing agents, dimethylaminopropylamine and diethylaminopropylamine, are preferred in applications such as in manufacture of castings, moldings, and other

TABLE 7 Trade Names and Suppliers of Epi-Bis Resins

Resin description	Shell (Epon)	Dow (D.E.R.)	Union Carbide (Bakelite)	Celanese (Epi-Rez)	Ciba (Araldite)	Reichhold (Epotuf)
Low-viscosity modified	817	...	2712	...	502	37–131
Low-viscosity modified	819H–13	321	2713	5077	507	37–128
Flexible	...	732–736	...	5021–5042	508	...
B.G.E.* modified low-viscosity	815	334–335	2795	5071–504	506	37–130
P.G.E.† modified low-viscosity	820	336	2600	...	...	37–135
Standard liquids, low to high viscosity‡	826	330	2772	509	5006	37–139
	828	331	2774	510	6010	37–140
Unmodified	830	...	...	...	6020	...
	834	337	3794	515		
	836					

* Butyl glycidyl ether.
† Phenyl glycidyl ether.
‡ Lowest viscosity.

TABLE 8 Properties of an Epi-Bis Resin Cured with Various Hardeners[3]

Hardener	phr*	Tensile strength, lb/in.² at 25°C	Tensile modulus, lb/in.² ×10⁻⁶	Tensile elongation, %	Dielectric constant, 60 Hz at			Dielectric strength, S/T, V/mil	Dissipation factor, 60 Hz at				Volume resistivity, MΩ-cm	Arc resistance, s (ASTM D 495)
					25°C	50°C	100°C		25°C	50°C	100°C	150°C		
Aliphatic amines:														
Diethyl amino propylamine	7	8,500	....	....	....	....	....	....	....	....	....	....	....	58
Aminoethyl piperazine	20	10,000	....	5-6	....	....	....	....	....	....	....	....	....	62
Aromatic amines:														
Mixture of aromatic amines	23	12,600	....	6-5	....	....	....	....	....	....	....	....	....	80
p,p'-Methylene dianiline	27	9,500	0.4	4-5	4.4	....	4.6	420	0.007	....	0.002	....	>1 × 10¹⁰	83
m-Phenylene diamine	14.5	13,000	0.5	4	4.4	....	4.7	483	0.007	....	0.003	....	>1 × 10¹⁰	78
Diaminodiphenyl sulfone	30	7,000	0.47	....	4.22	....	4.71	410	0.004	....	0.068	....	>2 × 10⁸	65
Anhydrides:														
Methyl tetrahydrophthalic anhydride	84	12,200	0.44	6-4	3.0	3-0	3.0	377	0.006	0.005	0.005	0.09	>2 × 10⁹	110
Phthalic anhydride	78	8,500	0.5	....	3.5	....	3.8	390	0.004	....	0.006	....	>5 × 10⁹	95
"Nadic" methyl anhydride	80	11,400	0.40	5-6	3.3	3-4	3.4	443	0.003	0.002	0.004	....	>4.7 × 10¹⁰	110
BF₃MEA	3	5,900	0.4	1-8	3.6	....	4.1	480	0.004	....	0.049	....	13 × 10⁹	120

* Parts hardener per hundred parts resin.

parts employing large resin masses. They often require heating for complete cure. The aliphatic amines produce cured epoxies that are useful to 80 to 100°C and have reasonably good electrical, physical, and chemical-resistant qualities up to 70°C. Table 13 is a listing of typical amine catalysts in order of their reactivity.

TABLE 9   Cast-Resin Data on Blends of Cycloaliphatic Epoxy Resins[4]

Resin ERL–4221,* parts..........	100	75	50	25	0
Resin ERR–4090,* parts..........	0	25	50	75	100
Hardener, hexahydrophthalic (HHPA), phr†................	100	83	65	50	34
Catalyst (BDMA), phr†..........	1	1	1	1	1
Cure, h/°C....................	2/120	2/120	2/120	2/120	2/120
Postcure, h/°C.................	4/160	4/160	4/160	4/160	4/160
Pot life, h/°C..................	>8/25	>8/25	>8/25	>8/25	>8/25
HDT (ASTM D 648), °C........	190	155	100	30	−25
Flexural strength (D 790), lb/in.² ..	14,000	17,000	13,500	5,000	Too soft
Compressive strength (D 695), lb/in.²........................	20,000	23,000	20,900	Too soft	Too soft
Compressive yield (D 695), lb/in.²........................	18,800	17,000	12,300	Too soft	Too soft
Tensile strength (D 638), lb/in.²...	8,000–10,000	10,500	8,000	4,000	500
Tensile elongation (D 638), %.....	2	6	27	70	115
Dielectric constant (D 150), 60 Hz:					
25°C........................	2.8	2.7	2.9	3.7	5.6
50°C........................	3.0	3.1	3.4	4.5	6.0
100°C.......................	2.7	2.8	3.2	4.9	Too high
150°C.......................	2.4	2.6	3.3	4.6	Too high
Dissipation factor:					
25°C........................	0.008	0.009	0.010	0.020	0.090
100°C.......................	0.007	0.008	0.030	0.30	Too high
150°C.......................	0.003	0.010	0.080	0.80	
Volume resistivity (D 257), Ω-cm........................	1×10^{13}	1×10^{12}	1×10^{11}	1×10^{8}	1×10^{6}
Arc resistance (D 495), s..........	>150‡	>150	>150	>150	>150

* Union Carbide Corp.
† Parts per 100 resin.
‡ Systems started to burn at 120 s.   All tests stopped at 150 s.

TABLE 10   Heat-Distortion Temperatures* of Blends of Novolac Epoxy and Epi-Bis Resins[7]

Hardener	D. E. N. 438†	75/25	50/50	25/75	D. E. R. 332‡
TETA..........	§	§	133	126	127
MPDA..........	202	192	180	. . .	165
MDA...........	205	193	190	186	168
5% BF₃MEA....	235	. . .	. . .	204	160
HET...........	225	213	205	203	196

* Heat-distortion temperature, °C (stoichiometric amount of curing agent—except BF₃ MEA—cured 15 h at 180°C).
† Novolac resin, Dow Chemical Co.
‡ Epi-bis resin, Dow Chemical Co.
§ The mixture reacts too quickly to permit proper mixing by hand.

Aromatic amines   Aromatic amines such as metaphenylenediamine (MPDA) and methylene dianiline (MDA) are solid, moderate-temperature epoxy curing agents. They are often used in liquid eutectic mixtures, because of the inconvenience of dissolving the solids in hot resin. Epoxy compositions containing aromatic amines are typically cured for 2 h, at 80°C, followed by 2 h at 150°C. The resistance to

TABLE 11   Comparison of Chemical Resistance for a Cured Epoxy Novolac and an Epi-Bis Resin[7]

Chemical	Weight gain,* %	
	D.E.N. 438†	D.E.R.331‡
Acetone...	1.9	12.4
Ethyl alcohol.................................	1.0	1.5
Ethylene dichloride...........................	2.6	6.5
Distilled water................................	1.6	1.5
Glacial acetic acid...........................	0.3	1.0
30% sulfuric acid.............................	1.9	2.1
3% sulfuric acid..............................	1.6	1.2
10% sodium hydroxide........................	1.4	1.2
1% sodium hydroxide.........................	1.6	1.3
10% ammonium hydroxide....................	1.1	1.3

One-year immersion at 25°C; cured with methylene dianiline; gelled 16 h at 25°C, postcured 4½ h at 166°C; D.E.N. 438 cured additional 3½ h at 204°C.
* Sample size ½ by ½ by 1 in.
† Novolac epoxy resin, Dow Chemical Co.
‡ Epi-bis resin, Dow Chemical Co.

chemicals and electrical properties is good, and the physical properties are among the highest of any epoxy system. The upper use temperature is 135°C. These systems will B-stage overnight; that is, the mixture will partially cure to a solid that is soluble and can be melted and cured on further heating. Molding compounds can be produced this way. The aromatic amines are used in potting composition, laminates, and chemical-resistant castings.

Acids and anhydrides Epoxies cured with acids and anhydrides are generally useful at higher temperature than those cured with other agents. Very large castings are produced with these anhydrides because the exotherms are generally low. Most cycloaliphatic epoxies are cured with them. The chief disadvantages of the anhydrides are the necessity to melt the materials that are room-temperature solids, and the long high-temperature cures. Tertiary amines are often used to shorten the cure cycles.

Phthalic anhydride (PA) is an inexpensive room-temperature solid. The epoxy must be kept hot to prevent the PA from coming out of solution. At the high curing temperature needed to cure this material, the PA sublimes out of the mix, creating a production inconvenience and safety hazard. Very large unfilled castings can be produced that are tough, readily machinable, and possess good electrical properties.

Methyl nadic anhydride (MNA) is used extensively in laminating and casting because it is a liquid at room temperature, and the cured systems have excellent electrical properties and very low weight loss at high temperature. The weight loss of epi-bis epoxies cured with various hardeners is shown in Fig. 3.

Chlorendic or HET anhydride imparts flame resistance to epoxies and produces materials with excellent high-temperature electrical properties. After an initial very low weight loss at high temperature, HET anhydride-cured epoxies abruptly decompose in a relatively short time (Fig. 4).

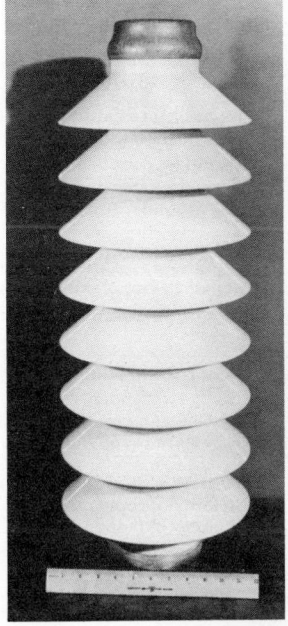

Fig. 1 High voltage bushing cast from a cycloaliphatic epoxy. (*Union Carbide Corporation.*)

TABLE 12   Comparison of a Chlorinated and a Nonchlorinated Epi-Bis Resin[8]

Property	Resin			
	A	B	C	D
Composition (parts by weight):				
Epi-Rez 5161*	100	. . .	100	
Epi-Rez 510†	. . .	100	. . .	100
m-Phenylenediamine	8.8	14.5		
70/30 HHPA/HET	. . .	. . .	64	116
DMP-10	. . .	. . .	0.5	0.5
Gel time at 250°F (min)	20	10	30	30
Physical properties:				
Ultimate flexural strength, lb/in.²	16,750	21,000	13,850	18,950
Izod impact (ft-lb/in. notch)	0.30	0.32	0.27	0.26
Hardness:				
Rockwell M	111	110	106	112
Shore D	80	80	80	80
Heat-distortion temp, °C	122	145	133	126
% water absorption, 24 h	0.06	0.17	0.06	0.09
% weight loss, 168 h at 350°F	5.76‡	1.11	1.45	1.22
% weight loss, 168 h at 300°F	1.01	0.51	0.22	0.27
Flame extinction§, s:				
Vertical	1	Keeps burning	2	Keeps burning
Horizontal	1	50	2	50

Each composition was cured at 250°F for 2 h, followed by a postcure of 3 h at 350°F.
* Chlorinated epi-bis resin, Celanese Chemical Co.
† Epi-bis resin, Celanese Chemical Co.
‡ Severe decomposition.
§ Description of test: sample 6 by ½ by ⅛ in. Test sample (¾ in.) is placed in the hottest part of a Meker-burner flame for 10 s. The flame is removed, and the time (in seconds) is measured for the sample to stop burning.

Pyromellitic dianhydride (PMDA) dissolves readily in hot epoxies. The mixture cures rapidly at 175°C, yielding a brittle, moderate-strength product. The high-temperature properties of this material are exceptional, particularly in adhesives.

Catalysts Tertiary amines such as benzyldimethylamine and tri-dimethyl amino-methyl phenol (Rohm and Haas's DMP-30) are sometimes used as sole curing agent,

TABLE 13   Approximate Order of Activity of Typical Amine Catalysts[9] (Arranged in Decreasing Order of Activity)

Type of amine	Source
DMP-30 (tri-dimethyl-aminomethyl-phenol)	Rohm & Haas Co.
Triethylenetetramine	Carbide & Carbon Chemical Co.
Diethylenetriamine	Carbide & Carbon Chemical Co.
Dimethylaminopropylamine	Carbide & Carbon Chemical Co.
Catalyst EC-1	Borden Co., Chemical Div.
DMP-10 (dimethyl-aminomethyl-phenol)	Rohm & Haas Co.
Diethylaminopropylamine	Carbide & Carbon Chemical Co.
Benzyldimethylamine	Rohm & Haas Co.
Piperidine	Matheson Co.
Diethylamine	Carbide & Carbon Chemical Co.
Dimethylaminopropionitrile	Matheson Co.
Shell catalyst D(2-ethyl hexoic acid salt of tri-dimethyl-aminomethyl phenol)	Shell Chemical Company
m-Phenylenediamine (Shell catalyst CL)	E. I. du Pont de Nemours & Co., National Aniline Division of Allied Chemical & Dye Corp.

but they are generally used to accelerate the curing of anhydride-epoxies mixtures. Piperidine has a high odor level and produces epoxy castings that have outstanding high-temperature stability even though the heat-deflection temperature is only 75°C.

Boron trifluoride complexes can also be classified as latent agents. They have adverse effects on wet electrical characteristics.

Latent curing agents   Latent curing agents such as dicyandiamide are used to produce one-part systems that have long room-temperature shelf lives but cure rapidly at elevated temperature. They are used in potting compounds, laminates, and adhesives. A new class of latent curing agent is being used that is insoluble in the epoxy at room temperature but goes into solution at moderate temperature to effect a rapid cure.

Flexibilizing curing agents   Flexibilizing curing agents are generally long aliphatic chains containing amino groups. Lancast A° and the Versamids† are members of this group. They cure slowly at room temperature to produce tough,

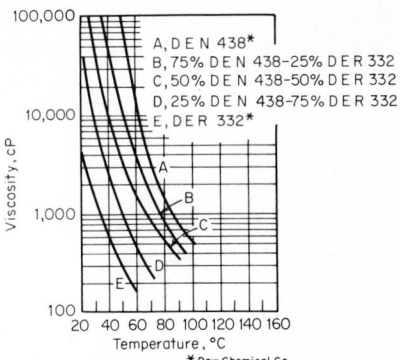

Fig. 2   Viscosity of blends of a novolac epoxy resin with an epi-bis resin as a function of temperature.[7]

water-resistant products. Adhesion is excellent, and the ratio of curing agent to resin is not critical. Emery's trimer acid is an example of another class of flexibilizer. It is viscous, is insoluble at room temperature, and has a long cure cycle. The electrical properties are better than those of other flexibilized systems, and the cured epoxies are usable over a wider temperature range.

Epoxy Modifiers

Reactive diluents   Reactive diluents, such as butyl glycidyl ether, reduce the viscosity of the epoxy mixture as shown in Fig. 5. There is relatively little change

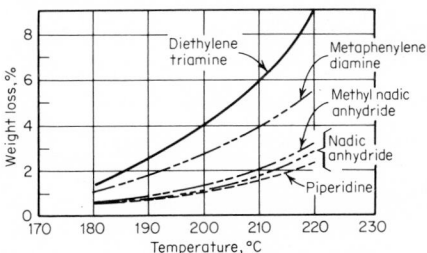

Fig. 3   1,000-h weight loss of an epi-bis resin cured with various hardeners.[10]

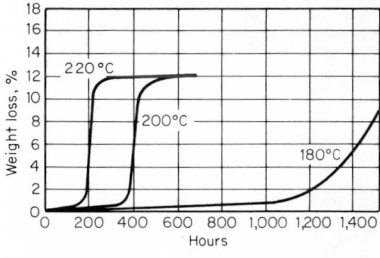

Fig. 4   Weight loss at various temperatures for a HET anhydride-cured epoxy resin.[10]

in cured properties when these diluents are used in concentrations up to 5 parts per hundred resins (phr). Above this concentration, there is a deterioration of electrical, mechanical, and chemical-resistance properties.

Reactive flexibilizers   Reactive flexibilizers, such as polysulfides, urethanes, polybutadienes, and polyesters, are copolymerized with epoxies to increase flexibility, improve adhesion, modify electrical properties, and/or improve resistance to chemicals. Table 2 gives properties of modified epoxies. Figures 34 and 35 show the

° Ciba-Geigy.
† General Mills.

dielectric constant and dissipation factor of a phthalic-anhydride-cured epoxy-poly-butadiene as a function of temperature and frequency. This is covered further in the section on butadiene.

Nonreactive extenders Nonreactive extenders such as thermoplastics, asphaltums, and waxes are used to modify properties and reduce the cost of the compounds. These find application in areas of high-volume usage, such as paving and in the construction area.

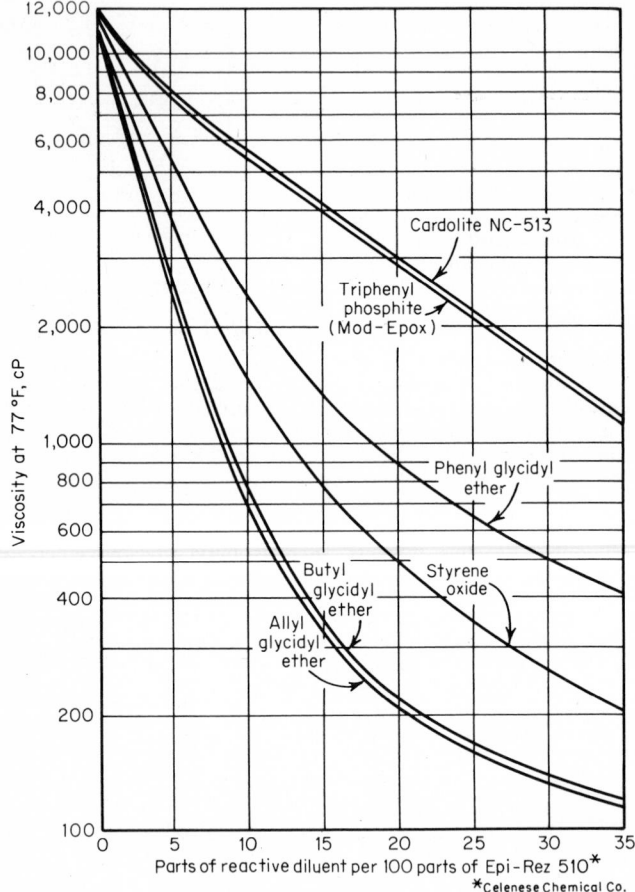

Fig. 5 Effect of various reactive diluents on the viscosity of an epoxy resin.[11]

Fillers

Most commercial epoxy compounds contain one or more fillers in the form of powders, fibers, or flakes. Fillers are used to modify viscosity, increase pot life, reduce exotherm, reduce cure shrinkage, modify density, improve heat resistance, modify thermal conductivity (usually to increase), reduce thermal coefficient of expansion, increase strength, improve machinability, increase hardness and wear resistance, modify electrical properties, increase chemical and solvent resistance, modify friction characteristics, improve thermal shock resistance, improve adhesion, and impart color. Generally, the fillers should be low in cost, reproducible in composition and particle size and shape, easy to disperse in the compound, and low in density, and should not increase the viscosity of the mixture excessively. The filler should stay in suspension or, at worst, be able to be resuspended with a minimum of stirring. Table 14 lists some of the more commonly used fillers and their characteristics. The effects of several fillers on the properties of epoxy and polyester

TABLE 14   Characteristics of Some Commonly Used Fillers

Filler	Characteristics
Silica	Inexpensive, hard, abrasive, lightweight, easily mixed in, good electrical properties, resistant to chemicals and weathering, difficult to machine, high loading possible
Calcium carbonate	Inexpensive, lightweight, improves machinability, high loadings possible with minimum viscosity increase, poor water and acid resistance, easily mixed in, poor electrical properties
Clay	Inexpensive, used as extender; some grades do not mix well
Aluminum oxide	Very hard, abrasive, castings are abrasion-resistant and must be ground to size, increases thermal conductivity, produces translucent compounds; some types used to help keep other fillers in suspension
Calcium silicate	Fibrous material increases impact strength
Glass spheres	Available in a variety of graded sizes, easily mixed in, good packing density, reduces thermal expansion
Hollow spheres	Glass, phenolic, thermoplastic available, reduces density, reduces thermal conductivity and dielectric constant
Fibers	Glass, asbestos, Dacron,* cotton, nylon, increases impact strength, high viscosity
Metal powders and particles	Heavy, settles rapidly, easy to mix in
	Aluminum—powder and pellets, increased thermal conductivity, easy to machine, castings are malleable
	Silver—flakes and powder, silver-coated copper less expensive, high electrical and thermal conductivity
	Copper, bronze, brass—flakes and powder, increased thermal and electrical conductivity, pigment
	Stainless steel—flakes, weather resistance, moisture barrier
Finely divided silica	Thickens compounds to reduce sagging
Pigments	To produce color and opacity, most easily incorporated from dispersions in epoxy resins or plasticizers

* Trademark of E. I. du Pont de Nemours & Company, Inc.

resins are given in Table 15. Figures 6 through 11 show the effect of filler concentration in exotherm, cure shrinkage, coefficient of thermal expansion, thermal conductivity, arc resistance, and viscosity. Figure 12 illustrates surface resistivity vs. relative humidity of epoxy castings containing various fillers.

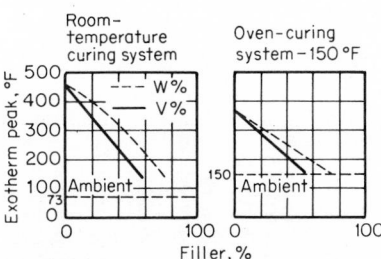

Fig. 6 Effect of filler concentration on exotherm of an epoxy resin.[13]

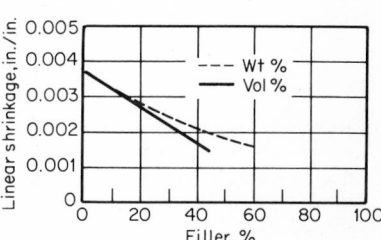

Fig. 7 Effect of filler concentration on shrinkage of an epoxy resin.[13]

TABLE 15   Effects of Several Fillers on Properties of Typical Epoxy and Polyester Resins[12]

Property	Epoxy resins				Polyester resins				
	Unfilled	Calcium carbonate	Mica	Glass	Alumina	Mica	Glass	Calcium carbonate	Unfilled
Coefficient of linear expansion	72	57	43	.....	70.5	65	70	84	97
Thermal conductivity, W/(in.²)(°C)(in.)	0.008	0.014	0.012	0.012	0.0109	0.0074	0.0073	0.0069	0.0075
Water absorption, mg	24	20	22	24	60	50	40	50	50
Specific gravity	1.2	15.8	14.1	15.2	18.7	13.8	15.5	15	12.3
Compressive strength, lb/in.²	15,900	7,540	5,700	.....	26,600	19,600	25,000	18,500	18,000
Tensile strength, lb/in.²	9,700	6,000	5,650	.....	3,700	5,000	5,000	4,250	7,000
Dielectric strength, V/mil	320	370	420	370	270	400	380	280	250–300
Brinell hardness (500-kg load, 10-mm ball)	.....	.....	.....	.....	76	49	49	65	49

Power factor, tan δ,									
MHz:									
1..........	0.029	0.026	0.035	0.026	0.0194	0.0213	0.0165	0.0173	0.018
10.........	0.029	0.026	0.034	0.026	0.0188	0.0204	0.0171	0.0171	0.031 (at 1 kHz)
20.........	0.028	0.026	0.032	0.026	0.0183	0.0197	0.0170	0.0170	0.045 (at 50 Hz)
50.........	0.026	0.025	0.030	0.025	0.0176	0.0185	0.0165	0.0167	0.016
100........	0.020	0.023	0.026	0.023	0.0177	0.0176	0.0160	0.0164	
Dielectric constant,									
MHz:									
1..........	3.9	4.45	4.05	4.05	5.05	3.62	4.33	4.75	3.13
10.........	3.75	4.2	3.9	3.9	4.92	3.53	4.4	4.63	3.18 (at 1 kHz)
20.........	3.7	4.2	3.85	3.85	4.9	3.5	4.35	4.60	3.91 (at 50 Hz)
50.........	3.65	4.15	3.75	3.8	4.85	3.47	4.29	4.55	4.3
100........	3.6	4.03	3.7	3.75	4.8	3.44	4.25	4.51	
Remarks.......	......	Best general characteristics	Best for high dielectric strengths	......	Highest thermal conductivity	Lowest coefficient of linear expansion			

* ×10⁻⁶ per °C; measured at 60–80°C for epoxy resins and at 60°C for polyester resins.

Applications

Epoxy resins have found wide use in adhesives, potting, art reproductions, construction, jigs and fixtures, laminates, etc.   Undersea pipes and tanks of epoxy composites are joined with epoxy cement;[15] large contoured jet nozzles have been cast.[16]

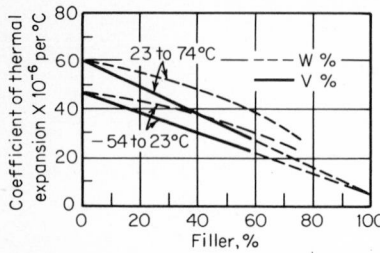

Fig. 8 Effect of filler concentration on coefficient of thermal expansion of an epoxy resin.[13]

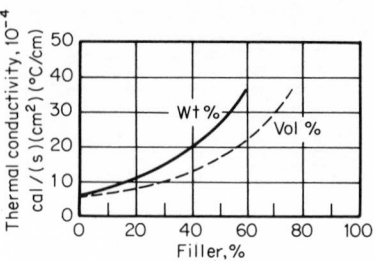

Fig. 9 Effect of filler concentration on thermal conductivity of an epoxy resin.[13]

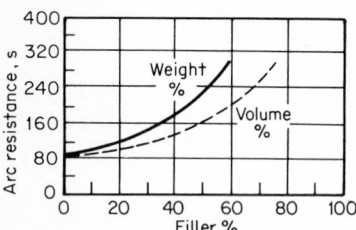

Fig. 10 Effect of filler concentration on arc resistance of an epoxy resin.[13]

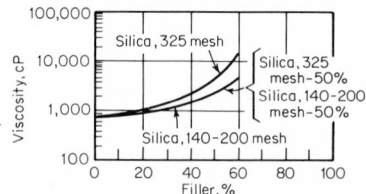

Fig. 11 Effect of fillers on viscosity of an epoxy resin.[13]

Syntactic foam forms the hull of deep-submergence vessels.   Binder for terrazzo flooring, heavy-duty floor retopping, missile casings, chemical-resistant paint, and laminates for the process industry are only a few of the increasing applications of the versatile epoxies.

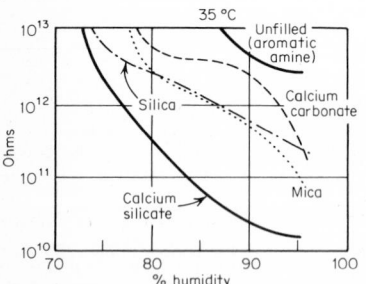

Fig. 12 Surface resistivity of an epoxy resin containing various fillers as a function of percent humidity.[14]

POLYESTERS

Polyesters are becoming increasingly popular because of their low cost, ease of use, and versatility. They find applications in laminates, castings, art objects, industrial construction, insulation, embedments, molding compounds, coatings, and adhesives.

Other advantages of polyesters are their high impact resistance, good weathering resistance, transparency, and good surface effects. Disadvantages include poor adhesion, high cure shrinkage, and inhibition of cure by air and some fillers.

There are three required components in a polyester formulation: (1) an unsaturated polyester, prepared from organic acids and polyols; (2) a monomer; and (3) a curing system. There are usually also fillers, reinforcement, and other modifiers.

Basic Types of Polyesters

Saturated and unsaturated organic acids such as maleic and phthalic acid are reacted with hydroxyl-containing chemicals such as ethylene glycol to produce unsaturated polyesters that are very thick or solid at room temperature. The general-purpose polyester, when blended with a monomer and cured, produces rigid, rapidly curing, transparent castings whose properties are shown in Table 16. A second type produces a flexible material with properties given in Table 17. This

TABLE 16   Properties of a Typical Rigid Polyester*[17]

Property	Value
Flexural properties:	
Ultimate strength, lb/in.^{2}	17,500
Elastic modulus, lb/in.^{2}	620,000
Tensile properties:	
Ultimate strength, lb/in.^{2}	9,000
Elastic modulus, lb/in.^{2}	590,000
Ultimate elongation, %	1.7
Compressive properties:	
Ultimate strength, lb/in.^{2}	24,100
Elastic modulus, lb/in.^{2}	660,000
Impact strength, Charpy unnotched, ft-lb/in.	2.6
Hardness:	
Rockwell (M scale)	115
Barcol	45
Heat-distortion temp, °C, 264 lb/in.^{2}, 2°C/min	78
Polymerization shrinkage, vol. %	7.0
Specific gravity, 25°C	1.235
Refractive index $N_1{}^{25°C}$	1.5564
Water absorption, % wt after 7 days immersion at 25°C	0.4
Dielectric constant:	
60 Hz	3.4
10^3 Hz	3.3
10^6 Hz	3.3
Power factor:	
60 Hz	0.009
10^3 Hz	0.004
10^6 Hz	0.014
Loss factor:	
60 Hz	0.029
10^3 Hz	0.013
10^6 Hz	0.048
Dielectric strength, V/mil, 100-mil cast specimen, 25°C	450

* Paraplex P-43, Rohm and Haas.

type of polyester is tougher and slower-curing, and it produces lower exotherms and less cure shrinkage. Flexible polyesters absorb more water and are more easily scratched, but show more abrasion resistance than the rigid. The two types can be blended to produce intermediate properties as shown in Figs. 13 through 16.

Polyesters that are inherently fire-resistant are produced from chlorine containing anhydrides such as chlorendic anhydride. Their properties are similar to the properties of other polyesters except for a higher density.

A group of polyesters are available that are used where resistance to chemicals is important. Some of these incorporate bisphenol A in their structures; others contain isophthalic anhydride. They also have improved impact resistance, flex strength, corrosion resistance, and crack resistance over the general-purpose polyesters.[19,20]

Monomers

Styrene is the most widely used monomer because of its low cost, high solvency, low viscosity, reactivity, and the desirable cured properties it imparts to the polyester. Abrasion and wear properties increase in most polyester laminates when the

TABLE 17   Properties of a Typical Flexible Polyester*[18]

Property	Value
Flexural property:	
Elastic modulus, $lb/in.^2$	6,400
Tensile properties:	
Ultimate strength, $lb/in.^2$	1,600
Ultimate elongation, %	220
Hardness, Shore A	80–85
Polymerization shrinkage, vol. %	9.0
Specific gravity, 25°C	1.122
Refractive index, $N_1^{25°C}$	1.5378
Water absorption, % wt after 7 days immersion at 25°C	0.4
Dielectric constant:	
60 Hz	4.2
10^3 Hz	4.2
10^6 Hz	4.0
Power factor:	
60 Hz	0.005
10^3 Hz	0.011
10^6 Hz	0.052
Loss factor:	
60 Hz	0.021
10^3 Hz	0.046
10^6 Hz	0.208
Dielectric strength, V/mil, 100-mil cast specimen, 25°C	370

* Paraplex P-13, Rohm and Haas.

styrene content is raised from 21 to 31 percent.   Figure 17 shows cure time and gel time vs. styrene content.   The effect of styrene content on mechanical properties is shown in Table 18.   The effect on wear and hardness characteristics is given in Figs. 18 and 19.   The choice of monomer is predicated on both curing characteristics and properties of the resulting polyester.   Figure 20 presents exotherm curves for

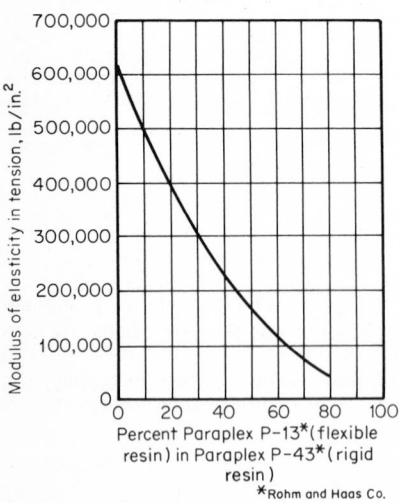

Fig. 13   Tensile modulus as a function of the composition of blends of rigid and flexible polyesters.[17]

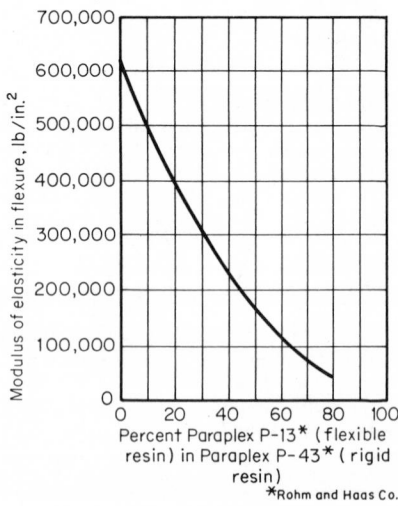

Fig. 14   Modulus of elasticity in flexure as a function of the composition of blends of rigid and flexible polyesters.[17]

the various monomers. The important properties of the monomers are listed below:
ALPHA METHYL STYRENE: used with styrene at concentration of 2 to 4 percent reduces exotherm without affecting cure time excessively; used in large castings; higher concentrations decrease hardness and tensile strength.

METHYL METHACRYLATE: used in combination with styrene, increases durability, color retention, resistance to fiber erosion; improves weathering resistance; proper

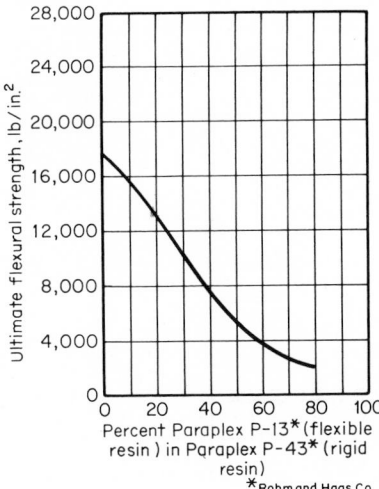

Fig. 15 Flexural strength as a function of the composition of blends of rigid and flexible polyesters.[17]

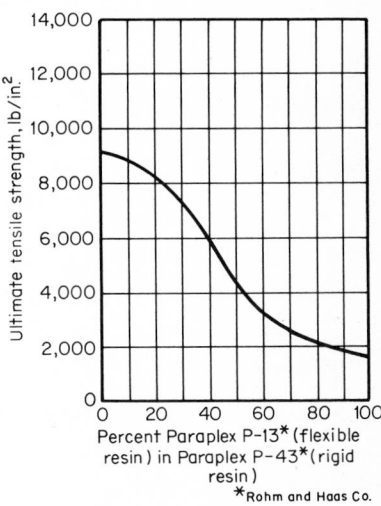

Fig. 16 Tensile strength as a function of the composition of blends of rigid and flexible polyesters.[17]

concentration will change index of refraction of the polyester to match that of glass and thus produce a nearly transparent laminate; lower boiling point, higher cost, and greater cure shrinkage than styrene.

VINYL TOLUENE: compared with styrene—shorter cure time, higher boiling point, lower cure shrinkage, higher viscosity.

DIALLYL PHTHALATE: low volatility, gives prepregs long shelf life, lower cure shrinkage, lower exotherms, reduced odor, increased flexural strength and Izod impact resistance, higher cost.

TRIALLYL CYANURATE: greatly improved high-temperature stability, greater cure shrinkage, more brittle.

DIVINYL BENZENE: used with styrene, castings are harder, more heat-resistant, brittle.

CHLOROSTYRENE: increased chemical resistance.

Curing Systems

The curing system does not greatly change the properties of the final polyester product, but the storage life, mold or press time, postcure if any, and maximum temperature developed are controlled. The reactivity of the polyester used, and the size and shape of the product affect the choice of systems. This

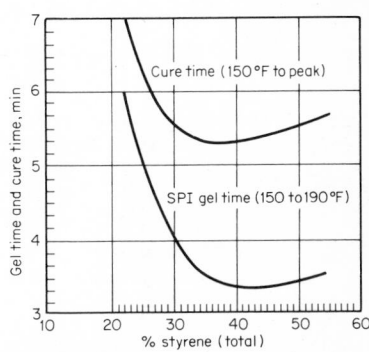

Fig. 17 Gel time and cure time as a function of total styrene content of a polyester resin.[21]

TABLE 18 Effect of Styrene Concentration on Propylene Glycol-Maleate-Phthalate Polyester Resins

Mole ratio maleate/ phthalate	Styrene, %	Peak exotherm, °F	Flexural strength, lb/in.²	Flexural modulus, lb/in.² × 10⁵	Tensile strength, lb/in.²	Tensile modulus, lb/in.² × 10⁵	HDT °F	Elongation, %	Water absorption, % after 24 h at 77°F
40/60.........	20	323	20,700	6.6	8,200	8.4	147	1.2	0.17
40/60.........	30	347	16,100	5.9	7,900	7.8	158	1.31	0.21
40/60.........	40	349	14,100	5.9	9,100	6.8	172	1.73	0.17
40/60.........	50	340	15,700	5.3	9,500	7.5	176	1.85	0.17
50/50.........	20	340	20,000	6.5	8,100	8.5	158	1.3	0.19
50/50.........	30	380	19,000	5.8	8,300	8.1	194	1.32	0.23
50/50.........	40	392	17,700	6.3	9,200	7.9	201	1.7	0.21
50/50.........	50	396	15,700	5.2	8,000	7.2	199	1.7	0.20
60/40.........	20	356	19,500	6.5	8,000	8.7	169	....	0.23
60/40.........	30	400	17,200	5.7	8,600	7.9	219	1.38	0.25
60/40.........	40	407	17,800	6.0	7,200	7.7	226	1.46	0.25
60/40.........	50	404	17,700	5.1	6,600	7.0	225	1.23	0.28

versatility of cure schedule from very rapid room-temperature cure to extended temperature cure is one of polyester's most attractive attributes. The catalyst is generally a peroxide such as benzoyl peroxide or methyl ethyl ketone peroxide. Promoters, used singly or in combination, and retarders complete the cure system.

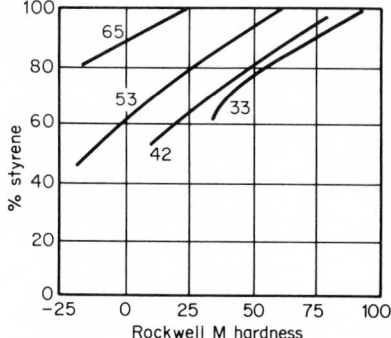

Fig. 18 Styrene content vs. Rockwell M hardness of triethylene maleate polyester castings.[23]

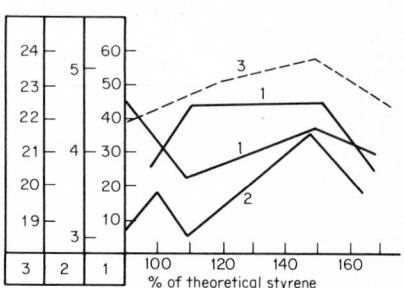

Fig. 19 Wear characteristics of a typical polyester resin as a function of styrene content.[24] Curve 1, Taber abrasion, milligrams loss per 1,000 cycles. Curve 2, diamond threshold tear. Curve 3, Taber scratch factor.

Fillers

The fillers used with polyesters are much the same as those used with epoxies. They are used to reduce shrinkage, modify processing properties, increase hardness, increase thermal conductivity, decrease thermal expansion, and increase wear resistance and chemical resistance. Greater care must be used in picking the correct filler with polyester, as some fillers will prematurely gel the mix, while others will reduce the curing rate or even inhibit the cure completely. Calcium carbonate is the most widely used filler for polyesters because of its low cost and compatibility and the high loadings possible. The effect of varying concentrations and types of calcium carbonate filler on the viscosity, chemical resistance, and flexural, impact, and compressive strength is shown in Tables 19 through 23.

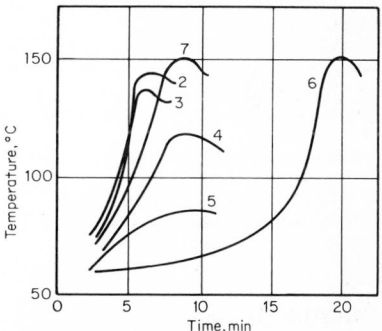

Applications of Polyesters

Polyesters are ideal for do-it-yourselfers, as well as in large industrial operations, because they are easy to formulate, mix, and cure, with a minimum of elaborate equipment. Polyesters are most widely used in glass-cloth laminates and with chopped glass fibers in molding compositions because they

Fig. 20 Exotherm curves for polyesters containing different monomers.[25] (1) Styrene. (2) Vinyl acetate. (3) Acrylonitrile. (4) Methyl methacrylate. (5) Diallyl phthalate. (6) Triallyl cyanurate.

wet the glass well, bonding to it and producing strong, attractive, long-lasting composites. For accurate reproduction of mold details, an additive has been developed that produces zero cure shrinkage in polyester-glass bulk compounds.[27] Polyester–fiber glass laminates are used in boats, automobile bodies, translucent building panels, bathtub and shower units, chemical-storage tanks and piping, chemical-resistant tank cars, fishing rods, and snowmobiles. Bowling balls, synthetic marble, and furniture parts are cast from polyesters. Water-extended polyester (WEP) has been widely used to make plaques, lamp bases, and furniture. In this process 50 to 80 percent water is incorporated in the mix before casting. This results in a lower cost, lighter weight, easier mixing, more controllable curing, and the ability to fabricate finished castings with conventional woodworking tools.[28, 29]

Hobbyists and artists use polyesters extensively to embed objects and produce works of art such as those shown in Figs. 21 and 22.

TABLE 19   Effect of Calcium Carbonate on the Viscosity[a] of Polyester Resins[26]

Polyester	% Surfex MM[b, c]						
	0	10	20	30	40	50	60
Selectron 5003:[d]							
Original viscosity....	1,000	1,100	1,300	1,600	2,400	3,800	10,000
24 h viscosity.......	1,100	1,600	2,200	2,700	4,600	10,000	20,000
48 h viscosity.......	1,300	1,900	2,700	3,800	7,100	13,300	Gelled
Permafil 2864:[e]							
Original viscosity....	300	400	400	500	600	1,000	2,600
24 h viscosity.......	300	400	400	600	800	1,500	5,900
48 h viscosity.......	400	500	500	600	1,000	2,200	10,200
Vibrin 114:[f]							
Original viscosity....	1,000	1,100	1,400	1,700	2,500	5,000	14,400
24 h viscosity.......	1,000	1,200	1,400	1,900	3,000	7,000	Gelled
48 h viscosity.......	1,200	1,400	1,600	2,000	3,700	Gelled	
Paraplex P-43:[g]							
Original viscosity....	3,900	4,600	5,100	5,800	7,600	15,000	36,000
24 h viscosity.......	5,000	6,000	7,500	9,100	16,000	177,000	Gelled
48 h viscosity.......	6,000	7,000	9,400	11,000	26,000	Gelled	
Laminac 4115:[h]							
Original viscosity....	1,000	1,000	1,200	1,500	2,200	3,600	12,000
24 h viscosity.......	1,000	1,200	1,700	2,400	5,400	13,000	Gelled
48 h viscosity.......	1,100	1,200	3,000	3,800	11,000	Gelled	

[a] Brookfield, 76°F, No. 3 spindle, 5 r/min, measured in centipoises.
[b] Diamond Shamrock Chemical Co.
[c] By weight on the resin.
[d] P. P. G. Industries, Inc.
[e] General Electric Corp.
[f] Uniroyal Inc., Chemical Div.
[g] Rohm and Haas Co.
[h] American Cyanamid Co.

POLYURETHANES

Polyurethanes can be formulated to produce a range of materials from elastomers as soft as a Shore A of 5 to tough solids with a Shore D of 90. Figure 23 shows the range of hardness used in various applications. Polyurethanes can have extremely high abrasion resistance and tear strength, excellent shock absorption, resistance to a broad spectrum of solvents, good electrical properties, and excellent resistance to oxygen aging. They have limited life in high-humidity and high-temperature applications. Figure 24 presents the effect of humidity aging on the hardness of polyurethanes derived from various polyols. The liquid components must be kept dry during storage and fabrication.

The polyurethanes are reaction products of an isocyanate, a polyol, and a curing agent. Because of the hazards involved in handling free isocyanate, prepolymers of the isocyanate and the polyol are generally used in the casting.

Types of Polyurethanes

The polyurethanes are classed according to the polyol used. The polyols are generally polyethers and polyesters. Polyester-based urethanes are more resistant to hydrolysis, and have higher resilience, good energy-absorption characteristics, good hysteresis characteristics, and good all-around chemical resistance. A polyester-based material will generally be stiffer and will have higher compression and tensile moduli, higher tear strength and cut resistance, higher operating temperature, lower compression set, optimum abrasion resistance, and good oil and fuel resistance.[32]

TABLE 20 Effect of Calcium Carbonates on Chemical Resistance of Polyesters[26]

Chemical	Control (unfilled), %	% wt changes		
		40% Surfex*	40% Suspenso*	40% Multiflex* MM
Water absorption:				
7 days....................	0.8	0.8	0.7	0.3
14 days.................	1.1	1.2	1.1	0.5
10% calcium chloride:				
7 days....................	0.4	0.4	0.4	0.3
14 days.................	0.6	0.6	0.6	0.4
10% nitric acid:				
7 days....................	0.5	−0.7	−2.5	−1.4
14 days.................	0.7	−0.9	−3.4	−1.6
5% acetic acid:				
7 days....................	0.5	0.1	−1.3	−1.9
14 days.................	0.8	0.2	−1.5	−2.2
30% sulfuric acid:				
7 days....................	0.3	0.3	0.2	0.2
14 days.................	0.5	0.7	0.5	0.4
3% sulfuric acid:				
7 days....................	0.4	0.7	0.3	−0.3
14 days.................	0.6	1.3	0.6	−0.1
10% hydrochloric acid:				
7 days....................	0.2	−0.4	−3.2	−3.8
14 days.................	0.3	−0.7	−4.8	−5.2
10% sodium hydroxide:				
7 days....................	0.5	−4.4	−2.6	−2.0
14 days.................	0.9	−2.1	3.6	−2.2
1% sodium hydroxide:				
7 days....................	0.9	0.1	0.4	−0.1
14 days.................	1.6	0.5	0.9	0.5
10% ammonium hydroxide:				
7 days....................	3.8	2.9	2.3	2.1
14 days.................	5.3	2.4	2.7	2.1

* Diamond Shamrock Chemical Co.

Table 24 summarizes the characteristics of the two materials. Table 25 lists properties of a family of polyester-based polyurethanes. Resistance to temperature exposure is given in Fig. 25.

Castor oil–based potting compounds are convenient to use. They have improved hydrolytic stability and poorer physical properties than polyether- or polyester-based urethanes. Polybutadiene–based polyurethanes have outstanding resistance to humidity, but they have higher viscosities and only moderate physical properties.

Curing Agents

The choice of curing agent influences the curing characteristics and final properties. Diamines are the best general-purpose curing agents; the highest physical properties

TABLE 21   Effect of Calcium Carbonate Loading on Flexural Strength, lb/in.^{2}, of Glass-Filled Polyesters[26]

Polyester	% filler*				
	0%	30%	40%	50%	60%
Selectron 5003,† Surfex MM:‡					
Flatwise, dry strength.........	26,000	27,000	27,000	24,000	27,000
Flatwise, wet strength.........	24,000	26,000	23,000	22,000	23,000
Edgewise, dry strength........	24,000	25,000	25,000	23,000	25,000
Edgewise, wet strength........	22,000	26,000	23,000	23,000	24,000
Vibrin 114,§ Surfex MM:					
Flatwise, dry strength.........	24,000	25,000	22,000	25,000	24,000
Flatwise, wet strength.........	15,000	18,000	15,000	16,000	16,000
Edgewise, dry strength........	22,000	26,000	20,000	23,000	24,000
Edgewise, wet strength........	14,000	19,000	16,000	17,000	18,000
Laminac 4115,¶ Surfex MM:					
Flatwise, dry strength.........	24,000	24,000	21,000	27,000	22,000
Flatwise, wet strength.........	21,000	21,000	18,000	21,000	20,000
Edgewise, dry strength........	26,000	24,000	20,000	24,000	23,000
Edgewise, wet strength........	19,000	19,000	19,000	21,000	22,000

Glass content: 20% by volume
  * By weight on the resin.
  † Diamond Shamrock Chemical Co.
  ‡ P. P. G. Industries, Inc.
  § Uniroyal, Inc., Chemical Div.
  ¶ American Cyanamid Co.

are produced with MOCA [4,4′ methylenebis (2-chloroaniline)]. As most useful diamines are room-temperature solids, they must be melted and mixed hot, reducing pot life and complicating the casting procedure. The other major class of curing agents, polyols, are more convenient to use, but the products have lower physical properties. The properties of these curing agents are summarized in Table 26.

TABLE 22   Effect of Calcium Carbonate Loading on Impact Strength, ft-lb/in. of notch, of Glass-Filled Polyesters[26]

Polyester	% filler*				
	0%	30%	40%	50%	60%
Selectron† 5003, Surfex:‡					
Edgewise, dry strength..................	21.1	16.2	21.0	20.9	15.5
Edgewise, wet strength..................	12.4	15.8	13.8	11.8	15.6
Vibrin 114,§ Surfex MM:					
Flatwise, dry strength...................	17.8	21.8	19.2	21.2	22.9
Flatwise, wet strength...................	16.4	18.1	15.8	19.9	19.3
Edgewise, dry strength..................	16.1	20.4	17.4	18.1	18.1
Edgewise, wet strength..................	13.1	17.5	13.7	16.1	14.2
Laminac 4115,¶ Surfex MM:					
Flatwise, dry strength...................	23.4	20.7	22.5	21.6	25.7
Flatwise, wet strength...................	20.6	21.2	18.6	20.0	20.8
Edgewise, dry strength..................	17.0	15.5	13.2	18.5	18.6
Edgewise, wet strength.................	14.0	15.4	11.7	15.0	14.8

Glass content: 20% by volume
  * By weight of resin.
  † P. P. G. Industries, Inc.
  ‡ Diamond Shamrock Chemical Co.
  § Uniroyal, Inc., Chemical Div.
  ¶ American Cyanamid Co.

TABLE 23   Compressive Strength of Polyester Resins Filled with Calcium Carbonate and Glass Fibers[26]

Polyester	% filler*				
	0%	30%	40%	50%	60%
Selectron 5003,† Surfex MM:‡					
Flatwise, dry strength.........	46,000	43,000	43,000	46,000	44,000
Flatwise, wet strength.........	36,000	37,000	34,000	42,000	35,000
Edgewise, dry strength........	27,000	21,000	23,000	27,000	24,000
Edgewise, wet strength........	20,000	22,000	20,000	25,000	19,000
Vibrin 114,§ Surfex MM:					
Flatwise, dry strength.........	25,000	28,000	25,000	27,000	32,000
Flatwise, wet strength.........	20,000	23,000	20,000	24,000	23,000
Edgewise, dry strength........	19,000	14,000	16,000	19,000	21,000
Edgewise, wet strength........	12,000	11,000	9,000	12,000	11,000
Laminac 4115,¶ Surfex MM:					
Flatwise, dry strength.........	35,000	32,000	25,000	24,000	37,000
Flatwise, wet strength.........	28,000	31,000	23,000	22,000	30,000
Edgewise, dry strength........	22,000	12,000	16,000	17,000	24,000
Edgewise, wet strength........	16,000	15,000	12,000	13,000	14,000

Glass content: 20% by volume
* By weight on resin.
† P. P. G. Industries, Inc.
‡ Diamond Shamrock Chemical Co.
§ Uniroyal, Inc., Chemical Div.
¶ American Cyanamid Co.

Fig. 21   "The Group"—polyester resin cast in an RTV silicone mold attached to an acrylic sheet.   (*Courtesy of Thelma R. Newman.*)

Fig. 22 "The Crowd"—polyester resin cast in an RTV silicone mold attached to an acrylic sheet. (*Courtesy of Thelma R. Newman.*)

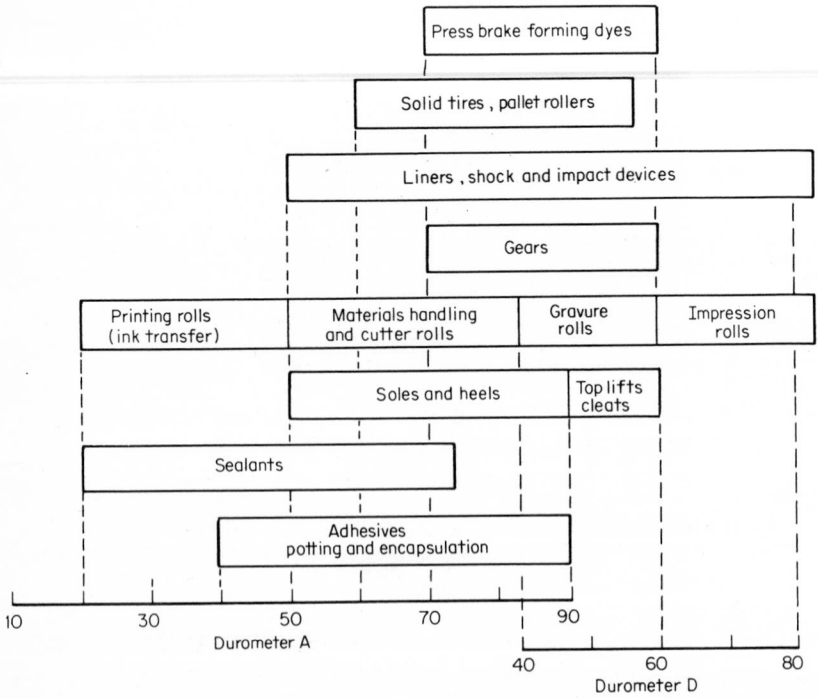

Fig. 23 Applications of urethane elastomers and their relation to elastomer hardness.[30]

Applications

Most polyurethanes make use of the good abrasion resistance and low coefficient of friction.   Polyurethanes are used in roller coatings and press pads, where they last ten times longer than the rubber they replace.   Other applications are gaskets, encapsulation, casting molds, timing belts, bumpers, wear strips, liners of equipment-handling abrasive materials, heels, and soles.   The outstanding wear resistance is shown in the truck wheels pictured in Fig. 26.   Figure 27 is a picture of a polyurethane impeller compared with a steel impeller that has had the same usage.

Polyurethanes are also used in flexible adhesives and wear-resistant coatings for bowling balls and pins, gymnasium floors, golf balls, etc.

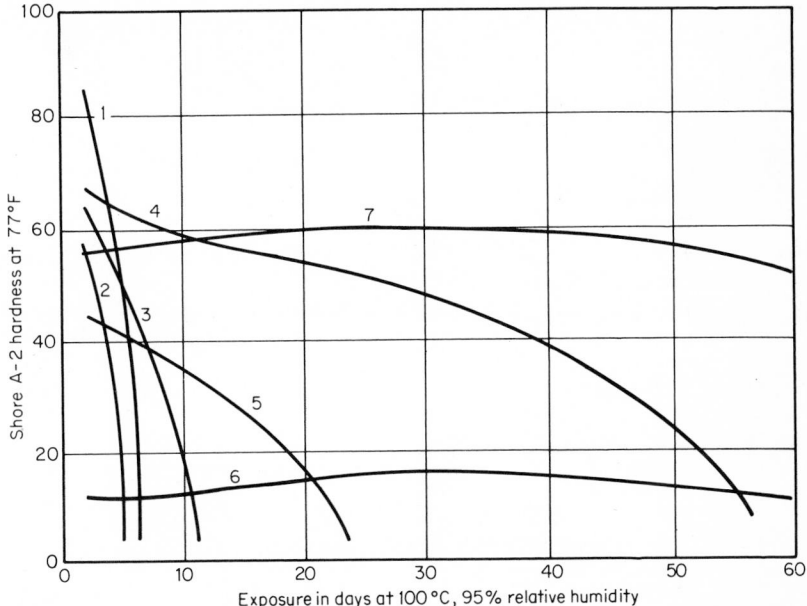

Fig. 24   Hardness of polyurethanes as a function of humidity exposure.[31]   (1) Polyesterurethane.   (2) Polyesterurethane.   (3) Polycaprolactone.   (4) Polyetherurethane.   (5) Polyetherurethane.   (6) Polybutadiene polyol urethane.   (7) Polybutadiene polyol urethane.

The largest use of polyurethanes is in rigid and flexible foams.   Rigid foams blown with halocarbons have the lowest thermal conductivity of any commercially available insulation.   They are used in refrigerators, picnic boxes, and building construction. Flexible foam is used in furniture, clothing, packaging, and shock and vibration mounts.   Foams are covered in detail in a separate chapter of this handbook.

Hydrolytic Stability

Water affects polyurethane in two ways: temporary plasticization and permanent degradation.   Moisture plasticization results in a slight reduction in hardness and tensile strength; the original properties are restored when the absorbed water is removed.

Hydrolytic (chemical) degradation causes a permanent reduction in the cured physical and electrical properties.   The functional groups present in the chain are hydrolyzed, resulting in both chain breaking and loss of cross linking.   Disastrous failures of polyurethane potting compounds in humid atmospheres have been reported.

TABLE 24 Typical Properties of Urethane Rubbers (High-Performance Castables)[32]

Properties	Ester-based			Ether-based		
Hardness ranges:						
Shore A	20-50	60-75	80-90	50-80	75-80	80-90
Shore D	......	......	......	......	......	......
Tensile strength, lb/in.²	80-3,500	4,500-6,500	6,500	4,800-10,000	8,000-9,500	4,400-5,000
Elongation, %	525-650	460-615	550-660	250-450	270-300	450-575
100% modulus, lb/in.²	25-150	265-410	530-1,000	1,100-5,500	4,100-5,000	700-1,050
300% modulus, lb/in.²	45-300	560-795	1,000-1,900	2,500-4,700	......	1,600-1,700
Tear strength, pli*						
Die C	30-128	168-370	420-650	450-2,000	......	550-675
Split	10-34	30-200	250-400	310-1,100	115-125	85-120
Compression set, % (22 h at 158°F)	6-1.7	1.4-25	25-38	0.8-33	......	28-35
Bashore rebound, % (78°F)	19-21	8-24	27-32	23-35	48-50	50-55

* Pounds per lineal inch.

TABLE 25 Properties of a Family of Polyether-Based Polyurethane Resins[33]

	Adiprene Product No.								
	L-100	L-100	L-42	L-83	L-100	L-167	L-200	L-213	L-315
Compound:									
Adiprene*	100	100	100	100	100	100	100	100	100
MOCA†			8.8	10.3	12.5	19.5	23.2	25	26
1,4-Butanediol	3.2								
Trimethylolpropane	0.9	9.6							
Methylene dianiline		50							
Dioctyl phthalate									
Mix temp, °F	212	150	212	212	212	158	185	175	170
Cure, h at °F	16/212	1/212	3/212	3/212	3/212	2/212	2/100	2/212	2/212
Physical properties:									
Hardness, Durometer A	60	75	80	83	90	95	58	75	75
Hardness, Durometer D					40	48			
100% modulus, lb/in.²	275	550	400	700	1,100	1,800	3,000	3,800	4,300
300% modulus, lb/in.²	500	800	625	1,200	2,100	3,400	7,800		
Tensile strength, lb/in.²	2,100	2,850	3,000	4,400	4,500	5,000	8,200	7,500	11,000
Elongation at break, %	500	625	800	575	450	400	315	250	270
Tear strength, Graves, lb/in.	150			400	500	600			725
Tear strength, split, lb/in.	12	70	70	85	75	150	135	145	115
Abrasion resistance, NBS index, %	25	600	110	160	175	285	370	350	435
Compression set, method B, 22 h at 158°F	9	27	45	35	27	45	40	50	
Compression set, method A, 22 h at 158°F, 1350 lb/in.²									
Resilience—Yerzley, %	72	70	70		9	9			9
Resilience—Bashore, %	60	60	60		45	39		50	48
Specific gravity	1.06	1.06	1.08	1.05	1.10	1.14	1.14	1.19	1.19

* Du Pont Co.
† Du Pont's registered trademark for its curing agent 4,4'-methylene-bis-(2-chloroaniline).

In an attempt to accelerate evaluation of these resin systems a number of aging tests have been developed. These include (1) exposure to an environment of 95% RH at 100°C, (2) exposure to an environment of 15 lb/in.² steam at 110°C, and (3) continuous immersion in boiling water. A minimum Shore A hardness of 30 after a 28-day exposure to 95% RH, 100°C is a widely accepted criterion for satisfactory performance. Figure 24 shows the change in hardness of polyurethanes derived from various polyols when they are exposed to 95% RH at 100°C. Ricinoleate–based polyurethane resins have stabilities between those of the polyether polyurethanes and polybutadiene polyol polyurethanes.

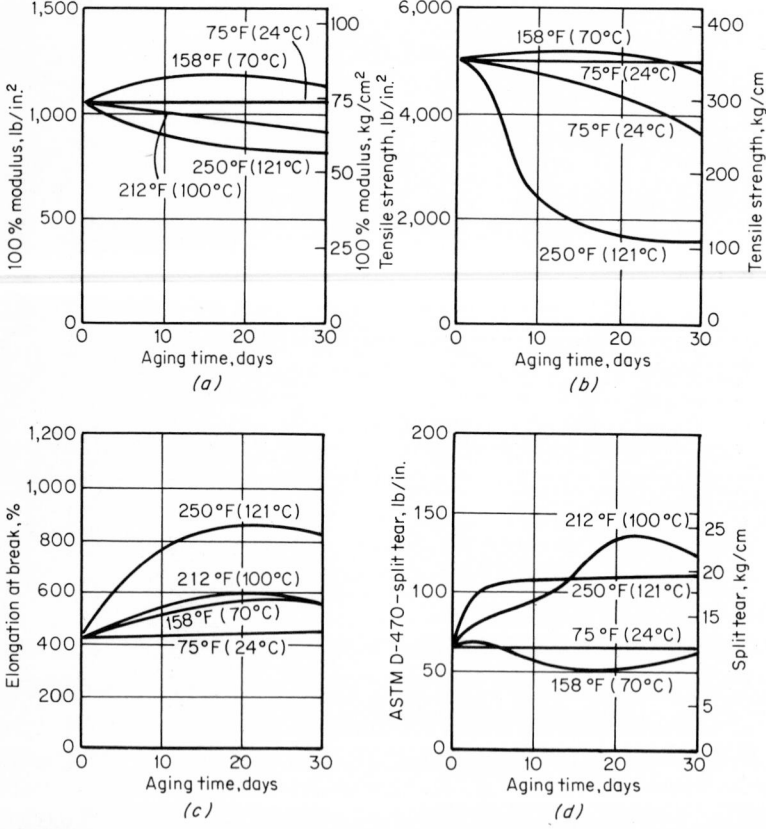

Fig. 25 Physical properties as a function of time at elevated temperature of a polyether polyurethane.[33] (a) 100 percent modulus vs. aging time at elevated temperature. (b) Tensile strength vs. aging time. (c) Elongation at break vs. aging time. (d) Split tear vs. aging time.

TABLE 26 Properties of Polyurethane Resins Cured with Diamines and Polyols[33]

Property	Diamine cures	Polyol cures
Resilience.....................	Medium	Medium to high
Reactivity....................	Medium to high	Low to medium
Modulus.....................	Medium to high	Low
Tensile strength..............	Medium to high	Low to medium
Ultimate elongation...........	Medium to high	High
Tear strength................	High	Low to medium
Abrasion resistance:		
Sliding....................	High to very high	Low to medium
Impact....................	Medium to high	High to very high
Compression set..............	Medium	Low
Hardness....................	High	Low to medium

SILICONES

Silicone synthetic polymers are partly organic and partly inorganic in nature. They have a combination of properties that make them the unique choice in many applications. Silicone elastomers remain flexible as low as −80°C and stable at temperatures as high as 300°C for extended periods of time. They are virtually unchanged after long weathering; their excellent electrical properties remain constant with temperature and frequency; they are strongly hydrophobic, but they readily transmit water vapor and other gases; and almost all castable materials readily release from silicone surfaces without pretreatment. Silicone rubber's use has been limited by its high cost, low strength, high thermal expansion, cure inhibition of some compounds, and occasionally poor cure of some types of silicones.

The principal companies supplying silicone resins are General Electric, Dow Corning Corporation, Stauffer Chemical Co., and Union Carbide Corp.

Types of Silicone and Their Applications

RTV silicone elastomers The RTV (room-temperature vulcanizing) silicones are available in a wide range of viscosities. Their cure time and temperature are readily adjustable by varying the type and amount of catalysts used. Most of the newer materials cure well in deep sections. The electrical properties are excellent and the physical properties are adequate for electrical encapsulating.

Fig. 26 Truck wheel cast from a polyurethane elastomer. (*E. I. du Pont de Nemours & Company, Inc.*)

Properties of a series of RTVs are given in Tables 27 and 28. Chemical resistance is shown in Table 29. The low modulus over the −65 to 200°C range places a minimum of stress on the electrical components during thermal cycling and shock. These RTVs are also used as molds for casting art objects, furniture parts, and low-melting metals. They are also used for formed-in-place gaskets.

Flexible resins Flexible resins (or vinyl silicones) are two-part systems used in much the same applications as the silicone elastomers. The encapsulating grades

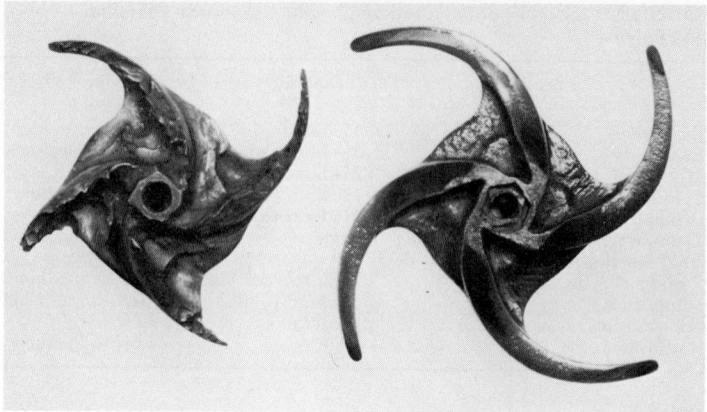

Fig. 27 Cast polyurethane impeller (*right*) compared with a steel impeller with equivalent service. (*E. I. du Pont de Nemours & Company, Inc.*)

TABLE 27   Viscosities and Consistencies of a Family of RTV Silicone Rubbers[34]

Material	Color	Viscosity, poises	Consistency	Solids, %
RTV-11*	White	120†	Easily pourable	100
RTV-20	Pink	300†	Pourable	100
RTV-40	White	450†	Pourable	100
RTV-60	Red	550†	Pourable	100
RTV-88	Red	10,000‡	Spreadable thixotropic§ paste	
RTV-90	Red	12,000‡	Stiff paste	100

* General Electric Corp.
† Measured with a Brookfield RVT viscometer, No. 6 spindle, 10 r/min.
‡ Measured with a Brookfield HBF viscometer, No. 7 spindle, 2 r/min.
§ Does not flow under static conditions, but acts like a liquid under pressure.

TABLE 28   Properties of a Family of RTV Silicone Rubbers*[34]

Property	RTV-11	RTV-20	RTV-40	RTV-60	RTV-88	RTV-90
Specific gravity	1.18	1.35	1.37	1.47	1.47	1.47
Durometer hardness, Shore A	45	50	55	60	65	60
Tensile strength, lb/in.²	350	450	550	650	750	750
Elongation, %	180	140	120	110	110	160
Tear strength, lb/in., die B	15	25	25	50	40	50
Peel strength, lb/in. (bonded to primed stainless steel)	3†	3.5†	4†	4.5†	5†	6.5†
Linear shrinkage, %	0.5	0.2	0.2	0.2	0.5	0.2

* General Electric Corp. Properties determined on cold-pressed ASTM sheets cured with 0.5% Thermolite-12 catalyst. Heat aging in circulating air oven.
† Cohesive failure.

include Dow Corning's Sylgards and General Electric's RTV 615 and RTV 616. These materials are water-white or pigmented. Properties of a series of these materials are given in Table 30. The excellent resistance to heat aging is shown in Fig. 28. The clarity of these materials allows easy identification and replacement of electrical parts. They are used in aircraft canopies as a clear, flexible buffer

TABLE 29   Typical Chemical Resistance of RTV Silicone Rubber[35]

Test fluid	Tensile-strength change, %	Volume change, %
MIL-L-7808 (70 h/300°F)	−40	+20
Silicone oil SF-96(100) (70 h/300°F)	−55	+30
ASTM No. 3 (70 h/300°F)	−50	+30
Skydrol 500 (70 h/212°F)	−65	+85
JP-4 (70 h/80°F)	−15	+80
5% NaCl in distilled water* (70 h/80°F)	0	+1

* Cured 144 h/80°F.

between the rigid plastic outer faces. These flexible resins are stronger than the other RTVs and can be cured very rapidly in thick sections at elevated temperatures. Their chief disadvantage is their cure inhibition by a variety of materials, including sulfur-vulcanized rubber, certain RTV silicone rubbers, certain clays, void fillers, some waxes, and pressure-sensitive adhesives. This problem can be overcome by coating the inhibiting surface with a barrier coating before casting.

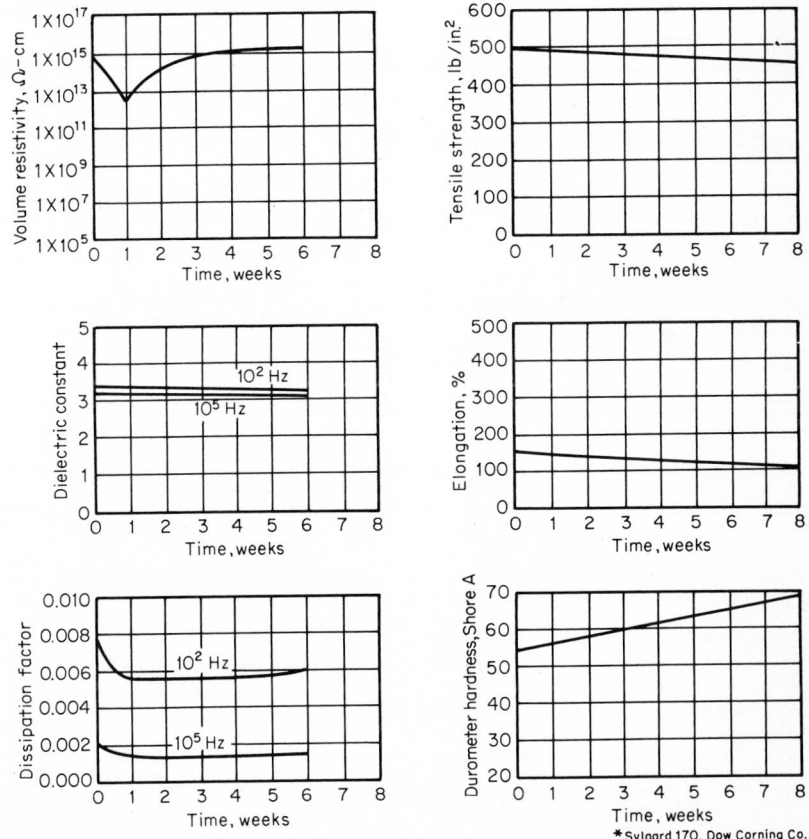

Fig. 28   Properties of a vinyl silicone aged at 200°C.[37]

TABLE 30 Typical Physical Properties of Vinyl Silicone Resins[36]

Property	ASTM test method	182	183	184	185
		As Supplied			
Specific gravity at 25°C (77°F)	D 702	1.05	1.22	1.05	1.23
Viscosity at 25°C, (77°F) cP		5,500	8,000	5,500	8,000
Viscosity, immediately after adding curing agent, cP		3,900	5,000	3,900	5,000
Pot life† at 25°C (77°F) (with curing agent added), h		8	4	2	2
		As Cured			
Color		Clear	Black	Clear	Black
Hardness, Shore A scale	D 676	40	45	35	50
Elongation, %	D 412	100	150	100	150
Specific gravity	D 702	1.05	1.22	1.05	1.23
Tensile strength, lb/in.²	D 412	900	875	900	800
Thermal conductivity, cal/(cm²)(s)(°C/cm)		3.5×10^{-4}	7.5×10^{-4}	3.5×10^{-4}	7.5×10^{-4}
Linear coefficient of thermal expansion, cm/cm/°C (−55 to 150°C) (−67 to 302°F)		3.0×10^{-4}	2.5×10^{-4}	3.0×10^{-4}	2.50×10^{-4}
Volume expansion, cm³/(cm³)(°C) from −55 to 150°C (−67 to 302°F)		9.6×10^{-4}	7.8×10^{-4}	9.6×10^{-4}	7.8×10^{-4}
MIL-I-16923 C, thermal shock resistance, from −55 to 155°F (−49 to 68°C)		Passes 10 cycles	Passes 10 cycles	Passes 10 cycles	Passes 10 cycles

Property	ASTM				
Weight loss,‡ %:					
After 1,000 h at 150°C (302°F)	......	......	......	1.6	2.55
After 1,000 h at 200°C (392°F)	......	1.6	2.35	4.0	4.10
After 168 h at 250°C (482°F)	......	3.2	3.95	.....	.....
Water absorption, % after 7 days immersion at 25°C (77°F)	D 570	0.10	0.12	Less than 0.10	Less than 0.10
Brittle point, °C, lower than	D 746	−70	−65	−120	−120
Refractive index	D 1218	1.430	DNA§	1.430	DNA§
Radiation resistance, cobalt-60 source	......	Still usable after exposure to 200 megarads; hard and brittle after 500 megarads	Still usable after exposure to 250 megarads; hard and brittle after 500 megarads	Still usable after exposure to 200 megarads; hard and brittle after 500 megarads	Still usable after exposure to 250 megarads; hard and brittle after 500 megarads
Flammability	D 635		Self-extinguishing		

* Dow Corning Corp., Sylgard resins.
† Time required for catalyzed viscosity to double at 25°C (77°F).
‡ Specimen size: 1 by 1½ by 1/16 in. thick.
§ DNA = does not apply.

Another group of related flexible resins is widely used in mold making because of high tear strength, low cure shrinkage, and good release properties. These materials are cure-inhibited by the same materials as the encapsulating flexible resins. Properties are given in Table 31 for one of these compounds.

Rigid solventless silicones This group is used where the excellent high-temperature properties of silicones are needed in a rigid structure such as impregnating coils. They are relatively weak, air-inhibited, and have high cure shrinkage. Properties of two of these materials are given in Table 32.

One-part silicones These materials are used primarily as adhesives and sealants. They cure by reacting with the moisture in the air, liberating such materials as acetic acid and methanol. The acetic acid–liberating compounds must be used with caution near copper or other materials that might be attacked. This cure mechanism makes it imperative that at least one of the adherents be permeable to water and that relatively thin layers are formed. The bond is excellent to most materials including metals, glass, ceramics, plastics, and silicone rubbers. The one-part silicones are available in a variety of colors, flame resistance, solvent resistance, and viscosity. They resist ozone, oxidation, moisture, and harsh chemicals, and have high dielectric strength. They are used in formed-in-place gaskets, high-vacuum feed-through seals, construction and swimming-pool sealants, plastic piping seals, on automobile windshields, and as printed-circuit-board coatings.

CAST PHENOLICS

Unlike most of the materials covered in this chapter, the phenolics cure by a condensation mechanism, generating water in the process. This water is finely dispersed in the casting, producing a brilliancy found in few other plastics. This and the high density of the unfilled castings have made this material desirable in beads, knife handles, paperweights, and billiard balls. These products are either cast directly to final shape or machined from cast rods and sheets. The phenolics are not widely used in electrical applications because of the corrosive effects of the acid catalyst and the water vapor given off in curing. Phenolic dip coatings have been successfully used for protecting electronic components. Properties of phenolic casting resins are given in Table 33.

ALLYLIC RESINS

Allylic resins have been used in optical parts because of their excellent clarity, hardness, abrasion resistance, and color stability. Their electrical properties are outstanding, but the casting difficulties and moderate mechanical properties have limited their use in potting. The allylics have never competed with acrylics for glazing because of their higher price and poorer weathering properties. Low vapor pressure of the monomers and long pot life have led to their use in prepregs and as impregnants.

Types Available

Diallyl phthalate and diallyl isophthalate (Dapon from the FMC Corporation) are the allylics produced in largest quantity. Diethylene glycol bis(allyl carbonate), used largely in optical castings, is marketed by PPG Industries as CR-39. Triallyl cyanurate (TAC) and other allylics are used as monomers in polyesters. Diallyl chlorendate (DAC) is used in flame-resistant compositions.

Processing Characteristics

The vapor pressure of all the allyl ester monomers is very low, as shown in Table 34. The low odor level is also a plus. The peroxide-catalyzed materials have a pot life extending to over a year, and cure rapidly at temperatures above 150°C. If castings are made directly from monomers, the cure shrinkage is about 12 percent. Incorporation of prepolymer in the casting composition reduces the shrinkage considerably but raises the viscosity, precluding electrical potting.

TABLE 31 Properties of a Family of Silicone Rubber Used in Mold Making[38]

Property	ASTM test method	Silastic A* RTV rubber	Silastic B RTV rubber	Silastic C RTV rubber	Silastic D RTV rubber	Silastic E RTV rubber	Silastic G RTV rubber
Before catalyzing:							
Color	MIL-S-7502	White	White	Red	Red	White	Pink
Flow (sag or slump)	....	Pourable	Pourable	Nil	Pourable	Pourable	Pourable
Specific gravity	....	1.14	1.38	1.47	1.50	1.12	1.1
Nonvolatile content, %	....	98	98	98	98	98	97
Viscosity, poise†	....	140	300	8,000	400	1,200	1,400
After vulcanizing for 72 h at 77° (25°C):							
Durometer hardness, Shore A scale	ASTM D 176	45	60	65	60	35	35
Tensile strength, lb/in.²	ASTM D 412	400	600	650	600	700	850
Elongation, %	ASTM D 412	180	120	180	100	400	300–350
Tear strength, die B, pli	ASTM D 624	15	30	40	30	90	110
Temp stability, upper limit, deg		500°F (260°C)	600°F (315°C)	600°F (315°C)	600°F (315°C)	500°F (260°C)	
Shrink, linear % after 24 h at 77°F (25°C)		0.3	0.2	0.3	0.2	Nil	0.3
After 7 days at 77°F (25°C)		0.6	0.4	0.6	0.4	0.1	0.6
Thermal conductivity cal/(cm)(°C)(s)		0.52×10^{-3}	7.5×10^{-4}	7.5×10^{-4}	7.5×10^{-4}		
Volume coefficient of thermal expansion per °C (0–100)		7.50×10^{-4}	3.25×10^{-4}	3.25×10^{-4}	3.25×10^{-4}		

* Dow Corning Corp.
† Brookfield viscometer, No. 5 spindle, 5 r/min.

TABLE 32 Properties of Solventless Silicone Resins[39]

Property	ASTM method	R-7501 resin*		R-7521 resin*	
		Unfilled	Filled†	Unfilled	Filled‡
Hardness, Barcol		10	90	10	90
Flexural strength, lb/in.²:	D 790-49T				
At 77°F (25°C)		5,900	7,900	7,600	7,900
At 77°F (25°C) after 200 h at 500°F (260°C)		......	7,500	......	8,400
Compressive strength, lb/in.²:	D 649-42T				
At 77°F (25°C)		17,000	12,000	18,000	13,000
At 77°F (25°C) after 200 h at 500°F (260°C)		......	13,000	......	15,000
Tensile strength, lb/in.² at 77°F	D 882-49T	2,800	4,300	3,500	4,000
Izod impact strength, ft-lb/in. notch	D 256-43T method A	0.20	0.30	0.20	0.30
Coefficient of linear expansion, in./(in.)(°C), from 25 to 150°C		150×10^{-6}	100×10^{-6}	125×10^{-6}	80×10^{-6}
Weight loss, %, 200 h at 250°C		3.5	1.2	5.4	1.6
Water absorption, % after 24 h immersion in water at 77°F (25°C)	B 570-54T	0.04	0.03	0.08	0.04
Electric strength, V/mil§:	D 149-55T				
In air		350	350	350	350
In oil		350	350	350	350
In oil, after 200 h at 250°C		......	370	......	350
Dissipation factor:	D 150-54T				
Dry, at 60 cycles		0.003	0.008	0.002	0.008
At 10³ cycles		0.003	0.004	0.002	0.004
At 10⁶ cycles		0.001	0.002	0.001	0.010
Wet,¶ at 60 cycles		0.0035	0.020	0.003	0.010
At 10³ cycles		0.003	0.010	0.002	0.006
At 10⁶ cycles		0.002	0.004	0.001	0.003

	ASTM method				
Dissipation factor, after aging 200 h at 480°F (249°C):	D 150-54T				
Dry, at 60 cycles		.....	0.004	.....	0.004
At 10^3 cycles		.....	0.002	.....	0.002
At 10^6 cycles		.....	0.001	.....	0.002
Wet,¶ at 60 cycles		.....	0.010	.....	0.010
At 10^3 cycles		.....	0.006	.....	0.005
At 10^6 cycles		.....	0.004	.....	0.003
Dielectric constant:	D 150-54T				
Dry, at 60 cycles		2.95	3.65	2.85	3.65
At 10^3 cycles		2.90	3.60	2.80	3.60
At 10^6 cycles		2.90	3.60	2.80	3.60
Wet,¶ at 60 cycles		2.95	3.75	2.85	3.70
At 10^3 cycles		2.90	3.70	2.80	3.65
At 10^6 cycles		2.90	3.60	2.80	3.60
Dielectric constant, after aging 200 h at 380°F (249°C):	D 150-54T				
Dry, at 60 cycles		.....	3.55	.....	3.80
At 10^3 cycles		.....	3.50	.....	3.75
At 10^6 cycles		.....	3.50	.....	3.75
Wet,¶ at 60 cycles		.....	3.65	.....	3.95
At 10^3 cycles		.....	3.60	.....	3.90
At 10^6 cycles		.....	3.55	.....	3.80
Volume resistivity, Ω-cm, dry	D 257-54T	7×10^{15}	7×10^{14}	5×10^{15}	4×10^{15}
Wet¶		6×10^{14}	3×10^{14}	4×10^{14}	3×10^{15}
Surface resistivity, Ω, dry	D 257-54	1×10^{15}	$>7 \times 10^{15}$	2×10^{15}	3×10^{15}
Wet¶		1×10^{13}	2×10^{14}	2×10^{14}	2×10^{15}

* Dow Corning Corporation. Catalyzed with 2% recrystallized peroxide, based on weight of resin solids. Curing schedule was 16 h at 125°C, plus an additional 3 h at 200°C.
† Formula: 100 parts R-7501 resin to 150 parts silica flour (Ottawa sand).
‡ Formula: 100 parts R-7521 resin to 230 parts silica flour (Ottawa sand).
§ Determined with $\frac{1}{4}$-in. electrodes on a section 0.125 in. thick.
¶ Condition C (96 h at 23°C and 96% relative humidity).

TABLE 33   Properties of Cured Cast-Phenolic Resins*[40]

Property	Test method	Test values
Inherent properties:		
Specific gravity..........................	D 792	1.30–1.32
Specific volume, in.^{3}/lb....................	D 792	21.3–20.9
Refractive index, n_D......................	D 542	1.58–1.66
Cured properties:		
Tensile strength, lb/in.^{2}..................	D 638–D 651	6,000–9,000
Elongation, %...........................	D 638	1.5–2.0
Modulus of elasticity in tension, 10^5 lb/in.^{2}...	D 638	4–5
Compressive strength, lb/in.^{2}..............	D 695	12,000–15,000
Flexural strength, lb/in.^{2}.................	D 790	11,00–17,000
Impact strength, ft-lb/in. notch		
($\frac{1}{2}$- × $\frac{1}{2}$-in. notched bar, Izod test)........	D 256	0.25–0.40
Hardness, Rockwell......................	D 785	M93–M120
Thermal conductivity,		
10^4 cal/(s)(cm^2)(°C)(cm)..................	C 177	3–5
Specific heat, cal/(°C)(g)..................	.....	0.3–0.4
Thermal expansion, 10^{-5} per °C.............	D 696	6–8
Resistance to heat (continuous), °F..........	.....	160
Heat-distortion temperature, °F.............	D 648	165–175
Volume resistivity (50% RH and 23°C),		
Ω-cm....................................	D 257	10^{12}–10^{13}
Dielectric strength ($\frac{1}{8}$-in. thickness), V/mil:		
Short time.............................	D 149	350–400
Step-by-step...........................	D 149	250–300
Dielectric constant:		
60 Hz.................................	D 150	5.6–7.5
10^5 Hz.................................	D 150	5.5–6.0
10^6 Hz.................................	D 150	4.0–5.5
Dissipation (power) factor:		
60 Hz.................................	D 150	0.10–0.15
10^3 Hz.................................	D 150	0.01–0.05
10^6 Hz.................................	D 150	0.04–0.05
Arc resistance, s..........................	D 495	200–250
Water absorption (24 h, $\frac{1}{8}$-in. thickness), %..	D 570	0.3–0.4
Burning rate.............................	D 635	Very low
Effect of sunlight........................	.....	Colors may fade
Effect of weak acids.....................	D 543	None to slight, depending on acid
Effect of strong acids....................	D 543	Decomposed by oxidizing acids; reducing and organic acids, none to slight effect
Effect of weak alkalies....................	D 543	Slight to marked, depending on alkalinity
Effect of strong alkalies..................	D 543	Decomposes
Effect of organic solvents.................	D 543	Attacked by some
Machining qualities......................	.....	Excellent
Clarity..................................	.....	Transparent, translucent, opaque

* Trade names and suppliers: Gemstone, Knoedler Chemical Co.; Marblette, Marblette Corp.; Syncast, Snyder Chemical Corp.

Properties

The properties of the allylic resins are tabulated in Tables 35 through 38. The low tensile strength and impact resistance preclude the use of DAP and DAIP in unfilled castings where physical abuse is likely. In most applications of CR-39, it is used in competition with glass or polymethyl methacrylate (PMMA), where its abrasion resistance (Fig. 29), high heat distortion, or impact resistance are needed.

TABLE 34 Properties of Allyl Ester Monomers[41]

Property	Diallyl phthalate	Diallyl isophthalate	Diallyl maleate	Diallyl chlorendate	Diallyl adipate	Diallyl diglycollate	Triallyl cyanurate	Diethylene glycol bis(allyl carbonate)
Density, at 20°C	1.12	1.12	1.076	1.47	1.025	1.113	1.113	1.143
Molecular wt	246.35	246.35	196	462.76	226.14	214.11	249.26	274.3
Boiling pt, °C, 4 mm Hg	160	181	111	......	137	135	162	160
Freezing pt, °C	−70	−3	−47	29.5	−33	......	27	−4
Flash pt, °C	166	340	122	210	150	146	>80	177
Viscosity at 20°C, cP	12	16.9	4.5	4.0	4.12	7.80	12	9
Vapor pressure at 20°C, mm Hg	27	......						
Surface tension at 20°C, dynes/cm	34.4	35.4	33	......	32.10	34.37	......	
Solubility in gasoline, %	24	100	23	......	100	4.8	80	35
Thermal expansion, in./(in.)(°C)	0.00076							

TABLE 35 Properties of Two Cured Allyl Esters[35]

Property by ASTM procedures	Diallyl phthalate	Diallyl isophthalate
Dielectric constant:		
25°C and 60 Hz	3.6	3.5
25°C and 10^6 Hz	3.4	3.2
Dissipation factor:		
25°C and 60 Hz	0.010	0.008
25°C and 10^6 Hz	0.011	0.009
Volume resistivity, Ω-cm at 25°C	1.8×10^{16}	3.9×10^{17}
Volume resistivity, Ω-cm at 25°C (wet)*	1.0×10^{14}	
Surface resistivity, Ω at 25°C	9.7×10^{15}	8.4×10^{12}
Surface resistivity, Ω at 25°C (wet)*	4.0×10^{13}	
Dielectric strength, V/mil at 25°C†	450	422
Arc resistance, s	118	123–128
Moisture absorption, %, 24 h at 25°C	0.09	0.1
Tensile strength, lb/in.^{2}	3,000–4,000	4,000–4,500
Specific gravity	1.270	1.264
Heat-distortion temp., °C at 264 lb/in.^{2}	155 (310°F)	238‡ (460°F)
Heat-distortion temp., °C at 546 lb/in.^{2}	125 (257°F)	184–211 (364–412°F)
Chemical resistance, % gain in wt.		
After 1 month immersion at 25°C in:		
Water	0.9	0.8
Acetone	1.3	−0.03
1% NaOH	0.7	0.7
10% NaOH	0.5	0.6
3% H_2SO_4	0.8	0.7
30% H_2SO_4	0.4	0.4

* Tested in humidity chamber after 30 days at 70°C (158°F) and 100% relative humidity.
† Step by step.
‡ No deflection.

Electrical properties are excellent. The variation of dissipation factor, dielectric constant, and dielectric strength with temperature and frequency is given in Figs. 30 through 33. The surface and volume resistivities remain high after prolonged exposure to high humidity. Resistance to solvents and acids is excellent, and resistance to alkalies is good.

TABLE 36 Properties of Cured Diethylene Glycol bis(Allyl Carbonate)[41]

Property	Value
Specific gravity, 25°C	1.32
Hardness, Rockwell M	100
Heat distortion, 10 mil, 264 lb/in.^{2}, °C	65
Izod impact, ft-lb/in.	0.3
Compressive strength, lb/in.^{2}	22,500
Flexural strength, lb/in.^{2}	9,000
Flexural modulus, lb/in.$^2 \times 10^{-6}$	0.30
Tensile strength, lb/in.^{2}	6,000
Volume resistivity, Ω-cm	4.1×10^{14}
Surface resistivity, Ω	3.4×10^{11}
Dielectric strength, V/mil	560
Dissipation factor, 1 MHz, %	5.6
Dielectric constant, 1 MHz	3.5
Arc resistance, s	249
Water absorption, 24 h, %	0.2

TABLE 37   Properties of Cast Alkyd–Triallyl Cyanurate Resin[41]

Property	Value
Specific gravity	1,346
Flexural strength, lb/in.^{2}	7,200
Flexural modulus, lb/in.^{2} $\times$ 10^{-6}	0.75
Heat distortion, °C	270
Refractive index	1.550
Dielectric constant:	
60 Hz	3.97
1 MHz	3.60
9,375 MHz	2.85
Dissipation factor, 60 Hz	0.016
Water absorption, % 24 h	0.39

TABLE 38   Properties of Cast Diallyl Orthophthalate–Diallyl Chlorendate Copolymer (80% DAP, 20% DAC)[41]

Property	Value
Hardness, Rockwell M	118
Heat distortion, °F	324
Impact, Izod notched, ft-lb/in	0.27
Dissipation factor, %:	
60 Hz	1.0
1 kHz	1.0
1 MHz	1.3
Dielectric constant:	
60 Hz	3.6
1 kHz	3.5
1 MHz	3.5
Volume resistivity, Ω-cm	8.9×10^{15}
Burning rate:*	
ASTM 635-56T	Self-extinguishing
ASTM 747-49, in./min	0.15

* 10% antimony oxide added.

Applications

Diallyl phthalate and diallyl isophthalate are seldom used as cast homopolymers except for small electrical parts.   CR-39 cast into sheet and other shapes has been used in glazing, safety-goggle lenses, face shields, sunglasses, tape take-up reels, and transparent plotting boards.

DAP is used for impregnation of ferrous and nonferrous castings because of its low viscosity, excellent sealant characteristics, low resin bleed-out, and ease of cleanup.   Wood impregnated with DAP has reduced water absorption and increased impact, compressive, and shear strengths.   Oppressive odors of wet lay-ups of styrene polyesters were eliminated by adding 3% DAP.[44]

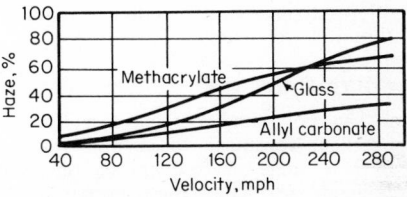

Fig. 29   Effect of falling emery on haze of transparent optics.[42]

DAP and DAIP-glass laminates have high-temperature electrical properties superior to most other structural laminates. The cure cycles are shorter, and little or no postcure is necessary to provide usable strength up to 250°C.

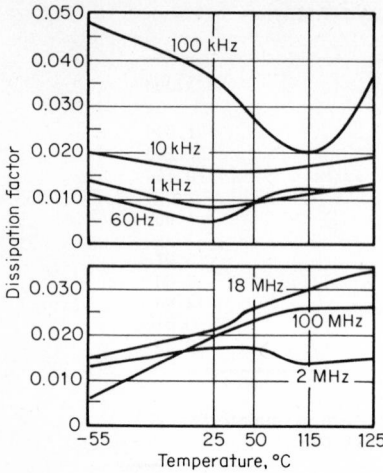

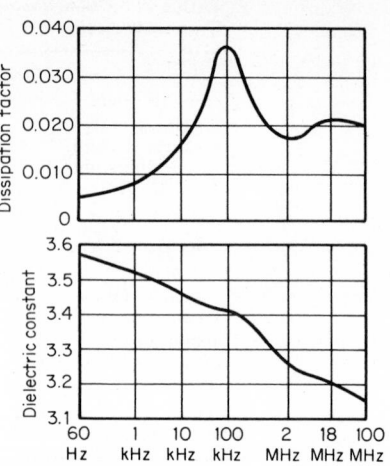

Fig. 30 Dissipation factor of cast diallyl phthalate as a function of frequency and temperature.[43]

Fig. 31 Dissipation factor and dielectric constant of cast diallyl phthalate as a function of frequency.[43]

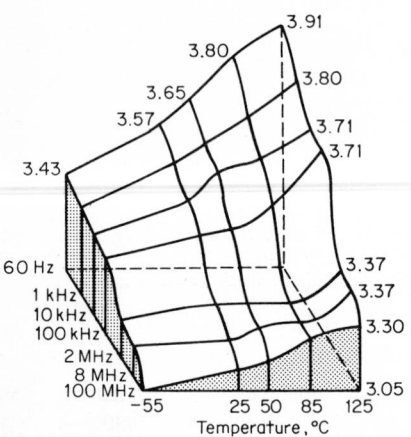

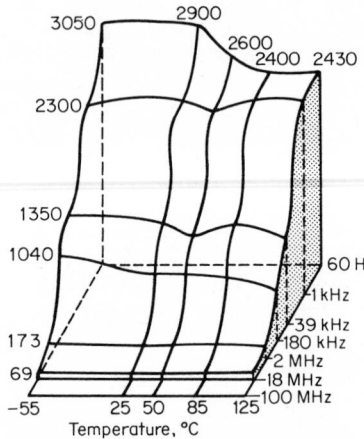

Fig. 32 Dielectric constant of cast diallyl phthalate as a function of temperature and frequency.[43]

Fig. 33 Dielectric strength of cast diallyl phthalate as a function of temperature and frequency.[43]

POLYBUTADIENES

The polybutadienes have a hydrocarbon structure that is responsible for the excellent resistance to chemicals and solvents, electrical properties that are good over a range of frequencies, temperatures, and humidity and resistance to high temperature. The basic materials are homopolymers or copolymers that react through terminal hydroxyl groups, terminal carboxyl groups, vinyl groups, or a combination of these. These materials have limited use as casting compounds because of their high viscosity.

Hydroxyl Terminated Resins

The Arco Chemical Company, division of Atlantic Richfield Company, markets Poly BD, a group of hydroxyl terminated butadiene homopolymers and copolymers with styrene or acrylonitrile. Properties of these liquid resins are shown in Table 39. These materials are generally reacted with isocyanates to produce polyurethanes that have excellent resistance to boiling water. Typical properties are shown in Table 40. Figure 24 demonstrates the water resistance of these polyurethanes compared with those derived from other polyols.

TABLE 39 Typical Properties of Hydroxyl Terminated Polybutadienes[31]

Property	Poly BD liquid resin		
	R-45	CS-15	CN-15
Polymer type............................	Homo-polymer	Styrene copolymer	Acrylonitrile copolymer
Butadiene, wt. %........................	100	75	85
Styrene, wt. %...........................	. . . .	25	
Acrylonitrile, wt. %.....................	. . . .	. . . .	15
Viscosity, poise, at 30°C................	50	225	525
Hydroxyl content, meq/g................	0.80	0.65	0.60
Moisture, wt. %.........................	0.05	0.05	0.05
Iodine No..............................	398	335	345
Polybutadiene microstructure (all polymers):			
Trans-1,4.............................		60%	
Cis-1,4..............................		20%	
Vinyl-1,2.............................		20%	

Carboxyl Terminated Resins

Homopolymers and copolymers with acrylonitrile are in the Hycar series, produced by the B. F. Goodrich Chemical Company. The incorporation of acrylonitrile increases the viscosity and imparts oil resistance, adhesion, and compatibility with epoxy resins. These materials are generally cured with epoxies employing standard epoxy curing agents. The epoxy content can be varied to produce materially ranging from elastomers to toughened rigid resins, as shown in Table 41. Some commercial and potential applications include electrical potting compounds (with transformer oil), sealants, and moisture-block compounds for telephone cables. The dissipation factor and dielectric constant of phthalic anhydride–cured epoxy butadienes are given in Figs. 34 and 35.

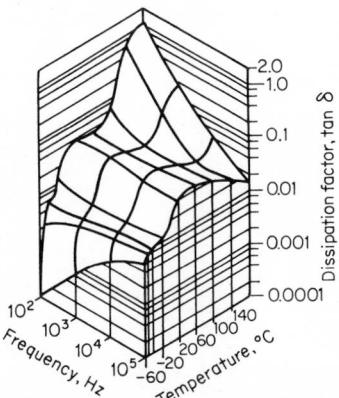

Fig. 34 Dissipation factor of phthalic-anhydride-cured epoxy polybutadiene as a function of temperature and frequency.[46]

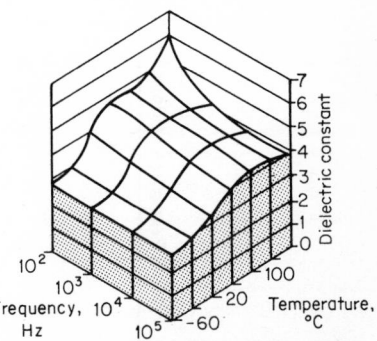

Fig. 35 Dielectric constant of phthalic-anhydride-cured epoxy polybutadiene as a function of temperature and frequency.[46]

TABLE 40 Typical Properties of Polybutadiene Polyurethanes[31]

Property	Soft Gumstock oil-extended	Medium Gumstock	Medium Mixed polyols, one-step	Hard Urea urethanes	Hard Urea urethanes two-step	Hard Mixed polyols
Hardness, Shore A...........	32	52	80	83	95	70 Shore D
Tensile strength, lb/in.^{2}.....	240	510	2,200	1,500	3,130	6,500
% elongation...............	520	420	315	230	475	30
Modulus, lb/in.^{2}, 100%......	100	190	765	930	1,410	
Tear strength, pli..........	32	42	175	235	275	

Other Typical Values:

Property	One-shot system	Prepolymer
Shrinkage, in./in............................	.....	0.012
Thermal conductivity, Btu/(ft^2)(h)(in./°F).....	1.95	1.50
Water absorption, %.........................	.....	0.38 (48 h)

TABLE 41 Properties of Blends of Carboxyl Terminated Polybutadienes and Epi-bis Epoxies[45]

Property	A	B	C	D	E	F	G	H
Hycar* CTBN	100	100	100	100	50	25	12.5	
DER-331†	12.5	25	50	100	100	100	100	100
Triethylene tetramine	0.75	1.5	3	6	6	6	6	6
Cured at room temp:								
Work life	>14 days	>6 days	>2 h	>2 h	2 h	2 h	70 min	45 min
Cure time		2 weeks	1 week	24 h	4 h	3 h	2½ h	2½ h
Cured 16 h at 105°C:								
Tensile strength, lb/in.²	133	555	1,060	2,250	>3,000	>3,000	>3,000	>3,000
Elongation at break, %	230	175	75	50	>10	>10	>10	<10
Hardness:								
Duro A	45	83	83	63	60	83	83	88
Duro D								
Flexural strength, lb/in.²				3,193	4,940	9,060	12,675	
Heat-distortion temp, ASTM D 648, °C:								
Temp.				44	52	68	69	73
180° bend	OK	OK	OK	OK	Fail	Fail	Fail	Fail
Pan cure hardness, inst./10 in.:								
Duro A	47/41	61/55	88/85	55/47	60/55	75/73	76/74	82/81
Duro D								

* Registered trademark of B. F. Goodrich Chemical Co.
† A bisphenol A type of epoxy resin supplied by the Dow Chemical Company.

Resins Containing Pendant Vinyl Groups

Ricons* are butadiene homopolymers and butadiene-styrene copolymers that cure through pendant vinyl groups. For electrical potting, a monomer such as vinyl toluene is added to reduce the viscosity and modify cured properties, and the mixture is cured with a peroxide catalyst. The excellent electrical properties, thermal stability, moisture and chemical resistance, and low density make these materials useful in potting and metal-protective coatings. Disadvantages of these materials are high cure shrinkage (7 to 10 percent), low strength, and high-temperature cures.

Three series of Hystl† resins are produced that have a high concentration of pendant vinyl. One series is carboxyl terminated, the second is hydroxyl terminated, and the third does not contain terminal functional groups. The combination of reactive sites allows these materials to be chain-extended and then cross-linked through the vinyl unsaturation.

BUTYL LM RUBBER

This material, produced by the Enjay Chemical Company, is similar to conventional butyl rubber. The important properties include the lowest moisture-vapor-transmission rate of any elastomer (Table 42), resistance to degradation in high-humidity, high-temperature environments (very little change after 250°F steam for

TABLE 42   Water-Vapor-Transmission Rates of Elastomeric Potting Compounds[47]

Elastomer	WVTR†	WVTR relative to butyl LM
Butyl LM:*		
2-part RT cure	0.04	1
2-part conformal coating	0.04	1
Silicone:		
1-part RT cure	8.31	207
2-part RT cure	8.30	210
Urethane, 2-part RT cure	2.16	54
Polysulfide, 2-part RT cure	2.80	70

* Registered trademark of Enjay Chemical Co.
† Grams of water per 24 h per 100 in.^{2} at 100°F and 95% RH normalized to 20-mil thickness.

1,000 h), excellent electrical properties, resistance to soil bacteria, excellent weathering, and resistance to chemicals and oxidation. The physical properties are poor when the butyl LM is compounded as a potting compound, as shown in Table 43.

Because of the high viscosity of the rubber, oil or solvent must be added to produce a pourable compound. This material is useful in applications where it can be applied under pressure and protected from physical abuse. Uses of butyl LM include roofing coating, reservoir liners, aquarium sealants, and conformal coatings.

POLYSULFIDE RUBBER

The Thiokol Corporation markets a series of liquid polysulfides that can be oxidized to rubbers. These rubbers have excellent resistance to most solvents, good water and ozone resistance, very low specific permeability to many highly volatile solvents and gases, objectionable odor, poor physical properties, and a service-temperature range of −55 to 150°C. A typical formulation and resulting properties are given in Table 44.

Polysulfides have been used extensively as flexibilizing coreactants with epoxy

* Richardson Company.
† Hystl Development Company.

TABLE 43 Composition and Properties of an Elevated-Temperature-Cured, Low-Molecular-Weight, Butyl-Rubber Compound[48]

Composition, parts/100 resins:
Enjay Butyl LM 430*	100
Univolt 33 (transformer oil)	75
ZnO	5
Stearic acid	1

Molecular sieve 4A	5
p-Quinone dioxime	3.5
Pb_3O_4	8.0
Altax	5.0

Physical properties, 75-mil pad cured 15 minutes at 300°F:
Hardness, Shore A	5
Tensile strength, lb/in.²	25
Ultimate elongation, %	300

Electrical properties at room temp:
Dielectric constant at 1 kHz	2.73
Dissipation factor at 1 kHz	0.005
Volume resistivity at 500 Vdc, Ω-cm	9×10^{15}
Dielectric strength on 40-mil pad 500 V/min rise, V/mil	375

* Enjay Chemical Company.

TABLE 44 Typical Physical Properties of a Polysulfide Converted by the p-Quinone dioxime-Diphenylguanidine and Trinitrobenzene Cures[49]

	GMF-DPG cure	TNB cure
Compounding Recipes, Parts:		
Thiokol LP-3*	100	100
Titanox A†	50	50
2,4,6-Trinitrobenzene	...	4
p-Quinone dioxime	7	
Diphenylguanidine	3	3
Sulfur	...	1
Vulcanizate properties (cured 16 h at 158°F):		
Shore A sheet hardness	40	40
300% modulus, lb/in.²	290	180
Tensile strength, lb/in.²	350	250
Elongation, %	400	600
Compression set, %‡ (2 h at 158°F, 25% comp.)	40	40
Low-temperature torsional flexibility,§ absolute modulus:		
G_{rt}, lb/in.²/100° twist	120	103
$G_{5,000}$, °F	−56	−56
$G_{10,000}$	−60	−60
$G_{20,000}$	−61	−63

* Thiokol Corp.
† Titanium pigments incorporated.
‡ ASTM method D 395-49T, method B.
§ Determined according to ASTM method D 1043-49T; G_{rt} denotes the torsional modulus at room temperature, whereas the values for $G_{5,000}$, $G_{10,000}$, and $G_{20,000}$ denote the temperature at which the torsional moduli were 5,000, 10,000, and 20,000 lb/in.², respectively.

resins. The effect of concentration of the polysulfide on the heat-distortion temperature of epoxy formulations is shown in Fig. 36; properties of one mixture are included in Table 1.

Applications of polysulfide rubbers include impregnation of leather to impart water and solvent resistance, protective coatings for metals, binder in gaskets, sealing off aircraft gasoline tanks, deck calking, patching hose, printers' rollers, and adhesives. In the electrical industry, flexible-cable connections are sealed with polysulfide compounds.

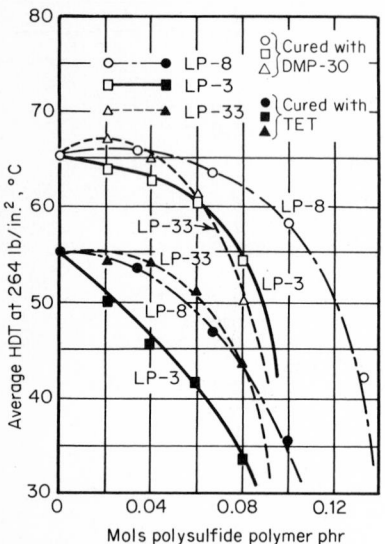

Fig. 36 Effect of flexibilizer (polysulfide polymer) content on the heat-distortion temperature of an epi-bis resin.[50]

CAST NYLON

There is no theoretical limitation to size of nylon parts that can be cast. Castings weighing as much as 1,500 lb have been produced. The process lends itself to the production of large, complex shapes where the reduced cost of the mold required at atmospheric pressures offsets the curing time required.[51] The nylon 6 is polymerized in place from caprolactam anionically. Castings have a 15 percent increase in tensile strength and a 17 percent increase in flexural modulus over the injection-molded product. Typical properties of cast nylon are shown in Table 45. Complex castings can be made having a minimum of internal stress, resulting in better dimensional stability and greater strength. Molybdenum disulfide can be added for increased permanent lubricity, and reinforcements for increased impact strength.

Cast-nylon gears are widely used because of their resistance to abrasion and galling, their light weight, and their superior bearing properties. Nylon's resistance to strong alkalies and most solvents makes it useful as tank linings. Other applications include gears, elevator buckets, crane bearings, wear plates, ball-valve seats, bushings, and roll covers, such as those shown in Fig. 37. The complexities and controls necessary in casting nylon and the patent situation limit the practitioners to a few specialized companies.

ACRYLICS

Castings of acrylics esters have excellent optical clarity; outstanding weather resistance; resistance to discoloring; resistance to acids, alkalies, and oils; and excellent electrical properties, especially arc resistance. Physical properties of three commonly used acrylic-ester monomers are given in Table 46. Methyl methacrylate is the most widely used monomer of the family. Properties of cured methyl methacrylate are given in Table 47. Forty-five percent of the polymethyl methacrylate produced is sold as cast sheets. It is used as glazing where breakage is a problem. Being a thermoplastic, it can be formed into a variety of shapes such as aircraft canopies, furniture, and decorative forms. Tubes, rods, and decorative embedments are also produced.

Ethyl methacrylate is sometimes used in place of the methyl ester because its higher boiling point reduces the possibility of bubbles during curing. Butyl methacrylate produces a permanently soft polymer which is easily sliced, allowing embedded specimens to be cut into thin sections. Polymethyl alpha chloroacrylate sheet has a higher heat-distortion temperature, hardness, and scratch resistance than

TABLE 45   Properties of Cast Nylon 6[52]

Property	ASTM No.	Value
Specific gravity	D 792	1.15–1.17
Tensile strength, lb/in.²	D 638	11,000–14,000
Elongation, %	D 638	10–60
Modulus of elasticity, lb/in.²	D 638	350,000–450,000
Compressive strength, lb/in.²	D 695	
At 0.1% offset	.....	9,000
At 1.0% offset	.....	12,000
Shear strength, lb/in.²	D 732	10,500–11,500
Hardness (Rockwell)	D 785	112–120
Tensile impact, ft-lb/in.²	.....	80–130
Deformation under load, %, 122°F, 2,000 lb/in.²	D 621	0.5–1.0
Stiffness, lb/in.²	D 747	200,000–400,000
Heat-distortion temp:		
66 lb/in.², °F	D 648	425
264 lb/in.², °F	D 648	200–425
Flammability	D 635	Self-extinguishing
Coefficient of linear thermal expansion, in./(in.)(°F)	D 696	5.0×10^{-5}
Dielectric strength (short-time 0.040 in. thick), V/mil	D 149	500–600
Dielectric constant:		
60 Hz	D 150	3.7
10^3 Hz	D 150	3.7
10^6 Hz	D 150	3.1
Power factor:		
60 Hz	D 150	0.02
10^3 Hz	D 150	0.02
10^6 Hz	D 150	0.02
Water absorption, 24 h, %	D 570	0.6–1.2
Water absorption, saturation, %	D 570	5.5–6.5

TABLE 46   Physical Constants of Three Commonly Used Acrylic-Ester Monomers[53]

Properties	Methyl acrylate	Ethyl acrylate	Methyl methacrylate
Molecular weight	86	100	100
Boiling point, °C	79.6–80.3	99.3–99.7	100.6–101.1
Freezing point, °C	< −75	< −75	−48.2
Refractive index (N_d at 25°C)	1.401	1.404	1.412
Specific gravity, 25/25°C	0.952	0.919	0.940
Flash point, °F:			
Closed cup, Tagliabue	26.6	48.2	50.0
Open cup, Cleveland	60	85	85
Viscosity, cSt at 25°C	0.503	0.596	0.569
Vapor pressure, mm Hg at °C:			
2	...	21	18
10	46	27	24
20	72	40	35
30	112	61	53
40	177	93	81
50	270	142	124
60	395	210	189
70	565	300	279
80	765	425	397
90	...	582	547
Solubility in water at 30°C, % by wt	5.2	1.82	1.50
Heat of polymerization, kcal/mole	19–20	...	13
Density, lb/gal	7.95	7.65	7.8

Fig. 37   Cast-nylon roll covers.   (*Polymer Corporation.*)

TABLE 47   Typical Properties of Clear Transparent Acrylic Sheeting[54]

Property	Continuously cast	Conventional cell-cast	Extruded
Mechanical:			
Tensile strength, lb/in.²	10,600	10,800	9,400
Tensile modulus ($\times 10^{-6}$), lb/in.²	0.47	0.47	0.45
Tensile elongation, %	5.0	5.5	4.0
Flexural strength, lb/in.²	16,500	16,900	14,900
Hardness, Barcol	50	50	47
Hardness, Rockwell M	98	98	89
Optical:			
Haze, %	1	1	3
Luminous transmittance, %	92	92	91
Light reflectance, %	96	97	91
Thermal:			
Heat-distortion temp, °F	204–209	204–217	162–183
Miscellaneous:			
Specific gravity	1.19	1.19	1.19
Water absorption (wt. gain), %	0.2	0.2	0.3

polymethyl methacrylate, but it is much more expensive. Methacrylate polymers modified with alpha methyl styrene have exceptional stability to ultraviolet light and a high heat-distortion point.[55]

Acrylics are cured with peroxide catalysts, slowly under very rigidly controlled conditions. Because of the skill and experience needed to produce high-quality castings, a few specialized companies produce most of the finished products.

Other uses for acrylics include dentures, fillings, and incorporation in polyesters. Wear-resistant flooring can be made by impregnating the wood with methyl methacrylate monomer and polymerizing it with radiation from a cobalt-60 source.[56]

VINYL PLASTISOLS

A plastisol is a liquid dispersion of a fine-particle-size polyvinyl chloride in a plasticizer. Plastisols are converted to flexible shapes by heating to 175 to 190°C. This solidification is produced by a purely physical change, the PVC absorbing the plasticizer. The property range shown in Table 48 is produced by varying the amount and type of resin and plasticizer used. If a more rigid part is needed, part or all of the plasticizer can be replaced by a cross-linking reactive monomer such as diallyl orthophthalate. A typical formulation for a rigidsol is given in Table 49; properties of this material are given in Table 50.

TABLE 48   Property Ranges of Plastisol Systems[57]

Viscosity	As required from 1,000 to more than 1 million cP
Tensile strength	From 500 to 2,700 lb/in.^{2}
Elongation	From 150 to 600%
Hardness	From 10 on the Shore A$_2$ durometer scale to 65 on the Shore D durometer scale
Heat resistance	Continuous exposure at 200°F for 2,000 h and intermittent exposure at 450°F for 2 h or less
Chemical resistance	Excellent for most acids, alkalies, detergents, oils, and solvents
Flexibility	Good to temperatures as low as −65°F
Flame resistance	Self-extinguishing
Dielectric strength	Min of 400 V/mil at room temp in 3-mil thicknesses; 800 V/mil in 10-mil thicknesses
Volume-resistivity range	At 25°C, 14.5 × 10^8 − 2.44 × 10^{16} Ω-cm

TABLE 49   Allylic Rigidsol Formulation[41]

	Parts by wt.	
Ingredient	For thick section parts	For thin section parts
Blacar 1716 (Cary Chemical)	70	70
Geon 202 (B. F. Goodrich Co.)	30	30
Diallyl orthophthalate (FMC Corp.)	60	60
Dyphos (National Lead)	10	5
Benzoyl peroxide	0.1	0.2
Dicumyl peroxide (Hercules Powder Co.)	0.2	1.0
Polyethylene glycol-400 monolaurate	1.0	1.0
Thermolite-13 (M & T Chemical Co.)	...	2.0

Plastisols are used where the extra expense of the finely divided PVC particles used is justified by the production conveniences of the material. Manufacturing processes used include dip coating, slush molding, and fabric coating. Advantages are ability to produce hollow parts, inexpensive molds, in-place forming, and large part capability.

TABLE 50   Properties of Allylic Vinyl Rigidsols[41]

Property	Value
Hardness, Shore D	84
Tensile strength, lb/in.^{2}	5,800
Elongation, %	7
Flexural strength, lb/in.^{2}	9,200
Heat-distortion temp, at 264 lb/in.^{2}, °F	126
Volatility, % loss	0.2
Kerosene extraction, % loss	0.2

Applications include beer-cap gasketing, coated glass bottles, coated battery holders, armrests, strippable coatings, prosthetic devices, crown seals,[58] die-coated yarn, storage-tank linings,[57] footballs, bottles, dolls' heads,[42] molds,[59] and mechanically frothed carpet backing.[60]

REFERENCES

1. Lundberg, C. V.: A Guide to Potting and Encapsulation Materials, *Mater. Eng.*, May 1960.
2. Sheard, E. A.: "Commercial Status of Liquid Elastomers," Division of Rubber Chemistry, American Chemical Society, April 1972.
3. Bakelite Liquid Epoxy Resins and Hardeners, *Union Carbide Corp. Tech. Bull.*
4. Bakelite Cycloaliphatic Epoxides, *Union Carbide Corp. Tech. Bull.*
5. Thomsen, F.: Centrifugal Casting, *Kunststoffe*, October 1969.
6. Delmonte, J.: Epoxies for Massive Castings, *Plast. Des. Processing*, January 1969.
7. Dow Epoxy Novolac Resins, *Dow Chemical Co. Tech. Bull.*
8. Epi-Rez 5161—Flame Resistant Resin, Celanese Resins Div., *Celanese Coatings Co. Tech. Bull.*
9. Thiokol Liquid Polymer/Epoxy Resin Casting Compounds, *Thiokol Chemical Corp. Tech. Bull.*
10. Buchoff, L. S., and W. R. Sherwin: Properties of High Temperature Epoxy Systems, *Rubber Plast. Age*, September 1962.
11. Epoxy Resins for Plastics, Celanese Resins Div., *Celanese Coatings Co. Tech. Bull.*
12. Dummer: Embedded and Printed Circuits, *Electrotechnol.*, May 1953.
13. Formo, J. F., and R. E. Isliefson: Producing Special Properties in Plastics with Fillers, *Reg. Tech. Conf.*, Society of Plastics Engineers, Fort Wayne, Ind., May 1959.
14. Buchoff, L. S.: Effect of Humidity on Surface Resistance of Filled Epoxy Resins, *Ann. Nat. Tech. Conf.*, Society of Plastics Engineers, Chicago, Ill., January 1960.
15. Marina Goes to Epoxy: 5900 Others to Follow, *Modern Plastics*, June 1971.
16. Glick, H., and Y. Tuan: Epoxy Casting of Large Contoured Nozzles, *SPE J.*, September 1969.
17. *Rohm and Haas Corp. Tech. Bull.* 442a.
18. *Rohm and Haas Corp. Tech. Bull.* 451.
19. In Gel Coats, the Iso's Have It, *Modern Plastics*, November 1970.
20. Sommer, J.: Isophthalic Polyesters, New Tool for Industry, *Australian Chem. Eng.*, September 1968.
21. *Inmont Corp. Tech. Bull.* 5-1.
22. Slone, N. C.: *SPE J.*, 1960.
23. Church, T. M., and C. Berenson: *Ind. Eng. Chem.*, 1955.
24. Sias, C. B.: *15th Ann. Conf.*, Reinforced Plastics Div., Society of the Plastics Industry, 1960.
25. Benhke, E.: *Kunststoffe Rundschau*, 1957.
26. The Application of Precipitated Calcium Carbonates with Polyester Resins, *Diamond Shamrock Chemical Co. Tech. Bull.*
27. Latest Development in Low Profile Polyesters: Zero Shrinkage, *Modern Plastics*, March 1970.
28. New Water Filled Polyesters, *Plast. World*, November 1969.
29. Leitheiser, R. H., et al.: *27th Ann. Tech. Conf.*, Water Extended Polyester Resin (WEP) New Low Cost Molding Material, Society of Plastics Engineers, May 1964.

30. Reference Issue, Plastics, *Machine Des.*, December 1968.
31. Poly bd, the Best of Two Worlds, *Arco Tech. Bull.*, Arco Chemical Co., Division of Atlantic Richfield Co.
32. Processing Versatility Comes to Urethane Elastomers and Plastics, *Mater. Eng.*, February 1972.
33. Adiprene, *Du Pont Tech. Bull.*, Elastomer Chemicals Dept.
34. Technical Data Book S-23, General Electric Corp., Silicone Products Dept.
35. Directory/Encyclopedia Issue, *Insulation/Circuits*, June/July 1972.
36. *Dow Corning Corp. Tech. Bull.* 67-005.
37. Information about Sylgard Elastomers, *Dow Corning Corp. Tech. Bull.*
38. Information about Mold Making Materials, *Dow Corning Corp. Tech. Bull.*
39. Dow Corning Corp., Silicone Notes Reference 10–261.
40. Plastics Property Chart, "*Modern Plastics* Encyclopedia," 1970–1971.
41. Raech, H., Jr., "Allylic Resins and Monomers," Van Nostrand Reinhold, New York, 1965.
42. CR-39 Allyl Diglycol Carbonate in Cast Sheets and Forms, *PPG Industries, Inc., Tech. Bull.*
43. Chapman, J. J., and L. J. Frisco: The Electrical Properties of Diallyl Phthalate Resin (Dapon), *Final Rept.*, Jan. 1, 1956, to Jan. 31, 1957.
44. DAP Suppresses Styrene Odors in Polyester Aircraft Parts, *Plast. World*, December 1971.
45. Hycar Reactive Liquid Polymers, *B. F. Goodrich Chemical Co. Tech. Bull.*
46. Fitzgerald, C. G., et al.: Epoxy-Butadiene Resins, *SPE J.*, January 1957.
47. Glazman, J.: Low Molecular Weight Butyl Rubber, *28th Ann. Tech. Conf.*, Society of Plastics Engineers, May 1970.
48. Properties and Applications of Butyl L M, *Enjay Chemical Co., Tech. Bull.*
49. Polymer LP-3, *Thiokol Chemical Corp. Tech. Bull.*
50. Breslau, A. J.: Heat Distortion Properties of Polysulfide Polymer-modified Epoxy Resins, *Plast. Technol.*, April 1957.
51. Hemmel, H., et al.: Activated, Anionic Polymerization of Lactams and Its Utilization in Pressureless Casting, *Kunststoffe*, July 1969.
52. Monocast and Nylatron GSM Nylon, *The Polymer Corp. Tech. Bull.*
53. Horn, M. B.: "Acrylic Resins," Reinhold, New York.
54. Now—Cast Acrylic in Rolls of Continuous Sheet, *Plast. Des. and Processing*, November 1965.
55. Methyl Methacrylate, *Chem. Purchasing*, September 1971.
56. Witt, A. E., and J. E. Morrissey: Economics of Making Irradiated Wood-Plastic Products, *Modern Plastics*, January 1972.
57. New Plastisol Markets Emerge, *Plast. World*, April 1969.
58. Hull, J. W.: Progress in Vinyl Dispersions, *Plast. Technol.*, October 1968.
59. Flexible Mold Material for Casting, *Plast. Des. and Processing*, November 1971.
60. Simoneau, E. T.: Silicone Surfactants in Mechanically Frothed Vinyl Plastisol Foams, *Rubber World*, August 1970.

Chapter 9

Plastic and Elastomeric Coatings

JOHN A. MOCK

Materials Engineering,
Reinhold Publishing Corporation,
Stamford, Connecticut

INTRODUCTION

Surface coatings have been used for many thousands of years. The Egyptians developed a weather-resistant coating based on beeswax which was applied in the molten state. Bitumen was used to preserve the wooden hulls of ships, as noted in the Biblical story of Noah's ark, which was treated within and without with pitch.

In the Middle Ages the use of casein solubilized by means of lime was discovered. Up until the twentieth century, most organic coatings were based on natural resins. One of the first plastic-base coatings to come into general use was rosin-modified phenolic, which was introduced in 1913. Nitrocellulose was introduced as a binder in lacquers in 1920, and alkyd, as well as 100 percent phenolic, were introduced in 1928. The 1930s saw the introduction of such important synthetic-resin binders as chlorinated rubber, vinyl, and modified urea.[1]

Synthetic resins enjoy wider use today than natural resins because of their more consistent quality, the generally improved properties of vehicles made with them, and the wide choice of unique qualities which they give to a surface coating. A modern finish can be individually tailored to a given application by selecting a given synthetic resin or combination of resins. Cost, of course, varies with type of resin. Table 1 lists prices of typical resins used in coatings.

TABLE 1   Prices of Some Resins Used in Finishes

Resin	Dollars per lb
Acrylic copolymer emulsions	$0.21\frac{1}{2}$–0.22
Acrylonitrile-styrene copolymer	0.44–1.00
Alkyd for high-grade flat finishes	0.2065
Butadiene-styrene lattices (dry)	$0.29\frac{1}{2}$–0.35
Epoxy liquid resin	0.65–0.78
Fluorocarbon resin:	
Aqueous dispersion	5.15–8.00
Maleic esters	0.21–$0.27\frac{1}{4}$
Maleic resin for lacquers	$0.23\frac{1}{2}$
Melamine-formaldehyde resin	$0.32\frac{3}{4}$–0.40
Nylon resins	1.18–1.28
Phenolic coating resin	0.30–0.58
Polyamide resin	0.52–0.90
Polyester coating resin base	0.36
Polyethylene, low-molecular-weight powder	0.30–0.37
Polystyrene emulsions	0.20
Saran latex–vinylidene chloride–acrylonitrile 52%	0.45
Styrene polymer, naturals, clear	$0.40\frac{1}{2}$–$0.43\frac{1}{2}$
Urea-formaldehyde resins	$0.20\frac{1}{2}$–$0.47\frac{1}{4}$
Vinyl acetate resin, solution in acetone, per gallon	5.10
Vinyl chloride resins	$0.23\frac{1}{2}$–$0.46\frac{1}{2}$
Vinyl chloride–acetate resins	$0.23\frac{1}{2}$–0.64

The basic purpose of a protective coating is to form a film over a substrate which will cohere to itself and adhere to the surface over which it is applied.

TYPES OF PLASTIC COATINGS

Film-forming resins may be thermoplastic, i.e., softened by heat but regaining their original properties upon cooling; and thermosetting, i.e., not softened by heat as long as the temperature is not high enough to cause thermal decomposition. (See also Chap. 1 for identification of plastic types.) Thermoplastic film-forming types of resins are acrylics, acetates, butyrates, polyamides, vinyls, etc. Thermosetting film-forming resins include epoxies, melamines, phenolics, polyesters, ureas, etc. Table 2 is a general guide to the major types of thermoplastic and thermosetting plastic-base coatings in use currently.[27]

Elastomeric coatings, on the other hand, are a class of materials based on rubber-like polymers which are cross-linked. Elastomeric coatings are characterized by the fact that they can be used in heavy thicknesses while still retaining film integrity despite vibration and other stresses. Most of the synthetic rubbers and natural rubber can be used in these types of coatings. Commonly used rubberlike materials include neoprene, chlorosulfonated polyethylene, urethane elastomers, polysulfide, and fluoroelastomer.[1]

Literally thousands of coating formulations are available today to do a multitude of jobs. These include specially formulated coatings which are applied by electrodeposition. These coatings are water-reducible systems and are commonly used as a primer or as a one-coat finish on automobiles, appliances, office equipment, and many other products. A chief advantage of an electrodeposited coating is that the material gives total coverage in that every corner, side, angle, and louver is covered and inaccessible holes and recesses are consistently, smoothly, and uniformly painted.

Powder-coating materials will become very popular in the years ahead because of federal, state, and local government regulations concerning air pollution. Powder coatings offer the advantage that no solvent system is used and thus harmful solvent emission is eliminated. These coatings are applied by several different methods, including fluidized bed, electrostatic spray, flame spraying, electrostatic fluidized bed, and powder flood coating. However, the two most frequently used coating processes for powders are fluidized bed and electrostatic spray. Table 3 shows plastics suitable for fluidized-bed coating. There is a wide choice of materials for powder coating. These include vinyl, nylon, polypropylene, cellulosics, epoxies, polyesters, and many others.[13]

Another development in plastic coatings is coil coating. Here the coating is applied to coils of steel or aluminum on a continuous basis. Coatings are permanently bonded to one or both sides of the metal in many combinations of color, design, or texture. Besides liquid coatings, plastic films are laminated to metal on a coil-coating line.

Curing Although impressive gains have been made in new-coating formulations, there is still the problem of curing by relatively slow methods. However, there is promise of breaking this barrier with radiation-curing systems. Radiation curing can be up to 200 times faster than conventional curing, with the added advantage of being at room temperature. Special radiation-sensitive coatings, now commercially available, are used. These coatings are based on monomers rather than polymers as in conventional coating systems. The paint formulations are solventless and can be chemically engineered to incorporate structures similar to those present in polyurethanes, silicones, unsaturated polyesters, melamine-modified alkyds, acrylic-base materials, and epoxies. However, many monomers are volatile, and radiation systems do not necessarily eliminate problems associated with solvent-base finishes.[16]

As mentioned, literally thousands of coating formulations are available. A typical example is fire-retardant paint.[20] Figure 1 shows how a fire-retardant paint reduces damage caused by a fire. This type of coating fights fire by intumescing when exposed to flames. A tough insulating layer resulting from the intumescence keeps damaging heat from flammable substrates like wood for a period of time. These coatings are increasingly popular in public institutions, such as nursing homes,

TABLE 2 Properties of Coatings by Polymer Type[27]

Coating type	Electrical properties				Physical characteristics				
	Volume resistivity, ohm-cm (ASTM D 257)	Dielectric strength, volts/mil	Dielectric constant	Dissipation factor	Maximum continuous service temperature, °F	Adhesion to metals	Flexibility	Approximate Sward hardness (higher number is harder)	Abrasion resistance
Acrylic	10^{14}–10^{15}	450–550	2.7–3.5	0.02–0.06	180	Good	Good	12–24	Fair
Alkyd	10^{14}	300–350	4.5–5.0	0.003–0.06	200 / 250 T.S.	Excellent	Fair to good / Low temperature—poor	3–13 (air dry) / 10–24 (bake)	Fair
Cellulosic (nitrate butyrate)		250–400	3.2–6.2		180	Good	Good / Low temperature—poor	10–15	
Chlorinated polyether (Penton*)	10^{15}	400	3.0	0.01	250	Excellent	Good		
Epoxy-amine cure	10^{14} at 30°C / 10^{10} at 105°C	400–550	3.5–5.0	0.02–0.03 at 30°C	350	Excellent	Fair to good / Low temperature—poor	26–36	Good to excellent
Epoxy-anhydride, Dicy		650–730	3.4–3.8	0.01–0.03	400	Excellent	Good to excellent / Low temperature—poor	20	Good to excellent
Epoxy-polyamide	10^{14} at 30°C / 10^{10} at 105°C	400–500	2.5–3.0	0.008–0.02	350	Excellent	Good to excellent / Low temperature—poor	20	Fair to good
Epoxy-phenolic	10^{12}–10^{13}	300–450			400	Excellent	Good / Low temperature—fair		Good to excellent
Fluorocarbon TFE	10^{18}	430	2.0–2.1	0.0002	500	Can be excellent; primers required.	Excellent		
FEP / CTFE	10^{18} / 10^{18}	480 / 500–600	2.1 / 2.3–2.8	0.0003–0.0007 / 0.003–0.004	400 / 400	Can be excellent; primers required	Excellent		
Parylene (polyxylylenes)	10^{16}–10^{17}	700	2.6–3.1	0.0002–0.02	240°F (air) / 510°F (inert atm.)	Good	Good		

Phenolics	10^9-10^{12}	100-300	4-8	0.005-0.5	350	Excellent	Poor to good Low temperature—poor	30-38	Fair
Phenolic-oil varnish					250	Excellent	Good Low temperature—fair		Poor to fair
Phenoxy	$10^{13}-10^{14}$	500	3.7-4.0	0.001	180	Excellent	Excellent		
Polyamide (nylon)	$10^{13}-10^{15}$	400-500	2.8-3.6	0.01-0.1	225-250	Excellent			
Polyester	$10^{12}-10^{14}$	500	3.3-8.1	0.008-0.04	200	Good on rough surfaces; poor to polished metals.	Fair to excellent	25-30	Good
Chlorosulfonated (polyethylene (Hypalon))†				0.03-0.07	250	Good	Elastomeric	Less than 10	
Polyimide	$10^{16}-10^{18}$	400 3,000 (10 mil)	6-10	0.003	500	Good	Fair to excellent		Good
Polystyrene	$10^{10}-10^{19}$	500-700	3.4-3.8	0.0001-0.0005	140-180		Poor to fair		
Polyurethane	$10^{12}-10^{13}$	450-500 3,800 (1 mil)	2.4-2.6	0.02-0.08	250	Often poor to metals. (Excellent to most non-metals.)	Good to excellent. Low temperature—poor.	10-17 (castor oil) 50-60 (polyester)	
Silicone	$10^{14}-10^{16}$	550	6.8 (1 kHz) 4.4 (1 MHz)	0.001-0.008	500	Varies, but usually needs primer for good adhesion.	Excellent Low temperature—excellent	12-16	Fair to excellent
Vinyl chloride (poly-)	$10^{11}-10^{15}$	300-800	3.0-4.2	0.04-0.14	150	Excellent, if so formulated.	Excellent Low temperature—fair to good	5-10	
Vinyl chloride (plastisol, organisol)	10^9-10^{16}	400	3-9	0.10-0.15	150	Requires adhesive primer.	Excellent Low temperature—fair to good	3-6	
Vinyl fluoride	$10^{13}-10^{14}$	260 1,200 (8 mil)	2.3-9	0.05-0.15	300	Excellent, if fused on surface.	Excellent Low temperature—excellent		
Vinyl formal (Formvar‡)	$10^{13}-10^{15}$	850-1,000	6.4-8.4	0.007-0.2	200	Excellent			

* Trademark of Hercules Powder Co., Inc., Wilmington, Del.
† Trademark of E. I. du Pont de Nemours & Co., Wilmington, Del.
‡ Trademark of Monsanto Co., St. Louis, Mo.

TABLE 2 Properties of Coatings by Polymer Type[27] (Continued)

Coating type	Resistance to environmental effects						Film formation			Typical uses
	Chemical and solvent resistance	Moisture and humidity resistance	Weather-ability	Resistance to micro-organisms	Flamma-bility	Repairability	Method of cure	Cure schedule	Application method	
Acrylic	Solvents—poor Alkalies—poor Dilute acids—poor to fair	Good	Excellent resistance to UV and weather	Good	Medium	Remove with solvent.	Solvent evaporation	Air dry or low-temperature bake	Spray, brush, dip	Coatings for circuit boards quick dry protection for markings and color coding.
Alkyd		Poor	Good to excellent	Poor	Medium	Poor	Oxidation or heat	Air dry or baking types	Most common methods	Painting of metal parts and hardware.
Cellulosic (nitrate butyrate)	Solvents—good Alkalies—good Acids—good	Fair		Poor to good	High	Remove with solvents.	Solvent evaporation	Air dry or low-temperature bake	Spray, dip	Lacquers for decoration and protection. Hot-melt coatings.
Chlorinate polyether (Penton*)		Good			Low		Powder or dispersion fuses	High temperature fusion	Spray, dip, fluid bed	Chemically resistant coatings.
Epoxy-amine cure	Solvents—good to excellent Alkalies—good Dilute acids—fair	Good	Pigmented—fair; clear—poor (chalks)	Good	Medium	No	Cured by catalyst reaction	Air dry to medium bake	Spray, dip, fluid bed	Coatings for circuit boards. Corrosion-protective coatings for metals.
Epoxy-anhydride, Dicy	Solvents—good Alkalies— Dilute acids—	Good		Good	Medium	No	Cured by chemical reaction	High bakes 300 to 400°F	Spray, dip, fluid bed, impreg.	High-bake, high-temperature-resistant dielectric and corrosion coatings.
Epoxy-polyamide	Solvents—fair Alkalies—good Dilute acids—poor	Good		Good	Medium	No	Cured by coreactant	Air dry or medium bake	Spray, dip	Coatings for circuit boards. Filleting coating.
Epoxy-phenolic	Solvents—excellent Alkalies—fair Dilute acids—good	Excellent	Pigmented—fair; clear—poor	Good	Medium	No	Cured by coreactant	High bakes 300 to 400°F	Spray, dip	High-bake solvent and chemical resistant coating.
Fluocarbon TFE	Solvents—excellent Alkalies—good Dilute acids—excellent	Excellent		Good	None	No	Fusion from water or solvent dispersion	Approx. 750°F	Spray, dip	High-temperature-resistant insulation for wire.
FEP CTFE		Excellent		Good	None	No	Fusion from water or solvent dispersion	500–600°F	Spray, dip	High-temperature-resistant insulation. Extrudable.
Parylene (polyxylylenes)		Excellent			None		Vapor phase deposition and polymerization requiring special license from Union Carbide.			Very thin, pinhole-free coatings, possible semi-conductable coating.

Phenolics	Solvents—good to excellent; Alkalies—poor; Dilute acids—good	Excellent	Fair	Poor to good	Medium	No	Cured by heat	Bake 300–500°F	Spray, dip	High-bake chemical and solvent-resistant icoatings.
Phenolic-oil varnish	Solvents—poor; Alkalies—poor; Dilute acids—good to excellent	Good	Good	Poor, unless toxic—additive	Medium	Poor	Oxidation or heat		Spray, brush, dip-impregnate	Impregnation of electronic modules, quick protective coating.
Phenoxy		Good		Good			Cured by heat			Chemical resistant coating.
Polyamide (nylon)		Fair				Fairly solderable				Wire coating.
Polyester	Solvents—poor; Alkalies—poor to fair; Dilute acids—good	Fair	Very good	Good	Medium	Poor	Cured by heat or catalyst	Air dry or bake 100–250°F	Spray, brush, dip	Wire coating.
Chlorosulfonated polyethylene (Hypalon†)	Solvents—poor; Alkalies—good; Dilute acids—good	Good		Good	Low		Solvent evaporation	Air dry or low temperature bake	Spray, brush	Moisture and fungus proofing of materials.
Polyimide	Solvents—excellent; Alkalies—; Dilute acids—	Good	Good	Good	Low	Poor	Cured by heat	High bake	Dip, impregnate, wire coater	Very high temperature resistant wire insulation.
Polystyrene	Solvents—good; Dilute alkalies—fair; Dilute acids—good	Good		Good	High	Dissolve with solvents.	Solvent evaporation	Air dry or low bake	Spray, dip	Coil coating, low dielectric constant, low loss in radar uses.
Polyurethane	Solvents—good; Dilute alkalies—fair; Dilute acids—good	Good		Poor to good	Medium	Excellent; melts, solder-through properties	Coreactant or moisture cure	Air dry to medium bake	Spray, dip	Conformal coating of circuitry, solderable wire insulation.
Silicone	Solvents—poor; Alkalies—good (dilute) poor (concentrated); Dilute acids—good	Excellent	Excellent	Good	Very low (except in O₂ atm.)	Fair to excellent. Cut and peel.	Cured by heat or catalyst	Air dry (RTV) to high bakes	Spray, brush, dip	Heat-resistant coating for electronic circuitry. Good moisture resistance.
Vinyl chloride (poly-)	Solvents—alcohol, good; Alkalies—good	Good	Pigmented—fair to good; Clear—poor	Poor to good (depends on plasticizer)	Very low	Dissolve with solvents.	Solvent evaporation	Air dry or elevated temperature for speed	Spray, dip, roller coat	Wire insulation. Metal protection (especially magnesium, aluminum).
Vinyl chloride (plastisol, organisol)		Good		Poor to good (depends on plasticizer)	Low	Poor	Fusion of liquid to gel	Bake 250–350°F	Spray, dip, reverse roll	Soft-to-hard thick coatings, electroplating racks, equipment.
Vinyl fluoride		Good	Excellent	Good	Very low	Poor	Fusion from water or solvent dispersion	Bake 400–500°F	Spray, roller coat	Coatings for circuitry. Long-life exterior finish.
Vinyl formal (Formvar‡)		Good	Good	Good	Medium	Poor	Cured by heat	Bake 350–500°F	Roller coat, wire coater	Wire insulation (thin coatings) coil impregnation.

* Trademark of Hercules Powder Co., Inc., Wilmington, Del.
† Trademark of E. I. du Pont de Nemours & Co., Wilmington, Del.
‡ Trademark of Monsanto Co., St. Louis, Mo.

TABLE 3   Summary of Plastics for Fluidized-Bed Coating[27]

	Fluidizing conditions			Fluidized-bed powder		
		Cure or fusion				
Resin	Preheat temperature, °F	Temperature, °F	Time, min	Maximum operating temperature, °F	Adhesion	Weather resistance
Epoxy..............	250–450	250–450	1–60	200–400	Excellent	Good
Vinyl..............	450–550	400–600	1–3	225	Poor	Good
Cellulose acetate butyrate............	500–600	400–550	1–3	225	Poor	Good
Nylon.............	550–800	650–700	1	300	Poor	Fair
Polyethylene.......	500–600	400–600	1–5	225	Fair	Good
Polypropylene......	500–700	400–600	1–3	260	Poor	Good
Penton............	500–650	450–600	1–10	350	Poor	Good
Teflon.............	800–1000	800–900	1–3	500	Poor	Good

Fig. 1   Illustration of protection given by fire-retardant coating (top) compared with conventional coating. (*Avco Systems.*)

TABLE 4   Specific Test Methods for Coatings[27]

Test	ASTM	Fed. STD. 141a, method	MIL-STD-202, method	Fed. STD. 406, method	Others
Abrasion	D 968	6191 (Falling Sand) 6192 (Taber)	....	1091	Fed. Std. 601, 14111
Adhesion	 D 2197	6301.1 (Tape Test, Wet) 6302.1 (Microknife) 6303.1 (Scratch Adhesion) 6304.1 (Knife Test)	....	1111	Fed. Std. 601, 8031
Arc resistance	D 495	....	303	4011	
Dielectric constant	D 150	....	301	4021	Fed. Std. 101, 303
Dielectric strength (breakdown voltage)	D 149 D 115	....	....	4031	Fed. Std. 601, 13311
Dissipation factor	D 150	....	....	4021	
Drying time	D 1640 D 115	4061.1			
Electrical insulation resistance	D 229 D 257	....	302	4041	MIL-W-81044, 4.7.5.2
Exposure (exterior)	D 1014	6160 (On Metals) 6161.1 (Outdoor Rack)			
Flash point	D 56, D 92 D 1310 (Tag Open Cup)	4291 (Tag Closed Cup) 4294 (Cleveland Open Cup)	....	....	Fed. Std. 810, 509
Flexibility	....	6221 (Mandrel) 6222 (Conical Mandrel)	....	1031	Fed. Std. 601, 11041
Fungus resistance	D 1924	....	....	....	MIL-E-5272, 4.8 MIL-STD-810, 508.1 MIL-T-5422, 4.8
Hardness	.... D 1474	6211 (Print Hardness) 6212 (Indentation)			
Heat resistance	D 115 D 1932	6051			
Humidity	D 2247	6071 (100% RH) 6201 (Continuous Condensation)	103 106A	....	MIL-E-5272, Proc. 1 Fed. Std. 810, 507
Impact resistance	....	6226 (G.E. Impact)	1074		
Moisture-vapor permeability	E 96 D 1653	6171	7032		
Nonvolatile content	....	4044			
Salt spray (fog)	B 117	6061	101C	6071	MIL-STD-810, 509.1 MIL-E-5272, 4.6 Fed. Std. 151, 811.1 Fed. Std. 810, 509
Temperature-altitude	....	....	....	....	MIL-E-5272, 4.14 MIL-T-5422, 4.1 MIL-STD-810, 504.1
Thermal conductivity	D 1674 (Cenco Fitch) C 177 (Guarded Hot Plate)	....	....	....	MIL-I-16923, 4.6.9
Thermal shock	....	....	107	....	MIL-E-5272, 4.3 MIL-STD-810, 503.1
Thickness (dry film)	D 1005 D 1186	6181 (Magnetic Gage) 6183 (Mechanical Gage)	....	2111, 2121, 2131, 2141, 2151	Fed. Std. 151, 520, 521.1
Viscosity	D 1545 D 562 D 1200 D 88	4271 (Gardner Tubes) 4281 (Krebs-Stormer) 4282 (Ford Cup) 4285 (Saybolt) 4287 (Brookfield)			
Weathering (accelerated)	D 822	6151 (Open Arc) 6152 (Enclosed Arc)	....	6024	

A more complete compilation of test methods is found in J. J. Licari, "Plastic Coatings for Electronics," McGraw-Hill Book Company, New York, 1970.

The major collection of complete test methods for coatings is "Physical and Chemical Examination of Paints, Varnishes, Lacquers, and Colors," by Gardner and Sward, Gardner Laboratory, Bethesda, Md.   This has gone through many editions.

hospitals, and schools. Another type of coating is one that provides a temporary surface on parts during handling, shipping, storage, or fabrication and then is peeled or stripped off.

Probably the chief reason for using a plastic or elastomeric coating is to provide protection against corrosion, stains, weather, abrasion, wear, and other service environments. Added to these is the fact that many modern plastic and elastomer coatings can provide other characteristics, such as electrical insulation, lubricity, temperature control, fire retardancy, control of marine fouling, and sound deadening.

For the sake of this discussion we will concentrate on the following types of plastic coatings: alkyd, acrylic, cellulosics, epoxy, fluorocarbon, melamine, nylon,

TABLE 5   Source of Specifications[27]

Abbreviation	Full name	Source of document
AMS	Aerospace Material Specifications	Society of Automotive Engineers 485 Lexington Ave., New York, N.Y.
ASTM	American Society for Testing and Materials	American Society for Testing and Materials 1916 Race St., Philadelphia, Pa.
ERF	Epoxy Resin Formulators	Society of Plastics Industry 250 Park Ave., New York, N.Y.
Fed. Std.	Federal Standard	General Services Administration 30 Church St., New York, N.Y. 300 N. Los Angeles St., Los Angeles, Calif. or 7th and D Sts., S.W., Washington, D.C., and Naval Supply Depot, 5801 Tabor Ave., Philadelphia, Pa.
MIL or MIL-STD	Military or Military Standard	Naval Supply Depot, 5801 Tabor Ave., Philadelphia, Pa.
CS	Commercial Standard	Superintendent of Documents, U.S. Government Printing Office, Washington, D.C.
AS	American Standard	American National Standards Institute 10 East 40th St., New York, N.Y.
AATCC	American Association of Textile Chemists and Colorists	American Association of Textile Chemists and Colorists, Lowell Technical Institute, Box 28, Lowell, Mass.
(Letter)-(Letter)-(No.)	....	Same as for Fed. Std.
IEC	International Electrotechnical Commission	American National Standards Institute 10 East 40th St., New York, N.Y.

phenolic, polyester, silicone, polyethylene, polyvinyl, and specialty coatings. Table 4 lists specific test methods for coatings, whereas Table 5 gives sources of coating specifications.

Alkyds

Alkyd coatings are the most popular type of plastic coating used currently. This type of coating offers low cost, ease of application, resistance to weathering, good gloss retention, good stability to sunlight, and good resistance to moisture.

Quite often, alkyds are combined with other synthetic resins to obtain enhanced properties. For example, alkyds often are combined with urea and/or melamine for use on appliances. Automobile enamel top coats traditionally consist of melamine with alkyd (see section on melamine), whereas primer for the automobile consists of urea with alkyd. Another advantage of alkyd coatings is their resistance to greases, many chemicals, moderate temperatures, and outdoor weather.

Chemistry   The term *alkyd* refers to a variety of resinous products. An important group in this classification is oil-modified-glyceryl phthalate resins. They are the reaction products of trihydric alcohol, glyceryl, and dibasic acid or phthalic acid modified with either drying or nondrying oils. Resins known as *resinous plasticizers,* like glycol or glycerol sebacate, also are alkyds. Another group is more generally known as polyesters.[7]

Alkyds are specified as being short, medium, or long. The greater the oil content in an alkyd resin, the more flexible the coating will be. Alkyds with shorter oil lengths have quicker drying times. However, they show less wetting ability and thus require careful surface preparation. Because oil-modified alkyds dry by both condensation and autoxidation, they are suitable for use in both baking and air-drying finishes. The straight and oil-modified alkyds constitute about half the total alkyds consumed by the coatings industry.[2]

Uses Alkyd coatings are more widely used on metal and wood products than any other coating system. Alkyds or modified alkyds have been used extensively as finishes for aluminum. Durability of these coatings generally is good, but they have limited formability and impact resistance. Their life expectancy is over 10 years if applied at 1 to 1.5 mils dried film thickness.

Alkyds are widely used on wood because they offer good all-round performance as fillers, sealers, and top coats at comparatively low cost. They may be formulated with low to high solids content and have good resistance to chemicals, stains, water, and abrasion.

Alkyds are used in electrodeposition applications. The most common water-soluble coatings in electrodeposition are those containing carboxyl groups along the polymer chain. The polymer backbone may be based on a variety of resins, including alkyds. Electrodeposited water-soluble alkyd coatings generally have high electrical resistance. They quickly coat all surfaces, and as these surfaces are coated, electric current approaches zero and the process stops for all practical purposes.[19]

Alkyd paints are applied by a variety of other methods, including brush, spray, or roller. Coverage is 350 to 400 ft^2/gal per coat.

Acrylic

Coatings produced from acrylic resins are characterized by excellent resistance to ultraviolet light and weather, clarity, and resistance to discoloration at high temperatures. They find use as automotive lacquers, airplane finishes, and clear coatings for bright metals, such as chromium electroplates. Because of their durable gloss during atmospheric exposure, acrylic coatings also find wide use as a finish coat for epoxies and vinyl to provide improved appearance for longer periods of time.[15]

Chemistry Acrylic resins used in coatings are derived from esters of acrylic and methacrylic acid by a vinyl type of polymerization. Included in this classification are methyl methacrylate and ethyl methacrylate. Two distinct types are in use: thermoplastic acrylics (lacquer) and thermosetting (enamel).[12] Coating formulations containing acrylate and methacrylate resins generally are baked and can be used for white enamels having excellent resistance to chemical fumes and to acid and alkali solutions. The acrylates in conjunction with vinyl chloride polymers have been used for clear coatings on a variety of metals. When properly baked, thermosetting acrylics, like lacquers, have exceptional exterior durability and exceptional gloss retention. In addition, they are harder and more resistant to detergents, grease stains, and chemicals than acrylic lacquers.

Nonaqueous acrylic dispersion coatings offer high application solids in a low-cost aliphatic hydrocarbon medium. Despite high solids content, the dispersion finishes have excellent reflow capabilities.[17] Experience to date indicates that these newer acrylics offer the best properties of thermosetting and thermoplastic acrylics combined in one.

Water-reducible acrylic coatings offer an excellent balance of postformability and durability in coil-coating operations. Water-reducible acrylic coatings can be used for exterior, as well as interior, applications where a hard, mar-resistant yet flexible finish, with excellent durability is required. Water-reducible coatings, like the acrylics, are designed to comply with air-pollution regulations, such as Rule 66 of Los Angeles County.

Acrylic coatings can be applied by most coating methods, including spraying, roller coating, coil coating, and electrodeposition.[5] Compared with conventional finishes, electrodeposited acrylic allows parts and assemblies to be produced twice as fast as before in many applications.

Acrylic resins are used in radiation-cured coating systems. In operation, radiation curing is done this way: electrons are generated by passing direct electric current through a filament at the top of an accelerator tube. As electrons are emitted and passed down a column through a series of accelerating electrodes, their energy is increased to the designed maximum. The method is fast and is done at room temperature. Equipment is expensive. Properties of radiation-cured acrylic coatings can be controlled by varying the polymer composition, polymer processing, monomer type, monomer-polymer ratio, and pigment and filler content. Because the monomers, which act as cross-linking agents, become an integral part of the cured film, the amount of material lost by evaporation is considerably less than with ordinary solvents. Radiation-cured acrylic coatings are used to coat ABS instrument panels in several automobiles.

Acrylic resins are used in electrically conductive coatings. The undersides and interiors of many consumer items, electronic musical instruments, and industrial equipment, such as glass-fiber-encased computer terminals, are coated with acrylic-base conductive coatings. The coatings block electrical interference, bleed static charges, and ensure complete electrical grounding. Also, acrylic conductive coatings are used on nonmetallic substrates, such as glass fiber and wood, for radar-signal reflection.

Cellulosics

Types of cellulosics   There are three major cellulose thermoplastics for coatings. These include cellulose nitrate, cellulose acetate butyrate, and ethyl cellulose. Cellulose coatings are modified with many different types of synthetic resins.[6]

Alkyds modified with either drying or nondrying types of oils are among resins widely used in high-solids-type cellulose lacquers. Also, phenolic resins are incorporated in nitrocellulose lacquers to improve resistance to moisture, acids, and alkalies.

Cellulose esters are obtained by replacing the alcoholic groups of cellulose with acid radicals. In the manufacture of nitrocellulose, cotton linters are treated with nitric and sulfuric acid mixtures until the desired degree of nitration is achieved. Ethyl cellulose, and ether, is produced by the reaction of alkali cellulose with ethyl chloride under heat and pressure. Ethyl cellulose is tasteless and nontoxic. Like nitrocellulose, it is soluble in many esters, ethers, ketones, and aromatic hydrocarbons. Ethyl cellulose is more resistant to heat, light, and electrical breakdown and is less flammable than nitrocellulose. Ethyl cellulose is used in hot-melt coatings, such as those used on army-ordnance hardware.

Nitrocellulose is popular as a coating material because it applies very easily, dries fast by solvent evaporation to a high gloss, and has excellent adhesion. Nitrocellulose lacquers are noted for their fine rubbing and buffing qualities, are fairly mar- and wear-resistant, and have good holdout when used as a base coating for printing inks.[9]

Uses   Nitrocellulose lacquers can be modified with urea, alkyd, and other resins to improve solids content, as well as solvent and stain resistance, and durability. They are used widely on wood furniture because they polish easily and can be retouched in case of damage. Cellulose lacquers are also commonly used on plastics. Striking decorative effects are achieved with lacquers by applying consecutive marks and multicolor spray coats. Also, dramatic increase in depth perspective is obtainable by lacquer coating the back of a transparent molded plastic part.

Cellulose lacquers are particularly good for thermoplastics because the film hardens quickly and completely without baking. Air drying eliminates heat shrinking and warping of molded plastic parts. Also, with lacquers plastic parts rejected because of blotches or errors in painting usually can be reclaimed by removing the lacquer with a combination of solvents similar to those used in the lacquer. Solvents are used to maintain proper lacquer viscosity during application. However, if too much solvent or unsuitable solvent is used, the lacquer may craze the plastic surface. This can be a critical problem. Crazing can occur on lacquered plastic parts whenever any portion of the part is subjected to greater tensile deformation or elongation than it would tolerate. Table 6 shows which solvents affect different thermoplastic surfaces.

Cellulose lacquers have a high surface luster and are pleasant to the touch. Color formulations range from crystal clear to color effects, from transparents and opaques to variegated colors. They have good dimensional stability, strength and toughness, low water absorption, low heat conductivity, and high dielectric strength. Also, they have good resistance to microorganisms (Table 7).

TABLE 6 Solvents That Affect Plastics

Resin	Heat-distortion point, °F	Solvents that affect surface
Acetal..............................	338	None
Methyl methacrylate.................	160–195	Ketones, esters, aromatics
Modified acrylic....................	170–190	Ketones, esters, aromatics
Cellulose acetate...................	110–209	Ketones, some esters
Cellulose propionate................	110–250	Ketones, esters, aromatics, alcohols
Cellulose acetate butyrate..........	115–227	Alcohols, ketones, esters, aromatics
Nylon...............................	260–360	None
Polyethylene:		
High density....................	140–180	
Med. density....................	120–150	None
Low density.....................	105–121	
Polypropylene.......................	210–230	None
Polycarbonate.......................	210–290	Ketones, esters, aromatics
Polystyrene (G.P. high heat)........	150–195	Some aliphatics, ketones, esters, aromatics
Polystyrene (impact, heat-resistant)....	148–200	Ketones, esters, aromatics, some aliphatics
ABS................................	165–225	Ketones, esters, aromatics, alcohol

Uses Cellulosic coatings are used on a broad range of industrial, as well as consumer metal products, including kitchen and bathroom fixtures, electrical, mechanical, and marine hardware, automotive parts, indoor and outdoor furniture, housings for outdoor electrical equipment, and many other metal products.[4]

Powder-coating materials based on cellulose acetate butyrate and cellulose acetate propionate are available in a range of particle sizes to produce films from 2 to 50 mils in an almost unlimited range of colors and hues. The butyrate ester results in a finish of extreme toughness and resists the elements of outdoor weathering. The propionate offers a finish which meets the specifications of the Food and Drug Administration (FDA), making it acceptable on containers that are in contact with food. A primer is required with cellulosic powder coatings. After application, the primer is flash-dried for several minutes and, if electrostatic powder application is to be used, is then passed to the coating operation. Some typical powder-coating costs, including those of cellulose acetate and butyrate, are shown in Table 8.

Epoxies

Chemistry Prepared by a condensation reaction between epichlorohydrin bisphenol A, epoxy resins are linear polymers that become cross-linked upon curing. The basic raw materials for epoxy resins come from petroleum. Chemically, epoxy resins are glycidyl ethers of a bishydroxyphenyl propane containing the epoxy carbon-oxygen-carbon ring as the significant functional group. They incorporate an ether-type linkage that contributes stability. Epoxies have well-spaced hydroxyl groups available for esterification and epoxide units that enhance reactivity. The curing or hardening epoxide resins can be obtained by reacting with amines or with polymers such as phenolic resins.[9]

Catalytic hardening agents may be either nonfunctional, secondary, or tertiary amines. Epoxy resins cured by reacting with phenolic or amino resins provide

TABLE 7   Resistance of Coating
Resins to Microorganisms[27]

Coating type	Resistance to attack
Acrylic	G
Alkyd	P
Cellulosics:	
Cellulose acetate	G to P
Cellulose acetate butyrate	G
Cellulose nitrate	P
Ethylcellulose	G
Chlorinated rubber	G
Chlorosulfonated polyethylene	G
Epoxy-amine	G
Epoxy-polyamide	G
Epoxy-phenol-formaldehyde	G
Epoxy-urea-formaldehyde	G
Melamine-formaldehyde	G to P
Phenol-formaldehyde	G to P
Phenoxy	G
Polyester	G
Polyethylene:	
Low molecular weight	P
High molecular weight	G
Polychlorofluoroethylene	G
Polyisobutylene	G
Polystyrene	G
Polyurethane	P
Polyvinyl acetate	P
Polyvinyl butyrate	P
Polyvinyl chloride	G*
Polyvinyl chloride–acetate	G
Polyvinylidene chloride	G to P
Polyvinylidene fluoride	G
Silicone	G
Silicone-alkyds	G to P
TFE fluorocarbon	G
Urea formaldehyde	G

* Except when plasticized.
G   Good
P   Poor

baking finishes of outstanding properties.   The phenolic blends display excellent chemical resistance and toughness.   Epoxy-urea coatings, although somewhat lower in chemical resistance, show improved color and gloss retention.

In common with other paints containing drying oils, epoxy paints are not suitable

TABLE 8   Comparison of Powder-coating Costs

Material	Cost per lb	Coverage, lb/ft² at 10 mils	Cost, cents/ft² at 10 mils
Vinyl	0.40–1.00	0.067–0.750	2.7–7.5
Polyethylene	0.25–0.70	0.050–0.055	1.3–3.5
Cellulose acetate butyrate	0.70–0.80	0.065–0.070	4.5–5.6
Polyester	1.00–1.30	0.065–0.070	6.5–9.1
Nylon	2.00–2.50	0.056–0.060	11.2–15.0
Chlorinated polyether	6.50–7.00	0.075	49.0–53.0

for water immersion. Adhesion is probably one of the prime reasons for using an epoxy coating. Epoxies also contribute exceptional corrosion-resistant qualities to coatings. Also, they are tough, abrasion-resistant, and flexible.

Availability Epoxy coatings are supplied in many forms, including solvent-base systems, powders, and water-base systems. Other types include two-package, solvent-free viscous liquids; two-package, amine-cured epoxy solution coatings; single-package, baking-type epoxy esters which can be either air-drying or baking; and other systems.

Epoxy coatings are popular as an electrical insulation. They have high dielectric strength over a wide temperature range, up to 1,200 V/mil in heavier thicknesses (see Table 9). Other electrical properties are given in Tables 10 to 12.

Preheat temperature requirements for epoxy-powder coatings are low (250 to 400°F) because of the low melt viscosity of uncured resins. No primers are needed and adhesion is excellent. Chemical resistance is good, but epoxy coatings tend to become brittle when thick. Epoxy-powder prices range from $1 to $1.50/lb.[13]

Epoxy coatings do not require special application techniques. They can be applied by brush, spray, flow coating, dipping, or rolling.

One unique coating is an acrylic-modified epoxy for radiation curing. Radiation curing is fast, allowing production-line speeds of up to 2,000 ft/min. Also, the technique takes place at room temperature, and thus, heat-sensitive materials like wood or hardboard can be given the equivalent of a baked finish at speeds never before possible. The only solvents which can be used in radiation-cured paint formulations are reactive monomers which become part of the final film. This feature reduces material cost, avoids air-pollution problems, and produces an unusually uniform film unflawed by vaporizing solvents in the coating. A typical radiation-curing system is shown in Fig. 2.

Fluorocarbons

The four major fluorocarbon polymers used for coatings, in order of degree of fluorination, are (1) polyvinyl fluoride, (2) polyvinylidene fluoride, (3) polytrifluorochloroethylene, and (4) polytetrafluoroethylene.

Although fluorine can be combined with many materials to yield fluorinated polymers, the major impact of the fluorine/carbon combination has been in the series of substituted ethylene polymers upon which fluorocarbon coatings are based. Physical, mechanical, and electrical properties of the four fluorocarbon polymers are given in Table 13. As can be seen, fluorocarbon finishes, besides offering nonstick properties, also provide heat and chemical resistance, cryogenic stability, low coefficient of friction, unique electrical properties, and nonwetting features.[1]

Polytrifluorochloroethylene and polytetrafluoroethylene are used as chemical-resistant coatings and linings where severe chemical environments justify the high cost of these extremely resistant materials. The coatings have exceptional release characteristics, and polytetrafluoroethylene (PTFE) commonly is used as a nonstick finish on cookware. Polyvinyl fluoride and polyvinylidene fluoride are used for long-term protection of metals outdoors. Typical applications include factory buildings, aircraft hangars, and other outdoor structures.[6]

Polytetrafluoroethylene (PTFE) has a strongly bonded molecular structure consisting of carbon and fluorine atoms only. Several clear and pigmented PTFE coating grades are available, and all of them are water-borne dispersions requiring a primer before application of top coat. A close relative to polytetrafluoroethylene (PTFE) is polyfluoroethylene propylene, or FEP resin. This coating material retains most of the properties of PTFE except maximum service temperature, which is 400°F, compared with about 550°F for conventional PTFE coatings. An advantage of an FEP coating is that it can be baked at 600 to 700°F, compared with about 725 to 750°F for regular PTFE. Also, FEP coatings are less porous and less permeable than PTFE coatings because FEP melt flows to form the coating film, whereas PTFE sinters by interparticulate contact.[1]

Uses Besides cookware, the nonstick feature of fluorocarbon finishes makes them useful for such products as putty knives, pruning saws, door-lock parts, fan and blower blades, sliding- and folding-door hardware, skis, and snow shovels (see

TABLE 9   Dielectric Strengths of Polymer Coatings[27]

Material	Dielectric strength, volts/mil	Comments*	Source of information
Polymer coatings:			
Acrylics..............	450–550	Short-time method	a
	350–400	Step-by-step method	a
	400–530	....	b
	1,700–2,500	2-mil-thick samples	Columbia Technical Corp., Humiseal Coatings
Alkyds..............	300–350	....	b
Chlorinated polyether	400	Short-time method	a
Chlorosulfonated poly- ethylene...........	500	Short-time method	a
Diallyl phthalate.....	275–450	....	b
	450	Step-by-step method	a
Diallyl isophthalate...	422	Step-by-step method	a
Depolymerized rubber (DPR)	360–380	....	H. V. Hardman Co., DPR Subsidiary
Epoxy..............	650–730	Cured with anhydride–castor oil adduct	Autonetics, Div. of North American Rockwell
Epoxy..............	1,300	10-mil-thick dip coating	
Epoxies, modified.....	1,200–2,000	2-mil-thick sample	Columbia Technical Corp., Humiseal Coatings
Neoprene...........	150–600	Short-time method	a
Phenolic............	300–450	....	b
Polyamide..........	780	106 mils thick	
Polyamide-imide......	2,700		
Polyesters...........	250–400	Short-time method	a
	170	Step-by-step method	a
Polyethylene.........	480	....	b
	300	60-mil-thick sample	c
	500	Short-time method	a
Polyimide...........	3,000	Pyre-M.L., 10 mils thick	d
	4,500–5,000	Pyre-M.L. (RC-675)	e
	560	Short-time method, 80 mils thick	a
Polypropylene........	750–800	Short-time method	a
Polystyrene..........	500–700	Short-time method	a
	400–600	Step-by-step method	a
	450	60-mil-thick sample	c
Polysulfide...........	250–600	Short-time method	a
Polyurethane (single component)........	3,800	1-mil-thick sample	f
Polyurethane (two components)/castor oil cured...........	530–1,010	....	g
Polyurethane (two components, 100% solids)	275	125-mil-thick sample	Products Research & Chem. Corp. (PR-1538)
	750	25 mils thick sample	
Polyurethane (single component)	2,500	2-mil-thick sample	Columbia Technical Corp., Humiseal 1A27
Polyvinyl butyral.....	400		
Polyvinyl chloride....	300–1,000	Short-time method	
	275–900	Step-by-step method	
Polyvinyl formal......	860–1,000	....	b
Polyvinylidene fluoride	260	Short-time, 500-volt/sec, ⅛-in. sample	h
	1,280	Short-time, 500-volt/sec, 8-mil sample	h
	950	Step by step (1-kv steps)	h

TABLE 9 Dielectric Strengths of Polymer Coatings[27] (Continued)

Material	Dielectric strength, volts/mil	Comments*	Source of information
Polyxylylenes:			
Parylene N	6,000	Step by step	Union Carbide Corp.
	6,500	Short time	Union Carbide Corp.
Parylene C	3,700	Short time	Union Carbide Corp.
	1,200	Step by step	Union Carbide Corp.
Parylene D	5,500	Short time	Union Carbide Corp.
	4,500	Step by step	Union Carbide Corp.
Silicone	500	Sylgard 182	Dow Corning Corp.
Silicone	550–650	RTV types	General Electric & Stauffer Chemical Co. bulletins
Silicone	800	Flexible dielectric gel	Dow Corning Corp.
Silicone	1,500	2-mil-thick sample	Columbia Technical Corp., Humiseal 1H34
TFE fluorocarbons	400	60-mil-thick sample	c
	480	Short-time method	a
	430	Step by step	a
Teflon TFE dispersion coating	3,000–4,500	1–4-mil-thick sample	E. I. du Pont de Nemours & Co.
Teflon FEP dispersion coating	4,000	1.5-mil-thick sample	E. I. du Pont de Nemours & Co.
Other materials used in electronic assemblies:			
Alumina ceramics	200–300	. . . .	b
Boron nitride	900–1,400		
Electrical ceramics	55–300	. . . .	b
Forsterite	250	. . . .	b
Glass, borosilicate	4,500	40-mil sample	c
Steatite	145–280	. . . .	b

* All samples are standard 125 mils thick unless otherwise specified.
 [a] *Insulation*, Directory Encyclopedia Issue, no. 7, June–July, 1968.
 [b] *Mater. Eng.*, Materials Selector Issue, vol. 66, no. 5, Chapman-Reinhold Publication, mid-October, 1967–1968.
 [c] Kohl, W. H.,: "Handbook of Materials and Techniques for Vacuum Devices," p. 586, Reinhold Publishing Corporation, New York, 1967.
 [d] Learn, J. R., and M. P. Seegers: Teflon-Pyre-M. L. Wire Insulation System, *13th Symp. on Tech. Progr. in Commun. Wire and Cables*, Atlantic City, N.J., Dec. 2–4, 1964.
 [e] Milek, J. T.: Polyimide Plastics: A State of the Art Report, *Hughes Aircraft Rep.* S-8, October, 1965.
 [f] *Hughson Chemical Co. Bull.* 7030A.
 [g] *Spencer-Kellogg* (Division of Textron, Inc.) *Bull.* TS-6593.
 [h] *Pennsalt Chemicals Corp. Prod. Sheet* KI-66a, Kynar Vinylidene Fluoride Resin, 1967.

Fig. 3). There are many automotive applications for nonstick fluorocarbon finishes. A typical example is a fluorocarbon-coated expender ring for pistons. Here, the fluorocarbon coating prevents carbon and sludge buildup, and sticking and clogging problems are eliminated. In another automotive use, carburetor throttle shafts are coated with fluorocarbon because of the slick finish, which is applied to a butterfly valve on top of the carburetor as well as to the throttle shaft underneath the carburetor, provides good dry lubrication, short break-in period, good solvent resistance, and good temperature resistance from −30 to 325°F.

Application Fluorocarbon coatings, because they have nonstick and nonwetting characteristics, are difficult to adhere to other materials. The first step in applying a fluorocarbon finish is grit blasting or mechanical abrading. The object is to achieve

TABLE 10   Volume Resistivities of Polymer Coatings[27]

Material	Volume resistivity at 25°C, ohm-cm	Source of information
Acrylics....	10^{14}–10^{15}	a
	$>10^{14}$	b
	7.6×10^{14}–1.0×10^{15}	Columbia Technical Corp., Humiseal
Alkyds........................	10^{14}	a
Chlorinated polyether............	10^{15}	b
Chlorosulfonated polyethylene.....	10^{14}	b
Depolymerized rubber...........	1.3×10^{13}	H. V. Hardman, DPR Subsidiary
Diallyl phthalate................	10^{8}–2.5×10^{10}	a
Epoxy (cured with DETA)........	2×10^{16}	c,d
Epoxy polyamide	1.1–1.5×10^{14}	
Phenolics......................	6×10^{12}–10^{13}	a
Polyamides....................	10^{13}	
Polyamide-imide................	7.7×10^{16}	e
Polyethylene...................	$>10^{16}$	
Polyimide.....................	10^{16}–10^{18}	f
Polypropylene..................	10^{10}–$>10^{16}$	a
Polystyrene....................	$>10^{16}$	b
Polysulfide....................	2.4×10^{11}	g
Polyurethane (single component)...	5.5×10^{12}	h
Polyurethane (single component)...	2.0×10^{12}	h
Polyurethane (single component)...	4×10^{13}	Columbia Technical Corp.
Polyurethane (two components)....	1×10^{13}	Products Research &
	$5 \times 10^{9}(300°\text{F})$	Chemical Corp. (PR-1538)
Polyvinyl chloride..............	10^{11}–10^{15}	b
Polyvinylidene chloride..........	10^{14}–10^{16}	a
Polyvinylidene fluoride...........	2×10^{14}	h
Polyxylylenes (parylenes).........	10^{16}–10^{17}	Union Carbide Corp.
Silicone (RTV).................	6×10^{14}–3×10^{15}	Stauffer Chemical Co., Si-O-Flex SS 831, 832, & 833
Silicone, flexible dielectric gel......	1×10^{15}	Dow Corning Corp.
Silicone, flexible, clear............	2×10^{15}	Dow Corning Corp.
Silicone.......................	3.3×10^{14}	Columbia Technical Corp. Humiseal 1H34
Teflon TFE	$>10^{18}$	b
Teflon FEP....................	$>2 \times 10^{18}$	b

[a] *Mater. Eng.*, Materials Selector Issue, vol. 66, no. 5, Chapman-Reinhold Publication, mid-October, 1967–1968.

[b] *Insulation*, Directory Encyclopedia Issue, no. 7, June–July, 1968.

[c] Lee, H., and K. Neville: "Epoxy Resins," McGraw-Hill Book Company, New York, 1966.

[d] Tucker, Cooperman, and Franklin: Dielectric Properties of Casting Resins, *Electron. Equip.*, July, 1956.

[e] Freeman, J. H.: A New Concept in Flat Cable Systems, *5th Ann. Symp. on Advan. Tech. for Aircraft Elec. Syst.*, Washington, D.C., October, 1964.

[f] Milek, J. T.: Polyimide Plastics: A State of the Art Report *Hughes Aircraft Rep.* S-8, October, 1965.

[g] Hockenberger, L.: *Chem.-Ing. Tech.*, vol. 36, 1964.

[h] *Hughson Chemical Co. Tech. Bull.* 7030A; *Pennsalt Chemicals Corp. Prod. Sheet* KI-66a, Kynar Vinylidene Fluoride Resin, 1967.

TABLE 11   Dielectric Constants of Polymer Coatings[27] (at 25°C)

Coating	60–100 Hz	10^6 Hz	>10^6 Hz	Reference source
Acrylic		2.7–3.2		a
Alkyd			3.8 (10^{10} Hz)	b
Asphalt and tars			3.5(10^{10} Hz)	b
Cellulose acetate butyrate		3.2–6.2		a
Cellulose nitrate		6.4		a
Chlorinated polyether	3.1	2.92		a
Chlorosulfonated polyethylene (Hypalon)	6.19 7–10(10^3 Hz)	~5		E. I. du Pont de Nemours & Co.
Depolymerized rubber (DPR)	4.1–4.2	3.9–4.0		H. V. Hardman, DPR Subsidiary
Diallyl isophthalate	3.5	3.2	3(10^8 Hz)	a,c
Diallyl phthalate	3–3.6	3.3–4.5		a,c
Epoxy-anhydride—castor oil adduct	3.4	3.1	2.9(10^7 Hz)	Autonetics, Division of North American Rockwell
Epoxy (one component)	3.8	3.7		Conap Inc.
Epoxy (two components)	3.7			Conap Inc.
Epoxy cured with methylnadic anhydride (100:84 pbw)	3.31			d
Epoxy cured with dodecenyl-succinic anhydride (100:132 pbw)	2.82			d
Epoxy cured with DETA	4.1	4.2	4.1	e
Epoxy cured with m-phenylenediamine	4.6	3.8	3.25(10^{10} Hz)	e
Epoxy dip coating (two components)	3.3	3.1		Conap Inc.
Epoxy (one component)	3.8	3.5		Conap Inc.
Epoxy-polyamide (40 % Versamid* 125, 60 % epoxy)	3.37	3.08		e
Epoxy-polyamide (50 % Versamid 125, 50 % epoxy)	3.20	3.01		e
Fluorocarbon (TFE, Teflon)	2.0–2.08	2.0–2.08		E. I. du Pont de Nemours & Co.
Phenolic		4–11		a
Phenolic	5–6.5	4.5–5.0		c
Polyamide	2.8–3.9	2.7–2.96		
Polyamide-imide	3.09	3.07		
Polyesters	3.3–8.1	3.2–5.9		c
Polyethylene		2.3		a
Polyethylenes	2.3	2.3		c
Polyimide-Pyre-M.L.† enamel	3.8	3.8		f
Polyimide–Du Pont RK-692 varnish	3.8			g
Polyimide–Du Pont RC-B-24951	3.0(10^3 Hz)			g
Polyimide–Du Pont RC-5060	2.8(10^3 Hz)			g
Polypropylene		2.1		a
Polypropylene	2.22–2.28	2.22–2.28		c
Polystyrene	2.45–2.65	2.4–2.65	2.5(10^{10} Hz)	b
Polysulfides	6.9			h
Polyurethane (one component)	4.10	3.8		Conap Inc.
Polyurethane (two components—castor oil cured)		2.98–3.28		i
Polyurethane (two components)	6.8(10^3 Hz)	4.4		Products Research & Chemical Corp. (PR-1538)
Polyvinyl butyral	3.6	3.33		a
Polyvinyl chloride	3.3–6.7	2.3–3.5		a
Polyvinyl chloride–vinyl acetate copolymer	3–10			a
Polyvinyl formal	3.7	3.0		a
Polyvinylidene chloride		3–5		a
Polyvinylidene fluoride	8.1	6.6		j
Polyvinylidene fluoride	8.4	6.43	2.98(10^9 Hz)	k

TABLE 11   Dielectric Constants of Polymer Coatings[27] (Continued)

Coating	60–100 Hz	10^6 Hz	$>10^6$ Hz	Reference source
p-Polyxylylene:				
Parylene N	2.65	2.65	....	Union Carbide Corp.
Parylene C	3.10	2.90	....	Union Carbide Corp.
Parylene D	2.84	2.80	....	Union Carbide Corp.
Shellac (natural, dewaxed)	3.6	3.3	2.75(10^9 Hz)	*l*
Silicone (RTV types)	3.3–4.2	3.1–4.0	....	General Electric and Stauffer Chemical Cos.
Silicone (Sylgard ‡ type)	2.88	2.88	....	Dow Corning Corp.
Silicone, flexible dielectric gel	3.0	....	....	Dow Corning Corp.
FEP dispersion coating	2 1(10^3 Hz)	....	....	E. I. du Pont de Nemours & Co.
TFE dispersion coating	2.0–2.2(10^3 Hz)	....	....	E. I. du Pont de Nemours & Co.
Wax (paraffinic)	2.25	2.25	2.22(10^{10} Hz)	*l*

* Trademark of General Mills, Inc., Kankakee, Ill.
† Trademark of E. I. du Pont de Nemours & Co., Wilmington, Del.
‡ Trademark of Dow Corning Corporation, Midland, Mich.
a Mater. Eng., Materials Selector Issue, vol. 66, no. 5, Chapman-Reinhold Publication, mid-October 1967–1968.
b Volk, M. C., J. W. Lefforge, and R. Stetson: "Electrical Encapsulation," Reinhold Publishing Corporation, New York, 1962.
c Insulation, Directory Encyclopedia Issue, no. 7, June-July, 1968.
d Coombs, C. F. (ed.): "Printed Circuits Handbook," McGraw-Hill Book Company, New York, 1967.
e Lee, H., and K. Neville: "Handbook of Epoxy Resins," McGraw-Hill Book Company, New York, 1967.
f Learn, J. R., and M. P. Seegers: Teflon-Pyre-M.L. Wire Insulation System, *13th Symp. of Tech. Progr. in Commun. Wire and Cable*, Atlantic City, N.J., Dec. 2–4, 1964.
g Du Pont Bull. H65-4, Experimental Polyimide Insulating Varnishes, RC-B-24951 and RC-5060, January, 1965.
h Hockenberger, L.: *Chem.-Ing. Tech.* vol. 36, 1964.
i Spencer-Kellogg (division of Textron, Inc.) *Bull.* TS-6593.
j Barnhart, W. S., R. A. Ferren, and H. Iserson: *17th ANTEC of SPE*, January, 1961.
k Pennsalt Chemicals Corp. Prod. Sheet KI-66a, Kynar Vinylidene Fluoride Resin, 1967.
l Von Hippel, A. R. (ed.): "Dielectric Materials and Applications," Technology Press of MIT and John Wiley & Sons, Inc., New York, 1961.

surface profile which is adequate to provide satisfactory tooth, but not so deep as to leave a surface which is too rough or easily damaged. The next step is application of a primer. After the primer has been baked and cooled, it is followed immediately by top-coat application. A top-coat coating thickness of 0.7 to 1.2 mil in one application is recommended. After the top coat is applied, it is flashed to remove water. Then, it is baked at a temperature over 725°F in order to obtain coalescence. Coalescence is obtained through softening of the particle surface and interparticulate contact. The above procedures apply to PTFE water-dispersion coatings and, although not used as such for other fluorocarbon-coating systems, do give an indication of how fluorocarbons are applied as a coating.

Melamine

Although melamine resins, including urea formaldehyde, are rarely used alone as finishes, they are an important group of resins as additives for coatings and are mentioned here separately. Melamine and other amino resins are used mostly in combinations with alkyds and other resins. Melamine resins offer very high hardness, resistance to detergents, high-arc tracking resistance, and premanency of color and are self-extinguishing. One of the most important contributions of melamine resins is rapid curing at lower temperatures.[7]

Urea and melamine resins basically are urea and formaldehyde condensation products. Resins recommended for use with alkyds are modified with butyl alcohol

TABLE 12   Dissipation Factors of Polymer Coatings[27] (at 25°C)

Coating	60–100 Hz	10^6 Hz	>10^6 Hz	Reference source
Acrylics	0.04–0.06	0.02–0.03		a
Alkyds	0.003–0.06			a
Chlorinated polyether	0.01	0.01		a
Chlorosulfonated polyethylene	0.03	0.07(10^3 Hz)		b
Depolymerized rubber (DPR)	0.007–0.013	0.0073–0.016		H. V. Hardman, DPR Subsidiary
Diallyl phthalate	0.010	0.011	0.011	b
Diallyl isophthalate	0.008	0.009	0.014(10^{10} Hz)	b
Epoxy dip coating (two components)	0.027	0.018		Conap Inc.
Epoxy (one component)	0.011	0.004		Conap Inc.
Epoxy (one component)	0.008	0.006		Conap Inc.
Epoxy polyamide (40 % Versamid 125, 60 % epoxy)	0.0085	0.0213		c
Epoxy polyamide (50 % Versamid 115, 50 % epoxy)	0.009	0.0170		c
Epoxy cured with anhydride–castor oil adduct	0.0084	0.0165	0.0240	Autonetics, North American Rockwell
Phenolics	0.005–0.5	0.022		a
Polyamide	0.015	0.022–0.097		
Polyesters	0.008–0.041	.....		a
Polyethylene (linear)	0.00015	0.00015	0.0004(10^{10} Hz)	d
Polymethyl methacrylate	0.06	0.02	0.009(10^{10} Hz)	d
Polystyrene	0.0001–0.0005	0.0001–0.0004		
Polyurethane (two component, castor oil cure)		0.016 –0.036		e
Polyurethane (one component)	0.038–0.039	0.068–0.074		Conap Inc.
Polyurethane (one component)	0.02	.....		Conap Inc.
Polyvinyl butyral	0.007	0.0065		
Polyvinyl chloride	0.08–0.15	0.04–0.14		a
Polyvinyl chloride, plasticized	0.10	0.15	0.01(10^{10} Hz)	d
Polyvinyl chloride–vinyl acetate copolymer	0.6–0.10			
Polyvinyl formal	0.007	0.02		
Polyvinylidene fluoride	0.049	0.17		f
	0.049	0.159	0.110	g
Polyxylylenes:				
Parylene N	0.0002	0.0006		Union Carbide Corp.
Parylene C	0.02	0.0128		Union Carbide Corp.
Parylene D	0.004	0.0020		Union Carbide Corp.
Silicone (Sylgard 182)	0.001	0.001		Dow Corning Corp.
Silicone, flexible dielectric gel	0.0005	.....		Dow Corning Corp.
Silicone, flexible, clear	0.001	.....		Dow Corning Corp.
Silicone (RTV types)	0.011–0.02	0.003–0.006		General Electric
Teflon FEP dispersion coating	0.0002–0.0007	.....		E. I. du Pont de Nemours & Co.
Teflon FEP	<0.0003	<0.0003		a
Teflon TFE	<0.0003			a
Teflon TFE	0.00012	0.00005		Union Carbide Corp.
Other materials:				
Alumina (99.5 %)		0.0001		h
Beryllia (99.5 %)		0.0003		h
Glass silica	0.0006	0.0001	0.00017(10^{10} Hz)	d
Glass, borosilicate		0.013–0.016		Corning Glass Works
Glass, 96 % silica		0.0015–0.0019		Corning Glass Works

a *Mach. Des.*, Plastics Reference Issue, vol. 38, no. 14, Penton Publishing Co., 1966.
b *Insulation*, Directory, Encyclopedia Issue, no. 7, June–July, 1968.
c Lee, H. and K. Neville: "Handbook of Epoxy Resins," McGraw-Hill Book Company, New York, 1967.
d Mathes, K.: Electrical Insulation Conference, 1967.
e *Spencer-Kellogg* (Division of Textron, Inc.) *Tech. Bull.* TS-6593.
f Barnhart, W. S., R. A. Ferren, and H. Iserson: *17th ANTEC of SPE*, January, 1961.
g *Pennsalt Chemicals Corp. Prod. Sheet* KI-66a, Kynar Vinylidene Fluoride Resin, 1967.
h *Mach. Des.*, Design Guide, Sept. 28, 1967.

to produce butylated amino resins. This imparts solubility in combinations with butyl alcohol and aromatic and aliphatic hydrocarbons, which are practical solvents for alkyds. Amino resins are used in baking finishes and occasionally in some air-drying coatings. Baked films of unmodified resins are hard and brittle and have poor adhesion to metals.

Nylon (Polyamide)

Nylon coatings on metal surfaces make it possible to combine the desirable frictional and wear characteristics of the material with the greater strength, dimensional

Fig. 2   Radiation curing is fast, allowing production-line speeds of up to 2,000 ft/min. The technique takes place at room temperature, and heat-sensitive materials like wood or hardboard can be given the equivalent of a baked finish at speeds never before possible. (*Radiation Dynamics.*)

stability, and lower cost of metals. The coefficient of friction of nylon and other coatings is shown in Table 14. Adherent coatings ranging in thickness from 2 to 30 mils have been applied to aluminum and many types of steel, including stainless. Also, ceramics have been coated successfully.

Nylon or polyamide resins are polymers made by the reaction of dibasic acids or their amide-forming derivatives with polyamines or by the polymerization of amino acids or lactams. The oldest and most widely used type of nylon is that based on adipic acid and hexamethylene diamine (called nylon 6/6 because both the acid and amine components have 6 carbon atoms). Another major type is nylon 6/10, in which sebacic acid is used instead of adipic acid to yield a softer, more flexible polymer with lower water absorption and greater stiffness at high humidities. A newer type, nylon 6, is made by heating caprolactam (a 6-carbon ring-type amide) under conditions in which the ring opens to form a linear polymer. This type has a lower softening temperature and a wider solubility than other types of nylons.[18]

The unique properties of nylon or polyamide coatings make them useful in a variety of applications. For example, they can be used to protect surfaces from abrasion, scratching, and mechanical forces; protect surfaces from the effects of weather, corrosive environments, and moisture; protect against the attack of solvents and chemicals; give improved functionality to a surface, such as reducing coefficient of friction; and enhance appearance. The properties of nylon coatings are given

TABLE 13   Comparison of Properties of Four Fluorocarbons

Property	Polyvinyl fluoride (PVF) $(CH_2\text{-}CHF)_n$	Polyvinylidene fluoride (PVF-2) $(CH_2\text{-}CF_2)_n$	Polytrifluorochloroethylene (PTFCl) $(CClF\text{-}CF_2)_n$	Polytetrafluoroethylene (PTFE) $(CF_2\text{-}CF_2)_n$
Physical properties:				
Density	1.4	1.76	2.104	2.17–2.21
Fusing temp, °F	300	460	500	750
Max continuous service and temp, °F	225	300	400	550
Coefficient of friction	0.16	0.16	0.15	0.1
Flammability	Burns	Non-flammable	Non-flammable	Non-flammable
Mechanical properties:				
Tensile strength, lb/in.^{2}	7,000	7,000	5,000	2,500–3,500
Elongation, %	115–250	300	250	200–400
Izod impact, ft-lb/in.	....	3.8	5	3
Durometer hardness	....	80	74–78	50–65
Yield strength at 77°F, lb/in.^{2}	6,000	5,500	4,500	1,300
Heat-distortion temp at 66 lb/in.^{2}, °F	NA	300	265	250
Coefficient of linear expansion	2.8×10^5	8.5×10^5	15×10^5	8×10^5
Modulus (tension) $\times 10^5$ lb/in.^{2}	2.5–3.7	1.2	1.9	0.6
Electrical properties:				
Dielectric strength, V/mil	....	260 (0.125)	500 (0.063)	600 (0.060)
Short time, V/mil, in.	3,400 (0.002)			
Dielectric constant, 10^3 Hz	8.5	7.72	2.6	2.1
Arc resistance (77°F) ASTM D 495	NA	60	300	300
Volume resistivity Ω-cm at 50% RH 77°F	10^{12}	10^{14}	10^{16}	10^{18}
Dissipation factor, 100 Hz	1.6	0.05	0.022	0.0003

in Table 15, which gives a comparison of three types of higher-molecular-weight nylon products.

Application Nylon coatings are applied in several ways, including flame spraying and the fluidized-bed process. In the fluidized process, nylon powder is held in suspension in a closed vessel by means of a cushion of air. The part or product to be coated is normally heated above the melting point of the resin and then dipped into the powder until is it coated to the desired thickness. Of several powder-coating types available, nylon 11 is particularly useful because of its low coefficient of friction and high resistance to abrasion, chipping, and scratching. Nylon 6/10 is a fair substitute for nylon 11 but is about 20 percent more expensive. Nylon 6/10 is more moisture-resistant than nylon 6 and 6/6, more flexible, and easier to work with. Nylon 12 is one of the newest nylons to be used in plastic-powder coatings. Nylon requires a primer. Price ranges from $2.50 to $3.00/lb.

Use of a primer is recommended where a nylon coating is intermittently exposed to steam or boiling water, as in steam sterilization. A primer should also be used for intermittent exposure to hot water or even to high humidities and warm temperatures for extended periods of time to retain adhesion. With a primer, nylon coatings can withstand boiling water without loss of adhesion for at least 8 h and can be steam-sterilized. Nylon coatings of 10 to 20 mils are preferred for most purposes. However, in some cases, thinner coatings down to 2 to 3 mils are desirable.[1]

Phenolic

Phenolic coatings are outstanding in their resistance to alkalies, chemicals, and moisture, and find their widest use in products subject to such exposures. These

Fig. 3 Nonstick feature of fluorocarbon finishes makes them useful for products like putty knives, pruning saws, door-lock parts, fan and blower blades, sliding- and folding-door hardware, skis, and snow shovels. (*E. I. du Pont de Nemours & Company, Inc.*)

finishes adhere well to metals, dry fast, and withstand high temperatures for short periods of time. Phenolic coatings are among the must durable finishes for outdoor uses.

Phenolic resins were the first truly synthetic resins to become commercially available (in 1909). These resins are made by the condensation of a phenyl with an aldehyde. A variety of types are available, depending on the nature and proportions of the phenyl and aldehyde, type of catalyst used, and reaction conditions. Formaldehyde, because of its high reactivity and low cost, is by far the most used aldehyde. Besides phenyl, many substituted phenyls are of wide commercial importance. Both heat-reactive and non-heat-reactive types of phenolic resins are available. Heat reactivity, with the associated capability for cross linking and insolubilization, is favored by high proportions of formaldehyde and by alkaline catalysts. Acid catalysts generally yield thermoplastic resins that are permanently fusible and soluble, although a high formaldehyde ratio can impart cross-linking characteristics to these resins also. Phenolic resins have a yellow color which tends to darken on aging. Hence, they are not usually used in white finishes. The darkening tendency

TABLE 14   Coefficients of Friction of Typical Coatings[27]

Coating	Coefficient of friction, μ	Information source
Polyvinyl chloride	0.4–0.5	a
Polystyrene	0.4–0.5	a
Polymethyl methacrylate	0.4–0.5	a
Nylon	0.3	a
Polyethylene	0.6–0.8	a
Polytetrafluoroethylene (Teflon)	0.05–0.1	a
Catalyzed epoxy air-dry coating with Teflon filler	0.15	b
Parylene N	0.25	c
Parylene C	0.29	c
Parylene D	0.31–0.33	c
Polyimide (Pyre-M.L.)	0.17	d
Graphite	0.18	d
Graphite–molybdenum sulfide:		
Dry-film lubricant	0.02–0.06	e
Steel on steel	0.45–0.60	e
Brass on steel	0.44	e
Babbitt on mild steel	0.33	e
Glass on glass	0.4	e
Steel on steel with SAE no. 20 oil	0.044	e
Polymethyl methacrylate to self	0.8 (static)	e
Polymethyl methacrylate to steel	0.4–0.5 (static)	e

[a] Bowder, F. P., *Endeavor*, vol. 16, no. 61, p. 5, 1957.
[b] Product Techniques Incorporated: Bulletin on PT-401 TE, Oct. 17, 1961.
[c] Union Carbide data.
[d] *Du Pont Tech. Bull.* 19, Pyre-M.L. Wire Enamel, August, 1967.
[e] Electrofilm, Inc. data.

is greatest with the resin-modified phenolic and least with the parasubstituted pure phenolics.[10]

Types   Phenolic resins used in coatings may be classified on the basis of their composition and properties into the following main categories: (1) modified phenolic resins, oil-soluble; (2) pure phenolic resins (100 percent phenolic); (3) phenolic dispersion resins. Quite often phenolic coatings are used in electrical and electronic

TABLE 15   Comparison of the Properties of Nylon Coatings

	Nylon 11	Nylon 6/6	Nylon 6
Elongation (73°F), %	120	90	50–200
Tensile strength (73°F), lb/in.^{2}	8,500	10,500	10,500
Modulus of elasticity (73°F), lb/in.^{2}	178,000	400,000	350,000
Rockwell hardness	R 100.5	R 118	R 112–118
Specific gravity	1.04	1.14	1.14
Moisture absorption, %, ASTM D 570	0.4	1.5	1.6–2.3
Thermal conductivity Btu/(ft^2)(h)(°F/in.)	1.5	1.7	1.2–1.3
Dielectric strength (short time), V/mil	430	385	440
Dielectric constant (10 Hz)	3.5	4	4.8
Effect of:			
Weak acids	None	None	None
Strong acids	Attack	Attack	Attack
Strong alkalies	None	None	None
Alcohols	None	None	None
Esters	None	None	None
Hydrocarbons	None	None	None

applications.   In such cases, the phenolic often is modified with another synthetic resin like polyester.

The fast drying and poor wetting characteristics of phenolic coatings necessitate very thorough surface preparation to obtain adequate adhesion.   Because phenolic coatings continue to cure and harden on again, failure from a lack of intercoat adhesion can occur if too much time elapses between coats.   Ordinarily, successive coats of phenolic paint should be applied within 24 h after the preceding coat.   Phenolic coatings generally are applied by brushing or spraying.

Polyester

The term *polyester* commonly is used to denote a linear-type alkyd having carbon-to-carbon double bond on saturation in the polymer chain.   These unsaturated polyesters may be cross-linked by reaction with a monomer such as styrene or diallyl phthalate to form an insoluble and infusible resin, without the formation of a by-product during the curing reaction.   Unsaturated polyesters are reacted to form solids by using catalysts, often MEK.   The resulting solids are thermosetting plastics.   Unsaturated polyesters are free of oil, fatty acids, or other natural products and are prepared with hydroxyl, carboxyl, or other reactive groups at the terminal sites and along the molecular chain.   Alkyd resins, although they are a form of reactive polyester, have long been prepared from naturally occurring oils.   Thus, alkyds differ from unsaturated polyesters by the presence of natural oils or resins.[12]

Good adhesion, high hardness, outstanding solvent and stain resistance, good over-bake resistance, outstanding flexibility and impact resistance, and excellent salt-spray resistance are some of the characteristics of polyester coatings.   Because of these features, polyesters are being used for transportation equipment, appliances, sporting goods, business machines, and other original equipment market (OEM) products.

Powders   Thermosetting and thermoplastic polyester powders are offered at moderate cost.   Some advantages of these powders are that they require no primer and can be applied in relatively thin films.   They are available in a wide range of colors and have excellent decorative properties, as well as outstanding electrical properties.   Thermoplastic polyester powder coatings can be used in food-handling equipment, such as bakery shelving and meat-processing equipment.   Coatings can be formulated which meet USDA and SDA regulations for food handling.   Other uses for thermoplastic polyester coating powders include hardware, toys, tennis rackets, outdoor-sign letters, and highway posts.[13]

Besides powders, polyester and polyester-type coatings are offered in both solvent-base and water-base formulations.   Solvent-base polyester enamel can be applied easily and can be used with existing solvent compositions.   These enamels offer short cure cycles when fast-curing melamine resins are used in the formulation.   In addition, the polyester enamels offer excellent flow-out with little tendency to crater and minimum solvent popping.   If desired, aromatic solvents can be used alone or in blends with active solvents during coating operations.

Water-base polyester coating systems are used in electrocoating applications such as automobile bodies and parts, electrical equipment, farm implements, metal furniture, structural-steel shelving, and aluminum extrusions.   Key properties of these water-base electrodeposited materials include film uniformity, reduced fire hazard, good adhesion, and complete coverage of complex shapes and sharp edges.

Ultraviolet-cured polyester coatings are used on wood surfaces as fillers and as top coats.   Properly cured and sanded, an ultraviolet-cured polyester filler produces a smooth surface which remains virtually undisturbed when overcoated.   Clear top coats, like the filler, are classed as 100 percent solids and require no catalyst additive.   They can be formulated for roller coating or spray application.   A mercury-vapor lamp or a plasma-arc radiation system can be used in ultraviolet curing.[15]

Typical uses   A major use for polyester and polyester-type coating material is as a gel coat on glass-fiber-reinforced plastic parts such as bathroom units and boats.   The gel coat is a resin formulation that is sprayed on a mold prior to molding.   Ultimately, the gel coat becomes the exterior layer of the reinforced-plastic part.   The gel coat features high gloss, good color retention, and a surface that resists scrubbing and cleaning agents.   Typically, a polyester gel coat is used to give a decorative and protective surface to tub/shower units which are made out of glass-fiber-reinforced plastics (see Fig. 4).[22]

Polyethylene

Polyethylene is the generic term for a variety of thermoplastic resins made by polymerization of pure ethylene. The small molecules of ethylene containing two carbon atoms and four hydrogen atoms are polymerized, or joined end to end, forming very long molecules which are called *polyethylene*. It is these giant molecules which give polyethylene its unusual and useful properties.

The most important application for polyethylene in the coatings field is in packaging. Extruded in a thin film onto a moving sheet of paper, aluminum foil, or

Fig. 4 Polyester gel coats are used to give a decorative and protective surface to tub/shower units which are made out of glass-fiber-reinforced plastics. (*Owens-Corning Fiberglas Corp.*)

other packaging material, the plastic provides a clear, strong, flexible, abrasion-resistant, heat-sealable coating of low permeability; it is highly resistant to strong acids and alkalies, solvents, greases, and oils. Polyethylene-coated aluminum foil is heat-sealable and completely moistureproof. Moisture-vapor transmission rates for polyethylene and other plastics are given in Table 16.

Besides extrusion coatings, polyethylene resins find use as a powder coating. Polyethylene powders can be applied by fluidized-bed or electrostatic-spray techniques. The resulting coatings combine low water absorption and excellent chemical resistance with good electrical insulation. Polyethylene-powder finishes are used on food-handling equipment, chemical-processing equipment, and fans. The cost of a polyethylene-powder coating ranges from 25 to 60 cents/lb.[1]

Polyimide

Coatings based on polyimide resins can be used over the temperature range −400 to 900°F. Coatings offer excellent long-term thermal stability and outstanding wear and friction properties, combined with reliability, very good electrical properties at high temperatures, and moderately good resistance to chemicals and water.

Tests show the polyimide finishes take 30 min at 800 to 850°F, 24 h at 600 to 650°F, and weeks at 500 to 550°F, maintaining excellent film properties and only moderate darkening of color. Some formulations have been subjected to temperatures over 1200°F for short periods and water-quenched without deterioration. When deterioration does occur, the finish vaporizes without leaving any apparent residue.[7]

TABLE 16   Moisture-vapor Transmission Rates (MVTR) of Plastic Coatings and Film   (in g/(mil)(in.^{2})(24 hr)[27]

Coating or film	MVTR	Information source
Epoxy-anhydride..................	2.38	Autonetics data (25°C)
Epoxy–aromatic amine.............	1.79	Autonetics data (25°C)
Neoprene........................	15.5	Baer[29] (39°C)
Polyurethane (Magna X-500).......	2.4	Autonetics data (25°C)
Polyurethane (isocyanate-polyester)	8.72	Autonetics data (25°C)
Olefane,* polypropylene...........	0.70	Avisun data
Cellophane (type PVD uncoated film)	134	Du Pont
Cellulose acetate (film).............	219	Du Pont
Polycarbonate....................	10	FMC data
Mylar†..........................	1.9	Baer[29] (39°C)
	1.8	Du Pont data
Polystyrene......................	8.6	Baer[29] (39°C)
	9.0	Dow data
Polyethylene film.................	0.97	Dow data (1-mil film)
Saran resin (F120).................	0.097 to 0.45	Baer[29] (39°C)
Polyvinylidene chloride............	0.15	Baer[29] (2-mil sample, 40°C)
Polytetrafluoroethylene (PTFE)......	0.32	Baer[29] (2-mil sample 40°C)
PTFE, dispersion cast.............	0.2	Du Pont cast
Fluorinated ethylene propylene (FEP)	0.46	Baer[29] (40°C)
Polyvinyl fluoride.................	2.97	Baer[29] (40°C)
Teslar..........................	2.7	Du Pont data
Parylene N......................	14	Union Carbide data (2-mil sample)
Parylene C......................	1	Union Carbide data (2-mil sample)
Silicone (RTV 521)................	120.78	Autonetics data
Methyl phenyl silicone.............	38.31	Autonetics data
Polyurethane (AB0130–002)........	4.33	Autonetics data
Phenoxy........................	3.5	Lee, Stoffey, and Neville[40]
Alkyd-silicone (DC-1377)...........	6.47	Autonetics data
Alkyd-silicone (DC-1400)...........	4.45	Autonetics data
Alkyd-silicone...................	6.16–7.9	Autonetics data
Polyvinyl fluoride (PT-207).........	0.7	Product Techniques Incorp.

* Trademark of Avisun Corporation, Philadelphia, Pa.
† Trademark of E. I. du Pont de Nemours & Co., Wilmington, Del.

Polyimides have excellent resistance to most chemicals, except hot and/or strong caustic solutions. They can be applied by standard or electrostatic spraying, as well as by roller coating to any material that can withstand a baking temperature of 550°F.

Properties   Polyimide resins usually are based on an aromatic dianhydride and an aromatic diamine. Outstanding characteristics include continuous service in air at 550°F, wear resistance, low friction, good strength, toughness, dimensional stability, dielectric strength, radiation resistance, and low outgassing. Polyimides are resistant to most dilute or weak acids and organic solvents. They are not affected by aliphatic or aromatic solvents, ethers, esters, alcohols, tertiary amine compounds, most hydraulic fluids, or jet fuels.

At elevated temperatures, polyimide coatings retain a high degree of strength and stiffness. However, prolonged exposure to air at high temperatures causes a gradual reduction in tensile strength in these resins. At temperatures up to at least 650°F, the loss of properties in inert atmospheres, such as nitrogen or a vacuum, is negligible. As mentioned earlier, the upper temperature limit for intermittent exposure is about 900°F.

Polyimide's combination of high-temperature resistance, strength, dimensional stability, and good electrical properties is unique among plastic resins. The material's high ratings in such areas as dielectric strength, dissipation factor, dielectric

Fig. 5 Polyimide coating is used as a protective finish on the inside of aluminum, stainless steel, and other metal cookware. (*Mirro Aluminum Co.*)

constant, volume resistivity, and arc resistance are excellent. Corona resistance is superior to that obtainable with fluorocarbons and polyethylenes.

Uses Polyimide coatings, because of their unique combination of properties, are used in many industries, including aircraft, appliance, hardware, electrical, and business machine. Typical applications include electrical motors, transformers, production jigs and fixtures, insulating varnishes, jet-engine parts, valves, chemical-process equipment, sporting equipment, and small appliances like hair dryers. The coating is used also as a decorative and protective finish for aluminum, stainless steel, and other metal cookware (see Fig. 5). All colors, except white and certain pastels, are possible. Table 17 gives NEMA standards for polyimide and other wire-insulating materials.[27]

Polyphenylene Sulfide

Polyphenylene sulfide is a nonflammable plastic which has outstanding high-temperature resistance up to 600°F and resistance to corrosive chemicals. Coatings are free of pinholes and have excellent thermal stability. The thermoplastic resin is a crystalline aromatic polymer with a symmetrical, rigid backbone chain consisting of para-substituted benzene rings connected by a single sulfur atom between rings.

Application Coatings can be applied by several techniques. Each of these requires a baking operation after the initial coating is applied. Multiple coatings

TABLE 17   NEMA Standards and Manufacturers' Trade Names for Wire Insulation[27]

Manufacturer	Plain enamel	Polyvinyl formal	Polyvinyl formal modified	Polyvinyl formal with nylon overcoat	Polyvinyl formal with butyral overcoat	Poly-amide	Acrylic	Epoxy
Thermal class..............	105°C	105°C	105°C	105°C	105°C	105°C	105°C	130°C
NEMA Standard[6].........	MW 1	MW 15	MW 27	MW 17	MW 19	MW 6	MW 4	MW 9
Anaconda Wire & Cable Co.	Plain enamel	Formvar	Hermetic Formvar	Nyform	Cement coated Formvar	....	....	Epoxy epoxy-cement coated
Asco Wire & Cable Co.......	Enamel	Formvar	....	Nyform	Formbond	Nylon	Acrylic	Epoxy
Belden Manufacturing Co...	Beld-enamel	Formvar	....	Nyclad	....	....	....	Epoxy
Bridgeport Insulated Wire Co.	....	Formvar	....	....	Quickbond	Quick-Sol	....	....
Chicago Magnet Wire Corp.	Plain enamel	Formvar	....	Nyform	Bondable Formvar	Nylon	Acrylic	Epoxy
Essex Wire Corp............	Plain enamel	Formvar	Formetex	Nyform	Bondex	....	Ensolex/ESX	Epoxy
General Cable Corp.........	Plain enamel	Formvar	Formetic	Formlon	Formeze	....	Solderable acrylic	Epoxy
General Electric Co........	....	Formex	....	....	....	Nylon	....	....
Haveg-Super Temp Div..... Hitemp Wires Co. Division Simplex Wire & Cable Co.	....	....	....	....	....	....	....	....
Hudson Wire Co............	Plain enamel	Formvar	....	Nyform	Formvar AVC	Ezsol	....	....
New Haven Wire & Cable, Inc.	Plain enamel	....	....	....	....	....	....	....
Phelps Dodge Magnet Wire Corp.	Enamel	Formvar	Hermeteze	Nyform	Bondeze	....	....	....
Rea Magnet Wire Co., Inc...	Plain enamel	Formvar	Hermetic Formvar special	Nyform	Koilset	Nylon	....	Epoxy
Viking Wire Co., Inc........	Enamel	Formvar	....	Nyform	F-Bondall	Nylon	....	....

Courtesy of Rea Magnet Wire Co., Inc.

may be applied if each coat is thoroughly baked before proceeding with the next coat. The material is supplied in several grades, choice being dictated by the particular method of application used.

In a liquid-slurry method, the polymer is mixed with reinforcing agents and, in some grades, polytetrafluoroethylene, and made into a slurry with water. The slurry is applied to metal by spraying, followed by baking 30 min at 700°F. Flat pieces can be coated with a doctor blade or coating rod instead of spraying. Slurries can be made with organic liquids such as toluene or ethylene glycol, instead of water. However, provision must be made for handling fumes evolved during spraying and curing.

In dry-powder spraying the formulation is sprayed onto a preheated metal surface. After spraying, the coating is cured in the usual manner, 30 to 60 min at 700°F in an oven with air circulation.

In fluidized-bed coating, the object to be coated is heated to 700°F and dipped immediately in a fluidized bed of polyphenylene sulfide. The hot coated object is

TABLE 17 NEMA Standards and Manufacturers' Trade Names for Wire Insulation[27] (Continued)

Teflon	Poly-urethane	Poly-urethane with friction surface	Poly-urethane with nylon overcoat	Poly-urethane with butyral overcoat	Poly-urethane with nylon and butyral overcoat	Polyester	Polyester with overcoat	Poly-imide	Polyester polyimide	Ceramic, ceramic-Teflon, ceramic-silicon
200°C MW 10	105°C MW 2	105°C	130°C MW 28	105°C MW 3 (PROP)	130°C MW 29 (PROP)	155°C MW 5	155°C MW 5	220°C MW 16	180°C	180°C+ MW 7
....	Analac	....	Nylac	Cement-coated analac	Cement coated nylac	....	Anatherm D, Anatherm 200	Al 220 M.L.	Anatherm N Anamid M (amide-imide)	
....	Poly	....	Nypol	Asco bond-P	Asco bond	Ascotherm	Isotherm 200	M.L.	Ascomid	
....	Beldure	....	Beldsol	....	....	Isonel	Polyther-maleze	M.L.		
....	Polyure-thane	Uniwind	Poly-nylon	Polybond	....	Isonel 200				
....	Soderbrite	....	Nysod	Bondable polyure-thane	....		Polyester 155			
....	Soderex	....	Soderon	Soder-bond	Soder-bond N	Thermalex F	Polyther-malex/PTX 200	Allex		
....	Enamel "G"	....	Genlon	....	....	Gentherm	Polyther-maleze 200			
....	....	....	....	....	....	Alkanex Isonel				
Teflon Temprite	....	....	....	....	....	Isonel				
....	Hudsol	Gripon	Nypoly	Hudsol AVC	Nypoly AVC	Isonel 200	Isonel 200-A	M.L.	Isomid	
....	Impsol	....	Impsolon	....	....		Imp-200			
....	Sodereze	Gripeze	Nyleze	....	S-Y Bondeze	....	Polyther-maleze 200 II	M.L.		
....	Solvar	....	Nylon solvar	Solvar koilset	....	Isonel 200	Polyther-maleze 200	Pyre M.L.	Isomid	Ceroc
....	Polyure-thane	....	Poly-nylon	P-Bondall	....	Isonel 200	Iso-poly	M.L.	Isomid Isomid-P	

immediately transferred to an oven and baked for 30 to 90 min at 700°F. Heavy objects which retain their heat during the dipping operation are more suitable for coating by this technique. Objects of steel, aluminum, titanium, Hastelloy C, brass, glass, quartz, and ceramics can be coated with the material in the fluidized-bed technique.

Typical uses Polyphenylene sulfide coatings are resistant to attack by organic acids, esters, aromatic hydrocarbons, aliphatic hydrocarbons, alcohols, inorganic salts, and a variety of inorganic bases. The coating is somewhat affected by some aqueous mineral acids and by a few chlorinated solvents.

Besides corrosion-resistant applications, the material is used for nonstick applications for cooking utensils as well as food-processing equipment. The scratch-resistant finish has been proved nontoxic in various studies.

Polyphenylene sulfide coatings well withstand forward and reverse impact of 160 in.-lb and a 3/16-in. mandrel bend of 180°. Elongation of the coating is greater than 32 percent (limit of test). Thus, toughness and extensibility are excellent. Such coatings have been exposed to air at 450°F for several months and 600°F for a few weeks without appreciable deterioration of the coating.

Polypropylene

While likely to find its widest use in molded parts and film, polypropylene resin can also be applied as a coating in much the same manner as polyethylene, as mentioned earlier. This resin can be applied to metals by dipping a heated object into a powder resin and then completing the fusion and flow-out process by heating in an oven at about 480°F.[23]

The coating obtained is heat-stable, clear, tough, chemically resistant, and highly impermeable to many vapors and gases. Polypropylene wire coatings, applied by extrusion, show excellent electrical and mechanical properties, with high resistance to heat, cracking, and abrasion. Polypropylene also can be extruded onto paper and fabrics, like polyethylene, to provide a flexible, durable coating in a variety of colors.

Polyvinyl Chloride

Polyvinyl chloride (PVC) is a thermoplastic polymer formed by polymerization of vinyl chloride. A variety of copolymers of vinyl chloride and other vinyl monomers, notably vinyl acetate and vinylidene chloride, are of commercial importance. For coating uses, vinyl resins are supplied in the form of solutions or as dispersions or emulsions. This versatility makes it possible to use a wide range of polymer compositions and molecular weights and, thus, to obtain coatings which vary in properties, thickness, and cost. The properties of various vinyl coatings including solution, plastisol, and organosol are given in Table 18.

Vinyl coatings have outstanding durability and excellent resistance to acids, alkalies, chemicals, and salt water. They are superior to chlorinated rubber in resistance to oils and fats. Like chlorinated rubber, vinyl paint will not support combustion. However, their adherence is poor unless special primers are used. A vinyl butyral wash primer is recommended to improve adherence. Wash primers contain phosphoric acid, which reacts with the metal surface to which it is applied, producing a phosphate surface overcoated with a tightly bonded, thin coat of vinyl butyral resin. Other primers formulated for increased adhesion combine vinyls with resins such as alkyds.

Plastisols are liquid dispersions of fine-particle-size emulsion polyvinyl chloride in plasticizers. Organosols are essentially the same as plastisols but contain a volatile liquid organic diluent to reduce viscosity and facilitate processing. In many coating operations, conventional paint-spraying equipment is used. In other techniques, parts can be dipped, knife, or roller-coated. Plastisols and organosols are used extensively for dip coating of wire products (see Fig. 6) such as household utensils and knife coating of fabrics and paper. Plastisol and organosol products can be varied from hard to very soft.[21]

The absence of odor or taste makes vinyl suitable for can coatings and for other types of food containers. Because of their high flexibility and durability vinyls are used for coil-coated metal. Vinyl's chemical resistance makes it useful for applications such as linings for chemical containers, vats, and pipes handling corrosive materials. The toughness feature of vinyl coatings makes them useful as decorative and functional coatings on metal furniture, signs, aluminum siding, and office machines.

Powders Besides solution, organosol, and plastisol formulations, polyvinyl chloride is supplied in the form of powder. The superior insulating properties of vinyl powders make them useful for coating electrical conduit and fittings. Their moisture resistance, equal to the highest-grade vinyl plastisol, provides protection for piping used in irrigation and in the oil and mining industries. In addition, vinyl-powder coatings can be applied in a wide range of thicknesses from 6 to 35 mils, making them suitable for small wire products, as well as large coated coils.[3]

Vinyl powders can be applied by high-speed production-line fluidized-bed techniques. Essentially, what happens is that a coating film is extruded at the point of fusion to the moving substrate, such as a metal coil. Postcuring is shortened or eliminated and finishing costs are lowered. Heating substrate to 400°F and exposure

TABLE 18 Comparison of the Properties of Vinyl Coatings[a]

Coating type	Outstanding characteristics	Mechanical properties					Color and gloss			Weathering properties	
		Hardness	Abrasion resistance	Adhesion	Flexibility	Toughness	Film color	Color retention	Gloss	Weather resistance	Gloss retention
Solution[b]	Excellent color, flexibility, chemical resistance; tasteless, odorless	F	E	F to G	E	E	E[f]	E	G	E[f]	E
Plastisol[c]	Toughness; resilience; abrasion resistance; can be applied without solvents	F	E	E to cloth[e]	E	E	E[f]	E	F	E[f]	F to G
Organosol[d]	High solids content; excellent color, flexibility; tasteless, odorless	F	E	E to cloth[e]	E	E	E[f]	E	P to G	E[f]	G

[a] E = excellent; G = good; F = fair; P = poor.
[b] Vinyl chloride acetate copolymers; resins vary widely in compatibility with other materials.
[c] Vinyl chloride acetate copolymer and vinyl chloride resins.
[d] Vinyl chloride acetate copolymers; require grinding for good dispersions.
[e] Requires primer for use on metal.
[f] Pigmented.

of the powder for 30 s is all that is required for almost instantaneous vinyl-powder application.

Another application method for vinyl powders is electrostatic spraying. With this method, preheating is eliminated for some applications. Selected areas can be coated, and automation is simplified. In electrostatic spraying, a valve-operated, injector-type air pump meters powder from a reservoir to a gun. Each powder particle receives a maximum electrical charge as it leaves a diffuser at the front of the gun. Movement of powder to the part to be coated is controlled by electro-static force.[5]

In many applications, vinyls are combined with alkyds to reduce raw-material cost and increase solids content. Chemical resistance of vinyl-alkyd coatings is

Fig. 6 Vinyl plastisols and organosols are used extensively for dip coating of wire products. The coatings can be varied from hard to very soft. (*M&T Chemicals.*)

superior to straight alkyds, and water resistance, adhesion to wood and other sub-strates, drying, and toughness are excellent. Although vinyl-alkyd formulations cost more than straight alkyd coatings, they cost less than vinyls while offering most of the superior performance of vinyls.

Silicone

Silicone resins are based on a highly stable, inorganic silicone-oxygen structure coupled with various organic radicals. Lower-molecular-weight linear polymers are heat- and oxidation-resistant fluids, whereas high-molecular-weight linear polymers are elastomers. The cross-linked types, in a range of molecular weights, are the silicone resins used as coatings. Cross linking generally is accomplished by intro-ducing trifunctional silicon without any atoms at intervals along the siloxane chains to form oxygen bridges between trifunctional atoms on different chains. Silicone coatings are quite expensive and can result in a paint cost averaging from about $15 to $20/gal.[1]

Uses Because of their exceptional heat resistance, silicone coatings find their chief use in heat-stable maintenance coatings and industrial finishes for severe high-temperature applications such as hot stacks, oven walls, combustion-engine mufflers, incinerator interiors, arc-light reflectors, wall and space heaters, and fireplace equip-ment (see Fig. 7). Although used primarily for their high-temperature resistance, silicone resins also perform well at low temperatures. Applied from dilute solutions to various surfaces, the silicones provide excellent water repellency.[9]

Where maximum heat resistance obtainable with pure silicone resins is not required, more economical coatings can be made with modified silicone resins that have been blended or copolymerized with organic resins such as alkyds, phenolics, epoxies, acrylics, and cellulose derivatives.

Baking-type silicone-modified alkyd enamels retain their color and gloss for long periods at temperatures up to 400°F. Air-drying aluminum paints based on silicone-modified alkyds have excellent corrosion resistance and weatherability and withstand

Fig. 7 Silicone coatings find their chief use in heat-stable finishes for severe high-temperature applications such as fireplace equipment (shown above), incinerator interiors, arc-light reflectors, and wall and space heaters. (*Copper Development Association.*)

temperatures up to 800°F for limited periods of time. Other silicone-modified resins include combinations of silicones with phenolic resins to produce weather- and water-resistant enamels; with acrylic resins for durable white enamels; with chlorinated rubber to make tough, heat- and corrosion-resistant paints; with epoxy resins for strong, heat-resistant electrical-insulating coatings; and with cellulose and cellulose nitrate to make heat- and weather-resistant lacquers. Thermal conductivity of silicone and other coatings is given in Table 19.

Application Silicone and silicone-modified coatings are applied in much the same way as other organic coatings, namely, by brushing, spraying, or dipping. Surface preparation is important, and loose paint should be scraped off and scale or rust removed from metal surfaces by sandblasting or wire brushing. A good organic primer should be used under silicone-modified paints for temperatures up to 300°F. A silicone-modified zinc-dust primer should be used for higher temperatures. No more than two coats of silicone coating should be applied, with a heat cure

TABLE 19   Thermal Conductivity Data for Polymer Coatings[27]

Material	k value,* cal/(sec)(cm²) (°C/cm) $\times 10^4$	Source of information
Unfilled plastics:		
Acrylic................................	4–5	*a*
Alkyd.................................	8.3	*a*
Depolymerized rubber..............	3.2	H. V. Hardman, DPR Subsidiary
Epoxy................................	3–6	*b*
Epoxy (electrostatic spray coating)....	6.6	Hysol Corp., DK-4
Epoxy (electrostatic spray coating)....	2.9	Minnesota Mining & Mfg., No. 5133
Epoxy (Epon† 828, 71.4% DEA, 10.7%)	5.2	
Epoxy (cured with diethylenetriamine)	4.8	*c*
Fluorocarbon (Teflon TFE)...........	7.0	Du Pont
Fluorocarbon (Teflon FEP)..........	5.8	Du Pont
Nylon................................	10	*d*
Polyester............................	4–5	*a*
Polyethylenes.......................	8	*a*
Polyimide (Pyre-M.L. enamel)........	3.5	*e*
Polyimide (Pyre-M.L. varnish)........	7.2	*f*
Polystyrene..........................	1.73–2.76	*g*
Polystyrene..........................	2.5–3.3	*a*
Polyurethane........................	4–5	*n*
Polyvinyl chloride...................	3–4	*a*
Polyvinyl formal....................	3.7	*a*
Polyvinylidene chloride..............	2.0	*a*
Polyvinylidene fluoride..............	3.6	*h*
Polyxylylene (Parylene N)...........	3	Union Carbide
Silicones (RTV types)..............	5–7.5	Dow Corning Corp.
Silicones (Sylgard types)............	3.5–7.5	Dow Corning Corp.
Silicones (Sylgard varnishes and coatings)............................	3.5–3.6	Dow Corning Corp.
Silicone (gel coating)................	3.7	Dow Corning Corp.
Silicone (gel coating)................	7 (150°C)	Dow Corning Corp.
Filled plastics:		
Epon 828/diethylenetriamine = A.....	4	*b*
A + 50% silica.....................	10	*b*
A + 50% alumina	11	*b*
A + 50% beryllium oxide..........	12.5	*b*
A + 70% silica....................	12	*b*
A + 70% alumina	13	*b*
A + 70% beryllium oxide..........	17.8	*b*
Epoxy, flexibilized = B..............	5.4	*i*
B + 66% by weight tabular alumina	18.0	*i*
B + 64% by volume tabular alumina	50.0	*i*
Epoxy, filled........................	20.2	Emerson & Cuming, 2651 ft
Epoxy (highly filled).................	15–20	Wakefield Engineering Co.
Polyurethane (highly filled)..........	8–11	International Electronic Research Co.
Other materials used in electronic assemblies:		
Alumina ceramic.....................	256–442 (20–212°F)	*a*
Aluminum............................	2767–5575	*a*
Aluminum oxide (alumina), 96%......	840	*i*
Beryllium oxide, 99%...............	5500	*j*
Copper...............................	8095–9334	*a*
Glass (Borosill, 7052)...............	28	*k*
Glass (pot-soda-lead, 0120)..........	18	*k*

TABLE 19   Thermal Conductivity Data for Polymer Coatings[27] (Continued)

Material	k value,* cal/(sec)(cm²) (°C/cm) × 10[h]	Source of information
Glass (silica, 99.8% SiO₂)............	40	*l*
Gold.............................	7104 (20–212°F)	*a*
Kovar...........................	395	*m*
Mica............................	8.3–16.5	*a*
Nichrome‡.......................	325	*m*
Silica...........................	40	*k*
Silicon nitride...................	359	*m*
Silver...........................	9995 (20–212°F)	*a*
Zircon..........................	120–149	*a*

* All values are at room temperature unless otherwise specified.
† Trademark of Shell Chemical Co., New York, N.Y.
‡ Trademark of Driver-Harris Co., Harrison, N.J.
a Mater. Eng., Materials Selector Issue, vol. 66, no. 5, Chapman-Reinhold Publication, mid-October, 1967.
b Wolf, D. C., *Proc. Nat. Electron. and Packag. Symp.*, New York, June, 1964.
c Lee, H., and K. Neville: "Handbook of Epoxy Resins," McGraw-Hill Book Company, New York, 1966.
d Davis, R.: *Reinf. Plast.*, October, 1962.
e Du Pont Tech. Bull. 19, Pyre-M.L. Wire Enamel, August, 1967.
f Du Pont Tech. Bull. 1, Pyre-M.L. Varnish RK-692, April, 1966.
g Teach, W. C., and G. C. Kiessling: Polystyrene, Reinhold Publishing Corporation, New York, 1960.
h Barnhart, W. S., R. A. Ferren, and H. Iserson: *17th ANTEC of SPE*, January, 1961.
i Gershman, A. J., and J. R. Andreotti: *Insulation*, September, 1967.
j American Lava Corp. Chart 651.
k Shand, E. B., "Glass Engineering Handbook," McGraw-Hill Book Company, 1958.
l Kingery, W. D.: Oxides for High Temperature Applications, *Proc. Int. Symp.*, Asilomar, Calif., October, 1959, McGraw-Hill Book Company, New York, 1960.
m Kohl, W. H.: "Handbook of Materials and Techniques for Vacuum Devices," Reinhold Publishing Company, New York, 1967.
n "Modern Plastics Encyclopedia," McGraw-Hill, Inc., New York, 1968.

recommended after the first coat.   Thin coats up to 1 mil thick per coat provide the best service.[4]

ELASTOMERIC COATINGS

Synthetic rubbers find important uses in coatings where resilience, abrasion resistance, distensibility, and elasticity are of prime importance.   Among the synthetic elastomers available are materials which combine excellent heat, chemical, and weathering resistance along with other properties.   Those of particular interest to the coatings field include chlorosulfonated polyethylene, fluoroelastomers, neoprene, polysulfide, and polyurethane.

Butyl

Butyl is a general-purpose synthetic rubber made by copolymerizing isobutylene with 1 to 3 percent isoprene at temperatures below 140°F in the presence of aluminum chloride catalyst.   Both isobutylene and isoprene are available in substantial quantities from petroleum-refinery operations.

Butyl is superior to natural rubber and styrene-butadiene rubber in resistance to aging and weathering, sunlight, heat, air, ozone and oxygen, chemicals, flexing and cut growth, and impermeability to gases and moisture.[2]

Chlorinated Rubber

Chlorinated rubber is produced by the chlorination of rubber dissolved in a suitable solvent. In the reaction, chlorine saturates the double bond and replaces part of the hydrogen of the rubber molecule. The resulting product is a hard resin which is soluble in hydrocarbon and lacquer solvents and compatible with plasticizers and certain resins.

Chlorinated rubber traces its origin back to World War II when natural rubber was not available. The material was produced by chlorinating natural rubber, and it contained about 67 percent chlorine. Chlorinated-rubber films dry by evaporation of the solvent. There are no cross-linkages formed by oxidation or polymerization. Consequently, the films remain soluble in solvents but are unaffected by nonsolvents. The chlorinated rubber is a thermoplastic resin rather than an elastomer and is used chiefly in coatings where excellent resistance to water and common corrosive chemicals is needed (see Table 20). The material has a high degree of impermeability to water vapor and a very low water absorption.

Chlorinated rubber is highly resistant to strong alkalies and acids, strong bleaches, corrosive vapors and fumes, soaps and detergents, mineral oils, mold, and mildew. However, chlorinated rubber does deteriorate by heat and ultraviolet light but can be stabilized to improve its resistance to these influences. It is odorless, tasteless, nontoxic, and nonflammable.

Uses   Chlorinated rubber is used in corrosion-resistant coatings and primers, industrial-maintenance coatings, traffic paints, swimming-pool paints, and concrete and masonry paints. Other uses include flame-retardant coatings, textile coatings, printing inks, and adhesives. The maximum dry-heat resistance of chlorinated-rubber coatings is 190°F, and it may be used on surfaces subjected to low temperatures down to −40°F. The coatings are used also in food-processing plants, breweries, sewage-disposal plants, and other operations where good resistance to moisture, fungus, mildew, and chemicals is required.[1]

Chlorosulfonated Polyethylene

Chlorosulfonated polyethylene is an elastomeric material produced by the simultaneous chlorination and sulfonation of already polymerized, relatively high molecular weight, low-density polyethylene, by treatment with gaseous chlorine and sulfur dioxide in the presence of a radical initiator (a peroxide) as catalyst. Chlorosulfonated polyethylene usually is derived from a polyethylene having a molecular weight of about 20,000 and contains about 27 percent chlorine and 1.5 percent sulfur. The cured polymer has excellent resistance to abrasion, heat, weathering, and cracking, and is completely resistant to ozone. It has good resistance to mineral oils and to corrosive chemicals, such as oxidizing acids, alkalies, chromium-plating baths, and hydrogen peroxide.

Solution coatings of chlorosulfonated polyethylene can be formulated in a variety of colors, including white. They can be applied to most other elastomeric compositions to provide protection against weathering and ozone. The coatings can be used without curing if a light stabilizer and detackifying resin are incorporated into the formulation. However, a heat cure is required for maximum resistance properties. Coatings that will cure at room temperature can be prepared by using special accelerators.

Typical applications   Unpigmented lacquers based on chlorosulfonated polyethylene are applied to rubber sporting goods and footwear to provide a durable, protective finish. Cotton and synthetic fabrics coated with this material show excellent weathering resistance. Chlorosulfonated-polyethylene coatings are used also on upholstery fabrics, awnings, tarpaulins, heavy-duty gloves, and as industrial-maintenance coatings in corrosive or abrasive environments. Chlorosulfonated-polyethylene coatings can be used for continuous service at temperatures up to 220°F.

Fluoroelastomers

Fluoroelastomers are rubberlike polymers which form useful coatings when compounded with other materials and cured. Fluoroelastomers are self-extinguishing, and the materials have excellent resistance to hydrocarbon solvents, fuels, and lubricants, as well as halogenated solvents.

Two principal types of fluoroelastomers are used in coatings. One is a copolymer of trifluorochloroethylene and vinylidene fluoride, and the other is a copolymer of hexafluoropropylene and vinylidene fluoride. The cost of these materials is very

TABLE 20   Water-Absorption Data for Polymer Coatings[27]

Material	% absorption in 24 hr*	Source of information
Chlorinated polyether (Penton).............	Negligible	Hercules Powder Co.
Depolymerized rubber....................	0.5	H. V. Hardman, DPR Subsidiary
Diallyl phthalate........................	0.09	a
Epoxy.................................	0.04	a
Epoxies...............................	0.11–1.84	b
Epoxy cured with DETA..................	0.11–0.12	c
Fluorocarbon (Teflon)....................	0.01	a
Fluorocarbon (Teflon TFE dispersion coating)	<0.1	Du Pont
Fluorocarbon (Teflon FEP dispersion coating)	<0.01 0.3 –0.4	Du Pont
Phenolic...............................	0.9	e
Phenoxy...............................	0.13	d
Polyesters (copolymerized with styrene)......	0.17–0.28	e
Polyethylene...........................	0.01	a
Polyimide..............................	0.7	Union Carbide
Polyimide-amide........................	0.5 –1.9 (in boiling water)	
Polystyrene............................	0.04	a
Polyvinyl fluoride.......................	0.04	Product Techniques, Inc., PT-207 Coating
Polyvinyl formal (Formvar)...............	0.75–1.1	
Polyvinylidene chloride–polyvinyl chloride (Saran)................................	0.05	
Polyvinylidene fluoride...................	0.04	f
Polyxylylene (Parylene C)................	0.06	Union Carbide
Polyxylylene (Parylene N)...............	0.01	Union Carbide
Silicone...............................	0.15	a

* At room temperature, unless otherwise specified.
a Piser, J.: *Electron. Des.* vol. 17, Aug. 16, 1967.
b *Celanese Corp. Tech. Bull.* 1164.
c Lee, H., and K. Neville: "Epoxy Resins," McGraw-Hill, New York, 1966.
d Lee, H., D. Stoffey, and K. Neville: "New Linear Polymers," McGraw-Hill, New York, 1967.
e Parker, E. E., and E. W. Moffett: *Ind. Eng. Chem.*, vol. 46, p. 1615, 1954.
f Barnhart, W. S., R. A. Ferren, and H. Iserson: *17th ANTEC of SPE*, January, 1961.
g "Modern Plastics Encyclopedia" (annual), McGraw-Hill, Inc., New York.

high, ranging from fifteen to twenty times more than that for other elastomeric materials.

Besides excellent solvent resistance, fluoroelastomers have excellent heat resistance, surpassing that of all other elastomers. Coatings based on these materials can be used at temperatures up to 600°F for short periods, and almost indefinitely at 450°F. Application of fluoroelastomer coatings generally is done by dipping, roller coating, or knife coatings.[6]

Neoprene

The excellent combination of resistance and durability properties inherent in neoprene has been proved in many applications since the material was introduced in 1932. Neoprene coatings are widely used to provide lasting protection against corrosion and abrasion of industrial equipment such as pipelines, storage tanks, coal chutes, pump and blower blades, ship-propeller shafts, tool handles, and machinery. Neoprene coatings on military aircraft have shown excellent resistance to abrasion and erosion, although the weight of the thick coatings used (about 10 mils) constitutes somewhat of a drawback in aircraft applications. Neoprene coatings are widely specified as maintenance finishes in oil refineries, chemical plants, gas stations, and aircraft installations.[2]

Neoprene or polychloroprene is prepared by the emulsion polymerization of chloroprene in an alkaline medium. The polymerization is halted when about 30 percent complete, while the polymer is still thermoplastic. The polymer is recovered by acidifying the emulsion to a point just short of coagulation and then free-coagulating the polymer onto a brine-cool drum, from which it is stripped, washed, and dried. Several types of neoprene, differing in uncured properties and manner of cure, are manufactured.

Neoprene-rubber coatings are usually supplied as solutions of the uncured polymer in aromatic solvents. Catalyst or accelerators are added immediately prior to use. Heat-curing types are usually syrupy liquids that are applied by brush or dip coating. Air-drying types can be applied by brush, dip, spray, roller, or flow. Depending on the nature of the formulation, thicknesses of 3 to 6 mils per coat are obtained easily by various coating methods. An adequate period (several hours) of air drying should precede baking to prevent blister formation from too rapid an escape of solvent. Generally, neoprene-rubber coatings are used at thicknesses of 10 to 15 mils. A primer is required for maximum adhesion, and chlorinated-rubber-base primers have proved to be very effective for this purpose. Neoprene-rubber coatings generally are available in gray or dark colors, chiefly black and green.

Nitrile

Nitrile rubber on butadiene-acrylonitrile copolymers are similar to butadiene-styrene rubbers in general appearance and physical properties. However, nitrile rubbers have better resistance to most solvents, greater resistance to the deteriorating effects of heat and sunlight, and excellent resistance to oils, greases, and waxes.

Nitrile rubber is used in coatings as nonvolatile, nonextractable plasticizers for phenolic resins and for polyvinyl chloride resins. In latex form, nitrile rubbers are used as coatings and impregnants for paper, fabrics, and leather products.[1]

Polysulfide

Coatings based on polysulfide rubber have low permeability to gases, good electrical properties, and excellent resistance to aging, sunlight, and ozone, and they afford good protection against fresh and salt water. Polysulfide rubbers have excellent resistance to organic solvents, including most aliphatic and aromatic hydrocarbons, alcohols, ketones, and esters. Their resistance to acids and alkalies is only fair.

Polysulfide-rubber coatings find use as solvent-resistant linings for chemical tanks and equipment (see Fig. 8) and as solvent-resistant coatings for paper, fabrics, leather, wood, and concrete.

Polysulfide rubbers are the products of a condensation polymerization in which one or more organic dihalides are reacted with an aqueous solution of sodium polysulfide, producing high-molecular-weight polymers. Polysulfide rubbers are available not only as solids but also as liquids. The liquids are 100 percent polymer, unadulterated with solvents or diluents, which also can be converted to highly elastic rubbers with properties closely approaching those of cured, solid polysulfides. The liquid polymers are basically reactive, low-molecular-weight synthetic rubbers which can be chemically cured to flexible, elastomeric coatings.[6]

Polysulfide coatings from liquid polymers show excellent resistance to aging, ozone, oxidation, and weathering. Polysulfide coatings have a service-temperature range of −65 to 250°F and withstand short exposure to 300°F. Properly compounded liquid polysulfide coatings do not have any corrosive effect on aluminum, copper, magnesium, iron, steel, silver, or gold.

A water-dispersed polysulfide polymer, which forms a film without cracking or checking, is used as a liner for concrete fuel-storage tanks. This film performs satisfactorily even in contact with high-octane gasoline. There are several water-dispersed polysulfide polymers, differing in properties obtained in the finished coating. Besides concrete, water-dispersed polysulfide coatings can be applied to wood, fabrics, felt, leather, and paper in applications where a tough, flexible film having good adhesion and excellent resistance to various types of fuels, oils, water, and mild acids and bases is required.

As with most elastomeric coatings, thick polysulfide-rubber films can be built up without sacrificing resiliency. Thus, films up to 100 mils thick are sometimes applied where erosive and corrosive conditions are severe.

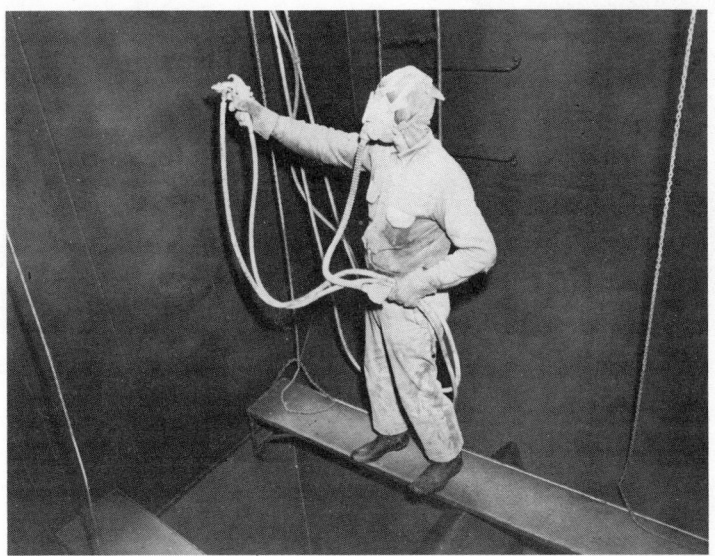

Fig. 8 Polysulfide-rubber coating is sprayed on the inside of railroad hopper car used to transport corrosive liquids. Coating's service life is roughly 5 years in this service. (*Thiokol Chemical Co.*)

Polyurethane

Polyurethane coatings offer an unusual combination of properties, such as adhesion, impact resistance, flexibility, gloss, and weatherability over a wide range of temperatures, humidity, and climatic conditions.

The coatings fall into two general categories: one-component and two-component reactive coatings. The one-component types include urethane oils, moisture-cured, and blocked or heat-cured coatings. The two-component types are isocyanate-terminated prepolymers cured with polyols and catalyst.[15]

Uses Polyurethane coatings find use on metal, plastic, and rubber substrates, such as low-profile reinforced polyesters, ABS, nylon, barrier-coated polystyrene, treated polyolefins, and various types of rubber products.

Polyurethane coatings have found many uses for appliances, electrical products, transportation equipment (see Fig. 9), recreational products, etc. In the automobile industry urethane coatings based on aliphatic-type polyisocyanates are being used in production to coat elastomeric bumpers, instrument pads, and polycarbonate headlight bezels. Also, the coatings are used, in some cases, on integral-skin foam parts, including horn buttons, steering wheels, arm and head rests, visors, airscoops, strip moldings, fender and skirt extensions, and specialty wheel covers.[2]

Weatherability and color retention have been problems with polyurethane coatings up until now. New grades have been developed which are eminently more stable toward discoloration and chalking; thus outdoor uses for polyurethane coatings are increasing.

The newer polyurethane coatings offer 40 to 50 percent more abrasion resistance and 20 percent more mar resistance compared with conventional alkyd or acrylic finishes. Polyurethane in recent tests came out to be 40 percent better for color and gloss retention compared with conventional polyester gel coats and alkyd finishes after exposure to Florida weather for 12 months.

Fig. 9 Polyurethane coatings find wide use on transportation equipment such as this cement-mixer truck. Coating is tough, resists scratches, and holds up well under rough service conditions. (*E. I. du Pont de Nemours & Company, Inc.*)

Properties Because of the variety of polyisocyanate and polyol compounds available and the almost infinite freedom in formulating polyurethane coatings, it is difficult to classify rigidly the properties of urethane films. However, Table 21 is an attempt at illustrating the average properties which can be expected from various two-package, one-package, and lacquer-type compositions. Figure 10 shows properties related to molecular weight for cross-link density.

Application Polyurethane coatings can be applied by dipping, spraying (the most popular), brushing, flowing, curtain, knife, and roller techniques. In the case of prereacted urethanes, i.e., lacquers and urethane oils, or blocked types, no special precautions are necessary, and these types offer the greatest application versatility.

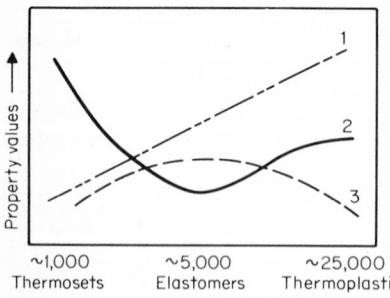

Fig. 10 Properties of polyurethane coatings related to molecular weight for cross-link density. Curve 1, solvent swelling and elongation. Curve 2, hardness. Curve 3, elasticity. (*Interscience Publishers.*)

Reactive, two-component types can be applied by various coating techniques also, but precautions should be taken because of the coating's pot life. For example, as the coating ages after premixing, viscosity builds. This condition usually results in poor flow and film control. Eventually, gelation occurs, and this results in clogged guns, troughs, tubes, etc.

If the work life of the coating is less than 2 h, common application techniques are not applicable and special equipment usually is required. On the other hand, coatings with a pot life greater than 3 h require only minor equipment modifications or procedural controls.

TABLE 21 Guide to Selecting Polyurethane Coatings

Property	One-component			Two-component	Lacquer
	Urethane oil	Moisture	Blocked		
Abrasion resistance	Fair–good	Excellent	Good–exc.	Excellent	Fair
Hardness	Medium	Med.–hard	Med.–hard	Soft–very hard	Soft–med.
Flexibility	Fair–good	Good–exc.	Good	Good–exc.	Excellent
Impact resistance	Good	Excellent	Good–exc.	Excellent	Excellent
Solvent resistance	Fair	Poor–fair	Good	Excellent	Poor
Chemical resistance	Fair	Fair	Good	Excellent	Fair–good
Corrosion resistance	Good	Fair	Good	Excellent	Good–exc.
Adhesion	Good	Fair–good	Fair	Excellent	Fair–good
Toughness	Poor	Excellent	Good	Excellent	Good–exc.
Elongation	Fair	Poor	Poor	Excellent	Excellent
Tensile	....	Good	Fair–good	Good–exc.	Excellent
Weatherability	....	....	....	....	....
Aliphatic	Good	Poor–fair	....	Good–exc.	Good
Conventional	Poor–fair	Poor–fair	Poor–fair	Poor–fair	Poor
Pigmented glass	High	High	High	High	Medium
Cure rate	Slow	Slow	Fast	Fast	None
Cure temp	Room temp	Room temp	300–390°F	212°F	150–225°F
Work life	Infinite	1 year	6 months	1 s–24 h	Infinite

Reactive-type urethane coatings should be stored and handled so that exposure to extreme temperatures, humidity, and/or reactive contaminants is avoided. Also, containers should be tightly sealed after use.[2]

Polyurethane coatings are characterized by very good toughness and abrasion resistance (see Table 22) combined with excellent chemical resistance depending on type. They have good to excellent abrasion resistance, good to excellent impact resistance, fair to excellent solvent resistance, and good to excellent adhesion to substrates.

TABLE 22   Abrasion Resistance of Organic Coatings[27]

Coating	Taber wear index, mg/kHz
Polyurethane Type 1	55–67
Polyurethane Type 2 (clear)	8–24
Polyurethane Type 2 (pigmented)	31–35
Polyurethane Type 5	60
Urethane oil varnish	155
Alkyd	147
Vinyl	85–106
Epoxy-amine-cured varnish	38
Epoxy-polyamide enamel	95
Epoxy-ester enamel	196
Epoxy-polyamide coating (1:1)	50
Phenolic spar varnish	172
Clear nitrocellulose lacquer	96
Chlorinated rubber	200–220
Silicone, white enamel	113
Catalyzed epoxy, air-cured (PT-401)	208
Catalyzed epoxy, Teflon-filled (PT-401)	122
Catalyzed epoxy, bake-Teflon-filled (PT-201)	136
Parylene N	9.7
Parylene C	44
Parylene D	305
Polyamide	290–310
Polyethylene	360
Alkyd TT-E-508 enamel (cured for 45 min at 250°F)	51
Alkyd TT-E-508 (cured for 24 hr at room temperature)	70

Styrene Butadiene

Styrene-butadiene copolymers of high styrene content have been of importance in developing latex paints. Styrene-butadiene copolymer elastomers used in coatings contain about 75 percent butadiene. However, copolymers ranging from 50 percent butadiene to 100 percent polybutadiene are manufactured and are available as solid polymers and as latexes. Styrene-butadiene copolymers are used in coatings for rug backing, paper, fabrics, and leathers. These coatings are tough, flexible, abrasion-resistant, and acid- and alkali-resistant, but are attacked by aromatic and chlorinated hydrocarbons and are swelled by oils and greases.

Styrene has found many uses in protective coatings. For example, styrene polymerized in the presence of water produces a high-molecular-weight polystyrene latex. Polystyrene is brittle but, suitably plasticized, is useful in protective-coating formulations. Polystyrene has good electrical properties and is resistant to acids and alkalies.

APPLICATION METHODS

The choice of a proper application method for painting depends on a number of factors, such as the nature of the coating material, type and texture of surface on which it is to be applied, shape and size of substrate, environmental consideration, available equipment and facilities, skills available, and fire and toxicity hazards.[5]

Generally, application outside a factory is limited to brushing, rolling, and conventional cold- or hot-spraying methods. Factory coating operations are generally accomplished with more elaborate types of spraying, including electrostatic, airless, or by special techniques such as fluidized-bed.[27] Advantages and limitations of various methods are given in Table 23.

Brushing   Painting by brush dates back centuries and is still used for a substantial portion of painting done outside a factory. Although brush application is considerably slower than spraying, it gives excellent results when done with skill using the proper brush and technique. A high-quality paintbrush is an expensive tool that deserves proper care to maintain its usefulness.

Roller   Roller application is a fast and convenient means for applying various types of pigmented coatings to large, flat surfaces that are reasonably free of interference from protrusions and mounted fixtures. This technique gives excellent results on porous or textured surfaces such as brick, cinder block, and plaster. Although it does not lend itself to the fine workmanship that can be achieved with skilled brush application, the roller is an excellent tool which is capable of good results.

Spray   Spray application is the most widely used method for applying paint. Spraying is convenient for overhead work, for large areas, and for reaching into areas inaccessible by brushing (see Fig. 11). Successful spray painting depends on skill developed through knowledge and experience. Although a number of different spray methods are available, all of them require clean paint and clean equipment. Correct lapping is very important in achieving a uniform thickness and appearance. The chief disadvantages of conventional spraying methods are those associated with the overspray caused by air and paint rebound from the surface being painted. The overspray wastes paint, necessitates the careful masking of adjacent areas, and may contaminate surrounding objects.[10]

Conventional cold spray is the application most widely used for outdoors. Although cold spraying has the disadvantage associated with overspray, the method is fast and economical compared with brushing, especially on large projects or rough surfaces. The equipment required is simple and inexpensive compared with that of more elaborate methods. A typical system consists of an air compressor capable of providing up to about 90 lb/in., an air transformer to regulate pressure, flexible housing, and a spray gun itself. Before spraying, the coating material should be thinned in accordance with the manufacturer's instructions. As a general guide, common paints and enamels are usually sprayed at air pressures of 60 to 70 lb/in.^{2} and fluid pressures of 15 to 20 lb/in.^{2} Lacquers and other low-solids liquids can be sprayed at air pressures of 40 to 50 lb/in.^{2} and fluid pressures of 10 to 15 lb/in.^{2}

Airless spraying is a recent development. Hydraulic pressure rather than air is used to accomplish the atomization of a coating material. High pressures, around 2,500 lb/in.^{2}, are required to force liquids of paint viscosity through a very small, precision orifice which is used in the technique. Both portable and fixed types of airless-spray equipment are available. Also, airless spray may be combined with hot-spray techniques to permit spraying at lower pressures through a triggered gun. The chief advantage of airless spray over conventional spray lies in the great reduction in overspray which is achieved by eliminating the atomizing air.[5]

Hot spraying, in which the coating material to be sprayed is first heated to a temperature of 150 to 175°F, can be applied to both airless and conventional spraying techniques. The paint is heated by passage through a hot-water heat exchanger. Heating reduces the viscosity of the material, lessens the amount of thinner required, and permits spraying at a higher solid content. Coatings for hot-spray application must be specially formulated for proper solvent balance and viscosity. With proper formulation, hot-spray coatings are less prone than conventionally sprayed coatings to such defects as running, sagging, and orange-peel surface.

Electrostatic   Electrostatic spraying uses the attraction between electrically charged particles of material of opposite polarity. The coating particles are charged to a high negative potential as they contact or pass through a suitably shaped

TABLE 23 Application Methods for Coatings[27]

Method	Advantages	Limitations	Typical applications
Spray	Fast, adaptable to varied shapes and sizes. Equipment cost is low.	Difficult to completely coat complex parts and to obtain uniform thickness and reproducible coverage.	Motor frames and housings, electronic enclosures, circuit boards, electronic modules.
Dip	Provides thorough coverage, even on complex parts such as tubes and high-density electronic modules.	Viscosity and pot life of dip must be monitored. Speed of withdrawal must be regulated for consistent coating thickness.	Small- and medium-sized parts, castings, moisture and fungus proofing of modules, temporary protection of finished machined parts.
Brush	Brushing action provids good "wetting" of surface, resulting in good adhesion. Cost of equipment is lowest.	Poor thickness control; not for precise applications. High labor cost.	Coating of individual components, spot repairs, or maintenance.
Roller	High-speed continuous process; provides excellent control on thickness.	Large runs of flat sheets or coil stock required to justify equipment cost and setup time. Equipment cost is high.	Metal decorating of sheet to be used to fabricate cans, boxes.
Impregnation	Results in complete coverage of intricate and closely spaced parts. Seals fine leaks or pores.	Requires vacuum or pressure cycling or both. Special equipment usually required.	Coils, transformers, field and armature windings, metal castings, and sealing of porous structures.
Fluidized bed	Thick coatings can be applied in one dip. Uniform coating thickness on exposed surfaces. Dry materials are used, saving cost of solvents.	Requires preheating of part to above fusion temperature of coating. This temperature may be too high for some parts.	Motor stators; heavy-duty electrical insulation on castings, metal substrates for circuit boards, heat sinks.
Screen-on	Deposits coating in selected areas through a mask. Provides good pattern deposition and controlled thickness.	Requires flat or smoothly curved surface. Preparation of screens is time-consuming.	Circuit boards, artwork, labels, masking against etching solution, spot insulation between circuitry layers or under heat sinks or components.
Electrocoating	Provides good control of thickness and uniformity. Parts wet from cleaning need not be dried before coating.	Limited number of coating types can be used; compounds must be specially formulated ionic polymers. Often porous, sometimes nonadherent.	Primers for frames and bodies, complex castings such as open work, motor end bells.
Vacuum deposition	Ultrathin, pinhole-free films possible. Selective deposition can be made through masks.	Thermal instability of most plastics; decomposition occurs on products. Vacuum control needed.	Experimental at present. Potential use is in microelectronics, capacitor dielectrics.
Electrostatic spray	Highly efficient coverage and use of paint on complex parts. Successfully automated.	High equipment cost. Requires specially formulated coatings.	Heat dissipators, electronic enclosures, open-work grills and complex parts.

electrode. The surface to be coated is grounded so that it has a high positive potential with respect to the negatively charged coating particles. Several different methods of electrostatic spraying are available. Although all use the basic principle of electrostatic attraction by giving the coating particles a high negative charge, they differ in the manner in which the coating material is atomized and directed. Also, the original use of electrostatic spray for applying liquid coating materials has recently been supplemented by an electrostatic method for applying dry-powder coatings.

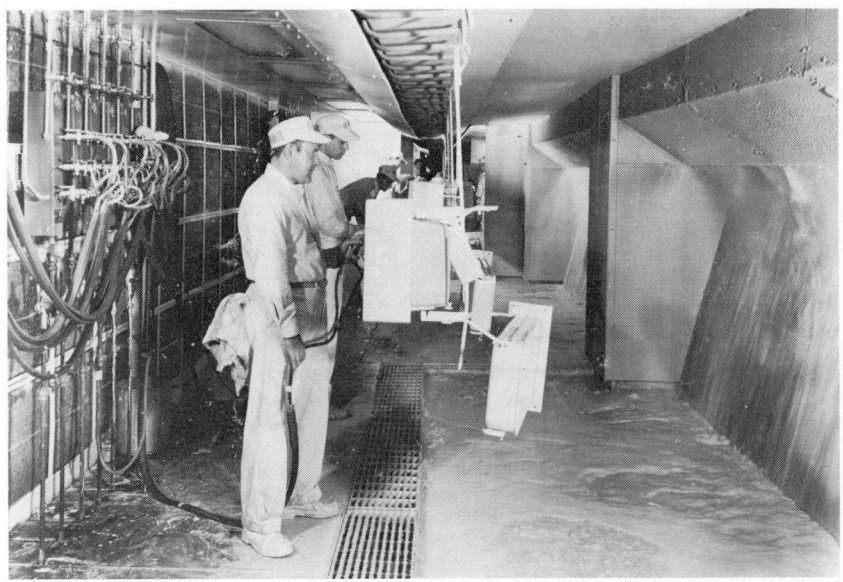

Fig. 11 Spray application is the most widely used method for applying plastic- and elastomeric-type coatings to industrial products such as appliances and cars. (*Courtesy of American Iron and Steel Institute.*)

Dipping Dipping is a fast and economical method of industrial finishing. The shape and surface texture of the object and the rate of its withdrawal from the dip tank have a direct effect on drainage and on the thickness and uniformity of the coating. The faster the object is withdrawn, the greater is the thickness and the less uniform the coating tends to be. In general, best results are obtained by with-drawing the object at a slow, uniform speed that permits most of the drainage to take place as the withdrawal proceeds.

Flow Flow coating is a process in which the coating material is flowed onto the object from a hose, after which excess material is allowed to drain off into a shallow tank. Flow coating is similar to dipping in its requirements for close control of the coating composition, viscosity, drying rate, and drainage. However, flow coating has the advantage of not requiring a large dip tank. The coating material is sup-plied from a central reservoir which can feed several hoses at a time, if desired.

Knife and roller coating are high-speed industrial coating methods for applying a wide variety of coatings to flat surfaces on a continuous basis.

Knife Knife coating is similar in principle to the doctor-blade technique used in the laboratory for applying paint films. Only one side can be coated at a time. The coating is flowed on and spread evenly over the surface of the substrate material. Coating thickness will vary with nonuniformity in the thickness of the substrate.

Roller Roller coating uses hard-rubber or steel rolls to apply the coating to flexible, flat sheet of metal, cloth, paper, and thin composition board. The coating is applied and spread evenly over the surface of the material by a roller rotating in the direction opposite to that of the strip of material being coated. Coating speeds of 50 ft/min are readily achieved. If desired, both sides of the strip can be coated at the same time.

Fluidized bed   Fluidized-bed coating is a rapid, convenient, and economical method for applying solventless coatings on objects of metal, glass, ceramic, plastics, and other materials. The technique is suitable for coating small objects and for applying thick, uniform coatings on complex shapes. In the fluidized process, the coating material in the form of a fine, dry powder is fluidized by suspending the coating particles in a controlled, upward stream of gas. The coating particles must be spherical in shape to present a uniform surface resistance in the gas stream. Both powder and gas must be clean and dry to ensure free mobility of the particles. The article to be coated is preheated to a temperature above the melting point of the resin and then dipped into the fluidized bed, where it acquires a uniform coating of the material as the particles melt onto it. Upon withdrawal, the coating congeals into a more or less continuous film which may be further annealed in an oven to ensure good leveling and a pinhole-free coating.[13]

The fluidized-bed method is most widely used for thermoplastic resins which are readily fused into a continuous film. However, it can also be used with thermosetting resins that can be liquefied by heat prior to being cured by heating. Thermoplastic resins which have been successfully applied by the fluidized-bed method include such materials as polyethylene, polypropylene, nylon, and the fluorocarbons, as well as polystyrene, vinyls, acrylics, and cellulosics.

Electrodeposition   Electrodeposition or electrocoating is a process in which paint is deposited from a water dispersion or solution on an electrode (substrate) when direct electric current is applied. Unlike conventional paint systems, an extremely clean substrate is required because scratches and surface defects are faithfully reproduced owing to the nature of the process.[19]

Water-reducible coating formulations used in electrodeposition are of two types, emulsion and water-soluble. As a rule, the coatings are formulated as simply as possible, since pigment and resin must move in the same direction and deposit at reproducible rates. Complications arise as the number of constituents in the formulation increases. A wide range of polymers are being used as primary resins in the electrodeposition process. These include maleinized drying oils, phenolic-modified maleinized oils, alkyds, epoxy esters, styrene–allyl alcohol copolymer esters, and acrylate copolymers.

Electrodeposited water-soluble coatings quickly coat all surfaces, and as these surfaces are coated, electric current approaches zero and the process stops for all practical purposes. Some coatings are 1 mil or less thick, and heavier coatings are as thick as 2 mils.

Electrodeposition is used to prime-coat automobile bodies, wheels, and other automobile parts. Also, it is used to coat metal furniture, appliances, toys, office furniture, electrical parts, tools, and other metal products.

The process lends itself well to coating ferrous and nonferrous metals. Although nonmetals have been coated by the process, there are problems relating to flow of electric current.

ECOLOGY

Rule 66, the Los Angeles County air-pollution regulation, has been in effect for several years, as has Regulation 3 in the San Francisco Bay area. These regulations are serving as prototypes for pollution-abatement legislation in other sections of the country. Rule 66 prohibits the sale and application of noncomplying products, whereas Regulation 3 is directed toward emission into the atmosphere rather than toward the products themselves. In both cases severe limitations have been placed on the classes and amounts of organic solvents used in coatings. Solvent restrictions for both Rule 66 and Regulation 3 are similar, and a formulation meeting Rule 66 requirements will most likely comply with Regulation 3, but the reverse is not necessarily true.

As a result many industrial-paint suppliers are formulating and reformulating paints to comply with these laws. Some coatings contain exempt solvents, whereas many others are water-reducible systems or 100 percent solids. Water-reducible coatings are available for flow coating, dipping, spraying, and electrodeposition. Some water-base coatings are epoxy, acrylic, phenolic, etc.

Historically, solvents and thinners have been necessary to allow coatings to be handled and applied. Economically, they have been costly wastes. Powder coatings offer a way around solvents. Fluidized bed, electrostatic spray, or combinations of

Fig. 12 Prelaminated metals offer a way out of air-pollution problems encountered in industrial finishing. Materials are sold to the user ready for fabrication. (*Litho-Strip Corp.*)

these two basic methods are used to apply the powders. Typical powders include vinyls, nylons, cellulosics, epoxies, polyolefins, and other materials.[15]

As in other paint-application methods, there are limitations to the fluidized process. For example, coatings less than 0.010 in. thick are difficult to attain and coatings less than 0.005 in. are practically unattainable, except by electrostatic processes. Another possible limitation is that the part to be coated must be able to withstand preheat temperatures of 450 to 650°F or higher. Surface preparation is very important in the fluidized-bed process, as it is in all coating applications.

Pretreatments Another potential problem regarding pollution is that of pretreat chemicals. Traditionally, these materials contain toxic substances in many instances. Cleaning chemicals, classically containing high percentages of phosphates, are being formulated phosphate-free. Most cleaning compounds now formulated contain surfactant systems which are classified as biodegradable. Biodegradable surfactants can be broken down by bacteria present in natural waters, sewage, or soils, into carbon dioxide in water. Rinses based on solutions of certain nontoxic materials compare in performance with toxic-chromium-bearing rinses. Normally these nontoxic rinses can be discharged into a sewer system without any waste treatment and can be handled with no danger.[14]

Prepainted metal Prepainted and prelaminated metal is another way around air- and water-pollution problems associated with industrial finishing. In this case, prepainted metals, also known as coil-coated metals, are prepainted and prelaminated by the supplier (see Fig. 12). They are sold to users ready for fabrication. Besides eliminating pollution problems, users do not have to set up a painting line to finish parts, production floor space is saved, and equipment, insurance, and labor costs often are reduced.

The coatings often can be formed and fabricated like uncoated metals. Naturally, some forming operations will damage the prepainted finish, but by and large, parts can be fabricated without surface damage by observing a few precautions.

A wide variety of coatings are available. These include acrylic, alkyds, silicone-alkyds, vinyl-alkyds, epoxies, phenolics, polyester, polyvinylidene fluoride and polyvinyl fluoride fluorocarbon coatings, vinyl solution, vinyl plastisols, and vinyl organosols. Prelaminated metals include those with a film of polyvinyl chloride or polyvinyl fluoride.

Prepainted metals find many uses, including transportation equipment, luggage, shelving, cabinets, appliance housings, metal furniture, baseboard-heating jackets, exterior siding, metal-container linings, drapery hardware, awnings, lighting fixtures, automobile parts, and mobile homes.

Probably one of the biggest drawbacks in using prepainted metals is the fact that conventional welding techniques cannot be used. However, magnetic-force welding or projection-stud welding are possible. In addition, other fastening methods can be used, such as lock seaming, snap fitting, riveting, and adhesive bonding.

REFERENCES

1. Roberts, A. G.: "Organic Coatings: Properties, Selection, and Use," National Bureau of Standards, 1968.
2. Burns, R. M., and W. W. Bradley: "Protective Coatings for Metals," 3d ed., Van Nostrand Reinhold, New York, 1970.
3. Sarvetnick, H. A.: "Plastisols and Organosols," Van Nostrand Reinhold, New York, 1970.
4. Murphy, J. A.: "Surface Preparation and Finishes for Metals," McGraw-Hill, New York, 1970.
5. Gross, W. F.: "Applications Manual for Paint and Protective Coatings," McGraw-Hill, New York, 1970.
6. Martens, C. R. (ed.): "Technology of Paints, Varnishes, and Lacquers," Van Nostrand Reinhold, New York, 1968.
7. Myers, R. R., and J. S. Lang: "Treaties on Coatings," vol. 1, parts 1 and 2; vol. 2, Marcel Dekker, Inc., New York, 1968, 1969.
8. Parker, D. H. (ed.): "Principles of Surface Coating Technology," Wiley, New York, 1965.
9. Payne, H. F.: "Organic Coating Technology," Wiley, New York, 1961.
10. Tysall, L. A.: "Basic Principles of Industrial Paints," Pergamon, New York, 1967.
11. Uhlig, H. H.: "Corrosion and Corrosion Control," Wiley, New York, 1971.
12. Simonds, H. R., and J. M. Church: "The Encyclopedia of Basic Materials for Plastics," Van Nostrand Reinhold, New York, 1967.
13. Ranney, M. W.: "Powder Coatings and Fluidized Bed Techniques," Noyes Data Corp., 1971.
14. Cleaning and Preparing Metal in Aircraft Production, *Oakite Products, Inc., Tech. Bull.*
15. Mock, J. A.: Guide to Organic Coatings, *Mater. Eng.*, August 1972.
16. Burnside, G. L., and G. E. F. Brewer: Ford Electrocoating Process: Corrosion Protection Standards, *Metal Finishing*, December 1967.
17. Wiesman, D. H., Sr., D. H. Wiesman, Jr., G. A. Hagan, and J. W. Hagan: Non-Aqueous Acrylic Dispersion Coatings, *SAE Congr.*, January 1968.
18. Pettigrew, C. K.: Plastic Coatings: A Unique Concept in Mechanical Design, *ASME Des. Eng. Conf.*, April 1968.

19. Ranney, M. W.: "Electrodeposition and Radiation Curing of Coatings," Noyes Data Corp., 1970.
20. Mock, J. A.: Intumescent Coatings Make Flammable Materials Safer, *Mater. Eng.*, July 1968.
21. McCormick, R. L., Jr.: Tough Vinyl Organosols Now Are Stain Resistant, *Mater. Eng.*, September 1972.
22. Gel Coat Handbook, *Ferro Corp. Tech. Bull.*
23. Powder Coating '72, *SME Symp.*, March 1972.
24. Waterborne Coating, *DeSoto Inc. Tech. Bull.*
25. "Paint, Varnish, Lacquer, and Related Products: Materials Specifications and Tests: Naval Stories; Aromatic Hydrocarbons," Part 20, ASTM Standards, 1972.
26. Federation of Societies for Paint Technology: "Thesaurus of Paint and Allied Technology," 1968.
27. Harper, Charles A. (ed.): "Handbook of Materials and Processes for Electronics," Chap. 5, by J. J. Licari and E. R. Brands, McGraw-Hill, New York, 1970.

Chapter **10**

Plastics and Elastomers as Adhesives

EDWARD M. PETRIE

Westinghouse Electric Corporation, Research and
Development Center, Pittsburgh, Pennsylvania

INTRODUCTION TO ADHESIVES

Definition

Modern structural adhesives are primarily synthetic materials based on plastic or elastomeric compounds. Adhesives were first used many thousands of years ago, and until relatively recently most adhesives were derived from vegetable, animal, or mineral substances. Synthetic polymeric adhesives displaced many of these products owing to their stronger adhesion and greater resistance to operating environments. Because of the importance and wide use of plastic- and elastomeric-based adhesives, they are the principal concern of this chapter.

An adhesive is defined as a substance capable of holding materials together by surface attachment. Unfortunately, a material conforming to this definition does not necessarily ensure success with an assembly problem. For an adhesive to be useful, it must not only hold materials together but also withstand operating loads and last the life of the product.

The successful application of an adhesive depends on many factors. Anyone intending to use adhesives faces the complex task of selecting the proper adhesive and correct time-temperature-pressure relationship that allows it to harden. He must also determine the substrate-surface treatment which will permit an acceptable degree of permanence and bond strength. The adhesive joint must be correctly designed to avoid stresses in the bond that could cause premature failure. Also, the physical and chemical stability of the bonded joint must be forecast with relation to its service environment. This chapter is intended to be a guide in the selection and use of adhesives so that adequate strength and service life may be realized in adhesive-bonded assemblies.

TABLE 1   Advantages and Disadvantages of Adhesive Bonding

Advantages	Disadvantages
1. Provides large stress-bearing area	1. Surfaces must be carefully cleaned
2. Provides excellent fatigue strength	2. Long cure times may be needed
3. Damps vibration and absorbs shock	3. Limitation on upper continuous operating temperature (generally 350°F)
4. Minimizes or prevents galvanic corrosion between dissimilar metals	4. Heat and pressure may be required to set adhesive
5. Joins all shapes and thicknesses	5. Jigs and fixtures may be needed
6. Provides smooth contours	6. Rigid process control usually necessary
7. Seals joints	7. Inspection of finished joint is difficult
8. Joins any combination of similar or dissimilar materials	8. Useful life depends on the environment
9. Often less expensive and faster than mechanical fastening	
10. Heat, if required, is too low to affect metal parts	
11. Provides attractive strength/weight ratio	

Advantages and Disadvantages of Adhesive Bonding

The science of adhesive bonding has been advanced to a degree where adhesives must be considered an attractive and practical alternative to mechanical fastening. Adhesive bonding presents several distinct advantages over conventional mechanical methods of fastening. There are also some disadvantages which may make adhesive bonding impractical. These pros and cons are summarized in Table 1.

The design engineer must consider and weigh these factors before deciding on a method of fastening. However, in many applications adhesive bonding is the only practical, high-strength method for solving assembly problems. In the aircraft industry, for example, adhesives make the use of thin metal and honeycomb structures feasible because stresses are transmitted more effectively by adhesives than by rivets or welds. Plastics, elastomers, and some metals such as aluminum and titanium can also be more reliably joined by adhesives than by any other method.

Mechanical advantages The most common methods of structural fastening are shown in Fig. 1. Welding or brazing, useful on heavy-gage metal, is expensive and requires great heat. Many light metals such as aluminum, magnesium, and titanium are difficult to weld and are weakened or distorted by the heat of welding. High-strength, reliable joints can be made on these metals with adhesives. Holes, needed for rivets or other fasteners, are not required for an adhesive bond, thereby allowing the stresses to be uniformly distributed over the entire bonded area.

The stress-distribution characteristics and inherent toughness of adhesives provide bonds with superior fatigue resistance, as shown in Fig. 2. Generally in well-designed joints, the adherends fail in fatigue before

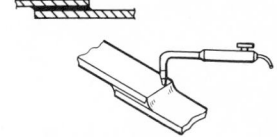

Brazing is an expensive bonding method. Requiring excessive heat, it often results in irregular, distorted parts. Adhesive bonds are always uniform.

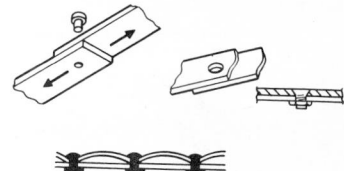

When joining a material with mechanical fasteners, holes must be drilled through the assembly. These holes weaken the material and allow concentration of stress.

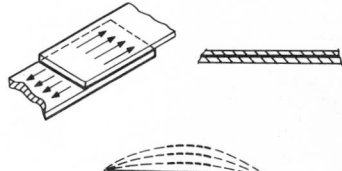

A high strength adhesive bond withstands stress more effectively than either welds or mechanical fasteners.

Fig. 1 Common methods of structural fastening.[1]

the adhesive. With the proper selection and use of adhesives, shear strengths as high as 7,000 lb/in.^{2} are available. Adhesives may also act as vibration dampers to reduce the noise and oscillation encountered in some assemblies.

Design advantages   Adhesives offer certain design advantages that are often valuable. Unlike rivets or bolts, adhesives produce smooth contours that are aerodynamically and cosmetically beneficial. Adhesives also offer a better strength/weight ratio than other methods of mechanical fastening.

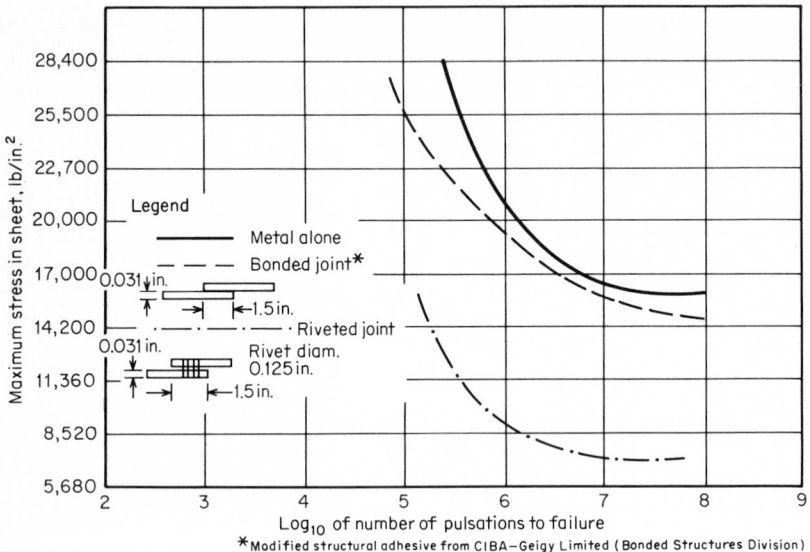

Fig. 2   Fatigue strengths of aluminum-alloy specimens under pulsating tensile load.[2]

Adhesives can join any combination of solid materials regardless of shape or thickness. Materials such as plastics, elastomers, ceramics, and wood can be joined more economically and efficiently by adhesive bonding than by other methods.

Adhesive bonding is frequently faster and less expensive than conventional fastening methods. It is especially well suited for high-volume production or assemblies requiring large bonded areas. As the size of the area to be joined increases, the time and labor saved by using adhesives instead of mechanical fasteners become progressively greater because the entire joint area can be bonded in one operation. Figure 3 shows the economy of metal-to-metal bonding compared with riveting.

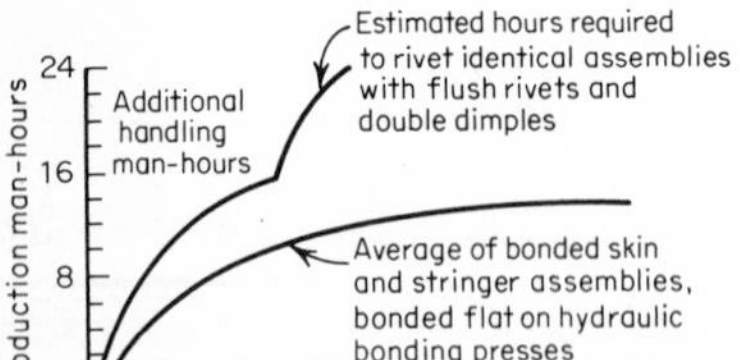

Other advantages   Another property of adhesives that may be advantageous is their ability to function as electrical and thermal insulators in a joint. The degree of insulation can be varied with different adhesive formulations and fillers. Adhesives can even be made electrically and thermally conductive with silver and boron nitride fillers, respectively. However, since adhesives generally do not conduct electricity, they prevent galvanic corrosion when dissimilar metals are bonded. Adhesives can also perform sealing functions, offering a barrier to the passage of fluids and gases.

Fig. 3   The economy of metal-to-metal bonding compared with conventional riveted structure.[3]

Frequently, adhesives may be called upon to do multiple functions. In addition to performing a mechanical fastening operation, an adhesive may also be used as a sealant, vibration damper, insulator, and gap filler.

Mechanical limitations The most serious limitation on the use of polymeric adhesives is their time-dependent strength in degrading environments such as moisture, high temperatures, or solvents.

There are organic adhesives that perform well between −60 and 350°F. But only a few adhesives can withstand operating temperatures outside that range. Adhesives are also degraded by chemical environments and outdoor weathering. The rate of strength degradation may be accelerated by continuous stress or elevated temperatures. Since nearly every adhesive application is somewhat unique, the adhesive manufacturers often do not have data concerning the aging characteristics of their adhesives in specific environments. Thus, before any adhesive is incorporated into production, a thorough evaluation should be made in a simulated operating environment.

With most structural adhesives, strength is more directional than with mechanical fasteners. Generally, adhesives perform better when stressed in shear or tension than when exposed to cleavage or peel forces.

Design limitations The adhesive joint must be carefully designed for optimum performance. Design factors must include the type of stress, environmental influences, and production methods to be used. Since the true "general-purpose" adhesive has not yet been developed, the design engineer should allow time to test candidate adhesives and bonding processes. Further time must be allowed to train personnel in what can be a rather complex and critical fastening technique.

Although the material cost of adhesives is relatively low, some systems may require metering, mixing, and dispensing units as well as curing fixtures such as ovens and presses. Capital equipment investment should be included in any economic evaluation of an adhesive system. The following items contribute to a "hidden cost" of using adhesives, and they also could contribute to serious production difficulties:

- The storage life of the adhesive may be unrealistically short; some adhesives require refrigerated storage.
- The adhesive may begin to solidify before the worker is ready.
- The cost of surface preparation and primers, if necessary, must be considered.
- Ease of handling, waste, and reliability can be essential cost factors.
- Cleanup is a cost factor, especially where misapplied adhesive may ruin the appearance of a product.
- Once bonded, samples cannot easily be disassembled; if misalignment occurs and the part cures, usually the part must be scrapped.

Many of these costs can be minimized by the proper choice and use of adhesives.

Production limitations Several problems must be considered when bonding operations are projected. All adhesives require clean surfaces to attain optimum results. Depending on the type and condition of the substrate and the bond strength desired, surface preparations ranging from a simple solvent wipe to chemical etching are necessary.

If the adhesive has multiple components, the parts must be carefully weighed and mixed. The setting operation often requires heat and pressure. Lengthy set time makes jigs and fixtures necessary for assembly. Rigid process controls are also necessary, because the adhesive properties are dependent on the curing parameters and surface preparations.

Adhesives are generally composed of material that may present personnel hazards, including flammability and dermatitis, in which case the necessary precautions must be considered. Finally, the inspection of finished joints for quality control is very difficult. This necessitates strict control over the entire bonding process to ensure uniform bond quality.

Nomenclature

Some of the more commonly used terms in the adhesives industry are defined in the glossary section of this handbook. A much more complete glossary of terms pertaining to adhesives can be found in the 1970 American Society for Testing and Materials "Book of Standards," Specification D 907-70.[4]

Theories of Adhesion

The actual mechanism of adhesive attachment is not yet very well defined. Various theories attempt to describe the phenomena of adhesion. Each theory contains points that are partially correct. No single theory explains adhesion in a general, all-encompassing way. A knowledge of adhesion theories can assist in understanding the basic requirements for a good bond.

Mechanical theory   At one time, adhesion was thought to occur by the adhesive flowing and filling microcavities on the substrate. When the adhesive then hardens, the substrates are held together mechanically. The surface of a solid material is never truly smooth but consists of a maze of peaks and valleys. According to the mechanical theory, in order to function properly the adhesive must penetrate the cavities on the surface and displace the trapped air at the interface.

Mechanical anchoring of the adhesive appears to be a prime factor in bonding many porous substrates. Adhesives also frequently bond better to nonporous abraded surfaces than to natural surfaces. This beneficial effect of mechanical roughening may be due to:

1. Mechanical interlocking
2. Formation of a clean surface
3. Formation of a highly reactive surface
4. Formation of a larger surface

It is widely believed that although the surface becomes rougher because of abrasion, it is a change in both physical and chemical properties of the surface layer which produces an increase in adhesive strength. While some adhesive applications can be explained by mechanical interlocking, it has been shown that mechanical effects are not always of great importance.

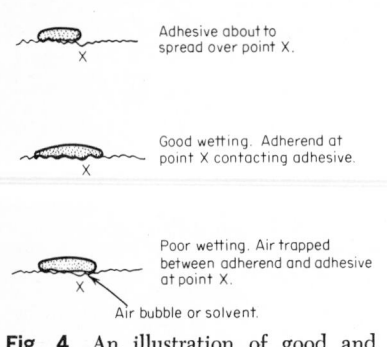

Adhesive about to spread over point X.

Good wetting. Adherend at point X contacting adhesive.

Poor wetting. Air trapped between adherend and adhesive at point X.

Air bubble or solvent.

Fig. 4   An illustration of good and poor wetting by an adhesive spreading over a surface.[5]

Adsorption theory   The adsorption theory states that adhesion results from molecular contact between two materials and the surface forces which develop.

The process of establishing continuous contact between an adhesive and the adherend is known as "wetting." For an adhesive to wet a solid surface, the adhesive should have a lower surface tension than the solid's critical surface tension. Figure 4 illustrates good and poor wetting of an adhesive spreading over a surface. Good wetting results when the adhesive flows into the valleys and crevices on the substrate surface; poor wetting results when the adhesive bridges over the valley. Poor wetting causes less actual area of contact between the adhesive and adherend, resulting in lower overall joint strength.

Most organic adhesives easily wet metallic solids. But many solid organic substrates have surface tensions less than those of common adhesives. The fact that good wetting requires the adhesive to have a lower surface tension than the substrate explains why organic adhesives such as epoxies have excellent adhesion to metals but offer weak adhesion on untreated polymeric substrates such as polyethylene, polypropylene, and the fluorocarbons.

After intimate contact is achieved between adhesive and adherend through wetting, it is believed that permanent adhesion results primarily through forces of molecular attraction. Four general types of chemical bonds are recognized as being involved in adhesion and cohesion: electrostatic, covalent, and metallic, which are referred to as *primary bonds,* and van der Waals forces, which are referred to as *secondary bonds.*

Electrostatic and diffusion theories   In the United States the electrostatic and diffusion theories of adhesion are not regarded as highly as the adsorption theory.

The electrostatic theory states that electrostatic forces in the form of an electrical double layer are formed at the adhesive-adherend interface. These forces account for resistance to separation. This theory gathers support from the fact that electrical discharges have been noticed when an adhesive is peeled from a substrate.

The fundamental concept of the diffusion theory is that adhesion arises through the interdiffusion of molecules in the adhesive and adherend. The diffusion theory is primarily applicable when both the adhesive and adherend are polymeric, having long-chain molecules capable of movement. The solvent or heat welding of thermoplastics is considered to be due to diffusion of molecules.

Weak-boundary-layer theory   According to the weak-boundary-layer theory as first described by Bikerman, when bond failure seems to be at the interface, usually a cohesive break of a weak boundary layer is the real event.[6] Weak boundary layers can originate from the adhesive, the adherend, the environment, or a combination of any of the three.

Weak boundary layers can occur on the adhesive or adherend if an impurity concentrates near the bonding surface and forms a weak attachment to the substrate. When failure occurs, it is the weak boundary layer that fails, although failure seems to occur at the adhesive-adherend interface.

Two examples of this effect are polyethylene and metal oxides. Polyethylene has a weak, low-molecular-weight constituent that is evenly distributed throughout the polymer. This weak boundary layer is present at the interface and contributes to low failing stress when polyethylene is used as an adhesive or adherend. Some metallic oxides are weakly attached to their base metals. Failure of adhesive joints made with these materials occurs cohesively within the oxide. It should also be noted that some oxides such as aluminum oxide are very strong and do not significantly impair joint strength. Weak boundary layers such as in polyethylene and metal oxides can be removed or strengthened by various surface treatments.

Weak boundary layers formed from the bonding environment, generally air, are very common. When the adhesive does not wet the substrate as shown in Fig. 4, a weak boundary layer of air is trapped at the interface, causing lowered joint strength.

Requirements for a Good Bond

The basic requirements for useful adhesive-bonded assemblies are cleanliness, wetting, solidification, and proper choice of adhesive and joint design.

Cleanliness   To achieve an effective adhesive bond, one must start with a clean surface. Foreign materials such as dirt, oil, moisture, and weak oxide layers must be removed from the substrate surface, or else the adhesive will bond to these weak boundary layers rather than the substrate in question. There are various surface preparations that remove or strengthen the weak boundary layer. These treatments generally involve physical or chemical processes or a combination of both. Surface-preparation methods for specific substrates will be discussed in a later section.

Wetting   While it is in the liquid state, the adhesives must "wet" the substrate. The term wetting refers to a liquid spreading over and intimately contacting a solid surface. Examples of good and poor wetting have been illustrated and explained. The result of good wetting is greater contact area between adherend and adhesive over which the forces of adhesion may act.

Solidification   The liquid adhesive, once applied, must be capable of conversion into a solid. The process of solidifying an adhesive can be completed in three ways: (1) chemical reaction by any combination of heat, pressure, and curing agents, (2) cooling from a molten liquid to a solid, and (3) drying due to solvent evaporation. The method by which solidification occurs depends on the choice of adhesive.

Adhesive choice   When determining which adhesives are suitable candidates for an application, a number of important considerations must be taken into account. Many of the factors most likely to influence adhesive selection are listed in Table 2. With regard to the controlling factors involved, the many adhesives available can usually be narrowed to a few candidates that are most likely to be successful. The general areas of concern to the design engineer when selecting adhesives should be the material to be bonded, service requirements, production requirements, and cost.

Joint design   The adhesive joint should be designed to take advantage of the good properties of adhesives and to minimize their shortcomings. Such design considerations will be discussed in the next section. Although adequate adhesive-bonded

TABLE 2   Factors Influencing Adhesive Selection[7]

Stress

Tension.............. Forces acting perpendicular to the plane of the adhesive. Not commonly encountered in bonding thin plastic or metal sheets, leather, cork compositions, etc.

Shear.............. Forces acting in the plane of the adhesive. Pure shear is seldom encountered in adhesive assemblies; substantial tension components are usually found

Impact.............. Minimum force required to cause the adhesive to fail in a single blow. May be determined in tension or shear. Measures brittleness

Peel............... Stripping of a flexible member fastened with adhesive to another flexible or rigid member. Stress is applied at a line; test loads are expressed in pounds per inch width. Commonly used angles of peel in tests are 90° for relatively stiff and 180° for flexible members

Cleavage............ Forces applied at one end of a rigid bonded assembly which tend to split the bonded members apart. Can be considered as "peel" of two rigid members

Fatigue............. Dynamic—alternate loading in shear or tension-compression. Static —maximum load sustained for long periods of time in tension or shear; tests are also used to determine creep

Chemical Factors

External............ Effect of chemical agents such as water, salt water, gasoline, by hydraulic fluid, acids, alkalies, etc.

Internal............ Effect of adherend on adhesive (i.e., exuded plasticizers in certain plastics and rubber); effect of adhesive on the adherend (crazing, staining, etc.)

Exposure

Weathering.......... Combined effect of rainfall, sunlight, temperature changes, type of atmosphere

Light.............. Important only with translucent adherends. Effect of artificial or natural light, or ultraviolet

Oxidation........... Usually tested by exposure to ozone with the joint either unstressed or stressed, in which case deterioration is faster

Moisture............ Either adhesive or adherend may be affected by high humidity or wet conditions. Cyclic testing with alternate moist and dry conditions can be valuable. May cause dimensional changes

Salt spray........... Important only in coastal or marine atmospheres. Possible corrosion of adherend should also be considered

Temperature

High............... Normal atmospheric variations may be encountered, or exceptional conditions. Bond strength may be affected by reactions in adhesive or adherend; decomposition or changes in physical properties of adhesive are important

Low............... May cause crystallization or embrittlement, detected by strength test. Cyclic testing with low or high temperatures may detect lack of durability

Biological Factors

Bacteria or mold..... Usually warm, humid tropical conditions. Can affect bond strength, and cause emission of odor or discoloration

Rodents or vermin.... Adhesives of animal or vegetable origin may be attacked by rats, cockroaches, etc.

Working Properties

Application.......... Brushing, spray, trowel, or knife-spreader application characteristics are usually determined by trial and error. Consistency or viscosity may be adequate indications. Mechanical stability of emulsions and dispersions, and foaming tendency, can be important for machine application

Bonding range....... Minimum drying or solvent-reactivation time before suitable bond can be obtained. Maximum allowable time before assembly. Permissible temperature range with heat-activated adhesives

TABLE 2   Factors Influencing Adhesive Selection[7] (Continued)

Working Properties (Continued)

Blocking............	Tendency of surfaces coated for storage before assembly to adhere under slight pressure, or changes in humidity or temperature
Curing rate.........	Minimum curing time, and effect of overcuring. May be determined as a shear or tensile-strength vs. curing-time curve at a specific curing temperature
Storage stability......	Physical and chemical changes in original unapplied state as a result of storage for extended time periods at representative storage temperatures
Coverage............	Area of bond that can be formed with unit weight or volume of adhesive; expressed as pounds per 1,000 ft of bond line, or square feet per gallon. Depends on method of application; dimensions of work or of adhesive-coated area in relation to part size may affect coverage

assemblies have been made from joints designed for mechanical fastening, the maximum benefits of the adhesive can be obtained only in assemblies specially designed for adhesive bonding.

Mechanism of Bond Failure

Adhesive joints may fail adhesively or cohesively.   Adhesive failure is interfacial bond failure between the adhesive and adherend.   Cohesive failure occurs when fracture allows a layer of adhesive to remain on both substrates.   When the adherend fails before the adhesive, it is known as a cohesive failure of the adherend.   The various modes of possible bond failures are shown in Fig. 5.

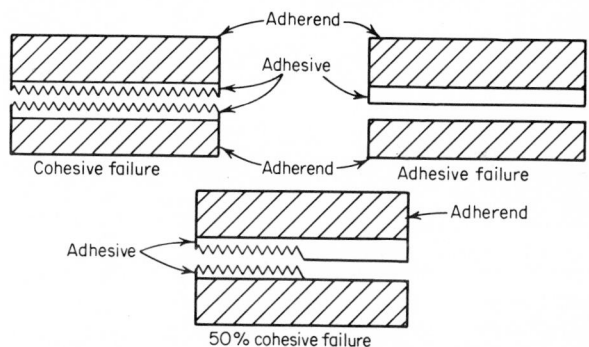

Fig. 5   Cohesive and adhesive bond failure.[3]

Cohesive failure within the adhesive or one of the adherends is the ideal type of failure, because the maximum strength of the materials in the joint has been reached. However, failure mode should not be used as a criterion for a useful joint.   Some adhesive-adherend combinations may fail in adhesion but exhibit greater strength than a similar joint bonded with a weaker adhesive that fails cohesively.   The ultimate strength of the joint is a more important criterion than the mode of joint failure. However, an analysis of failure mode can be an extremely useful tool to determine if the failure was due to a weak boundary layer or improper surface preparation.

The exact cause of premature adhesive failure is very hard to determine, because so many factors in adhesive bonding are interrelated.   However, there are certain common factors at work when an adhesive bond is made that contribute to the weakening of all bonds.   The influences of these factors are qualitatively summarized in Fig. 6.

If the adhesive does not wet the surface of the substrate, the joint will be inferior. It is also important to allow the adhesive enough time to wet the substrate effectively before gelation occurs.

Internal stresses occur in the adhesive joint during production because of a natural tendency of the adhesive to shrink during setting and because of differences in physical properties between adhesive and substrates. The coefficient of thermal expansion of adhesive and adherend should be as close as possible to limit stresses that may develop during thermal cycling or after cooling from an elevated-temperature cure. As shown in Fig. 7 polymeric adhesives generally have a thermal-expansion coefficient an order of magnitude greater than metals. Adhesives can be formulated with various fillers to modify their thermal-expansion characteristics and limit internal stresses. A relatively elastic adhesive capable of accommodating internal stress may also be useful when thermal-expansion differences are of concern.

Once an adhesive bond is made and placed in service, other forces are at work weakening the bond. The type of stress involved, its orientation to the adhesive, and

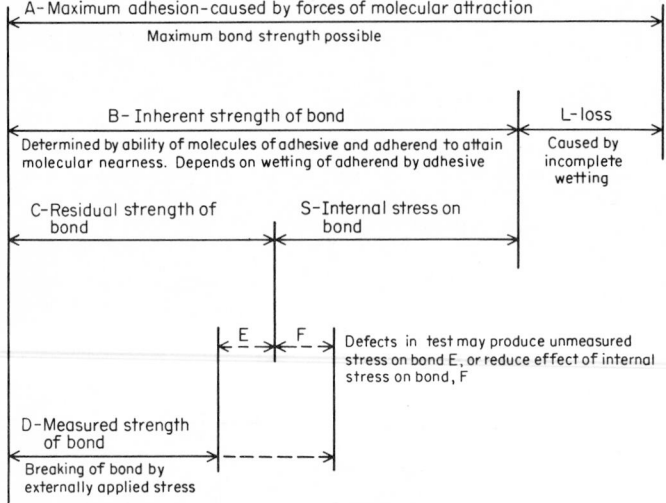

Fig. 6 Relations between the factors involved in adhesion.[8]

the rate in which the stress is applied are important. Sustained loads can cause premature failure in service even though similar unloaded joints may exhibit adequate strength when tested after aging. Some adhesives show very little resistance to dead load, especially during exposure to heat or moisture. The possibility of constant loads should be looked for in every application, and the adhesive should be tested accordingly. Most adhesives have poor strength when the stresses are acting to peel or cleave the adhesive from the substrate. Many adhesives are sensitive to the rate in which the joint is stressed. Rigid, brittle adhesives sometimes have excellent tensile or shear strength but stand up very poorly under an impact test.

Operating environments are capable of degrading an adhesive joint in various ways. The adhesive may have to withstand temperature variation, weathering, oxidation, moisture, and other exposure conditions. If more than one of these factors are present in the operating environment, their combined effect could cause a catastrophic decline in adhesive strength.

The possible combinations of adhesives, adherends, stress, and environments are so great that reliable adhesive strength and aging data are seldom available to the design engineer. Where time and funds permit, the candidate adhesive joints should always be evaluated under simulated operating loads in the intended environment.

Basic Adhesive Materials

Adhesives may be classified by many methods. The most common methods are by function, chemical composition, mode of application and setting, physical form, cost, and end use.

Function   The functional classification defines adhesives as being structural or nonstructural.   Structural adhesives are materials of high strength and permanence. Their primary function is to hold structures together and be capable of resisting high loads.

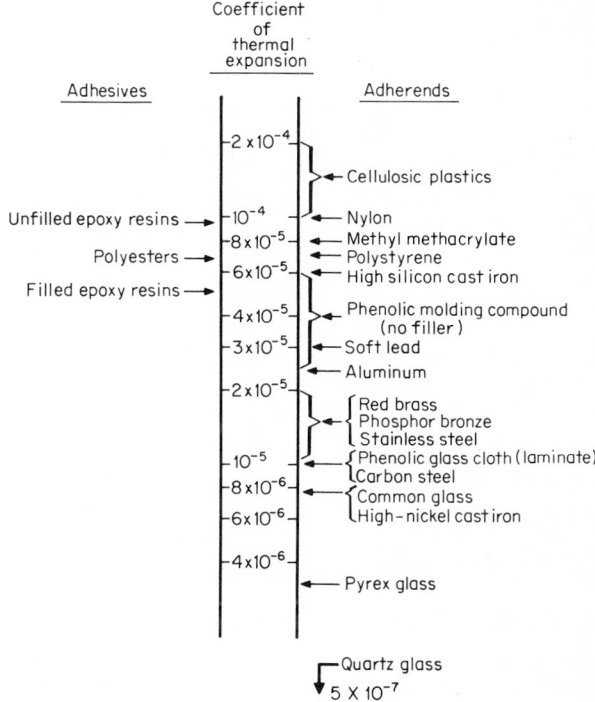

Fig. 7   Coefficients of thermal expansion, in./in./°C.[9]

Nonstructural adhesives are not required to support substantial loads but merely hold lightweight materials in place.   Nonstructural adhesives are sometimes referred to as holding adhesives.   Sealing adhesives usually have a nonstructural function. They are principally intended to fill a gap between adherends to provide a seal without having high degrees of strength.

Chemical composition   The composition classification describes synthetic adhesives as thermosetting, thermoplastic, elastomeric, or combinations of these.   They are described generally in Table 3.

Thermosetting adhesives are materials that cannot be heated and softened repeatedly after the initial cure.   Adhesives of this sort cure by chemical reaction at room or elevated temperatures, depending on the type of adhesive.   Substantial pressure may also be required with some thermosetting adhesives, and others are capable of providing strong bonds with only contact pressure.   Thermosetting adhesives are

TABLE 3 Adhesives Classified by Chemical Composition[10]

Classification	Thermoplastic	Thermosetting	Elastomeric	Alloys
Types within group...	Cellulose acetate, cellulose acetate butyrate, cellulose nitrate, polyvinyl acetate, vinyl vinylidene, polyvinyl acetals, polyvinyl alcohol, polyamide, acrylic, phenoxy	Cyanoacrylate, polyester, urea formaldehyde, melamine formaldehyde, resorcinol and phenol-resorcinol formaldehyde, epoxy, polyimide, polybenzimidazole, acrylic, acrylate acid diester	Natural rubber, reclaimed rubber, butyl, polyisobutylene, nitrile, styrene-butadiene, polyurethane, polysulfide, silicone, neoprene	Epoxy-phenolic, epoxy-polysulfide, epoxy-nylon, nitrile-phenolic, neoprene-phenolic, vinyl-phenolic
Most used form......	Liquid, some dry film	Liquid, but all forms common	Liquid, some film	Liquid, paste, film
Common further classifications.........	By vehicle (most are solvent dispersions or water emulsions)	By cure requirements (heat and/or pressure most common but some are catalyst types)	By cure requirements (all are common); also by vehicle (most are solvent dispersions or water emulsions)	By cure requirements (usually heat and pressure except some epoxy types); by vehicle (most are solvent dispersions or 100% solids); and by type of adherends or end-service conditions
Bond characteristics...	Good to 150–200°F; poor creep strength; fair peel strength	Good to 200–500°F; good creep strength; fair peel strength	Good to 150–400°F; never melt completely; low strength; high flexibility	Balanced combination of properties of other chemical groups depending on formulation; generally higher strength over wider temp range
Major type of use.....	Unstressed joints; designs with caps, overlaps, stiffeners	Stressed joints at slightly elevated temp	Unstressed joints on light-weight materials; joints in flexure	Where highest and strictest end-service conditions must be met; sometimes regardless of cost, as military uses
Materials most commonly bonded........	Formulation range covers all materials, but emphasis on nonmetallics—esp. wood, leather, cork, paper, etc.	For structural uses of most materials	Few used "straight" for rubber, fabric, foil, paper, leather, plastics films; also as tapes. Most modified with synthetic resins	Metals, ceramics, glass, thermosetting plastics; nature of adherends often not as vital as design or end-service conditions (i.e., high strength, temp)

sometimes provided in a solvent medium to facilitate application. They are also commonly available as solventless liquids, pastes, and solids.

Thermosetting adhesives may be sold as multiple and single-part systems. Generally the single-part adhesives require elevated-temperature cure, and they have a limited shelf life. Multiple-part adhesives have longer shelf lives, and some can be cured at room temperature or more rapidly at elevated temperatures. But they require metering and mixing before application. Once the adhesive is mixed, the working life is limited.

Because molecules of thermosetting resins are densely cross-linked, their resistance to heat and solvents is good, and they show little elastic deformation under load at elevated temperatures.

Thermoplastic adhesives do not cross-link during cure; so they can be resoftened with heat. They are single-component systems that harden upon cooling from a melt or by evaporation of a solvent or water vehicle. Hot-melt adhesives commonly used in packaging are examples of a solid thermoplastic material that is applied in a molten state, and adhesion develops as the melt solidifies during cooling. Wood glues are thermoplastic emulsions that are a common household item. They harden by evaporation of water from an emulsion.

Thermoplastic adhesives have a more limited temperature range than thermosetting types. It is not suggested to use thermoplastic adhesives over 150°F. Their physical properties vary over a wide range because many polymers are used in a single adhesive formulation.

Elastomeric-type adhesives are based on synthetic or naturally occurring polymers having great toughness and elongation. These adhesives may be supplied as solvent solutions, latex cements, dispersions, pressure-sensitive tapes, and single- or multiple-part solventless liquids or pastes. The curing requirements vary with the type and form of elastomeric adhesive. These adhesives can be formulated for a wide variety of applications, but they are generally noted for their high degree of flexibility and good peel strength.

Adhesive alloys are made by combining thermosetting, thermoplastic, and elastomeric adhesives. They utilize the most useful properties of each material. However, the adhesive alloy is usually never better than its weakest constituent. Thermosetting-resin additions provide a very strong, cross-linked structure. Impact, bending, and peel strength are provided by the addition of thermoplastic or elastomeric materials. Adhesive alloys are commonly available in solvent solutions and as supported or unsupported film.

Mode of application and setting Adhesives are often classified by their mode of application. Depending on viscosity, adhesives are sprayable, brushable, or trowelable. Heavily bodied adhesive pastes and mastics are considered extrudable; they are applied by syringe, caulking gun, or pneumatic pumping equipment.

Another distinction between adhesives is the manner in which they flow or solidify. As shown in Table 4, some adhesives solidify simply by losing solvent while others harden as a result of heat activation or chemical reaction. Pressure-sensitive systems flow under pressure and are stable when pressure is absent.

Physical form Another method of distinguishing between adhesives is by physical form. The physical state of the adhesive generally determines how it is to be applied. Liquid adhesives lend themselves to easy handling via mechanical spreaders or spray and brush. Paste adhesives have high viscosities to allow application on vertical surfaces with little danger of sag or drip. These bodied adhesives also serve as gap fillers between two poorly mated substrates. Tape and film adhesives are poor gap fillers but offer a uniformly thick bond line, no need for metering, and easy dispensing. Adhesive films are available as a pure sheet of adhesive or with cloth or paper reinforcement. Another form of adhesive is powder or granules which must be heated or solvent-activated to be made liquid and appliable. The characteristics and advantages of the various adhesive forms are summarized in Table 5.

Cost The cost of fastening with adhesives must include the material cost of the adhesive, the cost of labor, the cost of equipment, the time required to cure the adhesive, and the economic loss due to rejects of defective joints.

Adhesive price is dependent on development costs and volume requirements.

TABLE 4 Adhesives Classed by Activation and Cure Requirements[10]

Requirement	Types available	Forms used	Remarks
Heat	Room temp to 450°F types available; 250 to 350°F types most common for structural adhesives	Formulated in all forms; liquid most common	Applying heat will usually increase bond strength of any adhesive, even room temp types
Pressure	Contact to 500 lb/in.² types available; 25 to 200 lb/in.² types most common for structural adhesives	Formulated in all forms; liquid or powder most common	Pressure types usually have greater strength (not true of modified epoxies)
Time	Types requiring a few seconds to a week available; ½ to 24-h types most common for structural adhesives	Formulated in all forms; solvent solutions common	Time required varies with pressure and temp applied and immediate strength
Catalyst	Extremely varied in terms of chemical catalyst required; may also contain thinners, etc.	Two components—paste (or liquid) + liquid	Sometimes catalyst types may require elevated temp (<212°F) and/or pressure instead of, or in addition to, a chemical agent

Adhesives that have been specifically formulated to be resistant to adverse environments are more expensive than "general-purpose adhesives." Adhesive prices range from pennies a pound for inorganic and animal-based systems to over \$20/lb for some heat-resistant types. Common epoxy adhesives are generally about \$1.50 to \$3.50/lb in 5-gal lots. Adhesives in film or powder form require more processing than liquid or paste types and are more expensive.

Adhesive-materials cost should be calculated on a cost per bonded area basis. Since many adhesives are sold as dilute solutions, a cost per unit weight or volume basis may lead to erroneous comparisons.

Specific adherends or applications Adhesives may also be classified according to their end use. Thus, metal adhesives, wood adhesives, and vinyl adhesives refer to the substrates they bond; and acid-resistant adhesives, heat-resistant adhesives, and weatherable adhesives indicate the environments for which each is suited. The remainder of this chapter will principally classify adhesives according to their end use and chemical composition.

TABLE 5 Adhesives Classified by Form[10]

Type	Remarks	Advantages
Liquid	Most common form; practically every formulation available. Principally solvent-dispersed	Easy to apply. Viscosity often under control of user. Major form for hand application
Paste	Wide range of consistencies. Limited formulations; principally 100% solids modified epoxies	Lends itself to high-production setups because of less time wait. High shear and creep strengths
Powder	Requires mixing or heating to activate curing	Longer shelf life; mixed in quantities needed
Mastic	Applied with trowel	Void-filling, nonflowing
Film, tape	Limited to flat surfaces, wide range of curing ease	Quick and easy application. No waste or run-over; uniform thickness
Other	Rods, supported tapes, precoated copper for printed circuits, etc.	Ease of application and cure for particular use

Adhesive Composition

Modern-day adhesives are often fairly complex formulations of components that perform specialty functions.

The adhesive *base* or *binder* is the primary component of an adhesive that holds the substrates together. The binder is generally the component from which the name of the adhesive is derived. For example, an epoxy adhesive may have many components, but the primary material is epoxy resin.

A *hardener* is a substance added to an adhesive to promote the curing reaction by taking part in it. Two-part adhesive systems generally have one part which is the base and a second part which is the hardener. Upon mixing, a chemical reaction ensues which causes the adhesive to solidify. A *catalyst* is sometimes incorporated into an adhesive formulation to speed up the reaction between base and hardener.

Solvents are sometimes needed to disperse the adhesive to a spreadable consistency. Solvents used with synthetic resins and elastomers are generally organic in nature, and often a mixture of solvents is required to achieve the desired properties.

An ingredient added to an adhesive to reduce the concentration of binder is called a *diluent*. Diluents are principally used to lower the viscosity and modify the processing conditions of some adhesives. Reactive diluents react with the binder during cure, become part of the product, and do not evaporate as does a solvent.

Fillers are relatively nonadhesive substances added to the adhesive to improve its working properties, strength, permanence, or other qualities. Fillers are also used to reduce material cost. By selective use of fillers the properties of an adhesive can be changed tremendously. Thermal expansion, electrical and thermal conduction, shrink-age, viscosity, and thermal resistance are only a few properties that can be modified by use of fillers.

A *carrier* or *reinforcement* is usually a thin fabric or paper used to support the semicured adhesive composition to provide a tape or film. The carrier also serves as a bond-line spacer and reinforcement for the adhesive.

Adhesive-Bonding Process

A typical flow chart for the adhesive-bonding process is shown in Fig. 8. The bonding process is as important as the adhesive itself for a successful end product.

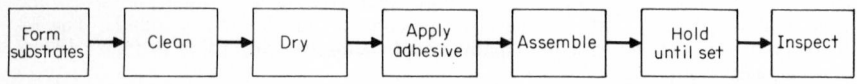

Fig. 8 Basic steps in the bonding process.[11]

Many of the adhesive problems that develop are not due to a poor choice of material or joint design but are directly related to faulty production techniques. The adhesive user must obtain the proper processing instructions from the manufacturer and follow them closely and consistently to ensure acceptable results. Adhesive production involves four basic steps:
1. Design of joints and selection of adhesive
2. Preparation of adherends
3. Applying and curing the adhesive
4. Inspection of bonded parts

The remainder of this chapter will discuss these steps in detail.

DESIGN AND TESTING OF ADHESIVE JOINTS

Types of Stress

Four basic types of loading stress are common to adhesive joints: tensile, shear, cleavage, and peel. Any combination of these stresses, illustrated in Fig. 9, may be encountered in an adhesive application.

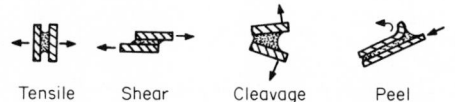

Tensile   Shear   Cleavage   Peel

Fig. 9 The four basic types of adhesive stress.[12]

Tensile stress develops when forces acting perpendicular to the plane of the joint are distributed uniformly over the entire bonded area. Adhesive joints show good resistance to tensile loading because all the adhesive contributes to the strength of the joint. In practical applications, unfortunately, loads are rarely axial, and cleavage or peel stresses tend to develop. Since adhesives have poor resistance to cleavage and peel, joints designed to load the adhesive in tension should have physical restraints to ensure axial loading.

Shear stress results when forces acting in the plane of the adhesive try to separate the adherends. Joints dependent upon the adhesive's shear strength are relatively easy to make and are commonly used. Adhesives are generally strongest when stressed in shear because all the bonded area contributes to the strength of the joint.

Cleavage and peel stresses are undesirable. Cleavage occurs when forces at one end of a rigid bonded assembly act to split the adherends apart. Peel stress is similar to cleavage but applies to a joint where one or both of the adherends are flexible. Joints loaded in peel or cleavage offer much lower strength than joints loaded in shear because the stress is concentrated on only a very small area of the total bond. Peel stresses particularly should be avoided where possible, since the stress is confined to a very thin line at the edge of the bond. The remainder of the bonded area makes no contribution to the strength of the joint.

Maximizing Joint Efficiency

Although adhesives have often been used successfully on joints designed for mechanical fastening, the maximum efficiency of bonded joints can be obtained only by designing the joint specifically for adhesive bonding. To avoid concentration of stress, the joint designer should take into consideration the following rules:

1. Keep the stress on the bond line to a minimum.
2. Design the joint so that the operating loads will stress the adhesive in shear.
3. Peel and cleavage stresses should be minimized.
4. Distribute the stress as uniformly as possible over the entire bonded area.
5. Adhesive strength is directly proportional to bond width. Increasing width will always increase bond strength; increasing the depth does not always increase strength.
6. Generally rigid adhesives are better in shear, and flexible adhesives are better in peel.

Brittle adhesives are particularly weak in peel because the stress is localized at only a thin line, as shown in Fig. 10. Tough, flexible adhesives distribute the peeling stress over a wider bond area and show greater resistance to peel.

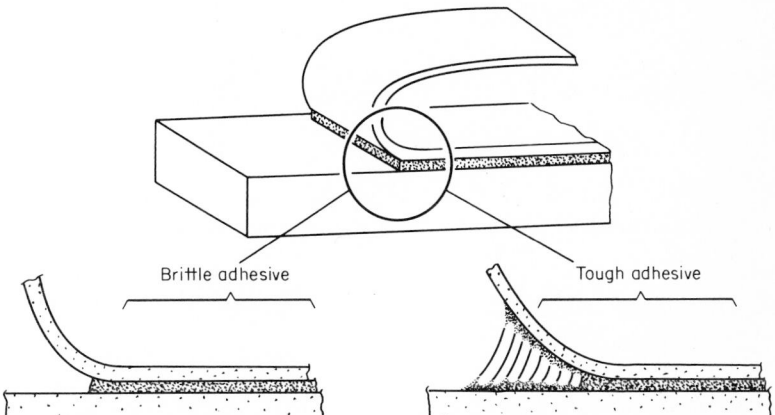

Brittle adhesive   Tough adhesive

Fig. 10   Tough, flexible adhesives distribute peel stress over a larger area.[11]

For a given adhesive and adherend, the strength of a joint stressed in shear depends primarily on the width and depth of the overlap and the thickness of the adherend. Adhesive shear strength is directly proportional to the width of the joint. Strength can sometimes be increased by increasing the overlap depth, but the relationship is not linear. Since the ends of the bonded joint carry a higher proportion of the load than the interior area, the most efficient way of increasing joint strength is by increasing the width of the bonded area.

In a shear joint made from thin, relatively flexible adherends, there is a tendency for the bonded area to distort because of eccentricity of the applied load. This distortion, illustrated in Fig. 11, causes cleavage stress on the ends of the joint, and the joint strength may be considerably impaired. Thicker adherends are more rigid, and the distortion is not as much a problem as with thin-gage adherends. Figure 12 shows the general interrelationship between failure load, depth of overlap, and adherend thickness for a specific metallic adhesive joint.

Since the stress distribution across the bonded area is not uniform and depends on joint geometry, the failure load of one specimen cannot be used to predict the failure load of another specimen with different joint geometry. The results of a particular shear test pertain only to joints that are exact duplicates. To characterize an adhesive more closely, the ratio of overlap length to adherend thickness l/t is plotted against shear strength. A set of l/t curves for aluminum bonded with a nitrile-rubber adhesive is shown in Fig. 13. From such a chart, the design engineer can determine

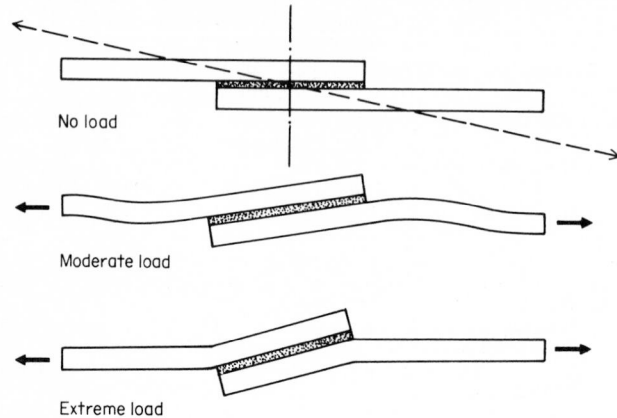

No load

Moderate load

Extreme load

Fig. 11 Distortion caused by loading can introduce cleavage stresses and must be considered in joint design.[10]

the overlap length required for a given adherend thickness to obtain a specific joint strength.

The strength of an adhesive joint also depends on the thickness of the adhesive. Thin adhesive films offer the highest shear strength provided that the bonded area does not have "starved" areas where all the adhesive has been forced out. Excessively

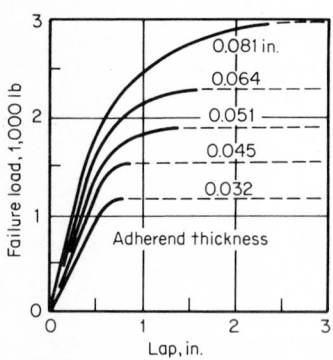

Fig. 12 Interrelation of failure loads, depth of lap, and adherend thickness for lap joints with a specific adhesive and adherend.[9]

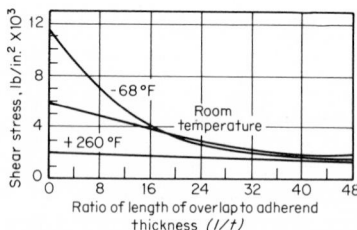

Fig. 13 l/t curves at three test temperatures for aluminum joints bonded with a nitrile-rubber adhesive.[13]

heavy adhesive-film thicknesses cause greater internal stresses during cure and concentration of stress under load at the ends of a joint. Optimum adhesive thickness for maximum shear strength is generally between 2 and 8 mils. Strength does not vary significantly with bond-line thickness in this range.

Joint Design

The ideal adhesive-bonded joint is one in which under all practical loading conditions the adhesive is stressed in the direction in which it most resists failure. A favorable stress can be applied to the bond by using proper joint design. Some joint designs may be impractical, expensive to make, or hard to align. The design engineer will often have to weigh these factors against optimum adhesive performance.

Joints for flat adherends   The simplest joint to make is the plain butt joint. Butt joints cannot withstand bending forces because the adhesive would experience cleavage stress. If the adherends are too thick to design simple overlap-type joints, the butt joint can be improved by redesigning in a number of ways, as shown in Fig. 14. All the modified butt joints reduce the cleavage effect caused by side loading. Tongue-and-groove joints also have an advantage in that they are self-aligning and act as a reservoir for the adhesive. The scarf joint keeps the axis of loading in line with the joint and does not require a major machining operation.

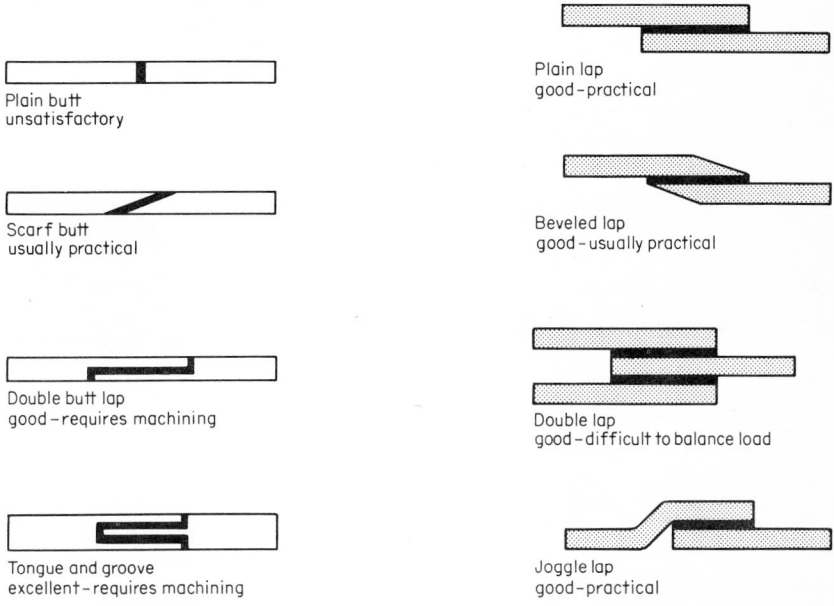

Plain butt
unsatisfactory

Scarf butt
usually practical

Double butt lap
good - requires machining

Tongue and groove
excellent - requires machining

Fig. 14   Butt connections.

Plain lap
good - practical

Beveled lap
good - usually practical

Double lap
good - difficult to balance load

Joggle lap
good - practical

Fig. 15   Lap connections.

Lap joints are the most commonly used adhesive joint because they are simple to make, are applicable to thin adherends, and stress the adhesive primarily in its strongest direction. Tensile loading of a lap joint causes the adhesive to be stressed in shear. However, the simple lap joint is offset and the shear forces are not in line, as was illustrated in Fig. 11. This factor results in cleavage stress at the ends of the joint, which seriously impairs its efficiency. Modifications of lap-joint design include

1. Redesigning the joint to bring the load on the adherends in line
2. Making the adherends more rigid (thicker) near the bond area (see Fig. 12)
3. Making the edges of the bonded area more flexible for better conformance, thus minimizing peel

Modifications of lap joints are shown in Fig. 15.

The joggle-lap-joint design is the easiest method of bringing loads into alignment. The joggle lap can be made by simply bending the adherends. It also provides a surface to which it is easy to apply pressure. The double-lap joint has a balanced construction which is subjected to bending only if loads on the double side of the lap are not balanced. The beveled lap joint is also more efficient than the plain lap joint. The beveled edges allow conformance of the adherends during loading, thereby reducing cleavage stress on the ends of the joint. Figure 16 illustrates the general effect of plain-lap and beveled-lap designs on joint strength.

Strap joints keep the operating loads aligned and are generally used where overlap joints are impractical because of adherend thickness. Strap-joint designs are shown in Fig. 17. Like the lap joint, the single strap is subjected to cleavage stress under bending forces. The double strap joint is more desirable when bending stresses are encountered. The beveled double strap and recessed double strap are the best joint designs to resist bending forces. Unfortunately, they both require expensive machining.

When thin members are bonded to thicker sheets, operating loads generally tend to peel the thin member from its base, as shown in Fig. 18a. The subsequent illustrations show what can be done to decrease peeling tendencies in simple joints.

Stiffening joints Often thin sheets of a material are made more rigid by bonding stiffening members to the sheet. When such sheets are flexed, the bonded joints are subjected to cleavage stress. Some design methods for reducing cleavage stress on stiffening joints are illustrated in Fig. 19. Resistance of stiffening members to bending forces is increased by extending the bond area, providing greater flange flexibility, and increasing the stiffness of the base sheet.

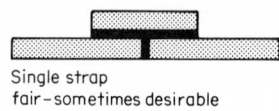

Single strap
fair-sometimes desirable

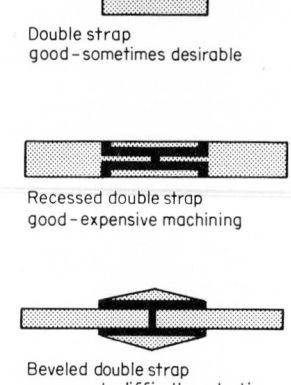

Double strap
good-sometimes desirable

Recessed double strap
good-expensive machining

Beveled double strap
very good-difficult production

Fig. 17   Strap connections.

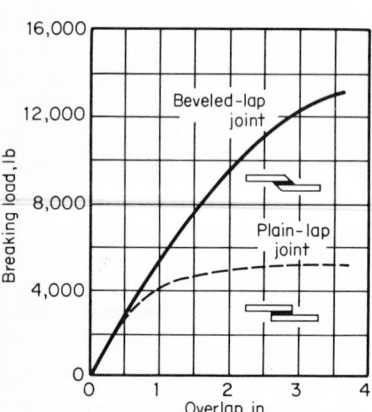

Fig. 16   The effect of beveled and plain lap-joint design on breaking load.[14]

Cylindrical-joint design Several recommended designs for rod and tube joints are illustrated in Fig. 20. These designs should be used instead of the simpler butt joint. Their resistance to bending forces and subsequent cleavage is much better, and the bonded area is larger. Unfortunately, most of these joint designs require a machining operation.

Angle- and corner-joint designs A butt joint is the simplest method of bonding two surfaces that meet at an odd angle. Although the butt joint has good resistance to pure tension and compression, its bending strength is very poor. Dado, L, and T angle joints, shown in Fig. 21, offer greatly improved properties. The T design is the preferable angle joint because of its large bonding area and good strength in all directions.

Corner joints for relatively flexible adherends such as sheet metal should be designed with reinforcements for support. Various corner-joint designs are shown in Fig. 22.

With very thin adherends, angle joints offer low strengths because of high peel concentrations. A design consisting of right-angle corner plates or slip joints offers the most satisfactory performance. Thick, rigid members such as rectangular bars and wood may be bonded with an end lap joint, but greater strengths can be obtained with mortise and tenon. Hollow members such as extrusions fasten together best with mitered joints and inner splines.

Plastic- and elastomer-joint design Design of joints for plastics and elastomers generally follows the same practice as for metal. However, the designer should be aware of certain characteristics for these materials that require special consideration.

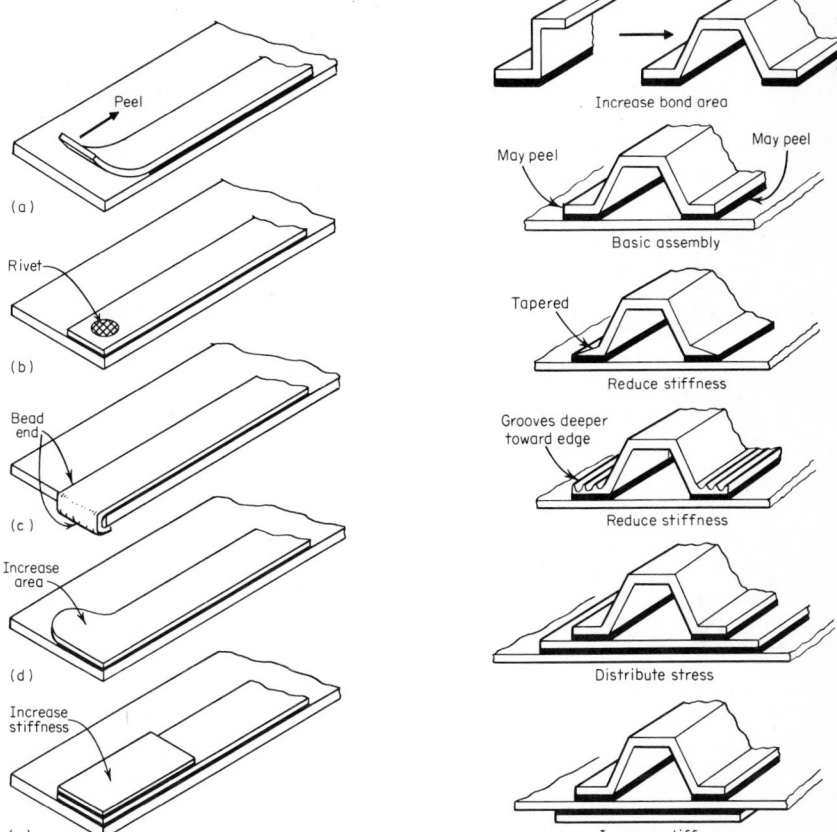

Fig. 18 Minimizing peel in adhesive joints.[7]

Fig. 19 Stiffener-assembly designs.[7]

Flexible Plastics and Elastomers. Thin or flexible polymeric substrates may be joined using a simple or modified lap joint. The double strap joint is best, but also the most time-consuming to fabricate. The strap material should be made out of the same material as the parts to be joined, or at least have approximately equivalent strength, flexibility, and thickness. The adhesive should have the same degree of flexibility as the adherends.

If the sections to be bonded are relatively thick, a scarf joint is acceptable. The length of the scarf should be at least four times the thickness; sometimes larger scarfs may be needed.

When bonding elastic material, forces on the elastomer during cure of the adhesive should be carefully controlled, since excess pressure will cause residual stresses at the bond interface.   Stress concentrations may also be minimized in rubber-to-metal joints by elimination of sharp corners and using metal thick enough to prevent peel stresses that may arise with thinner-gage metals.

As with all joint designs, polymeric joints should avoid peel stress.   Figure 23 illustrates methods of bonding flexible substrates so that the adhesive will be stressed in its strongest direction.

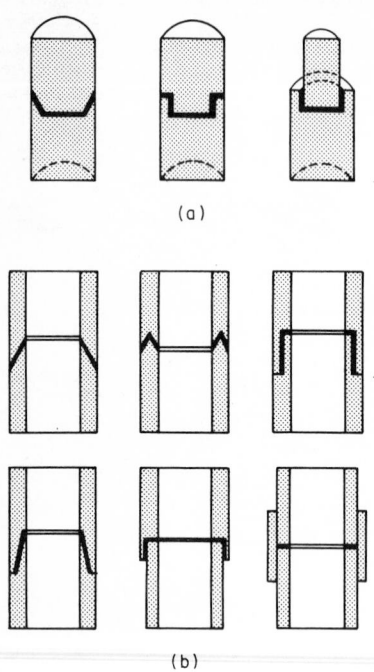

(a)

(b)

Fig. 20 Recommended designs for rod and tube joints.[15] (a) Three joint designs for adhesive bonding of round bars. (b) Six joint configurations useful in adhesive-bonding cylinders or tubes.

Rigid Plastics.   Reinforced plastics are often anisotropic materials.   This means their strength properties are directional. Joints made from anisotropic substrates should be designed to stress both the adhesive and adherend in the direction of greatest strength.   Laminates, for example, should be stressed parallel to the laminations. Stresses normal to the laminate may cause the substrate to delaminate.

Single and joggle lap joints are more likely to cause delamination than scarf or beveled lap joints.   The strap-joint variations are useful when bending loads may be imposed on the joint.

Wood-joint design   Wood is an anisotropic and dimensionally unstable material. Strength properties differ with grain direction.   Tensile strength of wood in the line of the grain is approximately ten times greater than in a direction perpendicular to the grain.   Adhesive joints should be designed to take this factor into account.

Wood changes dimensions with moisture absorption.   The degree of dimensional stability is also dependent on grain direction.   Joints should be made from similar types and cuts of wood to avoid stress concentrations.   When wood is bonded to a much more dimensionally stable material, the adhesive must be strong enough to withstand the loads caused by dimensional distortion of the wood.

A plain scarf joint with very flat slopes is the most efficient joint design for wood, but it requires the application of lateral or transverse pressure during the time the adhesive hardens.   Common wood joints that are self-aligning and require no pressure are shown in Figs. 24 and 25.

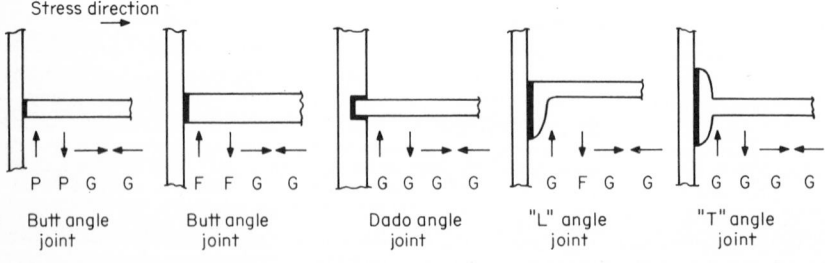

Butt angle joint   Butt angle joint   Dado angle joint   "L" angle joint   "T" angle joint

P = Poor   F = Fair   G = Good

Fig. 21 Types of angle joints, and methods of reducing cleavage.[7]

ASTM Test Methods

A number of standard tests for adhesive bonds have been specified by the American Society for Testing and Materials (ASTM). Selected ASTM standards are presented in Table 6. The properties usually reported by adhesive suppliers are ASTM tensile-shear and peel strength. Such data are sufficient to compare strengths of vari-

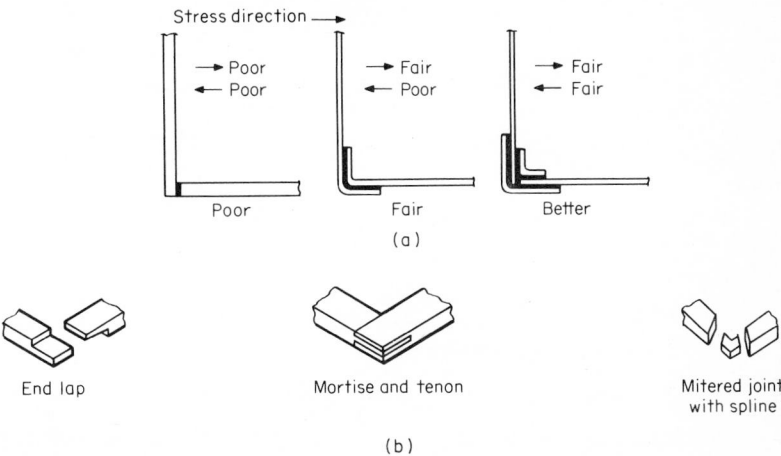

Fig. 22 Reinforcement of bonded corners.[7,12] (a) Corner joints for relatively thin adherends. (b) Corner joints for thick adherends.

ous bonding systems when stressed in pure shear or peel. But they cannot be readily translated into specific strength values for an actual production joint. The most reliable adhesion test is to measure the strength of an actual assembly under simulated operating conditions. Unfortunately, such tests are often impractical.

Lap-shear tests The lap-shear or tensile-shear test measures the strength of the adhesive in shear. It is the most common adhesive test because the specimens are inexpensive, easy to fabricate, and simple to test. This method is described in ASTM D 1002, and the standard test specimen in shown in Fig. 26a. The specimen is loaded in tension, causing the adhesive to be stressed in shear until failure occurs. Since the test calls for a sample population of five, specimens can be made and cut from large test panels illustrated in Fig. 26b.

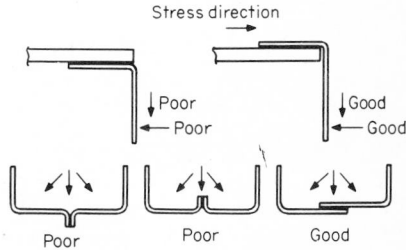

Fig. 23 Methods of joining flexible plastic or rubber.[7]

The test report should include the type of adhesive, bonding procedures, loading rate, bond-line thickness, area of the bond, test environment, pounds per square inch of bonded area to cause failure, and type of failure. An analysis of type (or mode) of failure is an extremely valuable tool to determine the cause of adhesive failure. The failed joint should be visually examined to determine where and to what extent failure occurred. Failure may be found to be 100 percent cohesive, 50 percent cohesive–50 percent adhesive, 50 percent cohesive–50 percent primer, or any one of a number of possible combinations. The purpose of this exercise is to establish the weak link in the joint so that remedial action may be taken.

Tensile tests   The tensile strength of an adhesive joint is seldom reported in the adhesive supplier's literature because pure tensile stress is not often encountered in actual production.   Tensile test specimens also require considerable machining to assure parallel surfaces.

ASTM tension tests are described in D 2095 and employ bar- or rod-shaped butt joints.   The maximum load at which failure occurs is recorded in pounds per square

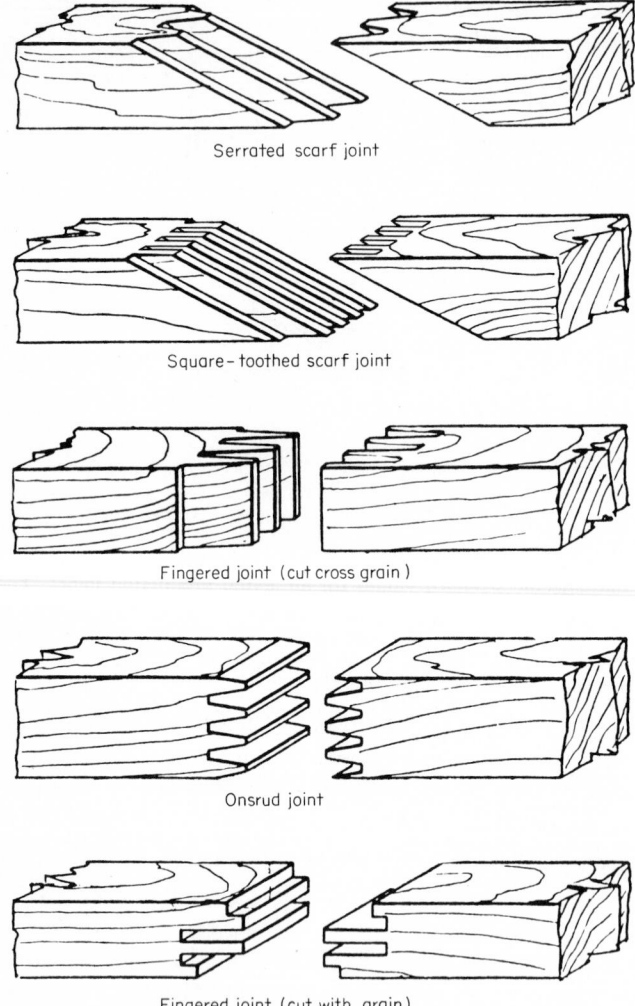

Serrated scarf joint

Square-toothed scarf joint

Fingered joint (cut cross grain)

Onsrud joint

Fingered joint (cut with grain)

Fig. 24   End joints for wood.[16]

inch of bonded area.   Test environment, joint geometry, and type of failure should also be recorded.

A simple cross-lap specimen to determine tensile strength is described in ASTM D 1344.   This specimen has the advantage of being easy to make, but grip alignment and adherend deflection during loading can cause irreproducibility.   A sample population of at least 10 is recommended for this test method.

Peel test A well-designed joint will minimize peel stress, but not all peel forces can be eliminated. Because adhesives are notoriously weak in peel, tests to measure peel resistance are very important. Peel tests involve stripping away a flexible adherend from another adherend that may be flexible or rigid. The specimen is usually peeled at an angle of 90 or 180°.

The most common types of peel test are the T-peel, Bell, and climbing-drum

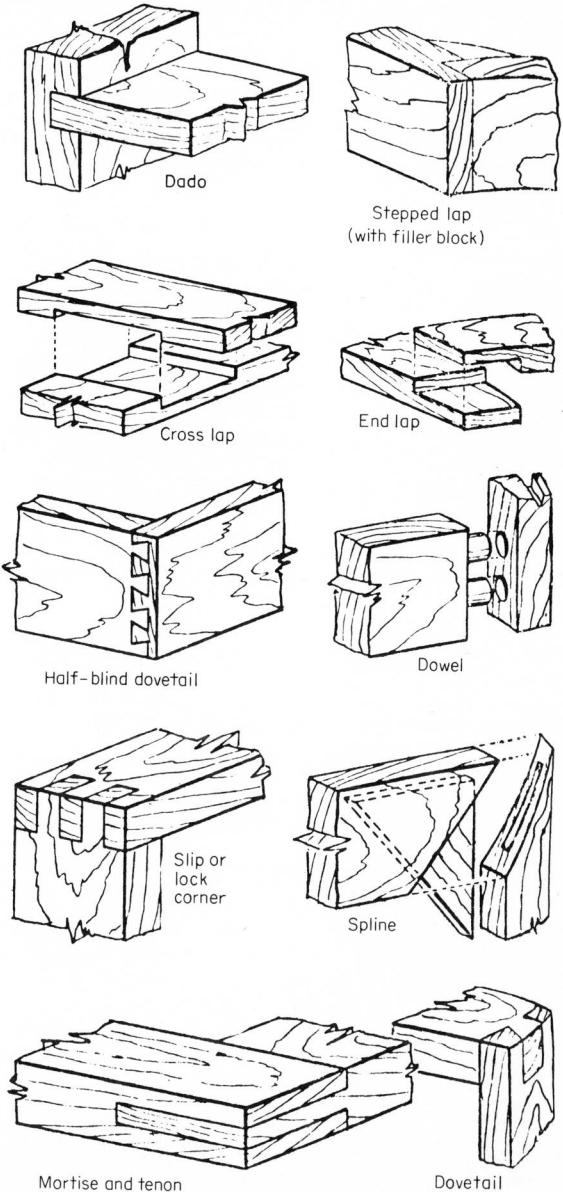

Dado

Stepped lap
(with filler block)

Cross lap

End lap

Half-blind dovetail

Dowel

Slip or
lock
corner

Spline

Mortise and tenon

Dovetail

Fig. 25 Construction adhesive joints for wood.[16]

TABLE 6   ASTM Adhesive Standards*

Adhesion

Adhesion of Bondable Silicone Rubber Tapes Used for Electrical Insulation, Testing (D 2148) 29

Bond of Oil and Resin-base Caulking Compounds, Test for (D 2450) 21

Bond Strength of Plastics and Electrical Insulating Materials, Test for (D 952) 27

Preparation of Bar and Rod Specimens for Adhesion Tests, Recommended Practice for (D 2094) 16

Preparation of Metal Surfaces for Adhesive Bonding, Recommended Practice for (D 2651) 16

Preparation of Surfaces of Plastics Prior to Adhesive Bonding, Recommended Practice for (D 2093) 16

Pressure-sensitive Adhesive Coated Tapes Used for Electrical Insulation, Testing (D 1000) 29

Pressure-sensitive Tack of Adhesives, Test for (D 1878) 16

Rubber Insulating Tape, Specification for (D 119) 28

Adhesives

Adhesives, Definitions of Terms Relating to (including Tentative Revision) (D 907) 16

Adhesives for Structural Laminated Wood Products for Use under Exterior (Wet Use) Exposure Conditions, Specification (D 2559) 16

Adhesives Relative to Their Use as Electrical Insulation, Testing (D 1304) 16

Apparent Viscosity of Adhesives Having Shear-rate-dependent Flow Properties, Test for (D 2556) 16

Applied Weight per Unit Area of Liquid Adhesive, Test for (D 899) 16

Applied Weight per Unit Area of Dried Adhesive Solids, Test for (D 898) 16

Blocking Point of Potentially Adhesive Layers, Test for (D 1146) 16

Consistency of Adhesives, Tests for (D 1084) 16

Density of Adhesives in Fluid Form, Test for (D 1875) 16

Flow Properties of Adhesives, Test for (D 2183) 16

Hydrogen Ion Concentration of Dry Adhesive Films, Test for (D 1583) 16

Penetration of Adhesives, Test for (D 1916) 16

Shear Strength and Shear Modulus of Structural Adhesives, Test for (E 229) 30

Storage Life of Adhesives by Consistency and Bond Strength, Test for (D 1337) 16

Working Life of Liquid or Paste Adhesives by Consistency and Bond Strength, Test for (D 1338) 16

Adhesives, Bonding Strength

Atmospheric Exposure of Adhesive Bonded Joints and Structures, Recommended Practice for (D 1828) 16

Cleavage Strength of Metal-to-Metal Adhesive Bonds, Test for (D 1062) 16

Climbing Drum Peel Test for Adhesives, (D 1781) 16

Cross-lap Specimens for Tensile Properties of Adhesives, Testing (D 1344) 16

Determining Strength Developments of Adhesive Bonds, Recommended Practice for (D 1144) 16

Effect of Bacterial Contamination on Permanence of Adhesive Preparations and Adhesive Bonds, Test for (D 1174) 16

Effect of Moisture and Temperature on Adhesive Bonds, Test for (D 1151) 16

Effect of Mold Contamination on Permanence of Adhesive Preparations and Adhesive Bonds, Test for (D 1286) 16

Exposure of Adhesive Specimens to High-energy Radiation, Recommended Practice for (D 1879) 16

Impact Strength of Adhesive Bonds, Test for (D 950) 16

Integrity of Glue Joints in Structural Laminated Wood Products for Exterior Use, Test for (D 1101) 16

Peel Resistance of Adhesives (T-peel Test), Test for (D 1876) 16

Peel or Stripping Strength of Adhesive Bonds, Test for (D 903) 16

Resistance of Adhesive Bonds to Chemical Reagents, Test for (D 896) 16

Resistance of Adhesives to Cyclic Laboratory Aging Conditions, Test for (D 1183) 16

Shear-block Test for Quality Control of Glue Bonds in Scarf Joints, Conducting (D 1759) 16

Strength of Adhesive Bonds on Flexural Loading, Test for (D 1184) 16

Strength Properties of Adhesives in Shear by Tension Loading in the Temperature Range of -267.8 to -55 C (-450 to -57 F), Test for (D 2557) 16

Strength Properties of Adhesives in Shear by Tension Loading at Elevated Temperatures (Metal-to-Metal), Test for (D 2295) 16

Strength Properties of Adhesive Bonds in Shear by Compression Loading, Test for (D 905) 16

Strength Properties of Adhesives in Shear by Tension Loading (Metal-to-Metal), Test for (D 1002) 16

Strength Properties of Metal-to-Metal Adhesive by Compressive Loading (Disk Shear), Test for (D 2182) 16

Tensile Properties of Adhesive Bonds, Test for (D 897) 16

Tensile Strength of Adhesives by Means of Bar and Rod Specimens, Test for (D 2095) 16

Adhesives, Creep Properties

Conducting Creep Tests of Metal-to-Metal Adhesives, Recommended Practice for (D 1780) 16

Creep Properties of Adhesives in Shear by Compression Loading (Metal-to-Metal), Test for (D 2293) 16

Creep Properties of Adhesives in Shear by Tension Loading (Metal-to-Metal), Test for (D 2294) 16

* Number following test method number is volume number of ASTM book, 1970 edition.

methods. Representative test specimens are shown in Fig. 27. The values resulting from each test method can be substantially different; hence it is important to specify the test method employed.

Peel-test reports should contain the same type of data recorded for shear tests. The rate of peel loading is more important than in lap-shear loading, and should be known and controlled as closely as possible. Adhesive thickness has a significant effect on peel-strength values, especially for the more elastic adhesives. Peel strength is generally reported as high, low, and average loads sustained while separating the substrates. Peel values are recorded in pounds per inch of width of the bonded specimen. They tend to fluctuate more than other adhesive test results because of the extremely small area at which the stress is localized during loading.

The T-peel test is described in ASTM D 1876 and is the most common of all peel tests. The T-peel specimen is shown in Fig. 28. Generally this test method is used when both adherends are flexible. The Bell peel test is used when one adherend is flexible and the other is rigid. The flexible member is peeled at a constant angle through a spool arrangement. Thus, the values obtained are generally more reproducible. The Bell peel test was developed by the Bell Aircraft Corp. and does not

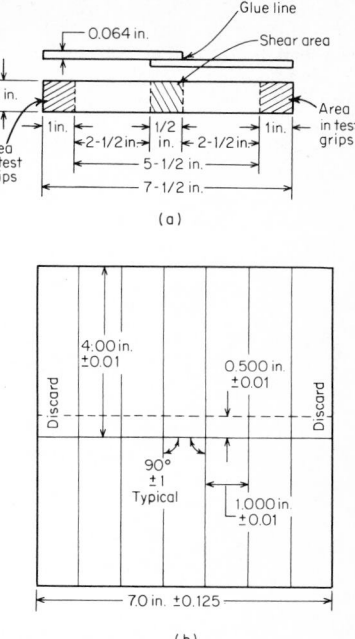

(a)

(b)

Fig. 26 Standard lap-shear test-specimen design.[4] (*a*) Form and dimensions of lap-shear test specimen. (*b*) Standard test panel of five lap-shear specimens.

have an ASTM designation. The climbing-drum peel specimen is described in ASTM D 1781. This test method is intended primarily for determining peel strength of thin metal facings on honeycomb cores, although it can be used for joints where at least one member is flexible.

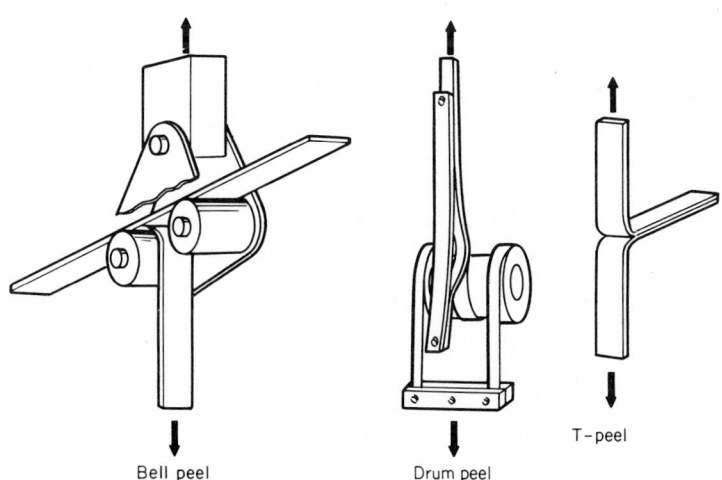

Bell peel   Drum peel   T-peel

Fig. 27 Common types of adhesive peel tests.[15]

A variation of the T-peel test is a 180° stripping test illustrated in Fig. 29 and described in ASTM D 903. This method is commonly used when one adherend is flexible enough to permit a 180° turn near the point of loading. This test offers more reproducible results than the T-peel test because the angle of peel is maintained constant.

Cleavage test Cleavage tests are conducted by prying apart one end of a rigid bonded joint and measuring the load necessary to cause rupture. The test method is

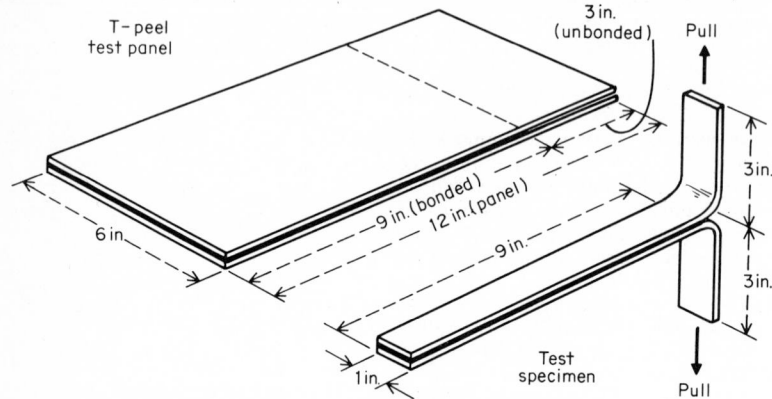

Fig. 28   Test panel and specimen for T peel.[4]

described in ASTM D 1062. A standard test specimen is illustrated in Fig. 30. Cleavage values are reported in pounds per inch of adhesive width. Because cleavage test specimens involve considerable machining, peel tests are usually preferred.

Fatigue test Fatigue testing places a given load repeatedly on a bonded joint. Standard lap-shear specimens are tested on a fatiguing machine capable of inducing cyclic loading (usually in tension) on the joint. The fatigue strength of an adhesive is reported as the number of cycles of a known load necessary to cause failure.

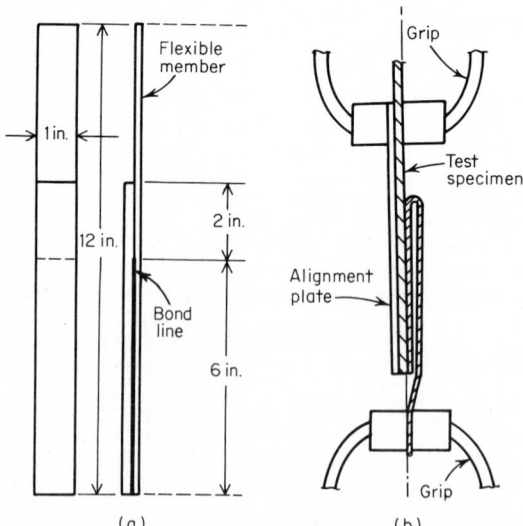

Fig. 29   180° peel test specimens.[4]   (a) Specimen design.   (b) Specimen under test.

Fatigue strength is dependent on adhesive, curing conditions, joint geometry, mode of stressing, magnitude of stress, and frequency of load cycling.

Impact test   The resistance of an adhesive to impact can be determined by ASTM D 950. The specimen is mounted in a grip shown in Fig. 31 and placed in a standard impact machine. One adherend is struck with a pendulum hammer traveling at 11 ft/s, and the energy of impact is reported in pounds per square inch of bonded area.

Creep test   The dimensional change occurring in a stressed adhesive over a long time period is called *creep*. Creep data are seldom reported in the adhesive supplier's literature because the tests are time-consuming and expensive. This is very unfortunate, since sustained loading is a common occurrence in adhesive applications. All adhesives tend to creep, some much more than others. With weak adhesives, creep may be so extensive that bond failure occurs prematurely. Adhesives have also been found to degrade

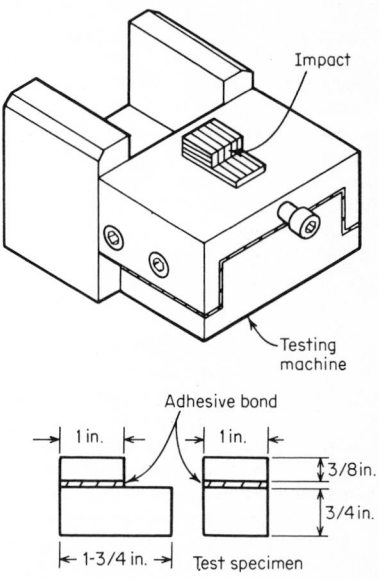

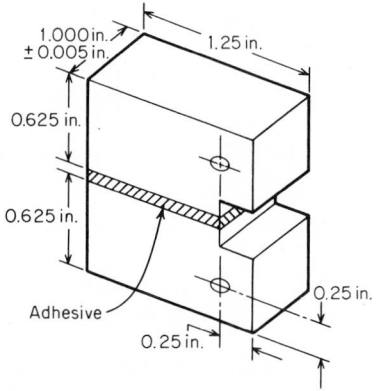

Fig. 30   Cleavage test specimen.[18]

Fig. 31   Impact test specimen and holding fixture.[4]

much more rapidly when aged in a stressed rather than an unstressed condition. This phenomenon will be investigated more closely in a later section.

Creep-test data are accumulated by loading a specimen with a predetermined stress and measuring the total deformation as a function of time or measuring the time necessary for complete failure of the specimen. Depending on the adhesive, loads, and testing conditions the time required for a measurable deformation may be extremely long. ASTM D 2294 defines a test for creep properties of adhesives utilizing a spring-loaded apparatus to maintain constant stress.

Environmental tests   Strength values determined by short-term tests do not give an adequate indication of an adhesive's efficiency during continuous environmental exposure. Laboratory-controlled aging tests seldom last longer than a few thousand hours. To predict the permanence of an adhesive over a 20-year product life requires accelerated test procedures and extrapolation of data. Such extrapolations are extremely risky because the causes of adhesive-bond deterioration are not well understood. Unfortunately no universal method has yet been established to estimate bond life accurately from short-term aging data.

Adhesives may experience many different and exotic environments. Environmental aging of adhesives is accomplished by exposing a stressed or unstressed joint to simulated operating conditions. The type of stress and environment should be selected to be a close approximation of real-life conditions. Two simultaneous environments such as heat and moisture may cause the adhesive to degrade much faster than when exposed in any single environment, because one condition could accelerate the effect of the other.

Federal Test Specifications

A variety of federal and military specifications describing adhesives and test methods have been prepared. Selected government specifications are described in Table 7.

SURFACE PREPARATION

Importance of Surface Preparation

Surface preparation of adherends prior to bonding is one of the most important factors in the adhesive-bonding process. Bond strength and permanence are greatly dependent on the type of surface that is in contact with the adhesive. Prebond treatments are intended to provide cohesively strong and easily wettable surfaces. As a general rule, all adherends must be treated in some manner prior to bonding.

The degree to which adherends must be prepared is related to the service environment and the ultimate joint strength required. Surface preparations can range from simple solvent wiping to a combination of mechanical abrading, chemical cleaning, and acid etching. In many low- to medium-strength applications extensive surface preparation may be unnecessary. But, where maximum bond strength, permanence, and reliability are called for, carefully controlled surface-treating processes are required. The following factors should be considered in the selection of a surface preparation:

1. The ultimate initial bond strength required
2. The degree of permanence necessary and the service environment
3. The amount of contamination initially on the adherend
4. The type of adherend

The strength of an adhesive joint is significantly increased when loose deposits such as rust, scales, flaking paint, and organic contaminants are removed from the surface so that the adhesive can more easily wet the substrate. Table 8 shows the effect of metallic-surface preparations on adhesive-joint strength.

Surface preparations enhance the quality of a bonded joint by performing one or more of the following functions: (1) remove contaminants, (2) control absorbed water, (3) control oxide formation, (4) poison surface atoms which catalyze polymer breakdown, (5) protect the adhesive from the adherend and vice versa, (6) match the adherend crystal structure to the adhesive molecular structure, and (7) control surface roughness.[24] Thus, surface preparations affect the permanence of the joint as well as its initial strength. Figure 32 illustrates the effect of outdoor aging on bonded aluminum joints treated with various processes prior to bonding.

Plastic and elastomeric adherends are even more dependent than metals on surface preparation. Most of these materials are contaminated with mold-release agents or are waxed to improve appearances. Such contaminants must be removed before bonding. Polytetrafluoroethylene, polyethylene, and some other polymeric materials are completely unsuitable for adhesive bonding in their natural state. The surfaces of these materials must be chemically altered prior to bonding.

General Cleaning Methods

Listed here are several methods of preparing substrates for adhesive bonding. They are listed in approximate order of increasing effectiveness. The chosen method will ultimately be the process which yields the necessary strength and permanence with the least time and labor.

Solvent wiping   Where loosely held dirt, grease, and oil are the only contaminants, simple solvent wiping will provide surfaces for weak- to medium-strength bonds. Solvent wiping is widely used, but it is the least effective substrate treatment. Volatile solvents such as toluene, acetone, methylethyl ketone, and trichloroethylene are acceptable. Trichloroethylene is often favored because of its nonflammability. A clean cloth should be saturated with the solvent and wiped across the area to be

TABLE 7   Government Adhesive Specifications

Military Specifications

MIL-A-928........	Adhesive; Metal to Wood, Structural
MIL-A-1154.......	Adhesive; Bonding, Vulcanized Synthetic Rubber to Steel
MIL-C-1219.......	Cement, Iron and Steel
MIL-C-3316.......	Adhesive; Fire Resistant, Thermal Insulation
MIL-C-4003.......	Cement; General Purpose, Synthetic Base
MIL-A-5092.......	Adhesive; Rubber (Synthetic and Reclaimed Rubber Base)
MIL-A-5534.......	Adhesive; High Temperature Setting Resin (Phenol, Melamine and Resorcinol Base)
MIL-C-5339.......	Cement; Natural Rubber
MIL-A-5540.......	Adhesive; Polychloroprene
MIL-A-8576.......	Adhesive; Acrylic Monomer Base, for Acrylic Plastic
MIL-A-8623.......	Adhesive; Epoxy Resin, Metal-to-Metal Structural Bonding
MIL-A-9117.......	Adhesive; Sealing, for Aromatic Fuel Cells and General Repair
MIL-C-10523......	Cement, Gasket, for Automobile Applications
MIL-S-11030.......	Sealing Compound, Noncuring Polysulfide Base
MIL-S-11031.......	Sealing Compound, Adhesive; Curing, Polysulfide Base
MIL-A-11238......	Adhesive; Cellulose Nitrate
MIL-C-12850......	Cement, Rubber
MIL-A-13554......	Adhesive for Cellulose Nitrate Film on Metals
MIL-C-13792......	Cement, Vinyl Acetate Base Solvent Type
MIL-A-13883......	Adhesive, Synthetic Rubber (Hot or Cold Bonding)
MIL-A-14042......	Adhesive, Epoxy
MIL-C-14064......	Cement; Grinding Disk
MIL-P-14536......	Polyisobutylene Binder
MIL-I-15126.......	Insulation Tape, Electrical, Pressure-Sensitive Adhesive and Pressure-Sensitive Thermosetting Adhesive
MIL-C-18726......	Cement, Vinyl Alcohol-Acetate
MIL-A-22010......	Adhesive, Solvent Type, Polyvinyl Chloride
MIL-A-22397......	Adhesive, Phenol and Resorcinol Resin Base
MIL-A-22434......	Adhesive, Polyester, Thixotropic
MIL-C-22608......	Compound Insulating, High Temperature
MIL-A-22895......	Adhesive, Metal Identification Plate
MIL-C-23092......	Cement, Natural Rubber
MIL-A-25055......	Adhesive; Acrylic Monomer Base, for Acrylic Plastics
MIL-A-25457......	Adhesive, Air-drying Silicone Rubber
MIL-A-25463......	Adhesive, Metallic Structural Honeycomb Construction
MIL-A-46050......	Adhesive, Special; Rapid Room Temperature Curing, Solventless
MIL-A-46051......	Adhesive, Room-Temperature and Intermediate-Temperature Setting Resin (Phenol, Resorcinol, and Melamine Base)
MIL-A-52194......	Adhesive, Epoxy (for Bonding Glass Reinforced Polyester)
MIL-A-9067C......	Adhesive Bonding, Process and Inspection Requirements for
MIL-C-7438.......	Core Material, Aluminum, for Sandwich Construction
MIL-C-8073.......	Core Material, Plastic Honeycomb, Laminated Glass Fabric Base, for Aircraft Structural Applications
MIL-C-21275......	Core Material, Metallic, Heat-Resisting, for Structural Sandwich Construction
MIL-H-9884.......	Honeycomb Material, Cushioning, Paper
MIL-S-9041A......	Sandwich Construction, Plastic Resin, Glass Fabric Base, Laminated Facings for Aircraft Structural Applications

Federal Specifications

MMM-A-181.......	Adhesive, Room-Temperature and Intermediate-Temperature Setting Resin (Phenol, Resorcinol and Melamine Base)
MMM-A-00185.....	Adhesive, Rubber
MMM-A-00187.....	Adhesive, Synthetic, Epoxy Resin Base, Paste Form, General Purpose
MMM-A-132.......	Adhesives, Heat Resistant, Airframe Structural, Metal-to-Metal
Federal Test........ Method 175	Adhesives; Methods of Testing
MIL-STD-401......	Sandwich Construction and Core Materials; General Test Methods

TABLE 8   Effect of Substrate Pretreatment on Strength of Adhesive-bonded Joints

Adherend	Treatment	Adhesive	Shear strength, lb/in.²	Ref.
Aluminum..........	As received	Epoxy	444	19
	Vapor degreased		837	
	Grit blast		1,751	
	Acid etch		2,756	
Aluminum..........	As received	Vinyl-phenolic	2,442	20
	Degreased		2,741	
	Acid etch		5,173	
Stainless steel.......	As received	Vinyl-phenolic	5,215	20
	Degreased		6,306	
	Acid etch		7,056	
Cold-rolled steel.....	As received	Epoxy	2,900	21
	Vapor degreased		2,910	
	Grit blast		4,260	
	Acid etch		4,470	
Copper.............	Vapor degreased	Epoxy	1,790	22
	Acid etch		2,330	
Titanium...........	As received	Vinyl-phenolic	1,356	20
	Degreased		3,180	
	Acid etch		6,743	
Titanium...........	Acid etch	Epoxy	3,183	23
	Liquid pickle		3,317	
	Liquid hone		3,900	
	Hydrofluorosilicic acid etch		4,005	

bonded until no signs of residue are evident on the cloth or substrate. Special precautions are necessary to prevent the solvent from becoming contaminated. For example, the wiping cloth should never touch the solvent container, and new wiping cloths must be used often. After cleaning, the parts should be air-dried in a clean, dry environment before being bonded.

Vapor degreasing   Vapor degreasing is a form of solvent cleaning that is attractive when many parts must be prepared. This method is also more reproducible than solvent wiping. It consists of suspending the adherends in a container of chlorinated solvent vapor such as trichloroethylene (boiling point 250°F). When the hot vapors come into contact with the relatively cool substrate, solvent condensation occurs which dissolves the organic contaminants from the surface. Vapor degreasing is preferred to solvent wiping because the surfaces are continuously being washed in distilled uncontaminated solvent. It is recommended that the vapor degreaser be cleaned and a fresh supply of solvent used when the contaminants dissolved in the old solvent lower the boiling point significantly.

Plastic and elastomeric material should not be treated by the vapor-degreasing process because the hot solvent may detrimentally affect the part. For these materials carefully selected solvents or detergent solutions are acceptable cleaning liquids.

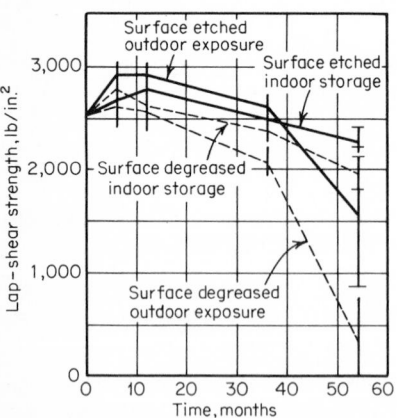

Fig. 32 Effect of time on lap-shear strength of aluminum specimens bonded with a bisphenol-A epoxy cured with a polyamine-modified polyamide.[25]

Abrasive cleaning Mechanical methods for surface preparation include sand-blasting, wire brushing, and abrasion with sandpaper, emery cloth, or metal wool. These methods are most effective for removing heavy, loose particles such as dirt, scale, tarnish, and oxide layers.

The parts should be degreased prior to abrasive treatment to prevent contaminants from being rubbed into the surface. Solid particles left on the surfaces after abrading can be removed by blasts of clean, dry air or solvent wiping. A wet-abrasive-blasting process can be employed which is more adaptable than conventional dry blasting. Wet-blasting units can handle a wider range of abrasive materials, and a spray rinse automatically removes the blasting residue.

Each metal and alloy reacts favorably with a specific range of abrasive sizes. It has been found that joint strength generally increases with the degree of surface roughness (Table 9).[19,26] However, excessively rough surfaces also increase the probability

TABLE 9   Effect of Surface Roughness on Butt Tensile Strength of DER 332-Versamid 140 (60/40)* Epoxy Joints[26]

Adherend		Adherend surface†	Butt tensile strength, $lb/in.^2$
6061	A1	Polished	$4,720 \pm 1,000$
6061	A1	0.005-in. grooves	$6,420 \pm 500$
6061	A1	0.005-in. grooves, sandblasted	$7,020 \pm 1,120$
6061	A1	Sandblasted (40–50 grit)	$7,920 \pm 530$
6061	A1	Sandblasted (10–20 grit)	$7,680 \pm 360$
304	SS	Polished	$4,030 \pm 840$
304	SS	0.010-in. grooves	$5,110 \pm 1,020$
304	SS	0.010-in. grooves, sandblasted	$5,510 \pm 770$
304	SS	Sandblasted (40–50 grit)	$7,750 \pm 840$
304	SS	Sandblasted (10–20 grit)	$9,120 \pm 470$

* 74°C/16 h cure.
† Adherend surfaces were chromate-etched.

TABLE 10   Recommended Abrasive Methods and Sizes for Various Metal Substrates[27]

Material	Method	Size
Steel	Dry blast	80–100 grit abrasive
Aluminum	Wet blast	140–325 grit abrasive
Brass	Dry blast	80–100 grit abrasive
Brass	Wet blast	140–325 grit abrasive
Stainless steel	Wet blast	140–325 grit abrasive

that voids will be left at the interface, causing stress risers which may be detrimental to the joint in service. Table 10 presents a range of abrasive sizes and methods that have been found favorable for abrasive cleaning of metallic substrates.

Hand sanding, wire brushing, and other methods that are highly related to the operator's skill and patience are not recommended as abrasive surface treatments. These methods provide irreproducibility and should be used only when no other method is possible.

Chemical cleaning Strong detergent solutions are used to emulsify surface contaminants on both metallic and nonmetallic substrates. These solutions are usually heated for more efficiency. Parts for cleaning are immersed in a well-agitated solution maintained at 150 to 210°F for approximately 10 min. The surfaces are then rinsed immediately with deionized water and dried. Chemical cleaning is generally used in combination with other surface treatments. Chemical cleaning by itself will not remove heavy or strongly attached contaminants such as rust or scale.

Alkaline detergents recommended for cleaning are combinations of alkaline salts such as sodium metasilicate and tetrasodium pyrophosphate with surfactants included. Many commercial detergents such as Oakite (Oakite Products, Inc., 19 Rector Street, New York) and Sprex (Du Bois Co., 1120 West Front, Cincinnati, Ohio) are available.

Other cleaning methods   Vapor-honing and ultrasonic cleaning are efficient treating methods for small, delicate parts.   In cases where the substrate is so delicate that usual abrasive treatments may be too rough, contaminants can be removed by vapor honing.   This method is similar to grit blasting except that very fine abrasive particles are suspended in a high-velocity water or steam spray.   Thorough rinsing after vapor honing is usually not required.

Ultrasonic cleaning employs a bath of cleaning liquid or solvent that is ultrasonically activated by a high-frequency transducer.   The part to be cleaned is im-

TABLE 11   Surface Preparation of Aluminum Substrates vs. Lap-Shear Strength[19]

Group treatment	$\bar{X}$, lb/in.^{2}	s, lb/in.^{2}	C_v, %
1. Vapor degrease, grit blast 90-mesh grit, alkaline clean, $Na_2Cr_2O_7$–H_2SO_4, distilled water.	3,091	103	3.5
2. Vapor degrease, grit blast 90-mesh grit, alkaline clean, $Na_2Cr_2O_7$–H_2SO_4, tap water.	2,929	215	7.3
3. Vapor degrease, alkaline clean, $Na_2Cr_2O_7$–H_2SO_4, distilled water.	2,800	307	10.96
4. Vapor degrease, alkaline clean, $Na_2Cr_2O_7$–H_2SO_4, tap water.	2,826	115	4.1
5. Vapor degrease, alkaline clean, chromic–H_2SO_4, deionized water.	2,874	163	5.6
6. Vapor degrease, $Na_2Cr_2O_7$–H_2SO_4, tap water.	2,756	363	1.3
7. Unsealed anodized.	1,935	209	10.8
8. Vapor degrease, grit blast 90-mesh grit.	1,751	138	7.9
9. Vapor degrease, wet and dry sand, 100 + 240 mesh grit, N_2 blown.	1,758	160	9.1
10. Vapor degrease, wet and dry sand, wipe off with sandpaper.	1,726	60	3.4
11. Solvent wipe, wet and dry sand, wipe off with sandpaper (done rapidly).	1,540	68	4
12. Solvent wipe, sand (not wet and dry), 120 grit.	1,329	135	1.0
13. Solvent wipe, wet and dry sand, 240 grit only.	1,345	205	15.2
14. Vapor degrease, aluminum wool.	1,478		
15. Vapor degrease, 15% NaOH.	1,671		
16. Vapor degrease.	837	72	8.5
17. Solvent wipe (benzene).	353		
18. As received.	444	232	52.2

$\bar{X}$ = average value. s = standard deviation. C_v = coefficient of variation.   Resin employed is EA 934 Hysol Division, Dexter Corp.; cured 16 h at 75°F plus 1 h at 180°F.

mersed in the liquid, which carries the sonic waves to the surface of the part.   High-frequency vibrations then dislodge the contaminants.   Commercial ultrasonic cleaning units are available from a number of manufacturers.

Chemical etch   Chemical treatment of substrate surfaces is the most effective method of surface preparation.   This type of treatment changes the physical and chemical properties of the surface to produce greater wettability and a stronger surface.   Specific chemical treatments are required for each substrate material.   The part or area to be bonded is usually immersed in an acid solution for a matter of minutes.   The parts are then immediately rinsed with deionized water and dried.

Chemical solutions must be changed regularly to prevent contamination and assure repeatable concentration.   Tank temperature and agitation must also be controlled. Personnel need to be trained in the handling and use of acidic solutions and must wear the proper clothing.

Some paste-type etching products are available that simultaneously clean and chemically treat surfaces. They react at room temperature and need only be applied to the specific area to be bonded. However, these paste etchants generally require much longer treatment time than acid-bath processes.

Combined cleaning methods More than one cleaning method is usually required for optimum adhesive properties. A three-step process that is recommended for most substrates consists of (1) degreasing, (2) mechanical abrasion, and (3) chemical treatment. Table 11 shows the effect of various combinations of aluminum-surface preparations on lap-shear strength. Optimum bond strength on aluminum occurs when a treatment consisting of vapor degreasing, abrading, alkaline cleaning, and acid etching is used.

Activated-gas surface treatment of polymers Recently it has been reported that the treatment of certain polymeric surfaces with excited inert gases greatly improves the bond strength of adhesive joints prepared from these materials.[28,29] With this technique, called "plasma treatment," a low-pressure inert gas is activated by an electrodeless radio-frequency discharge or microwave excitation to produce metastable excited species which react with the polymeric surface. The type of plasma gas can be selected to initiate a wide assortment of chemical reactions. Atoms are expelled

TABLE 12   Typical Adhesive-Strength Improvement with Plasma Treatment; Aluminum-Plastic Shear Specimen Bonded with Epon 828-Versamid 140 (70/30) Epoxy Adhesive[29]

Material	Strength of bond, lb/in.^{2}	
	Control	After plasma treatment
High-density polyethylene.................	315 + 38	Greater than 3,125 + 68
Low-density polyethylene..................	372 + 52	Greater than 1,466 ± 106
Nylon 6.................................	846 ± 166	Greater than 3,956 ± 195
Polystyrene.............................	566 ± 17	Greater than 4,015 ± 85
Mylar* A...............................	530 ± 51	1,660 ± 40
Mylar D................................	618 ± 25	1,185 ± 83
Polyvinylfluoride (Tedlar)*..............	278 ± 2	Greater than 1,280 ± 73
Lexan†.................................	410 ± 10	928 ± 66
Polypropylene...........................	370 ±	3,080 ± 180
Teflon* TFE............................	75	750

* Trademark of E. I. du Pont de Nemours & Company, Inc.
† Trademark of General Electric Co.

from the polymeric surface to produce a strong, wettable, cross-linked skin. Commercial instruments are available that can treat polymeric materials in this manner. Table 12 presents bond strength of various plastic-aluminum composites pretreated with activated gas and bonded with an epoxy adhesive.

Evaluation of Cleaned Parts Before and After Bonding

On many nonporous surfaces a useful and quick method for testing the effectiveness of the surface preparation is the "water-break test." If distilled water beads when sprayed on the surface do not wet the substrate, the surface-preparation steps should be repeated. If the water wets the surface in a uniform film, an effective surface operation has been achieved.

The objective of treating an adherend prior to bonding is to obtain a joint where the weakest link is the adhesive layer and not the interface. Thus, destructively tested joints should be examined for mode of failure. If failure is cohesive (within the adhesive layer or adherend), the surface treatment is optimum for that particular combination of adherend, adhesive, and testing condition. But it must be realized that specimens may exhibit cohesive failure initially and interfacial failure after aging. Both adhesive and surface preparations need to be tested with respect to the intended service environment.

Substrate Equilibrium

After the surface-preparation process has been completed, the substrates may have to be stored before bonding. Typical storage life for various metals subjected to different treatments is shown in Table 13. Because of the relatively short storage life of many treated materials, the bonding operations should be conducted as soon after the surface preparation as possible.

TABLE 13   Maximum Allowable Time between Surface Preparation and Bonding or Priming[27]

Metal	Surface	Time
Aluminum	Wet-abrasive-blasted	72 h
Aluminum	Sulfuric–chromic acid etched	6 days
Aluminum	Anodized	30 days
Stainless steel	Sulfuric acid etched	30 days
Steel	Sandblasted	4 h
Brass	Wet-abrasive-blasted	8 h

If prolonged storage is necessary, a compatible primer may be used to coat the treated substrates after preparation. The primer will protect the surface during storage and interact with the adhesive during bonding. Many primer systems are sold with adhesives for this purpose.

Specific Surface Treatments and Adherend Characteristics

Metallic adherends   Table 14 lists common recommended surface-treating procedures for metallic adherends. The general methods previously described are all applicable to metallic surfaces, but the processes listed in Table 14 have been specifically found to provide reproducible structural bonds and fit easily into the bonding operation. The metals most commonly used in bonded structures and their respective surface treatments are described more fully in the following sections.

TABLE 14   Surface Preparation for Metals (Based on Refs. 3, 17, 30, 31, 32)

Adherend	Degreasing solvent	Method of treatment	Remarks
Aluminum and aluminum alloys	Trichloro-ethylene	1. Sandblast or 100-grit emery cloth followed by solvent degreasing	Medium- to high-strength bonds, suitable for noncritical applications
		2. Immerse for 10 min at 77 ± 6°C in a commercial alkaline cleaner or	Optimum bond strength, specified in ASTM D 2651 and MIL-A-9067. Solvent degrease may replace alkaline cleaning

$$
\begin{array}{lr}
 & \textit{Parts} \\
 & \textit{by wt.} \\
\text{Sodium metasilicate} & 30.0 \\
\text{Sodium hydroxide} & 1.5 \\
\text{Sodium pyrophosphate} & 1.5 \\
\text{Nacconol NR (Allied} & \\
\quad \text{Chemical Co.)} & 0.5 \\
\text{Water (distilled)} & 128.0 \\
\end{array}
$$

Wash in water below 65°C and etch for 10 min at 68±3°C in

$$
\begin{array}{lr}
 & \textit{Parts} \\
 & \textit{by wt.} \\
\text{Sodium dichromate} & 1 \\
\text{Sulfuric acid (96\%,} & \\
\quad \text{sp. gr. 1.84)} & 10 \\
\text{Water (distilled)} & 30 \\
\end{array}
$$

Rinse in distilled water after washing in tap water, and dry in air

TABLE 14   Surface Preparation for Metals (Based on Refs. 3, 17, 30, 31, 32) (Continued)

Adherend	Degreasing solvent	Method of treatment	Remarks
		3. Vapor degrease or solvent wipe. Immerse for 5 min at 71–82°C in Sulfuric acid (96%, sp. gr. 1.84)......... 1 gal Chromic acid........ 45 oz Distilled water....... 9 gal Rinse in distilled water and dry in air	Alternative to sulfuric-dichromate etch
		4. Degrease with 50/50 solution of methyl ethyl ketone and chlorothene. Abrade lightly with mildly abrasive cleaner. Rinse in deionized water; wipe; or air dry. Etch 20 min at RT in *Parts by wt.* Sodium dichromate... 2 Sulfuric acid (96%, sp. gr. 1.84)......... 7 Rinse thoroughly in deionized water; dry at 70°C for 30 min	Room-temp etch
		5. Form a paste using sulfuric acid–dichromate solution and finely divided silica or fuller's earth. Apply; do not permit paste to dry. Time depends on degree of contamination (usually greater than 10 min at RT). Wash very thoroughly with deionized water, and air-dry	Paste form of acid etch useful when part cannot be immersed. ASTM D 2651
Beryllium	Trichloro-ethylene	1. Scrub with a nonchlorinated cleaner. Rinse with deionized water. Force dry at 250–300°F. Check for water break and repeat procedure if necessary	
		2. Immerse for 5–10 min at 20°C in *Parts by wt.* Sodium hydroxide..... 20–30 Water (distilled)...... 170–180 Rinse in distilled water after washing in tap water, and oven-dry for 10 min at 121–177°C	
		3. Same procedure as 2, except immerse for 3–4 min at 75–80°C	
Brass and bronze (see also Copper and copper alloys)	Trichloro-ethylene	1. Etch for 5 min at 20°C in *Parts by wt.* Zinc oxide........... 20 Sulfuric acid (96%, sp. gr. 1.84)........ 460	Temperatures must not exceed 65°C when washing and drying

TABLE 14   Surface Preparations for Metals (Based on Refs. 3, 17, 30, 31, 32) (Continued)

Adherend	Degreasing solvent	Method of treatment	Remarks
		Nitric acid (69 %, sp. gr. 1.41)............ 360 Rinse in water, below 65°C, and re-etch in the acid solution for 5 min at 49°C. Rinse in distilled water after washing, and dry in air	
Cadmium	Trichloroethylene	1. Abrasion. Grit or vapor blast, or 100-grit emery cloth, followed by solvent degreasing	Suitable for general-purpose bonding
		2. Electroplate with nickel or silver	For maximum bond strength
Chromium	Trichloroethylene	1. Abrasion. Grit or vapor blast, or 100-grit emery cloth, followed by solvent degreasing	Suitable for general-purpose bonding
		2. Etch for 1–5 min at 90–95°C in *Parts by wt.* Hydrochloric acid (37 %)............. 17 Water.............. 20 Rinse in distilled water after cold/hot-water washing, and dry in hot air	For maximum bond strength
Copper and copper alloys	Trichloroethylene	1. Abrasion. Sanding, wire brushing, or 100-grit emery cloth, followed by vapor or solvent degreasing	Suitable for general-purpose bonding. Use 320-grit emery cloth for foil
		2. Etch for 10 min at 66°C in *Parts by wt.* Ferric sulfate........ 1.0 Sulfuric acid (96 %)... 0.75 Water.............. 8.0 Wash in water at 20°C, and etch in cold solution of *Parts by wt.* Sodium dichromate... 5 Sulfuric acid (96 %)... 10 Water.............. 85 Etch until a bright clean surface has been obtained. Rinse in water, dip in ammonium hydroxide (sp. gr. 0.88), and wash in tap water. Rinse in distilled water, and dry in warm air	For maximum bond strength. Suitable for brass and bronze. ASTM D 2651
		3. Etch for 1–2 min at 20°C in *Parts by wt.* Ferric chloride (42 % w/w solution)....... 0.75 Nitric acid (sp. gr. 1.41).............. 1.5 Water.............. 10.0 Rinse in distilled water after cold-water wash and dry in air stream at 20°C	Room-temp etch. ASTM D 2651

TABLE 14   Surface Preparations for Metals (Based on Refs. 3, 17, 30, 31, 32) (Continued)

Adherend	Degreasing solvent	Method of treatment	Remarks
		4. Etch for 30 s at 20°C in Parts by wt. Ammonium persulfate 1 Water............... 4 Rinse in distilled water after cold-water wash, and dry in air stream at 20°C	Alternative etching solution to above where fast processing is required
		5. Solvent degrease. Immerse 30 s at 20°C in Parts by vol. Nitric acid (69%).... 10 Deionized water...... 90 Rinse in running water and transfer immediately to next solution; immerse for 1–2 min at 98°C in Ebonol C (Ethone, Inc., New Haven, Conn.) 24 oz and equivalent water to make 1 gal Rinse in deionized water, and air-dry	For copper alloys containing over 95% copper. Stable surface for hot bonding. ASTM D 2651
Germanium	Trichloro-ethylene	Abrasion. Grit or vapor blast followed by solvent degreasing	
Gold	Trichloro-ethylene	Solvent or vapor degrease after light abrasion with a fine emery cloth	
Iron		See Steel (mild)	
Lead and solders	Trichloro-ethylene	Abrasion. Grit or vapor blast, or 100-grit emery cloth, followed by solvent degreasing	
Magnesium and magnesium alloys	Trichloro-ethylene	1. Abrasion with 100-grit emery cloth followed by solvent degreasing	Apply the adhesive immediately after abrasion
		2. Vapor degrease. Immerse for 10 min at 60–70°C in Parts by wt. Deionized water...... 95 Sodium metasilicate.. 2.5 Trisodium pyrophosphate............. 1.1 Sodium hydroxide.... 1.1 Nacconal NR (Allied Chemical Co.)...... 0.3 Rinse in water and dry below 60°C	Medium to high bond strength. ASTM D 2651
		3. Vapor degrease. Immerse for 10 min at 71–88°C Parts by wt. Water............... 4 Chromic acid........ 1 Rinse in water, and dry below 60°C	High bond strength. ASTM D 2651

TABLE 14 Surface Preparation for Metals (Based on Refs. 3, 17, 30, 31, 32) (Continued)

Adherend	Degreasing solvent	Method of treatment	Remarks
		4. Vapor degrease. Immerse for 5–10 min at 63–80°C in *Parts by wt.* Water.............. 12 Sodium hydroxide.... 1 Rinse in water. Immerse for 5–15 min at RT in *Parts by wt.* Water.............. 123 Chromic acid........ 24 Calcium nitrate...... 1.8 Rinse in water, and dry below 60°C	Alternative to procedure 2
		5. Light anodic treatment and various corrosion-preventive treatments have been developed by magnesium producers (Dow 17 and Dow 7, Dow Chemical Co.)	Dow 17 preferred under extreme environmental conditions
Nickel	Trichloroethylene	1. Abrasion with 100-grit emery cloth followed by solvent degreasing	For general-purpose bonding
		2. Etch for 5 s at 20°C in nitric acid (69%, sp.gr.1.41). Wash in cold and hot water followed by a distilled-water rinse, and air-dry at 40°C	For general-purpose bonding
Platinum	Trichloroethylene	Solvent or vapor degrease after light abrasion with a 320-grit emery cloth	
Silver	Trichloroethylene	Abrasion with 320-grit emery cloth followed by solvent degreasing	
Steel (stainless)	Trichloroethylene	1. Abrasion with 100-grit emery cloth, grit or vapor blast, followed by solvent degreasing	
		2. Solvent degrease and abrade with grit paper. Degrease again. Immerse for 10 min at 71–82°C in *Parts by wt.* Sodium metasilicate.. 3 Tetrasodium pyrophosphate.......... 1.5 Sodium hydroxide.... 1.5 Nacconol NR (Allied Chemical Co.)...... 0.5 Distilled water....... 138.0 Rinse in deionized water; dry in air, at 93°C. Immerse for 10 min at 85–91°C in *Parts by wt.* Oxalic acid......... 1 Sulfuric acid (sp. gr. 1.84).............. 1 Distilled water....... 8 Rinse in deionized water; dry at 93°C for 10–15 min	Heat-resistant bond. Alkaline clean alone sufficient for general bonding. Commercial alkaline cleaners (Prebond 700, American Cyanamid) available

TABLE 14   Surface Preparation for Metals (Based on Refs. 3, 17, 30, 31, 32) (Continued)

Adherend	Degreasing solvent	Method of treatment	Remarks

		3. Etch for 15 min at 63°C in
		
		              Parts
		              by wt.
		Sodium dichromate
		  (saturated solution).. 0.30
		Sulfuric acid......... 10.0
		Remove carbon residue with nylon brush while rinsing. Rinse in distilled water and dry in warm air, at 93°C
			ASTM D 2651

		4. Etch for 2 min at 93°C in
		
		              Parts
		              by wt.
		Hydrochloric acid
		  (37%).............. 20
		Orthophosphoric acid
		  (85%).............. 3
		Hydrofluoric acid
		  (35%).............. 1
		Rinse in warm water with a final rinse in distilled water. Dry in air below 93°C
			For maximum resistance to heat and environment. ASTM D 2651

		5. Vapor degrease for 10 min and pickle for 10 min at 20°C in
		
		              Parts
		              by vol.
		Nitric acid (69%, sp.
		  gr. 1.41)........... 10
		Hydrofluoric acid
		  (48%).............. 2
		Water.............. 88
		Dry in air under 70°C
			Room-temperature etch. Treatment may be followed by passivation for 20 min in 5–10% w/v chromic acid (CrO_3) solution

Steel (mild), iron, and ferrous metals other than stainless	Trichloroethylene	1. Abrasion. Grit or vapor blast followed by solvent degreasing with water-free solvents	Xylene or toluene is preferred to acetone and ketone, which may be moist enough to cause rusting

		2. Etch for 5–10 min at 20°C in
		              Parts
		              by wt.
		Hydrochloric acid
		  (37%).............. 1
		Water.............. 1
		Rinse in distilled water after cold-water wash, and dry in warm air for 10 min at 93°C
			Bonding should follow immediately after etching treatment since ferrous metals are prone to rusting. Abrasion is more suitable for procedure where bonding is delayed

		3. Etch for 10 min at 60°C in
		
		              Parts
		              by wt.
		Orthophosphoric acid
		  (85%).............. 1
		Ethyl alcohol
		  (denatured)......... 2
		Brush off carbon residue with nylon brush while washing in running water. Rinse with distilled water, and heat for 1 h at 120°C
			For maximum strength

Tin	Trichloroethylene	Solvent or vapor degrease after light abrasion with a fine emery cloth (320-grit)	

TABLE 14   Surface Preparation for Metals (Based on Refs. 3, 17, 30, 31, 32) (Continued)

Adherend	Degreasing solvent	Method of treatment	Remarks
Titanium and titanium alloys	Trichloro-ethylene	1. Abrasion. Grit or vapor blast, or 100-grit emery cloth, followed by solvent degrease; or scour with a nonchlorinated cleaner, rinse, and dry 2. Etch for 5–10 min at 20°C in *Parts by wt.* Sodium fluoride...... 2 Chromium trioxide... 1 Sulfuric acid (96%, sp. gr. 1.84)........ 10 Water............... 50 Rinse in water and distilled water. Dry in air at 93°C	For general-purpose bonding
		3. Etch for 2 min at 20°C in *Parts by vol.* Hydrofluoric acid (60%)............. 63 Hydrochloric acid (37%)............. 841 Orthophosphoric acid (85%)............. 89 Rinse in water and distilled water. Dry in air at 93°C	Suitable for alloys to be bonded with polybenzimidazole adhesives. Bond within 10 min of treatment. ASTM D 2651
		4. Etch for 10–15 min at 38–52°C in *Parts by vol.* Nitric acid (69%).... 6 Hydrofluoric acid (60%)............. 1 Water............... 20 Rinse with water and distilled water. Dry in oven at 71–82°C for 15 min	Alternative etch for alloys to be bonded with polyimide adhesives is nitric: hydrofluoric: water in ratio 5:1:27 by wt. Etch 30 s at 20°C
		5. Commercial etching liquids and pastes (Plasa-Jell 107C available from Semco, South Hoover, Los Angeles, Calif. 18881)	
Tungsten and tungsten alloys	Trichloro-ethylene	Etch for 1–5 min at 20°C in *Parts by wt.* Nitric acid (69%, sp. gr. 1.41)............ 6 Hydrofluoric acid (60%)............. 1 Sulfuric acid (96%, sp. gr. 1.84)........ 10 Water............... 3 Add a few drops of hydrogen peroxide (20%). Rinse with water and distilled water. Dry in air at 71–82°C for 15 min	
Uranium	Trichloro-ethylene	Abrasion of the metal in a pool of liquid adhesive (epoxy resins have been used)	Prevents oxidation of metal

TABLE 14   Surface Preparation for Metals (Based on Refs. 3, 17, 30, 31, 32) (Continued)

Adherend	Degreasing solvent	Method of treatment	Remarks
Zinc and zinc alloys	Trichloro-ethylene	1. Abrasion. Grit or vapor blast, or 100-grit emery cloth, followed by solvent degreasing	For general-purpose bonding

		2. Etch for 2–4 min at 20°C in
		                 Parts by vol.
		 Hydrochloric acid
		   (37%)............... 10–20
		 Water............... 90–80
		 Rinse with warm water and distilled water. Dry in air at 66–71°C for 30 min
			Glacial acetic acid is an alternative to hydrochloric acid

		3. Etch for 3–6 min at 38°C in
		                 Parts by wt.
		 Sulfuric acid (96%,
		   sp. gr. 1.84)......... 2
		 Sodium dichromate
		   (crystalline)........ 1
		 Water............... 8
			Suitable for freshly galvanized metal

Aluminum and Aluminum Alloys.   Aluminum alloys have been bonded with adhesives more than any other substrate.   Adhesive-bonded aluminum joints provide important engineering properties such as a high strength/weight ratio, corrosion resistance, and low cost.

The effects of various aluminum surface treatments have been studied extensively. The most widely used process for high-strength, environment-resistant adhesive joints is the sodium dichromate–sulfuric acid etch which was developed by Forest Product Laboratories and is known as the FPL etch process.   Abrasion or solvent degreasing results in lower bond strengths, but these simpler processes are more easily placed in production.   Table 15 qualitatively lists the bond strengths which can be realized by various aluminum treatments.

TABLE 15   Surface Treatments for Adhesive-Bonding Aluminum[33]

Surface treatment	Type of bond
Solvent wipe (MEK, MIBK, trichloroethylene)...............	Low to medium strength
Abrasion of surface, plus solvent wipe (sandblasting, coarse sandpaper, etc.)..	Medium to high strength
Hot-vapor degrease (trichloroethylene).....................	Medium strength
Abrasion of surface, plus vapor degrease...................	Medium to high strength
Alodine treatment.......................................	Low strength
Anodize..	Medium strength
Caustic etch*..	High strength
Chromic acid etch (sodium dichromate–sulfuric acid)†.........	Maximum strength

* A good caustic etch is Oakite 164 (Oakite Products, Inc., 19 Rector Street, New York, N.Y.).

† Recommended pretreatment for aluminum to achieve maximum bond strength and weatherability:

1. Degrease in hot trichloroethylene vapor (160°F).
2. Dip in the following chromic acid solution for 10 min at 160°F:
   - Sodium dichromate ($Na_2Cr_2O_7 \cdot 2H_2O$)............... 1 part/wt.
   - Conc. sulfuric acid (sp. gr. 1.86).................... 10 parts/wt.
   - Distilled water...................................... 30 parts/wt.
3. Rinse thoroughly in cold, running, distilled, or deionized water.
4. Air-dry for 30 min, followed by 10 min at 150°F.

Copper and Copper Alloys. Surface preparation of copper alloys is necessary to remove a weak oxide layer attached to the copper surface. This oxide layer is especially troublesome because it forms very rapidly. Copper specimens must be bonded or primed as quickly as possible after surface preparation.

One of the better surface treatments for copper, utilizing a commercial product named Ebonol C,° does not remove the oxide layer but creates a deeper and stronger oxide formation. This process is commonly used when bonding requires elevated temperatures, for example, laminating copper foil to polyethylene.

Magnesium and Magnesium Alloys. Magnesium is the lightest metal used in structural bonding. The surface is very sensitive to corrosion, and chemical products are often formed at the adhesive-metal interface during bonding. Preferred surface preparations for magnesium develop a strong surface coating to prevent corrosion. Proprietary methods of producing such coatings have been developed by magnesium producers such as Dow Chemical Co. Table 16 summarizes the effect of surface preparation on joints made with various magnesium alloys.

TABLE 16   Surface-Preparation Effect on Various Magnesium Alloys[34]

Adhesive evaluated: alloy	Surface-preparation selection in order of greater bond capability (bond strength in parentheses, lb/in.²)
Polyamide–epoxy with:	
AZ 31	Hot chromic (2210), Ebonol C (2150), Dow 17 (2110), alkaline (2040)
ZK 60-T6	Hot chromic (2170), alkaline (1890), Dow 17 (1850), RT chromic (1830)
HM 21	Hot chromic (2718), RT chromic (2280), Dow 17 (2170), Ebonol C (2110)
HK 31	Hot chromic (2700), RT chromic (2200), Dow 17 (2808), Ebonol C (1780)
LA 141	Ebonol C (2360), RT chromic (2200), alkaline (2160)
Phenolic–epoxy with:	
AZ 31	Ebonol C (2100)
ZK 60-T6	Ebonol C (1310), hot chromic (1160), Dow 17 (1120)
HM 21	Ebonol C (1830), hot chromic (1450), Dow 17 (1430), RT chromic (1420)
HK 31	RT chromic (1510), hot chromic (1498), Dow 17 (1440); Ebonol C (1320)
LA 141	Hot chromic (1816), RT chromic (1770)

Steel and Stainless Steel. Steels generally are easy to bond provided that all rust, scale, and organic contaminants are removed. This may be accomplished easily by a combination of mechanical abrasion and solvent cleaning. Table 17 shows the effect of various surface treatments on the tensile shear strength of steel joints bonded with a vinyl-phenolic adhesive.

Prepared steel surfaces are easily oxidized. Once processed, they should be kept free of moisture and primed or bonded within 4 h. Stainless surfaces are not as sensitive to oxidation as carbon steels, and a slightly longer time between surface preparation and bonding is acceptable.

Titanium Alloys. Titanium is achieving prominence in supersonic-aircraft structures because of the high skin temperatures encountered and the metal's extremely high strength/weight ratio. Because of the intended use of titanium at high temperatures, most surface preparations have been devoted to improving the thermal resistance of titanium joints. Like magnesium, titanium can also react with the adhesive during cure and create a weak boundary layer.

Various titanium treatments and their effect on adhesive properties are listed in Table 18. A 10 percent solution of hydrofluorosilicic acid has also been reported

° Trademark of Enthone, Inc., New Haven, Conn.

as an excellent surface etchant. Titanium cleaned with hydrofluorosilicic acid and bonded with a modified epoxy adhesive (Metlbond 329) exhibited tensile shear strength of 4,005 lb/in.[23]

Plastic adherends Many plastics and plastic composites can be treated prior to bonding by simple mechanical abrasion or alkaline cleaning to remove surface contaminants. Figure 33 illustrates the effect of various methods of surface abrasion on the bond strength of an epoxy laminate joint. In some cases it is necessary that the polymeric surface be physically or chemically modified to achieve acceptable bonding. This applies particularly to crystalline thermoplastics such as the polyolefins, linear polyesters, and fluorocarbons. Methods used to improve the bonding characteristics of these surfaces include

1. Oxidation via chemical treatment or flame treatment
2. Electrical discharge to leave a more reactive surface
3. Ionized inert gas which strengthens the surface by cross linking and leaves it more reactive
4. Metal-ion treatment

TABLE 17 Effect of Pretreatment on the Shear Strength of Steel Joints Bonded with a Polyvinyl Formal Phenolic Adhesive[30]

Pretreatment	Martensitic steel		Austenitic steel		Mild steel	
	M	C	M	C	M	C
Grit blast + vapor degreasing..	5,120	13.1	4,100	4.7	4,360	7.8
Vapor blast + vapor degreasing	6,150	5.6	4,940	7.1	4,800	5.9
Following treatments were preceded by vapor degreasing:						
Cleaning in metasilicate solution................	4,360	5.7	3,550	7.8	4,540	6.1
Cleaning in proprietary alkaline solution..........	5,140	8.3	3,210	11.6	3,740	3.3
Acid-dichromate etch.......	5,780	5.8	2,150	22.5	4,070	4.0
Vapor blast + acid dichromate etch...........	6,180	4.1				
Hydrochloric acid etch + phosphoric acid etch.......	3,700	17.9	950	20.2	3,090	20.7
Nitric/hydrofluoric acid etch.....................	6,570	7.5	3,210	15.2	4,050	8.4

M = mean failing load, lb/in.². C = coefficient of variation, %.

Table 19 lists common recommended surface treatments for plastic adherends. These treatments are necessary when plastics are to be joined with adhesives. Solvent and heat welding are other methods of fastening plastics which do not require chemical alteration of the surface. Welding procedures will be discussed in another section of this chapter. As with metallic substrates, the effects of plastic surface treatments decrease with time. It is necessary to prime or bond soon after the surfaces are treated. Listed below are some common plastic materials that require special physical or chemical treatments to achieve adequate surfaces for adhesive bonding.

Fluorocarbons. Fluorocarbons such as polytetrafluoroethylene (TFE), polyfluoroethylene propylene (FEP), polychlorotrifluoroethylene (CFE), and polymonochloro-trifluoroethylene (Kel-F) are notoriously difficult to bond because of their low surface tension. However, epoxy and polyurethane adhesives offer moderate strength if the fluorocarbon is treated prior to bonding.

The fluorocarbon surface may be made more "wettable" by exposing it for a brief moment to a hot flame to oxidize the surface. The most satisfactory surface treatment is achieved by immersing the plastic in a sodium-naphthalene dispersion in tetrahydrofuran. This process is believed to remove fluorine atoms, leaving a carbonized surface

TABLE 18 Effect of Various Cleaning Methods on Titanium Joints Bonded with a Modified Epoxy Adhesive (Metlbond 329, Narmco Materials)[23]

Cleaning method	RT	Lap-shear strength, lb/in.² after various environment exposures				
		30-day salt, tested at RT	10 h at 350°F followed by 30-day salt, tested at RT	10 h at 350°F, tested at 350°F	10 h at 350°F followed by 30-day salt, tested at 350°F	30-day salt, followed by 10 h at 350°F, tested at 350°F
Liquid hone	3,900	2,290	1,442	1,808	506	1,267
Pasa Jell paste 107‡	3,183	3,083	2,517	1,734	416	1,264
Titanium pickle*	3,317	927	1,747	1,431	791	1,377
Liquid Pasa Jell 107-C7	3,590	2,633	2,387	1,927	709	1,278
P.S. 12035†	2,003	1,420	1,437	842	187	309
Pasa Jell paste 107/primer	2,923	2,966	2,950	1,411	838	1,217

* Hydrochloric–phosphoric–hydrofluoric acid etch.
† Alkaline clean followed by sulfuric acid–sodium dichromate etch.
‡ Product name of Semco, Los Angeles, Calif.

which can be wet easily.   Fluorocarbon films treated for adhesive bonding are available from most suppliers.   A formulation and description of the sodium-naphthalene process may found in Table 19.   Commercial chemical products for etching fluorocarbons are also listed.

Another process for treating fluorocarbons as well as other hard-to-bond plastics has been developed by Bell Telephone Laboratories.[36]   This process, called CASING, or plasma treatment, cross-links the polymeric surface by exposing it to an electrically activated inert gas such as neon or helium.   This forms a tough, cross-linked surface

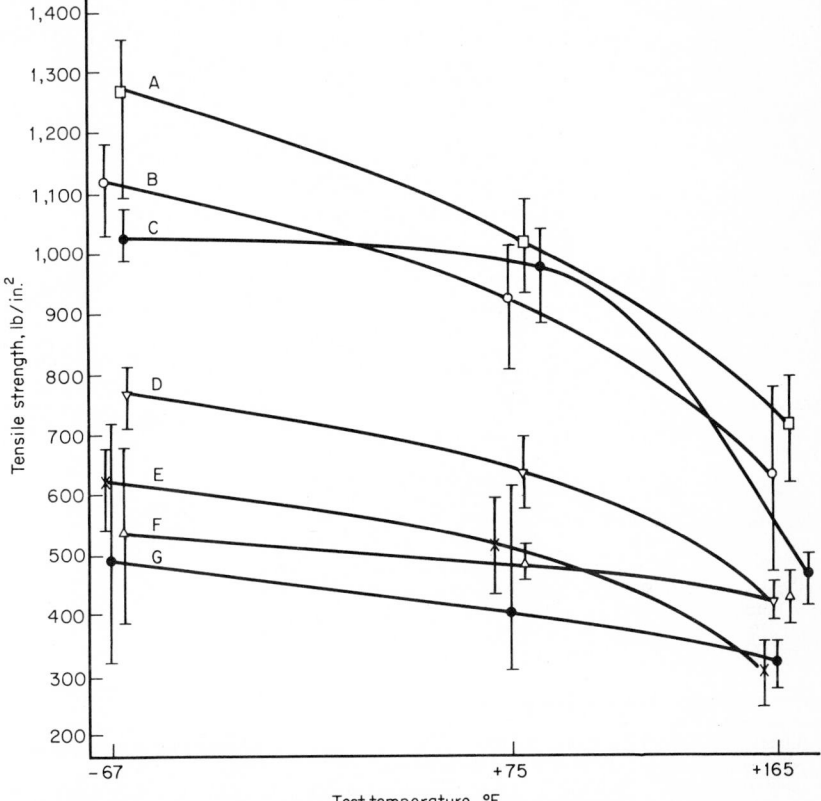

Fig. 33 Effect of plastic surface preparations on the tensile strength of an epoxy laminate bonded with a room-temperature-curing epoxy.[35]   (*A*) Vapor blasting.   (*B*) Grooving and sandblasting.   (*C*) Raised knurling with vapor blasting.   (*D*) Raised knurling.   (*E*) Grooving.   (*F*) Depressed knurling.   (*G*) Solvent wiping.

that wets easily and is adequate for printing and painting as well as bonding. Figure 34 shows the tensile-shear strength of aluminum-TFE joints after various surface preparations.

Polyethylene Terephthalate (Mylar).   A medium-strength bond can be obtained with polyethylene terephthalate plastics and films by abrasion and solvent cleaning. However, a stronger bond can be achieved by immersing the surface in a warm solution of sodium hydroxide or in an alkaline cleaning solution for 2 to 10 min.

Polyolefins, Polyformaldehyde, Polyether.   These materials can be effectively bonded only if the surface is first oxidized.   Polyethylene and polypropylene can be

TABLE 19   Surface Preparations for Plastics (Based on Refs. 3, 17, 30, 31, 32)

Adherend	Degreasing solvent	Method of treatment	Remarks
Acetal (co-polymer)	Acetone	1. Abrasion. Grit or vapor blast, or medium-grit emery cloth followed by solvent degreasing	For general-purpose bonding
		2. Etch in the following acid solution: 　　　　　　　Parts by wt. Potassium dichromate　75 Distilled water......　120 Concentrated sulfuric acid (96%, sp. gr. 1.84)..............　1,500 for 10 s at 25°C. Rinse in distilled water, and dry in air at RT	For maximum bond strength. ASTM D 2093
Acetal (homo-polymer)	Acetone	1. Abrasion. Sand with 280A-grit emery cloth followed by solvent degreasing	For general-purpose bonding
		2. "Satinizing" technique. Immerse the part in 　　　　　　　Parts by wt. Perchloroethylene....　96.85 1,4-Dioxane.........　3.00 p-Toluenesulfonic acid..............　0.05 Cab-o-Sil (Cabot Corp.)............　0.10 for 5–30 s at 80–120°C. Transfer the part immediately to an oven at 120°C for 1 min. Wash in hot water. Dry in air at 120°C	For maximum bond strength. Recommended by du Pont
Acrylonitrile butadiene styrene	Acetone	1. Abrasion. Grit or vapor blast, or 220-grit emery cloth, followed by solvent degreasing	
		2. Etch in chromic acid solution for 20 min at 60°C	Recipe 2 for methyl pentane
Cellulosics: Cellulose, cellulose acetate, cellulose acetate butyrate, cellulose nitrate, cellulose propionate, ethyl cellulose	Methanol, isopropanol	1. Abrasion. Grit or vapor blast, or 220-grit emery cloth, followed by solvent degreasing	For general bonding purposes
		2. After procedure 1, dry the plastic at 100°C for 1 h, and apply adhesive before the plastic cools to room temperature	
Diallyl phthalate, diallyl isophthalate	Acetone, methyl ethyl ketone	Abrasion. Grit or vapor blast, or 100-grit emery cloth, followed by solvent degreasing	Steel wool may be used for abrasion

TABLE 19   Surface Preparations for Plastics (Based on Refs. 3, 17, 30, 31, 32) (Continued)

Adherend	Degreasing solvent	Method of treatment	Remarks
Epoxy resins	Acetone, methyl ethyl ketone	Abrasion. Grit or vapor blast, or 100-grit emery cloth, followed by solvent degreasing	Sand or steel shot are suitable abrasives
Ethylene vinyl acetate	Methanol	Prime with epoxy adhesive and fuse into the surface by heating for 30 min at 100°C	
Furane	Acetone, methyl ethyl ketone	Abrasion. Grit or vapor blast, or 100-grit emery cloth, followed by solvent degreasing	
Ionomer	Acetone, methyl ethyl ketone	Abrasion. Grit or vapor blast, or 100-grit emery cloth, followed by solvent degreasing	Alumina (180-grit) is a suitable abrasive
Melamine resins	Acetone, methyl ethyl ketone	Abrasion. Grit or vapor blast, or 100-grit emery cloth, followed by solvent degreasing	
Methyl pentene	Acetone	1. Abrasion. Grit or vapor blast, or 100-grit emery cloth, followed by solvent degreasing	For general-purpose bonding
		2. Immerse for 1 h at 60°C in *Parts by wt.* Sulfuric acid (96%, sp. gr. 1.84)......... 26 Potassium chromate.. 3 Water.............. 11 Rinse in water and distilled water. Dry in warm air	
		3. Immerse for 5–10 min at 90°C in potassium permanganate (saturated solution), acidified with sulfuric acid (96%, sp. gr. 1.84). Rinse in water and distilled water. Dry in warm air	
		4. Prime surface with lacquer based on urea-formaldehyde resin diluted with carbon tetrachloride	Coatings (dried) offer excellent bonding surfaces without further pretreatment
Phenolic resins, phenolic melamine resins	Acetone, methyl ethyl ketone, detergent	1. Abrasion. Grit or vapor blast, or abrade with 100-grit emery cloth, followed by solvent degreasing	Steel wool may be used for abrasion. Sand or steel shot are suitable abrasives. Glass-fabric decorative laminates may be degreased with detergent solution
		2. Removal of surface layer of one ply of fabric previously placed on surface before curing. Expose fresh bonding surface by tearing off the ply prior to bonding	

TABLE 19   Surface Preparations for Plastics (Based on Refs. 3, 17, 30, 31, 32) (Continued)

Adherend	Degreasing solvent	Method of treatment	Remarks
Polyamide (nylon)	Acetone, methyl ethyl ketone, detergent	1. Abrasion. Grit or vapor blast, or abrade with 100-grit emery cloth, followed by solvent degreasing	Sand or steel shot are suitable abrasives
		2. Prime with a spreading dough based on the type of rubber to be bonded in admixture with isocyanate	Suitable for bonding polyamide textiles to natural and synthetic rubbers
		3. Prime with resorcinol-formaldehyde adhesive	Good adhesion to primer coat with epoxy adhesives in metal-plastic joints
Polycarbonate, allyl diglycol carbonate	Methanol, isopropanol, detergent	Abrasion. Grit or vapor blast, or 100-grit emery cloth, followed by solvent degreasing	Sand or steel shot are suitable abrasives
Fluorocarbons: Polychlorotrifluoroethylene, polytetrafluoroethylene, polyvinyl fluoride, polymonochlorotrifluoroethylene	Trichloroethylene	1. Wipe with solvent and treat with the following for 15 min at RT: Naphthalene (128 g) dissolved in tetrahydrofuran (1 liter) to which is added sodium (23 g) during a stirring period of 2 h. Rinse in deionized water, and dry in warm air	Sodium-treated surfaces must not be abraded before use. Hazardous etching solutions requiring skillful handling. Proprietary etching solutions are commercially available (see 2). PTFE available in etched tape. ASTM D 2093
		2. Wipe with solvent and treat as recommended in one of the following commercial etchants: Bond aid.....W. S. Shamban and Co. 11617 W. Jefferson Blvd. Culver City, Calif. Fluorobond...Joclin Mfg. Co. 15 Lufbery Ave. Wallingford, Conn. Fluoroetch....Action Associates 1180 Raymond Blvd. Newark, N.J. Tetraetch....W. L. Gore Associates 487 Paper Mill Rd. Newark, Del.	
		3. Prime with epoxy adhesive, and fuse into the surface by heating for 10 min at 370°C followed by 5 min at 400°C	
		4. Expose to one of the following gases activated by corona discharge: Air (dry) for 5 min Air (wet) for 5 min Nitrous oxide for 10 min Nitrogen for 5 min	Bond within 15 min of pretreatment
		5. Expose to electric discharge from a tesla coil (50,000 V ac) for 4 min	Bond within 15 min of pretreatment

TABLE 19   Surface Preparations for Plastics (Based on Refs. 3, 17, 30, 31, 32) (Continued)

Adherend	Degreasing solvent	Method of treatment	Remarks
Polyesters, polyethylene terephthalate (Mylar)	Detergent, acetone, methyl ethyl ketone	1. Abrasion. Grit or vapor blast, or 100-grit emery cloth, followed by solvent degreasing	For general-purpose bonding
		2. Immerse for 10 min at 70–95°C in *Parts by wt.* Sodium hydroxide.... 2 Water............. 8 Rinse in hot water and dry in hot air	For maximum bond strength. Suitable for linear polyester films (Mylar)
Chlorinated polyether	Acetone, methyl ethyl ketone	Etch for 5–10 min at 66–71°C in *Parts by wt.* Sodium dichromate...... 5 Water................ 8 Sulfuric acid (96%, sp. gr. 1.84).............. 100 Rinse in water and distilled water. Dry in air	Suitable for film materials such as Penton. ASTM D 2093
Polyethylene, polyethylene (chlorinated), polyethylene terephthalate (see polyesters), polypropylene, polyformaldehyde	Acetone, methyl ethyl ketone	1. Solvent degreasing	Low-bond-strength applications
		2. Expose surface to gas-burner flame (or oxyacetylene oxidizing flame) until the substrate is glossy	
		3. Etch in the following: *Parts by wt.* Sodium dichromate... 5 Water............. 8 Sulfuric acid (96%, sp. gr. 1.84)........ 100 Polyethylene 60 min at 25°C or and polypropylene 1 min at 71°C Polyformaldehyde......10 s at 25°C	For maximum bond strength. ASTM D 2093
		4. Expose to following gases activated by corona discharge: Air (dry)...........For 15 min Air (wet)...........For 5 min Nitrous Oxide.......For 10 min Nitrogen...........For 15 min	Bond within 15 min of pretreatment. Suitable for polyolefins.
		5. Expose to electric discharge from a tesla coil (50,000 V ac) for 1 min	Bond within 15 min of pretreatment. Suitable for polyolefins.
Polymethyl methacrylate, methacrylate butadiene styrene	Acetone, methyl ethyl ketone, detergent, methanol, trichloroethylene, isopropanol	Abrasion. Grit or vapor blast, or 100-grit emery cloth, followed by solvent degreasing	For maximum strength relieve stresses by heating plastic for 5 h at 100°C

TABLE 19   Surface Preparations for Plastics (Based on Refs. 3, 17, 30, 31, 32) (Continued)

Adherend	Degreasing solvent	Method of treatment	Remarks
Poly-phenylene	Trichloro-ethylene	Abrasion. Grit or vapor blast, or 100-grit emery cloth, followed by solvent degreasing	
Poly-phenylene oxide	Methanol	Solvent degrease	Plastic is soluble in xylene and may be primed with adhesive in xylene solvent
Polysty-rene	Methanol, isopro-panol, deter-gent	Abrasion, Grit or vapor blast, or 100-grit emery cloth, followed by solvent degreasing	Suitable for rigid plastic
Poly-sulfone	Methanol	Vapor degrease	
Polyure-thane	Acetone, methyl ethyl ketone	Abrade with 100-grit emery cloth and solvent degrease	
Polyvinyl chloride, polyvinyl-idene chloride polyvinyl fluoride	Trichloro-ethylene, methyl ethyl ketone	1. Abrasion. Grit or vapor blast, or 100-grit emery cloth followed by solvent degreas-ing 2. Solvent wipe with ketone	Suitable for rigid plastic. For maximum strength, prime with nitrile-phenolic adhesive Suitable for plasticized material
Styrene acrylo-nitrile	Trichloro-ethylene	Solvent degrease	
Urea for-malde-hyde	Acetone, methyl ethyl ketone	Abrasion. Grit or vapor blast, or 100-grit emery cloth, followed by solvent degreasing	

prepared for bonding by holding the flame of an oxyacetylene torch over the plastic until it becomes glossy, or else by heating the surface momentarily with a blast of hot air. It is important not to overheat the plastic, thereby causing deformation. The treated plastic must be bonded as quickly as possible after surface preparation.

Polyolefins, such as polyethylene, polypropylene, and polymethyl pentene, as well as polyformaldehyde and polyether, may be more effectively treated with a sodium dichromate–sulfuric acid solution. This treatment oxidizes the surface, allowing better wetting by the adhesive. Plasma treatment, described in the section on fluorocarbons, is also an effective treatment for these plastics. Table 20 shows the tensile-shear strength of bonded polyethylene pretreated by these various methods.

Elastomeric adherends Vulcanized-rubber parts are often contaminated with mold release and plasticizers or extenders that can migrate to the surface. As shown in Table 21, solvent washing and abrading are common treatments for most elas-tomers, but chemical treatment is required for maximum properties. Many synthetic and natural rubbers require "cyclizing" with concentrated sulfuric acid until hairline fractures are evident on the surface.

Fluorosilicone and silicone rubbers must be primed before bonding. The primer acts as an intermediate interface, providing good adhesion to the rubber and a more wettable surface for the adhesive.

TABLE 20 Effects of Surface Treatments on Bonding to Polyethylene with Various Types of Adhesives[38]

Specimen No.	Control	Flame treated	Sanded	Acid treated, oven-dried at 90°C	Acid treated, oven-dried at 71°C	Acid treated, wiped, air-dried at 22°C	Acid treated, acetone-dried	Plasma treatment			
								Helium (30 s)	Helium (30 min)	Oxygen (30 s)	Oxygen (30 min)
Epoxy											
1	40	464*	186	454	428*	500*	516*	468*	...	423*	490*
2	48	480*	166	480*	440*	524*	490*	470*	...	463*	424*
3	24	452*	182	472*	532*	500*	502*	470*	...	484*	439*
4	58	486*	220	462*	460*	524*	500*	450*	...	495*	424*
5	58	520*	216	506*	424*	448*	476*				
Avg lb/in².	46	480	195	475	457	499	497	464	...	466	445
Polyester											
1	74	502*	214	290	300	294	452*	196	284	264	480*
2	102	472*	146	290	288	416*	462*	230	396*	246	514*
3	70	430*	170	230	322	412	392	214	380	320	300
4	70	364*	188	268	318	426*	464*	160	400*	240	370
5	108	400*	178	308	256	236	200	148	372	244	484*
Avg lb/in².	85	434	175	277	297	357	394	190	346	263	430
Nitrile-Rubber-Phenolic											
1	42	196	54	102	100	124	106	...	166	...	210
2	38	120	54	100	92	128	136	...	110	...	170
3	46	88	52	88	64	120	102	...	276	...	220
4	46	120	64	96	158	88	54	...	112	...	110
5	48	166	56	96	124	88	164	...	224	...	170
Avg lb/in².	44	138	56	96	108	110	112	...	178	...	176

* Adherend failed rather than bond. All values in this table are based upon lap-shear strength calculated as lb/in².

Other adherends   Table 22 provides surface treatments for a variety of materials not covered in the preceding tables. Bonding to painted or plated parts presents a problem not encountered with other adherends. It is not recommended to bond to painted surfaces because the resulting bond is only as strong as the adhesion of the paint to the base material.

Primers and Adhesion Promoters

Primers are applied and cured onto the adherend prior to adhesive bonding. They serve four primary functions singly or in combination:

1. Protection of surfaces after treatment (primers can be used to extend the time between preparing the adherend surface and bonding)

2. Developing tack for holding or positioning parts to be bonded

3. Inhibiting corrosion during service

4. Serving as an intermediate layer to enhance the physical properties of the joint and improve bond strength

Primers developed to protect treated surfaces prior to bonding are generally proprietary formulations manufactured by the adhesive producer to match the adhesive. These primers are applied quickly after surface preparation and result in a dry or slightly tacky film less than 1 mil in thickness. During the bonding operation the primer and adhesive fuse together.

Some primers go through a tacky phase at room or elevated temperature during which the adhesive user may wish to assemble the parts to be bonded. Tacky primers do not alleviate the need for pressure during the bonding operation, but they can be used to hold the parts together before pressure is applied to the assembly.

Some primers have been found to provide corrosion resistance for the joint during service. The primer protects the adhesive-adherend interface and lengthens the service life of the bonded joint. Representative data are shown in Fig. 35.

Primers may also be designed to modify the characteristics of the joint. For example, elastomeric primers are used with rigid adhesives to provide greater peel or cleavage resistance. Primers can also chemically react with the adhesive and adherend to provide greater joint strengths. This type of primer is referred to as an *adhesion promoter*. The use of reactive silane to improve the adhesion of resin to glass fabric in polymeric laminates is well known in the plastic industry. Recently developed silane and silyl peroxide adhesion promoters are claimed to be capable of bonding a variety of polymers not only to glass reinforcement or inorganic fillers but also to metals as well as other polymers.[40]

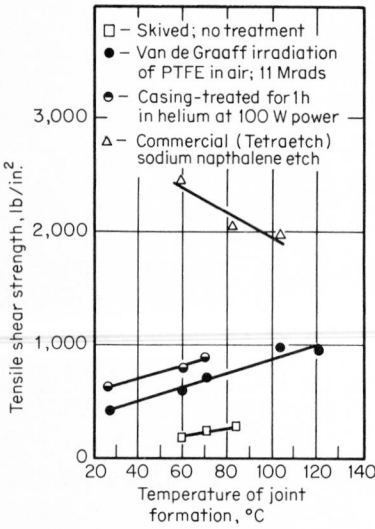

Fig. 34   The tensile-shear strength of a composite consisting of aluminum–epoxy adhesive–PTFE film–epoxy adhesive–aluminum plotted as a function of the ultimate temperature of joint formation.[37]

TYPES OF ADHESIVES

Adhesives, as noted before, are classified primarily as being thermoplastic, thermosetting, elastomeric, or an alloy blend. Thermoplastic adhesives revert to a fluid state when heated; thermosetting adhesives do not. Elastomeric adhesives have high elongation and are generally based on natural or synthetic rubbers. They soften with heat but do not become fluid as do the thermoplastics. Adhesive alloys are a

TABLE 21   Surface Preparations for Elastomers (Based on Refs. 3, 17, 30, 31, 32)

Adherend	Degreasing solvent	Method of treatment	Remarks
Natural rubber	Methanol, isopropanol	1. Abrasion followed by brushing. Grit or vapor blast, or 280-grit emery cloth, followed by solvent wipe	For general-purpose bonding
		2. Treat the surface for 2–10 min with sulfuric acid (sp. gr. 1.84) at RT. Rinse thoroughly with cold water/hot water. Dry after rinsing in distilled water. (Residual acid may be neutralized by soaking for 10 min in 10% ammonium hydroxide after hot-water washing)	Adequate pretreatment is indicated by the appearance of hairline surface cracks on flexing the rubber. Suitable for many synthetic rubbers when given 10–15 min etch at room temperature. Unsuitable for use on butyl, polysulfide, silicone, chlorinated polyethylene, and polyurethane rubbers
		3. Treat surface for 2–10 min with paste made from sulfuric acid and barium sulfate. Apply paste with stainless-steel spatula, and follow procedure 2, above	
		4. Treat surface for 2–10 min in <div style="text-align:right">*Parts by vol.*</div> Sodium hypochlorite.. 6 Hydrochloric acid (37%)............. 1 Water.............. 200 Rinse with cold water and dry	Suitable for those rubbers amenable to treatments 2 and 3
Butadiene styrene	Toluene	1. Abrasion followed by brushing. Grit or vapor blast, or 280-grit emery cloth, followed by solvent wipe	Excess toluene results in swollen rubber. A 20-min drying time will restore the part to its original dimensions
		2. Prime with butadiene styrene adhesive in an aliphatic solvent.	
		3. Etch surface for 1–5 min at RT, following method 2 for natural rubber.	
Butadiene nitrile	Methanol	1. Abrasion followed by brushing. Grit or vapor blast, or 280-grit emery cloth, followed by solvent wipe	
		2. Etch surface for 10–45 s at RT, following method 2 for natural rubber	
Butyl	Toluene	1. Solvent wipe	For general-purpose bonding
		2. Prime with butyl-rubber adhesive in an aliphatic solvent	For maximum strength
Chlorosulfonated polyethylene	Acetone or methyl ethyl ketone	Abrasion followed by brushing. Grit or vapor blast, or 280-grit emery cloth, followed by solvent wipe	General-purpose bonding
Ethylene propylene	Acetone or methyl ethyl ketone	Abrasion followed by brushing. Grit or vapor blast, or 280-grit emery cloth, followed by solvent wipe	General-purpose bonding

TABLE 21   Surface Preparations for Elastomers (Based on Refs. 3, 17, 30, 31, 32) (Continued)

Adherend	Degreasing solvent	Method of treatment	Remarks
Fluoro-silicone	Methanol	Application of fluorosilicone primer (A 4040) to metal where intention is to bond unvulcanized rubber	Primer available from Dow Corning
Polyacrylic	Methanol	Abrasion followed by brushing. Grit or vapor blast, or 100-grit emery cloth followed by solvent wipe	General-purpose bonding
Polybuta-diene	Methanol	Solvent wipe	General-purpose bonding
Polychloro-prene	Toluene, methanol, isopropanol	1. Abrasion followed by brushing. Grit or vapor blast, or 100-grit emery cloth, followed by solvent wipe 2. Etch surface for 5–30 min at RT, following method 2 for natural rubber	Adhesion improved by abrasion with 280-grit emery cloth followed by acetone wipe
Polysulfide	Methanol	Immerse overnight in strong chlorine water, wash and dry	
Polyure-thane	Methanol	1. Abrasion followed by brushing. Grit or vapor blast, or 280-grit emery cloth followed by solvent wipe 2. Incorporation of a chlorosilane into the adhesive-elastomer system. 1% w/w is usually sufficient	Chlorosilane is available commercially. Addition to adhesive eliminates need for priming and improves adhesion to glass, metals. Silane may be used as a surface primer
Silicone	Acetone or methanol	1. Application of primer, Chemlok 607, in solvent (dries 10–15 min) 2. Expose to oxygen gas activated by corona discharge for 10 min	Primer available from Hughson Chemical Company

TABLE 22   Surface Preparations for Materials Other Than Metals, Plastics, and Elastomers (Based on Refs. 3, 17, 30, 31, 32)

Adherend	Degreasing solvent	Method of treatment	Remarks
Asbestos (rigid)	Acetone	1. Abrasion. Abrade with 100-grit emery cloth, remove dust, and solvent degrease 2. Prime with diluted adhesive or low-viscosity rosin ester	Allow the board to stand for sufficient time to allow solvent to evaporate off
Brick and fired non-glazed building materals	Methyl ethyl ketone	Abrade surface with a wire brush; remove all dust and contaminants	
Carbon graphite	Acetone	Abrasion. Abrade with 220-grit emery cloth and solvent degrease after dust removal	For general-purpose bonding

TABLE 22   Surface Preparations for Materials Other Than Metals, Plastics, and Elastomers (Based on Refs. 3, 17, 30, 31, 32) (Continued)

Adherend	Degreasing solvent	Method of treatment	Remarks
Glass and quartz (nonoptical)	Acetone, detergent	1. Abrasion. Grit blast with carborundum and water slurry, and solvent degrease. Dry for 30 min at 100°C. Apply the adhesive before the glass cools to RT	For general-purpose bonding. Drying process improves bond strength
		2. Immerse for 10–15 min at 20°C in *Parts by wt.* Sodium dichromate... 7 Water.............. 7 Sulfuric acid (96%, sp. gr. 1.84)........ 400 Rinse in water and distilled water. Dry thoroughly	For maximum strength
Glass (optical)	Acetone, detergent	Clean in an ultrasonically agitated detergent bath. Rinse; dry below 38°C	
Ceramics and porcelain	Acetone	1. Abrasion. Grit blast with carborundum and water slurry, and solvent degrease	Suitable for unglazed ceramics such as alumina, silica
		2. Solvent degrease or wash in warm aqueous detergent. Rinse and dry	For glazed ceramics such as porcelain
		3. Immerse for 15 min at 20°C in *Parts by wt.* Sodium dichromate... 7 Water.............. 7 Sulfuric acid (96%, sp. gr. 1.84)........ 400 Rinse in water and distilled water. Oven-dry at 66°C	For maximum strength bonding of small ceramic (glazed) artefacts
Concrete, granite, stone	Perchloroethylene, detergent	1. Abrasion. Abrade with a wire brush, degrease with detergent, and rinse with hot water before drying	For general-purpose bonding
		2. Etch with 15% hydrochloric acid until effervescence ceases. Wash with water until surface is litmus-neutral. Rinse with 1% ammonia and water. Dry thoroughly before bonding	Applied by stiff-bristle brush. Acid should be prepared in a polyethylene pail 10–12% hydrochloric or sulfuric acids are alternative etchants. 10% w/w sodium bicarbonate may be used instead of ammonia for acid neutralization
Wood, plywood		Abrasion. Dry wood is smoothed with a suitable emery paper. Sand plywood along the direction of the grain	For general-purpose bonding
Painted surface	Detergent	1. Clean with detergent solution, abrade with a medium emery cloth, final wash with detergent	Bond generally as strong as the paint
		2. Remove paint by solvent or abrasion, and pretreat exposed base	For maximum adhesion

class of adhesives compounded from a mixture of polymers. Such hybrids are primarily thermosetting in character.

These four adhesive groups can be further subdivided by specific chemical composition as described in Tables 23 through 26. The types of resins that go into thermosetting and alloy adhesives are noted for high strength, creep resistance, and resistance to environments such as heat, moisture, solvents, and oils. Their physical properties are well suited for structural-adhesive applications. Elastomeric and thermoplastic adhesives are not used in joints requiring continuous loading because of their tendency to creep under stress. They are also degraded by many common service environments. These adhesives find greatest use in low-strength applications such as pressure-sensitive tape and packaging seals.

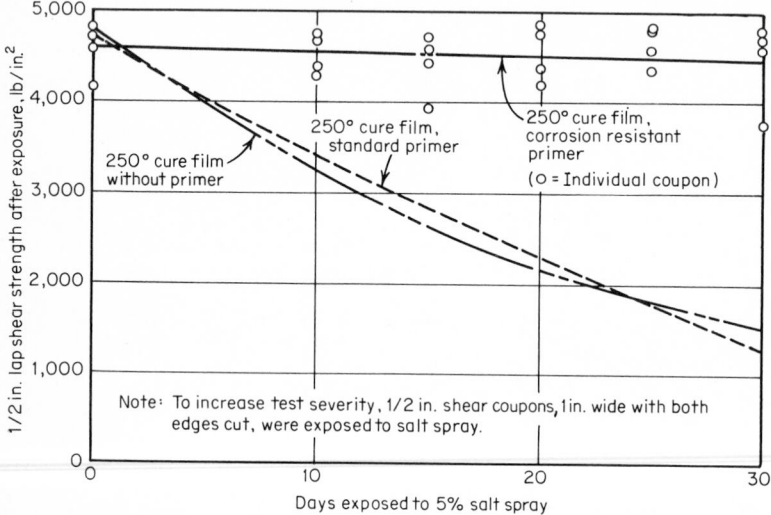

Fig. 35   Effect of primer on lap-shear strength of aluminum joints exposed to 5 percent salt spray.[39]

Structural Adhesives

Epoxies   Epoxy adhesives offer a high degree of adhesion to all substrates except some untreated plastics and elastomers. Cured epoxies have thermosetting molecular structures. They exhibit excellent tensile-shear strength but poor peel strength unless modified with a more resilient polymer. Epoxy adhesives offer excellent resistance to oil, moisture, and many solvents. Low cure shrinkage and high resistance to creep under prolonged stress are characteristic of epoxy resins.

Epoxy adhesives are commercially available as liquids, pastes, film, and solids. A list of selected epoxy adhesives and their properties is presented in Table 27. Commercial epoxy adhesives are primarily composed of an epoxy resin and curing agent. The curing agent may be incorporated into the resin to provide a one-component adhesive, or else it may be provided in a separate container to be mixed into the resin prior to curing. Secondary ingredients in epoxy adhesives include reactive diluents to adjust viscosity; mineral fillers to lower cost, adjust viscosity, or modify coefficient of thermal expansion; and fibrous fillers to improve cohesive strength. Epoxy adhesives are generally supplied as 100 percent solids, but some are available as sprayable solvent systems. Epoxy resins have no evolution of volatiles during cure and are useful in gap-filling applications.

Depending on the type of curing agent, epoxy adhesives can cure at room or elevated temperatures. Higher strengths and better heat resistance are generally obtained with the heat-curing types. Room-temperature-curing epoxies can harden in as little as 1 min at room temperature, but most systems require from 18 to 72 h. The curing time is greatly temperature-dependent, as shown in Fig. 36.

Epoxy resins are the most versatile of structural adhesives because they can be cured and coreacted with many different resins to provide widely varying properties. Table 28 describes the influence of curing agents on the bond strength of epoxy to various adherends. The type of epoxy resin used in most adhesives is derived from the reaction of bisphenol A and epichlorohydrin. This resin can be cured with amines or polyamides for room-temperature-setting systems; anhydrides for elevated-temperature cure; or latent curing agents such as boron trifluoride complexes for use in one-component, heat-curing adhesives. The characteristics of curing agents used with epoxy resins in adhesive formulations are summarized in Table 29. Polyamide curing agents are used in most "general-purpose" epoxy adhesives. They provide a room-temperature cure and bond well to many substrates including plastics, glass, and elastomers. The polyamide-cured epoxy also offers a relatively flexible adhesive with fair peel and thermal-cycling properties.

Epoxy alloys   A variety of polymers can be blended and coreacted with epoxy resins to provide certain desired properties. The most common of these are phenolic, nylon, and polysulfide resins. Properties and curing characteristics of commercially available epoxy-alloy adhesives are presented in Table 30.

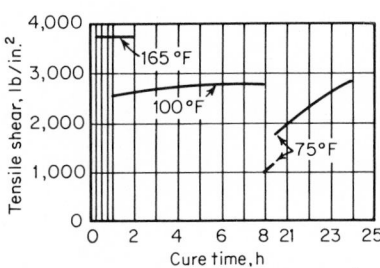

Fig. 36   Characteristics of a particular epoxy adhesive under different cure time and temperature relationships.[41]

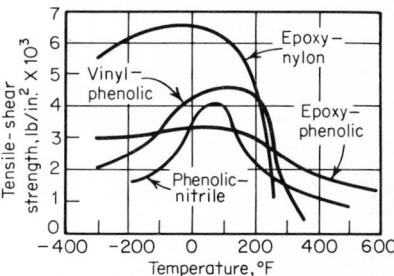

Fig. 37   The effect of temperature on the tensile-shear strength of adhesive alloys (substrate material is aluminum).[42]

Epoxy-Phenolic.   Adhesives based on epoxy-phenolic blends are good for continuous high-temperature service in the 350°F range or intermittent service as high as 500°F. They retain their properties over a very wide temperature range, as shown in Fig. 37. Shear strengths of up to 3,000 lb/in.² at room temperature and 1,000 to 2,000 lb/in.² at 500°F are available. Resistance to oil, solvents, and moisture is very good. Because of the rigid nature of the constituent materials, epoxy-phenolic adhesives have low peel strength and limited thermal-shock resistance.

These adhesives are available as pastes, solvent solutions, and film supported on glass fabric. Cure requires 350°F for 1 h under 15 to 50 lb/in.² pressure. Epoxy-phenolic adhesives were developed primarily for bonding metal joints in high-temperature applications.

Epoxy-Nylon.   Epoxy-nylon adhesives offer both excellent shear and peel strength. They maintain their physical properties at cryogenic temperatures but are limited to a maximum service temperature of 180°F.

Epoxy-nylon adhesives are available as unsupported film or solvent solutions. A pressure of 25 lb/in.² and temperature of 350°F are required for 1 h to cure the adhesive. Because of their excellent filleting properties and high peel strength, epoxy-nylon adhesives are often used to bond aluminum skins to honeycomb core in aircraft structures.

Epoxy-Polysulfide.   Polysulfide resins combine with epoxy resins to provide adhesives with excellent flexibility and chemical resistance. These adhesives bond well to many different substrates. Shear strength and elevated temperature properties are

TABLE 23 Thermosetting Adhesives

Adhesive	Description	Curing method	Special characteristics	Usual adherends	Price range	Selected suppliers
Cyanoacrylate	One-part liquid	Rapidly at RT in absence of air	Fast setting; good bond strength; low viscosity; high cost; poor heat and shock resistance; will not bond to acidic surfaces	Metals, plastics, glass	Very high	Eastman Chemical Loctite Corp.
Polyester	Two-part liquid or paste	RT or higher	Resistant to chemicals, moisture, heat, weathering. Good electrical properties; wide range of strengths; some resins do not fully cure in presence of air; isocyanate-cured system bonds well to many plastic films	Metals, foils, plastics, plastic laminates, glass	Low-med	Asland Chemical Cadillac Plastic John C. Dolph Co. PPG Industries USM Corp., du Pont
Urea formaldehyde	Usually supplied as two-part resin and hardening agent. Extenders and fillers used	Under pressure	Not as durable as others but suitable for fair range of service conditions. Generally low cost and ease of application and cure. Pot life limited to 1 to 24 h	Plywood	Low	American Cyanamid Borden Chemical Monsanto Pacific Resins & Chemical Weldwood Products Wilhold Glues
Melamine formaldehyde	Powder to be mixed with hardening agent	Heat and pressure	Equivalent in durability and water resistance (including boiling water) to phenolics and resorcinols. Often combined with ureas to lower cost. Higher service temp than ureas	Plywood, other wood products	Medium	American Cyanamid Monsanto Reichhold Pacific Resins & Chemical West Coast Adhesives
Resorcinol and phenol-resorcinol formaldehyde	Usually alcohol-water solutions to which formaldehyde must be added	RT or higher with moderate pressure	Suitable for exterior use; unaffected by boiling water, mold, fungus, grease, oil, most solvents. Bond strength equals or betters strength of wood; do not bond directly to metal	Wood, plastics, paper, textiles, fiberboard, plywood	Medium	Chrysler Chemical Inmont Corp. National Casein Pacific Resins & Chemical Weldwood Products

Type	Form	Cure	Properties	Substrates	Cost	Suppliers
Epoxy	Two-part liquid or paste; one-part liquid, paste, or solid; solutions	RT or higher	Most versatile adhesive available; excellent tensile-shear strength; poor peel strength; excellent resistance to moisture and solvents; low cure shrinkage; variety of curing agents/hardeners results in many variations	Metals, plastic, glass, rubber, wood, ceramics	Medium	Emerson & Cuming, Epoxylite Corp., H. B. Fuller, Hysol, Ren Plastics, Narmco, Hughson Chemical, 3M, National Starch
Polyimide	Supported film, solvent solution	High temp	Excellent thermal and oxidation resistance; suitable for continuous use at 550°F and short-term use to 900°F; expensive	Metals, metal foil, honeycomb core	Very high	American Cyanamid, Monsanto, Epoxylite, Narmco
Polybenzimidazole	Supported film	Long, high-temp cure	Good strength at high temperatures; suitable for continuous use at 450°F and short-term use at 1000°F; volatiles released during cure; deteriorate at high temperatures on exposure to air; expensive	Metals, metal foil, honeycomb core	Very high	Narmco
Acrylic	Two-part liquid or paste	RT	Excellent bond to many plastics, good weather resistance, fast cure, catalyst can be used as a substrate primer; poor peel and impact strength	Metals, many plastics, wood	Medium	Hughson Chemical, B. F. Goodrich
Acrylate acid diester	One-part liquid or paste	RT or higher in absence of air	Chemically blocked, anaerobic type; excellent wetting ability; useful temperature range −65 to 300°F; withstands rapid thermal cycling; high-tensile-strength grade requires cure at 250°F, cures in minutes at 280°F	Metals, plastics, glass, wood	Very high	Loctite Corp.

TABLE 24 Thermoplastic Adhesives

Adhesive	Description	Curing method	Special characteristics	Usual adherends	Price range	Selected suppliers
Cellulose acetate, cellulose acetate butyrate	Solvent solutions	Solvent evaporation	Water-clear, more heat resistant but less water resistant than cellulose nitrate; cellulose acetate butyrate has better heat and water resistance than cellulose acetate and is compatible with a wider range of plasticizers	Plastics, leather, paper, wood, glass, fabrics	Low	Adhesive Products Borden C. L. Hathaway Industrial Polychemicals National Starch
Cellulose nitrate	Solvent solutions	Evaporation of solvent	Tough, develops strength rapidly, water-resistant; bonds to many surfaces; discolors in sunlight; dried adhesive is flammable	Glass, metal, cloth, plastics	Low	Borden Chemical Chemical Dev. Corp. General Adhesives National Starch Polymer Chemical Upaco Adhesives
Polyvinyl acetate	Solvent solutions and water emulsions, plasticized or unplasticized, often containing fillers and pigments. Also dried film which is light-stable, water-white, transparent	On evaporation of solvent or water; film by heat and pressure	Bond strength of several thousand lb/in.^{2} but not under continuous loading. The most versatile in terms of formulations and uses. Tasteless, odorless; good resistance to oil, grease, acid; fair water resistance	Emulsions particularly useful with porous materials like wood and paper. Solutions used with plastic films, mica, glass, metal, ceramics.	Low	Adhesive Products Borden Chemical Peter Cooper Co. National Starch Swift Chemical H. B. Fuller Co.
Vinyl vinylidene	Solutions in solvents like methyl ethyl ketone	Evaporation of solvent	Tough, strong, transparent and colorless. Resistant to hydrocarbon solvents, greases, oils	Particularly useful with textiles; also porous materials, plastics	Medium	Atlantic Paste & Glue Imperial Adhesives Macco Adhesives Pierce & Stevens Uniroyal Inc.

Type	Forms	Method of cure	Properties	Materials bonded	Cost	Manufacturers
Polyvinyl acetals	Solvent solutions, film, and solids	Evaporation of solvent; film and solid by heat and pressure	Flexible bond; modified with phenolics for structural use; good resistance to chemicals and oils; includes polyvinyl formal and polyvinyl butaryl types	Metals, mica, glass, rubber, wood, paper	Medium	Borden Chemical Cadillac Plastics Imperial Adhesives Mace Adhesives National Starch Talon Adhesives
Polyvinyl alcohol	Water solutions, often extended with starch or clay	Evaporation of water	Odorless, tasteless and fungus-resistant (if desired). Excellent resistance to grease and oils; water soluble	Porous materials such as fiberboard, paper, cloth	Low	Adhesive Products Borden Chemical Franklin Glue Co. Imperial Adhesives H. B. Fuller Co. National Starch
Polyamide	Solid hot-melt, film, solvent solutions	Heat and pressure	Good film flexibility; resistant to oil and water; used for heat-sealing compounds	Metals, paper, plastic films	Medium	Adhesive Products Borden Chemical Chrysler Chemical General Mills National Starch USM Chemical
Acrylic	Solvent solutions, emulsions and mixtures requiring added catalysts	Evaporation of solvent; RT or elevated temp (two-part)	Good low-temperature bonds; poor heat resistance; excellent resistance to ultraviolet; clear; colorless	Glass, metals, paper, textiles, metallic foils, plastics	Medium	Adhesive Products Borden Chemical B. F. Goodrich National Starch USM Chemical
Phenoxy	Solvent solutions, film, solid hot-melt	Heat and pressure	Retain high strength from 40 to 180°F; resist creep up to 180°F; suitable for structural use	Metals, wood, paper, plastic film	Medium	Union Carbide

TABLE 25 Elastomeric Adhesives

Adhesive	Description	Curing method	Special characteristics	Usual adherends	Price range	Selected suppliers
Natural rubber	Solvent solutions, latexes and vulcanizing type	Solvent evaporation, vulcanizing type by heat or RT (two-part)	Excellent tack, good strength. Shear strength 30–180 lb/in.²; peel strength 0.56 lb/in. width. Surface can be tack-free to touch and yet bond to similarly coated surface	Natural rubber, masonite, wood, felt, fabric, paper, metal	Medium	Chicago Adhesive Prod. Hardman Inc. B. F. Goodrich Goodyear National Starch
Reclaimed rubber	Solvent solutions, some water dispersions. Most are black. some gray and red	Evaporation of solvent	Low cost, widely used. Peel strength higher than natural rubber; failure occurs under relatively low constant loads	Rubber, sponge rubber, fabric, leather, wood, metal, painted metal, building materials	Low	Borden Chemical General Adhesives B. F. Goodrich National Starch Stabond Corp.
Butyl	Solvent system, latex	Solvent evaporation, chemical cross linking with curing agents and heat	Low permeability to gases, good resistance to water and chemicals, poor resistance to oils, low strength	Rubber, metals	Medium	Acme Adhesives Can-Tite Rubber General Adhesives Intercoastal Corp. Permallastic Uniroyal
Polyisobutylene	Solvent solution	Evaporation of solvent	Sticky, low-strength bonds; co-polymers can be cured to improve adhesion, environmental resistance, and elasticity; good aging; poor thermal resistance; attacked by solvents	Plastic film, rubber, metal foil, paper	Low	Borden Chemical H. B. Fuller Co. General Adhesives 3M, B. F. Goodrich Goodyear
Nitrile	Latexes and solvent solutions compounded with resins, metallic oxides, fillers, etc.	Evaporation of solvent and/or heat and pressure	Most versatile rubber adhesive. Superior resistance to oil and hydrocarbon solvents. Inferior in tack range; but most dry tack-free, an advantage in pre-coated assemblies. Shear strength of 150–2,000 lb/in.², higher than neoprene, if cured	Rubber (particularly nitrile) metal, vinyl plastics	Medium	Chrysler Chemical B. F. Goodrich Goodyear, 3M National Starch

Type	Form	Setting action	Properties	Materials bonded	Cost	Manufacturers
Styrene butadiene	Solvent solutions and latexes. Because tack is low, rubber resin is compounded with tackifiers and plasticizing oils	Evaporation of solvent	Usually better aging properties than natural or reclaimed. Low dead load strength; bond strength similar to reclaimed. Useful temp range from −40 to 160°F	Fabrics, foils, plastics film laminates, rubber and sponge rubber, wood	Low	Borden Chemical, Chrysler Chemical, Goodyear, 3M, National Starch, H. B. Fuller
Polyurethane	Two-part liquid or paste	RT or higher	Excellent tensile-shear strength from −400 to 200°F; poor resistance to moisture before and after cure; good adhesion to plastics	Plastics, metals, rubber	Medium	Devcon, B. F. Goodrich, Conap Inc., Furane Plastics, USM Corp.
Polysulfide	Two-part liquid or paste	RT or higher	Resistant to wide range of solvents, oils, and greases; good gas impermeability; resistant to weather, sunlight, ozone; retains flexibility over wide temperature range; not suitable for permanent load-bearing applications	Metals, wood, plastics	High	Essex Chemical, Cadillac Plastic, H. B. Fuller Co., Hardman Inc., 3M, Permalastic
Silicone	Solvent solution: heat or RT curing and pressure-sensitive; and RT vulcanizing solventless pastes	Solvent evaporation, RT or elevated temp	Of primary interest is pressure-sensitive type used for tape. High strengths for other forms are reported from −100 to 500°F; limited service to 700°F. Excellent dielectric properties	Metals; glass; paper; plastics and rubber, including silicone and butyl rubber and fluorocarbons	High–very high	Dow Corning, General Electric, Emerson & Cuming
Neoprene	Latexes and solvent solutions, often compounded with resins, metallic oxides, fillers, etc.	Evaporation of solvent	Superior to other rubber adhesives in most respects—quickness; strength; max temp (to 200°F, sometimes 350°F); aging; resistant to light, weathering, mild acids, and oils	Metals, leather, fabric, plastics, rubber (particularly neoprene), wood, building materials	Medium	Armstrong Cork, 3M, National Starch, H. B. Fuller, Franklin Glue Co., Essex Chemical, USM Corp.

TABLE 26 Alloy Adhesives

Adhesive	Description	Curing method	Special characteristics	Usual adherends	Price range	Selected suppliers
Epoxy-phenolic	Two-part paste, supported film	Heat and pressure	Good properties at moderate cures; volatiles released during cure; retains 50% of bond strength at 500°F; limited shelf life; low peel strength and shock resistance	Metals, honeycomb core, plastic laminates, ceramics	Medium	American Cyanamid Emerson & Cuming Epoxy Technology Isochem Resins Narmco Reliable Mfg.
Epoxy-polysulfide	Two-part liquid or paste	RT or higher	Useful temperature range −70 to 200°F, greater resistance to impact, higher elongation, and less brittleness than epoxies	Metals, plastic, wood, concrete	Medium	American Cyanamid TraCon Adhesive Engineering Chrysler Chemical
Epoxy-nylon	Solvent solutions, supported and unsupported film	Heat and pressure	Excellent tensile-shear strength at cryogenic temperature; useful temperature range −423 to 180°F; limited shelf life	Metals, honeycomb core, plastics	Medium	Adhesive Engineering American Cyanamid B. F. Goodrich Hysol, 3M, Narmco Reliable Mfg.
Nitrile-phenolic	Solvent solutions, unsupported and supported film	Heat and pressure	Excellent shear strength; good peel strength; superior to vinyl and neoprene–phenolics; good adhesion	Metals, plastics, glass, rubber	Medium	American Cyanamid Chrysler Chemical H. B. Fuller Co. National Starch 3M, USM Corp.
Neoprene-phenolic	Solvent solutions, supported and unsupported film	Heat and pressure	Good bonds to a variety of substrates; useful temp range −70 to 200°F, excellent fatigue and impact strength	Metals, glass, plastics	Medium	Narmco
Vinyl-phenolic	Solvent solutions and emulsions, tape, liquid and coreacting powder	Heat and pressure	Good shear and peel strength; good heat resistance; good resistance to weathering, humidity, oil, water, and solvents, vinyl formal and vinyl butyral forms available, vinyl formal–phenolic is strongest	Metals, paper, honeycomb core	Low-medium	American Cyanamid Borden Chemical B. F. Goodrich National Starch Talon Adhesives Thiokol Chemical

low, but resistance to peel forces and low temperatures is very good. The epoxy-polysulfide alloy is supplied as a two-part, flowable paste that cures to a rubbery solid at room temperature.

Resorcinol and phenol resorcinols Resorcinol adhesives are primarily used for bonding wood structures, plastic skins to wood cores, and primed metal to wood. Adhesive bonds as strong as the wood itself are obtainable. Resorcinol adhesives are resistant to boiling water, oil, many solvents, and mold growth. Their service temperature ranges from -300 to $350°F$. Because of high cost, resorcinol resins are often modified by the addition of phenolic resins to form phenol resorcinol.

Resorcinol and phenol-resorcinol adhesives are available in liquid form and are mixed with a powder hardener before application. Generally, these adhesives are cold-setting, but they can also be hot-pressed.

Melamine formaldehyde and urea formaldehydes Melamine-formaldehyde resins are used as colorless adhesives for wood. Because of high cost, they are sometimes blended with urea formaldehyde. Melamine formaldehyde is usually supplied in powder form and reconstituted with water; a hardener is added at the time of use. Temperature of about $200°F$ is necessary for cure. Adhesive strength is greater than the strength of wood.

Urea-formaldehyde adhesives are not as strong or as moisture-resistant as the resorcinols. However, they are inexpensive, and both hot- and cold-setting types are available. Maximum service temperature of urea adhesive is approximately $140°F$. Cold-water resistance is good, but boiling-water resistance may be improved by the addition of melamine-formaldehyde or phenol-resorcinol resins. Urea-based adhesives are used mainly in plywood manufacture.

Modified phenolics Phenolic or phenol formaldehyde is primarily used as an adhesive for bonding wood. However, because of its brittle nature when cured, this resin is unsuitable alone for more extensive adhesive applications. By modifying phenolic resin with various synthetic rubbers and thermoplastic materials, flexibility is greatly improved. The modified adhesive is well suited for structural bonding of many materials, notably metals. Some commercially available phenolic-alloy adhesives are described in Table 31.

Nitrile-Phenolic. Certain blends of phenolic resins with nitrile rubber produce adhesives useful up to $300°F$ continuously. On metals, nitrile phenolics offer shear strength of up to 4,000 lb/in.^{2} along with excellent peel properties. Good bond strengths can also be achieved on rubber, plastics, and glass. These adhesives have high impact strength and resistance to creep and fatigue. Their resistance to solvent, oil, and water is also good.

Nitrile-phenolic adhesives are available as solvent solutions as well as supported and unsupported film. They require heat curing at 300 to $500°F$ under pressure of up to 200 lb/in.^{2} The nitrile-phenolic systems with the highest curing temperature have the greatest resistance to elevated temperatures during service. Because of their good peel strength and elevated-temperature properties, nitrile-phenolic adhesives are used commonly for bonding linings to brake shoes.

Vinyl-Phenolic. Vinyl-phenolic adhesives are based on a combination of phenolic resin with polyvinyl formal or polyvinyl butyral resins. Because of their excellent shear and peel strength, vinyl-phenolic adhesives are one of the most successful structural adhesives for metal. Room-temperature shear strength as high as 5,000 lb/in.^{2} is available. Maximum operating temperature, however, is only $200°F$ because the thermoplastic constituent softens at elevated temperatures. Chemical resistance and impact strength are excellent.

Vinyl-phenolic adhesives are supplied in solvent solutions and as supported and unsupported film. The adhesive cures rapidly at elevated temperatures under pressure. They are generally used to bond metals, rubbers, and plastics to themselves or each other. A major application of vinyl-phenolic adhesive is the bonding of copper sheet to plastic laminate in printed-circuit-board manufacture.

Neoprene-Phenolic. Neoprene-phenolic alloys are used to bond a variety of substrates. Normal service temperature is from -70 to $200°F$. Because of high resistance to creep and most service environments, neoprene-phenolic joints can withstand prolonged stress. Fatigue and impact strengths are also excellent. Shear strength, however, is lower than that of other modified phenolic adhesives.

TABLE 27 Suppliers for Selected Epoxy-based Adhesives

Adhesive supplier	Trade name	Form	Pot life*	Setting conditions†		Continuous-service temp, °F		Shear strength‡ per ASTM D 1002
				Temp, °F	Time	Min	Max	
Adhesive Engineering	Aerobond 3034	Sup. film	6 months at 0°F	350	10 min§	NA	300	4,200
Amicon	A-100	1-pt., paste	4 months	250	3 h	NA	250	3,900
	A-310	1-pt., paste	9 months	250	20 min	NA	300	2,900
	A-410	1-pt., paste	12 months	280	20 min	NA	300	4,500
Armstrong Products	A-36	2-pt., paste	1 min	77	5 min	−60	150	2,000
	A-297	2-pt., paste	8 h	300	15 min	−60	150	3,000
Emerson and Cuming	Eccobond 45	2-pt., paste	3 h	77	8 h	NA	200	3,200
	Eccobond 285	2-pt., paste	30 min	77	24 h	NA	200	2,100
	Eccobond 104	2-pt., paste	12 h	250	6 h	NA	400	2,500
Epoxylite	No. 211	2-pt., liquid	30 min	77	4 h	NA	220	3,200
	No. 810	3-pt., paste	15 min	77	24 h	NA	400	1,800
	No. 5302	2-pt., paste	4 days	400	1 h	NA	400	1,450
H. B. Fuller	Resiweld 7004	2-pt., liquid	30 min	77	24 h	−100	200	3,240
Furane	EPI-100-A	1-pt., powder	6 months	350	1 h	NA	200	5,000
	1331 A/B	2-pt., paste	10 min	77	3 h	NA	200	2,600
	8510/9863	2-pt., paste	60 min	250	1 h	NA	350	3,300
Hightemp Resins	No. 5504/122	2-pt., paste	8 h	300	3 h	NA	350	6,000

Hughson	Chemlok 304	2-pt., paste	90 min	77	16 h	−100	150	2,000
Hysol	EA 907	2-pt., paste	20 min	77	24 h	−67	180	3,300
	EA 9410	2-pt., paste	30 min	77	72 h	−67	160	4,750
	EA 934	2-pt., paste	40 min	77	24 h	−423	350	3,100
	EA 917	1-pt., powder	24 h	300	1 h	−67	300	2,400
Loctite	Epoxy 45	2-pt., paste	4 min	77	1 h	−67	150	2,000
3M Co.	2214 HiTemp	1-pt., paste	6 months at 40°F	250	40 min	−67	300	2,000
	EC2216 A/B	2-pt., paste	90 min	77	8 h	−67	180	4,000
	1751 A/B	2-pt., paste	45 min	77	48 h	−67	180	2,300
Marblette	Maraset 534B	1-pt., liquid	4 months	300	4 h	−67	300	3,500
	Maraset 537	2-pt., paste	4 h	77	48 h	−67	250	1,700
Narmco	Metlbond 329	Sup. film	30 days	350	60 min§§	−423	450	3,000
	Metlbond 326	Sup. film	45 days at 0°F	175	150 min§§	−67	160	2,600
National Starch	M777	2-pt., paste	45 min	77	24 h	−67	180	3,000
	M620	1-pt., paste	1 year	200	90 min	−300	350	2,500
REN	RP-142	1-pt., paste	90 days	300	20 min	−67	266	3,850
	RP-144/H-997	2-pt., paste	30 min	77	8 h	−67	140	4,250
	RP-134/H-997	2-pt., liquid	60 min	77	24 h	−67	176	3,555
	RP-135/H-15	2-pt., solvent system	30 days	300	120 min	−67	212	3,700
Tra-Con	Tra-Bond 2108	2-pt., liquid	90 min	200	2 h	−67	350	2,000

* Pot life at 77°F except where indicated.
† Contact pressure required except where indicated.
‡ Tensile-shear strength on etched aluminum adherends at 77°F.
§ Pressure of 50 lb/in.² required.

TABLE 28 The Influence of Epoxy Curing Agent on the Bond Strength Obtained with Various Base Materials[42]

Curing agent	Amount*	Cure cycle, h at °F	Tensile-shear strength, lb/in.²					
			Polyester glass-mat laminate	Polyester glass-cloth laminate	Cold-rolled steel	Aluminum	Brass	Copper
Triethylamine..........	6	24 at 75, 4 at 150	1,850	2,100	2,456	1,810	1,765	655
Trimethylamine..........	6	24 at 75, 4 at 150	1,054	1,453	1,385	1,543	1,524	1,745
Triethylenetetramine..........	12	24 at 75, 4 at 150	1,150	1,632	1,423	1,675	1,625	1,325
Pyrrolidine..........	5	24 at 75, 4 at 150	1,250	1,694	1,295	1,733	1,632	1,420
Polyamid amine equivalent 210–230......	35–65	24 at 75, 4 at 150	1,200	1,450	2,340	3,120	2,005	1,876
Metaphenylenidiamine..........	12.5	4 at 350	780	640	2,150	2,258	2,150	1,650
Diethylenetriamine..........	11	24 at 75, 4 at 150	1,010	1,126	1,350	1,420	1,135	1,236
Boron trifluoride monoethylamine.....	3	3 at 375	.....	.....	1,732	1,876	1,525	1,635
Dicyandiamide..........		4 at 350	530	432	2,680	2,785	2,635	2,550
Methyl nadic anhydride..........	85	6 at 350	600	756	2,280	2,165	1,955	1,835

Epoxy resin used was derived from bisphenol A and epichlorohydrin and had an epoxide equivalent of 180 to 195; the adhesives contained no filler.
* Per 100 parts by weight of resin.

TABLE 29 Characteristics of Curing Agents Used with Epoxy Resins in Adhesive Formulation[42]

Curing agent	Physical form	Amount required*	Cure temp, °F	Pot life at 75°F†	Complete cure conditions	Max use temp, °F
Triethylenetetramine	Liquid	11–13	70–275	30 min	7 days (75°F)	160
Diethylenetriamine	Liquid	10–12	70–200	30 min	7 days (75°F)	160
Diethylaminopropylamine	Liquid	6–8	83–300	5 h	30 min (75°F)	185
Metaphenylenediamine	Solid	12–14	150–400	8 h	1 h (185°F) 2 h (325°F)	300
Methylene dianiline	Solid	26–30	150–400	8 h	1 h (185°F), 2 h (325°F)	300
Boron trifluoride monoethylamine	Solid	1–4	275–400	6 months	3 h (325°F)	325
Methyl nadic anhydride	Liquid	80–100	250–100	5 days	3 h (320°F)	325
Triethylamine	Liquid	11–13	70–200	30 min	7 days (75°F)	180
Polyamides:						
Amine value 80–90	Semisolid	30–70	70–300	5 h	5 days (75°F)	‡‡
Amine value 210–230	Liquid	30–70	70–300	5 h	5 days (75°F)	‡‡‡
Amine value 290–320	Liquid	30–70	70–300	5 h	5 days (75°F)	‡

* Per 100 parts by weight; for an epoxy resin with an epoxide equivalent of 180 to 190.
† Five hundred grams per batch; with a bisphenol A-epichlorohydrin derived epoxy resin with an epoxide equivalent of 180 to 190.
‡ Highly dependent on concentration.

TABLE 30 Suppliers for Selected Epoxy-Alloy Adhesives

Adhesive supplier	Trade name	Form	Pot life	Setting conditions			Continuous-service temp, °F		Shear strength per ASTM D 1002
				Temp, °F	Time	Pressure, lb/in.²	Min	Max	
Epoxy-Phenolic–Based Systems									
Adhesive Engineering	Aerobond 422	Sup. film	NA	350	45	50	−300	350	3,000
American Cyanamid	HT-424	Sup. film or 2-pt., paste	6 months at <40°F	350	60	50	−423	350	3,550
Narmco	Metlbond 302	Sup. film	90 days at <40°F	350	60	50	−423	350	2,300
Epoxy-Polysulfide–Based Systems									
Adhesive Engineering	Concresive No. 2	2-pt., paste	1 h	77	24 h	Contact	−70	200	2,000
American Cyanamid	BR-93	2-pt., liquid	6 min	77	12 h	Contact	−100	180	300
Chrysler Chemical	Cycleweld C-14	2-pt., liquid	6 h	75	5 h	Contact	−20	220	3,500
Tra-Con	Tra-Bond 2132	2-pt., liquid	30 min	77	18 h	Contact	−100	100	2,500
Epoxy-Nylon–Based Systems									
American Cyanamid	FM-1000	Unsup. film	6 months	350	60 min	25	−423	180	7,090
Hysol	EA 951	Unsup. film	6 months at 40°F	350	60 min	50	−423	300	6,320
Narmco	Metlbond 400	Unsup. film	90 days at 40°F	350	60 min	50	−423	180	6,200
	Metlbond 332	1-pt., liquid	60 days	350	60 min	50	−423	180	5,000

TABLE 31 Suppliers for Selected Phenolic–Alloy Adhesives

Adhesive supplier	Trade name	Form	Pot life	Setting conditions			Continuous-service temp, °F		Shear strength per ASTM D 1002
				Temp, °F	Time, min	Pressure, lb/in.^{2}	Min	Max	
Vinyl-Phenolic—Based Systems									
Narmco...........	Narmtape 105	Sup. film	6 months	300	30	10-30	-67	180	2,830
American Cyanamid....	FM-47	Sup. or unsup. film	12 months	350	30	10-30	-400	225	5,000
Nitrile-Phenolic—Based Systems									
3M Co..........	AF-31	Unsup. film	6 months at 40°F	350	60	150	-67	300	4,200
	AF-15	Unsup. film	6 months at 40°F	350	60	150	-67	300	2,198
American Cyanamid....	FM-238	Unsup. film	NA	350	60	200	-70	350	4,400
Narmco...........	Metlbond 402	Unsup. film	6 months	350	60	100	-67	260	4,500
Neoprene-Phenolic—Based Systems									
Narmco...........	Narmtape 102	Sup. film	6 months	350	60	10-50	-67	180	1,920

Temperatures over 300°F and pressure greater than 50 lb/in.² are needed for cure. Neoprene-phenolic adhesives are available as solvent solutions and film. During cure these adhesives are quite sensitive to atmospheric moisture, surface contamination, and other processing variables.

Polyaromatics Polyimide and polybenzimidazole resins belong to the aromatic heterocycle polymer family, which is noted for its outstanding thermal resistance. These two adhesives are the most thermally stable systems commercially available. The polybenzimidazole (PBI) adhesive has shear strength on steel of 3,000 lb/in.² at room temperature and 2,500 lb/in.² at 700°F. The polyimide adhesive offers a shear strength of approximately 3,000 lb/in.² at room temperature but does not have the excellent strength at 700 to 1,000°F which is characteristic of PBI. Polyimide adhesives offer better long-term elevated-temperature aging properties than PBI. The maximum continuous operating temperature for a polyimide adhesive is 600 to 650°F, whereas PBI adhesives oxidize rapidly at temperatures over 500°F.

Both adhesives are available as supported film, and polyimide resins are also available in solvent solution. During cure, temperatures of 550 to 650°F and high pressure are required. Volatiles are released during cure which contribute to a porous, brittle bond line with relatively low peel strength. Table 32 lists physical properties and curing characteristics of these high-temperature adhesive systems.

Polyesters Polyesters are a large class of synthetic resins having widely varying properties. They may be divided into two distinct groups: saturated and unsaturated.

Unsaturated polyesters are fast-curing, two-part systems that harden by the addition of catalysts, usually peroxides. Styrene monomer is generally used as a reactive diluent for polyester resins. Cure can occur at room or elevated temperature depending on the type of catalyst. Accelerators such as cobalt naphthalene are sometimes incorporated into the resin to speed cure. Unsaturated polyester adhesives exhibit greater shrinkage during cure and poorer chemical resistance than epoxy adhesives. Some types of polyesters are inhibited from curing by the presence of air, but they cure fully when enclosed between two substrates. Depending on the type of resin, polyester adhesives can be quite flexible or very rigid. Uses include patching kits for the repair of automobile bodies and concrete flooring. Polyester adhesives also have strong bond strength to glass-reinforced polyester laminates.

Saturated polyester resins exhibit high peel strength and are used to laminate plastic films such as polyethylene terephthalate (Mylar). They also offer excellent clarity and color stability. These polyester types, in both solution and solid form, can be chemically cross-linked with curing agents such as the isocyanates for improved thermal and chemical stability.

Polyurethane Polyurethane-based adhesives form tough bonds with high peel strength. Generally supplied as a two-part liquid, polyurethane adhesives can be cured at room or elevated temperatures. They have exceptionally high strength at cryogenic temperatures, but maximum operating temperature is limited to 200°F. Like epoxies, urethane adhesives can be applied by a variety of methods and form strong bonds to most surfaces. Some polyurethane adhesives degrade substantially when exposed to high-humidity environments.

Polyurethane adhesives bond well to many substrates including hard-to-bond plastics. Since they are very flexible, polyurethane adhesives are often used to bond films, foils, and elastomers. The properties of a number of commercial polyurethane adhesives are presented in Table 33.

Anaerobic adhesives Acrylate acid diester and cyanoacrylate resins are called "anaerobic" adhesives because they cure when air is excluded from the resin. Table 33 lists properties and curing characteristics of these adhesives. Anaerobic resins are noted for being simple to use, one-part adhesives, having fast cure at room temperature and very high cost ($60/lb). However, the cost is equivalent to ¼ cent/in.² of bonded area because only a small amount of adhesive is required. Anaerobic adhesives do not cure when gaps between adherend surfaces are greater than ~ 10 mils.

The acrylate acid diester adhesives are available in various viscosities. They cure in minutes at room temperature when a special primer is used or in 3 to 10 min at 250°F without the primer. Without the primer the diester adhesive requires 3 to 4 h at room temperature to cure.

TABLE 32 Suppliers for Selected Polyaromatic Adhesives

| Adhesive supplier | Trade name | Form | Pot life | Setting conditions | | | Continuous-service temp, °F | | Shear strength per ASTM D 1002 |
				Temp, °F	Time	Pressure, lb/in.²	Min	Max	
Narmco.	Imidite 850	PBI	Sup. film	650	60	50	−423	450	3,000
American Cyanamid.	FM-34	PI	Sup. film	550	90	25	−423	600	3,300
Epoxylite.	No. 814	PI	1-pt., paste	500	90	10	−423	500	>1,000

TABLE 33 Suppliers for Selected Polyurethane, Anaerobic, and Thermosetting Acrylic Adhesives

Adhesive supplier	Trade name	Form	Pot life	Setting conditions		Continuous-service temp, °F		Shear strength per ASTM D 1002
				Temp, °F	Time	Min	Max	
Anaerobic Systems								
Eastman Chemical (cyanoacrylate)....	910	1-pt., liquid	Indef.	77	10 min	−20	100	1,920
	910 FS	1-pt., liquid	Indef.	77	10 min	−20	100	1,970
	910MHT	1-pt., liquid	Indef.	77	10 min	−20	225	1,350
Loctite (diester).........	306	1-pt., paste	Indef.	250	3 min	−65	300	1,800
	307	1-pt., liquid	Indef.	250	3 min	−65	200	3,500
	310	1-pt., paste	Indef.	250	3 min	−65	200	4,000
Polyurethane Systems								
American Cyanamid..........	Cybond 4010	2-pt., sol. sys.	8 h	300	1 h	−423	180	NA
Goodyear..........	Pliobond 6000	2-pt., paste	<3 min	77	24 h	−423	180	2,444
3M Co..........	PB-3543	2-pt., paste	50 s	77	2 h	NA	180	2,000
Thermosetting Acrylic Adhesive Systems								
B. F. Goodrich..........	Tame 100	2-pt., liquid	80 min	77	24 h	−40	200	4,430
	Tame 250	2-pt., liquid	16 h	77	24 h	−40	200	5,120
	Tame 300	2-pt., liquid	50 min	77	24 h	−40	200	3,250
Hughson..........	TS-500	2-pt., liquid	40 min	77	6 h	−40	180	3,900
	TS-520	2-pt., liquid	40 min	77	2 h	−40	180	3,600
	TS-540	2-pt., liquid	3 min	77	10 h	−40	180	2,600

The cyanoacrylate adhesives are more rigid and less resistant to moisture than diester adhesives. They are available only as low-viscosity liquids that cure in seconds at room temperature without the need of a primer. The cyanoacrylate adhesives bond well to a variety of substrates, as shown in Table 34, but have poor thermal resistance. Recently three modifications of the cyanoacrylate resin have been introduced to provide faster cures, higher strengths with some plastics, and greater thermal resistance.[43]

Phenoxy   Phenoxy adhesives are based on an epoxy-like polymer with creep resistance and good shear strength to a variety of surfaces.[44] Although a thermoplastic resin, phenoxy has a strength and heat resistance sufficient to be considered a structural adhesive. The adhesive softens at 180°F, but its chemical resistance and flexibility are good.

TABLE 34   Performance of Cyanoacrylate Adhesives on Various Substrates[43]

Substrate	Age of bond	Shear strength, lb/in.² of adhesive bonds
Steel–steel	10 min	1,920
	48 h	3,300
Aluminum–aluminum	10 min	1,480
	48 h	2,270
Butyl rubber–butyl rubber	10 min	150*
SBR rubber–SBR rubber	10 min	130
Neoprene rubber–neoprene rubber	10 min	100*
SBR rubber–phenolic	10 min	110*
Phenolic–phenolic	10 min	930*
	48 h	940*
Phenolic–aluminum	10 min	650
	48 h	920*
Aluminum–nylon	10 min	500
	48 h	950
Nylon–nylon	10 min	330
	48 h	600
Neoprene rubber–polyester glass	10 min	110*
Polyester glass–polyester glass	10 min	680
Acrylic–acrylic	10 min	810*
	48 h	790*
ABS–ABS	10 min	640*
	48 h	710*
Polystyrene–polystyrene	10 min	330*
Polycarbonate–polycarbonate	10 min	790
	48 h	950*

* Substrate failure.

The phenoxy resin can be dissolved and applied like a lacquer or solvent adhesive. It also can be extruded as a hot melt from pellets or applied in film form. Since phenoxy is a thermoplastic resin, it must be bonded at elevated temperature and pressures. Phenoxy resins are available from Union Carbide Corporation, New York.

Thermosetting acrylics   Thermosetting acrylic adhesives are newly developed two-part systems which provide high shear strength to many metals and plastics, as shown in Table 35. Acrylics retain their adhesion properties up to 200°F. They are relatively rigid adhesives with poor peel strength. These adhesives are particularly noted for their weather and moisture resistance as well as fast cure at room temperature. Selected thermosetting acrylic adhesives are listed in Table 33.

One manufacturer has developed an acrylic adhesive system where the hardener is applied to the substrate as a primer.[46] The substrate can then be stored for up to 6 months. When the parts are to be bonded, only the acrylic resin need be applied between the already primed substrates. Cure can occur in minutes at room temperature depending on the type of acrylic resin used. Thus, in effect this system offers a fast-reacting, one-part adhesive with indefinite shelf life.

Nonstructural Adhesives

Nonstructural adhesives are characterized by low strength and poor creep resistance at slightly elevated temperatures. The most common nonstructural adhesives are based on elastomers and thermoplastics. Although these systems have low strength, they usually are easy to use and very fast setting. Because of these characteristics most nonstructural adhesives are used in assembly-line fastening operations.

Rubber-based adhesives   Natural- or synthetic-rubber-based adhesives usually have excellent peel strength but low shear strength. Their resiliency provides good fatigue and impact properties. Temperature resistance is generally limited to 150 to 200°F, and creep under load occurs at room temperature.

The basic types of rubber-based adhesives used for nonstructural applications are shown in Table 36. These systems are generally supplied as solvent solutions, latex cements, and pressure-sensitive tapes. The first two forms require driving the solvent or water vehicle from the adhesive before bonding. This is accomplished by either

TABLE 35   Tensile-Shear Strength of Various Substrates Bonded with Thermosetting Acrylic Adhesives[45]

| | Avg lap shear, lb/in.^{2} at 77°F | | |
Substrate*	Adhesive A	Adhesive B	Adhesive C
Alclad aluminum, etched	4,430	4,235	5,420
Bare aluminum, etched	4,305	3,985	5,015
Bare aluminum, blasted	3,375	3,695	4,375
Brass, blasted	4,015	3,150	4,075
302 stainless steel, blasted	4,645	4,700	5,170
302 stainless steel, etched	2,840	4,275	2,650
Cold-rolled steel, blasted	2,050	3,385	2,135
Copper, blasted	2,915	2,740	3,255
Polyvinyl chloride, solvent wiped	1,375†	1,250†	1,250†
Polymethyl methacrylate, solvent wiped	1,550†	1,160†	865†
Polycarbonate, solvent wiped	2,570†	960	2,570†
ABS, solvent wiped	1,610†	1,635†	1,280†
Alclad aluminum–PVC	1,180†		
Plywood, ⅝-in. exterior glued (lb/in.)	802†	978†	
AFG-01 gap fill (¹⁄₁₆-in.) (lb/in.)	.....	1,083†	

* Metals solvent cleaned and degreased before etching or blasting.
† Substrate failure.

simple evaporation or forced heating. Some of the stronger and more environmental-resistant rubber-based adhesives require an elevated-temperature cure. Generally only slight pressure is required to achieve a substantial bond. Such pressure-sensitive adhesives are permanently tacky and flow under pressure to provide intimate contact with the adherend surface. Pressure-sensitive tapes are made by placing these adhesives on a backing material such as rubber, vinyl, canvas, or cotton cloth. After pressure is applied, the adhesive tightly grips the part being mounted as well as the surface to which it is affixed. The ease of application and the many different properties that can be obtained from elastomeric adhesives account for their wide use.

In addition to pressure-sensitive adhesives, elastomers go into mastic compounds that find wide use in the construction industry. Neoprene and reclaimed-rubber mastics are used to bond gypsum board and plywood flooring to wood-framing members. Often the adhesive bond is much stronger than the substrate. These mastic systems cure by evaporation of solvent through the porous substrates.

Elastomer-adhesive formulation is more complex than that of other adhesives because of the need for antioxidants and tackifiers. To make the situation even more complex, each formulation is generally designed to fulfill a specific set of properties. Commercial elastomeric adhesives have such widely differing properties that a list of a few specific adhesives would have little meaning.

Silicone Silicone resins have low shear properties but excellent peel strength and heat resistance.   Silicone adhesives are generally supplied as solvent solutions for pressure-sensitive application.   The adhesive reaches maximum physical properties after being cured at elevated temperature with an organic peroxide catalyst.   A lesser degree of adhesion can also be developed at room temperature.   Silicone adhesives retain their adhesive qualities over a wide temperature range, and after extended ex-

TABLE 36   Properties of Elastomeric Adhesives Used in Nonstructural Applications[42]

Adhesive	Application	Advantages	Limitations
Reclaimed rubber	Bonding paper, rubber, plastic and ceramic tile, plastic films, fibrous sound insulation and weather-stripping; also used for the adhesive on surgical and electrical tape	Low cost, applied very easily with roller coating, spraying, dipping, or brushing, gains strength very rapidly after joining, excellent moisture and water resistance	Becomes quite brittle with age, poor resistance to organic solvents
Natural rubber	Same as reclaimed rubber; also used for bonding leather and rubber sides to shoes	Excellent resilience, moisture and water resistance	Becomes quite brittle with age; poor resistance to organic solvents; does not bond well to metals
Neoprene rubber	Bonding weather stripping and fibrous sound-proofing materials to metal; used extensively in industry; bonding synthetic fibers, i.e., Dacron	Good strength to 150°F, fair resistance to creep	Poor storage life, high cost; small amounts of hydrochloric acid evolved during aging that may cause corrosion in closed systems; poor resistance to sunlight
Nitrile rubber	Bonding plastic films to metals, and fibrous materials such as wood and fabrics to aluminum, brass, and steel; also, bonding nylon to nylon and other materials	Most stable synthetic-rubber adhesive, excellent oil resistance, easily modified by addition of thermosetting resins	Does not bond well to natural rubber or butyl rubber
Polyiso-butylene	Bonding rubber to itself and plastic materials; also, bonding polyethylene terephthalate film to itself, aluminum foil and other plastic films	Good aging characteristics	Attacked by hydrocarbons; poor thermal resistance
Butyl	Bonding rubber to itself and metals; forms good bonding with most plastic films such as polyethylene terephthalate and polyvinylidene chloride	Excellent aging characteristics; chemically cross-linked materials have good thermal properties	Metals should be treated with an appropriate primer before bonding; attacked by hydrocarbons

posure to elevated temperature.   Table 37 shows typical adhesive-strength properties of pressure-sensitive tape prepared with aluminum-foil backing.

Room-temperature-vulcanizing (RTV) silicone-rubber adhesives and sealants form flexible bonds with high peel strength to many substrates.   These resins are one-component pastes which cure by reacting with moisture in the air.   Because of this unique curing mechanism, adhesive bond lines should not overlap by more than 1 in.

TABLE 37   Effect of Temperature and Aging on Silicone Pressure-sensitive Tape (Aluminum-Foil Backing)[47]

Testing temperature		Adhesive strength, oz/in.	
°C	°F	Uncatalyzed adhesive	Catalyzed adhesive (1% benzoyl peroxide)
−70	−94	Over 100*	Over 100*
−20	−4	Over 100*	Over 100*
100	212	60	48
150	302	60	45
200	392	Cohesive failure	40
250	482	Cohesive failure	35
Heat-aging cycle prior to testing (tested at 25°C, 77°F):			
No aging................................		60	52
1 h at 150°C (302°F).....................		55	50
24 h at 150°C (302°F)...................		60	50
7 days at 150°C (302°F).................		70	50
1 h at 250°C (482°F).....................		60	50
24 h at 250°C (482°F)...................		65	65
7 days at 250°C (482°F)................		Cohesive failure	65

* maximum limit of testing equipment.

RTV silicone materials cure at room temperature in about 24 h.   Fully cured adhesives can be used for extended periods up to 450°F and for shorter periods up to 500°F.   Figure 38 illustrates the peel strength of an RTV adhesive on aluminum as a function of heat aging.   With most RTV silicone formulations, acetic acid is released during cure.   Consequently corrosion of metals such as copper and brass in the bonding area may be a problem.   Silicone rubber bonds best to clean metal, glass, wood, silicone resin, vulcanized silicone rubber, ceramic, and many plastic surfaces.

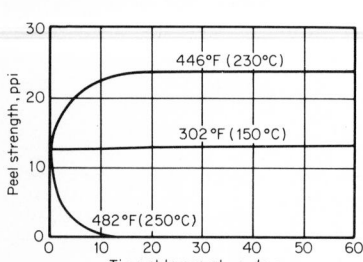

Fig. 38   Peel strength of RTV silicone rubber bonded to aluminum as a function of heat aging.[48]

Thermoplastic adhesives   Thermoplastic adhesives can be repeatedly melted and solidified without greatly hindering their properties.   Table 24 describes the most common types of thermoplastic adhesives.   These adhesives are useful in the −20° to 150°F temperature range.   Their physical properties vary with chemical type.   Some resins like the polyamides are quite tacky and flexible, while others are very rigid.

Thermoplastic adhesives are generally available as solvent solutions, water-based emulsion, and hot melts.   The first two systems are useful in bonding porous materials such as wood, plastic foam, and paper.   Water-based systems are especially useful for bonding foams that could be affected adversely by solvent-based adhesives.   Generally when hardened, thermoplastic adhesives are very nonresistant to the solvent in which they are originally supplied.

Hot-melt systems are usually flexible and tough.   They are used extensively for sealing applications involving paper, plastic films, and metal foil.   Table 38 offers a general comparison of hot-melt adhesives.   Hot melts can be supplied as (1) tapes or ribbons, (2) films, (3) granules, (4) pellets, (5) blocks, or (6) cards which are melted and pressed between the substrate.   The rate at which the adhesive cools and sets is dependent on the type of substrate and whether it is preheated or not.   Table 39 lists the advantages and disadvantages associated with the use of water-based, solvent-based, and hot-melt thermoplastic adhesives.

TABLE 38 General Comparison of Common Hot-Melt Adhesives[49]

Property	Ethylene vinyl acetate	Ethylene ethyl acetate	Ethylene acrylic acid	Ionomer	Phenoxy	Polyamide	Polyester	Polyethylene	Polyvinyl acetate	Polyvinyl butyral
Softening point, °C	40	60	70	75	100	100	.....	.....	65–195	130
Melting point, °C	95 (80–100)	90 (80–100)	.....	.....	.....	.....	267	137	.....	130
Crystallinity	L	L	M	L	L	L	H	H or L	L	L
Melt index	6 (2–20)	3 (2–20)	3 (0.5–400)	2	2.5	2	5	5 (0.5–20)	.....	.....
Tensile strength, lb/in.^{2}	2,750	2,000	3,300	4,000	9,500	2,000	4,500	2,000	5,000	6,500
% elongation	800	700	600	450	75	300	500	150	10	100
Lap shear on metal, lb/in.^{2}	.....	.....	1,700	.....	3,500	1,050	.....	.....	.....	.....
Film peel, lb/in.	1	4.5	12	12	.....	.....	5	2	.....	.....
Cost	M	M	M	M	H	M	H	L	M–L	M
Usage	H	M	L	L	L	H	H	L	M	H

H = high.   M = medium.   L = low.

TABLE 39 Advantages and Disadvantages of Thermoplastic-Adhesive Forms[50]

Form	Advantages	Disadvantages
Water base.......	Lower cost Nonflammable Nontoxic solvent Wide range of solids content Wide range of viscosity High concentration of high-molec- ular-weight material can be used Penetration and wetting can be varied	Poorer water resistance Subject to freezing Shrinks fabrics Wrinkles or curls paper Can be contaminated by some metals used for storage and appli- cation Corrosive to some metals Slow drying Poorer electrical properties
Solvent base......	Water-resistant Wide range of drying rates and open times Develop high early bond strength Easily wets some difficult surfaces	Fire and explosive hazard Health hazard Special explosion proof and venti- lating equipment needed
Hot melt.........	Lower package and shipping cost per pound of solid material Will not freeze Drying and equipment for drying are unnecessary Impervious surfaces easily bonded Fast bond-strength development Good storage stability Provides continuous water-vapor impermeable and water-resistant adhesive film	Special applicating equipment needed Limited strength because of vis- cosity and temperature limita- tions Degrades on continuous heating Poorer coating weight control Preheating of adherends may be necessary

SELECTING THE ADHESIVE

Factors Affecting Selection

There is no general-purpose adhesive. The best adhesive for a particular applica- tion will depend on the materials to be bonded, the service and assembly require- ments, and the economics of the bonding operation. By using these factors as criteria for selection, the many commercially available adhesives can be narrowed down to a few possible candidates. All the properties desired are often not available in a single adhesive system. In these cases, a compromise adhesive can usually be selected by deciding which properties are of major and minor importance.

Adhesive substrates The materials to be bonded are a prime factor in determining which adhesive to use. Some adherends such as stainless steel or wood can be suc- cessfully bonded with a great many adhesive types; other adherends such as nylon can be bonded with only a few. Typical adhesive-adherend combinations are listed in Table 40. A number in a given square indicates the particular adhesive which will bond to a cross-referenced substrate. This information is intended only as a guideline to show common adhesives which have been used successfully in various applications. The adhesive selections are listed without regard to strength or service requirements. Lack of a suggested adhesive for a particular substrate does not necessarily mean that a poor bond will result, only that information is not commonly available concerning that particular combination.

If two different materials are to be bonded, the recommended adhesives in Table 40 are those showing identical numbers under both substrates. Although adhesives may bond well to different substrates, joints may be inferior because of mismatch in thermal-expansion coefficients. In these cases a resilient adhesive should be carefully chosen to compensate for such stress.

Service requirements Adhesives must also be selected with regard to the type of stress and environmental conditions to which they will be exposed. These factors will further limit the number of candidate adhesives to be tested. Information on the environmental resistance of various adhesive classifications will appear in the next section.

TABLE 40 Selecting Adhesives for Use with Various Substrates (Based on Reference 30)

Adhesives (number key given in Table 41)

Adherend	Applicable adhesive numbers
Metals, Alloys	
Aluminum	1 2 3; 5; 9 10 11; 14 15; 21; 23 24 25; 27 28 29 30 31 32 33 34 35 36 37 38 39 40 41 42 43
Brass	1; 5; 9; 11; 14; 21; 23 24; 30; 32; 35 36; 39 40 41 42 43
Bronze	1; 5; 9; 11; 21; 23 24; 30; 32; 35 36
Cadmium	9 10
Chromium	1 2 3; 9 10 11; 21; 24
Copper	1 2; 5; 9 10 11; 14 15; 21; 23 24; 29; 32; 34 35 36; 39 40 41 42 43
Germanium	9; 10 11
Gold	2; 5; 10 11; 29 30; 32; 37
Lead	2 3; 9 10 11
Magnesium	1; 5; 9 10 11; 19; 21; 23 24; 27 28 29 30 31 32 33 34 35 36 37; 39 40 41 42 43
Nickel	5; 9 10 11; 29; 36
Platinum	9; 11; 23; 38
Silver	5; 9 10 11; 12 13 14 15 16 17 18; 21; 23 24 25; 27 28 29 30 31 32 33 34 35 36 37; 39 40 41
Steel (mild), iron	1 2 3; 5; 9 10 11; 12 13 14 15 16 17 18 19; 21; 23 24 25; 27 28 29 30 31 32 33 34 35 36 37; 39 40 41 43
Steel (stainless)	1 2 3; 5; 9 10 11; 12 13 14 15 16 17 18 19; 21; 23 24 25; 27 28 29 30 31 32 33 34 35 36 37; 39 40 41 43
Tin	1 2; 5; 9 10 11; 12 13; 23
Titanium	9 10 11; 29 30 31 32 33 34 35 36 37; 39 40 41 42 43
Tungsten	9 10
Uranium	1 2 3; 5; 9 10 11; 19; 23 24; 27 28 29 30 31 32 33 34 35 36; 40
Zinc	1 2; 11
Plastics	
Acetals	1 2; 11; 30; 36; 40
Acrylonitrile butadiene styrene	1 2 3; 9; 11; 14 15; 24; 27; 29 30; 32; 36; 40
Allyl diglycol carbonate	24
Cellulose	16 17 18; 26
Cellulose acetate	1 2; 16 17 18 19; 23; 26; 29 30; 36; 40
Cellulose acetate—butyrate	1 2; 16 17 18; 23; 26; 29 30; 36; 40
Cellulose nitrate	18; 23
Cellulose propionate	1 2; 16 17 18 19; 23; 26; 29 30; 36; 40
Diallyl isophthalate	4; 9; 31; 36; 40 42
Diallyl phthalate	4; 9; 31; 36; 40 42
Epoxy resins	2; 5; 9; 11; 27 28 29 30; 32; 35 36 37; 40
Ethyl cellulose	18
Methacrylate butadiene styrene	31
Melamine formaldehyde	4; 7; 9; 19; 30; 36; 40; 43
Methyl pentene	9
Phenol formaldehyde	4; 7; 9; 11; 19; 30; 36; 40; 43
Phenolic—melamine	2; 5; 7; 9 10 11; 14; 23; 27 28 29 30 31 32; 35 36 37; 39 41 43

TABLE 40 Selecting Adhesives for Use with Various Substrates (Based on Reference 30) (Continued)

Adherend	Adhesives (number key given in Table 41)
Plastics (Continued)	
Phenoxy	1 2, 7 8, 11
Polyamide (nylon)	1 2 3, 9, 11, 14 15, 23, 30, 32, 40
Polycarbonate	1 2 3, 9 10 11, 14 15, 32, 40
Polychlorotrifluoroethylene	9 10 11, 29, 40
Polyester (fiber composite)	2 3, 9, 11, 14, 24, 27 28 29, 34 35
Polyether (chlorinated)	
Polyethylene	9, 11, 23, 30, 32, 36
Polyethylene (film)	2, 23, 30, 36
Polyethylene (chlorinated)	2, 5, 23, 29, 30, 32, 35 36, 40
Polyethylene terephthalate (Mylar)	2, 23, 29, 30, 32, 35 36, 40
Polyformaldehyde	2, 5, 10, 19, 23, 29, 30, 32, 35
Polyimide	2, 9 10 11 12, 24, 30, 39
Polymethyl methacrylate	1, 5, 9, 11, 14 15, 19, 24, 29, 30, 32, 34 35 36 37, 40
Polyphenylene	11, 34 35 36
Polyphenylene oxide	2 3, 9, 11, 14, 29, 30, 32, 36, 40
Polypropylene	11, 30
Polypropylene (film)	2, 11
Polystyrene	1 3, 11, 14 15, 19, 23, 35
Polystyrene (film)	1, 11, 15, 24
Polysulfone	11, 14
Polytetrafluoroethylene	9 10 11, 29, 34 35, 40
Polyurethane	11, 30, 32, 36
Polyvinyl chloride	1 2, 9, 11, 14 15, 19, 24, 29, 30, 32, 35
Polyvinyl chloride (film)	2, 19, 24, 30, 40
Polyvinyl fluoride	2 3, 11, 30, 32, 33, 40
Polyvinylidene chloride	2, 23, 30, 32, 40
Silicone	2
Styrene acrylonitrile	11, 30, 34 35
Urea formaldehyde	4, 14
Foams	
Epoxy	1 2, 9, 11, 26 27
Latex	9, 11
Phenol formaldehyde	2, 5, 9, 11, 19, 27, 28, 30, 32, 36, 41
Polyethylene	9, 11, 29 30
Polyethylene–cellulose acetate	9, 11
Polyphenylene oxide	2, 9, 11, 21, 28, 30 31
Polystyrene	2, 9, 11, 14, 19, 28, 30, 32, 36
Polyurethane	2, 5, 9, 11, 14, 19, 24, 28, 30, 32, 35
Polyvinylchloride	2 3, 9, 11, 15, 19, 28, 30, 32, 33, 36
Silicone	4, 34
Urea formaldehyde	7, 43

Rubbers

Material	Cross-reference numbers
Butyl	1 2 · 7 9 · 15 · 28 · 35 · 40 43
Butadiene nitrile	1 2 · 9 10 · 15 · 32 · 35 · 40 43
Butadiene styrene	1 2 · 9 10 · 15 · 30 31 32 · 35
Chlorosulfonated polyethylene	11 · 30 32
Ethylene propylene	15 · 34 35
Fluorocarbon	1 · 9 10 11 · 34
Fluorosilicone	
Polyacrylic	24
Polybutadiene	2 · 8 10 11 · 15 · 31
Polychloroprene (neoprene)	1 2 · 7 9 10 11 · 15 · 26 · 30 31 32 33 34 35 36 · 40 43
Polyisoprene (natural)	1 2 · 7 9 10 11 · 15 · 26 27 · 30 31 32 33 34 35 · 40 43
Polysulfide	33
Polyurethane	1 2 · 11 · 15 · 30 32 33 · 35 36 · 40 41
Silicone	28 29 30 · 34 35 · 40 41

Wood, Allied Materials

Material	Cross-reference numbers
Cork	2 3 4 5 6 7 8 9 · 11 · 16 17 18 19 · 21 23 · 26 27 28 · 30 31 32 33 34 · 36 · 40 41
Hardboard, chipboard	4 · 6 7 · 31 33
Wood	2 3 4 5 6 7 8 9 · 11 · 14 15 16 17 18 19 · 21 23 · 25 26 27 28 29 · 30 31 32 33 34 · 36 · 40
Wood (laminates)	2 3 4 5 6 7 8 9 · 11 · 16 17 18 19 · 21 23 · 25 26 28 29 · 30 31 32 33 34 · 36 · 40

Fiber Products

Material	Cross-reference numbers
Cardboard	2 3 4 5 · 7 8 9 · 11 · 16 17 18 19 · 21 22 23 24 · 26 27 28 29 · 30 31 32 33 · 36 37 · 40 41
Cotton	2 3 4 5 · 9 · 11 · 16 17 18 19 · 23 24 · 26 27 · 30 31 32 33 34 35 36 · 40 41
Felt	3 4 5 · 7 9 · 11 · 16 17 18 19 · 23 24 · 26 27 · 30 31 32 33 34 36 · 40 41
Jute	2 3 · 7 9 · 11 · 16 17 19 · 24 · 26 27 · 30 32 33 34 36
Leather	2 3 · 7 9 · 11 · 16 17 18 19 · 21 23 24 · 26 27 · 30 31 32 33 34 36 · 40 41
Paper (bookbinding)	16 17 18 19 · 22 23 · 29 · 31 32 33 36
Paper (labels)	16 17 18 19 · 22 23 24 · 29 · 30 31 32 33 34 36
Paper (packaging)	2 · 16 17 18 19 · 21 22 · 26 27 28 29 · 30 31 32 33 36
Rayon	2 3 · 7 9 · 11 · 16 17 18 19 · 23 · 26 27 28 29 · 30 31 32 33 34 35 36 · 40 41
Silk	2 3 4 5 · 7 9 · 11 · 16 17 18 19 · 21 23 24 · 26 27 28 29 · 30 31 32 33 34 35 36 · 40 41 43
Wool	2 3 4 5 · 7 9 · 11 · 16 17 18 19 · 21 23 24 · 26 27 28 · 30 31 32 33 34 35 36 · 41

Inorganic Materials

Material	Cross-reference numbers
Asbestos	2 3 · 5 · 7 8 9 · 11 · 14 16 18 19 · 21 · 27 28 29 · 30 31 32 33 34 35 · 37 · 40 43
Carbon	9 · 11
Carborundum	1 2 3 4 · 7 8 9 10 11 · 14 15 16 17 18 19 · 21 · 27 28 29 · 30 31 32 33 34 35 36 37 · 40 41
Ceramics (porcelain, vitreous)	2 3 4 · 7 8 · 14 · 35
Concrete, stone, granite	9 · 11 · 14
Ferrite	9 10 11
Glass	1 2 3 · 9 10 11 · 14 15 16 17 18 · 28 · 30 31 32 · 34 35 · 37 · 40
Magnesium fluoride	
Mica	5 · 14 · 35 · 43
Quartz	9 · 11 · 14
Sodium chloride	9 · 11
Tungsten carbide	9 · 11

The chosen adhesive should have strength great enough to resist the maximum stress during any time in service with reasonable safety factors. Overspecifying may result in certain adhesives being overlooked which can do the job at lower cost and with less demanding curing conditions.

Assembly requirements   Adhesives require time, pressure, and heat singly or in combination to achieve hardening. Curing conditions are often a severe restricting factor in the selection of an adhesive. Typical production factors involved in assembly are the equipment available, allowable cure time, pressure required, necessary bonding temperature, degree of substrate preparation required, and physical form of the adhesive. Table 41 lists the available forms and processing requirements for various types of adhesives.

Cost requirements   Economic analysis of the bonding operation must consider not only the raw-materials cost but also the processing equipment necessary, time and labor required, and cost incurred by wasted adhesive and rejected parts. Raw-materials cost for the adhesive should not be based on price per weight or volume because of variations in solids and density of adhesives. Cost per bonded part is a more realistic criterion for selection of an adhesive.

Sources of Information

To select adhesive candidates for testing, it is necessary to have meaningful physical data for comparison. Literature from adhesive manufacturers and trade periodicals offers some adhesive design and test information. Usually the data are from reproducible ASTM test methods. General physical properties such as shear or peel strength can easily be compared. Unfortunately information concerning specific stresses, environments, or adherends is often lacking.

Adhesive manufacturer   The adhesive supplier can help to narrow the choice of adhesives, recommend a certain formulation, assist in designing the joint, and establish surface-preparation and bonding requirements. Before contacting the adhesive supplier or attempting to select candidates alone, the buyer must first clearly define his bonding problem. A questionnaire such as illustrated in Fig. 39 is often used by suppliers to determine adhesive-bond requirements.

Literature sources   Adhesive information and property data can be found in the technical literature. Two periodicals are published in the United States which deal specifically with adhesives: *Adhesives Age* (Palmerton Publishing Co.) and the *Journal of Adhesion* (Gordon and Breach Publishers). Other trade periodicals such as *Insulation/Circuits, Materials Engineering, Machine Design,* and *Product Engineering* often have adhesive articles relative to a specific application. Many excellent texts also exist on adhesive technology.

Adhesives for Metal

The chemical types of structural adhesives for metal bonding were described in the preceding section. Since organic adhesives readily wet most metallic surfaces, the adhesive selection does not depend as much on the type of metal substrate as on other bonding requirements.

Selecting a specific adhesive from a table of general properties is difficult because formulations within one class of adhesives may vary widely in physical properties. General physical data for structural metal adhesives are presented in Table 42. This table may prove useful in making preliminary selections or eliminating obviously unsuitable adhesives. Once the candidate adhesives are limited to only a few types, the designer can search more efficiently for the best bonding system.

Nonstructural adhesives for metals include elastomeric and thermoplastic resins. These are generally used as pressure-sensitive or hot-melt adhesives. They are noted for fast production, low cost, and low to medium strength. Typical adhesives for nonstructural-bonding applications were described previously. Most pressure-sensitive and hot-melt cements can be used on any clean metal surface and on many plastics and elastomers.

Bonding Plastics

Bonding of plastics is generally more difficult than metal bonding because of the lower surface energy of polymeric substrates. But most thermoplastics can be joined by solvent or heat welding as well as with adhesives.   Table 43 indicates joining methods which are suitable for common types of plastics.   Descriptions of these various bonding techniques are presented in Table 44.   The plastic manufacturer is generally the leading source of information on the proper methods of joining a particular plastic.

Direct-heat welding   Welding by direct application of heat provides an advantageous method of joining many thermoplastics that do not degrade rapidly at their melt temperatures.   The three principal methods of direct welding are heated-tool welding, hot-gas welding, and resistance-wire welding.

WHAT
MATERIALS
ARE
BEING
JOINED?

1

MY APPLICATION IS TO: ☐bond ☐coat ☐seal ☐pot

TO _____

Please be specific—say "aluminum to foamed styrene" NOT "metal to plastic." Submit samples if possible.

WHAT
ARE
THE
END-USE
SERVICE
REQUIREMENTS?

2

MY ASSEMBLY WILL BE EXPECTED TO WITHSTAND:

A. Temperature extremes of_____°F ☐continuously ☐intermittently

B. These chemicals or solvents _____

☐immersion ☐vapor ☐continuously ☐intermittently

C. These conditions: ☐moisture ☐sunlight ☐outdoor weathering

D. Color requirements: _____

E. These loads: (Describe amount and type of strength required. Sketch, if possible)_____

WHICH
**PRODUCTION
METHOD**
WILL
BE
USED?

3

A. I prefer the following form of adhesive: ☐liquid ☐paste ☐film ☐powder

B. This will be ☐hand application.
              ☐machine application.

C. My equipment makes it desirable to use:

D. These are my application requirements:

Viscosity desired: _____

Drying time required: _____

Curing temp. available: _____

Curing time available: _____

Pressure available: _____

E. If necessary, we ☐can use a 2-part adhesive.
                   ☐cannot

☐brush
☐extruder
☐knife or trowel
☐caulking gun
☐spray
☐roll coater
☐other _____

Estimated monthly volume:_____   Seasonal: ☐no ☐yes

Price limitations:_____

Present products used: (Samples if possible)_____

Why unsatisfactory: _____

Remarks: _____

Fig. 39   Problem-analysis form for selection of adhesives.[51]

Heated-Tool Welding.   In this method the surfaces to be fused are heated by holding them against a hot surface; then the parts are brought into contact and allowed to harden under slight pressure.   Electric-strip heaters, soldering irons, hot plates, and resistance blades are common methods of providing heat locally.

A simple hot plate has been used extensively with many plastics.   The parts are held on the hot plate until sufficient fusible material has been developed.   Table 45 lists typical hot-plate temperatures for a variety of plastics.   A similar technique involves butting flat plastic sheets on a table next to a resistance blade that runs the length of the sheet.   Once the plastic begins to soften, the blade is raised, and the sheets are pressed together and fused.

TABLE 41 Selecting Adhesives with Respect to Form and Processing Factors

Adhesive type	Common forms available						Method of cure			Processing conditions			
	Solid	Film	Paste	Liquid	Solvent sol, emulsion	Solvent release	Fusion on heating	Pressure sensitive	Chemical reaction	Room temp	Elevated temp	Bonding pressure required	Bonding pressure not required
Thermosetting Adhesives													
1. Cyanoacrylate				X					X	X		X	
2. Polyester + isocyanate		X			X	X			X	X	X	X	X
3. Polyester + monomer			X	X					X	X	X	X	
4. Urea formaldehyde	X			X					X		X	X	
5. Melamine formaldehyde	X								X		X	X	
6. Urea–melamine formaldehyde									X		X	X	
7. Resorcinol formaldehyde	X			X					X	X	X	X	
8. Phenol-resorcinol formaldehyde				X					X	X	X	X	
9. Epoxy (+ polyamine)	X	X	X	X					X	X	X		X
10. Epoxy (+ polyanhydride)			X	X					X	X	X		X
11. Epoxy (+ polyamide)		X	X	X					X		X		X
12. Polyimide		X			X				X		X	X	
13. Polybenzimidazole		X	X						X		X	X	
14. Acrylic				X					X	X			X
15. Acrylate acid diester			X	X					X	X	X	X	
Thermoplastic Adhesives													
16. Cellulose acetate					X	X				X			X
17. Cellulose acetate butyrate					X	X				X			X
18. Cellulose nitrate					X	X				X			X

19. Polyvinyl acetate	X	X*	X*	X			X*	X	X		X
20. Vinyl vinylidene	X			X				X	X		X
21. Polyvinyl acetal	X	X*	X*	X			X*	X	X		X
22. Polyvinyl alcohol	X	X	X	X				X	X		X
23. Polyamide					X						X
24. Acrylic		X	X	X			X	X	X		X
25. Phenoxy		X	X				X	X	X		X
Elastomer Adhesives											
26. Natural rubber		X	X	X	X	X		X	X		X
27. Reclaimed rubber		X		X	X	X		X	X		X
28. Butyl		X	X	X	X	X		X	X		X
29. Polyisobutylene		X		X	X	X		X	X		X
30. Nitrile	X	X		X	X	X		X	X		X
31. Styrene butadiene		X	X*	X	X	X	X*	X	X		X
32. Polyurethane	X			X	X			X	X	X	X
33. Polysulfide	X			X	X				X	X	X
34. Silicone (RTV)			X		X	X					
35. Silicone resin		X		X		X		X	X		X
36. Neoprene		X						X	X		X
Alloy Adhesives											
37. Epoxy-phenolic	X	X	X		X			X	X	X	X
38. Epoxy-polysulfide				X	X			X	X	X	X
39. Epoxy-nylon		X	X		X			X	X		X
40. Phenolic-nitrile		X	X		X			X	X		X
41. Phenolic-neoprene		X	X		X			X	X		X
42. Phenolic-polyvinyl butyral		X	X		X			X	X		X
43. Phenolic-polyvinyl formal		X	X		X			X	X		X

* Heat and pressure required for hot-melt types.

TABLE 42 Properties of Structural Adhesives Used to Bond Metals

Adhesive	Service temp, °F Max	Service temp, °F Min	Shear strength, lb/in.²	Peel strength	Impact strength	Creep resistance	Solvent resistance	Moisture resistance	Type of bond
Epoxy-amine	150	−50	3,000–5,000	Poor	Poor	Good	Good	Good	Rigid
Epoxy-polyamide	150	−60	2,000–4,000	Medium	Good	Good	Good	Medium	Tough and moderately flexible
Epoxy-anhydride	300	−60	3,000–5,000	Poor	Medium	Good	Good	Good	Rigid
Epoxy-phenolic	350	−423	3,200	Poor	Poor	Good	Good	Good	Rigid
Epoxy-nylon	180	−423	6,500	Very good	Good	Medium	Good	Poor	Tough
Epoxy-polysulfide	150	−100	3,000	Good	Medium	Medium	Good	Good	Flexible
Nitrile-phenolic	300	−100	3,000	Good	Good	Good	Good	Good	Tough and moderately flexible
Vinyl-phenolic	225	−60	2,000–5,000	Very good	Good	Medium	Medium	Good	Tough and moderately flexible
Neoprene-phenolic	200	−70	3,000	Good	Good	Good	Good	Good	Tough and moderately flexible
Polyimide	600	−423	3,000	Poor	Poor	Good	Good	Medium	Rigid
Polybenzimidazole	500	−423	2,000–3,000	Poor	Poor	Good	Good	Good	Rigid
Polyurethane	150	−423	5,000	Good	Good	Good	Medium	Poor	Flexible
Acrylate acid diester	200	−60	2,000–4,000	Poor	Medium	Good	Poor	Poor	Rigid
Cyanoacrylate	150	−60	2,000	Poor	Poor	Good	Poor	Poor	Rigid
Phenoxy	180	−70	2,500	Medium	Good	Good	Poor	Good	Tough and moderately flexible
Thermosetting acrylic	250	−60	3,000–4,000	Poor	Poor	Good	Good	Good	Rigid

Figure 40 illustrates a heated tool commonly used to bond plastic sheet and film. Care must be taken especially with thin film not to apply too much pressure or heat and melt through the plastic. Table 46 provides heat-sealing temperature ranges for common plastic films.

Hot-Gas Welding. An electrically or gas-heated welding gun with an orifice temperature of 425 to 700°F can be used to bond many thermoplastic materials. The pieces to be joined are beveled and positioned with a small gap between them. A welding rod made of the same plastic that is being bonded is laid in joint with a steady pressure. The heat from the gun is directed to the tip of the rod, where it fills the gap, as shown in Fig. 41.

For polyolefins the heated gas must be inert, since air will oxidize the surface of the plastic. After welding, the joint should not be stressed for several hours. This is particularly true for polyolefins, nylons, and polyformaldehyde. Hot-gas welding is not recommended for filled materials or substrates less than $\frac{1}{16}$ in. thick. Applications

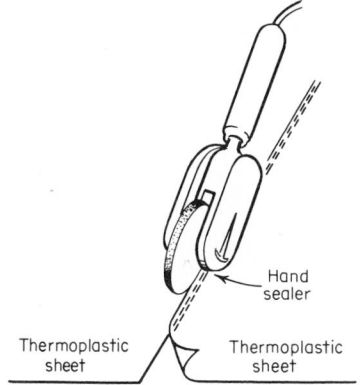

Fig. 40 Hand-operated heat-sealing tool.[55]

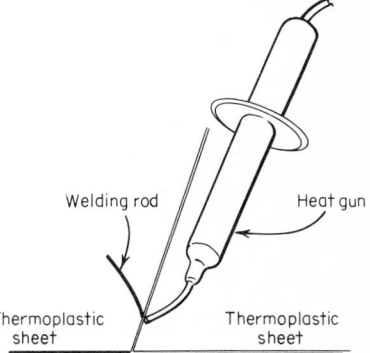

Fig. 41 Hot-gas-welding apparatus.[55]

are usually large structural assemblies. The weld is not cosmetically attractive, but tensile strengths 85 percent of the parent material are average.

Resistance-Wire Welding. This method employs an electrical-resistance-heating wire laid between mating substrates to generate heat of fusion (Fig. 42). After the bond has been made, the exterior wire is cut off. A similar type of process can be used to cure thermosetting adhesives when the heat generated by the resistance wire is used to advance the cure.

Indirect thermal welding Indirect heating occurs when some form of energy is applied to the joint which acts on the plastic to cause heating at the interface or in the plastic as a whole. The applied energy is generally in the form of friction, high frequency, induction, or ultrasonics.

Friction or Spin Welding. Spin welding uses the heat of friction to cause fusion at the interface. One substrate is rotated very rapidly while in touch with the other substrate so that the sufaces melt without damaging the entire part. Sufficient pressure is applied during the process to force out excess air bubbles. The rotation is then stopped, and pressure is maintained until the weld sets. Rotation speed and pressure are dependent on the thermoplastics being joined.

A wide variety of joints can be made by spin welding. Since the outer edges of the rotating substrate move considerably faster than the center, joints are generally designed to concentrate pressure at the center. Spin welding is a popular method of joining large-volume products such as packaging and toys.

Dielectric Heating. Dielectric sealing can be used on most thermoplastics except those which are relatively transparent to high-frequency electric fields. It is used mostly to seal vinyl sheeting such as automobile upholstery, swimming-pool liners, and

TABLE 43 Assembly Methods for Plastics[52]

Plastic	Adhesives	Dielectric welding	Induction bonding	Mechanical fastening	Solvent welding	Spin welding	Thermal welding	Ultrasonic welding
Thermoplastics								
ABS	X		X	X	X	X	X	X
Acetals	X		X	X	X	X	X	X
Acrylics	X		X	X	X	X		X
Cellulosics	X				X	X		
Chlorinated polyether	X			X				
Ethylene copolymers		X					X	
Fluoroplastics	X							
Ionomer		X					X	
Methylpentene								X
Nylons	X		X	X	X	X		X
Phenylene oxide–based materials	X			X	X	X	X	X
Polyesters	X			X	X	X		X
Polyamide-imide	X			X				X
Polyaryl ether	X			X				X
Polyaryl sulfone	X			X				X
Polybutylene								X

Polycarbonate	X	X	X	X	X	X		X
Polycarbonate/ABS	X	X	X	X		X		X
Polyethylenes	X	X			X	X	X	X
Polyimide								X
Polyphenylene sulfide	X	X			X	X		X
Polypropylenes	X	X	X	X	X	X	X	X
Polystyrenes	X	X		X	X	X		X
Polysulfone								X
Propylene copolymers		X	X	X	X	X	X	X
PVC/acrylic alloy	X				X	X		X
PVC/ABS alloys		X	X	X	X		X	X
Styrene acrylonitrile	X	X				X		
Vinyls		X	X	X	X	X	X	X
Thermosets								
Alkyds						X		X
Allyl diglycol carbonate						X		X
Diallyl phthalate								X
Epoxies								X
Melamines						X		X
Phenolics								X
Polybutadienes						X		X
Polyesters						X		X
Silicones								X
Ureas								X
Urethanes						X		X

Technique	Description	Advantages	Limitations	Processing considerations
Solvent cementing and dopes	Solvent softens the surface of an amorphous thermoplastic; mating takes place when the solvent has completely evaporated. Bodied cement with small percentage of parent material can give more workable material, fill in voids in bond area. Cannot be used for polyolefins and acetal homopolymers	Strength, up to 100% of parent materials, easily and economically obtained with minimum equipment requirements	Long evaporation times required; solvent may be hazardous; may cause crazing in some resins	Equipment ranges from hypodermic needle or just a wiping media to tanks for dip and soak. Clamping devices are necessary, and air dryer is usually required. Solvent-recovery apparatus may be necessary or required. Processing speeds are relatively slow because of drying times. Equipment costs are low to medium. The high range might be $1,000–$2,000
		Thermal Bonding		
Ultrasonics	High-frequency sound vibrations transmitted by a metal horn generate friction at the bond area of a thermoplastic part, melting plastics just enough to permit a bond. Materials most readily weldable are acetal, ABS, acrylic, nylon, PC, polyimide, PS, SAN, phenoxy	Strong bonds for most thermoplastics; fast, often less than 1 s. Strong bonds obtainable in most thermal techniques if complete fusion is obtained	Size and shape limited. Limited applications to PVCs, polyolefins	Converter to change 20 kHz electrical into 20 kHz mechanical energy is required along with stand and horn to transmit energy to part. Rotary tables and high-speed feeder can be incorporated. Cost: from $2,400 up for stand, converter; $65–$1,000 for horn
Hot-plate and hot-tool welding	Mating surfaces are heated against a hot surface, allowed to soften sufficiently to produce a good bond, then clamped together while bond sets. Applicable to rigid thermoplastics.	Can be very fast, e.g., 4–10 s in some cases; strong	Stresses may occur in bond area	Use simple soldering guns and hot irons, relatively simple hot plates attached to heating elements up to semiautomatic hot-plate equipment. Clamps needed in all cases. Costs run from $4.99 gun to $2,000–$20,000 range for commercial hot-plate units
Hot-gas welding	Welding rod of the same material being joined (largest application is vinyl) is softened by hot air or nitrogen as it is fed through a gun that is softening part surface simultaneously. Rod fills in joint area and cools to effect a bond	Strong bonds, especially for large structural shapes	Relatively slow; not an "appearance" weld	Requires a hand gun, special welding tips, an air source and welding rod. Regular hand-gun speeds run 6 in./min; high-speed hand-held tool boosts this to 48–60 in./min. Costs begin at under $100

Process	Description	Advantages	Limitations	Equipment and cost
Spin welding	Parts to be bonded are spun at high speed, developing friction at the bond area; when spinning stops, parts cool in fixture under pressure to set bond. Applicable to most rigid thermoplastics	Very fast (as low as 1–2 s); strong bonds	Bond area must be circular	Basic apparatus is a spinning device, but sophisticated feeding and handling devices are generally incorporated to take advantage of high-speed operation. Costs go up to the $30,000–$40,000 range
Dielectrics	High-frequency voltage applied to film or sheet causes material to melt at bonding surfaces. Material cools rapidly to effect a bond. Most widely used with vinyls	Fast seal with minimum heat applied	Only for film and sheet	Requires rf generator, dies, and press. Operation can range from hand-fed to semiautomatic with speeds depending on thickness and type of product being handled. 3–25 kW units are most common with costs running $500/kW and up
Induction	A metal insert or screen is placed between the parts to be welded, and energized with an electromagnetic field. As the insert heats up, the parts around it melt, and when cooled form a bond. For most thermoplastics	Provides rapid heating of solid sections to reduce chance of degradation	Since metal is embedded in plastic, stress may be caused at bond	High-frequency generator, heating coil, and inserts (generally 0.02–0.04 in. thick). Hooked up to automated devices, speeds are high. Cost: in the $1,000/kW range (1–5 kW used). Work coils, water cooling for electronics, automatic timers, multiple-position stations may also be required
Adhesives*				
Liquids solvent, water base, anaerobics	Solvent-and-water-based liquid adhesives, available in a wide number of bases—e.g., polyester, vinyl—in one- or two-part form fill bonding needs ranging from high-speed lamination to one-of-a-kind joining of dissimilar plastics parts. Solvents provide more bite, but cost much more than similar base water-type adhesive. Anaerobics are a group of adhesives that cure in the absence of air, with a minimum amount of pressure required to effect the initial bond. Adhesives are used for almost every type of plastic	Easy to apply; adhesives available to fit most applications	Shelf and pot life often limited. Solvents may cause pollution problems; water-base not as strong; anaerobics toxic	Application techniques range from simply brushing on to spraying and roller coating-lamination for very high production. Adhesive application techniques, often similar to decorating equipment, from hundreds (e.g., a glue pump for $400) to thousands of dollars with sophisticated laminating equipment costing in the tens of thousands of dollars. Anaerobics are generally applied a drop at a time from a special bottle or dispenser

TABLE 44   Bonding and Joining Techniques for Plastics: What Is Available and What They Offer[53] (Continued)

Technique	Description	Advantages	Limitations	Processing considerations
		Adhesives* (Continued)		
Pastes, mastics	Highly viscous single- or two-component materials which cure to a very hard or flexible joint depending on adhesive type	Does not run when applied	Shelf and pot life often limited	Often applied via a trowel, knife or gun-type dispenser; one-component systems can be applied directly from a tube. Various types of roller coaters are also used. Metering-type dispensing equipment in the $2,500 range has been used to some extent
Hot melts	100% solids adhesives that become flowable when heat is applied. Often used to bond continuous flat surfaces	Fast application; clean operation.	Virtually no structural hot melts for plastics	Hot melts are applied at high speeds via heating the adhesive, then extruding (actually squirting) it onto a substrate, roller coating, using a special dispenser or roll to apply dots or simply dipping
Film	Available in several forms including hot melts, these are sheets of solid adhesive. Mostly used to bond film or sheet to a substrate	Clean, efficient	High cost	Film adhesive is reactivated by a heat source; production costs are in the medium–high range depending on heat source used
Pressure-sensitive	Tacky adhesives used in a variety of commercial applications (e.g., cellophane too). Often used with polyolefins	Flexible	Bonds not very strong	Generally applied by spray with bonding effected by light pressure
Mechanical fasteners (staples, screws, molded-in inserts, snap fits and variety of proprietary fasteners)	Typical mechanical fasteners are listed on the left. Devices are made of metal or plastic. Type selected will depend on how strong the end product must be, appearance factors. Often used to join dissimilar plastics or plastics to nonplastics	Adaptable to many materials; low to medium costs; can be used for parts that must be disassembled	Some have limited pull-out strength; molded-in inserts may result in stresses	Nails and staples are applied by simply hammering or stapling. Other fasteners may be inserted by drill press, ultrasonics, air or electric gun, hand tool. Special molding—i.e., molded-in-hole—may be required

* Typical adhesives in each class are: Liquids: 1. Solvent—polyester, vinyl, phenolics acrylics, rubbers, epoxies, polamide; 2. Water—acrylics, rubber-casein; 3. Anaerobics—cyanoacrylate; mastics—rubbers, mastics; hot melts—polyamides, PE, PS, PVA; film—epoxies, polyamide, phenolics; pressure, sensitive—rubbers.

TABLE 45   Hot-Plate Temperatures to Weld Plastics[54]

Plastic	Temp, °F
ABS	450
Acetal	500
Phenoxy	550
Polyethylene LD	360
HD	390
Polycarbonate	650
PPO	650
Noryl*	525
Polypropylene	400
Polystyrene	420
SAN	450
Nylon 6, 6	475
PVC	450

* Trademark of General Electric Co.

rainwear.   An alternating electric field is imposed on the joint which causes rapid reorientation of polar molecules and heat generated by molecular friction.   Pressure is applied and held until the weld cools.   Variables in the bonding operation are the frequency generated, dielectric loss of the plastic, the power applied, pressure, and time.   Generally an increase in any of these factors will cause additional heating.   Dielectric heating can also be used to generate the heat necessary for curing polar, thermosetting adhesives.

Induction Heating. An electromagnetic-induction field can be used to heat a metal grid or insert placed between mating thermoplastic substrates. When the joint is positioned between induction coils, the hot insert causes the plastic to melt and fuse together.   Slight pressure is maintained as the induction field is turned off and the joint hardens.

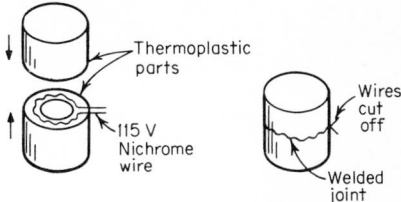

Fig. 42   Process setup for resistance-wire welding.[57]

Electromagnetic adhesives can be made from metal-filled thermoplastics.   These adhesives can be shaped into gaskets or film which will melt in an induction field. The advantage of this method is that stresses caused by large metal inserts are avoided.   Table 47 shows compatible plastic combinations for electromagnetic adhesives.

TABLE 46   Heat-sealing Temperatures for Plastic Films[56]

Film	Temp, °F
Coated cellophane	200–350
Cellulose acetate	400–500
Coated polyester	490
Poly(chlorotrifluoroethylene)	415–450
Polyethylene	250–375
Polystyrene (oriented)	220–300
Poly(vinyl alcohol)	300–400
Poly(vinyl chloride) and copolymers (nonrigid)	200–400
Poly(vinyl chloride) and copolymers (rigid)	260–400
Poly(vinyl chloride)–nitrile rubber blend	220–350
Poly(vinylidene chloride)	285
Rubber hydrochloride	225–350
Fluorinated ethylene–propylene copolymer	600–750

TABLE 47 Compatible Plastic Combinations for Bonding with Electromagnetic Adhesives[58]

	ABS	Acetal	Acrylic	Nylon	PC	PE	PP	PS	PVC	SAN
ABS.	X	..	X	..	..	..	..	X	X	
Acetal.	..	X								
Acrylic.	X	..	X	..	..	..	..	X	X	
Nylon.	..	..	..	X						
Polycarbonate. . .	..	..	..	..	X					
Polyethylene. . . .	..	..	..	..	..	X				
Polypropylene. . .	..	..	..	..	..	..	X			
Polystyrene.	X	..	X	..	..	..	..	X	X	
Polyvinyl chloride.	X	..	X	..	..	..	..	X	X	
SAN.	..	..	..	..	..	..	..	..	..	X

X = compatible combinations.

Ultrasonic Welding.   During ultrasonic welding, a high-frequency electrodynamic field is generated which resonates a metal horn.   The horn vibrates the substrate sufficiently fast to cause great heat at the interface.   With pressure and subsequent cooling a strong bond can be obtained with many thermoplastics.   Rigid plastics with a high modulus of elasticity are best.   Excellent results generally are obtainable with polystyrene, SAN, ABS, polycarbonate, and acrylic plastics.

Typical ultrasonic joint designs are shown in Fig. 43.   Ultrasonics can also be used to stake plastics to other substrates and for inserting metal parts.   Ultrasonic bonding is considered faster than methods of direct-heat welding.   Ultrasonic heating is also applicable to hot-melt and thermosetting adhesives.[59]

Solvent cementing   Solvent cementing is the simplest and most economical method of joining noncrystalline thermoplastics.   Solvent-cemented joints are less sensitive to thermal cycling than joints bonded with adhesives, and they are as resistant to degrading environments as the parent plastic.   Bond strength 85 to 100 percent of the

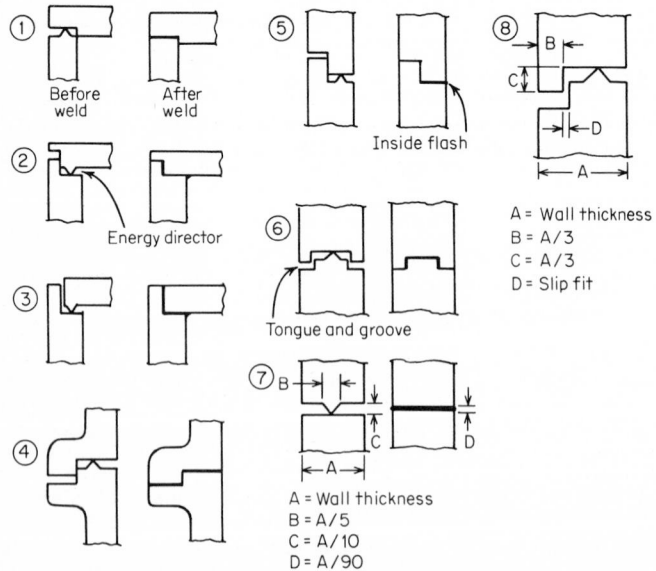

Fig. 43 Typical joint designs used in ultrasonic welding.[57]

parent plastic can be obtained. The major disadvantage of solvent cementing is the possibility of stress cracking. When two dissimilar plastics are to be joined, adhesive bonding is generally desirable because of solvent and polymer compatibility problems.

Solvent cements should be chosen with approximately the same solubility parameter as the plastic to be bonded. Table 48 lists typical solvents used to bond major plastics. It is common to use a mixture of fast-drying solvent with a less volatile solvent to prevent crazing. The solvent cement can be bodied up to 25 percent by weight with the parent plastic to fill gaps and reduce shrinkage and internal stress during cure.

The parts to be bonded should be unstressed and annealed if necessary. The solvent cement is generally applied to the substrate with a syringe or brush. In some cases the surface may be immersed in the solvent. After the area to be bonded softens, the parts are mated and held under pressure until dry. Pressure should be low and uniform so that the joint will not be stressed. After the joint hardens, the

TABLE 48   Typical Solvents for Solvent Cementing[53]

Plastic	Solvent
ABS	Methyl ethyl ketone, methyl isobutyl ketone, tetrahydrofuran, methylene chloride
Acetate	Methylene chloride, acetone, chloroform, methyl ethyl ketone, ethyl acetate
Acrylic	Methylene chloride, ethylene dichloride
Cellulosics	Methyl ethyl ketone, acetone
Nylon	Aqueous phenol, solutions of resorcinal in alcohol, solutions of calcium chloride in alcohol
PPO	Trichloroethylene, ethylene dichloride, chloroform, methylene chloride
PVC	Cyclohexane, tetrahydrofuran, dichlorobenzene
Polycarbonate	Methylene chloride, ethylene dichloride
Polystyrene	Methylene chloride, ethylene ketone, ethylene dichloride, trichloroethylene, toluene, xylene
Polysulfone	Methylene chloride

These are solvents recommended by the various resin suppliers. A key to the selection of solvents is how fast they evaporate: a fast-evaporating product may not last long enough for some assemblies; too slow evaporation could hold up production.

pressure is released, and an elevated-temperature cure may be necessary depending on the plastic and desired joint strength. The bonded part should not be packaged or stressed until the solvent has adequate time to escape from the joint.

Adhesive bonding of plastics   Adhesive bonding is the easiest and usually the best method of fastening one type of plastic to another type or to a different substrate. Thermoplastic and rubber-based contact adhesives are often used to bond plastics when the application is nonstructural. These adhesives are generally less sensitive to heat and solvents than the parent plastic. Thermosetting adhesives such as epoxies or polyurethanes are commonly used for structural applications. When bonding plastics to almost any other type of material, the mismatch in thermal-expansion coefficients can impose great stress on the bond during thermal cycling. Thus, adhesives for plastics are generally resilient materials to allow distribution of stress at temperature extremes.

To have acceptable joint strength, many plastic materials require surface treatments prior to bonding. Table 19 offers a guideline for which treatments to use. Recommended adhesive selections for plastics are listed in Table 40.

Recommended solvent and adhesive-bonding processes for common plastics
Some plastic materials act differently from others when solvent-welded or bonded with adhesives. The most common types of plastics and their joining procedures are described below.

ABS.   Solvents listed in Table 48 can be bodied up to 25 percent with resin. It is several days at room temperature before the effect of the solvent passes, although holding strength occurs in a few hours. Because of the rapid softening action of the solvent, pressure and the amount of solvent applied should be minimal. The best adhesives for ABS are epoxies, polyurethane, and thermosetting acrylics. Acrylic adhesives have shown strength greater than that of ABS; however, they form a very rigid bond.

Acetal.   A room-temperature-solvent adhesive for bonding acetal copolymer (Celcon*) to itself, nylon, and ABS is hexafluoroacetone sesquihydrate (Allied Chemical Corp.). Lower-strength bonds can also be made to polyester, cellulosics, and rubber. This solvent will not bond acetal to metal. Acetal homopolymer (Delrin†) cannot be solvent-cemented because of the plastic's good solvent resistance.

Acetal homopolymer can be effectively bonded with various adhesives after the surface is either sanded or "satinized." Performance of adhesives on acetal is described in Table 49.

Acrylics.   Acrylics must be annealed before solvent welding to avoid stress cracking and achieve maximum joint strength. Typical annealing temperatures are about 10°F below the distortion temperature. The following conditions have been found suitable in some applications:

Annealing Temperature

Acrylic (easy flow)...   140°F   2 h*
Acrylic (hard flow)...   170°F   2 h*

* Approximate time for thin sections; longer times required for heavy sections.

A polymerizable solvent cement for acrylics can be made from a mixture of 60 parts methylene chloride, 40 parts methyl methacrylate monomer, 0.2 part benzoyl peroxide, and sufficient resin for body. This is a very good solvent cement with gap-filling properties; but being a two-part adhesive with a definite working life, it is a difficult adhesive to use.

Adhesive joints usually give lower strength than solvent- or heat-welded joints. Cyanoacrylate, epoxy, and thermosetting acrylic adhesives offer good adhesion but poor resistance to thermal aging.

Cellulosics.   Either of the solvents listed in Table 48 is appropriate for welding cellulose-based plastics. Ethyl cellulose is usually bonded with ethyl acetate. Polyurethane, epoxy, and cyanoacrylate adhesives are commonly used to bond cellulosics. Cellulosic materials may contain plasticizers that are not compatible with the adhesive. The extent of plasticizer migration should be determined before an adhesive is selected.

Chlorinated Polyether.   These plastics can be bonded with epoxy, polyurethane, and polysulfide-epoxy adhesives after treatment with a hot chromic acid solution. Tensile-shear strength of 1,270 lb/in.² has been achieved with an epoxy-polysulfide adhesive.

Fluorocarbons.   Epoxies and polyurethane give good bond strengths to treated fluorocarbon surfaces. Table 50 shows the effect of surface treatments on the bondability of epoxy adhesives to polytetrafluoroethylene (Teflon) and polychlorotrifluoroethylene (Kel-F).

Nylon.   The recommended adhesives for bonding nylon to nylon are generally solvents, such as aqueous phenol, solutions of resorcinol in alcohol, and solutions of calcium chloride in alcohol. These solvents are sometimes bodied by adding nylon resin.

* Trademark of Celanese Plastics Co.
† Trademark of E. I. du Pont de Nemours & Co., Inc.

TABLE 49 Performance of Adhesives with Acetal Homopolymer[60]

Adhesive	Base	Color of adhesive	Tensile shear strength, lb/in.²						Adhesive curing conditions	
			Acetal to acetal		Acetal to aluminum		Acetal to steel		Room temp	Oven, °F
			Sanded	"Satinized"	Sanded	"Satinized"	Sanded	"Satinized"		
F&F 46950- and 2% RC 805[a]	Polyester	Amber	300	450	200	....	150	600	......	1 h at 275
F&F 4684[a]	Rubber	Amber	300	250	300	100	300	300	......	1 h at 275
F&F 46950[a]	Polyester	Amber	300	700	250	300	350	700	......	1 h at 275
Chemlock 220–203[b]	Rubber	Black	....	300	....	150	....	150	......	1 h at 250
Chemlock 220[b]	Rubber	Black	....	300	....	150	....	250	......	1 h at 250
3M–EC711[c]	Rubber	Tan	500	400	450	200	500	350	......	1 h at 250
Phenoweld No. 8[d]	Phenolic	Red	100	600	....	400	250	400	......	30 min at 300
Narmco 3135A[e] 3135B	Modified epoxy	Yellow	150	600	....	450	300	850	24 h or	1 h at 200
Narmco 3134[e] 3134C	Modified epoxy	Gray	150	450	....	400	250	830	24 h or	1 h at 200
Resiweld No. 7004[f]	Epoxy	Yellow	150	850	....	600	200	600	24–48 h	50 min at 200
Biggs R–363[g]	Epoxy	Amber	....	550	....	....	550	....	8 h or	30 min at 150
Biggs R–313[g]	Epoxy	Gray	500	600	....	60	300	50	8 h or	30 min at 150
Cycleweld C–14[h]	Epoxy	Straw yellow	....	400	....	150	....	200	......	1 h at 150
Cycleweld C–6[h]	Epoxy	Straw yellow	....	500	....	....	....	....	......	1 h at 180
CD 200[i]	Vinyl	Amber	300	500	....	550	400	700	......	1 h at 180
EA VI[j]	Epoxy	Red	500	500	....	600	500	600	48 h or	2 h at 165

a E. I. du Pont de Nemours & Co., Inc., Industrial Finishes Dept., Wilmington, Del.
b Hughson Chemical Company, Erie, Pa.
c 3M Corporation, 900 Bush Avenue, St. Paul, Minn.
d H. V. Hardman Co., 571 Cortland St., Belleville, N.J.
e Narmco Resins & Coatings Company, 600 Victoria St. Costa Mesa, Calif.
f H. B. Fuller Co., St. Paul, Minn.
g Carl H. Biggs, Inc., 1547 15 St., Santa Monica, Calif.
h Cycleweld Cement Product Co. 5437 W. Jefferson St., Trenton, Mich.
i Chem. Develop. Corp., Danvers, Mass.
j Hysol Corp., Olean, N.Y.

TABLE 50 Effect of Surface Treatment on Bondability of Epoxy Adhesives to Teflon and Kel-F[61]

Adhesive	Ratio	Cure	Fluorinated material	Tensile shear strength, lb/in.^{2}			
				No treatment	No treatment, abraded	Treated*	Abraded and treated
Armstrong A2/Act E........	100:6	4 h/165°F	Teflon	Fell apart	. . . . .	1,570	1,920
Epon 828/Versamid 125........	60:40	16 h RT, 4 h/165°F	Kel-F 270	380	1,120	2,820	3,010
Epon 828/Versamid 125........	60:40	16 h RT, 4 h/165°F	Kel-F 300	550	1,250	2,030	2,910
Epon 828/Versamid 125........	60:40	16 h RT, 4 h/165°F	Kel-F 500	780	1,580	1,780	2,840
Epon 828/Versamid 125........	60:40	16 h RT, 4 h/165°F	Teflon*	. . . . .	. . . . .	1,150	

* Treated with sodium naphthalene etch solution.

Aqueous phenol containing 12 percent water is the most generally used cement. After the solvent is applied, the parts are clamped together and after air drying they are placed in boiling water for 5 min to remove excess phenol. In cases where immersion is impractical, the joints may be cured by prolonged air exposure at ambient condition or at 150°F. Handling strength is achieved in 48 h and maximum strength in 3 to 4 days. Bond strength as high as 2,400 lb/in.^{2} in shear can be obtained with aqueous phenol.

A solution of resorcinol in ethanol provides equivalent bond strength, and the cement is less corrosive and hazardous than aqueous phenol. Equal parts by weight resorcinol and 95 percent ethanol make up the cement. The solution is brushed on both surfaces and allowed to dry for 30 s. The parts are then assembled and held firmly. The joint will cure at room temperature. Handling strength is achieved after 90 min, and maximum strength occurs after 4 to 5 days. If desired, the cure can be accelerated by a 30-min, 150°F oven exposure.

Calcium chloride (22.5 parts) and ethanol (67.5 parts) bodied with nylon produce bonds as strong as those obtained with phenol or resorcinol except when used with high-molecular-weight nylons. This adhesive also forms a less brittle bond. The

TABLE 51   Tensile-Shear Strength of Nylon-to-Metal Joints Bonded with Various Adhesives[62]

Adhesive	Tensile-shear strength, lb/in.^{2}			
	Aluminum	Brass	Stainless steel	Iron
Phenoweld No. 7*....................	900	1,200	1,500	1,600
NT 422† (primer) + Penacolite G-1124‡.........................	1,000	1,100	1,300	1,200
Cycleweld 6§.........................	500	500	500	500
Cycleweld 14§.........................	800	800	900	800

* H. V. Hardman Co., Belleville, N.J.
† Har-Lee Fishing Rod Co., Jersey City, N.J.
‡ Koppers Co., Pittsburgh, Pa.
§ Cycleweld Chemical Products Div., Trenton, Mich.

cement is painted on both surfaces, and after 60 s the parts are assembled and clamped. Bond strength develops at a rate similar to the resorcinol-ethanol cement.

Some epoxy, resorcinol formaldehyde, phenol-resorcinol, and rubber-based adhesives have been found to produce satisfactory joints between nylon and metal, wood, glass, and leather. Table 51 shows bond strength of various adhesives in nylon-to-metal joints.

Phenoxy. Phenoxy can be solvent-welded with a 40 to 50 percent solution of phenoxy resin in methyl ethyl ketone.

Polycarbonates. Methylene chloride is a very fast solvent cement for polycarbonate. This solvent is recommended only for temperate-climate zones and on small areas. A mixture of 60 percent methylene chloride and 40 percent ethylene chloride is slower-drying and the most common solvent cement used. Ethylene chloride is recommended alone in very hot climates. These solvents may be bodied with 1 to 5 percent polycarbonate resin where gap-filling properties are important. A pressure of 200 lb/in.^{2} is recommended.

Table 52 shows bond strength of solvent-welded polycarbonate joints compared with adhesive-bonded joints. Epoxy and polyurethane adhesives offer good bond strengths to polycarbonates.

Polyolefins. To obtain a usable adhesive bond with polyolefins, the surface must be treated in one of the ways listed in Table 19. Polyurethanes, epoxies, and some rubber-based adhesives provide acceptable strength to treated polyolefin surfaces.

TABLE 52　Bonding Data for Polycarbonate Plastics[63]

Method or materials	For bonds to — Polycarbonate	Metals	Rubber, wood, plastics	Hardener	Proportions, pph	Cure temp, °F	Time, h	Adhesion quality	Tensile-shear strength, lb/in.²	Service range, °F	Color of adhesive
Solvent cementing:											
Methylene chloride	X			...	...	Room	48	Excellent	4,500–6,500	−40 + 270	Clear
Ethylene dichloride	X			...	...	Room	48	Excellent	4,500–6,500	−40 + 270	Clear
One-part systems:											
Room temp. curing:											
Eastman 910	X	X	X	...	...	Room	2–24	Excellent	...	70–212	Clear
Central Coil HB-56		X	X	...	...	Room	4–24	Good	...	70–175	Red
Vulcalock R-266-T			X	...	...	Room	Air dry	Good	...		Amber
G-E RTV-102	X		X	...	...	Room	2–24	Excellent	350	−75–350	White
Hot-melt adhesive											
Terrell Corp. 10 × 104	X		X			Melts at 350	Instantaneous	Excellent	...	32–175	Dark
Two-part systems:											
Room temp curing:											
Uralane 8089	X	X	X	Part B	100:40	Room	24–48	Very good	150	−40 + 250	Translucent amber
Epon 828	X	X	X	Diethylene triamine	100:8	Room	2–48	Excellent	2,500		Clear
Epon 828	X	X		Versamid 115	100:100	Room	24–48	Fair*	600	32–160	Dark amber
Epon 828—Thiokol LP-3	X	X	X	DMP-30	100:100:10	Room	48	Fair*	1,800	70–160	Dark amber
EA IV	X	X	X	Cur. Ag. A	100:6	Room	24–6 days	Very good	2,900	−60 + 190	Red—tan
EA 913	X	X	X	Part B	100:12	Room	3–4 days	Very good	2,000	−60 + 250	Gray—black
Bondmaster M688	X	X	X	CH-8	100:16	Room	4–24	Very good		−40 + 212	Tan
Elevated temp curing:											
Epon 828		X	X	Methyl nadic anhydride DMP-30	100:84:2	175	8 (min)	Very good	1,200	32–250	Light amber
Epon 828			X	Phthalic anhydride	100:40	250	8 (min)	Very good	1,200	−40 + 250	Light amber
EA IV			X	Cur. Ag. A	100:6	200	45 min	Very good	1,800	−60 + 200	Red—tan
EA 913		X	X	Part B	100:12	180	3	Very good	2,900	−60 + 250	Gray—black
Bondmaster M688	X	X	X	CH-8	100:16	200	15 min	Very good	600	−40 + 250	Tan
Flexicast 4002		X	X	Part B	100:25	212	3	Very good		−90 + 250	Gray—black
Pressure and Heat-sensitive:											
Lepage 4050		X	X	...	...	Room		Excellent	25 lb/in.² at point of bond	Room	Clear
Nafco No. 101	X	X	X	...	...	Room				−40–400	Clear
Kleen—Stik 295	X	X	X			120–140†	<1 min	Good	...	−40 + 200	Amber
Kleen—Stik 298	X	X	X			120–140†	<1 min	Good	...	−40 + 200	Amber
G-P Transmount	X	X	X			150–250	10–15 min	Excellent		−60 + 250	Amber

Adhesive Sources

One-part system	Two-part systems	Pressure and heat-sensitive
Eastman Chemical Products, Kingsport, Tenn Central Coil Co, Indianapolis, Ind. B. F. Goodrich, Akron, Ohio Terrell Corp., Waltham, Mass. G.E. Co., Silicone Products Dept., Waterford, N. Y.	Furane Plastics Inc., Los Angeles, Calif. Hysol Corp, Olean, N. Y. National Starch Corp, Bloomfield, N. J. Industrial Tires, Ltd., Toronto, Canada	Lepage's Permacel, Inc., New Brunswick, N. J. Wesrep Corp., Los Angeles, Calif. Kleen-Stik Products, Inc., Newark, N. J. Girder Process, Rutherford, N. J.

* Both combinations make flexible joints, with the Epon-Thiokol system the more flexible.　Both fail in hot water.

† Also cures at room temperature.

Table 53 presents the tensile strength of epoxy-bonded polyethylene and polypropylene joints after various surface treatments.

Polyphenylene Oxides. Polyphenylene oxide joints must mate almost perfectly; otherwise solvent welding provides a weak bond. Very little solvent cement is needed. Best results have been obtained by applying the solvent cement to only one substrate. Optimum holding time has been found to be 4 min at approximately 400 lb/in.^{2}. Table 54 lists properties of solvent-welded PPO bonds. A mixture of 95 percent chloroform and 5 percent carbon tetrachloride is the best solvent system for general-purpose bonding, but very good ventilation is necessary. Ethylene dichloride offers a slower rate of evaporation for large structures or hot climates.

Various adhesives can be used to bond polyphenylene oxide to itself or other substrates; however, etching is required for optimum bond strength. Table 55 lists tensile-shear strength of PPO bonded with a variety of adhesives.

Polystyrenes. Many solvents will bond polystyrene to itself. The selection is usually based on the time required for the joint to harden. The faster-drying solvents, as shown in Table 56, cause crazing. Slower-drying solvents are often mixed with these for optimum properties. A 50:50 mixture of ethyl acetate and toluene bodied with polystyrene is an excellent general-purpose adhesive. Bond strengths up to 100 percent of the parent material are common.

Polyurethane or cyanoacrylate adhesives offer bonds to polystyrene that are often stronger than the substrate itself. Unsaturated polyester and epoxy adhesives give lower bond strengths but are acceptable in many applications.

Polysulfone. A 5 percent solution of polysulfone resin in methylene chloride is recommended as a solvent cement. A minimum amount of adhesive should be used. The assembled pieces should be held for 5 min under 500 lb/in.^{2}. The strength of the joint will improve over a period of several weeks as the residual solvent evaporates. Strength of solvent-welded joints has surpassed the strength of the polysulfone part.

A number of adhesives have been found suitable for joining polysulfone to itself or other materials. An epoxy adhesive, EC-2216, and a pressure-sensitive adhesive, EC-880, both from 3 M Co., offer good bonds at temperatures to 180°F. EC-880 is not considered a structural adhesive. For higher-temperature applications, BR 89 and BR 92 (American Cyanamid) are recommended. These adhesives require heat cure but are useful to 300 to 350°F.

Polyurethane. Urethane, epoxies, and nitrile adhesives give generally good bonds to urethane polymers. A resorcinol-formaldehyde adhesive offers excellent adhesion but also brittleness and higher rigidity, which can cause the polyurethane to fail at relatively low loads.

Polyvinyl Chloride. PVC and its acetate copolymers can be bonded using solvent solutions containing ketones. Methyl ethyl ketone is commonly used as a cement for low- and medium-molecular-weight polyvinyl chloride copolymers. High-molecular-weight copolymers require cyclohexane or tetrahydrofuran. The following formulation has worked satisfactorily to bond unsupported flexible PVC to itself:

Parts by weight	*Material*
100	Polyvinyl chloride resin (medium molecular weight)
100	Tetrahydrofuran
200	Methyl ethyl ketone
1.5	Tin organic stabilizer
20	Dioctylphthalate (a plasticizer)
25	Methyl isobutyl ketone

This cement also works well on rigid PVC.

Plasticizer migration from vinyls to the adhesive bond line can cause difficulty. Adhesives must be tested for their ability to resist the plasticizer. Nitrile-rubber adhesives are very resistant to plasticizers. Also used are polyurethane and neoprenes. PVC can be made with a wide variety of plasticizers. An adhesive suitable on a certain PVC formulation may not be compatible on a PVC from another supplier.

TABLE 53 Tensile Strength of Polyethylene and Polypropylene Bonds after Surface Treatment[61]

Adhesive	Ratio	Cure	Adherend materials	Tensile shear strength, lb/in.^{2}			
				No treatment	No treatment, abraded	Unabraded, treated	Abraded, treated
Epon 828/Versamid 125......	50:50	2 h RT +6 h 165°F	Aluminum to polypropylene	Fell apart (3)	393 (3)	760 (3)	695 (3)
Epoxy 1001/Versamid 125....	50:50	16 h 165°F	Aluminum to polyethylene	.....	366 (4)	1,000 (4)	
Epon 828/Versamid 125......	60:40	RT 7 days	Aluminum to polyethylene	83 (5)	.....	580 (5)	

The number of specimens tested is indicated in parentheses. Polyolefin treatment was a sulfuric acid–sodium dichromate etch.

TABLE 54 Tensile-Shear Strength of Solvent-cemented PPO Bonds[64]

Solvent	Cure temp	Cure time, h	Tensile-shear strength, lb/in.^{2}
95% chloroform + 5% carbon tetrachloride	100°C	2	4,000
	RT	100	4,100
Chloroform..........................	100°C	1	2,350
	RT	48	1,150
	RT	200	2,650
Ethylene dichloride....................	100°C	1	2,000
	RT	48	1,000
	RT	200	2,700
Ethylene dichloride + 2% PPO.........	100°C	2	2,150
	RT	200	2,700
Methylene chloride + 2% PPO.........	100°C	3	2,100
	RT	48	1,050
	RT	200	2,650

TABLE 55 Tensile-Shear Strength of PPO Bonded with Various Adhesives[64]

Adhesive type and manufacturer	Breaking strength, lb/in.^{2}	
	Sanded surface	Etched surface*
Epoxy:		
J-1151 (E-9) (Armstrong Cork)......................	880 (1,180)	1,315 (2,447)
Armstrong A-701 (Armstrong Prod.).................	1,355 (3,628)	1,398 (3,013)
Chemlock 305 (Hughson Chem.).....................	704 (940)	2,145 (2,817)
Ray-Bond R-86009 (Raybestos-Manhatten)..........	600 (1,052)	1,545 (3,060)
EA 907 (Hysol Corp.).............................	1,322 (722)	2,167 (2,186)
Ecco Bond 285 (HV) (Emerson-Cuming).............	....	1,450
Ecco Bond 285 (HVA) (Emerson-Cuming)............	....	1,340
Polyester:		
46930 (6% RC-805) (E. I. du Pont de Nemours).......	1,340 (1,180)	1,594 (1,557)
46950 (9% RC-805)...............................	1,086 (1,182)	903 (1,127)
46971 (6% RC-805)...............................	924 (1,460)	1,225 (1,190)
46960 (8% RC-805)...............................	738 (387)	951 (432)
Silicone:		
RTV 102 (General Electric)........................	49 (72)	82 (148)
RTV 108 (General Electric)........................	46 (73)	106 (132)
RTV 103 (General Electric)........................	53 (84)	59 (97)
Elastomer and resin-modified elastomer:		
Pliobond 20 (Goodyear Tire & Rubber)..............	545 (1,108)	470 (1,285)
Pliobond HT 30 (Goodyear Tire & Rubber)...........	280 (375)	255 (440)
D-220 (Armstrong Cork)..........................	225 (645)	208 (500)
Ray-Bond R-81005 (Raybestos-Manhatten)..........	475 (700)	935 (1,703)
Ray-Bond R-82015 (Raybestos-Manhatten)..........	184 (140)	140 (145)
Ray-Bond R-82016 (Raybestos-Manhatten)..........	343 (507)	330 (488)
Cyanoacrylate:		
Eastman 910 (Eastman Chem. Products).............	1,160	
Polyurethane:		
Uralane 8089 (Furane Plastics).....................	....	748 (1,036)

Values in parentheses have been obtained on samples exposed for 1 week to 100°C.
* Chromic acid etch, conditions not specified.

Thermosetting Plastics. Thermosetting plastics cannot be heat- or solvent-welded. They are easily bonded with many adhesives, some of which have been listed in Table 40. Abrasion is generally recommended as a surface treatment.

Reinforced Plastics. Adhesives giving satisfactory results on the resin matrix alone may also be used to bond reinforced plastics. Some reinforced thermoplastics may be solvent-cemented, but adhesive bonding is usually recommended.

Plastic Foams. Some solvent cements and solvent-containing pressure-sensitive adhesives will collapse thermoplastic foam. Water-based adhesives based on SBR or polyvinyl acetate and 100 percent solid adhesives are often used. Butyl, nitrile, and polyurethane adhesives are often used for flexible polyurethane foam. Epoxy adhesives offer excellent properties on rigid polyurethane foam.

TABLE 56   Solvent Cements for Polystyrene[65]

Solvent	Boiling point, °C	Tensile strength of joint, lb/in.^{2}
Fast-drying:		
Methylene chloride............................	40	1,800
Carbon tetrachloride...........................	77	1,350
Ethyl acetate.................................	77	1,500
Benzene......................................	80	
Methyl ethyl ketone..........................	80	1,600
Ethylene dichloride...........................	84	1,800
Trichloroethylene.............................	87	1,800
Medium-drying:		
Toluene......................................	111	1,700
Perchloroethylene............................	121	1,700
Ethyl benzene................................	136	1,650
Xylenes.....................................	138–144	1,450
Diethyl benzene..............................	185	1,400
Slow-drying:		
Mono-amyl benzene..........................	202	1,300
Ethyl naphthalene............................	257	1,300

Adhesives for Bonding Elastomers

Vulcanized elastomers   Bonding of vulcanized elastomers to themselves and other materials is generally completed by using a pressure-sensitive adhesive derived from an elastomer similar to the one being bonded. Flexible thermosetting adhesives such as epoxy-polyamide or polyurethane also offer excellent adhesive strength to most elastomers. Surface treatment consists of washing with a solvent, abrading, or acid cyclizing as described in Table 21.

Elastomers vary greatly in formulation from one manufacturer to another. Fillers, plasticizers, antioxidants, etc., may affect the adhesive bond. Adhesives should be thoroughly tested for a specific elastomer and then reevaluated if the elastomer manufacturer or formulation is changed.

Unvulcanized elastomers   Unvulcanized elastomers may be bonded to metals and other rigid adherends by priming the adherend with a suitable air- or heat-drying adhesive before the elastomer is molded against the adherend. The most common elastomers to be bonded in this way include nitrile, neoprene, urethane, natural rubber, SBR, and butyl rubber. Less common unvulcanized elastomers such as the silicones, fluorocarbons, chlorosulfonated polyethylene, and polyacrylate are more difficult to bond. However, recently developed adhesive primers improve the bond of these elastomers to metal. Surface treatment of the adherend before priming should be according to good standards. Figure 44 illustrates a bondability index of common and specialty elastomers.

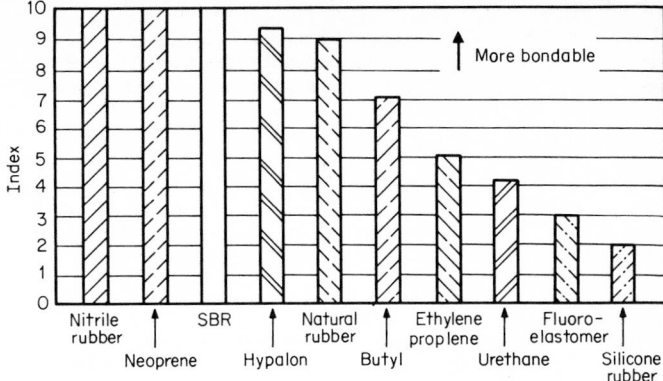

Fig. 44   Bondability index of common and specialty elastomers.[66]

Adhesive Bonding of Wood

Resorcinol-formaldehyde resins are cold-setting adhesives for wood structures. Urea-formaldehyde adhesives, commonly modified with melamine formaldehyde, are used in the production of plywood and in wood veneering for interior applications. Phenol-formaldehyde and resorcinol-formaldehyde adhesive systems have the best heat and weather resistance.

Polyvinyl acetates are quick-drying, water-based adhesives commonly used for the assembly of furniture. This adhesive produces bonds stronger than the wood itself but is not resistant to moisture or high temperature. Table 57 describes common adhesives used for bonding wood.

TABLE 57   Properties of Common Wood Adhesives[67]

Resin type used	Resin solids in glue mix, %	Principal use	Method of application	Principal property	Principal limitation
Urea formalde-hyde	23–30	Wood to wood interior	Spreader rolls	Bleed-through-free; good adhesion	Poor durability
Phenol formalde-hyde	23–27	Plywood ex-terior	Spreader rolls	Durability	Comparatively long cure times
Melamine for-maldehyde	68–72	Wood to wood, splicing, patching, scarfing	Sprayed, combed	Adhesion, color, dura-bility	Relative cost; poor wash-ability; needs heat to cure
Melamine urea 2/1	55–60	End and edge gluing exte-rior	Applicator	Colorless, durability and speed	Cost
Resorcinol for-maldehyde	50–56	Exterior wood to wood (lam-inating)	Spreader rolls	Cold sets durability	Cost, odor
Phenol-resorcinol 10/90	50–56	Wood to wood exterior (lam-nating)	Spreader rolls	Warm-set durability	Cost, odor
Polyvinyl-acetate emulsion	45–55	Wood to wood interior	Brushed, sprayed, spreader rolls	Handy	Lack of H_2O and heat resistance

Adhesives for Glass

Glass adhesives are generally transparent heat-setting resins that are water-resistant to meet the requirements of outdoor applications. Commercial adhesives generally used to bond glass and their physical characteristics are presented in Table 58.

EFFECT OF ENVIRONMENT

For an adhesive bond to be useful, it not only must withstand the mechanical forces acting on it, but must also resist the service environment. Adhesive strength is influenced by many common environments including temperature, moisture, chemical fluids, and outdoor weathering. Table 59 summarizes the relative resistance of various adhesive types to common environments.

TABLE 58   Commercial Adhesives Most Desirable for Glass[68]

Trade name	Chemical type	Bond characteristics		
		Strength, lb/in.^{2}	Type of failure	Weathering quality
Butacite, Butvar........	Polyvinyl butyral	2,000–4,000	Adhesive	Fair
Bostik 7026, FM-45, FM-46...............	Phenolic butyral	2,000–5,500	Glass	Excellent
EC826, EC776.........	....	....	Adhesion and glass	
N-199, Scotchweld.......	Phenolic nitrile	1,000–1,200	....	Excellent
Pliobond M-20, EC847...	Vinyl nitrile	1,200–3,000	Adhesion and glass	Fair to good
EC711, EC882.........	....	....	....	....
EC870...............	Neoprene	800–1,200	Adhesion and cohesive	Fair
EC801, EC612.........	Polysulfide	200– 400	Cohesive	Excellent
EC526, R660T, EC669...	Rubber base	200– 800	Adhesive	Fair to poor
Silastic...............	Silicone	200– 300	Cohesive	Excellent
Rez-N-Glue, du Pont 5459...............	Cellulose vinyl	1,000–1,200	Adhesive	Fair
Vinylite AYAF, 28-18....	Vinyl acetate	1,500–2,000	Adhesive	Poor
Araldite, Epon L-1372, ERL-2774, R-313, C-14, SH-1, J-1152..........	Epoxy	600–2,000	Adhesive	Fair to good

In applications where possible degrading elements exist, candidate adhesives must be tested under simulated service conditions. Table 60 lists common ASTM environmental tests that are often reported in the literature. Time and economics generally allow only short-term tests. As yet, no decisive test has been developed to determine life of an adhesive based on short-term data. It is tempting to try to accelerate service environment in the laboratory by increasing temperature or humidity, for example, and to extrapolate the results to actual conditions. However, often too many variables are affected to achieve a reliable estimate of life.

High Temperature

All polymeric materials are degraded to some extent by exposure to elevated temperatures. Not only are physical properties lowered at high temperatures, but they also degrade during thermal aging. Newly developed polymeric adhesives have been found to withstand 500 to 600°F continuously. To use these materials, the designer must pay a premium in adhesive cost and also be capable of providing long, high-temperature cures.

For an adhesive to withstand elevated-temperature exposure, it must have a high melting or softening point and resistance to oxidation. Materials with a low melting point, such as many of the thermoplastic adhesives, may prove excellent adhesives at room temperature. However, once the service temperature approaches the glass transition point of these adhesives, plastic flow results in deformation of the bond and degradation in cohesive strength. Thermosetting materials, exhibiting no melting point, consist of highly cross-linked networks of macromolecules. Many of these materials are suitable for high-temperature applications. When considering thermosets, the critical factor is the rate of strength reduction due to thermal oxidation and pyrolysis.

Thermal oxidation initiates a progressive chain scission of molecules resulting in losses of weight, strength, elongation, and toughness within the adhesive. Figure 45 illustrates the effect of oxidation by comparing adhesive joints aged in both high-temperature air and inert-gas environments. The rate of strength degradation in air depends on the temperature, the adhesive, the rate of airflow, and even the type of adherend. Some metal-adhesive interfaces are chemically capable of accelerating the rate of oxidation. It has been found that nearly all types of structural adhesives exhibit better thermal stability when bonded to aluminum than when bonded to stainless steel or titanium (Fig. 45).

High-temperature adhesives are usually characterized by a rigid polymeric structure, high softening temperature, and stable chemical groups. The same factors also

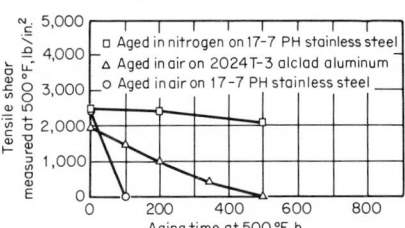

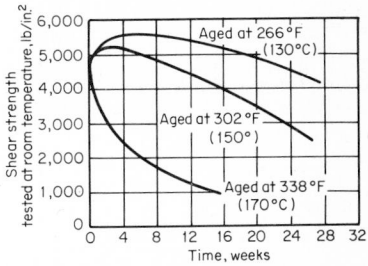

Fig. 45 The effect of 500°F aging in air and nitrogen on an epoxy-phenolic (HT 424) adhesive.[70]

Fig. 46 Effect of temperature aging on typical epoxy adhesive in air. Strength measured at room temperature.[71]

make these adhesives very difficult to process. Only epoxy-phenolic-, polyimide-, and polybenzimidazole-based adhesives can withstand long-term service temperatures greater than 350°F.

Epoxy   Epoxy adhesives are generally limited to applications below 250°F. Figure 46 illustrates the aging characteristics of a typical epoxy adhesive at elevated temperatures. Certain epoxy adhesives have been able to withstand short terms at 500°F and long-term service at 300 to 350°F. These systems were formulated especially for thermal environments by incorporation of stable epoxy coreactants or high-temperature curing agents into the adhesive.

One of the most successful epoxy coreactant systems developed thus far is an epoxy-phenolic alloy. The excellent thermal stability of the phenolic resins is coupled with the adhesion properties of epoxies to provide an adhesive capable of 700°F short-term operation and continuous use at 350°F. The heat-resistance and thermal-aging properties of an epoxy-phenolic adhesive are compared with those of other high-temperature adhesives in Fig. 47.

Anhydride curing agents give unmodified epoxy adhesives greater thermal stability than most other epoxy curing agents. Phthalic anhydride, pyromellitic dianhydride, and chlorendic anhydride allow greater cross linking and result in short term-heat resistance to 450°F. Long-term thermal endurance, however, is limited to 300°F.

TABLE 59 Relative Resistance of Synthetic Adhesives to Common Service Environments[69]

Adhesive type	Shear	Peel	Heat	Cold	Water	Hot water	Acid	Alkali	Oil, grease	Fuels	Alcohols	Ketones	Esters	Aromatics	Chlorinated solvents
Thermosetting Adhesives															
1. Cyanoacrylate	2	6	5	..	6	6	6	6	3	3	5	5	5	4	4
2. Polyester + isocyanate	2	2	3	2	1	3	3	2	2	2	3	2	2	6	2
3. Polyester + monomer	2	6	5	3	3	6	3	6	2	2	3	6	6	6	6
4. Urea formaldehyde	2	6	3	3	2	6	2	2	2	2	2	2	2	2	2
5. Melamine formaldehyde	2	6	2	2	2	5	2	1	2	2	2	2	2	2	2
6. Urea–melamine formaldehyde	2	6	2	2	2	2	1	1	2	2	2	2	2	2	2
7. Resorcinol formaldehyde	2	6	2	2	2	2	2	2	2	2	2	2	2	2	2
8. Phenol-resorcinol formaldehyde	2	6	3	2	2	2	2	2	2	2	2	2	2	2	2
9. Epoxy (+ polyamine)	2	5	3	5	2	2	2	2	2	3	1	6	6	1	
10. Epoxy (+ polyanhydride)	2	5	1	4	3	3	2	2	..	2	2	6	6	2	
11. Epoxy (+ polyamide)	2	2	6	2	2	6	3	6	2	2	1	6	6	3	
12. Polyimide	2	4	1	1	2	4	2	2	2	2	2	2	2	2	2
13. Polybenzimidazole	2	4	1	1	2	4	2	2	2	2	2	2	2	2	2
14. Acrylic	2	6	5	3	1	3	2	2	2	2	2	2	2	2	2
15. Acrylate acid diester	2	5	3	3	4	4	6	6	3	3	5	5	5	4	4
Thermoplastic Adhesives															
16. Cellulose acetate	2	6	2	3	1	6	1	2	..	2	4	6	6	6	6
17. Cellulose acetate butyrate	2	3	3	3	2	..	3	2	..	..	6	6	6	6	6
18. Cellulose nitrate	2	6	3	3	3	3	3	6	2	2	6	6	6	6	6
19. Polyvinyl acetate	2	6	6	..	3	6	3	3	2	2	6	6	6	6	6
20. Vinyl vinylidene	2	3	3	3	3	3	..	..	2	2	2	2	2		

21. Polyvinyl acetal	2	6	5	2	2	...	6	3	2	2	3	3	6	3	2
22. Polyvinyl alcohol	...	2	3	...	6	6	5	5	2	1	3	1	1	1	1
23. Polyamide	2	3	5	...	5	6	6	2	2	2	6	2	2	2	6
24. Acrylic	2	2	4	3	3	3	...	...	2	...	...	4	4	...	4
25. Phenoxy	2	3	4	3	3	4	3	2	3	5	5	...	...	6	
Elastomer Adhesives															
26. Natural rubber	2	3	3	...	3	...	3	3	6	6	2	4	4	6	6
27. Reclaimed rubber	2	3	3	...	2	...	3	3	6	6	2	4	4	6	6
28. Butyl	3	6	6	3	2	6	1	2	6	6	2	2	2	6	6
29. Polyisobutylene	6	6	6	3	2	6	2	2	6	6	2	2	2	6	6
30. Nitrile	2	3	3	3	2	5	5	6	2	2	3	6	6	3	6
31. Styrene butadiene	3	6	3	3	1	...	3	2	...	5	2	6	6	6	6
32. Polyurethane	2	3	3	2	2	3	3	3	2	2	2	5	5	...	5
33. Polysulfide	3	2	6	2	1	6	2	2	2	2	2	6	6	2	6
34. Silicone (RTV)	3	5	1	1	2	2	3	3	2	3	3	3	3	3	3
35. Silicone resin	2	2	1	2	2	2	...	2	2	2	2	4	4	3	6
36. Neoprene	2	3	3	3	2	...	2	2	2	2	3	6	6	6	6
Alloy Adhesives															
37. Epoxy-phenolic	1	6	1	3	2	2	2	2	3	3	2	6	6	2	
38. Epoxy-polysulfide	2	2	6	2	1	6	2	2	2	2	2	6	6	2	6
39. Epoxy-nylon	1	1	6	2	2	6	...	...	...	2	3	6	6	6	6
40. Phenolic-nitrile	2	2	2	3	2	2	2	2	2	2	2	6	6	6	6
41. Phenolic-neoprene	2	3	3	2	2	...	3	2	2	2	3	6	6	6	6
42. Phenolic–polyvinyl butyral	2	3	3	3	2	3	4	2	2	2	4	6	6	6	6
43. Phenolic–polyvinyl formal	2	3	6	6	2	6	4	2	2	2	4	6	6	6	6

Key: 1. Excellent 2. Good 3. Fair 4. Poor 5. Very poor 6. Extremely poor

Table 61 shows the high-temperature properties of a typical epoxy formulation cured with pyromellitic dianhydride.

Advantages of epoxy-based systems include relatively low cure temperatures, no volatiles formed during cure, low cost, and a variety of formulating and application possibilities. The higher-temperature adhesives lose these advantages in favor of improved thermal-aging characteristics.

Modified phenolics Of the common modified phenolic adhesives, the nitrile-phenolic blend has the best resistance to elevated temperatures. As shown in Table 62, nitrile phenolic adhesives have high shear strength up to 250 to 350°F, and the strength retention on aging at these temperatures is very good. The nitrile phenolic adhesives are extremely tough and provide high peel strength.

Silicone Silicone adhesives have very good thermal stability but low strength. Their chief application is in nonstructural uses such as high-temperature pressure-sensitive tape. Attempts have been made to incorporate silicones with other resins such as epoxies and phenolics, but long cure times and low strength have limited their use.

TABLE 60   ASTM Test Methods for Determining Environmental Resistance of Adhesives

Title	Method
Atmospheric Exposure of Adhesive Bonded Joints and Structures, Recommended Practice for	ASTM D 1828
Exposure of Adhesive Specimens to High Energy Radiation, Recommended Practice	ASTM D 1879
Integrity of Glue Joint in Structural Laminated Wood Products for Exterior Use, Test for	ASTM D 1101
Resistance of Adhesives to Cyclic Laboratory Aging Conditions, Tests for	ASTM D 1183
Effect of Bacteria Contamination on Permanence of Adhesive Preparations and Adhesive Bonds	ASTM D 1174
Effect of Moisture and Temperature on Adhesive Bonds, Tests for	ASTM D 1151
Effect of Mold Contamination on Permanence of Adhesive Preparations and Adhesive Bonds, Test for	ASTM D 1286
Resistance of Adhesive Bonds to Chemical Reagents, Test for	ASTM D 896
Strength Properties of Adhesives in Shear by Tension Loading in the Temperature Range of −450 to −57°F, Test for	ASTM D 2557
Strength Properties of Adhesives in Shear by Tension Loading at Elevated Temperatures, Test for	ASTM D 2295

Polyaromatics The polyaromatic resins, polyimide and polybenzimidazole, offer greater thermal resistance than any other commercially available adhesive. The rigidity of their molecular chains decreases the possibility of chain scission caused by thermally agitated chemical bonds. The aromaticity of these structures provides high bond-dissociation energy and acts as an "energy sink" to the thermal environment.

Polyimide. The strength retention of polyimide adhesives for short exposures to 1000°F is slightly better than that of an epoxy-phenolic alloy. However, the thermal endurance of polyimides at temperatures greater than 500°F is unmatched by other commercially available adhesives.

Polyimide adhesives are usually supplied as a glass-fabric-reinforced film having a limited shelf life. A cure of 90 min at 500 to 600°F and 15 to 200 lb/in.^{2} pressure is necessary for optimum properties. High-boiling volatiles are released during cure, which causes a somewhat porous adhesive layer. Because of the inherent rigidity of this material, peel strength is low.

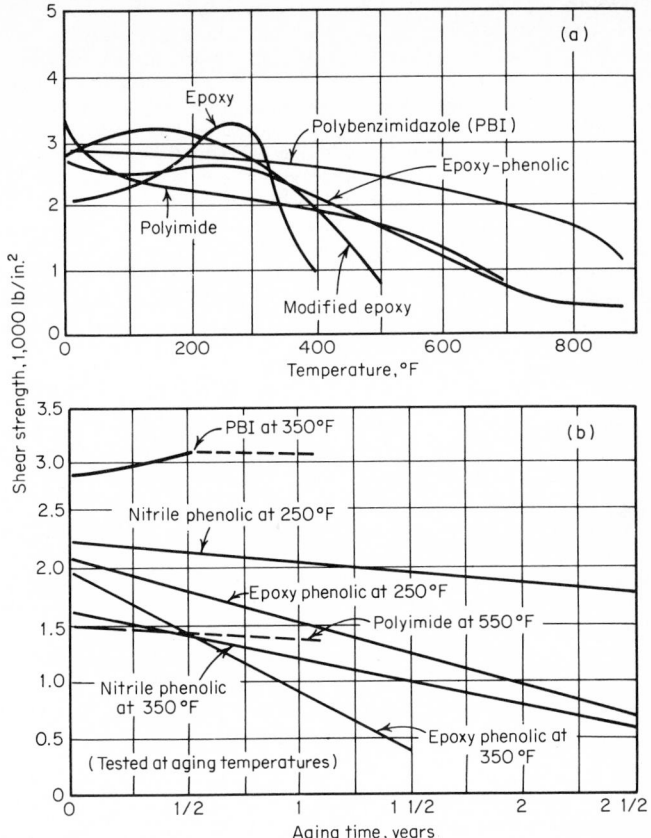

Fig. 47 Comparison of (*a*) heat resistance and (*b*) thermal aging of high-temperature structural adhesives.[72]

Polybenzimidazole. As illustrated in Fig. 47, polybenzimidazole (PBI) adhesives offer the best short-term performance at elevated temperatures. However, PBI resins oxidize rapidly and are not recommended for continuous use at temperatures over 450°F.

PBI adhesives require a cure at 600°F. Release of volatiles during cure contributes to a porous adhesive bond. Supplied as a very stiff, glass-fabric-reinforced film, this adhesive is very expensive; and applications are limited by a long, high-temperature curing cycle.

TABLE 61 Tensile-Shear Strength of Anhydride-cured Epoxy Adhesives at Elevated Temperatures[73]

Substrate	Treatment	Tensile-shear strength, lb/in.², at		
		75°F	300°F	450°F
Aluminum..............................	Etched	2,300	2,600	900
Aluminum..............................	Untreated	1,800	1,400	500
Steel, cold-rolled........................	Etched	1,700	2,000	1,100
Steel, cold-rolled........................	Untreated	1,200	1,200	900

Low Temperature

The factors which determine the strength of an adhesive at very low temperatures are (1) the difference in coefficient of thermal expansion between adhesive and adherend, (2) the elastic modulus, and (3) the thermal conductivity of the adhesive. The difference in thermal expansion is very important, especially since the elastic modulus of the adhesive generally decreases with falling temperature. It is necessary that the adhesive retain some resiliency if the thermal-expansion coefficients of adhesive and adherend cannot be closely matched. The adhesive's coefficient of thermal conductivity is important in minimizing transient stresses during cooling. This is why thinner bond thicknesses have better cryogenic properties than thicker ones.

Low-temperature properties of common structural adhesives used for cryogenic applications are illustrated in Fig. 48. Epoxy-polyamide adhesive systems can be made serviceable at very low temperatures by the addition of appropriate fillers to control thermal expansion. But the epoxy-based systems are not as attractive as some others because of brittleness and corresponding low peel and impact strength at cryogenic temperatures.

TABLE 62   Shear Strength of High-Temperature Structural Adhesives[73]

Property	Modi-fied epoxy*	Epoxy-phenolic	Nitrile-phenolic	Poly-imide*	Poly-benzimid-azole*	Vinyl phenolic
Initial room-temp shear strength, lb/in.²	3,800	3,240	3,200–4,100	3,300	2,920	3,800–4,500
% of original shear strength after 10-min soak:						
At 180°F	83	94	35–70	.....	99	35–106
At 250°F	74	81	31–42	.....	99	12–79
At 350°F	.....	66	....	64	98	6–11
At 500°F	28	54	....	61	83	
At 650°F	.....	43	....	45	74	
At 800°F	.....	15	....	.....	68	
After 1,000-h soak:						
At 180°F	.....	96	62–78	.....	.....	60–122
At 250°F	.....	81	52–66	.....	.....	33–82
At 350°F	.....	43	27–56	.....	110	13–30
At 500°F	.....	.....	....	63	6–7	
At 600°F	.....	.....	....	61		

* Stainless steel used instead of aluminum as adherend material.

Epoxy-phenolic adhesives are exceptional in that they have good adhesive properties at both elevated and low temperatures. Vinyl-phenolic adhesives maintain fair shear and peel strength at −423°F, but strength decreases with temperature because of increasing brittleness in the thermoplastic constituent. Nitrile-phenolic adhesives do not have high strength at low service temperatures because of rigidity.

Polyurethane and epoxy-nylon systems offer outstanding cryogenic properties. Polyurethane adhesives are easily processible and bond well to many substrates. Peel strength ranges from 22 lb/in. at 75°F to 26 lb/in. at −423°F, and the increase in shear strength at −423°F is even more dramatic. Epoxy-nylon adhesives also retain flexibility and yield 5,000 lb/in.² shear strength in the cryogenic-temperature range.

Heat-resistant polyaromatic adhesives also have shown promising low-temperature properties. The shear strength of a polybenzimidazole adhesive on stainless-steel substrates is 5,690 lb/in.² at a test temperature of −423°F,[74] and polyimide adhesives have exhibited shear strength of 4,100 lb/in.² at −320°F.[75] These adhesives have not been extensively evaluated at low temperatures. Success in this area would indicate the applicability of polyaromatic adhesives on structures seeing both very high and low temperatures such as the proposed reusable space shuttle.

Humidity and Water Immersion

Moisture can affect adhesive strength in two significant ways. Some polymeric materials, notably ester-based polyurethanes, will "revert," i.e., lose hardness, strength, and in the worst cases turn fluid during exposure to warm, humid air. Water can also permeate the adhesive and preferentially displace the adhesive at the bond interface. This mechanism is the most common cause of adhesive-strength reduction in moist environments.

The rate of reversion or hydrolytic instability depends on the chemical structure of the base adhesive, the type and amount of catalyst used, and the flexibility of the adhesive. Certain chemical linkages such as ester, urethane, amide, and urea can be hydrolyzed. The rate of attack is fastest for ester-based linkages. Ester linkages are present in certain types of polyurethanes and anhydride-cured epoxies. Generally amine-cured epoxies offer better hydrolytic stability than anhydride-cured types. Figure 49 illustrates the hydrolytic stability of various polymeric materials determined by a hardness measurement. This technique of measuring hydrolytic stability is known as the *Gahimer and Nieske procedure*.[71] The reversion rate also depends on

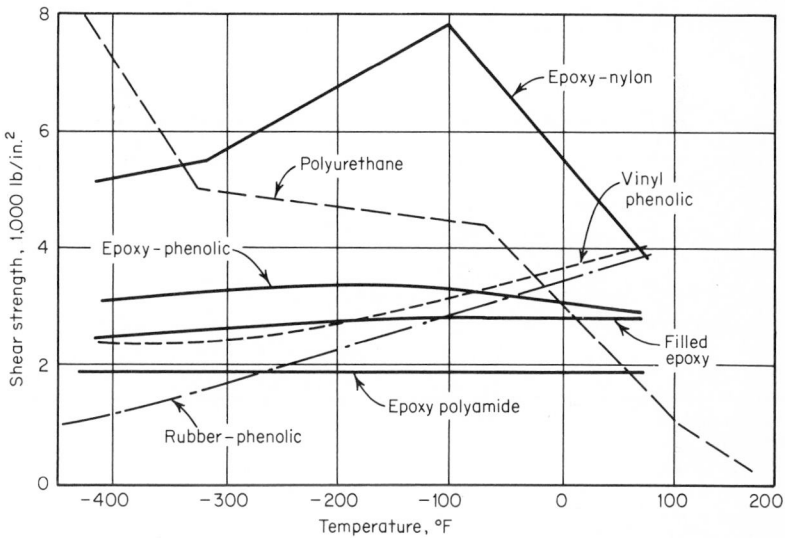

Fig. 48 Properties of cryogenic structural-adhesive systems.[72]

the amount of catalyst used in the formulation. Best hydrolytic properties are obtained when the proper stoichiometric ratio of base material to catalyst is used. Reversion is usually much faster in flexible materials because water permeates it more easily.

Structural adhesives not susceptible to the reversion phenomenon are also likely to lose adhesive strength when exposed to moisture. The degradation curves shown in Fig. 50 are typical for an adhesive exposed to moist, high-temperature environments. The mode of failure in the initial stages of aging usually is truly cohesive. After 5 to 7 days the failure becomes one of adhesion. It is expected that water vapor permeates the adhesive through its exposed edges and concentrates in weak boundary layers at the interface. This effect is greatly dependent on the type of adhesive. Table 63 illustrates that a nitrile-phenolic adhesive does not succumb to failure through the mechanism of preferential displacement at the interface. Failures occurred cohesively within the adhesive even when tested after 24 months of immersion in water. A nylon-epoxy adhesive degraded rapidly under the same conditioning owing to permeability and preferential displacement by moisture. Adhesive strength deteriorates more quickly in an aqueous-vapor environment than in liquid water because of more rapid permeation of the vapor.

Because of the importance of the interface, primers and surface treatments tend to hinder adhesive-strength degradation in moist environments. A fluid primer that easily wets the interface presumably tends to fill in minor discontinuities on the

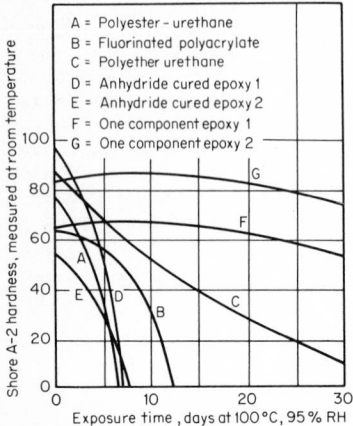

Fig. 49 Hydrolytic stability of potting compounds. Materials showing rapid hardness loss soften similarly after 2 to 4 years of service at ambient temperatures in high-humidity climates such as in Vietnam.[76]

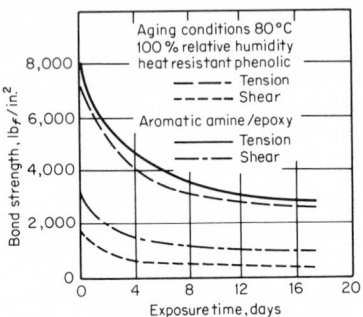

Fig. 50 Effect of humidity on adhesion of two structural adhesives to stainless steel.[78]

surface. Chemical etching, which removes surface flaws, also results in improved resistance to high humidity.

Stress accelerates the effect of environments on the adhesive joint. Few data are available on this phenomenon because of the time and expense associated with stress-aging tests. However, it is known that moisture, as an environmental burden, markedly decreases the ability of an adhesive to bear prolonged stress. Figure 51 illustrates the effect of stress aging on specimens exposed to humidity cycling from 90 to 100 percent and simultaneous temperature cycling from 80 to 120°F. The loss of load-bearing ability of a flexibilized epoxy adhesive (Fig. 51b) is exceptionally strong. The stress on this particular adhesive had to be reduced to 13 percent of its original strength in order for the joint to last a little more than 44 days.

TABLE 63 Effect of Humidity and Water Immersion on the Shear Strength (Lb/in.^{2}) of Two Structural Adhesives[79]

Exposure time, months	Nylon/epoxy adhesive		Nitrile/phenolic adhesive	
	Humidity cycle*	Water immersion	Humidity cycle*	Water immersion
0......................	4,370	4,370	3,052	3,052
2......................	1,170	2,890	2,180	2,740
6......................	950	1,700	2,370	2,280
12.....................	795	500	2,380	2,380
18.....................	1,025	200	2,350	2,640
24.....................	850	120	2,440	2,390

Substrate aluminum treated with a sulfuric acid–dichromate etch.
* Humidity cycle of 93% RH between 149 and 89°F with a cycle time of 48 h.

Outdoor Weathering

By far the most detrimental factors influencing adhesives aged outdoors are heat and humidity. Thermal cycling, ultraviolet radiation, and cold are relatively minor factors. The reasons why warm, moist climates degrade adhesive joints were presented in the last section.

When exposed to weather, structural adhesives rapidly lose strength during the first 6 months to 1 year. After 2 to 3 years the rate of decline usually levels off at 25 to 30 percent of the initial joint strength depending on the climate zone, adherend, adhesive, and stress level. Figure 52 shows the weathering characteristics of unstressed epoxy adhesives to the Richmond, Va., climate.

The following generalizations are of importance in designing a joint for outdoor service:

1. The most severe locations are those with high humidity and warm temperatures.
2. Stressed panels deteriorate more rapidly than unstressed panels.
3. Stainless-steel panels are more resistant than aluminum panels because of corrosion.
4. Heat-cured adhesive systems are generally more resistant than room-temperature-cured systems.
5. Using the better adhesives, unstressed bonds are relatively resistant to severe outdoor weathering, although all joints will eventually exhibit some strength loss.

MIL-STD-304 is a commonly used accelerated-exposure technique to determine the effect of weathering and high humidity on adhesive specimens. Only comparisons can be made with this type of test. In this procedure bonded panels are exposed to alternating cold (−65°F), dry heat (160°F), and heat and humidity (160°F, 95% RH) for 30 days. The effect of MIL-STD-304 conditioning on the joint strength of common structural adhesives is presented in Table 64.

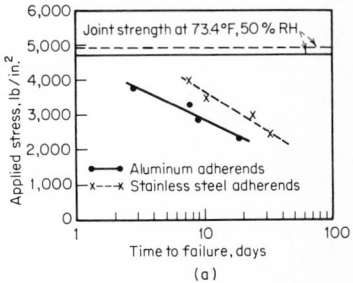

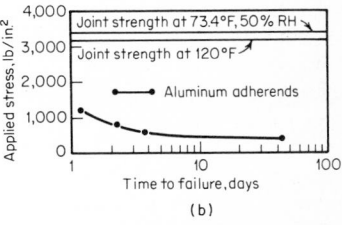

Fig. 51 Time to failure vs. stress for two adhesives in a warm, high-humidity environment.[80] (*a*) Adhesive, one-part, heat-curing, modified epoxy. (*b*) Adhesive, flexibilized amine-cured epoxy.

Chemicals and Solvents

Most organic adhesives tend to be susceptible to chemicals and solvents, especially at elevated temperatures. Standard test fluids and immersion conditions used by adhesive suppliers and MMM-A-132 are listed in Table 65. Unfortunately exposure tests lasting less than 30 days are not applicable to many service-life requirements. Practically all adhesives are resistant to these fluids over short time periods and at room temperatures. Some epoxy adhesives even show an increase in strength during aging in fuel or oil (Fig. 53). This effect is possibly due to a postcuring or plasticizing of the epoxy by oil.

Epoxy adhesives are generally more resistant to a wide variety of liquid environments than other structural adhesives. However, the resistance to a specific environment is greatly dependent on the type of epoxy curing agent used. Aromatic amine, such as metaphenylene diamine, cured systems are frequently preferred for long-term chemical resistance.

There is no universally "best adhesive" for all chemical environments. As an example, maximum resistance to bases almost axiomatically means poor resistance to acids. It is relatively easy to find an adhesive that is resistant to one particular chemical environment. It becomes much more difficult to find an adhesive that will not degrade in two widely differing chemical environments. Generally, adhesives which are most resistant to high temperatures have the best resistance to chemicals and solvents.

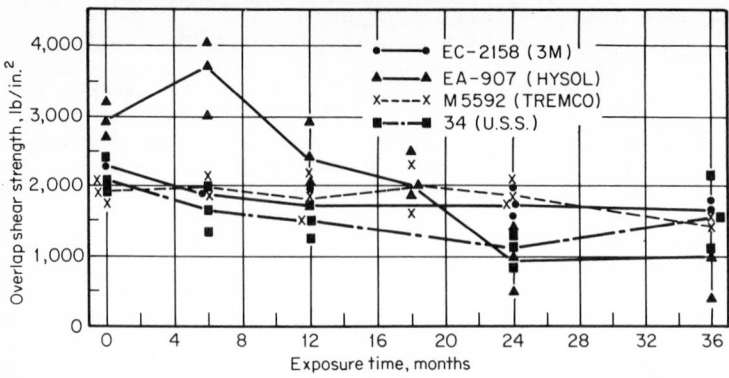

Fig. 52 Effect of outdoor weathering on typical aluminum joints made with four different two-part epoxies cured at room temperature.[33]

The temperature of the immersion medium is a significant factor in the aging properties of the adhesive. As the temperature increases, more fluid is generally adsorbed by the adhesive, and the degradation rate increases.

From the rather meager information reported in the literature, it may be summarized that:

1. Chemical-resistance tests are not uniform, in concentrations, temperature, time, or properties measured.

TABLE 64   Effect of MIL-STD-304 Aging on Bonded Aluminum Joints[81]

Adhesive	Shear, lb/in.², 73°F		Shear, lb/in.², 160°F	
	Control	Aged	Control	Aged
Room-temp. cured:				
Epoxy–polyamide.....................	1,800	2,100	2,700	1,800
Epoxy–polysulfide....................	1,900	1,640	1,700	6,070
Epoxy–aromatic amine................	2,000	Failed	720	Failed
Epoxy–nylon.........................	2,600	1,730	220	80
Resorcinol epoxy–polyamide...........	3,500	3,120	3,300	2,720
Epoxy–anhydride.....................	3,000	920	3,300	1,330
Polyurethane........................	2,600	1,970	1,600	1,560
Cured 45 min at 330°F, epoxy–phenolic......	2,900	2,350	2,900	2,190
Cured 1 h at 350°F:				
Modified epoxy......................	4,900	3,400	4,100	3,200
Nylon–phenolic......................	4,600	3,900	3,070	2,900

TABLE 65   Standard Test Fluids and Immersion Conditions for Adhesive Evaluation

Conditions, 77°F after
30 days in RT water
30 days in 110°F humid cabinet (100% RH)
30 days in 95°F salt-spray cabinet (5% salt)
7 days in JP-4 jet-engine fuel
7 days in anti-icing fluid (isopropyl alcohol)
7 days in hydraulic oil (MIL-H-5606)
7 days in HC test fluid (70/30 v/v isooctane/toluene)

2. Generally, chlorinated solvents and ketones are severe environments.

3. High-boiling solvents such as dimethylformamide, dimethyl sulfoxide, and Skydrol* are severe environments.

4. Acetic acid is a severe environment.

5. Amine curing agents for epoxies are poor in oxidizing acids.

6. Anhydride curing agents are poor in caustics.

Vacuum

The ability of an adhesive to withstand long periods of exposure to a vacuum is of primary importance for materials used in space travel. The degree of adhesive evaporation is a function of its vapor pressure at a given temperature. Loss of low-molecular-weight constituents such as plasticizers or diluents could result in hardening and porosity of adhesives or sealants.

Since most structural adhesives are relatively high-molecular-weight polymers, exposure to pressures as low as 10^{-9} torr is not harmful. However, high temperatures, nuclear radiation, or other degrading environments may cause the formation

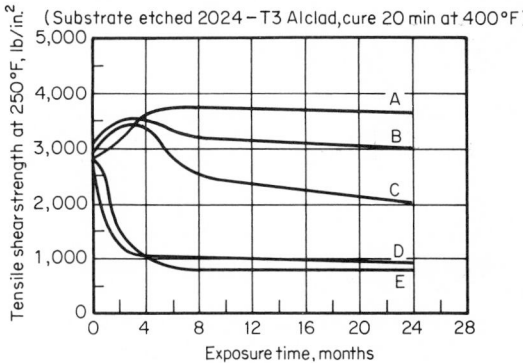

Fig. 53 Effect of immersion in various chemical environments on a one-part, heat-curing epoxy adhesive (EA 929, Hysol Corp.)[82] (A) Gasoline at 75°F. (B) Gear oil at 250°F. (C) Distilled water at 75°F. (D) Tapwater at 212°F. (E) 38% Shellzone† at 220°F.

of low-molecular-weight fragments which tend to bleed out of the adhesive in a vacuum.

Epoxy and polyurethane adhesives are not appreciably affected by 10^{-9} torr for 7 days at room temperature. However, polyurethane adhesives exhibit significant outgassing when aged under 10^{-9} torr at 225°F.[83]

Radiation

High-energy particulate and electromagnetic radiation including neutron, electron, and gamma radiation have similar effects on organic adhesives. Radiation causes molecular-chain scission of polymers used in structural adhesives, which results in weakening and embrittlement of the bond. This degradation is worsened when the adhesive is simultaneously exposed to elevated temperatures and radiation.

Figure 54 illustrates the effect of radiation dosage on the tensile-shear strength of structural adhesives. Generally, heat-resistant adhesives have been found to resist radiation better than less thermally stable systems. Fibrous reinforcement, fillers, curing agents, and reactive diluents affect the radiation resistance of adhesive systems. In epoxy-based adhesives, aromatic curing agents offer greater radiation resistance than aliphatic-type curing agents.

* Trademark of Monsanto Corp.
† Trademark Shell Oil Co.

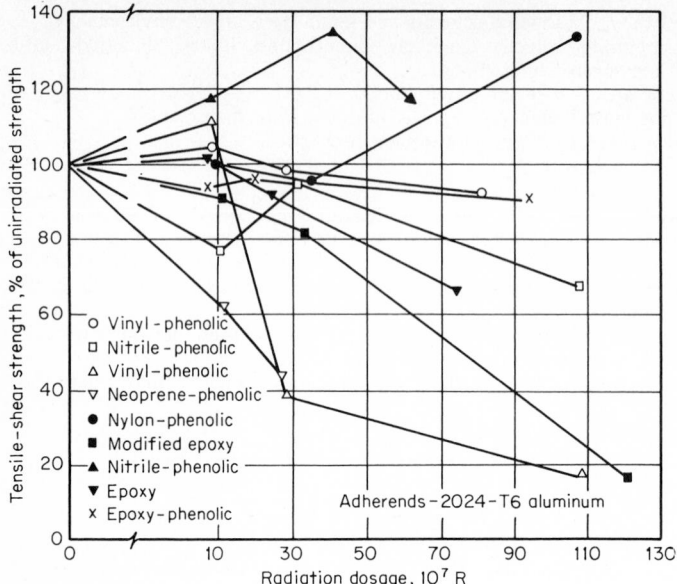

Fig. 54 Percent change of initial tensile-shear strength caused by nuclear radiation dosage.[84]

PROCESSING AND QUALITY CONTROL OF ADHESIVE JOINTS

Processing and quality control are usually the final considerations in the design of an adhesive-bonding system. These decisions are very important, however, because they alone may (1) restrict the degrees of freedom in designing the end product, (2) widen or narrow the types of adhesives that can be considered, (3) affect the quality and reproducibility of the joint, and (4) affect the total assembly cost.

Measuring and Mixing

When a multiple-part adhesive is used, the concentration ratios have a significant effect on the quality of the joint. Strength differences caused by varying curing-agent concentration are most noticeable when the joints are tested at elevated temperatures or after exposure to water or solvents. Exact proportions of resin and hardener must be weighted out on an accurate balance or in a measuring container for best adhesive quality and reproducibility.

The weighted-out components must be mixed thoroughly. Mixing should be continued until no color streaks or density stratifications are noticeable. Caution should be taken to prevent air from being mixed into the adhesive through over-agitation. This can cause foaming of the adhesive during heat cure, resulting in porous bonds. If air does become mixed into the adhesive, vacuum degassing may be necessary before application.

Only enough adhesive should be mixed to work with before the adhesive begins to cure. Working life of an adhesive is defined as the period of time during which an adhesive after mixing with catalyst remains suitable for use. Working life is decreased as the ambient temperature increases and the batch size becomes larger. One-part and some heat-curing, two-part adhesives have very long working lives at room temperature, and application and assembly speed or batch size are not critical.

For a large-scale bonding operation, hand mixing is costly, messy, and slow; repeatability is entirely dependent on the operator. Equipment is available that can meter, mix, and dispense multicomponent adhesives on a continuous or shot basis. Figure 55 illustrates a standard adhesive-dispensing machine for multicomponent systems. An important advantage for using such equipment is that it allows the use of short-working-life materials by mixing only the smallest practical quantity continuously. This offers shorter curing cycles, resulting in faster assembly times.

Application of Adhesives

The selection of an application method depends primarily on the form of the adhesive: liquid, paste, powder, or film. Table 66 describes the advantages and limitations realized in using each of the four basic forms. Other factors influencing the application method are the size and shape of parts to be bonded, the areas where the adhesive is to be applied, and production volume and rate.

Liquids Liquids, the most common form of adhesive, can be applied by a variety of methods. Brush, simple rollers, and glue guns are manual methods that provide simplicity, low cost, and versatility. Spray, dipping, and mechanical roll coaters are generally used on large production runs. Mechanical-roller methods are commonly used to apply a uniform layer of adhesive to a continuous roll or coil. Such automated systems are used with adhesives that have a long working life and low viscosity. Spray methods can be used on both small and large production runs. The spray adhesive is generally in solvent solution, and sizable amounts of adhesive may be lost from overspray. Two-component adhesives are usually mixed prior to placing in the spray-gun reservoir. Application systems are available, however, that meter and mix the adhesive in the spray-gun barrel. This is ideal for fast-reacting systems.

Pastes Bulk adhesives such as pastes or mastics are the simplest and most reproducible adhesive to apply. These systems can be troweled on or extruded through a caulking gun. Little operator skill is required. Since the thixotropic nature of the paste prevents it from flowing excessively, application is usually clean, and not much waste is generated.

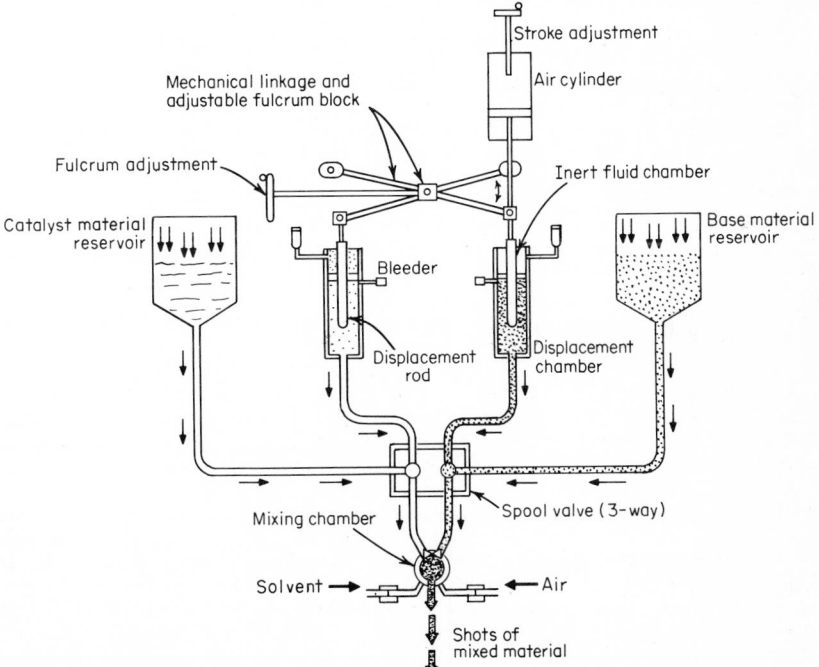

Fig. 55 Operating schematic of an automatic metering-mixing-dispensing machine for adhesive systems.[85]

TABLE 66 Characteristics of Various Adhesive-Application Methods[86]

Application method	Viscosity	Operator skill	Production rate	Equipment cost	Coating uniformity	Material loss
Liquid						
Manual, brush or roller..........	Low to medium	Little	Low	Low	Poor	Low
Roll coating, reverse, gravure........	Low	Moderate	High	High	Good	Low
Spray, manual, automatic, airless, or external mix........	Low to high	Moderate to high	Moderate to high	Moderate to high	Good	Low to high
Curtain coating........	Low	Moderate	High	High	Good to excellent	Low
Bulk						
Paste and mastic........	High	Little	Low to moderate	Low	Fair	Low to high
Powder						
Dry or liquid primed........	......	Moderate	Low	High	Poor to fair	Low
Dry Film........	......	Moderate to high	Low to high	Low to high	Excellent	Lowest

Powders   Powder adhesives can be applied in three ways.   They may be sifted onto a preheated substrate.   The powder which falls onto the substrate melts and adheres.   The assembly is then mated and cured according to recommended processes.   A preheated substrate could also be dipped into the powder and then extracted with an attached coating of adhesive.   This method helps to assure even powder distribution.   Lastly, the powder can be melted into a paste or liquid and applied by conventional means.   Powder adhesives are generally one-part, epoxy-based systems that require heat and pressure to cure.   They do not require metering and mixing but often must be refrigerated for extended shelf life.   Because coating uniformity is poor, large variations of joint strength may result with powder adhesives.

Films   The use of dry adhesive films is expanding more rapidly than other forms because of their following advantages:

1. High repeatability—no mixing or metering, constant thickness.
2. Easy to handle—low equipment cost, relatively hazard-free; clean operating.
3. Very little waste—preforms can be cut to size.
4. Excellent physical properties—wide variety of adhesive types available.

TABLE 67   Characteristics of Adhesive Films[87]

Weight:
From 0.030 to greater than 0.100 lb/ft^2

Handleability:
From dry and stiff to very tacky and drapable

Costs:
From 12 cents/ft^2 to greater than $1/ft^2

Strength:
From 1,500 to greater than 6,000 lb/in.^{2} shear
From 1 to greater than 100 lb/in. peel
From no strength above 200°F to high strength above 600°F
From brittle at −67°F to tough at −400°F

Mobility and support:
From unsupported film through multiple nonstructural carriers to fully structural carriers
From highly fluid during cure conditions to little or no flow during cure

Chemical variations:
From one polymeric base to a blend of six or more polymers of widely varying characteristics
From one curing agent to a blend of four or more curing and catalytic agents
From compatible solutions to multiphase systems
From unfilled to highly filled systems

Curing variations:
From room temp cure to 500° cures requiring long postcure times
From 1 to 4 h out time at room temperature to 30 or more days at room temp

Films are limited to flat surfaces or simple curves.   Application requires a relatively high degree of care to ensure nonwrinkling and removal of separator sheets.   Characteristics of available film adhesives (Table 67) vary widely depending on the type of adhesive used.

Film adhesives are made in both unsupported and supported types.   The carrier for supported films is generally fibrous fabric or mat.   Film adhesives are supplied as heat-activated, pressure-sensitive, or solvent-activated forms.

Solvent-activated adhesives are made tacky and pressure-sensitive by wiping with solvent.   They are not as strong as the other types but are well suited for contoured, curved, or irregularly shaped parts.   Manual solvent-reactivation methods should be closely monitored so that excessive solvent is not used.   Chemical formulations available in solvent-activated films include neoprene, nitrile, and butyral phenolics.   Decorative trim and nameplates are usually fastened onto a product with solvent-activated adhesives.

Hot-melt adhesives   Hot-melt adhesives are available in a variety of solid forms. Usually they are supplied as billets and applied to the substrate through a hot-melt glue gun. The hot-melt application system is equipped with a temperature-regulated glue reservoir and insulated hosing to the gun. Once out of the gun, the adhesive solidifies quickly. A high degree of operator skill is necessary to make the assembly as quickly as possible.

Bonding Equipment

After the adhesive is applied, the assembly must be mated as quickly as possible to prevent contamination of the adhesive surface. The substrates are held together under pressure and heated if necessary until cure is achieved. The equipment required to perform these functions must provide adequate heat and pressure, maintain constant pressure during the entire cure cycle, and distribute pressure uniformly over the bond area. Of course, many adhesives cure with simple contact pressure at room temperature, and extensive bonding equipment is not necessary.

Pressure equipment   Pressure devices should be designed to maintain constant pressure on the bond during the entire cure cycle. They must compensate for thickness reduction from adhesive flow-out or thermal expansion of assembly parts. Thus, screw-actuated devices like C clamps and bolted fixtures are not acceptable when constant pressure is important. Spring pressure can often be used to supplement clamps and compensate for thickness variations. Dead-weight loading may be applied in many instances; however, this method is sometimes impractical, especially when heat cure is necessary.

Pneumatic and hydraulic presses are excellent tools for applying constant pressure. Steam or electrically heated platen presses with hydraulic rams are often used for adhesive bonding. Some units have multiple platens, thereby permitting the bonding of several assemblies at one time.

Large bonded areas such as on aircraft parts are usually cured in an autoclave. The parts are mated first and covered with a rubber blanket to provide uniform pressure distribution. The assembly is then placed in an autoclave, which can be pressurized and heated. This method requires heavy capital-equipment investment.

Vacuum-bagging techniques can be an inexpensive method of applying pressure to large parts. A film or plastic bag is used to enclose the assembly, and the edges of the film are sealed airtight. A vacuum is drawn on the bag, enabling atmospheric pressure to force the adherends together. Vacuum bags are especially effective on large areas because size is not limited by equipment.

Heating equipment   Many structural adhesives require heat as well as pressure. Most often the strongest bonds are achieved by an elevated-temperature cure. With many adhesives, trade-offs between cure times and temperature are permissible. But generally, the manufacturer will recommend a certain curing schedule for optimum properties.

If, for example, a cure of 60 min at 300°F is recommended, this does not mean that the assembly should be placed in a 300°F oven for 60 min. It is the bond line that should be at 300°F for 60 min. Total oven time would be 60 min plus whatever time is required to bring the adhesive up to 300°F. Large parts act as a heat sink and may require substantial time for an adhesive in the bond line to reach the necessary temperature. Bond-line temperatures are best measured by thermocouples placed very close to the adhesive. In some cases, it may be desirable to place the thermocouple in the adhesive joint for the first few assemblies being cured.

Oven heating is the most common source of heat for bonded parts, even though it involves long curing cycles because of the heat-sink action of large assemblies. Ovens may be heated with gas, oil, electricity, or infrared units. Good air circulation within the oven is mandatory to prevent nonuniform heating.

Heated-platen presses are good for bonding flat or moderately contoured panels when faster cure cycles are desired. Platens are heated with steam, hot oil, or electricity and are easily adapted with cooling-water connections to further speed the bonding cycle.

Induction and dielectric heating are the fastest heating methods because they focus heat at or near the adhesive bond line. Workpiece-heating rates greater than

100°F/s are possible with induction heating. For induction heating to work, the adhesive must be filled with metal particles or the adherend must be capable of conducting electricity or being magnetized as shown in Fig. 56. Dielectric heating is an effective way of curing adhesives if at least one substrate is a nonconductor. Metal-to-metal joints tend to break down the microwave field necessary for dielectric heating. This heating method makes use of the polar characteristics of the adhesive and adherend to cause molecular movement and frictional heating within polymeric materials. Both induction and dielectric heating involve relatively expensive capital-equipment outlays, and the bond area is limited. Their most important advantages are assembly speed and the fact that an entire assembly does not have to be heated to cure only a few grams of adhesive.

Adhesive-thickness control   It is highly desirable to have a uniformly thin (2- to 10-mil) adhesive bond line. Starved adhesive joints, however, will yield exceptionally poor properties. Three basic methods are used to control adhesive thickness. The first method is to use mechanical shims or stops which can be removed after the curing operation. Sometimes it is possible to design stops into the joint.

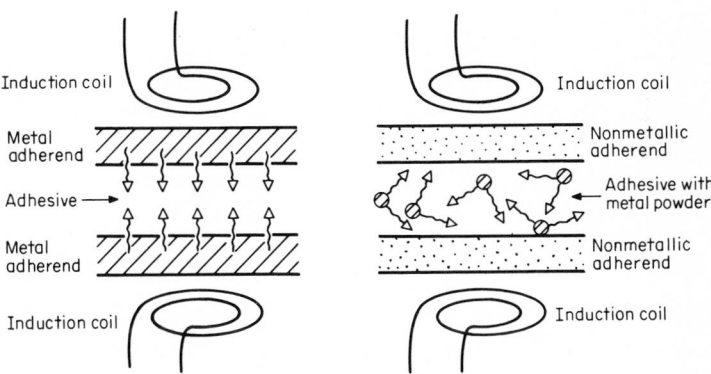

Fig. 56   Left, induction-heated metal parts heat adhesive by conduction; right, parts not heated by induced electric currents are bonded with adhesive that is heated because it contains metal particles.[88]

The second method is to employ a film adhesive that becomes highly viscous during the cure cycle preventing excessive adhesive flow-out. With supported films, the adhesive carrier itself can act as the "shims." Generally the cured bond-line thickness will be determined by the original thickness of the adhesive film. The third method of controlling adhesive thickness is to use trial and error to determine the correct pressure-adhesive viscosity factors that will yield the desired bond thickness.

Quality Control

A flow chart of a quality-control system for a major aircraft company is illustrated in Fig. 57. This system is designed to ensure reproducible bonds and, if a substandard bond is detected, to make suitable corrections. Quality control should cover all phases of the bonding cycle from inspection of incoming material to the inspection of the completed assembly. In fact, good quality control will start even before receipt of materials.

Prehandling conditions   The human element enters the adhesive-bonding process more than in other fabrication techniques. An extremely high percentage of defects can be traced to poor workmanship. This generally prevails in the surface-preparation steps but may also arise in any of the other steps necessary to achieve a bonded assembly. This problem can be largely overcome by proper motivation and education. All employees from design engineer to laborer to quality-control inspector should be somewhat familiar with adhesive-bonding technology and be aware of the circumstances that can lead to poor joints. A great many defects can also be traced to poor design engineering. Table 68 lists the polled replies of experts to a question: "Where do design engineers most often err in designing bonded joints?"

The plant's bonding area should be as clean as possible prior to receipt of materials. The basic approach to keeping the assembly area clean is to segregate it from the other manufacturing operations either in a corner of the plant or in isolated rooms. The air should be dry and filtered to prevent moisture or other contaminants from gathering at a possible interface. The cleaning and bonding operations should be separated from each other. If mold release is used to prevent adhesive flash from sticking to bonding equipment, it is advisable that great care be taken to assure that the release does not contaminate the adhesive or adherends. Spray mold releases, especially silicone release agents, have a tendency to migrate to undesirable areas.

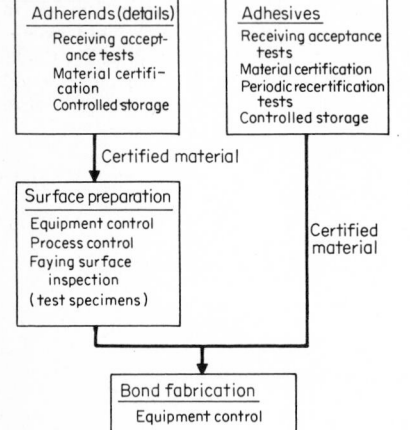

Quality control of adhesive and surface treatment   Acceptance tests on adhesives should be directed toward assurance that incoming materials are identical from lot to lot. The tests should be those which can quickly and accurately detect deficiencies in the adhesive's physical or chemical properties. ASTM lists various test methods that are commonly used for adhesive acceptance. Actual test specimens should also be made to verify strength of the adhesive. These specimens should be stressed in directions that are representative of the forces which the bond will see in service, i.e., shear, peel, tension, or cleavage. If possible, the specimens should be prepared and cured in the same manner as actual production assemblies. If time permits, specimens should also be tested in simulated service environments, e.g., high temperature, humidity.

Surface preparations must be carefully controlled for reliable production of adhesive-bonded parts. If a chemical surface treatment is required, the process must be monitored for proper sequence, bath temperature, solution concentration, and contaminants. If sand or grit blasting is employed, the abrasive must be changed regularly. An adequate supply of clean wiping cloths for solvent cleaning is also mandatory. Checks should be made to determine if cloths or solvent containers may have become contaminated.

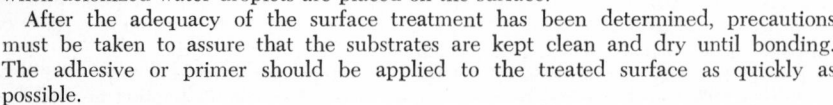

Fig. 57   Flow chart of a quality-control system for adhesive bonding.[89]

The specific surface preparation can be checked for effectiveness by the water-break free test. After the final treating step the substrate surface is checked for a continuous film of water that should form when deionized water droplets are placed on the surface.

After the adequacy of the surface treatment has been determined, precautions must be taken to assure that the substrates are kept clean and dry until bonding. The adhesive or primer should be applied to the treated surface as quickly as possible.

Quality control of the bonding process   The adhesive metering and mixing operation should be monitored by periodically sampling the mixed adhesive and testing it for adhesive properties. A visual inspection can also be made for air entrapment and degree of mixing. The quality-control engineer should be sure that the oldest adhesive is used first and that the specified shelf life has not been exceeded.

During the actual assembly operation, the cleanliness of the shop and tools should be verified. The shop atmosphere should be controlled as closely as possible. Temperature in the range of 65 to 90°F and relative humidity from 20 to 65 percent is best for almost all bonding operations.

The amount of the applied adhesive and the final bond-line thickness must also be monitored because they can have a significant effect on joint strength. Curing conditions should be monitored for heat-up rate, maximum and minimum temperature during cure, time at the required temperature, and cool-down rate.

Bond inspection   After the adhesive is cured, the joint area can be inspected to detect gross flaws or defects. This inspection procedure can be either destructive or nondestructive in nature. Destructive testing generally involves placing samples of the production run in simulated or accelerated service and determining if it has

TABLE 68   Where Design Engineers Err[90]

Adhesive technology	Design consideration	Production-line problems
Low peel strength	Using a butt joint when lap joint would be stronger	Lack of careful surface preparation
Overlooking such factors as pot life, curing time, operating temperatures	Loads causing unsuspected cleavage forces	Expecting prototype performance from bonds made on assembly line
Failure to get technical help from supplier in selecting an adhesive	Overlooking effect of increased service temp in decreasing resistance to chemicals	Failure to keep surfaces clean until adhesive is applied
Assuming that strongest adhesive is always the best without considering cost or processing	Failure to check coefficients of expansion when unlike materials are bonded	Failure to consider the application method and equipment when designing joint
Lack of care in test procedures	Calling for heat-curing adhesive on a part that will not stand the heat	
	Overdesigning by asking for more strength or heat resistance than is needed	

similar properties to a specimen that is known to have a good bond and adequate service performance. The causes and remedies for faults revealed by such mechanical tests are described in Table 69. Nondestructive testing (NDT) is far more economical, and every assembly can be tested if desired. Great amounts of energy are now being devoted to improve NDT techniques.

Nondestructive-testing procedures   *Visual Inspection.*   A trained eye can detect a surprising number of faulty joints by close inspection of the adhesive around the bonded area. Table 70 lists the characteristics of faulty joints that can be detected visually. The most difficult defect to be found by any way are those related to improper curing and surface treatments. Therefore, great care and control must be given to surface-preparation procedures and shop cleanliness.

Sonic Inspection.   Sonic and ultrasonic methods are at present the most popular NDT technique for use on adhesive joints. Simple tapping of a bonded joint with a coin or light hammer can indicate an unbonded area. Sharp clear tones indicate that adhesive is present and adhering to the substrate in some degree; dull hollow tones indicate a void or unattached area.

Ultrasonic testing basically measures the response of the bonded joint to loading by low-power ultrasonic energy. Comparative investigation by several sources has shown that the Fokker Bond Tester is the most reliable method of quantitatively estimating bond strength. Table 71 indicates the degree of correlation between

TABLE 69   Faults Revealed by Mechanical Tests

Fault	Cause	Remedy
Thick, uneven glue line	Clamping pressure too low	Increase pressure. Check that clamps are seating properly
	No follow-up pressure	Modify clamps or check for freedom of moving parts
	Curing temperature too low	Use higher curing temperature. Check that temperature is above the minimum specified throughout the curing cycle
	Adhesive exceeded its shelf life, resulting in increased viscosity	Use fresh adhesive
Adhesive residue has spongy appearance or contains bubbles	Excess air stirred into adhesive	Vacuum-degas adhesive before application
	Solvents not completely dried out before bonding	Increase drying time or temperature. Make sure drying area is properly ventilated
	Adhesive material contains volatile constituent	Seek advice from manufacturers
	A low-boiling constituent boiled away	Curing temperature is too high
Voids in bond (i.e., areas that are not bonded), clean bare metal exposed, adhesive failure at interface	Joint surfaces not properly treated	Check treating procedure; use clean solvents and wiping rags. Wiping rags must not be made from synthetic fiber. Make sure cleaned parts are not touched before bonding. Cover stored parts to prevent dust from settling on them
	Resin may be contaminated	Replace resin. Check solids content. Clean resin tank
	Uneven clamping pressure	Check clamps for distortion
	Substrates distorted	Check for distortion; correct or discard distorted components. If distorted components must be used, try adhesive with better gap-filling ability
Adhesive can be softened by heating or wiping with solvent	Adhesive not properly cured	Use higher curing temperature or extend curing time. Temperature and time must be above the minimum specified throughout the curing cycle. Check mixing ratios and thoroughness of mixing. Large parts act as a heat sink, necessitating larger cure times

predicted and actual bond strength of the Fokker Bond Tester compared with two other NDT instruments.

The Fokker Bond Tester is a portable electronic unit consisting of a power drive, meter, and small probe. The instrument makes a comparison between the ultrasonic response of a joint with no adhesive and joints with varying degrees of voids at the interface or within the adhesive. Of course, the instrument's readings must

TABLE 70   Visual Inspection for Faulty Bonds

Fault	Cause	Remedy
No appearance of adhesive around edges of joint or adhesive bond line too thick	Clamping pressure roo low Starved joint Curing temperature too low	Increase pressure. Check that clamps are seating properly Apply more adhesive Use higher curing temperature. Check that temperature is above the minimum specified
Adhesive bond line too thin	Clamping pressure too high Curing temperature too high Starved joint	Lessen pressure Use lower curing temperature Apply more adhesive
Adhesive flash breaks easily away from substrate	Improper surface treatment	Check treating procedure; use clean solvents and wiping rags. Make sure cleaned parts are not touched before bonding
Adhesive flash is excessively porous	Excess air stirred into adhesive Solvent not completely dried out before bonding Adhesive material contains volatile constituent	Vacuum-degas adhesive before application Increase drying time or temperature Seek advice from manufacturers
Adhesive flash can be softened by heating or wiping with solvent	Adhesive not properly cured	Use higher curing temperature or extend curing time. Temperature and time must be above minmum specified. Check mixing

be correlated with laboratory-prepared specimens having different strengths to yield a reliable relationship between actual and predicted values.

The Fokker method is most sensitive to properties which physically affect adhesion such as voids, porosity, and incomplete wetting. Unfortunately, it is not capable of detecting incomplete cure, poor surface preparation, or contamination at the interface.

Other NDT Methods. Radiography (x-ray) inspection can be used to detect voids or discontinuities in the adhesive bond. This method is more expensive and requires more skilled experience than ultrasonic methods.

TABLE 71   Correlation between Predicted and Actual Bond Strengths Made by Three Ultrasonic NDT Instruments[89]

Type of bond	Adhesive material	Nondestructive test instrument		
		Fokker, lb/in.^{2}	Coinda scope, lb/in.^{2}	Stubmeter, lb/in.^{2}
Overlap..................	FM-47	± 950	$\pm 1,300$	$\pm 1,400$
Overlap..................	HT-424	± 360	$\pm 1,100$	± 950
Overlap..................	FM-58	± 740	$\pm 1,175$	$\pm 1,050$
Overlap..................	Metlbond 4021	± 730	± 600	$\pm 1,150$
Honeycomb..............	FM-47	± 250		
Honeycomb..............	HT-424	± 200		
Overlap..................	FM-47	± 750		
Honeycomb..............	Epoxy–phenolic	± 20 to ± 150		
Overlap..................	Redux 775	± 570		

95% confidence limits.

Thermal-transmission methods are relatively new techniques for adhesive inspection. Liquid crystals applied to the joint can make voids visible if the substrate is heated. This test is simple and inexpensive, although materials with poor heat-transfer properties are difficult to test, and the joint must be accessible from both sides. An experimental infrared inspection technique has also been developed for detection of internal voids and nonbonds.[91] This technique is expensive, but it can accurately determine the size and depth of the flaw.

The science of holography has also been used for NDT of adhesive bonds. Holography is a method of producing photographic images of flaws and voids using coherent light such as that produced by a laser. The major advantage of holography is that it photographs successive "slices" through the scene volume. A true three-dimensional image of a defect or void can then be reconstructed.

REFERENCES

1. Structural Adhesives, Minnesota Mining and Manufacturing Co., Adhesives, Coatings and Sealers Division, *Tech. Bull.*
2. Powis, C. N.: Some Applications of Structural Adhesives, in D. J. Alner (ed.), "Aspects of Adhesion," vol. 4, University of London Press, Ltd., London, 1968.
3. Cagle, C. V.: "Adhesive Bonding Techniques and Applications," McGraw-Hill, New York, 1968.
4. "Book of Standards," Part 16, American Society for Testing and Materials, 1970.
5. Schneberger, G. L.: Chemical Aspects of Adhesive Bonding, Part II, Physical Properties, *Adhesives Age,* March 1970.
6. Bikerman, J. J.: Causes of Poor Adhesion, *Ind. Eng. Chem.,* September 1967.
7. Koehn, G. W.: Design Manual on Adhesives, *Machine Des.,* April 1954.
8. Reinhart, F. W.: Survey of Adhesion and Types of Bonds Involved, in J. E. Rutzler and R. L. Savage (eds.), "Adhesion and Adhesives Fundamentals and Practices," Society of Chemical Industry, London, 1954.
9. Perry, H. A.: Room Temperature Setting Adhesives for Metals and Plastics, in J. E. Rutzler and R. L. Savage (eds.), "Adhesion and Adhesives Fundamentals and Practices," Society of Chemical Industry, London, 1954.
10. Merriam, J. C.: Adhesive Bonding, *Mater. Des. Eng.,* September 1959.
11. Rider, D. K.: Which Adhesives for Bonded Metal Assembly, *Prod. Eng.,* May 25, 1964.
12. Facts about Use of Adhesives and Adhesive Joints, Hysol Division, Dexter Corporation, *Bull. Gl-700.*
13. Lunsford, L. R.: Design of Bonded Joints, in M. J. Bodnar (ed.), "Symposium on Adhesives for Structural Applications," Interscience, New York, 1962.
14. Adhesive Bonding of Metals: What Are Design Factors? *Iron Age,* June 15, 1961.
15. "Adhesive Bonding Alcoa Aluminum," Aluminum Co. of America, 1967.
16. Sharpe, L. H.: Assembling with Adhesives, *Machine Des.,* Aug. 18, 1966.
17. DeLollis, N. J.: "Adhesives for Metals Theory and Technology," Industrial Press, New York, 1970.
18. Parker, R. S., and R. Taylor: "Adhesion and Adhesives," p. 41, Pergamon, New York, 1966.
19. Chessin, N., and V. Curran: Preparation of Aluminum Surfaces for Bonding, in M. J. Bodnar (ed.), "Structural Adhesive Bonding," Interscience, New York, 1966.
20. Muchnick, S. N.: Adhesive Bonding of Metals, *Mech. Eng.,* January 1956.
21. Vazirani, H. N.: Surface Preparation of Steel for Adhesive Bonding and Organic Coatings, *J. Adhesion,* July 1969.
22. Vazirani, H. N.: Surface Preparation of Copper and Its Alloys for Adhesive Bonding and Organic Coatings, *J. Adhesion,* July 1969.
23. Walter, R. E., D. L. Voss, and M. S. Hochberg: Structural Bonding of Titanium for Advanced Aircraft, *Nat. SAMPE Tech. Conf. Proc.,* vol. 2, Aerospace Adhesives and Elastomers, 1970.
24. Thelen, E.: Adherend Surface Preparation, in M. J. Bodnar (ed.), "Symposium on Adhesives for Structural Applications," Interscience, New York, 1962.
25. Scott, J. A.: Durability of Adhesive—Bonded Aluminum Joints Exposed to Outdoor Weathering, Durability of Adhesive Joints, *ASTM Spec. Tech. Pub.* 401, 1966.
26. Jennings, C. W.: Surface Roughness and Bond Strength of Adhesives, *J. Adhesion,* May 1972.

27. Rogers, N. L.: Surface Preparation of Metals, in M. J. Bodnar (ed.), "Structural Adhesive Bonding," Interscience, New York, 1966.
28. Hall, J. R., et al.: Effects of Activated Gas Plasma Treatment Time on Adhesive Bondability of Polymers, *J. Appl. Polymer Sci.*, vol. 16, pp. 1465–1477, 1972.
29. Bersin, R. L.: How to Obtain Strong Adhesive Bonds via Plasma Treatment, *Adhesives Age*, March 1972.
30. Schields, J.: "Adhesives Handbook," CRC Press, 1970.
31. Guttmann, W. H.: "Concise Guide to Structural Adhesives," Reinhold, New York, 1961.
32. Preparing the Surface for Adhesive Bonding, Hysol Division, Dexter Corp., *Bull.* Gl–600.
33. "Adhesive Bonding Aluminum," Reynolds Metals Co., 1966.
34. Jackson, L. C.: Effects of Surface Preparations on Bond Strength of Magnesium, in M. J. Bodnar (ed.), "Structural Adhesive Bonding," Interscience, New York, 1966.
35. Jackson, L. C.: Preparing Plastic Surfaces for Adhesive Bonding, *Adhesives Age*, April 1961.
36. Schonhorn, H., and R. H. Hansen: Surface Treatment of Polymers for Adhesive Bonding, *J. Appl. Polymer Sci.*, vol. 11, p. 1461, 1967.
37. Schonhorn, H., and F. W. Ryan: Adhesion and Polytetrafluoroethylene, *J. Adhesion*, January 1969.
38. Devine, A. T., and M. J. Bodnar: Effects of Various Surface Treatments on Adhesive Bonding of Polyethylene, *Adhesives Age*, May 1969.
39. Krieger, R. B.: Advances in the Corrosion Resistance of Bonded Structures, *Nat. SAMPE Tech. Conf. Proc.*, vol. 2, Aerospace Adhesives and Elastomers, 1970.
40. Fan, Y. L., and R. G. Shaw: Silyl Peroxides—A Novel Family of Adhesion Promoters, *25th Ann. Tech. Conf.*, Society of Plastics Industry, 1970.
41. Austin, J. E., and L. C. Jackson: Management: Teach Your Engineers to Design Better with Adhesives, *SAE J.*, October 1961.
42. Burgman, H. A.: Selecting Structural Adhesive Materials, *Electrotechnol.*, June 1965.
43. Three New Cyanoacrylate Adhesives, Eastman Chemical Co., Leaflet R-206A.
44. Bugel, T. E., et al.: Phenoxy Resin—A New Thermoplastic Adhesive, Adhesion, *ASTM Spec. Tech. Pub.* 360, 1964.
45. TAME, A New Concept in Structural Adhesives, *B. F. Goodrich General Products Co. Bull.* GPC-72-AD-3.
46. Hughson Acrylic Structural Adhesives, *Hughson Chemical Co. Tech. Bull.*
47. Information about Pressure Sensitive Adhesives, *Dow Corning Bull.* 02-032.
48. Information about Silastic RTV Silicone Rubber, *Dow Corning Tech. Bull.* 61-015a.
49. Bruno, E. J. (ed.): "Adhesives in Modern Manufacturing," p. 29, Society of Manufacturing Engineers, 1970.
50. Lichman, J.: Water-based and Solvent-based Adhesives, in I. Skeist (ed.), "Handbook of Adhesives," Reinhold, New York, 1962.
51. Problem Analysis Form, *National Starch Corp. Bull.*
52. Engineer's Guide to Plastics, *Mater. Eng.*, May 1972.
53. Trauernicht, J. O.: Bonding and Joining: Weigh the Alternatives, Part 1, Solvent Cements, Thermal Welding, *Plast. Technol.*, August 1970.
54. Gentle, D. F.: Bonding Systems for Plastics, in D. J. Alner (ed.), "Aspects of Adhesion," vol. 5, University of London Press, London, 1969.
55. Bodnar, M. J.: Bonding Plastics, in I. Skeist (ed.), "Handbook of Adhesives," Reinhold, New York, 1962.
56. H. F. Mark, N. G. Gaylord, N. M. Bikales (eds.): "Encyclopedia of Polymer Science and Technology," vol. 1, p. 536, Wiley, New York, 1964.
57. How to Fasten and Join Plastics, *Mater. Eng.*, March 1971.
58. Electromagnetic Bonding—It's Fast, Clean, and Simple, *Plast. World*, July 1970.
59. Hauser, R. L.: Ultra Adhesives for Ultrasonic Bonding, *Adhesives Age*, March 1969.
60. Delrin Acetal Resins Design Handbook, *E. I. du Pont de Nemours & Co., Tech. Bull.*
61. DeLollis, N. J., and O. Montoya: Surface Treatment for Difficult-to-Bond Plastics, *Adhesives Age*, January 1963.
62. King, A. F.: Adhesives for Bonding DuPont Plastics, E. I. du Pont de Nemours & Co., TR 152, August 1966.
63. Lexan Polycarbonate Resins Fabrication Data, General Electric Plastics Dept., TR CDC-446.

64. Abolins, V., and J. Eickert: Adhesive Bonding and Solvent Cementing of Polyphenylene Oxide, *Adhesives Age*, July 1967.
65. Cementing, Welding, and Assembly of Plastics, in "Plastic Engineers Handbook," chap. 17, p. 489, Society of the Plastics Industry, 1960.
66. Cox, D. R.: Some Aspects of Rubber to Metal Bonding, *Rubber J.*, April–May 1969.
67. Hemming, C. B.: Wood Gluing, in I. Skeist (ed.), "Handbook of Adhesives," Reinhold, New York, 1962.
68. Moser, F.: Bonding Glass, in I. Skeist (ed.), "Handbook of Adhesives," Reinhold, New York, 1962.
69. Weggemans, D. M.: Adhesive Charts, in "Adhesion and Adhesives," vol. 2, Elsevier, Amsterdam, 1967.
70. Krieger, R. B., and R. E. Politi: High Temperature Structural Adhesives, in D. J. Alner, (ed.), "Aspects of Adhesion," vol. 3, University of London Press, London, 1967.
71. Burgman, H. A.: The Trend in Structural Adhesives, *Machine Des.*, Nov. 21, 1963.
72. Kausen, R. C.: Adhesives for High and Low Temperatures, *Mater. Eng.*, August–September 1964.
73. Licari, J. J.: High Temperature Adhesives, *Prod. Eng.*, Dec. 7, 1964.
74. Litvak, S.: Polybenzimidazole Structural Adhesives for Bonding Stainless Steel, Beryllium and Titanium Alloys, AFML-TR-65-426, March 1966.
75. FM-34 Adhesive Film, *American Cyanamid Tech. Bull.* January 1968.
76. Bolger, J. C.: New One-Part Epoxies Are Flexible and Reversion Resistant, *Insulation*, October 1969.
77. Gahimer, F. H., and F. W. Nieske: Navy Investigates Reversion Phenomers of Two Elastomers, *Insulation*, August 1968.
78. Falconer, D. J., et al.: The Effect of High Humidity Environments on the Strength of Adhesive Joints, *Chem. Ind.*, July 4, 1964.
79. DeLollis, N. J., and O. Montoya: Mode of Failure in Structural Adhesive Bonds, *J. Appl. Polymer Sci.*, vol. 11, pp. 983–989, 1967.
80. Sharpe, L. H.: Aspects of the Permanence of Adhesive Joints, in M. J. Bodnar (ed.), "Structural Adhesive Bonding," Interscience, New York, 1966.
81. Tanner, W. C.: Adhesives and Adhesion in Structural Bonding for Military Material, in M. J. Bodnar (ed.), "Structural Adhesive Bonding," Interscience, New York, 1966.
82. Aerospace Adhesive EA929, Hysol Division, *Dexter Corp. Tech. Bull.* A5-129.
83. Roseland, L. M.: Structural Adhesives in Space Applications, in M. J. Bodnar (ed.), "Structural Adhesive Bonding," Interscience, New York, 1966.
84. Arlook, R. S., and D. G. Harvey: Effects of Nuclear Radiation on Structural Adhesive Bonds, Wright Air Development Center, *Rept.* WADC-TR-46-467, February 1957.
85. Pyles Industries, Inc., Form 1-810-100.
86. Carroll, K. W.: How to Apply Adhesives, *Prod. Eng.*, Nov. 22, 1965.
87. King, H. A.: Taking a Look at Tape and Film Adhesives, *Adhesives Age*, February 1972.
88. Bolger, J. C., and M. J. Lysaght: New Heating Methods and Cures Expand Uses for Epoxy Bonding, *Assembly Eng.*, March 1971.
89. Smith, D. F., and C. V. Cagle: A Quality Control System for Adhesive Bonding Utilizing Ultrasonic Testing, in M. J. Bodnar (ed.), "Structural Adhesive Bonding," Interscience, New York, 1966.
90. Hiler, M. J.: Adhesive Failures: Reasons and Preventions, *Adhesives Age*, January 1967.
91. Apple, W. R.: Infrared Inspection of Adhesive Bonds, *Adhesives Age*, July 1970.

Chapter 11

Commercial and Government Specifications and Standards

ARTHUR H. LANDROCK
and
NORMAN E. BEACH*

Plastics Technical Evaluation Center,
Picatinny Arsenal, Dover, New Jersey

* Mr. Beach, who prepared the tabular material used in the section on Identification of Materials Specifications, Test Methods, and Standards beginning on p. 11-62, passed away some months after submitting this material. Mr. Beach had retired from the Federal Service a few years earlier.

INTRODUCTION

The purpose of this chapter is to provide the reader with a working knowledge of specifications, standard practices, test methods, and other standardization documents involving plastics and elastomers and issued by standardizing agencies. Such agencies may be government, quasi-government, trade societies, and professional and technical societies. A recent government publication[1] lists 486 American organizations that consider standardization to be a major or important part of their work. An even more recent compilation[2] tabulates over 9,000 United States national and international standards on plastics and related materials in effect at the end of 1970. These standards are published by technical societies, trade associations, government agencies, and military organizations.

Standardization provides industry with economies through mass production, production of uniform goods, and savings in time and materials through standard designs, equipment, procedures, and testing. National technical and trade associations are in a position to represent entire industries and to bring the requirements of the consumer into line with current engineering and manufacturing practices. Some organizations are concerned with the preparation of standards, while others develop test methods or certification and labeling plans to guarantee that products meet the specified standards.[1]

The emphasis in this chapter is on standardization documents involving plastics and elastomers. Chapter 10 considers specifications involving adhesives. Nevertheless, the user of this handbook who is interested in adhesives specifications will do well to become acquainted with the contents of this chapter, for many of the sources described in detail are sources of adhesives specifications. The emphasis in the treatment to follow is on specifications and other documents involving plastic and elastomeric materials, rather than end items. Intermediate stages such as plastic laminates, films, and sheeting will be considered. On the other hand, no attempt will be made to consider the myriad of specifications involving end items fabricated from plastics and elastomers.

The standardization documents discussed in this chapter change frequently. The reader is therefore advised to consult the latest revision of any referenced document for up-to-date information.

HOW TO USE SPECIFICATION LITERATURE

There are a number of ways in which specifications and standards information in plastics and elastomers are used. The Plastics Technical Evaluation Center (PLASTEC) is the recipient of a number of requests covering all aspects of this area. Probably the most common request is: "What is Type I resin under MIL-P-XXXX?" Usually the caller is bidding on a defense contract and is not familiar with the specifications pertinent to the contract. The easiest way to answer this type of question is to refer to *PLASTEC Note* 6B,[3] Part 5, the numerical index, and determine the appropriate reference number, which can then be tracked down in Parts 1 and 2, where the complete title and other descriptive material are available. If a copy of the specification is on hand, either in hard copy or in microfiche form, it is a simple matter to scan it to determine the required information. The well-known periodically updated DOD Index of Specifications and Standards[4] (DODISS) can also be used, in this case by searching the numerical listing. This index covers all subjects, not just plastics and elastomers, and is therefore much bulkier than specialized sources. Because of its magnitude, the type size is considerably reduced, making the use of a magnifying glass almost a necessity.

Another common request is: "What specifications are there for low-density polyethylene resins, or for silicone resins, etc.?" It is convenient in questions of this type to go to *PLASTEC Note* 6B[3] and check the alphabetical subject index, or simply scan the entries themselves, which are listed alphabetically according to plastic family groupings, "Acetal, Acrylic, etc." This source has only a very limited number of specifications involving elastomers, however. The DODISS[4] may or may not be helpful. In this case, the alphabetical listing would be used. Only a few specifications are listed under "Polyethylene," but the applicable specifications are found under "Plastic Molding Materials." It should be noted that nowhere does the DODISS provide an alphabetical breakdown of all key words used in the specification titles. The alphabetization is based solely on the first significant word in the title, such as "Plastic Molding Material, Polyethylene, Low and Medium Density," L-P-390B(1), or "Plastic Molding and Extrusion Material, Ionomer Resin," MIL-P-46124. In the latter case, as in all Military Specifications, the letter after MIL- refers to the first word of the title.

A recently published source that can be very helpful in tracking down specifications and standards in particular subject areas is the National Bureau of Standards *World Index of Plastics Standards.*[2] This book is organized as a key-word-in-context (KWIC) index. To use it, one would look at the vertical center of the page for the subject word of interest, such as "Polyethylene," and then check the rest of the titles for the appropriate specification. Words following the subject word are used in alphabetical order. Details on this source are given below. The main advantage of this book is that it covers a large number of sources of specifications and standards (22 U.S. technical associations, 26 federal and state agencies, and 22 foreign standardization bodies). Only enough information is given to identify the standardization document. More information must be obtained from other sources.

SPECIFICATION SOURCES

U.S. Government Sources

U.S. Department of Commerce, National Bureau of Standards (NBS)

Index of U.S. Voluntary Engineering Standards.[5] This computer-produced index of 1,000 pages contains the permuted titles of more than 19,000 voluntary engineering and related standards, specifications, test methods, and recommended practices in effect as of the end of 1969, published by some 360 U.S. technical societies, professional organizations, and trade associations. It does not include federal and military, or company standards and specifications. The information contained in the index was compiled by the Information Section of the Bureau's Office of Engineering Standards Services. This is a key-words-in-context (KWIC) index. The title of each standard can be found under all the significant key words which it contains. These key words are arranged alphabetically down the center of each page, together with their surrounding context. The date of publication or last revision, the standard number, and an acronym designating the standards-issuing organization appear as part of each entry. A list of these acronyms and the names and addresses of the organizations which they represent are found at the beginning of the index.

When applicable, cross references to other organizations besides the group originally issuing the standard are provided. An example of the type of entries provided is shown in Table 1, where nine representative specifications or standards involving plastics are tabulated as they appear in the subject document. The reader will note that SAE 3842 has an AMS designation, meaning that this is an Aeronautical Material Specification published by the Society of Automotive Engineers.

PLASTEC has on hand only the microfiche version of this source. It is the writers' opinion that this less expensive form should be purchased only for initial screening to determine whether or not the hard copy should be procured. It should be remembered that this is a general, rather than a specialized compilation, and in all probability, documents of interest in plastics and elastomers covered in this source would be listed in NBS's World Index of Plastics Standards.[2]

World Index of Plastics Standards.[2] This computer-produced index of 458 pages contains the printed titles of more than 9,000 national and international standards on plastics and related materials, including adhesives and elastomers, which were in effect as of the end of December 1970. These standards are published by technical societies, trade associations, government agencies, and military organizations. The title of each standard can be found under all the significant key words which it contains. These key words are arranged alphabetically down the center of each page, together with their surrounding context. The date of publication or last revision designating the standard-issuing organization appears as part of each entry.

TABLE 1 Examples of Typical Entries in NBS Special Publication 329 Involving Plastics

	Conditioning	PI* 1P-51
	Plastic Films for Testing (1951)	MCA SG5
	Plastic Foams—Storage, Handling and Fabrication (1960)	SAE 1327
	Plastic Materials (1968)	NSA 3438
Test Method for Determining Cold Cracking of Flexible Vial, Plastic (1967)		FME 8-17
	Storage of Expanded Plastics and Rubber (1966)	AST D1708
Standard Method of Test for Tensile Properties of Plastics by Use of Microtensile Specimens (1966)		USC CS252
TFE-Fluorocarbon (Polytetrafluoroethylene) Resin Electrical Tubing (1963)		SAE 3842AMS
reinforced-Polytetrafluoroethylene Sheet, TFE Fluorocarbon Resin (1961) Asbestos Fiber		SPI ERF17
Method of Test for Deflection Temperature of Cured Epoxy Resins under Load (1964)		

* Acronyms listed in this table are for the following organizations:

PI	Packaging Institute, Inc.	FME	Factory Mutual System
MCA	Manufacturing Chemists Association	USC	U.S. Department of Commerce, Office of Engineering Standards Services, NBS
SAE	Society of Automotive Engineers, Inc.		
NSA	National Standards Association	SPI	Society of the Plastics Industry, Inc.
AST	American Society for Testing and Materials		

A list of those acronyms and addresses (even telephone numbers in most cases) is found at the beginning of the index.

A total of 124 standardizing agencies are listed. Those of perhaps the greatest interest and utility are the following:

American National Standards Institute, Inc.

American Society for Testing and Materials

British Standards Institution

Factory Mutual System

General Services Administration (for Federal Specifications)

International Organization for Standardization

Manufacturing Chemists' Association

Naval Publications and Forms Center (for all Military Specifications, standards, etc.)

National Electrical Manufacturers Association

National Fire Protection Association

National Flexible Packaging Association

National Standards Association

Packaging Institute, Inc.

Rubber Manufacturers Association

Society of Automotive Engineers, Inc.

Society of the Plastics Industry, Inc.

Underwriters' Laboratories, Inc.

U.S. Department of Commerce

The reader will find that this cloth-bound index will be very helpful in giving him a direct entry into most of the standards and specifications involving plastics and elastomers in which he may have any interest. The specific documents found by searching this source can then be tracked down by reference to the literature published by the standardizing agencies, much of which is described below. Table 2 shows representative sections from the actual index.

List of Voluntary Product Standards, Commercial Standards, and Simplified Practice Recommendations.[6] This brief publication, obtainable on a no-charge basis from the National Bureau of Standards, is intended to assist all interested parties in obtaining copies of Voluntary Product Standards, Commercial Standards, and Simplified Practice Recommendations published by the U.S. Department of Commerce. It contains subject and numerical indexes of the standards available, and instructions for ordering them.

In 1965, the Department of Commerce announced a change in the names of their voluntary standards developed cooperatively with industry groups. The two types of voluntary standards published by the Department prior to 1966 were referred to as "Commodity Standards" and were identified as Simplified Practice Recommendations (R) and Commercial Standards (CS). It was felt that it was desirable to combine Simplified Practice Recommendations and Commercial Standards into a single type document and to provide a new name. The term "Voluntary Product Standard" was chosen, and is being used to identify all new standards, as well as revisions of existing standards. The abbreviation "PS" is being used to identify the "Voluntary Product Standards." Existing Commercial Standards and Simplified Practice Recommendations will be identified by the abbreviations (CS) and (R), respectively, until such time as they are revised.

Perhaps the following description of how these documents are developed and used will be helpful. It should be remembered that Commercial Standards and Simplified Practice Recommendations, although still in existence, are being phased out by Voluntary Product Standards.

Voluntary Product Standards. Voluntary Product Standards are standards developed under procedures established by the Department of Commerce. The standards may include (1) dimensional requirements for standard sizes and types of various products, (2) technical requirements, and (3) methods of testing, grading, or marking. The objective of a Voluntary Product Standard is to establish requirements which are in accordance with the principal demands of the industry and, at the same time, are not contrary to the public interest.

The adoption and use of a Voluntary Product Standard is, as the title implies, voluntary. However, when reference to a Voluntary Product Standard is made in contracts, labels, invoices, or advertising literature, the provisions of the standard are enforceable through use of legal channels as a part of the sales contract. NBS has the following role in the development of Voluntary Product Standards. It:

TABLE 2 Examples of Typical Entries in NBS Special Publication 352

Polyvinyl Acetate	Adhesive (1970)	CNS	2676
Silicone Rubber	Adhesive, Air Drying (1964)	MIL	A25457
Tape, Pressure Sensitive	Adhesive, Aluminum Backed (1965)	FED	L780
Tape, Pressure Sensitive	Adhesive, Aluminum Foil Backed, for High Temperature Application (1966)	MIL	TN1287
	Adhesive, Bonding Vulcanized Synthetic Rubber to Steel (196 6)	FED	MMMA121
Tape, Pressure Sensitive	Adhesive, Contact, Chloroprene (1964)	FED	MMMA130
Tape, Pressure Sensitive	Adhesive, Corrosion Resistant (1962)	MIL	T4053
	Adhesive, Double Coated, Cellophane, Cellulose Acetate (196 2)	FED	UUT91
	Flexible Pipe Coupling (1966)	FME	1920
Rigid and	Flexible Plastic Cellular Foam Nomenclature and Test Conditions (1968)	UNI	6347
Rigid and	Flexible Plastic Foam Test for Apparent Density (1968)	UNI	6349
Rigid and	Flexible Plastic Foam Test for Linear Dimensions (1968)	UNI	6348
Mounting Bases,	Flexible Plastic Foam (1968)	MIL	M81288
Styrene Butadiene	Molding Material (1968)	FED	LP398
ABS Rigid	Molding Material (1968)	FED	LP1183
Urea Formaldehyde	Molding Material (1968)	FED	LP401
Plastic	Molding Material, Acetal, Glass Fiber Reinforced (1970)	MIL	P46137
Polyethylene	Molding Material, Low and Medium Density (1966)	FED	LP390
Polypropylene	Molding Material,injection and Extrusion (1964)	FED	LP394
Plastic Type of	Molding Material,polyester,compression Molded (1968	DIN	16911
Plastic Type of Standard Injected Molded Specimen (1/	Molding Material,polymethyl Methacrylate,acrylic Properties	DIN	7745
	Nylon Litter,nonrigid ,polcless (1952)	MIL	L16942
	Nylon Molded or Extruded (1955)	MIL	N18352
	Nylon Molded Parts Weather Resistant (1955)	MIL	M19098
Specification for Reinforced and Filled	Nylon Molding and Extrusion Materials (1970)	AST	D2897
	Nylon Molding Material Glass Filled (1964)	FED	LP395
	Nylon Molding Material (1966)	MIL	M20693
	Nylon Mountaineering Ropes (1959)	BSI	3104
Silicone Viscosity, Short Pot Life (1965)	Rubber Compound, Room Temp. Vulcanizing, 50,000 Centipoises	SAE	3364AAMS
Silicone Viscosity (1965)	Rubber Compound, Room Temp. Vulcanizing, 55,000 Centipoises	SAE	3366BAMS
Test for Staining of Polyvinyl Chloride Compositions by	Rubber Compounding Ingredients (1968()	AST	D2151
Test for	Rubber Compounds Chemical Analysis (1970)	FED	60116000
Nonsilver Staining Natural	Rubber Compounds (1959)	BSI	3106
Vulcanized Natural	Rubber Compounds (1962)	BSI	1154
Vulcanized Butadiene,acrylonitrile	Rubber Compounds (1962)	BSI	2751
Vulcanized Chloroprene	Rubber Compounds (1962)	BSI	2752
Low Compression Set Acrylonitrile Butadiene Vulcanized	Rubber Compounds (1962)	BSI	3222

1. Provides editorial assistance in the preparation of the standard.
2. Supplies such assistance and review as are required to assure the technical soundness of the standard.
3. Acts as an unbiased coordinator in the development of the standard.
4. Sees that the standard is representative of the views of producers, distributors, and users or consumers.
5. Seeks satisfactory adjustment of valid points of disagreement.
6. Determines the compliance with the criteria established in the Department's procedures cited above.
7. Publishes the standard.

Industry customarily:

1. Initiates and participates in the development of a standard.
2. Provides technical counsel on a standard.
3. Promotes the use of, and support for, the standard.

The Index of Standards provided in this pamphlet covers 89 subject areas. Those that appear to be of interest in plastics and elastomers are tabulated in Table 3. It is recognized that these standards involve end items to a considerable degree, rather than raw materials. Printed standards available from the Superintendent of Documents, Government Printing Office, Washington, D.C. 20402, are shown in the index with the actual cost in the price column. The prices quoted are for delivery to addresses in the United States and its territories and possessions and certain countries which extend the franking privilege. In the case of all other countries, one-fourth the cost of the publication should be added to cover postage. Remittances should be made either by coupons (obtainable from the Superintendent of Documents in denominations of 5, 10, 25, and 50 cents, and good until used) or by check or money order payable to the Superintendent of Documents.

If the symbol M appears in the price column, the standard is out of print. Requests for single copies of these standards or inquiries regarding their status should be sent to the Office of Engineering Standards, NBS, Washington, D.C. 20234.

U.S. Department of Defense (DOD)

Department of Defense Index of Specifications and Standards (DODISS).[4] This index and supplements may be purchased from the Superintendent of Documents. It is also available for reference at most military installations and procurement offices and at all Small Business Administration (SBA) field offices. The DODISS is comprised of two parts as follows:

Part I—Alphabetical Listing (514 pages for 1 July 1972)

Part II—Numerical Listing (598 pages + 68 pages for July 1, 1972)

Part II—Appendix—A cumulative listing of canceled documents formerly listed in the DODISS. This appendix, published on a triennial basis, is kept current by adding to it cumulative cancellation listings of the DODISS Part II Supplement (green pages) and subsequent canceled listings appearing in the basic DODISS, Part II. This appendix will not be distributed. Military activities desiring copies can obtain them by submitting DD Form 1425 to: Commanding Officer, Naval Publications and Forms Center, 5801 Tabor Avenue, Philadelphia, Pa. 19120.

Private industry and individuals may purchase the Appendix from: Superintendent of Documents, Government Printing Office, Washington, D.C. 20402.

The price is $2.00 for domestic mailing and $2.50 for foreign.

The DODISS lists the unclassified federal, military, and departmental specifications, standards, and related standardization documents, and those industry documents which have been adopted by the Department of Defense. Other standardization documents developed by industry must be obtained from other sources.

An example of the DODISS Part I is shown in Table 4.

In this listing L indicates a limited coordination document (see Military Specifications below for explanation) and Q indicates that a Qualified Products List (list of manufacturers' products tested and known to meet the requirements of the specification) has been issued. The Military Specifications are those beginning with MIL and the Federal Specifications are those with letter combinations, such as L-P-383 and ZZ-R-001415. The FSC number is the Federal Supply Classification Code listing. PREP is the code for the preparing agency, and those designations under

TABLE 3 List of Voluntary Product Standards, Commercial Standards, and Simplified Practice Recommendations Apparently Involving Plastics and Elastomers from NBS List of Publications 53

Standard	Price
Voluntary Product Standards:	
PS 3-66 TFE-fluorocarbon (polytetrafluoroethylene) resin skived tape	$0.10
PS 10-69 Polyethylene (PE) plastic pipe (schedule 40—inside diameter dimensions)	0.15
PS 11-69 Polyethylene (PE) plastic pipe (SDR)	M
PS 12-69 Polyethylene (PE) plastic pipe (schedules 40 and 80—outside diameter dimensions)	0.15
PS 13-69 Uncored slab urethane foam for bedding and furniture cushioning	0.10
PS 15-69 Custom contact-molded reinforced-polyester chemical-resistant process equipment	0.20
PS 17-69 Polyethylene sheeting (construction, industrial, and agricultural applications)	0.15
PS 18-69 Acrylonitrile-butadiene-styrene (ABS) plastic pipe (schedules 40 and 80)	0.15
PS 19-69 Acrylonitrile-butadiene-styrene (ABS) plastic pipe (standard dimension ratio)	0.15
PS 21-70 Poly(vinyl chloride) (PVC) plastic pipe (schedules 40, 80, and 120)	0.15
PS 22-70 Poly(vinyl chloride) (PVC) plastic pipe (standard dimension ratio)	M
PS 24-70 Melamine dinnerware (alpha-cellulose-filled) for household use	0.10
PS 25-70 Heavy-duty alpha-cellulose-filled melamine tableware	0.10
PS 26-70 Rigid poly(vinyl chloride) (PVC) profile extrusions	0.10
PS 28-70 Glass stopcocks with polytetrafluoroethylene (PTFE) plugs	0.25
PS 29-70 Plastic heat-shrinkable film	0.15
PS 31-70 Polystyrene plastic sheet	0.15
PS 33-70 Polytetrafluoroethylene (PTFE) plastic lined steel pipe and fittings	0.10
PS 34-70 Fluorinated ethylene-propylene (FEP) plastic lined steel pipe and fittings	0.10
PS 35-70 Poly(vinylidene fluoride) (PVF$_2$) plastic lined steel pipe and fittings	0.10
PS 43-71 Fluorinated ethylene-propylene (FEP) plastic tubing	0.10
PS 47-71 Heat-shrinkable fluorocarbon plastic tubing	0.20
PS 52-71 Polytetrafluoroethylene (PTFE) plastic tubing	0.20
PS 53-72 Glass-fiber reinforced polyester structural plastic panels (supersedes CS 214-57)	0.20
PS 55-72 Rigid poly(vinyl chloride) (PVC) plastic siding	M

Commercial Standards:

CS 182-51	Latex foam mattresses for hospitals (under revision)	$ M
CS 192-53	General-purpose vinyl plastic film (amendments 1 and 2)	M
CS 201-55	Rigid polyvinyl chloride sheets (reprinted 1963)	M
CS 209-57	Vinyl chloride plastics garden hose	0.10
CS 227-59	Polyethylene film (amended 1963) (under revision)	M
CS 230-60	Vinyl plastic weatherstrip	M
CS 239-63	TFE fluorocarbon (polytetrafluoroethylene) resin sheet	M
CS 245-62	Vinyl-metal laminates	M
CS 247-62	TFE-fluorocarbon (polytetrafluoroethylene) resin flexible hose (wire braid reinforced)	M
CS 248-64	Vinyl-coated glass fiber insect screening and louver cloth	0.10
CS 257-63	TFE-fluorocarbon (polytetrafluoroethylene) resin molded basic shapes (errata 1963)	M
CS 258-63	Expanded vinyl fabrics for apparel use	M
CS 273-65	Expanded vinyl fabric for furniture upholstery use	0.10
CS 274-66	TFE-fluorocarbon (polytetrafluoroethylene) resin sintered thin coatings for dry film lubrication	M

Simplified Practice Recommendations:

R 242-53	Vinyl- and pyroxylin-coated fabrics (amended June 1962)	M

TABLE 4 Typical Section of DODISS Part I—Alphabetical Listing

ALPHABETICAL LISTING

	TITLE	DOCUMENT NUMBER	FSC	PREP	DATE	CUSTODIAN
L	Plastic Material, Heat Resistant, Low Pressure Laminated Glass Fiber Base, Polyester Resin	MIL-P-25395A	9330	AS	23 Dec 63	11
Q	Plastic Material, Laminated Phenolic, For Bearings (Water Or Grease Lubrication)	MIL-P-18324D (2)	9330	SH	14 Jun 68	MR SH 84
	Plastic Material, Laminated Thermosetting Sheets, Nylon Fabric Base, Phenolic-resin	MIL-P-15047C	9330	SH	11 Jun 71	EL SH 11
Q	Plastic Material, Laminated, Thermosetting, Electric Insulation, Sheets, Glass Cloth, Silicone Resin	MIL-P-997D (2)	5970	SH	14 Jun 68	EL SH 11
	Plastic Material, Molding, Acrylic, Colored And White, Heat Resistant, For Lighting Fixtures	MIL-P-19735B	9330	SH	21 Mar 62	MR AS 11
	Plastic Material, Phenolic Resin, Glass-fiber Base, Laminated	MIL-P-25515C	9330	AS	04 Dec 68	MR AS
	Plastic Material, Polyester Resin, Glass Fiber Base, Filament Wound Tube	MIL-P-82540 (1)	9330	OS	03 Sep 70	OS 11
	Plastic Material, Polyester Resin, Glass Fiber Base, Low Pressure Laminated	L-P-383	9330	/AS	24 Oct 67	MR AS 11
L	Plastic Material, Pressure Sensitive, For Aircraft Identification And Marking	MIL-P-33477	9330	11	27 Jul 66	11
	CHANGE 1 28 JUNE 1967					
	Plastic Material, Silicone Resin, Glass Fiber Base, Low-pressure Laminated (Asg)	MIL-P-25518A (1)	9330	AS	20 Oct 64	AS 11
L	Plastic Material, Unicellar, Sheet, Elastomeric	MIL-P-24333 (2)	9330	SH	17 Jun 70	SH
	Plastic Material, Unicellular (Sheets And Tubes)	MIL-P-15280E INT AMD 1 (SH)	9330	SH	17 Jun 68	EL SH 11
	Plastic Materials, Asbestos Base. Phenolic Resin, Low Or High Pressure Laminates (Asg)	MIL-P-25770A	9330	AS	27 Nov 61	AS 11
	Plastic Mix, Refractory	HH-P-00430	9350	‡	01 Mar 65	
L	Plastic Mix, Refractory (Water-wall, Boiler, Chrome Ore Plastic)	MIL-P-15384E	9350	SH	22 Apr 71	SH
Q	Plastic Mix, Refractory, Fire Clay, Super Duty	MIL-P-15731D	9350	SH	25 May 71	SH 84
	Plastic Molding (And Extrusion) Material, Nylon, Glass Fiber Reinforced	L-P-395B	9330	/MR	13 Nov 69	MR SH 11
	Plastic Molding And Extrusion Material, Cellulose Acetate Butyrate	L-P-349C NOTICE 2 (11)	9330	*EL	20 Jan 71	EL AS
L	Plastic Molding And Extrusion Material, Ethyl Cellulose	MIL-P-22985A (1)	9330	MR	23 Aug 68	MR OS 11
L	Plastic Molding And Extrusion Material, Ionomer Resins	MIL-P-46124	9330	MR	15 Aug 67	
L	Plastic Molding And Extrusion Material, Phenoxy	MIL-P-46088A	9330	MR	10 Jul 68	
L	Plastic Molding And Extrusion Material, Poly(Aryl Sulfone Ether) Resin, Thermoplastic	MIL-P-46133	9330	MR	30 Aug 68	
L	Plastic Molding And Extrusion Material, Polyphenylene Oxide	MIL-P-46115A	9330	MR	07 May 71	
L	Plastic Molding And Extrusion Material, Polyphenylene Oxide, Modified	MIL-P-46129 (1)	9330	MR	05 Aug 68	
L	Plastic Molding And Extrusion Material, Polyphenylene Oxide, Modified, Glass Fiber Reinforced	MIL-P-46131	9330	MR	15 Nov 68	
L	Plastic Molding And Extrusion Material, Polysulfone	MIL-P-46120	9330	MR	17 Mar 67	

CUSTODIAN represent the Army, Navy, and Air Force custodians of the documents. These agencies are responsible for coordinating the documents. Full details on codes are given in the front of the compilation. Letters such as A and B after the specifications refer to revisions, the original document having no letter designation. Numbers in parentheses after specifications are changes, usually minor revisions, additions, or deletions. Details on the specifications themselves will be given later in this chapter.

An example of the DODISS Part II is shown in Table 5.

This table is taken from Part II, Section A, which takes up 598 pages in the July 1, 1972, issue. This section is a cumulative listing of documents numerically by document number. Section A includes the additions and changes to standardization documents listed in Section B.

Part II, Section B (68 pages) is a ready-reference listing of additions, changes, and cancellations to standardization documents since the previous bimonthly supplement. This list will be issued with supplements effective Sept. 1, Nov. 1, Jan. 1, and May 1. The supplements include changes in entries and cancellations.

The cumulative collection of canceled documents, formerly included in Section A of the Part II—Numerical Listing basic document, is no longer published. Instead, Section A will contain an annual compilation of documents canceled during the previous year. The cumulative collection of canceled documents is now published on a triennial basis as an Appendix to the Part II basic document. DODISS holders having a need for a complete list of canceled documents should have the following cancellation lists:

DODISS Part II, July 1, 1972, or the current issue
DODISS Part II Appendix, July 1, 1972, or the current triennial issue

The cancellation listing may be kept current by adding to it the cumulative cancellation listing of DODISS Part II Supplement (A Section) until the next basic revision of DODISS, in this case July 1, 1973, or later.

The organization of the Numerical Listing (Part II) is as follows:

Military Specifications
Military Standards
Federal Specifications
Federal Standards
Qualified Products List
Industry Documents*
International Standardization Documents
Military Handbooks
Federal Handbooks
Air Force–Navy Aeronautical Standards
Air Force–Navy Aeronautical Design Standards
Air Force–Navy Aeronautical Specifications
U.S. Air Force Specifications
Other Departmental Documents
Air Force–Navy Aeronautical Bulletins
U.S. Air Force Specifications Bulletins

As indicated above, the Q designation in both the alphabetical and numerical listings indicates that a Qualified Products List exists for the specifications. This list, found toward the back of Part II, gives the official designation of the QPL for each specification. Thus, MIL-P-18324D (2), as shown in Table 4, has a Q designation and is listed as QPL-18324-21 in the left-hand Document Number column. The title of the specification is then given, along with the FSC Category, Preparing Agency, and Date of Issue of the QPL. The number 21 in the QPL number means that this is the 21st revision of QPLs issued for this specification.

An example of the DODISS FSC Listing is shown in Table 6.

This section of the listing was chosen to show the transition between the FSC

* Include AIA, AMS, ANSI, ASTM, AWS, IEEE, MCA, MMPA, NEMA, NFPA, and RWMA.

TABLE 5 Typical Sections of DODISS Part II—Numerical Listing

NUMERICAL LISTING
MILITARY SPECIFICATIONS

DOCUMENT NUMBER	TITLE	FSC	PREP	DATE	CUSTODIAN	
MIL-R-43024C (1)	Refrigerator, Mechanical, Commercial, Portable, Reach-in, 70 Cubic Foot, less Mechanical Unit	4110	GL	20 Oct 69	GL	YD 82
MIL-S-43026C	Sewing Machines, Industrial, Complete With Carrying, Shipping And Storage Cases	3530	GL	24 Apr 70	GL	MC 84
MIL-P-43027C	Paper, Chart, Map And Book, Lithographic Finish, Coated And Uncoated	9310	GL	21 Mar 69	GL	84
MIL-L-43028A	Loading Standard Assembly, Liquid Transfer, Bulk	3990	ME	05 Sep 68	ME	SA 84
MIL-R-43031D	Refrigeration Units, Mechanical, Panel Type, 3000 Btu/hr Capacity	4110	GL	31 Mar 70	GL	YD 82
MIL-S-43032A	Saw, Hand, Crosscut, 24 Inch Blade	5110	GL	28 Feb 67	MU	
MIL-C-43033	Cloth, Cotton, Water Repellant	1395	MU	10 Jul 61	MU	
MIL-S-43034A	Skid, Platform	3990	ME	03 Dec 65	ME	
MIL-C-43035A (1)	Chair, Straight, Kitchen	7105	GL	07 Jun 68	MU	
MIL-T-43036A (2)	Tape, Pressure Sensitive Adhesive, Plastic Film, Fildment Reinforced (For Sealing Fiber Containers And Cans)	8135	MU	16 Mar 67	MU	AS 84
MIL-P-43038A	Plastic Molding Material, Polyester, Low Pressure Laminating, High Temperature Resistant	9330	MR	10 May 68	MR	
MIL-R-43040B	Removal Tools, Connector Electrical Contact	5120	GL	02 Mar 72	GL	SH
MIL-P-43043A	Plastic, Polyester, Glass Fiber Filled, Pre-mix Molding Compound	9330	MR	22 Mar 67	MR	

NUMERICAL LISTING
FEDERAL SPECIFICATIONS

DOCUMENT NUMBER	TITLE	FSC	PREP	DATE	CUSTODIAN	
L-P-00370B	Plastic Film (Polyvinyl Chloride And Copolymer Of Vinylidene Chloride And Vinyl Chloride)	8135	†	24 Jun 70	SH	
L-P-375C (1)	Plastic Film, Flexible, Vinyl Chloride	8135	/GL	23 Jul 71	GL	SH
L-P-377B	Plastic Sheet And Strip, Polyester	*8135	/GL	28 Feb 66	GL	SA
L-P-378C	Plastic Sheet And Strip, Thin Gauge, Polyolefin	8135	/GL	30 Oct 69	GL	SA 69
INT AMD 2						
L-P-380B	Plastic Molding Material, Methacrylate	9330	/MR	05 Jan 68	MR	11
L-P-383	Plastic Material, Polyester Resin, Glass Fiber Base, Low Pressure Laminated	9330	/AS	24 Oct 67	MR	AS 11
L-P-385B	Plastic Molding Material, Polychlorotrifluoroethylene	9330	/MR	15 Oct 68	MR	SH 11
L-P-00386A	Plastic Material, Cellular Urethane (Flexible)	9330	†	15 May 68		11
L-P-387A	Plastic Sheet, Laminated, Thermosetting(For Designation Plates)	9330	*EL	29 Oct 63	EL	SH 11
INT AMD 2 (SH)						
L-P-389A (1)	Plastic Molding Material, Fep Fluoro-carbon, Molding And Extrusion	9330	/EL	11 Feb 65	EL	SH 11
L-P-390C	Plastic, Molding And Extrusion Material, Polyethylene And Copolymers (Low, Medium, And High Density)	9330	*EL	10 Aug 71	EL	SH 11
L-P-391C	Plastic, Sheet, Rods And Tubing, Rigid, Multiapplication (Methacrylate, Cast)	9330	/MR	28 May 71	MR	SH 11
L-P-392A	Plastic Molding Material, Acetal, Injection And Extrusion	9330	*EL	29 Jan 64	EL	SH 11
L-P-393A (2)	Plastic Molding Material, Polycarbonate, Injection And Extrusion	9330	/EL	22 Jun 70	EL	SH 11

11-12

TABLE 6 Typical Section of DODISS FSC Listing of DOD Standardization Documents

FSC LISTING

FSC	TITLE	DOCUMENT NUMBER	PREP	DATE	CUSTODIAN
9320	Testing Hard Rubber Products USER-AT MC OS REV-AV GL ME EL AS YD 82 GS	ASTM-D530 / D530-68	MR	20 Jan 70	MR SH
9320	Testing Hard Rubber Products. USER-AT OS MC REV-EL ME GL AV YD AS 82 GS	ASTM-530 / D530-68	MR	20 Jan 70	MR SH 85
9320	Testing Rubber O-rings USER-AT MC OS REV-AV GL EL ME GS	ASTM-D1414 / D1414-68	MR	20 Jan 70	MR SH
9320	Tile, Rubber, Underwater Acoustic Decoupler	MIL-T-24180	SH	01 Dec 65	SH
9320	Visual Inspection Guide For Rubber Sheet Material REV-WC 82	MIL-STD-289A	SH	02 Dec 68	MR SH 11
9330	Allyl Plastic Sheet, Rod, And Tube Validat may 68 USER-MS REV-EL SH YD	L-A-499	/AS	27 Aug 52	MR AS
9330	Asbestos Felting (B) Stage Phenolic Resin Impregnated Low Pressure Molding	AMS-3858 / 3858	AS	29 Jan 71	MR AS 11
9330	Asbestos, Polytetrafluoroethylene Impregnated Sintered	AMS-3840 / 3840	AS	29 Jan 71	MR AS 11
9330	Baffle Material, Aircraft Fuel Tank REV-82	MIL-B-83054 (1)	11	25 Feb 71	11
9330	Boron Monofilament, Continuous, Vapor Deposited REV-84	MIL-B-83353	11	21 Dec 71	11
9330	Cast Polymethyl Methacrylate Thick Windows	MIL-C-24449	SH	01 Mar 71	SH
9330	Cellulose - Nitrate - Plastic (Celluloid Or Pyroxylin Type) (For Use In Ammunition)	MIL-C-15567 (1)	OS	17 Mar 67	OS
9330	Cellulose Acetate Sheet (For Inhibitors)	MIL-C-17576B	OS	10 Sep 71	MU OS
9330	Cloth, Coated, Viny' Coated (Artificial Leather)	MIL-HDBK 700 / CHANGE 1 / 1 JUN 1966	MR	01 Nov 65	MR
9330	Coating, Conformal, Resin Validat jan 71	MIL-C-22627 (1)	AS	19 Nov 64	AS
9330	Coating, Pipe, Thermoplastic Resin Or Thermosetting Epoxy USER-SH REV-MU CE MR MD	L-C-5308	/ME	04 Jun 70	ME YD 84
9330	Core Material, Cellular Cellulose Acetate Validat apr 68 REV-SH	MIL-C-18345A	AS	21 Nov 62	AS
9330	Core Material, Foamed-in-place, Urethane Type REV-MI 84	MIL-C-8087C	AS	24 Apr 68	AS 11
9330	Ethyl Cellulose Molded Plastics	L-E-210	‡	05 Feb 54	
9330	Fabric, Wire Reinforced Asbestos Polytetrafluoroethylene Impregnated, Sintered	AMS-3839 / 3839	AS	29 Jan 71	MR AS 11
9330	Fiber Sheet, Vulcanized USER-EL MI SM REV-MR MD 45	MIL-F-10336D	GL	30 Dec 71	GL
9330	Film, Elastomeric, Fluorescent, For Weapons Systems (Asg) Validat nov 70 REV-11	MIL-F-22735B	AS	25 Feb 65	AS 11

categories for rubber and plastics fabricated materials. The Defense Standardization Directory (SD-1)[7] and a similar source[8] listing FSC Class and Area Assignments show the following breakdown for Nonmetallic Fabricated Materials:

Group 93GP Nonmetallic Fabricated Materials
9310   Paper and Paperboard
9320   *Rubber Fabricated Materials*
9330   *Plastics Fabricated Materials*
9340   Glass Fabricated Materials
9350   Refractories and Fire Surfacing Materials
9390   Miscellaneous Fabricated Nonmetallic Materials

The term "fabricated" may be misleading to the reader since, by glancing through the index, it can be seen that plastic resins are included in the 9330 category.

The use of the DODISS is mandatory on all military activities. This means that the Federal and Military Specifications, standards, and related standardization documents must be considered in identifying items for procurement actions.

Military Handbooks. These handbooks are generally guides to a subject area and are often concerned with covering the salient properties and intended uses of current Federal and Military Specifications. Military activities desiring copies can obtain them by submitting DD Form 1425 to the Naval Publications and Forms Center, 5801 Tabor Avenue, Philadelphia, Pa. 19120. Others may obtain copies by sending a check for the purchase price to the Superintendent of Documents. Unfortunately this price is not listed in the DODISS listing of Military Handbooks. If the preparing agency is known (see the DODISS listing), contact with the agency will usually be wise to obtain the price.

The following is a list of handbooks obviously of interest in the fields of plastics and elastomers:

MIL-HDBK-17A   Plastics for Aerospace Vehicles, Part I—Reinforced Plastics
MIL-HDBK-17   Plastics for Flight Vehicles, Part II—Transparent Glazing Materials
MIL-HDBK-23A   Structural Sandwich Composites
MIL-HDBK-132   Protective Finishes
MIL-HDBK-139   Plastic (sic), Processing of
MIL-HDBK-149A   Rubber and Rubber-like Materials
MIL-HDBK-212   Gasket Materials (Nonmetallic)
MIL-HDBK-223   Coded List of Materials
MIL-HDBK-304   Package Cushioning Design
MIL-HDBK-691A   Adhesives
MIL-HDBK-692   A Guide to the Selection of Rubber O-Rings
MIL-HDBK-695   Rubber Products, Shelf Storage Life
MIL-HDBK-696   Plastic Coating Compound, Strippable, Hot and Cold Dip
MIL-HDBK-699   A Guide to the Specifications for Flexible Rubber Products
MIL-HDBK-700   Plastics (listing in DODISS July 1, 1972, issue is incorrect)
MIL-HDBK-725   Adhesives—A Guide to Their Properties and Uses as Described by Federal and Military Specifications

Examples of typical pages in two particular Military Handbooks are given in Tables 7 and 8. These formats are not typical of all Military Handbooks, however. MIL-HDBK-699 is completely in tabular form, as shown in the table. On the other hand, MIL-HDBK-17A presents information in a number of different ways, including discussion, tables of information and other data, and graphical presentations, including plots of data. Table 8 shows part of the coverage on epoxy resins. This handbook has a great deal of information on laminates made with epoxy resin, including stress-strain curves.

Military Standards. Military Standards, like Federal Standards, are issued to eliminate waste and improve efficiency in operations. They are available in two forms, those listed as MIL-STD-XXXX, which are essentially general in nature and are issued in book format, and those listed as MS-XXXX, which are specific to particular end items or parts and are issued in sheet format. The book-format standards are used for the comprehensive presentation of engineering practices (including test methods), procedures, processes, codes, safety requirements, symbols, abbreviations, nomenclature, type designations, and characteristics for standard equipments,

TABLE 7 Typical Page from MIL-HDBK-699, A Guide to Specifications for Flexible Rubber Products

MIL-HDBK-699(MR)

MULTIPURPOSE RUBBERS

Environment and Test	Temp. °F	Time, Hrs.	Specification Requirements	QPL Issued	Elastomer Commonly Used	FSC Class	Specification Title	Specification Number
Air, original properties	Ambient	–	T, 1000-3500 psi; E, 100-600%; H, 30-90		Natural rubber	9320	Rubber Composition, Vulcanized General Purpose, Solid (Symbols and Tests)	MIL-STD-417 Type R, Class RN
Cooling medium, ASTM D746	-67	–	F2 suffix, no cracks					
Cooling medium, ASTM D746	-40	–	F1 suffix, no cracks					
Air, compression set	158	22	Basic cmpd., 50% max; B suffix, 25% max.					
Air, original properties	Ambient	–	T, 1000-2500 psi; E, 100-400%; H, 30-90		SBR or Butyl	9320	Rubber Composition, Vulcanized General Purpose, Solid (Symbols and Tests)	MIL-STD-417 Type R, Class RS
Cooling medium, ASTM D746	-67	–	F2 suffix, no cracks					
Cooling medium, ASTM D746	-40	–	F1 suffix, no cracks					
Air, oven aging	212	70	A1 suffix, max. change: T, -25%; E, -35%					
Air, original properties	Ambient	–	T, 500-2000 psi; E, 100-400%; H, 40-90		Polysulfide	9320	Rubber Composition, Vulcanized General Purpose, Solid (Symbols and Tests)	MIL-STD-417 Type S, Class SA
Cooling medium, ASTM D746	-67	–	F2 suffix with H, 40-60, no cracks					
Cooling medium, ASTM D746	-40	–	F2 suffix, no cracks					
Air, oven aging	212	70	A1 suffix, max. change: T, -10%; E, -30%					
ASTM #3 petroleum oil	212	70	Volume change, 0 to +10%					
Air, original properties	Ambient	–	T, 500-2000 psi; E, 100-450%; H, 40-90		Nitrile	9320	Rubber Composition, Vulcanized General Purpose, Solid (Symbols and Tests)	MIL-STD-417 Type S, Class SB
Cooling medium, ASTM D746	-67	–	F2 suffix, no cracks					
Cooling medium, ASTM D746	-40	–	F2 suffix, no cracks					
Air, compression set	212	70	Basic cmpd., 60% max; B suffix, 35% max.					
ASTM #3 petroleum oil	212	70	Volume change, 0 to +40%					
Air, original properties	Ambient	–	T, 500-2500 psi; E, 200-400%; H, 30-90		Chloroprene	9320	Rubber Composition, Vulcanized General Purpose, Solid (Symbols and Tests)	MIL-STD-417 Type S, Class SC
Cooling medium, ASTM D746	-67	–	F2 suffix with H, 30-60, no cracks					
Cooling medium, ASTM D746	-40	–	F2 suffix, no cracks					
Air, compression set	158	22	50% max.					
ASTM #3 petroleum oil	212	70	Volume change, 0 to +120%					
Air, original properties	Ambient	–	T, 500-1200 psi; E, 50-500%; H, 40-80		Silicone	9320	Rubber Composition, Vulcanized General Purpose, Solid (Symbols and Tests)	MIL-STD-417 Type T, Class TA
Cooling medium, ASTM D746	-67	–	No cracks					
Cooling medium, ASTM D746	-103	–	F3 suffix, T, -50%; H, +15					
Air, oven aging	437	22	Max. change: T, -50%; E, -50%; H, +10					
ASTM #1 petroleum oil	302	70	E1 suffix, volume change, 0 to +20%					
Air, original properties	Ambient	–	T, 500-1500 psi; E, 100-200%; H, 40-80		Polyacrylate	9320	Rubber Composition, Vulcanized General Purpose, Solid (Symbols and Tests)	MIL-STD-417 Type T, Class TB
Air, oven aging	347	70	Max. change: T, -30%; E, -50%; H, +10					
Air, compression set	302	70	60% max.					
ASTM #3 petroleum oil	302	70	Volume change, 0 to +20%					
(All Grades Covered by this Document are also Covered Under MIL-STD-417 and ASTM D2000								
(Many grades available. Following is a typical example) Air, original properties	Ambient	–	T, 2000 psi. min; E, 500% min; H, 45 ± 5		Natural		Elastomeric Compounds for Automotive Applications	ASTM D735 (SAE J14)
Air, oven aging	158	70	Min values: T, 1500 psi; E, 400%					
(Many grades available. Following is a typical example) Air, original properties	Ambient	–	T, 1500 psi. min; E, 400% min; H, 55 ± 5		Chloroprene		Latex Dipped Goods and Coatings for Automotive Applications	ASTM D1764 (SAE J19) Type LR
Air, oven aging	212	70	Min values: T, 1200 psi; E, 300%				Latex Dipped Goods and Coatings for Automotive Applications	ASTM D1764 (SAE J19) Type LS, Class LSC
ASTM #2 petroleum oil	212	22	Volume change % max: Basic cmpd, 80; E2 suffix, 50					
Air, original properties	Ambient	–	T, 500-3500 psi; E, 75-600%; H, 30-90		Natural rubber, SBR or butyl		Elastomeric Materials For Automotive Applications	ASTM D2000 (SAE J200) Material AA
Cooling medium, ASTM D746	-40	–	F17 suffix, pass ± 30%					
Air, oven aging	158	70	Max. change: T, ± 30%; E, -50%; H, ± 15					
Air, compression set	158	22	Basic cmpd, 50% max; B13 suffix, 25% max.					

TABLE 8   Typical Page from MIL-HDBK-17A, Plastics for Aerospace Vehicles, Part I, Reinforced Plastics

CHAPTER 2

MATERIALS AND PROCESSES

2.1 INTRODUCTION

Information on resin systems, reinforcements and fabrication processes is presented in this chapter. The resin systems which are used for reinforced composites include the epoxies, phenolics, polyesters, polyimides and silicones. Of these, the epoxies, which exhibit superior mechanical properties, are of major importance.

Fiberglass, boron and graphite are the principal reinforcing materials for aerospace structural composites. Fiberglass is utilized in various forms such as woven fabrics, unidirectional tapes, nonwoven broadgoods, or roving. Boron and graphite are utilized as continuous high modulus filaments in nonwoven forms. Other reinforcements which find use in applications such as heat shields or other ablative or heat resistant components are listed herein and briefly described.

The fabrication processes of specific interest for aerospace parts are the hand-layup bag-molding techniques such as vacuum bagging, autoclave molding and combined vacuum/autoclave processes. Press molding with matched dies is used to a lesser extent. The filament winding process for handling roving is used extensively for such aerospace components as rocket motor cases, pressure vessels, radomes and others.

2.2 RESINS

2.2.1 EPOXY RESINS

2.2.1.1 *Definition.* Epoxy resins are polymers containing one or more epoxide groups ($-CH-CH-$) and which can be converted to a thermoset stage by reaction with appropriate curing agents. This definition is applicable to a broad class of resins which differ in properties and end use. The epoxy resins referred to in the handbook are therefore identified by specific chemical type and are limited to resins for laminating and filament winding.

2.2.1.2 *Epoxy Types.* The more common epoxy resins for aerospace applications are conventional epoxy (diglycidyl ether of bisphenol A) and epoxy novolacs (glycidyl ethers of novolacs). Other types, which are used to a lesser extent, include cycloaliphatics based on diepoxide carboxylates or adipates, glycidyl ethers of various phenols or bisphenols, and resins based on cyclopentyl ethers. Conventional epoxy is frequently blended with a solid polyfunctional resin (tetraglycidyl ether of tetraphenylethane).

2.2.1.3 *Specifications.* Performance requirements and classifications of epoxy resins are given in MIL-R-9300, Resin, Epoxy, Low-Pressure Laminating. This specification classifies epoxy resins as either general purpose or heat resistant type. Both types are further classified into electrical grades depending on the dielectric constant. MIL-R-9300 resins may be qualified as liquids or solids suitable for impregnating and wet layups or as impregnated fabrics for dry layup.

2.2.1.4 *Epoxy Resin Systems.* A mixture of a base resin (or blend of resins) and a curing agent (or combination of curing agents and accelerators) is generally referred to as the epoxy resin system. Some systems may also contain diluents, fillers or modifiers. Diluents serve mainly to reduce resin viscosity prior to cure and are not recommended for aerospace use. Small amounts of fillers (approximately 2 to 3 percent) are added to some formulations to prevent excessive resin flow during cure. Modifiers are usually either polyamides or nitriles, and are added to adhesive-type systems to increase the flexibility of the cured resin.

2-1

TABLE 8   Typical Page from MIL-HDBK-17A, Plastics for Aerospace Vehicles, Part I, Reinforced Plastics (Continued)

2.2.1.5 *Curing of Epoxy Resins.* Epoxy resins are converted to a cured thermoset stage by chemical combination with curing agents or by the action of catalysts. Both curing mechanisms normally take place without the evolution of by-products. Most aerospace systems are cured by direct reaction with curing agents.

The temperatures at which curing reactions occur depend primarily on the specific curing agent or catalyst. Reactions may be initiated at ambient temperature or they may require the addition of heat. Nearly all aerospace resin formulations employ elevated temperature curing systems. For general purpose systems, cures are typically at about 250F, while heat resistant systems are cured from 300F to 350F with postcures at 400F or higher.

Optimum cure cycles for any system are determined empirically. The variables to be considered include the concentration of curing agents, accelerators or catalysts; the time-temperature schedules for B-staging, initial cure and postcure; and the interaction of these factors. Normally each resin property is optimized independently, and the usual practice is to establish conditions which yield a desirable combination of properties.

2.2.1.6 *Curing Agents.* Various curing agents for epoxy resins are listed below, and are classified according to chemical type (Ref. 2-1):

Aromatic Amines: These agents, which require elevated temperature cures, are widely used in laminating and filament winding formulations. The more common diamines are metaphenylene diamine (MPDA), 4-4' methylenedianiline (MDA) and diaminodiphenylsulfone (DADPS). The ability of these materials to cure to a dry B-stage makes them useful for prepregs. The commercial aromatic diamines are solids at ambient temperature, and require melting prior to mixing with the resin. Liquid versions are available as proprietary eutectic mixtures, adducts or solutions. In some cases accelerators are added to reduce cure time.

Aliphatic Primary Amines: Most curing agents in this group are reactive with glycidyl ether resins at ambient temperatures and consequently are not suitable for preimpregnated systems. Aerospace use is infrequent and is limited chiefly to formulations for repair of damaged parts. The principal examples are diethylenetriamine (DETA), triethylenetetramine (TETA) and tetraethylenepentamine (TEPA).

Tertiary Amines: Tertiary amines act as sole curing agents, cocuring agents or accelerators. As sole curing agents, they react with liquid ether resins upon the addition of heat, but are reactive at ambient temperatures with higher molecular weight resins. Benzyldimethylamine (BDMA) is frequently used as an accelerator in laminating formulations.

Amides: Dicyandiamide (DICY) is used with solid ethers of bisphenol A where it is latent at ambient temperatures and gives rapid cures at elevated temperatures. Fatty polyamides, substituted nylons and soluble nylons are frequently employed in adhesive formulations.

Latent Curing Agents: Borontrifluoromonoethylamine (BF_3MEA) is the most common curing agent (catalyst) in this group. Normally the catalyzed resin is gelled at about 250F to reduce exotherm and is given post cure to develop optimum properties.

Acid Anyhydrides: Most anhydrides require elevated temperature cures and postcures with glycidyl ether resins. They generally provide improved thermal stability compared to amine curing agents, and add increased pot-life. and lower viscosity to the laminating formulation. Accelerators are usually required to effect cure. Typical anyhdrides are maleic (MA), succinic (SA), pyromellitic dianhydride (PMDA), hexadrophthalic (HHPA) and nadic methyl anhydride (NMA). Chlorendic anhydride (HET) is used in flame retardant formulations.

2.2.1.7 *Properties of Epoxy Resins.* Properties obtained from cast resin samples are used in some instances to compare or optimize resin systems and to estimate composite properties by means of micromechanics analysis. It is cautioned that evaluations based on cast resin properties do not insure that similar performances will be attained in a composite. Other factors such as resin distribution, cure cycles, void content, resin shrinkage and processing conditions can materially affect the results. Verification by testing of composites is therefore essential.

Mechanical Properties: Tensile, compressive and flexural properties of representative epoxy systems are listed in Table 2-1; and tensile stress-strain relations in Figure 2-1.

Electrical Properties: The dielectric constant and loss tangent for conventional epoxy with two curing agents are shown in Table 2-2. As indicated, both properties increase as the test temperature is increased. When fiberglass is used as a reinforcement, the composite dielectric strength is higher and the loss tangent lower than for resin alone. Laminate electrical properties as specified in MIL-R-9300 are summarized in Table 2-3. (These values also apply to MIL-R-7575 and MIL-R-25042 polyester laminates).

2-2

either singly or in families. Military standards in the book format are also used to cover overall characteristics of a family of end items or major components. In the DODISS Part II the MIL-STD entries take up 12 pages, compared with 77 pages for the MS listings.

Military Standards are further classified into Coordinated Standards and Limited Coordination Standards. The latter have the letter L preceding the MIL-STD or MS number designation, and the former, constituting the bulk of the standards, are unmarked. The Coordinated Standards have been concurred in by all interested activities. The Limited Coordination Standards are issued by one department to satisfy an immediate need for covering an area unique to that department.

Military Standards are changed by either complete revisions or revised pages issued by Notices. Complete revisions are indicated by the addition of a capital letter following the designation number, and revised pages by the date of the Notice.

Military Standards of the MIL-STD category believed to be of interest in plastics and elastomers are listed below:

MIL-STD-190B   Identification Marking of Rubber Products
MIL-STD-297   Visual Inspection Guide for Hard Rubber (Ebonite) Items
L   MIL-STD-298   Visual Inspection Guide for Rubber Extruded Goods
MIL-STD-401B   Sandwich Constructions and Core Materials, General Test Methods
L   MIL-STD-417   Rubber Composition, Vulcanized General Purpose, Solid (Symbols and Tests)
MIL-STD-698A   Quality Standards for Aircraft Pneumatic Tires and Inner Tubes
MIL-STD-768A, Part 1   Instructions for Repair of Aircraft and Weapons Reinforced Plastic and Sandwich Structures, Part 1, All Plastic Construction
MIL-STD-768, Part 2   Instructions for Repair of Aircraft and Weapons Sandwich Structures, Part 2, Metal Constructions (ASG)
MIL-STD-845   Tetrafluoroethylene Hose High Temperature, Medium Pressure, Classification of Defects
L   MIL-STD-860   Fokker Ultrasonic Adhesive Bond Tester
L   MIL-STD-1256A   Rubber Coated Parts for Machine Gun, 7.62 MM M60
L   MIL-STD-1514   Fitting, Tetrafluoroethylene Hose High Temperature High Pressure (3000 Psi) Hydraulic and Pneumatic, Classification of Defects

Military Specifications. These specifications, which are widely used and referred to, are administered by the Department of Defense and developed by a military department (Army, Navy, or Air Force), and each is assigned a custodian from that department. The custodian may prepare the specification, or have it assigned to a joint agency, such as the Aeronautical Standards Group (ASG). After preparation, Military Specifications are coordinated and approved by the interested departments and activities. When so coordinated, they are called Coordinated Military Specifications, meaning that they have been coordinated through the Army, Navy, and Air Force. When a specification is of interest to only one department, or if insufficient time is available for full coordination by all military departments, it can be released as a Limited Coordination Specification. However, it is released as approved for use by all departments and agencies of the Department of Defense.

Military specifications are designated by an identification system consisting of three parts. The first part, MIL, is an abbreviation for military. The second part is a single letter which is the first letter of the title of the specification. The third part is the serial number. Thus, MIL-P-22985A (1), which is listed in Table 4, consists of 12 pages, including 1 page for Amendment 1 and 11 pages for the basic document. The amendment page for this specification, which has the title "Plastic Molding and Extrusion Material, Ethyl Cellulose," consists of a brief section indicating a deletion on page 7, table VI, and an addition to page 9, par. 4.3.10. In the upper right-hand corner of the Amendment page, which always precedes the rest of the specification, is the designation:

MIL-P-22985A
AMENDMENT 1
23 August 1968

The designation in the upper right-hand corner of the basic specification is as follows:

MIL-P-22985A
16 August 1967
SUPERSEDING
MIL-P-22985
11 August 1961

Note that the DODISS entry shown in Table 4 lists the complete document as MIL-P-22985A (1), with the preparation date as 23 August 68. Thus, the date indicated is that of the latest revision (Amendment 1). Note that the A revision in 1967 superseded the original specification in 1961. In some cases, a B revision will supersede an A revision, etc. Since this specification is fully coordinated, there is no L designation in Table 4 for the entry, and no parenthetical suffix in the specification number designation on the actual document. The designations in the lower left-hand corner of the last page list the custodians, review activities, and user activities for the Army, Navy, and Air Force. These are given in code form, the code being given in the front of the DODISS Part II.

For comparison, the specification MIL-P-46120 shown at the bottom of Table 4, covering "Plastic Molding and Extrusion Material, Polysulfone," is a Limited Co-ordination Specification, as shown by the designation L before the document number. This entry notes that MR is the preparing agency and sole custodian. Reference to the code in the front of the DODISS Part II shows that MR refers to the U.S. Army Materials and Mechanics Research Center (AMMRC) at Watertown, Mass. The designation in the upper right-hand corner of the first page of the specification is:

MIL-P-46120 (MR)
17 March 1967

Note that this specification has no amendment and no revisions. It is correctly cited as MIL-P-46120 (MR), the parenthetical suffix indicating that it is a limited coordination document, the MR designating the sole coordinating agency. The last page, lower left, also lists codes for Review Activities and User Activities, in this case all Army agencies.

If this Limited Coordination Specification should become coordinated, the preparing activity suffix is dropped. Whenever a military specification becomes obsolete and revision is delayed, a "used in lieu of" military specification can be released by any of the military departments to cover the necessary requirements. These specifications carry the same number as the basic specification, except that two zeroes are placed in front of the serial number. Additionally, there is a suffix in parentheses after the serial number indicating the activity issuing the document, and the notation "Use in lieu of" is added. The authors know of no military specification relating to plastics and elastomers in this category, although there may be some. A typical example is MIL-I-0043705A, "Ice Cream Maker, etc.," listed in the DODISS.

Perhaps some discussion of the organization of military specifications will be helpful. Military specifications are broken down into six sections as follows:

1. Scope
2. Applicable Documents
3. Requirements
4. Quality Assurance Provisions
5. Preparation for Delivery
6. Notes

Under 1. *Scope* is a simple statement about what the specification covers. Usually this is followed by an outline of the classification of the various subtypes. In the case of MIL-P-22985A, there are three types, *General Purpose, Improved Resistance to Impact,* and *Electrical,* each type having one or more classes. In this case minimum tensile strength values are given for each type.

Under 2. *Applicable Documents,* all specifications and standards referenced in Sections 3, 4, and 5 are tabulated. Under 3. *Requirements,* the basic plastic material is further described and property values are given, largely in tabular form, for each type and class. Some general requirements are given in paragraph form. Under 4. *Quality Assurance Provisions,* detailed instructions are given on responsibility, sampling plans, inspection techniques, etc. An important part of this section from the technical point of view is a tabulation of test methods used. Usually standard methods, such as ASTM, are specified, with additional details or deviations noted. Under 5. *Preparation for Delivery,* details of packaging and packing are given for various "levels," depending on the military requirements. Level A is the most severe requirement and Level C the least severe. Under 6. *Notes,* a paragraph is given on *Intended Use* of the specification. Ordering data are also given to show what procurement documents should specify. Additional notes, such as, in this case, *Suitability for Use with Explosives,* are also given.

The lower right corner of each military specification gives the Federal Supply Classification Code (FSC) number, usually in a box. In the case of MIL-P-22985A, this is FSC 9330.

Actual samples of two pages of another Military Specification, MIL-R-83330, are shown in Tables 9 and 10. Note in Par. 3.4, Table I that references are given to both ASTM and Federal Test Method Standard (FTMS) 406. Both these sources are discussed later in this chapter.

As indicated above, some Military Specifications have *Qualified Products Lists* (QPL). Perhaps a brief explanation of what these lists are and how they are used will help the reader. As a means of expediting contract awards and deliveries of certain products, manufacturers are sometimes required to have their products pretested, either by a government or a designated independent laboratory, to determine if the products meet the specific requirements of government specifications. If the products meet specification requirements, the manufacturer's product may be placed on a Qualified Products List. These lists, which may be either federal or military, are prepared for only those specifications which require prequalification tests.[9] The initial inclusion of a manufacturer's product on a QPL does not guarantee the continued qualification of the product. Assurance that the proper quality is being maintained is achieved by adequate testing of deliveries under each contract. Testing is usually limited to normal acceptance-type inspection of selected samples of a shipment.

The cost of testing a manufacturer's product for inclusion on a QPL is ordinarily chargeable to the manufacturer. Products which require qualification testing are generally confined to items in such commodity groups as metals, paints, pressure-sensitive tapes, cleaning compounds, tools, ball bearings, and some electrical equipment. Military QPLs include other commodity groups, however. At the time a specification is developed, interested manufacturers are given the opportunity to submit any products of the general type covered by the specification for qualification tests.

Manufacturers are not permitted to advertise or publicize the fact that a product appears on a QPL. Considering this limitation, no sample QPL list will be shown in this exposition. Some idea of the limited scope of products covered by QPLs may be obtained from the fact that fewer than 50 of the over 5,000 Federal and Interim Federal Specifications incorporate Qualified Product Lists as part of their requirements. Only 10 percent of the approximately 40,000 specifications in the DODISS contain qualification requirements.[9]

Naval Ordnance Systems Command "Index of Ordnance Specifications and Weapons Specifications."[10] The Naval Ordnance Systems Command, Washington, D.C., issues this document and supplement, which should be of interest and use to the reader. NOSC general policy with respect to the use of specifications in design and construction is described in MIL-STD-143B, "Specifications and Standards, Order of Precedence for the Selection of." This publication provides a list of NAVORD OS (Ordnance Specifications) and WS (Weapons Specifications) specifications, descriptions, and requirements, covering materials, articles, devices, explosives, parts or assemblies, and processes used by the Naval Ordnance Systems Command.

Part 1 of this document is a numerical listing of NAVORD OS numbers issued, including security classification, document number, latest revision, title, and Federal

Supply Class (FSC) number when known. Part 2 is a listing of NAVORD WS numbers, and is otherwise similar to Part 1. Part 3 is an alphabetical listing of active documents listed in Parts 1 and 2. Part 4 is a numerical listing of canceled OS documents, while Part 5 is a similar listing of canceled WS documents.

TABLE 9   First Page of Typical Military Specification on Plastics

```
                                                    MIL-R-83330
                                                    15 July 1971

                              MILITARY SPECIFICATION

                          RESIN, HEAT-RESISTANT, LAMINATING

                    This specification is mandatory for use by all Depart-
                    ments and Agencies of the Department of Defense.

       1.  SCOPE

       1.1  Scope.  This specification covers one type of resin used in
    fabricating reinforced plastic structural parts for aircraft, missiles,
    and other applications in which extreme heat resistance is essential.

       1.2  Classification.  The resin shall be of the following type and
    classes (see 6.2):

          Type I  -  Polyimide (PI)

          Class 1  -  10 - 50 psi laminating pressure.
          Class 2  -  50 - 300 psi !aminating pressure.

       2.  APPLICABLE DOCUMENTS

       2.1  The following documents of the issue in effect on date of invita-
    tion for bids or request for proposal, form a part of this specification
    to the extent specified herein.

    SPECIFICATIONS

       Federal

       PPP-C-96      Cans, Metal, 28 Gage and Lighter
       PPP-D-705     Drum: Metal Shipping, Steel (Over 12 and Under 55 Gallon)
       PPP-D-729     Drums, Metal 55-Gallon (for Shipment of Noncorrosive
                     Material)
       PPP-P-704     Pails, Shipping, Steel (1 through 12 Gallon)

    STANDARDS

       Federal

       Federal Test Method Standard No. 406    Plastics: Methods of Testing


                                                            FSC 9330
```

Examples of Part 3 listings covering subjects of interest to the readers of this handbook are shown in Table 11. These are only partial listings for the particular subjects covered (based on the first words of the specification as listed). It is readily apparent that there are a considerable number of specifications available from NAVORD that cover plastics and elastomers.

Plastics Technical Evaluation Center (PLASTEC). This center, which is the Department of Defense (DOD) information analysis center with responsibility in plastics, has issued two documents that involve standardization. The first is *Government Specifications and Standards for Plastics, Covering Defense Engineering Materials and Applications (Revised).*[3] This document lists the specifications and

TABLE 10   Page of Typical Military Specification on Plastics Showing Requirements

---

MIL-R- 83330

considered requirements of this specification.

   3.4 <u>Mechanical and physical properties</u>. The mechanical and physical properties of laminates of the resin shall conform to the values given in table I. These properties were developed using an aminosilane finish and avoiding content of less than 2 percent.

Table I. Mechanical and physical properties.

Property	Test [1] Condition	Type I Requirement Class 1 & 2	Methods	
			ASTM	FTMS 406
Flexural strength (psi) ultimate	1 2 3	65,000 28,000 15,000	D 790	1031
Initial modulus of elasticity	1 2 3	3.3×10^6 2.5×10^6 1.2×10^6		
Tensile strength (psi)	1 2	50,000 20,000	D 638	1011
Compressive strength (psi)	1 2	55,000 25,000	D 695	1021
Interlaminar shear strength (psi)	1	2,000		1042
Flammability (in/min)	1	Self- extinguishing	D 635	2021

  [1] Test conditions shall be as follows:
     1. Ambient conditions (see 4.4.1)
     2. @ 550°F after 1000 hr @ 550°F (+5°F forced air oven).
     3. @ 700°F after 100 hr @ 700°F (+5°F forced air oven).

   3.5 <u>Electrical properties of the laminate</u>. The electrical properties of a glass cloth base laminate, fabricated as specified in 4.3.1.2, shall conform to the values given in table II.

4

---

standards for those plastic materials and plastics applications which are considered to be of interest to engineers concerned with the design, development, production, and handling of defense hardware. Included are specifications for the basic or raw materials, composite materials, and the items and applications of potential defense concern. Excluded are specifications on life-situation items: clothing, utensils, and furniture, and decorating or preservative coatings.

The body of the material is presented in five parts: specifications for or involving specific plastic materials, title—stated or otherwise identified; general reference documents; specifications for or involving unspecified plastics; a subject index; a numerical index. Complete citations for 1,750 specifications are given, covering: number, date, title, the existence of a Qualified Products List, the status of coordination among the services (whether limited to one, or not limited), the agency which prepared the specification, and the custodian or custodians of the specification. Included is a discussion of government specifications, and instructions on their procurement.

The intended use of the material or end item covered by a government specification is usually stated under *Notes* in Section 6. To make this report truly authoritative, it would have been necessary to call in all specifications pertaining to plastics (or expected to be so), for evaluation. This would have been extremely time-consuming. The authors felt that since specifications are subject to change, a report involving them is timely only if it is produced with speed. Thus, the selection of items to enter was made on editorial judgment, based on two considerations: the wording of the title of the specification, and the "hardware status" of the preparing agency and the custodian or custodians of the specification.

Coverage of specifications which mention a plastic material or the word "plastic" in the title is quite complete. On the other hand, coverage of specifications without such title-stated guides to their plastics relationship is of indefinite completeness. All known cases of this type have been evaluated for inclusion. The trade literature has been combed for leads to the unknown, and hundreds of specifications of possible interest have been called in for checking. It is felt, however, that many other such "hidden-plastics" specifications exist.

The first part of the report covers those specifications for which the plastic material is title-stated or known. These specifications are listed under the heading of the particular material. Thus the engineer can see at a glance all the defense-related specifications on acrylics, epoxies, fluorocarbons, phenolics, polyesters, silicones, vinyls, etc. With the engineer still in mind, the within-group listings of the specifications are subdivided. First in order are listed those specifications pertaining to materials (basic, raw, or composite) and prime forms (sheets, rods, tubes). These are followed, after a dividing space, by specifications on applications of the particular material. These are alphabetized by application (which is presented in all capitals) rather than by first word of title, as in the DODISS. Thus: "COATING system, epoxy-polyamide, chemical and solvent resistant"; "OBTURATING BANDS, plastic, rigid, high impact, polyvinyl chloride (molded)."

An example of Part 1 is shown in Table 12. The letters LC refer to Limited Coordination Document, and those items marked with an X are limited coordination documents. The letters A, N, and AF refer to Army, Navy, and Air Force, respectively. The codes are given in the front of the document. It is interesting to note Ref. 286, listed as JAN-E-480. JAN refers to joint Army-Navy, and this specification goes back to 1947. Very few of these specifications are still in existence. Reference 288 is for a Military Standard. The reference numbers are assigned serially, and these are referred to in the subject index (Part 4) and the numerical index (Part 5).

The second part of the report covers specifications and standards considered of reference interest to engineers involved in plastics. These pertain to tests, methods, tolerances, and similar general information and requirements. An example of this part is shown in Table 13. The third part of the report covers those specifications for which the particular plastic material is not known or specified. At the beginning of this part is a list of "material" specifications in which the material is not identified. This list is followed by a listing of "application" specifications involving adhesives, plastics, and elastomers. These are alphabetized, but again, the preference is for grouping by the application involved (cited in all capitals), rather than the first word of the title. An example of the "materials" section of Part 3 is shown in Table 14. This section was chosen because it listed both plastics and elastomers.

It should be noted that in all three parts just described the identification of a

TABLE 11 Samples of Alphabetical Listing (Part 3) in NAVORD OS O
of Particular Interest in Plastics and Elastomers

✕ NOMENCLATURE	NAVORD NUMBERS
ADHESIVE, EPOXY RESIN, STRUCTURAL BONDING.	WS 8035
ADHESIVE, EPOXY RESIN, STRUCTURAL BONDING.	WS 8993
ADHESIVE, EPOXY RESIN, STRUCTURAL BONDING.	WS 8994
ADHESIVE, EPOXY RESIN, STRUCTURAL BONDING.	WS 9124
ADHESIVE, EPOXY RESIN, STRUCTURAL BONDING.	WS 9125
ADHESIVE, EPOXY RESIN SYSTEM, WITH FILLERS (POLARIS FBM).	OS 10502
ADHESIVE, EPOXY RESIN SYSTEM WITH INORGANIC FILLERS (POLARIS FBM).	OS 10505
ADHESIVE, EPOXY RESIN SYSTEM, POLARIS FLEET BALLISTIC MISSILE.	WS 3283
ADHESIVE, EPOXY RESIN, TWO COMPONENT.	WS 3490
ADHESIVE, EPOXY RESIN, TWO COMPONENT, FILLED.	WS 11540
ADHESIVE, EPOXY RESIN, TWO COMPONENT, UNFILLED.	WS 11539
ADHESIVE, EPOXY RESIN, TWO PARTS POLARIS FLEET BALLISTIC MISSILE.	WS 3298
ADHESIVE, EPOXY RESIN, UNFILLED; LOW TEMPERATURE CURE.	WS 5140
ADHESIVE, FILLE, THERMOSETTING.	WS 11773
ADHESIVE FILM SEE ALSO FILM, ADHESIVE	
ADHESIVE, FLUOROCARBON RUBBER, HIGH-TEMPERATURE, HALOCARBON FLUID RESISTANT, POLARIS FLEET BALLISTIC MISSILE, GENERAL SPECIFICATION FOR.	WS 3295
ADHESIVE, GLASS SUPPORTED EPOXY.	XWS 3889
ADHESIVE, HEAT CONDUCTIVE.	WS 8872
ADHESIVE, FOR HEAT SHRINKABLE MATERIALS, POLARIS FBM.	OS 11881
ADHESIVE, INORGANIC, HIGH-TEMPERATURE INSULATION-BONDING POLARIS FLEET BALLISTIC MISSILE, GENERAL SPECIFICATION FOR.	OS 9304
ADHESIVE, LIQUID, THERMOSETTING.	OS 11098
ADHESIVE, LIQUID, VINYL-PHENOLIC RESIN.	WS 4523
ADHESIVE MIXING.	WS 8486
ADHESIVE MIXING.	WS 8491
ADHESIVE MIXING.	WS 8492
COATING, APPLICATION OF CONFORMAL, GENERAL SPECIFICATION FOR.	WS 12521
COATING, BALLISTIC ROD.	OS 11642
COATING, CHLOROSULFONATED POLYETHYLENE, MOISTURE AND FUNGUS RESISTANT, FLEET BALLISTIC MISSILE, GENERAL SPECIFICATION FOR.	WS 3568
COATING COMPOUND, METAL PRETREATMENT.	WS 8814
COATING COMPOUND, NON-SLIP PROTECTIVE.	WS 5114
COATING COMPOUND, POLYURETHANE, INSULATING ELECTRICAL (ROOM TEMPERATURE CURE).	OS 10252
COATING COMPOUND, PROTECTIVE, SILICONE BASE.	WS 5164
COATING COMPOUND, SYNTHETIC MODIFIED RESIN.	OS 9815
COATING, CONDUCTIVE, HIGH-TEMPERATURE POLARIS FLEET BALLISTIC MISSILE, GENERAL SPECIFICATION FOR.	OS 9312
COATING, CONDUCTIVE SILVER.	OS 12725
COATING, CONFORMAL.	WS 13297
COATING, CONFORMAL PROTECTIVE (FOR PRINTED CIRCUIT ASSEMBLIES).	OS 8549
COATING FOR DETONATING CORD POLARIS FLEET BALLISTIC MISSILE, GENERAL SPECIFICATION FOR.	OS 9457
COATING, ELASTOMERIC, FOR MARKING OF BUTYL RUBBER POLARIS FLEET BALLISTIC MISSILE, GENERAL SPECIFICATION FOR.	WS 3556
COATING, ELECTRICALLY INSULATING EPOXY, FLEET BALLISTIC MISSILE.	WS 11749
COATING FOR ELECTRODEPOSITION FLEET BALLISTIC MISSILE, GENERAL SPECIFICATION FOR.	WS 11755
COATING, EPOXY CONDUCTIVE.	WS 13227
COATING, EPOXY, ELECTRICAL INSULATION, POLARIS FBM.	WS 1365
COATING, EPOXY RESIN.	WS 3235
COATING, EPOXY RESIN BASE, CLEAR.	OS 9911
PLASTIC, ACETAL, RIGID: MOLDED PARTS, RODS AND SHEETS, POLARIS FBM.	WS 1362
PLASTIC COATING, POLYAMIDE (NYLON), FUSION BONDED.	WS 5120
PLASTIC COMPOUND, COATING, MOLDING, AND POTTING.	WS 2059
PLASTIC COMPOUND, GLASS STRAND-EPOXY RESIN.	WS 3218
PLASTIC COMPOUND, LOW PRESSURE MOLDED.	WS 3217
PLASTIC COMPOUND, MOLDING AND EXTRUDING, ACETAL RESIN.	OS 11781
PLASTIC COMPOUND, MOLDING AND EXTRUDING, POLYCARBONATE RESIN.	OS 12531
PLASTIC COMPOUND, MOLDING AND EXTRUDING, POLYCARBONATE RESIN, GLASS REINFORCED, PROCUREMENT OF.	WS 6136
PLASTIC COMPOUND, MOLDING AND EXTRUDING, POLYPHENYLENE OXIDE RESIN, GLASS-FILLED, MODIFIED.	WS 6268
PLASTIC COMPOUND, POLYVINYL CHLORIDE, CONDUCTIVE, PROCUREMENT OF.	OS 12859
PLASTIC FILM METALLIC COATED.	XWS 3923
PLASTIC FILM, POLYETHYLENE TEREPHTHALATE, ALUMINIZED.	OS 8939
PLASTIC, FLUOROCARBON, GLASS FIBER FILLED, DESCRIPTION AND REQUIREMENTS.	WS 4145
PLASTIC, FLUOROCARBON, MOS 2 FILLED, DESCRIPTION AND REQUIREMENTS.	WS 4146
PLASTIC FOAM, POLYURETHANE, FLEXIBLE, PRESSURE-SENSITIVE ADHESIVE BACKED.	WS 5141
PLASTIC FOAM, POLYVINYL CHLORIDE, RIGID.	WS 5133
PLASTIC FOAM SHEET AND PARTS-TYPE U-1 MACHINED FINISH.	WS 3214
PLASTIC, FOAMED-IN-PLACE, POLYISOCYANATE RESIN.	WS 8946
PLASTIC, IRRADIATED POLYETHYLENE, MOLDED AND EXTRUDED SHAPES, SHEET AND TUBING.	OS 7477
PLASTIC, IRRADIATED, SIZE-AND FORM-RECOVERY TYPE.	WS 1647
PLASTIC LACQUER, FOR LOW-TORQUE LOCKING SEAL.	OS 7647
PLASTIC, LOW PRESSURE, MOLDED, LAMINATED, REINFORCED SANDWICH STRUCTURES.	WS 3221
PLASTIC MATERIAL, ACETAL RESIN FOR INJECTION MOLDING AND EXTRUSIONS.	OS 12702
PLASTIC MATERIALS, ASBESTOS BASE-PHENOLIC, LOW PRESSURE LAMINATES.	OS 7405
PLASTIC MATERIAL, CELLULAR POLYSTYRENE FOAM, POLARIS FLEET BALLISTIC MISSILE.	WS 1094
PLASTIC MATERIAL, COMPOSITE POLARIS FLEET BALLISTIC MISSILE, GENERAL SPECIFICATION FOR.	WS 3470
POTTING COMPOUND.	WS 13245
POTTING COMPOUND, (ALUMINUM SILICATE FILLED, POLYESTER TYPE).	OS 8916
POTTING COMPOUND FOR CIRCUIT BLOCKS.	XWS 2624
POTTING COMPOUND, CORK-FILLED, ROOM-TEMPERATURE-CURING.	WS 9138
POTTING COMPOUND E-360.	WS 13248
POTTING COMPOUND, EPOXY RESIN, SILICA FILLED, LOW VISCOSITY.	WS 8687
POTTING COMPOUND, FOAMED-IN-PLACE, POLYURETHANE, DESCRIPTIONS AND REQUIREMENTS FOR.	WS 3985
POTTING COMPOUND ISOCYANATE FOAM.	OS 9950
POTTING COMPOUND, MIXING AND CURING OF (FOAMED-IN-PLACE).	WS 4056
POTTING COMPOUND, POLYURETHANE (POLARIS FBM).	OS 11924
POTTING COMPOUND, POLYURETHANE RUBBER.	OS 7505
POTTING COMPOUND, RESILIENT.	WS 5685
POTTING COMPOUND, SEMI-RIGID.	WS 5686
POTTING COMPOUND SILICONE RUBBER ROOM TEMPERATURE VULCANIZING (FOR BT-3 TERRIER AND TARTAR MISSILES).	OS 7527
POTTING COMPOUND, URETHANE.	WS 13249
POTTING ELECTRICAL SYSTEM COMPONENTS.	WS 5687
POTTING AND ENCAPSULATING WITH EPOXY RESIN.	WS 13243

TABLE 11   Samples of Alphabetical Listing (Part 3) in NAVORD OS O of Particular Interest in Plastics and Elastomers (Continued)

```
RUBBER, ADHESIVE, (RUBBER TO METAL), POLARIS FLEET BALLISTIC MISSILE.            WS 3308
RUBBER, ADHESIVE, (RUBBER TO METAL), POLARIS FBM, GENERAL REQUIREMENTS FOR.      WS 1056
RUBBER, ADHESIVE, (RUBBER TO METAL), TWO-PART.                                   WS 8072
RUBBER, ADHESIVE, (RUBBER TO METAL), TWO PARTS, POLARIS FLEET BALLISTIC MISSILE. WS 3311
RUBBER, ADHESIVE, SOLVENT CONTAINING TYPE.                                       WS 11969
RUBBER, ADHESIVE, SYNTHETIC.                                                     WS 13231
RUBBER, AGE CONTROL, FLEET BALLISTIC MISSILE, GENERAL SPECIFICATION FOR.         WS 6758
RUBBER, AGE CONTROL, POLARIS FLEET BALLISTIC MISSILE.                            OS 10674
RUBBER-TO-ALUMINUM CLOSURE, FABRICATION AND BONDING OF.                          WS 8094
RUBBER-BASE INSULATING COMPOUND, PURCHASE DESCRIPTION.                           WS 6529
RUBBER BUNA-N.                                                                   OS 8291
RUBBER, BUTADIENE ARCYLONITRILE, ASBESTOS FILLED, UNVULCANIZED.                  WS 8467
RUBBER, BUTADIENE-ACRYLONITRILE, SILICONE MODIFIED, PURCHASE DESCRIPTION.        WS 7166
RUBBER, BUTYL COMPOUND 70 SHORE A HARDNESS.                                      WS 6260
RUBBER, BUTYL-SILICONE MODIFIED FILLING, PURCHASE DESCRIPTION.                   WS 7165
RUBBER - CELLULARS, SILICONE SPONGE.                                             OS 7906
RUBBER, CHLOROBUTYL, SHEET.                                                      WS 8007
RUBBER COMPOUND CHLOROPRENE TYPE; FOR COMPRESSION MOLDING, SYNTHETIC.            WS 4864

☆ FOR EXACT TITLE, COGNIZANT SYMBOL, AND REMARKS, CONSULT THE NUMERICAL PORTION OF THIS INDEX.
```

TABLE 12   Example of Part 1—Specifications for or Involving Specific Plastic Materials, Title-Stated or Otherwise Identified (from *PLASTEC Note* 6B)

Ref	Title	Spec Number	Date	QPL	LC	Prepared by	Custodian "A"	"N"	"AF"
	PHENOLIC (Continued)								
286	ENAMEL, baking, phenol- or urea-formaldehyde	JAN-E-480	24 Jul 47	-	-	MR	MR	-	-
287	INSULATING compound, asbestos phenolic	MIL-I-81255	23 Mar 65	-	X	AS	-	AS	-
288	SPRING, Belleville ((phenolic))	MS-75044	12 Jan 68	-	X	EL	EL	-	-
289	VARNISH, phenolic, baking	MIL-V-12276	8 Sep 65	-	X	MR	MR	-	-
290	VARNISH, phenol-formaldehyde, clear, and aluminum pigmented	MIL-V-13750	4 Nov 66	-	X	MR	MR	-	-
291	VARNISH (mixing, phenolic)	MIL-V-15218	1 Nov 67	-	-	SH	MR	SH	-
	PHENOXY								
292	Plastic molding and extrusion material, phenoxy	MIL-P-46088	10 Jul 68	-	X	MR	MR	-	-
	------------------------								
	PLASTISOL								
293	Plastic, plastisol (for coating metallic objects)	MIL-P-20689	24 Jun 63	X	-	MR	MR	SH	ASD
294	Resin, plastisol grade, polyvinyl chloride	MIL-R-23444	16 Nov 66	-	X	OS	-	OS	-
	------------------------								
	POLYALKENE								

295	WIRE, electric, crosslinked polyalkene insulated, copper *QPL in preparation	MIL-W-81044 (/1 to /15)	17 Jul 68	-*	X	AS	-	AS	-
	POLYAMIDE (NYLON)								
296	Plastic molding material, polyamide (nylon), glass fiber filled	L-P-395	10 Apr 64	-	-	MR	MR	SH	ASD
297	Nylon plastic, flexible, molded or extruded	MIL-N-18352	14 Jan 55	-	X	AS	-	AS	-
298	Molding plastic, polyamide (nylon) and molded and extruded polyamide plastic parts - weather resistant	MIL-M-19098	11 Nov 55	-	X	SH	-	-	-
299	Molding plastic, polyamide resin (nylon) glass-fiber filled and molded polyamide resin glass-fiber filled plastic parts	MIL-M-19887	27 Jul 62	-	X	SH	-	-	-
300	Molding plastic, polyamide (nylon), rigid	MIL-M-20693	21 Dec 66	-	-	EL	EL	SH	ASD

TABLE 13   Example of Part 2—General Reference Documents (from *PLASTEC Note* **6B)**

Ref	Title	Spec Number	Date	QPL	LC	Pre-pared by	Custodian "A"	"N"	"AF"
785	ADHESIVES	MIL-HDBK 691	17 May 65	-	-	MR	MR	-	-
786	ADHESIVES, methods of testing	FED-STD-175	27 Sep 67	-	-	MR	MR	SH	ASD
787	AGE CONTROLS of age-sensitive elastomeric items	ANA 438	15 Feb 65	-	-	ED	-	-	ED
788	Procedure for AIRFRAME contractor's purchase of supply items requiring qualification tests	ANA 437	10 Sep 58	-	-	ASD	-	-	ASD
789	Visual inspection standards and inspection procedures for inspection of packaging, packing and marking of small arms AMMUNITION	MIL-STD-644	3 Jun 65	-	-	MU	MU	OS	OAM
790	Packaging, packing and marking for shipment of inert AMMUNITION components	MIL-STD-1169	30 Dec 66	-	X	MU	MU	-	-
791	Ballistic acceptance test method for personal ARMOR material	MIL-STD-662	15 Jun 64	-	-	MC	WC	MC	ASD

plastic material has sometimes been added, in double parentheses, to the title. Thus, in Ref. 288 in Table 12, phenolic is identified in this way.

The fourth part is a subject index of the materials covered by the specifications cited, their characteristics and uses. An example is shown in Table 15. This listing has been carried out in depth, so that the engineer can trace back (through the within-report reference system) to all specifications pertaining to his particular interest, whether they involve a material (polyester), a form (sheet), or a property (solar-heat reflection).

The fifth part is a numerical index of all the specifications cited in this report. An example is shown in Table 16. The numbers are listed serially by logical groupings, such as military specifications, military standards, military handbooks, federal

TABLE 14   Example of Part 3—Specifications for or Involving Unidentified Plastics (Sometimes Noted but Unconfirmed) (from *PLASTEC Note* **6B)**

Ref	Title	Spec Number	Date	QPL	LC	Pre-pared by	Custodian "A"	"N"	"AF"
	MATERIALS								
961	Plastic sheet, laminated, thermosetting (for designation plates) ((unspecified resin))	L-P-387	29 Oct 63	-	-	EL	EL	SH	ASD
962	Plastic sheet, laminated, decorative and non-decorative ((unspecified resin))	L-P-508	23 Feb 68	-	-	GS	-	-	-
963	Rubber sheet, strip, extruded, and molded shapes, synthetic, oil resistant	MIL-R-2765	9 Jul 65	-	-	SH	WC	SH	-
964	Rubber, synthetic, sheet, strip and molded	MIL-R-3533	16 Nov 67	-	-	OS	MR	OS	KEL
965	(Not used)								
966	Synthetic rubber sheets, strips, molded or extruded shapes	MIL-S-6855	29 Mar 68	-	-	AS	MR	AS	ASD
967	Rubber, synthetic, solid, sheet and fabricated parts, synthetic oil resistant	MIL-R-7362	2 May 67	-	-	ASD	MR	AS	ASD
968	Rubber sheets and molded shapes, cellular, synthetic, exploded cell (foamed latex)	MIL-R-20092	17 May 63	-	-	SH	MR	SH	KEL
969	Rubber, high-temperature, fluid resistant	MIL-R-25897	4 Oct 65	-	X	ASD	-	AS	ASD
970	Rubber, hard (Eponite), natural or synthetic, sheet, strip, rod, tubing, and molded parts	MIL-R-45036	3 Jul 67	-	X	MR	MR	-	-
971	Rubber, synthetic, heat shrinkable	MIL-R-46846	30 Mar 66	-	X	MI	MI	-	-
972	Rubber, synthetic, shock absorbing, high compressive strength, ion resistant	MIL-R-81297	27 Jul 65	-	X	AS	-	AS	-

TABLE 15   Example of Part 4—Subject Index (in Terms of Reference Number)
(from *PLASTEC Note* 6B)

ABS - see "acrylonitrile-butadiene-styrene"	Adhesive, heat sealing tape - 1654
Acetal - 1, 2	Adhesive, pressure sensitive tape - 1659 to 1661, 1663, to
Acetate - 81, 82, 94, 99; see also "cellulose acetate"	1666, 1668 to 1670, 1672, 1675 to 1677, 1679 to 1683
Acetate, saponified, impregnated twine - 1726	Adhesive, sealing compound, polysulfide - 619
Acid resistant battery liners - 1041	Aeronautical lighting, colors - 816, 819
Acid resistant cloth - 57	Aeronautical lighting, covers - 825
Acid resistant coating, polyester - 508	Age controls, elastomers - 787
Acid resistant coating, polyester-epoxy - 145	Aircraft arresting, webbing for, nylon - 452
Acrylic - 3 to 30	Aircraft identification and marking, polyester - 517
Acrylic, coating - 1157	Aircraft structure sealing compound - 1595
Acrylic fiber filled molding material - 116	Airfield surfacing - 1455
Acrylic film - 1245	Airway, pharyngeal - 996
Acrylic nitrocellulose - 26, 27, 29	Alkyd - 34 to 44
Acrylic nitrocellulose, impregnant - 1357	Alkyd phenol - 37, 248
Acrylic, protractors - 1541	Alkyd, varnish - 1729
Acrylic, scales - 1575	Allyl - 45, 46
Acrylic sheet, coating for - 1147	Aluminum-backed tape - 1655, 1681

specifications, or federal standards, although such headings are not used. The reference numbers listed for each subject heading lead back to the appropriate specifications or other documents cited in Parts 1, 2, and 3.

No attempt has been made by the author to carry through the identification of "revision" and "amendment" for the many specifications listed, since (1) this information may be outdated in some instances even before the report is printed, (2) when information on the exact status of a specification is required, it will have to be obtained from a holder of the latest DODISS and supplement, and (3) at any request through proper channels for copies of either military or federal specifications, the latest revision and amendment are automatically provided. A similar attitude has been taken toward the existing method of numbering specifications

TABLE 16   Example of Part 5—Numerical Index (in Terms of Reference Number)
(from *PLASTEC Note* 6B)

MIL-M-14 — 116, 226, 251, 665, 479	MIL-F-1148 — 973	MIL-C-3432 — 1085
MIL-C-17 — 1082	MIL-G-1149 — 1260	MIL-D-3464 — 831
MIL-T-27 — 946	MIL-G-1366 — 1277	MIL-I-3505 — 1375
MIL-P-79 — 227, 258	MIL-S-1491 — 1619	MIL-I-3511 — 999
MIL-W-80 — 30	MIL-R-1688 — 400	MIL-R-3533 — 964
MIL-P-116 — 901	MIL-B-1860 — 1075	MIL-A-3562 — 984
MIL-B-117 — 1035	MIL-L-1910 — 1439	MIL-C-3608 — 1169
MIL-B-121 — 1034	MIL-T-1956 — 1687	MIL-C-3643 — 821
MIL-B-131 — 1036	MIL-H-1987 — 1316	MIL-P-3671 — 1528
MIL-P-149 — 1144	MIL-C-1997 — 1198	MIL-C-3702 — 1086
MIL-V-173 — 1727	MIL-R-2033 — 1559	MIL-R-3745 — 246
MIL-B-197 — 1046	MIL-H-2217 — 1334	MIL-C-3787 — 1277
MIL-G-252 — 1299	MIL-T-2283 — 420	MIL-G-3820 — 1261
MIL-P-265 — 752	MIL-M-2301 — 1444	MIL-I-3825 — 1376
MIL-C-333 — 82	MIL-S-2329 — 1445	MIL-C-3849 — 1189
MIL-A-388 — 78	MIL-S-2412 — 1576	MIL-C-3883 — 1190

which are under a major change or are being revised to meet an emergency. Thus L-P-0065 is listed as L-P-65. The date given for each specification is the date of the latest action on that specification, as determined from the latest supplement to the DODISS in effect at the writing of the report.

Any future revision of the report will be in a considerably different form. The section of this chapter on Identification of Materials Specifications, Test Methods, and Standards is a partial attempt at such a revision. It was prepared by Mr. Beach. The second standardization document issued by PLASTEC is *Guide to Test Methods for Plastics and Related Materials.*[11] This document lists test methods pertaining to plastic materials and processes. The methods listed are identified from lists of existing standards and test methods promulgated by various government and nongovernment standardizing groups, many of whose listings are described in this chapter, and from a study of specifications relating to plastics. Standardized test methods which constitute standardization documents on their own are covered, as are test methods which constitute only parts of materials specifications. The citations are presented in alphabetical order, by test method subject ("CREEP") rather than by material tested ("cushioning material"). The work covers 1,000 test subjects and over 4,000 citations. Standardizing sources covered include:

AATCC (American Association of Textile Chemists and Colorists)
AMS (Aerospace Materials Specifications—SAE)
ANSI (American National Standards Institute)
ASTM (American Society for Testing and Materials)
CS (Commercial Standards)
ERF—SPI (Epoxy Resin Formulators, Society of the Plastics Industry)
Federal Standards
IEC (International Electrotechnical Commission)
ISO (International Organization for Standardization)
Military Specifications
Military Standards
NASA MSFC (Marshall Space Flight Center)
NEMA (National Electrical Manufacturers Association)
PI (Packaging Institute)
REA (USDA Rural Electrification Administration)
TAPPI (Technical Association of the Pulp and Paper Industry)

Examples of typical entries in this document are shown in Table 17. Note that paragraph numbers or method numbers are given so that the reader can go directly to the test method described in the cited documents.

Department of Defense Single Stock Point for Specifications and Standards.[12] This brochure, the cover of which is shown in Fig. 1, is a brief outline of the Specifications and Standards Program and is offered to private industry as an aid to facilitate the obtaining of specifications and standards required to do business with the Department of Defense. Index subjects are as follow:

DOD-SSP (Single Stock Point) Highlights
How to Order Specifications and Standards
Subscription Service (Automatic Distribution)
Redistribution of Specifications
DODISS (Department of Defense Index of Specifications and Standards)
Inquiries and Complaints

The Naval Publications and Forms Center (NPFC), Philadelphia, is the Department of Defense Single Stock Point (DOD-SSP) for unclassified specifications and standards throughout the Department of Defense for military procurement. It has the responsibility of indexing, determining requirements, procuring, receiving, distributing, issuing, and controlling those current specifications and standards listed in the DODISS. The following types of documents are stocked and distributed:

Military Specifications
Military Standards
Federal Specifications
Federal Standards
Qualified Products Lists
Industry Documents
Military Handbooks
Air Force–Navy Aeronautical Standards
Air Force–Navy Aeronautical Design Standards
Air Force–Navy Aeronautical Specifications
U.S. Air Force Specifications
Other Departmental Documents
Air Force–Navy Aeronautical Bulletins
U.S. Air Force Specifications Bulletins

Customers of the DOD-SSP include all military services, federal, state, and municipal agencies, foreign governments, and industrial and commercial firms, both

DEPARTMENT OF DEFENSE
SINGLE STOCK POINT
(DOD-SSP)
for
SPECIFICATIONS AND STANDARDS

A GUIDE FOR
PRIVATE INDUSTRY

NAVAL PUBLICATIONS AND FORMS CENTER
5801 TABOR AVENUE
PHILADELPHIA, PA. 19120

Fig. 1 DOD Guide for Private Industry on Specifications and Standards, showing cover page.

national and foreign.   Specifications and standards are issued to private industry on an "as ordered basis" without charge.

Requests for copies of specifications and standards can be obtained from this source[13] by telephone, telegraph, mail in any form, or bearer.   However, it is preferred that private industry use a simplified order form, DD Form 1425, which includes a self-addressed label.   Once a customer orders specifications and standards, he will automatically be provided by NPFC with sufficient blank forms to reorder continually in the preferred manner.   To submit a request by telephone, the requester should call (215) 697-3321 during the hours 8 A.M. to 4:30 P.M. Philadelphia time, Monday through Friday.   Off duty telephone requests will be serviced by an automatic answering device 7 days a week.

Department of Defense Standardization Manual 4120.3-M.[13]   This very useful manual was developed by, and is maintained in cooperation with the Departments

TABLE 17   Examples of Typical Listings in PLASTEC's Guide to Test Methods for Plastics and Related Materials (*PLASTEC Note* 17)

TEST SUBJECT	TEST METHOD
DIMENSIONAL STABILITY, glass filled plastic	MIL-P-19833, para 4.4.2.7
DIMENSIONAL STABILITY, insulation board	HH-I-530, para 4.4.4
DIMENSIONAL STABILITY, molded parts	L-P-410, para 4.4.10
DIMENSIONAL STABILITY, pipe and fittings	CS 228, para 7.10
DIMENSIONAL STABILITY, post forming grades, CF sheet	NEMA LI-1, para 12.11
DIMENSIONAL STABILITY, tubes and rods	MIL-P-22296, para 4.6.2.4
DIMENSIONAL STABILITY, tubing	FD-110, para 8.5
DIMENSIONAL STABILITY, high temperature	MIL-M-14, para 4.5.2.6
DIMENSIONAL STABILITY, low temperature thermal block and pipe insulation	ASTM C 548
DIMENSIONAL STABILITY, rods, tubes, sheets, shapes	AMS 3651, para 4.2.11
DIMENSIONAL STABILITY, post-forming sheet	AMS 3605, para 4.3.4
Hygroscopic DIMENSIONAL STABILITY, cartographic sheet	L-P-505, para 4.4.2.4
Thermal DIMENSIONAL STABILITY, cartographic sheet	L-P-550, para 4.4.2.3
Thermal DIMENSIONAL STABILITY, sheet	MIL-P-55010, para 4.4.4
IMPACT, shower receptor	CS 222, para 5.4
IMPACT, spoon, knife and fork (picnic)	MIL-S-676, para 4.4.2
IMPACT, syntactic material	MIL-S-24167, para 4.7.4.4
Tensile IMPACT energy to break plastics and electrical insulating materials	ASTM D 1822
Tensile IMPACT tests, tank backing mtl	MIL-P-8045, para 4.6.5
Low-temperature IMPACT, coating compound	MIL-C-16555, para 4.5.9
Low-temperature IMPACT, film	CS 192, para 4.9
Low temperature IMPACT, hydraulic parts	MIL-P-5517, para 4.2.2.5
IMPACT PUNCTURE RESISTANCE of films and barriers	FED-STD-101, method 313; MIL-B-131; MIL-F-22191; MIL-F-23712
IMPACT, buried distribution wire	(REA) PE-24, para 6.044; (REA) PE-44, para 6.044; (REA) PE-45, para 10.055
IMPACT, buried service wire	(REA) PE-43, para 5.054
IMPACT RESISTANCE of plastics and electrical insulating materials	ASTM D 229, para 21; ASTM D 256; NEMA LD-1, para 2.15; (AS)C59.11
IMPACT RESISTANCE of plastics at subnormal and supernormal temperatures	ASTM D 758
IMPACT RESISTANCE, primer coating	MIL-P-46056, para 4.4.16
IMPACT RESISTANCE, battery containers	FED-STD-601, method 11261
Resistance to IMPACT by falling ball	FED-STD-406, method 1074; ERF 23

of the Army, Navy, and Air Force and the Defense Supply Agency in accordance with Department of Defense Directive 4120.3.   It is mandatory for use by all defense activities.   It sets forth the scope, purpose, and basic objectives of the Defense Stabilization Program and provides a statement of the principles under which this program is to be carried out.   The manual includes detailed instructions for preparation of standardizing documents, including specifications, standards, and handbooks, both federal and military.

General Services Administration, Federal Supply Service   For a comprehensive discussion of the role of GSA-FSC in standardization documents involving plastics and elastomers the reader should refer to the Directory of United States Standardization Activities.[1]   There is an "Index of Federal Specifications and Standards,"[14] published by GSA-FSC, but the reader should be able to find any relevant document in the DODISS[4] described above.   The Federal Index is easier to read and use than the DODISS, however, and lists prices so that the user will be able to send a check for the price of specifications and other standardization documents he purchases. This source provides general information on the Federal Standardization Document System, on how to use the Index, and on how to obtain Federal Standardization Documents. Tables of alphabetical listings are given for Federal and Interim Federal Specifications, Federal and Interim Standards, and Federal Handbooks. Also given are numerical listings for the same sources and also for the Federal Qualified Product Lists. Other tables include Federal Supply Classification (FSC) listings and Federal Specifications and Standards canceled or superseded during the previous year. Finally, a complete list of specifications and standards canceled and superseded through the previous year is given. An example of entries in the Federal Index is given below. Note that prices of the documents are provided. If a QPL existed, the letter Q would be inserted in the QPL column.

ALPHABETICAL LISTING OF FEDERAL AND INTERIM FEDERAL SPECIFICATIONS							59
TITLE	DOCUMENT NUMBER	QPL	FSC	PREP	DATE	PRICE	
Plastic Molding Material Phenolic Resin	L-P-1125		9330	FSS	25 Apr 67	.05	
Plastic Molding Material, Acrylonitrile-butadiene-styrene (Abs), Rigid	L-P-1183A		9330	MR	23 Jul 71	.15	
Plastic Rod, Solid; Plastic Tubes And Tubing,heavy Walled: Polyvinyl Chloride, Rigid	L-P-1036(1)		9330	MR	29 Jan 68	.05	
Plastic Sheet And Strip, Polyester	L-P-377B		*8135	GL	28 Feb 66	.10	
Plastic Sheet And Strip, Thin Gauge, Polyolefin	L-P-378C(1)		8135	GL	16 Mar 71	.20	
	INT AMD 2				8 Oct 71		

Federal Specifications and Standards are administered by the General Services Administration, and are developed by either civilian or military agencies, depending on the interest and technical assistance available for their preparation. After preparation, the specifications are forwarded to other interested government agencies for comments and approval, and are then released. In general, Federal Specifications and other related documents are written to cover requirements that are not essentially military in nature. But in use, there appears to be no sharp dividing line. Many military agencies are required to procure materials and items called out in Federal Specifications. It is quite probable too that, on the other hand, many nonmilitary and even nongovernment agencies specify that Military Specification requirements must be met.

Federal Specifications. Federal Specifications are promulgated by GSA for mandatory use by all Federal Agencies. They describe the essential and technical requirements for items, materials, or services used by, or for partial use in two or more Federal Agencies, including the procedures for determining that the requirements have been met.[6] In appearance, these documents very closely resemble Military Specifications. Indeed, many of them are written by individuals who also prepare Military Specifications. Their organization is essentially identical to that of Military Specifications, as described above. About the only difference noticeable is that Section 1 is listed as Scope and Classification, instead of just Scope, as in the military form. This extended title appears to be more descriptive of the actual use of the section than the simpler term.

Federal Specifications are designated by an identification system consisting of three parts. The first part is comprised of from one to three capital letters identifying the Federal procurement group to which the items belong. Those of interest in plastics and elastomers are the following:

L   Cellulose Products and Synthetic Resins
ZZ   Rubber and Rubber Goods
MMM Adhesives
PPP   Packaging and Packing

The second part is a single letter which formerly designated the first letter of the specification title. This letter is no longer significant.[9] The third part is the serial number determined by its alphabetical location within the procurement group. Thus, L-P-395B covers "Plastic Molding (and Extrusion) Material, Nylon, Glass Fiber Reinforced." This particular Federal Specification is also listed in Table 4, as shown in the DODISS alphabetical listing. It would, of course, also be found in the numerical listing in the DODISS, and in the Federal Index.[14] As indicated above for Military Specifications, the B refers to the second revision of the document.

Interim Federal Specifications are potential Federal Specifications issued in interim form for optional use by all Federal Agencies. They may be released when there is an immediate need and insufficient time to process the specification through concerned departments. These are designated in the same manner as Regular Federal Specifications, except that two zeroes precede the serial number, and in parentheses after the serial number there is a code giving the agency responsible for the actual development and preparation of the specification. Thus L-P-00524 (GSA-FSS) covers "Plastic Sheet, Polyethylene, Laminated, Nylon Reinforced," issued by the General Services Administration and prepared by the Federal Supply Service. An Interim Federal Specification with a military sponsor is MMM-A-00131 (Army-MR), covering "Adhesive: Glass-to-Metal (for Bonding of Optical Elements)," issued by the Army and prepared by the Army Materials and Mechanics Research Center at Watertown, Mass.

Emergency Specifications are also designated in the same manner as Regular Federal Specifications, except that one zero precedes the serial number.

Federal Specifications are modified by *amendments* or by complete *revisions.* Amendments are used only when minor changes are made; each carries the specification symbol, the amendment number in arabic numbers, and the date. Only one amendment is in existence at any one time. Subsequent minor changes are made by superseding amendments; these include changes still in effect at the date of issue and are indicated by arabic numerals. *Revisions* of Federal Specifications are made when significant changes are introduced. They are indicated by the addition of a letter A, B, etc., to the basic symbol, as discussed in the example above.

Federal Standards. A Federal Standard is a standard promulgated by GSA for mandatory use by all Federal Agencies. It establishes engineering and technical limitations and applications for items, materials, processes, methods, designs, and engineering practices. Federal Standards are designed to achieve overall savings and related benefits through such means as (1) assuring the highest practical degree of uniformity, (2) providing for interchangeability of parts, (3) reducing the number of qualities, sizes, colors, varieties, and types, or (4) standardizing on test methods and engineering practices. Federal Standards are tabulated in the DODISS[4] and the Index of Federal Specifications and Standards.[14] The latest listings show approximately 255 Federal Standards. These standards are identified by arabic numerals and the date of issue. Interim Federal Standards show a double zero before the serial number. The date of issue is given in each document. Each revision is indicated by a capital letter, and each change notice and amendment by an arabic numeral. Change notices used to amend Federal Standards are issued in loose-leaf form.

Federal Standards are of five types: (1) Supply Standards, (2) Test Method Standards, (3) Material Standards, (4) Engineering Standards, and (5) Procedural Standards. The following Federal Standards taken from the DODISS[4] appear to be of interest to the readers of this handbook. Note that no indication is given in the DODISS as to which of the five types the standards represent.

FED-STD-1   Standard for Laboratory Atmospheric Conditions for Testing
FED-STD-67A   Finger Pads, Rubber
FED-STD-0067B   Finger Pads, Rubber
FED-STD-68A   Rubber Band
° FED-STD-101B   Preservation, Packaging, and Packing Materials,
    Test Procedures
FED-STD-00160   Rubber Products, Definitions and Terms for Visible Defects of
FED-STD-162   Hose, Rubber, Visual Inspection Guide for
° FED-STD-175A   Adhesives, Methods of Testing
FED-STD-00308   Inner Tube, Pneumatic Tire
° FED-STD-406   Plastics: Methods of Testing
FED-STD-501A   Floor Coverings, Resilient, Nontextile: Sampling and Testing
° FED-STD-601   Rubber: Sampling and Testing

The Federal Index[14] does indicate which standards are Test Method Standards, and these are indicated in the listing above with an asterisk (*). There are more than a dozen of these Test Methods Standards. They appear as large compilations of widely coordinated test methods covering many different product or commodity areas. They adopt ASTM methods to a great extent. The loose-leaf arrangement of the individual test methods permits ready revision. This type of standard reflects agreement and uniformity among government, suppliers, and industrial users of standard test methods. The individual standards involving test methods are designated as Federal Test Method Standards, and the designations incorporate the words Test Methods, as: Fed. Test Method Std. No. 601: Rubber: Sampling and Testing.

Contrast this designation with the tabulated listings in the DODISS[4] and Index of Federal Specifications and Standards,[14] where the term "Test Method" is not used in the title. This particular Federal Test Method Standard has some 450 loose-leaf pages. It describes the general physical and chemical methods for testing natural and synthetic rubber products for conformance with requirements with Federal Specifications. It was prepared in order to eliminate unnecessary and undesirable variations in the general sampling and testing procedures. This standard does not include special test methods applicable to certain natural and synthetic products that are described in the appropriate detail specifications, nor does it include all the test methods for natural and synthetic rubber materials used in industry. Where conflict exists between the provisions of these methods and those of the individual test procedures in specifications for particular materials, the latter take precedence.

The organization of FED TEST METHOD STD No. 601, which may be regarded as typical of FED TEST METHOD STDS, is as follows:

Section 1. Scope and Contents
2. Alphabetical Index of Test Methods
3. Numerical Index of Test Methods
4. Subject Index
5. Definitions of Terms
6. Sampling and Acceptance Procedures
7. Temperature and Humidity of Conditioning Room
8. Notes
9. Test Methods (this section, of course, constitutes the great bulk of this document)

In this standard, as in the others, similar tests have been grouped together (in Section 3, Numerical Index of Test Methods), as shown in Tables 18 and 19.

Another important FED TEST METHOD STD is No. 406, Plastics: Methods of Testing. The test method groupings of this standard are shown in Table 20.

Table 21 shows the methods listed under one typical Group. The number of times any given test method has been revised is indicated by the number following the decimal point.

TABLE 18   Group Headings—FED TEST METHOD STD No. 601

Group 1000	Preparation of Materials and Samples (3 methods)
Group 2000	Geometric Measurements (17 methods)
Group 3000	Rheological Tests (11 methods)
Group 4000	Tension Tests (10 methods)
Group 5000	Thermal Tests (10 methods)
Group 6000	Liquid Treatment Tests (15 methods)
Group 7000	Accelerated Aging Tests (10 methods)
Group 8000	Adhesion Tests (6 methods)
Group 9000	Electrical Tests (2 methods)
Group 10000	Hydrostatic Tests (13 methods)
Group 11000	Hard Rubber (18 methods, as shown in Table 8)
Group 12000	Cellular Rubber (16 methods)
Group 13000	Tape (13 methods)
Group 14000	Miscellaneous Physical Tests (4 methods)
Group 15000	Chemical Analysis of Synthetic Rubber Compounds
Group 16000	Chemical Analysis of Rubber Compounds and Packings

**TABLE 19 Test Methods under Group 11000—Hard Rubber—
FED TEST METHOD STD No. 601**

Title	Method No.
Hard Rubber, General	11001
Tensile Strength, Hard Rubber	11011
Elongation, Hard Rubber	11021
Flexural Strength, Hard Rubber	11041
Deflection, Hard Rubber	11051
Cold Flow, Hard Rubber	11121
Impact Resistance, Hard Rubber, General	11211
Impact Resistance, Hard Rubber, Cantilever	11221
Impact Resistance, Hard Rubber, Simple Beam	11231
Impact Resistance, Hard Rubber, Variable Size Ball	11241
Impact Resistance, Hard Rubber, Fixed Size Ball	11251
Impact Resistance, Battery Containers	11261
Softening Point, Hard Rubber	11311
Bulge Test, Battery Containers	11321
Weight Change, Hard Rubber in Battery Acid	11411
Dimensional Changes, Hard Rubber in Battery Acid	11421
Penetration, Acid, Hard Rubber	11431
Voltage Withstand, Battery Containers	11521

TABLE 20 Group Headings—FED TEST METHOD STD No. 406

Group 1000	Mechanical Test Methods (28 methods)
Group 2000	Thermal Test Methods (8 methods, as shown in Table 21)
Group 3000	Optical Test Methods (6 methods)
Group 4000	Electrical Tests (6 methods)
Group 5000	Miscellaneous Physical Tests (5 methods)
Group 6000	Permanence Tests (15 methods)
Group 7000	Chemical Tests (9 methods)

**TABLE 21 Test Methods under Group 2000—Thermal Test Methods—
FED TEST METHOD STD No. 406**

Title	Method No.	ASTM No.
Deflection Temperature under Load	2011	D 648–56,* Procedure 6(a)
Flammability of Plastics over 0.050 Inch in Thickness	2021	D 635–56T
Flammability of Plastics 0.050 Inch and under in Thickness	2022	D 568–56T
Flame Resistance	2023	
Linear Thermal Expansion (Fused-Quartz Tube Method)	2031	D 696–44
Thermal Expansion Test (Strip Method)	2032	
Flow Temperature Test for Thermoplastic Molding Materials	2041	D 569–59, Method A
Brittleness Temperature of Plastics by Impact	2051	D 746–57T

* ASTM designations are given as of the time of writing of the above cited document. Certainly the year designations, as given above, are no longer current. The latest designations can be found in the current "Annual Book of ASTM Standards."[25] ASTM designations in FED TEST METHOD STDS are given for reference only. There may be differences between the ASTM methods and the FED TEST METHOD STD.

FED TEST METHOD STD No. 101B, Jan. 15, 1969, is a recent revision of 101A, dated Oct. 21, 1959. This standard, on Preservation, Packaging, and Packing Materials: Test Procedures, has considerable interest in plastics and elastomers. The test method groupings of this standard are shown in Table 22.

Examples of some of the test methods in 101B are given in Table 23.

FED TEST METHOD STD No. 501a, June 15, 1966, supersedes an earlier Interim FED TEST METHOD STD. It has an obvious interest to the readers of this handbook, for most of the floor coverings considered are plastics, elastomers, and elasto-

TABLE 22   Group Headings—FED TEST METHOD STD No. 101B

Group 1000	Dimensional Measurements (3 methods)
Group 2000	Strength and Elastic Properties of Materials (48 methods)
Group 3000	Resistance Properties of Materials (30 methods)
Group 4000	General Physical Properties (43 methods)
Group 5000	Properties of Containers, Packages, Packs, and Packaging Materials (27 methods)
Group 6000	Chemical Analyses

TABLE 23   Representative Test Methods under Various Groups in FED TEST METHOD STD No. 101B

Title	Method No.
Creep Properties of Package Cushioning Materials under Compression	2013
Tensile Tear Test of Plastic Film and Thin Sheet	2046
Deterioration Resistance of Materials in Soil	3008
Oil Transmission Test for Packaging Materials	3017
Leakage of Gaskets for Floating Bag Barriers	4024
Water Absorption Test of Plastics (Other than Cushioning Materials)	4036
Water Vapor Permeability of Containers or Packages	5021
Penetration of Packaging Materials by Water	5026
Ash and Loss on Ignition of Solid Materials	6005

TABLE 24   Group Headings—FED TEST METHOD STD No. 501a

Group 1000	Preparation of Material for Test (4 methods)
Group 2000	Geometrical Measurements (12 methods)
Group 3000	Rheological Tests (13 methods)
Group 4000	Tension Tests (5 methods)
Group 5000	Accelerated Aging Tests (5 methods)
Group 6000	Thermal Tests (6 methods)
Group 7000	Miscellaneous Physical Tests (9 methods)
Group 8000	Electrical Tests (4 methods)
Group 9000	Chemical Tests (5 methods)

plastics, with or without additives. The test method groupings of this standard are shown in Table 24.

Some examples of the test methods in 501a are given in Table 25.

Details on the makeup of FED TEST METHOD STD No. 175A, Adhesives: Methods of Testing, will not be given here, since adhesives are not really within the scope of this chapter.

National Aeronautics and Space Administration   NASA specifications and other standardization documents are not nearly as well known to the technical world as Military and Federal Specifications and Standards. Their field of interest is much more limited than is the case with military and federal standardization documents.

The following documents will give the reader a good idea of the type of documentation available through NASA. It will be apparent that NASA also uses a considerable number of standardization documents issued by other agencies.

NASA Specifications and Standards (NASA SP-9000).[15] This volume, published in 1967, is the first effort to assemble in one publication an up-to-date listing of all specifications and standards originated by NASA, its centers, and the Jet Propulsion Laboratory (JPL) through December 1966. It is intended to help determine what specifications and standards have been generated, where they were developed, and the organizational component responsible. It is arranged in three sections.

Section 1 is a listing of specifications and standards, displayed by the alphanumeric designation assigned by the originating organizations. The title appears below the alphanumeric designation, followed by the date of issue plus the date or dates of any amendments. To the right of the alphanumeric designation is the code for the NASA Center at which the specification and/or standard was generated, followed by the four-digit code (e.g., 00–01) for the organizational component serving as source. These origin codes and source codes are identified in the front part of the volume.

TABLE 25   Representative Test Methods under Various Groups in FED TEST METHOD STD No. 501a

Title	Method No.
Flexibility, Folding Method	3121
Deflection	3131
Durometer Hardness	3511
International Hardness of Elastomers	3531
Flame Spread Index	6421
Dynamic Coefficient of Friction	7121
Scratch Resistance	7711
Resistance to Detergents	9341

An example of an entry from the alphanumeric section is as follows:

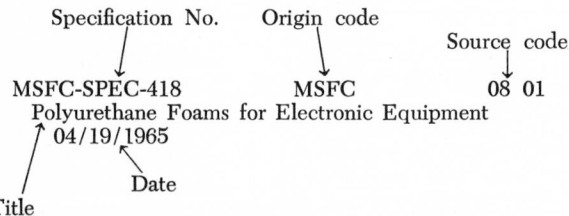

In this case MSFC means George C. Marshall Space Flight Center, Huntsville, Ala., and 08 01 refers to the Documentatation Repository Branch at that center.

Section 2 is a subject index containing subject headings, arranged alphabetically, derived from the significant terms in the titles of the items listed in this publication. Under each term are listed the titles of those specifications and standards containing that term. The titles are followed by the alphanumeric designation and the source code.

An example of the above entry from the Subject Index is as follows:

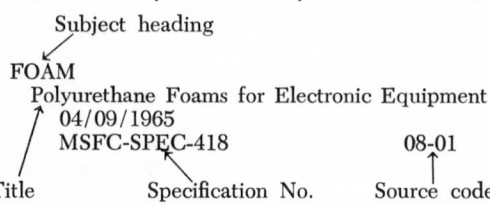

Other subject headings that will be of interest to the readers of this handbook are as follows:

Coating	Polymeric
Bonding	Polytetrafluoroethylene
Adhesive	Polyurethane
Elastomeric	Potting
Foams	Rubber
Plastic	Silicone
Packaging	Polyethylene

This section is by far the most extensive section in the document.

Section 3 is an Originator Index, which is comparable to Section 1, except that the listing of specifications and standards is grouped by NASA Center of origin. These centers are as follows:

Ames Research Center	Jet Propulsion Laboratory
Electronics Research Center	Kennedy Space Center
Flight Research Center	Manned Spacecraft Center
Goddard Space Flight Center	Marshall Space Flight Center
Headquarters	

It should be noted that this document includes only specifications and standards originated by NASA. Contractor-generated specifications and standards which NASA may use from time to time are not included.

TABLE 26 Example of Section I—Numerical Listing, Microfiche Reference Index for MSFC Specifications, Standards, and Procedures

MICROFICHE REFERENCE INDEX FOR MSFC SPECIFICATIONS, STANDARDS, AND PROCEDURES				PAGE 23
		NOVEMBER 1972		
DOCUMENT			TYPE OF	
NUMBER	REV.	TITLE	DOCUMENT	DATE
411-1		QUALIFIED PRODUCTS LIST-1	SPEC.	MAR 66
412		MODULES, ELECTRONIC, ENCAPSULATING	PROC.	MAR 65
AMD-2				SEP 65
417	A	VEHICLE CONFIGURATION SYSTEM DATA REQUIREMENTS	STD.	APR 66
NOTICE-1				DEC 66
* 418		POLYURETHANE FOAMS FOR ELECTRONIC EQUIPMENT	SPEC.	APR 65
AMD-2				FEB 70
* 418-1		QUALIFIED PRODUCTS LISTS-1	SPEC.	APR 65

MSFC Specifications, Standards, and Procedures—Microfiche Reference File Index.[16] This index, dated November 1972, has been prepared to identify information contained on microfiche in the A&PS-MS-D Documentation Repository. It is divided into two sections. Section I is a listing by document number in numerical sequence. Table 26 shows a sample of Section I. The entries shown include the polyurethane foam specification cited above from the entry in NASA SP-9000 (this particular entry is marked with an asterisk (*)). Table 27 shows an example of Section II, the listing by title in alphabetical sequence, again including the same polyurethane foam specification. The entries in Table 26 show that a Qualified Products List exists for this specification. It also shows that Amendment 2 was issued in February 1970.

George C. Marshall Space Flight Center Technical Specification Listing (MSFC-TSL-488B).[17] The purpose of this document is to identify the specifications and standards listed by user and subject, provide a brief scope of the document listed, and delineate pertinent use data (limitations, unique characteristics, etc.) relative to multiapplication technical specifications used by design and procuring activities of MSFC and its associated contractors. The technical specifications listed are for use by MSFC and its associated contractor designers in selecting economically and technically suitable specifications, standards, etc., suitable for call-outs in engineering drawings and contracts applicable to space programs and projects.

TABLE 27 Example of Section II—Alphabetical Listing, Microfiche Reference Index for MSFC Specifications, Standards, and Procedures

MICROFICHE REFERENCE INDEX FOR MSFC SPECIFICATIONS, STANDARDS, AND PROCEDURES PAGE 54
NOVEMBER 1972

TITLE	DOCUMENT NUMBER	REV.	TYPE OF DOCUMENT	DATE
PLATING, CADMIUM-COVER SHEET FOR QQ-P-416AG-AMD 1	1041996C	A	SPEC.	NOV 64
PLATING, MECHANICAL, ZINC	40MC038C		SPEC.	JUL 66
PLATINUM, IRIDIUM, ALLOY, PT85IR15, WIRE, STRIP,RCD	10MC176C		SPEC.	DEC 63
PLATIUM, RUTHENIUM ALLOY, PT90-RU10, WIRE, SHEET	10MC175E		SPEC.	DEC 63
PLUS OR MINUS 5 VOLT INVERTER + TEMPCIRCUIT	40M38752	A	SPEC.	FEB 72
POLYTETRAFLUOROETHYLENE, PARTS, RODS + TUBING	236		SPEC.	JUL 62
POLYURETHANE FOAM, FILLING S-1 AND S-18 TORQUE BOX	10MC1778		SPEC.	MAR 64
* POLYURETHANE FOAMS FOR ELECTRONIC EQUIPMENT	418		SPEC.	APR 65

Revisions to specifications listed in this listing are identified by a letter suffix to the basic number of the technical specification (e.g., MSFC-SPEC-750B). Amendments are identified by AMDT and an arabic number in parentheses following the specification number, including revision letter when applicable [e.g., MSFC-SPEC-750B (AMDT 2)].

The organization of this massive volume (576 pages) is as follows:
Section I. Introduction
Section II. Technical Specification Listing (the bulk of the document)
Section III. Specification Indices
Section II, the Technical Specification Listing, is by far the bulkiest section in

TABLE 28 Typical Page from Technical Specification Listing, MSFC-TSL-488B

CATEGORY & SPECIFICATION NO.	TITLE	TYPE	GSE ONLY	USER	SCOPE & REMARKS
17.1 MATERIALS (CONTINUED)					
* MIL-STD-670B	CLASSIFICATION SYSTEM AND TESTS FOR CELLULAR ELASTOMERIC MATERIALS	S		CDI	ESTABLISHES TESTS AND A CLASSIFICATION SYSTEM FOR ELASTOMERIC (CELLULAR) MATERIALS.
17.1.1 FOAM					
BMS 8-38D	FOAM-URETHANE, RIGID FLUOROCARBON-BLOWN	M		B	SPECIFIES SPECIAL FOAM WITH A FLUOROCARBON BLOWING AGENT.
BMS 8-39D	FLEXIBLE URETHANE FOAM	M		B	PROVIDES MATERIALS TO BE USED FOR CUSHIONING, SHOCK ABSORBING, AND GELLED WATER HOLDER APPLICATIONS. THESE MATERIALS ARE FROM THREE TYPES (APPLICATION) FIVE GRADED (DENSITY), AND FOUR CLASSES (COMPRESSION-DEFLECTION) AND ARE USED IN TWO FORMS (SHEET STOCK AND MOLDED PARTS). INCLUDES A QPL DATED 2-10-66.
BMS 8-61B	FOAM-URETHANE, FLEXIBLE	M		B	SPECIFIES A FLEXIBLE URETHANE FOAM TO BE USED AS A PROTECTIVE CUSHIONING MATERIAL. THE MATERIAL, SUPPLIED AS TWO UNREACTED LIQUID COMPONENTS, IS TO BE MOLDED IN THE 4 TO 6 LBS. PER CUBIC FOOT DENSITY RANGE. INCLUDES A QPL DATED JUNE 2, 1966.
60B32013B	STRUCTURAL POLYURETHANE FOAM	M		B	DESCRIBES ROOM TEMPERATURE CURING, CLOSED CELL, RIGID POLYURETHANE FOAM, PRIMARILY FOR STRUCTURAL APPLICATIONS.
1P20002	FOAM, POLYESTER-URETHANE, FLEXIBLE OPEN CELL	M		D	PROVIDES REQUIREMENTS FOR FLEXIBLE FOAMED POLY-URETHANE IN SHEET OR ROLL FORM.

NOTE: USERS MAY BE USING DIFFERENT REVISION.
*RECOMMENDED FOR NEW DESIGN

MSFC-TSL-488B
September 30, 1969

this document. A typical page is shown in Table 28 for Plastic Foams. As shown below, 17.0 covers Plastics, and 17.1 Plastic Materials. About two more pages similar to this are given for the plastic foam category. Under Type the code S refers to Standard and M to Material. The code CDI under User indicates that the MIL STD is used by Chrysler, McDonnell-Douglas Astronautics, and IBM. The single D means that the specification is used by McDonnell-Douglas. The letter B refers to The Boeing Company. "BMS" means Boeing Materials Specification.

Under Section III—Specification Indices are three major groupings, a Specification Category Index, a Subject Index, and a Numerical Index. The Specification Category Index is broken down as follows, with the categories of particular interest marked by asterisks (*):

* 1.0 Adhesives
2.0 Brazing
3.0 Chemical Machining
4.0 Cleaning
5.0 Conductors, Electrical
6.0 Configuration and Design Control
7.0 Electro-Magnetic Interference/ Telemetry
8.0 Fabric
9.0 Finishes
10.0 Gases and Liquefied Gases
11.0 Heat Treatment
12.0 Hydraulic/Pneumatic

13.0 Identification
*14.0 Insulation
15.0 Lubricants
16.0 Metals
*17.0 Plastics
*18.0 Potting/Molding/Encapsulation
*19.0 Preservation/Packaging/Handling
20.0 Propellants/Explosives
21.0 Quality/Reliability
*22.0 Rubber
23.0 Soldering
24.0 Testing
25.0 Welding

Those categories of particular interest to the readers of this handbook are broken down further as follows:

1.0 ADHESIVES
  1.1 Materials
    1.1.1 General Purpose
    1.1.2 Structural
  1.2 Bonding and Cementing Procedures
    1.2.1 Structural

14.0 INSULATION
  14.1 Materials
    14.1.1 Electrical
    14.1.2 Thermal
    14.1.3 Tape
  14.2 Procedures

17.0 PLASTICS
  17.1 Materials
    17.1.1 Foam
    17.1.2 Honeycomb
    17.1.3 Laminates
    17.1.4 Printed Circuit Materials
    17.1.5 Tape
    17.1.6 Thermoplastic Materials
    17.1.7 Thermosetting Materials
  17.2 Procedures

18.0 POTTING/MOLDING/ENCAPSULATION
  18.1 Compounds
  18.2 Procedures

19.0 PRESERVATION/PACKAGING/HANDLING
  19.1 Materials
  19.2 Procedures

22.0 RUBBER
  22.1 Materials
  22.2 Procedures
  22.3 Standards

Examples of the Subject Index under Section III are shown as follows:

Subject	Section	Category
Acrylic-plastic	17.1.3	Plastics
Extrusions-plastic	17.1	Plastics
Extrusions-rubber	22.1	Rubber
Filler	18.1	Potting-molding-encapsulation
Polyurethane	17.1	Plastics
Printed circuit boards—materials	17.1.4	Plastics
Teflon	17.1	Plastics

The Numerical Index lists document number, date, index (really section), and page number for all listed documents for 18 different sources. A typical entry is as follows:

     AMS 2243D   January 15, 1962   21.2   381

The 18 sources covered are as follows:

3.1   Aerospace/Aeronautical Material Specifications
3.2   American Society for Testing and Materials
3.3   Army Ballistic Missile Agency
3.4   The Boeing Company
3.5   McDonnell-Douglas Astronautics
3.6   Federal Specifications
3.7   Federal Standards
3.8   General Electric
3.9   International Business Machines
3.10   Marshall Space Flight Center Drawings
3.11   Marshall Space Flight Center
3.12   Manned Space Flight Center
3.13   Military Specifications
3.14   Military Standards
3.15   National Aeronautics and Space Administration
3.16   North American Rockwell
3.17   Ordnance Corps
3.18   Miscellaneous

The typical entry shown just before this listing is for a document from the Aeronautical Material Specifications (AMS).

This document may be used in a number of ways. To find documents in a particular field, the reader will have to refer to the Subject Index, which will then lead him directly to the listing of applicable documents in Section II. If he wants information on a specification the number of which he knows, he must look in the Numerical Index, and then follow into Section II.

George C. Marshall Space Flight Center Specifications and Standards for Space Systems and Related Equipment.[18] This manual has been published to provide the MSFC with a means of realizing performance, cost, or schedule benefits through the use of approved design standards and material, process, and parts specifications that establish requirements for system/equipment definition and acquisition. It provides a brief abstract, implementation media, typical implementation wording, and information notes to facilitate the use of requirements reflected by specifications and standards listed in it. The use of this document will provide a common base line for uniformly specifying common engineering requirements and will assure compatibility between the procuring activity and contractor and between the contractor and subcontractor.

The manual is arranged in a general section and four annexes. The general section states the purpose, scope, referenced documents, policy, definitions and abbreviations, background, general provisions, and responsibilities. Annex A contains the basic specifications and standards applicable to space systems and ground support equipment (GSE). Annex B contains documentation unique to GSE. Annexes C and D contain contract management publications and information pub-

---

lications, respectively. Annex A subject headings are exactly as listed above for MSFC-TSL-488B where the Specification Category Index is listed (page 39), except that a twenty-sixth heading is listed, Flying Qualities. An alphabetical index of subjects covered is also provided. An example of two items listed under Appendix A is given in Table 29. It should be noted that the 18.f entry is for the same polyurethane foam specification whose entry in NASA SP-9000 is described above on page 36. Standards and specifications tabulated in this document come from the following sources:

NASA Headquarters
NASA Marshall Space Flight Center (specifications and drawings)
NASA Manned Space Center
Federal Standards and Specifications
Military Specifications
American National Standards Institute
Flight Research Center

TABLE 29   Sample of Appendix A in MSFC MM 8070.2

18.f MSFC-SPEC-418(2) - POLYURETHANE FOAMS FOR ELECTRONIC EQUIPMENT

(1) Abstract. - This specification covers the requirements for foam-in-place liquid resin and catalyst capable of producing rigid, thermosetting, unicellular, polyurethane foam used for embedment of electronic components and circuits, and for void filling to provide lightweight reinforcement. This specification also describes qualification and acceptance requirements for preparation for delivery.

(2) Implementation. - Through contract statement, SOW, or CEI.

(3) Typical Implementation Wording. - "Encapsulation materials shall conform to the requirements of MSFC-SPEC-418."

(4) Notes/Cautions/Warnings. - Quality Control Plan. Supplier shall submit a copy of inspection plan to MSFC with bid or proposal. Inspection and test records shall be available to the procuring agency.

18.g 50M60441 - PLASTICS FOR ELECTRICAL PACKAGING FOR SATURN V AND ATM

(1) Abstract. - Document is a listing of qualified packaging plastics (potting, molding and encapsulating materials) that have been used on the Saturn V project and ATM.

(2) Implementation. - Through contract statement or SOW.

(3) Typical Implementation Wording. - "Plastics for electrical or electronic packaging shall be selected from 50M60441."

(4) Notes/Cautions/Warnings. - Some materials in the listing have not been qualified for ATM use.

Manned Spacecraft Center Nonmetallic Materials Design Guidelines and Test Data Handbook (MSC-02681 Revision B).[19] This voluminous document (number of pages not known because of layout) provides information on the selection of nonmetallic materials for application in manned spacecraft. Offgassing and combustion test requirements, test data, and the interpretation of test results are included. In addition, the document contains the latest Manned Spacecraft Center criteria for

the selection of nonmetallic materials based on flammability and offgassing requirements. This is not really a standardization document, but it will be of considerable
interest to those involved with standardization of plastics and elastomers, however.
Its distribution is restricted.

*Office of Manned Space Flight, NASA Headquarters, NHB 8060.1 Flammability,
Odor, and Offgassing Requirements and Test Procedures for Materials in Environments That Support Combustion.*[20] This document is issued to establish uniform
material evaluation and control criteria for all materials that are under consideration
for use in and around manned spacecraft during flight and test operations.   Its
contents, based on the materials technology developed during the Apollo and Skylab
programs, is intended for implementation on all future contracts and programs as
applicable. The document provides:

■ Standard requirements for control of flammability, odor, and offgassing of materials to be used in the design and development of manned space vehicles.

■ Guidelines and directions for material selection.

■ Testing procedures for the candidate materials used in and around manned
space vehicles during flight and test operations.

The provisions of this handbook are applicable to the NASA installations responsible for the hardware design and development of manned space vehicles.   The
appropriate section of the handbook will be referenced in all future contracts and
programs involving manned space vehicles.

Table 30 shows the contents of this source.

A sample page of this handbook is shown in Table 31. The reader will note the
references to test methods presented elsewhere in the publication.

Rural Electrification Administration (REA)   This agency is a unit of the U.S.
Department of Agriculture. It issues a number of specifications involving telephone
equipment installation, many of which involve plastics and elastomers. Its mission
is not listed in the NBS Directory,[1] nor are its specifications tabulated in the World
Index of Plastics Standards[2] and the DODISS.[4] Its specifications and standards are
listed in the following reference, and those specifications known to involve plastics
and elastomers are described.

Current REA Publications—REA Telephone Borrowers.[21] Table 32 is a list of
specifications published by REA that are known to be concerned with plastics and
elastomers.[21]

Single copies of the above documents may be obtained directly from REA. Orders
for more than a single copy can be obtained by sending a check or money order
for the proper amount to the Government Printing Office, Washington, D.C. 20402.
Unfortunately the reference cited does not give the prices. The numbers in parentheses in the above list are the actual specification numbers. The bulletin numbers
are for the bulletins announcing and commenting on the specifications.

REA Specifications are organized somewhat like Military and Federal Specifications, although there are differences. PE-220 on Flexible and Semirigid Polyvinyl
Chloride Raw Material is, perhaps, one of their more interesting specifications from
the point of view of plastic materials and, as such, will be outlined briefly here.

1. SCOPE

    1.1   This specification covers the requirements for polyvinyl chloride raw materials
to be supplied in two formulations, flexible jacketing grade and semirigid
insulating grade.

    1.2   All polyvinyl chloride material covered by this specification shall be free
from excessive moisture and shall not be contaminated by foreign matter
such as dirt or metallic particles.

    1.3   The products using the two polyvinyl chloride compounds described herein
include, but are not limited to, the following:

    1.31 *Flexible Polyvinyl Chloride:* Jacketing material for telephone drop wires,
station wires, special equipment cables, and switchboard cables.

    1.32 *Semirigid Polyvinyl Chloride:* Primary insulation for special equipment cables,
switchboard cables and load coil stub cables.

TABLE 30   Contents of NASA NHB 8060.1

TABLE 31   Sample Page from NASA NHB 8060.1

CHAPTER 2   REQUIREMENTS

200   MATERIALS SELECTION

1. GENERAL. All materials considered for use in a manned spacecraft including the materials of the spacecraft, stowed equipment, or experiments shall be evaluated for flammability, odor, and off-gassing characteristics and shall meet the applicable requirements specified in this publication. Materials shall be evaluated in accordance with the applicable test methods specified herein. A minimum of three specimens (excluding black boxes) of the materials in the thickness intended for use shall be subjected to each applicable test at the maximum use environment. Material applications previously reviewed and approved should be considered to minimize additional testing.

2. MATERIALS GROUPING AND RESTRICTIONS. Materials considered for use shall be grouped as follows based upon the results of screening tests and shall be approved for specific applications by the NASA Center responsible for the spacecraft/hardware development.

GROUP I.- Materials that are noncombustible or self-extinguishing before 6 inches of the sample is consumed in accordance with Test No. 1 (Upward Propagation Test) using silicone or equivalent igniter or are noncombustible in accordance with Tests No. 4 (Electrical Insulation), and No. 5 (Potting Compounds) as applicable. Group I materials will be verified in end item configuration and/or full scale test in accordance with paragraph 205 unless it can be shown by the rationale resulting from an engineering evaluation and analyses of test data that configuration tests are not required.

Note: There are no restrictions on Group I Materials in design. The maximum burn time for propagation tests is 10 minutes maximum.

GROUP II - Materials that fail to self-extinguish when subjected to Test No. 1. The flame propagation rate of all materials shall be determined in accordance with Test No. 2. Group II materials shall not be used unless functional requirements preclude the use of Group I materials and the materials are specifically approved by the cognizant NASA Center on an individual basis. Should it be imperative to use Group II materials in habitable environments, the functionally acceptable material with the lowest flame propagation rate and the lowest production rate for offgassed products shall be used. Group II materials shall be evaluated according to Test No. 3 and shall be acceptable for design if they exhibit a flash or fire point above 450°F. All pertinent test and design data including drawings, photographs, specifications, environmental and functional requirements, and one (1) square foot samples of materials in the proper thickness shall be submitted to the

2-1

2. FORMULATIONS

2.1 The flexible jacketing grade compounds should contain the ingredients in the proportions shown below unless substitutions are approved by REA:

	Formulation (percent)	
	Outdoor use	Indoor use
Polyvinyl Chloride*...............................	46.0	52.0
Normal-Octyl Normal-Decyl o-Phthalate............	32.0	
Di-2-Ethylhexyl o-Phthalate,.....................	....	27.9
etc.		etc.

* Resin GP-$\bar{5}$-00003 in accordance with ASTM D 1755–66.

2.11 The flexible compounds shall be capable of meeting the performance requirements contained in Appendix I.

TABLE 32   REA Specifications Involving Plastics and Elastomers

Source	Title
REA Bull. 345–14	Specification for Cable, Fully Color-Coded, Polyethylene-Insulated, Double Polyethylene-Jacketed for Direct Burial (PE-23)
REA Bull. 345–13	Specification for Cable, Fully Color-Coded, Polyethylene-Insulated, Polyethylene-Jacketed (PE-22)
REA Bull. 345–51	Specification for Crystalline Propylene/Ethylene Copolymer Raw Material (PE-210)
REA Bull. 345–49	Specification for Fiberglass Buried Plant Terminal Housings (PE-51)
REA Bull. 345–29	Specification for Figure 8 Cable (PE-38)
REA Bull. 345–58	Specification for Flexible and Semi-Rigid Polyvinyl Chloride Raw Material (PE-220)
REA Bull. 345–21	Specification for Polyethylene Raw Material (PE-200)
REA Bull. 345–24	Specification for Spindle-Threaded Steel Communication Insulator Pins and Associated Plastic Bushings (PE-34)
REA Bull. 345–31	Specification for Wire, Bridle, Polyethylene Insulated (PE-19)
REA Bull. 345–42	Specification for Low Loss Buried Distribution Wire (PE-44)
REA Bull. 345–17	Specification for Wire, Plastic Insulated Conductor (PE-21)
REA Bull. 345–18	Specification for Wire, Plastic Insulated, Plastic Jacketed (PE-20)
REA Bull. 345–48	Specification for Buried Distribution Wires (PE-50)

2.2   The semirigid insulating grade compounds shall contain the ingredients in the proportions shown below unless substitutions are approved by REA:

Formulation (percent)

Polyvinyl Chloride*	70.00
Di-2-Ethylhexyl *o*-Phthalate, ...	25.00
etc.	etc.

* Resin GP-5̄-00003 in accordance with ASTM D 1755-66.

2.21   The semirigid polyvinyl chloride compounds shall be capable of meeting the performance requirements contained in Appendix II.

3. MOLDING PROCEDURE

3.1   Where molded test specimens are required they shall be prepared from compression molded sheets which shall be molded according to the following procedure:

(details given)

4. COLORS

4.1   Color concentrates may be added to natural resin by the wire and cable manufacturers. However, material so colored shall meet all the property requirements of this specification. Table 1 gives the color tolerance limits based on the Munsell Color Standards.

Table 1. Tolerance Limits—Hue, Value and Chroma for the component colors.

APPENDIX I   Flexible Polyvinyl Chloride Raw Material Requirements (Jacketing Grade)

APPENDIX II   Semirigid Polyvinyl Chloride Raw Material Requirements (Insulating Grade)

Nongovernment Sources

American National Standards Institute (ANSI)   This is a nonprofit membership organization whose bylaws provide for membership from national trade, technical, and professional groups, firms from commerce and industry, government agencies and departments, consumer groups, and similar organizations. It is described in the Directory of United States Standardization Activities[1] under the name United

States of America Standards Institute, by which it was known from 1966 to 1969. Previously it was known as the American Standards Association. The Institute serves as the national clearinghouse for standards and provides the machinery for developing and approving standards which are supported by a national consensus. Technical societies, trade associations, consumer groups, and the like make up the Member Bodies of the Institute. As of December 1972 there were 5,000 American National Standards on the organization's books. Many of these have resulted from submittals by competent organizations of standards which they have developed through their own procedures, together with evidence as to the existence of consensus in support of such standards. The balance have come through the work of sectional committees. A few—mostly simple standards—have been approved after acceptance by a General Conference. Many of the standards on the current list have been revised a number of times. Standards approved by the Institute, formerly called USA Standards, are now called American National Standards.

The American National Standards Institute, Inc., is now located at 1430 Broadway, New York, N.Y. 10018. ANSI is the United States member of the *International Organization for Standardization (ISO)*, which in 1972 consisted of the national standards bodies of 70 countries. The United States' viewpoints to be presented in the technical work of the ISO are developed through a committee specially organized as a USA Technical Advisory Group for an ISO Technical Committee. The work of the ISO technical committees results eventually in ISO Standards which may be embodied in the national standards of the ISO's Member Bodies. Before 1972, ISO documents were described as Recommendations. Since 1972, new documents are issued as Standards. Many earlier Recommendations are being transformed into Standards. A number of ISO Recommendations or Standards have been embodied in American National Standards.[1,22]

ISO publishes several information documents which are described in the ANSI Catalog.[23] Of greatest interest is the ISO Catalog (1972), which contains a full list of ISO Recommendations (and Standards) classified according to subject and number. In the 1972 catalog the sections on rubber and plastics run to 10 pages of English and French text, and list approximately 80 Recommendations originated by Technical Committee (TC), 45 on Rubber and 110 Recommendations by TC 61 on Plastics. The Catalog is issued annually, and a supplement listing revisions and new standards is published at 3-month intervals. The 1972 issue of the ISO Catalog costs $9.00.[22]

Through an exchange system with the national standards bodies of all countries where such exist, ANSI has available national standards and catalogs of national standards of many countries. Most catalogs issued by other countries, as well as most standards, are available only in the language of origin. However, some catalogs and standards are available in an English version.[22]

Every 2 weeks the Institute publishes the *ANSI Reporter* and *Standards Action*. The *Reporter* is a newsletter describing the Institute's major activities on both the national and international levels. *Standards Action* solicits comments on standards that are being developed and lists those which have recently been approved or revised. *Standards Action* and the *Reporter* are available to nonmembers on a subscription basis.[22]

The reader will find it profitable to obtain a copy of the current ANSI catalog, which is available free of charge. The 1972 issue[23] lists some 463 American National Standards and 138 ISO Recommendations involving plastics and rubber. The count, by the authors of this chapter, is reasonably accurate. The relationships with international publications (i.e., between American National Standards and ISO Recommendations) are shown by cross reference enclosed in italic parentheses. Since cross references must be brief, the following system has been used in the catalog: Where there is complete agreement with an international recommendation, the words "Agrees with" precede the number designations of the standard or international recommendation; where there is partial agreement, only the number designation is given. The actual relationships can be determined only through careful comparison of the documents involved. The American National Standards listings also include cross referencing to ASTM standards, the number designations of which

TABLE 33   Examples of Subject Index Entries Involving Plastics and Rubber (from ANSI Catalog)

Plastics
  abbreviations, ISO R1043
  absorption, water K65.16; ISO R62; R117
  acetic acid yield, ISO R1597
  acetone extraction, K65.12
  aqueous dispersion, ISO R1147; R1148; R1625
  bathtubs, Z124.1
  brittleness temperature, ISO R974
  carbon black, K65.89
  cellulose, K64.6; K65.22-K65.27; ISO R1598; R1600
  cellulose acetate, ISO R1061; R1157; ISO R585
  chlorine content, K65.9
  color change resistance, ISO R877-R879
  coloring materials, ISO R183
  compressive properties, K65.1; ISO R604
  conditioning, C59.28
  definition of terms, ISC R472
  deformation under load, K65.4; ISO R75
  density, apparent, K65.19; K65.87
  dissipation factor, C59.60
  ethyl cellulose molding compounds, K64.1
  ethylene, K65.47
  film, K65.67; K65.90
  flammability, K65.21; K65.28; K65.110; K65.111; K65.120, ISO R1210
  flatness, measurement, K65.18

Rubber (see also Latex)
  atmospheric conditions, testing, ISO R471
  cables, insulated, J8.11; CEE 2
  carbon black, ISO R1124-R1126; R1138; R1306; R1310; R1435; R1437; R1866-R1868
  chemical analysis, J10.1
  classification, J8.9
  elastomeric, J8.15
  friction tape, J8.13
  hose, J6.8; ISO R1307; R1401-R1403; R1823; R1825; R1746
  manganese, determination, ISO R1397
  mills and calenders, safety code, B28.1
  nomenclature, J9.1; ISO R1629
  products, test methods, J1.1; J1.2; J1.3; J8.1
  protective equipment, J6
  raw natural, ISO R247; R248; R249; R250; R1434; ISO R1654-R1656; R1795
  sealing rings, ISO R1398
  tape, insulating, C59.6
  viscosity, ISO R289; R1652
  vulcanized test methods, J2; J3; J4; J5.1; J7; ISO R33; R35; R36; R37; R48; R132; R133; R188; R814; R816; R1396; R1399; R1400; R1432; R1653; R1747; R1767; R1817; R1818; R1826; R1827

are listed after the titles. An excellent subject index to titles of Standards and International Specifications (including ISO) is published at the end of the catalog. Examples of entries for Plastics and Rubber are shown in Table 33.

As mentioned above, American National Standards and ASTM standards are cross-referenced in the ANSI Catalog. ASTM standards will be discussed further in this chapter, but the example of the first page of an ASTM specification in Fig. 2 will

 Designation: D 2220 – 68

American National Standard C8 46 1970
American National Standards Institute

Standard Specification for

VINYL CHLORIDE PLASTIC INSULATION FOR WIRE AND CABLE, 75 C OPERATION[1]

This Standard is issued under the fixed designation D 2220; the number immediately following the designation indicates the year of original adoption or, in the case of revision, the year of last revision. A number in parentheses indicates the year of last reapproval.

Fig. 2 ASTM Specification showing ANSI cross reference; copied from "1972 Annual Book of ASTM Standards," Part 28.[25]

TABLE 34 Examples of Listings from ANSI Catalog of American National Standards on Plastics and Rubber and ISO Recommendations Involving Plastics and Rubber

●K65—Plastics: (Continued)

K65.36-1967 Evaluation of Welding Performance for Poly (Vinyl Chloride) Structures, Method of Test for (ASTM D1789-65) 1.50

K65.37-1967 Alcohol-Benzene Soluble Matter in Cellulose, Method of Test for (ASTM D1794-62) 1.50

K65.38-1967 Intrinsic Viscosity of Cellulose, Method of Test for (ASTM D1795-62) 1.50

K65.39-1967 Chromatographic Analysis of Chemically Refined Cellulose, Method of Test for (ASTM D1915-63) 1.50

K65.40-1967 Carboxyl Content of Cellulose, Methods of Test for (ASTM D1926-63) 1.50

K65.41-1967 Melting Point of Semicrystalline Polymers, Method of Test for (ASTM D2117-64) (ISO R1218) 1.50

K65.42-1967 Solubility of Cellulose in Sodium Hydroxide, Method of Test for (ASTM D1696-61) 1.50

K65.43-1970 Ethylcellulose, Method of Testing (ASTM D914-69) 1.50

◆K65.44-1971 Polybutylene Plastics, Specification for (ASTM D2581-69) 1.50

◆K65.45-1971 Lot-To-Lot Stabilization Uniformity of Propylene Plastics, Method of Test for (ASTM D2342-68) 1.50

◆K65.46-1971 Propylene Molding and Extrusion Materials, Specification for (ASTM D2146-69) 1.50

◆K65.47-1971 Degree of Crosslinking in Crosslinked Ethylene Plastics as Determined by Solvent Extraction, Methods of Test for (ASTM D2765-68) 1.50

◆K65.48-1971 Use of a Melt Index Strand for Determining Density of Polyethylene, Practice for (ASTM D2839-69) 1.50

◆K65.49-1971 Sampling of Plastics, Practice for (ASTM D1898-68) 1.50

◆K65.50-1971 Statistical Design in Interlaboratory Testing of Plastics, Practice for (ASTM D2188-69) 1.50

◆K65.51-1971 Nomenclature Relating to Plastics (ASTM D883-69a) 1.50

Specifications for Rubber

J1.1-1970 Compound and Sample Preparation for Physical Testing of Rubber Products, Methods of (ASTM D15-69) 1.50

◆J1.2-1971 Conditioning of Elastomeric Materials for Low-Temperature Testing, Practice for (ASTM D832-64 (1970)) 1.50

◆J1.3-1971 Test Temperatures for Rubber and Rubber-Like Materials, Practice for (ASTM D1349-62 (1968)) 1.50

J2.1-1969 Tension Testing of Vulcanized Rubber, Method of (ASTM D412-68 (ISO R37) 1.50

◆J2.2-1971 Low-Temperature Stiffening of Rubber and Rubber-Like Materials by Means of a Torsional Wire Apparatus, Method of Measuring (ASTM D1053-65 (1970)) 1.50

◆J2.3-1971 Air-Pressure Heat Test of Vulcanized Rubber, Method of (ASTM D454-53 (1970)) 1.50

◆J2.4-1971 Sponge and Expanded Cellular Rubber Products, Specifications and Methods of Testing (ASTM D1056-68) 1.50

◆J2.6-1971 Ply Separation and Cracking of Rubber Products, Methods of Dynamic Testing for (ASTM D430-59 (1970)) 1.50

◆J2.7-1971 Change in Properties of Elastomeric Vulcanizates Resulting from Immersion in Liquids, Method of Test for (ASTM D471-68) 1.50

◆J2.8-1971 Resistance to Surface Cracking of Stretch Rubber Compounds, Method of Test for (ASTM D518-61 (1968)) 1.50

◆J2.9-1971 Automotive Hydraulic Brake Hose, Methods of Testing (ASTM D571-55 (1970)) 1.50

◆J2.10-1971 Automotive Air Brake and Vacuum Brake Hose, Methods of Testing (ASTM D622-65 (1970)) 1.50

◆J2.11-1971 Compression Fatigue of Vulcanized Rubber, Methods of Test for (ASTM D623-67) 1.50

◆J2.12-1971 Coated Fabrics, Methods of Testing (ASTM D751-68) 1.50

◆J2.13-1971 Young's Modulus in Flexure of Natural and Synthetic Elastomers at Normal and Subnormal Temperatures, Method of Test for (ASTM D797-64 (1970)) 1.50

TABLE 34   Examples of Listings from ANSI Catalog of American National Standards on Plastics and Rubber and ISO Recommendations Involving Plastics and Rubber (Continued)

ISO No.

→ **R117-1959** Plastics, Determination of Boiling Water Absorption.... 2.80
→ **R118-1959** Plastics, Determination of Methanol-Soluble Matter in Polystyrene..2.80
→ **R119-1959** Plastics, Determination of Free Phenols in Phenol-Formaldehyde Mouldings.................................. 2.80
→ **R120-1959** Plastics, Determination of Free Ammonia and Ammonium Compounds in Phenol-Formaldehyde Mouldings (K66.1-1969)..2.80
♦ **R121-1971** (2nd edition) Composition of Magnesium-Aluminium-Zinc Alloy Castings and Mechanical Properties of Sand Cast Reference Test Bars.................................... 2.80
R122-1959 Composition of Magnesium-Aluminum-Zinc Alloy Ingots for Casting Purposes............................... 2.45
→ **R123-1968** Sampling of Latex................................. 2.80
→ **R124-1966** Determination of Total Solids of Latex.............. 2.45
→ **R125-1966** Determination of Alkalinity of Latex................ 2.45
→ **R126-1959** Determination of Dry Rubber Content of Latex........2.45
→ **R127-1959** Determination of Koh Number of Latex............. 2.80
R128-1959 Engineering Drawing, Principles of Presentation (Y14.2-1957 and Y14.3-1957)............................. 6.65
R129-1959 Engineering Drawing, Dimensioning (Y14.5-1966)..... 7.00
R130-1959 Colour Identification of Mechanical Control Circuits for Aircraft... 3.85
R131-1959 Expression of the Physical and Subjective Magnitudes of Sound or Noise (Agrees with S1.1-1960) (Including Supplement R357-1963)................................. 3.50
→ **R132-1959** Determination of Resistance to Flex Cracking of Vulcanized Natural or Synthetic Rubber (de Mattia Type Machine).. 4.90
→ **R133-1959** Determination of Resistance to Crack Growth of Vulcanized Natural or Synthetic Rubber (de Mattia Type Machine).. 3.85

▲ not yet available; † quantity and member discounts apply;

♦ approved since last price list; R, reaffirmed

serve to show how American National Standards are cross-referenced in ASTM standards:

Table 34 shows the types of listings that are found in the ANSI Catalog. Plastics are listed under Category K—Chemical Industry, while Rubber has its own Category J. The oldest standards listed for plastics go back to 1959, and the oldest for rubber to 1969. Note that all the plastics standards shown have cross references to ASTM standards. One standard, K65.41-1967, also has a cross reference to an ISO Recommendation. A similar situation prevails with the rubber entries. Standard J2.1-1969 also has a cross reference to an ISO Recommendation. The figures in the right-hand column are prices. Full details on how to order are given in the front of the catalog. In the case of ISO Recommendations the reader will see from Table 34 that listings on plastics and rubber are mixed in with other subjects. Apparently serial numbers are assigned on a continuous basis, regardless of the subject area. The portion of the tabulation shown in Table 34 for ISO Recommendations was chosen because it had a relatively large number of entries on plastics and rubber, as indicated by the arrows. Again, the figures in the right-hand column are prices. Note that the entry for R131-1959 indicates that the Recommendation

"Agrees with S1.1-1960" (an American National Standard). The subject area of this standard is not in the fields of plastics and elastomers, but nevertheless it serves as an example of this type of entry, which is discussed briefly above.

The authors have recently received a copy of a brochure[24] outlining ANSI's role. This publication was received too late to incorporate its points in this chapter, but it is well worth obtaining.

American Society for Testing and Materials (ASTM)   The American Society for Testing and Materials, founded in 1899, is a scientific and technical organization formed for "the development of standards on characteristics and performance of materials, products, systems and services; and the promotion of related knowledge." It is the world's largest source of voluntary consensus standards. The Society operates through more than 109 main technical committees. These committees function in prescribed fields under regulations that insure balanced representation among producers, consumers, and general-interest participants. The Society in 1972 had about 22,000 active members, of whom approximately 14,000 serve as technical experts on committees. The Society's work concerns primarily standardization and research in materials. It is, and has been for more than 70 years, specifically interested in the quality and test of materials and only indirectly does it become involved in design problems, dimensional standards, and related matters.[1] It is the purpose of this presentation not to discuss the many publications issued by the Society, but rather to cover only standardization documents. For ease of reference the standards are published in collective form, and each is also issued in separate pamphlet form. Of predominant interest is the "Annual Book of ASTM Standards,"[25] published each year in 33 "parts." The 33 parts or volumes, as they really are, contain 32,000 pages and over 4,600 standards (in the 1973 edition). Those parts which are of probable interest to the readers of this handbook are listed and described in the list of references below.[25] It should be noted that the different parts are published throughout the year. Thus at any given month the most up-to-date versions may represent one calendar year for one part and a different calendar year for another. Probably the most useful parts are Numbers 26, 27, and 28, at least as far as plastics and rubbers are concerned.

Table 35 shows the titles of a number of standards published in Parts 26, 27, and 28. These examples are taken from the List by Subjects, which is really a subject index. Each volume also has a "Contents" listing, which is a numerical listing of standards. In most cases it is not necessary to use the latter index, since the standards are printed in the volumes in numerical order.

The first page of ASTM D 2133-66, as published in the "1972 Annual Book of ASTM Standards," Part 26, is shown in Table 36. Note the reference to an American National Standard* in the upper right portion of the page. ASTM always lists the cognizant ASTM Committee, and, in some cases, the subcommittee, as Footnote No. 1. Space is not available to show the entire standard, but the paragraph headings are listed below to show what type of coverage is given in the specification. D 2133 is actually one of the shorter ASTM standards, with only three pages. Continuing from 6.1.3 on the sample page, the paragraphs are:

  6.1.4 Melting Point
  6.1.4.1 Apparatus
  6.1.4.2 Procedure
  6.1.5 Melt Flow Rate
  6.1.6 Tensile Yield Strength, Ultimate Elongation, and Modulus (Tensile) of Elasticity
  7. Number of Tests
  7.1 Duplicate determinations using etc. . . .
  8. Retest and Rejection
  8.1 If any failure occurs, etc. . . .
  9. Packaging and Marking
  9.1 Packaging
  9.2 Marking

* Joint American National Standards—ASTM Standards were actually developed by ASTM. ANSI does not develop its own standards.

Until September 1972, copies of ASTM individual standards could be obtained from the Society at $1.25 each. Now individual standards in lots less than 50 may be purchased by coupon only, each coupon redeemable for one standard. A five-coupon book is available at $6 and a 25-coupon book at $30.

It is interesting to point out that there is considerable supplemental material in the later pages of Part 28. This is presented in pages outlined in a gray border. This material includes an "Elastomers Manual" and has a useful cross-reference listing of ISO Recommendations and ASTM Methods on Rubber.

TABLE 35   Examples from Lists by Subjects from Annual Book of ASTM Standards

---

LIST BY SUBJECTS

1972 ANNUAL BOOK OF ASTM STANDARDS, PART 26
PLASTICS—SPECIFICATIONS; METHODS OF TESTING PIPE, FILM,
REINFORCED AND CELLULAR PLASTICS: FIBER COMPOSITES

Since the standards in this book are arranged in numeric sequence, no page numbers are given in this list by subjects.

PLASTICS

Olefin Plastics

Specifications for:

D 2647 – 70	Crosslinkable Ethylene Plastics
D 2952 – 71	Ethylene Plastics
D 2581 – 71	Polybutylene Plastics
D 1248 – 72	Polyethylene Molding and Extrusion Materials
*D 2146 – 69	Propylene Plastic Molding and Extrusion Materials

Methods of Test for:

*D 1693 – 70	Environmental Stress Cracking of Ethylene Plastics
*D 2552 – 69	Environmental Stress Rupture of Type III Polyethylenes Under Constant Tensile Load
*D 2342 – 68	Lot-to-Lot Stabilization Uniformity of Propylene Plastics
*D 1928 – 70	Preparation of Compression-Molded Polyetylene Test Samples
*D 2445 – 65 T	Thermal Oxidative Stability of Propylene Plastics
D 3012 – 72	Thermal Oxidative Stability of Propylene Plastics, Using a Biaxial Rotator

Recommended Practice for:

*D 2839 – 69	Use of Melt Index Strand for Determining Density of Polyethylene

Thermoplastic Materials

Specifications for:

*D 2133 – 66 (1972)	Acetal Resin Injection Molding and Extrusion Materials
*D 1788 – 68	Acrylonitrile - Butadiene - Styrene (ABS) Plastics, Rigid
*D 706 – 63	Cellulose Acetate Molding and Extrusion Compounds

1972 ANNUAL BOOK OF ASTM STANDARDS, PART 27

PLASTICS—GENERAL METHODS OF TESTING, NOMENCLATURE

Since the standards in this book are arranged in numeric sequence, no page numbers are given in this list by subjects.

PLASTICS

Mechanical Properties

Methods of Test for:

*D 1044	56 (1969)	Abrasion Resistance, Surface, of Transparent Plastics
*D 1242	56 (1969)	Abrasion Resistance of Plastic Materials
*D 1602	60 (1969)	*Bearing Load of Corrugated Plastics Panels (see Part 26)*
*D 953	54 (1969)	Bearing Strength of Plastics
*D 1893	67 (1972)	Blocking of Plastic Film
*D 952	51 (1969)	Bond Strength of Plastics and Electrical Insulating Materials
D 1180	57 (1972)	Bursting Strength of Round Rigid Plastic Tubing
D 3028	72	Kinetic Coefficients of Friction of Plastics
*D 1894	63 (1972)	Coefficients of Friction of Plastic Film
*D 695	70	Compressive Properties of Rigid Plastics
*D 1621	64	*Compressive Strength of Rigid Cellular Plastics (see Part 26)*
*D 2143	69	*Cyclic Pressure Strength of Reinforced, Thermosetting Plastic Pipe (see Part 26)*
*D 2236	70	Dynamic Mechanical Properties of Plastics by Means of a Torsional Pendulum
*D 1604	63 (1969)	*Flatness of Plastics Sheet or Tubing, Measuring (see Part 26)*
D 790	71	Flexural Properties of Plastics
D 671	71	Flexural Fatigue of Plastics by Constant-Amplitude-of-Force

TABLE 35   Examples from Lists by Subjects from Annual Book of ASTM Standards (Continued)

1972 ANNUAL BOOK OF ASTM STANDARDS, PART 28

Rubber, Carbon Black, Gaskets

Since the standards in this book are arranged in alphanumerical sequence, no page numbers are given in this list.

RUBBER AND RUBBER-LIKE MATERIALS
Processability Tests

Methods of Test for:

D 2704 – 68 T	Curing Characteristics with a Compression Modulus Cure Meter, Measurement of (Discontinued 1972†)
D 2705 – 68 T	Curing Characteristics with the Oscillating Disk Rheometer Cure Meter, Measurement of (Discontinued 1972†)
D 2706 – 68 T	Curing Characteristics with the Viscurometer Cure Meter, Measurement of (Discontinued 1972†)
D 2230 – 68	Extrudability of Unvulcanized Elastomeric Compounds
D 2084 – 71 T	Measurement of Curing Characteristics with the Oscillating Disk Cure Meter
*D 926 – 67	Plasticity and Recovery of Rubber and Rubber-Like Materials by the Parallel Plate Plastometer
D 1917 – 67	Shrinkage of Styrene-Butadiene Rubbers (SBR) or Compounds of SBR
*D 1646 – 72	Viscosity and Curing Characteristics of Rubber by the Shearing Disk Viscometer

Recommended Practice for:

*D 1349 – 62(1968)	Standard Test Temperatures for Rubber and Rubber-Like Materials

Interlaboratory Tests
Recommended Practice for:

*D 1421 – 68	Interlaboratory Testing of Rubber and Rubber-Like Materials

* Approved as **American National Standard** by the American National Standards Institute.

Currently there is a move on the part of the Federal Government to use ASTM standards and specifications in purchasing documents. The National Bureau of Standards (NBS) has recently adopted the policy of withdrawal of a number of Voluntary Product Standards for commercial products. Requests for many standards are channeled to ASTM.[26] The Code of Federal Regulations, which sets procedures for the Bureau's development of Voluntary Product Standards, now asserts the philosophy that the private sector has the primary responsibility for the development of voluntary standards, and the Bureau should develop only those standards of broad public concern which cannot be effectively developed by a private standards-writing body. The Bureau's policy now is not to compete with private standard-writing bodies, but to cooperate with them and supplement their activities when necessary.

It should be noted that the Federal Government has a rather direct influence in the development of ASTM standards, since many of ASTM's committees have members active in writing standards who are also representatives of Federal Agencies. Even when the federal representatives have no direct hand in the writing or other preparation of a standard, they have an opportunity to participate in balloting for or against the standard, and during the balloting can make suggestions for modifications of the standards to make them acceptable.

As an example, the senior author of this chapter is, at the time of the writing of this chapter, chairman of the Editorial and Definitions Subcommittee of ASTM Committee G-3 on Deterioration of Nonmetallic Materials. As such, he is responsible for approving the actual write-ups of all standards developed by this committee and, in practice, is very active in the later stages of their preparation.

National Electrical Manufacturers Association (NEMA) This is a trade association of manufacturers of equipment and apparatus for the generation, transmission, distribution, and utilization of electric power.[1] The membership consists of over

TABLE 36   First Page of Typical ASTM Specification in Plastics*

 Designation: D 2133 – 66 (Reapproved 1972)   American National Standard K65.200-1971
Approved May 20, 1971
By American National Standards Institute

Standard Specification for
ACETAL RESIN INJECTION MOLDING AND EXTRUSION MATERIALS[1]

This Standard is issued under the fixed designation D 2133; the number immediately following the designation indicates the year of original adoption or, in the case of revision, the year of last revision. A number in parentheses indicates the year of last reapproval.

1. Scope

1.1 This specification covers two grades of acetal resin materials suitable for injection molding or extrusion, or both.

NOTE 1—The values stated in U.S. customary units are to be regarded as the standard.

2. Grades

2.1 This specification covers two grades of acetal resins, classified according to melt flow, representing materials generally processed by injection molding (Grade I) or extrusion (Grade II).

3. General Requirements

3.1 The material shall be of uniform composition and so prepared as to conform to the requirements of this specification.

3.2 All resins described in this specification shall be free of foreign matter to such a contamination level as may be agreed upon between the purchaser and the seller.

4. Detail Requirements

4.1 The acetal resins covered by this specification shall conform to the requirements prescribed in Tables 1 and 2 when tested by the procedures specified in Section 6. Table 2 lists those tests requiring a molded specimen.

5. Sampling

5.1 Unless otherwise agreed upon between the purchaser and the seller, the materials shall be sampled in accordance with ASTM Recommended Practice D 1898, for Sampling of Plastics.[2] Adequate statistical sampling shall be considered an acceptable alternative.

6. Methods of Test

6.1 The mechanical properties of specimens molded or extruded from acetal resins are influenced by a number of factors, such as cylinder temperature, die temperature, rate of filling the die, and preconditioning prior to testing. It is essential that the manufacturer's recommendations on molding and extrusion conditions be followed. The properties enumerated in these specifications shall be determined in accordance with the following methods:

6.1.1 *Conditioning*—For tests for tensile properties, the molded test specimens shall be conditioned in accordance with Procedure A of ASTM Methods D 618, Conditioning Plastics and Electrical Insulating Materials for Testing.[2] The other tests require no conditioning.

6.1.2 *Test Conditions*—Tests for tensile properties shall be conducted in the Standard Laboratory Atmosphere of 23 ± 2 C (73.4 ± 3.6 F) and 50 ± 5 percent relative humidity as defined in Procedure A of Methods D 618. Tests for specific gravity, melting point, and melt flow rate may be conducted under ordinary laboratory conditions.

6.1.3 *Specific Gravity*—Determine the specific gravity in accordance with ASTM Methods D 792, Tests for Specific Gravity and Density of Plastics by Displacement.[2]

---

[1] This specification is under the jurisdiction of ASTM Committee D-20 on Plastics and is the direct responsibility of subcommittee D-20.15 on Thermoplastic Materials.
Current edition effective Sept. 30, 1966. Originally issued 1962. Replaces D 2133 – 64 T.
[2] *Annual Book of ASTM Standards*, Part 27.

311

---

* From "1972 Annual Book of ASTM Standards," Part 26.

500 of the major electrical manufacturing companies in the country. The membership is limited to corporations, firms, and individuals actively engaged in the manufacture for sale in the open market of products included within the product scope of one or more of the many NEMA product subdivisions. NEMA may be considered an aggregate of product sections, each representing a group of manufacturers of certain classes of products, such as Conduits, Ducts, and Fittings; Insulating Materials; and Wires and Cables. Sections with related interests are organized into divisions.[1]

NEMA has published some 210 separate standards publications in a number of classifications. A considerable amount of NEMA standardization activity is in cooperation with other organizations engaged in standardization, such as ASTM, Edison Electrical Institute, NFPA, Underwriters' Laboratories, IEEE, and other associations, laboratories, or government bodies.

Many electrical product standards originating within NEMA or initiated by other organizations are of such national significance as to make desirable their adoption as American National Standards under the procedures established by ANSI, which provides for participation by and consideration of the views of all interested groups. NEMA participates in the International Electrotechnical Commission (IEC), the international body for the development and approval of recommendations for electrical standards.

NEMA, upon invitation, furnishes the military and other governmental standardizing bodies information and recommendations for use in preparation of their initial drafts or revisions of various Federal and Military Specifications affecting products within its scope.[1]

Table 37 shows examples of NEMA Standards believed to involve plastics or elastomers. They are taken from the current NEMA Catalog.[27]

Many, but not all, of the publications listed in the NEMA Catalog are provided with an automatic revision service. These books are shipped complete with all interim revisions available at the time of shipment. To secure subsequent revisions, the purchaser should fill out and return the postal card which accompanies each of these books.

Rubber Manufacturers Association (RMA) The technical committees of this organization's several product groups collaborate with federal, state, and municipal agencies, domestic and foreign associations, and individual commercial users in the development of new or improved specifications for all kinds of rubber products.[1] Emphasis is placed on the practicality of physical requirements and minimization of sizes and types of products, endeavoring to achieve the utmost simplification of sizes and types, with greatest potential economic benefit to producers, wholesalers, retailers, and consumers in processing, distribution, and inventory costs. Most efficient use of raw materials and labor is another objective.[1]

The products covered by these standardization activities are many and varied, and include both military and civilian requirements. Typical are motor vehicle and agricultural tires; transmission, conveyor and elevator belts; suction, discharge, air, steam, gasoline and fire hose; rubber-covered rolls; molded and extruded rubber products; sheet packing, mats and matting; O rings and shaft seals; rubber protective and rubber-and-canvas footwear; druggists' sundries; surgical, hospital, and industrial rubber gloves, tubing, sheeting, and other items; sponge rubber and foam latex mattresses, pillows, cushions, furniture upholstery, carpet and rug underlays; rubber heels, solding materials, and other rubber shoe products; vinyl and rubber tile, sheet flooring and cove base; natural-rubber-type descriptions; and standard samples. Some of the publications of interest taken from RMA's catalog[28] are listed in Table 38. Single copies of the publications listed are free, unless otherwise noted.

Society of Automotive Engineers (SAE) Standardization has been an important activity in SAE from its inception.[1] It now carries on technical standardization work for the motor-vehicle, aircraft, airline, space-vehicle, farm-tractor, earth-moving and road-building machinery, and other manufacturing industries using internal-combustion engineers. The standards of the SAE are published annually in the SAE Handbook,[29] except for the standards of the aerospace industry, which are published in loose-leaf form as Aerospace Materials Specifications (AMS).[30]

The society's standardization work is carried out under the general direction of the SAE Technical Board, which organizes such technical committees as may be necessary to carry on the work. Most of these technical committees are of a permanent nature, but some are appointed to handle specific projects and are disbanded upon completion. The society's standards and technical committee activities also include active advisory cooperation with the armed forces and numerous other

TABLE 37 Examples of NEMA Standards Related to Plastics and Elastomers

Standard	Price
Conduits, Ducts, and Fittings	
Corrugated Polyethylene Coilable Electrical Plastic Duct, TC 5-1971	$0.85
Electrical Plastic Tubing (EPT), Conduit (EPC-40 and EPC-80) and Fittings, TC 2-1970	1.30
Plastic Utilities Duct for Underground Installation, TC 6-1971	1.00
Polyethylene Fittings for Use with Rigid Polyethylene Conduit, TC 4-1969	1.25
PVC Fittings for Use with Rigid PVC Conduit and Tubing, TC 3-1967	0.85
Styrene and PVC Ducts and Duct Fittings for Underground Installations, TC 1-1970	1.00
Insulating Materials	
Films, Test Methods for Plastic, MX 3-1971	2.00
Flexible Electrical Insulation	
B-Stage Materials, Glass-Polyester, VF 10-1964 (R 1971)	0.50
Composite Slot and Phase Insulation, VF 20-1963 (R 1971)	0.60
Fabric, Glass	
Continuous-filament Woven, VF 40-1963 (R 1971)	0.70
Heat-resistant Flexible Resin-coated, VF 11-1966	0.50
Silicone-elastomer (Rubber) Coated, VF 8-1961 (R 1971)	0.25
Fabric, Polyester-glass Coated, MX 1-1967	0.65
Fabric, Varnished	
Glass, Silicone, VF 5-1956 (R 1971)	0.90
Polyester, Nonwoven, MX 2-1967	0.65
Laminated Thermosetting Decorative Sheets, LD 1-1971	4.50
Laminates, Industrial, LI 1-1971	8.00
High-temperature Properties, LI 3-1961 (R 1971)	0.35
Temperature Indices, LI 5-1969	3.00
Filament-wound Tubes, LI 5.1-1972	In press
Polyester Glass Laminates, LI 5.2-1972	In press
Sleeving, Coated Electrical, VS 1-1962	0.50
Wires and Cables	
Cross-linked Thermosetting-Polyethylene-Insulated, WC 7-1971((IPCEA S-66-524)	5.00
Ethylene-Propylene-Rubber-Insulated, Ozone-Resistant	
Cables Rated 0-35 000 Volts, Interim Standard #1 to NEMA WC 8 and IPCEA S-68-516	0.50
Cables Rated 5000 Volts and Less with Integral Insulation and Jacket, Interim Standard #2 to NEMA WC 8 and IPCEA S-68-516	0.50
Primary Underground Residential Distribution Cables, Interim Standard #3 to NEMA WC 8 and IPCEA S-68-516	0.50
Concentric-neutral Underground Distribution Cables, Interim Standard #4 to NEMA WC 8 and IPCEA S-68-516	0.50
Rubber-insulated, WC 3-1969 (IPCEA S-19-81 Fifth Edition)	8.00
Thermoplastic-insulated, WC 5-1968 (IPCEA S-61-402 Second Edition)	4.50

federal and state government agencies. The SAE is active in international standardization, and sponsors several U.S. National Committees for ISO projects under the auspices of ANSI. In addition, many SAE standards have found international usage by being voluntarily adopted by foreign industry or government people.

All standards, specifications, and reports developed by the society are made available for industry, government, or other usage on a voluntary basis. Frequent checks

are made to determine the use of each document. Unused documents are canceled.
The SAE has been influential in aiding government agencies, both civil and defense,
in the development of sound technical documents. SAE standards are recognized
in government publications as sources for establishing minimum technical requirements in areas where government regulatory control has been established.[1]

Table 39 lists 31 standardization documents published by SAE in its current
handbook[29] that appear to involve plastics and elastomers. The reader will note
that a few of these are also ASTM standards. In those cases, ASTM makes this
recognition in the upper right-hand corner of the first page of their standards, just
as with ANSI.

SAE Aerospace Materials Specifications (AMS)   As indicated above, the standards
and specifications published by SAE for the aerospace industry are issued in loose-
leaf form. Individual specifications are sold at $3 each and are listed in the current

TABLE 38   RMA Rubber Product Publications (Partial List)

Publication	Price
Industrial Products	
Hose Handbook, 3rd Edition	$3.00
Sheet Rubber Handbook, 2nd Edition	1.25
Handbook—Specifications and Tolerances—Rubber-Covered Rolls, 2nd Edition	2.00
Protective Linings Technical Information—A series of related bulletins	
Industrial Rubber Glossary	
Molded and Extruded Products	
Rubber Handbook, 3rd Edition—Specifications for Rubber Products—	
Molded · Extruded · Lathe-Cut · Cellular	1.50
Vendor's Identification Guide	
Natural Rubber	
International Standards of Quality and Packing for Natural Rubber Grades (The Green Book)	
Copy International Samples (See Part III—Green Book) Sample Books Covering 29 of the 35 Separate Grades	35.00

AMS catalog.[30]   Currently there are some 216 specifications involving plastics,
adhesives, and elastomers, according to the authors' count, not counting those that
are concerned solely with fibers. The breakdown under appropriate subject areas
is as follows:

Quality controls and processes	1
Nonmetallics	202
Accessories, fabricated parts, and assemblies	13
	216

The single entry under Quality Control and Processes is AMS 2810 Identification—
Natural and Synthetic Rubber Materials. Examples of entries under the other two
categories are shown in Table 40. Note the entries where "similar specifications"
(i.e., Military and Federal, in these cases) exist.

Aerospace Materials Specifications are also listed in the DODISS.[4] Titles and
AMS numbers are given for many of the AMS specifications in Table 43 presented
under Identification of Materials Specifications, Test Methods and Standards.

Until recently, Aerospace Material Specifications took the following form, or a
form fairly close to this:

1. Acknowledgment (a vendor shall mention this specification number in all
quotations and when acknowledging purchase orders)
2. Form
3. Application

TABLE 39 SAE Standards, Recommended Practices, and Information Reports Involving Plastics and Rubbers*

Standard	Title	Designation
SAE J14	Specifications for Elastomer Compounds for Automotive Applications (formerly same as ASTM D 735, no longer an ASTM document).	SAE Standard
SAE J200f	Classification System for Elastomeric Materials for Automotive Applications (same as ASTM D 2000).	SAE Information Report
SAE J15	Flexible Foams Made from Polymers or Copolymers of Vinyl Chloride (same as ASTM D 1565).	SAE Recommended Practice
SAE J16	Classification of Elastomer Compounds for Automotive Resilient Mountings (same as ASTM D 1207).	SAE Recommended Practice
SAE J17	Latex Foam Rubbers (substantially same as ASTM D 1055)	SAE Recommended Practice
SAE J18a	Sponge- and Expanded Cellular-Rubber Products (substantially same as ASTM D 1056).	SAE Recommended Practice
SAE J815	Load Deflection Testing of Urethane Foams for Automotive Seating.	SAE Recommended Practice
SAE J19	Latex Dipped Goods and Coatings for Automotive Applications (substantially same as ASTM D 1764).	SAE Standard
SAE J369	Flammability of Automotive Interior Trim Materials—Horizontal Test Method	SAE Recommended Practice
SAE J882	Test Method for Measuring Thickness of Automotive Textiles and Plastics.	SAE Recommended Practice
SAE J388	Dynamic Flex Fatigue Test for Slab Polyurethane Foam.	SAE Recommended Practice
SAE J954	Urethane for Automotive Seating	SAE Recommended Practice
SAE J322	Nonmetallic Trim Materials—Test Method for Determining the Staining Resistance to Hydrogen Sulfide Gas	SAE Recommended Practice
SAE J323	Test Method for Determining Cold Cracking of Flexible Plastic Materials	SAE Recommended Practice
SAE J20d	Coolant System Hoses.	SAE Standard
SAE J30a	Fuel and Oil Hoses.	SAE Standard
SAE J1401	Hydraulic Brake Hose.	SAE Standard
SAE J1402b	Air Brake Hose.	SAE Standard
SAE J1403	Vacuum Brake Hose.	SAE Standard
SAE J188	Power Steering Pressure Hose—High Volumetric Expansion Type.	SAE Standard
SAE J189	Power Steering Return Hose—Low Pressure.	SAE Standard
SAE J190	Power Steering Pressure Hose—Wire Braid.	SAE Standard
SAE J191	Power Steering Pressure Hose—Low Volumetric Expansion Type.	SAE Standard
SAE J50a	Windshield Wiper Hose.	SAE Standard
SAE J151b	Automotive Air Conditioning Hose.	SAE Standard
SAE J60a	Rubber Cups for Hydraulic Actuating Cylinders	SAE Standard
SAE J65a	Specifications and Methods of Test for Rubber Boots for Use on Drum Type Hydraulic Brake Wheel Cylinders.	SAE Standard
SAE J80	Automotive Rubber Mats.	SAE Standard
SAE J946	Guide to the Application and Use of Radial Lip Type Oil Seals.	SAE Recommended Practice
SAE J120a	Rubber Rings for Automotive Applications.	SAE Recommended Practice
SAE J130	Nonrigid Thermoplastic Compounds for Automotive Applications.	SAE Recommended Practice
SAE J250	Synthetic Resin Plastic Sealers, Nondrying Type.	SAE Recommended Practice
SAE J855a	Test Method of Stretch and Set of Textiles and Plastics.	SAE Recommended Practice

* Excerpted with permission, copyright, Society of Automotive Engineers, 1972, all rights reserved.

TABLE 40 Examples of Aerospace Material Specifications (AMS) Involving Plastics and Elastomers*

AMS 3588 THRU 3894/2		
AMS	**TITLE OF SPECIFICATION**	**SIMILAR SPECIFICATION****

NON-METALLICS (continued)

AMS	TITLE OF SPECIFICATION	SIMILAR SPECIFICATION
3588	Plastic Tubing, Electrical Insulation – Irradiated Polyolefin, Clear, Very Flexible, Heat Shrinkable (Low Recovery Temperature) 2 to 1 Shrink Ratio	
3589	Plastic Tubing, Electrical Insulation, Resin Bonded – Polyethylene Terephthalate, Heat Shrinkable, 1.3 to 1 Shrink Ratio	
3590A	Plastic Sheet, Copper Faced–Paper Reinforced Phenol-Formaldehyde	
3591	Plastic Tubing, Electrical Insulation, Thermally Welded - Polyethylene Terephthalate, Heat Shrinkable, 1.6 to 1 Shrink Ratio	
3598A	Plastic Sheet, Copper Faced – Glass Fabric Reinforced Polytetrafluoroethylene	
3601B	Plastic Sheet, Copper Faced–Glass Fabric Reinforced Epoxy Resin	
3605D	Plastic Sheet–Post Forming, Cotton Fabric Reinforced Phenol-Formaldehyde	MIL-P-8655
3607C	Plastic Sheet and Plate–Cotton Fabric Reinforced Phenol-Formaldehyde	L-L-31, MIL-P-15035, Type FBM
3608	Plastic Sheet–Methyl Methacrylate, General Purpose	L-P-391, Type I
3609	Plastic Sheet–Methyl Methacrylate, Heat Resistant	L-P-391, Type II
3610C	Sheet–Water Vapor Resistant, Flexible, Transparent	MIL-B-131
3611B	Plastic Sheet–Polycarbonate	
3612	Polyester Film, Electrical Grade, General Purpose	MIL-I-631, Type G
3613	Polyester Film, Copper Clad – For Electronic Applications	
3615B	Plastic Tubing–Cotton Fabric Reinforced Phenol-Formaldehyde	MIL-79, Type FBE
3617	Plastic Moldings & Extrusions–Polyamide (Nylon)	MIL-P-17091, MIL-M-20693
3618	Resin, Polyimide, Thermosetting – High Heat (290C) Resistant	
3620B	Plastic Moldings & Extrusions – Polystyrene	L-P-416, Type I
3622A	Plastic Moldings & Extrusions–Cellulose Acetate, General Purpose	
3623	Elastomeric Tubing – Electrical Insulation, Irradiated Polychloroprene, Flexible, Heat Shrinkable, 1.750 to 1 Shrink Ratio	
3624A	Plastic Moldings & Extrusions–Cellulose Acetate Butyrate	L-P-349, Type II
3625	Elastomeric Tubing – Electrical Insulation, Crosslinked Silicone, Pigmented, Flexible, Heat Shrinkable, 1.750 to 1	
3626C	Plastic Moldings & Extrusions–Methyl Methacrylate	L-M-500
3627	Plastic Moldings & Extrusions–Methyl Methacrylate, Heat Resistant	L-M-500
3628A	Plastic Moldings & Extrusions–Polycarbonate	
3629	Tubing–Extruded–Polyvinyl Chloride, High Temperature, Electrical Insulation	
3630C	Plastic Extrusions–Flexible–Polyvinyl Chloride	MIL-I-7444, MIL-I-631, Ty.F
3631	Plastic Extrusions–Flexible, High Temperature, Polyvinyl Chloride	MIL-I-631, Type F
3632B	Plastic Tubing–Electrical Insulation, Irradiated Polyvinylidene Fluoride, Heat Shrinkable, Semi-Rigid, 2 to 1 Shrink Ratio ———Kynar	

ACCESSORIES, FABRICATED PARTS AND ASSEMBLIES (continued)

AMS	TITLE OF SPECIFICATION	SIMILAR SPECIFICATION
* 7273A	Rings, Sealing–Fluorosilicone Rubber, High Temperature Fuel and Oil Resistant (70-80)	
7274E	Rings, Sealing–Synthetic Rubber, Oil Resistant (65-75)	
7277B	Rings, Sealing–Synthetic Rubber, Phosphate Ester Hydraulic Fluid Resistant, Butyl Type (70-85)	
7278D	Rings, Sealing–Synthetic Rubber, High Temp. Fluid Resist. Fluorocarbon Type (70-80)	MIL-R-25897, Type I
7279D	Rings, Sealing–Synthetic Rubber, High Temp. Fluid Resist. Fluorocarbon Type (85-95)	
7280	Rings, Sealing, Synthetic Rubber – High Temp. Fluid Resist, Low Compression Set, FPM Type, 70-80 ———FPM	

***Indicates New and Revised specifications issued November 15, 1972.**

Revisions are marked with an "A" or "B", etc., following the specification number.

4. Composition
5. Technical requirements
6. Quality
7. Reports
8. Packing
9. Approval
10. Rejections

Currently Aerospace Material Specifications are taking a form more closely resembling Military and Federal Specifications, as follows, as used in AMS 3894 Graphite Fiber Tape and Sheet—Epoxy Resin Impregnated for Hand Layup, 15 November 1972:

1. Scope (including form, application, and classification)
2. Applicable documents
3. Technical requirements
4. Quality assurance provisions
5. Preparation for delivery
6. Acknowledgment
7. Rejections
8. Notes

Society of the Plastics Industry (SPI)   This society has a Thermosetting Resin Formulators Division, which is comprised of members of the Epoxy Resin Formulators (ERF) and the Epoxy Molding Materials Institute (EMMI), which were formerly two separate and distinct divisions of SPI. This division has published over 30 test procedures in the ERF series and 20 test methods in the EMMI series. An SPI brochure[31] outlines all the purposes of this division. As far as specifications and test procedures are concerned, the purpose of the division is to: "develop in cooperation with other groups, technical specifications, chemical and physical test procedures, and performance standards on thermosetting resin compounds for adoption by the industry through the American Society for Testing and Materials, the U.S. Department of Commerce or American National Standards Institute procedures, and also to carry out testing programs as may be necessary to develop such performance standards covering these resin compounds used for industrial and consumer applications."

A partial list of test procedures in both series is given in Tables 41 and 42.

SPI is also very active in the preparation of Voluntary Product Standards and Commercial Standards, as might be expected of a major society representing the plastics industry. These standards are described above.

Underwriters' Laboratories   The Underwriters' Laboratories (UL) has its main office at 207 East Ohio Street, Chicago, Ill. 60611, and laboratories at Chicago and Northbrook, Ill., Melville, L.I., N.Y., and Santa Clara, Calif. UL is an organization established to maintain and operate laboratories for the examination and testing of devices, systems, and materials.[1] It was founded in 1894 and is operated as a nonprofit service organization. The object of the UL is to conduct scientific investigations, studies, experiments, and tests to determine the relation of various materials, devices, constructions, and methods to life, fire, and casualty hazards, and to ascertain, define, and publish standards, classifications, and specifications for materials, devices and constructions, and methods affecting such hazards, and other information tending to reduce and prevent loss of life and property from fire, crime, and casualty.[1]

Each of the several engineering departments in UL deals with distinct and separate subjects, including fire protection. Each department has prepared standards providing specifications and requirements for construction and performance under test and actual use of systems, materials, and appliances of numerous classes submitted to the Laboratories.

The UL has issued more than 230 of these standards and sets of requirements. The requirements of a standard are so stated that, if correctly applied, there is no discrimination between the products of two or more manufacturer submitters. They are published so that others may know the basis for the Laboratories' opinions, and the standards must necessarily justify the opinions.

A number of Federal Specifications covering materials or appliances of classes which are under the supervision of Underwriters' Laboratories recognize the Laboratories' UL label as evidence of compliance with the applicable requirements of such specifications. In its work with standardization, the Underwriters' Laboratories cooperates with many other organizations, such as ANSI, NFPA, and ASTM.[1]

TABLE 41   SPI Epoxy Resin Formulators Division Test Methods

Method	Title
ERF 1-61	Method of Test for Viscosity of Epoxy Compounds and Related Components
ERF 2-61	Method of Test for Gel Time and Peak Exothermic Temperature of Epoxy Compounds
ERF 3-61	Tentative Method of Test for Specific Gravity of Epoxy Compounds and Components
ERF 4-62	Method of Test for Long-Time Creep of Epoxy Compounds (same as ASTM D 674-56, with additions)
ERF 5-62	Method of Test for Flexural Properties of Epoxy Compounds (same as ASTM D 790-61)
ERF 6-62	Method of Test for Tensile Properties of Epoxy Compounds (same as ASTM D 638-60T)
ERF 7-62	Method of Test for Flammability of Hardened Epoxy Compounds over 0.050 In. in Thickness (same as ASTM D 635-56T with additions and changes)
ERF 8-62	Method of Test for Compressive Properties of Epoxy Casting or Molding Compounds (same as ASTM D 695-61T with additions)
ERF 9-63	Method of Test for Hardness of Cured Epoxy Materials
ERF 10-63	Method of Test for Peel Strength of Epoxy Adhesive (same as ASTM D 1781-62)
ERF 11-63	Method of Test for Coefficient of Linear Thermal Expansion of Epoxy Casting Systems
ERF 12-64	Method of Test for Linear Shrinkage of Epoxy Casting Resins during Cure
DRF 13-64	Method of Test for Pot Life of Epoxy Compounds
ERF 14-64	Method of Test for Dimensional Stability of Epoxy Casting Compounds
ERF 15-64	Method of Test for Tensile Shear Strength of Adhesives (Metal-to-Metal) (same as ASTM D 1002 with additions and changes)
ERF 16-64	Method of Test for Coverage of Laminating Systems
DRF 17-64	Method of Test for Deflection Temperature of Cured Epoxy Resins under Load (same as ASTM D 648-56 with additions and changes)
ERF 18-64	Method of Test for Edgewise Compressive Properties of Laminating System (Fed. Test Method Std. No. 406, Method 1021, which is based on ASTM D 695-61T)
ERF 19-64	Method of Test for Resistance of Plastics to Chemical Reagents (same as ASTM D 543-60T with addition)
ERF 20-66	Method of Test for Shear Strength of Epoxy Compounds (same as ASTM D 732-46)
ERF 21-66	Method of Test for Bearing Strength of Epoxy Compounds (same as ASTM D 953-54)
ERF 22-66	Method of Test for Determination of the Thermal Conductivity of Cured Epoxy Resin Systems
ERF 23-66	Method of Test for Resistance to Impact by Falling Ball

Miscellaneous Sources

Adhesives   A very useful source on adhesives specifications is a book by Katz,[32] which is really a compilation of the essential data and other information contained in adhesive-material specifications. This fine book, recently revised by Cagle, has a brief introductory section on "Adhesives Selection" covering a number of technical subjects related to adhesives types and applications. Of particular interest is an excellent summary of "specification technology," which is essentially a tabulation of specifications, some of which involve polymeric adhesives, and others which deal with inorganic materials.   The bulk of the book is the second part, Adhesives Properties Data, which consists of abstracts or abridgments of over 250 federal, military, commercial, and industrial standards and specifications.   Many of the specifications are so abridged that they are given on only a single page of the text. Properties given in tabular form are not abridged.   The specification entries take the following form:

Title
Scope
Classification
Properties
Form
QPL (yes or no as to existence)
Notes (not always used)

An index of the specifications covered in this book is given in an appendix.

TABLE 42   SPI Epoxy Molding Materials Institute Test Methods

Method	Title
EMMI 1-66	Method of Test for Spiral Flow
EMMI 2-67	Method of Test for Bulk Factor of Epoxy Molding Materials (same as ASTM D 1895-65T)
EMMI 3-67	Method of Testing for Specific Gravity of Epoxy Molding Compounds (same as ASTM D 792-64T)
EMMI 4-67	Method of Test for Water Absorption of Epoxy Molding Compounds (same as ASTM D 570-63)
EMMI 5-67	Method of Test for Flammability of Epoxy Molding Compounds (same as ASTM D 635-63)
EMMI 6-67	Method of Test for Hardness of Epoxy Molding Compounds (ASTM Methods D 785-65 and D 1706-61 are used)
EMMI 7-67	Method of Test for Heat Deflection Temperature of Epoxy Molding Compounds (same as ASTM D 648-56)
EMMI 8-67	Method of Test for Tensile Strength of Epoxy Molding Compounds (same as ASTM D 638-64T, Type I)
EMMI 9-67	Method of Test for Impact Strength of Epoxy Molding Compounds (same as ASTM D 256-56, Procedure A—notched Izod method)
EMMI 10-67	Method of Test for Flexural Strength of Epoxy Molding Compounds (same as ASTM D 790-66)
EMMI 11-67	Method of Test for Compressive Strength of Epoxy Molding Compounds (same as ASTM D 795-63T)
EMMI 12-67	Method of Testing for Volume Resistivity of Epoxy Molding Compounds (same as ASTM D 257-61)
EMMI 14-67	Method of Test for Arc Resistance of Epoxy Molding Compounds (same as ASTM D 495-61)
EMMI 16-68	Method of Test for Dielectric Constant and Dissipation Factor of Epoxy Molding Compounds (same as ASTM D 150-65T with changes)
EMMI 17-68	Method of Test for Coefficient of Linear Thermal Expansion of Epoxy Molding Compounds (same as ASTM D 696-44)
EMMI 18-68	Method of Test for Heat Resistance of Epoxy Molding Compounds (same as ASTM D 794-64T)

Plastic-Laminate Materials   The second book in the series started by Katz[32] and discussed above under Adhesives is one by Beach[33] on plastic-laminate materials. Abstracted are Military and Federal Specifications, Commercial Standards, ASTM, NEMA, and AMS specifications.   Reference is also made to Military and Federal Standards and to other pertinent documents.

IDENTIFICATION OF MATERIALS SPECIFICATIONS, TEST METHODS, AND STANDARDS

This section has been prepared by using the World Index of Plastics Standards,[2] the PLASTEC tabulation of specifications and standards,[3] and the DODISS.[4] It is organized alphabetically by materials into four tables (Tables 43 to 46), one each for plastics, elastomers, coatings, and insulations. Only specifications on materials are given, covering resins, forms, shapes, sheets, films, etc. End items using polymeric materials in whole or in part are not covered, since this would require too large a listing. No guarantee is made that all pertinent documents are listed, but the tables should be quite complete. In some cases the title was general and the actual type of resin used had to be determined by examination of the document itself, or by other means.

Note that notations as to amendments, revisions, notes, changes, etc., are not given for Military and Federal Specifications, and T designations (for Tentative) and dates are not given for ASTM documents. These can be looked up in the appropriate documents cited elsewhere in this chapter. Specifications listed include Military, Federal, AMS, NEMA, and ASTM.

The key words in the specifications which identify the plastic or elastomeric material or type of material are shown in CAPITALIZED form as an aid to the reader. The alphabetization is on the basis of the key word, as ACETAL, ACRYLIC. Closely related specifications are grouped together, but there is no attempt to set up any other pattern of listing, such as listing specifications in numerical order within Military or Federal Specifications. There will necessarily be a certain amount of confusion in alphabetical listings, since specification titles are frequently inconsistent. Thus, polystyrene materials may be found under POLYSTYRENE and under STYRENE, polyvinyl chloride may also be found under VINYL CHLORIDE, etc. In cases where the specification title gives no indication of the material or type of material covered, this is given in a double parentheses, as ((POLYESTER)). The information as to the actual material was obtained in a number of different ways, the most common of which was direct examination of the document.

TABLE 43   Guide to Specifications for Plastic Materials (Molding, Extrusion, Films, Foams, Laminates; Sheets, Rods, Tubes, Foams)

Material	Specification
Plastic molding material, ACETAL, injection and extrusion	L-P-392
ACETAL resin for injection molding and extrusion materials	ASTM D 2133
Plastic molding material, ACETAL, glass-fiber-reinforced	MIL-P-46137
Glass-reinforced ACETAL parts for molding and extrusion	ASTM D 2948
ACRYLIC moldings and extrusions, METHYL METHACRYLATE	AMS 3626
ACRYLIC METHACRYLATE molding and extrusion	ASTM D 788
ACRYLIC castings, METHYL METHACRYLATE, general-purpose	AMS 3580
ACRYLIC cast METHACRYLATE plastic sheet, rods, tubes, and shapes	ASTM D 702
ACRYLIC castings, METHYL METHACRYLATE, heat-resistant	AMS 3581
Plastic sheet, extruded ACRYLIC	L-P-507
ACRYLIC sheet, METHYL METHACRYLATE, general-purpose	AMS 3608
ACRYLIC sheet, METHYL METHACRYLATE, heat-resistant	AMS 3609
Extruded ACRYLIC plastic sheet	ASTM D 1547
Plastic sheet, ACRYLIC, heat-resistant	MIL-P-5425
Plastic sheet, ACRYLIC, modified	MIL-P-8184
Plastic sheet, ACRYLIC, modified, laminated	MIL-P-25374
Plastic sheet, corrugated translucent, glazing ((ACRYLIC, glass-reinforced))	L-P-505
Plastic sheet, cast, ACRYLIC, shipboard application (illumination and signal lighting)	MIL-P-24191
Plastic sheets and parts, modified ACRYLIC base, monolithic, crack-propagation-resistant	MIL-P-25690
Plastic material, molding, ACRYLIC, colored and white, heat-resistant, for lighting fixtures	MIL-P-19735

TABLE 43   Guide to Specifications for Plastic Materials (Molding, Extrusion, Films, Foams, Laminates; Sheets, Rods, Tubes, Forms) (Continued)

Material	Specification
Rigid ABS plastics ((ACRYLONITRILE BUTADIENE STYRENE))	ASTM D 1788
Plastic molding material, ACRYLONITRILE BUTADIENE STYRENE (ABS), rigid	L-P-1183
Plastic sheets, ACRYLONITRILE BUTADIENE STYRENE copolymer, rigid	MIL-P-19904
Plastic sheet and plastic rod, thermosetting, cast ((ALKYD copolymer))	L-P-516
ALLYL molding compounds	ASTM D 1636
ALLYL plastic sheet, rod, and tube	L-A-499
Plastic sheet and plastic rod, thermosetting, cast ((ALLYL))	L-P-516
Plastic molding material, CELLULOSE ACETATE	L-P-397
CELLULOSE ACETATE molding and extrusion compounds	ASTM D 706
Plastic moldings and extrusions, CELLULOSE ACETATE	AMS 3622
Core material, cellular CELLULOSE ACETATE	MIL-C-18345
CELLULOSE ACETATE plastic sheet	ASTM D 786
Plastic sheet, flexible, reinforced ((CELLULOSE ACETATE))	L-P-1196
Plastic sheet (sheeting), pressure-sensitive, adhesive-coated, CELLULOSE ACETATE, transparent	MIL-P-46140
Plastic sheet, adhesive-coated, paper-backed ((CELLULOSE ACETATE))	L-P-514
Plastic sheet and film, CELLULOSE ACETATE	L-P-504
CELLULOSE ACETATE sheet and film	ASTM D 1202
Plastic sheet, CELLULOSE ACETATE, optical quality	MIL-P-21094
CELLULOSE ACETATE BUTYRATE	MIL-C-5537
Plastic molding and extrusion material, CELLULOSE ACETATE BUTYRATE	L-P-349
CELLULOSE ACETATE BUTYRATE molding and extrusion compounds	ASTM D 707
Plastic molding and extrusion, CELLULOSE ACETATE BUTYRATE	AMS 3624
CELLULOSE ACETATE BUTYRATE film and sheeting for fabrication and forming	ASTM D 2411
Plastic sheet, flexible, reinforced ((CELLULOSE ACETATE BUTYRATE))	L-P-1196
CELLULOSE NITRATE (pyroxylin) plastic sheets, rods, and tubes	ASTM D 701
Electronic-grade CELLULOSE NITRATE	ASTM F 63
Plastic molding material, CELLULOSE PROPIONATE	MIL-P-46074
CELLULOSE PROPIONATE molding and extrusion compounds	ASTM D 1562
Plastic sheet and film, CHLOROTRIFLUOROETHYLENE (CTFE) copolymer	L-P-1174
Plastic sheet, rods, tubes, and disks, CHLOROTRIFLUOROETHYLENE polymer	MIL-P-46036
Molding plastics and molded plastic parts, thermosetting ((DIALLYL PHTHALATE, filled))	MIL-M-14
Plastic molding material and plastic molded parts, glass-fiber filler, DIALLYL PHTHALATE RESIN	MIL-P-19833
EPOXY resins	ASTM D 1763
EPOXY molding compounds	ASTM D 3013
Resin, EPOXY	MIL-R-21931
Molding material, glass, EPOXY impregnated	MIL-M-56862
Molding material, glass mat, EPOXY coated	MIL-M-46861
Resin, casting, EPOXY, aluminum-filled	MIL-R-22628
Molding material, plastic, EPOXY compounds, thermosetting	MIL-M-24325
Molding plastic, glass/EPOXY premix	MIL-M-46069
Laminating resin ((EPOXY))	MIL-L-52696
Resin, EPOXY, low-pressure laminating	MIL-R-9300
Plastic, laminating, EPOXY base	MIL-P-43398
Plastic sheet, laminated, thermosetting, paper-base, EPOXY resin	MIL-P-22324
Plastic sheet, laminated, thermosetting, glass-fiber-base, EPOXY resin	MIL-P-18177
Plastic laminates, glass-fabric base, high-strength ((EPOXY))	MIL-P-22693
Plastic material, glass-fiber base, EPOXY resin, low-pressure laminated	MIL-P-25421
Glass-fiber-base laminate, EPOXY resin	MIL-G-21729
Plastic sheet, rod, and tube, laminated, thermosetting ((EPOXY))	L-P-509
Glass roving, EPOXY resin-impregnated, type S glass	AMS 3832

TABLE 43   Guide to Specifications for Plastic Materials (Molding, Extrusion, Films, Foams, Laminates; Sheets, Rods, Tubes, Forms) (Continued)

Material	Specification
Glass roving, EPOXY resin-impregnated, type E glass	AMS 3828
Cloth, type E glass, B stage EPOXY resin-impregnated, style 181-75E	AMS 3822
Plastic embedding compound, EPOXY, rigid	MIL-P-46067
Plastic filler compound, EPOXY, for honeycomb panels	MIL-P-46094
Syntactic buoyancy material for high hydrostatic pressures ((such as EPOXY))	MIL-S-24154
Plastic sheet, vibration-damping (type ML-02)	MIL-P-22581
Potting compound, EPOXY, filled, 9–12 CTE, 250 HDT	AMS 3739
Potting compound, EPOXY, filled, 10–15 CTE, 225 HDT	AMS 3735
Potting compound, EPOXY, filled, 15–20 CTE, 180 HDT	AMS 3738
Potting compound, EPOXY, filled, 15–20 CTE, 225 HDT	AMS 3740
Potting compound, EPOXY, filled, 25–30 CTE, 175 HDT, room-temperature cure	AMS 3736
Potting compound, EPOXY, unfilled, general-purpose, room-temperature cure	AMS 3734
Potting compound, EPOXY, unfilled, 35–40 CTE, 160 HDT, room-temperature cure	AMS 3737
Potting compound, EPOXY, flexible, durometer 75–85	AMS 3750
Potting compound, foamed EPOXY type, amine-hardened	AMS 3730
ETHYL CELLULOSE	MIL-E-10853
ETHYL CELLULOSE molded plastics	L-E-210
Plastic molding and extrusion material, ETHYL CELLULOSE	MIL-P-22985
ETHYL CELLULOSE molding and extrusion compounds	ASTM D 787
ETHYLENE plastics	ASTM D 2952
Cross-linked ETHYLENE plastics	ASTM D 2647
FEP-FLUOROCARBON molding and extrusion materials	ASTM D 2116
Plastic molding and extrusion material, IONOMER resins	MIL-P-46124
ISOBUTYL METHACRYLATE polymer	MIL-I-10903
MELAMINE molded plastics	L-M-181
MELAMINE-FORMALDEHYDE molding compound	ASTM D 704
Molding plastics and molded plastic parts, thermosetting ((MELAMINE-filled))	L-M-181
Plastic moldings, MELAMINE FORMALDEHYDE, mineral-filled	AMS 3640
Plastic sheet, laminated, thermosetting, glass cloth, MELAMINE resin	MIL-P-15037
Plastic sheet, rod, and tube, laminated, thermosetting ((MELAMINE))	L-P-509
Plastic rod and tube, thermosetting, laminated ((MELAMINE))	MIL-P-79
Plastic laminates, glass-fabric base, high-strength ((MELAMINE))	MIL-P-22693
METHACRYLATE molding and extrusion compounds	ASTM D 788
Plastic molding material, METHACRYLATE	L-P-380
Cast METHACRYLATE plastic sheets, rods, tubes, and shapes	ASTM D 702
Plastic sheet, rods, and tubing, rigid, multiapplication (METHACRYLATE, cast)	L-P-391
METHYL METHACRYLATE polymer	MIL-M-10851
Plastic moldings and extrusion, METHYL METHACRYLATE, heat-resistant	AMS 3627
NYLON injection-molding and extrusion materials	ASTM D 789
Reinforced and filled NYLON injection-molding and extrusion materials	ASTM D 2897
Reinforced OLEFIN polymers for injection molding or extrusion	ASTM D 2853
Plastic sheet and coating material, PARA-XYLENE polymers	MIL-P-46121
Resin, PHENOLIC, laminating	MIL-R-9299
Plastic sheet, rod, and tube, laminated, thermosetting ((PHENOLIC))	L-P-509
Plastic rod and tube, thermosetting, laminated ((PHENOLIC))	MIL-P-79
Plastic materials, asbestos-base, PHENOLIC resin, low- or high-pressure laminates	MIL-P-25770
Plastic sheet, laminated, thermosetting, cotton-fabric base, PHENOLIC resin, postforming	L-P-511
Plastic sheet, laminated, thermosetting, cotton-fabric base, PHENOLIC resin	MIL-P-15035
Plastic laminates, glass-fabric base, high-strength ((PHENOLIC))	MIL-P-22693
Plastic material, PHENOLIC resin, glass-fiber base, laminated	MIL-P-25515

TABLE 43   Guide to Specifications for Plastic Materials (Molding, Extrusion, Films, Foams, Laminates; Sheets, Rods, Tubes, Forms) (Continued)

Material	Specification
Plastic material, laminated thermosetting sheets, nylon-fabric base, PHENOLIC resin	MIL-P-15047
Plastic sheet, laminated, thermosetting, paper-base, PHENOLIC resin	L-P-513
Tape, asbestos felt, PHENOLIC resin-impregnated	MIL-T-60330
Tape, asbestos felt, impregnated with PHENOLIC	MIL-T-81141
Asbestos felting, B stage PHENOLIC resin-impregnated, low-pressure molding	AMS 3858
Cloth, silica, B stage PHENOLIC resin-impregnated, high-pressure molding	AMS 3830
Cloth, silica, PHENOLIC impregnated	MIL-C-81251
Silica fabric, PHENOLIC impregnated	MIL-S-81256
Asbestos felt or mat, resin-impregnated ((PHENOLIC))	MIL-A-81264
PHENOLIC molding materials	ASTM D 700
Resin, PHENOLIC (alkyd resin), oil-soluble	MIL-R-15189
Plastic molding material, PHENOLIC resin	L-P-1125
Molding plastics and molded plastic parts, thermosetting ((PHENOLIC, filled))	MIL-M-14
Plastic molding compound, glass-fiber-reinforced PHENOLIC resin, thermosetting	MIL-P-23943
Molding plastic and molded plastic parts, asbestos-fiber-filled, arc- and flame-resistant PHENOLIC resin	MIL-M-21556
Plastic, PHENOLIC, graphited, sheets, rods, tubes, and shapes	MIL-P-5431
Plastic, filled PHENOLIC, uncured	MIL-P-13436
Plastic sheet and plastic rod, thermosetting, cast ((PHENOLIC))	L-P-516
Plastic, thermosetting, pulp-filled preforms ((PHENOLIC))	MIL-P-16617
Resin, PHENOL FORMALDEHYDE, laminating	MIL-R-3745
Plastic moldings, thermosetting, PHENOL FORMALDEHYDE, macerated-fabric-filled	AMS 3641
Plastic sheet and plate, cotton-fabric-reinforced, PHENOL FORMALDEHYDE	AMS 3607
Plastic sheet, postforming, cotton-fabric-reinforced, PHENOL FORMALDEHYDE	AMS 3605
Plastic tubing, cotton-fabric-reinforced, PHENOL FORMALDEHYDE	AMS 3615
Plastic molding and extrusion material, PHENOXY	MIL-P-46088
PHENOXY plastics molding and extrusion materials	ASTM D 2531
Plastic, POLYAMIDE (nylon), rigid, rods, tubes, flats, molded and cast parts	L-P-410
Plastic moldings and extrusions, POLYAMIDE (nylon)	AMS 3617
Resin, POLYAMIDE, thermosetting, high heat (290°C) resistant	AMS 3618
Molding plastic, POLYAMIDE (nylon) rigid	MIL-M-20693
Plastic, POLYAMIDE (nylon), flexible molding and extrusion material	MIL-P-22096
Nylon plastic, flexible, molded or extruded ((POLYAMIDE))	MIL-N-18352
Plastic molding (and extrusion) material, nylon, glass-fiber-reinforced ((POLYAMIDE))	L-P-395
Molding plastic, POLYAMIDE resin (nylon) glass-fiber-filled and molded POLYAMIDE resin glass-fiber-filled plastic parts	MIL-P-19887
Polytetrafluoroethylene parts and coating and POLYAMIDE parts	MIL-P-24074
Plastic sheet, vibration-damping (type ML-02) ((POLYAMIDE))	MIL-P-22581
Plastic molding and extrusion material, POLY(ARYL SULFONE ETHER) resin, thermoplastic	MIL-P-46133
Plastic potting and impregnating material, POLYBUTENE	MIL-P-43045
POLYBUTYLENE plastics	ASTM D 2581
POLYBUTYLENE (PB) plastic tubing	ASTM D 2666
Plastic molding material, POLYCARBONATE, injection and extrusion	L-P-393
POLYCARBONATE plastic molding, extrusion and casting materials	ASTM D 2473
Plastic moldings and extrusions, POLYCARBONATE	AMS 3628
Plastic molding material, POLYCARBONATE, glass-fiber-reinforced	MIL-P-81390
Reinforced POLYCARBONATE injection-molding and extrusion materials	ASTM D 2848
Plastic sheet, POLYCARBONATE	AMS 3611

TABLE 43   Guide to Specifications for Plastic Materials (Molding, Extrusion, Films, Foams, Laminates; Sheets, Rods, Tubes, Forms) (Continued)

Material	Specification
Plastic sheet (sheeting), POLYCARBONATE	MIL-P-46144
Plastic sheet, POLYCARBONATE, transparent	MIL-P-83310
Plastic molding material, POLYCHLOROTRIFLUOROETHYLENE	L-P-385
POLYCHLOROTRIFLUOROETHYLENE (PCTFE) plastics	ASTM D 1430
Plastic material, molding, rigid thermoplastic, POLYDICHLORO-STYRENE, for use in electronic, communications, and allied electrical equipment	MIL-P-3409
Resin, POLYESTER, bisphenol-A type	MIL-R-46068
Resin, POLYESTER, low-pressure laminating	MIL-R-7575
Resin, POLYESTER, low-pressure laminating, fire-resistant	MIL-R-21607
Resin, POLYESTER, high-temperature-resistant, low-pressure laminating	MIL-R-25042
POLYESTER molding compounds	ASTM D 1201
Molding plastics and molded plastic parts, thermosetting ((POLYESTER, filled))	MIL-M-14
Molding plastics and molded plastic parts, thermosetting ((POLYESTER, diallyl phthalate cross-linked, filled))	MIL-M-14
Plastic molding material, POLYESTER, low-pressure laminating, high-temperature resistant	MIL-P-43038
Plastic material, heat-resistant, low-pressure laminate, glass-fiber base, POLYESTER resin	MIL-B-25395
Plastic material, POLYESTER resin, glass-fiber base, low-pressure laminated	L-P-383
Plastic material, POLYESTER resin, glass-fiber base, filament-wound tube	MIL-P-82540
Plastic laminates, fibrous-glass-reinforced, marine structural ((POLYESTER))	MIL-P-17549
Plastic laminates, glass-fabric base, high-strength ((POLYESTER))	MIL-P-22693
Plastic sheet, laminated, thermosetting, electrical insulation sheet, glass mat ((POLYESTER))	MIL-P-24364 (24364/1 to 24634/3)
Plastic film, nonrigid, transparent ((POLYESTER))	MIL-P-14591
POLYESTER film, electrical-grade, general-purpose	AMS 3612
Plastic sheet, thermosetting, transparent ((POLYESTER))	MIL-P-8257
Plastic sheet and strip, POLYESTER	L-P-377
Plastic sheet, corrugated translucent, glazing ((POLYESTER resin, glass-reinforced))	L-P-505
Plastic sheet, rod, and tube, laminated: thermosetting ((POLYESTER))	L-P-509
Tubing, plastic, glass-fabric base, low-pressure laminated ((POLYESTER))	MIL-T-10652
Glass POLYESTER B-stage materials	NEMA VF 10
Plastic, POLYESTER, glass-fiber-filled, premix molding compound	MIL-P-43043
Sandwich construction, plastic resin, glass-fabric base, laminated facing and honeycomb core for aircraft structural applications ((POLYESTER))	MIL-S-9041
Core material, plastic honeycomb, laminated, glass-fabric base, for aircraft structural applications ((POLYESTER))	MIL-C-8073
Syntactic material, rigid, pour-in-place, structural-void-filling ((POLYESTER resin with phenolic microspheres))	MIL-S-24167
Sealing compound, thread, polymerizing, room-temperature ((POLYESTER))	MIL-S-3927
POLYETHER cushioning material, foam-in-place, flexible	MIL-P-43226
POLYETHYLENE plastic molding and extrusion materials	ASTM D 1248
Plastic molding and extrusion material, POLYETHYLENE and copolymers (low-, medium-, and high-density)	L-P-390
Plastic molding material (potting and impregnating), POLYETHYLENE, low molecular-weight	MIL-P-43081
POLYETHYLENE film and sheeting	ASTM D 2103
Plastic rods and tubes, POLYETHYLENE	MIL-P-21922
POLYETHYLENE (PE) plastic tubing	ASTM D 2737
Plastic sheet (sheeting), POLYETHYLENE	L-P-512
Plastic sheet, POLYETHYLENE, laminated, nylon-reinforced	L-P-524

TABLE 43   Guide to Specifications for Plastic Materials (Molding, Extrusion, Films, Foams, Laminates; Sheets, Rods, Tubes, Forms) (Continued)

Material	Specification
POLYETHYLENE plastic sheets, virgin and borated....................	MIL-P-23536
Cushioning material, unicellular, POLYETHYLENE foam (for packaging purposes)...	MIL-C-46842
Plastic sheet, POLYETHYLENE TEREPHTHALATE..................	MIL-P-55010
POLYFLUOROETHYLENEPROPYLENE film and sheet...............	AMS 3647
Resin, heat-resistant, laminating ((POLYIMIDE)).......................	MIL-P-83330
Plastic sheet and strip, POLYIMIDE.................................	MIL-P-46112
Plastic sheet and strip, thin-gage, POLYOLEFIN......................	L-P-378
Plastic molding and extrusion material, POLYPHENYLENE OXIDE.....	L-P-378
POLYPHENYLENE OXIDE molding and extrusion materials...........	ASTM D 2874
Plastic molding and extrusion material, POLYPHENYLENE OXIDE, modified..	MIL-P-46129
Plastic molding and extrusion material, POLYPHENYLENE OXIDE, glass-fiber-reinforced...	MIL-P-46131
Plastic molding material, POLYPROPYLENE, injection and extrusion....	L-P-394
POLYPROPYLENE plastic molding and extrusion materials.............	ASTM D 2146
Plastic molding material, POLYPROPYLENE, glass-fiber-reinforced......	MIL-P-46109
Oriented POLYPROPYLENE film....................................	ASTM D 2673
Nonoriented POLYPROPYLENE plastic film..........................	ASTM D 2530
Cushioning material, sheet form, low-density, unicellular, POLYPROPYLENE foam...	MIL-C-81823
Plastic molding material and plastic extrusion material, POLYSTYRENE..	L-P-396
POLYSTYRENE molding and extrusion materials......................	ASTM D 703
Plastic moldings and extrusions, POLYSTYRENE.....................	AMS 3620
Reinforced and filled POLYSTYRENE, STYRENE-ACRYLONITRILE, and ACRYLONITRILE-BUTADIENE-STYRENE for molding and extrusion..	ASTM D 3011
Plastic molding material, POLYSTYRENE, glass-fiber-reinforced........	MIL-P-21347
Plastic sheet and film, POLYSTYRENE, oriented......................	L-P-506
Plastic molding material (POLYSTYRENE foam, expanded bead)........	MIL-P-19644
Plastic material, cellular, POLYSTYRENE (for buoyancy applications)....	MIL-P-40619
Plastic molding and extrusion material, POLYSULFONE................	MIL-P-46120
POLYTETRAFLUOROETHYLENE.................................	AMS 3651
Plastic molding material, POLYTETRAFLUOROETHYLENE (TFE fluorocarbon)...	L-P-403
POLYTETRAFLUOROETHYLENE moldings, premium-grade, as sintered..	AMS 3668
POLYTETRAFLUOROETHYLENE moldings, as sintered, general-purpose grade..	AMS 3660
POLYTETRAFLUOROETHYLENE extrusions, premium-strength, stress-relieved...	AMS 3659
POLYTETRAFLUOROETHYLENE extrusions, radiographically inspected, premium strength, as sintered............................	AMS 3657
POLYTETRAFLUOROETHYLENE parts and coatings and polyamide parts...	MIL-P-24074
Plastic rods, POLYTETRAFLUOROETHYLENE, molded and extruded..	MIL-P-19468
Plastic sheet, laminated, glass cloth, POLYTETRAFLUOROETHYLENE resin..	MIL-P-19161
POLYTETRALFUOROETHYLENE sheet, molded, as sintered, general-purpose grade..	AMS 3667
POLYTETRAFLUOROETHYLENE sheet, molded, premium grade, as sintered..	AMS 3669
POLYTETRAFLUOROETHYLENE sheet, glass-fabric-reinforced.......	AMS 3666
Plastic tubes and tubing, POLYTETRAFLUOROETHYLENE (TFE fluorocarbon resin), heavy walled.................................	MIL-P-22296
Plastic film, POLYTETRAFLUOROETHYLENE (TFE fluorocarbon resin)...	MIL-P-22241
POLYTETRAFLUOROETHYLENE film, premium-grade..............	AMS 3661
POLYTETRAFLUOROETHYLENE film, noncritical-grade.............	AMS 3652
POLYTETRAFLUOROETHYLENE film, general-purpose grade.........	AMS 3662

TABLE 43   Guide to Specifications for Plastic Materials (Molding, Extrusion, Films, Foams, Laminates; Sheets, Rods, Tubes, Forms) (Continued)

Material	Specification
Fabric, asbestos, wire-reinforced, POLYTETRAFLUOROETHYLENE impregnated, sintered	AMS 3839
Asbestos, POLYTETRAFLUOROETHYLENE impregnated, sintered	AMS 3840
Asbestos-fiber-reinforced, POLYTETRAFLUOROETHYLENE sheet, TFE fluorocarbon resin, high-compressibility	AMS 3843
POLYTETRAFLUOROETHYLENE sheet, TFE fluorocarbon resin, asbestos-fiber-reinforced	AMS 3842
POLYTRIFLUOROCHLOROETHYLENE, unplasticized	AMS 3650
POLYTRIFLUOROCHLOROETHYLENE, compression-molded, heavy sections, unplasticized (Kel-F)	AMS 3645
POLYTRIFLUOROCHLOROETHYLENE film, unplasticized	AMS 3649
POLYTRIFLUOROCHLOROETHYLENE sheet, molded, unplasticized	AMS 3646
POLYTRIFLUOROCHLOROETHYLENE tubing, unplasticized	AMS 3648
Molding and potting compound, chemically cured, POLYURETHANE (polyester-based)	MIL-M-24041
Plastic, POLYURETHANE, flexible, potting and molding compound	MIL-P-46076
Foam, plastic, flexible, open-cell, polyester-type POLYURETHANE	MIL-F-81334
POLYURETHANE foam, flexible, open-cell, medium-flexibility, 2.5 lb/ft^3	AMS 3570
POLYURETHANE foam, rigid or plastic, for packaging	MIL-P-26514
Plastic foam, POLYURETHANE (for use in aircraft)	MIL-P-46111
Plastic material, foamed POLYURETHANE for encapsulating electronic components	MIL-P-46847
Plastic material, cellular POLYURETHANE, foam-in-place, rigid, 2 and 4 lb/ft^3	MIL-P-21929
Plastic material, cellular POLYURETHANE, rigid, void filler, pour-in-place, large-scale, and installation of	MIL-P-24249
Plastic sheet and plastic tubing (flexible, POLYURETHANE film)	MIL-P-43604
Sandwich construction, plastic resin, glass-fabric-base, laminated facings and POLYURETHANE foamed-in-place core. For aircraft structural applications	MIL-S-25392
POLYVINYL ALCOHOL, granular	MIL-P-265
POLY(VINYL CHLORIDE) resins	ASTM D 1755
Resin, plastisol grade, POLYVINYL CHLORIDE	MIL-R-23444
Plastic extrusions, flexible, POLYVINYL CHLORIDE	AMS 3630
Plastic extrusions, flexible, high temperature, POLYVINYL CHLORIDE	AMS 3631
Rigid POLY(VINYL CHLORIDE) plastic sheets	ASTM D 1927
POLY(VINYL CHLORIDE) compounds and chlorinated POLY(VINYL CHLORIDE) compounds	ASTM D 1784
Plastic film (POLYVINYL CHLORIDE and copolymer of vinylidene chloride and vinyl chloride)	L-P-370
Plastic rod, solid, plastic tubes and tubing, heavy-walled, POLYVINYL CHLORIDE, rigid	L-P-1036
Cushioning material, POLYVINYL CHLORIDE, plasticized, cellular	MIL-C-40010
Sponge made from closed-cell POLY(VINYL CHLORIDE) or copolymers	ASTM D 1667
Rigid POLYVINYL CHLORIDE VINYL ACETATE compounds	ASTM D 2114
Rigid POLY(VINYL CHLORIDE–VINYL ACETATE) plastic sheet	ASTM D 2123
Plastic sheets and strips (POLYVINYL FLUORIDE)	L-P-1040
Plastic molding material and plastic extrusion material, POLYVINYLIDENE FLUORIDE polymer and copolymer	MIL-P-46122
RESORCINOL	MIL-R-22578
Resin, SILICONE, low-pressure laminating	MIL-R-25506
Molding plastics and molded plastic parts, thermosetting ((SILICONE filled))	MIL-M-14

TABLE 43 Guide to Specifications for Plastic Materials (Molding, Extrusion, Films, Foams, Laminates; Sheets, Rods, Tubes, Forms) (Continued)

Material	Specification
Plastic moldings, thermosetting, glass-roving-filled SILICONE, heat-resistant	AMS 3643
Plastic sheet, rod, and tube, laminated, thermosetting ((SILICONE))	L-P-509
Plastic material, laminated, thermosetting, electric insulation, sheets, glass cloth, SILICONE resin	MIL-P-997
Plastic material, SILICONE resin, glass-fiber-base, low-pressure laminated	MIL-P-25518
Biaxially oriented STYRENE plastics sheet	ASTM D 1463
Plastic sheet and plastic rod, thermosetting, cast ((STYRENE copolymer))	L-P-516
STYRENE-ACRYLONITRILE copolymer molding and extrusion material	ASTM D 1431
Plastic molding and extrusion material, STYRENE ACRYLONITRILE copolymers	L-P-399
STYRENE ACRYLONITRILE copolymer molding and extrusion materials	ASTM D 1431
Plastic sheet, STYRENE ACRYLONITRILE	L-P-526
STYRENE-BUTADIENE molding and extrusion materials	ASTM D 1892
Plastic molding material, STYRENE BUTADIENE	L-P-398
Compression molding of specimens of STYRENE BUTADIENE molding and extrusion materials	ASTM D 2292
Plastic sheet, STYRENE BUTADIENE	L-P-527
Plastic molding material, FEP fluorocarbon, molding and extrusion ((TETRAFLUOROETHYLENE-HEXAFLUOROPROPYLENE copolymer))	L-P-389
Plastic sheet and film, FEP fluorocarbon, extruded ((TETRAFLUOROETHYLENE-HEXAFLUOROPROPYLENE copolymer))	L-P-523
TFE-FLUOROCARBON resin molding and extrusion materials	ASTM D 1457
TFE-FLUOROCARBON rod	ASTM D 1710
Plastic molding material, UREA FORMALDEHYDE	L-P-401
UREA-FORMALDEHYDE molding compound	ASTM D 705
Rigid URETHANE foam	ASTM D 2341
Foam, URETHANE	MIL-F-81254
Slab flexible URETHANE foam	ASTM D 1564
Core material, foamed-in-place, URETHANE type	MIL-C-8087
Plastic sheets, VINYL, flexible, transparent, optical-quality	MIL-P-18080
Plastic sheet, cellular, shock-absorbing (closed-cell, foamed, modified VINYL sheet)	AMS 3635
Plastic sheet and film, VINYL copolymers	MIL-P-6264
Plastic molding material, VINYL CHLORIDE polymer and vinyl chloride vinyl acetate copolymer, rigid	L-P-1035
VINYL CHLORIDE copolymer resins	ASTM D 2474
Nonrigid VINYL CHLORIDE polymer and copolymer molding and extrusion compounds	ASTM D 2287
Plastic film, flexible, VINYL CHLORIDE	L-P-375
Flexible foams made from polymers or copolymers of VINYL CHLORIDE	ASTM D 1565
Nonrigid VINYL CHLORIDE plastic sheet	ASTM D 1593
Plastic sheet (sheeting), plastic strip, VINYL CHLORIDE polymer and vinyl chloride vinyl acetate copolymer, rigid	L-P-535
Plastic sheet, flexible, reinforced ((VINYL CHLORIDE))	L-P-1196
VINYLIDENE CHLORIDE molding compound	ASTM D 729
Plastic molding material, VINYLIDENE CHLORIDE, copolymer	L-P-1041
Plastic film (polyvinyl chloride and copolymer of VINYLIDENE CHLORIDE and VINYL CHLORIDE	L-P-370

TABLE 44   Guide to Specifications for Elastomeric Materials: Rubbers, Synthetics

Material	Specification
Rubber, fabricated parts	MIL-R-3065
Rubber, high-temperature, fluid-resistant	MIL-R-25897
Rubber, electrically conductive, fuel-resistant	MIL-R-81090
Rubber sheet, cellular, hardboard	ZZ-R-785
Rubber sheet, solid and molded shapes, aromatic and nonaromatic fuel-resistant	MIL-R-7582
Rubber, cellular, chemically blown	MIL-R-6130
Sponge and expanded cellular rubber products	ASTM D 1056
Rubber, anodic	MIL-R-6891
Synthetic rubber compound, BUTADIENE-STYRENE type, ozone-resistant, for low-temperature service	MIL-S-21923
Synthetic rubber, phosphate-ester-resistant, BUTYL type, 35-45	AMS 3237
Synthetic rubber, phosphate-ester-resistant, BUTYL type, 65-75	AMS 3238
Synthetic rubber, phosphate-ester-resistant, BUTYL type, 85-95	AMS 2339
Rubber, CARBOXY-NITROSO, nitrogen tetroxide (N_2O_4)	MIL-R-83322
Rubber, CHLORINATED, natural, power	MIL-R-60671
Synthetic rubber, electrical-resistant, CHLOROPRENE type, 65-75	AMS 3210
Synthetic rubber, flame-resistant, CHLOROPRENE type, 65-75	AMS 3244
Synthetic rubber sheet, cotton-fabric-reinforced, weather-resistant, CHLOROPRENE type	AMS 3270
Synthetic rubber sheet, nylon-fabric-reinforced, weather-resistant, CHLOROPRENE type	AMS 3273
Synthetic rubber—sponge, CHLOROPRENE type, soft	AMS 3197
Synthetic rubber—sponge, CHLOROPRENE type, medium	AMS 3198
Synthetic rubber—sponge, CHLOROPRENE type, firm	AMS 3199
Synthetic rubber, weather-resistant, CHLOROPRENE type, 25-35	AMS 3207
Synthetic rubber, weather-resistant, CHLOROPRENE type, 35-45	AMS 3240
Synthetic rubber, weather-resistant, CHLOROPRENE type, 45-55	AMS 3208
Synthetic rubber, flame-resistant, CHLOROPRENE type, 55-65	AMS 3243
Synthetic rubber, weather-resistant, CHLOROPRENE type, 55-65	AMS 3241
Synthetic rubber, weather-resistant, CHLOROPRENE type, 65-75	AMS 3209
Synthetic rubber, weather-resistant, CHLOROPRENE type, 75-85	AMS 3242
Rubber, CHLOROSULFONATED POLYETHYLENE elastomer, sheet and molded shapes, ozone-resistant	MIL-R-81828
Rubber, hard (EBONITE), natural or synthetic, sheet, strip, tubing, and molded parts	MIL-R-45036
Plastic material, cellular, ELASTOMERIC	MIL-P-12420
Plastic material, unicellular, sheet, ELASTOMERIC	MIL-P-24333
Cellular ELASTOMERIC materials, fabricated parts	MIL-C-3133
Sealing compound, ELASTOMERIC, accelerator required, aircraft structural	MIL-S-7124
Rubber, ETHYLENE-PROPYLENE, general-purpose	MIL-R-83285
Synthetic rubber, phosphate-ester-resistant, ETHYLENE PROPYLENE type, 55-65	AMS 3248
Rubber, FLUOROCARBON elastomer, high-temperature, fluid, and compression-set-resistant	MIL-R-83248
Rubber, FLUOROSILICONE elastomer, oil- and fuel-resistant, sheets, strips, molded parts, and extruded shapes	MIL-R-25988
Plastic film (RUBBER HYDROCHLORIDE base)	L-P-1167
Latex foam rubbers	ASTM D 1055
Rubber sheets and molded shapes, cellular, synthetic open cell (foamed latex)	MIL-R-20092
Rubber, cellular sheet, molded and hand-built shapes, latex foam	MIL-R-5001
Sealing compound, noncuring, POLYBUTENE	MIL-S-12158
Sealing compound, noncuring, POLYSULFIDE base	MIL-S-11030
Sealing compound, adhesive, curing (POLYSULFIDE base)	MIL-S-11031
Rubber, SILICONE	ZZ-R-765
SILICONE rubber, general-purpose, 35-45	AMS 3301
SILICONE rubber, general-purpose, 45-55	AMS 3302
SILICONE rubber, general-purpose, 55-65	AMS 3303
SILICONE rubber, general-purpose, 65-75	AMS 3304

TABLE 44   Guide to Specifications for Elastomeric Materials: Rubbers, Synthetics (Continued)

Material	Specification
SILICONE rubber, general-purpose, 75–85	AMS 3305
SILICONE resin, elastomeric, transparent, elevated-temperature cure, Sylgard 182	AMS 3368
SILICONE resin, elastomeric, transparent, room-temperature cure, Sylgard 184	AMS 3370
SILICONE resin, elastomeric, opaque, elevated-temperature cure, Sylgard 185	AMS 3369
SILICONE resin, elastomeric, opaque, room-temperature cure, Sylgard 185	AMS 3371
SILICONE resin, elastomeric, high tear strength, elevated-temperature cure, Sylgard 187	AMS 3372
SILICONE rubber, extreme-low-temperature-resistant, 15–30	AMS 3332
SILICONE rubber, extreme-low-temperature-resistant, 45–55	AMS 3335
SILICONE rubber, 1,000 lb/in.^{2}, 45–55	AMS 3345
SILICONE rubber, 1,000 lb/in.^{2}, 55–65	AMS 3346
SILICONE rubber, 1,100 lb/in.^{2}, high resilience, 65–75	AMS 3349
SILICONE rubber, 1,150 lb/in.^{2} tensile strength, high-resiliency, 25–35	AMS 3348
SILICONE rubber, 1,200 lb/in.^{2}, high modulus, 45–55	AMS 3347
SILICONE rubber, 1,800 lb/in.^{2}, 45–55	AMS 3344
SILICONE rubber, extreme-low-temperature-resistant, 35–45	AMS 3334
SILICONE rubber, fuel- and oil-resistant, 50–65	AMS 3326
SILICONE rubber, extreme-low-temperature-resistant, 55–65	AMS 3336
SILICONE rubber, fuel- and oil-resistant, 55–65	AMS 3325
SILICONE rubber, lubricating-oil- and compression-set-resistant, electrical-grade, 55–65	AMS 3356
SILICONE rubber, high- and extreme-low-temperature-resistant, 65–75	AMS 3337
SILICONE rubber, lubricating-oil- and compression-set-resistant, 65–75	AMS 3357
SILICONE rubber, low compression set, non-oil-resistant, 65–75	AMS 3307
SILICONE rubber, high-temperature fuel- and oil-resistant, 70–80	AMS 3327
SILICONE rubber, extreme-low-temperature-resistant, 75–85	AMS 3338
SILICONE rubber compound, room-temperature-vulcanizing, 15,000 cP viscosity, durometer 35–55	AMS 3362
SILICONE rubber compound, room-temperature-vulcanizing, 35,000 cP viscosity, durometer 40–45	AMS 3365
SILICONE rubber compound, room-temperature-vulcanizing, 50,000 cP viscosity, durometer 30–45	AMS 3363
SILICONE rubber compound, room-temperature-vulcanizing, 50,000 cP viscosity, short pot life, durometer 35–55	AMS 3364
SILICONE rubber compound, room-temperature-vulcanizing, 55,000 cP viscosity, durometer 55–70	AMS 3366
SILICONE rubber compound, room-temperature-vulcanizing, 1,200,000 cP viscosity, durometer 55–70	AMS 3367
Sealing compound, electrical, SILICONE rubber, accelerator required	MIL-S-23586
Rubber, sponge, SILICONE, closed-cell	MIL-R-46089
SILICONE rubber sponge, closed-cell, firm	AMS 3196
SILICONE rubber sponge, closed-cell, medium	AMS 3195
SILICONE rubber sponge, closed-cell, medium, extreme low temperature	AMS 3193
SILICONE rubber sponge, closed-cell, firm, extreme low temperature	AMS 3194
SILICONE rubber sheet, glass-fabric-reinforced	AMS 3315
SILICONE rubber sheet, glass-fabric-reinforced, heat- and weather-resistant, 60–80	AMS 3320
Potting compound, SILICONE rubber, room-temperature-vulcanizing	MIL-P-46838
Rubber, SILICONE, encapsulating compound	MIL-R-46092
Rubber, SYNTHETIC, heat-shrinkable	MIL-R-46846
Rubber, SYNTHETIC, sheet, strip, and molded	MIL-R-3533
SYNTHETIC rubber sheets, strips, molded or extruded shapes	MIL-S-6855
Rubber sheet, strip, extruded and molded shapes, SYNTHETIC, oil-resistant	MIL-R-2765
Rubber, SYNTHETIC, solid, sheet and fabricated parts, synthetic oil-resistant	MIL-R-7632

TABLE 44   Guide to Specifications for Elastomeric Materials: Rubbers, Synthetics (Continued)

Material	Specification
Asbestos and SYNTHETIC rubber sheet, hot-oil-resistant	AMS 3232
SYNTHETIC rubber sheet, nylon-fabric-reinforced, aromatic-fuel-resistant	AMS 3274
Rubber, SYNTHETIC, shock-absorbing, high compressive strength, oil-resistant	MIL-R-81297
SYNTHETIC rubber, low-temperature-resistant, 25–35	AMS 3204
SYNTHETIC rubber, aromatic-fuel-resistant, 35–45	AMS 3214
SYNTHETIC rubber, dry-heat-resistant, 35–45	AMS 3201
SYNTHETIC rubber, hot-oil-resistant, high-swell, 45–55	AMS 3222
SYNTHETIC rubber, hot-oil- and coolant-resistant, low-swell, 45–55	AMS 3226
SYNTHETIC rubber, low-temperature-resistant, 45–55	AMS 3205
SYNTHETIC rubber, 55–65	AMS 3220
SYNTHETIC rubber, aromatic-fuel-resistant, 55–65	AMS 3212
SYNTHETIC rubber, dry-heat-resistant, 55–65	AMS 3202
SYNTHETIC rubber, hot-oil- and coolant-resistant, low-swell, 55–65	AMS 3227
SYNTHETIC rubber, hydraulic-fluid (petroleum-base) resistant, 55–65	AMS 3200
SYNTHETIC rubber, aromatic-fuel-resistant, 65–75	AMS 3215
SYNTHETIC rubber, hot-oil- and coolant-resistant, low-swell, 65–75	AMS 3228
SYNTHETIC rubber, aromatic-fuel-resistant, 75–85	AMS 3213
SYNTHETIC rubber, hot-oil-resistant, low-swell, 75–85	AMS 3229
Sealing compound, SYNTHETIC rubber, electrical connections and electric systems, accelerator required	MIL-S-8516

TABLE 45   Guide to Specifications for Coating Materials

Material	Specification
Coating compound, ACRYLIC, clear	MIL-C-17504
Plastic coating compound, strippable (hot dipping) ((CELLULOSE ACETATE BUTYRATE))	MIL-P-149
Plastic coating compound, strippable ((CELLULOSE ACETATE BUTYRATE))	MIL-P-23242
Plastic coating compound, strippable, cold-dipping, 120°F (49°C) ((CELLULOSE ACETATE BUTYRATE, modified))	MIL-P-45021
Coating systems, ELASTOMERIC, thermally reflective and rain-erosion-resistant	MIL-C-27315
Primer coating, EPOXY	MIL-P-52192
Coating, EPOXY, for steel structures	MIL-C-82407
Coating, EPOXY, baking type, for magnesium castings	MIL-C-46079
Resin compound, thermosetting, room-temperature-curing, for metal coating ((EPOXY))	MIL-C-22627
Coating compound, (EPOXY) thermal-insulating (intumescent)	MIL-C-46081
Coating, EPOXY-POLYAMIDE	MIL-C-22750
Primer coating, EPOXY-POLYAMIDE, chemical- and solvent-resistant	MIL-P-23377
Plastic compound, EPOXY-POLYAMIDE, marine-splash-zone application	MIL-P-28579
Plastic coating compound, strippable (hot dipping) ((ETHYL CELLULOSE))	MIL-P-149
Coating, PHENOLIC resin base, baking	MIL-C-18467
Coating, PHENOLIC resin base, air-drying	MIL-C-18468
PHENOLIC resin coating material, TFE pigmented, 300°F cure	AMS 3136
Plastic, PLASTISOL (for coating metallic objects)	MIL-P-20689
Primer coating, POLYAMIDE, EPOXY	MIL-P-46056
Resin compound, thermosetting, room-temperature cure, for metal coating ((POLYESTER))	MIL-R-23461
Coating compound, elastomeric, POLYSULFIDE	MIL-C-38713
POLYTETRAFLUOROETHYLENE resin coating, low build, 700–750°F (371–399°C) fusion	AMS 2515

TABLE 45   Guide to Specifications for Coating Materials (Continued)

Material	Specification
POLYTETRAFLUOROETHYLENE resin coating, high build, 700–750°F (371–399°C) fusion	AMS 2516
Coating, POLYURETHANE	MIL-C-46057
Coating, POLYURETHANE, for aircraft application	MIL-C-27227
Coating, POLYURETHANE, aliphatic, weather-resistant	MIL-C-81773
Coating, POLYURETHANE, clear, linseed-oil-modified	TT-C-540
SILICONE resin coating material, 400°F (204°C) cure	AMS 3135
Compound, SILICONE, soft film	MIL-C-21567
Resin, conformal coating, URETHANE	AMS 3137
Coating compound, strippable, sprayable ((VINYL))	MIL-C-16555

TABLE 46   Guide to Specifications for Insulation Materials; Electrical, Thermal

Material	Specification
Insulating compound, molded, electrical, heat-shrinkable BUTYL RUBBER, flexible, cross-linked	MIL-I-81765/5
Plasticized CELLULOSE ACETATE sheet and film for primary insulation	ASTM D 1202
Preformed flexible ELASTOMERIC cellular thermal insulation in sheet and tubular form	ASTM C 534
Insulating compound, molded, electrical, heat-shrinkable FLUOROELASTOMER, flexible, cross-linked	MIL-I-81765/4
Insulation batt, NYLON, thermal	MIL-I-52172
Insulating compound, asbestos PHENOLIC	MIL-I-81255
Insulating compound, molded, electrical, heat-shrinkable, POLYCHLOROPRENE, flexible, cross-linked	MIL-I-81765/2
Elastomeric tubing, electrical insulation, irradiated POLYCHLOROPRENE, flexible, heat-shrinkable, 1.750:1 shrink ratio	AMS 3623
Plastic tubing, electrical insulation, thermally welded, POLYETHYLENE TEREPHTHALATE, heat-shrinkable, 1.6:1 shrink ratio	AMS 3591
Insulation, electrical, dielectric barrier, laminated, plastic film and synthetic fiber mat ((POLYETHYLENE TEREPHTHALATE))	MIL-I-22834
Insulating compound, molded, electrical, heat-shrinkable POLYOLEFIN, cross-linked, semirigid, and flexible	MIL-I-81765/1
Plastic moldings, electrical insulation, cross-linked POLYOLEFIN, pigmented, flexible, heat-shrinkable	AMS 3674
Plastic tubing, electrical insulation, irradiated POLYOLEFIN, heat-shrinkable, pigmented, flexible, 2:1 shrink ratio (RNF-100, type 1)	AMS 3636
Plastic tubing, electrical insulation, irradiated POLYOLEFIN, clear, flexible, heat-shrinkable, 2:1 shrink ratio (RNF-100, type 2)	AMS 3637
Plastic tubing, electrical insulation, irradiated POLYOLEFIN, semirigid, pigmented, heat-shrinkable, 2:1 shrink ratio	AMS 3638
Plastic tubing, electrical insulation, irradiated POLYOLEFIN, clear, semirigid, heat-shrinkable, 2:1 shrink ratio	AMS 3639
Plastic tubing, electrical insulation, irradiated POLYOLEFIN, heat-shrinkable	AMS 3633
Plastic tubing, electrical insulation, POLYOLEFIN, dual-wall, semirigid, heat-shrinkable	AMS 3634
Plastic tubing, electrical insulation, irradiated POLYOLEFIN, pigmented, very flexible, heat-shrinkable (low recovery temperature), 2:1 shrink ratio	AMS 3587
Plastic tubing, electrical insulation, irradiated POLYOLEFIN, clear, very flexible, heat-shrinkable (low recovery temperature), 2:1 shrink ratio	AMS 3588
Preformed, block-type cellular POLYSTYRENE thermal insulation	ASTM C 578
Extruded POLYTETRAFLUOROETHYLENE tubing, electrical insulation, standard wall (Teflon)	AMS 3653
Extruded POLYTETRAFLUOROETHYLENE tubing, electrical insulation, light wall (Teflon)	AMS 3654

TABLE 46   Guide to Specifications for Insulation Materials; Electrical, Thermal (Continued)

Material	Specification
Extruded POLYTETRAFLUOROETHYLENE tubing, electrical insulation, thin wall (Teflon)	AMS 3655
Insulation sleeving, electrical, POLYTETRAFLUOROETHYLENE resin, nonrigid	MIL-I-22129
Insulation tape, electrical, high-temperature, POLYTETRAFLUORO-ETHYLENE, pressure-sensitive	MIL-I-23594
Insulation tape, glass fabric, POLYTETRAFLUOROETHYLENE coated	MIL-I-18746
Plastic tubing, electrical insulation, POLYTETRAFLUOROETHYLENE, heat-shrinkable, overexpanded	AMS 3584
Plastic tubing, electrical insulation, POLYTETRAFLUOROETHYLENE, heat-shrinkable, thin wall	AMS 3585
Plastic tubing, electrical insulation, POLYTETRAFLUOROETHYLENE, heat-shrinkable, standard wall	AMS 3586
Insulation, plastic, cellular POLYURETHANE, rigid, preformed, and foam-in-place	MIL-I-24172
Plastic tubing, electrical insulation, non-cross-linked POLYVINYL CHLORIDE, flexible, heat-shrinkable. 1.6:1 shrink ratio	AMS 3579
Plastic tubing, electrical insulation, cross-linked POLYVINYL CHLORIDE, flexible, heat-shrinkable, 2:1 shrink ratio	AMS 3582
Plastic tubing, electrical insulation, cross-linked POLYVINYL CHLORIDE, semirigid, heat-shrinkable, 2:1 shrink ratio	AMS 3583
Tubing, extruded, POLYVINYL CHLORIDE, high-temperature, electrical-nisulation	AMS 3629
Plastic tubing, electrical-insulation, irradiated POLYVINYL FLUORIDE, heat-shrinkable, semirigid, 2:1 shrink ratio (Kynar)	AMS 3632
SILICONE compound (for electrical insulating and sealing)	MIL-S-8660
Insulating compound, molded, electrical, heat-shrinkable SILICONE RUBBER, flexible, cross-linked	MIL-I-81765/3
Insulation sheets, electrical, pasted mica, SILICONE bonded	MIL-I-19526
Insulation sheet, electrical, mica paper, SILICONE bonded	MIL-I-19917
Insulation sleeving, electrical, flexible, glass fiber, SILICONE RUBBER treated	MIL-I-18057
Insulation tape, electrical, self-adhering, unsupported SILICONE RUBBER	MIL-I-46852
Insulation tape, electrical, self-bonding SILICONE RUBBER treated, bias-weave or sinusoidal-weave glass, cable splicing, naval standard	MIL-I-22444
Elastomeric tubing, electrical insulation, cross-linked SILICONE, pigmented, flexible, heat-shrinkable, 1.750:1 shrink ratio	AMS 3625
Insulation, SYNTHETIC, rubberlike, chemically expanded, cellular (sheet form)	MIL-I-16562
Insulation, electrical, SYNTHETIC RUBBER composition, nonrigid	MIL-I-631
Rigid preformed cellular URETHANE thermal insulation	ASTM C 591
Insulation sleeving, electrical, flexible ((VINYL))	MIL-I-7444
Insulation tubing, electrical, nonrigid, VINYL, very-low-temperature grade	MIL-I-22076
Insulation sleeving, electrical, flexible, glass fiber, VINYL treated	MIL-I-21557
Insulation tape, electrical, glass fiber (resin-filled) and cord, fibrous glass ((VINYL))	MIL-I-3158

REFERENCES

1. Hartmann, J. E.: Directory of United States Standardization Activities, *Nat. Bur. Std. Misc. Pub.* 288, Aug. 1, 1967. (For sale by the Superintendent of Documents, Government Printing Office, Washington, D.C. 20402 at $2.)
2. Breden, L. H. (ed.): World Index of Plastics Standards, *Nat. Bur. Std. Spec. Pub.* 352, December 1971. (For sale by the Superintendent of Documents, Government Printing Office, Washington, D.C. 20402. Order by SD Catalog No. C13.10:352 at $5.50.)

3. Beach, N. E.: Government Specifications and Standards for Plastics, Covering Defense Engineering Materials and Applications (Revised), Plastics Technical Evaluation Center, Picatinny Arsenal, Dover, N.J., *PLASTEC Note* 6B, June 1969. [Available from the National Technical Information Service (NTIS), U.S. Department of Commerce, 5285 Port Royal Road, Springfield, Va. 22161 at $4 prepaid or $4.50 if billed. Order as No. AD 697 181.]

4. Department of Defense (U.S.): Index of Specifications and Standards (DODISS), Part I, Alphabetical Listing; Part II, Numerical Listing and the Federal Supply Classification Listing. (Parts I and II are available at $20.00/year, including supplements and quarterlies, from the Superintendent of Documents, Government Printing Office, Washington, D.C. 20402. The Federal Supply Classification Listing is available from the same source at $10/year. This listing does not have a part number.)

5. Slattery, W. J. (ed.): An Index of U.S. Voluntary Engineering Standards, *Nat. Bur. Std. Spec. Pub.* 329, March 1971. (For sale by the Superintendent of Documents, Government Printing Office, Washington, D.C. 20402. Order by SD Catalog No. C13.10:329 at $9.90 for hard copy; microfiche, 15 pages, available from National Technical Information Service, U.S. Department of Commerce, Springfield, Va. 22151, as COM 71-50172 at $0.95.)

6. "List of Voluntary Product Standards, Commercial Standards, and Simplified Practice Recommendations," National Bureau of Standards, U.S. Department of Commerce, Office of Engineering Standards, Washington, D.C. 20234, NBS List of Publications 53, revised September 1972, 16 pages.

7. Department of Defense (U.S.): "Standardization Directory" SD-1 (FSC Class and Area Assignments), Defense Standardization Program, Oct. 1, 1972. This directory is issued on a quarterly basis as of January, April, July, and October.

8. "Federal Supply Groups and Classes," Cataloging Handbook H2-1. (Available from the Superintendent of Documents, Government Printing Office, Washington, D.C. 20402 at $0.30 per copy. This handbook has not been seen by the authors.)

9. "Guide to Specifications and Standards of the Federal Government," General Services Administration, Washington, D.C., 24 pages, June 1969 (reprinted December 1971). (Copies available without charge from GSA Business Service Centers and field offices of the Small Business Administration and the U.S. Department of Commerce.)

10. Naval Ordnance Systems Command: "Index of Ordnance Specifications and Weapons Specifications," NAVORD OS O, 10th Revision, Jan. 31, 1972, 434 pages. [This document is issued annually. Supplements listing additions, changes, and cancellations are issued each midyear. The latest issue is July 31, 1972. Requests for copies of this index, documents listed therein, or requests that contractors or activities be placed on distribution list for this index, should be addressed to: Commanding Officer, Naval Ordnance Station, Central Technical Documents Office, Louisville, Ky. 40214, ATTN: Code 80121. Emergency calls for information on OS's and WS's may be made to Miss Emma C. Karrer, Code V2, by calling (502) 361-2641 or Autovon 726-1550, Ext. 323.]

11. Plastics Technical Evaluation Center (PLASTEC): Guide to Test Methods for Plastics and Related Materials, *PLASTEC Note* 17, by Norman E. Beach, August 1967, 73 pages. (Available from the National Technical Information Service, U.S. Department of Commerce, Springfield, Va. 22161 at $6 prepaid or $6.50 if billed. Order as AD 662 049.

12. "Department of Defense Single Stock Point (DOD-SSP) for Specifications and Standards—A Guide for Private Industry," Naval Publications and Forms Center, 5801 Tabor Avenue, Philadelphia, Pa. 19120, Form 4ND-NPFC-4120/3 (rev. September 1972, 8 pages.)

13. Department of Defense (U.S.): "Defense Standardization Manual 4120.3-M, Standardization Policies, Procedures and Instructions," Office of the Assistant Secretary of Defense (Installations and Logistics), Washington, D.C., January 1972, over 600 pages. [Copies available to the general public at $2.50 from the Superintendent of Documents, Government Printing Office, Washington, D.C. 20402. (This issue supersedes the 4120.3-M issue dated April 1966, at one time designated M200B)].

14. "Index of Federal Specifications and Standards," General Services Administration, Federal Supply Service, FPMR 101-29.1 (41 CFR 101-29.1). Jan. 1, 1972 issue is 336 pages plus a 59-page appendix listing canceled and superseded documents. (For sale by the Superintendent of Documents, Government

Printing Office, Washington, D.C. 20402. Subscription price is $7.50/year and $9.25/year for foreign mailing, including the basic index and monthly cumulative supplements.)

15. "NASA Specifications and Standards," National Aeronautics and Space Administration, Scientific and Technical Information Division, Washington, D.C., NASA SP-9000, May 1967, 380 pages. (Available to organizations registered with NASA to receive documents without charge through the NASA Scientific and Technical Information Facility, P.O. Box 33, College Park, Md. 20740. Others may purchase copies at $3 per copy from the National Technical Information Service, U.S. Department of Commerce, Springfield, Va. 22161.)

16. "MSFC Specifications, Standards and Procedures—Microfiche Reference File Index," National Aeronautics and Space Administration, George C. Marshall Space Flight Center, Huntsville, Ala. 35812, File 005, November 1972, 104 pages. Prepared by Hayes International Corporation, published by Documentation Repository Division, Management Services Office, MSFC.

17. "Technical Specification Listing," National Aeronautics and Space Administration, George C. Marshall Space Flight Center, Huntsville, Ala. 35812, MSFC-TSL-488B, Sept. 30, 1969, 576 pages. Prepared by Requirements Division, Central Systems Engineering, S&E-CSE-C for the MSFC Specification Coordination Board.

18. "Specifications and Standards for Space Systems and Related Equipment," National Aeronautics and Space Administration, George C. Marshall Space Flight Center, Huntsville, Ala. 35812, MM 8070.2, Sept. 15, 1972.

19. "Nonmetallic Materials Design Guidelines and Test Data Handbook," National Aeronautics and Space Administration, Manned Spacecraft Center, Houston, Tex., MSC-02681 Revision B, August 1971. (Distribution of this handbook is limited to individuals and/or organizations having a need to know its contents. Distribution is further limited to the minimum number of copies necessary to support the immediate requirements of each contractor. New requests for this document on a one-time or continuous basis must be fully justified in writing by the contractor's government technical monitor. Requests by him should be addressed to: Chief, Reliability Division, Code NB, Manned Spacecraft Center, Houston, Tex. 77058.)

20. "Flammability, Odor, and Offgassing Requirements and Test Procedures for Materials in Environments That Support Combustion," National Aeronautics and Space Administration, Office of Manned Space Flight, Washington, D.C. 20546, NHB 8060.1, November 1971. (For sale by the Superintendent of Documents, Government Printing Office, Washington, D.C. 20402 at $1.)

21. "Current REA Publications" (to REA Telephone Borrowers), Rural Electrification Administration, U.S. Department of Agriculture, Washington, D.C. 20250, Feb. 12, 1971, 37 pages.

22. Resnick, I.: American National Standards Institute, private communication, January 1973.

23. "ANSI Catalog 1972," American National Standards Institute, Inc., 1430 Broadway, New York, N.Y. 10018, Mar. 1, 1972, 142 pages.

24. "The Role of the American National Standards Institute," American National Standards Institute, December 1972, 8 pages.

25. "1973 Annual Book of ASTM Standards," American Society for Testing and Materials, 1919 Race Street, Philadelphia, Pa. 19103, to be published in the following parts (Parts listed are only those relevant to plastics and elastomers, including adhesives):

Part 15   Paper; Packaging; Cellulose; Casein; *Flexible Barrier Materials;* Business Copy Products; Leather; 1,154 pages; 251 standards. ($36.50; available April 1973.)

Part 16   *Structural Sandwich Construction;* Wood; *Adhesives;* 998 pages; 174 standards. ($31.50; available July 1973.)

Part 26   *Plastics—Specifications; Methods of Testing Pipe, Film, Reinforced and Cellular Plastics; Fiber Composites;* 1,240 pages; 209 standards. ($39; available July 1973.)

Part 27   *Plastics—General Methods of Testing, Nomenclature;* 1,040 pages; 170 standards. ($32.75; available July 1973.)

Part 28   *Rubber;* Carbon Black; Gaskets; 1,348 pages; 212 standards. ($42.25; available July 1973.)

Part 29   *Electrical Insulating Materials;* 1,350 pages; 220 standards. ($42.25; available July 1973.)

Part 30   General Test Methods; Radioisotopes and Radiation Effects; Statistical Methods; Appearance of Materials; Emission, Molecular and Mass Spectroscopy; Sensory Evaluation of Materials and Products; Chromatography; Temperature Measurement; Space Simulation; Resinography; Microscopy; *Deterioration of Nonmetallic Materials;* 1,716 pages; 209 standards. ($54; available July 1973.)

Part 33   Index to Standards; 240 pages. ($4.25; available November 1973.) The 1972 edition of these parts will be supplied until the 1973 version becomes available. The prices will probably be somewhat lower for the earlier versions, and the number of pages may be less. The 1972 issue contains 4,500 standards, of which 1,300 are new or revised. Part 33, the Index of Standards, can be used to locate the specific parts in which standards are presented. It is available at $4.

26. Staff: NBS Policy on Voluntary Product Standards, *Materials Research and Standards* (ASTM), November 1972.
27. "NEMA Standards Publications," National Electrical Manufacturers Association, 155 East 44th Street, New York, N.Y. 10017, March 1972 catalog, 7 pages.
28. "RMA Rubber Product Publication Catalog—Technical and Educational Materials," Rubber Manufacturers Association, 444 Madison Avenue, New York, N.Y. 10022, August 1971, 1 page.
29. "1972 SAE Handbook," Society of Automotive Engineers, 2 Pennsylvania Plaza, New York, N.Y. 10001, 1,372 pages, $40.
30. "Index of Aerospace Material Specifications," Society of Automotive Engineers, 2 Pennsylvania Plaza, New York, N.Y. 10001, Nov. 15, 1972, 30 pages, published semiannually in May and November.
31. Society of the Plastics Industry, Inc., Thermosetting Resin Formulators Division, 250 Park Avenue, New York, N.Y. 10017; untitled brochure covers: Purposes of this Division. What Has the TRF Done? What Is Being Accomplished? Who May Belong? How May I Benefit?
32. Katz, I. (revised by C. V. Cagle): "Adhesive Materials—Their Properties and Usage," Foster Publishing Company, 1602 Pattiz Ave., Long Beach, Calif. 90815, 1971. (Also available from Palmerton Publishing Co., Inc., 101 West 31st St., New York, N.Y. 10001.)
33. Beach, N. E.: "Plastic Laminate Materials—Their Properties and Usage," Foster Publishing Company, 1602 Pattiz Ave., Long Beach, Calif. 90815, 1967.

Chapter **12**

Design and Fabrication of Plastic Parts

PERCY GRAFTON
Boonton Molding Company,
Boonton, New Jersey

INTRODUCTION

The versatility of plastic materials, mold designs, molding techniques, and finishing methods influences the product's design. Properties favorable to versatile and economical product design are the variety of surface textures, colors, and complex shapes (without secondary operations) and the wide range of physical, electrical, mechanical, and thermal characteristics.

Thermoset Molding

Thermoset-molding materials, when molded, will undergo chemical reaction under heat and pressure. After the chemical reaction during the mold cycling, thermoset materials cannot be reused. Method of processing is by feeding cool or preheated material into a hot mold under pressure and holding for a predetermined cure time.

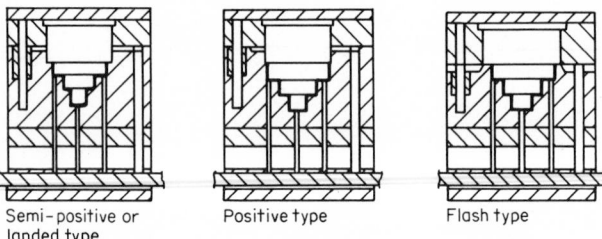

Semi-positive or   Positive type   Flash type
landed type

Fig. 1 Compression-type molds for thermosets.

Possible materials are alkyds, epoxies, phenolics, polyesters, melamines, silicones, and ureas. These materials can be filled with wood flour, cotton, rayon, mica, glass, and other fillers in order to meet special requirements. There are numerous trade names for most of these plastics (see Chap. 1).

Compression molds Both positive and semipositive compression molds (Fig. 1) are used when a molded part (requiring strength) is made of plastic material filled with long fibers. Compression molds are not recommended for delicate mold sections or long pins with small diameters. The flash-type mold is used only in isolated cases, as this design loses material from the cavity and may produce parts with insufficient density.

Transfer molds Transfer molds (Fig. 2) can be used for delicate mold sections and small pins with some reservations. The transfer method has a tendency to break up the fibers in the material when transferring the plastic through the mold gates, thereby reducing the strength of the molded parts contrary to the material specifications.

Injection molds for thermosets Injection molds for thermosets (Fig. 2) are similar in method to the transfer types, except that they are fed through the center of the mold by a reciprocating screw in a screw-type press.

Thermoplastic Molding

Thermoplastics are materials which will repeatedly soften when heated and harden when cooled. Method of processing is by feeding hot plastic material under pressure into a cool mold and holding for a predetermined set time. These materials can be reused, eliminating some possible waste. A few of these materials are ABS resins, acrylics, nylons, phenoxy, polyethylene, acetal, and vinyl. Many of these materials can be supplied with glass, asbestos, or other fillers. There are many trade names for most of these plastics (see Chap. 1).

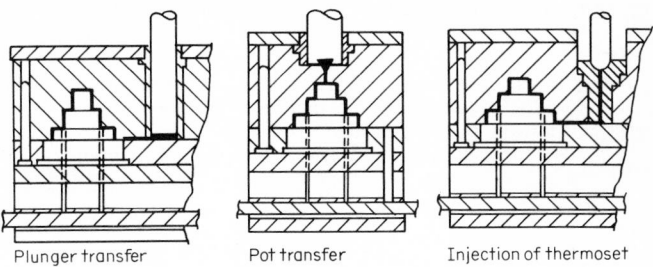

Plunger transfer   Pot transfer   Injection of thermoset

Fig. 2 Plunger, transfer, and injection molds for thermosets.

Injection molds for thermoplastics For the injection process there are several types of molds (Fig. 3). The difference is mainly in the runner system (hot runner or insulated runner) and the method of gating (edge, tunnel, center, and disk).

Material Selection

There are approximately 40 different basic materials and many variations of each type. It can be said that no one material has all the qualities that may be desired, but there is a material that will satisfy a singular or combination of requirements. In selecting a material from a supplier specification sheet, it should be taken into consideration that the data were obtained from test bars under laboratory conditions, whereas in practice the geometric shapes may create variations from material specification values.

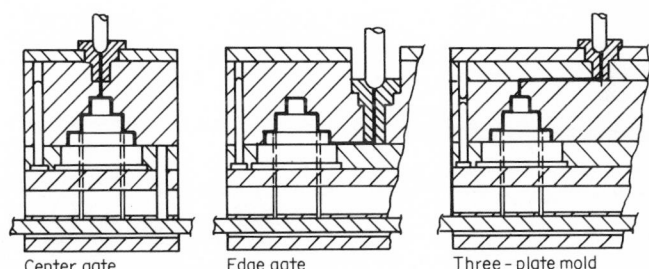

Center gate   Edge gate   Three - plate mold

Fig. 3 Injection-type molds for thermoplastics.

Qualitative specifications Many materials will be eliminated from consideration by comparison of material specifications with qualitative requirements, such as FDA approval, UL approval, weather-exposure modifications (sunlight fading, freezing brittleness, moisture absorption, etc.) transparency, buoyancy, fungus resistance, flame resistance (nonburning, slow-burning, self-extinguishing, etc.), high-temperature limitations, high-impact limitations, and live hinges.

Quantitative specifications Further rejection of possible materials will occur by referring to material specifications for quantitative requirements, such as the temperature range, mechanical attributes (density, etc.), electrical attributes (dielectric, arc resistance, etc.), and chemical attributes (acid resistance, solvent resistance, etc.).

Applications and finishing operations   Final selection of a material will be based on the product's use and manufacturer's secondary operations (drilling, tapping, machining, sonic welding, insert assembling, plating, painting, hot stamping, etc.).

The material manufacturer should be consulted to confirm the applicability of the material selected.

DESIGN FEATURES AND GUIDELINES

Flow Patterns

Flow and weld lines   The flow pattern is the path of soft material flowing around an obstruction, forming a pattern as the plastic material meets and welds together. Flow patterns are always present in some form as material moves around openings and slots (Fig. 4), holes, bosses, and ribs (Fig. 5).   To obtain maximum weld

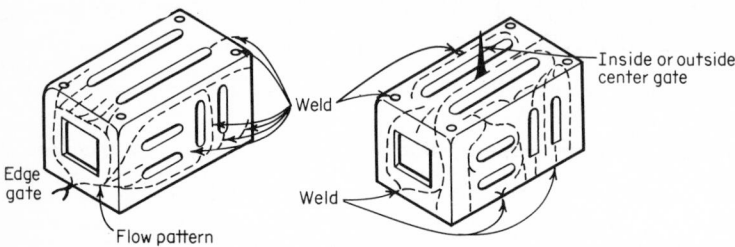

Fig. 4   Flow patterns around holes, slots, and openings.

strength, gate the part so as to obtain the minimum distance of flow around an obstruction and at the same time direct the flow to allow the formation of welds at points that will assure maximum strength with acceptable and unsightly weld lines. Many factors affect the flow pattern, such as the geometric design, location of gates, vents, and the type of plastic used.   Wall thicknesses (Fig. 6) will affect flow patterns in thermoset material.   Parts may be porous in transfer-type moldings and injection of thermosets unless correct wall thickness and gate locations are used.

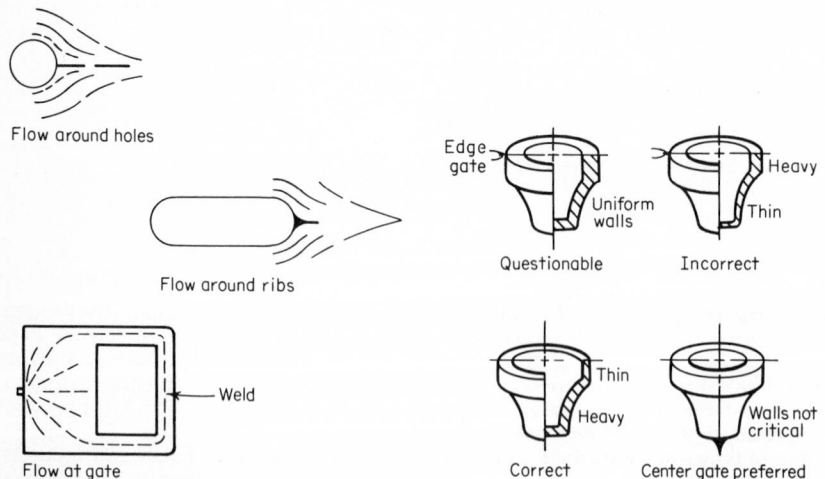

Fig. 5   Flow patterns around holes, bosses, ribs, and windows.

Fig. 6   Flow patterns affected by wall thickness.

Compression-molded parts will not be affected because of the method of loading plastic into cavities.   Injection of thermoplastics will be affected by improper wall thickness and gate locations (Fig. 6), which could cause gas pockets and surface burns.

Resin-rich section   High-impact materials do not always assure maximum overall strength.   Delicate thin sections (Fig. 7) in the center may be resin-rich and without fibers for strength.   The resin may flow off the long fibers and into thin sections before the part is completely filled.   This is especially true of thermoset material, depending upon the geometric design and method of molding.

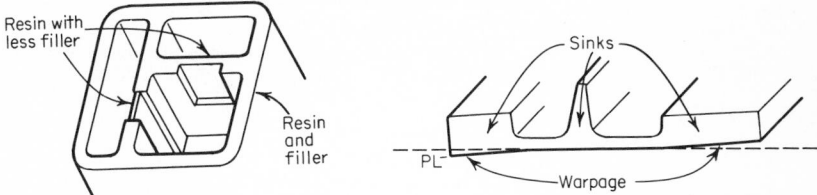

Fig. 7 Flow patterns in and around thin delicate sections.

Fig. 8 Uneven sections.

Heavy and Uneven Sections

Undesirable sections   In thermoplastic products, heavy and uneven sections (Fig. 8) cause sink marks, warpage, longer molding cycles, and dimensional difference due to distortion.   In thermoset products heavy and/or uneven sections (Fig. 8) cause longer molding cycle, distortion, cracks, blisters, and strains.   Coring to form uniform thin walls (Fig. 9) will eliminate the undesirable conditions mentioned above and reduce manufacturing cost caused by long cycle time and excessive rejects.

Delicate Sections

Delicate-edged parts   Products with sharp edges (Fig. 10) should be avoided, especially with brittle molding material, as parts will be subjected to chipping, breakage in handling, and undesirable parting-line flashing.

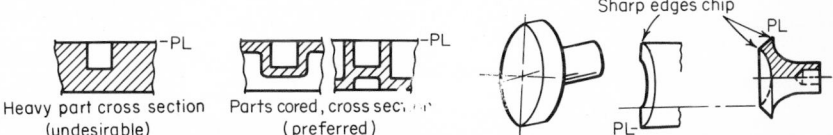

Fig. 9 Heavy sections and coring.

Fig. 10 Undesirable edges.

Delicate mold sections   Hole locations must be away from sharp edges of thin walls, as proper flow is prevented around mold pins.   Also thin walls (Fig. 11) will chip near the hole's edge.   For reduced maintenance cost, molds with laminated sections and replaceable blades should be used when delicate and tight-toleranced dimensions are necessary.   The shape of holes (Fig. 12) may be altered from round to square when laminated-type molds are required.

Designed Undercuts

Undercuts should be avoided wherever possible, as they require tricky mold designs and special molding techniques.   In using thermoplastic materials, stripper-type molds can be used in many cases, depending upon the plastic and depth of the undercut.

Internal undercuts   Internal undercuts (Fig. 13) can be produced by cams, wedges, and ejector pins.

External undercuts   External undercuts (Fig. 14) and outside holes can be made by cams.

Collapsible cores   In cases of round internal undercuts (Fig. 15) retractable internal sectional wedges (collapsible cores) can be used for rigid materials. In some thermoplastic materials stripper-type molds may be used. In cases of movable mold members, the product will be marked with visible seam-line scars.

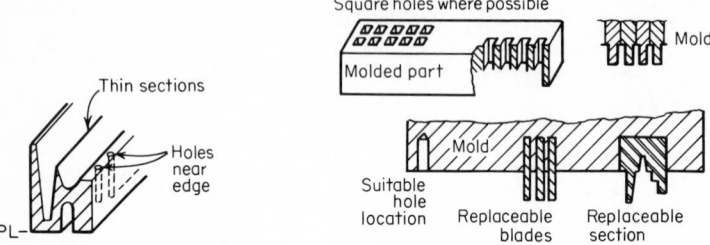

Fig. 11   Delicate edges and holes.   **Fig. 12**   Delicate sections.

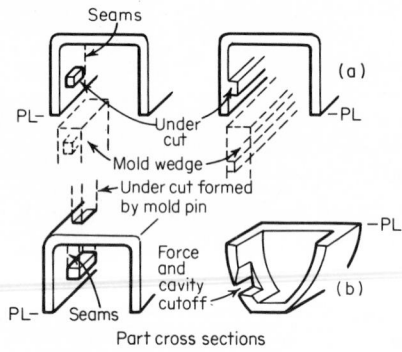

Fig. 13   Inside undercuts.

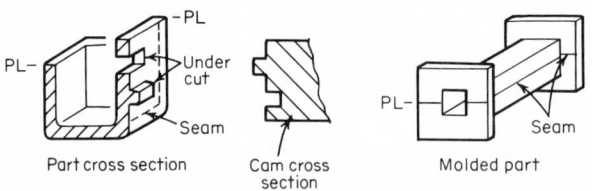

Fig. 14   Outside undercuts.

Fig. 15   Stripped undercuts.

Parting Lines

Possibilities   The shape of the part will determine the contour of the parting line. A straight-line type (Fig. 16a) is preferred, although the odd-step type (Fig. 16b) is required in many cases.

Designed mismatch A 0.010-in. mismatch or bead (Fig. 17) is intentionally designed to disguise an undesirable mismatch in the mold parting line.

Parting-line wear Cylindrical, conical, or spherical surfaces (Fig. 18) may be flattened along the path of the parting lines in order to adjust for objectionable

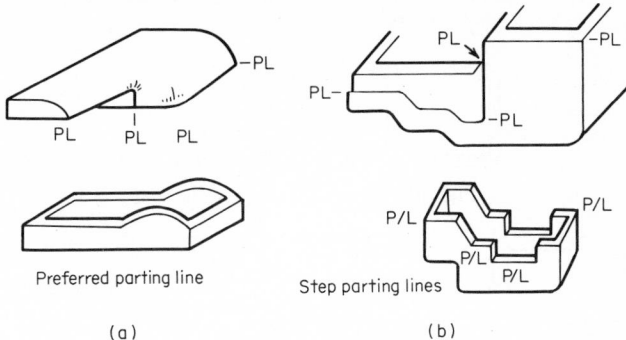

Preferred parting line

Step parting lines

(a)

(b)

Fig. 16 Parting-line locations.

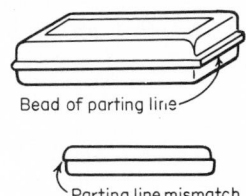

Bead of parting line

Parting line mismatch

Fig. 17 Parting-line mismatch.

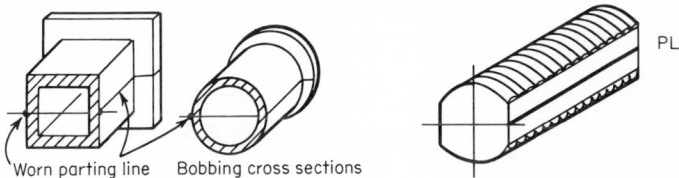

Worn parting line   Bobbing cross sections

Fig. 18 Parting-line wear.

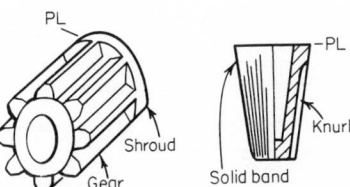

Shroud

Gear

Knurl

Solid band

Fig. 19 Parting-line shroud.

burrs or seams. Flattening the curved surface along the parting line could make possible the functioning of a compatible part (such as a bolt and nut).

Flashing in teeth Many items, such as gears or round parts with knurled periphery, require shrouds, beads, or bands (Fig. 19) to prevent flashing within the teeth.

Designed Mismatch

Controlled mismatch   Utilizing a bead to disguise a possible mismatch (Fig. 20) will eliminate molded rejects.   When molding boxes and covers or similar items, a designed mismatch at the parting line (Fig. 21) will improve the appearance. Mismatch can be due to the difference in material shrinkage and/or warpage set up by strains during the molding cycle.

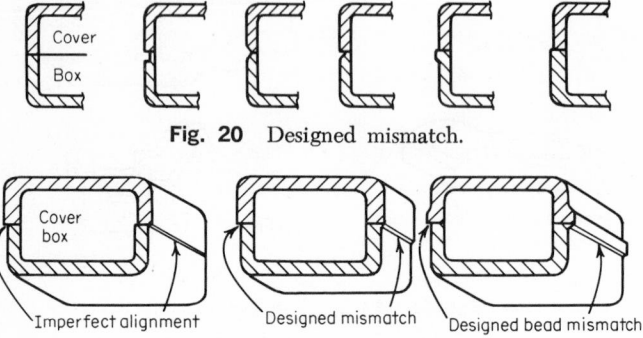

Fig. 20   Designed mismatch.

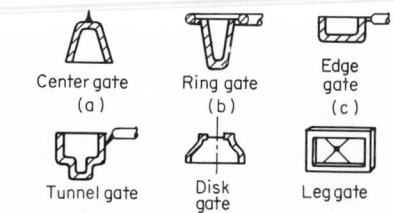

Fig. 21   Mismatch parting lines.

Gates

Variety of gates   Standard gates used are ring, center, tunnel, disk, and leg gates (Fig. 22).   These gates are used for both thermoset and thermoplastic materials, and each has its advantage as well as disadvantage when used.   Center gating (Fig. 22*a*) is preferred in most cases.   Ring gating (Fig. 22*b*) is used for tubular parts where the material enters the open end and flows uniformly along all walls.

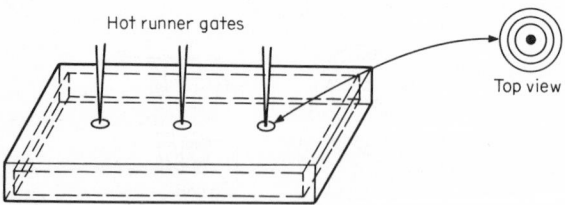

Fig. 22   Types of available gates.

Fig. 23   Multiple gates.

Edge gating (Fig. 22*c*) is used in many cases, depending upon the geometric configuration of the parts.   Hot and insulated runners (Fig. 23) are used for large parts.   Multiple gating, which usually consists of hot and insulated runners, is to keep molded parts flat.   This type of gating is also used for multicavity-type molding of small parts.

Hand-finished gates   For thermoset material, gating (Fig. 24a) is designed to be broken from the part whenever possible.   Other gates are machined or sanded flush (Fig. 24b) to a given dimension.

Step-type design   Here (Fig. 25) center gating is preferred when molding thermoplastic materials.   If edge gating is used for aesthetic reasons, molding conditions

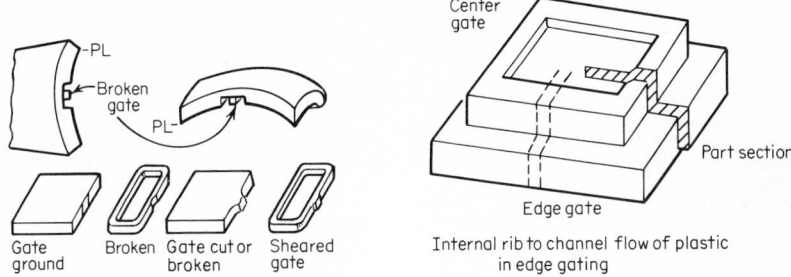

Fig. 24   Gates for thermoset.

Fig. 25   Problem gating.

will be improved by adding an internal rib.   The rib will channel the plastic into the upper sections before filling the sidewalls, thus preventing gas from being trapped within the top surface.   Molding of thermoset materials by the straight compression method will produce a product without gas pockets.

Hinges

Hinge variations   The conventional pin-type hinges (Fig. 26) can be used with either metal or plastic hinge pins.   The straight draw-molded hinge (Fig. 27) needs no cam slides or drilling but requires hinge pins.

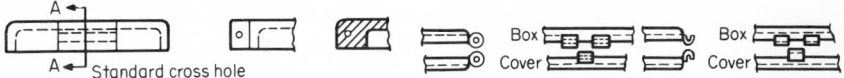

Fig. 26   Standard-type hinge with pin.

Fig. 27   Straight draw-type hinge.

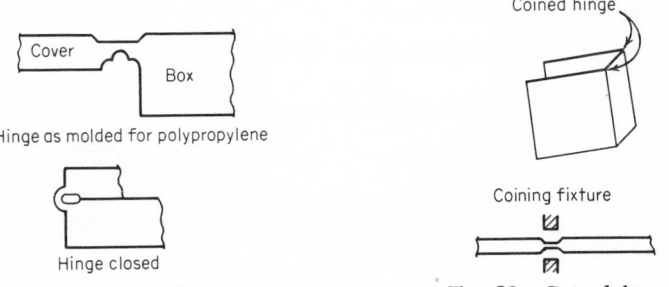

Fig. 28   Live hinge.

Fig. 29   Coined hinge.

Live hinges   In order to ensure good hinge life, live hinges (Fig. 28) are made of polypropylene material and must be flexed or bent before the plastic is cooled or permanently set.   Live hinges have the advantage of being complete without secondary operations.

Coined-type hinge   Coined-type hinges (Fig. 29) are cold-formed and limited to certain materials.

Flexing motion   A few materials can be used to accomplish oscillating and flexing motions (Fig. 30).

Ejector Pins and Vents

All types of molds must have vents (Fig. 31) and need ejector pins to remove parts from the mold. In most cases the ejector pins are altered (by making flat areas along the pins' surface) to facilitate venting.

Thermoset venting In thermoset compression or transfer molding, venting is necessary because both air (originally trapped within the preformed material) and

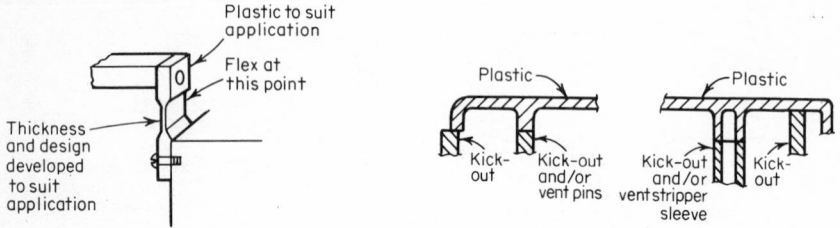

Fig. 30   Flexing member.   Fig. 31   Design for ejector-pin venting.

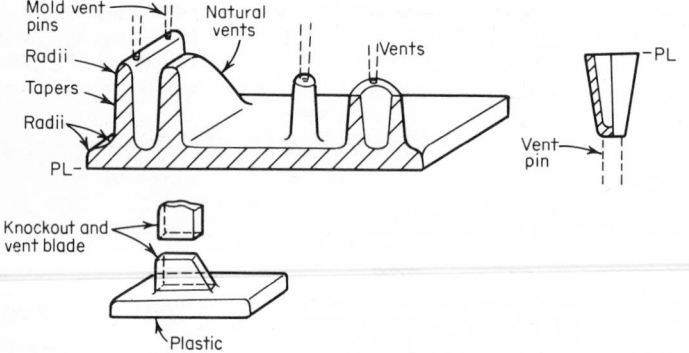

Fig. 32   Ejector pins and vents.

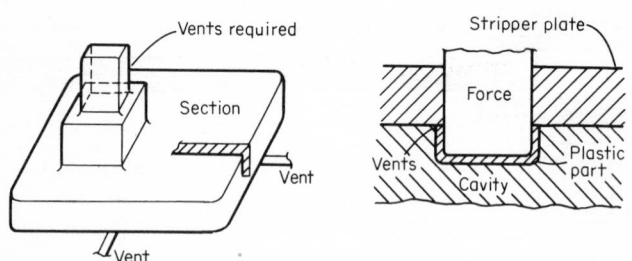

Fig. 33   Parting-line vents.

gas (created by preheating the material) will produce defective (porous areas) molded parts.

Thermoplastic venting In injection molding of thermoplastics, venting (Fig. 32) releases gas, which if trapped, would appear as a burned spot at the edges and a hole near center sections.

Sharp corners, vertical surfaces, and improper venting (Fig. 33) impede the flow of plastic materials during the molding cycle.

Holes and Threads

Molded holes The location of holes should be designed to allow proper flow of plastic and minimize weld lines. The depth of molded blind holes should not exceed 2.5 times their diameter. Molded through holes (Fig. 34) in plastic require a butt pin or a through pin which is anchored at both ends in the mold steel.

Drilled holes The drilling of most thermoset materials requires high-speed or carbide drills run at high speeds. Thermoplastic materials are drilled at slower speeds, as the material will melt from the frictional heat. Holes impractical to mold must be drilled. Holes must not be close to edges because chips and cracks will result. Since a long hole (Fig. 35) with a small diameter is difficult to drill along the intended direction, the most practical approach in many cases is to mold partway and drill the remainder of the distance. Center distances (Fig. 36) between mounting holes should, if possible, be designed to allow for the variations due to plastic shrinkage. Elongation of one hole is helpful.

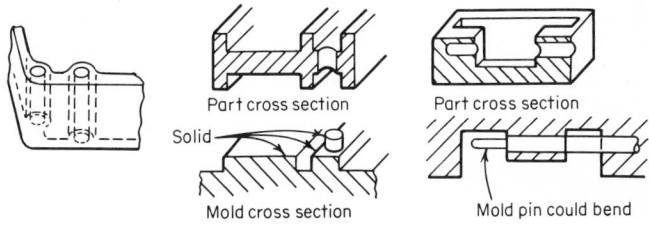

Part cross section

Part cross section

Solid

Mold cross section

Mold pin could bend

Fig. 34 Molded holes.

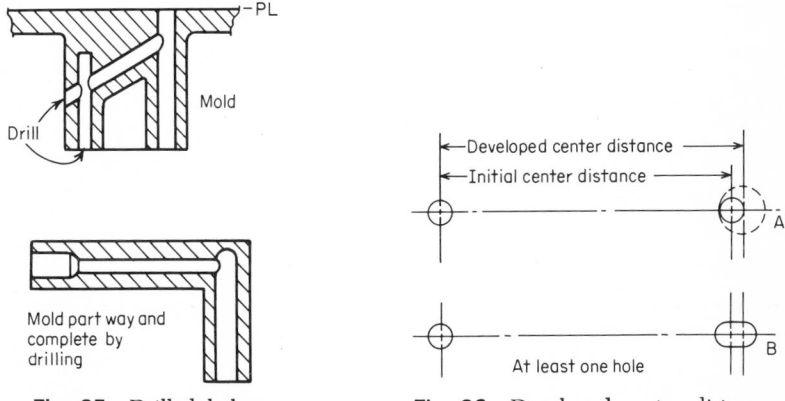

PL

Mold

Drill

Mold part way and complete by drilling

Fig. 35 Drilled holes.

Developed center distance

Initial center distance

A

At least one hole

B

Fig. 36 Developed center distance.

Threaded holes Threaded holes should be countersunk at least one thread to prevent brittle material from chipping. In cases where automatic unscrewing of a part from the mold is desired, the design should prevent the molded part from revolving during the unscrewing operation. Threads can be both molded or cut; however, the material selected and thread size will have a bearing on the thread strength (as shown in Fig. 37).

Warpage

At times a product will have molded-in strains which will warp (Fig. 38) the part, and the molder will try to compensate by using a cooling fixture to hold the piece while it is cooling. This procedure may not always succeed as it may overbend the part and at a later date the part will again distort because of the memory of the particular plastic.   Warpage allowance and direction (Fig. 39) should be indicated on the design drawing so that cooling fixtures, grade of materials, and molding techniques can be determined in order to achieve the best results in the finished product.   In long box-type shapes, add ribs if possible to maintain flat surfaces with the correct angles for adjacent walls.

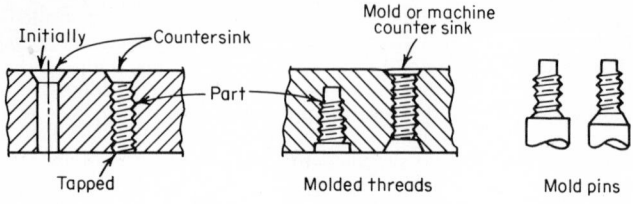

Fig. 37   Molded threads.

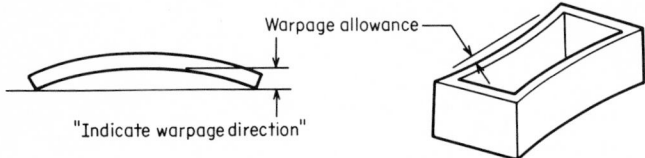

Fig. 38   Warpage allowances.

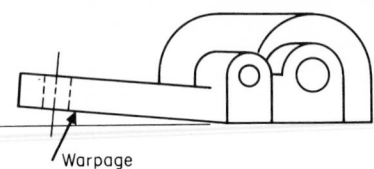

Fig. 39   Product warpage.

Sinks

Thicknesses of adjacent walls and ribs (Fig. 40) in thermoplastic parts should be approximately 60 percent of the main body to reduce the possibility of sink marks. A method for illustrating sinks (Fig. 41) is to roll a ball down the wall thickness.

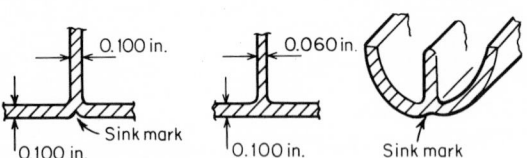

Fig. 40   Sink marks.

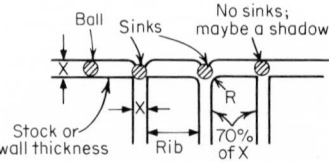

Fig. 41   The path of the ball indicates the possible location and depth of sink marks.

If the ball does not follow the outer wall contour, the part design will result in sink marks in the molded product.

Lettering and Etching

Molded characters Engraving or stamping makes possible the molding of raised monograms, lettering, etc. (Fig. 42), in plastic parts. Depressed backgrounds or lettering in molded parts require the construction of a removable engraved pad or the removal of background steel from the mold. Lettering, etc., should consist of smooth rounded characters to prevent the plastic material from sticking in the

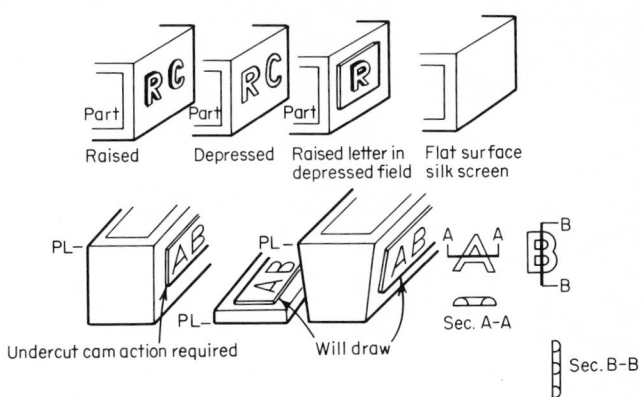

Fig. 42 Characters on cam splits.

mold. Also characters should be positioned in the direction of the mold draw. Wood-grain finishes, stipples, etchings, and other mold finishes (Fig. 43) should be 0.001 to 0.004 in. deep. On conventionally molded parts a 4 to 6° taper is necessary. Characters forming an undercut on walls, which are parallel to the mold draw, require cam action.

Postmolding decorating Silk screening, hot stamping, spray painting, and offset printing can be used to decorate flat surfaces.

Three-dimensional effects (Fig. 44) can be realized by molding characters and/or backgrounds on different levels within transparent or translucent plastic and then

Fig. 43 Wood-grain finish.

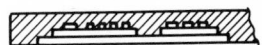

Fig. 44 Three-dimensional effects.

hot stamping, silk screening, and/or filling the characters with paint which is different in color from the background.

Inlay decorating For thermoset materials, inlay decorations (impregnated with an appropriate resin) are applied to plastic parts (dishware, etc.) during the molding cycle.

In use of thermoplastics, inlay decorations (printed on or impregnated with an appropriate resin) are held in the mold by a static charge prior to injecting plastic material into the mold.

Tolerances

Variables to be considered when stating tolerances are: (1) moldmaker's tolerances, (2) plastic-material shrinkage variances (including after shrinkages), and (3) molding-process or technique adjustments.

Variations produced by molding process   In compression molding, parting-line flash and buildup dimensions have the greatest influence upon the final dimensions (Fig. 45).

In transfer molding, the method of gating and direction of material flow will vary the ultimate dimensions differently depending upon the direction considered.

For thermoplastics, the temperature, pressure, and other variables will affect the dimensions of the plastic product.

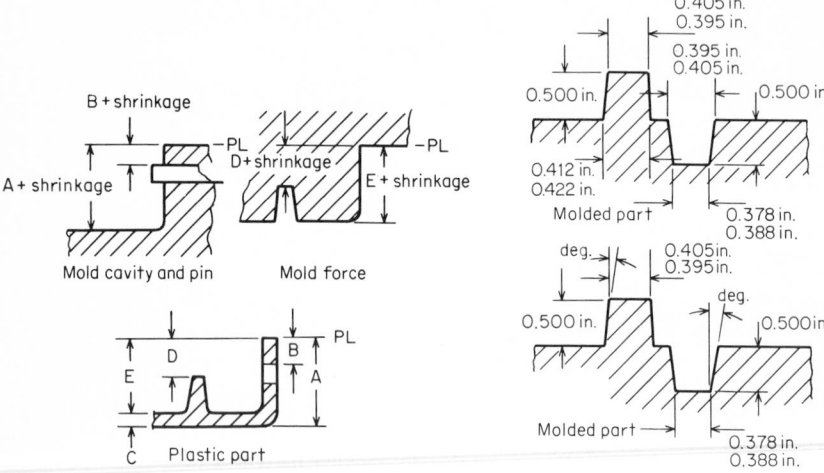

Fig. 45 Dimensions and tolerances over parting lines.   Dimensions *A*, *B*, and *C* will be increased by an amount equal to the parting line flash.

Fig. 46 Dimensioning tapers.

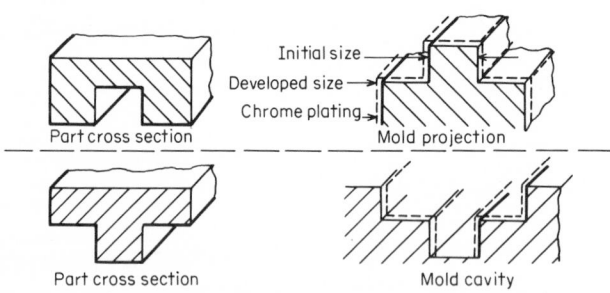

Fig. 47 Mold adjustments to meet tolerances.

Identification of critical specifications   General indication of the overall amount of allowable tolerance or permissible taper is insufficient.   Each critical dimension should be labeled individually (Fig. 46).

Mold adjustments to meet tolerances   Critical dimensions with tight tolerances may be compensated for in mold design (Fig. 47) by chrome plating or electrical-discharge machining (EDM) to develop correct dimensions.

Material shrinkage   Mold shrinkage is the difference in dimension on the drawing of the product and drawing of the mold.   Normally the shrinkage of most materials occurs up to the time the part is cooled, but in some materials the shrinkage (referred to as "after shrinkage") continues long after the part has cooled.

Inserts

Insert molded into plastic   Inserts should be made of soft metals such as brass or aluminum. Harder metals such as steel could accidentally topple over in the mold and cause expensive damage. Inserts with grooves and knurls (Fig. 48) will withstand considerable torque, but square or hexagon-shaped inserts will withstand greater torque. All hollow inserts (Fig. 48) should have a $\frac{1}{32}$- by $\frac{1}{32}$-in. recess which is designed to fit the mold to prevent plastic from entering the insert. If possible, hollow inserts which have an end molded into the plastic should be designed with a blind hole to prevent plastic from flowing into the insert.

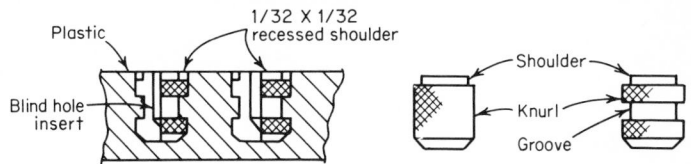

Fig. 48   Molded-in inserts.

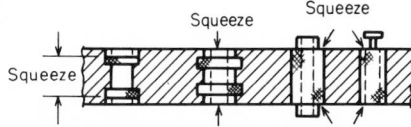

Fig. 49   Through inserts.

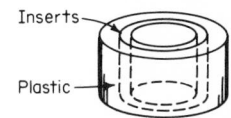

Fig. 50   Plastic around large insert.

Hollow inserts (Fig. 49) which go all the way through a plastic part should be 0.002 to 0.004 in. longer than necessary to permit squeezing by the closed mold. In this way, plastic is prevented from entering into the insert or flowing onto areas of the insert which must be kept clean.

When molding plastic around large metal inserts (Fig. 50), the plastic wall thickness must be sufficient to withstand cracking due to the shrinkage of plastic around insert.

Plastic which is molded inside a large metal insert (Fig. 51) will shrink away from the metal unless anchored by metal projections from the inserts.

Plastic will shrink along the length of a long insert (Fig. 52), exposing the insert's end.

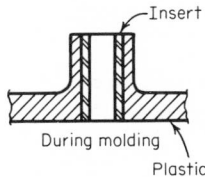

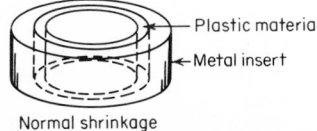

Fig. 51   Plastic inside large insert.

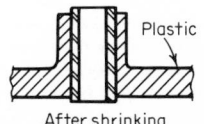

Fig. 52   Shrinkage along an insert.

Irregular-shaped inserts (Fig. 53) which must fit into a mold member and which protrude from the plastic part will necessitate costly deflashing operations.

Blistering will occur in thermoset material where thin walls of plastic (Fig. 54) are adjacent to the metal insert. Thermoplastic materials may not flow into some areas (creating pockets of trapped gas) when inserts are covered by thin walls (Fig. 54) of plastic.

Inserts which are assembled or molded (Fig. 55) must be designed to avoid jacking out of the insert from the plastic.

Insert assembled into plastic Inserts with accurate dimensions can be pushed into a molded hole while the plastic is still hot. The shrinkage of the plastic will grip

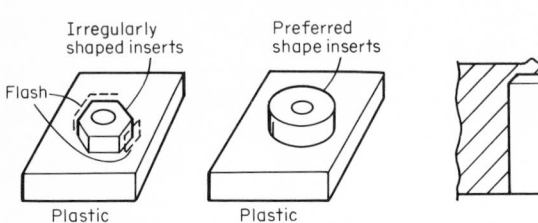

Fig. 53 Hex insert protruding from plastic.

Fig. 54 Thin plastic wall over inserts.

the metal insert. Inserts can be pushed into thermoplastic material (with or without a hole) while applying ultrasonic vibrations. Inserts can be assembled into place with an appropriate solvent or cement.

Special-design inserts (Fig. 56) with fishhook-type projections can be pushed into plastic with a hole whose dimensions are a few thousandths of an inch smaller than the insert.

Many threaded expandable-type inserts (Fig. 57) are designed to push into a molded or drilled hole. When the insert (Fig. 58) expands, it grips into the plastic.

Fastener product design In expansion-control insert (Fig. 59), as downward

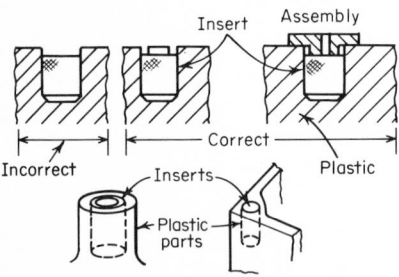

Fig. 55 Anchorage of inserts.

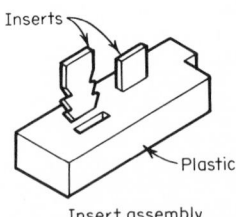

Fig. 56 Sawtooth-type insert.

pressure is applied, the diamond knurl section breaks off, slides over the cone section, and expands into the hole wall. The amount of expansion is easily controlled by using slight variations in hole depth. The expansion-control feature provides greater flexibility in meeting individual requirements for pull-out strength in a wide variety of materials. Screw sizes: Nos. 2–56 to ¼–20. Pull-out strength for a No. 6–32 in phenolic will be approximately 250 lb, with similar proportions for others.

For push-insert (Fig. 60), as downward pressure is applied, the straight knurl becomes firmly locked in the hole wall to prevent rotation. As the knurled spreader section contacts the bottom of the hole, it expands and becomes permanently anchored in the hole wall to prevent pull-out. Screw sizes: Nos. 2–56 to 10–32. Pull-out strength in phenolic will be approximately 175 lb, with similar proportions for other sizes. Torque strength is greater than the breaking point of the mating screw.

By the use of exclusive tapered threads (Fig. 61), installation of the mating screw expands and firmly anchors the diamond knurl into the hole wall to prevent rotation and pull-out. The straight knurl prevents the insert from turning during initial

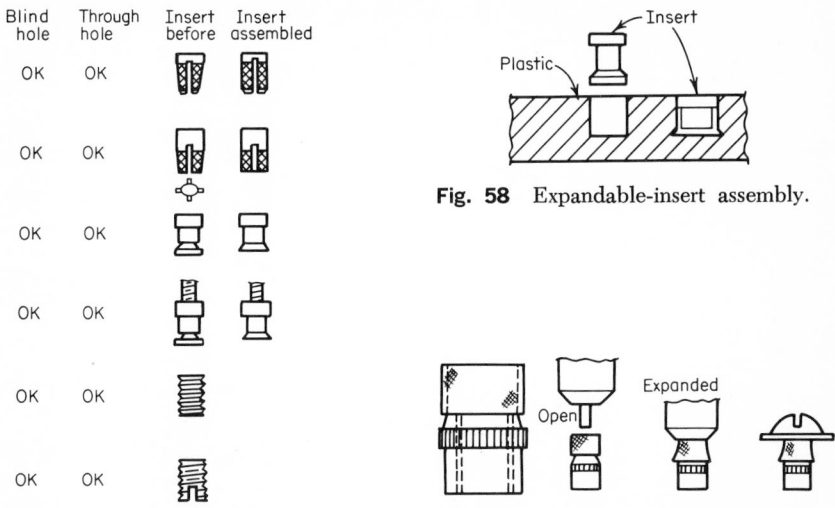

Blind hole	Through hole	Insert before	Insert assembled
OK	OK		
OK	OK		
OK	OK		
OK	OK		
OK	OK		
OK	OK		

Fig. 57 Expandable inserts.

Fig. 58 Expandable-insert assembly.

Fig. 59 Expansion-control insert. (*Fastener Products, Inc.*)

screw installation. It also holds the insert flush with the base-material surface prior to expansion of the diamond knurl. By changing the hole diameter, required torque strength and the amount of screw-locking action are controlled. To ensure maximum expansion, the mating screw should be long enough to protrude to the bottom of the insert or beyond if possible. Screw sizes: Nos. 2-56 to ¼-20. Pull-out strength for a No. 6–32 in phenolic will be 250 lb, with similar proportions for other sizes. Torque strength will be in excess of the breaking point of the mating screw.

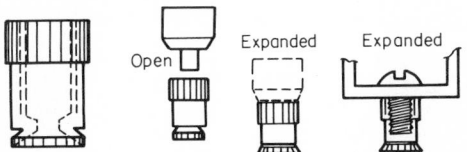

Fig. 60 Push-insert. (*Fastener Products, Inc.*)

The Fin-Set self-tapping insert* (Fig. 62) is installed quickly and easily by tapping its own threads as it is driven into a drilled or cored hole, either blind or through holes. It becomes permanently anchored in base material to resist high pull-out forces or backing-out due to prolonged vibration. Screw sizes: Nos. 2–56 to ¼-20. For installation of the Fin-Set self-tapping insert (Fig. 63) the breakaway drive tool should be used with a drill press and reversible tapping head. An electric or air gun may also be used as long as care is taken concerning proper alignment.

Dodge expansion inserts† Expansion inserts, designed primarily for use in plastics, are installed quickly and easily after molding.

* Trademark, Fastener Products, Inc., Southport, Conn.
† Trademark, Heli-Coil Products, Danbury, Conn.

The standard insert (Fig. 64) is used widely to provide strong brass threads in plastic parts of all types after molding. Highly versatile, standard inserts provide the answers for the majority to applications where long-life threads are required in plastic parts. Screw sizes: Nos. 4-40 to $\frac{5}{16}$-18.

The wedge insert (Fig. 65) is designed to put strong threads in soft plastics. The "expansion principle" forces the large wedge configurations of the insert body deeply into the wall of the drilled hole, permanently anchoring the insert and giving the installation much greater strength. Screw sizes: Nos. 6-32 to $\frac{1}{4}$-20.

The Flange Insert (Fig. 66) features a flange with a large surface for effective electrical contact. The flange may also be used to hold down mating parts. When

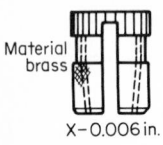
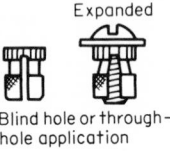
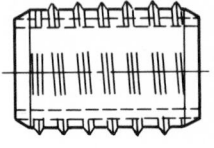

Fig. 61 Taper-thread insert. (*Fastener Products, Inc.*)

Fig. 62 Fin-Set self-tapping insert. (*Fastener Products, Inc.*)

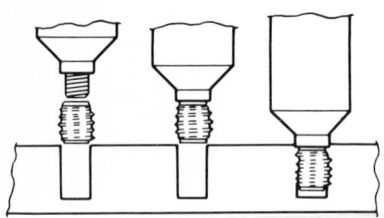

Fig. 63 Installation of Fin-Set self-tapping insert. (*Fastener Products, Inc.*)

Fig. 64 Standard insert. (*Heli-Coil Products.*)

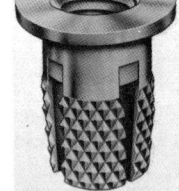

Fig. 65 Wedge insert. (*Heli-Coil Products.*)

Fig. 66 Flange insert. (*Heli-Coil Products.*)

used to attach separate parts permanently, the Flange Insert is placed through a hole in one part and into the cored hole in the parent material. This insert is especially well suited for thin-section applications, since the downward force of installation is absorbed by the flange. Screw sizes: Nos. 4-40 to $\frac{1}{4}$-20.

The Clinch Insert (Fig. 67) features a pilot and a flange. The pilot may be clinched over a terminal, eyelet fashion, while the flange offers a broad surface for good electrical contact. Since the downward force of installation is absorbed by the flange, the Clinch Insert may be used in relatively thin-section applications. Screw sizes: Nos. 4-40 to $\frac{5}{16}$-18.

* Trademark, Heli-Coil Products, Danbury, Conn.

*The Cone-Spread insert** (Fig. 68) is designed for use in miniaturized parts. Like the Dodge expansion inserts, the Cone-Spread insert is used to provide strong, reusable threads in all plastic parts after molding. Screw size: No. 0-3.

*The Ultrasert II** (Fig. 69) is installed into thermoplastic materials using equipment (usually ultrasonic) that develops frictional heat between the insert and the plastic, thus remelting a narrow zone around the insert which, when resolidified, provides the high-performance-strength values of a molded-in insert while still retaining all the economical advantages of installation after molding.

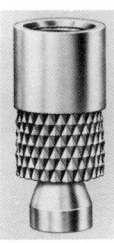

Fig. 67 Clinch insert. (*Heli-Coil Products.*)

Fig. 68 Cone-Spread insert. (*Heli-Coil Products.*)

Fig. 69 Ultrasert II. (*Heli-Coil Products.*)

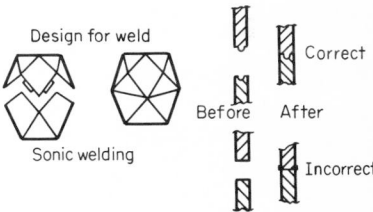

Fig. 70 Assembly by screws.

Fig. 71 Ultrasonic welding of joints.

ASSEMBLY AND MACHINING GUIDELINES

Assembly Methods

Conventional methods (Fig. 70) can be used along with ultrasonic welding (Fig. 71), spin welding (Fig. 72), hot-air welding (Fig. 73), solvent cementing (Fig. 74), and hot staking (Fig. 75). Ultrasonic welding requires specially designed points of contact. The ultrasonic vibrations travel through the plastic and release their energy as heat which melts and welds the plastic at the joint. Care must be taken in solvent cementing, as strains may be relieved, causing plastic to crack and craze.

* Trademark, Heli-Coil Products, Danbury, Conn.

In spin welding, the rotating plastic parts create frictional heat which welds the joint. Spin welding is limited to a few materials.

Flameless compressed air conveys concentrated heat to a welding rod of a compatible plastic material which is applied under pressure to the joint being welded.

Joint design for ultrasonic welding A tongue-and-groove joint (Fig. 76) usually has the capability of providing greatest strength. The need to maintain clearance on both sides of the tongue, however, makes this more difficult to mold. Draft

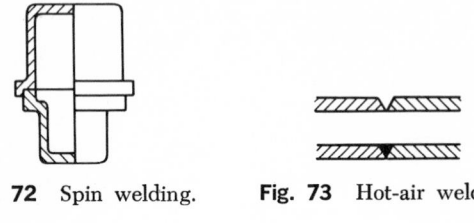

Fig. 72 Spin welding.   **Fig. 73** Hot-air welding.

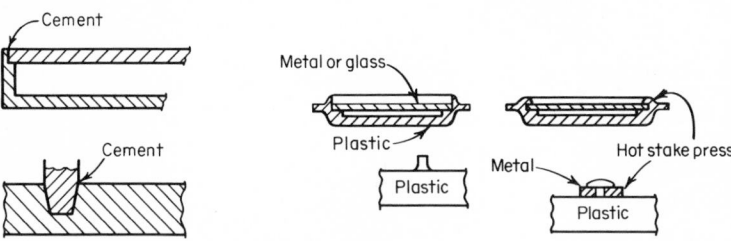

Fig. 74 Assembly by solvent cement.   **Fig. 75** Hot-stake assembly.

angles can be modified concurrently with good molding practices, but interference between elements must be avoided. Figure 77 illustrates basic joint variations suitable for ultrasonic welding. These are suggested guidelines for typical joint proportions. Specific applications may require slight modification. Practical considerations suggest a minimum height of 0.005 in. for the energy director. Where height greater than 0.020 in. is indicated, two or more directors should be provided with

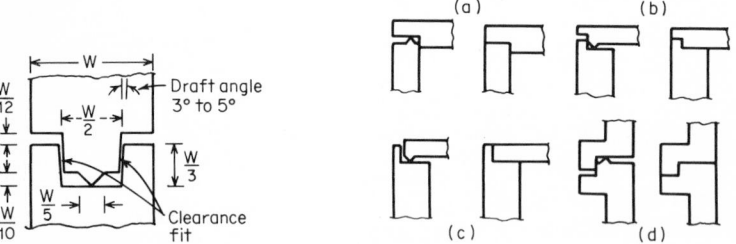

Fig. 76 Tongue-and-groove joint.
(*Branson Sonic Power Co.*)

Fig. 77 Basic joint variations suitable for ultrasonic welding. (*Branson Sonic Power Co.*)

the sum of heights equaling the formula dimension. Figure 78 shows the interference joint used when a hermetic joint is needed for the crystalline thermoplastics (nylon, acetal, polyethylene, polypropylene). Since crystalline resins (such as acetal and nylon) have a tendency to be watery in the molten state, the adjoining surfaces remain cool when the energy director has become molten, resulting in little or no interaction of melted and unmelted surfaces. When the interference joint is used, weld strength of crystalline materials approaches 95 percent of parent-material strength, as opposed to 40 to 70 percent when energy director is used. The interference joint permits interaction between the two surfaces during the entire melt cycle by exposing more and more surface area as the two surface planes interfere under ultrasonic and clamp force. Good fixturing should be used when the interference joint is utilized, as the outer walls of the part may flex or distort if not contained by the nest.

Ultrasonic inserting and staking For inserting (Fig. 79) a hole is premolded in the plastic slightly smaller in diameter than the insert to be received. Such a hole provides a certain degree of interference fit and guides the insert into place.

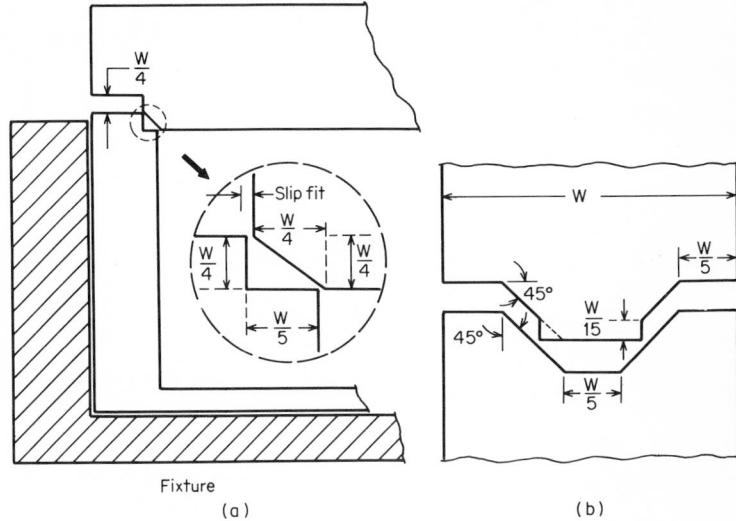

Fig. 78   Step-joint variation/tongue-and-groove variation. (*Branson Sonic Power Co.*)

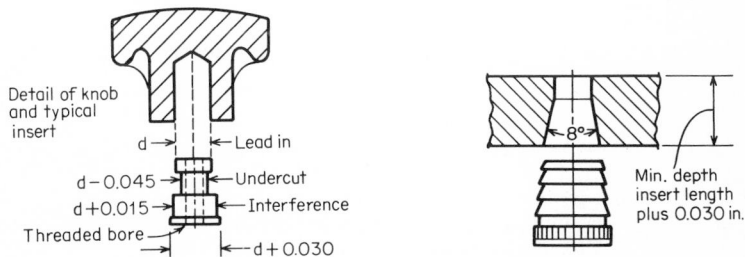

Fig. 79   Ultrasonic inserting. (*Branson Sonic Power Co.*)

Fig. 80   Tapered insert. (*Branson Sonic Power Co.*)

For a final interlocking assembly, the metal insert is usually knurled, undercut, or shaped to resist the loads imposed on the finished assembly. Ultrasonic vibrations travel through the driven part until they meet the joining area between the metal and plastic. At this joint (or interference) the energy of the ultrasonic vibrations is released as heat. The intensity of heat created by the vibration between the plastic and the metal is sufficient to melt the plastic momentarily, permitting the inserts to be driven into place. Ultrasonic exposure time is usually less than 1 s, but during this brief contact the plastic reforms itself around knurls, flutes, undercuts, or threads to encapsulate the insert. Tapered inserts (Fig. 80) show the relationship of dimensions between a typical undercut insert and the diameter of the premolded hole. For tapered inserts, the hole should also be tapered as shown. This permits accurate positioning of the insert and reduces installation time. The choice of contacting the plastic or metal surface with the horn will depend upon the configuration of the part, and the ability of the assembly to accept the required vibratory intensity. As with ultrasonic welding and inserting, ultrasonic staking (Fig. 81) employs the same principle of creating localized heat through the application of high-frequency vibrations. Most staking applications involve the assembly of metal and plastic. In staking, a hole in the metal receives a plastic stud. Ultrasonic staking requires the release of vibratory energy only at the surface of the plastic stud; therefore, the contact area between horn and plastic must be kept as small

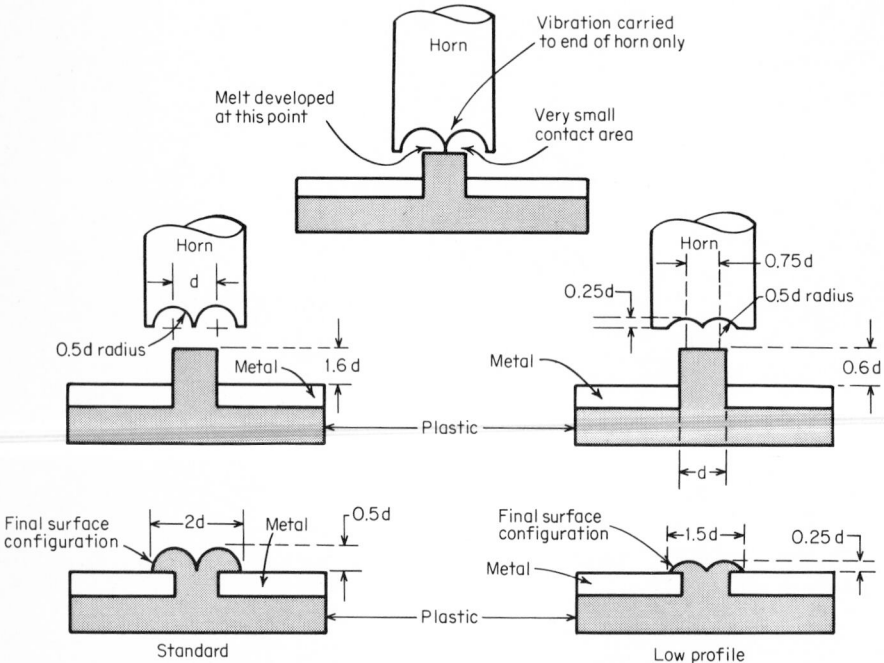

Fig. 81   Ultrasonic staking variations.   (*Branson Sonic Power Co.*)

as possible. The horn is usually contoured to meet the specific requirements of the application. With the introduction of ultrasonic vibrations, the stud melts and reforms to create a locking head over the metal.

There are two head forms that will satisfy the requirements of a majority of applications (Fig. 81). The first, generally considered standard, produces a head having twice the diameter of the original stud, with a height one-half the stud diameter. The second, referred to as a low-profile head, has a head diameter 1½ times the stud diameter, with a head height one-fourth the size of the head diameter.

Unlike ultrasonic plastic welding, staking requires that out-of-phase vibrations be generated between horn and plastic surfaces. Light initial contact pressure is therefore a requirement for out-of-phase vibratory activity within the limited contact area, as shown in Fig. 77. It is the progressive melting of plastic under continuous but light pressure that forms the head. Adjustment of flow-control valve and trigger switch may be required to reduce pressure to the desired level.

Techniques for Machining and Secondary Operations

Most plastics can be machined with accuracy if the proper tools and cutting conditions are applied. Tool geometry, cutting speeds, and coolants are critical in machining plastics. Machining operations will remove the luster from most molded surfaces.

Mechanical finishing of thermoplastics Most thermoplastics have a low heat-distortion point, thus requiring appropriate cutting speeds and feeds to prevent damage from frictional heat. If more accuracy and longer life are to be attained, carbide tools should be used. If a mirrorlike finish is expected, the cutters should be diamond-tipped. Diamond tools must be fed uniformly along the traverse of the cut. Sharp tools carefully ground with minimum cutting edge should be used. Removal of the molded surface layer during machining operations will affect the physical properties of the plastic material. Adding water-soluble agents to the air jet improves the cooling system but requires a subsequent cleaning operation. Coolant should be selected with care, as some liquids may craze, crack, or dissolve the plastic. Glass-filled material presents the problem of abrasiveness, which in turn reduces the cutter life.

The spiral plastic cuttings, which leave the cutting edge, create problems of entanglement. These problems usually can be resolved by air jets and vacuum attachments directing the chips away from the cutter. However, the type of thermoplastic to be machined will vary the techniques used.

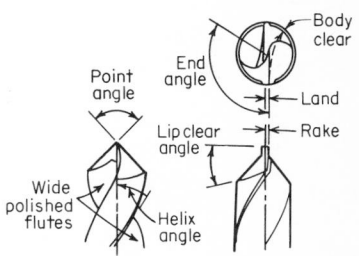

Fig. 82 Characteristics of a drill.

Drills for thermoplastics Drills which are not made of high-speed steel or carbide should have carbide or diamond tips. Also drills should have highly polished flutes and chrome-plated or nitrided surfaces. The drill design should have the land with conventional or narrower dimensions ($\frac{1}{16}$), the spiral with regular or higher helix angle (22 to 40°), the rake with positive angle (0 to +5°), the point with conventional or sharper angle (small drills 60 to 90° and large drills 90 to 118°), the end angle with conventional values (120 to 135°), and the lip clearance angle with conventional values (12 to 18°). (See Fig. 82.) Refer to Table 1 for consecutive listing of drill sizes.

Drills are fed at the rate of about 0.005 in. per revolution. Cutting speed will vary with type of plastic. Slower speeds will prevent overheating. Drills specifically designed for thermoplastics are available with high spiral helix angles (34 to 45°).

Suggested drill speeds:

Small to 0.125 in. diam	5,000 r/min
0.126–0.177 in. diam	3,000 r/min
0.180–0.228 in. diam	2,500 r/min
0.232–0.312 in. diam	1,700 r/min
0.315–0.375 in. diam	1,300 r/min
0.377–0.500 in. diam	1,000 r/min

For larger holes, use fly cutter at high speeds. The type of plastic to be drilled will vary the technique.

Taps for thermoplastics For tapping thermoplastics, taps with a slightly negative rake and with two or three flutes are preferred. Some plastics, such as nylon, may require slightly oversized taps because of the resiliency of the plastic after the tap is removed.

Solid carbide taps and standard taps of high-speed steel with flash-chrome-plate or nitrided surfaces are necessary. During tapping, the tap should be backed out of the hole periodically to clear the threads of chips. Refer to the formula for tapping thermoset material given in the following section on Tapping and Threading Thermosets. Thermoplastics may require one size larger tap drill for a given tap. Refer to Table 2.

TABLE 1   Consecutive Listing of Regular Drill Sizes*

Catalog designation	Decimal equivalents	Catalog designation	Decimal equivalents	Catalog designation	Decimal equivalents
No. 80	0.0135	1/16	0.0625	3.30 mm	0.1299
0.35 mm	0.0138	1.60 mm	0.0630	3.40 mm	0.1339
No. 79	0.0145	No. 52	0.0653	No. 29	0.1360
1/64	0.0156	1.65 mm	0.0650	3.50 mm	0.1378
0.40 mm	0.0158	1.70 mm	0.0669	No. 28	0.1405
No. 78	0.0160	No. 51	0.0670	9/64	0.1406
0.45 mm	0.0177	1.75 mm	0.0689	3.60 mm	0.1417
No. 77	0.0180	No. 50	0.0700	No. 27	0.1440
0.50 mm	0.0197	1.80 mm	0.0709	3.70 mm	0.1457
No. 76	0.0200	1.85 mm	0.0728	No. 26	0.1470
No. 75	0.0210	No. 49	0.0730	3.75 mm	0.1476
0.55 mm	0.0217	1.90 mm	0.0748	No. 25	0.1495
No. 74	0.0225	No. 48	0.0760	3.80 mm	0.1496
0.60 mm	0.0236	1.95 mm	0.0768	No. 24	0.1520
No. 73	0.0240	5/64	0.0781	3.90 mm	0.1535
No. 72	0.0250	No. 47	0.0785	No. 23	0.1540
0.65 mm	0.0256	2.00 mm	0.0787	5/32	0.1562
No. 71	0.0260	2.05 mm	0.0807	No. 22	0.1570
0.70 mm	0.0276	No. 46	0.0810	4.00 mm	0.1575
No. 70	0.0280	No. 45	0.0820	No. 21	0.1590
No. 69	0.0292	2.10 mm	0.0827	No. 20	0.1610
0.75 mm	0.0295	2.15 mm	0.0846	4.10 mm	0.1614
No. 68	0.0310	No. 44	0.0860	4.20 mm	0.1654
1/32	0.0312	2.20 mm	0.0866	No. 19	0.1660
0.80 mm	0.0315	2.25 mm	0.0886	4.25 mm	0.1673
No. 67	0.0320	No. 43	0.0890	4.30 mm	0.1693
No. 66	0.0330	2.30 mm	0.0906	No. 18	0.1695
0.85 mm	0.0335	2.35 mm	0.0925	11/64	0.1719
No. 65	0.0350	No. 42	0.0935	No. 17	0.1730
0.90 mm	0.0354	3/32	0.0938	4.40 mm	0.1732
No. 64	0.0360	2.40 mm	0.0945	No. 16	0.1770
No. 63	0.0370	No. 41	0.0960	4.50 mm	0.1772
0.95 mm	0.0374	2.45 mm	0.0965	No. 15	0.1800
No. 62	0.0380	No. 40	0.0980	4.60 mm	0.1811
No. 61	0.0390	2.50 mm	0.0984	No. 14	0.1820
1.00 mm	0.0394	No. 39	0.0995	No. 13	0.1850
No. 60	0.0400	No. 38	0.1015	4.70 mm	0.1850
No. 59	0.0410	2.60 mm	0.1024	4.75 mm	0.1870
1.05 mm	0.0413	No. 37	0.1040	3/16	0.1875
No. 58	0.0420	2.70 mm	0.1063	4.80 mm	0.1890
No. 57	0.0430	No. 36	0.1065	No. 12	0.1890
1.10 mm	0.0433	2.75 mm	0.1083	No. 11	0.1910
1.15 mm	0.0453	7/64	0.1094	4.90 mm	0.1929
No. 56	0.0465	No. 35	0.1100	No. 10	0.1935
3/64	0.0469	2.80 mm	0.1102	No. 9	0.1960
1.20 mm	0.0472	No. 34	0.1110	5.00 mm	0.1968
1.25 mm	0.0492	No. 33	0.1130	No. 8	0.1990
1.30 mm	0.0512	2.90 mm	0.1142	5.10 mm	0.2008
No. 55	0.0520	No. 32	0.1160	No. 7	0.2010
1.35 mm	0.0531	3.00 mm	0.1181	13/64	0.2031
No. 54	0.0550	No. 31	0.1200	No. 6	0.2040
1.40 mm	0.0551	3.10 mm	0.1220	5.20 mm	0.2047
1.45 mm	0.0571	1/8	0.1250	No. 5	0.2055
1.50 mm	0.0591	3.20 mm	0.1260	5.25 mm	0.2067
No. 53	0.0595	3.25 mm	0.1280	5.30 mm	0.2087
1.55 mm	0.0610	No. 30	0.1285	No. 4	0.2090

* Ace Drill Corporation, Adrian, Mich.

TABLE 1   Consecutive Listing of Regular Drill Sizes* (Continued)

Catalog designation	Decimal equivalents	Catalog designation	Decimal equivalents	Catalog designation	Decimal equivalents
5.40 mm	0.2126	O	0.3160	13.5 mm	0.5315
No. 3	0.2130	8.10 mm	0.3189	35/64	0.5469
5.50 mm	0.2165	8.20 mm	0.3228	14.0 mm	0.5512
7/32	0.2188	P	0.3230	9/16	0.5625
5.60 mm	0.2205	8.25 mm	0.3248	14.5 mm	0.5709
No. 2	0.2210	8.30 mm	0.3268	37/64	0.5781
5.70 mm	0.2244	21/64	0.3281	15.0 mm	0.5906
5.75 mm	0.2264	8.40 mm	0.3307	19/32	0.5938
No. 1	0.2280	Q	0.3320	39/64	0.6094
5.80 mm	0.2283	8.50 mm	0.3346	15.5 mm	0.6102
5.90 mm	0.2323	8.60 mm	0.3386	5/8	0.6250
A	0.2340	R	0.3390	16.0 mm	0.6299
15/64	0.2344	8.70 mm	0.3425	41/64	0.6406
6.00 mm	0.2362	11/32	0.3438	16.5 mm	0.6496
B	0.2380	8.75 mm	0.3445	21/32	0.6562
6.10 mm	0.2402	8.80 mm	0.3465	17.0 mm	0.6693
C	0.2420	S	0.3480	43/64	0.6719
6.20 mm	0.2441	8.90 mm	0.3504	11/16	0.6875
D	0.2460	9.00 mm	0.3543	17.5 mm	0.6890
6.25 mm	0.2461	T	0.3580	45/64	0.7031
6.30 mm	0.2480	9.10 mm	0.3583	18.0 mm	0.7087
1/4	0.2500	23/64	0.3594	23/32	0.7188
E	0.2500	92.0 mm	0.3622	18.5 mm	0.7283
6.40 mm	0.2520	9.25 mm	0.3642	47/64	0.7344
6.50 mm	0.2559	9.30 mm	0.3661	19.0 mm	0.7480
F	0.2570	U	0.3680	3/4	0.7500
6.60 mm	0.2598	9.40 mm	0.3701	49/64	0.7656
G	0.2610	9.50 mm	0.3740	19.5 mm	0.7677
6.70 mm	0.2638	3/8	0.3750	25/32	0.7812
17/64	0.2656	V	0.3770	20.0 mm	0.7874
6.75 mm	0.2657	9.60 mm	0.3780	51/64	0.7969
H	0.2660	9.70 mm	0.3819	20.5 mm	0.8071
6.80 mm	0.2677	9.75 mm	0.3839	13/16	0.8125
6.90 mm	0.2717	9.80 mm	0.3858	21.0 mm	0.8268
I	0.2720	W	0.3860	53/64	0.8281
7.00 mm	0.2756	9.90 mm	0.3898	27/32	0.8438
J	0.2770	25/64	0.3906	21.5 mm	0.8465
7.10 mm	0.2795	10.00 mm	0.3937	55/64	0.8594
K	0.2810	X	0.3970	22.0 mm	0.8661
9/32	0.2812	Y	0.4040	7/8	0.8750
7.20 mm	0.2835	13/32	0.4062	22.5 mm	0.8858
7.25 mm	0.2854	Z	0.4130	57/64	0.8906
7.30 mm	0.2874	10.50 mm	0.4134	23.0 mm	0.9055
L	0.2900	27/64	0.4219	29/32	0.9062
7.40 mm	0.2913	11.00 mm	0.4331	59/64	0.9219
M	0.2950	7/16	0.4375	23.5 mm	0.9252
7.50 mm	0.2953	11.50 mm	0.4528	15/16	0.9375
19/64	0.2969	29/64	0.4531	24.0 mm	0.9449
7.60 mm	0.2992	15/32	0.4688	61/64	0.9531
N	0.3020	12.00 mm	0.4724	24.5 mm	0.9646
7.70 mm	0.3031	31/64	0.4844	31/32	0.9688
7.75 mm	0.3051	12.50 mm	0.4921	25.0 mm	0.9843
7.80 mm	0.3071	1/2	0.5000	63/64	0.9844
7.90 mm	0.3110	13.0 mm	0.5118	1	1.0000
5/16	0.3125	33/64	0.5156		
8.00 mm	0.3150	17/32	0.5312		

* Ace Drill Corporation, Adrian, Mich.

Table 2 Tap Sizes*

To minimize tapping problems and lengthen tool life, use the largest drill possible to produce a minor diameter that will result in the lowest percentage of full thread consistent with adequate strength. A minor diameter that provides a 55 to 65% thread is sufficient for most requirements, but in some cases a higher percentage of thread may be necessary to conform with the minor-diameter limits of the thread class specified.

Drills generally cut holes larger than their diameters. In the table below, the probable percentages of full thread were determined by the average amount oversize the various drills are expected to cut. Reaming becomes necessary when closer control of the hole size is required.

Suggested Percentage of Full Thread in Tapped Holes

Material	Deep-hole tapping†	Average commercial work	Thin sheet stock or stampings
Free-cutting: Aluminum, brass, bronze, cast iron, copper, mild steel, tool steel..........	60–70%	65–70%	75–85%
Hard or tough: Cast steel, drop forging, monel metal, nickel steel, stainless steel........	55–65%	60–70%	75–85%

† Generally, deeper than 1½ times the hole diameter.

Tap drill sizes: Probable percentage of full thread produced in tapped hole using stock sizes of drill

Tap	Tap drill	Decimal equivalent of tap drill	Theoretical % of thread	Probable oversize (mean)	Probable hole size	% of thread
0–80	56	0.0465	83	0.0015	0.0480	74
	3/64	0.0469	81	0.0015	0.0484	71
1–64	54	0.0550	89	0.0015	0.0565	81
	53	0.0595	67	0.0015	0.0610	59
1–72	53	0.0595	75	0.0015	0.0610	67
	1/16	0.0625	58	0.0015	0.0640	50
2–56	51	0.0670	82	0.0017	0.0687	74
	50	0.0700	69	0.0017	0.0717	62
	49	0.0730	56	0.0017	0.0747	49
2–54	50	0.0700	79	0.0017	0.0717	70
	49	0.0730	64	0.0017	0.0747	56
3–48	48	0.0760	85	0.0019	0.0779	78
	5/64	0.0785	77	0.0019	0.0800	70
	47	0.0781	76	0.0019	0.0804	69
	46	0.0810	67	0.0019	0.0829	60
	45	0.0820	63	0.0019	0.0839	56
3–56	46	0.0810	78	0.0019	0.0829	69
	45	0.0820	73	0.0019	0.0839	65
	44	0.0860	56	0.0020	0.0880	47
4–40	44	0.0860	80	0.0020	0.0880	74
	43	0.0890	71	0.0020	0.0910	65
	42	0.0935	57	0.0020	0.0955	51
	3/32	0.0938	56	0.0020	0.0958	50
4–48	42	0.0935	68	0.0020	0.0955	61
	3/32	0.0938	68	0.0020	0.0958	60
	41	0.0960	59	0.0020	0.0980	52

Tap	Tap drill	Decimal equivalent of tap drill	Theoretical % of thread	Probable oversize (mean)	Probable hole size	% of thread
10–24	24	0.1520	70	0.0032	0.1552	64
	23	0.1540	67	0.0032	0.1572	61
	5/32	0.1563	62	0.0032	0.1595	56
	22	0.1570	61	0.0032	0.1602	55
10–32	5/32	0.1563	83	0.0032	0.1595	75
	22	0.1570	81	0.0032	0.1602	73
	21	0.1590	76	0.0032	0.1622	68
	20	0.1610	71	0.0032	0.1642	64
	19	0.1660	59	0.0032	0.1692	51
12–24	11/64	0.1719	82	0.0035	0.1754	75
	17	0.1730	79	0.0035	0.1765	73
	16	0.1770	72	0.0035	0.1805	66
	15	0.1800	67	0.0035	0.1835	60
	14	0.1820	63	0.0035	0.1855	56
12–28	16	0.1770	84	0.0035	0.1805	77
	15	0.1800	78	0.0035	0.1835	70
	14	0.1820	73	0.0035	0.1855	66
	13	0.1850	67	0.0035	0.1885	59
1/4–20	3/16	0.1875	61	0.0035	0.1910	54
	9	0.1960	83	0.0038	0.1998	77
	8	0.1990	79	0.0038	0.2028	73
	7	0.2010	75	0.0038	0.2048	70
	13/64	0.2031	72	0.0038	0.2069	66
	6	0.2040	71	0.0038	0.2078	65
	5	0.2055	69	0.0038	0.2093	63
	4	0.2090	63	0.0038	0.2128	57

Tap	Tap drill	Decimal equivalent of tap drill	Theoretical % of thread	Probable oversize (mean)	Probable hole size	% of thread
1/2–13	27/64	0.4219	78	0.0047	0.4266	73
1/2–20	7/16	0.4375	63	0.0047	0.4422	58
	29/64	0.4531	72	0.0047	0.4578	65
9/16–12	15/32	0.4688	87	0.0048	0.4736	82
	31/64	0.4844	72	0.0048	0.4892	68
9/16–18	1/2	0.5000	87	0.0048	0.5048	80
	33/64	0.5156	65	0.0048	0.5204	58
5/8–11	17/32	0.5313	79	0.0049	0.5362	75
	35/64	0.5469	66	0.0049	0.5518	62
5/8–18	9/16	0.5625	87	0.0049	0.5674	80
	37/64	0.5781	65	0.0049	0.5831	58
3/4–10	41/64	0.6406	84	0.0050	0.6456	80
	21/32	0.6563	72	0.0050	0.6613	68
3/4–16	11/16	0.6875	77	0.0050	0.6925	71
7/8–9	49/64	0.7656	76	0.0052	0.7708	72
	25/32	0.7812	84	0.0052	0.7864	79
	51/64	0.7969	67	0.0052	0.8021	62
7/8–14	13/16	0.8125	77	0.0052	0.8177	61
1–8	55/64	0.8594	67	0.0059	0.8653	83
	7/8	0.8750	77	0.0059	0.8809	73
	57/64	0.8906	67	0.0059	0.8965	64
1–12	29/32	0.9063	77	0.0059	0.9122	54
	59/64	0.9219	87	0.0059	0.9279	81
1–14	15/16	0.9375	72	0.0060	0.9435	67
	59/64	0.9219	58	0.0060	0.9279	52
			84	0.0060		78

Tap Drill Sizes

Tap	Drill	Dec. equiv.	% full thread	Diff.	Dec. equiv.	% full thread
5–40	40	0.0980	83	0.0023	0.1003	76
	39	0.0995	79	0.0023	0.1018	71
	38	0.1015	72	0.0023	0.1038	65
	37	0.1040	65	0.0023	0.1063	58
5–44	38	0.1015	79	0.0023	0.1038	72
	37	0.1040	71	0.0023	0.1063	63
	36	0.1065	63	0.0023	0.1088	55
6–32	37	0.1040	84	0.0023	0.1063	78
	36	0.1065	78	0.0023	0.1088	72
	7/64	0.1094	70	0.0026	0.1120	64
	35	0.1100	69	0.0026	0.1126	63
	34	0.1110	67	0.0026	0.1136	60
	33	0.1130	62	0.0026	0.1156	55
6–40	34	0.1110	83	0.0026	0.1136	75
	33	0.1130	77	0.0026	0.1156	69
	32	0.1160	68	0.0026	0.1186	60
8–32	29	0.1360	69	0.0029	0.1389	62
	28	0.1405	58	0.0029	0.1434	51
8–36	29	0.1360	78	0.0029	0.1389	70
	28	0.1405	68	0.0029	0.1434	57
10–24	9/64	0.1406	68	0.0029	0.1435	57
	27	0.1440	85	0.0032	0.1472	79
	26	0.1470	79	0.0032	0.1502	74
	25	0.1495	75	0.0032	0.1526	69

Tap	Drill	Dec. equiv.	% full thread	Diff.	Dec. equiv.	% full thread
1/4–28	3	0.2130	80	0.0038	0.2168	72
	7/32	0.2188	67	0.0038	0.2226	59
	2	0.2210	63	0.0038	0.2248	55
5/16–18	F	0.2570	77	0.0038	0.2608	72
	G	0.2610	71	0.0041	0.2651	66
	17/64	0.2656	65	0.0041	0.2697	59
	H	0.2660	64	0.0041	0.2701	59
5/16–24	H	0.2660	86	0.0041	0.2701	78
	I	0.2720	75	0.0041	0.2761	67
	J	0.2770	66	0.0041	0.2811	58
3/8–16	5/16	0.3125	77	0.0044	0.3169	72
	O	0.3160	73	0.0044	0.3204	68
	P	0.3230	64	0.0044	0.3274	59
3/8–24	21/64	0.3281	87	0.0044	0.3325	79
	Q	0.3320	79	0.0044	0.3364	71
	R	0.3390	67	0.0044	0.3434	58
7/16–14	T	0.3580	86	0.0046	0.3626	81
	23/64	0.3594	84	0.0046	0.3640	79
	U	0.3680	75	0.0046	0.3726	70
	3/8	0.3750	67	0.0046	0.3796	62
	V	0.3770	65	0.0046	0.3816	60
7/16–20	W	0.3860	79	0.0046	0.3906	72
	25/64	0.3906	72	0.0046	0.3952	65
	X	0.3970	62	0.0046	0.4016	55

Tap	Drill	Dec. equiv.	% full thread	Diff.	Dec. equiv.	% full thread
1 1/8–7	15/16	0.9375	67	0.0060	0.9435	61
	31/32	0.9688	84	0.0062	0.9750	81
	63/64	0.9844	76	0.0067	0.9911	72
	1	1.0000	67	0.0070	1.0070	64
	1 1/64	1.0156	59	0.0070	1.0226	55
1 1/8–12	1 1/32	1.0313	87	0.0071	1.0384	80
	1 3/64	1.0469	72	0.0072	1.0541	66
1 1/4–7	1 3/32	1.0938	84		Reaming recommended	
	1 7/64	1.1094	76			
	1 1/8	1.1250	67			
1 1/4–12	1 5/32	1.1563	87			
	1 11/64	1.1719	72			
1 3/8–6	1 3/16	1.1875	87			
	1 13/64	1.2031	79			
	1 7/32	1.2188	72			
	1 15/64	1.2344	65			
1 3/8–12	1 9/32	1.2813	87			
	1 19/64	1.2969	72			
1 1/2–6	1 5/16	1.3125	87			
	1 21/64	1.3281	79			
	1 11/32	1.3438	72			
	1 23/64	1.3594	65			
1 1/2–12	1 13/32	1.4063	87			
	1 27/64	1.4219	72			

Formula: tap drill size

$$\text{Major diam of tap} - \frac{0.01299 \times \% \text{ of full thread}}{\text{No. of threads per inch}} = \text{drill size}$$

Example: Determine drill size for 2 in.—12N Tap, 70% Full Thread.

Basic major diam of tap = 2.0000 in.

$0.01299 \times 70 = 0.0993 \div 12 = 0.0758$ in.

Theoretical drill size = 1.9242 in.

Nearest standard drill size (1 59/64) = 1.9219 in.

Formula: percentage of full thread

$$\text{Threads per inch} \times \frac{\text{major diam of tap} - \text{drill diam}}{0.01299} = \% \text{ of full thread}$$

Example: Determine % of full thread for 2 in.—12N tap using 1.9219 in. drill.

$$\frac{2.000 - 1.9219 = 0.0781 \div 0.01299 = 6.012}{\text{Threads per inch} = 12}$$

Theoretical % of full thread = 72.144

Suggested Pipe Tap Drill Sizes

Tap size	1/16	1/8	1/4	3/8	1/2	3/4	1	1 1/4	1 1/2	2	2 1/2	3	3 1/2	4
Drill size:														
Taper pipe tap‡	D	R	7/16	37/64	45/64	59/64	1 5/32	1 1/2	1 47/64	2 7/32	2 5/8	3 1/4	3 3/4	4 1/4
Straight pipe tap§	1/4	11/32	7/16	37/64	23/32	59/64	1 5/32	1 1/2	1 3/4	2 7/32	2 21/32	3 1/4	3 3/4	4 1/4

‡ Sizes given permit direct tapping without reaming the hole, but only give a full thread for the first two or three threads.

§ For Dryseal Straight Pipe Threads suggested drill sizes are as shown, except 1/4" pipe, 0.444 drill size and 3/4" pipe, 1.955 drill size are recommended.

Table 2 Tap Sizes (Continued)

Tap Terminology

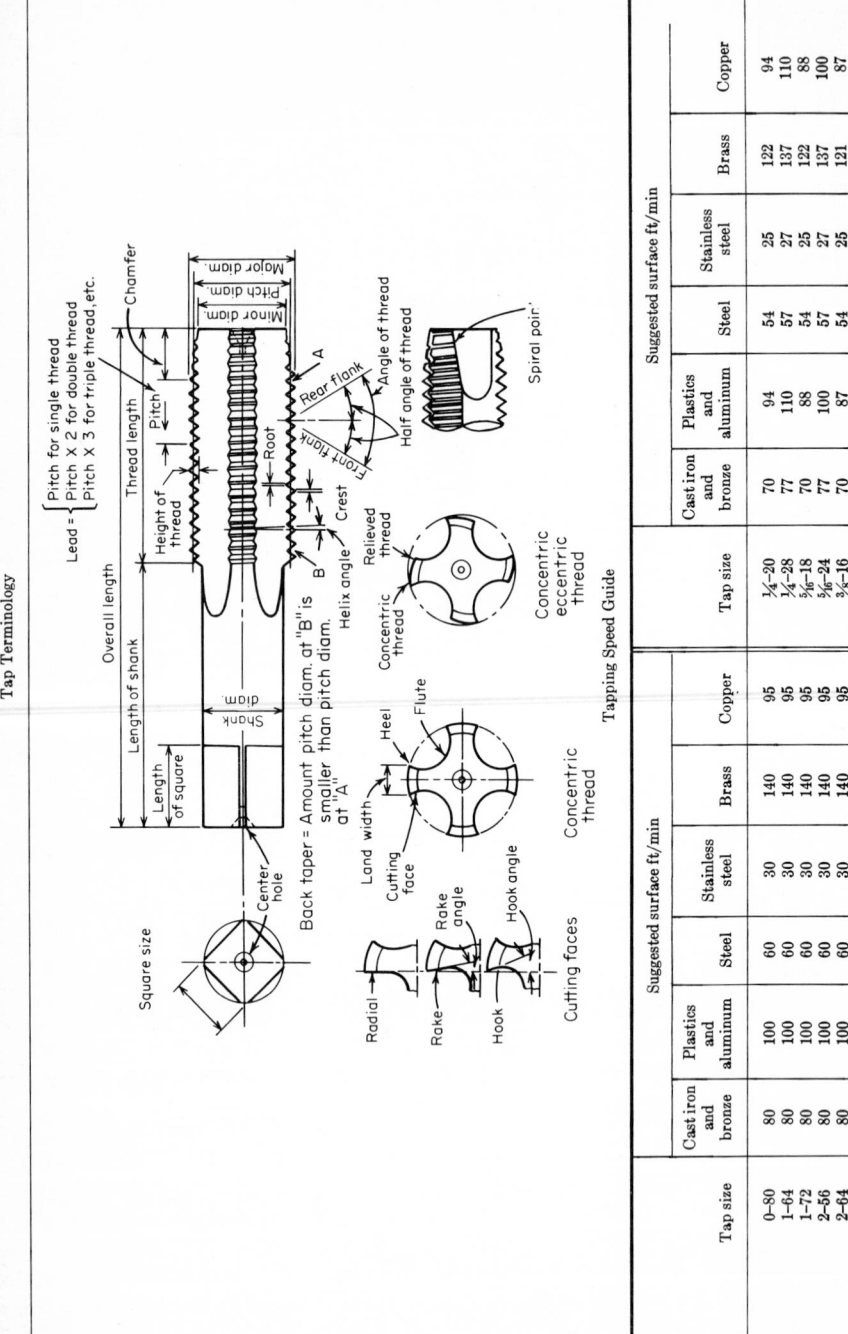

Tapping Speed Guide

Tap size	Suggested surface ft/min					
	Cast iron and bronze	Plastics and aluminum	Steel	Stainless steel	Brass	Copper
0–80	80	100	60	30	140	95
1–64	80	100	60	30	140	95
1–72	80	100	60	30	140	95
2–56	80	100	60	30	140	95
2–64	80	100	60	30	140	95

Tap size	Suggested surface ft/min					
	Cast iron and bronze	Plastics and aluminum	Steel	Stainless steel	Brass	Copper
¼–20	70	94	54	25	122	94
¼–28	77	110	57	27	137	110
5/16–18	70	88	54	25	122	88
5/16–24	77	100	57	27	137	100
3/8–16	70	87	54	25	121	87

Top chart

Tap size	20	25	30	40	50	60	Tap size	90	100	110	120	130	140
3-48	80	100	60	30	140	95	3/8-24	77	100	56	27	132	100
3-56	80	100	60	30	140	95	7/16-14	68	87	53	25	116	87
4-40	80	100	60	30	140	95	7/16-20	74	100	56	27	131	100
4-48	80	100	60	30	140	95	1/2-13	67	86	51	24	114	86
5-40	80	100	60	30	140	95	1/2-20	74	100	56	27	130	100
5-44	80	100	60	30	140	95	9/16-12	67	86	51	24	114	86
6-32	80	100	60	30	140	95	9/16-18	73	96	56	27	129	96
6-40	80	100	60	30	140	95	5/8-11	67	82	51	24	114	82
8-32	80	100	60	30	140	95	5/8-18	73	95	56	27	129	95
8-36	80	100	60	30	140	95	3/4-10	66	81	51	24	113	80
10-24	75	95	55	25	125	90	3/4-16	72	92	55	26	129	92
10-32	80	100	60	25	140	95	7/8-9	66	81	45	20	112	80
12-24	75	95	55	25	125	90	7/8-14	69	92	54	25	125	92
12-28	80	110	60	30	140	100	1 - 8	66	78	45	20	112	78
							1 -14	69	92	54	25	125	92

Conversion Table. Surface Feet per Minute to Revolutions per Minute

Surface ft/min	20	25	30	40	50	60	70	80	90	100	110	120	130	140	150
Tap size															
0	1273	1592	1910	2546	3183	3820	4456	5093	5730	6366	7003	7639	8276	8913	9549
1	1047	1308	1570	2093	2617	3140	3663	4186	4710	5233	5756	6279	6808	7326	7849
2	888	1110	1333	1777	2221	2665	3109	3554	3999	4442	4886	5330	5774	6218	6662
3	772	964	1157	1543	1929	2315	2701	3086	3472	3858	4244	4629	5015	5401	5787
4	682	853	1023	1364	1705	2046	2387	2728	3069	3411	3751	4092	4434	4775	5116
5	611	764	917	1222	1528	1833	2139	2445	2750	3056	3361	3667	3973	4278	4584
6	553	691	829	1106	1382	1658	1934	2211	2487	2764	3040	3316	3592	3869	4145
8	466	583	699	932	1165	1398	1631	1864	2097	2330	2563	2796	3029	3262	3495
10	402	502	603	804	1005	1205	1406	1607	1808	2009	2210	2411	2612	2813	3014
12	354	442	531	707	884	1061	1238	1415	1592	1769	1945	2122	2300	2476	2653
1/4	306	382	458	611	764	917	1070	1222	1375	1528	1681	1833	1986	2139	2292
5/16	245	306	367	489	611	733	856	978	1100	1222	1345	1467	1589	1711	1833
3/8	204	255	306	407	509	611	713	815	917	1019	1120	1222	1324	1426	1528
7/16	175	219	262	349	437	524	611	698	786	873	960	1048	1135	1222	1310
1/2	153	191	229	306	382	458	535	611	688	764	840	917	993	1070	1146
9/16	137	172	206	275	344	412	481	550	619	687	756	825	893	963	1031
5/8	122	153	183	244	306	367	428	489	550	611	672	733	794	856	917
3/4	102	128	153	203	255	306	357	407	458	509	560	611	662	713	764
7/8	87	109	131	175	218	262	306	350	392	437	480	524	568	611	655
1	76	96	115	153	191	230	268	306	344	382	420	468	497	535	573

* John Bath & Company, Inc.

Reaming for thermoplastics High-speed or carbide-steel machine reamers will ream accurately sized holes in thermoplastic. It is advisable to use a reamer 0.001 to 0.002 in. larger than the desired hole size to allow for the resiliency of plastic. Tolerances as close as ±0.0005 in. can be held in through holes ¼ in. in diameter. Fluted reamers are best for obtaining a good finished surface. Reamer speeds should approximate those used for drilling. The amount of material removed per cut will vary with the hardness of the plastic. Reaming can be done dry, but water-soluble coolants will produce better finishes.

Turning and milling Use tungsten carbide or diamond-tipped tools with negative back rake and front clearance. Milling cutters and end mills will remove undesirable material, such as flush gates, on a molded article. Refer to Table 3 for cutting-edge design. Turning speeds 200 to 700 sfpm (surface feet per minute) and feed rate of 0.02 to 0.1 ipr (inches per revolution) give the best results.

Mechanical finishing of thermosets The machining of thermoset plastics must consider the abrasive interaction with tools but does not involve the problems of melting from high-speed frictional heat. Although high-speed steel tools may be used, carbide and diamond tools will perform much better with a longer tool life. Higher cutting speeds improve machined finishes, but the high-speed wearing reduces the tool life. Since the machining of thermosets produces cuttings in powder form, vacuum hoses and air jets adequately remove the abrasive chips. To prevent grabbing, tools should have an O rake, which is similar to the rake of tools for machining brass. Adding a water-soluble coolant to the air jet will necessitate a secondary cleaning operation. The type of plastic will vary the machining technique.

Drilling thermosets Refer to Table 1 for consecutive listing of drill sizes. Drills which are not made of high-speed steel or solid carbide should have carbide or diamond tips. Also, drills should have highly polished flutes and chrome-plated or nitrided surfaces. The drill design should have the conventional land, the spiral with regular or slower helix angle (16 to 30°), the rake with positive angle (0 to +5°), the point angle conventional (90 to 118°), the end angle with conventional values (120°–135°), and the lip clearance angle with conventional values (12°–18°). See Fig. 82 for drill characteristics. Because of the abrasive material, drills should be slightly oversize, 0.001 to 0.002 in.

Suggested drill sizes and speeds:

Small to 0.093 in. diam	5,000 r/min
0.094–0.140 in. diam	3,000 r/min
0.141–0.191 in. diam	2,500 r/min
0.192–0.250 in. diam	1,700 r/min
0.252–0.312 in. diam	1,300 r/min
0.315–0.375 in. diam	1,000 r/min
0.377–0.500 in. diam	700 r/min

Tapping and threading thermosets The data in Table 2 on sizes of tap drills for plastics are based on experience and are specifically for thermosetting plastics.[*] The sizes given are the nearest available drill sizes. If one does not have the specified type, the nearest larger size should be used. Specification of percentage thread is based upon diameters in small sizes and upon pitch in larger sizes as indicated below.

The formula used is $D = T - n \times 2d$, in which D = drill diameter, T = outside diameter of thread or tap, d = depth of thread, n = percentage of thread depth desired (expressed as a decimal).

Percentages of depth used in preparing Table 2 were as follows: C indicates standard National Coarse Series (NCS) size; F indicates standard National Fine Series (NFS) size.

50 percent below No. 6
60 percent No. 6 through 14 } smaller sizes, based on diameters
70 percent No. 15 through 30

70 percent ¼ through ½ in. NCS
70 percent ¼ through 1 in. NFS } larger sizes, based on pitch
75 percent ⁹⁄₁₆ through 1 in. NCS

---

[*] For thermoplastics, use the same drill size or one size larger.

TABLE 3 Cutting-Tool Designs

Plastic	A	A-3	B	C	D	E	F	G	R1	R2	R3	Surface ft/min
Thermoplastic................	8°	8°	8°	0°	8°	20°	10°	20°	0.0–0.030	0.0–0.030	0.093	500–900
Thermoset..................	8°	8°	8°	0°	8°	20°	10°	20°	0.0–0.030	0.0–0.030	.....	1,400–1,800

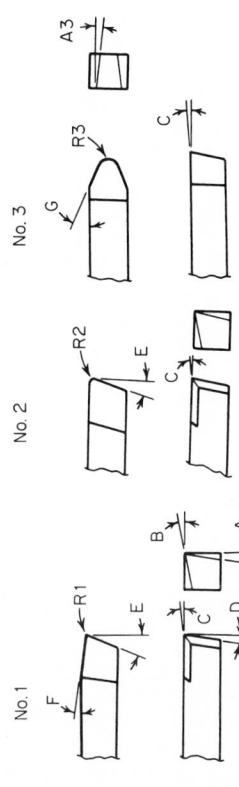

Cutting-tool geometry.   No. 1: conventional turning.   No. 2: turning and facing.   No. 3: facing.

Taps for thermosets   Solid carbide taps and standard taps of high-speed steel with flash-chrome-plate or nitrided surfaces are necessary. When tapping thermosets, taps should be oversize by 0.002 to 0.003 in. and have two or three flutes. Water-soluble lubricants and coolants are preferred. Refer to Table 2 for tap sizes.

Machining operations will remove the luster from molded samples. Turning and machining tools should be high-speed steel, carbide- or diamond-tipped. Refer to Table 3 for tool-bit design. Polishing, buffing, waxing, or oiling will return the luster to the machined part, where this is allowable.

PLASTIC MOLDING AND DECORATING

Thermoplastics

Thermoplastic materials may be modified or lubricated at times to suit a given molding condition, and many problems may be handled by a change in technique. Materials, in most cases, are dried in a circulating oven prior to placing in the press hopper, and in some cases infrared lights maintain heat in the hopper. Automatic hopper-heater dryers are also used (material is forced by air from the drum into the hopper and then fed into the press).

Brief description of thermoplastic press action and equipment   The mold is clamped between the press platens (this can be by direct hydraulic pressure or toggles hydraulically operated). The plunger or screw may be either hydraulic or electrically operated. Presses are specified by clamp pressure in tons and the maximum volume of material in ounces per shot displaced from the cylinder. Thus, the volume of the material and the projected area of the part being molded determine press size, provided that physical mold size fits within the press tie bars. Note that injection presses are used for thermoplastic, thermoset, and some rubber products by a change in cylinder and screw.

The injection press (Fig. 83) has a hydraulic clamping unit (Fig. 84) which is a low-pressure pump controlled by a low-pressure pilot head that takes over on the advancing cylinder to close the clamp. This low pressure protects the mold against any foreign objects that might get caught in the mold. Clamp tonnage is built up rapidly by a high-volume, high-pressure pump acting on both the advancing cylinder and the main ram. There is also a hydraulically operated screw-type direct-drive injection unit (Fig. 85) and optional electronic controls for its operation. The machine is designed for fast cycles, low-pressure mold protection, precise shot control, etc.

Some other presses available are a standard injection press (Fig. 86); a press with double-toggle clamp (Fig. 87) which is hydraulically operated to provide fast, smooth clamp action; and a reciprocating-screw injection unit (Fig. 88) driven by an electrically operated motor in which, regardless of the material being molded, the basic principle of the reciprocating-screw injection system is the same. A complete cycle involves these basic steps: (1) material is fed from the hopper to the front of the screw during the plasticizing operation; (2) the accumulated material forces the screw to reciprocate to the rear; (3) when a sufficient quantity of material has accumulated in front of the screw, the screw stops; (4) the screw is moved forward by the hydraulic injection cylinder; (5) the material is injected into the mold; and (6) the screw starts plasticizing again and preheats the required amount of material for the next shot as the completed part is removed from the mold. Then there is a press with computer control system (Fig. 89) which is capable of control and management of a large number of various types of plastic-processing machines. It is particularly useful in the control of injection-molding machines where the discontinuous nature of the process and the large number of variables which affect product quality make it difficult to control quality consistently by manual means. The digital computer, with its tremendous computing and control capacity, provides the means for a control and management system which can control the critical processing parameters of each machine and ensure optimum output of the equipment.

Fig. 83 Injection press. (*Cincinnati Milacron Co.*)

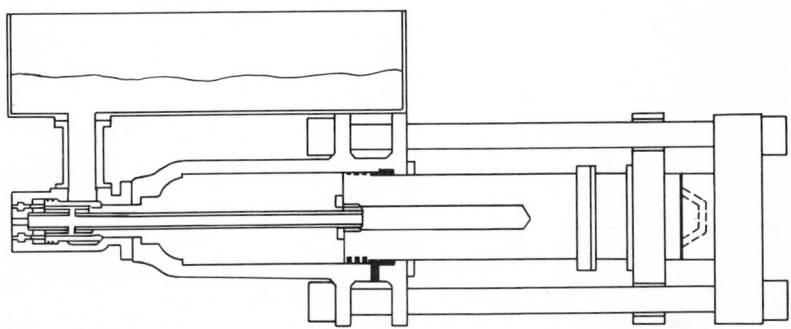

Fig. 84 Hydraulic clamping unit. (*Cincinnati Milacron Co.*)

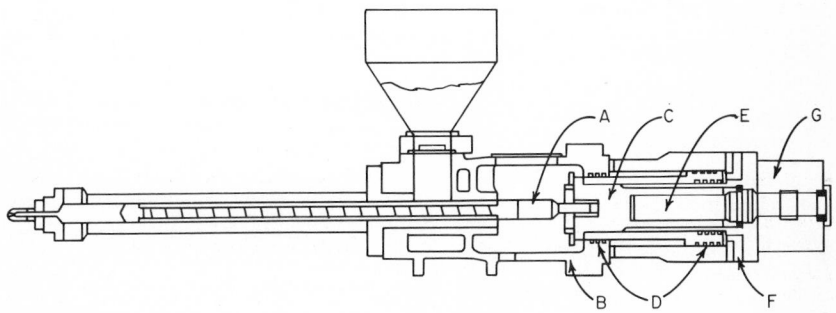

Fig. 85 Direct-drive injection unit. (*A*) Plasticizing screw. (*B*) Injection unit housing. (*C*) Piston component. (*D*) Piston rings. (*E*) Drive shaft. (*F*) Oil port. (*G*) Hydraulic-motor area. (*Cincinnati Milacron Co.*)

Fig. 86   Standard injection press.   (*New Britain Machine Co.*)

Fig. 87   Double-toggle clamp.   (*New Britain Machine Co.*)

Thermosets

Thermoset materials are available in many forms, such as powder, granular, and modular, with short or long glass fibers and with high-bulk fabric fillers. Material can be processed as powder or shaped preforms. In case of bulky material, hand weighing and cold pressing are required. Material can be used at room temperature or preheated. Methods of preheating are air-circulating oven, infrared lights, or electronic preheaters. Molds may be heated by steam, hot oil, or electric cartridges.

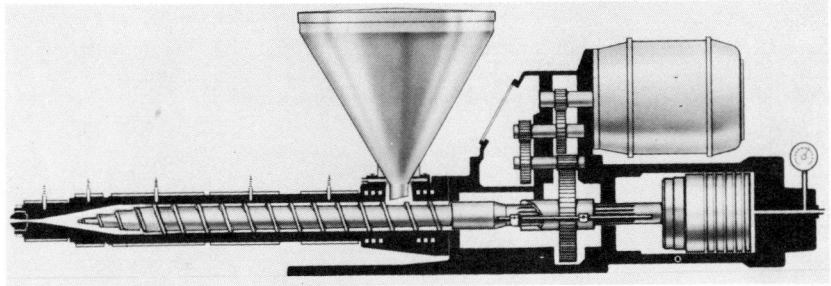

Fig. 88   Reciprocating-screw injection unit.   (*New Britain Machine Co.*)

Fig. 89   Press with computer control system.   (*New Britain Machine Co.*)

Press requirements are based on total square inches of projected area of the parts to be molded, multiplied by the required pounds per square inch whose total tonnage must be within the clamping pressure of the press in tons. Depending upon the plastic and the fillers to be molded, straight compression molding may require from 1 to 4 tons/in.^{2} on the total projected area.

Transfer molding may require from fractions of tons to 6 tons/in.^{2} at the transfer ram, but the total clamping pressure of the press in tons must exceed the total transfer tonnage by at least 10 tons to prevent the press from opening and flashing during the molding cycle. (Total transfer tonnage equals projected area in square inches of the part or parts to be molded, plus the area of the transfer ram and runner system, multiplied by the pounds per square inch produced at the ram on the plastic.)

Brief description of thermoset press equipment   The older hydraulic presses operate from central accumulator systems using both high- and low-pressure water or oil.   The later presses are equipped as self-contained units and operate with high and low hydraulic oil pressure.   There are presses for compression and transfer molding that require hand feeding of the molding compound, and also transfer-type presses that will automatically feed the powder, preform, preheat, transfer, and complete the molding cycle.   Presses are available, both vertical and horizontal, in many sizes for thermosets in semiautomatic compression and/or transfer (Fig. 90), automatic transfer (Fig. 91), automatic compression (Fig. 92), and automatic injection of thermosets (Fig. 93).   Note that by changing the heating cylinder and screw, the injection press may be used for thermoplastic materials.

Fig. 90   Semiautomatic transfer and/or compression press.   (*Hull Corp.*)

Roll-Leaf Stamping Process

Hot-leaf stamping is one of the oldest methods of marking and decorating.   At one time, pure gold was hammered by hand into extremely thin sheets.   This gold leaf was applied to buildings, books, etc., for decorative effects.   Later, methods were devised for plating gold onto a "carrier" strip of thin paper, acetate, or Cellophane.   This material was then supplied in roll form.   The process of applying the leaf in roll form was called roll-leaf stamping.   The development of imitation gold and silver, and pigmented foils and improved stamping equipment made the roll-leaf process an economical, high-speed method of marking and decorating.

The roll-leaf process is a branding technique which involves the use of an engraved metal die, metal type, or a flat silicone-rubber pad, a stamping press, the roll-leaf material, and at times a nest, jig, or fixture which both supports and positions the item.

The stamping die is mounted in the heated head of the press. The item to be marked is positioned on the worktable of the press, directly below the die. A roll of stamping leaf is mounted so that the carrier strip passes between the die and the item. When the die is brought into contact with the article under pressure, the heat transfers the roll-leaf coating into depressions made in the item by the die. After each impression, the strip of roll leaf is advanced to an unused portion by an automatic roll-leaf attachment. Since roll-leaf stamping is a "dry" method, the object can be handled immediately without fear of smearing the mark. This complete marking cycle normally will take from ½ to 5 s, depending upon the degree of hardness of the item, the item's resistance to heat, the boldness of the mark, the stamping-die temperature, and the pressure applied by the die.

Fig. 91 99-D automatic transfer press. (*Hull Corp.*)

Plastic items that have been hot-stamped will have the metallic or pigment coating "inlaid" into the depressions made by the heated metal die, or raised areas molded in relief will be coated with roll leaf using a silicone-rubber pad.

One of the advantages of hot stamping is that both flat and contoured surfaces can be marked. In order to mark a contoured surface, a die must be accurately shaped to match the part.

The roll-leaf process is used to mark items made of paper, cloth, leather, wood, plastics, and most other materials except glass, soft rubber, and uncoated metal.

An air-operated roll-leaf stamping press (Fig. 94), the Kensol 50, was specifically designed for hot stamping, creasing, and light embossing on flat items. It is available with one or two main cylinders for 6½ tons of pressure. This high-speed press will greatly increase production on work that formerly had to be stamped on

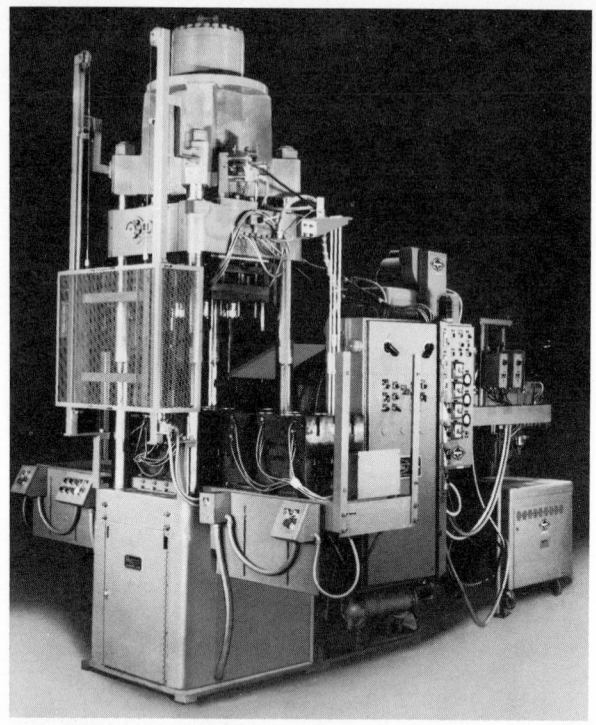

Fig. 92   200-ton compression shuttle with preheat.   (*Hull Corp.*)

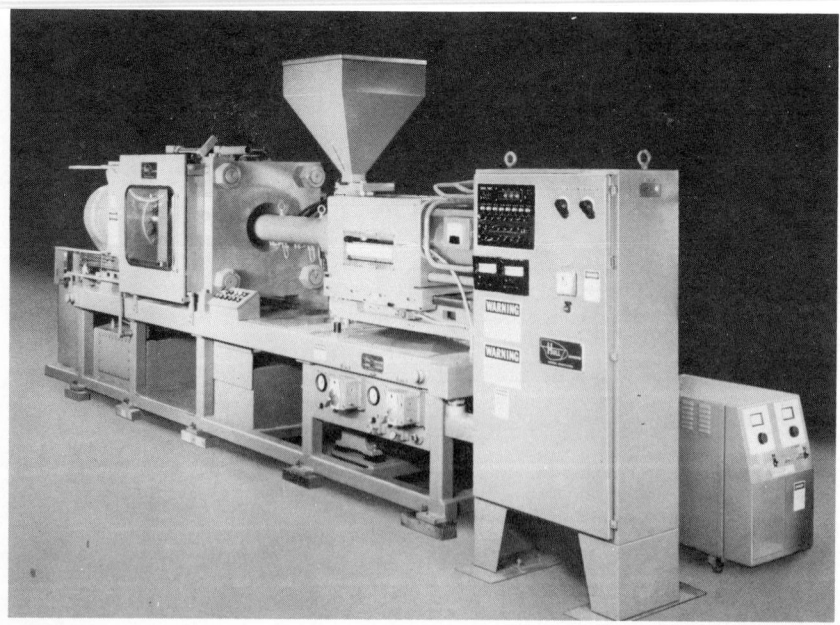

Fig. 93   Automatic injection press.   (*Hull Corp.*)

slow, expensive motor-driven equipment. The stamping heat action can be adjusted to give a powerful "squeeze" impression or a fast "kiss" impression that is required on materials with a plastic finish. Rotary machines for decorating the periphery of cylindrical objects are available along with units to produce multicolor heat transfer.

Dual-purpose rotary decorating machines (Fig. 95) are available with a photocell registration mechanism for preprinted multicolor Kensol Letraset heat transfers.

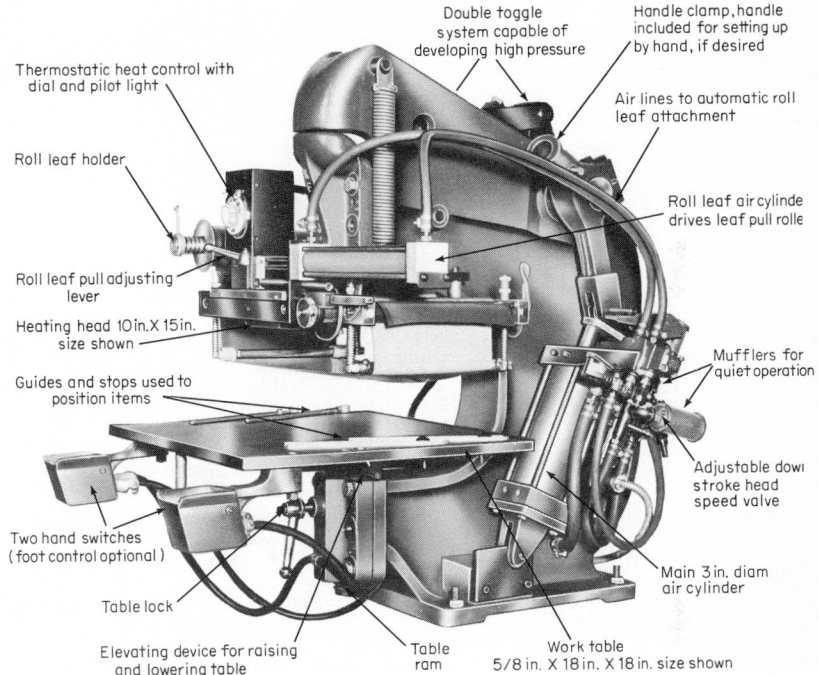

Fig. 94   Air-operated roll-leaf stamping press.   (*Kensol-Olsenmark, Inc.*)

With the Letraset heat-transfer process, it is possible to imprint intricate designs in up to six colors in one simple operation. Transfers with any design, pattern, or full-color picture can be supplied to satisfy individual requirements. Silk-screened or gravure-printed onto a clear carrier film, the transfers actually fuse with the container being marked. Thus, the poor abrasion resistance and loss of vivid color often experienced with other transfers are eliminated.

In general, two types of machine are used in roll-leaf stamping, namely, reciprocating and rotary. With a reciprocating press, the complete area of the die contacts the item to be marked instantaneously in one "stroke." With a rotary press, a cylindrical die rotates across an object, or a cylindrical object is rotated across a flat die. Higher pressures are required with a reciprocating press, since the complete die contacts the items to be decorated. On a rotary machine, low pressures are required, since only "line" contact is made between the die and the object being marked.

Reciprocating machines are recommended for decorating flat items, and items of irregular shapes, where an engraved stamping die can be made to conform to the shape. A reciprocating press *cannot* decorate more than 90° of a complete cylinder.

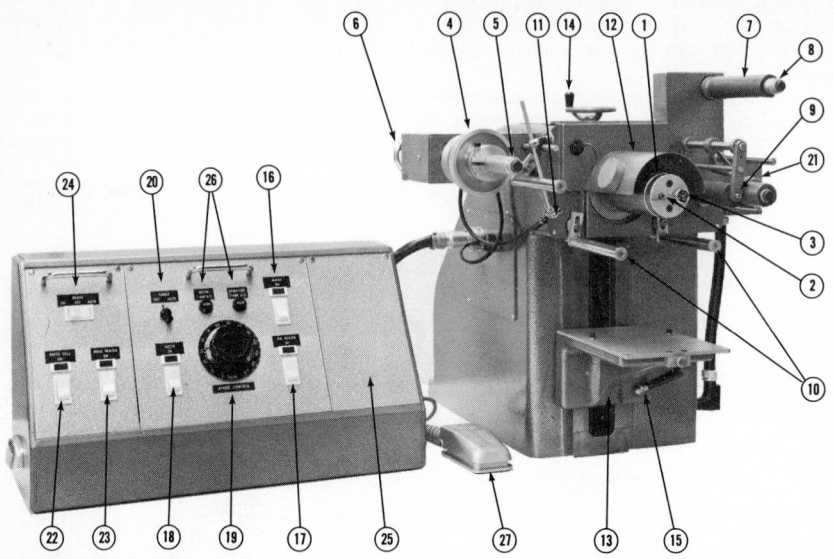

Fig. 95 Dual-purpose rotary decorating machine. (*Kensol-Olsenmark, Inc.*)

1. Rotating, heated die sleeve.
2. Thermostatic heat control.
3. Thermometer.
4. Unwind roller.
5. Unwind roller friction-drive adjustment.
6. Alignment adjusting handwheel.
7. Rewind roller.
8. Rewind roller friction-drive adjustment.
9. Drive and pinch roller assembly.
10. Adjustable guide rollers.
11. Photocell for registering transfers.
12. Heated hood.
13. Knee table.
14. Handwheel for adjusting knee table.
15. Knee-table lock screw.
16. Main on-off switch and pilot light.
17. Die-heater on-off switch and pilot light.
18. Drive-motor on-off switch.
19. Die speed-control knob.
20. "Selector" switch.
21. Interrupt button (partially shown).
22. Photocell on-off switch and pilot light.
23. Heated-hood on-off switch and pilot light.
24. Brake on-off-auto switch.
25. Spare panel.
26. Fuses.
27. Foot switch.

Rotary machines are recommended for decorating more than 90° of a cylinder (a full 360° of a cylinder can be decorated by rotary). They are also recommended for applying a continuous decoration on long lengths of molding, hose, and other extrusions. Perhaps the newest rotary application is coating of large areas with roll leaf (wood grain, on sides of plastic TV sets, coating large solid areas on displays, etc.).

Glossary of Terms

ABS plastics An abbreviated phrase used to refer to acrylonitrile-butadiene-styrene copolymers; an elastomer-modified styrene.

accelerator A chemical used to speed up a reaction or cure. For example, cobalt naphthanate is used to accelerate the reaction of certain polyester resins. The term accelerator is often used interchangeably with the term promoter. An accelerator is often used along with a catalyst, hardener, or curing agent.

adhere To cause two surfaces to be held together by adhesion.

adherend A body which is held to another body by an adhesive.

adhesion The state in which two surfaces are held together by interfacial forces which may consist of valence forces or interlocking action, or both.

adhesive A substance capable of holding materials together by surface attachment.

adhesive, contact An adhesive that is apparently dry to the touch and which will adhere to itself instantaneously upon contact; also called contact-bond adhesive or dry-bond adhesive.

adhesive, heat-activated A dry adhesive film that is rendered tacky or fluid by application of heat or heat and pressure to the assembly.

adhesive, pressure-sensitive A viscoelastic material which in solvent-free form remains permanently tacky. Such material will adhere instantaneously to most solid surfaces with the application of very slight pressure.

adhesive, room-temperature-setting An adhesive that sets in the temperature range of 20 to 30°C (68 to 86°F), in accordance with the limits for Standard Room Temperature specified in ASTM Methods D 618, Conditioning Plastics and Electrical Insulating Materials for Testing.

adhesive, solvent An adhesive having a volatile organic liquid as a vehicle.

adhesive, solvent-activated A dry adhesive film that is rendered tacky just prior to use by application of a solvent.

aging The change in properties of a material with time under specific conditions.

alcohols Characterized by the fact that they contain the hydroxyl (–OH) group, they are valuable starting points for the manufacture of synthetic resins, synthetic rubbers, and plasticizers.

aldehydes In general, aldehydes are volatile liquids with sharp, penetrating odors and are slightly less soluble in water than corresponding alcohols. The group (—CHO) which characterizes all aldehydes contains the most active form of the carbonyl radical, and makes the aldehydes important as organic synthetic agents. They are widely used in industry as chemical building blocks in organic synthesis.

aliphatic hydrocarbon See Hydrocarbon.

ambient temperature The temperature of the surrounding cooling medium such as gas or liquid, which comes into contact with the heated parts of the apparatus.

amine adduct Products of the reaction of an amine with a deficiency of a substance containing epoxy groups.

anhydrides Anhydrides result from the removal of water from organic acids. In general, they are liquids that boil at higher temperatures than the corresponding acids. They are particularly valuable for reactions with the alcohol groups found in cellulose, phenols, sugars, and vegetable oils, yielding esters which are less water-soluble than the original alcohol.

antioxidant A chemical used in the formulation of plastics to prevent or slow down the oxidation of material exposed to the air.

antistatic agents Agents which, when added to the molding material or applied onto the surface of the molded object, make it less conducting (thus hindering the fixation of dust).

arc resistance The time required for an arc to establish a conductive path in a material.

aromatic hydrocarbon See Hydrocarbon.

assembly A group of materials or parts, including adhesive, which has been placed together for bonding or which has been bonded together.

atactic When the radical groups are arranged heterogeneously around the carbon chain.

autoclave A closed vessel for conducting a chemical reaction or other operation under pressure and heat.

autoclave molding After lay-up, the entire assembly is placed in a steam autoclave at 50 to 100 lb/in.^{2}. Additional pressure achieves higher reinforcement loadings and improved removal of air.

B stage An intermediate stage in the curing of a thermosetting resin. In this stage, resins can be heated and caused to flow, thereby allowing final curing in the desired shape. The term A stage is used to describe an earlier stage in the curing reaction, and the term C stage is sometimes used to describe the cured resin. Most molding materials are in the B stage when supplied for compression or transfer molding.

bag molding A method of applying pressure during bonding or molding, in which a flexible cover, usually in connection with a rigid die or mold, exerts pressure on the material being molded, through the application of air pressure or drawing of a vacuum.

basket weave In this type of weave, two or more warp threads cross alternately with two or more filling threads. The basket weave is less stable than the plain weave but produces a flatter and stronger fabric. It is also a more pliable fabric than the plain weave and a certain degree of porosity is maintained without too much sleaziness, but not as much as the plain weave.

blister A raised area on the surface of a molding caused by the pressure of gases inside it on its incompletely hardened surface.

blowing agent Chemicals that can be added to plastics and that generate inert gases upon heating. This blowing or expansion causes the plastic to expand, thus forming a foam. Also known as foaming agent.

blow molding A method of fabrication of thermoplastic materials in which a parison (hollow tube) is forced into the shape of the mold cavity by internal air pressure.

bond The union of materials by adhesives. To unite materials by means of an adhesive.

bond strength The unit load applied in tension, compression, flexure, peel, impact, cleavage, or shear, required to break an adhesive assembly with failure occurring in or near the plane of the bond.

boron fibers High-modulus fibers of elemental boron deposited onto a thin tungsten wire. Supplied as single strands or tapes.

boss   Projection on a plastic part designed to add strength, to facilitate alignment during assembly, to provide for fastenings, etc.

bulk density   The density of a molding material in loose form (granular, nodular, etc.) expressed as a ratio of weight to volume (e.g., g/cm^3 or lb/ft^3).

bulk factor   Ratio of the volume of loose molding powder to the volume of the same weight of resin after molding.

bulk rope molding compound   Molding compound made with thickened polyester resin and fibers less than $\frac{1}{2}$ in. Supplied as rope, it molds with excellent flow and surface appearance.

capacitance (capacity)   That property of a system of conductors and dielectrics which permits the storage of electricity when potential difference exists between the conductors. Its value is expressed as the ratio of quantity of electricity to a potential difference. A capacitance value is always positive.

cast   To embed a component or assembly in a liquid resin, using molds that separate from the part for reuse after the resin is cured. Curing or polymerization takes place without external pressure. (See Embed and Pot.)

catalyst   A chemical that causes or speeds up the cure of a resin, but that does not become a chemical part of the final product. Catalysts are normally added in small quantities. The peroxides used with polyester resins are typical catalysts.

cavity   Depression in a mold, which usually forms the outer surface of the molded part; depending on the number of such depressions, molds are designated as single-cavity or multicavity.

centrifugal casting   Fabrication process in which the catalyzed resin is introduced into a rapidly rotating mold where it forms a layer on the mold surfaces and hardens.

chlorinated hydrocarbon   An organic compound having hydrogen atoms and, more importantly, chlorine atoms in its chemical structure. Trichloroethylene, methyl chloroform, and methylene chloride are chlorinated hydrocarbons.

coat   To cover with a finishing, protecting, or enclosing layer of any compound (such as varnish).

coefficient of expansion   The fractional change in dimension of a material for a unit change in temperature.

cohesion   The state in which the particles of a single substance are held together by primary or secondary valence forces. As used in the adhesive field, the state in which the particles of the adhesive (or the adherend) are held together.

cold flow (creep)   The continuing dimensional change that follows initial instantaneous deformation in a nonrigid material under static load.

cold-press molding   Molding process where inexpensive plastic male and female molds are used with room-temperature-curing resins to produce accurate parts. Limited runs are possible.

cold pressing   A bonding operation in which an assembly is subjected to pressure without the application of heat.

composite   A homogeneous material created by the synthetic assembly of two or more materials (a selected filler or reinforcing elements and compatible matrix binder) to obtain specific characteristics and properties.

compound   Some combination of elements in a stable molecular arrangement.

compression molding   A technique of thermoset molding in which the molding compound (generally preheated) is placed in the heated open mold cavity and the mold is closed under pressure (usually in a hydraulic press), causing the material to flow and completely fill the cavity, with pressure being held until the material has cured.

compressive strength   The maximum compressive stress a material is capable of sustaining. For materials which do not fail by a shattering fracture, the value is arbitrary, depending on the distortion allowed.

condensation   A chemical reaction in which two or more molecules combine with the separation of water, or some other simple substance.

condensation resins   Any of the alkyd, phenol-aldehyde, and urea-formaldehyde resins.

copolymer   See Polymer.

crazing   Fine cracks which may extend in a network on or under the surface or through a layer of a plastic material.

creep   The dimensional change with time of a material under load, following the initial instantaneous elastic deformation; the time-dependent part of strain resulting from force. Creep at room temperature is sometimes called "cold flow". See Recommended Practices for Testing Long-Time Creep and Stress-Relaxation of Plastics under Tension or Compression Loads at Various Temperatures, ASTM D 674.

cross-linking   When chemical links set up between molecular chains, the plastic is said to be cross-linked. In thermosets, cross linking makes one infusible super-molecule of all the chains, contributing to strength, rigidity, and high-temperature resistance. Thermoplastics (e.g., polyethylene) can also be cross-linked (e.g., by irradiation or chemically through formulation) to produce three-dimensional structures that are thermoset in nature and offer improved tensile strength and stress-crack resistance.

crow's foot   In this type of weave there is a 3 by 1 interlacing. That is, a filling thread floats over the three warp threads and then under one. This type of fabric looks different on one side than the other. Fabrics with this weave are more pliable than either the plain or basket weave and, consequently, are easier to form around curves.

crystalline melting point   The temperature at which the crystalline structure in a material is broken down.

crystallinity   A state of molecular structure referring to uniformity and compactness of the molecular chains forming the polymer and resulting from the formation of solid crystals with a definite geometric pattern. In some resins, such as polyethylene, the degree of crystallinity indicates the degree of stiffness, hardness, environmental stress-check resistance, and heat resistance.

cull   Material remaining in a transfer chamber after the mold has been filled. Unless there is a slight excess in the charge, the operator cannot be sure the cavity is filled. The charge is generally regulated to control the thickness of the cull.

cure   To change the physical properties of a material (usually from a liquid to a solid) by chemical reaction, by the action of heat and catalysts, alone or in combination, with or without pressure.

curing agent   See Hardener.

curing temperature   The temperature at which a material is subjected to curing.

curing time   In the molding of thermosetting plastics, the time it takes for the material to be properly cured.

cycle   One complete operation of a molding press from closing time to closing time.

damping   In a material, the ability to absorb energy to reduce vibration.

decorative laminates   High-pressure laminates consisting of a phenolic-paper core and a melamine-paper top sheet with a decorative pattern.

deflashing   Covers the range of finishing techniques used to remove the flash (excess unwanted material) from a plastic part; examples are filing, sanding, milling, tumbling, and wheelabrating. (See Flash.)

deflection temperature   Formerly called heat-distortion temperature (HDT).

diallyl phthalate   Ester polymer resulting from reaction of allyl alcohol and phthalic anhydride.

dielectric constant   (permittivity or specific inductive capacity). That property of a dielectric which determines the electrostatic energy stored per unit volume for unit potential gradient.

dielectric loss   The time rate at which electric energy is transformed into heat in a dielectric when it is subjected to a changing electric field.

dielectric-loss angle   (dielectric-phase difference)   The difference between 90° and the dielectric-phase angle.

dielectric-loss factor   (dielectric-loss index)   The product of dielectric constant and the tangent of dielectric-loss angle for a material.

dielectric-phase angle   The angular difference in phase between the sinusoidal alternating potential difference applied to a dielectric and the component of the resulting alternating current having the same period as the potential difference.

dielectric power factor The cosine of the dielectric-phase angle (or sine of the dielectric-loss angle).

dielectric strength The voltage which an insulating material can withstand before breakdown occurs, usually expressed as a voltage gradient (such as volts/mil).

diluent An ingredient usually added to a formulation to reduce the concentration of the resin.

diphenyl oxide resins Thermosetting resins based on diphenyl oxide and possessing excellent handling properties and heat resistance.

dissipation factor (loss tangent, tan δ, approximate power factor) The tangent of the loss angle of the insulating material.

DODISS Abbreviation for Department of Defense Index of Specifications and Standards.

dry To change the physical state of an adhesive on an adherend through the loss of solvent constituents by evaporation or absorption, or both.

ductility The ability of a material to deform plastically before fracturing.

eight harness satin A fabric with this type of weave has a 7 by 1 interlacing in which a filling thread floats over seven warp threads and then under one. Like the crowfoot weave, it looks different on one side than on the other. This weave is more pliable than any of the others and is especially adaptable where it is necessary to form around compound curves, such as on radomes.

elastic limit The greatest stress a material is capable of sustaining without any permanent strain remaining when the stress is released.

elasticity That property of a material by virtue of which it tends to recover its original size and shape after deformation. If the strain is proportional to the applied stress, the material is said to exhibit Hookean or ideal elasticity.

elastomer A material which at room temperature can be stretched repeatedly to at least twice its original length and, upon release of the stress, will return with force to its approximate original length. Plastics with such or similar characteristics are known as elastomeric plastics. The expression is also used when referring to a rubber (natural or synthetic) reinforced plastic, as in "elastomer-modified" resins.

electric strength (dielectric strength or disruptive gradient) The maximum potential gradient that a material can withstand without rupture. The value obtained for the electric strength will depend on the thickness of the material and on the method and conditions of test.

electrode A conductor of the metallic class through which a current enters or leaves an electrolytic cell, at which there is a change from conduction by electrons to conduction by charged particles of matter, or vice versa.

elongation The increase in gage length of a tension specimen, usually expressed as a percentage of the original gage length. (See also Gage length.)

embed To encase completely a component or assembly in some material—a plastic, for instance. (See Cast and Pot.)

encapsulate To coat a component or assembly in a conformal or thixotropic coating by dipping, brushing, or spraying.

environmental stress cracking The susceptibility of a thermoplastic article to crack or craze under the influence of certain chemicals or aging, or weather, and stress. Standard ASTM test methods that include requirements for environmental stress cracking are indexed in Index of ASTM Standards.

epoxy Thermosetting polymers containing the oxirane group. Mostly made by reacting epichlorohydrin with a polyol like bisphenol A. Resins may be either liquid or solid.

eutectic A mixture whose melting point is lower than that of any other mixture of the same ingredients.

exotherm The characteristic curve of resin during its cure, which shows heat of reaction (temperature) vs. time. Peak exotherm is the maximum temperature on this curve.

exothermic Chemical reaction in which heat is given off.

extender An inert ingredient added to a resin formulation chiefly to increase its volume.

extrusion The compacting of a plastic material and the forcing of it through an orifice.

fabric   A planar structure produced by interlacing yarns, fibers, or filaments.

failure, adhesive   Rupture of an adhesive bond, such that the separation appears to be at the adhesive-adherend interface.

fiber glass   An individual filament made by attenuating molten glass. A continuous filament is a glass fiber of great or indefinite length; a staple fiber is a glass fiber of relatively short length (generally less than 17 in.).

filament winding   A process for fabricating a composite structure in which continuous reinforcements (filament, wire, yarn, tape, or other) either previously impregnated with a matrix material or impregnated during the winding are placed over a rotating and removable form or mandrel in a previously prescribed way to meet certain stress conditions. Generally the shape is a surface of revolution and may or may not include end closures. When the right number of layers are applied, the wound form is cured and the mandrel removed.

filler   A material, usually inert, that is added to plastics to reduce cost or modify physical properties.

film adhesive   Thin layer of dried adhesive. Also describes a class of adhesives provided in dry-film form with or without reinforcing fabric, which are cured by heat and pressure.

fish paper   Electrical-insulation grade of vulcanized fiber in thin cross section.

flash   Extra plastic attached to a molding along the parting line; under most conditions it would be objectionable and must be removed before the parts are acceptable.

flexibilizer   A material that is added to rigid plastics to make them resilient or flexible. Flexibilizers can be either inert or a reactive part of the chemical reaction. Also called a plasticizer in some cases.

flexural modulus   The ratio, within the elastic limit, of stress to corresponding strain. It is calculated by drawing a tangent to the steepest initial straight-line portion of the load-deformation curve and calculating by the following equation:

$$E_B = \frac{L^3 m}{4bd^3}$$

where E_B = modulus
b = width of beam tested
d = depth of beam
m = slope of the tangent
L = span, in.

flexural strength   The strength of a material in bending expressed as the tensile stress of the outermost fibers of a bent test sample at the instant of failure.

fluorocarbon   An organic compound having fluorine atoms in its chemical structure. This property usually lends chemical and thermal stability to plastics.

four harness satin   This fabric is also named crowfoot satin because the weaving pattern when laid out on cloth design paper resembles the imprint of a crow's foot. In this type of weave there is a 3 by 1 interlacing. That is, a filling thread floats over the three warp threads and then under one. This type of fabric looks different on one side than on the other. Fabrics with this weave are more pliable than either the plain or basket weave and, consequently, are easier to form around curves.

gage length   The original of that portion of the specimen over which strain is measured.

gate   In injection and transfer molding, the orifice through which the melt enters the cavity.

gel   The soft rubbery mass that is formed as a thermosetting resin goes from a fluid to an infusible solid. This is an intermediate state in a curing reaction, and a stage in which the resin is mechanically very weak. *Gel point* is defined as the point at which gelation begins.

gel coat   A resin applied to the surface of a mold and gelled prior to lay-up. The gel coat becomes an integral part of the finished laminate, and is usually used to improve surface appearance, etc.

glass transition point   Temperature at which a material loses its glasslike properties and becomes a semiliquid.

glue line (bond line)   The layer of adhesive which attaches two adherends.

glue-line thickness   Thickness of the fully dried adhesive layer.

glycol   An alcohol containing two hydroxyl (–OH) groups.

graphite fibers   High-strength, high-modulus fibers made by controlled carbonization and graphitization of organic fibers, usually rayon, acrylonitrile, or pitch.

hand lay-up   The process of placing (and working) successive plies of reinforcing material or resin-impregnated reinforcement in position on a mold by hand.

hardener   A chemical added to a thermosetting resin for the purpose of causing curing or hardening.   Amines and acid anhydrides are hardeners for epoxy resins. Such hardeners are a part of the chemical reaction and a part of the chemical composition of the cured resin.   The terms hardener and curing agent are used interchangeably.   Note that these can differ from catalysts, promoters, and accelerators.

hardness   See Indentation hardness.

heat-distortion point   The temperature at which a standard test bar deflects 0.010 in. under a stated load of either 66 or 264 lb/in.^{2}.   See Standard Method of Test for Deflection Temperature of Plastics under Load, ASTM D 648.

heat sealing   A method of joining plastic films by simultaneous application of heat and pressure to areas in contact.   Heat may be supplied conductively or dielectrically.

high-frequency preheating   The plastic to be heated forms the dielectric of a condenser to which is applied a high-frequency (20 to 80 MHz) voltage.   Dielectric loss in the material is the basis.   The process is used for sealing vinyl films and preheating thermoset molding compounds.

hot-melt adhesive   A thermoplastic adhesive compound, usually solid at room temperature, which is heated to a fluid state for application.

hydrocarbon   An organic compound having hydrogen and carbon atoms in its chemical structure.   Most organic compounds are hydrocarbons.   Aliphatic hydrocarbons are straight-chained hydrocarbons, and aromatic hydrocarbons are ringed structures based on the benzene ring.   Methyl alcohol, trichloroethylene, etc., are aliphatic; benzene, xylene, toluene, etc., are aromatic.

hydrolysis   Chemical decomposition of a substance involving the addition of water.

hydroxyl group   A chemical group consisting of one hydrogen atom plus one oxygen atom.

hygroscopic   Tending to absorb moisture.

impregnate   To force resin into every interstice of a part.   Cloths are impregnated for laminating, and tightly wound coils are impregnated in liquid resin using air pressure or vacuum as the impregnating force.

indentation hardness   Hardness evaluated from measurements of area or indentation depth caused by pressing a specified indentation into the material surface with a specified force.

inhibitor   A chemical added to resins to slow down the curing reaction.   Inhibitors are normally added to prolong the storage life of thermosetting resins.

injection molding   A molding procedure whereby a heat-softened plastic material is forced from a cylinder into a cavity which gives the article the desired shape. Used with all thermoplastic and some thermosetting materials.

inorganic chemicals   Chemicals whose chemical structure is based on atoms other than the carbon atom.

insulation resistance   The ratio of the applied voltage to the total current between two electrodes in contact with a specific insulator.

isotactic   When molecules are polymerized in parallel arrangements of radicals on one side of the carbon chain.

Izod impact strength   A measure of the toughness of a material under impact as measured by the Izod impact test.

joint   The location at which two adherends are held together with a layer of adhesive.

joint, lap   A joint made by placing one adherend partly over another and bonding together the overlapped portions.

joint, scarf   A joint made by cutting away similar angular segments of two adherends and bonding the adherends with the cut areas fitted together.

joint, starved   A joint that has an insufficient amount of adhesive to produce a satisfactory bond.

laminate   To unite sheets of material by a bonding material usually with pressure and heat (normally used with reference to flat sheets); a product made by so bonding.

latent curing agent   Curing agent that produces long-time stability at room temperature but rapid cure at elevated temperature.

lay-up   As used in reinforced plastics, the reinforcing material placed in position in the mold; also the resin-impregnated reinforcement. The process of placing the reinforcing material in position in the mold.

leno weave   A leno weave is a locking type weave in which two or more warp threads cross over each other and interlace with one or more filling threads. It is used primarily to prevent shifting of fibers in open fabrics.

limited-coordination specification (or Standard)   A specification (or standard) which has not been fully coordinated and accepted by all the interested activities. Limited-coordination specifications and standards are issued to cover the need for requirements unique to one particular department. This applies primarily to military agency documents.

liquid injection molding   Fabrication process in which catalyzed resin is metered into closed molds.

macerate   To chop or shred fabric for use as a filler for a molding resin. The molding compound obtained when so filled.

mat   A randomly distributed felt of glass fibers used in reinforced plastics lay-up molding.

matched metal molding   Method of molding reinforced plastics between two close-fitting metal molds mounted in a hydraulic press.

mechanical properties   Material properties associated with elastic and inelastic reactions to an applied force.

melamines   Thermosetting resins made from melamine and formaldehyde and possessing excellent hardness, clarity, and electrical properties.

micron   A unit of length equal to 10,000 Å, 0.0001 cm, or approximately 0.000039 in.

mock leno weave   An open type weave that resembles a leno and is accomplished by a system of interlacings that draws a group of threads together and leaves a space between the next group. The warp threads do not actually cross each other as in a real leno and, therefore, no special attachments are required for the loom. This type of weave is generally used when a high thread count is required for strength and at the same time the fabric must remain porous.

modifier   A chemically inert ingredient added to a resin formulation that changes its properties.

modulus of elasticity   The ratio of unidirectional stress to the corresponding strain (slope of the line) in the linear stress-strain region below the proportional limit. For materials with no linear range a secant line from the origin to a specified point on the stress-strain curve or a line tangent to the curve at a specified point may be used.

moisture resistance   The ability of a material to resist absorbing moisture, either from the air or when immersed in water.

mold   A medium or tool designed to form desired shapes and sizes. To process a plastics material using a mold.

mold release   A lubricant used to coat a mold cavity to prevent the molded piece from sticking to it, and thus to facilitate its removal from the mold. Also called "release agent."

mold shrinkage   The difference in dimensions, expressed in inches per inch, between a molding and the mold cavity in which it was molded, both the mold and the molding being at room temperature when measured.

molecular weight The sum of the atomic masses of the elements forming the molecule.

monomer A small molecule which is capable of reacting with similar or other molecules to form large chainlike molecules called polymers.

multilayer printed circuits Electric circuits made on thin copper-clad laminates, stacked together with intermediate prepreg sheets and bonded together with heat and pressure. Subsequent drilling and electroplating through the layers result in a three-dimensional circuit.

necking Localized reduction of the cross-sectional area of a tensile specimen which may occur during loading.

NEMA standards Property values adopted as standard by the National Electrical Manufacturers Association.

notch sensitivity The extent to which the sensitivity of a material to fracture is increased by the presence of a surface in homogeneity such as a notch, a sudden change in section, a crack, or a scratch. Low notch sensitivity is usually associated with ductile materials, and high notch sensitivity with brittle materials.

nylon The generic name for all synthetic polyamides. These are thermoplastic polymers with a wide range of properties.

olefin A family of unsaturated hydrocarbons with the formula C_nH_n named after the corresponding paraffins by adding "ene" or "ylene" to the stem, i.e., ethylene. Paraffins are aliphatic hydrocarbons. (See Hydrocarbon.)

organic Composed of matter originating in plant or animal life, or composed of chemicals of hydrocarbon origin, either natural or synthetic. Used in referring to chemical structures based on the carbon atom.

paste An adhesive composition having a characteristic plastic-type consistency, that is, a high order or yield value, such as that of a paste prepared by heating a mixture of starch and water and subsequently cooling the hydrolyzed product.

penetration The entering of one part or material into another.

permanence The resistance of a given property to deteriorating influences.

permittivity Dielectric constant.

phenolic A synthetic resin produced by the condensation of an aromatic alcohol with an aldehyde, particularly of phenol with formaldehyde.

phenysilane Thermosetting copolymer of silicone and phenolic resins; furnished in solution form.

plain weave The plain weave is the most simple and commonly used. In this type of weave, the warp and filling threads cross alternately. Plain woven fabrics are generally the least pliable, but they are also the most stable. This stability permits the fabrics to be woven with a fair degree of porosity without too much sleaziness.

plastic A material containing an organic substance of large molecular weight, which is solid in its final condition and which at some earlier time was shaped by flow. (See Resin and Polymer.)

plastic deformation A change in dimensions of an object under load that is not recovered when the load is removed; opposed to elastic deformation.

plasticity A property of plastics which allows the material to be deformed continuously and permanently without rupture upon the application of a force that exceeds the yield value of the material.

plasticize To soften a material and make it plastic or moldable, by means of either a plasticizer or the application of heat.

plasticizer A material incorporated in a resin formulation to increase its flexibility, workability, or distensibility. The addition of a plasticizer may cause a reduction in melt viscosity, lower the temperature of second-order transition, or lower the elastic modulus of the solidified resin.

plastisols Mixtures of vinyl resins and plasticizers which can be molded, cast, or converted to continuous films by the application of heat. If the mixtures contain volatile thinners, they are also known as organosols.

Poisson's ratio The absolute value of the ratio of transverse strain to axial strain resulting from a uniformly applied axial stress below the proportional limit of the material.

polyesters Thermosetting resins, produced by reacting unsaturated, generally linear, alkyd resins with a vinyl-type active monomer such as styrene, methyl styrene, or diallyl phthalate. Cure is effected through vinyl polymerization using peroxide catalysts and promoters, or heat, to accelerate the reaction. The resins are usually furnished in liquid form.

polyimide High-temperature resins made by reacting aromatic dianhydrides with aromatic diamines.

polymer A compound formed by the reaction of simple molecules having functional groups which permit their combination to proceed to high molecular weights under suitable conditions. Polymers may be formed by polymerization (addition polymer) or polycondensation (condensation polymer). When two or more monomers are involved, the product is called a copolymer.

Also, any high-molecular-weight organic compound whose structure consists of a repeating small unit. Polymers can be plastics, elastomers, liquids, or gums and are formed by chemical addition or condensation of monomers.

polymerize To unite chemically two or more monomers or polymers of the same kind to form a molecule with higher molecular weight.

pot To embed a component or assembly in a liquid resin, using a shell, can, or case which remains as an integral part of the product after the resin is cured. (See Embed and Cast.)

pot life The time during which a liquid resin remains workable as a liquid after catalysts, curing agents, promoters, etc., are added; roughly equivalent to gel time. Sometimes also called working life.

power factor The cosine of the angle between the voltage applied and the current resulting.

preform A pill, tablet, or biscuit used in thermoset molding. Material is measured by volume, and the bulk factor of powder is reduced by pressure to achieve efficiency and accuracy.

preheating The heating of a compound prior to molding or casting in order to facilitate the operation, reduce cycle, and improve product.

premix A molding compound prepared prior to and apart from the molding operations and containing all components required for molding: resin, reinforcement fillers, catalysts, release agents, and other compounds.

prepolymers As used in polyurethane production, reaction product of a polyol with excess of an isocyanate.

prepreg Ready-to-mold sheet which may be cloth, mat, or paper impregnated with resin and stored for use. The resin is partially cured to a B stage and supplied to the fabricator, who lays up the finished shape and completes the cure with heat and pressure.

pressure-bag molding A process for molding reinforced plastics, in which a tailored flexible bag is placed over the contact lay-up on the mold, sealed, and clamped in place. Fluid pressure, usually compressed air, is placed against the bag, and the part is cured.

primer A coating applied to a surface, prior to the application of an adhesive, to improve the performance of the bond.

printed-circuit laminates Laminates, either fabric- or paper-based, covered with a thin layer of copper foil and used in the photofabrication process to make light-weight circuits.

promoter A chemical, itself a weak catalyst, that greatly increases the activity of a given catalyst.

proportional limit The greatest stress a material can sustain without deviating from the linear proportionality of stress to strain (Hooke's Law).

pultrusion Reversed "extrusion" of resin-impregnated roving in the manufacture of rods, tubes, and structural shapes of a permanent cross section. The roving, after passing through the resin dip tank, is drawn through a die to form the desired cross section.

qualified products list (QPL) A list of commercial products which have been pretested and found to meet the requirements of a specification, especially government specifications.

reactive diluent As used in epoxy formulations—a compound containing one or more epoxy groups which functions mainly to reduce the viscosity of the mixture.

refractive index The ratio of the velocity of light in a vacuum to its velocity in a substance. Also the ratio of the sine of the angle of incidence to the sine of the angle of refraction.

reinforced molding compound A plastic to which fibrous materials such as glass or cotton has been added to improve certain physical properties like flexural strength.

reinforced plastic A plastic with strength properties greatly superior to those of the base resin, resulting from the presence of reinforcements in the composition.

reinforced thermoplastics Reinforced molding compounds in which the plastic is thermoplastic.

relative humidity The ratio of the quantity of water vapor present in the air to the quantity which would saturate it at any given temperature.

release agent See Mold Release.

resin High-molecular-weight organic material with no sharp melting point. For general purposes, the terms resin, polymer, and plastic can be used interchangeably. (See Polymer.)

resistivity The ability of a material to resist passage of electric current either through its bulk or on a surface. The unit of volume resistivity is the ohm-centimeter, and the unit of surface resistivity is ohm/square.

rigidsol Plastisol having a high elastic modulus, usually produced with a cross-linking plasticizer.

Rockwell hardness number A number derived from the net increase in depth of impression as the load on a penetrator is increased from a fixed minimum load to a higher load and then returned to minimum load. Penetrators include steel balls of several specified diameters and a diamond-cone penetrator.

rotational casting (or molding) A method used to make hollow articles from thermoplastic materials. Material is charged into a hollow mold capable of being rotated in one or two planes. The hot mold fuses the material into a gel after the rotation has caused it to cover all surfaces. The mold is then chilled and the product stripped out.

roving The term roving is used to designate a collection of bundles of continuous filaments either as untwisted strands or as twisted yarns. Rovings may be lightly twisted, but for filament winding they are generally wound as bands or tapes with as little twist as possible.

rubber An elastomer capable of rapid elastic recovery; usually natural rubber, Hevea. (See Elastomer.)

set (mechanical) Strain remaining after complete release of the load producing the deformation.

set (polymerization) To convert an adhesive into a fixed or hardened state by chemical or physical action, such as condensation, polymerization, oxidation, vulcanization, gelation, hydration, or evaporation of volatile constituents.

shape factor For an elastomeric slab loaded in compression, the ratio of the loaded area to the force-free area.

shear strength The maximum shear stress a material is capable of sustaining. In testing, the shear stress is caused by a shear or torsion load and is based on the original specimen dimensions.

sheet molding compound Compression-molding material consisting of glass fibers longer than $\frac{1}{2}$ in. and thickened polyester resin. Possessing excellent flow, it results in parts with good surfaces.

shelf life An expression to describe the time a molding compound can be stored without losing any of its original physical or molding properties.

Shore hardness A procedure for determining the indentation hardness of a material by means of a durometer. Shore designation is given to tests made with a specified durometer instrument.

silicones Resinous materials derived from organosiloxane polymers, furnished in different molecular weights including liquids and solid resins.

sink mark A depression or dimple on the surface of an injection-molded part due to collapsing of the surface following local internal shrinkage after the gate seals. May also be an incipient short shot.

slush molding Method for casting thermoplastics, in which the resin in liquid form is poured into a hot mold where a viscous skin forms. The excess slush is drained off, the mold is cooled, and the molding is stripped out.

solvent Any substance, usually a liquid, which dissolves other substances.

specific heat The ratio of a material's thermal capacity to that of water at 15°C.

spiral flow test A method for determining the flow properties of a thermosetting resin in which the resin flows along the path of a spiral cavity. The length of the material which flows into the cavity and its weight give a relative indication of the flow properties of the resin.

spray-up Techniques in which a spray gun is used as the processing tool. In reinforced plastics, for example, fibrous glass and resin can be simultaneously deposited in a mold. In essence, roving is fed through a chopper and ejected into a resin stream which is directed at the mold by either of two spray systems.

stabilizers Chemicals used in plastics formulation to assist in maintaining physical and chemical properties during processing and service life. A specific type of stabilizer, known as an ultraviolet stabilizer, is designed to absorb ultraviolet rays and prevent them from attacking the plastic.

storage life See Shelf life.

strain The deformation resulting from a stress, measured by the ratio of the change to the total value of the dimension in which the change occurred; the unit change, due to force, in the size or shape of a body referred to its original size or shape. Strain is nondimensional but is frequently expressed in inches per inch or centimeters per centimeter, etc.

strength, dry The strength of an adhesive joint determined immediately after drying under specified conditions or after a period of conditioning in the standard laboratory atmosphere.

strength, wet The strength of an adhesive joint determined immediately after removal from a liquid in which it has been immersed under specified conditions of time, temperature, and pressure.

stress The unit force or component of force at a point in a body acting on a plane through the point. Stress is usually expressed in pounds per square inch.

stress relaxation The time-dependent decrease in stress for a specimen constrained in a constant strain condition.

substrate A material upon the surface of which an adhesive-containing substance is spread for any purpose, such as bonding or coating. A broader term than adherend.

surface preparation A physical and/or chemical preparation of an adherend to render it suitable for adhesive joining.

surface resistivity The resistance of a material between two opposite sides of a unit square of its surface. Surface resistivity may vary widely with the conditions of measurement.

syntactic foams Lightweight systems obtained by the incorporation of prefoamed or low-density fillers in the systems.

tack The property of an adhesive that enables it to form a bond of measurable strength immediately after adhesive and adherend are brought into contact under low pressure.

tensile strength The maximum tensile stress a material is capable of sustaining. Tensile strength is calculated from the maximum load during a tension test carried to rupture and the original cross-sectional area of the specimen.

thermal conductivity The ability of a material to conduct heat; the physical constant for the quantity of heat that passes through a unit cube of a material in a unit of time when the difference in temperatures of two faces is 1°C.

thermal stress cracking Crazing and cracking of some thermoplastic resins which result from overexposure to elevated temperatures.

thermoplastic Plastics capable of being repeatedly softened by increases in temperature and hardened by decreases in temperature. These changes are physical rather than chemical.

thermoset A material that will undergo or has undergone a chemical reaction by the action of heat, catalysts, ultraviolet light, etc., leading to a relatively infusible state.

thermosetting A classification of resin which cures by chemical reaction when heated and, when cured, cannot be resoftened by heating.

thinner A volatile liquid added to an adhesive to modify the consistency or other properties.

thixotropic Describing materials that are gel-like at rest but fluid when agitated.

time, assembly The time interval between the spreading of the adhesive on the adherend and the application of pressure or heat, or both, to the assembly.

time, curing The period of time during which an assembly is subjected to heat or pressure, or both, to cure the adhesive.

transfer molding A method of molding thermosetting materials, in which the plastic is first softened by heat and pressure in a transfer chamber, then forced by high pressure through suitable sprues, runners, and gates into a closed mold for final curing.

twill weave A basic weave characterized by a diagonal rib, or twill line. Each end floats over at least two consecutive picks enabling a greater number of yarns per unit area than a plain weave, while not losing a great deal of fabric stability.

UV stabilizer (ultraviolet) Any chemical compound which, when admixed with a thermoplastic resin, selectively absorbs ultraviolet rays.

vacuum-bag molding. A process for molding reinforced plastics in which a sheet of flexible transparent material is placed over the lay-up on the mold and sealed. A vacuum is applied between the sheet and the lay-up. The entrapped air is mechanically worked out of the lay-up and removed by the vacuum, and the part is cured.

vacuum injection molding Molding process using a male and female mold wherein reinforcements are placed in the mold, a vacuum is applied, and a room-temperature-curing liquid resin is introduced which saturates the reinforcement.

Vicat softening temperature A temperature at which a specified needle point will penetrate a material under specified test conditions.

viscoelastic A characteristic mechanical behavior of some materials which is a combination of viscous and elastic behaviors.

viscosity A measure of the resistance of a fluid to flow (usually through a specific orifice).

volume resistivity (specific insulation resistance) The electrical resistance between opposite faces of a 1-cm cube of insulating material, commonly expressed in ohm-centimeters. The recommended test is ASTM D 257-54T.

vulcanization A chemical reaction in which the physical properties of an elastomer are changed by causing it to react with sulfur or other cross-linking agents.

vulcanized fiber Cellulosic material which has been partially gelatinized by action of a chemical (usually zinc chloride), then heavily compressed or rolled to required thickness, leached free from the gelatinizing agent, and dried.

warpage Dimensional distortion in a plastic object after molding.

water absorption The ratio of the weight of water absorbed by a material to the weight of the dry material.

water-extended polyester Casting formulation in which water is suspended in the polyester resin.

wetting Ability to adhere to a surface immediately upon contact.

working life The period of time during which a liquid resin or adhesive, after mixing with catalyst, solvent, or other compounding ingredients, remains usable. (See Pot life.)

woven roving A heavy glass-fiber fabric made by the weaving of roving.

yield value (yield strength) The lowest stress at which a material undergoes plastic deformation. Below this stress, the material is elastic, above it, viscous. Also, the stress at which a material exhibits a specified limiting deviation from the proportionality of stress to strain.

Index